THE BIOLOGY PROBLEM SOLVER®

REGISTERED TRADEMARK

Staff of Research and Education Association,

Dr. M. Fogiel, Director

Research and Education Association
61 Ethel Road West
Piscataway, New Jersey 08854

THE BIOLOGY
PROBLEM SOLVER®

Printed in the United States of America

Library of Congress Control Number 00-112209

International Standard Book Number 0-87891-514-1

IV-1

WHAT THIS BOOK IS FOR

For as long as biology has been taught in schools, students have found this subject difficult to understand and learn because of the broad scope of this subject, and the large number of complex interrelationships among life structures and their functions. Despite the publications of hundreds of textbooks in this field, each one intended to provide an improvement over previous textbooks, biology remains particularly perplexing and the subject is often taken in class only to meet school/departmental requirements for a selected course of study.

In a study of the problem, REA found the following basic reasons underlying students' difficulties with biology taught in schools:

(a) No systematic rules of analysis have been developed which students may follow in a step-by-step manner to solve the usual problems encountered. This results from the fact that the numerous different conditions and principles which may be involved in a problem, lead to many possible different methods of solution. To prescribe a set of rules to be followed for each of the possible variations, would involve an enormous number of rules and steps to be searched through by students, and this task would perhaps be more burdensome than solving the problem directly with some accompanying trial and error to find the correct solution route.

(b) Textbooks currently available will usually explain a given principle in a few pages written by a professional who has an insight in the subject matter that is not shared by students. The explanations are often written in an abstract manner which leaves the students confused as to the application of the principle. The explanations given are not sufficiently detailed and extensive to make the student aware of the wide range of applications and different aspects of the principle being studied. The numerous possible variations of principles and their applications are usually not discussed, and it is left for the students to discover these for themselves while doing exercises. Accordingly, the average student is expected to rediscover that which has been long known and practiced, but not published or explained extensively.

(c) The examples usually following the explanation of a topic are too few in number and too simple to enable the student to obtain a thorough grasp of the principles involved. The explanations do not provide sufficient basis to enable a student to solve problems that may be subsequently assigned for homework or given on examinations.

The examples are presented in abbreviated form which leaves out much material between steps, and requires that students derive the omitted material themselves. As a result, students find the examples difficult to understand--contrary to the purpose of the examples.

Examples are, furthermore, often worded in a confusing manner. They do not state the problem and then present the solution. Instead, they pass through a general discussion, never revealing what is to be solved for.

Examples, also, do not always include diagrams/graphs, wherever appropriate, and students do not obtain the training to draw diagrams or graphs to simplify and organize their thinking.

(d) Students can learn the subject only by doing the exercises themselves and reviewing them in class, to obtain experience in applying the principles with their different ramifications.

In doing the exercises by themselves, students find that they are required to devote considerably more time to biology than to other subjects of comparable credits, because they are uncertain with regard to the selection and application of the theorems and principles involved. It is also often necessary for students to discover those "tricks" not revealed in their texts (or review books), that make it possible to solve problems easily. Students must usually resort to methods of trial-and-error to discover these "tricks", and as a result they find that they may sometimes spend several hours to solve a single problem.

(e) When reviewing the exercises in classrooms, instructors usually request students to take turns in writing solutions on the boards and explaining them to the class. Students often find it difficult to explain in a manner that holds the interest of the class, and enables the remaining students to follow the material written on the boards. The remaining students seated in the class are, furthermore, too occupied with copying the material from the boards, to listen to the oral explanations and concentrate on the methods of solution.

This book is intended to aid students in biology to overcome the difficulties described, by supplying detailed illustrations of the solution methods which are usually not apparent to students. The solution methods are illustrated by problems selected from those that are most often assigned for class work and given on examinations. The problems are arranged in order of complexity to enable students to learn and understand a particular topic by reviewing the problems in sequence. The problems are illustrated with detailed step-by-step explanations, to save the students the large amount of time that is often needed to fill in the gaps that are usually found between steps of illustrations in textbooks or review/outline books.

The staff of REA considers biology a subject that is best learned by allowing students to view the methods of analysis and solution techniques themselves. This approach to learning the subject matter is similar to that practiced in various scientific laboratories, particularly in the medical fields.

In using this book, students may review and study the illustrated problems at their own pace; they are not limited to the time allowed for explaining problems on the board in class.

When students want to look up a particular type of problem and solution, they can readily locate it in the book by referring to the index which has been extensively prepared. It is also possible to locate a particular type of problem by glancing at just the material within the boxed portions. To facilitate rapid scanning of the problems, each problem has a heavy border around it. Furthermore, each problem is identified with a number immediately above the problem at the right-hand margin.

To obtain maximum benefit from the book, students should familiarize themselves with the section, "How To Use This Book," located in the front pages.

To meet the objectives of this book, staff members of REA have selected problems usually encountered in assignment and examinations, and have solved each problem meticulously to illustrate the steps which are usually difficult for students to comprehend.

The following persons also have contributed a great deal of support and much patient work to achieve the objectives of the book:
Claudia Chernov, David Eskreis, Carl Fuchs, Khaw Huathin, Mitchell Kahn, Steve Koevary, Joseph Krick, Debbie Levbarg, Robert Mossack, Vincent Patrone, Dianna Pintus, Margaret Polaneczky, Anthony Polizzi, Steven Schwartzberg, Lyle Walsh, and Alice Wong. Thanks are, furthermore, due to several contributors who devoted brief periods of time to this work.

Gratitude is also expressed to the many persons involved in the difficult task of typing the manuscript with its endless changes, and to the REA art staff who prepared the numerous detailed illustrations together with the layout and physical features of the book.

Finally, special thanks are due to Helen Kaufmann for her unique talents to render those difficult border-line decisions and constructive suggestions related to the design and organization of the book.

Max Fogiel, Ph.D.
Program Director

HOW TO USE THIS BOOK

This book can be an invaluable aid to biology students as a supplement to their textbooks. The book is divided into 31 chapters, each dealing with a separate topic. The subject matter is developed beginning with the molecular basis of life, the organization of cells and tissues, and cellular metabolism and extending through bacteria and viruses, plants, invertebrates, immunology, respiration, nutrition, homeostasis, locomotion, coordination, hormones, reproduction, and embryonic development. Extensive sections are also included on genetics, evolution, ecology, and animal behavior.

Each chapter of the book is accompanied by a series of short-answer questions to help in reviewing the study material and as an added tool for exam preparation.

HOW TO LEARN AND UNDERSTAND A TOPIC THOROUGHLY

1. Refer to your class text and read the section pertaining to the topic. You should become acquainted with the principles discussed there. These principles, however, may not be clear to you at the time.

2. Locate the topic you are looking for by referring to the Table of Contents in the front of this book.

3. Turn to the page where the topic begins and review the problems under each topic, in the order given. For each topic, the problems are arranged in order of complexity, from the simplest to the more difficult. Some problems may appear similar to others, but each problem has been selected to illustrate a different point or solution method.

To learn and understand a topic thoroughly and retain its contents, it will generally be necessary for students to review the problems several times. Repeated review is essential in order to gain experience in recognizing the principles that should be applied, and to select the best solution technique.

HOW TO FIND A PARTICULAR PROBLEM

To locate one or more problems related to particular subject matter, refer to the index. In using the index, be certain to note that the numbers given there refer to problem numbers, not to page numbers. This arrangement of the index is intended to facilitate finding a problem more rapidly, since two or more problems may appear on a page.

If a particular type of problem cannot be found readily, it is recommended that the student refer to the Table of Contents, and then turn to the chapter which is applicable to the problem being sought. By scanning or glancing at the material that is boxed, it will generally be possible to find problems related to the one being sought, without consuming considerable time. After the problems have been located, the solutions can be reviewed and studied in detail.

For the purpose of locating problems rapidly, students should acquaint themselves with the organization of the book as found in the Table of Contents.

In preparing for an exam, it is useful to find the topics to be covered in the exam from the Table of Contents, and then review the problems under those topics several times. This should equip the student with what might be needed for the exam.

CONTENTS

CHAPTER 1

THE MOLECULAR BASIS OF LIFE

UNITS AND MICROSCOPY

• PROBLEM 1-1

A biologist deals with things on a microscopic level. To describe cellular dimensions and the amount of materials present at the cellular level, units of an appropriately small size are needed. What are these units of measurements?

Solution: In biology, the units of length commonly employed include the micron (abbreviated by μ) and the Ångstrom (abbreviated by Å). A micron is equivalent to 10^{-3} millimeters (mm). An Ångstrom is equivalent to 10^{-4} μ or 10^{-7} mm.

Weights are expressed in milligrams (10^{-3} grams) micrograms (10^{-6} grams), and nanograms (10^{-9} grams). The unit of molecular weight employed is the dalton. A dalton is defined as the weight of a hydrogen atom. For example, one molecule of water (H_2O) weighs about 18 daltons . One dalton weighs 1.674×10^{-24} grams.

• PROBLEM 1-2

Why are the elements carbon, hydrogen, oxygen, and nitrogen (C, H, O, N) of immense importance in living matter?

Solution: Carbon, hydrogen, oxygen and nitrogen are the four most abundant elements in living organisms. In fact, they make up about 99 percent of the mass of most cells. These four elements must possess some unique molecular ability which caused them to be selected to be the major constituents of life. This can be seen by comparing the relative abundance of the major chemical elements in the earth's crust to those in the human body.

Examining the table shows that the only close correspondence is for oxygen. Thus, we can assume that there is some reason why these 4 elements have been selected for their role in life.

Earth's crust		human body	
element	%	element	%
O	47.0	H	63.0
Si	28.0	O	25.5
Al	7.9	C	9.5
Fe	4.5	N	1.4
Ca	3.5	Ca	0.31
Na	2.5	P	0.22
K	2.5	Cl	0.08
Mg	2.2	K	0.06
Ti	0.46	S	0.05
H	0.22	Na	0.03
C	0.19	Mg	0.01

Relative abundance of the major chemical elements in earth's crust and in the human body as the percent of total number of atoms.

One property that makes these elements special is that they can readily form covalent bonds by electron pair sharing. To complete their outer shell, hydrogen needs one electron, oxygen needs two, nitrogen needs three and carbon needs four. Thus, these four elements react with themselves and each other to form a large number of covalent compounds. Furthermore, carbon, oxygen, and nitrogen can also form double bonds by sharing two electron pairs. Even further, carbon can form triple bonds. This gives these elements a lot of versatility in forming chemical bonds.

Another property that makes carbon, oxygen, hydrogen and nitrogen uniquely fit for their role in living matter is that they are the lightest elements that can form covalent bonds. The strength of a covalent bond is inversely related to the atomic weights of the bonded atoms. Thus these four elements are capable of forming very strong covalent bonds.

• PROBLEM 1-3

By photographic and optical means, an image can be enlarged almost indefinitely. However, most light microscopes offer magnifications of only 1000 to 1500 times the actual size, while electron microscopes offer magnifications of 100,000 times or more. Explain.

Solution: All microscopes are characterized by limits of resolution. Resolution refers to the clarity of the image. Objects lying close to one another can not be distinguished (resolved) as separate objects if the distance between them is less than one half the wavelength of the light being used. The average wave-length of visible light is 550 nanometers (or 5500 A). Thus, for light microscopes, objects can be distinguished only if they lie farther apart than about 275 nanometers.

Objects closer together than 275 nm are not resolved and appear to be one object. Increasing the size of the image, or the magnification, will not give meaningful information unless resolution is also increased. Increasing magnification without increasing resolution results in a larger image that is still blurred.

Electron microscopes offer resolution of details separated by .1 to .5 nanometers. Electrons, rather than light, are the radiation used in electron microscopes. Recall that electrons have a wave property in addition to a particle property and may be regarded as a radiation of extremely short wavelength. Since the wavelength of an electron in motion is so much shorter than the wavelength of light, resolution is more than a thousandfold better. Structures such as the plasma membrane, endoplasmic reticulum, ribosomes, microtubules and microfilaments were not visible until the advent of the electron microscope. These structures are all less than 275 nm in width. The plasma membrane has a thickness of 7.5 to 10 nm (or 75 to 100 Å). The ribosome is 15 to 25 nm in diameter (or 150 to 250 Å). Microtubules are 20 to 30 nm in diameter, and microfilaments range from 5 to 10 nm. Electron microscopy has also made possible the visualization of the nuclear envelope and the internal membranes of mitochondria and chloroplasts.

• PROBLEM 1-4

Distinguish between the terms "in situ," "in vitro," and "in vivo."

Solution: These terms are all used to refer to where a biochemical reaction or process takes place. "In situ" is Latin for "in place"; it refers to a reaction that occurs in its natural or original position. "In vitro" is Latin for "in glass"; it refers to a reaction that does not occur in the living organism, but instead occurs in a laboratory such as a test tube. "In vivo" is Latin for "in the living" and refers to a reaction that occurs in the living organism.

PROPERTIES OF CHEMICAL REACTIONS

• PROBLEM 1-5

Define the following terms: atom, isotope, ion. Could a single particle of matter be all three simultaneously?

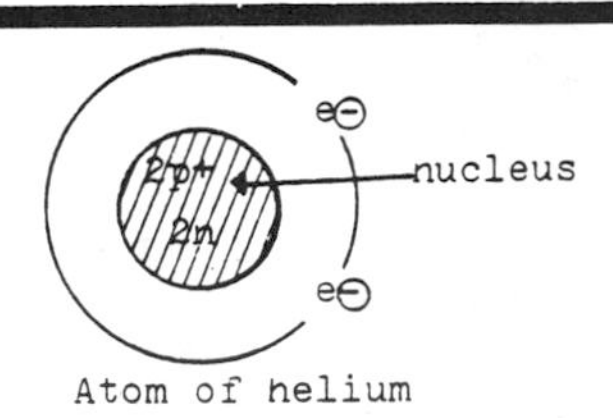

Atom of helium

Solution: An atom is the smallest particle of an element that can retain the chemical properties of that element. It is composed of a nucleus, which contains positively charged protons and neutral neutrons, around which negatively charged electrons revolve in orbits. For example, a helium atom contains 2 protons, 2 neutrons, and 2 electrons.

An ion is a positively or negatively charged atom or group of atoms. An ion which is negatively charged is called an anion, and a positively charged ion is called a cation.

Isotopes are alternate forms of the same chemical element. A chemical element is defined in terms of its atomic number, which is the number of protons in its nucleus. Isotopes of an element have the same number of protons as that element, but a different number of neutrons. Since atomic mass is determined by the number of protons plus neutrons, isotopes of the same element have varying atomic masses. For example, deuterium (^2H) is an isotope of hydrogen, and has one neutron and one proton in its nucleus. Hydrogen has only a proton and no neutrons in its nucleus.

A single particle can be an atom, an ion, and an isotope simultaneously. The simplest example is the hydrogen ion H^+. It is an atom which has lost one electron and thus developed a positive charge. Since it is charged, it is therefore an ion. A cation is a positively charged ion (i.e. H+) and an anion is a negatively charged ion (i.e. Cl^-). If one compares its atomic number (1) with that of deuterium (1), it is seen that although they have different atomic masses, since their atomic numbers are the same, they must be isotopes of one another.

• **PROBLEM 1-6**

Describe the differences between an element and a compound.

Solution: All substances are composed of matter in that they have mass and occupy space. Elements and compounds constitute two general classes of matter. Elements are substances that consist of identical atoms (i.e., atoms with identical atomic numbers). This definition of an element includes all isotopes of that element. Hence O^{18} and O^{16} are both considered to be elemental oxygen. A compound is a substance that is composed of two or more different kinds of atoms (two or more different elements) combined in a definite weight ratio. This fixed composition of various elements, according to law of definite proportions, differentiates a compound from a mixture. Elements are the substituents of compounds. For example, water is a compound composed of the two elements hydrogen and oxygen in the ratio 2:1, respectively. This compound may be written as H_2O, which is the molecular formula of water. The subscript "2" that appears after the hydrogen (H) indicates that in every molecule of water there are two hydrogen atoms. There is no subscript after the oxygen (O) in the molecular formula of water, which indicates that there is only one oxygen atom per molecule of water. Hence water is a compound whose molecules are each made up of two hydrogen atoms and one oxygen atom.

• PROBLEM 1-7

What are the three laws of thermodynamics? Discuss their biological significance.

Solution: The first law of thermodynamics states that energy can be converted from one form into another, but it cannot be created or destroyed. In other words, the energy of a closed system is constant. Thus, the first law is simply a statement of the law of conservation of energy.

The second law of thermodynamics states that the total entropy (a measure of the disorder or randomness of a system) of the universe is increasing. This is characterized by a decrease in the free energy, which is the energy available to do work. Thus, any spontaneous change that occurs (chemical, physical, or biological) will tend to increase the entropy of the universe.

The third law of thermodynamics refers to a completely ordered system, particularly, a perfect crystal. It states that a perfect crystal at absolute zero (0 Kelvin) would have perfect order, and therefore its entropy would be zero.

These three laws affect the biological as well as the chemical and physical world. Living cells do their work by using the energy stored in chemical bonds. The first law of thermodynamics states that every chemical bond in a given molecule contains an amount of energy equal to the amount that was necessary to link the atoms together. Thus, living cells are both transducers that turn other forms of energy into chemical bond energy and liberators that free this energy by utilizing the chemical bond energy to do work. Considering that a living organism is a storehouse of potential chemical energy due to the many millions of atoms bonded together in each cell, it might appear that the same energy could be passed continuously from organism to organism with no required extracellular energy source. However, deeper consideration shows this to be false. The second law of thermodynamics tells us that every energy transformation results in a reduction in the usable or free energy of the system. Consequently, there is a steady increase in the amount of energy that is unavailable to do work (an increase in entropy). In addition, energy is constantly being passed from living organisms to nonliving matter (e.g., when you write you expend energy to move the pencil, when it is cold out your body loses heat to warm the air, etc.). The system of living organisms thus cannot be a static energy system, and must be replenished by energy derived from the non-living world.

The second law of thermodynamics is also helpful in explaining the loss of energy from the system at each successive trophic level in a food pyramid. In the following food pyramid,

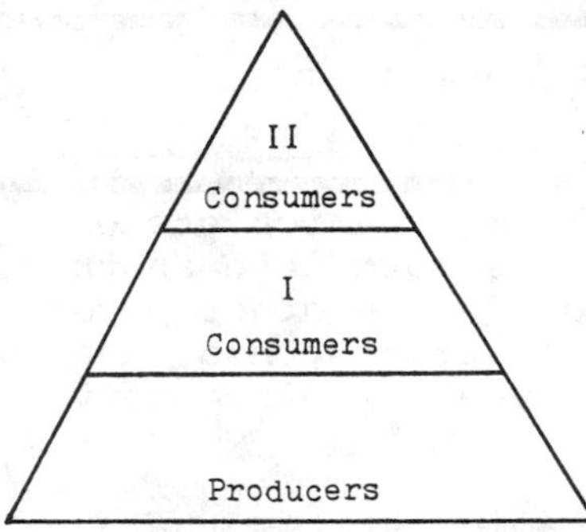

The energy at the producer level is greater than the energy at the consumer I level which is greater than the energy of the consumer II level. Every energy transformation between the members of the successive levels involves the loss of usable energy, and this loss serves to increase the entropy. Thus, this unavoidable loss causes the total amount of energy at each trophic level to be lower than at the preceding level.

• PROBLEM 1-8

What is a chemical reaction? What is meant by the rate-limiting step of a chemical reaction?

Solution: A necessary requirement of a chemical reaction is that the products be chemically different from the reactants. Therefore a chemical reaction is one in which bonds are broken and/or formed. Some chemical reactions have only a single step, with only one bond being formed or broken. In multistep reactions, a certain sequence of bond formation and bond breakages proceed. Usually the formation of a chemical bond releases energy, and the products formed are more stable than the reactants, in that they have a lower potential energy. The breakage of a bond requires some form of energy which is absorbed by the bond, giving products that, together, will exist at a higher potential energy. Thus, a reaction that requires heat to proceed and usually involves breaking bonds is termed an endothermic reaction. A reaction that proceeds with the release of heat and usually involves forming bonds is termed an exothermic reaction. Aside from bond changes, there is a change in potential energy or enthalpy and there can be a change in entropy, or orderliness that accompanies the reaction.

There is a tendency for chemical systems to go toward minimum energy or enthalpy and thus achieve maximum stability. There is also a tendency toward maximum disorder or entropy. These two concepts can be combined in a new function called free energy (G), which is the energy that is available for doing work.

$$G = \Delta H - T\Delta S \text{ , where}$$

ΔH represents the change in enthalpy
ΔS represents the change in entropy
T is the temperature in Kelvin
ΔG is the spontaneous change in free energy

In order for a chemical reaction to proceed, ΔG must be negative. A loss in enthalpy will be represented by a negative ΔH and a gain in entropy by a positive ΔS. If ΔG is positive, energy would have to be added to the system to drive the reaction forward.

Most chemical reactions occurring in the living cell are not simple, one-step reactions. Many contain a series of consecutive steps coupled, or linked, by common intermediates with the product of the first step being the reactant in the second step, and the product of the second step is the reactant in the third, and so on. The ΔG of such reactions is the sum of the ΔG's of the individual steps. Each step of the reaction has its own activation energy, and the one requiring the highest activation complex is kinetically the most unfavorable one. In other words, it will have the slowest rate and will for this reason determine the rate of the entire reaction. This step is better known as the rate-determining (or limiting) step of the reaction.

• PROBLEM 1-9

What is meant by the equilibrium of a chemical reaction? What factors influence the position of the equilibrium?

Solution: Chemical equilibrium is the state in which two opposing reactions (called forward and reverse) are proceeding at the same rate. Not all chemical reactions are reversible. Only if the products of a chemical reaction (the forward one) can recombine to form the original substances will it be possible to have the two opposing reactions necessary to reach a state of equilibrium. This state is reached when the rate of the forward reaction equals the rate of the reverse reaction. For example, the union of hydrogen and nitrogen to form ammonia (equation a) and the reverse or opposing reaction, which is the decomposition of ammonia to nitrogen and hydrogen (equation b) can be rewritten as a reaction at chemical equilibrium (equation c):

(a) $3H_2 + N_2 \xrightarrow{R_a} 2NH_3$ where R_a = rate of reaction a

(b) $2NH_3 \xrightarrow{R_b} 3H_2 + N_2$ where R_b = rate of reaction b

(c) $3H + N_2 \underset{R_a}{\overset{R_b}{\rightleftarrows}} 2NH_3$ where $R_a = R_b$.

It should be noted that the concentrations of reactants and products in reaction (c) or in any reaction at equilibrium may vary greatly. The amounts of substances present at equilibrium is dependent on the nature of the reactants, and on the pressure and the temperature at which equilibrium is established. The presence or absence of a catalyst has no influence on the relative amounts of products or reactants present at equilibrium.

The nature of the reactants affects the equilibrium concentrations. This is due to the difference in strengths of bonding forces. Consider the general equations:

(d) $A_2 + B_2 \rightarrow 2AB$

(e) $2AB \rightarrow A_2 + B_2$

If the energy needed to bond A_2 and B_2 in reaction (d) is greater that the energy needed to break the A-B bond in reaction (e), then reaction (e) has a greater tendency to occur than reaction (d). In this case, the equilibrium would favor one side:

$$A_2 + B_2 \rightleftharpoons 2AB$$

The effect of pressure on an equilibrium mixture is summed up best by Le Chatelier's principle which states that if the pressure on a system at equilibrium is changed, there is a tendency to diminish the change in pressure. The reaction that produces the fewer number of molecules and tends to lower the pressure is the reaction favored by the application of high pressure to the system. Thus if the equilibrium mixture:

$$N_2 + 3H_2 \rightleftarrows 2NH_3$$

(1 molecule + 3 molecules)
= 4 molecules 2 molecules

is established at a relatively low pressure, the equilibrium mixture contains a smaller proportion of ammonia and a larger proportion of nitrogen and hydrogen than it would at a high pressure. It should be realized that this only applies to a gaseous system. Increasing the pressure of a liquid or a solid would not appreciably increase the concentrations, and thus would have little effect on the equilibrium concentrations. A change in pressure has no effect on a gaseous mixture at equilibrium if neither the forward nor the reverse reaction diminishes the change in pressure; for example:

$$H_2 + Cl_2 \rightleftarrows 2HCl$$

(1 molecule + 1 molecule)
= 2 molecules 2 molecules

The influence of temperature is similar to that of pressure. A rise in temperature increases the rate of all reactions. Thus, both the rates of the forward and the backward reactions would be increased. But, one will probably be increased more than the other. To determine which reaction's rate will be increased and which side will be favored in equilibrium we can use a modified version of Le Chatelier's principle using temperature instead of pressure as the stress. If the temperature is increased, an endothermic reaction will be favored over an exothermic reaction because it will take in heat and thus, lessen the amount of heat present. An exothermic reaction would be favored by a decrease in temperature, because this reaction liberates heat to the surroundings and does so more readily if the surroundings are at a relatively low temperature.

MOLECULAR BONDS AND FORCES

• PROBLEM 1-10

Electrons are located at definite energy levels in orbitals around the nucleus of an atom. Sometimes electrons move from one orbital to another. What happens to cause such a change?

Solution: When an electron is in its ground state in the atom, it is occupying the orbital of lowest energy for that electron. There may be other electrons in the atom that are located in orbitals of even lower energy. The further the electron is from the nucleus, the greater the energy associated with that electron. An electron in the ground state may absorb energy from an external source (e.g. heat) , and be promoted to a higher-energy orbital. The electron would now be considered to be in an excited state. In a similar fashion, an excited electron may drop down to a lower-energy orbital (not necessarily the ground state), and in the process will decrease its energy content. This energy difference will be released in the form of a photon of light.

• PROBLEM 1-11

How is the bonding capacity of an atom related to its electron configuration? Illustrate with an example.

Solution: The electrons of an atom are distributed so that they occupy concentric atomic orbitals with each orbital containing two electrons. With increasing distance from the nucleus, there is a corresponding increase in the energy level of the orbital. Each atomic orbital has a characteristic energy level, hence the orbitals are said to be quantized. The quantum theory describes the characteristics of electrons in an atom by assigning four quantum numbers to each electron.

The principal quantum number, n, denotes the main energy level or shell in which the electron resides. The main energy levels of an atom are assigned integral values, starting with 1 as the shell closest to the nucleus, and continually increasing as the main energy levels become more distant from the nucleus. Hence, $n = 1, 2, 3, \ldots$. The angular momentum quantum number, ℓ, determines the shape of the volume in which the electron is most likely to be found. The shape of this volume can be illustrated with electron cloud diagrams, and depends upon the magnetic and electrostatic interactions of the electron with the nucleus and other electrons in neighboring orbitals. These interactions will be different for varying distances of the orbital to the nucleus. Consequently, the angular momentum of the electron depends upon the main energy level in which it resides. It has been determined that the quantum numbers ℓ and n are related as follows: $\ell = n - 1$. Therefore, $\ell = 0, 1, 2, 3, \ldots$, depending on n.

A moving charge generates a magnetic field. Since an electron is an elementary particle with a negative charge, and it moves within some volume described by ℓ, it will propagate a magnetic field.This is accounted for in quantum theory by the magnetic quantum number m, which depends upon ℓ in the following way: $m = -\ell, 0, \ell$. I.e., when $\ell = 0$, $m = 0$; when $\ell = 1$, $m = -1, 0, 1$; when $\ell = 2$, $m = -2, -1, 0, 1, 2$. One final characteristic of an electron is its rotational motion. An electron spins on its own axis. There are only two possible directions (clockwise and counterclockwise) that an electron can spin about its axis. The spin quantum number, s, accounts for this phenomenon, and the arbitrary values $+\frac{1}{2}$ and $-\frac{1}{2}$ are used to describe the spin of an electron. It is important to realize that no two paired electrons in the same atomic orbital can have the same spin quantum numbers. This would be a violation of the Pauli Exclusion Principle, which states that no two paired electrons in an atom may have identical sets of quantum numbers. Electrons in the same orbital have the same principal, angular momentum, and magnetic quantum numbers. Therefore their spin quantum number must differ to be in accordance with the Pauli Exclusion Principle.

The various atomic orbitals are labeled s, p, d and f. When $\ell = 0$, the orbitals are of the s variety; when $\ell = 1$ they are p orbitals; when $\ell = 2$ they are d orbitals; and when $\ell = 3$ they are f orbitals. The number of values of m will determine how many orbitals of each kind will be contained in each main energy level. For $\ell = 0$, $m = 0$ and this means that there is only one s orbital contained in each main energy level. For $\ell = 1$, $m = -1, 0, 1$ and hence there are three p orbitals contained in the main energy levels starting with $n = 2$. It is similarly found that there are five d orbitals and seven f orbitals in the main energy orbitals starting with $n = 3$ and $n = 4$, respectively.

With this basic knowledge of quantum theory, we can now assign an electron configuration to an atom. Electron configurations show how the electrons within an atom are distributed. From the electron configuration we can deduce the valence of the atom, and hence determine its bonding capacity. An important note should be made here: in determining the electron configuration of an atom, one uses the Aufbau process of filling atomic orbitals. By this process, the electrons fill the orbitals in order of increasing energy, and when there is more than one orbital in an energy sublevel (e.g. p orbitals - there are three of them in a sublevel), the electrons will fill the orbitals so that each one will have one electron before any orbital receives a pair of electrons. The relative energies of the orbitals in a main shell is given as $s < p < d < f$.

We will now assign an electron configuration for oxygen (atomic number = 8) and deduce its bonding capacity from its valence. There are eight electrons in the oxygen atom. In the first main energy level, $n = 1$, $\ell = 0$ and $m = 0$. There is just one s orbital in this level. Since an orbital can accommodate two electrons, the first main energy level will hold two electrons, leaving us with six electrons to account for. In the second main energy level, $n = 2$, $\ell = 0, 1$ and $m = -1, 0, 1$. There is one s orbital ($\ell = 0$) and three p orbitals ($\ell = 1$) in this level. Since s orbitals are lower in energy than p orbitals, they will be filled first, by two electrons. There are four electrons remaining to be placed in orbitals. Because there are three p orbitals in this level, the electrons will fill them so that, each orbital has one electron before any orbital has a complete pair. Hence the four electrons will occupy the three p orbitals so that one orbital will have a pair of electrons and the other two orbitals will each have one electron. The electron configuration of oxygen can be represented as:

$$\begin{array}{l} \qquad\qquad\qquad\qquad \underline{\uparrow\downarrow}_{2p_x}\ \underline{\uparrow}_{2p_y}\ \underline{\uparrow}_{2p_z} \\ \text{energy}\uparrow \qquad \underline{\uparrow\downarrow}_{2s} \\ \qquad \underline{\uparrow\downarrow}_{1s} \end{array}$$

Above: The numbers represent the principal quantum number n. Electrons with opposite spins are depicted as arrows ($\uparrow$ and $\downarrow$). The three p orbitals are labeled p_x, p_y, and p_z for differentiation.

The valence of an atom is the number of electrons that can be accommodated via bonding by the atom. The Octet Rule states that all atoms that can obtain an octet of electrons in the valence shell (outermost shell), will try to do so. By looking at the electron configuration of oxygen, we can see that its valence (the number of electrons needed to fill the valence shell)is two. Oxygen complies with the Octet Rule in that it seeks to obtain two electrons to fill up its second main energy level. Hence the bonding capacity (or the unpaired electrons available for bonding) of oxygen is two.

Let's now look at the electron configuration of carbon (atomic number = 6) which is one of the most biologically important elements. There are six electrons in a carbon atom. As in oxygen, two electrons will occupy the lone s orbital in the first main energy level. This leaves us with four electrons to account for it. If these remaining electrons filled the orbitals in the second main energy level as was true for oxygen, we would get the electron configuration for carbon:

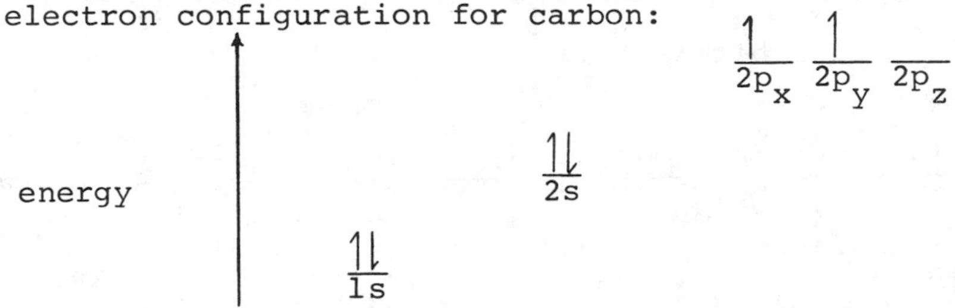

In this case the valence of carbon is four, and the bonding capacity is two. This is because there are two unpaired electrons in the 2p orbitals. It should be noted, however, that this representation of carbon's electron configuration does not comply with the Octet Rule. In this case, a fully bonded carbon atom would contain only six electrons in its valence shell, and has one empty orbital ($2p_z$). This is an energetically unstable condition. If it is at all possible for carbon to complete its valence shell, it will do so. It so happens that carbon does comply with the Octet Rule, and has all its orbitals in the first two main energy levels filled to capacity when it bonds with other atoms. This is facilitated by the hybridization of atomic orbitals. The electrons in the second main energy level occupy an orbital that is intermediate in energy between the 2s and 2p orbitals. This orbital is called an sp^3 hybrid orbital because it is an orbital formed by one s and three p orbitals. There are four sp^3 orbitals, each containing an unpaired electron. The electron configuration of carbon is shown as:

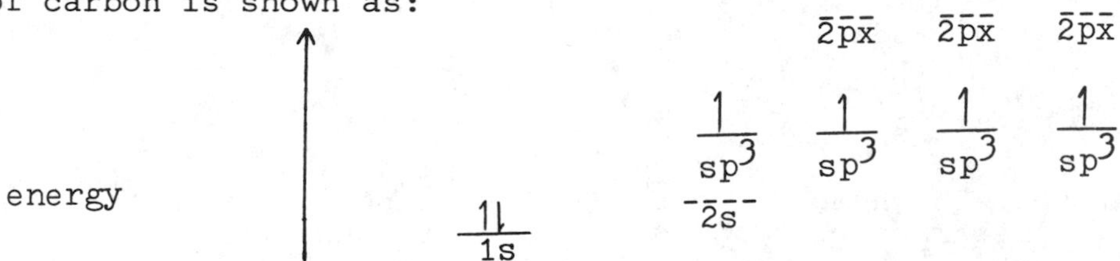

Now it can be seen that carbon has a valence of four, and a bonding capacity of four. In this case the Octet Rule is satisfied.

• PROBLEM 1-12

All organic matter is made up entirely or mostly of the basic elements carbon and hydrogen. In view of this, why is there such a diversity of organic compounds present?

Solution: The diversity of organic compounds is so vast that these organic compounds have been divided into families, such as alkanes, alkynes and aromatic compounds, which have no counterparts among inorganic compounds. The tremendous variety of these compounds is made possible by the unique properties of carbon. To be able to understand how carbon can form such a huge number of compounds with a great variety of properties, the way in which the atoms in these molecules are bonded together must be looked at.

Carbon has a valence number (the number of bonds an atom of an element can form) of four. This means that each carbon atom always has four bonds which can be either bonded to four other atoms, as in methane; to three other atoms, as in formaldehyde; or to 2 other atoms as in hydrogen cyanide. In other words carbon is capable of forming single, double and triple bonds:

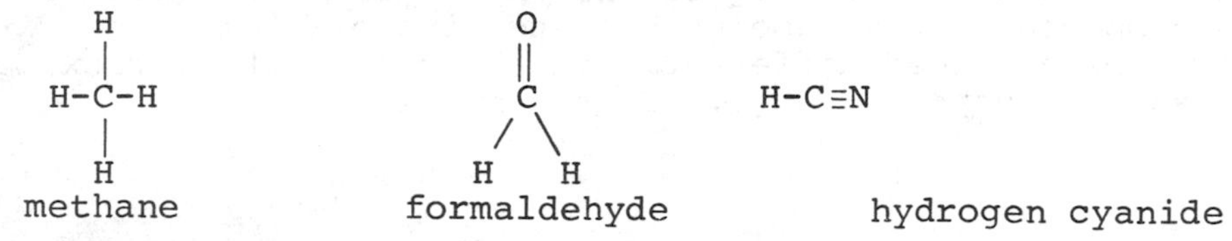

But the diversity this element possesses does not stop here. It is capable of bonding to other carbon atoms in an almost unique variety of chain and ring structures. This property is called catenation and accounts for the tremendous variety seen in the following compounds:

```
                                                   H
                                                   |
                                              H    C    H
                                               \  // \  /
  H H        H       H                           C     C
  | |         \     /                            |     ||
H-C-C-H        C=C          H-C≡C-H              C     C
  | |         /     \                           / \\  / \
  H H        H       H                         H    C    H
                                                    |
                                                    H
ethane      ethylene       acetylene              benzene
```

These examples show the use of a minimum number of carbon atoms. The diversity possible is apparent when one considers the fact that these different types of bond can be joined together in infinite numbers. To exemplify this, look at the compound:

```
                    H
                    |
                  H-C-H
                    |
              H     |    H
              |     |    |
            H-C-----C=C-H
              |
          H   C
           \ / \\
            C   C-H
            ||  |
     H      C   C-H
     |     / \ /
   H-C-C=C    C
     | | |    |
     H H H    Cl
```

As it can be seen, a multitude of different variations could be made with the only constraining factor being that carbon must have four bonds. The freedom allowed carbon is apparent when contrasted with the halogens (iodine, chlorine, bromine, and flourine) which are monovalent and thus can only form one bond, for example:

```
        H-Cl                   Br-Br

  hydrochloric acid          bromine
```

Another feature of the compounds of carbon that contributes to their variety is the existence of isomers (compounds composed of the same number and kind of atoms, but with the atoms arranged differently in space). Look, for example, at the compound C_4H_{10}. This can exist as:

```
  H H H H                H H H
  | | | |                | | |
H-C-C-C-C-H      OR    H-C-C-C-H
  | | | |                | | |
  H H H H                H | H
                           |
n-butane                 H-C-H
                           |
                           H   2-methyl propane
```

or look at C_5H_{10}. It can exist as:

```
               H                    H     H
               |                     \   /
             H C-H                    C=C   H H
             |/|                     /   \  | |
    H        C |                 H-C-H    C-C-H
     \      /| |                   |      | |
     H C=C   H H       or          H      H H
     |/   \
   H-C     H
     |
     H

  trans 2-pentene                 cis 2-pentene
```

Although these compounds have the same chemical formulas they have different chemical, physical and biological properties due to their arrangement in space.

• PROBLEM 1-13

Distinguish between covalent and ionic bonds.

Solution: A covalent bond is a bond in which two atoms are held together by a shared pair of electrons. An ionic bond is a bond in which oppositely charged ions are held together by electrical attraction.

In general, the electronegativity difference between two elements influences the character of their bond (see Table 1).

Table 1: Electronegativities of main groups of elements

IA	II A	III A	IV A	VA	VI A	VII A
H 2.1						
Li 1.0	Be 1.5	B 2.0	C 2.5	N 3.0	O 3.5	F 4.0
Na 0.9	Mg 1.2	Al 1.5	Si 1.8	P 2.1	S 2.5	Cl 3.0
K 0.8	Ca 1.0		Ge 1.8	As 2.0	Se 2.4	Br 2.8
Rb 0.8	Sr 1.0		Sn 1.8	Sb 1.9	Te 2.1	I 2.5
Cs 0.7	Ba 0.9		Pb 1.7	Bi 1.8		

Electronegativity measures the relative ability of an atom to attract electrons in a covalent bond. Using Pauling's scale, where fluorine is arbitrarily given the value 4.0 units and other elements are assigned values relative to it, an electronegativity difference of 1.7 gives the bond 50 percent ionic character and 50 percent covalent character. Therefore, a bond between two atoms with an electronegativity difference of greater than 1.7 units is mostly ionic in character. If the difference is less than 1.7 the bond is predominately covalent.

• PROBLEM 1-14

What are hydrogen bonds? Describe fully the importance of hydrogen bonds in the biological world.

Solution: A hydrogen bond is a molecular force in which a hydrogen atom is shared between two atoms. Hydrogen bonds occur as a result of the uneven distribution of electrons in a polar bond, such as an O-H bond. Here, the bonding electrons are more attracted to the highly electronegative oxygen atom, resulting in a slight positive charge (δ^+) on the hydrogen and a slight negative charge (δ^-) on the oxygen. A hydrogen bond is formed when the relatively positive hydrogen is attracted to a relatively negative atom of some other polarized bond. For example:

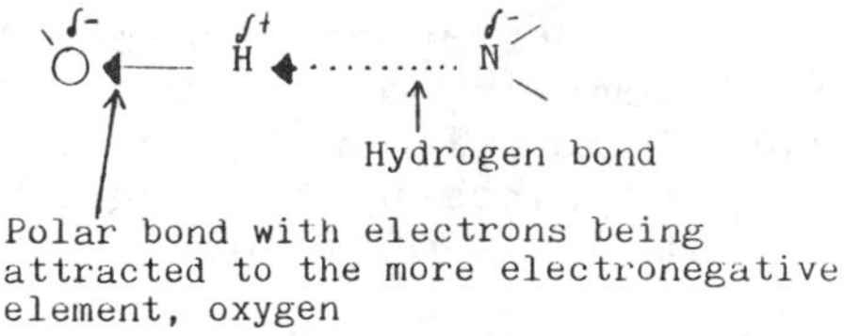

$$\overset{\delta-}{O} \leftarrow \overset{\delta+}{H} \leftarrow \cdots\cdots \overset{\delta-}{N}$$

Hydrogen bond

Polar bond with electrons being attracted to the more electronegative element, oxygen

The atom to which the hydrogen is more tightly linked or specifically the atom with which it forms the polar bond, is called the hydrogen donor, while the other atom is the hydrogen acceptor. In this sense, the hydrogen bond can be thought of as an intermediate type of acid-base reaction. Note, however, that the bond is an electrostatic one - no electrons are shared or exchanged, between the hydrogen and the negative dipole of the other molecule of the bond.

Hydrogen bonds are highly directional (note the arrows in the figure), and are strongest when all three atoms are colinear.

Bond energies of hydrogen bonds are in the range of about 3 to 7 kcal/mole. This is intermediate between the energy of a covalent bond and a van der Waals bond. However, only when the electronegative atoms are either F, O, or N, is the energy of the bond enough to make it important. Only these three atoms are electronegative enough for the necessary attraction to exist.

Hydrogen bonds are responsible for the structure of water and its special properties as a biological solvent. There is extensive hydrogen bonding between water molecules, forming what has been called the water matrix. This structure has profound effects on the freezing and boiling points of water, and its solubility properties. Any molecule capable of forming a hydrogen bond can do so with water, and thus a variety of molecules will dissociate and be soluble in water.

Hydrogen bonds are also most responsible for the maintenance of the shape of proteins. Since shape is crucial to their function (as both enzymes and structural components), this bonding is extremely important. For example, hydrogen bonds maintain the helical shape of keratin and collagen molecules and gives them their characteristic strength and flexibility.

It is hydrogen bonds which hold together the two helices of DNA. Bonding occurs between the base pairs. The intermediate bond strength of the hydrogen bond is ideal for the function of DNA - it is strong enough to give the molecule stability, yet weak enough to be broken with sufficient ease for replication and RNA synthesis.

• PROBLEM 1-15

What properties of water make it an essential component of living matter?

Solution: The chemistry of life on this planet is essentially the chemistry of water and carbon. Water is the most abundant molecule in the cell as well as on the earth. In fact, it makes up between 70 and 90 percent of the weight of most forms of life.

Life began in the sea and the properties of water shape the chemistry of all living organisms. Life developed as a liquid phase phenomenon because reactions in solution are much more rapid than reactions between solids, and complex and highly structural molecules can behave in solution in a way that they cannot behave in a gas. Water is an excellent solvent for living systems. It can stay in the liquid stage throughout a very wide range of temperature variation. Almost all chemicals present in living matter are soluble in it.

Water serves many functions in the living organism. It dissolves waste products of metabolism and assists in their removal from the cell and the organism. Water functions in heat regulation almost as an insulator would. It has a high heat capacity or specific heat in that it has a great capacity for absorbing heat with only minimal changes in its own temperature. This is because water molecules bond to one another by hydrogen bonds. Excess heat energy is dissipated by breaking these bonds, thus the living material is protected against sudden thermal changes. In addition, plants and animals utilize water loss to cool their bodies. When water changes from a liquid to a gas, it absorbs a great deal of heat. This enables the body to dissipate excess heat by the evaporation of water. In animals this process is sweating. Also, the good conductivity properties of water makes it possible for heat to be distributed evenly throughout the body tissues. Water serves as a lubricant, and is present in body fluids wherever one organ rubs against another, and in the joints where one bone moves on another. Water serves in the transport of nutrients and other materials within the organism. In plants, minerals dissolved in water are taken up by the roots and are transported up the stem to the leaves.

Water is also very efficient in dissolving ionic salts and other polar compounds because of the polar physical properties of water molecules. The proper concentration of these salts is necessary for life processes and it is important to keep them at extremely constant concentrations under normal conditions. These salts are important in maintaining osmotic relationships.

• PROBLEM 1-16

What are van der Waals forces? What is their significance in biological systems?

Solution: Van der Waals forces are the weak attractive forces that molecules of non-polar compounds have for one another. These are the forces that allow non-polar compounds to liquefy and/or solidify. These forces are based on the existence of momentary dipoles within molecules of non-polar compounds. A dipole is the separation of opposite charges (positive and negative). A non-polar compound's average distribution of charge is symmetrical, so there is no net dipole. But, electrons are not static, they are constantly moving about. Thus, at any instant in time a small dipole will probably exist. This momentary dipole will affect the distribution of charge in nearby non-polar molecules, inducing charges in them. This induction happens because the negative end of the temporary dipole will repel electrons and the positive end attracts electrons. Thus, the neighboring non-polar molecules will have oppositely oriented dipoles:

+ − + − + −
− + − + − +
+ − + − + −

These momentary, induced dipoles are constantly changing, short range forces. But, their net result is attraction between molecules.

The attraction due to van der Waals forces steadily increases when two non-bonded atoms are brought closer together reaching its maximum when they are just touching. Every atom has a van der Waals radius. The atoms are said to be touching when the distance between their nuclei is equal to the sum of their van der Waals radii. If the two atoms are then forced closer together, van der Walls attraction is replaced by van der Waals repulsion (the repulsion of the positively charged nuclei). The atoms then try to restore the state in which the distance between their two nuclei equals the sum of their van der Waals radii.

Both attractive and repulsive van der Waals forces play important roles in many biological systems. It is these forces, acting between non-polar chains of phospholipids, which serve as the cement holding together the membranes of living cells.

• PROBLEM 1-17

What is an anhydro bond? Give examples of these bonds in common biological molecules.

Solution: In biology, the most important covalent bonds joining molecules together are anhydro bonds. These are formed by removing a molecule of water to join two other molecules together. In other words, an -OH group is removed from one molecule and an -H group from the other to leave the two molecules bonded to each other usually by an oxygen or nitrogen atom.

Examples of anhydro bonds in common biological molecules include: glycoside bonds, peptide bonds, and ester bonds (see figure). Glycoside bonds are the anhydro bonds of carbohydrates and are formed by removing an H from an alcohol group (-OH) on one sugar and an OH from the other sugar. Peptide bonds are the anhydro bonds of proteins and are formed by removing an -OH from the carboxyl group (-COOH) of one amino acid and an H from the amino group ($-NH_2$) of another amino acid. Ester bonds are anhydro bonds of fats, formed by removing an OH from the carboxyl group of a fatty acid and an H from an alcohol group of glycerol. Other ester bonds that have great biological significance are phosphate esters [formed by removing an OH from phosphoric acid ($HO-PO_3H_2$) and an H from a sugar] and thioesters (formed by removing an OH from the carboxyl group of an acid and an H from a -SH group).

Some examples of anhydro bonds

Glycoside bond

α- glucose + α- glucose → maltose + H_2O

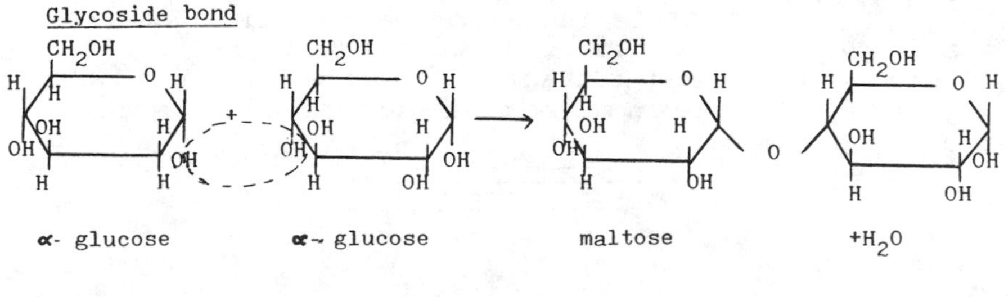

Peptide bond

$$CH_3-\underset{NH_2}{\overset{H}{C}}-\overset{O}{\overset{\|}{C}}-OH \quad H-\underset{H}{N}-\underset{H}{\overset{H}{C}}-\overset{O}{\overset{\|}{C}}OH \longrightarrow CH_3-\underset{NH_2}{\overset{H}{C}}-\overset{O}{\overset{\|}{C}}-\underset{H}{N}-CH_2\overset{O}{\overset{\|}{C}}OH + H_2O$$

Ester bond

$$CH_3-CH_2-\overset{O}{\overset{\|}{C}}-OH + HO-CH_3 \longrightarrow CH_3-CH_2-\overset{O}{\overset{\|}{C}}-O-CH_3 + H_2O$$

ACIDS AND BASES

• PROBLEM 1-18

Differentiate between acids, bases and salts. Give examples of each.

Solution: There are essentially 2 widely used definitions of acids and bases: the Lowry-Bronsted definition and the Lewis definition. In the Lowry-Bronsted definition, an acid is a compound with the capacity to donate a proton, and a base is a compound with the capacity to accept a proton. In the Lewis definition, an acid has the ability to accept an electron pair and a base the ability to donate an electron pair.

Salts are a group of chemical substances which generally consist of positive and negative ions arranged to maximize attractive forces and minimize repulsive forces. Salts can be either inorganic or organic. For example, sodium chloride, NaCl, is an inorganic salt which is actually best represented with its charges Na^+Cl^-; sodium acetate, CH_3COONa or $CH_3COO^-Na^+$ is an organic salt.

Some common acids important to the biological system are acetic acid (CH_3COOH), carbonic acid (H_2CO_3), phosphoric acid (H_3PO_4), and water. Amino acids, the building blocks of protein, are compounds that contain an acidic group (-COOH). Some common bases are ammonia (NH_3), pyridine (C_5H_5N), purine

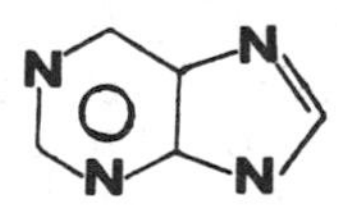

and water. The nitrogenous bases important in the structure of DNA and RNA carry the purine or pyridine functional group. Water has the ability to act both as an acid $\left[H_2O \xrightarrow{-H^+} OH^-\right]$ and as a base ($H_2O + H^+ \rightarrow H_3O^+$) depending on the conditions of the reaction, and is thus said to exhibit amphiprotic behavior.

• PROBLEM 1-19

What are acid-base reactions? In what way are they analogous to oxidation-reduction reactions?

Solution: An acid-base reaction involves proton transfer between two species: the proton donor and the proton acceptor. The proton donor is better known as a Bronsted acid and the proton acceptor as a Bronsted base. The general equation for an acid-base reaction can be written as:

$$\text{Proton donor} \longrightarrow H^+ + \text{proton acceptor}.$$

For example, the following reaction occurs between acetic acid (CH_3COOH) and water (H_2O):

$$CH_3COOH + H_2O \longrightarrow CH_3COO^- + H_3O^+$$

Acetic acid is the Bronsted acid because it donates a proton to become the acetate anion, and water is the Bronsted base since it accepts a proton to form the hydronium ion. The acetate anion now has the capacity to accept a proton, and therefore is a Bronsted base by definition. The acetic acid and the acetate anion together constitute a conjugate acid-base pair. The generalized formula of these types of reactions would be:

$$\text{Bronsted acid}_J + \text{Bronsted base}_K \longrightarrow \text{conjugate Bronsted base}_J + \text{conjugate Bronsted acid}_K$$

where the compounds written with the subscript, J are on conjugate acid-base pair and the compounds written with the subscript K are the other conjugate acid-base pair.

An oxidation-reduction reaction includes the transfer of electron(s), rather than proton(s), between two species. We can write the following general equation for oxidation-reduction (also known as redox) reactions, analogous to the one we wrote for acid-base reactions:

$$\text{Electron donor} \longrightarrow e^- + \text{electron acceptor}$$

The electron donor is the species that is undergoing oxidation (loss of electrons), and is also called the reducing agent (i.e., one that reduces another species). The electron acceptor is the species undergoing reduction (gain of electrons), and is also known as the oxidizing agent (i.e., one species that oxidizes another). The electron donor and acceptor together constitute a conjugate redox pair or couple. In some oxidation-reduction reactions, the transfer of one or more electrons is made via the transfer of hydrogen. Dehydrogenation is thus equivalent to oxidation. One example of a redox reaction is the one between NAD^+ and $NADH + H^+$:

$$NAD^+ + H_2 \underset{\text{oxidation}}{\overset{\text{reduction}}{\rightleftharpoons}} NADH + H^+$$

(NAD^+, nicotinamide adenine dinucleotide, will be discussed at length in a later chapter.)

As different acids and bases differ in their tendency respectively to dissociate and accept protons, different reducing and oxidizing agents differ in their tendency to lose and gain electrons or hydrogen, respectively. The acidity of a species is measured by its pH, while the electron-donating ability of a reducing agent is measured by its standard oxidation-reduction potential.

• PROBLEM 1-20

What does the "pH" of a solution mean? Why is a liquid with a pH of 5 ten times as acidic as a liquid with a pH of 6?

Solution: The pH (an abbreviation for "potential of hydrogen") of a solution is a measure of the hydrogen ion (H^+) concentration. Specifically, pH is defined as the negative log of the hydrogen ion concentration. A pH scale is used to quantify the relative acid or base strength. It is based upon the dissociation reaction of water: $H_2O \rightarrow H^+ + OH^-$. The dissociation constant (K) of this reaction is 1.0×10^{-14} and is defined as:

$$K = \frac{[H^+][OH^-]}{[H_2O]}$$ where $[H^+]$ and $[OH^-]$ are

the concentrations of hydrogen and hydroxide ions, respectively and $[H_2O]$ is the concentration of water (which is equal to one). The pH of water can be calculated from its dissociation constant k:

$$K = 1.0 \times 10^{-14} = \frac{[H^+][^-OH]}{[H_2O]} = [H^+][^-OH]$$

Since one H^+ and one ^-OH are formed for every dissociated H_2O molecule, $[H^+] = [^-OH]$.

$$1.0 \times 10^{-14} = [H^+]^2 \quad ; \quad [H^+] = 1.0 \times 10^{-7}$$

$$pH = -\log[H^+] = -\log(1.0 \times 10^{-7}) = 7$$

A pH of 7 is considered to be neutral since there are equal concentrations of hydrogen and hydroxide ions. The pH scale ranges from 0 to 14. Acidic compounds have a pH range of 0 to 7 and basic compounds have a range of 7 to 14.

• PROBLEM 1-21

Vinegar is classified as an acid. What does this tell you about the concentration of the hydrogen ions and the hydroxide ions?

Solution: Compounds can be classified as acidic

or basic depending upon the relative concentrations of hydrogen and hydroxide ions formed in aqueous solution. In an aqueous solution, acidic compounds have an excess of hydrogen ions, and basic compounds have an excess of hyroxide ions. Since vinegar is considered to be an acid, it will have an excess of hydrogen ions with respect to hydroxide ions when in solution. Thus, it can be deduced that the pH of an aqueous solution of vinegar will be within the range of 0 to 7.

Acetic acid is the constituent that gives vinegar its acidic properties. Acetic acid has a pH of 2.37 at a one molar concentration.

• PROBLEM 1-22

Calculate the hydrogen ion (H^+) concentrations of the following fluids: (a) blood plasma, (b) intracellular fluid of muscle, (c) gastric juice (pH - 1.4), (d) tomato juice, (e) grapefruit juice, (f) sea water. Use the accompanying table.

pH of some fluids

	pH
Seawater	7.0 - 7.5
Blood plasma	7.4
Interstitial fluid	7.4
Intracellular fluids:	
Muscle	6.1
Liver	6.9
Gastric juice	1.2 - 3.0
Pancreatic juice	7.8 - 8.0
Saliva	6.35-6.85
Cow's milk	6.6
Urine	5 - 8
Tomato juice	4.3
Grapefruit juice	3.2
Soft drink (cola)	2.8
Lemon juice	2.3

Solution: pH is defined as the logarithm ($\log_{10}$) of the reciprocal of the proton concentration, or log $\frac{1}{[H^+]}$. This expression can be rewritten a $-\log[H^+]$. Rearranging the equation $pH = -\log [H^+]$ gives us the means to solve for $[H^+]$: $[H^+] = 10^{-pH}$

(a) To find the $[H^+]$ of blood plasma:

The pH of blood plasma is given in the table to be 7.4. After inserting this into the equation, the $[H^+]$ is $10^{-7.4}$ or 3.98×10^{-8} Molar.

(b) $[H^+]$ in intracellular fluid of muscle:

The pH of intracellular fluid of muscle = 6.1

$\therefore\ [H^+] = 10^{-pH} = 10^{-6.1} = 7.94 \times 10^{-7}$ M.

(c) H^+ in gastric juice

pH of gastric juice = 1.4

$\therefore\ [H^+] = 10^{-1.4} = 3.98 \times 10^{-2}$ M.

(d) $[H^+]$ in tomato juice

pH of tomato juice = 4.3

$\therefore\ [H^+] = 10^{-4.3} = 5.01 \times 10^{-5}$ M.

(e) $[H^+]$ in grapefruit juice

pH of grapefruit juice = 3.2

$\therefore\ [H^+] = 10^{-3.2} = 6.31 \times 10^{-4}$ M.

(f) $[H^+]$ in seawater

pH of seawater = 7.0

$\therefore\ [H^+] = 10^{-7} = 1.0 \times 10^{-7}$ M.

• PROBLEM 1-23

Draw the general structure of an amino acid and discuss why it is both an acid and a base.

Solution: The general structure of an amino acid can be drawn as:

$$R - \underset{\underset{NH_3^{\oplus}}{|}}{\overset{\overset{H}{|}}{C}} - COO^-$$

where R represents a side group which can be changed to create any amino acid. Note that the molecule contains two polar groups (COO^- and NH_3^+) and for this reason, amino acids are called dipolar ions or zwitterions. It is these two polar groups which give amino acids the ability to function as both acids and bases.

Depending on the pH, amino acids can exist in one of three ionic forms:

$$\underset{\text{I}}{R\text{-}\underset{\underset{NH_3^+}{|}}{CH}\text{-}COOH} \underset{OH^-}{\overset{H^+}{\rightleftharpoons}} \underset{\text{II}}{R\text{-}\underset{\underset{NH_3^+}{|}}{CH}\text{-}COO^-} \underset{OH^-}{\overset{H^+}{\rightleftharpoons}} \underset{\text{III}}{R\text{-}\underset{\underset{NH_2}{|}}{CH}\text{-}COO^-}$$

The pH at which most of the amino acids exist in the zwitterion (dipolar) form is called its isoelectric point. When the pH is made more acidic, the dipolar ion is converted to the cation I. In this form, the amino acid will act as an acid with the carboxyl group capable of donating its proton. When the pH is made alkaline, the anion III is formed and predominates in solution. In this form, the amino acid will act as a base, with the amino group as the proton acceptor. At its isoelectric point, the amino acid is capable of acting as both an acid and a base, with both ionic forms existing in exact balance with each other.

The relative acidity and basicity of an amino acid is strongly influenced by the structure and properties of its particular R group. In addition, the R group can itself act as an acid or base. Thus, each amino acid will have its own characteristic isoelectric point, and its own particular K_a and K_b for its amino and carboxylic acid groups. At a given pH, some amino acids will be acidic (protonated or cationic), while others will be basic (anionic), and others will be neutral (zwitterionic). At physiological pH, most amino acids exist in the dipolar form, with acidity and basicity being determined by each particular R group. In the dipolar form, however, the amino acids can react with each other, via their polar groups, to form a covalent peptide linkage:

$$R_1 - \underset{NH_3^+}{\overset{H}{C}} - COO^- + {}^+NH_3 - \underset{COO^-}{CH} - R_2 \xrightarrow{-H_2O} R_1 - \underset{NH_3^+}{CH} - \overset{O}{\overset{\|}{C}} - NH - \underset{COO^-}{CH} - R_2$$

This peptide bond is the means by which large protein molecules are constructed from amino acids.

PROPERTIES OF CELLULAR CONSTITUENTS

• PROBLEM 1-24

Discuss some properties and functions of (a) carbohydrates (b) lipids, (c) proteins and (d) nucleic acids.

Solution: (a) Carbohydrates are made up of carbon, oxygen and hydrogen, and have the general formula $(CH_2O)_n$. Carbohydrates can be classified as monosaccharides, disaccharides, oligosaccharides or polysaccharides. The monosaccharides ("simple sugars") are further categorized according to the number of carbons in the molecule. Trioses contain 3 carbons: pentoses contain 5 carbons (i.e. ribose, dexyribose): and hexoses contain 6 carbons (i.e. glucose, fructose, galactose). The hexoses, which exist as straight chains or rings, are important building blocks for disaccharides and the more complex carbohydrates.

Disaccharides, important in nutrition, are chemical combinations of two monosaccharides:

Lactose = glucose + galactose
Sucrose (table sugar) = glucose + fructose
Maltose = glucose + glucose

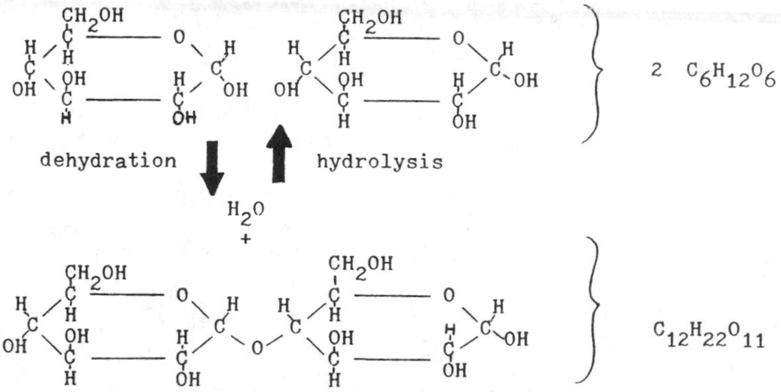

2 $C_6H_{12}O_6$

dehydration ↓ ↑ hydrolysis

H_2O +

$C_{12}H_{22}O_{11}$

As can be seen by the double arrows, the reverse of this reaction, hydrolysis, is also possible. Oligosaccharide means "few sugars" and is arbitrarily defined as compounds which upon hydrolysis yield 2-10 monosaccharides. Polysaccharides are complex carbohydrates made up of many monosaccharides bonded by glycosidic linkages. These long chains are formed by dehydration synthesis. They also can be broken down into monosaccharide units by hydrolysis. There are many complex polysaccharides that are of great biological significance. Their primary functions include both storage and structural properties. Examples of these are: starches (principal storage product of animals), and cellulose (major supporting material in plants). These are all polymers of glucose.

(b) Lipids are also composed principally of carbon, hydrogen and oxygen. However, they can also contain other elements, particularly phosphorus and nitrogen, and typically contain a much smaller proportion of oxygen than do carbohydrates. Lipids are insoluble in water. There are many known lipids, the most common include the fats, the phospholipids and the steroids. Fats are composed of two different types of compounds: glycerol (an alcohol) and fatty acids (organic compounds with a carboxyl group $-C(=O)OH$). Each molecule of fat contains one glycerol molecule and three fatty acids joined together by dehydration reactions.

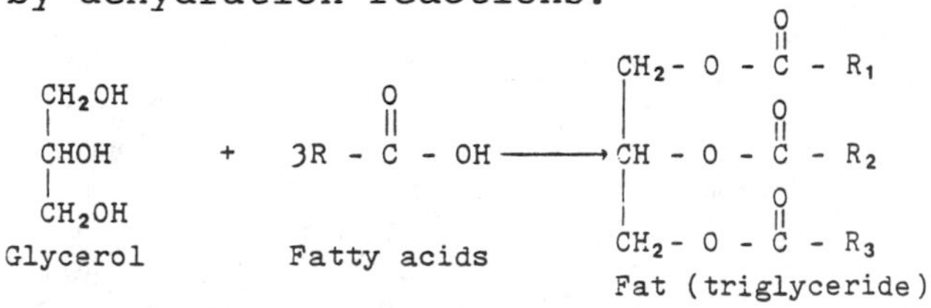

```
                                           O
                                           ||
                               CH2 - O  -  C  - R1
CH2OH              O            |
 |                 ||           |          O
CHOH     +   3R -  C - OH  ---> CH  - O  - C  - R2
 |                              |          O
CH2OH                           |          ||
                               CH2 - O  -  C  - R3
Glycerol       Fatty acids     Fat (triglyceride)
```

There are basically two groups of fats: the saturated (those fats which have the maximum possible number of hydrogen atoms and therefore have no carbon to carbon double or triple bonds) and the unsaturated (those which have at least one carbon to carbon double or triple bond). Unsaturated fats are an important part of our diet due to their storage capabilities. Phospholipids are composed of glycerol, fatty acids, phoshoric acid and usually a nitrogenous compound. They are important components of many cellular membranes. Steroids are classified as lipids because their solubility characteristics are similar to those of fats and phospholipids - they are all insoluble in water, but soluble in ether. However, steroids differ structurally from fats and phospholipids - they are not based on the bonding together of fatty acids and an alcohol. Instead, they are composed of four interlocking rings of carbon atoms with various side groups attached to the rings:

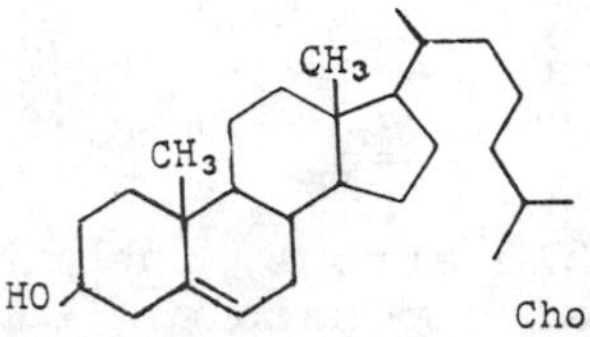

CH_3

CH_3

HO

Cholesterol

They are very important biologically. Steriods include the sex hormones,various other hormones and some vitamins. In addition they are also important structural elements in living cells, especially plants. c) Proteins are much more complex than either carbohydrates or lipids. They are made up of the four essential elements: carbon, hydrogen, oxygen and nitrogen bonded together to form compounds called amino acids. Some amino acids contain sulfur. The amino acid is represented as:

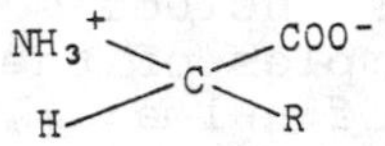

NH_3^+ — C — COO^-
H — C — R

where R is a chain which can be very simple (as in the amino acid glycine where R=H) or it can be very complex (as in tryptophan where R contains two ring structures). Proteins are long and complex polymers of varying combinations of the twenty amino acids which are formed by condensation reactions between the $-COO^-$ and $-NH_3^+$ groups of the amino acid building blocks. The bonds formed by these reactions are called peptide bonds. A dipeptide is a molecule with two amino acids joined together by one peptide bond. A tripeptide refers to three linked amino acids. Oligopeptides, which includes tripeptides, is the term for a short chain of amino acids. Polypeptides are the polymer of amino acids and may contain 1000 amino acids. Finally, a protein is one or more polypeptide chains coiled or folded into complex three-dimensional configurations. Often a metal ion or organic molecule is an integral part of the protein structure. Proteins are found in every part of the cell and are an integral part in both the structure and function of living things. They play an important part in many of the chemical reactions that occur within cells because the enzymes that catalyze these reactions are proteins themselves. Proteins also function as the structural and binding materials of organisms. Hair, fingernails, muscle, cartilage, tendons and ligaments are all structures which contain large amounts of proteins.

d) Nucleic acids, as their name implies, are found primarily in the nucleus. There are two types of nucleic acid: deoxyribonucleic acid (DNA) and ribonucleic acid (RNA). DNA and RNA molecules are very long chains composed of repeating subunits called nucleotides. A nucleotide is composed of any one of the following five nitrogenous bases: adenine, guanine, cytosine, thymine (only in DNA) or uracil (only in RNA), a five carbon sugar (ribose in RNA, deoxyribose in DNA) and a phosphate group. Both DNA and RNA are composed of many nucleotides linked together and are called polynucleotides or nucleic acids. In a polynucleotide, any two nucleotides are linked together by a dehydration reaction between the phosphate group of one nucleotide and the sugar group of another. The bases are attached to the sugars. Nucleic acids primarily function in heredity and governing the synthesis of many different kinds of proteins and other substances present in organisms. Chromosomes and genes are predominantly composed of DNA. Some DNA is also found in the mitochondria and the chloroplasts. Large quanitites of RNA are present in the nucleoli, the cytoplasm, and the ribosomes of most cells.

• PROBLEM 1-25

Draw the open-chain structure of glucose ($C_6H_{12}O_6$) and number the carbon atoms. Is this the correct structure of glucose in actuality?

Solution: The open-chain aldehyde form of a glucose molecule has the following structure and numbering system:

```
    O     H
     \\  /
      C1
      |
  H - C2 - OH
      |
 HO - C3 - H
      |
  H - C4 - OH
      |
  H - C5 - OH
      |
      C6H2OH
```

Numbering starts with the carbon of the carbonyl (—C(=O)—) functionality. Until the early 1890's this open-chain form had been widely accepted as the only structure of glucose.

This structure became questionable as it could not account for numerous experimental observations, among which was the fact that glucose fails to undergo the reactions typical of an aldehyde. In 1895 a cycle structure for glucose was proposed:

```
  H - C1 - OH  ---+
      |           |
  H - C2 - OH     |
      |           |
 HO - C3 - H      O        Or
      |           |
  H - C4 - OH     |
      |           |
  H - C5 ---------+
      |
      C6H2OH
```

```
        6CH2OH
          |
    H     C5 ------------- O
     \   / H5               \   H
      C4                     C1
     /   \ OH          H    /   \
   HO     C3 ----------- C2      OH
          H3             OH
```

This cyclic form can be looked upon as the result of an intramolecular reaction involving the aldehyde group at C_1 and the hydroxyl group at C_5 of the open-chain structure.

Actually, in an equilibrium mixture, there exist two forms of glucose which differ with respect to the configuration at C_1. These two forms, called anomers, are designated α (alpha) and β (beta):

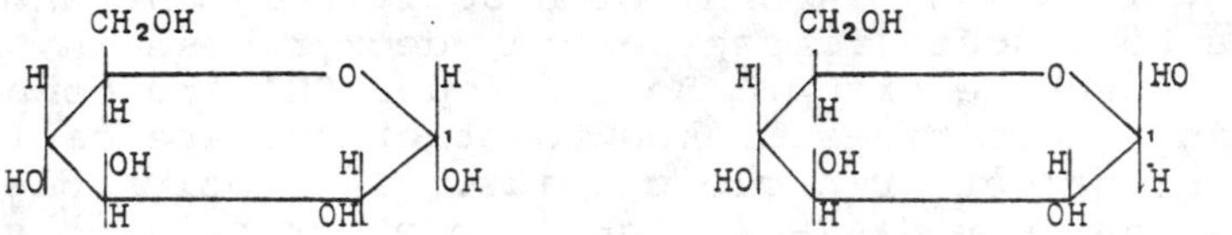

α- Glucose   β Glucose

These cyclic anomers of glucose exist in equilibrium with the open-chain form, but the latter occurs to a very minor extent. The following expression depicts their relationship:

$$\alpha\text{-glucose} \rightleftharpoons \begin{array}{c}\text{open-chain}\\ \text{aldehyde form}\end{array} \rightleftharpoons \beta\text{-glucose}$$

Experimentally it has been found that an equilibrium solution of glucose contains about 36% of the α form and 64% of the β form, with the open-chain form occurring less than 0.5% of the time.

• PROBLEM 1-26

What is the significance of hydrophobic and hydrophilic moieties in biological systems?

Solution: All cells possess a cell membrane, which serves to regulate the materials entering and leaving the cell, and to give the cell shape and protection. The regulation of the movement of materials across the cell membrane (which materials and how much of each) is referred to as selective permeability. Since the cell's internal and external environment are mainly aqueous, the cell membrane must have a unique structure to be able to allow for the passage of water-soluble as well as water-insoluble materials. Water-soluble substances are called hydrophilic ("love of water") and water-insoluble substances are called hydrophobic ("fear of water"). The cell membrane is composed of a double layer of phospholipid with protein molecules dispersed throughout the membrane. Phospholipids and proteins are molecules that have both a hydrophilic and a hydrophobic portion. It is precisely these qualities which bestow the selective permeability property to the cell membrane. The hydrophilic portions (the polar or charged portions) are arranged so that they face the inside or outside of the cell. The hydrophobic portions are arranged so that they lie within the cell membrane. The polar phosphate moiety of the phospholipid is hydrophilic and thus faces the inner or outer surface. The fatty acid tails, the hydrophobic regions of the phospholipid, are oriented towards the middle of the bilayer. Membrane proteins are classified as peripheral (extrinsic) or integral (intrinsic). Peripheral proteins are on the inner or outer surface, exposed to the aqueous solution. Integral proteins may span the entire membrane or just the bilayer. Only the polar sections face the surfaces.

Phospholipids are represented as: **Proteins are represented as:**

hydrophilic head **peripheral**

hydrophobic tail **integral**

Cell membrane model:

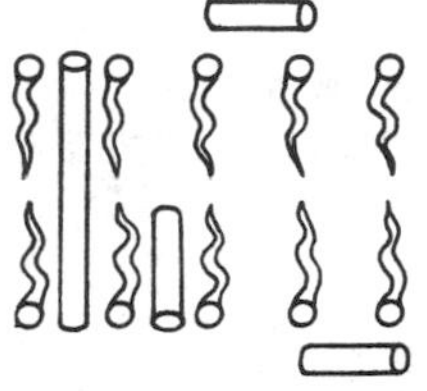

Note: The above model is not a monomeric unit of the cell membrane. Cell membranes have diversity both externally and internally.

• PROBLEM 1-27

Define a buffering system. What significance does it have in the living cell?

Solution: A buffering system is one that will prevent significant changes in pH, upon addition of excess hydrogen or hydroxide ion to the system. Buffering systems are of great importance in the maintenance of the cell. For example, the enzymes within a cell have an optimal pH range, and outside this range, enzymatic activity will be sharply reduced. If the pH becomes too extreme, the enzymes and proteins within the cell may be denatured, which would cause cellular activity to drop to zero and the cell would die. Hence a buffering system is essential to the existence of the cell. The average pH of a cell is 7.2, which is slightly on the basic side.

A buffer will prevent significant pH changes upon variation of the hydrogen ion concentration by abstracting or releasing a proton. Relatively weak diprotic acids (Ex: H_2CO_3, carbonic acid) are good buffers in living systems. When carbonic acid undergoes its first dissociation reaction, it forms a proton and a bicarbonate ion: $H_2CO_3 \rightleftarrows H^+ + HCO_3^-$. The pH of the system is due to the concentration of H^+ formed by the dissociation of H_2CO_3. Upon addition of H^+, the bicarbonate ion will become protonated so that the total H^+ concentration of the medium remains about the same. Similarly, upon addition of ^-OH, the bicarbonate ion releases a hydrogen to form water with the ^-OH. This maintains the H^+ concentration and hence keeps the pH constant. These reactions are illustrated below:

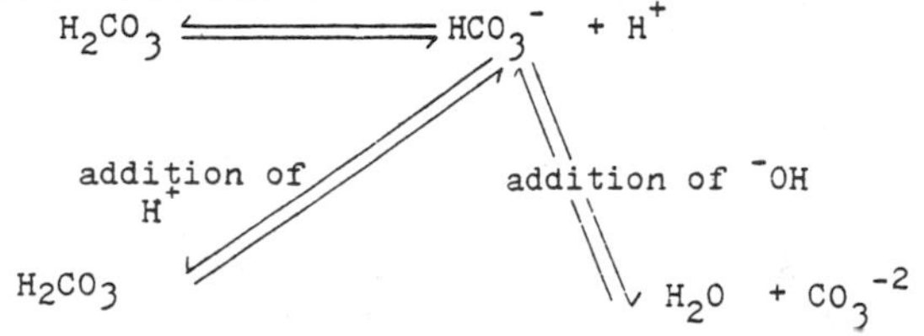

• **PROBLEM 1-28**

What are the properties of a colloidal system? Why does it afford a particularly good medium for chemical reactions? Give some examples of colloidal solutions.

Solution: A colloid is a stable suspension of particles that do not settle out of solution upon standing. The particles or dispersed phase, of a colloid are between 10^{-1} and 10^{-4} microns in size. The medium which holds these particles is called the dispersing phase.

Collodial systems display properties which are unique to them, such as the Tyndall effect and Brownian movement. The Tyndall effect is due to colloidal particles scattering light at all angles from the direction of a primary beam. This scattering of light can be observed, for example, when a beam of light enters a darkened room through a small opening. The particles of dust suspended in the air in the room cannot be seen if they are colloidal in size, but the light scattered by them, however, can be seen, and thus the particles appear as bright points in the beam. This phenomenon can only be observed at right angles to the light source.

Brownian movement is the random motion of colloidal particles in their dispersing medium. To observe this, an optical microscope should be focused on a colloidal solution. Due to the light-scattering properties just discussed, the dispersed substance will then be observed not as particles with definite outlines, but as small, sparkling specks traveling a random zig-zag path.

The reason colloidal solutions provide a particularly good medium for chemical reactions is another special property of colloids. Due to their physical characteristics, they have an enormous total surface area. All surfaces contain energy which can contribute to a chemical reaction. This is best exemplified by absorption (the process which accounts for the fact that solid substances can hold appreciable quantities of gases and liquids on their surface). The capacity for absorption and surface area are approximately proportional. The particles suspended in the dispersing phase can absorb ions, molecules and atoms of other substances. Many colloids have a preference for certain substances which permits the use of absorption in such processes as: removing undesirable colors and odors from certain materials, separation of mixtures, concentration of ores and various purification methods.

Examples of various colloidal systems are given in the following table:

dispersed substance	dispersing substance	general name	examples
gas	liquid	foam	whipped cream, beer froth
gas	solid	solid foam	pumice, marshmallow
liquid	gas	liquid aerosol	fog, clouds
liquid	liquid	emulsion	mayonnaise, milk
liquid	solid	solid emulsion	cheese, butter
solid	gas	solid aerosol	smokes, dust
solid	liquid	sol	most paints, jellies
solid	solid	solid sol	many alloys, black diamonds

• PROBLEM 1-29

How can radioactive isotopes be used in biological research such as tracing the pathways of compounds synthesized in the living cell?

Solution: Labeling substance with radioactive isotopes is a powerful method for studying specific metabolic pathways, as well as being applicable to many other areas of biological research. Radioactive labeling has been used, for example, in determining the rates of metabolic processes in intact organisms, and also in determining whether a given metabolic pathway is the predominant route, for a given metabolite. It is used in establishing whether a pathway which has been studied in a test tube is the same reaction that occurs in vivo. Labeling is done by taking the compound in question and substituting in a radioactive isotope for one of the naturally occuring elements comprising the compound. (see Table).

The radioactive isotope, when incorporated into a biological system will undergo spontaneous emmision of radiation while taking part in the same biological activities as its non-radioactive counterpart. Although this is not observable by eye, there are many known methods by which these emissions can be detected. The methods commonly employed include Geiger counters, photographic methods, fluorescent methods, and cloud chambers. In this way, the course of the reaction can be followed.

The value of radioactive tracers in biological research can be seen by considering the photosynthetic equation:

$$6\ CO_2 + 6\ H_2O \xrightarrow[\text{chlorophyll}]{\text{sunlight}} C_6H_{12}O_6 + 6\ O_2.$$

Although in the equation the photosynthetic process appears to be a simple reaction, it actually follows a very complicated course. This complex reaction, like most others, proceeds through a series of changes involving one or two molecules at a time. Researchers were able to use $^{14}_{6}C$, which was incorporated into CO_2, to determine the course of this reaction. They monitored the radioactivity through the use of a Geiger counter.

Some isotopes useful as radioactive tracers in biological systems.

naturally occuring element	radioactive isotopes	half - life
$^{1}_{1}H$	$^{3}_{1}H$	12.1 years
$^{12}_{6}C$	$^{14}_{6}C$	5,700 years
$^{23}_{11}Na$	$^{24}_{11}Na$	15 hours
$^{31}_{15}P$	$^{32}_{15}P$	14.3 days
$^{32}_{16}S$	$^{35}_{16}S$	87.1 days
$^{35}_{17}Cl$	$^{36}_{17}Cl$	3.1×10^5 years
$^{39}_{19}K$	$^{42}_{19}K$	12.5 hours
$^{40}_{20}Ca$	$^{45}_{20}Ca$	152 days
$^{56}_{26}Fe$	$^{59}_{26}Fe$	45 days
$^{127}_{53}I$	$^{131}_{53}I$	8 days

SHORT ANSWER QUESTIONS FOR REVIEW

Answer

To be covered when testing yourself

Choose the correct answer.

1. Two objects separated by a distance of 3000 Å are viewed under a light microscope. The light source has a wavelength of 0.7 μ. An observer looking through the eyepiece of the microscope would see (a) two distinct and separate objects (b) a single object. (c) nothing, because the objects will cancel out. — b

2. In a closed system, for the following reaction: HCl + Na $\rightleftharpoons$ ½ H_2 + NaCl, H_2 is a gas. If high pressure is applied to the system, the reaction will (a) cease to occur. (b) favor the formation of H_2 and NaCl. (c) favor the formation of HCl and Na. (d) decrease its rate. — c

3. The electron configuration of a certain neutral atom is:

   1s: ↑↓ 2s: ↑↓ 2px: ↑ 2py: ↑ 2pz: ↑

   The bonding capacity and the number of protons in this atom are respectively, (a) 3 and 7. (b) 1 and 4. (c) 7 and 3. (d) none of the above — a

4. What type of bond links the carbon to the amino group ($-NH_2$) in an amino acid? (a) hybridized (b) metallic (c) hydrogen (d) covalent — d

5. The helical structure of nucleic acids are maintained by (a) anhydride bonds. (b) phosphodiester bonds. (c) H bonds (d) covalent bonds — c

6. In the following neutralization reaction: HCl + ^{-}OH $\rightleftharpoons$ H_2O + Cl^- , the weaker conjugate base is (a) HCl. (b) Cl^-. (c) ^{-}OH. (d) H_2O. — b

7. During a titration of egg albumin (a protein enzyme) the net charge of the enzyme drops to zero at a hydrogen ion concentration of 2.51×10^{-5} molar. What is the pH at the isoelectric point of egg albumin? (a) 5.4 (b) 5.6 (c) 4.6 (d) 4.2 — c

8. Which of the following is not a protein or within a protein? (a) peptide bonds (b) biochemical catalysts (c) amino acids (d) nucleotides — d

Answer

To be covered when testing yourself

9. The reaction: $2\ C_6H_{12}O_6 \rightarrow H_2O + C_{12}H_{22}O_{11}$ is one (a) that results in the formation of a disaccharide. (b) example of a hydrolysis reaction. (c) that results in the formation of a lipid. (d) that summarizes the photosynthetic process. — a

10. The anomeric carbon of pure, crystallized β-D-Glucose is labeled with the radioactive isotope ^{14}C. When the sample is dissolved in water, radioactive α-and β-D-glucose is detected. One may infer that β-D-glucose underwent an isomerization to α-D-glucose. This reaction proceeds through (a) an open-chain protein. (b) an open-chain aldehyde. (c) a closed-chain (cyclic) polysaccharide. (d) an open-chain ketone. — b

11. Cell membranes are generally composed of (a) a double layer of phospholipids with proteins dispersed throughout the membrane. (b) a double layer of phosphoproteins with glucose dispersed throughout the membrane. (c) a double layer of nucleic acids. (d) a double layer of proteins with phospholipids dispersed throughout the membrane. — a

12. The phosphorylation of ADP to form ATP requires energy. What happens to most of this energy?
(a) It is lost as heat to the surroundings.
(b) It is converted into electrical energy.
(c) It is stored as chemical bond energy.
(d) It is destroyed during the course of the reaction. — c

13. A buffer serves to (a) prevent significant temperature variation. (b) prevent significant pH variation. (c) prevent significant [H+] variation. (d) b and c — d

14. When light is passed through a test tube containing an unknown liquid, there is no scattering of light observed when looking straight into the solution (the line of vision is parallel to the direction of light). The observer may conclude (a) that the liquid is a colloid. (b) nothing since there is not sufficient data. (c) that the liquid is a solution. (d) that the liquid is a suspension. — b

Fill in the blanks.

15. Resolution refers to the _____ of the image. — clarity

Answer

To be covered when testing yourself

16. The electron microscope uses _____ as the source of radiation. — electrons

17. The _____ states that a perfect crystal at absolute zero would have zero entropy. — third law of thermodynamics

18. A reaction that occurs with the release of heat is an _____ . — exothermic reaction

19. Consider the reaction A + B $\rightleftarrows$ C + Energy. The formation of C is favored by a _____ in temperature. — decrease

20. When an electron moves from an excited state to the ground state in an atom, energy is released in the form of a _____ . — photon of light

21. There are four quantum numbers to describe the characteristics of an electron. These are the _____ , _____ , _____ and _____ quantum numbers. — principal, angular momentum, magnetic, spin

22. _____ is the phenomenon that occurs when the momentary dipole of a molecule will affect the charge distribution of nearby non-polar molecules. — Induction

23. A Lewis acid has the ability to _____ an electron pair, and a Bronsted base has the ability to _____ a proton. — accept, accept

24. Oxidation can involve the removal of _____ or _____ , as well as the addition of _____ . — hydrogen, electrons, oxygen

25. The amphiprotic behavior of amino acids is attributable to the presence of the _____ and _____ functional groups. — amino, carboxyl

26. The final step in the aerobic respiratory process is the reduction of _____ to _____ . — oxygen, water

27. A colloid is composed of a _____ phase and a _____ phase. — dispersed, dispersing

Determine whether the following statements are true or false.

28. The resolution of an electron microscope is much better than a light microscope because of the greater magnification in the former. — False

Answer

To be covered when testing yourself

29. A professor performed an experiment in vitro and deduced the mechanism for a certain biochemical reaction. He can conclude that this is the mechanism by which the reaction occurs within a cell. — False

30. Ammonia (NH_3) is a compound because it is composed of more than one element, which occur in definite proportions. — True

31. When a liter of water at 100°C is combined with a liter of water at 0°C in a closed system, the resultant temperature is 50°C. In this process, entropy has increased and usable energy has decreased. — True

32. In a given reaction, if the change in enthalpy and entropy are both positive, and the product of temperature and entropy change is greater than the change in enthalpy, the reaction will proceed spontaneously. — True

33. The addition of a catalyst will increase the rate of reaction, and hence increase the equlibrium concentration of the products. — False

34. The lavender flame observed upon the burning of a sample of potassium is explained by the fact that electrons are dropping from orbitals of higher energy to ones of lower energy. — True

35. Trans-3-hexene and cis-3-hexene are organic isotopes of each other. — False

36. If the electronegativity difference between two atoms is zero, the bond is predominantly covalent. — True

37. Van der Waals forces are based on the existence of momentary dipoles, which facilitate the intermolecular attractions of the compound. — True

38. An oxidizing agent is one which reduces a substance, and is itself oxidized. — False

39. At a pH of 6.0, an amino acid whose isoelectric pH is 5.1 has a net negative charge. — True

40. Complete hydrolysis of a polysaccharide will result in the breakdown of the molecule into the individual monosaccharides. — True

41. Adenosine triphosphate contains three high energy phosphate bonds. — False

CHAPTER 2

CELLS AND TISSUES

CLASSIFICATION OF CELLS

• PROBLEM 2-1

What is a cell?

Solution: A cell is the fundamental organizational unit of life. One of the most important generalizations of modern biology is the cell theory. There are two components of the cell theory. It states: (1) that all living things are composed of cells and (2) that all cells arise from other cells. Living things are chemical organizations of cells and capable of reproducing themselves.

There are many types of cells, and just as many classifications to go with them. There are plant cells, animal cells, eucaryotic cells, procaryotic cells and many others. Also within each of these divisions, there are smaller subdivisions pertaining to the specific properties or functions of the cells. Cells exhibit considerable variation in properties based on different arrangements of components. Cells also vary in size, although most of them fall in the range of 5 to 20 mμ.

• PROBLEM 2-2

Even though there is no such thing as a "typical cell" - for there are too many diverse kinds of cells - biologists have determined that there are two basic cell types. What are these two types of cells?

Solution: Cells are classified as either procaryotic or eucaryotic. Procaryotes are strikingly different from eucaryotes in their ultrastructural characteristics. A key difference between the two cell types is that procaryotic cells lack the nuclear membrane characteristic of eucaryotic cells. Procaryotic cells have a nuclear region, which consists of nucleic acids. Eucaryotic cells have a nucleus, bounded by a double-layered membrane. The eucaryotic nucleus consists of DNA which is bound to proteins and organized into chromosomes.

Bacteria and blue-green algae are procaryotic unicellular organisms. Other organisms, for example, protozoa, algae, fungi, higher plants and animals are eucaryotic. Within eucaryotic cells are found discrete regions that are usually delimited from the rest of the cell by membranes. These are called membrane-bounded subcellular organelles. They perform specific cellular functions, for example, respiration and photosynthesis. The enzymes for these processes are located within membrane-bounded mitochondria and chloroplasts, respectively. In procaryotic cells, there are no such membrane-bounded organelles. Respiratory and photosynthetic enzymes are not segregated into discrete organelles although they have an orderly arrangement. Procaryotic cells lack endoplasmic reticulum, Golgi apparatus, lysosomes, and vacuoles. In short, procaryotic cells lack the internal membranous structure characteristic of eucaryotic cells.

There are other differences between procaryotic cells and eucaryotic cells. The ribosomes of bacteria and blue-green algae are smaller than the ribosomes of eucaryotes. The flagella of bacteria are structurally different from eucaryotic flagella. The cell wall of bacteria and blue-green algae usually contains muramic acid, a substance that plant cell walls and the cell walls of fungi do not contain.

• PROBLEM 2-3

What are the chief components and structures of a eucaryotic cell?

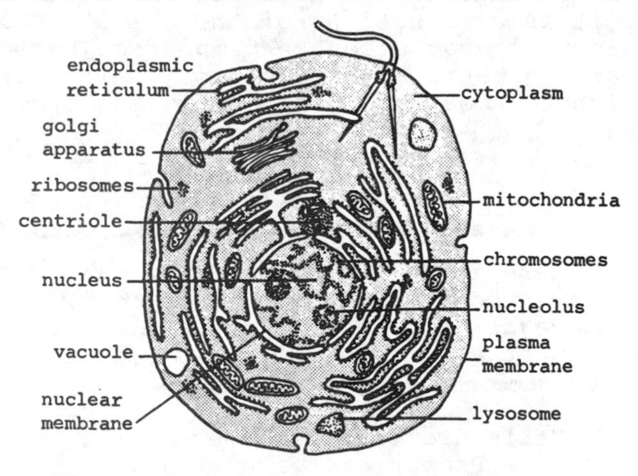

Fig. 1 Typical Animal Cell

Solution: Membranes are a crucial component of both animal and plant cells. The plasma membrane surrounds the cell and serves to separate the internal living matter from the external environment. Plant cells have in addition a cell wall external to the plasma membrane. It is composed mainly of cellulose and is fairly rigid but permeable. Selectivity of materials

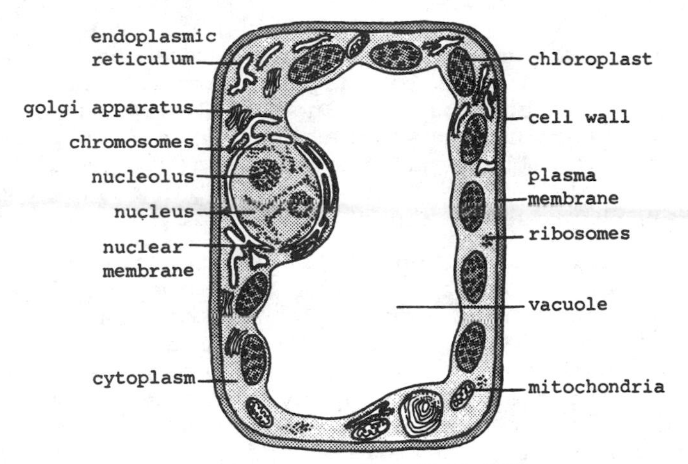

Fig. 2 Typical Plant Cell

entering the plant cells is not a function of the cell wall but of the plasma membrane. Membranes are also present inside the cell, dividing the cell into compartments distinctive in both form and function. All membranes are composed basically of lipids and proteins. For this reason the plasma membrane and the membranes inside the cell are termed unit membranes. Subcellular structures in the cytoplasm are known as organelles. When these are surrounded by membranes, they are termed membrane-bounded organelles. Each organelle performs some specific functions.

The nucleus is the controlling center of the cell. The nucleus is surrounded by two layers of unit membrane which form the nuclear envelope. Within the nucleus are found the chromosomes, the substances of inheritance. Chromosomes are composed chiefly of deoxyribonucleic acid (DNA) and protein. Genes located within the chromosomes, direct cellular function and are capable of being replicated in nuclear division. The nucleolus is a specialized region in the nucleus involved in the synthesis of ribosomal RNA, the material making up the ribosomes.

Mitochondria and chloroplasts are membrane-bounded organelles involved in energy production in the cell. Since energy can neither be created nor destroyed, these organelles actually convert one form of energy to another. Chloroplasts found in plant cells, convert solar energy to chemical energy contained in organic substances; the latter are oxidized in the mitochondria to yield energy useful to the cell in the form of ATP.

Cells which utilize food for energy must take food into the cell and degrade it. Lysosomes are membrane-bound organelles whose digestive enzymes break down organic substances to simpler forms, which can be used by the cell to yield energy for its life-sustaining process.

Most cells contain membrane-delimited bodies called vacuoles. Small vacuoles may be termed vesicles. These structures may contain the ingested materials taken from the cell exterior or materials to be released by the cell. Mature plant cells usually contain a single large fluid-filled vacuole. This vacuole aids the cell in maintaining an internal pressure and thus rigidity.

Organelles involved in the synthesis and transport of cellular components are the ribosomes, endoplasmic reticulum, and Golgi apparatus. Ribosomes are involved in protein synthesis. The endoplasmic reticulum is a system of membranes providing transport channels within the cell. It transports the synthesized protein to other parts of the cell. The Golgi apparatus is involved in the packaging of cellular products before they can be released by the cell to the outside.

Organelles involved in the maintenance of cellular shape and in movement are the microfilaments and microtubules, also known as the skeleton of the living cell. Microfilaments are involved in the connection of adjacent cells for intercellular communication. They also function in the transport of products within the cell. The microtubules are the basic substance in the cilia and flagella of motile cells. By the contracting action of microtubules, flagella and cilia beat and propel the cell. The microtubules have also been identified as the fundamental substance in the spindle apparatus during cell division. Both microfilaments and microtubules are protein structures.

• PROBLEM 2-4

What are the chief differences between plant and animal cells?

Solution: A study of both plant and animal cells reveals the fact that in their most basic features, they are alike. However, they differ in several important ways. First of all, plant cells, but not animal cells, are surrounded by a rigid cellulose wall. The cell wall is actually a secretion from the plant cell. It surrounds the plasma membrane, and is responsible for the maintenance of cell shape. Animal cells, without a cell wall, cannot maintain a rigid shape.

Most mature plant cells possess a single large central fluid sap, the vacuole. Vacuoles in animal cells are small and frequently numerous.

Another distinction between plant and animal cells is that many of the cells of green plants contain chloroplasts, which are not found in animal cells. The presence of chloroplasts in plant cells enable green plants to be autotrophs, organisms which synthesize their own food. As is generally known, plants are able to use sunlight, carbon dioxide and water to generate organic substances. Animal cells, devoid of chloroplasts, cannot produce their own food. Animals, therefore, are heterotrophs, organisms that depend on other living things for nutrients.

Some final differences between plant and animal cells are in the process of cell division. In animal cells, undergoing division, the cell surface begins to constrict, as if a belt were being tightened around it, pinching the old cell into two new ones. In plant cells, where a stiff cell wall interferes with this sort of pinching, new cell membranes form between the two daughter cells. Then a new cell wall is deposited between the two new cell membranes. During cell division, as the mitotic spindle apparatus forms, animal cells have two pairs of centrioles attached to the spindles at opposite poles of the cell. Even though plant cells form a spindle apparatus, most higher plants do not contain centrioles.

• **PROBLEM 2-5**

Most cells fall within the range of 5-20 microns. What factors keep the size of the cell within these boundaries?

Solution: Normally, a cell cannot attain a size greater than 20 microns due to the limits imposed by the size inter-relationships of its components. For example, there is a relationship between the amount of nuclear material and the size of the cell; this limits cells to a size where there is an optimal proportion of nuclear material to the rest of the cell. A cell can have only as much material as its number of genes can handle. And since the nucleus controls the entire cell, there is a limit to how far away parts can be located from the nucleus. However, some larger cells have solved these problems by having more than one nucleus.

Because materials must be transported throughout the cell and among the organelles, the relationship between the amount of cytoplasm and the amount of cellular organelles is also important in limiting cell size. If there is too much cytoplasm, transportation and communication between various areas of the cell would not be efficient. This transport problem is slightly minimized by processes such as cytoplasmic streaming.

The effect of the amount of membrane surface area as compared to cell volume is the most important factor in limiting cell size. The intake of nutrients, the exchange of gases, and the excretion of wastes are all dependent on the surface area available for diffusion of these materials. How this is affected by increases in cell size can be seen through the following mathematical relationships: volume (sphere) $= \frac{4}{3} \pi r^3$ (r = radius)

surface area $= 4\pi r^2$

proportionally: $\frac{\text{volume}}{\text{surface area}} = \frac{4/3\ \pi r^3}{4\pi r^2} = \frac{1}{3} r.$

It can be seen that the volume of the cell increases with the cube of the radius while the surface area increases only with the square of the radius. In other words, the surface area increases at a rate 1/3 r slower than the volume. Thus, if a cell's size increases, its surface-area may not increase proportionally enough to satisfy the needs of the larger cell.

There is also a limit as to how small a cell can be. The minimum size can be predicted by determining how many macromolecules a cell must have in order for it to keep up its metabolism. Mathematical analysis of the cellular content of the PPL organism (the smallest known single-celled organism) will supply the answer.

It is estimated that there are 1600 macromolecules (nucleic acids, proteins, etc.) in the PPL organism. Assuming that the cell must perform approximately 500 reactions in order to survive, we can determine that $\frac{\text{1600 macromolecules}}{\text{500 reactions}}$ are required on the average, or 3 to 4 macromolecules are required for each reaction. Taking into account the average size of the macromolecules necessary, the minimum size that any cell could be can be approximated.

FUNCTIONS OF CELLULAR ORGANELLES

• PROBLEM 2-6

Describe the structure and functions of the plasma membrane.

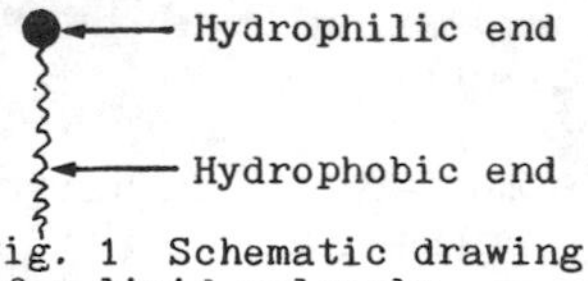

Hydrophilic end

Hydrophobic end

Fig. 1 Schematic drawing of a lipid molecule.

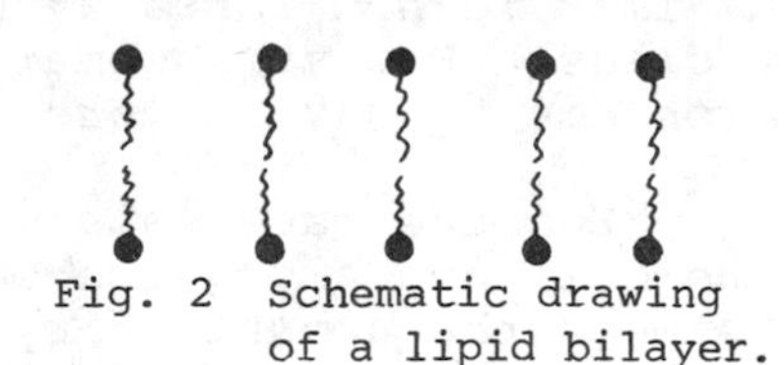

Fig. 2 Schematic drawing of a lipid bilayer.

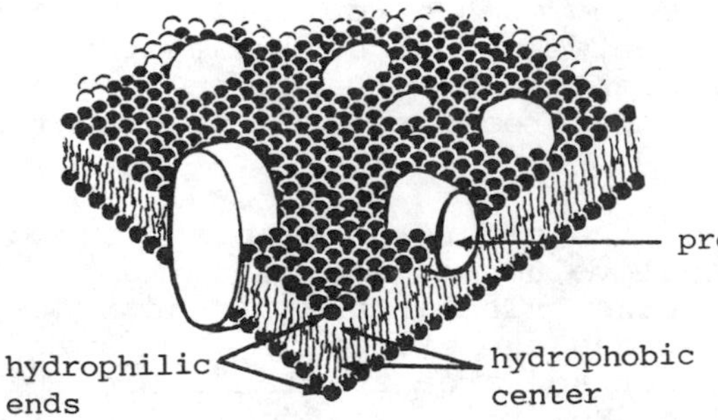

protein globule

hydrophilic ends

hydrophobic center

FIGURE 3. MODEL OF THE UNIT MEMBRANE

Solution: Each cell is surrounded by a selective membrane, a complex elastic covering that separates the cell protoplasm from the external environment. The structure of this covering, called the plasma membrane, has been under major investigation for many years. Studies of membrane permeability, electron microscopy, and biochemical analysis have enabled biologists to better understand the structure and composition of the plasma membrane. The plasma membrane contains about 40 percent lipid and 60 percent protein by weight, with considerable variation from cell type to cell type. The different types and amounts of lipids and proteins present determine to a great extent the characteristics of different membranes. As seen in electron micrographs, all membranes appear to have a similar fundamental structure. The plasma membrane is revealed by electron microscopy to resemble a railroad track in cross-section - two dark lines bordering a central lighter line. The membranes of cellular organelles also display this characteristic. The two dark lines were suggested to correspond to two layers of protein and the light middle layer to lipid. It was soon revealed that the lipid actually exists in two layers.

The lipid molecules of the plasma membrane are polar, with the two ends of each molecule having different electric properties. One end is hydrophobic ("fear of water"), which means it tends to be insoluble in water. The other end is hydrophilic("love of water"), which means it has an affinity for water (see Figure 1). The lipid molecules arrange themselves in two layers in the plasma membrane so that the hydrophobic ends are near each other, and the hydrophilic ends face outside toward the water and are stabilized by water molecules (see Figure 2). In this bilayer, individual lipid molecules can move laterally, so that the bilayer is actually fluid and flexible.

Protein molecules of the plasma membrane may be arranged in various sites but embedded to different degrees in relationship to the bilayer. Some of them may be partially embedded in the lipid bilayer, some may be present only on the outer surfaces, and still others may span the enitire lipid bilayer from one surface to the other (see Fig. 3). The different arrangements of proteins are determined by the different structural, conformational, and electrical characteristics of various membrane proteins. Like the lipid bilayer, the protein molecules tend to orient themselves in the most stable way possible. The proteins are usually naturally folded into a globular form which enables them to move laterally within the plane of the membrane at different rates. Certain proteins can actually move across the membrane. Thus membrane proteins are not static but dynamic.

The functions of the plasma membrane are highly specific and directly related to its structure, which is in turn, dependent on the specific types and amounts of proteins and lipids present. The discriminating permeability of the membrane is its primary function. It allows certain substances to enter or leave the cell, and prevents other substances from crossing it. Whether or not a molecule can cross a membrane depends on its size after hydration, electric charge, shape, chemical properties, and its relative solubility in lipid as compared to that in water. This selective permeability of the plasma membrane gives the cell the ability to keep its interior environment both chemically and physically different from the exterior environment.

The plasma membrane is also found to be particularly important in cell adherence. Because of the specificity of protein molecules on the membrane surface, cells can recognize each other and bind together through some interaction of their surface proteins. Surface proteins are believed to provide communication and linkage between cells in division so that cells divide in an organized plane, rather than in random directions giving rise to an amorphous mass of cells as in cancer. Surface proteins of the plasma membrane are also proposed to recognize foreign substance due to their remarkable specificity; they can bind with the foreign substance and inactivate it. Membrane proteins are further suggested to interact with hormones, or convey hormonal messages to the nucleus so that a physiological change can be effected. The plasma membrane also is involved in the conduction of impulse in nerve cells. The axon of nerve cells transmits impulse by a temporary redistribution of ions inside and outside the cell, with a subsequent change in the distribution of charges on the two surfaces of the membrane.

• PROBLEM 2-7

Why is the mitochondria referred to as the "powerhouse of the cell"?

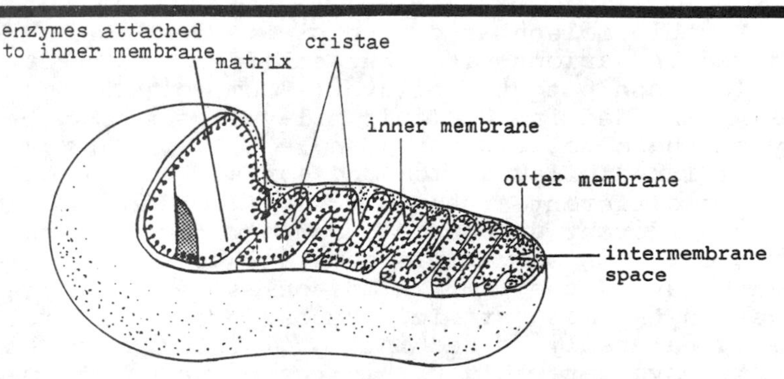

Diagram showing the internal structures of a mitochondrion through a cutaway view.

Solution: Mitochondria are membrane-bounded organelles concerned principally with the generation of energy to support the various forms of chemical and mechanical work carried out by the cell. Mitochondria are distributed throughout the cell, because all parts of it require energy. Mitochondria tend to be most numerous in regions of the cell that consume large amounts of energy and more abundant in cells that require a great deal of energy (for example, muscle and sperm cells).

Mitochondria are enclosed by two membranes. The outer one is a continuous delimiting membrane. The inner membrane is thrown into many folds that extend into the interior of the mitochondrion. These folds are called cristae. Enclosed by the inner membrane is the ground substance termed the matrix (see accompanying diagram). Many enzymes involved in the Krebs cycle are found in the matrix. Enzymes involved in the generation of ATP by the oxidation of $NADH_2$, or, the electron transport reactions, are tightly bound to the inner mitochondrial membrane. The enzymes for the specific pathways are arranged in sequential orders so that the products of one reaction do not have to travel far before they are likely to encounter the enzymes catalyzing the next reaction. This promotes a highly efficient energy production.

The reactions that occur in the mitochondria are all related in that they result in the production of ATP (adenosine triphosphate), which is the common currency of energy conversion in the cell. Some ATP is produced by reactions that occur in the cytoplasm, but about 95 percent of all ATP produced in the cell is in the mitochondria. For this reason the mitochondria are commonly referred to as the powerhouse of the cell.

• **PROBLEM 2-8**

Explain the importance and structure of the endoplasmic reticulum in the cell.

Solution: The endoplasmic reticulum is responsible for transporting certain molecules to specific areas within the cytoplasm. Lipids and proteins are mainly transported and distributed by this system. The endoplasmic reticulum is more than a passive channel for intracellular transport. It contains a variety of enzymes playing important roles in metabolic processes.

The structure of the endoplasmic reticulum is a complex of membranes that traverses the cytoplasm. The membranes form interconnecting channels that take the form of flattened sacs and tubes. When the endoplasmic reticulum has ribosomes attached to its surface, we refer to it as rough endoplasmic reticulum and when there are no ribosomes attached, it is called smooth endoplasmic reticulum. The rough endoplasmic reticulum functions in transport of cellular products; the role of the smooth endoplasmic reticulum is less well known, but is believed to be involved in lipid synthesis (thus the predominance of smooth end reticulum in hepatocytes of the liver).

In most cells, the endoplasmic reticulum is continuous and interconnected at some points with the nuclear membrane and sometimes, with the plasma membrane. This may indicate a pathway by which materials synthesized in the nucleus are transported to the cytoplasm. In cells actively engaged in protein synthesis and secretion (such as acinar cells of the pancreas), rough endoplasmic reticulum is abundant. By a well regulated and organized process, protein or polypeptide chains are synthesized on the ribosomes. These products are then transported by the endoplasmic reticulum to other sites of the cell where they are needed. If they are secretory products, they have to be packaged for release. They are carried by the endoplasmic reticulum to the Golgi apparatus, another organelle system. Some terminal portions of the endoplasmic reticulum containing protein molecules bud off from the membranes of the reticulum complex, and move to the Golgi apparatus in the form of membrane-bounded vesicles. In the Golgi apparatus, the protein molecules are concentrated, chemically modified, and packaged so that they can be released to the outside by exocytosis. This process is necessary because some proteins may be digestive enzymes which may degrade the cytoplasm ans lyse the cell if direct contact is made.

• PROBLEM 2-9

In microscopy, small spherical bodies are often seen attached to the network of endoplasmic reticulum. What are these bodies? What function do they serve in the cell?

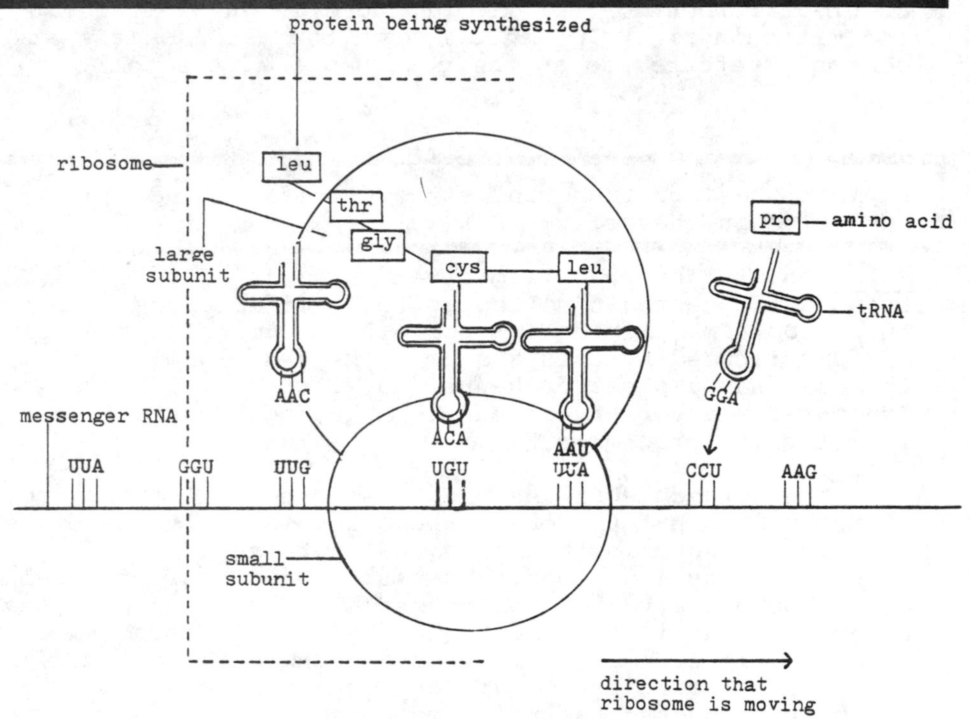

The role of the ribodome in protein synthesis.

Solution: The small spherical bodies that we see studding the endoplasmic reticulum - more accurately, the rough endoplasmic reticulum - are the ribosomes. The rough endoplasmic reticulum owes its rough appearance to the presence of ribosomes. The smooth endoplasmic reticulum appears smooth because it lacks ribosomes. Ribosomes consist of two parts, a large subunit and a small subunit. Both the large and small subunits are made of proteins and ribonucleic acid (RNA). However, the two subunits differ both in the number and in the type of proteins and RNA they contain. The large subunit contains large and more varied RNA molecules than the small subunit. It also has more protein molecules than the smaller one. An interesting point to note is that when we put together all the chemical components of a ribosome, under favorable conditions, these parts will rearrange themselves and come together, without direction from pre-existing ribosomes, to form a functional assembly. This ability to self-assemble may provide us with a clue to the origin of living things.

Ribosomes are the sites of protein synthesis in the cell. Messenger RNA (mRNA), which carries genetic information from the nucleus, associates with the small ribosomal subunit first and then binds to the large subunit as a prelude to protein synthesis. This association of mRNA to ribosomes holds the components of the complex system of protein synthesis together in a specific manner for greater efficiency than if they were dispersed freely in the cytoplasm. The mRNA then pairs with complementary molecules of transfer RNA (tRNA), each carrying a specific amino acid. The linking up of tRNA molecules into a chain complementary to mRNA brings together amino acids which bind with each other to form a highly specific protein molecule. Thus ribosomes are the sites where proteins are synthesized under genetic control.

• PROBLEM 2-10

How is the Golgi apparatus related to the endoplasmic reticulum in function?

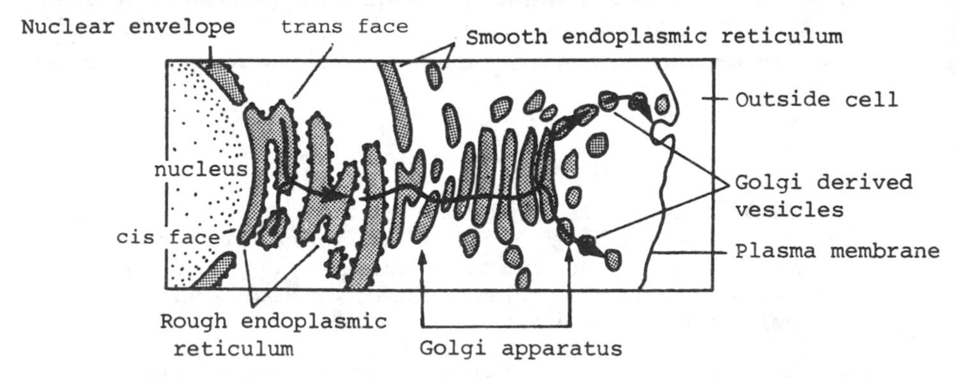

Schematic representation of the secretion of a protein in a typical animal cell. The solid arrow represents the probable route of secreted proteins.

Solution: The Golgi apparatus is composed of several membrane-bound, flattened sacs (or cisternae) arranged in parallel array about 300 Å apart. The sacs are disc-like and often slightly curved. Note the concavity on the trans face near the plasma membrane and the convexity of the cis face are thinner (more like reticulum membrane than like plasma membrane).

The function of the Golgi apparatus is best understood in cells involved in protein synthesis and secretion. The protein to be secreted is synthesized on the rough endoplasmic reticulum. Vesicles containing small quantities of the synthesized protein bud off from the endoplasmic reticulum. These vesicles carry the protein to the convex face of the Golgi apparatus.In the Golgi apparatus, the protein is concentrated bythe removal of water. In addition, chemical modifications of the protein, such as glycosylation (addition of sugar) occur. The modified protein is released from the concave surface in the form of secretory granules. The secretory granules containing the protein fuse with the plasma membrane and its contents (protein in this case) are expelled from the cell. This a process known as exocytosis.

• PROBLEM 2-11

What are the functions of the nucleus? What is the evidence that indicates the role of the nucleus in cell metabolism and heredity?

Solution: If the cell is thought of as a miniature chemical plant designed to carry out all the processes of life, then the nucleus can be compared to a central computer that controls a network of sophisticated and highly complicated biochemical machinery. This is because the nucleus contains the chromosomes, which bear the genes, the ultimate regulator of life.

The genes comprise a library of programs stored in the nucleus: programs that specify the precise nature of each protein synthesized by the cell. The nucleus monitors changing conditions both inside the cell and in the external environment, and responds to input of information by either activating or inhibiting the appropriate genetic programs.

Control of protein synthesis is the key to controlling the activities and responses of the cell, since a tremendeous array of important biological and biochemical processes are regulated by enzymatic proteins. By switching particular genes on and off, the cell controls not only the kinds of enzymes that it produces, but also the amounts. Both qualitative and quantitative control of

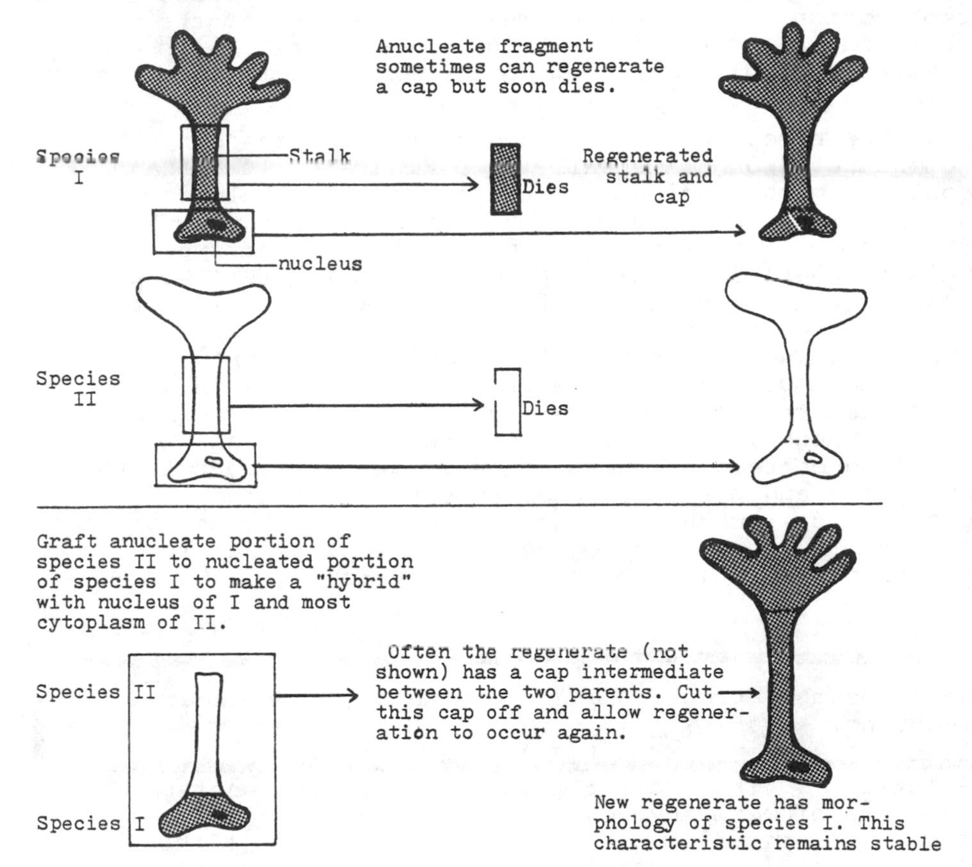

Experiments with the large single-celled alga Acetabularia.

enzyme synthesis are crucial to the proper functioning of the cell and the whole organism.

Cells without nuclei have a very limited range of function. The mammalian red blood cell does not contain a nucleus and is unable to reproduce; it is functional only for a relatively short period. Egg cells, from which nuclei have been experimentally removed, may divide for a while, but the products of division never differentiate into specialized cell types, and eventually die. Fragments without a nucleus, cut from such large unicellular organisms as the amoeba or the unicellular algae Acetabularia, survive temporarily; but ultimately they die, unless nuclei from other cells are transplanted into them. These experiments demonstrate that the nucleus is essential to long term continuation of life processes, and to structural and functional differentiation of cells.

The nucleus is of central importance also in the transmission of hereditary information. The nucleus carries the information for all the characteristics of the cell. This information is found in the chromosomes, which consist of protein and deoxyribonucleic acid (DNA). DNA is the hereditary material which contains all the information for the growth and reproduction of the cell. When the cell divides, the nuclear information is transmitted in an orderly fashion to the daughter cells by replication of the chromosomes and division of the nucleus. Thus, in this type of division, the daughter cells each possess a single complement of genetic information identical to that of the mother cell.

The hereditary importance of the nucleus can be demonstrated experimentally (see Figure). If a fragment containing the nucleus is cut from an Acetabularium of one species, characterized by a given appearance, the fragment will regenerate a whole cell of that species. This regenerative ability permits experiments of the type shown in the accompanying figure, in which nuclei of one species are grafted onto the cytoplasm of a different species. The observation from these experiments shows that the appearance of the regenerated cell eventually resembles that of the organism from which the nucleus is taken. Hence, we conclude that the nucleus, as a controlling center, produces information that controls the cytoplasm and participates in the regulation of cell growth and structure. Since the crucial information comes from the nucleus, the regenerated cell will resemble the organism from which the nucleus is taken.

• PROBLEM 2-12

Describe the structure and function of the nuclear membrane.

Solution: The nuclear membrane actually consists of two leaflets of membrane, one facing the nucleus and the other facing the cytoplasm. The structure of each of these two leaflets is fundamentally similar to the structure of the plasma membrane, with slight variations. However, the two leaflets differ from each other in their lipid and protein compositions. The nuclear membrane is observed under the electron microscope to be continuous at some points with the membranes of the endoplasmic reticulum.

Nuclear pores are the unique feature of the nuclear membrane. They are openings which occur at intervals along the nuclear membrane, and appear as roughly circular areas where the two membrane leaflets come together, fuse, and become perforated. Many thousands of pores may be scattered across a nuclear surface.

The nuclear pores provide a means for nuclear-cytoplasmic communication. Substances pass into and out of the nucleus via these openings. Evidence that the pores are the actual passageways for substances through the nuclear membrane comes from electron micrographs which show substances passing through them. The mechanisms by which molecules pass through the pores are not known. At present, we know that the pores are not simply holes in the membrane. This knowledge comes from the observation that in some cells, small molecules and ions pass through the nuclear membrane at rates much lower than expected if the pores were holes through which diffusion occurs freely.

• PROBLEM 2-13

Why do cells contain lysosomes?

Solution: Lysosomes are membrane-bounded bodies in the cell. All lysosomes function in, directly or indirectly, intra cellular digestion. The material to be digested may be of extracellular or intracellular origin. Lysosomes contain enzymes known collectively as acid hydrolases. These enzymes can quickly dissolve all of the major molecules that comprise the cell, and presumably would do so if they were not confined in structures surrounded by membranes.

One function of lysosomes is to accomplish the self-destruction of injured cells or cells that have outlived their usefulness. Lysosomes also destroy certain organelles that are no longer useful. Lysosomes are, in addition, involved in the digestion of materials taken into the cell in membranous vesicle. Lysosome fuse with the membrane of the vesicle so that their hydrolytic enzymes are discharged into the vesicle and ultimately digest the material. Lysosome play a part in the breakdown of normal cellular waste products, and in the turnover of cellular constituents.

Peroxisomes (or microbodies) are other membrane-bound vesicles containing oxidative enzymes. Peroxisomes play a role in the decomposition of some compounds.

• PROBLEM 2-14

Describe the structure of a centriole. What hypothetical function does it serve?

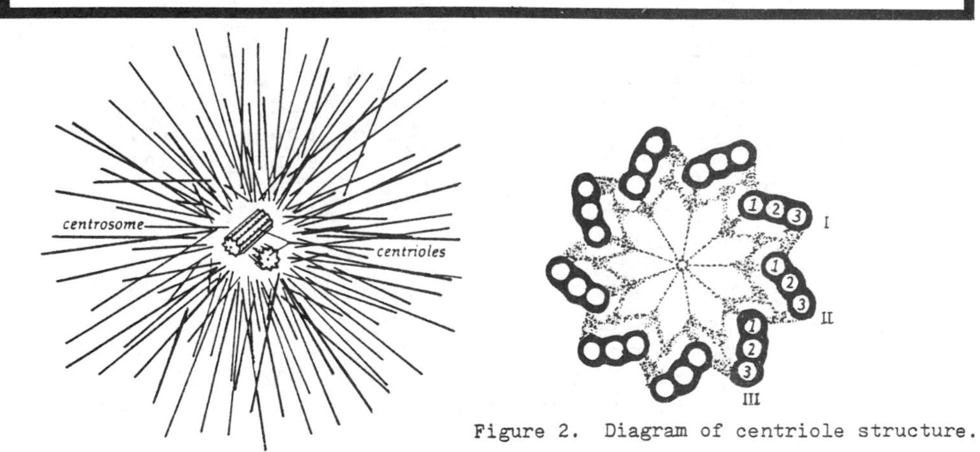

Figure 2. Diagram of centriole structure.

Figure 1. Diagram showing the relationship of the centrioles and centrosome.

Solution: Centrioles are small bodies located just outside the nucleus of most animal cells and some plant cells in a specialized region of cytoplasm that has been known to play a role in cell division. Centrioles usually occur as a pair in each cell oriented at right angles to each other. Each centriole of the pair is composed of nine groups of tubules arranged longitudinally in a ring to form a hollow cylinder. Each group is a triplet composed of three closely associated tubular elements called microtubules. (See Figures.) The space immediately surrounding a pair of centrioles is called the centrosome. The centrosome appears to be clear or empty in the light microscope.

Centrioles do not occur in most higher plant cells, although they are found in some algae and fungi and in a few reproductive cells of higher plants. The centrioles seem to play some part in directing the orderly distribution of genetic material during cell division. At the beginning of cell division the centrioles replicate, and the two pairs of centrioles that result move to opposite poles of the dividing cell. Under the electron microscope each pair is seen to send out spindle fibers, structures involved in separating and moving chromosomes to the opposite ends of the cell. This observation leads to the hypothesis that centrioles are needed in the formation of the spindle fibers. However, the existence of cells without centrioles yet capable of spindle formation seems to refute this hypothesis.

• **PROBLEM 2-15**

Microtubules and microfilaments both appear to be involved in intracellular motion and in intracellular structural support. How do the two organelles differ in structure and function and how are they similar?

Solution: Microtubules are thin, hollow cylinders, approximately 200 to 300 Angstroms in diameter. Microfilaments are not hollow, and are 50 to 80 Angstroms in diameter. Both microtubules and microfilaments are composed of proteins. The protein of microtubules is generally termed tubulin. Tubule proteins can be made to assemble into microtubules in a test tube, if the proper reagents are present. Some of the narrower microfilaments have been shown to be composed of proteins similar to actin. Actin is a protein involved in muscle cell contration. The composition of thicker microfilaments has not been completely determined.

Microtubules are often distributed in cells in arrangement that suggest their role in maintaining cell shape (a "cytoskeletal" role). For example, microtubules are arranged longitudinally within the elongated processes of nerve cells (axons) and the cytoplasmic extensions (pseudopods) of certain protozoa. Microtubules are arranged in a circular band in the disc-shaped fish red blood cells. Microfilaments may also play a structural role. They are often associated with some specialized region of the plasma membrane, such as the absorptive plasma membrane of the intestinal cells, or the portions of the plasma membrane which serve to anchor adjacent cells together so that they can intercommunicate.

Microtubules are components of cilia and flagella, and participate in the rapid movements of these structures. Microtubules are involved in directional movement within the cell during cell division, and are involved in certain types of oriented rapid intracellular movement. Microfilaments seem to be involved in many different types of cytoplasmic movement. They are believed to play a role in amoeboid motion of cells, and in the rapid streaming of the cytoplasm of plant cells about a central vacuole. Close associations between microfilaments and membrane-bound organelles suggests that microfilaments assist in intracellular transport and exchange of materials.

• PROBLEM 2-16

Explain the structural and functional aspects of cilia and flagella.

Solution: Some cells of both plants and animals have one or more hair-like structures projecting from their surfaces. If there are only one or two of these appendages and they are relatively long in proportion to the size of the cell, they are called flagella. If there are many that are short, they are called cilia. Actually, the basic structure of flagella and cilia is the same. They resemble centrioles in having nine sets of microtubules arranged in a cylinder. But unlike centrioles, each set is a doublet rather than than triplet of microtubules, and two central singlets are present in the center of the cylinder. At the base of the cylinders of cilia and flagella, within the main portion of the cell, is a basal body. The basal body is essential to the functioning of the cilia and flagella. From the basal body fibers project into the cytoplasm, possibly in order to anchor the basal body to the cell.

Both cilia and flagella usually function either in moving the cell, or in moving liquids or small particles across the surface of the cell. Flagella move with an undulating snake-like motion. Cilia beat in coordinated waves. Both move by the contraction of the tubular proteins contained within them.

• PROBLEM 2-17

Describe the composition and structure of the plant cell wall.

Solution: Because plant cells must be able to withstand high osmotic pressure differences, they require rigid cell walls to prevent bursting. This rigidity is provided primarily by cellulose, the most abundant cell wall component in plants. Cellulose, a polysaccharide, is a long, linear, unbranched polymer of glucose molecules. In the cell wall cellulose molecules are organized in bundles of parallel chains to form fibrils. The fibrils are often arranged in criss-cross layers reinforced and held together by three other polymeric materials: pectin and hemicellulose (complex polysaccharides) and extensin (a complex glycoprotein). Lignin is an additional polymeric substance found in the wood of trees. Together these substances provide a matrix capable of withstanding enormous stress.

A layer known as the middle lamella lies between and is shared by adjacent cells. The lamella is composed primarily of pectin. By binding the cells together, it provides additional stiffness to the plant.

The cell wall is usually interrupted at various locations by plasmodesmata. These are tiny holes in the cell walls through which run protoplasmic connections between adjacent cells. This provides for intercellular exchange of such materials as water and dissolved substances. Being extremely small, the plasmodesmata do not prevent the cell wall from exerting pressure on a swollen cell.

• PROBLEM 2-18

Plant cells are able to withstand much wider fluctuation in the osmotic pressure of the surrounding medium than animal cells. Explain why.

Solution: A plant cell is able to withstand much greater fluctuations in the makeup of the surrounding fluids than can an animal cell. This can be best understood by examining the role played by the plants' cell wall, a structure absent in animal cells.

First, take the case in which the cell is placed in a hypotonic medium. This means that the surrounding fluid has a lower concentration of solute molecules and a higher concentration of solvent (water) than the cell. Thus, the cell has a higher osmotic pressure than the medium, which means that the net movement of water should be into the cell. This is exactly what happens in an animal cell (see Figure 1). An animal cell will take in water by osmosis causing it to swell. It will take in enough water so that the osmotic pressure of the cell is equal to the osmotic pressure of the medium, thus they will be isotonic to each other. However, if the original difference in pressure was great, the animal cell may have to take in more water than its membrane can allow. Then the animal cell would burst. This is called lysis. If the original cell was a red blood cell, this is hemolysis. A plant cell placed in a hypotonic medium would also have water enter into it, causing it to swell, (see Figure 2). However, an upper limit as to how much water can enter is imposed by the cell wall. As the cell swells, its plasma membrane exerts pressure on the cell wall which is called turgor pressure. The wall exerts an equal and opposing pressure on the swollen membrane. When the pressure

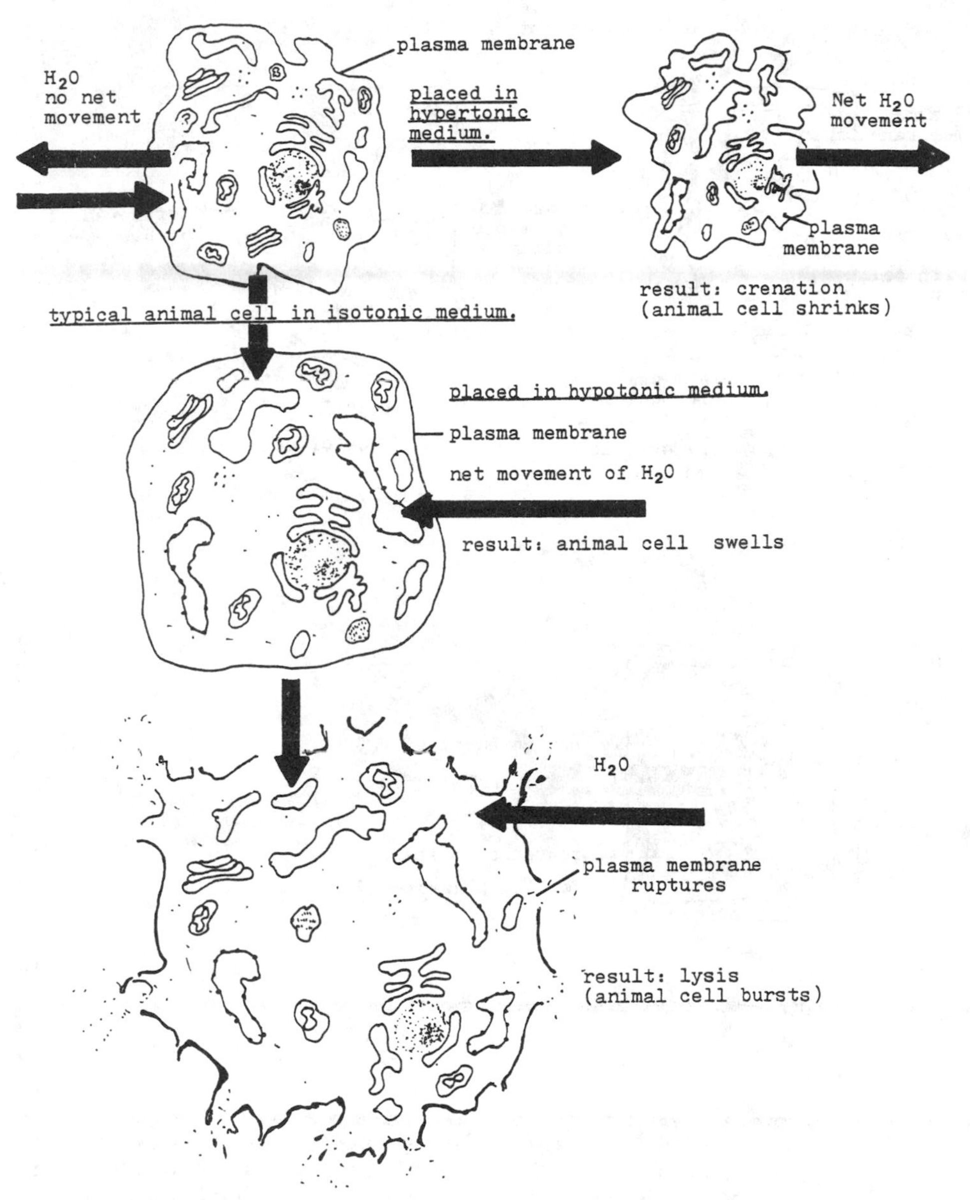

Figure 1. A typical animal cell in varying environments.

exerted by the cell wall is so great that further increase in cell size is not possible, water will cease to enter the cell. Thus, plant cells will only absorb a certain amount of water, even in an extremely dilute medium.

Now consider a cell placed in a hyperosmotic medium. In this case, the medium has a greater osmotic pressure than the cell. Thus, the net movement of water will be out of the cell and into the medium. Since the cell is losing water it will shrink. In an animal cell this phenomenon is called crenation (see Figure 1). In a plant cell, the shrinkage will cause the plasma membrane to pull away from the cell wall. The cell is said to be plasmolyzed, and the phenomenon is plasmolysis (see Figure 2).

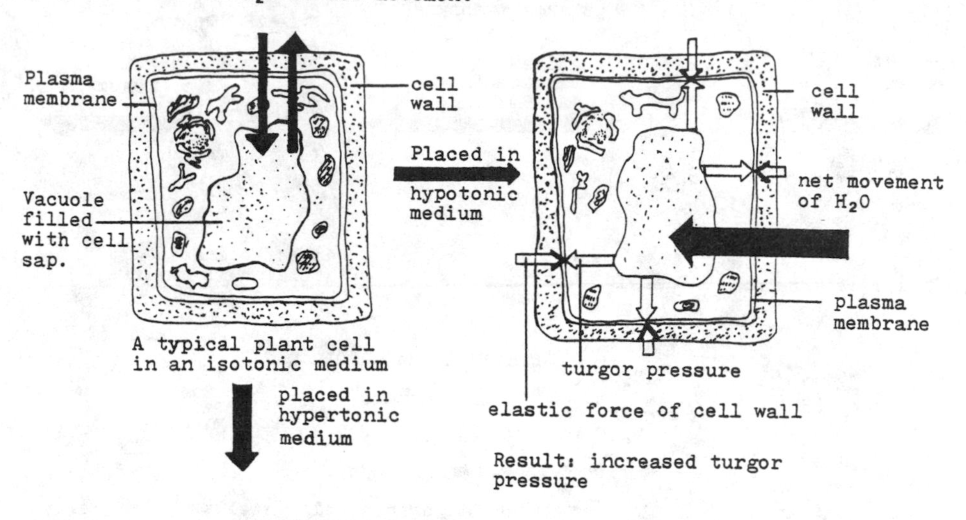

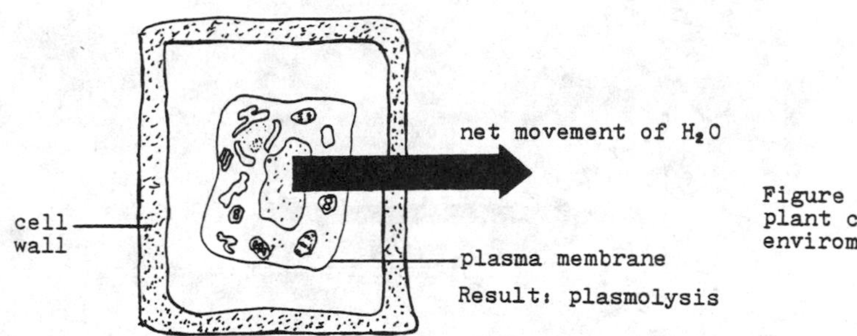

Figure 2. A typical plant cell in varying enviroments

• PROBLEM 2-19

What is the structure and function of the chloroplasts in green plants?

Solution: The chloroplasts have the ability to transform the energy of the sun into chemical energy stored in bonds that hold together the atom of such foods (fuel) as glucose. By the process of photosynthesis, carbon dioxide and water react to produce carbohydrates with the simultaneous release of oxygen. Photosynthesis, which occurs in the chloroplasts, is driven forward by energy obtained from the sun.

Photosynthesis involves two major sets of reactions, each consisting of many steps. One set depends on light and cannot occur in the dark; hence, this set is known as the light reaction.

It is responsible for the production of ATP from sunlight and the release of oxygen derived from water. This process is known as photophosphorylation. The other set, referred to as the dark reaction, is not dependent on light. In the dark reaction, carbon dioxide is reduced to carbohydrates using the energy of ATP from the light reactions.

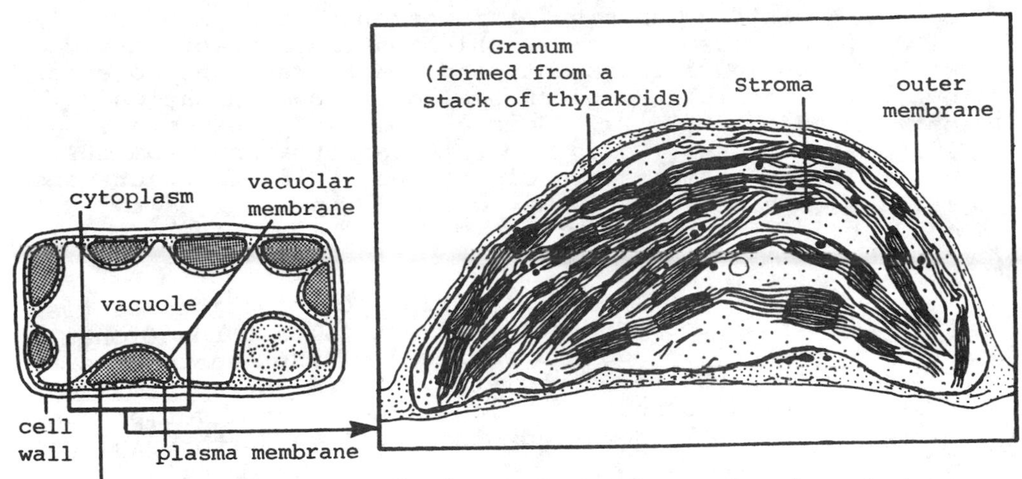

The internal structures of a choroplast.

Chlorophyll,the pigment contained in chloroplasts which gives plants their characteristic green color, is the molecule responsible for the initial trapping of light energy. Chlorophyll transforms light energy into chemical energy; then it passes this energy to a chain of other compounds involved in the light and dark reactions.

Plastids are organelles which contain pigments and/or function in nutrient storage. Chloroplasts are but one example os a plastid; they give the green color to plants. Each chloroplast is bounded by two layers of membranes; the outer one is not connected with the other membranes of the cell. The inner membrane gives rise to the complex internal system of the chloroplast. Surrounding the internal membranes is a ground substance termed the stroma. The stroma contains the enzymes which carry out the dark reactions of photosynthesis. Granules containing starch are also found in the stroma. The internal membrane system of the chloroplast are usually in the form of flattened sacs called thylakoids (see accompanying diagram). A stack of thylakoids is called a granum (plural - grana). The lamellae contain the chlorophyll molecules and other pigments involved in photosynthesis. The layered structure of the granum is extremely crucial for the efficient transfer of energy from one pigment molecule to another without the dissipation of a great deal of energy in the process.

TYPES OF ANIMAL TISSUES

• PROBLEM 2-20

Ordinarily, the bodies of multicellular organisms are organized on the basis of tissues, organs, and systems. Distinguish between these terms.

<u>Solution:</u> A tissue may be defined as a group or layer of similarly specialized cells which together perform a

certain specific function or functions. Each kind of tissue is composed of cells which have a characteristic size, shape and arrangement, and which are bound together by an intracellular substance allowing communication between adjacent cells. Some examples of tissues are epithelial tissue, which separates an organism from its external environments; muscle tissue, which contracts and relaxes; and nervous tissue, which is specialized to conduct information.

An organ is composed of various combinations of tissues, grouped together into a structural and functional unit. An organ can do a specific job: the heart pumps blood through out the circulatory system; or an organ can have several different functions: the human liver produces bile salts, breaks down red blood cells and stores glucose as glycogen.

A group of organs interacting and cooperating as a functional complex in the life of an organism is termed an organ system. In the human and other vertebrates, the organ systems are as follows:

(1) the circulatory system, which is the internal transport system of animals;
(2) the respiratory system, which provides a means for gas exchange between the blood and the environment;
(3) the digestive system, which functions in procurring and processing nutrients;
(4) the excretory system, which eliminates the waste products of metabolism;
(5) the integumentary system, which covers and protects the entire body;
(6) the skeletal system, which supports the body and provides for body shape and locomotion;
(7) the muscular system, which functions in movement and locomotions;
(8) the nervous system, which is a control system essential in coordinating and integrating the activities of the other systems with themselves and with the external environment;
(9) the endocrine system, which serves as an additional coordinator of the body functions; and
(10) the reproductive system, which functions in the production of new individuals for the continuation of the species.

• PROBLEM 2-21

List and compare the tissues that support and hold together the other tissues of the body.

Solution: Connective tissue functions to support and hold together structures of the body. They are classified into four groups by structure and/or function: bone, cartilage, blood, and fibrous connective tissue. The cells of these tissues characteristically screte a large amount of noncellular material, called matrix. The nature and function of each kind of connective tissue is determined largely by the nature of its matrix. Connective tissue cells are actually quite seperate from each other, for most of the connective tissue volume is made up of matrix. The cells between them function indirectly, by secreting a matrix which performs the actual functions of connection or support or both.

Blood consists of red blood cells, white blood cells, and platelets in a liquid matrix called the plasma. Blood has its major function in transporting alnost any substance that is needed, anywhere in the body.

The fibrous connective tissues have a thick matrix composed of interlacing protein fibers secreted by and surrounding the connective tissue cells. These fibers are of three types: collagenous fibers, which are flexible but resist stretching and give considerable strength to the tissues containing them; elastic fibers which can easily be stretched, but return to their normal length like a rubber band when released, and reticular fibers which branch and interlace to form complex networks. These fibrous tissues occur throughout the body, and hold skin to the muscle, keep glands in position, and bind together many other structures. Tendons and ligaments are specialized types of fibrous connective tissue. Tendons are not elastic but are flexible, cable-like cords that connect muscles to each other or to bones. Ligaments are semi-elastic and connect bones to bones.

The supporting skeleton of vertebrates is composed of the connective tissues cartilage and bone. Cartilage cells secrete a hard rubbery matrix around themselves. Cartilage can support great weight, yet it is flexible and somewhat elastic. Cartilage is found in the human body at the tip of the nose, in the ear flaps, the larynx and trachea, intervertebral discs, surfaces of skeletal joints and ends of ribs, to name a few places.

Bone has a hard, relatively rigid matrix. This matrix contains many collagenous fibers and water, both of which prevent the bone from being overly brittle. Bone is impregnated with calcium and phosphorus salts. These give bone its hardness. Bone cells that secrete the body matrix containing the calcium salts are widely seperated and are located in specialized spaces in the matrix. Bone is not a solid structure, for most bones have a large marrow cavity in their centers. Also, extending through the matrix are Haversian canals, through which blood vessels and nerve fibers run in order to supply the bone cells.

• PROBLEM 2-22

Describe the various types and functions pertaining to the epithelial tissues of animals.

Solution: Epithelial tissues form the covering or

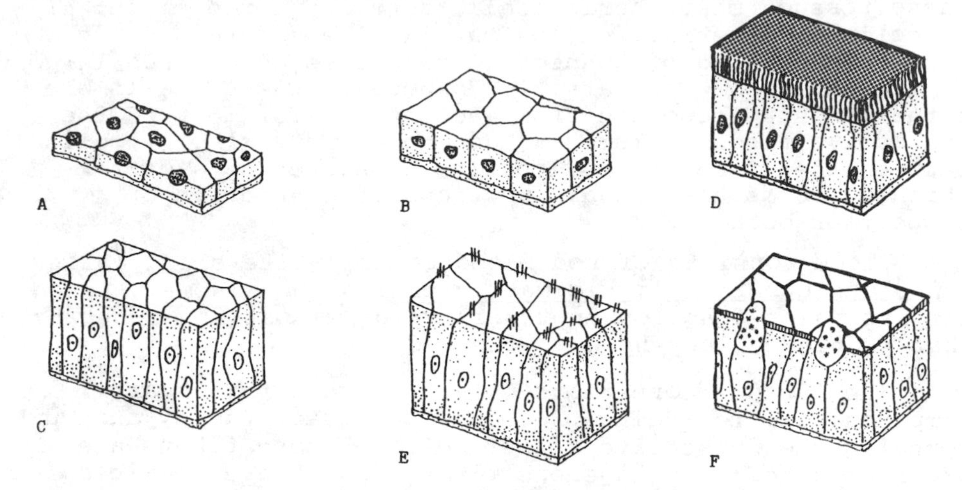

Types of epithelial tissue
A, squamous epithelium B,cuboidal epithelium
C, columnar epithelium D,ciliated columnar epithelium
E, sensory epithelium F,glandular epithelium

lining of the internal and external body surfaces. Epithelial tissues include, for example, the outer part of the skin, the linings of the digestive tract, the lungs, the urogenital tract, the covering of the body cavity and so forth. The cells that make up the epithelial tissues are packed closely together, thus providing a continuous protective barrier between the underlying cells and the outside world. This protection keeps the body safe from mechanical injury, from harmful chemicals and bacteria, and from drying. Epithelial cells may have other functions, such as absorption, secretion, and sensation.

The surface of the epithelial cell that is exposed to air or fluid often becomes specialized. This surface commonly bears cilia, hairs or fingerlike processes. It may also be covered with waxy or mucous secretions. The opposite surface rests upon other cell layers.

Epithelial cells are generally grouped into four categories according to their shape and function: squamous, cuboidal, columnar, and glandular epithelial cells.

Squamous cells are much broader than they are thick and have the appearance of thin, flat plates. They are found on the surface of the skin and the lining of the mouth, esophagus and vagina. Cuboidal epithelium (cube-shaped cells) is found in kidney tubules. The cells of columnar epithelium are long and thin, resembling pillars or columns. The stomach and intestines are lined with columnar eoithelium. Some columnar cells have cilia on their surfaces. The function of the cilia is to move substances past the cell. The respiratory tact is characterized by ciliated columnar epithelium. Finally, a gland may consist of one or many epithelical cells. Exocrine glands have ducts, while endocrine glands do not. Examples of secretions from glandular epithelium include wax, sweat and milk.

Epithelial tissue may further be classified as simple or stratified. Simple epithelium is one cell layer thick. Stratified epithelium consists of several layers of cells. Both simple and stratified epirhelium may be squamous, cuboidal, or columnar.

• PROBLEM 2-23

How are the types of muscle tissues differentiated?

Solution: The cells of muscle tissue have great capacity for contraction. Muscles are able to perform work by the summed contractions of their individual cells. The individual muscle cells are usually elongate, cylindrical or spindle-shaped cells that are bound together into sheets or bundles by connective tissue.

Three principal types of muscle tissue are found in vertebrates. Skeletal or striated muscle is responsible for most voluntary movements. Smooth muscle is involved in most involuntary movements of internal organs, such as the stomach. Cardiac muscle is the tissue of which much of the heart wall is composed; it is also involuntary.

Most skeletal muscle, as the name implies, is attached to the bones of the body, and its contraction is responsible for the movements of parts of the skeleton. Skeletal muscle contraction is also involved in other activities of the body, such as the voluntary release of urine and feces. Thus the movements produced by skeletal muscle are primarily involved with interactions between the body and the external environment.

Skeletal muscle reveals a striated appearance, and is therefore also referred to as striated muscle. These striations are actually due to the regular arrangement of thick and thin myofilaments in individual muscle fiber cells. Skeletal muscle is an exception to the common observation that each cell contains only one nucleus: each skeletal fiber cell is multinucleated: it has many nuclei. Skeletal muscle can contract cery rapidly but cannot remain contracted; the fibers must relax before the next contraction can occur.

Smooth muscle can be classified as visceral smooth muscle. It is found in the walls of hollow visceral organs, such as the uterus, urinary bladder, bronchioles and much of the gastrointestinal tract. Vascular smooth muscle refers to that smooth muscle in the walls of blood vessels. Unlike skeletal muscle cells that are cylindrical, multinucleate, and striated, smooth muscle cells are spindle-shaped, uninucleate, and lack striations. This lack of striations accounts for its smooth appearance. They have a slower speed of contraction, but can remain contracted for a longer period of time than the striated muscle cells.

Cardiac muscle has properties similar to those of both skeletal and smooth muscles. Like skeletal muscle cells, cardiac muscle cells are striated. Like smooth muscle fibers, cardiac muscle fibers are uninucleated and

Comparision of the types of muscle tissue

	Skeletal	Smooth	Cardiac
Location	Attached to skeleton	Walls of visceral organs, Walls of blood vessels	Walls of heart
Shape of fiber	Elongate, cylindrical, blunt ends.	elongate, spindle shaped, pointed ends.	Elongate, cylindrical, fibers branch and fuse.
Number of nuclei per fiber	Many	One	Many
Position of nuclei	Peripheral	Central	Central
Cross striations	Present	Absent	Present
Speed of contractions	Most rapid	Slowest	Intermediate
Ability to remain contracted	Least	Greatest	Intermediate
Type of control	Voluntary	Involuntary	Involuntary

are designed for endured contraction rather than speedy or strong contraction. Cardiac and smooth muscles are not voluntarily controlled but have spontaneous activities, and are regulated by the autonomic nervous system. The skeletal muscle, being voluntary, is controlled by the somatic nervous system.

• PROBLEM 2-24

Which tissue in the body is responsible for the rapid transmission of information?

Solution: To some extent, all cells have the property of irritability, the ability to respond to stimuli. Nervous tissue, however, is highly specialized not only for receiving and responding to such stimuli, but also for the transmission of stimuli. Nerve cells are easily stimulated and can transmit impulses very rapidly. Each nerve cell is specific for the type of information it transmits and the impulse is directed and coordinated to specific areas in the body.

Nervous tissue consists of neurons, cells that conduct electrochemical nerve impulses. Each neuron has an enlarged cell body, which contains the nucleus and two or more thin, hairlike processes extending from the cell body. There are two distinct types of these processes; they differ in the direction they normally conduct a nerve impulse. Axons conduct nerve impulses away from the cell body, while dendrites conduct impulses toward the cell body. The neurons are connected together in chains or networks in order to relay impulses for long distances to different parts of the body. The junction between the terminals of the axon of one neuron and the dendrite of the next neuron in line is called a synapse. The axon and dendrite do not actually touch at the synapse. There is a small gap between the two processes. An impulse can travel across the synapse only in one direction, from an axon to a dendrite. In this way the synapse functions in preventing impulses from backflowing in the wrong direction. A group of axons bound together by connective tissue constitutes a nerve. The functional combination of nerve and

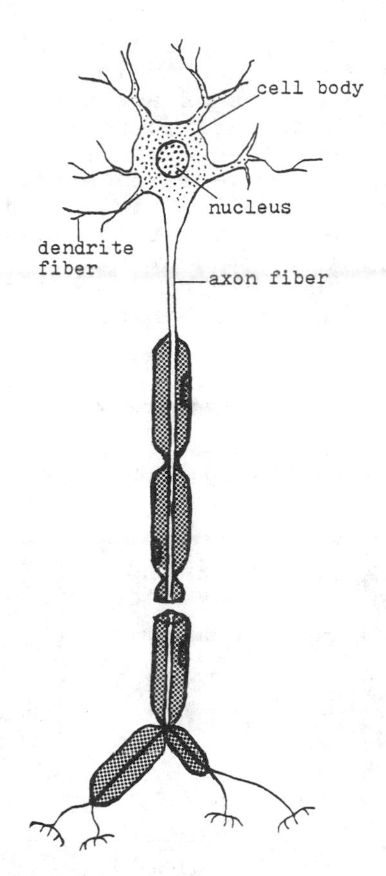

Figure. Nerve cell. The dendrites carry impulses toward the cell body. The axon carries impulses away from the cell body.

muscle tissue is fundamental to all multicellular animals except sponges. These tissues give animals their characteristic ability to move rapidly in response to stimuli. In other words, muscle contraction and thus movement are initiated and controlled by nervous tissue.

TYPES OF PLANT TISSUES

• PROBLEM 2-25

What are the difficulties in the classification of tissues of higher plants?

Solution: The characteristics of plant cells themselves make it difficult for botanists to agree on any one classification system. The different types of cells intergrade and a given cell can change from one type to another during the course of its life. As a result, the tissues formed from such cells also intergrade and can share functional and structural characteristics. Some plant tissues contain cells of one type while others consist of a variety of cell types. Plant tissues cannot be fully characterized or distinguished on the basis of any one single factor such as location, function, structure or evolutionary heritage. Plant tissues can be divided into two major categories: meristematic tissues, which are composed of immature cells and are regions of active cell division; and permanent tissues, which are composed of more mature, differentiated cells. Permanent tissues can be subdivided into three classes of tissues - surface, fundamental and vascular. However, the classification of plant tissues into categories based purely on their maturity runs into some difficulties. Some permanent tissues may change to meristematic activity under certain conditions. Therefore, this classification is not absolutely reliable.

• PROBLEM 2-26

Plant tissues that are neither considered surface tissues nor vascular tissues, are referred to as fundamental tissue. Describe the various types of fundamental tissues.

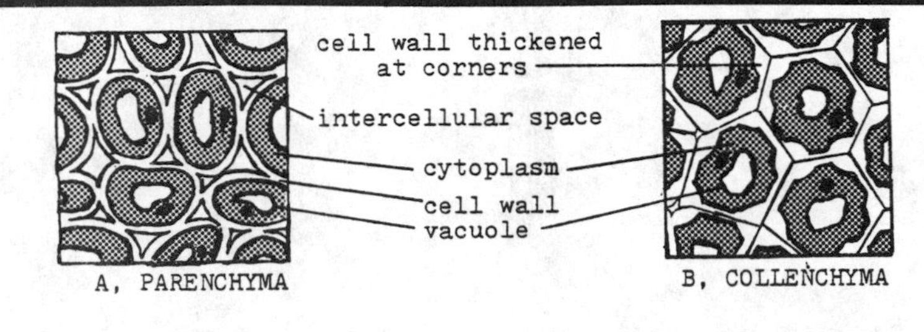

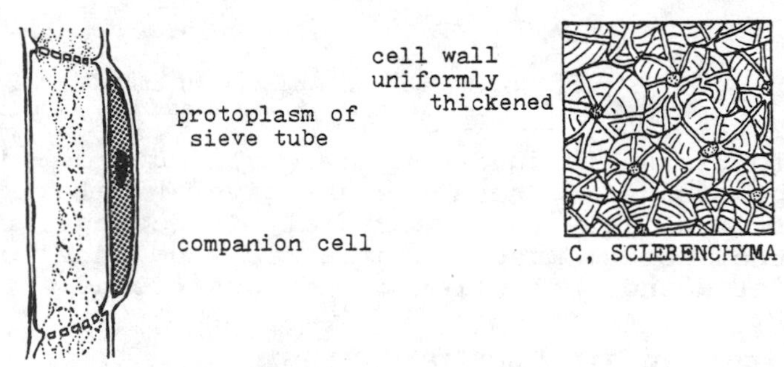

Some types of plant tissues: parenchyma, collenchyma, sclerenchyma, and endodermis.

Solution: Fundamental tissues make up most of the plant body. Examples include the soft parts of the leaf, the pith (central core of stems and roots), the cortex (outer area of stems and roots), and the soft parts of flowers and fruits. The chief functions of the fundamental tissues are the production and storage of food. Some fundamental tissues may also function in physical support of the plant body. There are four types of fundamental tissues, each composed of a single type of cell.

One type is parenchyma tissue, which occurs in roots, stems and leaves. Parenchyma consists of small cells with a thin cell wall and a thin layer of cytoplasm surrounding a large vacuole. The cells are loosely packed, resulting in abundant spaces in the tissue for gas and nutrient exchange. Most of the chloroplasts of leaves are found in these cells.

A second type is collenchyma tissue. It contains cells that are generally more elongate and their cell walls are irregularly thickened compared to the cell walls of parenchyma. The corners of their cell walls are thickened to provide the plant with support. These tissues occur just beneath the epidermis of stems and leaf stalks.

Sclerenchyma tissue, a third type, also functions in support. Most mature cells of this tissue are dead. They have very thick cells walls to provide support and mechanical strength. The cell wall is usually impregnated with an additional tough substance called lignin. The cell wall may be so thick that the internal space of the cell is nearly obliterated. Sclerenchyma is found in many stems and roots. Sclerenchyma cells are divided into two categories, fibers and sclereids. Fibers are elongate cells with tapered ends. Fiber cells are tough and strong, but flexible. Sclereids are more irregularly shaped cells. Stone cells are rounded sclereids. They are found in the hard shells of nuts and in seeds.

Endodermis, the fourth type of fundamental tissue, occurs in a layer surrounding the vascular core of roots. Endodermal cells have cell walls thickened with lignin and suberin which are chemical substances that make the cells waterproof. The endodermal layer is one cell thick, and the cells are compactly arranged. This cell layer is called the Casparian strip. The Casparian strip regulates the entry of water into the vascular tissues of roots.

• **PROBLEM** 2-27

In plants, which tissue is responsible for new growth?

Solution: Meristematic tissue is responsible for new growth in plants. There are several regions in a growing plant where meristematic tissues are found. The apical meristem near the tips of roots, the stem, and branches is important in the growth and differentiation in these areas of a plant. The meristems in the buds of a stem are responsible for outgrowths of the stem. Another meristematic tissue is found in the cambium; this tissue, called the vascular cambium, is where thickening of the stem occurs.

The cells of the meristematic tissue are often small, thin-walled, more or less rounded, each with a large nucleus and few or no vacuoles. Furthermore, meristematic cells generally are closely packed together, and thus few intercellular spaces are found. These cells are embryonic, relatively undifferentiated cells that are capable of rapid cell division. The resultant new cells grow and transform into specific types of plant tissue. The meristem of the root, for example, gives rise to cells that eventually differentiate into all of the cell types present in the root.

A plant embryo is composed entirely of meristematic cells and consequently is capable of rapid growth. As the plant develops, most of the meristem undergoes changes and differentiates into specific tissues. Some of the meristem continues to function in certain parts of the adult plant, and thus provides for continued growth. This gives some plants such as the perennials, the unique ability to continue growth throughout their lives.

• **PROBLEM 2-28**

Which tissue of plants most resembles, in function, the skin of animals?

Solution: Surface tissues form the protective outer covering of the plant body. The epidermis is the principal surface tissue of roots, stems, and leaves. It resembles the skin of animals in having a protective function. This tissue can be one cell thick, although some plants that live in very dry habitats, where protein from water loss is critical, have a thick epidermis consisting of many layers of cells. Most epidermal cells are relatively flat and contain a very large vacuole with only a thin layer of cytoplasm. Their outer walls are thicker than their inner walls. Epidermal cells often secrete a waxy, water-resistant substance, called cutin, on their outer surface. The thick outer cell wall and the cuticle aid in protection of the plant against loss of water.

Some epidermal cells on the surface of the leaves are specialized as guard cells. These regulate the size of small holes in the epidermis, the stomata, through which gases can move into or out of the leaf interior. Epidermal cells of roots have no cuticle which may interfere with their function of absorption. They are, however, characterized by hair-like outgrowths, called root hairs, that increase the absorptive surface for the intake of water and dissolved minerals from the soil.

Periderm is the surface tissue that constitutes the corky outer bark of tree trunks or the outer cork layer of large roots. Cork cells are dead when they are mature, and their cell walls contain another waterproof material, suberin, for additional protection.

MOVEMENT OF MATERIALS ACROSS MEMBRANES

• **PROBLEM 2-29**

Differentiate clearly between diffusion, dialysis and osmosis.

Solution: Diffusion is the general term for the net movement of the particles of a substance from a region where the substance is at a high concentration to regions where the substance is at a low concentration. The particles are in constant random motion with their speed being directly related to their size and the temperature. When the movements of all the particles are considered jointly, there is a net movement away from the region of high concentration towards regions of low concentration. This results in the particles of a given substance distributing themselves with relatively uniform density or concentration within any available space. Diffusion tends to be faster in gases than in liquids, and much slower in solids.

The movement or diffusion of water or solvent molecules through a semipermeable membrane is called osmosis. The osmotic pressure of a solution is a measure of the tendency of the solvent to enter the solution by osmosis. The more concentrated a solution, the more water will tend to move into it, and the higher is its osmotic pressure.

The diffusion of a dissolved substance through a semipermeable membrane is termed dialysis. Dialysis is the movement of the solute, while osmosis is the movement of the solvent through a semipermeable membrane. Dialysis and osmosis are just two special forms of diffusion.

• PROBLEM 2-30

Why is the phenomenon of diffusion important to movement of materials in living cells?

Solution: In a living cell, chemical reactions are constantly taking place to produce the energy or organic compounds needed to maintain life. The reacting materials of chemical reactions must be supplied continuously to the actively metabolizing cell, and the products distributed to other parts of the cell where they are needed or lower in concentration. This is extremely important because if the reactants are not supplied, the reaction ceases, and if the products are not distributed but instead accumulate near the site of reaction, LeChatelier's Principle of chemical reactions operates to drive the reversible reaction backward, diminishing the concentration of the products. Thus, in order to maintain a chemical reaction, the reactants must be continuously supplied and the products must move through the cell medium to other sites. Diffusion is how these processes occur.

When a certain chemical reaction is operating in the cell, some reacting substance will be used up. The concentration of this substance is necessarily lower in regions closer to the site of reaction than regions farther away from it. Under this condition, a concentration gradient is established. The concentration gradient causes the movement of molecules of this substance from a region of higher concentration to a region of lower concentration, or the reaction site. This movement is called diffusion. Thus, by diffusion, molecules tend to move to regions in the cell where they are being consumed. The products of the reaction travel away from the reaction site also by this process of diffusion. At the reaction site, the concentration of the products is highest, hence the products tend to move away from this region to ones where they are lower in concentration. The removal of products signals the reaction to keep on going. When the product concentration gets too high, the reaction is inhibited by a built-in feedback mechanism.

Thus diffusion explains how movement of chemical substances occurs into or out of the cell and within the cell. For example, oxygen molecules are directed by a concentration gradient to enter the cell and move toward the mitochondria. This is because oxygen concentration is necessarily the lowest in the mitochondria where oxidation reactions continually consume oxygen. Carbon dioxide is produced when an acetyl unit is completely oxidized in the citric acid cycle. The CO_2 will then travel away from the mitochondria, where it is produced, to other parts of the cell, or out of the cell into the bloodstream where it is lower in concentration.

• PROBLEM 2-31

Diffusion is too slow a process to account for the observed rate of transport of materials in active cells. Explain why.

Solution: The cytoplasm of a living cell is a watery medium separated by the plasma membrane from the external environment. The cytoplasm resembles a loose gel which is somewhat denser than pure water. In order to move from one part of a cell to another, materials have to move through this relatively dense gel. Thus diffusion of materials through a cell is necessarily a much slower process than diffusion of materials through water. Diffusion is a relatively slow method of transport of materials also because the molecules in motion are constantly colliding with other molecules. This causes them to bounce off in other directions and to take a zig-zag path instead of a straight one. Therefore molecules have to travel a longer distance, which requires more time, in order to get to their destinations.

Diffusion is only a passive process: it does not require energy. Its rate depends on the viscosity (thickness) of the medium, the difference in concentrations of molecules in the two regions, the distance between the two regions, the surface area of the two regions if they are in contact, and the temperature. Often the transport of materials in living cells occurs at so rapid a rate that it is not possible to be accounted for by simple (passive) diffusion alone. Instead, some active processes involving expenditure of energy are believed to be responsible.

• **PROBLEM 2-32**

The concentration of sodium ions (Na^+) inside most cells is lower than the concentration outside the cells. Why can't this phenomenon be explained by simple diffusion across the membrane and what process is responsible for this concentration difference?

Solution: Since the cell membrane is somewhat permeable to sodium ions, simple diffusion would result in a net movement of sodium ions into the cell, until the concentrations on the two sides of the membrane became equal. Sodium actually does diffuse into the cell rather freely, but as fast as it does so, the cell actively pumps it out again, against the concentration difference.

The mechanism by which the cell pumps the sodium ions out is called active transport. Active transport requires the expenditure of energy for the work done by the cell in moving molecules against a concentration gradient. Active transport enables a cell to maintain a lower concentration of sodium inside the cell, and also enables a cell to accumulate certain nutrients inside the cell at concentrations much higher than the extracellular concentrations.

The exact mechanism of active transport is not known. It has been proposed that a carrier molecule is involved, which reacts chemically with the molecule that is to be actively transported. This forms a compound which is soluble in the lipid portion of the membrane and the carrier compound then moves through the membrane against the concentration gradient to the other side. The transported molecule is then released, and the carrier molecule diffuses back to the other side of the membrane where it picks up another molecule. This process requires energy, since work must be done in transporting the molecule against a diffusion gradient. The energy is supplied in the form of ATP.

The carrier molecules are thought to be integral proteins;proteins which span the plasma membrane. These proteins are specific for the molecules they transport.

SPECIALIZATION AND PROPERTIES OF LIFE

• **PROBLEM 2-33**

What advantages do multicellular organisms have over single-celled organisms?

Solution: Single-celled organisms represent one of the great success stories of evolution. They probably comprise more than half the total mass of living organisms, and have successfully colonized even the harshest environments. Biochemically, many unicellular organisms, such as bacteria, are far more versatile than man being able to synthesize virtually everything they need from a few simple nutrients.

With the development of eukaryotic cells, certain evolutionary breakthroughs occurred. Eukaryotic cells are not only capable of being much larger than prokaryotes, they have the ability to aggregate into multicellular functional units. The cells of multicellular organisms are specialized for a variety of functions, and interact in a way that make the organism more than the sum of its parts.

Multicellular organisms employ a high order of complexity. The constituent cells are specialized for the division of labor. The cells that make up the body of a man are not all alike; each is specialized to carry out certain functions. For example, red blood cells carry oxygen from the lungs to the various parts of the body while nerve cells are involved in the transmission of impulses. This specialization allows each cell to function more efficiently at its own task, and also allows the organism as a whole to function more efficiently. Also, injury or death to a portion of the organism does not necessarily inhibit the functioning and survival of the individual as a whole.

Multicellular organisms are better adapted to survive in environments that are totally inaccessible to unicellular forms. This is most striking in the adaptation of multicellular organisms to land. Whereas unicellular animals and plants, such as the protozoans and the one-celled algae, survive primarily in a watery environment, the higher, multicellular organsisms, such as the mammals and angiosperms, are predominantly land-dwellers. Mutlicellularity also carries the potential for diversity. Millions of different shapes, specialized organ systems, and patterns of behavior are found in multicellular organsisms. This diversity greatly increases the kinds of environments that organisms are able to exploit.

• **PROBLEM 2-34**

If all the chemical components found in a living cell were to be put together in the right proportion, the mixture would not function as living matter. What characteristics would enable us to consider something living?

Solution: Living things are characterized by a specific, complex organization: the simplest functional unit of organization is the cell. The cell possesses a membrane,called the plasma membrane, which delimits its living substances from the surroundings. It also has a nucleus which is a specialized region of the cell controlling many important life processes.

Living things have to consume energy in order to stay alive. This energy must come from food; plants (with some exceptions) have the ability to manufacture their own food from sunlight and inorganic materials in the soil and air. Animals, lacking this ability, have to depend ultimately on plants for food. Complex food substances are taken in and degraded into simpler ones; and in the process, energy is liberated to carry out life sustaining functions.

Living things are characterized by their ability to move. The movement of most animals is quite obvious - they crawl, swim, climb, run or fly. The movements of plants are much slower and less apparent, but nonetheless do occur. A few animals, for example sponges and corals, are sedentary - they do not move from place to place. To procure food and other necessities of life, these animals rely on the beating of hair-like projections, called cilia, to move their surroundings (from which they obtain their food) past their body.

The ability to respond to physical or chemical changes in their immediate surrounding is another characteristic of living things. Higher animals have specialized cells or complex organs to respond to certain types of stimuli. Plants also show irritability; the Venus Flytrap is a remarkable example. The presence of an insect on the leaf of this plant stimulates the leaf to fold; the edges come together and ensnare the fly. The fly provides a source of nitrogen for the Venus Flytrap.

Living things grow, that is, increase in cellular mass. This growth can be brought about by an increase in the size of individual cells, by an increase in the number of cells, or both. Some organisms, such as most trees, will grow indefinitely, but most animals have a definite growth period which terminates in an adult form of a characteristic size or age.

Another characteristic of living things is the ability to reproduce. No organism can live forever; hence there must be some way for it to leave descendants that can continue its kind on earth. There are a variety of reproduction mechanisms. Some simple organisms split into two after having attained a certain size. Higher organisms have complex reproductive patterns involving partners of the opposite sex.

The ability of a plant or animal to adapt to its environment is another characteristic of living things. It enables an organism to survive the exigencies of a changing world. Each kind of organism can adapt by seeking out a favorable habitat or mode of life or by undergoing changes to make itself better fit to live in its present environment.

• PROBLEM 2-35

Although it is often possible to see unicellular organisms if a drop of pond water is placed on a microscope slide, it is impossible to see cuboidal epithelium if a rat kidney is placed under the microscope. Describe the steps necessary for the preparation of tissues for light microscopy.

Solution: There are four basic steps in the preparation of tissues for light microscopy.

(1) Fixing. The tissue is killed and stabilized with chemical agents, so that structures are preserved with a life-like appearance. Fixing prevents post-mortem changes of tissue structure caused by autolysis (self-digestion).

(2) Embedding. The fixed tissue is embedded in a hard material that provides firm support. This material is often paraffin or wax. The tissue is dehydrated and soaked in molten wax which penetrates the tissues and then is cooled to form a hard matrix.

(3) Sectioning. The support provided by the embedding material enables the tissue to be cut into very thin slices without being crushed by the knife edge. Sectioning tissue permits study of complex structures by providing an unimpeded view of deep layers.

(4) Staining. The cells are stained with dyes that color only certain organelles. This provides contrast between different cellular structures, for example, between nucleus and cytoplasm or between mitochondria and other cytoplasmic structures. Most cellular structures are colorless. If unstained, it is difficult and often impossible to distinguish different cellular regions by light microscopy.

Cytochemistry refers to staining procedures which specifically color certain cellular substances. For example, the Feulgen procedure specifically dyes DNA. DNA reacts with dilute hydrochloric acid to give aldehyde groups; these then react with the colorless Schiff reagent to change the dye to red. No other cellular components react in the right way with HCl to cause the Schiff reagent to become colored, so only DNA is stained. Other cytochemical techniques permit visualization of RNA, of the acid hydrolases of lysosomes, and of specific lipids, proteins, and carbohydrates.

SHORT ANSWER QUESTIONS FOR REVIEW

Answer

To be covered when testing yourself

Choose the correct answer.

1. Procaryotic cells differ from eucaryotic cells in that the former lack (a) ribosomes. (b) a plasma membrane. (c) endoplasmic reticulum. (d) a cell wall. — c

2. The nucleolus is an organelle that functions in (a) protein synthesis. (b) the production of ribosomal RNA. (c) formation of the mitotic spindle. (d) secretion. — b

3. Unlike plant cells, animal cells possess (a) a cell wall. (b) centrioles. (c) chloroplasts (d) a nuclear membrane. — b

4. Due to their autotrophic nature, plant cells can undergo a certain process which animal cells cannot. This process is (a) aerobic respiration. (b) photosynthesis. (c) protein synthesis. (d) oxidative phosphorylation. — b

5. A factor that is not involved in determining the size of a cell is (a) the number of nuclei per cytoplasmic volume. (b) the volume to surface area ratio. (c) the rate of gas exchange. (d) the rate of DNA replication. — d

6. Sperm cells are highly motile cells and require a great deal of energy to maintain their activity. An organelle that would be found in great abundance in this cell is the (a) mitochondrion. (b) ribosome. (c) lysosome. (d) testosterone. — a

7. A great deal of rough endoplasmic reticulum is observed in a cell. This would lead one to conclude that the cell is actively involved in (a) chromosomal replication. (b) triglyceride metabolism. (c) protein synthesis. (d) pinocytosis. — c

8. The Golgi apparatus primarily functions in (a) packaging protein for secretion. (b) synthesizing protein for secretion. (c) packaging protein for hydrolysis. (d) both a and b — a

9. In animal cells, centrioles play an important role in (a) mitosis. (b) thrombosis. (c) pinocytosis. (d) protein synthesis. — a

Answer

To be covered
when testing
yourself

10. The proteins that comprise microdilaments include (a) tubulin. (b) pseudopods. (c) myosin. (d) actin. — d

11. Intercellular protoplasmic channels in plant cells are known as (a) cellulose. (b) gap junctions. (c) plasmodesmata. (d) desmosomes. — c

12. The chloroplasts in green plants (a) are energy producers. (b) are energy transducers. (c) synthesize ATP from carbon dioxide. (d) result in ADP formation only in the presence of light. — b

13. Spectrin, the extrinsic endoprotein located immediately beneath the cell membrane of a red blood cell, serves (a) to synthesize hemoglobin. (b) to uncouple oxygen from oxyhemoglobin. (c) to provide the tensile strength to the erythrocyte cell membrane. (d) as a hormone receptor. — c

14. Unlike collenchyma and sclerenchyma tissues, parenchyma tissue does not function in (a) support. (b) gas exchange. (c) nutrient exchange. (d) b and c — a

15. The apical meristem is responsible for growth and differentiation in the (a) vascular cambium. (b) stem, root tips and branches. (c) stem, sepal and root tips. (d) blastula — b

16. A normal human erythrocyte contains 0.154 molar of sodium salts; the sodium ion is a non-penetrating particle to the erythrocyte membrane. When placed in a solution of 0.2 molar sodium chloride, the erythrocyte will (a) hemolyze (burst) due to an influx of water. (b) hemolyze due to an influx of sodium. (c) crenate (shrink) due to an efflux of water. (d) crenate due to an efflux of sodium. — c

17. A cell is placed in a solution of dye. After a while, the intracellular concentration of dye becomes much greater than the extracellular concentration. Upon addition of a metabolic inhibitor to the solution, the dye equilibrates across the cell membrane until the intra- and extracellular concentrations are equal. A possible role for this metabolic inhibitor might be to (a) inhibit protein synthesis. (b) delay chromosomal replication. (c) accelerate aerobic respiration. (d) inhibit ATP production. — d

	Answer To be covered when testing yourself
18. In preparing tissues for light microscopy, the sample must be fixed in order to (a) have a firm support. (b) distinguish subcellular regions from each other. (c) afford an unimpeded view of deep layers. (d) prevent autolysis.	d
Fill in the blanks.	
19. _____ serve as cytoskeletal elements in that they give support and shape to the cell.	Micro-tubules
20. Animals are _____ organisms because they depend on other living things for nutrients.	hetero-trophic
21. The inner membrane of the mitochondria is, by a series of folds called _____.	cristae
22. The nucleoprotein complexes within the nucleus comprise the _____, within which is situated the _____, the chief regulators of protein synthesis.	chromatin, genes
23. The pores of the nuclear membrane allow communication between the _____ and _____.	nucleus, cytoplasm
24. The type of enzymes that lysosomes possess are called _____.	acid hydrolases
25. The middle lamella is an intercellular layer in plants, and is mainly composed of _____, a complex polysaccharide.	pectin
26. The analogue of plasmolysis in animal cells is _____.	crenation
27. The inner membrane of a chloroplast is composed of flattened sacs called _____; collectively these sacs are termed _____.	thylakoids, grana
28. Smooth muscle has a lower rate of contraction and hence a _____ refractory period than skeletal muscle.	greater
29. Nerve impulses travel from the _____ to the dendrite across a _____.	axon, synapse
30. _____ is the surface tissue that comprises the outer bark of a tree.	Periderm
31. The trend of evolution has been one of increasing _____ of organs and organ systems.	special-ization

	Answer
	To be covered when testing yourself
Determine whether the following statements are true or false.	
32. Procaryotic cell walls contain muramic acid, whereas eucaryotic cell walls contain celluloid.	False
33. The last stage of mitosis in plant cells is characterized by the formation of a cell plate.	True
34. Surface proteins are believed to be responsible for the property of contact inhibition, which all normal cells exhibit.	True
35. It is believed that smooth endoplasmic reticulum is involved in lipid synthesis as well as ATP production.	False
36. Secretory granules have their contents released from the cell by exocytosis.	True
37. Lysosomes and peroxisomes are similar in that they contain enzymes that function in cellular transport.	False
38. Microfilaments are involved in the cytoplasmic streaming of plant cells.	True
39. The basal body of a cilium is composed of microtubules.	True
40. Turgor pressure will increase when a plant cell is placed in a hypertonic solution.	False
41. Photophosphorylation refers to the phosphorylation of ADP to ATP, using light energy as the driving force of the reaction.	True
42. A tissue is defined as a group of similarly specialized cells which together perform generalized functions.	False
43. Tendons connect muscle to bone whereas ligaments connect bone to bone.	True
44. Plasma is the acellular constituent of blood, and comprises the fluid matrix of this tissue type.	True
45. The tissue lining the kidney tubules is referred to as cuboidal epithelium.	True
46. Skeletal and smooth muscle are primarily responsible for most voluntary movements, whereas striated muscle primarily directs involuntary movements.	False

Answer

To be covered when testing yourself

Question	Answer
47. The Casparian strip is an endodermal plant cell layer which regulates the entrance of water into the vascular tissues of roots.	True
48. The cuticle of a plant cell helps prevent the loss of cytoplasm.	False
49. Guard cells regulate the size of the stomata in order to control the amount of gas exchange.	True
50. Dilute hydrochloric acid reacts with DNA to produce Schiff groups, and subsequent reaction with an aldehyde reagent will turn the solution to red.	False

CHAPTER 3

CELLULAR METABOLISM

PROPERTIES OF ENZYMES

• PROBLEM 3-1

Define the term catalyst. Give an example of an inorganic catalyst and of an organic catalyst.

Solution: A catalyst is a substance which affects the speed of a chemical reaction without affecting its final equilibrium point. The addition of a catalyst enables the reaction to reach equilibrium at a faster rate than it otherwise would. A catalyst is not incorporated into the products of the reaction and hence is capable of being regenerated.

To understand how a catalyst speeds up a chemical reaction, it is first necessary to define the activation energy of a reaction. There is usually an energy barrier to every chemical reaction which prevents the reaction from spontaneously occurring, even if the reaction is exergonic, that is, one with a negative ΔG (a decrease in free energy). This energy barrier is called the activation energy(Ea). One way to overcome the energy barrier, is through the addition of heat. This increases the internal energy content and therefore the frequency of collisions and reactions. A catalyst can also overcome the energy barrier by lowering the activation energy, and thereby increasing the number of molecules with a sufficient energy content to react and form a product. The catalyst acts by forming an unstable intermediate complex with the reactants. This complex is subsequently transformed into the product with the release of the catalyst.

$$\underset{\text{(catalyst+reactant)}}{(C + R)} \rightleftharpoons \underset{\text{(complex)}}{C \cdot R} \rightleftharpoons \underset{\text{(catalyst+product)}}{(C + P)}$$

The catalyst is therefore free to complex with a second molecule of the reactant. The intermediate complex formed provides an alternative pathway of reaction with a correspondingly lower activation energy, thus enabling the reaction to proceed at a faster rate.

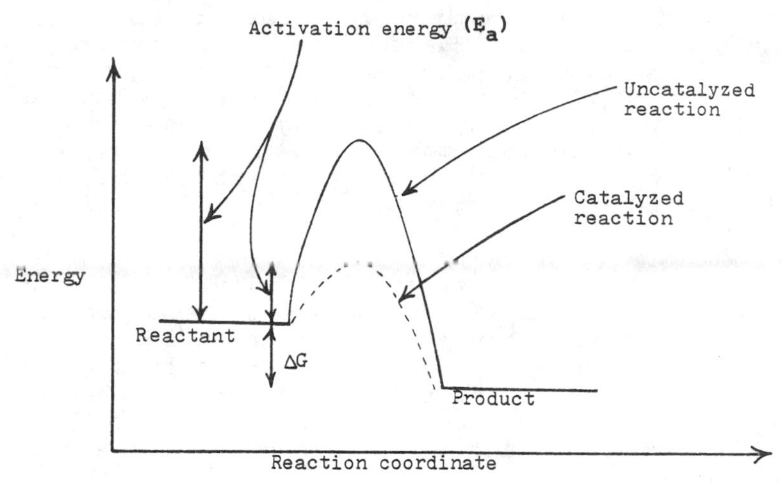

Figure showing a higher activation energy for an uncatalyzed reaction than for a catalyzed reaction, note that ΔG remains the same for both types of reactions.

There are both inorganic catalysts and organic catalysts. Most inorganic catalysts are non-specific, i.e., they will catalyze various similar reactions. For example, platinum black will catalyze any reaction involving H_2, since it serves to weaken the bond between the hydrogen atoms. Other inorganic catalysts include water, iron, nickel, and palladium. On the other hand, organic catalysts such as enzymes are highly selective as to the reactions they will catalyze. The specific molecules which an enzyme will bind to and react with are called substrates. There are many enzymes synthesized by living cells, all with specific functions and names. An example of an organic catalyst is sucrase, which is the enzyme that hydrolyzes sucrose into glucose and fructose. Other important enzymes include lysozyme, chymotrypsin, trypsin, and elastase.

• PROBLEM 3-2

What are some of the important properties and characteristics of enyzmes?

Solution: An important property of enzymes is their catalytic ability. Enzymes control the speed of many chemical reactions that occur in the cell. To understand the efficiency of an enzyme, one can measure the rate at which an enzyme operates – also called the turnover number. The turnover number is the number of molecules of substrate which is acted upon by a molecule of enzyme per second. Most enzymes have high turnover numbers and are thus needed in the cell in relatively small amounts. The maximum turnover number of catalase, an enzyme which decomposes hydrogen peroxide, is 10^7 molecules/sec. It would require years for an iron atom to accomplish the same task.

A second important property of enzymes is their

specificity, that is, the number of different substrates they are able to act upon. The surface of the enzyme reflects this specificity. Each enzyme has a region called a binding site to which only certain substrate molecules can bind efficiently. There are varying degrees of specificity: urease, which decomposes urea to ammonia and carbon dioxide, will react with no other substance; however lipase will hydrolyze the ester bonds of a wide variety of fats.

Another aspect of enzymatic activity is the coupling of a spontaneous reaction with a non-spontaneous reaction. An energy-requiring reaction proceeds with an increase in free energy and is non-spontaneous. To drive this reaction, a spontaneous energy-yielding reaction occurs at the same time. The enzyme acts by harnessing the energy of the energy-yielding reaction and transferring it to the energy-requiring reaction.

The structure of different enzymes differ significantly. Some are composed solely of protein (for example, pepsin). Others consist of two parts: a protein part (also called an apoenzyme) and a non-protein part, either an organic coenzyme or an inorganic cofactor, such as a metal ion. Only when both parts are combined can activity occur.

There are other important considerations. Enzymes, as catalysts, do not determine the direction a reaction will go, but only the rate at which the reaction reaches equilibrium. Enzymes are efficient because they are neeeded in very little amounts and can be used repeatedly. As enzymes are proteins, they can be permanently inactivated or denatured by extremes in temperature and pH, and also have an optimal temperature or pH range within which they work most efficiently.

• PROBLEM 3-3

Describe the mode of action of enzymes. What factors affect enzyme activity?

Solution: An enzyme (E) combines with its substrate (S) to form an intermediate enzyme-substrate complex(ES), which then decomposes into reaction products(P) and the free enzyme, as seen in the equation below.

$$E + S \longrightarrow \underset{\text{enzyme-substrate complex}}{ES} \longrightarrow E + P$$

An enzyme causes the substrate upon which it is acting to be much more reactive than when it is free. One postulate accounting for this is that the enzyme holds the substrate in a position which strains and weakens the substrate's molecular bonds. This weakening of the bonds within the substrate makes them easier to cleave and results in a general lowering of the energy of activation of the reaction. This postulate is extremely simplistic - the actual forces at work are much more numerous and complex.

When the substrate binds to the enzyme, it combines with only a relatively small part of the enzyme molecule the active site. Information about the active site, such as its location and the nature and sequence of amino acids in it, provides an indication of the mechanism of binding and catalysis. The binding of the substrate to the enzyme's active site depends on many forces: hydrogen bonding, the interaction of hydrophobic (water-repelling) groups, and the electrostatic interaction between charged groups on the amino acids. Many active sites also contain metal ions which aid in binding the substrate or expediting the catalytic reaction by withdrawing or stabilizing electrons. For example, the enzyme carboxypeptidase, which hydrolyzes polypeptide bonds of proteins in food, contains a zinc atom in its active site. The electrophilic (electron-attracting) zinc atom coordinates electrons from the carbonyl of the peptide bond, weakening the bond for attack by a specific amino acid of the enzyme at the active site. Such a mechanism, however, is beyond the scope of elementary biology and one would require a good course in biochemistry to understand fully.

Some enzymes, the regulatory or allosteric enzymes, have two binding sites: an active site and a regulatory site. Regulatory enzymes are a key controlling factor in metabolic pathways. If the end product of a pathway is in excess, it inhibits the action of the regulatory enzyme by binding to its regulatory site. The end product shuts off the catalytic activity of the active site by altering the arrangement of the enzyme's polypeptide chains, thus deforming and inactivating the enzyme (see diagram below).

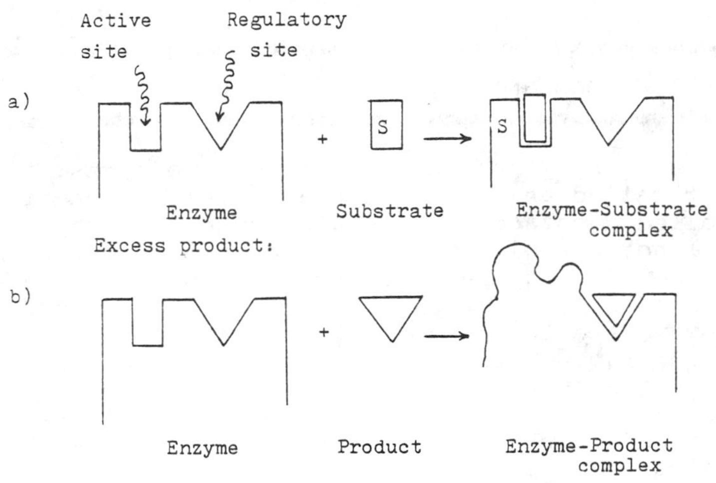

Schematic diagram showing binding at the active site (a) and regulatory site (b) of an enzyme. Note the change in enzyme conformation accompanying binding of product to the regulatory site.

This feedback mechanism is known as end-product inhibition and is important in preventing the accumulation of unwanted substances.

There are several factors which affect enzyme activity, one of which is temperature. High temperatures of 50°C or above can inactivate or denature most enzymes. Upon denaturation, the structure of the enzyme is permanently altered resulting in an irreversible loss of activity. When most organisms are exposed to high temperatures, death occurs due to enzyme inactivation and the resulting loss of metabolic activity. Enzymes are usually not denatured by freezing, but their activity is decreased or disappears. This loss of activity is temporary and the activity reappears upon exposure to normal temperatures. Most enzymes have an optimal temperature range. At temperatures below 50°C, enzymatic reactions double for each temperature rise of 10°.

Enzymes are also affected by and can be denatured by changes in pH. Although most enzymes have an optimum pH around neutrality, (pH 7) some require an acidic medium and others require an alkaline medium. For example, although both pepsin and trypsin works optimally at pH 8.5. The dependence of enzyme activity on pH is explained by the presence of ionizable groups on the protein molecules of the enzyme. pH controls, in part, the number of positive and negative charges on the enzyme molecule, which consequently affect activity.

A final factor affecting enzyme activity is the presence of enzyme poisons. Cytochrome oxidase, an enzyme involved in respiration, is inactivated by minute amounts of cyanide. Death from cyanide poisoning thus results from the inhibition of cytochrome enzymes. Other enzyme poisons include iodoacetic acid, flouride and lewisite.

• PROBLEM 3-4

Distinguish between apoenzymes and cofactors.

Solution: Some enzymes consist of two parts: a protein constituent called an apoenzyme and a smaller non-protein portion called a cofactor. The apoenzyme cannot perform enzymatic functions without its respective cofactor. Some cofactors, however, are able to perform enzymatic reactions without an apoenzyme, although the reactions proceed at a much slower rate than they would if the cofactor and apoenzyme were joined together to form what is called a holoenzyme.

The cofactor may be either a metal ion or an organic molecule called a coenzyme. A coenzyme that is very tightly bound to the apoenzyme is called a prosthetic group. Often, both a metal ion and a coenzyme are required in a holoenzyme.

Metalloenzymes are enzymes that require a metal ion; catalase, for example, which rapidly catalyzes the degradation of H_2O_2 to H_2O and O_2, needs Fe^{+2} or Fe^{+3}. Iron salts can catalyze the same reaction on their own, but the reaction proceeds much more quickly when the iron combines with the apoenzyme. Coenzymes usually function as intermediate carriers of the functional groups, atoms, or electrons that are transferred in an overall enzymatic transfer reaction.

• PROBLEM 3-5

Outline a method, giving the precautions to be aware of, by which an enzyme could be used to measure the amount of a given substrate.

Solution: An enzyme catalysed reaction can be represented by the equation:

$$E + S \rightleftharpoons ES \rightleftharpoons E + P$$

By increasing the concentration of substrate and keeping the concentration of enzyme constant, we can increase the rate of catalysis: the above reaction will shift toward the right. We can react a fixed amount of enzyme with different substrate concentrations to determine the corresponding reaction velocity. The reaction velocity must be an initial velocity, that is, the rate must be measured very soon after the start of the reaction. If the rate is measured over a long period, the measured rate may not be the true value since the reaction would tend to slow down because the amount of substrate becomes used up as the reaction proceeds. The plot of reaction velocity V, as a function of the substrate concentration [S], for a constant amount of enzyme is shown below:

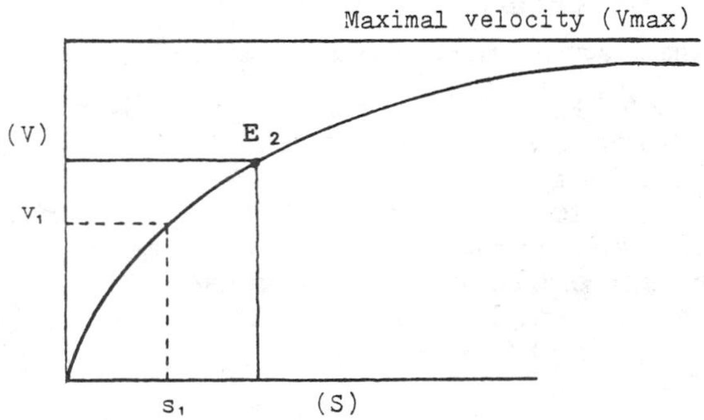

We now know that a specific substrate concentration corresponds to a specific rate. We can then react an unknown amount of substrate with the same amount of enzyme used to determine our original plot. By measuring the initial rate of the reaction, we can then read the corresponding concentration of substrate from our plot. For example, if we measured a rate V_1, we know that the 'unknown' concentration of substrate is S_1. An important precaution must be noted: at very large substrate concentrations, the plot is no longer linear. The rate increases very slowly at large [S] until it reaches a maximum velocity. The slow-down of rate increase occurs when the amount of enzyme is limiting. At this point, (point E_2 on the plot) most of the catalytic sites on the enzyme molecules are filled, so increasing the amount of substrate present has little effect on increasing the rate. Therefore, at very high [S], the rate becomes independent of the amount of substrate present. It is therefore critical that the unknown concentration of substrate be low enough to validly correspond to a specific rate.

One can measure the initial rate of reaction by stopping the reaction after a short period of time, usually by adding certain inhibitors or poisons. To the reaction mixture may then be added a substance which forms a colored complex with one of the reaction products. The denser the color, the more product formed within the measured time interval, and therefore, the higher the reaction rate. The density is related to how much light the colored sample absorbs when a beam of light is shone on the sample. (An instrument called a spectrophotometer is used to measure absorbance of light by a liquid sample.) The measured absorbance is compared to the absorbance of known amounts of product. Hence, when the absorbance measured is known, we will know how much product was formed by the reaction and since we know the time the reaction was stopped, we know how much product is produced per unit time, that is, the rate. The substrate concentration can then be read off the graph from its corresponding velocity.

TYPES OF CELLULAR REACTIONS

• PROBLEM 3-6

What is meant by cellular metabolism? How does metabolism differ from anabolism and catabolism?

Solution: Cellular metabolism includes the following processes that transform substances extracted from the environment: degradation, energy production and biosynthesis. All heterotrophic organisms break down materials taken from their environment, and utilize the product to synthesize new macromolecules. When materials are broken down, energy is released or stored in the cell; when the products are used in syntheses, energy is expended.

There are two general types of metabolism. That part of metabolism by which new macromolecules are synthesized with the consumption of energy is termed anabolism (Greek, ana = up, as in build up). The degradation reactions, which decompose ingested material and release energy, are collectively termed catabolism (Greek, cata = down, as in break down). The degradation of a glucose molecule to carbon dioxide and water during aerobic respiration is an example of catabolism. In the process, 38 molecules of ATP are formed for later use if needed. The degradation of fats is also an example of catabolism. The biosynthesis of proteins (from amino acids) and of carbohydrates like starch or glycogen, (from simple sugars) are two important anabolic processes.

• PROBLEM 3-7

What is the evidence that the chemical compounds of a cell are in a "dynamic state"?

Solution: The term "dynamic state" is meant to indicate that the chemical compounds of a cell are constantly being broken down and synthesized. At one time, it was thought that since an adult organism fails to change in outward appearance over a short period of time, its molecular constitution is also unchanging. That is, the cellular compounds are stable and remain in the cell for long periods of time without being degraded and replaced, the only exception being those compounds used as "fuel" for energy, which by necessity must undergo catabolism.

However, experimental evidence has shown that this belief in the static state of the cell's compounds is false. The cell's compounds are constantly changing - being broken down and replaced by newly synthesized compounds. Radioactively labelled amino acids, fats, and carbohydrates can be injected into laboratory animals in order to demonstrate the dynamic nature of the cell's environment. If the cell were static, the labelled molecules would only be broken down for fuel.

However, it has been observed that the labelled amino acids are incorporated into proteins and that the labelled fatty acids into fats(triglycerides). The marked proteins and fats are then degraded by the body to be replaced by the incorporation of new amino acids and fatty acids. All these metabolic processes occur without any visible change in body size. The only exception to the dynamic state of the cell is its DNA molecules, which contain genes, the body's basic units of genetic information. DNA is extremely stable in order that the hereditary characteristics, for which it codes, have little or no chance of changing.

Not only is the cell dynamic on the molecular level, but on the cellular level as well. Indeed, 2.5 million red blood cells alone are produced and destroyed per second in the human body.

• PROBLEM 3-8

How are oxidation and reduction reactions related? How does an oxidase enzyme differ from a dehydrogenase enzyme?

Solution: An oxidation reaction involves the removal of electrons from an atom or molecule. The reverse process, called a reduction reaction, involves the addition of electrons to an atom or molecule. Every oxidation reaction must be accompanied by a corresponding reduction reaction. The electrons given off by the substance oxidized (also called the reductant or reducing agent), must be accepted by the substance reduced (also called the oxidant or oxidizing agent). A simple example of an oxidation reaction is the oxidation of iron:

$$Fe^0 \longrightarrow Fe^{2+} + 2e^-$$

This involves the actual removal of electrons from the iron atom. The reverse process is a reduction reaction.

$$Fe^{2+} + 2e^- \longrightarrow Fe^0$$

A not-so-obvious oxidation reaction is the oxidation of methane (CH_4):

$$CH_4 + 2O_2 \longrightarrow CO_2 + 2H_2O$$

This involves the formation of bonds in which electrons are drawn toward a more electronegative (electron attracting) atom or atoms which is essentially the loss of electrons required in oxidation. Both carbon and hydrogen are considered to be oxidized, since in the new

$$\overset{\displaystyle O}{\overset{\|}{C}} = O \quad \text{and} \quad H - \overset{\displaystyle H}{\overset{|}{O}} \quad \text{bonds,}$$

the electrons are removed by oxygen farther from the C and H atoms then they were in the C-H bonds of methane. Oxygen is strongly electronegative and is a common agent in oxidation reactions.

Oxidation and reduction processes are important as a means of liberating and storing energy. This can be seen in the oxidation of succinate to fumarate by succinic dehydrogenase. The removal of two hydrogen atoms (with their corresponding electrons) from succinate forms fumarate. Succinate is therefore oxidized. The two hydrogen atoms are transferred to a molecule of flavin adenine dinucleotide (FAD) to form $FADH_2$. FAD is therefore reduced. This coupling of an oxidation to a reduction reaction traps the energy given off during the oxidation of succinate and uses it to synthesize $FADH_2$, an energy-requiring reduction process. $FADH_2$ can subsequently undergo an oxidation reaction and release this temporarily stored energy to another molecule. Oxidation processes thus liberate energy, while reduction processes act to store chemical energy.

Most organisms obtain energy by enzymatic reactions which involve the flow of electrons from NADH (reduced nicotinamide adenine dinucleotide) to oxygen, the final electron acceptor. The electrons are transferred by a system of enzymes located in the mitochondria called the electron transport system. An oxidase is an enzyme which removed electrons, while a dehydrogenase is an enzyme which removes hydrogen ions (H+) and their associated electrons. Cytochrome oxidase, the final enzyme in the electron transport chain, transfers electrons directly to O_2 to form H_2O. Succinate dehydrogenase, in the citric acid cycle, transfers electrons and hydrogen ions from succinate to FAD, yeilding fumarate and $FADH_2$ as a result.

ENERGY PRODUCTION IN THE CELL

• PROBLEM 3-9

Explain the events which take place during glycolysis.

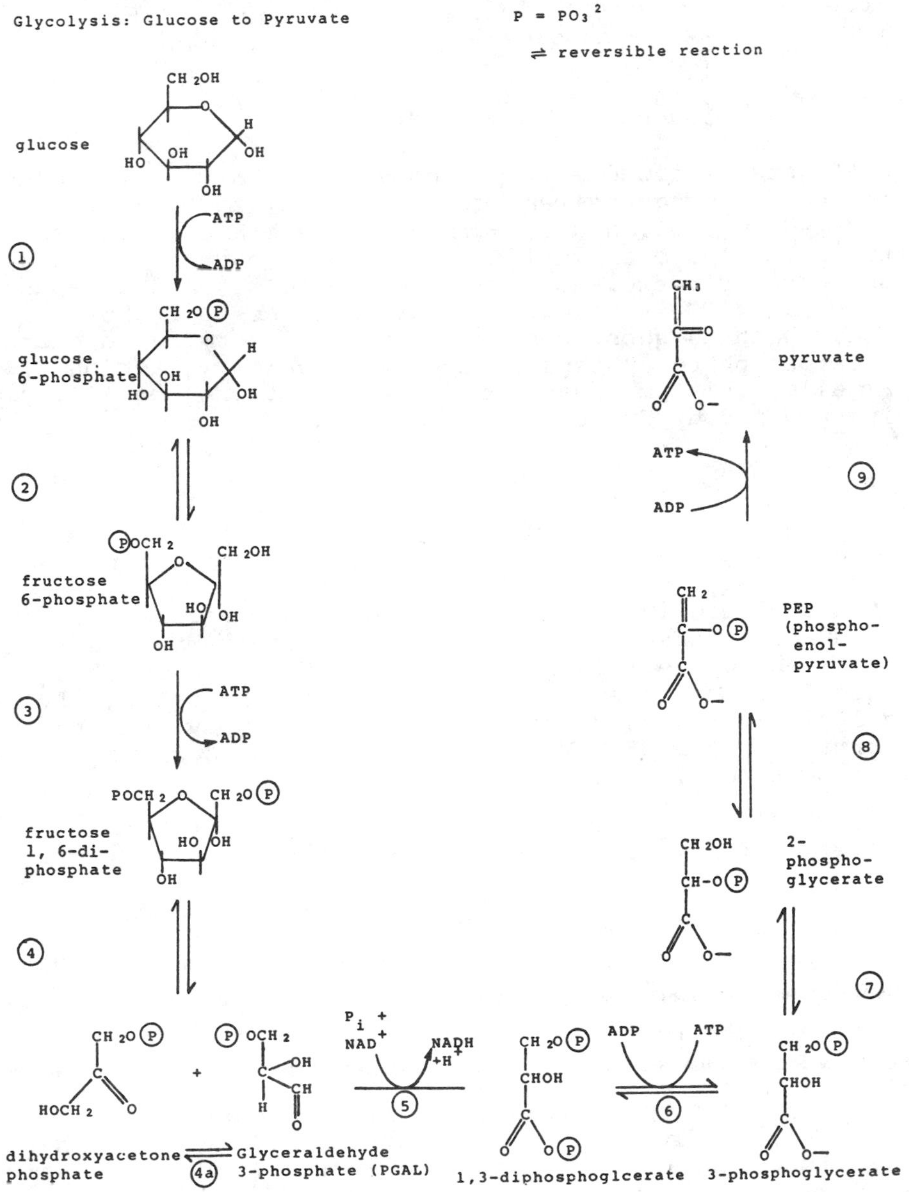

Glycolysis: Glucose to Pyruvate

P = $PO_3{}^2$

⇌ reversible reaction

glucose

① ATP → ADP

glucose 6-phosphate

②

fructose 6-phosphate

③ ATP → ADP

fructose 1, 6-diphosphate

④

dihydroxyacetone phosphate ⇌ (4a) Glyceraldehyde 3-phosphate (PGAL)

⑤ P_i + NAD^+ → NADH + H^+

1,3-diphosphoglcerate

⑥ ADP → ATP

3-phosphoglycerate

⑦

2-phosphoglycerate

⑧

PEP (phosphoenolpyruvate)

⑨ ADP → ATP

pyruvate

Solution:

Glycolysis is the series of metabolic reactions by which glucose is converted to pyruvate, with the concurrent formation of ATP. Glycolysis occurs in the cytoplasm of the cell, and the presence of oxygen is unnecessary. Glucose is first "activated," or phosphorylated by a high-energy phosphate from ATP (See reaction 1). The product, glucose-6-phosphate, undergoes rearrangement to fructose-6-phospate, which is subsequently phosphorylated by another ATP to yield fructose-1, 6-diphosphate (reactions 2 and 3). This hexose is then split into two three-carbon sugars, glyceraldehyde-3-phosphate (also called PGAL) and dihydroxyacetone, phosphate (see reaction 4). Only PGAL can be directly degraded in glycolysis; dihydroxyacetone phosphate is reversibly converted into PGAL by enzyme action (reaction 4a).

Since two molecules of PGAL are thus produced per molecule of glucose oxidized, the products of the subsequent reactions can be considered "doubled" in amount

$$1 \text{ glucose} \longrightarrow 2 \text{ PGAL}$$

PGAL gets oxidized and phosphorylated. NAD^+, the coenzyme in the dehydrogenase enzyme which catalyzed this step, gets reduced to NADH, and 1,3-diphosphoglycerate is formed (reaction 5). The energy-rich phosphate at carbon 1 of 1,3 diphosphoglycerate reacts with ADP to form ATP and 3-phosphoglycerate. This undergoes rearrangement to 2-phosphoglycerate, which is subsequently dehydrated, forming an energy-rich phosphate: phosphoenolpyruvate (PEP) (reaction 8). Finally, this phosphate group is transferred to ADP, yielding ATP and pyruvate. (reaction 9)

Since two molecules of PGAL are formed per molecule of glucose, four ATP molecules are produced during glycolysis. The net yield of ATP is only 2, since 2 ATP were utilized in initiating glycolysis (reactions 1 and 3). Pyruvate is then converted to acetyl coenzyme A which enters the Krebs cycle. In addition, two molecules of NADH are produced per molecule of glucose. Hence, the net result of glycolysis is that glucose is degraded to pyruvate with the net formation of of 2 ATP and 2 NADH. The process of glycolysis can be summarized as follows:

$$\text{glucose} + 2\text{ADP} + 2\text{Pi} + 2\text{NAD}^+ \longrightarrow 2 \text{ pyruvate} + 2\text{ATP} + 2\text{NADH} + 2\text{H}^+$$

• PROBLEM 3-10

What are the roles of coenzyme A in metabolism?

Solution: Coenzyme A is a large, complex organic molecule involved in a number of key reactions in the cell. The functions of coenzyme A are several. In glycolysis, glucose is converted to pyruvate. To enter the Krebs acid cycle (also known as the tricarboxylic acid, or TCA cycle and the citric acid cycle), pyruvate must be converted to another form; such a conversion first involves decarboxylation of pyruvate (loss of carbon dioxide), then a reaction with coenzyme A, and finally, a dehydrogenation reaction, to yield acetyl coenzyme A. The reactions are represented as follows:

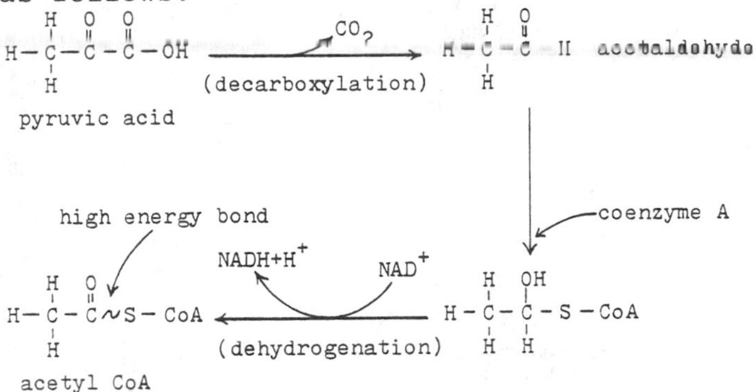

The acetate group is bonded to the coenzyme A molecule via a high energy bond. The energy of this bond in acetyl coenzyme A is utilized to transfer the acetyl group to oxaloacetic acid, the first reaction of the TCA cycle.

Coenzyme A is also involved in a reaction farther along in the cycle. An intermediate (derived from α-Ketoglutarate) combines with coenzyme A to yield succinyl coenzyme A. The bond connecting coenzyme A to succinate is also of high energy. The energy in this bond is converted to an energy-rich phosphate bond, ~P, through the addition of an inorganic phosphate. The phosphate group is subsequently transfered to GDP to produce GTP and free succinate. The reactions are diagrammed below:

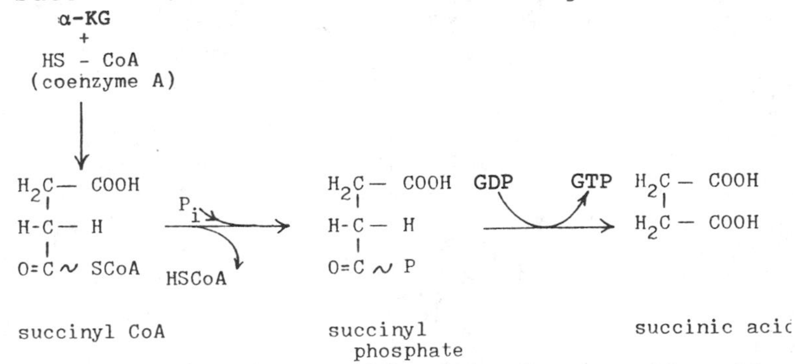

Coenzyme A is also utilized in fatty acid oxidation. The metabolism of a fatty acid initially involves its activation by ATP and coenzyme A to fatty acyl coenzyme A. Through a repeating series of reactions, fatty acyl coenyzme A is converted to acetyl coenzyme A, which, in turn, goes into the TCA cycle

As a general rule, coenzyme A is involved in reactions which utilize its high-energy bond in order to more effectively facilitate metabolism.

The sequence of reactions, wherein Acyl CoA is

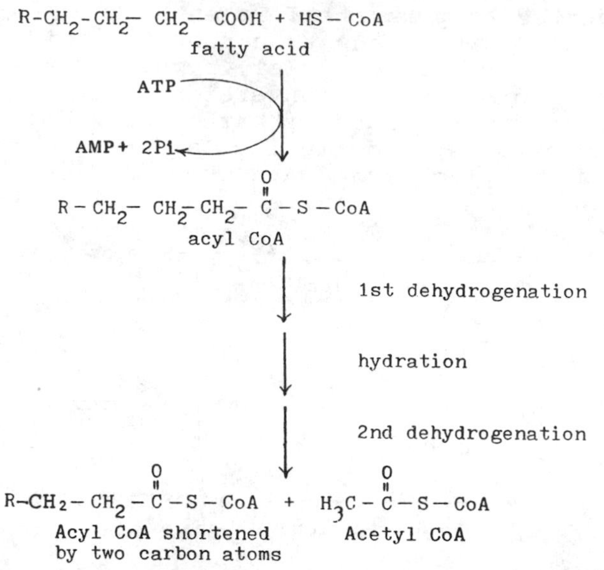

shortened by two carbon atoms, repeats itself until the Acyl CoA is shortened to Acetyl CoA.

• PROBLEM 3-11

> Discuss the citric acid cycle as the common pathway of oxidation of carbohydrates, fatty acids and amino acids.

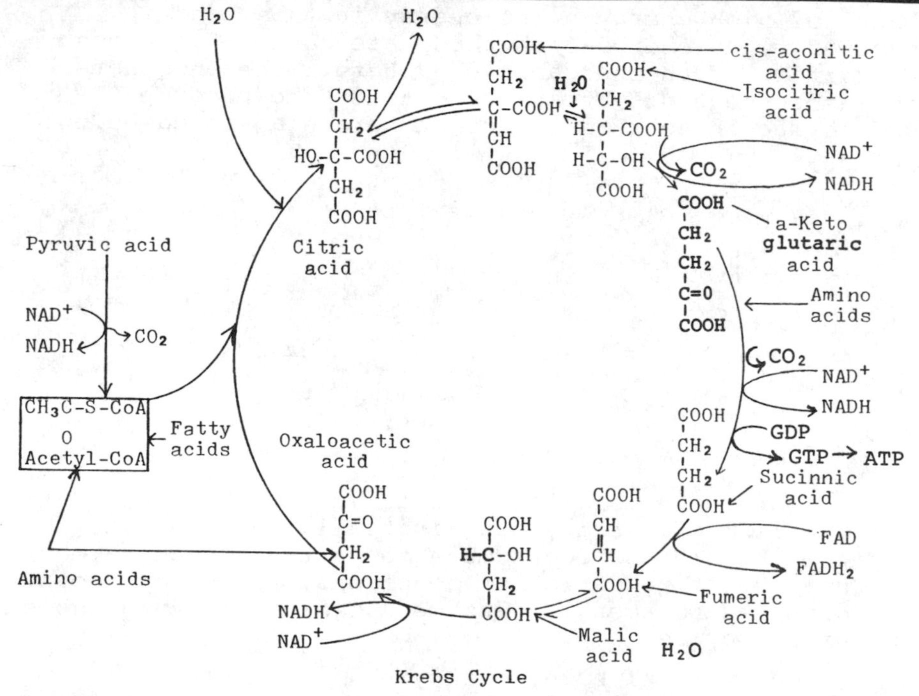

Krebs Cycle

Solution: The oxidation of the above substances all yield products which either enter or are intermediates of the citric acid (also known as Krebs and TCA, tricarboxylic acid) cycle. Glucose, a carbohydrate, undergoes a series of reactions in glycolysis to be oxidized to pyruvic acid. Pyruvic acid reacts with coenzyme A to yield acetyl coenzyme A. It is acetyl coenzyme A which enters the citric acid cycle (see Figure).

Acetyl coenzyme A is also formed from the oxidation of fatty acids. A repeating series of reactions cleaves the long carbon chain of a fatty acid into molecules of acetyl coenzyme A.

Finally, amino acids are metabolized by one of several reactions to products capable of participating in the reactions of the TCA cycle. The first step in the oxidation of all amino acids is a deamination reaction by which an amino group ($-NH_2$) is removed from the amino acid. For three amino acids, this is the only reaction required for conversion into a compound entering the TCA cycle directly. Alanine undergoes deamination to form pyruvate which is converted into acetyl coenzyme A. The deamination of glutamic acid yields α-ketoglutaric acid,while that of aspartic acid yields oxaloacetic acid. Both these compounds are intermediates of the TCA cycle. The TCA cycle is thus the final common pathway by which carbohydrates, fatty acids and amino acids are metabolized.

• PROBLEM 3-12

Does a yeast cell metabolize more efficiently in the presence or in the absence of oxygen? Explain.

Solution: Under aerobic conditions (in the presence of oxygen), glucose undergoes glycolysis to form pyruvate, which after conversion to acetyl CoA, enters the TCA cycle. The NADH produced is oxidized in the electron transport system provided that oxygen is available as the ultimate electron acceptor. Let us calculate the maximum amount of ATP produced from the oxidation of one mole of glucose. A net gain of 2 ATP is obtained directly in glycolysis (see figure). In addition, 2 NADH are generated. This occurs in the cytoplasm. To enter the electron transport system, the NADH must be transported into the mitochondrion, a process which requires 1 molecule of ATP for every NADH which crosses the mitochondrial membrane. So intead of three ATP being produced per NADH, each cytoplasmic NADH yields only 2 net ATP; that is, a total of 4 ATP are produced by the 2 NADH generated in glycolysis.

Recall that two 3-carbon molecules are obtained in glycolysis. Each of these two 3-carbon compounds can eventually enter the TCA cycle as acetyl CoA, and later, their electrons enter the electron transport system. Thus ATP production from the TCA cycle is twice that of a single turn. Two ATP are produced in the cycle via the formation of 2 molecules of GTP from 2 molecules of succinyl CoA. The conversion of pyruvate to acetyl CoA yields one NADH while the Krebs cycle yields 3 NADH. Thus 8 (4 x 2) NADH are produced per glucose at this point. The 8 (4 per turn) NADH formed in the cycle yield 24 ATP (3 ATP per NADH) by means of the electron transport chain. The 2 $FADH_2$ (1 per turn) produce 2 ATP each via electron transport, giving a total of 4 ATP. Summation of all the ATP from glycolysis, the citric acid cycle, and electron transport leads to a net production of 36 ATP per mole of glucose fully oxidized.

Under anaerobic conditions, oxygen is not available as a terminal electron acceptor, and the reactions of the electron transport system cease as the reduced intermediates build up (refer to questions 3-20 and 3-26). This leads to an accumulation of the TCA cycle intermediates and, since it cannot enter the cycle, an accumulation of pyruvate.Under these conditions in yeast cells, pyruvate is then decarboxylated to acetaldehyde, which is in turn reduced to ethyl alcohol (ethanol), regenerating oxidized NAD^+ for further use in glycolysis. This process is called alcoholic fermentation

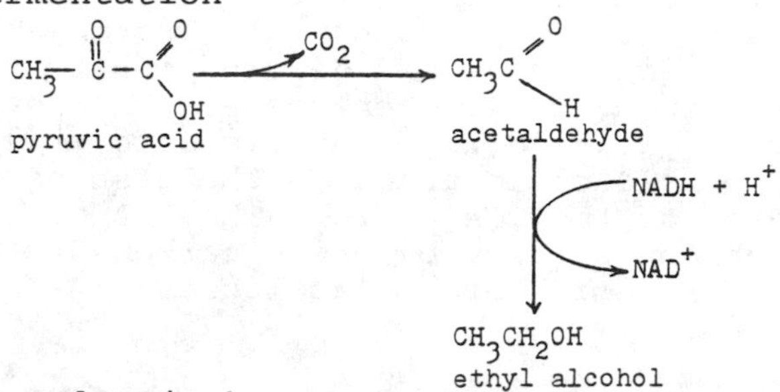

The analog in human muscle cells is the reduction of pyruvic acid to lactic acid. (Pyruvate and lactate are the ionized forms of these acids: i.e. Pyruvic acid is $CH_3\underset{O}{\underset{\|}{C}}-COOH$, while pyruvate is $CH_3\underset{O}{\underset{\|}{C}}-COO^-$). Under anaerobic conditions, this too regenerates NAD^+ in order to continue glycolysis for energy production: Pyruvic acid gets reduced to lactic acid as NADH gets oxidized back to NAD^+. (see next question).

Under anaerobic conditions then, only 2 ATP molecules are produced, as opposed to the 36 ATP produced when glycolysis is supplemented with aerobic respiration Thus, a yeast cell metabolizes 18 times more efficiently in the presence of oxygen than in the absence of oxygen.

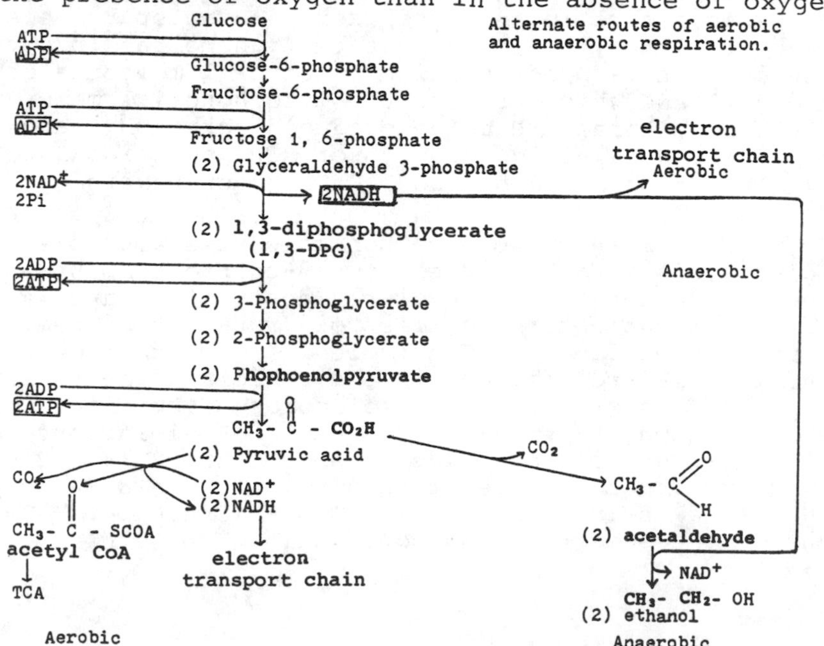

Alternate routes of aerobic and anaerobic respiration.

ATP yield from the complete oxidation of glucose

Reaction sequence	ATP yield per glucose
GLYCOLYSIS: GLUCOSE TO PYRUVATE (in the cytoplasm)	
Phosphorylation of glucose	-1
Phosphorylation of fructose 6-phosphate	-1
Dephosphorylation of 2 molecules of 1, 3-DPG	+2
Dephosphorylation of 2 molecules of phosphoenolpyruvate	+2
2 NADH are formed in the oxidation of 2 molecules of glyceraldehyde 3-phosphate	
CONVERSION OF PYRUVATE TO ACETYL CoA (inside mitochondria)	
2 NADH are formed	
CITRIC ACID CYCLE (inside mitochondria)	
Formation of 2 molecules of guanosine triphosphate from 2 molecules of succinyl CoA	+2
6 NADH are formed in **three oxidation steps.**	
2 $FADH_2$ are formed in 1 oxidation step.	
OXIDATION PHOSPHORYLATION (inside mitochondria)	
2 NADH formed in glycolysis; each yields 2 ATP (not 3 ATP each, because of the cost of the shuttle)	+4
2 NADH formed in the oxidative decarboxylation of pyruvate; each yields 3 ATP	+6
2 FADH formed in the citric acid cycle; each yields 2 ATP	+4
6 NADH formed in the citric acid cycle; each yields 3 ATP	+18
	36 ATP/glucose

• PROBLEM 3-13

Explain how the metabolism of a fermenting yeast cell differs from that of a fatigued muscle cell.

Solution: We shall first examine the anaerobic metabolism of yeast. When yeast is placed in an environment devoid of oxygen, their cells are able to extract energy from glucose in the absence of molecular oxygen via glycolysis, which occurs in both aerobic and anaerobic organisms. There are no hydrogen acceptors due to the absence of molecular oxygen, the final acceptor. When O_2 is not available, all intermediates of the respiratory chain are reduced, and the TCA cycle and electron transport system are turned off. Pyruvate, the degradation product of glucose, cannot therefore be oxidized to acetyl CoA and enter the TCA cycle for further energy extraction. Instead, pyruvate itself accepts the electrons (via acetaldehyde) and is reduced to ethyl alcohol (ethanol).

$$\underset{\text{pyruvate}}{CH_3-\overset{O}{\overset{\|}{C}}-C\overset{\nearrow O}{\underset{\searrow OH}{}}} \xrightarrow{CO_2} \underset{\text{acetaldehyde}}{CH_3-\overset{O}{\overset{\|}{C}}-H} \xrightarrow{NADH+H^+ \quad NAD^+} \underset{\text{ethanol}}{CH_3CH_2OH}$$

This process, known as alcoholic fermentation, converts each glucose molecule to two ethanol molecules and two CO_2. Alcoholic fermentation also serves to regenerate oxidized NAD^+, so that glycolysis can continue. Two net ATP's are formed per glucose molecule in alcoholic fermentation.

In man and other organisms, anaerobic respiration can supplement aerobic respiration for short periods of time. In a fatigued muscle cell, pyruvic acid, produced via glycolysis, again serves as an electron acceptor since oxygen is unavailable. However, unlike yeast, human cells produce lactic acid instead of ethanol and carbon dioxide:

$$\underset{\text{pyruvic acid}}{CH_3-\overset{O}{\overset{\|}{C}}-C\overset{\nearrow O}{\underset{\searrow OH}{}}} \xrightleftharpoons[]{NADH+H^+ \curvearrowright NAD^+} \underset{\text{lactic acid}}{CH_3-\underset{H}{\overset{OH}{C}}-C\overset{\nearrow O}{\underset{\searrow OH}{}}}$$

This process, sometimes called homolactic fermentation, occurs during strenuous exercise when muscle cells cannot obtain enough oxygen to support aerobic respiration. Instead, the energy expended comes from glycolysis (2 ATP per glucose molecule) with the build-up of lactic acid, the cause of muscle fatigue.

Although both processes, alcoholic and homolactic fermentation, enable energy production, they are not as efficient as aerobic respiration. A yeast cell or skeletal muscle cell can theoretically obtain 18 times as much energy by oxidizing a glucose molecule completely to carbon dioxide and water, instead of stopping at ethanol, if it employs the aerobic type of respiration.

• PROBLEM 3-14

If a living frog muscle is dissected out from a frog's body and placed in a closed container without oxygen, it can contract for a time when stimulated. Then it stops. But if the container is opened, it soon regains the ability to contract. Comment.

Solution: The important clue in this question is the presence or absence of oxygen and how it may affect metabolism. When the muscle is still intact within the frog, the muscle cells should be undergoing aerobic respiration, since oxygen is available to each cell (by diffusion from the blood). However, once the muscle is removed and placed in an anaerobic environment, aerobic respiration is replaced by homolactic fermentation. Recall that this process involves the reduction of pyruvic acid to lactic acid due to the absence of any hydrogen acceptors. (see previous question) The build-up of lactic acid produces fatigue and muscle cramps. Since glycolysis produces only 2 net ATP per glucose molecule consumed, the ATP produced thus is not sufficient to replenish the ATP expended in muscle contraction. Eventually, the cell's ATP storage is depleted and conraction can no longer occur. However, when the container is opened, oxygen is reintroduced and is now available to accept hydrogen. Lactic acid is first converted back to pyruvic acid which eventually enters the TCA cycle and the respiratory system to produce more ATP. The presence of oxygen enables the muscle cells to undergo aerobic respiration with the concomitant increase in ATP production and oxidation of lactic acid back to pyruvic acid. This oxidation occurs in the liver. Under these conditions, the muscle regains its ability to contract.

The same process occurs in humans after vigorous exercise. For example, once exercise is finished, heavy breathing continues. Such is the body's reaction in order to carry oxygen to the muscles as quickly as possible to repay the oxygen debt. Often this term is misleading, for oxygen debt does not mean the amount of oxygen the cells owe. Rather, it indicates the need of oxygen to oxidize the lactic acid accumulated when the cells are respiring anaerobically. The pyruvate formed as a result of the oxidation of lactate can enter into aerobic respiration to generate a larger amount of ATP. However, some of the lactate, transported to the liver, gets oxidized to pyruvate and then converted back to glucose when oxygen becomes available.

ANAEROBIC AND AEROBIC REACTIONS

• PROBLEM 3-15

How does a facultative anaerobe differ from an obligatory anaerobe?

Solution: Organisms that can live anaerobically are divided into two groups. The obligatory, or strict, anaerobe cannot use oxygen and dies in the presence of oxygen. These include denitrifying bacteria of the soil, which are responsible for reducing nitrate to nitrogen, and methane-forming bacteria, which produce marsh gas. Some obligatory anaerobes are pathogenic to man; these include *Clostridium botulinum*, responsible for botulism, a fatal form of food poisoning; *Clostridium perfringens*, which causes gas gangrene in wound infections; and *Clostridium tetani*, which causes the disease tetanus.

The facultative anaerobes can live either in the presence or absence of oxygen. Under anaerobic conditions, they obtain energy from a fermentation process; under aerobic conditions, they continue to degrade their energy source anaerobically (via glycolysis) and then oxidize the products of fermentation using oxygen as the final electron acceptor. Yeast will grow rapidly under aerobic conditions but will still continue to live when oxygen is removed. It reproduces more slowly but maintains itself by fermentation. Winemakers take advantage of this behavior by first aerating crushed grapes to allow the yeasts to grow rapidly. They then let the mixture stand in closed vats while the yeasts convert the grape sugar anaerobically to ethanol.

• PROBLEM 3-16

How does energy release in combustion differ from energy release in aerobic respiration?

Solution: Since glucose serves as the primary energy source, we will consider the oxidation of glucose. Combustion refers to the rapid oxidation of a substance with the liberation of heat. When a mole of glucose is burned outside of a living cell, approximately 690 kilocalories are released as heat. This great amount of heat would destroy a cell if released in one burst. Instead, the cell oxidizes glucose by a series of reactions (in aerobic respiration), each catalyzed by a specific enzyme. The release of energy is sequential, and in small packets thus preventing damage to the cell. During oxidative phosphorylation, the energy is transformed and stored in the form of ATP. The complete degradation of one mole of glucose yields 36 ATPs. Since each high energy phosphate bond ($\sim P$) holds about 7 kilocalories of energy, 252 (36×7) kilocalories are biologically useful, while the other 438 (690 252) kilocalories are released as heat. Aerobic respiration, although only 36.5% efficient, is able to harness the energy from the oxidation of glucose to provide an energy source for later needs, without resulting in trauma to the organism.

THE KREBS CYCLE AND GLYCOLYSIS

• PROBLEM 3-17

What structural advantages enable mitochondria to be efficient metabolic organelles?

Solution: Mitochondria are called the "powerhouses" of the cell because they are the major sites of ATP production in the cell. They contain the enzymes involved in both the Krebs (citric acid) cycle and the electron transport system, which work together to furnish 34 of the 36 ATPs produced by the complete oxidation of glucose. Another 2 ATPs are produced by glycolysis, which occurs in the cytoplasm, making the total 36 ATPs /oxidized glucose.

Pyruvate from glycolysis traverses both the outer and inner mitochondral membranes easily because the membranes are totally permeable to pyruvate. Once inside the mitochondrion's matrix (see figure) pyruvate is converted by the enzyme pyruvate dehydrogenase to acetyl coenzyme A. The coenzyme can then enter the Krebs Cycle which occurs in the matrix where all the enzymes that catalyze the cycle's reactions are present (except succinic dehydrogenase which is located on the inner membrane). Conveniently bound to the surface of the cristae, the greatly folded inner membrane of the mitochondria, are the enzymes of the electron transport system. Since the cristae jut into the matrix, the distance needed to travel by $FADH_2$ and NADH, generated by the TCA cycle, to the enzymes of the electron transport system is reduced, and, since the surface area of the inner membrane is so greatly enhanced by its cristae, the probabilities of NADH and $FADH_2$ encountering the electron transport system's enzymes which are concerned with ATP production become sharply increased. Also contributing to the efficiency of this organelle's ATP producing ability is the nature of the cristae's enzymes: they are actually multi-enzyme complexes which group together a number of sequentially acting enzymes responsible for electron transport and oxidative phosphorylation. These assemblies enhance the efficiency of respiration, since the product of one reaction is located near the enzyme of the subsequent reaction.

The Mitochondrion in the cell:

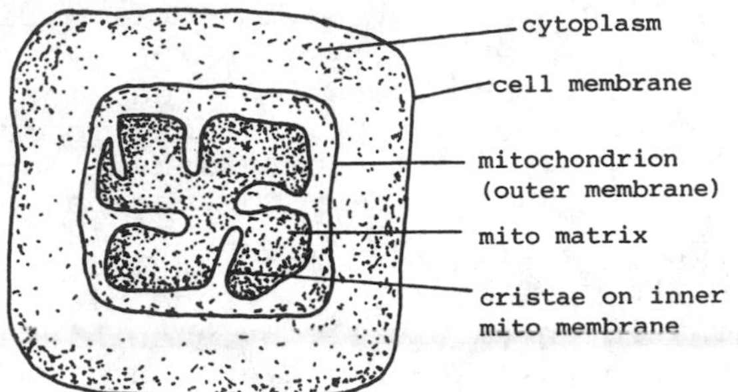

• PROBLEM 3-18

> Compare the reactions involved in the oxidation of lactic and succinic acids.

$$\underset{\text{Lactic acid}}{CH_3-CHOH-COOH} \xrightarrow{} 2H^+ + 2e^- + \underset{\text{Pyruvic acid}}{CH_3-CO-COOH}$$

$$\underset{\text{Pyruvic acid}}{CH_3-CO-COOH} \longrightarrow CO_2 + \underset{\text{Acetaldehyde}}{CH_3-CHO}$$

$$\underset{\text{Acetaldehyde}}{CH_3-CHO} + HS-CoA \longrightarrow CH_3-CHOH-S-CoA$$

$$CH_3-CHOH-S-CoA \longrightarrow 2H^+ + 2e^- + \underset{\text{Acetyl CoA}}{H_3C-\overset{O}{\overset{\|}{C}}\sim S-CoA}$$

Figure 1. The oxidation of lactic acid.

$$\underset{\text{Succinic acid}}{COOH-CH_2-CH_2-COOH} \longrightarrow 2H^+ + 2e^- + \underset{\text{Fumaric acid}}{COOH-CH=CH-COOH}$$

$$\underset{\text{Fumaric acid}}{COOH-CH=CH-COOH} + H_2O \longrightarrow \underset{\text{Malic acid}}{COOH-CH(OH)-CH_2-COOH}$$

$$\underset{\text{Malic acid}}{COOH-CH(OH)-CH_2-COOH} \longrightarrow 2H+2e^- + \underset{\text{Oxaloacetic acid}}{COOH-CO-CH_2-COOH}$$

$$\underset{\text{Oxaloacetic acid}}{COOH-CO-CH_2-COOH} \longrightarrow CO_2 + \underset{\text{Pyruvic acid}}{COOH-CO-CH_3}$$

$$\underset{\text{Oxaloacetic acid}}{COOH-CO-CH_2-COOH} + \text{acetyl CoA} \longrightarrow HSCoA + \underset{\text{Citric acid}}{HO-C(CH_2COOH)_2-COOH}$$

Figure 2. The oxidation of succinic acid.

Solution: Lactic acid is first dehydrogenated enzymatically by lactate dehydrogenase to form pyruvic acid. [Refer to Fig. 1.] The two hydrogen atoms and two electrons are transferred to a pyridine nucleotide, nicotinamide adenine dinucleotide (NAD^+). NAD^+ serves as an electron acceptor in dehydrogenation reactions involving substances (in this case, lactate) with an arrangement as follows:

$$\begin{matrix} | \\ H-C-OH. \\ | \end{matrix}$$

secondary alcohol

(Secondary alcohols, such as in lactic acid, get oxidized to ketones, such as in pyruvic acid).

Following this dehydrogenation reaction is the decarboxylation (loss of CO_2) of pyruvic acid to form acetaldehyde. Addition of coenzyme A to acetaldehyde forms a compound which undergoes another dehydrogenation to form acetyl coenzyme A.

The oxidation of succinic acid is part of the TCA cycle. [Refer to Fig. 2.] Succinic acid is first dehydrogenated enzymatically by succinic dehydrogenase to form fumaric acid. The two hydrogen atoms and their electrons are not transferred to NAD^+, as in lactic acid oxidation, but are transferred to flavin adenine dinucleotide (FAD). FAD is an electron acceptor in dehydrogenation reactions involving substances (in this case, acid. In this case, NAD^+ accepts the electrons (with a H) to form NADH. Oxaloacetic acid will either combine with acetyl coenzyme A to form citric acid in the TCA cycle, or be decarboxylated to form pyruvic acid. In most cases though, the oxoloacetata will combine with acetyl coenzyme A and continue the Krebs cycle. The reader should take note that in both the oxidation of lactate and succinate, dehydrogenation reactions occur, but different electron acceptors are used in each case: NAD^+ for lactate and FAD for the direct oxidation of succinate. The reducing equivalent NADH and $FADH_2$ are formed.

• PROBLEM 3-19

Discuss feedback control involved in glycolysis and the TCA cycle.

Solution: Glycolysis and the TCA cycle are both regulated by means of allosteric control of the enzymes involved. Allostery involves regulation of the activity of an enzyme by the binding of a molecule at a site other than the active site. The molecule that is bound changes the conformation of the enzyme, thereby changing its activity. If the molecule regulates an enzymatic reaction involved in its own synthesis, this regulation is called feedback control.

In glycolysis, one control point occurs at the conversion of fructose-6-phosphate to fructose 1,6-diphosphate, catalyzed by the enzyme phosphofructokinase. This enzyme can be allosterically inhibited by either ATP or citrate. This explains why the rate of glycolysis decreases as soon as aerobic respiration begins in yeast; once the electron transport chain begins to operate, much more ATP is formed per glucose molecule. This causes an accumulation of ATP in more than sufficient amounts for cellular metabolism. This large amount of ATP inhibits phosphofructokinase and therefore inhibits glycolysis. Since glycolysis is inhibited, so is ATP production, for the end products of glycolysis would normally enter the Krebs cycle in order to produce more ATP. This makes sense, since the cell no longer needs to utilize glucose at such a high rate in order to generate sufficient energy. The cell needs only to allow glycolysis to proceed at a rate such that the amount of glycolytic products entering the Krebs cycle will yield an appropriate supply of ATP.

Feedback inhibition provides the mechanism for this control. The inhibition by citrate, a Krebs cycle intermediate, works much the same way.

In the Krebs cycle, a major control point occurs at the conversion of isocitrate to α-ketoglutarate, catalyzed by the enzyme isocitrate dehydrogenase. This enzyme is allosterically inhibited by either ATP or NADH. If too much ATP is being synthesized, or too much NADH produced to be oxidized by the electron transport chain, the reaction is blocked and the whole cycle ceases. Some citrate is accumulated which acts to allosterically inhibit phosphofructokinase, thus shutting down glycolysis as well. With glycolysis shut down, pyruvate will not build up or be converted to lactate or ethanol (which would normally occur if only the TCA cycle were stopped).

Allosteric control allows for a high degree of efficiency and cooperation in glycolysis and the TCA cycle. In addition, there are other control points in both these pathways and in other pathways as well, all serving to keep the cell's metabolic processes in balance.

ELECTRON TRANSPORT

• PROBLEM 3-20

Describe the pathway of electron transport in the respiratory chain.

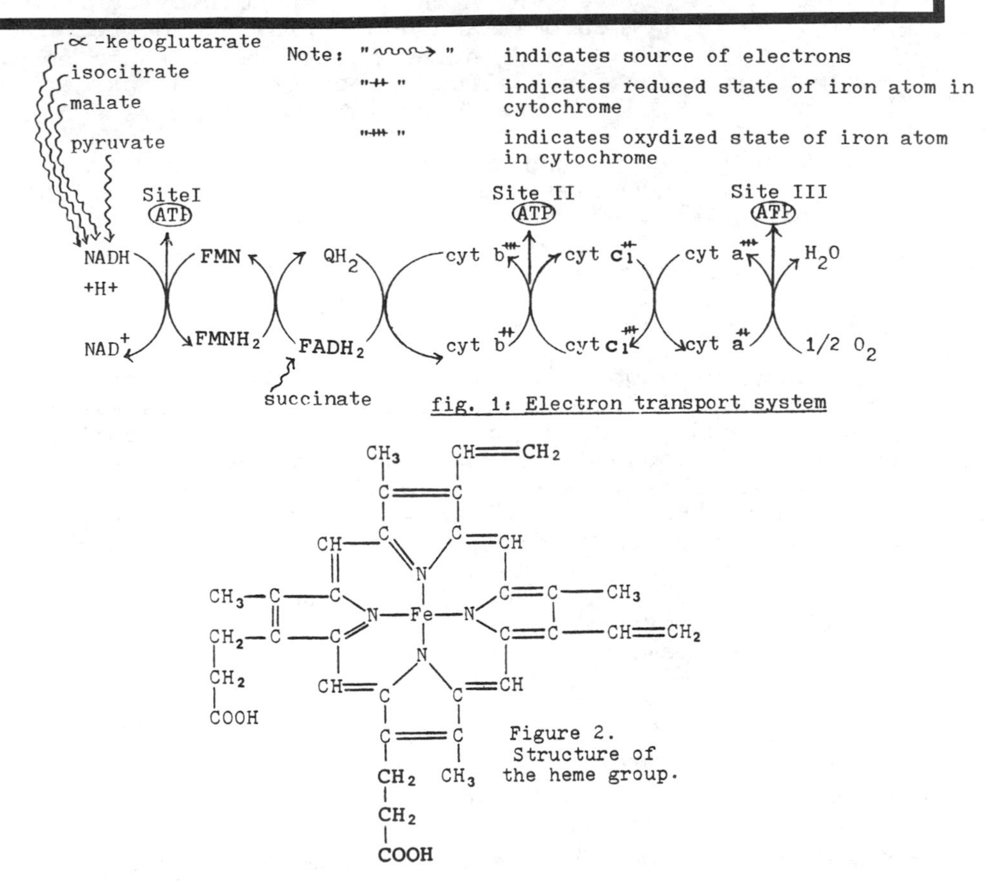

fig. 1: Electron transport system

Figure 2. Structure of the heme group.

Solution: A chain of electron carriers is responsible for transferring electrons through a sequence of steps from molecules such as NADH (reduced nicotinamide adenine dinucleotide) to molecular oxygen. NADH collects electrons from many different substrates through the action of NAD^+-linked dehydrogenases. For example, NADH is produced from the oxidation of glyceraldehyde-3-phosphate in glycolysis, and by several dehydrogenations in the citric acid cycle. In the electron transport system, NADH is oxidized to NAD^+ with the corresponding reduction of flavin mononucleotide (FMN) to $FMNH_2$. (see Figure 1). Some of the energy released is used for the production of one ATP molecule. Then $FMNH_2$ is oxidized back to FMN, and the electrons of $FMNH_2$ are transferred to ubiquinone or coenzyme Q. It is at this site where the electrons of $FADH_2$ are funneled into the electron transport chain.

As ubiquinone is oxidized, the first of a series of cytochromes, cytochrome b, is reduced. The cytochromes are large heme proteins (see Fig. 2). A heme group consists an iron atom surrounded by a flat organic molecule called a porphyrin ring.

It is the iron atom of the cytochrome which is oxidized (Fe^{3+}) or reduced (Fe^{2+}). Cytochrome b reduces cytochrome c_1; the energy released is used for the formation of a second ATP molecule. Cytochrome c_1 reduced to cytochrome c (not shown in figure) which in turn reduces cytochrome a (a complex of cytochromes a and a_3, plus two copper atoms). cytochrome a is oxidized in the last step, with molecular oxygen being the final electron acceptor. As oxygen is reduced to H_2O, a third and last ATP molecule is synthesized.

Thus for each NADH molecule entering the respiratory chain, 3 ATP molecules are produced. Therefore, eight NADH molecules give rise to 24 ATP per glucose molecule in the citric acid cycle. Only 4 ATP are produced by the 2 $FADH_2$ molecules, since they enter the electron transport system at ubiquinone, thus bypassing the first site of ATP synthesis.

The two cytoplasmic NADH from glycolysis produce only four ATP instead of six because one ATP is expended per cytoplasmic NADH in order to actively transport NADH across the mitochondrial membrane. By means of the respiratory chain, 32 ATP are produced per glucose molecule. The other 4 ATP molecules are produced during substrate-level phosphorylation in glycolysis and the TCA cycle. Thus there is a net yield of 36 ATP per glucose molecule oxidized. (Recall that "substrate-level" indicates reactions not involving the electron transport system.)

• PROBLEM 3-21

Small amounts of cyanide can kill a person. Why is cyanide such a deadly poison?

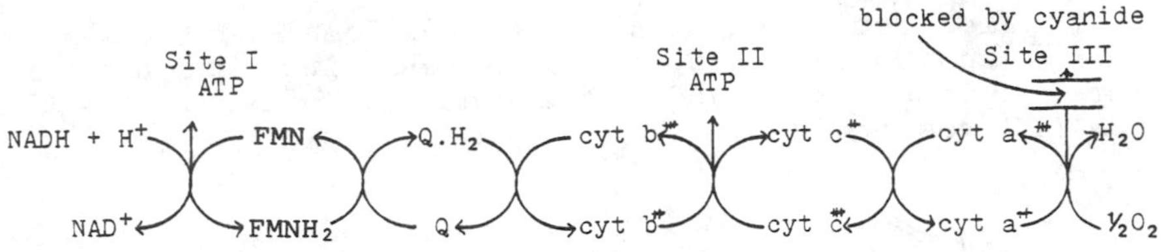

Electron transport chain. Note that all three ATP producing sites will eventually be blocked by cyanide poisoning.

Solution: Cyanide acts by inhibiting the final reaction of the respiratory chain, shutting it off and preventing the energy extraction process from taking place. Normally, reduced cytochrome a, the final element in the respiratory chain, would be reoxidized by oxygen, but since cyanide inhibits this from occuring, the whole chain is shut off as the intermediate acceptors are all converted to the reduced condition. Consequently, there is a severe decrease in the amount of ATP that can be produced for life processes, leading to the death of the victim. Hydrogen sulfide and carbon monoxide have similar effects at this site. Other poisons block either of the first two sites.

Looking at the electron transport chain it may not be obvious how the inhibition of the reoxidation of reduced cytochrome a can shut off the entire respiratory chain.Cyanide blocks ATP formation of the third site of ATP synthesis in the chain. However, this does not mean that ATP production still goes on as usual at the first and second site. Some ATP may be produced immediately after cyanide enters the electron transport chain, but eventually production stops. ATP production ceases because of the inability of the intermediate acceptors to oxidize NADH. Inhibition by cyanide generates an accumulation of reduced cytochrome a by preventing its oxidation. Since cytochrome a is in a reduced state, it cannot oxidize its neighbor, reduced cytochrome c. This leads to an accumulation of reduced cytochrome c and then c_1, which, being unable to oxidize cytochrome b, prevents the formation of ATP at the second site. Of more importance is the "backing up" of reduced states towards the beginning of the chain leading eventually to the reduction of all intermediate acceptors. These reduced intermediates cannot accept electrons; ATP formation in the respiratory chain requires intermediates to receive electrons which are liberated going from a reduced state to an oxidized state. Therefore, ATP formation cannot proceed normally and, in fact, finally stops if any step in the electron transport chain is blocked.

• PROBLEM 3-22

NAD^+ and $NADP^+$ play important roles in reactions associated with metabolism. Discuss these roles.

Solution: In a dehydrogenation reaction, the electrons removed from a substance cannot exist freely but must be transferred to another compound called an electron acceptor. Both nicotinamide adenine dinucleotide (NAD^+) and nicotinamide adenine dinucleotide phosphate ($NADP^+$)

are electron aceeptors. The functional part of both NAD^+ and $NADP^+$ is the nicotinamide ring, which accepts one hydrogen atom and two electrons to become NADH and NADPH respectively

$$NAD^+ + 2H^+ + 2e^- \longrightarrow NADH + H^+$$

Reduction of NAD^+ to NADH

NAD^+ and $NADP^+$ serve as electron acceptors in the dehydrogenation of substances with the secondary alcohol (H-C-OH) arangement. Although $NADP^+$ only differs from NAD^+ in having three rather than two phosphate groups on the R group, most dehydrogenases require either NAD^+ or $NADP^+$ to be present as a coenzyme and will not operate with the other one.

NAD^+ is used as an electron acceptor in glycolysis, the TCA cycle, and fatty acid oxidation. $NADP^+$ is used as an electron acceptor in the light reaction of photosynthesis, and in the pentose phosphate pathway (the source of pentoses, necessary constituents of nucleotides). We thus see that the function of NAD^+ or $NADP^+$ is to accept hydrogen atoms and electrons and to store energy released from oxidized substances in the form of NADH or NADH. The reverse is true for NADH or NADPH; that is, their function is to give up hydrogen atoms and electrons and energy to reducible compounds. For example, the hydrogen atoms of NADH in the electron transport chain are ultimately transferred to oxygen, producing high energy phosphate bonds in the form of ATP along the chain. The oxidation of one molecule of NADH yields 3 phosphate bonds; thus the 8 NADHs produced in the TCA cycle yield 24 phosphate bonds.

NADPH is formed from the reduction of $NADP^+$ in the light reaction of photosynthesis. It is subsequently utilized in the dark reactions of carbohydrate synthesis, in which carbon dioxide is used in the production of hexose molecules. To produce one hexose molecule (a 6-carbon sugar), 12 NADPH and 18 ATP are required, both of which are the products of the light reactions. Thus, both NAD^+, NADH and $NADP^+$, NADPH have major importance in reactions associated with metabolic and catabolic functions.

REACTIONS OF ATP

• PROBLEM 3-23

Compare the processes of oxidative phosphorylation and substrate level phosphorylation. How can certain evolutionary changes be explained by these differences?

Solution: In the complete oxidation of glucose to carbon dioxide and water, NADH (reduced nicotinamide adenine dinucleotide) is produced. High energy electrons are transferred from NADH to oxygen via the electron transport system. This transfer of electrons involves a series of enzyme catalyzed oxidation - reduction reactions in which the energy released is used in the production of high energy phosphate bonds. Three high energy phosphate bonds are made for each pair of electrons transferred to oxygen from NADH. This process of harnessing the energy released in the electron transport chain to synthesize ATP from ADP and inorganic phosphate is termed oxidative phosphorylation. The electron transport chain is not the only system that produces high energy bonds. Atp is produced in both glycolysis and the Krebs cycle.

Any reaction which synthesizes high energy phosphate bonds but is not involved in the electron transport system, is called a substrate level phosphorylation. In glycolysis, a net total of 2 ∿P s (high energy phosphate bonds) are produced at the substrate level; in the TCA cycle, one ∿P is synthesized at the substrate level. Most of the ATP produced during aerobic respiration, however, is synthesized by oxidative phosphorylation. Indeed, therein lies the source of the differences between aerobic and anaerobic organisms, and the reason that anaerobic organisms must be very small, and almost always unicellular. The high energy requirements of a larger organism could not be accomodated through the 3 ATP produced via substrate level phosphorylation.

• PROBLEM 3-24

What are the roles of ATP in the cell? How is ∿P produced, stored, and utilized?

Solution: One of the roles of ATP in the cell is to drive all of the energy-requiring reactions of cellular metabolism. Indeed, ATP is often referred to as the "energy currency" of the cell. The hydrolysis of one mole of ATP yields 7 kcal of energy:

$$\text{ATP} \xrightarrow{H_2O} \text{ADP} + \text{Pi} \qquad \Delta G^{o'} = -7 \text{ kcal/mole}$$

Because of their larger free energies of hydrolysis, the first and second bonds broken in ATP are called high energy phosphate bonds and can be written:

adenosine - P ∿ P ∿ P.

A second role of ATP is to activate a compound prior to its entry into a particular reaction. For example, the biosynthesis of sucrose has the following equation:

$$\text{Glucose} + \text{fructose} \rightleftharpoons \text{sucrose} + H_2O$$

The forward reaction is very unfavorable in a plant cell because of a preponderance of H_2O compared to glucose and fructose (recall Le Chatelier's principle). A great deal of energy would be needed to help the forward reaction occur. The cell alleviates this problem by first activating glucose with ATP:

glucose + ATP ———>glucose -1- phosphate + ADP

The phosphate bond attached to glucose is a high energy bond; the cleavage of this bond yields enough energy to enable the following reaction to proceed with the formation of sucrose:

glucose -1-phosphate + fructose ———>sucrose + phosphate

This series of reactions shows how activation by ATP permits a thermodynamically unfavored anabolic process to occur.

To produce energy-rich phosphate groups (∿P), energy from the complete oxidation of glucose in the process of respiration is utilized. The energy is used to add inorganic phosphate (P_i) to ADP to form ATP. The energy-rich phosphate groups are stored in the form of ATP. They are utilized during the hydrolysis of ATP to ADP and P_i. This energy-yielding reaction is coupled to energy-requiring reactions, allowing the latter to occur.

The amount of ATP present in a cell is usually small, so muscle cells, which require much energy during a brief period of time (during contraction), may run short of it. Therefore, an additional substance, creatine phosphate, serves as a reservoir of ∿P. The terminal phosphate of ATP is transferred by an enzyme to creatine to yield creatine phosphate and ADP. The phosphate bond is also high energy but the ∿P must be transferred back to ADP to form ATP in order for it to be used in an energy-requiring reaction. That is, creatine phospate replenishes the cell's ATP supply by donating its ∿P to ADP.

• PROBLEM 3-25

Compare the types of reactions in which the energy of ATP is made available for biosynthetic processes.

Solution: ATP has two high energy phosphate bonds and can transfer energy to another molecule in a variety of ways. The most common way is the transfer of the terminal phosphate group to the energy-requiring molecule while liberating ADP (adenosine diphosphate). This is known as a phosphorylation reaction, and is usually catalyzed by enzymes called kinases. The hydrolysis of ATP, in which the terminal phosphate group is transferred to water, releases energy ($\Delta G^{o'}$ = -7 kilocalories/mole). By coupling the energy-yielding hydrolysis of ATP to the reaction of a phosphate group with a given compound, one can "capture" part of the energy of hydrolysis in the

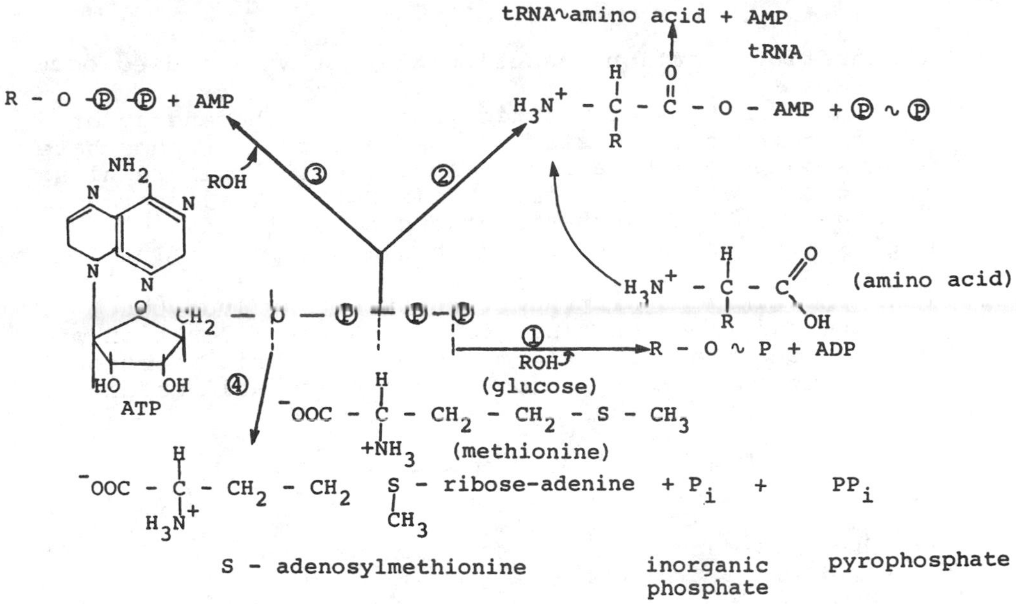

Four major types of ATP reactions: (1) the transfer of phosphate, (2) the transfer of pyrophosphate, (3) the transfer of AMP, and (4) the transfer of an adenosyl group.

phosphorylated compound, as seen below:

$$ATP + H_2O \rightleftharpoons ADP + phosphate \qquad \Delta G^{o\prime} = -7 \text{ kcals/mole}$$

$$phosphate + glucose \rightleftharpoons glucose\ 6\text{-}phosphate + H_2O \qquad \Delta G^{o\prime} = +3 \text{ kcals/mole}$$

$$ATP + glucose \rightleftharpoons glucose\ 6\text{-}phosphate + ADP \qquad \Delta G^{o\prime} = -4 \text{ kcals/mole}$$

The phosphorylated compound is now "activated," that is, it is of a higher energy content than before, and as such, is able to undergo reactions that were previously unfeasible.

A second type of reaction involves the transfer of AMP with the liberation of the last two phosphate groups as pyrophosphate (PPi). The energy released when ATP is split in this manner (-10 kcal/mole) is substantially greater than when a single phosphate group is cleaved off. A compound activated by AMP transfer can thus be of a significantly higher energy than its unactivated form. An example of this type of activation is the activation of an amino acid. Here, AMP is transferred to the COO^- group of the amino acid. The activated amino acid is then transferred to the terminus of a transfer RNA molecule for use in protein synthesis, forming a new high energy bond to the tRNA and releasing AMP. Usually, the phyrophosphate group initially generated is quickly cleaved to form two orthophosphate groups:

$$PPi + H_2O \longrightarrow Pi + Pi \qquad \Delta G^{o\prime} = -4.6 \text{ kcal/mole}$$

Note that the reaction releases a good deal of energy. Although this may appear wasteful, the reaction is highly useful in that it provides an exergonic "push" to the initial reaction, assuring that it will indeed occur to completeness.

A third type of reaction utilizing the energy of ATP involves the transfer of the last two pyrophosphate groups and the liberation of AMP. Again, this results in a highly activated compound (R∿P∿P). The intermediate compounds in the synthesis of cholesterol and its derivatives are activated in this manner.

The final major type of reaction involves the transfer of the adenosine group with the liberation of pyrophosphate and inorganic phosphate. This reaction is important in the activation of the amino acid methionine

$$\overset{+}{H_3N}-\underset{\displaystyle H}{\overset{\displaystyle COO^-}{C}}-CH_2-CH_2-S-CH_3$$

so that its methyl (CH_3) group can be transferred to other compounds.

• PROBLEM 3-26

How do we know that ATP is generated at three sites in the electron transport chain? What is the advantage of having a sequential transfer of electrons rather than one single transfer?

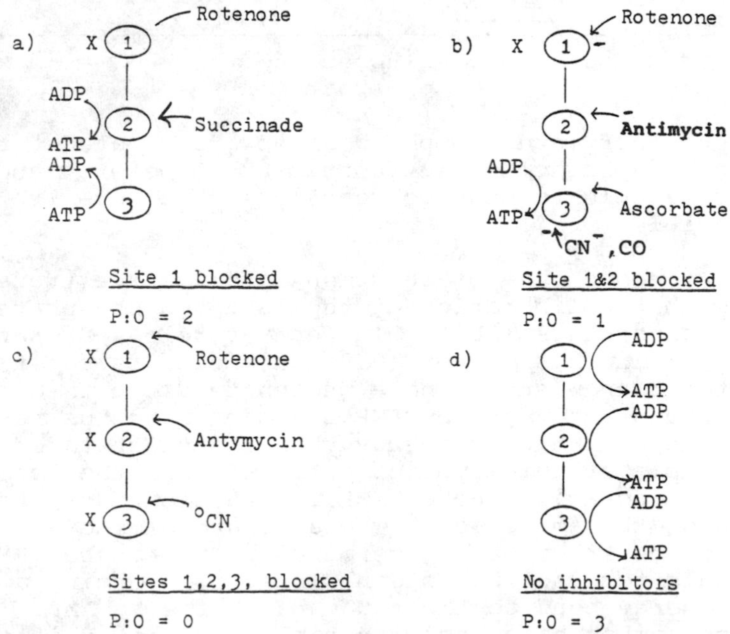

Solution: The NADH and $FADH_2$ formed in glycolysis, fatty acid oxidation and the citric acid (TCA) cycle are energy-rich molecules. When each of these molecules transfers its pair of electrons to molecular oxygen, a large amount of energy is released. This released energy can be used to generate ATP in oxidative phosphorylation. As electrons flow through the electron transport chain from NADH to O_2, ATP is formed at three sites along the chain. How do we know that three separate sites exist instead of one site where 3 ATP are produced?

The first evidence is the ATP yield from the oxidation of different substrates. The oxidation of NADH yields 3 ATP but the oxidation of succinate, which forms $FADH_2$, only yields 2 ATP. The electrons from $FADH_2$ enter the chain at coenzyme Q, which is past the first phosphorylation site. Only 1 ATP is formed when ascorbate (an artificial substrate also called Vitamin C) is oxidized, since its electrons enter at cytochrome c, which is past the first two phosphorylation sites. The P:O ratio (the ratio of organically bound phosphorous produced per mole of oxygen, which is often used as an index of oxidative phosphorylation) is 3:2:1 for NADH:succinate: ascorbate, respectively. Hence, by knowing the P:O ratios and that each substrate enters the process at different steps, we see that one mole of ATP is produced at each site.

The steps involved can be determined by the free energy change, ΔG, released during electron transfer. The ΔG for electron transfer from NADH to FMN is -12 kcal/mole, from cytochrome b to c, -10 kcal/mole, and from cytochrome a to O_2, -24 kcal/mole. These reactions are sufficiently exergonic to drive the synthesis of ATP, whereas the other electron transfer reactions do not release sufficient energy to allow ATP synthesis to take place.

There also exist inhibitors which specifically block certain steps in the chain. Rotenone specifically inhibits electron transfer from NADH to coenzyme Q, preventing ATP synthesis at the first site. Antimycin A inhibits electron flow at the second site, while cyanide and carbon monoxide block electron transfer at the third site. Each inhibitor has been observed to block all successive steps of the electron transport system. However it is possible to bypass the action of the inhibitor. For instance, the action of rotenone can be avoided through the addition of either succinate or ascorbate. Similarly, the action of antimycin A on the second site can be bypassed through the addition of ascorbate. By inhibiting the various sites and measuring the resulting P:O ratios it can be deduced that each site in the electron transport chain is responsible for the production of one molecule of ATP (see figure).

It should be noted that the addition of antimycin A will result in the blockage of site 1 as well as site 2. This is due to a backflow of electrons which results in the reduction of site 1, thus inactivating it. Likewise, the addition of cyanide results in the reduction of sites 1 and 2 as well as 3, thus rendering them all incapable of ATP formation. Hence cyanide has the well deserved reputation as one of the world's strongest toxins.

The advantage of the sequential transfer over one single transfer is that the sequential reactions divide up the free energy change of the oxidation of NADH, which is highly exergonic. The free energy of oxidation of

$$NADH + \frac{1}{2} O_2 + H^+ \rightleftharpoons H_2O + NAD^+$$

is about -53 kcal/mole. If all this energy were released at once, much of it would be wasted as heat since only 7 kcal are needed to phosphorylate ADP to form ATP. If all the energy from the oxidation of NADH is released at once to form 1 ATP, the efficiency of this system is

$$7/53 \times 100\% = 13\%.$$

However, if the same energy is released in a stepwise process to form 3 ATP, the efficiency is

$$21/53 \times 100\% = 40\%.$$

Thus, stepwise reactions increase the efficiency of oxidative phosphorylation from 13% to 40%.

In addition, the large amount of heat liberated would very easily destroy enzyme activity and thus is harmful to the cell if released in one burst.

The three reactions involved are "coupled" with the phosphorylation of ADP so that the liberated energy is used immediately to drive this otherwise non-spontaneous reaction.

The exact mechanism of oxidative phosphorylation, that is, how the energy liberated is transferred so as to form the high energy phosphate bond of ATP, is unknown.

The exact mechanism of oxidative phosphorylation, that is, how the energy liberated is transferred as so to form the high energy phosphate bond of ATP, is unknown. Mitchell proposed a chemiosmotic hypothesis in 1961. The three sites of "ATP synthesis" along the respiratory chain are actually sites where, at each, a pair of protons (2 H+) are pumped from the matrix to the intermembrane space. This creates a proton gradient. The protons flow back onto the matrix through special "lollypop" enzymes. As a pair of H+ flow back in down the proton gradient, one ATP is synthesized. Three pairs of H+ are extruded into the intermembrane space per NADH, so three pairs of H+ can re-enter and thus produce 3 ATPs. The mechanisms is similar for $FADH_2$, except since $FADH_2$ donates their electrons at the second site, only two pairs reenter and two ATPs are synthesized.

ANABOLISM AND CATABOLISM • PROBLEM 3-27

> Discuss how the carbons of a glucose molecule may be converted to the carbons of a fat molecule.

Solution: A fat molecule is actually a molecule of triacylglycerol (once called triglyceride), a compound formed when the three hydroxyl groups of glycerol are bonded to three fatty acids. A reaction of triacylglycerol synthesis can be written as follows:

$$\begin{array}{c} CH_2OH \\ | \\ CHOH \\ | \\ CH_2OH \end{array} \quad + \quad \begin{array}{c} R_1-COOH \\ R_2-COOH \\ R_3-COOH \end{array} \quad \begin{array}{l} CH_2-O-\underset{\underset{O}{\|}}{C}-R_1 \\ | \\ CH\ -O-\underset{\underset{O}{\|}}{C}-R_2 + 3H_2O \\ | \\ CH_2-O-\underset{\underset{O}{\|}}{C}-R_3 \end{array}$$

glycerol   **3 fatty acids**   **triacylglycerol**

R represents the hydrocarbon chain of the fatty acid, and will usually vary in its length and number of double or triple bonds. Thus, it is the fatty acid chain of fats which determine the nature of the fat. Triacylglycerols that are solid at room temperature are called "fats" and those which are liquid are called "oils."

Glucose is degraded in glycolysis to fructose -1,6-diphosphate, which is split into two three-carbon sugars, glyceraldehyde -3-phosphate (G-3-P) and dihydroxyacetone phosphate (DHAP). DHAP is reduced to α-glycerophosphate (glycerol-3-phosphate) with oxidation of NADH. α-glycerophosphate is dephosphorylated to glycerol with concomitant production of ATP.

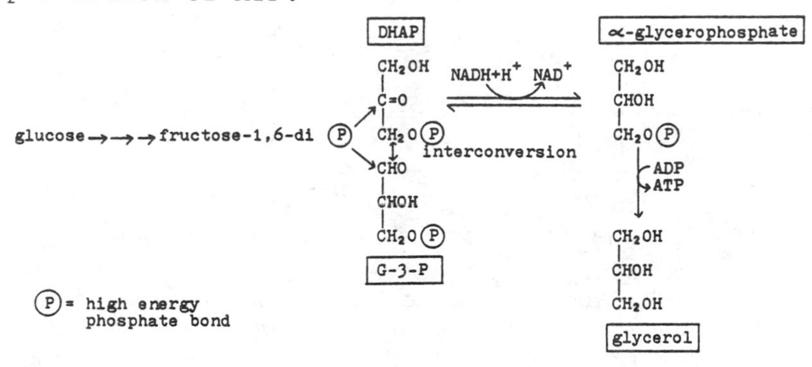

The synthesis of fatty acids requires molecules of acetyl coenzyme A as precursors of all the carbon atoms in the fatty acid chain. Chain growth during fatty acid synthesis starts at the carboxyl (-COOH) group of acetyl-CoA and proceeds by successive addition of acetyl

$$CH_3-\overset{\overset{O}{\|}}{C}-$$

residues at the carboxyl end of the growing chain. Acetyl-CoA is formed from oxidation of pyruvate, the breakdown product of glucose in glycolysis. Thus, both the glycerol and fatty acid carbon atoms of a fat molecule can originate from those of glucose molecules. Although this synthesis of fats from glycerol and fatty acids can occur in vitro (Latin, in glass), a different pathway is used in vivo (Latin, in the living). All naturally occurring triacylglycerols are formed by reaction of dihydroxyacetone phosphate and fatty acyl coenzyme A (a fatty acid "activated" by coenzyme A). Both these reactants can originate from glucose, as explained earlier.

• PROBLEM 3-28

Compare the processes by which glucose and fatty acids are activated.

Solution: Before glucose and fatty acids can be oxidized, they must first be activated in the first steps of their respective pathways. A fatty acid is activated by reaction with ATP and coenzyme A to yield a fatty acyl CoA:

$$\underset{\text{fatty acyl}}{\text{R-COOH}} + \text{ATP} + \text{CoA - SH} \longrightarrow \underset{\text{fatty acyl CoA}}{\text{R-CO}{\sim}\text{S-CoA}} + \text{AMP} + \text{PPi}$$

A series of enzymes, called acyl-CoA-synthetases are involved in the catalysis of this reaction. The bond formed between the fatty acid and coenzyme A is a high energy thioester bond. The energy for its formation comes from the hydrolysis of ATP to AMP and pyrophosphate. While a fatty acyl-AMP intermediate is formed, the final activated compound is a thioester.

The activated form of glucose, on the other hand, is phosphorylated.

Only ATP is involved in the activation, which yields glucose-6-phosphate and ADP

$$\text{glucose} + \text{ATP} \xrightleftharpoons{\text{hexokinase}} \text{glucose - 6 - P} + \text{ADP} + H^+$$

The enzyme involved is termed hexokinase.

Unlike fatty acid activation, ATP is hydrolyzed to ADP, not AMP, and only after glucose is converted to pyruvate does coenzyme A serve as an activator, enabling pyruvate to enter the TCA cycle.

• PROBLEM 3-29

How can glycogen (a polymer of glucose) be converted into fats (triacylglycerols) in the body?

Solution: A fat, or triacylglycerol molecule, consists of glycerol bonded to three fatty acids. Therefore, in order for glycogen to be converted into fats, it must first be broken down and converted into glycerol or an activated form of glycerol and fatty acids. Glycogen is broken down by the enzyme glycogen phosphorylase, to give glucose 1-phosphate (G1P). G1P is next converted to glucose 6-phosphate (G6P) by the enzyme phosphoglucomutase. G6P is then isomerized to fructose 6-phosphate by an isomerase enzyme. Further reaction of fructose 6-phosphate result in the formation of two triose phosphates: dihydroxyacetone phosphate and glyceraldehyde 3-phosphate. At this point, the conversion branches. Dihydroxyacetone-P (where P indicates phosphate) is reduced to glycerol 3-P which then gets hydrolyzed by a phosphatase to glycerol. Glyceraldehyde 3-P is converted to pyruvate by a series of reactions.

Pyruvate now undergoes oxidative decarboxylation in the mitochondria and is activated by Coenzyme A to become acetyl-CoA. The latter enters the citric acid cycle where it combines with oxaloacetate to form citrate. Citrate, unlike acetyl-CoA, can pass through the mitochondrial membrane to the cytoplasm, where it is reconverted to acetyl-CoA. Acetyl-CoA is then converted in the cytoplasm into malonyl-CoA, and finally, via a complex sequence of reactions, into fatty acids. Three fatty acids then react with one glycerol molecule to form a triacylglycerol.

This sequence of events, discussed only briefly above, is summarized below:

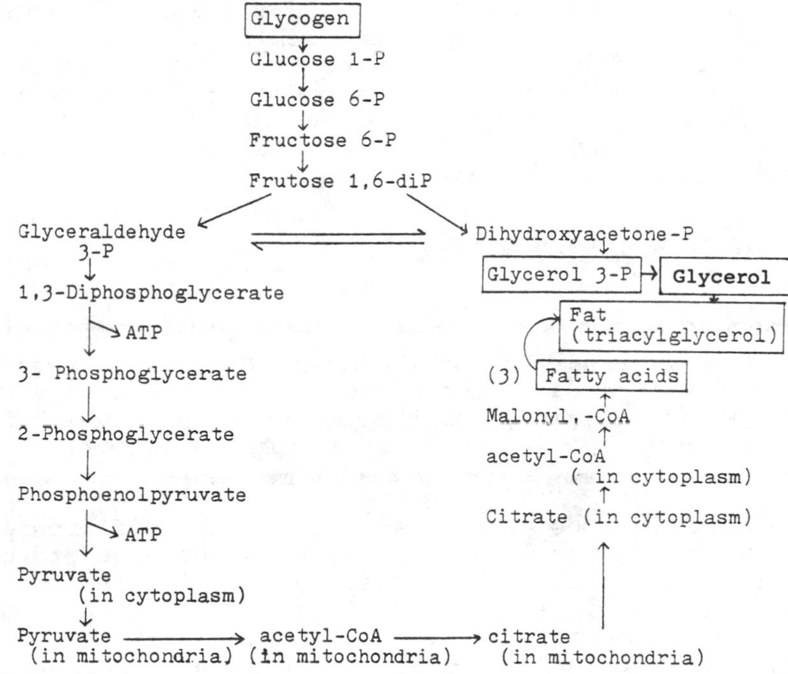

Note that pyruvate can only be converted to acetyl-CoA in the mitochondria. Fatty acids, however, are synthesized in the cytoplasm. Since acetyl-CoA cannot directly leave the mitochondria, it must first be converted to citrate, which can leave the mitochondria, and then be reconverted to acetyl-CoA in the cytoplasm. Although the enzymes for each reaction step are not indicated, it is important to realize that each step is catalyzed by an enzyme or complex of enzymes.

• PROBLEM 3-30

How can one demonstrate that amino acids are used to build proteins in cells, while the proteins already in the cell are used in catabolism?

Solution: One way of following the path of the amino acids is by a process called autoradiography. We can make amino acids radioactive by the incorporation of radioactive isotopes such as tritium (^3H) or carbon-14. We can then inject these labelled amino acids into cells in tissue culture, and note changes in the radioactive level. This is done by exposing the cells to an autoradiographic emulsion, which upon exposure to any radioactivity in the cells, produces small grains visible microscopically. If we give labelled amino acids to the cells in culture and we look for radioactivity at successive intervals, we should initially see an increase in the number of grains as the labelled amino acids are used to build proteins. As time progresses, the amount of radioactive protein should remain dynamically stable: as protein is broken down, new protein is synthesized. If we cease providing the cells with labelled amino acids and give them unlabelled amino acids instead, the amount of radioactive protein in the cells will decrease and eventually disappear as radioactive protein is degraded and excreted. The disappearance of radioactive protein indicates that protein is being catabolized. (Note: the new protein synthesized is not radioactive since unlabelled amino acids are involved.)

ENERGY EXPENDITURE

• PROBLEM 3-31

When estimating the total number of Calories required by a person in a day, we consider such factors as sex, amount of activity, and the temperature in which the person lives. Why are these factors necessary?

Solution: One must first clarify the difference between the terms "calorie" and "Calorie." One standard calorie is defined as the heat necessary to raise the temperature of one gram of water by 1°C from 14°C to 15°C. However, the nutritionist's Calorie is equal to 1000 of these small calories, or what a physicist would call a kilocalorie.

The energy yield of food measured in Calories is the amount of heat energy that can be extracted from a given amount of food. When one mole (180 g) of glucose is burned (oxidized) in a calorimeter, 690 Calories are given off. Fats have a higher caloric value than carbohydrates, yielding about 9 Calories per gram. This gives a value of 1620 Calories for 180 g of fats.

A minimum amount of Calories each day is needed to supply the energy requirements of the body. If we ingest more food than we really need the excess energy is usually stored as fat. Each person's Caloric requirement differs, however, according to sex, amount of activity and environmental temperature.

A man usually requires more Calories per day than does a woman. A woman's metabolism is usually lower than that of a man, so that women generally have to eat less per body weight than men. A lower metabolic rate indicates a lower Caloric need since less energy is required to drive the reactions of the body.

The Caloric need of a person also varies with the

amount of physical activity. More Calories are needed for greater amounts of activity. A highly active person requires more energy for body processes than does a sedentary person. A jogger needs ATP for muscle contraction during running; his cardiac and respiratory rates increase to speed the delivery of O_2 to his cells and the removal of CO_2 from his cells - these processes involve further ATP need. A sedentary person needs just enough Calories to sustain normal metabolic activity.

The temperature of a person's surroundings also affects Caloric requirements. A person working in a cold environment needs energy, not only for his regular metabolic needs, but also for heat production to maintain his normal body temperature. A person in a warmer climate has no such difficulties.

• PROBLEM 3-32

Discuss the various ways in which energy is expended within the human body.

Solution: Energy is primarily expended during the anabolic reactions of the cell; that is, energy is required for the synthesis of macromolecules. Some organisms ingest simple chemical substances and combine them to form the complex molecules needed for the building of new protoplasm. Others, the higher organisms, ingest complex organic materials, degrade them, and use the simple products to synthesize the complex organic compounds that they need. In both cases, growth results.

Energy is also expended during motion. Motion includes the contraction of muscle tissue, which allows an animal to either move itself (locomotion) or part of its body (the bending of a limb). Muscle contraction also occurs in the diaphragm during breathing, in the heart and in the walls of the digestive organs.

Energy is also required during movement of materials across the cell membrane. During active transport, molecules move from a region of low concentration to one of high concentration,a process which requires energy. Active transport is thought to involve a carrier molecule driven by the hydrolysis of ATP, the energy-releasing source.

Another energy-requiring process is heat production. During metabolic activity, the heat given off is important to "warm-blooded" animals (homeotherms) in keeping body temperature constant despite changes in evironmental temperatures. Even "cold-blooded" animals(poikilotherms) whose body temperature is determined by the environmental temperature, must release heat in order to survive extremely cold conditions. Homeotherms, which include birds and mammals, generally eat more food per body weight than do poikilotherms, which include fish, amphibians, reptiles, and invertebrates. This fact reflects the need to furnish more energy to maintain body temperature.

Both anabolism and muscle contraction require the energy obtained from the splitting of the energy-rich terminal phosphate group of ATP, which has the structure:

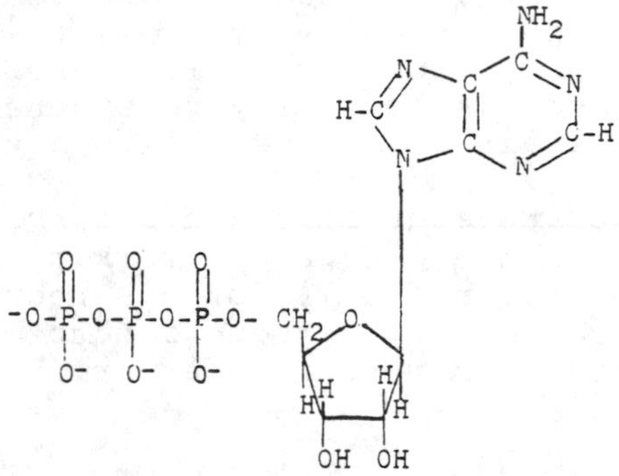

(at physiological pH)

ATP is produced during oxidative phosphorylation - the process by which the flow of electrons in the electron transport system of respiration transfers the released energy to be stored in ATP. However, certain compounds, such as the hormone thyroxine, can uncouple phosphorylation from the electron flow so that the energy is not trapped as ATP, but is released as heat. This mechanism is important when more heat is needed by an organism than is provided by its normal metabolic activity.

• PROBLEM 3-33

Explain three ways in which a person's basal metabolic rate may be determined.

Solution: A person's body is said to be at the basal metabolic rate when it is using energy at a rate just sufficient to maintain the life of its cells. A person's basal metabolic rate (BMR) usually approximates his metabolic rate immediately after awakening, when he is not physically active and only the necessary metabolic processes to maintain life are occurring. The BMR varies with age, height, weight and sex, as well as with certain abnormalities such as thyroid disease. The thyroid gland produces the hormone thyroxine, which regulates the rate of cellular metabolism. An abnormal BMR may indicate abnormal thyroxine levels. BMR has been measured by placing a person in a chamber and measuring the amount of heat given off (since heat is a by-product of metabolism), but such is an awkward and difficult method. The BMR is more easily determined by indirectly measuring the amount of oxygen used in a given period of time. Since any oxygen used will serve as the final electron acceptor in respiration, the amount of oxygen used is proportional to the metabolic rate. Another common method entails measuring the amount of thyroxine present in a person's blood. As mentioned previously, a high BMR could indicate an excess of thyroxine while a low BMR could indicate a deficiency. Each thyroxine molecule contains iodine and when released, is bound to a protein in the blood. This protein is the only protein in the blood which contains iodine. Thus, by measuring the amount of protein-bound iodine in a blood sample, the amount of thyroxine may be determined. This, in turn, gives an indication of the rate of metabolism.

BMR may be increased through administration of thyroid pills, which contain thyroxine produced by animals. The increased concentration of this hormone in the blood increases the metabolic rate. The underlying reason for this is that thyroxine acts to uncouple oxidative phosphorylation; i.e. it dissociates the formation of ATP from the electron transport chain in the cell. Much of the energy released during the oxidation of NADH will be lost as heat rather than trapped in the form of ATP. The ATP/ADP ratio declines, and the relatively heightened level of ADP acts as a positive or stimulatory allosteric modulator for the TCA cycle (through its action on an enzyme - isocitrate dehydrogenase). Thus the TCA cycle will function actively producing a continuous supply of NADH, the oxidation of which will fail to be coupled to ATP formation. Oxygen, the final electron acceptor, will be consumed rapidly, and a higher BMR measurement will result.

To decrease the BMR, thyroxine-suppressing drugs may be administered, or part of the thyroid gland may be surgically removed. Previously, radioactive iodine was injected in order to kill the cells of the thyroid gland. However, this treatment has recently been found to be extremely carcinogenic (cancer-causing).

• PROBLEM 3-34

A person whose diet is deficient in B vitamins is likely to be listless and lacking in energy for normal activities. Explain.

Solution: The vitamin B complex includes (among others) thiamine (B_1), riboflavin (B_2) pyridoxine (B_6), biotin, niacin or nicotinic acid, cobalamin (B_{12}) and pantothénic acid. All these vitamins function as parts of coenzymes used in the oxidation of glucose to form ATP.

Thiamine is an essential constituent of thiamine pyrophosphate, the coenzyme used in the oxidative decarboxylation of pyruvic and α-ketoglutaric acids in the TCA cycle. Characteristics of mild deficiencies are: fatigue, loss of appetite and weakness. Severe deficiencies enhance these symptoms and cause degeneration of nerves and muscle atrophy, a condition known as beriberi.

Niacin is a component of two coenzymes, NAD^+ and $NADP^+$, which serve as hydrogen acceptors in reactions of glycolysis and the TCA cycle. Deficiency of niacin causes stunted growth and the disease pellagra.

Riboflavin forms part of flavin adenine dinucleotide (FAD), a coenzyme of succinic dehydrogenase in the TCA cycle. Deficiency results in stunted growth, cracks in the corner of the mouth, and cornea damage.

Pantothenic acid is especially important since it forms part of coenzyme A, and thus used in a number of reactions in the metabolism of carbohydrates, fats and proteins. An expected symptom of pantothenic acid deficiency is growth failure.

A person with any of these vitamin deficiencies would show signs of slow ATP production since these vitamins are necessary providers of coenzymes involved in metabolic pathways leading to ATP formation. Since ATP is needed for energy-requiring reactions, a lack of ATP would cause the vitamin-deficient person to be sluggish and to exhibit slower growth.

• PROBLEM 3-35

How is the respiratory quotient obtained and how can it be used to determine when a person is in an advanced state of starvation?

Solution: The respiratory quotient (RQ) is a quantitative measure of the type of foods being used by the body. If we measure both the oxygen consumption and carbon dioxide release by the body, we can determine whether carbohydrates, fats, or proteins are being used for respiratory needs. This measure, called the respiratory quotient, is the ratio of carbon dioxide emitted to the amount of oxygen taken in:

$$RQ = \frac{CO_2 \text{ produced}}{O_2 \text{ consumed}}$$

During aerobic respiration, one molecule of glucose utilizes six molecules of oxygen and releases six molecules of carbon dioxide. This 1:1 ratio gives carbohydrates a respiratory quotient of 1.0. Fats, however, have a lower respiratory quotient. This indicates that more oxygen is necessary to break down a fat than a carbohydrate molecule. Oxygen is needed to convert the fat (or fatty acid) into intermediate products before it can enter the respiratory process. Free fatty acids are first formed from fats enzymatically, and are then oxidized to acetyl-CoA, which enters the TCA cycle. The overall equation for palmitic acid (a 16-carbon fatty acid) oxidation can be written:

$$\text{palmitate} + 23\ O_2 + 129\ Pi + 129\ ADP \longrightarrow 16\ CO_2 + 145\ H_2O + 129\ ATP$$

The respiratory quotient is therefore $\frac{16}{23} = 0.70$

This is typical for most fats.

The RQ of proteins lies between those of carbohydrates and fats - usually about 0.80. The oxidation of two molecules of alanine requires six molecules of oxygen and liberates five molecules of carbon dioxide according to the following equation:

$$2\ C_3H_7O_2N + 6\ O_2 \longrightarrow (NH_2)_2CO + 5\ CO_2 + 5\ H_2O.$$

This gives a respiratory quotient of 5/6, or 0.80 for proteins.

Normally, if an individual is consuming all three types of food, his RQ will be about .85. After a meal heavy in carbohydrates, however, a person's RQ will approach 1.0. If he stops eating, his RQ will decrease to .70 as his body uses its fatty deposits as an energy source. Further starvation (about 2 weeks) will tend to increase his RQ to 0.80 as the fatty deposits become exhausted and the proteins of the body are used. This is dangerous and can result in consumption and atrophy of muscles and organs. Therefore, a person in an advanced state of starvation will have an RQ of about 0.80.

Problem 3-36

Summarize the complete oxidation of glucose to CO_2 and H_2O

Solution:

See accompanying figure. The complete oxidation is written as:

$$\text{Glucose} + 6\ O_2 \dashrightarrow 6\ CO_2 + 6\ H_2O + 36\ \text{ATP}$$

Glucose enters the cell. In the cell cytoplasm, glycolysis breaks glucose down to pyruvate. Pyruvate enters the mitochondrion. In the matrix, pyruvate is converted to acetyl-CoA. Acetyl-CoA enters the Krebs cycle by combining with oxaloacetic acid (OAA), a four carbon molecule. Acetyl-CoA is 2 carbons, so when it joins with OAA it forms a 6-c citrate molecule. By a series of oxidation-reduction steps, OAA is regenerated to keep the cycle spinning, 2 CO_2s are produced per turn of the cycle, and the reducing equivalents NADH and $FADH_2$ are produced. The reducing equivalents go to the inner mitochondrial membrane and pass their electrons along. At three sites, a pair of protons are extruded into the intermembrane space (only two pairs for $FADH_2$). The last enzyme of the respiratory chain is cytochrome oxidase: oxygen accepts the electrons, picks up a pair of H+s, and forms water. Concomitant with the flow of electrons is the synthesis of ATP: The extruded H+s flow into the matrix via channels on the "lollypop" enzymes. One ATP is produced per pair of H+s.

Note on the figure that the only substances entering the cell are glucose and oxygen (circled) and the only substances leaving the cell are water, carbon dioxide and ATP (circled). This is in agreement with the equation above.

When glucose is oxidized completely, 36 ATPs are produced. Fatty acids and amino acids also feed into this scheme at acetyl-CoA, but different amounts of ATP are produced depending on the molecule.

Answer

To be covered when testing yourself

SHORT ANSWER QUESTIONS FOR REVIEW

Choose the correct answer.

1. Which of the following statements is incorrect? (a) catalysts can be organic, such as an enzyme, or inorganic, such as the metal nickel. (b) the enzyme chymotrypsin is not specific, as it will catalyze many different reactions in the body. (c) a catalyst can be used over and over again, as it is neither destroyed in the reaction, nor incorporated into the final product. (d) the "intermediate complex" consists of the substrate of the reaction attached to the catalyst. — b

2. Which of the following accounts for the increased reactivity of a substrate when it is attached to its specific enzyme? (a) bond strengthening (b) bond weakening (c) bond inactivity (d) none of the above — b

3. Which of the following is not correct concerning regulatory enzymes? (a) they usually have two binding sites one for the substrate and one for the "regulator". (b) they are often found in key metabolic pathways, such as the Krebs cycle. (c) the end-product of a pathway is never a regulator. (d) a product half-way down a pathway can act as a regulator. — c

4. Distinguish between the following: holoenzyme, coenzyme, metalloenzyme, and prosthetic group. (a) A holoenzyme is a cofactor joined to the apoenzyme. A coenzyme is a cofactor in the form of a metal ion. A metalloenzyme is a metal that acts as an enzyme, and a prosthetic group is a tightly bound cofactor. (b) A holoenzyme is the combined cofactor and apoenzyme. A coenzyme is an enzyme that requires a metal ion for activity. A metalloenzyme is an enzyme that can act as a metal, and a prosthetic group is a carbohydrate combined with an enzyme. (c) A holoenzyme is the combined cofactor and apoenzyme. A coenzyme is an organic molecule acting as a cofactor. A metalloenzyme is an enzyme that requires a metal for activity, and a prosthetic group is a coenzyme tightly bound to the apoenzyme. (d) A holoenzyme is an enzyme without its apoenzyme. A coenzyme is a fatty acid acting as an enzyme. A metallo-enzyme is an enzyme that requires a metal for its activity, and a prosthetic group is any organic molecule attached to an enzyme. — c

Answer

To be covered when testing yourself

5. Which of the following is incorrect? (a) The removal of electrons is called reduction. This process can be performed by a dehydroganase enzyme, such as the one used to transform succinate into fumarate. (b) An enzyme that tranfers electrons directly to oxygen is called cytochrome oxidase. When this occurs, oxygen is then reduced to H_2O.
(c) The addition of electrons is called reduction. When electrons are removed from succinate to form fumarate, the substance reduced is the coenzyme FAD (flavin adenine dinucleotide)
(d) When one compound is reduced, another compound must be oxidized. The substance reduced is therefore called the oxidizing agent, while the substance oxidized is called the reducing agent. — a

6. How many ATPs are derived from the oxidation of 1 molecule of pyruvate via the Krebs cycle and the electron transport system? (a) 12 (b) 14 (c) 15 (d) 20 — c

7. Which of the following contributes to the "powerhouse" properties of the mitochondria? (a) the presence on the mitochondria's inner membrane of the enzymes necessary for the functioning of the electron transport system (b) the presence of cristae, which increase the surface area of the mitochondria's inner membrane (c) the ease with which pyruvate can travel through the outer and inner membrane into the mitochondria's inner matrix (d) all of the above — d

8. Which of the following is incorrect concerning feedback regulation? (a) In the same way that high quantities of ATP and NADH are inhibitors of the Krebs cycle, high quantities of ADP and NAD^+ are stimulators of the Krebs cycle.
(b) Control of feedback regulation in both the glycolytic pathway and the Krebs cycle is due to the presence of allosteric enzymes. (c) Since many more ATPs are produced by the Krebs cycle and electron transport system than in glycolysis, once the former begin to function, they will inhibit the latter. (d) The only control point in the glycolytic pathway is at the conversion of fructose-6-phosphate to fructose-1, 6-diphosphate by the enzyme phosphofructokinase. It is inhibited by high quantities of ATP. — d

9. Which of the following is incorrect concerning the production and use of ATP? (a) ATP can be produced by either oxidative or substrate

Answer

To be covered when testing yourself

phosphorylation. Inorganic phosphate must be available for either of these reactions to occur. (b) the energy released upon hydrolysis of the terminal phosphate group is enough to drive many endergonic metabolic reactions, the use of an intermediate compound (one with a phosphate attached to it) never being necessary. (c) Not only does the terminal phosphate bond of ATP contain much energy, but both the terminal and second phosphate bonds can be used to drive endergonic reactions. (d) In the reaction forming sucrose from glucose and fructose, the energy needed originally comes form ATP, but must pass through a phosphorylated intermediate before it can ultimately be used. — b

10. If a metabolic reaction requires 5000 calories of energy in order to occur, will the hydrolysis of the terminal phosphate bond of ATP supply enough energy to drive the reaction? Will there be any energy left over? (a) No, the hydrolysis will not supply enough energy. (b) Yes, hydrolysis will supply exactly the energy needed. (c) Yes, the hydrolysis will supply the energy needed, plus some will be left over. (d) None of the above — c

11. Which of the following does not provide substantial proof of the location of the sites of oxidative phosphorylation during the electron transport system? (a) the use of specific inhibitors, such as cyanide and antimycin A, and measurement of the subsequent P:O ratio after their administration (b) the presence or absence of inorganic phosphate and ADP at specific sites on the cristae of the mitochondria. (c) measuring the ΔG (free energy change) from one step to another, and showing which sites generated enough energy (via electron transfer)to phosphorylate ADP. (d) feeding into the electron transport system different substrates such as succinate and ascorbate, and then measuring the subsequent P:O ratio. — b

12. Concerning the "activation" of glucose and fatty acids in preparation for oxidation, which of the following is incorrect? (a) a molecule of coenzyme A is needed for both the activation of glucose and fatty acids. (b) In fatty acid activation, a molecule of AMP is bound with the release of pyrophosphate; in glucose activation a phosphate is bound with the release of ADP. (c) Although both

Answer

To be covered when testing yourself

reactions are achieving the same results, namely activation, they are performed by completely different enzymes. (d) a thioester bond is involved only with fatty acid activation.

a

13. Which of the following does not contain any incorrect statements concerning triacylglycerol synthesis using glycogen? (a) glycogen is broken down to glucose-1-phosphate by the enzyme glycogen synthetase, once acetyl- CoA is formed, it must pass into the cytoplasm where it goes through a series of reactions to form fatty acids via malonyl-CoA (b) glycogen is broken down to glucose-1-phosphate by the enzyme glycogen phosphorylase; the acetyl- CoA eventually formed will be used to form citrate, which will pass from the mitochondria to the cytoplasm: the citrate is then reconverted to acetyl CoA; the fatty acids that are eventually synthesized from acetyl- CoA will combine with glycerol, formed ultimately from dihydroxyacetone-phosphate, to form a triacylglycerol.
(c) glycogen is broken down to glucose-1-phosphate by the enzyme glycogen phosphorylase; the glucose-1-phosphate is then directly converted to fructose-1, 6-diphosphate and then to glyceraldehyde-3-phosphate and dihydroxyacetone phosphate; both of the latter are then used in the final synthesis of a triacylglycerol.
(d) glycogen is broken down to glucose-1-phosphate by the enzyme glycogen phosphorylase; glucose-1-phosphate is then converted to glucose-6-phosphate, and that is eventually broken down to acetyl CoA. A molecule of glycerol-3-phosphate then enters the mitochondria and combines with the acetyl CoA, giving rise to a precursor of a triacylglycerol.

b

14. Which of the following B vitamins is incorrectly coupled with its function? (a) thiamine- decarboxylation (b) niacin - part of NAD^+ associated with accepting electrons during oxidation (c) pantothenic acid - assists in blood coagulation. (d) riboflavin - part of FAD associated with accepting electrons during oxidation.

c

15. Which of the following is incorrect? (a) one's basal metabolic rate can be severely disturbed by a thyroidectomy. (b) A normal respiratory quotient is about .85. This is, of course, if the individual is consuming normal amounts of all three foodstuffs. (c)

Answer

To be covered when testing yourself

a deficiency of thiamine (vitamin B_1) could lead to a condition called pellegra. (d) In latter states of starvation, one's respiratory quotient might be around .80, signifying the use of body proteins for energy.
energy. | c

Fill in the blanks.

16. The number of molecules of a substrate transformed to a product by one molecule of an enzyme per second is called the enzyme's _____ .

turnover number

17. Some enzymes consist of two parts: a protein part, called the _____, and a non-protein part, called the _____ . Together, they have the ability to use the energy of an energy-yielding reaction for a reaction that is _____ .

apoenzyme, coenzyme, energy-requiring

18. The portion of an enzyme where the substrate binds is called the _____ . This binding is dependent on many factors, including _____ bonding, electrostatic and _____ interactions between amino acids on the substrate and the enzyme.

active site, hydrogen, hydrophobic

19. The process by which macromolecules are broken down and energy gained is called _____ . The process by which macromolecules are synthesized, which requires energy, is called _____ . Both of these, working simultaneously, control one's _____ , an always dynamic process.

catabolism, anabolism, metabolism

20. The sequence of the glycolytic pathway is as follows: glucose ⟶ _____ , using 1 molecule of ATP, ⟶ fructose-6-phosphate ⟶ _____ , with use of another molecule of ATP; ⟶ glyceraldehyde-3-phosphate + _____ ⟶ 2 (_____) ⟶ 2(phosphoglyceric acid), with electrons being accepted by a molecule of _____ ; ⟶ 2(1,3 diphosphoglyceric acid,) with the new phosphate groups coming from _____ and the electrons being accepted by a molecule of _____; ⟶ 2(3-phosphoglyceric acid), the lost phosphate groups combining with _____ to form _____; ⟶ 2(_____) ⟶ 2(phospoenol pyruvic acid), ⟶ 2(pyruvic acid), the lost phosphate groups combining with _____ to form _____.

glucose-6-phosphate , fructose-1, 6-diphosphate, dihydroxy acetone phosphate, glyceraldehyde-3-phosphate, inorganic phosphate, NAD^+, ADP, ATP, 2-phosphoglyceric acid, ADP, ATP

		Answer
		To be covered when testing yourself
21.	The ratio of ATP produced aerobically to anaerobically by the oxidation of one molecule of glucose is _____ .	18 : 1
22.	In the process of winemaking, the fact that yeast is a _____ , allows it to first grow in the presence of oxygen; when the oxygen is removed, it will produce ethanol by the process of _____ .	facultative anaerobe, fermentation
23.	The correct sequence of electron acceptors in the electron transport chain with sites of ATP production noted, is as follows: $NAD^+ \longrightarrow$ _____ (the energy released used to synthesize ATP) $\longrightarrow$ Coenzyme Q $\longrightarrow$ _____ $\longrightarrow$ _____ (the energy released used to synthesize an ATP), $\longrightarrow$ cytochrome $a+a_3$ $\longrightarrow$ _____ (the energy released used to synthesize an ATP).	FMN, cytochrome b, cytochrome C_1 & C O_2
24.	When the energy released from the transfer of electrons from one acceptor to another acceptor is captured in a high energy bond containing phosphorus, this is called _____ . This occurs during _____ .	oxidative phosphorylation, electron transport
25.	The P:O ratio is a measure of the amount of organically bound phosphorus formed per mole of _____ reduced. By knowing this ratio, the number of _____ derived from the electron transport chain can be determined.	oxygen, ATPs,
26.	A triacylglycerol consists of a molecule of _____ bound to 3 _____ . Glucose, in its breakdown form of _____ , can serve as the starting point for synthesis of the fatty acid, while the glycolytic intermediate _____ serves as the starting point for the synthesis of the glycerol "backbone."	glycerol, fatty acids, acetyl CoA, dihydroxyacetone phosphate
27.	In the process called autoradiography, amino acids are made radioactive by incorporating into them radioactive isotopes such as _____ and _____ . By tracing the amount of radioactivity in the proteins, it can be seen that _____ and _____ of proteins are going on at a constant equal rate as long as amino acids are fed into the cell.	tritium, carbon 14, catabolism anabolism, amino acids
28.	A calorie (small "c") is defined as the _____ necessary to raise the _____ of _____ gram of _____ by _____ , from 14°C to 15°C. A calorie (uppercase "c") is _____ times that.	heat, temperature one, H_2O, 1°C, 1000

Answer

To be covered when testing yourself

29. The respiratory quotient is a measure of how much _____ is released per _____ consumed. Its numerical value will depend on the consistency of one's _____ , whether it be high in _____ , _____ , or _____ .

    carbon dioxide, oxygen, diet, carbohydrates, protein, fat

Determine whether the following statements are true or false.

30. A catalyst will speed up the rate of a chemical reaction by raising the activation energy, without affecting the reaction's final equilibrium point.

    False

31. Enzymes have specific pH levels and temperatures at which they are most effective; at extremes above and below these values, an enzyme can be inactivated.

    True

32. As with extremes of heat, extremes of cold will permanently destroy an enzyme's activity.

    False

33. It is important to remember that at high substrate concentrations, it is not accurate to measure the rate of a reaction using a Michaelis-Menten graph, as at those concentrations, the enzyme itself is rate-limiting.

    True

34. Coenzyme A is necessary for reactions such as the conversion of pyruvate to acetyl-CoA, the conversion of α-ketoglutarate to succinate, and the oxidation of fatty acids.

    True

35. All three foodstuffs, carbohydrate, proteins, and fats, can ultimately be broken down to funnel into the Krebs cycle.

    True

36. "Oxygen debt" is the amount of oxygen needed to oxidize the lactate built-up when exercising, in order to convert it back to pyruvate, so the latter can enter the Krebs cycle.

    True

37. Although occuring at different points in our metabolic cycle, the oxidation of both lactate and succinate, have in common the same electron acceptor.

    False

38. The coenzyme nicotinamide adenine dinucleotide (NAD^+) is used as the electron acceptor in the light reaction of photosynthesis, while nicotinamide adenine dinucleotide phosphate ($NADP^+$) is used as the electron acceptor in glycolysis and the Krebs cycle.

    False

	Answer
	To be covered when testing yourself
39. In muscle, creatine phosphate is used as the storage form for the high-energy phosphate bond. The energy derived from the hydrolysis of this bond is used directly to drive energy-requiring reactions in muscle.	False
40. The energy stored in a molecule of ATP can be used in only 3 ways; by cleavage of the terminal phosphate and its subsequent attachment to a compound, by cleavage of the last two phosphates and their subsequent attachment to a compound, and by cleavage of the last two phosphates and the subsequent attachment of the remaining AMP molecule to a compound.	False
41. When referring to the electron transport system; by not releasing all of the energy at one time, (that is, the energy released when electrons are moved from NADH to O_2), the smaller packets that are released are used much more efficiently	True
42. When synthesizing fatty acids from the breakdown products of glucose, it is essential to realize that this synthesis must take place in the cytoplasm, not the mitochondria.	True
43. Regardless of one's metabolic rate, the amount of calories needed to sustain one's life is always the same from one individual to another.	False
44. The theory behind altering one's basal metabolic rate through manipulations of thyroid gland functioning lies at the level of oxidative phosphorylation during the electron transport system.	True

CHAPTER 4

THE INTERRELATIONSHIP OF LIVING THINGS

TAXONOMY OF ORGANISMS

• PROBLEM 4-1

What is the basis of classification of living things used today and why is it better than some of the older methods?

Solution: The modern day basis of the classification of living things was developed by Linnaeus. Linnaeus based his system of classification upon similarities of structure and function between different organisms. Previously, a system based on the similarities of living habitats was used. Before Linnaeus, animals were categorized into three groups: those that lived in water, on the land, or in the air. We use structure and function today as a basis for grouping, since similar characteristics may indicate evolutionary relationships. For example, porpoises and alligators both live in the water, whereas cows and lizards both live on the land. Using Linnaeus's system of classification, porpoises are grouped with cows, and lizards with alligators. Porpoises and cows both give birth to live young, and maintain constant body temperature. Structurally, they both possess mammary glands, from which milk is obtained to feed their young. Alligators and lizards both have scales on their skins, and have similar respiratory and circulatory systems. Cows and porpoises are members of the class of mammals, even though their habitats are different. Alligators and lizards are members of the class of reptiles. This systematic method of biological classification based on evolutionary relationship is termed taxonomy. At present, both genetics and evolution are important in understanding taxonomy. Because of similarities in structure and function, both cows and porpoises are believed to have a common ancestor in the very distant past. A separate ancestor was probably shared by alligators and lizards.

• PROBLEM 4-2

The following are all classification groups: family, genus, kingdom, order, phylum, species, and class. Rearrange these so that they are in the proper order of sequence from the smallest grouping to the largest. Explain the scientific naming of a species.

Solution: Closely related species are grouped together into genera (singular-genus), closely related genera are grouped into families, families are grouped into orders, orders into classes, classes into phyla (singular - phylum), and phyla into kingdoms. Classes and phyla are the major divisions of the animal and plant kingdoms.

In order to give the scientific name for a certain species, the genus name is given first, with its first letter capitalized, the species name, given second, is entirely in lower case. The entire name is underlined or italicized. For example, the scientific name of the cat is Felis domestica and that one of the dog is Canis familaris. The cat and dog both belong to the class of mammals and the phylum of chordates.

An example of a complete taxonomic classification for a Manx cat is:

Kingdom - Animalia

Phylum - Chordata

Subphylum - Vertebrata

Class - Mammalia

Subclass - Eutheria

Order - Carnivora

Family - Felidae

Genus - Felis

Species - domestica

Variety - manx

As this example shows, important divisions may exist within a class or phylum, and the use of subclasses and subphyla is an aid to classification.

Note that the group "variety" allows us to refer exactly to the type of domesticated cats we are considering, not Siamese cats, not Persian cats, but Manx cats.

• PROBLEM 4-3

What are the major structural and functional differences between plants and animals?

Solution: The most important difference between plants and animals is in their methods of obtaining the energy that they need to grow and maintain life. Plants convert the light energy of sunlight into chemical energy via the process of photosynthesis. This is accomplished by the green pigment chlorophyll. Animals do not have this pigment and must obtain chemical energy directly. They do this by ingesting organic material from plants or from other animals.

In order to obtain food, most animals move about within their environment. Plants, in general, are stationary. The larger plants send roots into the soil in order to obtain water and minerals, and expose their leaves, which contain the photosynthetic machinery, to the sunlight.

At the microscopic level, a major structural difference between plants and animals is the outer covering of the cell. Plants have a rigid cell wall, external to the cell membrane, which is composed mainly of cellulose. Animal cells have no cell walls outside their cell membranes, and their shape is not rigid.

A second ultrastructural difference is the presence of organelles called chloroplasts. These are present only in plants. Chloroplasts contain chlorophyll and the enzymes necessary for photosynthesis.

Another difference between plants and animals is the ability for growth. Most plants are able to continue to grow throughout their lives. The body size of most animals, however, after a certain definite period of growth, remains fixed.

• **PROBLEM** 4-4

List some organisms which are difficult to assign to either the plant or animal kingdom. Does classifying these as members of a third kingdom, the Protista, solve the problems raised by attempts to divide the living world into "plants" and "animals"?

Solution: Many one-celled organisms are difficult to classify as either animals or plants. Some have characteristics of both animals and plants. Euglena are an example of this. Like plants, they contain chlorophyll and are capable of photosynthesis; like animals they are able to move about within the environment, ingesting and absorbing organic nutrients. Other organisms have characteristics strikingly different from either plants or animals. Bacteria, like plants, have a cell wall, however, unlike plants, their cell wall is not composed of cellulose. Certain bacteria are capable of photosynthesis, while other bacteria must obtain organic nourishment. Amongst the bacteria which do not rely on organic nourishment, some, like plants, obtain energy from sunlight, and some, unlike any other organism, obtain energy from the oxidation of inorganic compounds.

The formation of a third kingdom solves some of the problems which arise in attempting to classify these organisms. However, a problem still exists with the classification Protista since classification is based upon structural and functional similarities within groupings of organisms.

Many diverse organisms are included within the Protista which bear no closer relationship to each other than to plants or animals. For this reason a fourth and a fifth kingdom have also been proposed, the Monera and the Fungi. The Monera kingdom includes the bacteria and blue-green algae. These have cell walls different from plant cell walls, and lack nuclear membranes. Their ultrastructure differs from that of organisms in the other kingdoms. The members of the Fungi kingdom lack photosynthetic pigments but have cell walls and multiply in a manner more like plants or protists than animals.

The problems inherent in any classification system can never be fully resolved. In attempting to classify bacteria or Euglena, the problem is not in determining what the organism is, but in determining what other organisms it bears the closest resemblance to. As more is learned about the structure and function of different organisms, and about the relationships between different organisms through fossil records and biochemical analysis of their genetic compositions, more complete taxonomic systems can be established.

• PROBLEM 4-5

Discuss the relevant arguments for classifying the blue-green algae as monerans, as protists, and as plants.

Solution: The classification of blue-green algae depends primarily upon the system of classification which is used. The blue-green algae show similarities to members of the three kingdoms: Monera, Protista, and Plantae. It depends upon the opinion of the taxonomist as to how many kingdoms there should be and to which classification of organisms the blue-green algae bear the closest resemblance.

If five kingdoms were used in taxonomy, then blue-green algae would usually be grouped together with bacteria as monerans. Blue-green algae lack a nuclear membrane and have a single 'naked' chromosome, as do bacteria. Both blue-green algae and bacteria lack membrane-bounded subcellular organelles such as mitochondria and chloroplasts. Their ribosomes are unique.

Thus, on an ultrastructural level, the blue-green algea are most closely related to the monerans. Monerans are prokaryotes: they lack membrane-bounded nuclei; all other organisms are eukaryotes: they have membrane-bounded nuclei and membrane-bounded organelles.

If only three kingdoms are used, i.e., the Protista, plant and animal kingdoms, then more general characteristics for classification of blue-green algae must be used. Since the blue-green algae have characteristics of both plants and animals, they are placed in the kingdom for organisms with intermediate or hard to place traits: the Protista. A reason against doing this is that the kingdom becomes merely a dumping ground for misfit organisms.

The members of the kingdom do not necessarily share any structural or functional resemblances to each other - which is supposed to be the very basis for grouping organisms into kingdoms and its classifying subdivisions.

As 'plants,' algae have little internal differentiation, that is, they have no structures such as roots, leaves, or stems. For this reason, algae are often classified as protists rather than plants. However, if all organisms which are capable of photosynthesis are grouped with plants, then blue-green algae as well as other algae could be in this group.

NUTRITIONAL REQUIREMENTS AND PROCUREMENT

• PROBLEM 4-6

Why are autotrophic organisms necessary for the continuance of life on earth? Do all autotrophs require sunlight?

Solution: Autotrophic organisms have the capacity to generate all needed energy from inorganic sources. Heterotrophic organisms can only utilize the chemical energy present in organic compounds. There are two main types of autotrophs - photosynthetic organisms and chemosynthetic organisms. Photosynthetic autotrophs obtain energy from sunlight, and convert the radiant energy of sunlight to the chemical energy stored in the bonds of their organic compounds. Green plants obtain CO_2 from the atmosphere and minerals and water from the soil. Algae and photosynthetic bacteria absorb dissolved CO_2, water, and minerals through their cell membranes. Using energy from sunlight, the photosynthetic autotrophic organism converts CO_2, water, and minerals into all the constituents of the organism. Chemosynthetic organisms are much less common than photosynthetic organisms, and are always bacteria. Chemosynthetic bacteria do not require sunlight, and obtain energy by oxidizing certain substances. Two examples are the nitrifying bacteria which oxidize ammonia to nitrate (NO_2^-) or nitrates (NO_3^-) ultimately, and the sulfate bacteria, which oxidize sulfer to sulfates. The energy released from these chemical reactions is converted to a form of chemical energy utilized by the organism.

All the organisms which carry on respiration, that is, oxidize organic compounds to carbon dioxide and water, require oxygen. Respiration is the process by which heterotrophs obtain energy. Chemosynthetic bacteria also require oxygen in order to carry out oxidations of inorganic substances. The only source of oxygen on earth is the photosynthetic autotrophs. These organisms convert CO_2 and water to organic compounds, utilizing sunlight to provide energy, and generate O_2 in the process. If there were no green plants or photosynthetic marine organisms, the oxygen present in the atmosphere would quickly be used up by animals, bacteria and fungi.

The autotrophs are also responsible for providing the heterotrophs with organic nourishment. Sunlight is the most important source of energy on the earth, and it is only the photosynthetic autotrophs which can utilize this energy, converting it to chemical energy in organic compounds. Heterotrophs utilize the organic compounds produced by the autotrophs. Heterotrophs which obtain organic nourishment from other heterotrophs are also ultimately dependent upon autotrophs for nourishment, because the animals which are the prey have either directly or indirectly (through another animal) utilized the organic material of plants or algae. Photosynthetic autotrophs provide the earth with an energy source for living organisms and with oxygen. If photosynthetic autotrophs were not present, all life on earth would eventually cease as the food and O_2 would become depleted.

• **PROBLEM 4-7**

Differentiate between the several types of heterotrophic nutrition and give an example of each.

Solution: The different types of heterotrophic nutrition are defined according to either the type of food source used or the methods employed by the organism in utilizing the food to obtain energy. Holozoic nutrition is the process employed by most animals. In this process, food that is ingested as a solid particle is digested and absorbed. Holozoic nutrition can be further classified as to the food source: herbivores, such as cows, obtain food from plants; carnivores, such as wolves, obtain nutrients from other animals; omnivores, like man, utilize both plants and animals for food.

Saprophytic nutrition is utilized by yeasts, fungi, and most bacteria. These organisms cannot ingest solid food; instead they must absorb organic material through the cell membrane. They live where there are decomposing bodies of animals or plants, or where masses of plant or animal by-products are found. A saprophyte obtains nutrients from nonliving organic matter. An example of a saprophyte is given by yeasts which produce ethanol. Utilizing grape sugar as their energy source, they ferment glucose to carbon dioxide and ethanol. These yeasts are used to produce wine.

In parasitic nutrition, organisms called parasites obtain nourishment from a living host organism. Most parasites absorb organic material and are unable to digest a solid particle. This is true also of saprophytes, however, saprophytes do not require a living host in order to supply them with nourishment. Parasites are found in many classes of the plant and animal kingdoms, and frequently are bacteria, fungi, or protozoa. All viruses are parasites, requiring the host not only for nutrition, but also for synthetic and reproductive machinery. Some parasites exist in the host causing little or no harm. Parasites causing damage to the host are well known to man, and are termed pathogenic. Examples of these are the tapeworm, which lives in the intestine and prevents the host from obtaining adequate nutrition from the food which is eaten, and the tubercle bacillus which causes tuberculosis. Certain organisms are saprophytic in their natural habitat, but are capable of living in a host organism and causing disease. Clostridium tetani is an example. In forests, these bacteria obtain nutrients from decaying plant and animal material. When the bacteria enter a wound in a human, the toxic substances they release cause the disease tetanus.

• PROBLEM 4-8

Why are parasites usually restricted to one or a very few host species?

Solution: Parasites usually have extremely specific requirements for growth. They may grow only within a narrow range of pH and temperature, a certain oxygen concentration and may require a large number of different organic nutrients. The specific combination of optimal growing conditions which is required by a particular parasite can be found only in one host species, or in several closely related host species. Usually the parasite can live only in certain locations within the host. The tapeworm, for example, can live only in the intestine of a the human and will not infect the kidneys or the bones. Saprophytes, on the other hand, will grow within a broad range of temperature, pH and O_2 concentration, and they require very few organic nutrients. Yeasts, an example of a saprophyte, are able to grow at many different oxygen concentrations, oxidizing glucose to CO_2 and water if oxygen is present, and fermenting glucose to CO_2 and ethanol if oxygen is not present in sufficient amounts. Parasites, in general, are unable to switch from oxidative to fermentative metabolism, and must utilize either one or the other. In addition, yeasts can synthesize all their constituent proteins, nucleic acids, and other components if they are supplied with glucose. Parasitic bacteria lack the enzymes needed for this and must be supplied with amino acids, vitamins, and a mixture of sugars.

These complex growth requirements make it difficult to culture parasitic organisms in the laboratory. Many parasites can grow only if supplied with extracts from animal tissues. Some parasites, such as viruses and rickettsias, can grow only in the presence of living cells.

ENVIRONMENTAL CHAINS AND CYCLES

• PROBLEM 4-9

Animals at the top of a food pyramid may be larger in individual size than animals more immediate to the primary source of food, yet they represent a smaller total weight (biomass) in the aggregate. Explain.

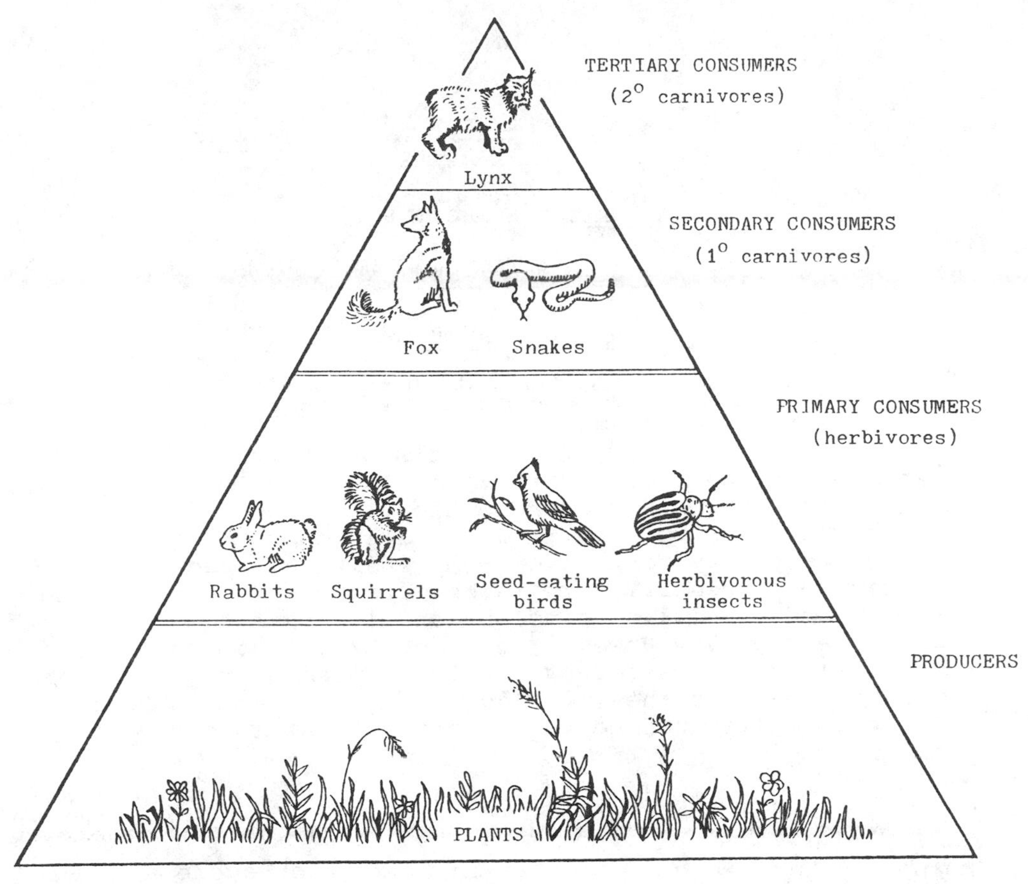

Figure 1. A hypothetical food pyramid. The amount of energy found at each level can be represented as: energy at producer level$\gg$ energy at primary consumer level$>$ energy at secondary consumer level $>$ energy at tertiary consumer level. The decrease in the amount of biomass (the mass of living organisms) going up the pyramid is also evident.

Solution: The food pyramid is the term that describes the successive transfer of the radiant energy trapped by plants through a series of consumers. In its simplest form, plants are eaten by primary consumers (herbivores), which are in turn eaten by secondary consumers (carnivores). Ultimately, decomposers (bacteria and fungi) degrade unused plant and animal matter into components useful to themselves or the green plants. At each level there is a decrease in the amount of usable energy because in the process of each transfer, energy is lost as heat. There is more energy available to plants than there is available to herbivores, more energy available to herbivores than to carnivores, more energy available to primary carnivores (carnivores that eat only herbivores) than to secondary carnivores (carnivores that also eat other carnivores). With each step, the animal must oxidize the food it eats to obtain energy to synthesize its own cellular constituents and to perform other life functions. The energy that is liberated is not all converted to the animal's cellular constitutes, rather a large proportion is lost as heat. The amount of available energy limits the number and mass of each organism which can utilize that energy to support life.

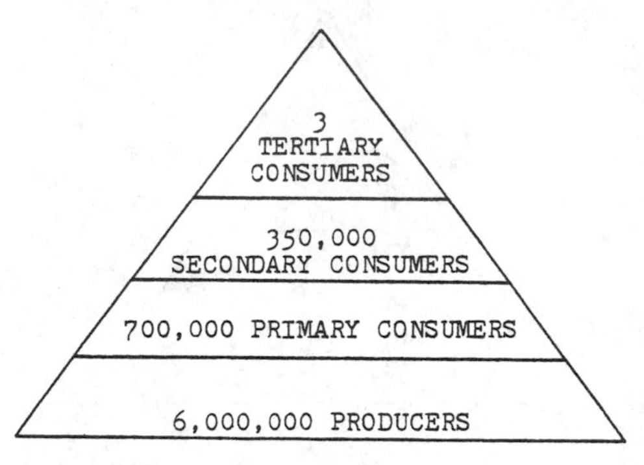

Figure 2. Pyramid of the biomass in a bluegrass field with respective approximate numbers.

The total mass of the animals at the top of the food pyramid, the secondary carnivores, is less than the total mass of the animals closer to the plant source of food, the herbivores, because there is less energy available to the secondary carnivores. An individual secondary carnivore is usually very large. Large body size is useful to these animals, since it enables them to capture and kill their prey. However, the number of such carnivores is small. In considering the total mass of the secondary carnivores, both the individual body mass and the number of individuals must be taken into account (See Figures 1 and 2).

• PROBLEM 4-10

Discuss the role of bacteria in the carbon cycle and the nitrogen cycle.

Solution: Cycling of the earth's resources is a process by which life is able to continue on earth. The carbon and nitrogen atoms that are present on the earth today are the same atoms that were present three billion years ago, and have been used over and over again. They are the fundamental constituents of organic compounds, and are present in large quantities in all organisms. Carbon and nitrogen atoms that were present in a dinosaur might now be found in an oak tree since matter is not destroyed, and cannot be created. The same carbon and nitrogen atoms are constantly being recycled.

In the carbon cycle, atmospheric carbon dioxide is converted into organic material by plants. Animals and bacteria convert some of this organic matter into CO_2 by respiration, however, most of the carbon remains fixed as organic matter in the bodies of plants and animals. Bacteria play a crucial role in the carbon cycle. Bacteria, as well as fungi, convert the carbon atoms in the organic matter of decaying plants, animals, and other bacteria and fungi to CO_2, which returns to the atmosphere (see Fig. 1).

Plants obtain nitrates from the soil, taking them up through their roots, and convert the nitrate into organic

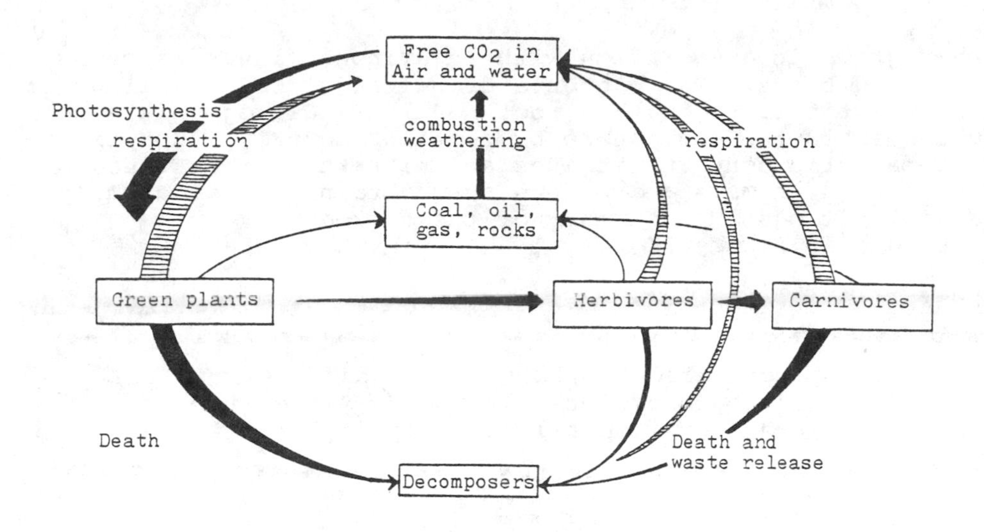

Figure 1. The carbon cycle.

compounds, mainly proteins. Animals obtain nitrogen from plant proteins and amino acids and they excrete nitrogen-containing wastes. These nitrogenous wastes are excreted in one of the following forms, depending on the species - urea, uric acid, creatinine and ammonia. Certain bacteria in the soil convert nitrogenous waste and the proteins of dead plants and animals into ammonia. Another type of bacteria present in the soil is able to convert ammonia to nitrate. These are termed nitrifying bacteria. They obtain energy from chemical oxidations. There are two types of nitrifying bacteria: nitrite bacteria, which convert ammonia into nitrite, and nitrate bacteria, which convert nitrite into nitrate. Nitrogen is thus returned to the cycle. Atmospheric nitrogen, N_2, cannot be utilized as a nitrogen source by either animals or plants. Only some blue-green algae and certain bacteria can convert N_2 to organic compounds. This process is termed nitrogen fixation. One genus of bacteria, Rhizobium, is able to utilize N_2 only

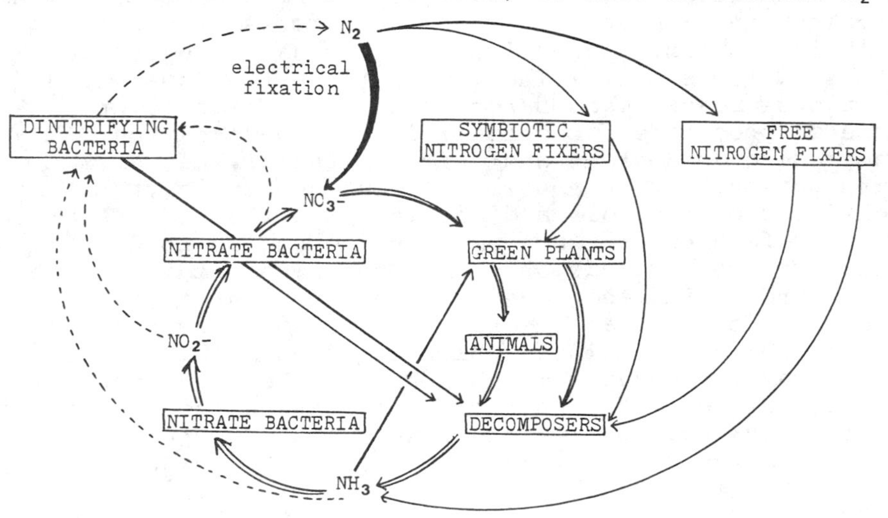

Figure 2. The nitrogen cycle.

when grown in association with leguminous plants, such as peas and beans. The bacteria grow inside tiny swellings of the plant's roots, called root nodules. Nitrogen is also returned to the atmosphere by certain bacteria. Denitrifying bacteria convert nitrites and nitrates to N_2, thus preventing animals and plants from obtaining biologically useful nitrogen. A summary of the nitrogen cycle is provided in Figure 2.

• PROBLEM 4-11

In what way does the phosphorus cycle differ from the carbon and nitrogen cycles? In what way does the energy cycle differ from all other cycles?

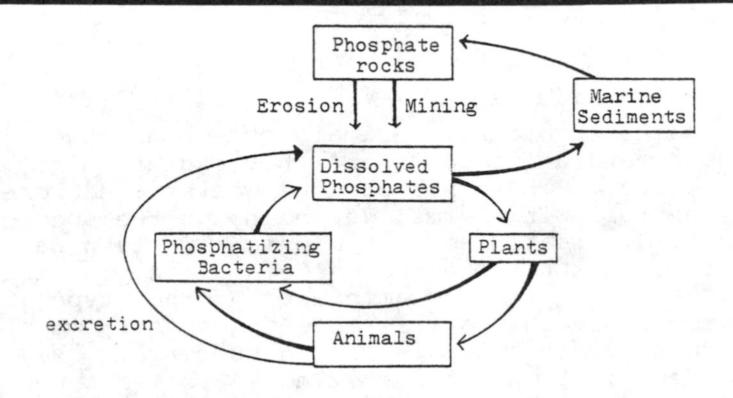

A simple version of the phosphorus cycle.

Solution: The phosphorus cycle differs from the carbon and nitrogen cycles in that it is not completely balanced, and relies less on living organisms to maintain it. The phosphorus cycle is an example of a sedimentary cycle in which the mineral cycles between land and sediments in the sea. Inorganic phosphate is found in rocks at the earth's available to plants and animals as inorganic phosphate in their water supply. The phosphate is then converted to various organic phosphates, which are an additional source of phosphate for animals in the food they eat. In degrading dead plant and animal material, bacteria return phosphate to the soil. The phosphate is then leached out of the soil and carried into the sea. Sea birds and fish return some of this phosphorus to the cycle. Marine birds deposit phosphorus-containing wastes on the land. Until the advent of chemical fertilizers, these deposits were an important source of phosphorus for agricultural use. Animals which eat fish and marine invertebrates recover some phosphorus. Despite these recovery methods, more phosphate is lost to the sea bottom than is returned. Over time, however, phosphorus may be returned to the cycle by geologic upheavals which bring the sea bottom up to the earth's surface, creating mountains and new land masses.

The energy cycle differs from other cycles in

that no cycling of useful energy occurs. The energy is converted into other forms of energy. However, in the process energy is lost as heat, which is not biologically useful. The energy cycle, thus, is not actually a cycle. All energy is derived from sunlight. Plants convert solar energy to chemical energy. Only three percent of the sun's radiation energy is trapped by the photosynthetic process. The chemical energy that is stored in plants is utilized by animals or bacteria. Since the process is not completely efficient, not all of the energy stored within the plant can be converted into useful energy by the animal; about 40 percent of the energy is lost as heat. If the animal is eaten by another animal, a further reduction in the amount of useful energy occurs, as additional heat is generated and lost when the chemical energy of the first animal is converted to the cellular constituents of the second animal. Eventually all the energy trapped by green plants is converted to heat, and all the carbon of organic compounds is converted to CO_2 and fossil fuels.

DIVERSIFICATION OF THE SPECIES

• PROBLEM 4-12

Suppose you were given a cage full of small, rat-like animals from the Rocky Mountain region and another cage full of similar animals from the Appalachian Mountains. How would you determine if the two groups were of the same or different species?

Solution: The species is the fundamental unit of biological classification. A species is defined as a group of organisms that is closely related structurally and functionally, which in a natural environment interbreeds and produces fertile offspring, but which seldom, if ever, breeds with organisms of another species. A species is reproductively isolated. The dog and the cat are examples of two different species. Dogs and cats cannot breed with one another.

Closely related species may interbreed, however, the offspring are rarely fertile. The horse and the donkey are examples of closely related species which can mate and produce viable offspring. This offspring is the mule, which is unable to reproduce. A mule can only result from the union of a horse and a donkey. The fact that mules are sterile helps to establish the fact that the horse and the donkey are of different species.

Species may be subdivided into subspecies, varieties, or races. The varieties are kept distinct either through geographic isolation, seasonal differences in reproductive patterns, or by the intervention of man. Dogs offer an example of controlled breeding. If unrestrained by man, the different varieties interbreed to produce mongrels. In certain cases, interbreeding between different subspecies is very difficult. The greatest problem in the case of dogs would be size; however, sexual attraction between different varieties exists and it can be proven that both varieties belong to the same species because they are structurally and functionally similar. Both the small and large varieties could breed with a medium sized variety, proving that the varieties are not reproductively isolated. If interbreeding between varieties is prevented by seasonal variation of reproductive patterns, careful regulation of laboratory conditions could enable two different subspecies to be ready for reproduction at the same time.

In the case of the rat-like animals, the two groups should be studied to see if they are similar structurally and functionally. If so, it must be determined whether the animals could breed with one another. If they could and if they produce fertile offspring, the two groups would be classified as members of the same species. If the offspring were all sterile, it would mean that the two groups of animals belonged to two different, closely related species. If the two groups seemed similar after studying their structural and functional characteristics, but did not breed with one another, an examination of their natural reproductive patterns might offer information as to why breeding did not occur under laboratory conditions. If conditions were altered and fertile offspring were produced, the two groups of animals could be members of the same species. If breeding did not occur under any conditions, or through an intermediate species, one would conclude that the two were members of different species.

• PROBLEM 4-13

Practically all of the corn planted in America today has been produced by the laborious task of hybridization of different varieties. Tell why this is done and explain in detail why the process is worthwhile.

Solution: A hybrid is an offspring of parents expressing different traits. Hybridization is the term referring to a mating that produces hybrids. The parents may be of the same species or of different species. A cross between an Irish setter and a collie is an example of a cross between different subspecies within the same species. A cross between a horse and a donkey is an example of a cross between different species. Offspring from such a cross are sterile.

Hybridization is commonly practiced by plant and animal breeders. Hybridization produces new varieties, which may show a greater strength and vigor than either of the parents, a characteristic termed hybrid vigor. However, not every hybrid will exhibit hybrid vigor. Those that show such quality have dominant desirable traits. Hybrids which have been passed on undesirable dominant traits occur just as frequently; when undesirable traits are dominant, a hybrid will not show any advantageous characteristics.

Newer and better varieties of roses, orchids, and garden vegetables are constantly being developed by selectively interbreeding plants with desirable dominant traits. Varieties of corn with hybrid vigor are obtained by artificially crossing two different parent varieties to obtain seeds. This is done by cutting the male reproductive parts from one variety of corn and dusting the female reproductive parts from the second variety of corn with the male's pollen. The seeds produced are crosses between the two varieties. When cultivated, the hybrid plants give a much higher yield than either of the parent varieties, and are more resistant to disease.

In order to understand why hybrid vigor occurs, it is necessary to understand Mendelian genetics. In these hybrids, the desirable traits which lead to increased strength, high yield, or long life are dominant over the less desirable traits of both parents, which are recessive. The hybrid tends to show the desirable traits of both parents, and the recessive traits are not expressed. In the case of hybrid varieties of corn, the desirable traits of the parent varieties combine so as to produce an offspring which gives a higher yield than either parent.

• **PROBLEM 4-14**

Why does the yield from hybrid corn go down when seeds from the hybrid are planted over the next several generations?

Solution: Hybrids of different varieties of corn will give a greater yield than either of the parent varieties if the desirable traits in both parents are carried on dominant genes. The less desirable parental traits would not appear in the first generation hybrids, because these traits are carried on recessive genes. When the hybrids reproduce, however, some of the crosses yield offspring in which both recessive genes come together, and the second generation expresses the recessive, less desirable characteristics. Future hybrid generations will continue to express recessive traits, and the future generations will cease to be exclusively high-yielding.

Hybridization between the parent varieties must be done every year in order to obtain the high-yielding hybrid seed. It is important to note that these crosses are between different varieties of the same species. Artificial crosses between the two parent varieties of corn give a hybrid variety which is capable of reproduction, and also has hybrid vigor. Crosses between different species may give an offspring which shows hybrid vigor, but the offspring will be almost invariably sterile.

• **PROBLEM 4-15**

Explain how a new species may arise after it has been separated from its ancestors for a considerable period of time.

Solution: Gene mutation and natural selection are two processes which enable a certain group of organisms to become a distinct species, different from its ancestral species. As a certain species distributes itself over a large geographical area, different subgroups become isolated in different environments. As genes mutate to give rise to different traits, certain traits are proven to be more valuable in certain environments and less valuable in others. The traits that enable the organism to be better adapted to its environment will tend to be conserved. This is because the organisms which express these traits will be more likely to survive and reproduce than organisms not so well adapted to their environment. Eventually, the genetic composition of an isolated group becomes so different from the genetic composition of the ancestral species, that breeding can no longer occur between the two groups of organisms. In addition, the isolated group will have different structural and functional characteristics. The new group has thus become a distinct species.

An example of this type of evolutionary divergence from a common ancestor is given by the camels and their relatives. Through study of fossil records, it has been determined that the ancestral species was similar to the South American llama. At some time in the distant past, some of these animals migrated across a land bridge to Africa. These animals, the immediate ancestors of the camels, evolved adaptive mechanisms that enabled them to survive in the desert environment. For example, the camel's hump can store food and water for a long time over great distances. This is an adaptive structure which llamas and their American relatives do not possess.

• PROBLEM 4-16

It is rather easy for botanists to create a new species of plant by hybridization of two existing species, yet such a technique is generally not possible for animals. Explain why, giving examples to illustrate.

Solution: A hybrid cross between two species yields an offspring with one set of chromosomes from each parent. This may create no problem in development or viability of the hybrid, but its ability to reproduce is likely to be impaired. During meiosis, irregularities in homologous pairing occur due to the differences between the two sets of chromosomes. Other meiotic irregularities follow, leading to nonviable gametes. If however, one were able to somehow overcome these meiotic irregularities, the offspring of interspecies crosses would be capable of reproduction. A method for producing viable gametes has been found for plants; this method, however, cannot be used for animals.

By applying the chemical colchicine, it is possible to induce a doubling of the chromosome number. If a certain species is diploid, and colchicine is applied, the species becomes tetraploid. In the case of a hybrid between two species, a doubling of the chromosome number results in the hybrid having two complete sets of chromosomes from each parent species. Proper pairing can now occur in meiosis, since the gamete will have one complete set of chromosomes from each parent species. A male and female gamete can combine to produce an embryo having two complete chromosome sets, with each set being diploid. The hybrid cross is no longer sterile. The polyploid hybrids will be incapable of breeding with the diploid parent species, but will be capable of breeding among themselves. Since the polyploid hybrids are structurally and functionally distinct from the parent species and since they are reproductively isolated, they are a new and separate species.

The first new species produced in this way was a cross between a cabbage and a radish. This plant had little agricultural value, but other new plant species have since been produced. Triticale, a cross between wheat and rye, is a high-yielding grain that has recently been developed. Triticale will not breed with wheat or with rye, but only with other triticale. Such hybrid species are termed allopolyploid, "allo" meaning different. The term indicates that the hybrids have a multiplication of the normal chromosome number and that the chromosomes are composed of complete sets from different species. Allopolyploid plants have also been found in nature. This method of creating new species, by doubling the chromosome number in a hybrid between two species, seems to have occurred in cotton, tobacco, and many other agriculturally important plants.

In plants, polyploidy often bestows added strength and vigor to the hybrid. In animals, however, polyploidy is very rarely found, and when it is, the organism is usually weak or malformed. This is probably due to the rather delicate sex balance in animals, in which any departure from the normal XY distribution is critical. Thus, even if breeding between different species of animals could be made to occur, and if colchicine could be applied to the developing hybrid offspring - both of which conditions are much more difficult in animals than in plants - it is unlikely that a polyploid hybrid would be viable.

Answer

SHORT ANSWER QUESTIONS FOR REVIEW

To be covered when testing yourself

Choose the correct answer.

1. Blue green algae can be classified as (a) monerans, (b) plants, (c) protists, (d) all of the above — d

2. The kingdom Monera includes: (a) only bacteria. (b) bacteria and blue-green algae. (c) blue-green algae and red algae. (d) all fungi. — b

3. The classification system used today is based upon (a) similarities of habitats between different organisms, (b) similarities of phenotypes between different organisms, (c) similarities of structures and functions between different organisms. (d) similarities of ecological niches between different organisms. — c

4. Which of the following classification groups are in the proper sequence from the largest grouping to the smallest? (a) phylum, class, order, family (b) kingdom, family, class, phylum (c) family, order, genus, species (d) kingdom, class, species, genus — a

5. The genus and species of a fruit fly is correctly written: (a) Drosophila melanogaster. (b) Drosophila Melanogaster. (c) drosophila melanogaster. (d) drosophila Melanogaster. — a

6. Organisms which absorb organic material through the cell membrane because they cannot ingest solid food, and live where there are decomposing bodies of animals or plants are: (a) holozotes, (b) omnivores, (c) parasites, (d) saphrophytes. — d

7. The carbon and nitrogen atoms found on earth today (a) were made by plants during photosynthesis. (b) are the same as those that were present on earth three billion years ago. (c) are synthesized during the reproduction process. (d) are produced by mitosis. — b

8. Nitrogenous wastes are excreted by different species of animals in all of the following forms, except: (a) creatinine, (b) uracil, (c) ammonia, (d) urea. — b

	Answer
	To be covered when testing yourself
9. Structurally, plant and animal cells differ in all of the following except (a) the outer cover of the cell. (b) the presence or absence of chloroplasts, (c) the enzymes present. (d) the presence or absence of mitochondria.	d
10. The greatest similarity in structure occurs between members belonging to the same (a) species, (b) genus, (c) family, (d) class.	a
11. An organism that makes its own food from carbon dioxide and water is (a) a green mold. (b) a green plant. (c) mushroom. (d) yeast.	b
12. In order to obtain high yielding hybrid seeds, hybridization between the two parental varieties must be done (a) once. (b) every year. (c) every five years. (d) every ten years.	b
13. In order for plants to obtain the necessary nitrogen, the free nitrogen in the air must be converted to nitrates. This is done by (a) legumes. (b) bacteria of decay. (c) nitrogen fixing bacteria. (d) denitrifying bacteria.	c
Fill in the blanks.	
14. The binomial system of classification which is being used today is based on the model developed by _____ .	Linnaeus
15. Some organisms are difficult to assign to either the plant or animal kingdom. Thus three more kingdoms have been proposed, the _____ , _____ and _____ .	Protista, Monera, Fungi
16. _____ organisms can produce energy by using inorganic sources.	Autotrophic
17. Most animals employ _____ nutrition in ingesting their food as a solid particle and then digesting and absorbing it.	holozoic
18. In a food pyramid, the amount of energy available to a deer is _____ than the amount of energy available to plants and _____ than that available to a cougar.	less, greater
19. Atmospheric nitrogen cannot be utilized as a nitrogen source by plants or animals. It must first be converted to _____ and/or _____ in order for it to be usable.	nitrites, nitrates

Answer

To be covered when testing yourself

20. _____ is the process in which plants can convert the light energy of sunlight to chemical energy. — photo-synthesis

21. Blue green algae can be classified in three different groups depending upon the taxonomy used. They can be called _____ because they lack a nuclear membrane and have a single 'naked' chromosome, or they can be placed in the _____ kingdom because they possess intermediate and hard to place traits. Finally, they can be grouped with the _____ because of their ability to photosynthesize. — Monerans, Protista, plants

22. Viable offspring are obtained from a cross between a male and a female of two different species and have a greater strength and vigor than either of the parents.This is an example of _____ . — hybrid vigor

23. The chemical _____ induces the doubling of the chromosome number. — colchicine

Determine whether the following statements are true or false.

24. Rodent-like creatures were collected from three different locations and brought back to a laboratory where they could be studied. They mated and produced viable young. This proved they were all of the same species. — False

25. A parasite can usually grow in many varied environments. — False

26. The reason that the amount of energy available to organisms is less as you proceed up a food pyramid is because energy is lost to the environment as heat. — True

27. Plants, like animals,grow during a certain definite period and then remain fixed at that size throughout their lives. — False

28. Green plants change inorganic materials into carbon dioxide. — False

29. The yield from corn with hybrid vigor will decrease when the seeds from the hybrid are planted over several generations. — True

30. Taxonomy is the science of classification. — True

31. A group of related species is called a genus. — True

CHAPTER 5

BACTERIA AND VIRUSES

BACTERIAL MORPHOLOGY AND CHARACTERISTICS

• PROBLEM 5-1

Discuss the major morphological features of bacteria. Explain why bacteria can produce changes in their environment so quickly.

Solution: Among the major characteristics of bacterial cells are their shape, arrangement, and size. These characteristics constitute the morphology of the bacterial cell.

Although there are thousands of different species of bacteria, the cells of most bacteria have one of three fundamental shapes: (1) spherical or ellipsoidal, (2) cylindrical or rodlike, and (3) spiral or helically coiled.

Spherical bacterial cells are called cocci (singular, coccus). Many of these bacteria form patterns of arrangement which can be used for identification. These patterns can be explained by peculiarities in the multiplication processes of the different bacteria. For example, diplococcal cells divide to form pairs. Streptococcal cells remain attached after dividing and form chains. Staphylococci divide three dimensionally to form irregular clusters of cocci, resembling bunches of grapes. Each of the cells in a diplococcal, streptococcal, or staphylococcal aggregate is an independent organism.

Cylindrical or rodlike bacterial cells are called bacilli (singular, bacillus). These do not form as wide a variety of arrangements as do cocci, but occasionally they are found in pairs or chains. These patterns do not arise from the multiplication process, but only from the particular stage of growth or growth conditions present, and hence bacilli usually appear as single, unattached cells. There are many variations in the thickness and length of these rodlike bacteria.

Spiral-shaped bacterial cells are called spirilla (singular, spirillum). Like the bacilli, they usually occur as unattached, individual cells. The spirilla exhibit considerable differences in length and in the frequency and amplitude of the spirals.

The average bacterial cell has dimensions of approximately 0.5 to 1.0 μm by 2.0 to 5.0 μm. (μm is the abbreviation for micrometer, which is 1/1000 of a millimeter or 10^{-6} meter.)

An important consequence of the very small size of a bacterial cell is that the ratio of surface area to volume is extremely high. This ratio allows a very large portion of the bacterial cell to be in contact with its environment. The result is that bacteria are able to rapidly ingest nutrients and growth factors and excrete wastes; their metabolic rate is correspondingly high. This high metabolic rate enables bacteria both to adjust to and to introduce changes in their environment in very short periods of time. These changes may be beneficial to the bacteria producing them, or to other species of bacteria. For instance, the release of carbon dioxide by certain bacteria increases the acidity of the growth medium and favors the growth of bacteria requiring a low pH environment.

The most important factor involved in the ability of bacteria to alter their environment is their ability to multiply rapidly. Within the multitudes of bacteria produced by a newly settled bacterium there will be a handful of mutants. Natural selection may allow one of these mutants to be most fit for survival in the changing environment and numerous offspring will be derived from it. This cycle may be repeated, demonstrating the capability of the bacterial species to survive under changing conditions.

• PROBLEM 5-2

What cell components typically found in a eucaryotic cell are missing from a bacterial (procaryotic) cell?

Solution: Cells are classified as eukaryotic or prokaryotic. The former means "true nucleus" while the latter means "before the nucleus". Prokaryotes include all bacteria including the cyanobacteria (blue-green algae), whereas eukaryotes include the protistan, fungal, animal and plant cells. The most obvious difference between bacteria and eucaryotes is the absence of the nuclear membrane in bacteria. The DNA of the bacterial cell is generally confined to a specific area but this region is not enclosed by a membrane. In addition, proteins, such as the histones of eucaryotic cells, are not found in association with bacterial DNA. Another distinctive feature is that during replication of the bacterial nuclear region, the mitotic spindle apparatus is not seen.

Bacterial cells also lack Golgi apparati, mitochondria, and endoplasmic reticulum. This means that the ribosomes found in the cytoplasmic region are free, i.e. they are not bound to endoplasmic reticulum. Membranous systems are not completely absent in bacteria. The plasma membrane folds inward at various points to form mesosomes. These membranes may be involved with the origin of other intracellular structures, and the compartmentalization and integration of biochemical systems. For example, although the electron transport system is located in the plasma membrane of bacteria, most of the respiratory enzymes are located in the mesosomes. The plasma membrane and mesosomes are the bacterial counterpart of the eucaryotic cells' mitochondria, serving to compartmentalize the respiratory enzymes. Those bacteria that contain chlorophyll do not contain plastids to "house" the chlorophyll. Instead, the chlorophyll is associated with membranous vesicles arising from mesosomes.

• PROBLEM 5-3

A biology student is presented with two jars. One is labelled "bacterial flagella" and the other is tagged "protozoan flagella". He is told they may be improperly labelled. How can he conclusively differentiate the flagella? Also discuss the role of flagella in bacteria.

Solution: The extremely thin hairlike appendages that protrude through the cell wall of some bacterial cells are called flagella. The length of a flagellum is usually several times the length of the cell, but the diameter of a flagellum is only a fraction of the diameter (the width) of the cell. Baccilli and spirilla usually have flagella, which are rarely found on cocci. Flagella occur singly, in tufts, or equally distributed around the periphery of the cell.

Flagella are responsible for the motility of bacteria. Not all bacteria have flagella, and thus not all are motile. The mechanism by which the flagella move the bacterial cell is not completely understood; however, it is proposed that movement requires rotation of the semi-stiff flagellum. Whatever the mechanism, the flagellum moves the cell at a very high speed and it enables the bacterium to travel many times its length per second.

Bacterial flagella are structurally different from the flagella of eucaryotic cells. This difference can be used to distinguish the bacterial (procaryotic) flagella from the protozoan (eucaryotic) flagella. By examining a cross section of the flagellum under an electron microscope, one can conclusively determine its identity. The eucaryotic flagellum contains a cytoplasmic matrix, with ten groups of tubular fibrils embedded in the matrix: nine pairs of fibrils around the periphery of the eucaryotic flagellum surround two central single fibrils. (see below).

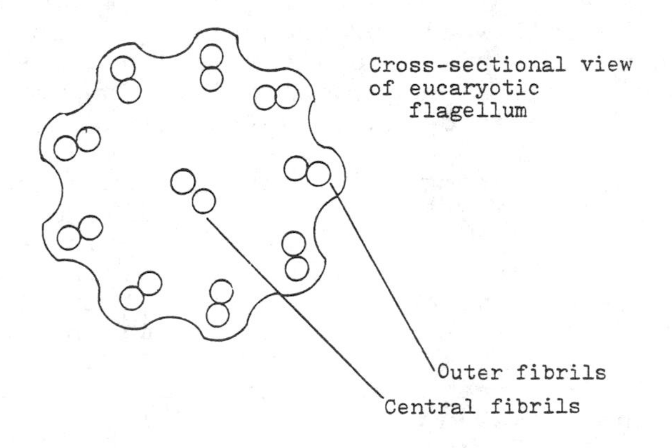

Bacterial flagella consist of only a single fibril and lack the "9+2" structural organization of the eucaryotic flagellum. Most bacterial flagella have the same diameter as a single fibril from a eucaryotic flagellum and hence they are usually thinner.

• PROBLEM 5-4

A biology student observes filamentous appendages on a bacterium and concludes they are flagella. However, he is told that the bacterium is non-motile. He therefore concludes that the flagella are non-operational. Why are both of his conclusions incorrect?

Solution: It is rare for an organism to have a structure which serves no function. The filamentous appendages observed by the student are not flagella, but pili (or fimbriae). Pili are hairlike structures found on some Gram-negative bacteria. They are smaller, shorter, and more numerous than flagella. They are found on non-motile as well as motile bacteria, but do not function in motility.

There are several different kinds of pili. The sex pilus (singular) is involved in bacterial conjugation (the transfer of part of a chromosome from one bacterial cell to another). Most pili aid the bacterium in adhering to the surfaces of animal and plant cells, as well as to inert surfaces such as glass. The pili enable the bacteria to fix themselves to tissues from which they can derive nutrients.

• PROBLEM 5-5

Explain why the bacterial cell wall is the basis for gram staining.

Solution: Gram staining is one of the most important differential staining techniques used to determine differences between bacterial cells. A bacterial cell is fixed to a microscope slide, and the slide is covered with the following solutions in the order listed: crystal violet, iodine solution, alcohol, and safranin. Bacteria stained by this method fall into two groups: gram-positive, which retain the crystal violet-iodine complex and are dyed a deep violet color, and gram-negative, which lose the crystal violet-iodine complex when treated with alcohol, and are stained by the safranin, giving them a red color.

The cell walls of bacteria are composed of peptidoglycan. This layer is thick in gram-positive cells, but thin in gram-negative cells. However, the latter have lipopolysaccharide layer surrounding the thin peptidoglycan layer.

The difference in staining is due to the high lipid content (about 20%) of the cell walls of gram-negative bacteria. (Lipids include fats, oils, steroids and certain other large organic molecules.) During the staining procedure of gram-negative cells, the alcohol treatment extracts the lipid from the cell wall, resulting in increased permeability of the wall. The crystal violet-iodine complex is thus leached from the cell in the alcohol wash. The decolorized cell then takes up the red safranin. The cell walls of gram-positive bacteria have a lower lipid content, and thus become dehydrated during alcohol treatment. Dehydration causes decreased permeability of the wall, so that the crystal violet-iodine complex cannot leave the cell, and the cell is thus violet-colored. In addition thick peptidoglycan layer of the gram + cell walls prevents decolorization by alcohol. Most bacteria fall into one of these two staining groups and gram-staining in an important means of bacterial classification.

• PROBLEM 5-6

A typical bacterial cell has a cell wall and plasma membrane. If the cell wall is removed by treatment with lysozyme, an enzyme that selectively dissolves the cell wall material, will the bacteria live? Explain. What is the biochemical difference between the cell walls of bacteria and plants?

Solution: The main function of the bacterial cell wall is to provide a rigid framework or casing which supports and protects the bacterial cell from osmotic disruption. Most bacteria live in a medium which is hypotonic relative to the bacterial protoplasm: i.e., the bacterial protoplasm is more concentrated than the medium, and water tends to enter the cell. If intact, the cell wall provides a rigid casing, which prevents the cell from swelling and bursting. If the cell wall were destroyed, water would enter the cell and cause osmotic lysis (bursting). If the bacterium is placed in a medium which has the same osmotic pressure as the bacterial cell contents, osmotic disruption would not occur when the cell wall was dissolved. A bacterium devoid of its cell wall is called a protoplast. Protoplasts can live in an isotonic medium,(medium having equal osmotic pressure as the bacterial protoplast). If the bacterium was placed in a hypertonic medium, water would leave the cell and plasmolysis (shrinkage) would ensue.

The cell wall varies in thickness from 100 to 250 Å (an angstrom, Å, is 10^{-10} meter) and may account for as much as 40% of the dry weight of the cell. While the cell wall in eucaryotes is composed of cellulose, in bacteria, the cell wall is composed of insoluable peptidoglycan. Peptidoglycan consists of sugars (N-aceylglucosamine and N-acetylmuramic acid) and amino acids, including diaminopinelic acid, an amino acid unique to bacteria. In gram-negative bacteria, the peptidoglycan constitutes a much smaller fraction of the wall component than it does in gram-positive bacteria. The higher lipid content in the cell walls of gram-negative bacteria accounts for the differences in gram-staining.

The plasma membrane, a thin covering immediately beneath the cell wall, is too fragile to provide the support needed by the cell. Instead, this semipermeable membrane controls the passage of nutrients and waste products into and out of the cell.

• **PROBLEM 5-7**

Explain why penicillin is effective only against actively growing bacteria. Describe the mode of action of some other antimicrobial agents.

Solution: An antimicrobial agent is one that interferes with the growth and activity of microorganisms. Knowledge of the mode of action of a particular agent makes it possible to determine the conditions under which it will act most effectively. Among the known sites of action of antimicrobial agents are the cell wall, cell membrane, protein structure and synthesis and enzyme activity.

The cell walls of some gram-positive bacteria are attacked by the enzyme lysozyme, which is normally found in tears, mucous secretions, and leucocytes (white blood cells). Lysozyme breaks down the peptide linkages in the cell wall complex. Some bacteria secrete enzymes which degrade cell walls of other bacteria or prevent cell wall formation. Without a cell wall to provide support for the bacterium, it will soon lyse and die. The antimicrobial effect of penicillin is attributed to its inhibition of cell wall synthesis. Penicillin prevents the incorporation of the amino sugar, N-acetylmuramic acid into the mucopeptide structure that comprises the cell wall. This is why penicillin only works on actively growing bacteria. If cell wall formation is complete, penicillin has no effect.

The cell membrane helps contain the cellular constituents and provides for selective transport of nutrients into the cell. Damage to this membrane thus inhibits growth or causes death. The bactericidal (kills bacteria) action of phenolic compounds, such as hexachlorophene, is attributed to their effect on cell permeability. This results in leakage of cellular constituents and eventual death.

Proteins are essential to the cell for both structure and enzymatic activity. Protein denaturation (alteration of their natural configuration) causes irreparable damage to the cell. High temperatures, acidity, and alcohol denature proteins. Streptomycin combines with the ribosomes of sensitive bacteria and disturbs protein synthesis.

Many agents inhibit enzymes involved in the energy-supplying reactions of the cell. For example, cyanide inhibits cytochrome oxidase in the electron transport chain, fluoride inhibits glycolysis, and dinitrophenol uncouples oxidative phosphorylation. All of these inhibit ATP synthesis.

A familiar antibiotic, sulfanilamide, works by blocking the synthesis of folic acid, a necessary substrate for certain reactions in the cell. A precursor of folic acid is para-aminobenzoic acid (PABA) whose structure is very similar to that of sulfanilamide:

Sulfanilamide, by effectively competing with this precursor for the binding site on an enzyme involved in the pathway, inhibits folic acid synthesis. Any compound, such as sulfanilamide, that interrupts synthetic processes by substituting itself for a natural metabolite is called an antimetabolite or metabolic analogue.

• PROBLEM 5-8

Why does milk "spoil" when kept in a refrigerator?

Solution: Even though bacteria may be provided with the proper nutrients for cultivation, it is necessary to determine the physical environment in which they will grow best. Bacteria exhibit diverse reactions to the temperature of their environment. The process of growth is dependent on chemical reactions, and the rates of these reactions are influenced by temperature. Temperature therefore affects the rate of growth of bacteria.

Most bacteria grow optimally within a temperature range of 25 to 40°C. (The normal temperature of the human body is 37°C). These bacteria are termed mesophiles. There are bacteria that grow best at temperatures between 45 to 60°C. These bacteria are termed thermophiles. Some thermophiles will not grow at temperatures in the mesophilic range. At the opposite end of this thermal spectrum are the psychrophiles, bacteria which are able to grow at 0°C or lower. Most psychrophiles grow optimally at higher temperatures of about 15-20° C. The psychrophiles are responsible for the spoilage of milk in the cool temperatures of a refrigerator (about 5°C). After a week or so, pasteurized milk will begin to "spoil." The accumulation of metabolic products of psychrophilic bacteria will impart an abnormal flavor or odor to the milk. The milk might become viscous, which is a condition referred to as "ropy" fermentation. The viscosity is caused by the accumulation of a gumlike material which normally forms a capsule around each bacterium. Sweet curdling may also occur, caused by the coagulation of casein, a milk protein.

• PROBLEM 5-9

Besides temperature, what other physical conditions must be taken into account for the growth of bacteria?

Solution: Although all organisms require small amounts of carbon dioxide, most require different levels of oxygen. Bacteria are divided into four groups according to their need for gaseous oxygen.

Aerobic bacteria can only grow in the presence of atmospheric oxygen. Shigella dysenteriae are pathogenic bacteria (causing dysentery) which require the presence of oxygen.

Anaerobic bacteria grow in the absence of oxygen. Obligate anaerobes grow only in environments lacking O_2. Clostridium tetani are able to grow in a deep puncture wound since air does not reach them. These bacteria produce a toxin which causes the painful symptoms of tetanus (a neuromuscular disease).

Facultative anaerobic bacteria can grow in either the presence or absence of oxygen. Staphylococcus, a genera commonly causing food poisoning, is a facultative anaerobe.

Microaerophilic bacteria grow only in the presence of minute quantities of oxygen. Propionibacterium, a genus of bacteria used in the production of Swiss cheese, is a microaerophile.

The growth of bacteria is also dependent on the acidity or alkalinity of the medium. For most bacteria, the optimum pH for growth lies between 6.5 and 7.5, although the pH range for growth extends from pH 4 to pH 9. Some exceptions do exist, such as the sulfer-oxidizing bacteria: Thiobacillus thiooxidans grow well at pH 1. Often the pH of the medium will change as a result of the accumulation of metabolic products. The resulting acidity or alkalinity may inhibit further growth of the organism or may actually kill the organism. This phenomenon can be prevented by addition of a buffer to the original medium. Buffers are compounds which act to resist changes in pH. During the industrial production of lactic acid from whey by Lactobacillus bulgaricus, lime, $Ca(OH)_2$, is periodically added to neutralize the acid. Otherwise, the accumulation of acid would retard fermentation.

BACTERIAL NUTRITION

• PROBLEM 5-10

What basic nutritional requirements do all living organisms have in common? Compare phototrophs and chemotrophs, autotrophs and heterotrophs.

Solution: Organisms, ranging from bacteria to man, share a set of nutritional requirements necessary for normal growth. These requirements must be known in order to cultivate microorganisms in the laboratory, and pure cultures may be obtained through the preparation of appropriate selective growth media.

All organisms require a source of energy. Green plants and some bacteria can utilize radiant energy (from the sun) and are thus called phototrophs. Animals and non-photosynthetic bacteria must rely on the oxidation of chemical compounds for energy and are thus called chemotrophs.

All organisms require a carbon source. Plants and many bacteria require only carbon dioxide as their carbon source. They are termed autotrophs. Animals and other bacteria require a more reduced form of carbon, such as an organic carbon compound. Sugars and other carbohydrates are examples of organic carbon compounds. Organisms which have this requirement are termed heterotrophs. They depend upon autotrophs for their organic form of carbon, which they use as both a carbon source and an energy source.

All organisms require a nitrogen source. Plants utilize nitrogen in the form of inorganic salts such as potassium nitrate (KNO_3), while animals must rely on organic nitrogen-containing compounds such as amino acids. Most bacteria utilize nitrogen in either of the above forms, although some bacteria can use atmospheric nitrogen.

All organisms require sulfur and phosphorous. While phosphorous is usually supplied by phosphates, sulfur may be supplied by organic compounds, by inorganic compounds, or by elementary sulfur.

All organisms need certain metallic elements, and many require vitamins. The metallic elements include sodium, potassium, calcium, magnesium, manganese, iron, zinc, and copper. While vitamins must be furnished to animals and to some bacteria, there are certain bacteria capable of synthesizing the vitamins from other nutrient compounds.

Finally, all organisms require water for growth. For bacteria and plants, all the above nutrients must be in solution in order to enter the organism.

Bacteria show considerable variation in the specific nutrients required for growth. For example, all heterotrophic bacteria require an organic form of carbon, but they differ in the kinds of organic compounds they can utilize. Different bacteria utilize nitrogen in its various forms. Some require several kinds of amino acids and vitamins, while others require only inorganic elements.

• **PROBLEM** 5-11

Design an experiment which would select for a nutritional mutant.

Solution: Organisms possess structures called genes which determine their characteristics. A gene is capable of changing, or mutating, to a different form so that it determines an altered characteristic. Any organism that has a specific mutation is called a mutant. Among the large variety of bacterial mutants are those which exhibit an increased tolerance to inhibitory agents such as antibiotics. There are also mutants that exhibit an altered ability to produce some end product, and mutants that are nutritionally deficient (unable to synthesize or utilize a particular nutrient.) These nutritional mutants are called auxotrophs because they require some nutrient not required by the original cell type or prototroph.

The first step in isolating a nutritional mutant is to

increase the spontaneous mutation rate of the bacteria, which usually ranges from 10^{-6} to 10^{-10} bacterium per generation. (This means that only 1 bacterium in 1 million to 1 in 10 billion is likely to undergo a mutational change.) The mutation rate can be significantly increased by exposing a bacterial culture (e.g. Escherichia coli), to ultraviolet radiation or x-rays.

A portion of the irradiated E.coli culture is placed on the surface of a Petri dish, and spread over the surface to ensure the isolation of colonies. This Petri dish must contain a "complete" medium, such as nutrient agar, so that all bacteria will grow, including the nutritional mutants. Nutrient media contain all the essential nutrients needed for growth. After incubation of this "complete" medium plate, the exact position of the colonies on the plate is noted. A replica plating device is then gently pressed to the surface of the complete plate, raised, and then pressed to the surface of a "minimal" media plate. A replica plating device consists of a sterile velveteen (a cotton fabric woven like velvet) cloth. Cells from each colony adhere to the cloth in specific locations so that each colony remains isolated. These cells serve as an inoculum when the cloth is pressed to another plate. The replica plating device tranfers the exact pattern of colonies from plate to plate. The positioning of the cloth on the "minimal" agar plate must be identical to its positioning when originally pressed to the complete plate. Colony locations will then be comparable on each of the two plates, which are termed "replicas." The "minimal" medium consists only of glucose and inorganic salts, which are nutrients which normally permit the growth of E. coli. From these basic nutrients, normal E. coli can synthesize all required amino acids, vitamins, and other essential components.

After incubation, colonies appear on the minimal plate at most of the positions corresponding to those on the complete plate. Those missing colonies on the minimal plate are assumed to be nutritional mutants, because they cannot grow on a glucose inorganic salts medium. These missing colonies can be located on the complete media plate by comparing the location of colonies on the replicas. If they had not been irradiated, and if mutations did not occur, all the colonies would have been able to grow on this minimal medium. The colonies that did grow were not affected by the irradiation, and were non-mutants.

To determine the exact nutritional deficiency of the auxotrophs, one can plate them on media which contain specific vitamins or amino acids in addition to glucose and inorganic salts. These compounds are normally synthesized by prototrophic E. coli. The mutation might have affected a gene which controls the formation of an enzyme which, in turn, regulates a step in the biosynthesis of one of these nutrients. By plating the mutant on several plates with each plate containing an additional specific nutrient, we could determine which nutrient the mutant cannot synthesize. The mutant will grow on that plate containing the specific nutrient which the mutant is unable to synthesize.

• PROBLEM 5-12

There are two bacterial species growing in a particular nutrient broth. One species can use maltose as its carbon source, while the other cannot and is capable only of utilizing glucose. Each species has a characteristic appearance. How can one separate and identify the two species?

Solution: Nutrient broth is a liquid medium which can support the growth of most heterotrophs. Nutrient broth is not a medium of any particular chemical composition but is simply a solution of complex raw materials which has been found to support the growth of many microorganisms. The point is to find a method to isolate each bacterial type. One way is to use a selective medium, which is a medium that either contains or lacks a particular substance which is necessery for the growth of an organism. To achieve separation, one must make one kind of medium which contains only maltose as its carbon source, and another medium which contains only glucose. (Both media would contain other essential nutrients) Maltose is a dissacharide which is composed of two glucose molecules. The maltose-utilizing bacteria have an enzyme that can break down maltose into glucose. They can therefore grow when presented with either maltose or glucose. But the glucose-utilizing bacteria lack this enzyme, and can use only glucose as a carbon source. The maltose plate thus serves as the selective medium.

Agar (used as a solidification agent) is added to the media and poured into separate Petri dishes (shallow, circular plates used for growing microorganisms). By means of a transfer loop (also called an inoculum loop- a long metal instrument with a wire loop at one end), a portion of the bacterial suspension is then placed on the surface of the maltose solid medium and streaked over the surface. The same "inoculation" procedure is performed on the glucose plate. This streaking thins out the bacteria on the agar surface, so that individual bacterial colonies are separated. Each colony develops from a single bacterial cell and is therefore genetically pure (barring mutations). All the colonies on the maltose plate consist of maltose-utilizing bacteria, since only this species can grow. The glucose-utilizing bacteria will not be able to grow on this plate. Both types of bacteria should be present on the glucose plate, since maltose-utilizing bacteria also use glucose (the product of maltose breakdown).

To select for glucose-utilizing bacteria, bacteria are transferred from any one colony on the glucose plate onto a maltose plate. If colonies form on the maltose plate, maltose-utilizing bacteria were selected. If no colony arises, it can be inferred that only glucose-requiring bacteria are present. Therefore, the colony on the glucose plate which corresponds to an area of no growth on the maltose plate will be the stock of glucose-requiring bacteria.

Most bacteria are heterotrophic, requiring an organic form of carbon, such as glucose, which they oxidize to obtain energy. How do chemotrophic bacteria which can utilize carbon dioxide as a sole carbon source acquire energy?

Solution: Autotrophs are organisms which require only carbon dioxide as a carbon source, from which they can construct the carbon skeletons of all their organic biomolecules. The carbon dioxide is reduced to glyceraldehyde-3-phosphate in order to synthesize carbohydrates. This reductive process requires much energy. Since glucose is not available to be oxidized for energy, these autotrophic bacteria must oxidize inorganic compounds. They are called chemoautotrophs, since they obtain their energy by oxidizing chemical compounds (as opposed to photoautotrophs, which obtain their energy from light). The inorganic compounds oxidized by the various bacteria include molecular hydrogen, ammonia, nitrite, sulfur (sulfide ions), and iron (ferrous ions). These oxidations result in electrons which enter the respiratory chain, with the concomitant production of ATP. ATP is then used as a source of energy in the reduction of CO_2.

Bacteria of the Hydrogenomonas genus obtain energy through the oxidation of hydrogen gas. They possess an enzyme, hydrogenase, which catalyzes the following reaction:

$$H_2 + 1/2\ O_2 \longrightarrow H_2O + 2e^-$$

The electrons are transferred to NAD to form $NADH_2$, which is then oxidized in the respiratory chain, yielding ATP.

The bacterium Nitrosomonas obtains energy by the oxidation of the ammonium ion:

$$2NH_4^+ + 3O_2 \longrightarrow 2NO_2^- + 2H_2O + 4H^+ + 2e^-$$

The electrons produced enter the respiratory chain where they are passed down to O_2, forming ATP along the chain.

Bacteria of the Nitrobacter genus obtain their energy through the oxidation of nitrite ions into nitrate ions:

$$2NO_2^- + O_2 \longrightarrow 2NO_3^- + 2e^-$$

Again the electrons produced are used in the formation of ATP.

The oxidation of ammonia into nitrate is called nitrification. It is one of the most important activities of autotrophic bacteria since it provides the form of nitrogen most available to plants. Nitrification is carried out in two stages, with the first stage involving Nitrosomonas and the second stage featuring the oxidation of nitrites by Nitrobacter. The opposite process, called nitrate respiration, is the reduction of nitrate to ammonia. This process is carried out by several heterotrophic bacteria under anaerobic conditions. The oxygen of the nitrate serves as the hydrogen acceptor (under aerobic conditions, molecular oxygen would normally serve as the final electron or hydrogen acceptor). The overall reaction is as follows:

$$HNO_3 + 4H_2 \longrightarrow NH_3 + 3H_2O$$

Nitrification should not be confused with nitrogen fixation or denitrification. Nitrogen-fixing microorganisms in the soil use molecular nitrogen in the atmosphere as their source of nitrogen and convert it into ammonia. The ammonia is then used in the synthesis of proteins and other nitrogenous substances. Denitrification is the reduction of nitrates to molecular nitrogen carried out by certain bacteria. such as Pseudomonas. The different kinds of reactions are outlined as follows:

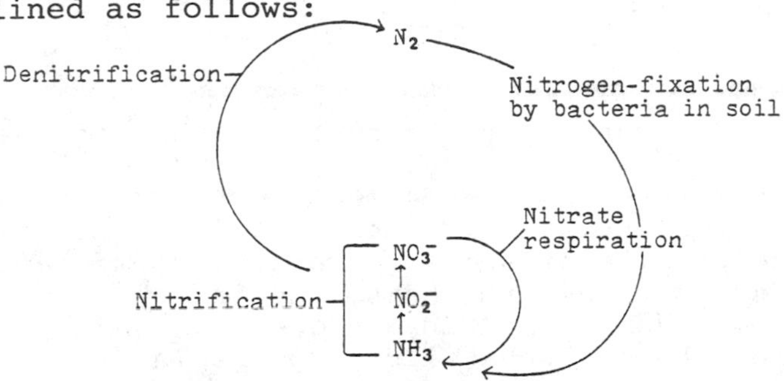

Reactions involving nitrogen-containing compounds.

BACTERIAL REPRODUCTION

• PROBLEM 5-14

Describe the most common process of bacterial reproduction.

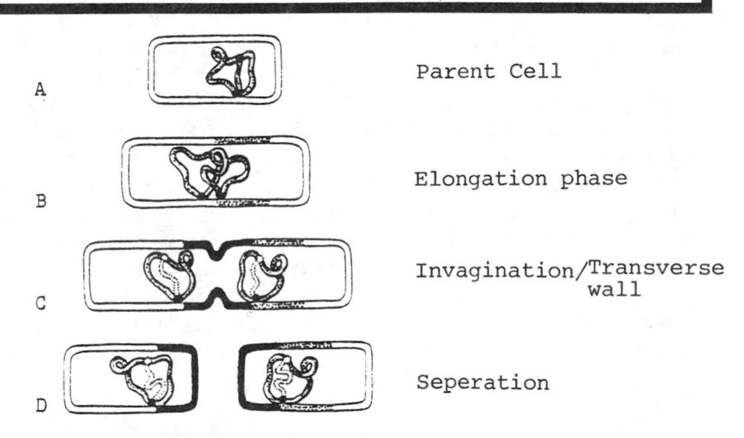

Binary fision of a Prokaryotic cell.

Solution: The most important process in the growth of bacterial populations is binary (transverse) fission. This is a type of cell division in which two identical daughter cells are produced as a result of the division of the parent cell. New cell wall material begins to form on the inner surface of the wall of the parent cell at a point midway along its length and this new wall material invaginates, dividing the cellular material of the parent cell evenly into two halves. The cell is then separated in two by the completion of the transverse wall. Each daughter cell possesses a complete set of genetic information. The genes, the units of inheritance, are arranged in sequence along a single, circular chromosome composed of DNA. During reproduction, the DNA is replicated and the two chromosomes move apart into separate nuclear areas in each half of the parent cell before the completion of the transverse cell wall. Each daughter cell has genetic information identical to the parent cell. This process of giving rise to new individuals by cell division is termed asexual reproduction.

Although binary fission is the major method of bacterial reproduction, another form of asexual reproduction budding is observed in some bacteria. Budding involves an outgrowth of the parental cell, which enlarges and separates to form a new cell.

• **PROBLEM** 5-15

Is endospore formation in bacteria a method of reproduction? Explain.

Solution: To answer this question, it is necessary to examine the results of spore formation. Some bacteria have the capacity to transform themselves into highly resistant cells called endospores. In a process known as sporulation, bacteria form these intracelluar spores in order to survive adverse conditions, such as extremely dry, hot, or cold environments. Each small endospore develops within a vegetative cell. The vegetative cell is the form in which these bacteria grow and reproduce. Each endospore contains DNA in addition to essential materials derived from the vegetative cell. All bacterial spores contain dipicolinic acid, a substance not found in the vegetative form. It is believed that a complex of calcium ion, dipicolinic acid, and peptidoglycan forms the cortex or outer layer of the endospore. This layer or coat helps the spore to resist the destructive effects of both physical and chemical agents. The dipicolinic acid/calcium complex may play a role in resuming metabolism during germination. The spore is generally oval or spherical in shape and smaller than the bacterial cell. Once the endospore is mature, the remainder of the vegetative cell may shrink and disintegrate. When the spores are transferred to an environment favorable for growth, they germinate and break out of the spore wall, and the germinating spore develops into a new vegetative cell. The endospore is incapable of growth or multiplication.

Endospore formation is neither a kind of reproduction nor a means of multiplication, since only one spore is formed per bacterial cell. During spore formation, a single endospore is present within the bacterium. The remaining portion of the vegetative cell dies off, while the endospore remains to later germinate into a new vegetative cell. This new vegetative cell is identical to the old one because it contains the same DNA. Spores only represent a dormant phase during the life of the bacterial cell. This phase is initiated by adverse environmental conditions.

Bacteria in the genera Bacillus and Clostridium are partially characterized by their ability to form endospores. Spores of Bacillus anthracis, the bacteria causing anthrax (primarily a disease of grazing animals), can germinate 30 years after they were formed.

• PROBLEM 5-16

How does the term "growth" as used in bacteriology differ from the same term as applied to higher plants and animals?

Solution: When a small number of bacteria are transferred into the proper medium and incubated under the appropriate physical conditions, a tremendous increase in the number of bacteria results in a short time. As applied to bacteria and microorganisms, the term "growth" refers to an increase in the entire population of cells. When we speak of the growth of plants and animals, we usually refer to the increase in size of the individual organism. The growth of bacteria involves the increase in numbers of cells over the initial quantity used to start the culture (called the inoculum). Some species of bacteria require only a day to reach their maximum population size, while others require a longer period of incubation. Growth can usually be determined by measuring cell number, cell mass, or cell activity.

• PROBLEM 5-17

Draw a typical bacterial growth curve and label the different phases. Discuss the factors for the existence of each phase.

Solution: If one inoculates a flask of nutrient broth with a given number of bacteria and follows the rate of growth during the incubation period, there is found a series of different growth rates. A plot of the logarithms of the number of cells versus time is used to illustrate the growth curve, which is shown in the following diagram:

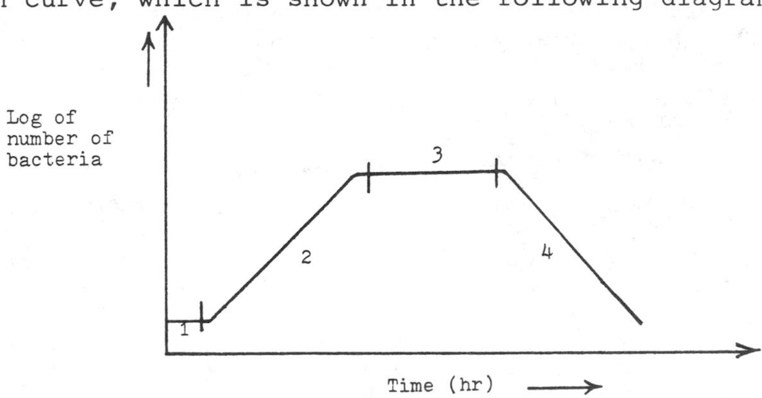

1 Lag phase 2 Exponential or logarithmic phase 3 Stationary phase

4 Death phase

There is an initial period of no growth, followed by one of rapid growth, and then there is a leveling off. The last period is one of decline in the population. The curved portions designates the transitional period between phases. These represent periods where the bacteria enter a new phase.

Following the addition of inoculum into the new medium, the bacterial population does not increase but each individual bacterium increases in size. During this period, called the lag phase, the bacteria are physiologically active and are adapting to the new environment, but the bacterial population number remains constant. In addition, the bacteria may alter the environment for their means (eg. exretion of CO_2 to lower pH, so that more favorable growth conditions are achieved.) At the end of the lag phase, the bacteria start to divide.

The lag phase is followed by the logarithmic or exponential phase where the cells divide at a constant and maximal rate according to their generation time. Most of the bacteria during this phase are uniform in terms of metabolic activity, unlike the other phases.

After the log phase, growth begins to level off in the stationary phase. The population remains constant because of a cessation of reproduction or an equalization of growth and death rates. The decreased growth is usually due to the exhaustion of nutrients or the production of toxic or inhibitory products.

The final phase is called the death phase. Here, the bacteria die faster than they are being produced, if any reproduction is occurring at all. Depletion of nutrients and accumulation of inhibitory products, such as acid, cause the increased death rate. The number of bacteria increases exponentially in the log phase, and the number of bacteria decreases exponentially in the death phase. Some species die rapidly, so that few living cells remain after about three days, while other species die after several months.

• PROBLEM 5-18

A biology student places one bacterium in a suitable medium and incubates it under appropriate conditions. After 3 hours and 18 minutes, he determines that the number of bacteria now present is 1000. What is the generation time of this bacteria?

Solution: Bacterial growth usually occurs by means of binary fission with one cell dividing into two and these two cells dividing into four, etc. The population at any time can be represented by a geometric progression (1, 2, 4, 8...). The time interval required for each division (or for the entire population to double) is called the generation time. Different bacteria have different generation times, ranging from 20 minutes for Escherichia coli to 33 hours for Treponema pallidum (the bacterium responsible for syphilis). The generation time also differs for the same bacterial species under different environmental conditions.

To determine the generation time (G) of a bacterial population, we must know the number of bacteria present initially (N_0), the number of bacteria present at the end of a given time period (N), and the time period (t). We can determine the generation time by using some simple mathematical expressions. If we start with a single bacterium, the total population (N) after the nth generation is 2^n:

$$N = 1 \times 2^n$$

For example, at the end of three generations, we would have eight bacteria, ($2^3 = 8$). However, since we usually start with many bacteria, we must modify the formula to account for more than one parental bacterium:

$$N = N_0 \times 2^n.$$

Solving for n, the number of generations, we get

$$\log N = \log N_0 + n \log 2.$$

By substituting .301 for the log 2 and rearranging,

$$n = \frac{\log N - \log N_0}{.301} = 3.3\ (\log N - \log N_0).$$

Since the difference of the logarithms of two numbers is the logarithm of the quotient of the two numbers,

$$n = 3.3 \log \frac{N}{N_0}\ .$$

The generation time G is simply equal to the time elapsed between N_0 and N divided by the number of generations:

$$G = \frac{t}{n} = \frac{t}{3.3 \log \frac{N}{N_0}}\ .$$

In our example: N = 1000, $N_0 = 1$ and t = 198 minutes. Substituting into the general formula,

$$G = \frac{198}{3.3 \times \log 1000}\ .$$

$$G = \frac{198}{3.3 \times 3}$$

$$= 20 \text{ minutes}$$

The time interval required for this particular bacterial species to divide is twenty minutes. If we plot the number of bacterial cells against time, we get a bacterial growth curve. In our case, with a generation time of 20 minutes, we would obtain this curve:

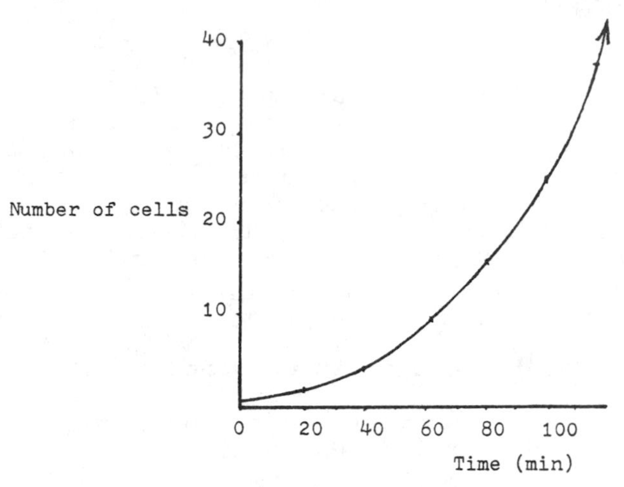

This curve does not represent the complete pattern of growth, but rather one selected portion of the normal growth curve. This portion is called the exponential or logarithmic phase of growth. Only at this phase does the population double at regular time intervals. Eventually the reproduction rate is checked because of the lack of nutrients or the accumulation of waste products. In our example, we assumed that the bacteria were growing at a logarithmic rate where both N_0 and N should be measured during the exponential growth phase.

• PROBLEM 5-19

Compare steady-state bacterial growth with synchronous growth. How can one obtain these growth conditions?

Solution: When bacteria are cultured in any closed system, such as a tube, flask, or tank, the pattern of growth includes four phases: lag, log, stationary, and death. However, one can set up an environment in which the bacterial population is maintained in the log (exponential) phase. This condition is called steady-state growth. Two devices, the turbidostat and the chemostat, are used to maintain the bacteria in a constant environment and promote steady-state growth. The culture is kept at a constant volume by adding fresh media at the same rate old media is removed. This serves to provide a continual replenishment of nutrients and a steady removal of any inhibitory products, maintaining the bacteria in the log phase of growth.

The turbidostat involves the use of a photoelectric eye that monitors the turbidity of the culture. The turbidity is proportional to the density of the bacterial cells. The turbidostat maintains the turbidity at a specific constant value by regulating the flow rate of media (fresh input and old output). The chemostat regulates the level of growth by maintaining a constant, limiting amount of an essential nutrient and draining waste medium. This amount is just sufficient to keep the bacteria in the log phase. A steady-state growth is usually desired for experiments studying metabolism (since the log phase cells are nearly uniform in metabolic activity).

Synchronous growth is irrelevant to the steady-state growth in the continuous culture of bacteria. In a population, all the bacterial cells do not normally divide at the same time (synchronous growth). If they did divide simultaneously there would be an instantaneous periodic doubling of the bacterial population instead of the constant increase. One can manipulate the culture to produce synchronous growth but it lasts only a few generations, since some bacteria reproduce out of phase with the others. One may obtain a synchronized culture by inoculating the medium at freezing temperatures. The bacteria will metabolize slowly and become ready to divide. When the temperature is raised, they should all initiate reproduction simultaneously. Another common method is to filter out the smallest cells of a log phase culture. Since these should be the ones that have just divided, they are rather well synchronized, and will eventually undergo division at approximately the same time.

Graphically, synchronous growth appears as follows:

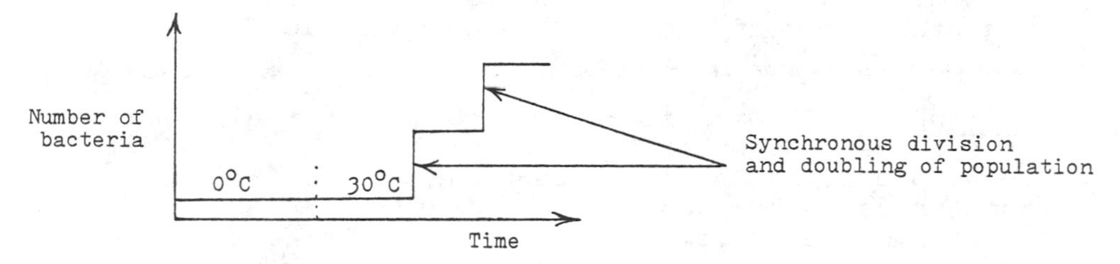

BACTERIAL GENETICS

• PROBLEM 5-20

A male bacterium conjugates with a female bacterium. After conjugation, the female becomes a male. Account for this "sex-change".

Solution: Conjugation occurs between bacterial cells of different mating types. Maleness in bacteria is determined by the presence of a small extra piece of DNA, the sex factor, which can replicate itself and exist autonomously (independent of the larger chromosome) in the cytoplasm. Male bacteria or donors having the sex factor, also known as the F factor, are termed F^+ if the sex factor exists extrachromosomally. F^+ bacteria can only conjugate with F^-, the femalecounterparts or recipients which do not posses the F factor. Genes on the F factor determine the formation of hairlike projections on the surface of the F^+ bacterium, called F or sex pili. The pili form cytoplamic bridges through which genetic material is transferred and aids the male bacterium in adhering to the female during conjugation. During conjugation of an F^+ with an F^- bacterium, the DNA that is the most likely to be transferred to the female is the F factor. Prior to transfer, the F factor undergoes replication. The female thus becomes a male by receiving one copy of the F factor, and the male retains its sex by holding on to the other copy of the sex factor. The DNA of the male chromosome is very rarely transferred in this type of conjugation.

If this were the only type of genetic exchange in conjugation, all bacteria would become males and conjugation would cease. However, in F^+ bacterial cultures, a few bacteria can be isolated which have the F factor incorporated into their chromosomes. These male bacteria that conjugate with F^- cells are called Hfr (high frequency of recombination) bacteria. They do not transfer the F factor to the female cells during conjugation, but they frequently transfer portions of their chromosomes. This process is unidirectional, and no genetic material from the F^- cell is transferred to the Hfr cell.

• PROBLEM 5-21

Genetic variation occurs in bacteria as a result of mutations. This would seem to be the only process giving rise to variation, for bacteria reproduce through binary fission. Design an experiment to support the existence of another process that results in variation.

Solution: It was originally thought that all bacterial cells arose from other cells by binary fission, which is the simple division of a parent bacterium. The two daughter cells are genetically identical because the parental chromosome is simply replicated, with each cell getting a copy. Any genetic variation was thought to occur solely from mutations. However, it can be shown that genetic variation also results from a mating process in which genetic information is exchanged.

One can show this recombination of genetic traits by using two mutant strains of E. coli, which lack the ability to synthesize two amino acids. One mutant strain is unable to synthesize amino acids A and B, while the other strain is unable to synthesize amino acids C and D. Both these mutant strains can be grown only on nutrient media, which contain all essential amino acids. When both strains are plated on a selective medium lacking all four amino acids in question, some colonies of prototrophic cells appear which can synthesize all four amino acids (A, B, C, and D). When the two strains are plated on separate minimal medium plates, no recombination occurs, and no prototrophic colonies appear.

These results may be proposed to be actually a spontaneous reversion of the mutations back to wild type (normal prototroph) rather than recombination. If a single mutation reverts to wild-type at a frequency of 10^{-6} mutations per cell per generation, two mutations would simultaneously revert at a frequency of $10^{-6} \times 10^{-6} = 10^{-12}$ mutations per cell per generation. If one plates about 10^9 bacteria, no mutational revertants for both deficiencies should occur. Recombination occurs at a frequency of

10^{-7}, and thus if 10^9 bacteria are plated, prototrophic colonies should be found.

This recombination of traits from the two parent mutant strains is brought about through a process called conjugation. During conjugation, two bacterial cells lie close to one another and a cytoplasmic bridge forms between them. Parts of one bacterium's chromosome are transferred through this tube to the recipient bacterium. The transferred chromosomal piece may or may not get incorporated into the recipient bacterium's chromosome.

• PROBLEM 5-22

There are genes on the E. coli chromosome which determine amino acid synthesis. For one particular strain of E. coli, how can one determine the order of genes for the synthesis of threonine (thr^+), methionine (met^+), histidine (his^+) and arginine (arg^+)? Use your knowledge of conjugation.

Solution: To become an Hfr bacterial cell, the F factor must become integrated into the chromosome. When an Hfr cell and an F^- cell begin conjugation, the F factor portion of the circular Hfr chromosome initiates synthesis of a linear chromosome. This linear chromosome carries a fragment of F factor on both its ends, with one end acting as the origin for the transfer of the chromosome. Since the other end of the chromosome contains the remaining portion of the F factor, and since the whole chromosome rarely gets transferred, the F^- cell usually does not receive a complete F factor to become F^+. In a particular Hfr strain, the F factor inserts at the same place on the chromosome, so that the origin of transfer is at the same site in all bacteria of this strain.

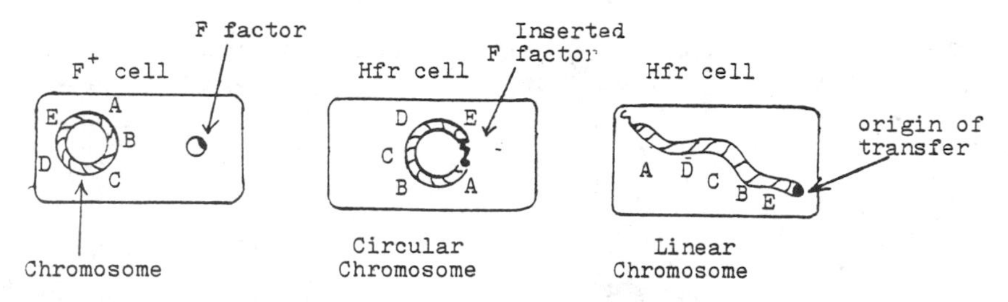

Figure 1. Insertion of F factor into chromosome.

During conjugation, the origin is the first part to travel through the conjugation tube and enter the F^- cell. Usually the conjugating cells separate before the complete transfer of the chromosome. The part that enters the F^- cell may or may not be incorporated into its chromosome. If it does become incorporated, recombination occurs and the F^- cell acquires new traits that are determined by the newly transferred Hfr genes.

To determine the order of the genes on the chromosome

of a particular Hfr strain, one can use the interrupted mating technique. We must first acquire an F^- strain that is auxotrophic (demonstrating a nutritional requirement) for the four amino acids but is resistant to streptomycin, an antibiotic which kills *E. coli*. The F^- genotype in consideration would be as follows: thr^-, met^-, his^-, arg^-, str^+. The Hfr strain must be prototrophic (can synthesize the amino acids in question) and be sensitive to streptomycin. It is also necessary that in the Hfr strain selected, the gene for streptomycin sensitivity be located far from the origin of transfer of the linear chromosome to avoid its transfer. The Hfr genotype would be thr^+, met^+, his^+, arg^+, str^-. The reason for selecting streptomycin sensitive and resistant strains will be made evident later.

The two cultures of F^- and Hfr cells are then mixed and incubated. At specific time intervals (e.g. every ten minutes), samples are removed from the conjugating mixture and agitated in a blender. This separates the conjugating bacteria to prevent genetic transfer. Since the linear Hfr chromosome always enters the F^- cell in a regular sequence, we can order the genes according to the length of time it took for the Hfr cells to transfer the genes to the F^- cells. For example, in the first ten minutes, only the thr^+ gene might be transferred. In the first fifty minutes, both the thr^+ and his^+ genes get transferred, the arg^+ gene after 80 minutes. The genes can then be located sequentially on the chromosome by noting their time of transfer. For this particular Hfr strain, the chromosome map is as follows:

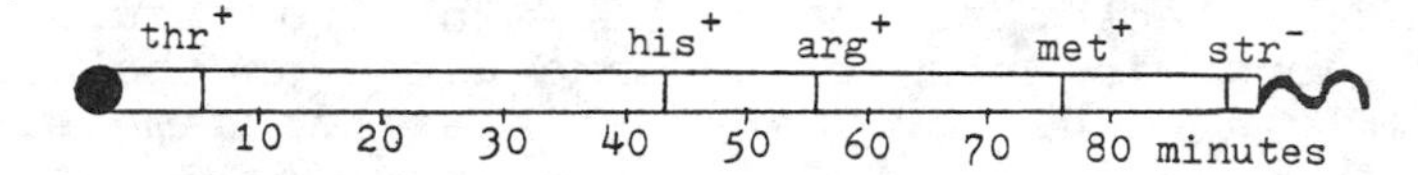

Figure 2.

In order to find out which gene is transferred after a certain period of time, the bacterial mixture is plated on four special media:

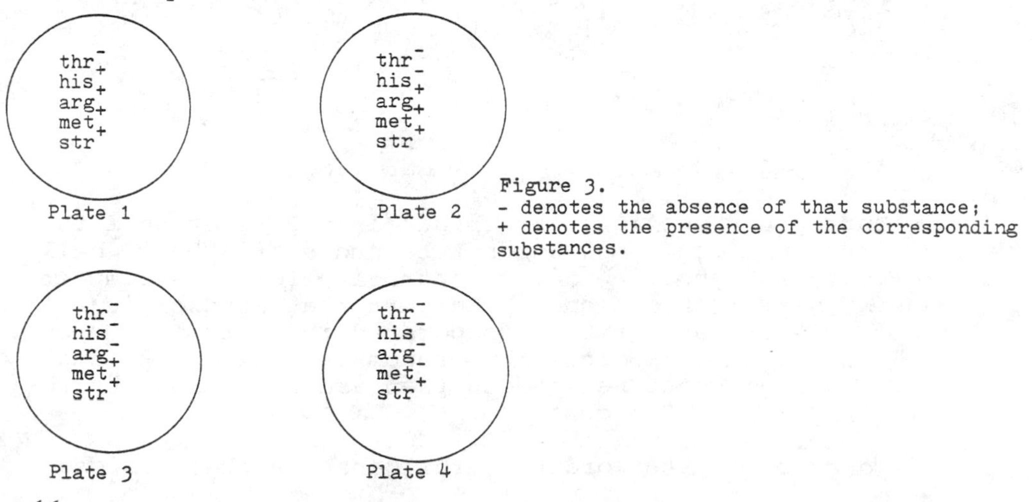

Figure 3.
- denotes the absence of that substance;
+ denotes the presence of the corresponding substances.

The sample produced after ten minutes will only grow on plate I since the only gene transferred is the thr^+ gene. It cannot grow on the others since it is still auxotrophic for the other three amino acids. The sample after the fifty minutes time interval would grow on both plates I and II. We thus know that the his^+ gene follows the thr^+ gene. The reason why all the plates contain streptomycin is that any Hfr bacteria remaining in the sample placed on the special media plates will be killed. Although F^- cells are streptomycin resistant, the only F^- cells that will grow on the special plates are those that have conjugated with Hfr cells and have received and incorporated the genes for determining amino acid synthesis. The streptomycin thus acts as a control to allow only newly-made recombinants to live. Another relevant point is that in an actual experiment, more media plates with different combinations of amino acid supplements would be required. For example, Plate I, which selected for thr^+, should also check for his^+, or for arg^+, or for met^+, since we would not know which gene is really first.

PATHOLOGICAL AND CONSTRUCTIVE EFFECTS OF BACTERIA

• PROBLEM 5-23

Distinguish between infection and infestation and between virulence and pathogenicity.

Solution: Although these terms are used interchangeably, the meanings of these words are rather distinct.

The term infection implies an interaction between two living organisms. The host and the parasitic microorganism compete for superiority over the other. If the microorganism prevails, disease results. If the host is dominant, immunity or increased resistance to the disease may develop.

Infestation indicates the presence of animal (nonmicrobial) parasites in or on the hosts' body. Lice, fleas and flat-worms are infesting organisms. They may transmit an infection (e.g. a louse carries typhus - a disease caused by microorganisms called Rickettsia) to man.

Parasitism is a type of antagonism in which one organism, the parasite, lives at the expense of the other - the host. Infection is a type of parasitism.

Pathogenicity refers to the ability of a parasite to gain entrance to a host and produce disease. (A pathogen is any organism capable of producing disease.) The degree of pathogenicity, or ability to cause infection, is called virulence. The virulence of a microorganism is not only determined by its inherent properties but also by the host's ability to resist the infection. A pathogen may be virulent for one host and nonvirulent for another. For example, streptococci, although found in the throats of some healthy individuals, can be pathogenic under different conditions or in other individuals.

Resistance is the ability of an organism to repel infection. Immunity is resistance (usually to one type of microorganism) developed through exposure to the pathogen by natural or artificial means. Lack of resistance is called susceptibility.

• PROBLEM 5-24

What are the bacterial factors influencing virulence?

Solution: Virulence is the degree of ability in microorganisms to produce disease. Some organisms are more virulent than others. There are several factors influencing virulence.

One of these is the production of toxins which are poisonous substances produced by some microorganisms. Both the ability to produce the toxin and the potency of the toxin affect the organism's capability to produce disease. Toxins which are secreted into the surrounding medium during cellular growth are called exotoxins, and toxins retained in the cell during growth and released upon cell death and lysis are called endotoxins. Exotoxins are released into the surrounding medium, e.g., into a can of vegetables containing Clostridium botulinum (the bacterium causing botulism) or into damaged tissue infected with Clostridium tetani (the bacterium causing tetanus). Both these types of bacteria exhibit difficulty in penetrating the host. However, these bacteria are virulent because of their toxins. When the spatial configuration of the amino acids in a toxin molecule is altered, the toxicity is lost, and the resulting substance is called a toxoid. Both toxins and toxoids are able to stimulate the production of antitoxins, substances made by the host which are capable of neutralizing the toxin. Endotoxins are liberated only when the microorganism disintegrate and are generally less toxic than exotoxins. They do not form toxoids and are usually pyrogenic, inducing fever in the host (some exotoxins produce fever also). Exotoxins are usually associated with gram-positive bacteria while endotoxins are associated with the gram-negative ones.

Another factor influencing virulence is the ability of bacteria to enter the host and penetrate host tissue. Specific bacterial enzymes are involved. One enzyme that is produced by some of the clostridia and cocci is hyaluronidase. This enzyme facilitates the spread of the pathogen by aiding its penetrance into host tissues. It hydrolyzes hyaluronic acid, an essential tissue "cement", and thus increases the permeability of tissue spaces to both the pathogen and its toxic products.

The virulence of pathogens is also influenced by their ability to resist destruction by the host. In certain bacteria, this resistance is due to the presence of a nontoxic polysaccharide material which forms a capsule surrounding the bacterial cell. For example, pneumococci are Streptococcus pnemoniae virulent when capsulated but are often avirulent when not capsulated. Capsulated pnemococci are resistant to phagocytosis (the ingestion and destruction of microorganisms by host phagocytes), while those without capsules are ingested by leucocytes and destroyed.

• PROBLEM 5-25

A woman opens a can of ham and beans and suspects food poisoning. She boils the contents for half an hour. A few hours after eating the ham and beans, nausea, vomiting, and diarrhea set in. Explain.

Solution: The woman probably suspected that the can of ham and beans was contaminated with Clostridium botulinum, the cause of the microbial food poisoning known as botulism. These bacteria sometimes, but not always, produce foul odor. Clostridium botulinum are usually found in canned, low-acid products since they are anaerobic, growing without the presence of oxygen. The poisoning is due to an extremely potent toxin produced by the bacteria. The toxin acts upon nerves and causes the paralysis of the pharynx and diaphragm, thus causing respiratory failure and sometimes death (65% of all cases are fatal). The toxin is heat-labile (unstable) and is destroyed by boiling for 15 minutes. Therefore, the woman denatured the botulinus toxin and her food poisoning cannot have been botulism. She also did not show the symptoms of botulism which affects the neuromuscular junction. There is a flaccid paralysis which may precede cardiac or respiratory failure. Other symptoms include difficulty in swallowing and speech. Double vision is often present as well.

The woman was probably poisoned by the toxin produced by Staphylococcus aureus, the most common agent causing food intoxication in the U.S. today. The bacteria may have come from food handlers who harbored the organism in the nose, throat, or sores on the skin (boils). The organism is a facultative anaerobe, able to live in the anaerobic environment of the can. Staphylococcal food poisoning usually involves food such as ham and dairy products. The important point in this type of poisoning is that the toxin is heat-stable and would be unaffected by boiling for 30 minutes. Therefore, the woman's precautionary action was not successful. Fortunately, death is rarely caused by this toxemia.

• PROBLEM 5-26

"A virulent organism is as good as dead if it is not communicable." Explain.

Solution: A virulent pathogen will eventually die if it is restricted to its original host. The host produces substances which either prevent the growth and spread of the pathogen or actually destroy it. Even if the host fails to do this, a virulent pathogen will bring about its own destruction by killing the host that sustains it. For the pathogen to survive and cause disease in a number of organisms, it must find new hosts to infect.

Communicable pathogens are transferred from one host to another by a number of means. Pathogens of the respiratory tract, such as Diplococcus pneumoniae, can leave the body in discharges from the mouth, nose and throat. Sneezing and coughing expedite the spread of these organisms. Enteric (intestinal) pathogens, such as those which cause typhoid fever, leave the host in fecal excretions and sometimes in the urine. The bacteria in these wastes contaminate the food and water which a subsequent host may ingest. Other pathogens cannot live long outside a host and are transmitted by direct contact, most commonly through breaks in the skin or contact of mucous membranes. The final common means of transmittance is by certain insect vectors, although these insects may be carried further by other organisms such as rats and man. For example, ticks transmit Rickettsiae ricketsii to man, causing Rocky Mountain spotted fever.

Some pathogens are transmitted by specific means, and enter the new host by a specific portal of entry. For example, enteric bacteria have a special affinity for the alimentary tract and are able to survive both the enzymatic activity of the digestive juices and the acidity of the stomach.

We thus see that the success of a pathogen depends on its successful transmittance to a new host via air, food, water, insects, or by contact.

• PROBLEM 5-27

A newborn infant is examined by a doctor, who determines that the child has syphilis. Explain how the child contracted syphilis.

Solution: Syphilis is a venereal disease of man, that is included in the more general category of contact diseases. Contact diseases of man usually result from the entry of an infectious agent into the individual through the skin or mucous membranes. Contact may be direct (through wounds and abrasions) or indirect (through vectors - mediating transmitters such as insects). In the case of veneral diseases, transmittance is usually through direct genital contact during sexual activity. The organism involved is a spirillum called Treponema pallidum.

However, direct sexual contact is not the only way in which syphilis is transmitted. An infected mother can transmit the organism by placental transfer to the fetus during the first four months of pregnancy. Contraction of the disease is also possible as the fetus passes through the infected vagina during birth.

The disease usually requires an incubation period of

3 to 6 weeks after infection. The disease produces lesions called chancres which resemble ulcerous sores. In the late stages of the disease, the cardiovascular and central nervous systems may be affected, with possible paralysis. The disease, if promptly detected, can be treated with penicillin. There is no known means for immunization against T. pallidum infection. Persons who have recovered from a syphilitic infection are just as likely to contract it upon subsequent exposure to the organism. Preventive measures include avoiding carriers and using local prophylactic measures such as condoms.

• PROBLEM 5-28

A microbiologist takes a stool specimen from you and a few days later tells you that Escherichia coli and Salmonella typhi are growing in your intestinal tract. Which type should you be concerned about?

Solution: Escherichia coli, or E. coli, is a species of bacteria which normally inhabits the intestine of man and other animals. Only rarely can E. coli be pathogenic. For example, they may cause diarrheal disease in infants, and are occasionally found in infections of the urogenital tract. The presence of E. coli in the stool should not cause any alarm, since it is one of the many types of bacteria that normally inhabit our intestine.

The genus Salmonella, however, includes several species which are pathogenic to man and other animals. Salmonella typhi, for example, causes the acute infectious disease known as typhoid fever. Typhoid fever is characterized by a fever, inflammation of the intestine, intestinal ulcers, and a toxemia (presence of toxins in the blood). The presence of Salmonella is therefore ample reason for concern.

Typhoid fever, like most intestinal infections, is transmitted from one person to another through food and water. Transmission may be indirect. For example, wastes from an infected person can pollute drinking water or food, or the infected person may handle food at some point in its processing or distribution, and contaminate it, affecting the consumer. The common housefly can also transmit Salmonella from wastes to food. Typhoid fever occurs in all parts of the world. In locations where good sanitation (proper disposal and treatment of biological waste, and water purification) is practiced, typhoid fever incidence is very low. Carriers are people who are infected with Salmonella typhi, but who have had only a slight intestinal infection, and hence do not know thay harbor the pathogenic organism. Carriers should not be allowed to handle or prepare food.

• PROBLEM 5-29

Suppose you discovered a new species of bioluminescent worm.

How could you prove that it was the worm itself and not some contaminating bacteria that was producing the light?

Solution: There are many bioluminescent organisms but sometimes it is difficult to distinguish whether it is the actual organism that is emitting light or whether the light is due to luminous bacteria living in the organism. Some species of fish have light organs under their eyes where light-emitting bacteria live. One possible way of determining if a bacterium is the source of the light is to take some of the light-producing substance and to place it in complete growth media (containing all possible nutrients). If the luminescent material proliferates, then a bacterium is most likely the causal factor. Another way would be to physically examine the light-emitting substance under a microscope. This could be done by scraping out some light-emitting substance from a bioluminescent organism and transferring it to a glass slide. If we see individual bacterium emitting light then we can conclude that the bioluminescence is caused by bacteria.

• **PROBLEM** 5-30

It was once thought that living things could form directly out of non-living matter. People believed that discarded food would be transformed into bacteria and molds. An experiment was prepared as illustrated in the figure below. Tube A was left untouched. Tubes B, C and D were prepared as illustrated and then boiled for 15 minutes in a beaker of water. Sample A spoiled in one day, B in two days, C in one week, and D remained fresh until the cork and the "S" tube were removed. How do these results disprove the following statements:

(a) Meat soup turns into bacteria

(b) The process of boiling meat soup alters it to prevent its change into bacteria

(c) Boiling the soup alters the air in the tube so that bacteria will not form from the soup.

Why does the soup in C eventually spoil while that in D remains fresh? How do these results help to show that bacteria are living creatures which can exist floating in the air around us?

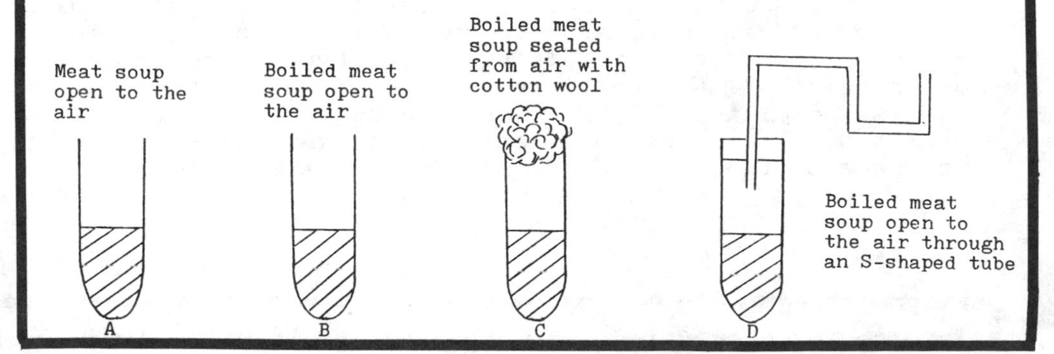

Solution: Bacteria are microorganisms found everywhere around us. They are responsible for turning food sour either by way of fermentation or putrefaction. Fermentation is the decomposition of carbohydrates by bacteria while putrefaction is the decay of proteins by the same agents.

Although the result in tube A suggests that meat soup spontaneously generates bacteria, the outcome in tube D shows this to be incorrect because it shows no sign of decay even though it contains boiled meat soup. The process of boiling meat soup does not prevent the soup from spoiling as shown by tube B. The soup in tube B does spoil, disproving statement (b). The result in tube C disproves the claim that boiling the soup alters the air because although the tube is sealed after boiling, the content spoils after a week.

It should be noted that the process of boiling kills the bacteria already present in the soup. Tube A spoils in a day because of the activity of the bacteria already present in it.

Tube B shows that although the soup is free from bacteria after boiling, it still spoils because it is exposed to the bacteria in the air. Tube C, with a cotton seal across its mouth, restricts the exposure of the soup to the air, thus reducing the chances of the bacteria getting into the soup and causing it to decay. But the cotton seal is not airtight and air does get into the soup gradually. That is why the soup spoils in a week.

Tube D has a special "S" tube attached to the mouth of it. Although the tube has direct access to the air, all the bacteria present in the air are trapped in the bend of the "S" tube. In this way, tube D can remain fresh indefinitely, because bacteria cannot get into the soup.

By this series of experiments, we conclude that exposed food does not spoil spontaneously. It decays because the bacteria that are present in the atmosphere have settled on and interacted with the food and produced decay.

• PROBLEM 5-31

In what ways are bacteria useful to the dairy industry?

Solution: Bacteria are a major cause of disease, but beneficial bacterial species outnumber harmful ones. There are many useful applications microorganisms in the dairy industry. Fermented milk products are made by encouraging the growth of the normal lactic acid-producing bacteria that are present in the culture. Specific organisms, or mixtures of them, are used to produce buttermilk, yogurt, and other fermented milk products. The principal bacteria used are species of Streptococcus, Leuconostoc, and Lactobacillus, which produce only lactic acid. Butter is made by churning pasteurized sweet or sour cream with the latter being fermented by streptococci and leuconostocs. Different types of cheeses are made by providing conditions that favor the development of selected types of bacteria and molds. The quality and characteristics of a cheese are determined by the biochemical activities of selected microorganisms. For example, Swiss cheese is produced through the fermentation of lactic acid by bacteria of the genus Propionibacterium. The products of the fermentation are propionic acid, acetic acid, and carbon dioxide. The acids give the characteristic flavor and aroma to the cheese, and the accumulation of carbon dioxide produces the familiar holes.

Bacteria are also useful to the dairy industry by providing the means by which certain milk-giving animals acquire nutrients. The ruminants (cattle, sheep, goats, and camels) are a group of herbivorous mammals whose digestive system has a compartmentalized stomach, the first section being termed the rumen. In the rumen, there exist certain bacteria and protozoa, which provide enzymes necessary to break down the cellulose acquired from the ingestion of plant material. The mammal does not normally have the enzymes needed to degrade cellulose into nutrients which the animal can use (the lack of these enzymes in humans also explains why man cannot break down plant material).

Bacteria are also important to the food industry in other ways. Pickles, sauerkraut, and some sausages are produced in whole or part by microbial fermentations.

VIRAL MORPHOLOGY AND CHARACTERISTICS

• PROBLEM 5-32

There is a disease of tobacco plants called tobacco mosaic. Explain how one can demonstrate the causative agent.

Solution: Tobacco mosaic disease causes the leaves of the tobacco plant to become wrinkled and mottled. By grinding these leaves, one can extract the juice from the infected plant. If this juice is rubbed onto the leaves of a healthy plant, it becomes infected. The agent can therefore be transmitted in the juice.

If the juice is boiled before being rubbed on a healthy plant, no disease develops. The agent might therefore be a bacterium, since most bacteria cannot survive temperatures above 70°C. To isolate the bacteria, we could use filters with known pore diameters. Since most bacteria are larger than .5 μ (a micron, μ, is equal to 10^{-6} meter), we could use a filter with a pore size slightly less than this. The liquid is passed through this very fine filter in order to remove the bacteria, and the filtrate is checked for the absence of bacteria. When the filtrate is rubbed on the leaves of a healthy plant, it still causes infection. The causative agent could therefore be a toxin produced by some bacteria or a bacterium smaller than any known. It can be demonstrated that the agent is not a toxin by showing that the agent can reproduce. If the filtrate is used to infect a healthy plant, and this plant is then ground up to obtain a new filtrate, this new filtrate could be used to infect another plant. If this procedure is repeated, and if the extent of mottling becomes decreased with subsequent infections, the agent is most probably a toxin. The attenuation of the disease can be attributed to the dilution of the toxin. However, in tobacco mosaic disease the extent of the mottling does not decrease with repetitions of filtration. One can assume that the causative agent does not become diluted. Another process must be occurring, since the agent is reproducing. Further experiments can show that the causative agent reproduces itself only inside the living plant; it cannot grow on artificial media. It is therefore not a bacterium, since bacteria do not require living host cells to reproduce.

The microbial agent of this disease is termed a virus: the tobacco mosaic virus, or TMV. It can be isolated and crystallized and observed using the electron microscope. It is a rod-shaped virus, much smaller than any known group of bacteria: .28 μ in length, .015 μ in diameter.

• PROBLEM 5-33

Why do some viruses contain only one type of protein in their protective coat?

Solution: The nucleic acid in a virus particle is surrounded by a protein coat, called a capsid, made up of protein subunits called capsomeres. A complete virus particle is called a virion and it may be covered by an envelope. Those that do not have envelopes are termed naked virions, while those virions that do have envelopes are termed enveloped. A typical naked and an enveloped virion are shown in Fig. 1.

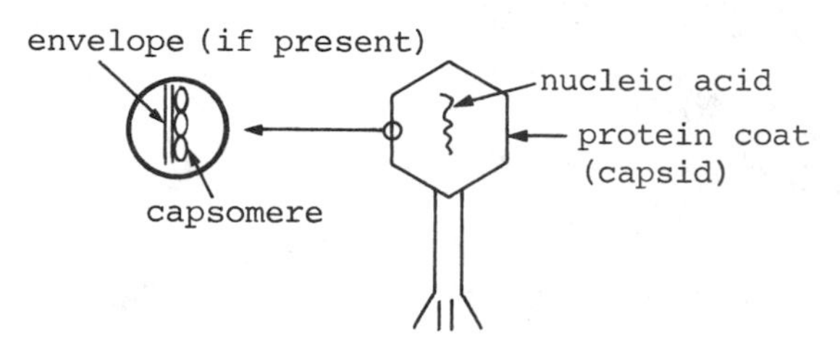

Fig. 1 Virus Particle (virion)

Some of the protein coats are complex, containing numerous layers and various different proteins. Some also include lipid and carbohydrate molecules. Others, however, contain only one or a few types of protein molecules in their coats. The use of a large number of identical proteins in the protective coat of viruses is often mandatory.

The extremely small viruses have a limited nucleic acid content which restricts the total number of amino acids that can be used to synthesize its protein. For example, there are approximately 6000 nucleotides in a TMV RNA chain. The TMV RNA can therefore code for a protein 2000 amino acids in length. (Three nucleotides are needed to code for one amino acid.) Since each amino acid weighs about 125 daltons (a dalton is a unit of weight equal to the weight of a single hydrogen atom), this protein molecule of 2000 amino acids should weigh 2.5×10^5 daltons. However, since we know the protein coat of TMV weighs 3.5×10^7 daltons, about 140 $(3.5 \times 10^7 \div 2.5 \times 10^5)$ of these protein molecules would be needed. The virus does not use all its nucleic acids to code for a single protein. Since only a very small part of the RNA chain codes for the coat protein, the proteins consist of fewer amino acids. Therefore, many smaller identical protein molecules are used to construct the protein coat. Since the number of identical protein molecules used to form the TMV protein coat is about 2150, each protein molecule consists of about 128 amino acids

$$\left(\frac{3.5 \times 10^7 \text{ daltons}}{2150 \text{ proteins}} \times \frac{1 \text{ amino acid}}{125 \text{ daltons}} = \frac{128 \text{ amino acids}}{\text{proteins}}\right).$$

The protein subunits either possess helical symmetry or cubical (or quasi-cubical) symmetry. The TMV shell consists of identical protein molecules helically arranged around a central RNA molecule (see Fig. 2). The Adenovirus is an icosahedral virion, exhibiting quasi-cubical symmetry (see Fig. 3). An icosahedron is a twenty-sided geometric figure.

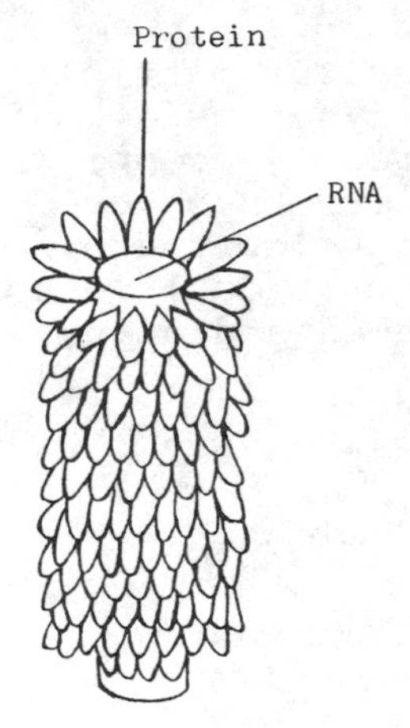

Figure 2. Structure of the TMV particle.

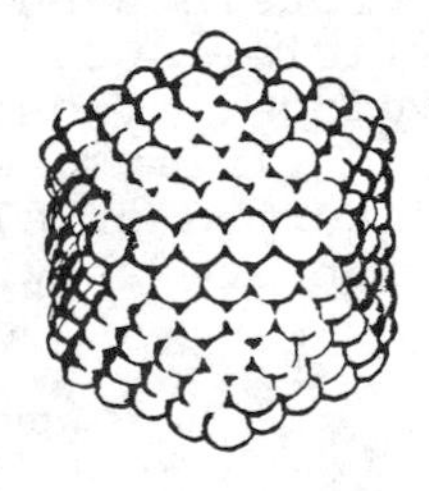

Figure 3. Adenovirus structure. These DNA-containing viruses, which multiply in animal cells, have very regular structures.

• PROBLEM 5-34

Viruses do not have any ribosomes which are the structures needed for protein synthesis. How are they able to synthesize their protein coats during their replication? In your explanation, describe the life cycle of a typical virus.

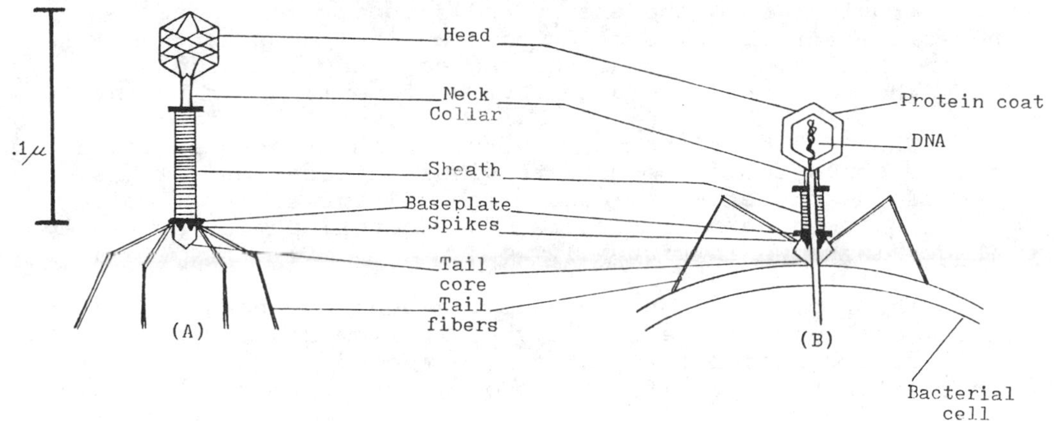

Figure 1. Structure of the T-even (2, 4, 6) phage particle. (A) shows the normal configuration. (B) shows the cutaway view of (A).

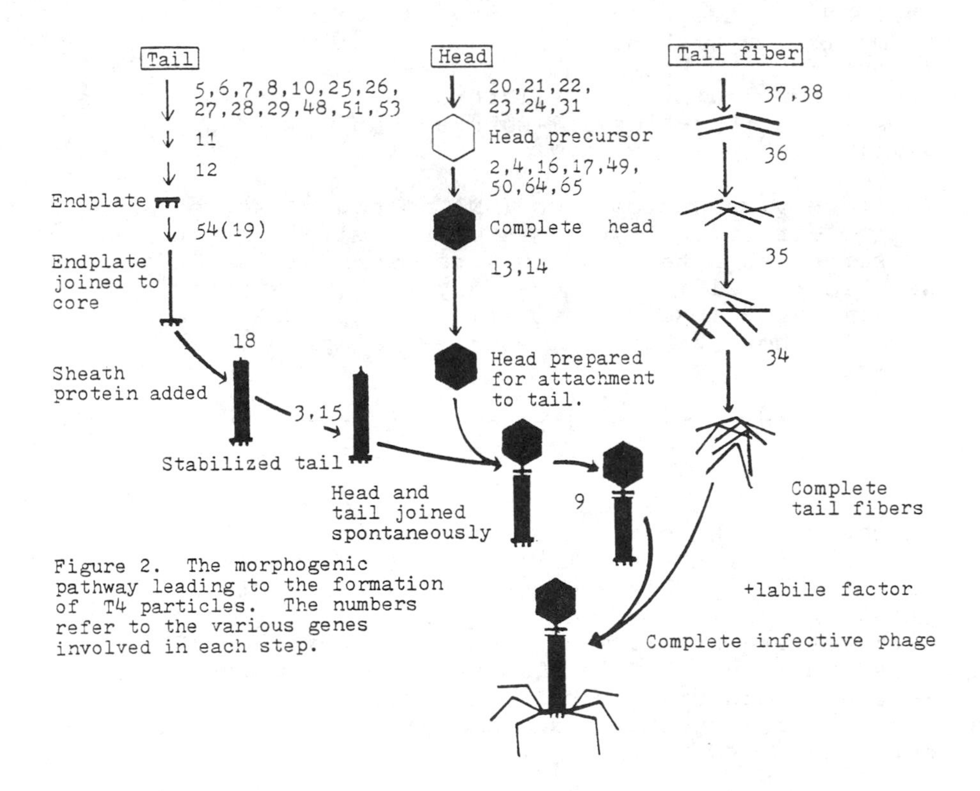

Figure 2. The morphogenic pathway leading to the formation of T4 particles. The numbers refer to the various genes involved in each step.

Solution: Viruses consist of a nucleic acid core (either DNA or RNA single or double stranded) surrounded by a protective protein coat, which may be further enveloped. There are no mitochondria, ribosomes, or any of the other organelles normally found in eucaryotic cells. How does the virus acquire the energy and the structural machinery (ribosomes) for making proteins? Since viruses have no independent metabolic activity, they are incapable of reproduction by fission, budding, or other simple means of reproduction. Instead, they replicate by inserting their nucleic acid into a functional host cell. The host cell provides both the energy and structural machinery necessary for the production of new viral components. After these components are assembled, they are released from the host cell and a new cycle of replication begins.

The T_4 bacteriophage (a virus which attacks bacteria) may be employed to demonstrate viral replication. T_4 is one of the bacterial viruses which multiplies in E. coli cells. The general structure of a bacteriophage, or simply phage, is shown in Figure 1.

The replication process begins when the phage particle collides with a sensitive bacterium and the phage tail fibers specifically attach to the bacterial cell wall. Host cells have specific receptor sites to which the phage attaches. An enzyme in the phage tail then breaks down a small portion of the cell wall, creating a small hole in the bacterium. The spikes are believed to help anchor the phage to the bacterium. The tail sheath then contracts, driving the core of the tail into the cell (see Figure 1B). The viral DNA is then injected into the bacterium through the enzymatically weakened cell wall. The protein coat is left outside the bacterium. (Most animal viruses are completely engulfed by the cell through a process known as pinocytosis, so that no viral substance remains outside the cell.)

The free viral chromosome (DNA molecule) serves as a template to direct the synthesis of new viral components. The DNA serves as a template for its own replication and for the viral-specific RNA necessary for the synthesis of its specific proteins. The viral-specific mRNA molecules attach to the host ribosomes, where viral protein synthesis takes place.

Sometimes viral infection does not affect normal cell metabolism, but more often, most of the host cell's metabolism is directed toward the synthesis of new molecules needed for new viral particles. In some cases, all DNA and RNA synthesis on the host chromosome is inhibited and preexisting bacterial RNA templates (mRNA) are broken down, so that all protein synthesis occurs only on new viral RNA templates. E. coli ribosomes may become modified, so that they work better with T_4 mRNA than with their own mRNA. This causes more viral molecular synthesis and less bacterial molecular synthesis. During infection, therefore, both host nucleic acid synthesis and protein synthesis are usually suppressed.

Once all the viral products are synthesized (there are over 40 of them), they interact in a definite sequence or morphogenetic pathway to produce the mature viruses.

The complete infective phages that are produced must now be released from the bacterium, which has a rigid cell wall. To ensure release, many phages have a gene which codes for lysozyme, an enzyme which degrades bacterial cell walls. Lysozyme begins to be synthesized when the coat proteins appear, so that its accumulation causes rupture of the bacterial cell wall when the phages are mature and ready for release. Usually several hundred viral particles are released from each bacterial cell.

These new viral particles then infect other E. coli, replicating themselves and lysing the bacteria. The life cycle of a T_2 phage is summarized in Figure 3.

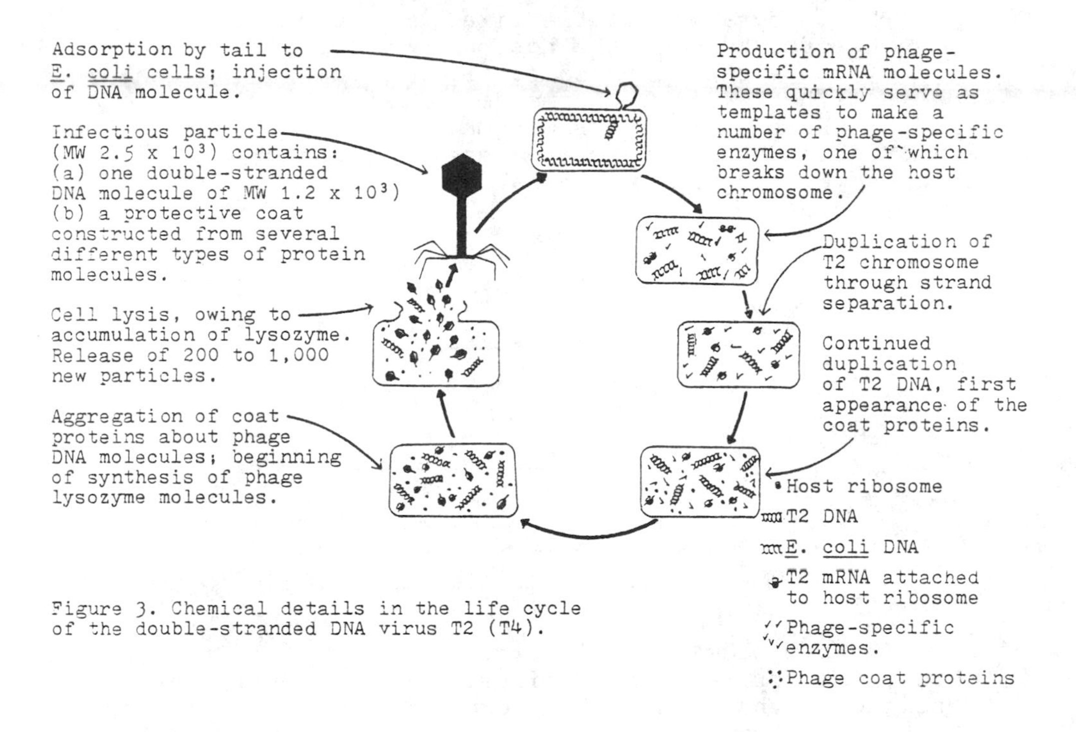

Figure 3. Chemical details in the life cycle of the double-stranded DNA virus T2 (T4).

• PROBLEM 5-35

A particular type of bacteriophage attaches to and penetrates a number of bacterial cells. After 20 minutes, some bacterial cells lyse and release many new viruse. Other bacteria remain intact and reproduce normally. After exposure to ultraviolet light, the remaining bacteria lyse within an hour. Explain.

Solution: Viruses such as the T_2 bacteriophage always multiply when they enter a host bacterial cell. Lysis eventually occurs and the newly synthesized progeny are released from the cell. Some viruses such as the λ (lambda) phage do not always multiply and lyse their host. They are referred to as the temperate phages, in contrast to the virulent phages which always kill their host. The temperate virus must be present in some inactive state within the bacterial cell. In this special relationship between the virus and the bacterium, called lysogeny, the viral DNA is incorporated into a specific section of the host's chromosome. The viral chromosome becomes an integral part of the host chromosome, and is duplicated along with it during each cell division. The phage DNA can be transferred from one cell to another during bacterial conjugation. When the viral chromosome is integrated within the host chromosome, it is called the prophage. Bacteria containing prophages are called lysogenic bacteria and viruses whose chromosomes can become prophages are called lysogenic viruses (in contrast to lytic viruses).

The integration of the viral chromosome occurs by a crossing over between the host chromosome and a circular form of the viral chromosome. Both chromosomes break and rejoin with the viral chromosome attaching to the broken ends of the bacterial chromosome. The point of attachment and crossing over occurs at a specific region of the chromosome. The insertion can be shown as follows:

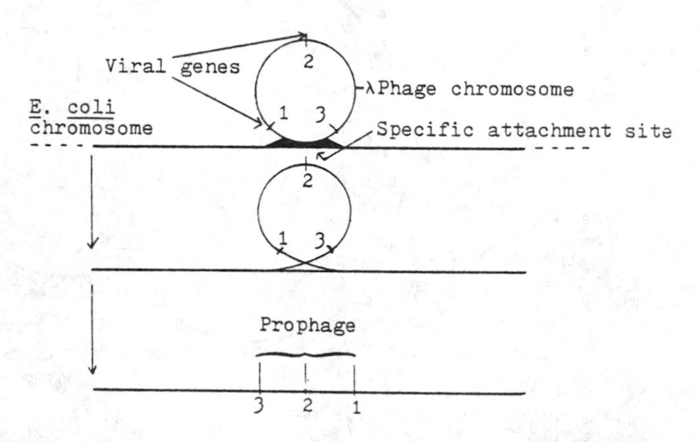

The prophage can remain integrated within the bacterial chromosome for many generations. Under certain natural conditions, the viral DNA becomes active and is removed from the host's chromosome. Once independent of the host chromosome, the lysogenic virus replicates and eventually lyses the cell. This entrance into the lytic cycle can be artificially induced by ultraviolet light or certain chemicals.

It is thought that while the viral DNA is inserted as a prophage, the transcription (the process where DNA is used to code for a complementary sequence of bases in an RNA chain, i.e. production of mRNA) of almost all phage genes is blocked. The only protein made by the λ phage DNA is the λ repressor, which causes the blockage of viral-specific mRNA transcription. Since no viral components can be made, the viral DNA remains inactive within the host chromosome. However, when the λ repressor is inactivated by an agent such as UV light, mRNA transcription occurs and the virus begins replication. Lysis eventually results with the release of new λ phages.

When a lysogenic virus is in the prophage state, its genes can produce phenotypic effects on the host bacterium. It can modify the cell wall, affect the production of enzymes and antigens, and possibly confer toxigenicity upon the lysogenic bacteria. Certain lysogenic phages enable the bacterium *Corynebacterium diphtheriae* to produce the toxin that causes the disease diphtheria.

The lytic and lysogenic cycles can be summarized in the following diagram:

Extrachromosomal Inheritance

Figure 2. Lytic and lysogenic cycles of bacteriophage

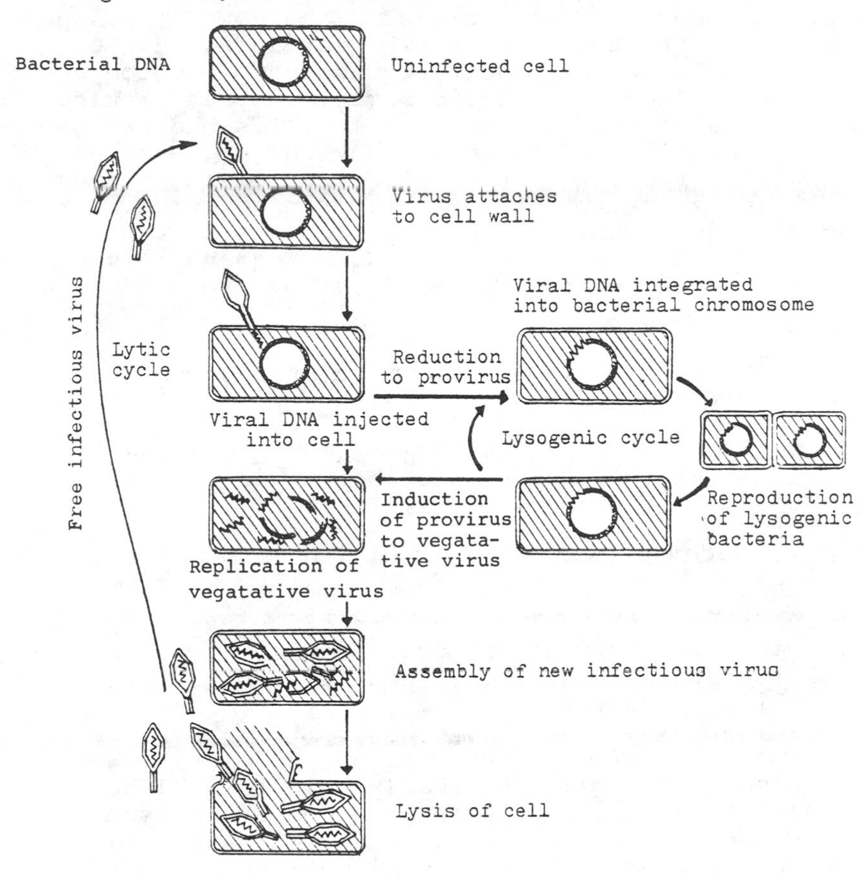

• PROBLEM 5-36

In what respects are viruses living things? How are they unlike living things?

Solution: All living things possess organization. The fundamental unit of organization is the cell. Living cells are variably bounded by a membrane, which regulates the movement of substances into and out of the cell. Viruses do not have any membranes because they have no need to take in or expel material. Viruses lack all metabolic machinery while cells possess this machinery in order to extract energy from the enviroment to synthesize their components. All cells produce ATP but viruses do not. Cells utilize ATP to build complex materials and to sustain active interactions with their environment. Viruses do not perform energy-requiring processes. Cells are capable of growth in size but viruses do not have this capacity. Most living things respond in complex ways to physical or chemical changes in their environment. Viruses do not.

Most importantly, living things possess the cellular machinery necessary for reproduction. They have a complete system for transcribing and translating the messages coded in their DNA. Viruses do not have this system. Although they do possess either RNA or DNA (cells possess both) they cannot independently reproduce, but must rely on host cells for reproductive machinery and components. Viruses use their host's ribosomes, enzymes, nucleotides, and amino acids to produce the nucleic acids and proteins needed to make new viral particles. They cannot reproduce on their own. However, viruses are unlike non-living things in that they possess the potentiality for reproduction. They need special conditions, such as the presence of a host, but they are able to duplicate themselves. Since non-living things do not contain any nucleic acids, they cannot duplicate themselves.

This is the critical argument of those who propose viruses to be living things.

The difficulty in deciding whether viruses should be considered living or non-living reflects the basic difficulty in defining life itself.

VIRAL GENETICS

• PROBLEM 5-37

Using a virus, how can one transform E. coli bacteria unable to utilize galactose (gal-mutants) into those that can utilize galactose (gal^+).

Solution: E. coli are usually able to utilize the monosaccharide galactose as a coarbon and energy source. However, mutations arise which affect an enzyme necessary for galactose utilization. These E. coli mutants are unable to utilize this sugar, and are called gal^- mutants. To transform some of these gal^- mutants into the wild type, gal^+, one can use a lysogenic virus such as the λ (lambda) bacteriophage.

One can infect a culture of prototrophic E. coli (gal^+) with the lysogenic phages. The viruses will inject their DNA into the bacterial cells. In the host, the viral DNA molecule changes from a linear structure to a circular one. At a specific attachment site on the bacterial chromosome, the viral DNA pairs with the bacterial chromosome and integrates into it after recombination. The viral DNA, now called a prophage, remains incorporated within the E. coli chromosome, until conditions are favorable for the excision of the prophage and its induction to a vegetative virus which replicates and lyses the cell.

The gene necessary for galactose utilization is very near the incorporated prophage. In the case of the λ phage, its proximity to the bacterial gal^+ gene allows rare errors to occur in which excision of the prophage includes the gal^+ gene. The λ prophage may coil in such a way that the recombination event leading to excision of the circular viral DNA includes the gal gene. To remain approximately the same size, the circular viral DNA leaves behind some of its own genes, which were replaced by the substituted bacterial region. Since it now lacks some necessary genes, the virus, now called a transducing particle or λ gal, is considered to be defective. These transducing particles can lyse the bacterial cell, but they cannot establish a lysogenic relatonship with, or cause lysis of, subsequent bacterial cells. The excision of the prophage which has the incorporated gal^+ gene can be illustrated as follows:

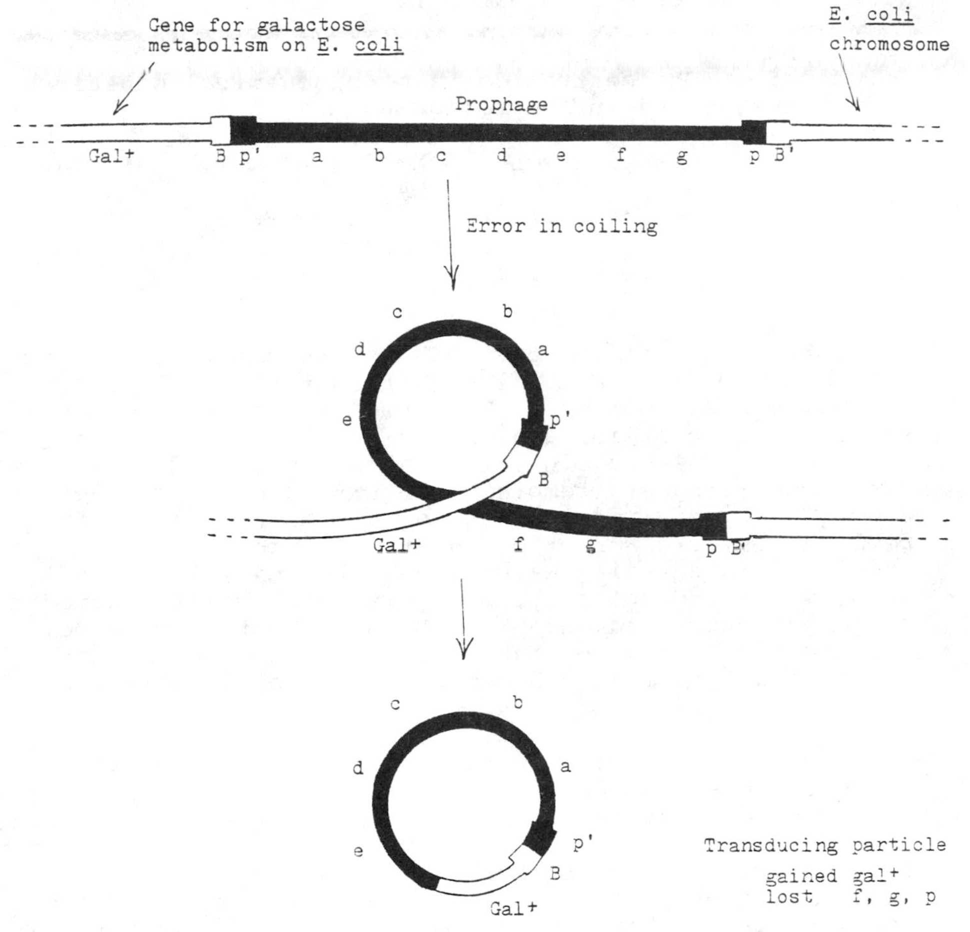

One can then harvest these transducing particles and add them to a culture of gal^- E. coli mutants. These phages attach to the mutant bacteria and inject their DNA (containing the gal^+ gene). The gal^+ region of the viral DNA may then recombine with the gal^- region of the bacterial chromosome. The gal^+ gene may be incorporated into the E. coli through the recombination process, transforming them to the wild type, which are able to utilize galactose. The whole viral DNA rarely becomes incorporated, since the transducing phage lacks a complete genome.

The viral chromosome, containing the gal^- gene, cannot be inserted into the bacterial chromosome and so is lost.

This process by which a virus transfers genes from one bacterial cell to another is called transduction. In specialized transduction, as opposed to generalized transduction, only one gene is transferred. Transduction thus serves as a mechanism for recombination in bacteria.

• **PROBLEM 5-38**

The central dogma of biochemical genetics is the basic relationship between DNA, RNA, and protein. DNA serves as a template for both its own replication and synthesis of RNA, and RNA serves as a template for protein synthesis. How do viruses provide an exception to this flow scheme for genetic information?

Solution: The central dogma of biochemical genetics can be summarized in the following diagram:

```
replication     ↻ 1
                DNA
transcription    ↓ 2
                RNA
  translation    ↓ 3
              PROTEIN
```

Arrow 1 signifies that DNA is the template for its own replication. Arrow 2 signifies that all cellular RNA molecules are made on DNA templates. All amino acid sequences in proteins are determined by RNA templates (arrow 3). However, in the viral infection of a cell, RNA sometimes acts as a template for DNA, providing an exception to the unidirectionality of the scheme involving replication, transcription and translation. The agents involved are certain RNA viruses. For example, an RNA tumor virus undergoes a life cycle in which it becomes a prophage integrated into the DNA of a host chromosome. But how can a single-stranded RNA molecule become incorporated into the double-helical structure of the host DNA? It actually does not.

An RNA tumor virus first adsorb to the surface of a susceptible host cell, then penetrates the cell by an endocytotic (engulfment) process, so that the whole virion (viral particle) is within the host cell. There, the particle loses its protein coat (probably by the action of cellular proteolytic enzymes). In the cytoplasm, the RNA molecule becomes transcribed into a complementary DNA strand. The enzyme mediating this reaction is called reverse transcriptase and is only found in viruses classified as retroviruses. (Thus, cells do not have the capacity to transcribe DNA fron an RNA template.) Cellular DNA polymerase then converts the virally produced single stranded DNA into a double-stranded molecule. This viral DNA forms a circle and then integrates into the host chromosome, where it is transcribed into RNA needed for new viral RNA and also for viral-specific protein synthesis. The RNA molecules and protein coats assemble, and these newly made RNA tumor viruses are enveloped by sections of the cell's outer membrane and detach from the cell surface to infect other cells. The release of the new virions does not require the lysis of the host cell and is accomplished by an evagination of the outer membrane. Other cells are infected through the fusion of the envelope and the potential host cell's outer membrane, thereby releasing the virion into the cytoplasm. Unlike the λ phage infection of an _E. coli_ cell, the RNA tumor virus does not necessarily interfere with normal cellular processes and cause death.

Thus, the enzyme reverse transcriptase of RNA viruses provides the basis for the single known exception to the standard relationship between DNA, RNA and proteins by catalyzing the transcription of DNA from RNA. The virus causing AIDS (Acquired Immune Deficiency Syndrome) is a retrovirus and thus utilizes the reverse transcriptase enzyme.

• PROBLEM 5-39

If DNAase is added to a bacterial cell, the DNA is hydrolyzed, the cell cannot make any more proteins and eventually dies. If DNAase is added to RNA viruses, they continue to produce new proteins. Explain.

Solution: By means of electron microscopy and x-rays diffraction studies, much has been discovered about the structure and composition of viruses. The infective ability of viruses is due to their nucleic acid composition. An individual virus contains either DNA or RNA but not both as is true for cells. Therefore since the virus proposed in the question contains RNA, it is not affected by DNAase. The RNA replicates, forming a complementary RNA strand which acts as messenger RNA in order to code for the synthesis of new viral proteins. To produce new viral RNA, the viral RNA first synthesizes a complementary strand and thus becomes double-stranded. The double-stranded RNA serves as a template for synthesis of new viral RNA. The virus could have been tobacco mosaic virus, influenza virus or poliomyelitis virus. These are all viruses which contain single-stranded RNA as their genetic material. There are at least two groups of RNA viruses in which the RNA is normally double-stranded, and assumes a double-helical form. DNA viruses, such as the smallpox virus, SV 40 (a tumor-inducing virus), and certain bacterial viruses, such as bacteriophages T_2, T_4, and T_6, contain double-stranded DNA. Yet there are some bacteriophages which have a single-stranded DNA molecule. It does not matter whether the genetic information is contained in DNA or RNA, or if it exists as a single strand or as a double helix. For viruses, the important point is that the genetic message is present as a sequence of nucleotide bases.

• PROBLEM 5-40

If antibodies can be brought into contact with a virus before it attaches itself to the host cell, the virus will be inactivated. Once infection has occurred, the production of interferon may protect the host from extensive tissue destruction. Explain.

Solution: Many human diseases have a viral etiology (cause). Among the more common viral diseases are smallpox, chicken pox, mumps, measles, yellow fever, influenza, rabies, poliomyelitis, viral pneumonia, fever blisters (cold sores), and the common cold. Although most viral infections do not respond to treatment with many of the drugs effective against bacterial infection, many are preventable by means of vaccines.

Buildup of an adequate supply of antibodies requires some time. During this period, extensive viral infection may occur. Most recoveries from viral diseases occur before full development of the immune response (production of antibodies). A factor that is important in the recovery from a viral infection is interferon. Interferon is a protein produced by virus-infected cells, which spreads to uninfected cells and makes them more resistant to viral infection. Thus, extensive viral production and resultant tissue damage are prevented.

Upon infection, the host cell is induced to synthesize interferon, a small protein which is then released into the extracellular fluid and affects the surrounding cells. Interferon binds to surface receptors of uninfected cells. This triggers these cells to synthesize a cytoplasmic enzyme which prevents viral multiplication. Note that the antiviral protein does not prevent entrance of the viral nucleic acid.

Interferon produced by a cell infected with a particular virus can prevent healthy cells from becoming infected by almost all other types of viruses. Interferon is therefore not virus-specific. However, interferon produced by one host species is not effective if administered to a different species. It is therefore host species specific. Since interferon produced by birds and other mammals cannot be used in treating human beings, it is difficult to obtain large enough quantities of interferon to provide effective chemotherapy for viral diseases. (Human donors cannot provide the needed amount of interferon.)

Interferon is a more rapid response to viral infection than antibody response. Interferon production is initiated within hours after viral infection, reaching a maximum after two days. Within three or four days, interferon production declines as antibodies are produced.

Prevention of viral infection by interferon production must be distinguished from another phenomenon - viral interference. Viral interference is observed when an initial viral infection prevents a secondary infection of the same cell by a different virus. The initial virus somehow prevents reproduction of the second virus or inactivates receptors on the cell membrane (there are specific sites for viral attachment). It may also stimulate production of an inhibitor of the second virus. Viral interference does not prevent uninfected cells from becoming infected.

Vaccination involes administration of an attenuated (live, yet weakened) or inactivated strain of virus (or other microorganism), which cannot produce the disease, but which stimulates antibody production and thus prevents infection by the pathogen.

VIRAL PATHOLOGY

• PROBLEM 5-41

We know that radiation and chemicals can cause cancer. Can viruses be a cause of cancer?

Solution: Cancer involves an abnormality in the control of cell division and cell function. Cancer cells usually produce cell masses called tumors. These cells lack the normal control systems that shut off unwanted cell division. The factors which give cancer cells their characteristic property of unregulated growth are passed on from parent to progeny cancer cells. For this reason, it has been suggested that genetic changes occur within chromosomal DNA. These changes lead to a heritable cancer phenotype. Somatic mutations (mutations in cells that are not destined to become gametes) may cause a cancer, if the mutation upsets a normal device controlling cell regulation. Some indication that somatic mutations can induce cancer comes from the fact that many carcinogens (agents that cause cancer) are also strong mutagens (agents that cause mutations.) Ultraviolet and ionizing forms of radiation are powerful carcinogens and are also highly mutagenic.

Exposure of skin to ultraviolet light frequently results in skin cancer, and x-rays applied to the thyroid are associated with increased incidence of thyroid cancer. Many chemicals are also highly carcinogenic and mutagenic. For example, nitrites and nitrates by themselves cause no genetic or carcinogenic changes, but in cells they may become converted to powerful carcinogens- the nitrosamines. Nitrosamines are also known to be mutagens. Nitrates and nitrites are found in cured meats (frankfurters, bacon, ham) and are used to inhibit growth of anaerobic bacteria.

In contrast to the somatic mutation proposal is the hypothesis that most cancers are induced by viruses. Viruses which produce tumors in animals are called oncogenic viruses. While somatic mutations are suspected to cause a loss of functional genetic material, viruses are vehicles into a cell: they introduce new genetic material that may transform the cell into a cancerous type. In addition to multiplying and lysing their host cell, certain viruses can insert their chromosomes into the host chromosome. In some animal species, this process can transform the cell into a morphologically distinguishable cancer cell. At this point, the virus is in the prophage state. Therefore, absence of detectable viruses does not provide ample evidence for or against a viral cause of a cancer.

In 1911, an RNA virus, called the Rous sarcoma virus, was shown to be the causal agent of a sarcoma (a tumor of connective tissue) in chickens. Other RNA viruses cause sarcomas and leukemias (uncontrolled proliferation of leucocytes) in both birds and mammals. One group of DNA viruses causes warts on the skin of mammals. Other DNA tumor viruses is a mouse virus called polyoma and a monkey virus called SV 40. A Herpesvirus, EB (Epstein-Barr) causes infectious mononucleosis and is probably involved involved in Burkitt's lymphoma, a cancer prevalent in humid tropical regions. Research is being conducted to find viral causes of cancer in humans.

• **PROBLEM** 5-42

Why is the organism that causes smallpox not used in the vaccine against smallpox?

Solution: Smallpox (variola) is an acute infectious disease that is caused by a virus. The disease is spread by droplet infection or by handling articles infected by a smallpox patient. The virus is thought to lodge in the nasopharynx (the part of the pharynx just dorsal to the soft palate), where it proliferates and spreads to the blood (viremia), enabling it to infect the skin and other tissues. The disease is characterized by an initial fever followed by the appearance of pustules (small fluid-filled eruptions) on the skin, which regress and leave the scars characteristic of smallpox.

It has long been known that contracting smallpox protects one against a second attack. For centuries, the Chinese inoculated the skin of a healthy person with the material from lesions on a smallpox patient. The variola virus, present in this material, would stimulate antibody production in the inoculated individual to give the person immunity to smallpox. However, the use of the actual virulent variola virus may be dangerous if too much is administered. In 1796, it was discovered by Jenner that inoculation with the vaccinia virus from cowpox lesions would confer immunity to smallpox as well as to cowpox. This immunization procedure is safe because smallpox is never produced. This vaccination procedure is practised today, using vaccinia virus taken from cows, sheep, or chick embryos.

Although the United States and many other countries have eliminated smallpox, outbreaks can occur as long as reservoirs of the variola virus still exist (e.g., Africa, India, and Southeast Asia). However, today the smallpox virus is basically confined to research laboratories. People are usually vaccinated very early in life then revaccinated every 3 to 5 years. Prevention is vital since there is no specific treatment for smallpox.

• **PROBLEM** 5-43

Discuss the etiology, epidemiology and prophylaxis of tuberculosis.

Solution: It is first necessary to define the terms used in the question before one can answer it. Etiology refers to the cause of a disease. Epidemiology deals with the cause and control of epidemics. An epidemic is the unusual prevalence or sudden appearance of a disease in a community. Prophylaxis is the preventive treatment used to protect against disease.

Many of the important infections of man are air-borne and cause diseases of the respiratory tract. Tuberculosis is an endemic respiratory disease; it is peculiar to a locality or a people. Although evidence of its existence has been found in Egyptian mummies 3000 years old, it is still a leading cause of death. In the nineteenth century, Robert Koch isolated the causative agent of human tuberculosis - Mycobacterium tuberculosis. Koch proved that pure cultures of these bacilli would produce the infection in experimental animals, and he later recovered the bacilli form these animals. There are several types of tubercle bacilli (the common name): the human, bovine, avian and other varieties. One strain is almost exclusively a human parasite, and is responsible for over ninety percent of all cases of tuberculosis. The less common bovine strain can cause tuberculosis in man through ingestion of infected beef or milk from an infected cow.

It is characteristic of tuberculosis and other respiratory diseases to occur in epidemic form, attacking many people within a short time. Their incidence usually increases during fall and winter, when many people frequently remain indoors.

The tubercle bacillus is transmitted by association with infected people through the secretions of the nose and throat, which is spread in droplets by coughing and sneezing. It may be transmitted by articles which have been used by an infected person, such as eating and drinking utensils and handkerchiefs. The lungs are the most commonly affected tissue, often exhibiting tubercles or small nodules. These are areas of destroyed lung tissue in which the tubercle bacilli grow. Symptoms of tuberculosis include pleurisy (inflamation of the pleural membrane which line the chest and cover the lungs), chest pains, coughing, fatigue and loss of weight. Treatment consists of bed rest, nourishing diet, and sometimes chemotherapy. The most effective drug, INH (isonicotinic acid hydrazide) is used both for chemotherapy and chemopropylaxis. For actual prevention, the drug is administered before infection to highly susceptible individuals.

High death rates in developing countries are due to substandard housing, overcrowding, and malnutrition. Prophylaxis for a community is best achieved when living and working conditions are not overcrowded, diet is adequate, and proper sanitation is available.

SHORT ANSWER QUESTIONS FOR REVIEW

Answer

To be covered when testing yourself

Choose the correct answer.

1. Which bacterial type appears spiral or helically coiled? (a) spirillum (b) bacillus (c) coccus (d) all of the above (e) none of the above — a

2. What may be a reason for the relative success of bacterial forms? (a) a large ratio of surface area to volume (b) rapid metabolic rates (c) rapid multiplication in number of offspring (d) all of the above (e) none of the above — d

3. Which of the following structures is found in a bacterial cell? (a) Golgi apparatus (b) endoplasmic reticulum (c) ribosomes (d) mitochondrion (e) nuclear membrane — c

4. The gram stain, which is used to differentiate bacterial cells, is based primarily on (a) the protein content in the respective bacterial cell wall. (b) the carbohydrate content in the respective bacterial cell wall. (c) the lipid content in the respective bacterial cell wall. (d) the diffusion rate of staining fluid through the bacterial cell wall. (e) none of the above. — c

5. How does penicillin inhibit bacterial proliferation? (a) inhibition of bacterial wall synthesis (b) inhibition of ribosomal function in the bacteria (c) inhibition of the bacterial glycolytic pathway (d) inhibition of the electron transport in the bacterium (e) inhibition of active sites in vital enzymes — a

6. Which bacteria would function best in hot temperatures (45 - 60°C)? (a) psychrophiles (b) thermophiles (c) mesophiles (d) all would do equally well. (e) none would survive at these temperatures. — b

7. What is the optimum pH for growth in most bacteria? (a) 2.5 - 3.5 (b) 3.5 - 4.5 (c) 4.5 - 5.5 (d) 5.5 - 6.5 (e) 6.5 - 7.5 — e

8. What is the correct sequence in the phases of the bacterial growth curve? (a) lag, exponential, stationary, death (b) exponential, stationary, lag, death (c) exponential, lag, stationary, death (d) stationary, exponential, lag, death (e) stationary, lag, exponential, death — a

Answer

To be covered when testing yourself

9. Which of the following is not considered a measure of growth of bacteria? (a) the size of the bacterium cell in the culture (b) the number of bacterial cells in the culture (c) the cell mass of the bacterial culture (d) the amount of cell activity in the culture (e) none of the above — a

10. After incubation of two bacteria cultures for 132 minutes, the number of bacteria present is 20,000. What is the generation time? (a) 2 minutes (b) 5 minutes (c) 10 minutes (d) 30 minutes (e) greater than 30 minutes — c

11. The determination of generation time is based on the premise that the culture number doubles in each generation. During which growth phase does this premise hold true? (a) lag phase (b) stationary phase (c) exponential phase (d) death phase (e) all of the above — c

12. A mutant bacterial strain which requires a supplement of nutrients not required by the original cell type is known as a(n) (a) prototroph (b) chemotroph (c) phototroph (d) auxotroph (e) heterotroph — d

13. During conjugation,what is transferred from the Hfr bacterium to the F^- bacterium? (a) the sex factor (F factor) (b) portions of the Hfr chromosome (c) the sex factor and portions of the Hfr chromosome (d) nothing is transferred (e) none of the above — b

14. Which statement correctly describes one of the phases in viral infection of bacteria? (a) Host cells have no specific receptor sites to which bacteriophages(viruses) attach. (b) The protein coat and DNA of the bacteriophage enter through the enzymatically weakened bacterial cell wall. (c) Viral specific mRNA produced within the host bacterium attach to the host ribosome, where viral protein synthesis takes place. (d) The extent of destruction of host tissues is always a good judge of a successful viral infection. (e) The lysozymes needed for rupture of the bacterial cell wall following infection are provided by the host cell. — c

15. Which of the following constituents are needed by the RNA virus to synthesize proteins in the host? (a) viral DNA (b) reverse transcriptase and amino acids (c) viral RNA and host ribosomes (d) host DNA polymerase and host ribosomes (e) all of the above — c

Answer

To be covered when testing yourself

In questions 16-25 match the following:

Question	Term	Answer
16. Salmonella	(a) hexachlorophene	16-d,
17. lysozyme	(b) prophage	17-g,
18. spherical chains	(c) male	18-j,
19. lysogenic bacteria	(d) typhoid	19-b,
20. bactericide	(e) food poisoning	20-a,
21. resistance	(f) mesophile	21-i,
22. Staphylococcus	(g) peptidase	22-e,
23. F^+	(h) pili	23-c,
24. adhesive structure	(i) immunity	24-h,
25. body temperature bacteria	(j) streptococcus	25-f,

(II)

Match question 26-35 with their corresponding terms:

Question	Term	Answer
26. preventive treatment	(a) hypotonic	26-i,
27. <u>Treponema pallidum</u>	(b) syphilis	27-b,
28. sulfanilamide	(c) faculative anaerobe	28-j,
29. exponential	(d) anaerobe	29-f,
30. T_4	(e) eucaryotic flagellum	30-c,
31. <u>Clostridium tetani</u>	(f) logarithmic	31-d,
32. auxotroph	(g) nutritional mutant	32-g,
33. sex factor	(h) F factor	33-h,
34. protoplasmic medium	(i) prophylaxis	34-a,
35. 9+2 arrangement	(j) folic acid synthesis inhibition	35-e.

(III)

For questions 36-45, match the corresponding terms.

Question	Term	Answer
36. mesosomes	(a) cell wall	36-f,
37. bacterial stain	(b) cell wall synthesis inhibitor	37-d,
38. osmotic stabilizer	(c) aerobe	38-a,
39. peptidoglycan component	(d) gram	39-i,
40. penicillin	(e) female	40-b,
41. <u>Shigella dysenteriae</u>	(f) procaryotic mitochondria	41-c,
42. <u>Propionibacterium</u>	(g) conjugation	42-h,
43. F^-	(h) Swiss cheese	43-e,
44. recombination	(i) N-acetylmuramic acid	44-g,
45. capsid	(j) protein coat	45-j,

Fill in the blanks.

46. Chains of spherical cells are classified as _____ bacteria.

Answer: strepto-coccal

47. The plasma membrane and mesosomes are the bacterial counterpart of the eucaryotic cell's _____, serving to compartmentalize the respiratory enzymes.

Answer: mito-chondria

	Answer
	To be covered when testing yourself
48. Bacteria that grow both under aerobic or anaerobic conditions depending on the environment are known as _____ .	facultative anaerobes
49. Those plants and bacteria that can utilize radiant energy (sunlight) are classified nutritionally as _____ .	photo-trophs
50. Autotrophs require only _____ from which they can construct the carbon skeletons of all their organic biomolecules.	carbon dioxide
51. There are many bacteria which are involved in the nitrogen cycle. _____ bacteria convert molecular nitrogen into ammonia while _____ bacteria reduce nitrates back into molecular nitrogen.	nitrogen fixing, denit-rifying
52. One can set up an environment in which the bacterial population is maintained in the log phase. This condition is called _____ .	steady-state growth
53. Conjugation in bacteria occurs across cytoplasmic bridges which are known as _____ .	sex pili
54. The degree of pathogenicity (the ability to cause disease) is called _____ .	virulence
55. Pathogens, such as those that cause typhoid fever, which inhibit the gastrointestinal tract are known as _____ pathogens.	enteric
56. Those organisms that carry a disease from one animal to another causing widespread infection are known as _____ .	vectors
57. _____ is a species of bacteria which normally inhabits the intestine of man and rarely is pathogenic	Escherichia coli
58. Certain viruses such as the λ (lambda) phage do not always multiply and lyse their host. They are referred to as _____ phages.	temperate
59. In RNA tumor viruses, RNA can be transcribed into a complementary DNA strand when mediated by the enzyme _____ .	reverse transcrip-tase
60. The sudden appearance and rapid spreading of a disease in a community is known as a(n) _____ .	epidemic

CHAPTER 6

ALGAE AND FUNGI

TYPES OF ALGAE

• PROBLEM 6-1

Algae may be microscopic single-celled organisms or enormous multicellular seaweeds hundreds of feet long. The body of one of these organisms is termed a thallus. How does a thallus differ from a plant body?

Solution: A plant body shows differentiation of parts, with its roots, stems, and leaves all varying greatly in structure and in function. Multicellular algae show very little, if any, tissue differentiation. For this reason, the body is termed a thallus (from the Greek thallos: a young, undifferentiated shoot or sprout). There is no anatomical basis for distinguishing the leaves or roots of multicellular alga, though in some species there may be a functional basis.

The plant kingdom has traditionally been divided into two groups: the Thallophyta (algae) and the Embryophyta (land plants). The Embryophyta, in addition to showing tissue differentiation, have a life cycle different from that of the Thallophyta.

The reproductive structures of the Thallophytes are usually unicellular, and the reproductive cells lack a protective surrounding wall or jacket of non-reproductive cells. Thallophyte zygotes do not develop into embryos until after they are released from the parent thallus. In Embryophyta, the reproductive structures are multicellular and are surrounded by non-reproductive cells. The early stages of embryonic development occur while the zygote is still contained within the parent plant.

The body of a fungus is also termed a thallus. Fungi, like algae, show little differentiation of parts, and lack multicellular sex organs with jacket cells. The embryo of a fungus develops outside the parent fungus.

• PROBLEM 6-2

Describe the kingdoms within which algae are classified.

Solution: Algae fall into the Plant, Protistan, and Moneran kingdoms. The three phyla divisions of the Thallophytes (a major plant division) includes the Chlorophyta (green algae), Rhodophyta (red algae), and Phaeophyta (brown algae). The Protistan algae include the Euglenophyta (photosynthetic flagellates), Chrysophyta (golden algae and diatoms), and the Pyrrophyta (dinoflagellates). The Moneran kingdom includes the Cyanophyta (blue-green algae).

• PROBLEM 6-3

All of the algae have the green pigment chlorophyll, yet the different phyla of algae show a great variety of colors. Explain.

Solution: The pigments found in the different algae phyla are extremely varied, and their concentrations result in different colors. The earliest classifications of algae were based on color. Fortunately, later study of algae showed that algae of similar pigmentation also shared other important characteristics and that the older classifications were still valid.

In addition to the green pigment chorophyll, most algae possess pigments of other colors called accessory pigments. The accessory pigments may play a role in absorbing light of various wavelenghts. The energy of these light wavelengths is then passed on to chorophyll. This absorption widens the range of wavelengths of light that can be used for photosynthesis.

The accessory pigments phycocyanin and phycoerythrin serve this function in the red algae, the Rhodophyta. These pigments give the Rhodophyta their characteristic red color, although occasionally, they may be black. The red algae often live at great depths in the ocean. The wavelengths absorbed by chlorophyll a do not penetrate to the depths at which the red algae grow. The wavelengths that do penetrate deep enough are mostly those of the central portion of the color spectrum. These wavelengths are readily absorbed by phycoerythrin and phycocyanin. The energy trapped by these pigments is then passed on to chlorophyll a, which utilizes this energy for photosynthesis.

In the green algae, the Chlorophyta, the chlorophyll pigments predominate over the yellow and orange carotene and xanthophyll pigments. The predominance of carotene pigments imparts a yellow color to the golden algae, members of the phylum Chrysophyta. The diatoms, the other important class of Chrysophyta, possess the brown pigment fucoxanthin. The Pyrrophyta (dinoflagellates) are yellow-green or brown, due to the presence of fucoxanthin and carotenes. Some red dinoflagellates are poisonous, containing a powerful nerve toxin. The blooming of these algae are responsible for the "red tides" that kill millions of fish.

The brown algae, the Phaeophyta, have a predominance of fucoxanthin. These algae range in color from golden brown to dark brown or black. The procaryotic cyanophyta (blue-green algae) have the blue pigment phycocyanin as well as phycoerythrin xanthophyll and carotene. The Euglenophyta contain chlorophyll a and b and some carotenoids.

• PROBLEM 6-4

What are the importance of desmids, diatoms and dino-flagellates to man?

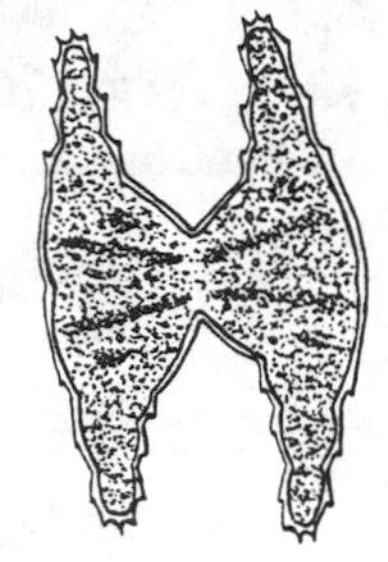

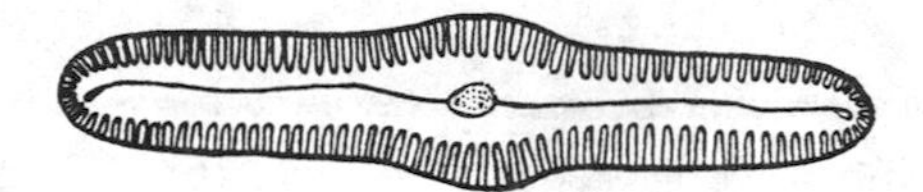

b. Typical diatom

a. Typical desmid

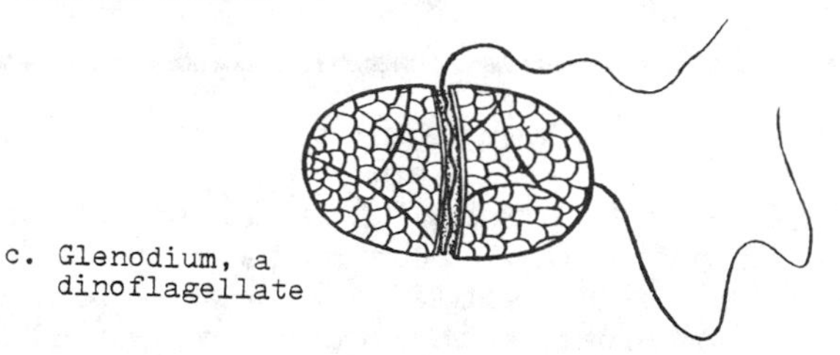

c. Glenodium, a dinoflagellate

Solution: Desmids, diatoms and dinoflagellates are all unicellular algae. Desmids are nonmotile, fresh-water green algae, commonly found in lakes and rivers. Desmids have symmetrical, curved, spiny or lacy bodies with a constriction in the middle of the cell. Under the microscope they look like snowflakes. Diatoms, members of Chrysophyta, are found in fresh and salt water. They have two shelled, siliceous cell walls and store food as leucosin and oil. The cell walls are ornamented with fine ridges, lines, and pores that are either radially symmetrical or bilaterally symmetrical along the long axis of the cell. Diatoms lack flagella, but are capable of slow, gliding motion. Dinoflagellates, the majority of the Pyrrophyta, are surrounded by a shell consisting of thick, interlocking plates. All are motile and have two flagella. A number of species lack chlorophyll. These are the heterotrophs which feed on particulate organic matter. Dinoflagellates are mainly marine organisms.

All of these organisms play extremely important roles in aquatic food webs. Plankton is defined as small aquatic organisms floating or drifting near the surface. Phyto-plankton are photosynthetic autotrophs and the zooplankton are heterotrophs. Desmids are important freshwater phyto-plankton. Diatoms are the most abundant component of marine plankton (a gallon of sea water often contains one or two million diatoms). Diatoms are also found in abundance in many rivers. Dinoflagellates are second only to the diatoms as primary producers of organic matter in the marine en-vironment. Probably three quarters of all the organic matter in the world is synthesized by diatoms and dino-flagellates. In addition to the production of organic material, these algae have the primary responsibility for the continued production of molecular oxygen via photosyn-thesis. Respiration by animals utilizes oxygen. If the supply were not constantly being replenished, the oxygen on earth would be exhausted.

Phytoplankton are essential to aquatic food chains; they are eaten by zooplankton, by invertebrates, and by some fish. The organisms which eat phytoplankton are in turn eaten by other organisms. Ultimately all aquatic life depends on the phytoplankton. Since terrestrial animals rely ultimately on the oceans, rivers, and lakes for a large part of their food source, the desmids, diatoms, and dinoflagellates are of crucial importance to land life as well.

Diatoms are also important to man by virtue of their glasslike cell walls. When the cells die, their silica-impregnated shells sink to the bottom of the sea, and do not decay. These shells accumulate in large quantities and geologic uplifts bring the diatomaceous earth to the surface, where it is mined and used commercially.

Diatomaceous earth is used as a fine abrasive in detergents, toothpastes, and polishes. It is also used as a filtering agent, and as a component in insulating bricks and soundproofing products.

Diatoms utilize oils as reserve material, and it is widely believed that petroleum is derived from the oil of diatoms that lived in past geologic ages.

• PROBLEM 6-5

List the structural characteristics of the blue-green algae. Where may they be found?

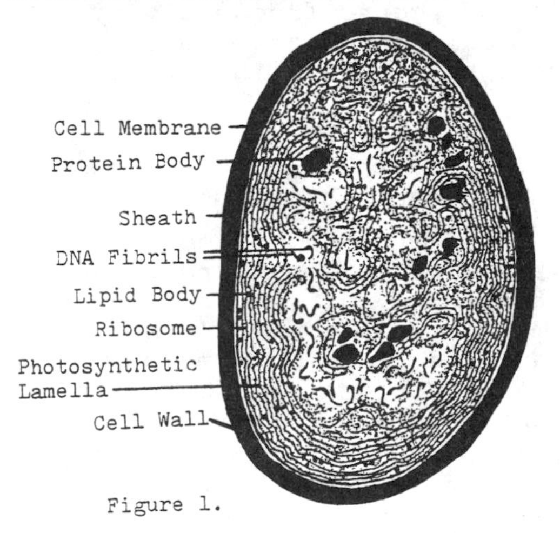

Figure 1.

Structure of a blue-green algal cell as seen under the electron microscope. Note its similarity to bacterial cell structure.

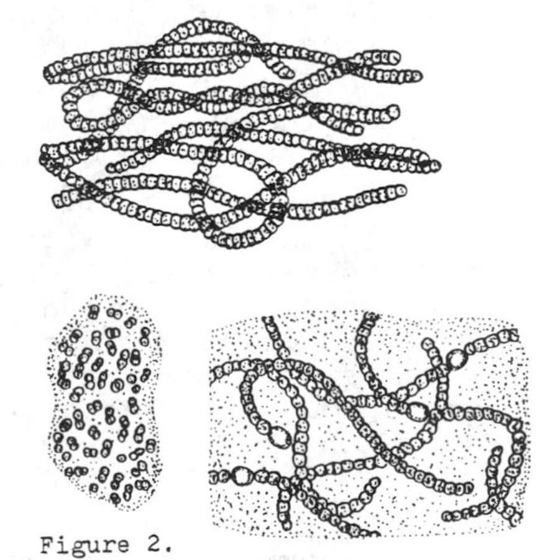

Figure 2.

Some common species of blue-green algae.

Solution: The blue-green algae, (Cyanophyta or cyanobacteria) are prokaryotic unicellular or filamentous organisms. The filamentous forms are strings of individual cells held together by fused walls. (See Figure 1.) In most species, ther are no protoplasmic connections, such as plasmodesmata, between the adjacent cells. The unicellular forms exist as either single rods or spheres, and each individual cell is capable of carrying out all necessary life processes.

The Cyanophyta are the most primitive chlorophyll-containing autotrophic organisms now living. The Cyanophyta are quite different from other algae, being structurally similar to bacteria. These similarities have formed the basis for grouping them with the bacteria in the kingdom Monera.

All blue-green algae possess photosynthetic pigments, which are located in folds or convolutions of the cell membrane extending into the interior of the cell. These photosynthetic pigments are chlorophyll a, carotenoids, phycocyanin (blue pigment) and sometimes phycoerythrin (red pigment). Like the bacteria, the blue-green algae lack mitochondria, Golgi apparatus, a nuclear membrane, endoplasmic reticulum and the large cell vacuole characteristic of higher plants. Ribosomes are present along with many proteinaceous granules and granules of a stored carbohydrate material known as cyanophycean starch, which is very similar to glycogen. The nuclear region consists of a single circular chromosome composed of double-stranded DNA.

The cell walls of blue-green algae contain some muramic acid and cellulose. Outside this cell wall is a sticky gelatinous sheath composed of pectic materials. It covers the entire cell. The cytoplasm of blue-green algae is unusually viscous, being composed of a very dense colloidal material.

Cell division in blue-green algae is accomplished by binary fission. Reproduction also occurs frequently by fragmentation of filaments. Sexual reproduction has never been observed in the Cyanophyta but is thought to occur infrequently.

No blue-green algae possess flagella, yet many species are capable of movement. The filamentous forms exhibit a peculiar slow gliding or oscillatory motion.

The blue-green algae are found in many varied habitats. Most are found in fresh water pools and ponds. A few species are found in hot springs, at temperatures up to 85°C. Some species are marine while others are common in soils, the banks of trees and the sides of damp rocks. Some species of blue-green algae exhibit a symbiosis with fungi, in which both act together to form a lichen.

CHARACTERISTICS OF FUNGI

• PROBLEM 6-6

How are fungi important to man?

Solution: Fungi are divided into four groups: Oomycetes (Oomycota)- the egg fungi, Zygomycetes (Zygomycota)- Zygo-spore-forming fungi, Ascomycetes (Ascomycota)- the sac fungi, and Basidiomycetes (Basidiomycota)- the club fungi.

Fungi are both beneficial and detrimental to man. Beneficial fungi are great importance commercially. Ascomycetes, or sac fungi, are used routinely in food production. Yeasts, members of this group, are utilized in liquor and bread manufacture. All alcohol production relies on the ability of yeasts to degrade glucose to ethanol and carbon dioxide, when they are grown in the absence of oxygen. Yeasts used in alcohol production continue to grow until the ethanol concentration reaches about 13 percent. Wine, champagne, and beer are not concentrated any further. However, liquors such as whiskey or vodka are then distilled, so that the ethanol concentration reaches 40 to 50 percent. The different types of yeasts used in wine production are in part responsible for the distinctive flavors of different wines. Bread baking relies on CO_2 produced by the yeasts which causes the dough to rise. Yeasts used in baking and in the brewing of beer are cultivated yeasts, and are carefully kept as pure strains to prevent contamination. Sac fungi of genus Penicillium are used in cheese production. They are responsible for the unique flavor of cheeses such as Roquefort and Camembert. The medically important antibiotic penicillin is also produced by members of this genus. Certain Ascomycetes are edible. These include the delicious morels and truffles.

The club fungi, or Basidiomycetes, are of agricultural importance. Mushrooms are members of this group. About 200 species of mushrooms are edible while a small number are poisonous. The cultivated mushroom, Agaricus campestris, differs from its wild relatives, and is grown commercially.

Fungi are often of agricultural significance in that they can seriously damage crops. Members of the Oomycetes, also known as water molds, cause plant seedling diseases, downy mildew of grapes, and potato blight (this was the cause of the Irish potato famine). Mildew is a water mold that grows parasitically on damp, shaded areas.

Rhizopus stolonifer, a member of the Zygomycetes, is known as black bread mold. Once very common, it is now controlled by refrigeration and by additives that inhibit mold growth. The Aascomycete Claviceps purpurea causes the disease ergot, which occurs in rye and other cereal plants and results in ergot poisoning of humans and livestock. This type of poisoning may be fatal. The disease caused in humans is called St. Vitus's dance. Visual hallucinations are a common symptom of this disease. Lysergic acid is a constituent of ergot and is an intermediate in the synthesis of LSD. The "dance macabre" of the Middle Ages is now believed to have been caused by ergot poisoning.

Basidiomycetes are also responsible for agricultural damage. Certain club fungi are known as smuts and rusts. Smuts damage crops such as corn, and rusts damage cereal crops such as wheat. Bracket fungi, another type of club fungi, cause enormous economic losses by damaging wood of both living trees and stored lumber.

Fungi are also important to man because of the diseases they cause in man and livestock. Candida albicans causes a throat and mouth disease, "thrush", and also infects the mucous membranes of the lungs and genital organs. Many skin diseases are caused by fungi, including ringworm and athlete's foot.

• PROBLEM 6-7

What advantages and disadvantages do the fungi have in comparison with chlorphyll-bearing plants in terms of survival?

Solution: The chlorophyll-bearing plants require no organic food source. Their sole needs are light, carbon dioxide, water, and inorganic minerals. Plants are able to grow in any environment where these needs are met. The ocean is a habitat very favorable to green plants (e.g. the green algae), but not to fungi. Since fungi do not photosynthesize, they are able to grow in the dark. Green plants cannot do this. Fungi do not require light penetration and they can survive with very thick, tough cell walls. The cell walls of fungi sometimes contain cellulose, but chitin is usually their most important constituent. Chitin, a polysaccharide of acetyl glucosamine, is also the principal constituent of the exoskeleton of arthropods-insects, lobsters, shrimp, and crabs. The strong cell wall of fungi permits them to grow in environments where no plants or other organisms can grow. Certain fungi are extremely resistant to plasmolysis (cell shrinkage in hypertonic medium), and can grow in concentrated salt solutions and sugar solutions. Certain fungi are able to withstand high concentrations of toxic substances. An example is the ethanol-producing yeasts, which grow at extremely high concentrations of ethanol. Other organisms are killed at these ethanol concentrations.

Fungi are less adept at surviving than green plants in that fungi require an organic food source, and are only able to grow where large quantities of foodstuffs are present. Fungi cannot capture their food, they must obtain nutrition by growing directly on or within their food source. For these reasons, they are not commonly found in ocean habitats.

• PROBLEM 6-8

What is the difference between a hypha and a mycelium and between an ascus and a basidium?

Solution: A few fungi are unicellular, but most have multicellular bodies made up of tubular branching filaments called hyphae. A mass of hyphae is called a mycelium. The hyphae of the algai fungi, or Phycomycetes (which contain the Oomycetes and the Zygomycetes), are not divided by cross walls between adjacent nuclei - they are coenocytic hyphae. These fungi are thus multinucleate. The Ascomycetes and Basidiomycetes have hypae divided by cross walls - they are septate; these fungi are multicellular. The mycelium may appear as a cobweb-like mass of fibers, as in bread mold, or may be fleshy and compact, as in truffles.

A basidium and an ascus are both reproductive structures. Ascus is Latin for sac, and fungi whose life

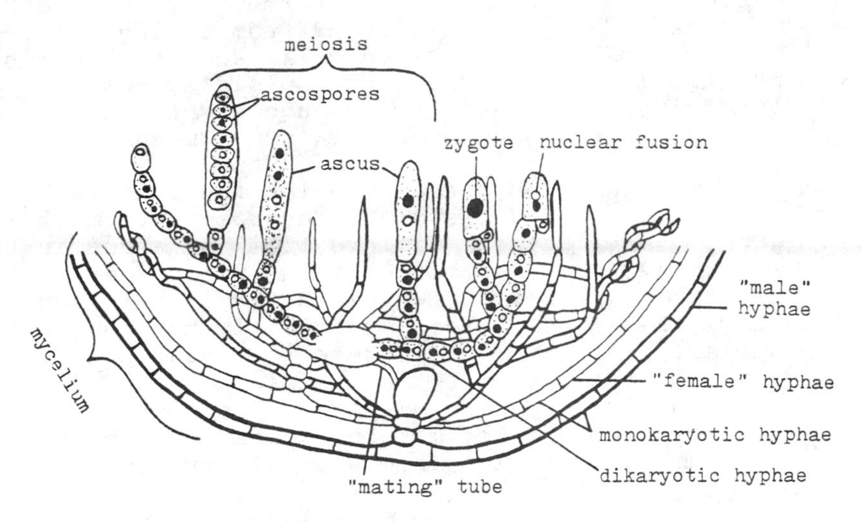

Figure 1. Sexual reproduction in the Ascomycetes. The "male" hyphae fuse with the "female" hyphae to produce dikaryotic hyphae, from which an ascus is produced. Meiosis occurs within the ascus to produce ascospores.

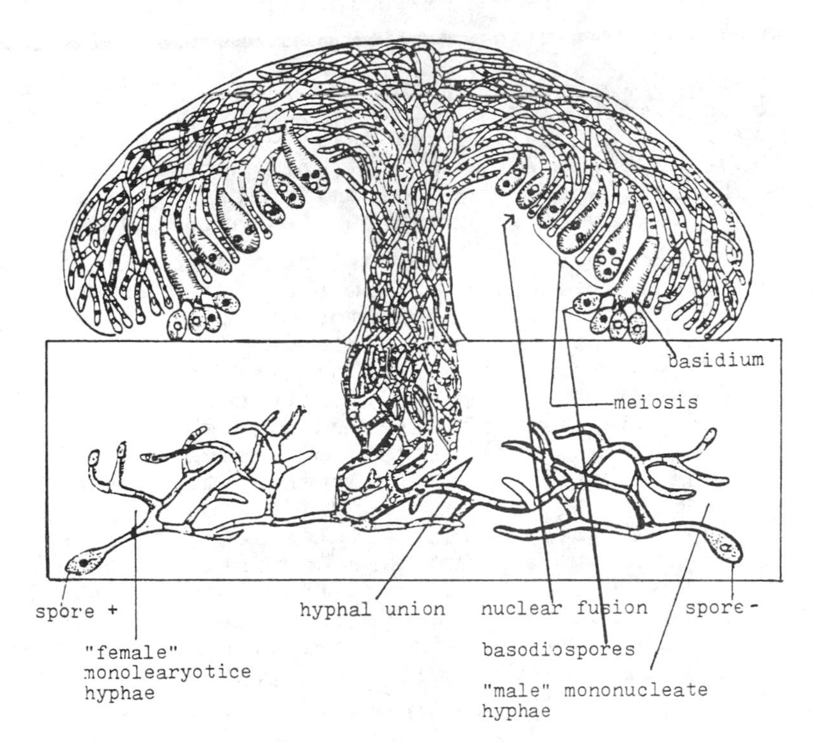

Figure 2. Sexual reproduction in the Basidiomycetes.

Diagram of section through a mushroom. Hyphae from two uninucleate mycelia--one of the plus, the other of the minus, mating type--unite and give rise to binucleate hyphae, which then develop into the above-ground part of the mushroom. The entire stalk and cap are composed of these hyphae tightly packed together. Spores are produced by basidia on the lower surface of the cap.

cycle includes an ascus are termed sac fungi or ascomycetes. The ascus is formed from a single parent cell. The zygote is an elongated cell, and its nucleus divides meiotically to produce four haploid spores. These usually divide mitotically to produce eight small spore cells. Spores are released when the ascus ruptures.

The basidiomycetes, or club fungi, have reproductive structures called basidia. Mushrooms are members of this group. The basidium differs from the ascus in that spores are formed on the outside of the parent cell rather than within it. The zygote nucleus within the elongated, club-shaped basidium divides meiotically to produce four haploid nuclei. These nuclei migrate to protuberances which develop at the tip of the basidium. Each protuberance then buds off to form a spore. The spores may fall from the basidium or they may be ejected. Both the ascus and the basidium develop from a single zygote.

• PROBLEM 6-9

In what ways are slime molds like true fungi? In what ways do they resemble animals?

Solution: Both cellular slime molds (acrasiomycota) and true slime molds (myxomycota) have unusual life cycles containing fungus-like and animal-like stages. Slime molds have membrane-bound nuclei, are heterotrophic, ingest food, lack cell walls and produce fruiting bodies. They belong to the kingdom Protista in the phylum Gymnomycota.

The true slime mold's adult vegetative stage is decidedly animal-like. At this stage, the slime mold is a large, diploid, multinucleated amoeboid mass called a plasmodium. It moves about slowly and feeds on organic material by phagocytosis. Plasmodium growth continues as long as an adequate food supply and moisture are available. When these run short, the plasmodium becomes stationary and develops organs specialized to produce haploid spores, known as the fruiting bodies. At this stage, the true slime mold is similar to the fungi. Meiosis occurs in the fruiting body, and spores with cellulose cell walls are released. Spores are a resistant and dormant form of a slime mold. When thw spores germinate, they lose their cell walls and become flagellated gametes. Gametes fuse to become zygotes. The zygotes lose their flagella, become amoeboid, and grow into multinucleated plasmodial slime molds.

Cellular slim molds differ from the true (acellular) slime molds in being haploid and in that the amoeboid cells, on swarming together, retain their identity as individual cells. The cellular slime molds resemble amoebas throughout most of their life cycle: they lack cell walls, move about and ingest particulate matter. Under certain conditions, the amoebas aggregate to form a multicellular pseudoplasmodium, called a slug. The pseudoplasmodium becomes stationary and fruiting bodies are formed. Spores are not produced by cellular division, but by the formation

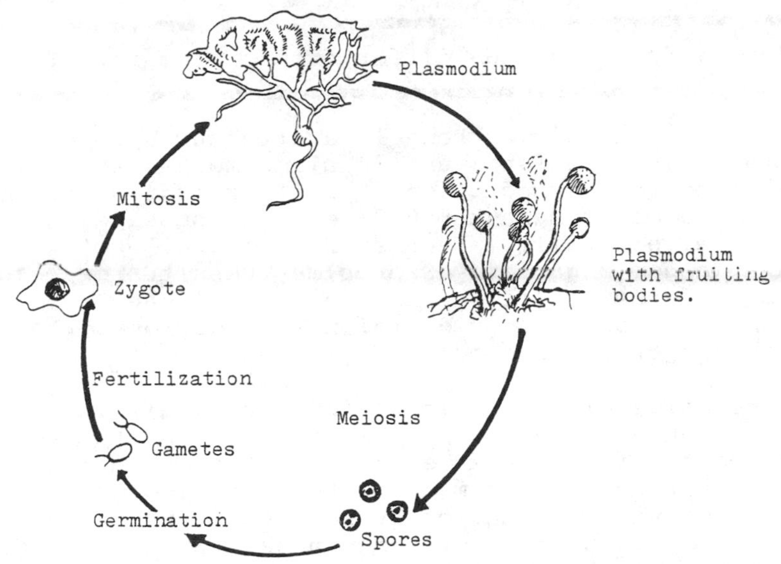

Figure 1. Life cycle of the true slime mold.

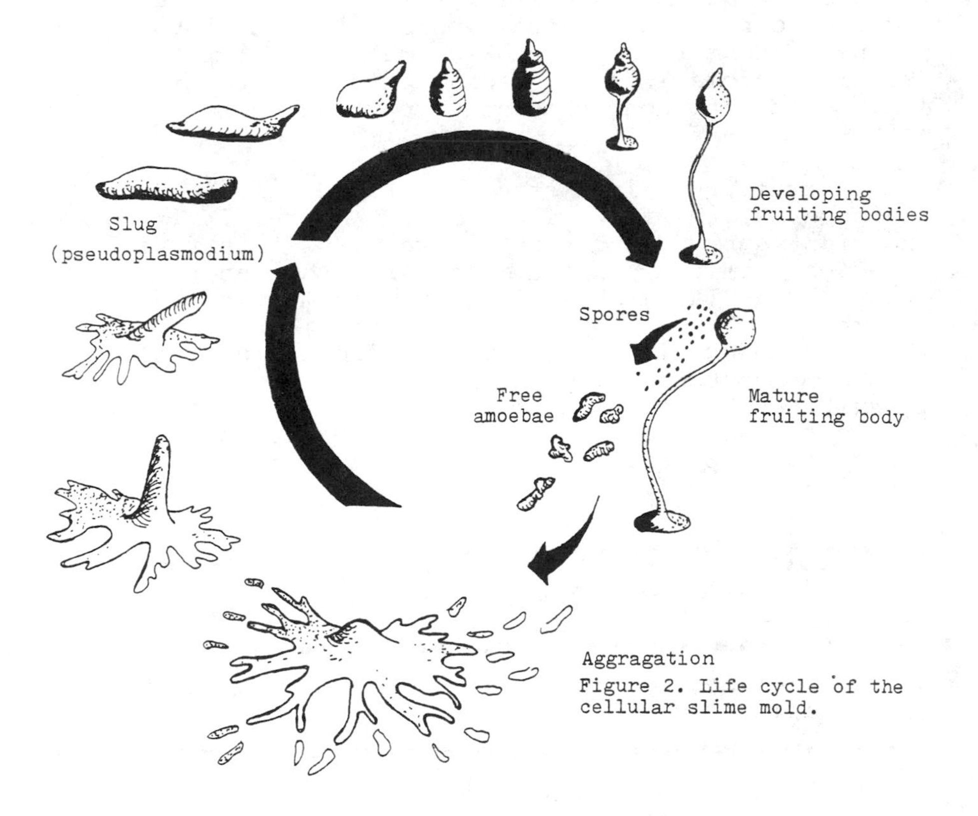

Figure 2. Life cycle of the cellular slime mold.

of cell walls around the individual amoeboid cells. Each spore becomes a new amoeboid cell when it germinates. Cellular slime molds are haploid throughout their life cycles. The formation of a fruiting, spore-forming body is characteristic of fungi.

• **PROBLEM** 6-9

How does a fruiting body differ from a true fruit?

Solution: A fruiting body is a structure, found in certain species of fungi and in slime molds, that is specialized for reproduction. Fruiting bodies usually release spores, single haploid cells possessing thick protective cell walls. Each spore is capable of becoming a new organism. Spores may result from meiotic division of the zygote, as in the case of the Basidiomycetes, or from meiotic division in a diploid organism, as with the true slime molds.

The cells of the fruiting body and their internal organization are usually similar to those of the vegetative parts of the organism. For example, the mushroom is the fruiting body of many of the basidiomycetes. The mushroom is composed of hyphae, as is the underground nonreproductive portion of the fungus. In mushrooms, these hyphae are compactly arranged, and may form a stem and a cap, but they are still similar in cellular morphology to the underground hyphae. "Fruiting body" is a general term referring to any multicellular structure which is specialized for reproduction and remains attached to the nonreproductive portion while releasing spores.

A true fruit is a structure found only in the Angiosperms, the flowering plants. A true fruit contains the plant embryo within a seed coat. A true fruit develops from the plant's maternal sex organ, the ovary. The plant embryo is a diploid multicellular organism which will develop into a diploid plant. Although there are many different types of fruits (grapes, tomatoes, peaches, chestnuts, and pea pods), the term "true fruit" is quite specific. The fruit differs from the nonreproductive portions of the plant in both its arrangement of cells and in the biochemical and morphological characteristics of the cells.

DIFFERENTIATION OF ALGAE AND FUNGI

• **PROBLEM** 6-10

How do the algae differ from the fungi in their method of obtaining food?

Solution: Algae are autotrophs, while the fungi are heterotrophs. Certain algae contain the green pigment chlorophyll,others have the accessory pigments. These pigments enable them to utilize solar energy to synthesize organic compounds from CO_2 and H_2O. Fungi, lacking photosynthetic pigments, must obtain organic compounds directly. The eucaryotic algae (all algae except the blue-green) contain chlorophyll in chloroplasts, subcellular organelles where photosynthesis occurs. Solar energy is trapped by the chlorophyll and converted to ATP, the ATP is used by the chloroplast to convert carbon dioxide and water to organic compounds, and oxygen is liberated. Algae obtain water and CO_2 from their environment, as well as the other minerals needed for their organic constituents. The CO_2 and minerals are dissolved in the aquatic environment, and are absorbed through the cell wall and cell membrane of the algae.

Fungi live where organic compounds are present. Some fungi are animal or plant parasites but most are saprophytes. Most sarophytic fungi excrete digestive enzymes into their environment, and break down food material extracellularly. The products of digestion are then absorbed through the cell wall and cell membrane by structures called the hautoria. Saprophytic fungi decompose vast quatities of dead organic material. Some cause spoilage of foodstuffs (bread, fruit, vegetables) and deterioration of leather goods, paper, fabrics, and lumber. Parastic fungi may carry out extracellular digestion or they may directly absorb organic materials produced by the body of their host. However, many fungi are benefical to man.

• PROBLEM 6-11

What are lichens? Describe the relationship that exists in a lichen. How do they reproduce?

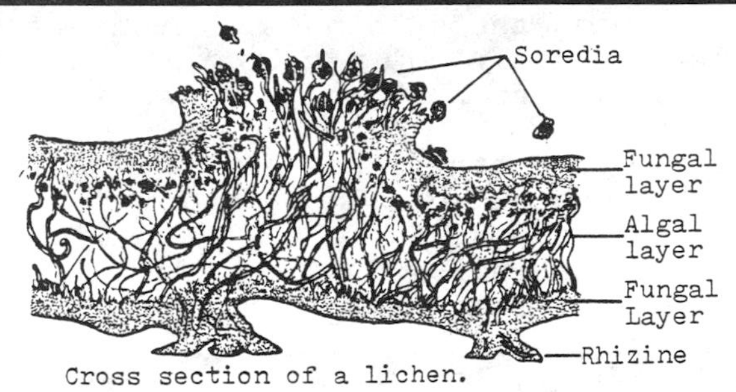

Cross section of a lichen.

Solution: Lichens are composite organisms consisting of algae and fungi. They grow on tree bark, rocks and other substrates not suitable for the growth of plants. Lichens may be found in low-temperature environments characteristic of polar regions and very high altitudes.

Structurally, a lichen can be likened to a fungal "sandwich" whose hyphae entwine a layer of algal cells (See figure). Structures known as rhizoids, which are short twisted strands of fungal hyphae, serve to attach the bottom layer to the substrate.

Not all species of algae or all species of fungi can enter into a lichenlike relationship. Most of the fungi found in lichens are Ascomycetes, although there are a few of the Basidiomycetes. Algae of the lichens are from the Cyanophyta (blue-green algae) or the Chlorophyta (green algae) phylum. Each lichen thallus consists of a single species of fungus associated with a single species of alga.

There are two types of lichens: fruticose and crustaceous. The fruticose have an erect, shrublike morphology. The crustaceous are very closely attached to the substrate or may even grow within its surface.

Lichens are the product of a relationship called symbiosis, in which each partner of the association derives something useful from the other for its survival. The algae provides the fungus with food, especially carbohydrates produced by photosynthesis, and possibly vitamins as well. The fungus absorbs, stores and supplies water and minerals required by the alga, as well as providing protection and a supporting framework for the alga.

Lichens appear to reproduce in a variety of ways. Fragmentation may occur, in which bits of the thallus are broken off from the parent plant and produce new lichens when they fall on a suitable substrate. Lichens may produce "reproductive bodies" called soredia which are knots of hyphae containing a few algal cells. In addition, the algal and fungal components of a lichen may reproduce independently of each other. The fungal component produces ascospores and the algal component reproduces by cell division or infrequently via sporulation. Some species of lichens in the Arctic have been alive for 4,500 years which suggests a very well balanced association between the symbionts. Lichen grow very slowly because of their low metabolic rate. They are also very resistant to heat and desiccation.

Lichens produce many interesting organic products. Unusual fats and phenolic compounds make up from two to twenty percent of the dry weight of the lichen body. Litmus, the pigment indicator, and essential oils used in perfumes are obtained from lichens.

EVOLUTIONARY CHARACTERISTICS OF UNICELLULAR AND MULTICELLULAR ORGANISMS

• PROBLEM 6-12

In what respects is Euglena like a plant and in what respects is it like an animal?

Solution: Euglena is a single-celled organism. Like plants, Euglena contains chlorophyll in chloroplasts and is capable of photosynthesis. Like animals, Euglena lacks a cell wall, is capable of active motion and ingests organic compounds as food. Botanists classified Euglena and its relatives as algae (phylum: Euglenophyta). Zoologists classified the euglenoids in the phylum Protozoa (class Phytomastigophora). Euglenoids are also often classified as a separate phylum in the kingdom Protista.

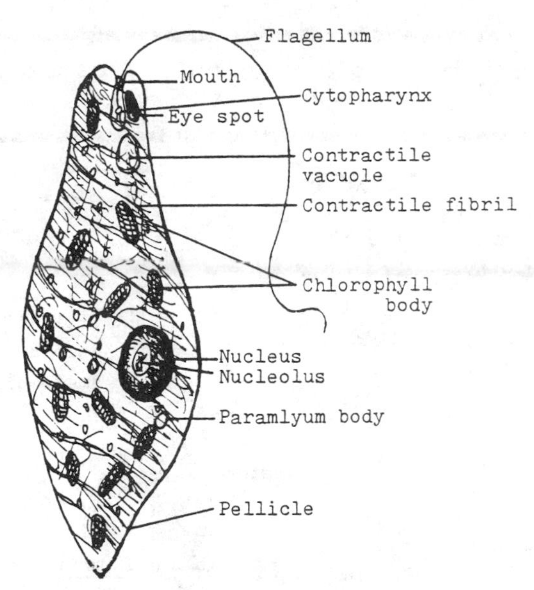

Diagram of Euglena.

Most euglenoids are capable of photosynthesis, but require certain organic nutrients as well as inorganic salts. Some are completely autotrophic, and some do not possess chlorophyll and are completely heterotrophic. Of the heterotrophs, some are saprophytic, feeding on dead organic materials, and others are holozoic, capturing and ingesting other organisms.

The euglenoids all have two flagella, by means of which they are capable of active motion. The euglenoid cell is delimited by a cell membrane within which there is a series of flexible, interlocking proteinaceous strips termed the pellicle. The flexible pellicle permits Euglena to change its shape, providing an alternate means of locomotion for mud-dwelling forms. As is true for protozoans, but not for plants, an area of the plasma membrane is specialized for the ingestion of food. This area is near the base of the flagella and is called the gullet. Unlike higher plants, but similar to many algae, the euglenoids have a stigma or an eyespot that functions as a light detector. The non-photosynthetic euglenoid species lack the stigma. Unique to the euglenoids is the storage product paramylum. This compound is a carbohydrate polymer unlike plant starch or animal glycogen. The paramylum is produced in an organelle called a pyrenoid. The cell division cycle of Euglena is also different from that of plants or animals. During division the nucleolus does not disappear, and the nuclear membrane remains intact.

The euglenoids have a mixture of plant and animal characteristics, in addition to having their own unique features.

They have no close relatives among the algae and few among the protozoans.

Define the terms haploid, dipolid, gamete, zygote, sporophyte and gametophyte.

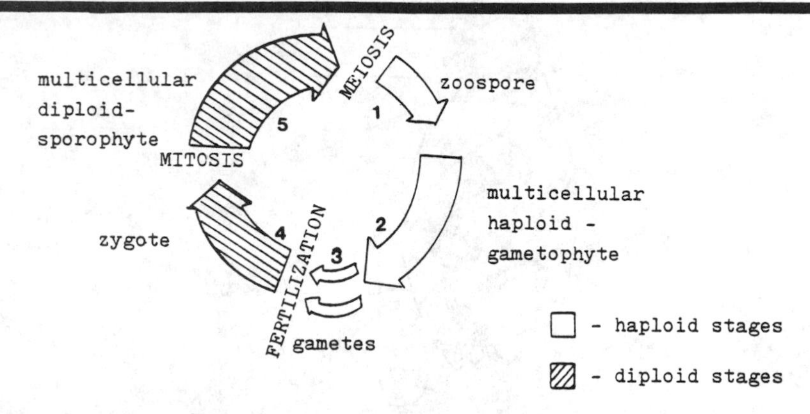

Life Cycle of Ulva and Ectocarpus

Solution: These terms refer to the life cycle and reproductive processes of algae, fungi, and higher plants. The chromosome number is termed haploid or diploid. A haploid organism is one which possesses a single complete set of chromosomes while a diploid organism has two sets of chromosomes.

A gamete is a cell specialized for sexual reproduction, and is always haploid. Two gametes fuse to form a diploid cell, which may give rise to a new multicellular individual via mitotic division, or which may itself undergo meiosis to form new haploid cells. The diploid cell that results when two gametes fuse is termed a zygote. The zygote is always diploid and is always a single cell.

Life cycles of algae and fungi show much more variability than life cycles of higher plants and animals. The different species within a phylum often have very different life cycles, and most algae and fungi are capable of both sexual and asexual reproduction. Sexual reproduction occurs when two haploid gametes fuse to form a zygote. Asexual reproduction occurs by fragmentation, with each fragment growing into a new plant or thallus. Asexual reproduction also involves mitotic division of a single cell of a thallus to produce daughter cells which are released from the parent thallus to grow into new algae or fungi. Daughter cells which are motile and specialized for asexual reproduction are termed zoospores.

In certain species, alternation of generations occurs. This type of life cycle includes both a multicellular haploid plant and a multicellular diploid plant, and both sexual and asexual reproduction. In algae, fungi, and most other lower plant forms, the dominant generation is the haploid one, and the reverse is true in the higher plants. Ulva (sea lettuce), a green alga, and Ectocarpus (a brown alga) are representative of organisms which show alternation of generations. The entire life cycle can be summarized as follows:

Haploid zoospores (stage 1) divide mitotically to produce a haploid multicellular thallus (stage 2). These haploid thalli may reproduce asexually by means of zoospores or sexually by means of gametes (stage 3). Gametes fuse to produce a diploid zygote (stage 4). The zygote divides mitotically to produce a diploid multicellular thallus (stage 5). Certain reproductive cells of the diploid thallus divide by meiosis to produce zoospores (stage 1). The zoospores are released, and begin the cycle anew. The haploid multicellular stage is called a gametophyte because it is a plant that produces gametes. The diploid multicellular stage is called a sporophyte because it is a plant that produces spores.

• PROBLEM 6-14

In the change from unicellular organisms to multicellular organisms, a number of different events occur. Describe these changes as they are evidenced by the motile-colony series of green algae.

Figure 1 : Clamydomonas, a single-celled alga.

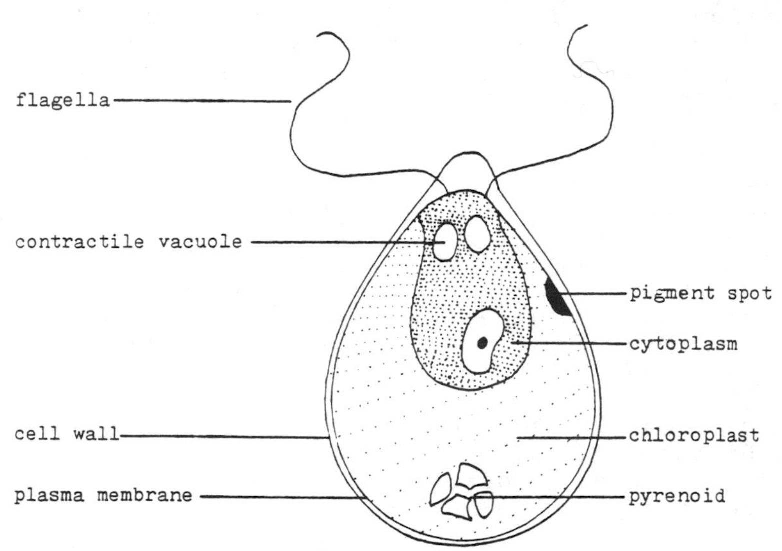

Solution: The motile-colony series, or the volvocine series, of Chlorophyta shows a gradual progression from the unicellular condition of Chlamydomonas to elaborate colonial organization. Gonium is an example of the simplest colonial stage. The cells of a Gonium colony are morphologically similar to the unicellular Chlamydomonas. Each cell has a single large cup-shaped chloroplast, containing a pyrenoid which functions

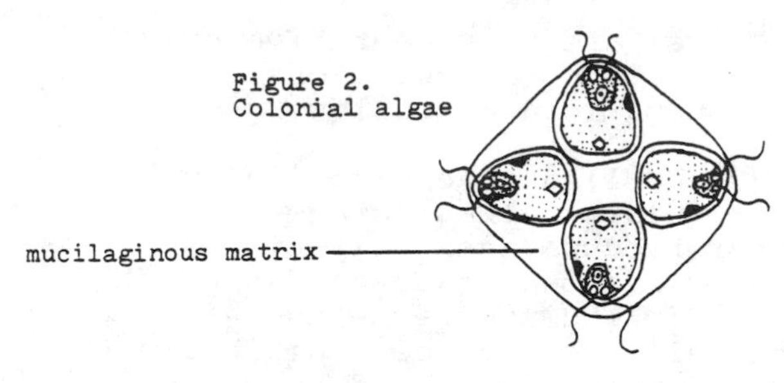

Figure 2.
Colonial algae

(a) Gonium sociale

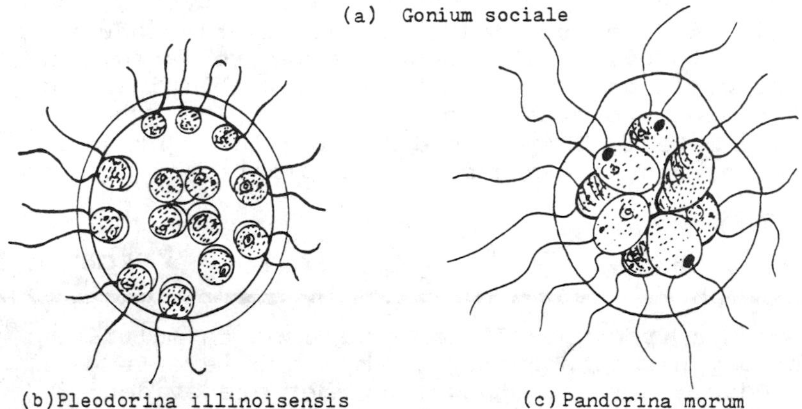

(b) Pleodorina illinoisensis (c) Pandorina morum

in starch synthesis, and an eye spot or stigma which functions in light detection. There are two flagella at the anterior end of each cell, and near the base of the flagella two small contractile vacuoles, which discharge rhythmically, expelling excess water from the cell. The Gonium cells are embedded in a mucilaginous matrix. There is coordination of activity between the cells, since the flagella of all the cells beat together, enabling the colony to swim as a unit.

Pandorina is at a more advanced stage of development than Gonium. The colony has anterior and posterior halves, and there is regional differentiation. The colony swims in an oriented fashion, with a definite anterior half. The stigmas are larger in the anterior cells. In addition, the vegetative cells of the colony are so dependent on one another that they cannot live apart from the colony, and the colony cannot survive if broken up.

The genus Pleodorina is further advanced with a considerable division of labor among the cells. The anterior cells are purely vegetative, and the posterior cells are larger and function in both sexual and asexual reproduction.

Each Gonium colony consists of 4, 16 or 32 cells and the Pandorina colony contains 8, 16, or 32 cells, and the Pleodorina colony has 32 to 128 cells. Volvox is the most advanced of the volvocine series. The colony consists of 500 to 50,000 cells. All the anterior cells are exclusively vegetative, and the cells in the posterior half of the colony are much larger than the vegetative cells, and are specialized for asexual reproduction. This is the production of male gametes, or of non-motile egg cells.

The changes represented by the volvocine series are as follows: (1) increasing size of colonies, (2) increasing coordination of activity among the cells, (3) increasing interdependence among the cells, so that vegetative cells cannot survive apart from the colony, and (4) increasing division of labor, particularly between vegetative and reproductive cells.

• PROBLEM 6-15

What trends in the evolution of sexual reproduction are evident among the green algae? Have these trends been continued in higher plants?

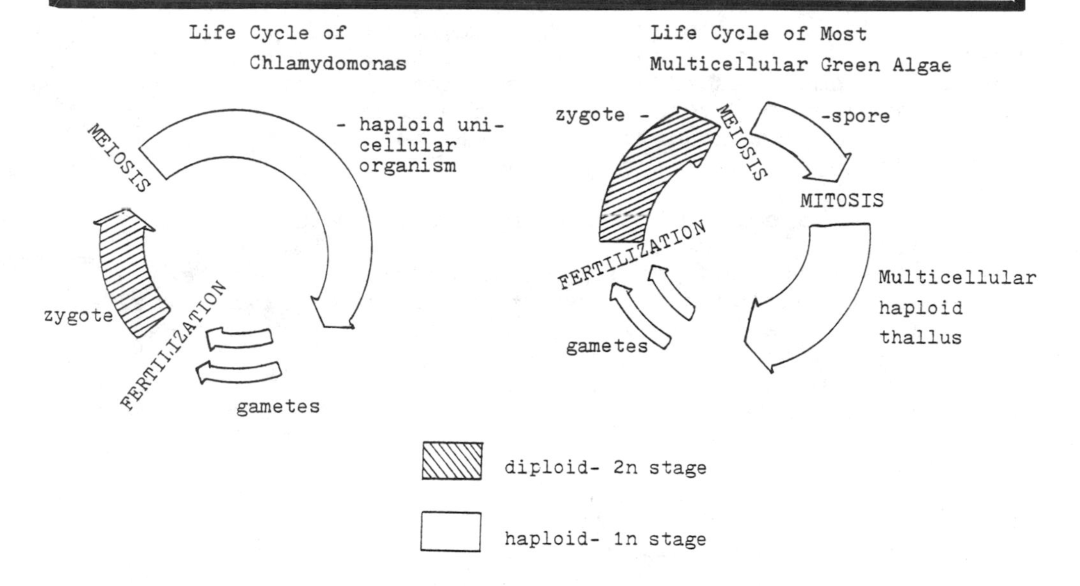

GREEN ALGAE: EVOLUTIONARY DEVELOPMENT

<u>Solution</u>: Among the green algae, the major lines of evolutionary development in sexual reproduction are: (1) the change from isogamy to heterogamy to oogamy, (2) the division of labor between the cells of a thallus so that vegetative and reproductive cells perform separate functions, (3) the development of conjugation processes which enable gene exchange through recombination, and (4) the presence of a diploid multicellular organism in the life cycle.

The unicellular green algae, Chlamydomonas, and the colonial genus Gonium have the simplest type of life cycles. The haploid adult organism is capable of asexual reproduction and of sexual reproduction. Usually Chlamydomonas reproduces asexually by zoospores. Under certain conditions, Chlamydomonas reproduces sexually. The mature cell divides mitotically several times to produce gametes. These gametes are all alike and they cannot be separated into male and female types. This condition is termed isogamy. In Gonium, individual cells are released from the colony and function as gametes. These gametes are also alike. Two gametes of one species fuse to form a zygote. The zygote divides meiotically to yield either individual haploid cells in Chlamydomonas, or a colony of haploid cells in Gonium. The zygote is the only diploid stage in these life cycles. In these two groups, the cells are not specialized and any cell is capable of either asexual or sexual reproduction.

Pandorina is a genus of colonial forms slightly more complex than Gonium. Sexual reproduction is heterogamous with small male gametes fusing with larger female gametes. Both types of gametes are free-swimming, and fertilization (fusion of gametes) occurs outside the parent colony.

Pleodorina is an even more advanced genus. In sexual reproduction, the female gametes are not flagellated and remain embedded in the parent colony. A flagellated male gamete swims to the nonmotile egg cell and fertilization occurs. This type of heterogamy is termed oogamy.

Volvox, a colonial genus, has an oogamous sexual reproductive life cycle, and also shows cellular specialization. Most of the cells of the large spherical colonies are exclusively vegetative, incapable of sexual reproduction. A few cells are specialized for reproduction. The female cell is large and non-motile and is produced by a specialized cell - an oogonium. The male cells are small and free-swimming, and are produced by sperm producing cells termed antheridia. Fertilization occurs inside the parent colony.

The change from isogamy to oogamy and the specialization of vegetative and reproductive cells are trends that have been continued among higher plants. All of the algae discussed so far have a dominant haploid stage and the only diploid stage is the one-celled zygote. Ulva, or sea lettuce, is a multicellular algae whose life cycle includes both multicellular haploid and multicellular diploid stages. Its life cycle exhibits alternation of generations. This alternation of generations is a major evolutionary event in the development of plants.

In the Ulva, reproductive cells differ from vegetative cells, but any vegetative cell is capable of becoming a reproductive cell.

Ulothrix is a multicellular filamentous algae which exhibits a type of isogamous sexual reproduction termed heterothallism. Isogametes are of two different mating types termed plus and minus. A plus gamete can only fuse with a minus type, and not with another plus gamete. Heterothallic isogamy is fairly common in algae and fungi, and may be considered an evolutionary stage between homothallic isogamy (where any gamete may fertilize any other gamete) and heterogamy. The sexual reproductive cells of Ulothrix differ from the usual vegetative cells and from zoospores. The zygote is the only diploid stage.

Spirogyra is a filamentous green algae with a rather odd sexual reproductive pattern. The gametes are not released from the parent thallus. Instead two filaments come to lie side by side and protuberances develop on cells next to one another. The protuberances enlarge, fuse, and the cell wall disintegrates between the two cells. One cell then travels through this conjugation tube and the two cells fuse to form a zygote. Any cell in a filament is capable of uniting with a cell in a neighboring filament. The two fusing cells are similar (isogamy). Although in higher plants, specialized tubes develop to allow fertilization (pollen tubes), this is not analogous to the conjugation tube of Spyrogyra. In Spirogyra, both haploid cells develop protuberances that will form the conjugation tube. In higher plants only the haploid male gametes form pollen tubes. These enable the male nucleus to unite with the nucleus of the female egg cell.

In all the life cycles discussed, both sexual and asexual reproduction occur. In filamentous algae, asexual reproduction occurs by fragmentation or by zoospores.

• PROBLEM 6-16

It is generally believed that higher plants evolved from the green algae. What is the reason for this belief?

Solution: The green algae share a number of characteristics with the higher plants. The other algal phyla have unique characteristics which are not shared with higher plants.

There are two types of chlorophyll present in most algae and plants. All photosynthetic organisms, except the photosynthetic bacteria, have chlorophyll a. The green algae and the higher plants have chlorophyll b in addition to chlorophyll a. Neither the brown algae nor the red algae have chlorophyll b. Instead, the brown algae have chlorophyll c and and the red algae possess chlorophyll d.

The principal reserve material of the Chlorophyta (green algae) and of higher plants is usually starch. The dinoflagellates utilize starch and oils. The diatoms and the golden algae store food as oils and as the polysaccharide leucosin, which is a white substance. Brown algae utilize a variety of unusual carbohydrates as food reserves,one of which is laminarin. Red algae do not utilize starch, but a polysaccharide similar to it called

Comparitive Summary of Characteristics in the Six Phyla of Eucaryotic Algae

Phylum	Number of Species	Photosynthetic pigments	Carbohydrate Food Reserve	Flagella	Cell wall components	Remarks
Chlorophyta (green algae)	7,000	Chlorophyll a and b, carotenoids	Starch	2 (or more) apical or sub equal, whiplash	Cellulose (xylan or mannan in a few)	Mostly fresh-water, but some marine.
Phaeophyta (brown algae)	1,500	a,c, caroteniods fucoxanthin	Laminarin	2, lateral, foward, tinsel, trailing whiplash; in reproductive cells only.	Cellulose matrix with alginic acids (polysaccharides)	Almost all marine, mostly temperate
Rhodophyta (red algae)	4,000	a, caroteniods, phycobilins	Floridean starch	None	Cellulose pectic materials common, xylan in Porphyra	Most marine, but some fresh-water; complex sexual cycle; many species tropical
Chrysophyta (golden algae and diatoms)	6-10,000	a, often c, carotiniods	Leucosin	1 or 2, apical, whiplash or tinsel, equal or unequal.	Pectic compounds with siliceous materials	Mostly marine
Pyrrophyta (golden brown algae)	1,100	a,c, caroteniods	Starch	2, lateral	Cellulose	Marine and freshwater, sexual repro-duction rare
Eugenophyta (euglenoids)	450	a,b, caroteniods	Paramylon	1 (to 2), apical, tinsel (one row of hairs)	Protein	Mostly fresh-water; sexual reproduction unknown.

the Floridean starch. The Euglenoids utilize the carbohydrate polymer paramylan as their reserve material.

The cell walls of Chlorophyta and the higher plants are composed chiefly of cellulose, with very little other materials. The Euglenoids lack a cellulose cell wall, and have instead a proteinaceous pellicle, which is internal to the plasma membrane. The cell walls of the Chrysophyta (diatoms, yellow-green, and golden algae) usually consist of pectic materials impregnated with silica. These siliceous cell walls consist of two halves, which fit together like the two parts of a pill box. Some unicellular dinoflagellates may lack cell walls, but most have thick cellulose cell walls that fit together as interlocking plates to form a shell. The cell walls of red algae have large quantities of mucilaginous material.

The higher plants are biochemically similar to the green algae and biochemically different from other algae. This leads biologists to propose that the ancestor of higher plants was similar to the present green algae. Also, among the Chlorophyta, there are many multicellular non-motile forms. Higher plants are always multicellular and non-motile. The Phaeophyta and Rhodophyta (brown and red algae) are also mostly multicellular and non-motile, but this is not true of the Euglenophyta, Chrysophyta, and dinoflagellates.

Answer

To be covered when testing yourself

SHORT ANSWER QUESTIONS FOR REVIEW

Choose the correct answer.

1. Which of the following is incorrect concerning the reproductive structures of thallophytes versus those of embryophytes? (a) both contain unicellular reproductive structures (b) embryophyta reproductive structures are surrounded by a "wall" or protective non-reproductive cells, while those of thallophytes have no such covering. (c) thallophyte zygotes differentiate into embryos only when they are released from the parent thallus. Embryophyte maturation takes place within the parent plant. (d) embryophyte reproductive structures are multicellular, while those of thallophytes are unicellular. — a

2. Which of the following correctly explains why the phylum rhodophyta exhibit a red color? (a) Since most rhodophyta grow at great depths, the chlorophyll can only absorb light in the red area of the spectrum, as only these wavelengths penetrate to great depths. (b) The wavelenghts of light that are absorbed by chlorophyll are passed to phycoerythrin, a red pigment. This imparts the red color to the algae. (c) Chlorophyll does not exist in the phylum Rhodophyta. The red pigment, phycoerythrin, absorbs all the light waves, but the algae appear red only. (d) In very deep ocean waters, only light waves from the central portion of the spectrum can penetrate that far down. The red pigment, phycoerythrin, can absorb this light, while chlorophyll can not. Rhodophyta contain this pigment, light is absorbed by it, and the energy derived is passed to chlorophyll for photosynthesis. Hence, rhodophyta are red in color. — d

3. Which of the following is incorrect concerning Desmids, Diatoms, and Dinoflagellates? (a) all three are unicellular algae (b) all three are motile organisms. (c) Desmids are commonly found in lakes and rivers and posess a snowflake-like appearance. Diatoms are usually radially or bilaterally symmetrical. Dinoflagellates are heterotrophs, usually with two flagella. (d) Desmids are an integral part of phytoplankton, while some Dinoflagellates are part of zooplankton and others are part of phytoplankton. — b

4. Which of the following is incorrect concerning blue-green algae. (a) they reproduce via

Answer

To be covered when testing yourself

binary fission. (b) they are capable of movement via flagella. (c) they are capable of storing carbohydrate in a form quite similar to glycogen. (d) when living associated with a fungus the two together form a lichen. — b

5. Which of the following correctly compares the hyphae structure of phycomycetes, ascomycetes, and basidiomycetes? (a) all three contain cross walls between successive nuclei of the hyphae, giving each hypha a multicellular appearance.

(b) Ascomycetes contain cross walls between successive nuclei, while Phycomycetes and Basidiomycetes contain no cross walls between successive nuclei, making them multinucleate. (c) all three contain no cross walls between successive nuclei, giving all three hyphae multinucleate appearance. (d) Phycomycetes contain no cross walls between successive nuclei of their hyphae, while Ascomycetes and Basidiomycetes contain cross walls between nuclei. — d

6. Which of the following correctly compares a basidium and an ascus? (a) A basidium is the reproductive structure for the Club fungi, while an ascus serves the same purpose for Sac fungi. Both form spores inside the parent cell. (b) A basidium is the asexual reproductive structure for Basidiomycetes, while an ascus is the sexual reproductive structure for Ascomycetes. They differ as to the location of spore maturation. (c) A basidium and an ascus are both sexual reproductive structures, the former for Basidiomycetes, the latter for Ascomycetes. With basidia, spores are formed outside the parent cell, while with asci, spores are formed within the parent cell. Both develop from a single zygote. (d) A basidium is another name for the mycelium of a basidiomycete, while an ascus refers to the same structure in an Ascomycete. — c

7. Characteristics of a "true fruit" include all of the following except: (a) it is a structure found only in the flowering plants (Angiosperms). (b) the cellular structure of the fruit is indistinguishable from the rest of the plant. (c) a true fruit develops from the maternal sex organ, the ovary. (d) inside a true fruit is a seed, which contains the plant embryo — b

8. Which of the following best describes a lichen? (a) It is any species of algae living para-

Answer

To be covered when testing yourself

sitically with any species of fungus. (b) a lichen is a special strain of fungus that can only grow near an abundant supply of algae. (c) a lichen consists usually of an Ascomycete fungus living in a mutualistic relationship with blue-green or green algae. (d) a lichen is always a Basidiomycetes fungus living off dead fungal matter.

c

9. Lichens can reproduce by which of the following methods? (a) a thallus from the parent lichen can fragment, attach to a suitable surface, and grow into a new lichen. (b) lichens are capable of producing reproductive bodies called soredia, which contain hyphae with a few intermingled algal cells, capable of growing into a new lichen. (c) both the fungus and the algae can reproduce independently of the other. (d) lichens can reproduce by any one of the three methods mentioned.

d

10. Which of the following is not a feature of euglenoids? (a) a non-distinct nucleolus and nuclear membrane during cell division. (b) a specialized area of the cell membrane for food ingestion called a gullet. (c) an eyespot for light detection in photosynthetic euglenoids. (d) a carbohydrate storage form called paramylum, produced in an organelle called a pyrenoid.

a

11. Which of the following correctly describes the sequence of events during alternation of generations? (a) the sporophyte stage gives rise to zoospores that fuse to form a zygote. It divides, giving rise to a diploid thallus. Then, certain cells divide, producing gametes. The cycle then starts again, (b) Zoospores produced from the previous sporophyte stage divide to form a haploid thallus. Gametes produced by the thallus fuse to form a diploid zygote. This zygote divides forming a diploid thallus. Then, certain cells of the thallus divide meiotically to form zoospores. The cycle begins again. (c) zoospores produced from the previous gametophyte stage divide to form a diploid thallus. Gametes produced by the thallus fuse to form a haploid zygote. The zygote then gives rise to a haploid thallus. Certain cells of the thallus then divide forming zoospores. The cycle can then begin again. (d) None of the previous explanations correctly describes alternation of generations.

b

	Answer
	To be covered when testing yourself
12. When going from the Pandorina stage to the Pleodorina stage in the Volvocine series, which of the following is the most distinctive, new, characteristic? (a) increased cell size. (b) increased colony size (c) increased division of labor (d) increased sexual reproduction.	c
13. Which of the following correctly distinguishes the three terms isogamy, heterogamy, and oogamy? (a) all three terms mean the same thing, with each one describing asexual reproduction in three different genera of algae. (b) isogamy refers to sexual reproduction in which the gametes are of different sizes. Heterogamy refers to sexual reproduction in which the gametes are the same size. Oogamy is a specific type of heterogamy. (c) isogamy refers to sexual reproduction in which the gametes that fuse are indistinguishable from each other. Heterogamy refers to sexual reproduction in which both gametes are free swimming, but are different sizes. Oogamy is sexual reproduction in which one gamete is non-motile and larger than the other free swimming "male" gamete. (d) isogamy is a type of asexual reproduction in which the species involved are of different classes, while oogamy is a type of sexual reproduction in which the species involved are of the same class. Heterogamy and oogamy are synonymous.	c
14. The green algae, spirogyra, utilize a special structure for reproductive purposes. Which of following is the correct name for this structure? (a) pollen tube (b) conjugation tube (c) plasmid (d) antheridia	b
15. Which of the following is incorrect concerning green algae and higher plants? (a) both are usually multicellular and motile (b) both capture energy by utilizing the same pigment. (c) the cell walls of both are mostly composed of cellulose. (d) both usually store glucose as starch.	a
Fill in the blanks.	
15. The plant kingdom can be divided into two subkingdoms: the _____ , which include algae and fungi, and the _____ , consisting of the higher plants. They differ in their relative amounts of _____ .	thallophytes, embryophytes, differentiation

	Answer
	To be covered when testing yourself
16. Chlorophyta have a larger quantity of the green pigment _____ than other pigments, so they appear green. Chrysophyta, on the other hand, have a predominance of the pigment _____ giving them a yellowish color.	chlorophyll, carotene
17. The difference between phytoplankton and zooplankton is that the former are _____, while the latter are _____ . Among the important jobs performed by these organisms is the production of _____ . They also serve as the main _____ source for many invertebrates and fish in our oceans.	autotrophs, heterotrophs, molecular oxygen, food
18. Blue-green algae are associated with bacteria for a number of reasons. They both _____ mitochondria and have _____ nuclear membrane. They are thus both _____ , while other algae and fungi are _____ .	lack, no, procaryotes, eucaryotes
19. Yeast, a member of the fungi class _____ , is used in the production of _____ . Another member of this class is used in the production of the antibiotic _____ .	Ascomycetes, alcohol, penicillin
20. The cell wall of fungi is usually composed of one of two subtances; one of them is _____, the most abundant consituent of all plant cell walls, and the other is _____, the same constituent that makes the _____ of some insects, all lobsters, and shrimp.	cellulose, chitin, exoskeleton
21. In its animal-like stage, a true slime mold is a diploid, ameoboid-like mass called a _____. It feeds via _____. When its food supply runs short, if reproduces by haploid spores contained in a _____. This type of reproduction is similar to that of _____ . Cellular slime molds differ from the animal-like stage of true slime molds only in the fact that the former are _____ .	plasmodium, phagocytosis, fruiting body, fungi, haploid
22. Digestive products are absorbed by the cell membrane and cell wall of fungi by structures known as the _____ .	haustoria
23. Lichens can be divided into two classes according to their appearance. _____ lichens have a shrublike, erect appearance, while the _____ lichens grow very close to the surface, appearing almost flat. The structures that attach a lichen on to the surface to which it is growing are specialized hyphae called _____ .	Fruticose, crustaceous rhizoids

	Answer To be covered when testing yourself
24. The special feature of the euglenoid cell membrane that allows it to change shape and be mobile, even in mud, is called the _____ . Its chemical make-up is _____ in nature.	pellicle, protein
25. An organism which has a single set of chromosomes is termed _____ , while one with two sets of chromosomes is _____ . A _____ is a reproductive cell and is always _____ . When two of the aforementioned cells fuse, a _____ is formed and is always _____ .	haploid, diploid, gamete, haploid, zygote, diploid
26. In the volvocine series of chlorophyta, the correct sequence of increasing differentiation, by stages, is as follows: _____ , _____ , _____ , and, _____ .	gonium, pandorina, pleodorina, volvox
27. At the level of the genus Volvox, occurs the first appearance of two cell types; the _____ cells are incapable of reproduction, while other cells are ____ for reproductive purposes. The female gametc is produced by a cell called an _____ , while the free-swimming male gametes are produced by a cell called an _____. Fertilization occurs _____ the parent colony.	vegetative, specialized oogonium, antheridium, outside
28. The filamentous, multicellar algae ulothrix, reproduces by a method known as _____ . Although this is a form of _____ , the fusion of gametes can only take place between two different "mating types" of gametes termed _____ and _____ . In this way, it can also be considered a primitive form of _____ .	hetero-thallism, isogamy, plus, minus, heterogamy
29. Among the similarities between chlorophyta and higher plants, is the fact that besides posessing the pigment _____ , they both utilize _____ . Brown algae, on the other hand, posess _____ , while rhodophyta only posess and utilize _____ .	chlorophyll a, chlorophyll b, chlorophyll c, chlorophyll a

Determine whether the following statements are true or false.

30. Thallophytes are considered "lower" forms of plant life. This is a reflection not of their lack of adaptation to their environment, but rather a reflection of their similarity to primitive forms of plant life.	True
31. Although the original classification of algae was done according to color, later studies have revealed that this type of classification is no longer valid.	False

	Answer To be covered when testing yourself
32. After diatoms die, their shells form a sediment near ocean's bottom that serve no purpose and can, if in excess, become a nuisance.	False
33. Some fungi posess the unique characteristic of causing disease, while others can prevent disease.	True
34. Fungi rarely grow in the ocean because in that particular environment, it is difficult for them to obtain their food, as they must get it directly by growing on or inside a food source.	True
35. Fungi that are multicellular are made up of branching filaments called hyphae. A mass of these filaments together is called a mycelium.	False
36. During its life cycle, a cellular slime mold might be difficult to distinguish from an amoeba.	True
37. An example of a "fruiting body" is a mushroom. Once it is formed, it detaches completely from the non-reproductive part of the basidiomycetes.	False
38. All algae, with the exception of the blue-green, are autotrophs, and obtain their energy via sunlight captured by chlorophyll contained in chloroplasts. Fungi, on the other hand, are heterotrophs, and must obtain their energy from the breakdown of either live organic matter, or dead organic matter.	True
39. A single euglena can be either heterotrophic or autotrophic, but never both at the same time.	False
40. Algae and fungi are both capable of sexual and asexual reproduction. Sexual reproduction involves fusion of haploid gametes, while asexual reproduction involves production of daughter cells called zoospores.	True
41. When referring to the volvocine series of chlorophyta, it can be stated that with increasing differentiation, there is an increasing dependence of one cell type on another cell type.	True
42. Sexual reproduction occurs outside the parent colony in all the genera of green algae.	False

	Answer To be covered when testing yourself
43. Referring to the multicellular algae ulva, although there is separation of vegetative and reproductive cells, a vegetative cell can become a reproductive cell.	True
44. On a strictly biochemical level, chlorophyta are more similar to rhodophyta and chrysophyta than they are to higher plants.	False

CHAPTER 7

THE BRYOPHYTES AND LOWER VASCULAR PLANTS

ENVIRONMENTAL ADAPTATIONS

• PROBLEM 7-1

Discuss the adaptations for land life evident in the bryophytes.

Solution: The bryophytes are considered one of the early invaders of the land. In order to survive on land, a plant must have certain structures which will enable it to exploit a terrestrial environment for food, water, gases and to afford it protection against environmental hazards. Although the bryophytes are far from being truly terrestrial and retain a strong dependency on a moist surrounding, they have accumulated important adaptations enabling them to live successfully on land.

Because land plants are removed from a water environment, they face the potential danger of dehydration due to evaporation. To reduce water loss, most bryophytes have an epidermis. In the mosses, it is thickened and waxy, forming a cuticle. Unlike aquatic plants, which obtain and excrete gases dissolved in water, land plants require an efficient means to exchange gases with the atmosphere. The epidermis of the bryophytes is provided with numerous pores for the diffusion of carbon dioxide and oxygen. Diffusion through pores is a much faster and more efficient process for gaseous exchange than simple diffusion through membranes.

Besides gases, land plants need to obtain water, which may be a limiting factor in a terrestrial environment. Most bryophytes absorb water and minerals directly and rapidly through their leaves and plant axis, which may have a central strand of thin-walled conducting cells. Generally, the plant is attached to the substrate by means of elongated single cells or filaments of cells called rhizoids. In those mosses having a cuticle on their leaves, the rhizoids may function to some extent in water absorption. In most bryophytes, however, the rhizoids are not true roots and serve only in anchoring the plant.

In addition to structures, the bryophytes have evolved a life cycle in which the developing zygote is protected within the female gametophyte. Here the zygote obtains food and water from the surrounding gametophytic tissue and is protected from drought and other physical hazards present in a terrestrial environment. The bryophyte sperm cell is flagellated and requires a moist medium for its transport. Fertilization, however, can occur after a rain or in heavy dew. The sperm cell swims, in a film of moisture, to the female gametophyte (archegonia) where gametic union occurs.

Because the bryophytes have acquired rhizoids, cutinized epidermis, a porous surface, and a reproductive style in which the embryo is protected within the female gametophyte, they are able to succeed in a terrestrial environment.

• PROBLEM 7-2

Why is a moss plant restricted to a height of less than about 15 cm?

Solution: A moss plant is rarely more than 15 cm tall due to the inefficiency of the rhizoids in absorbing water and minerals and the absence of true vascular and supporting tissues. The rhizoids of the moss are small, slender filaments of cells. Because of their small size and simplicity, they can provide anchorage and absorb enough water and salts only for a small plant. Moreover, the moss plant, like other bryophytes, lacks a vascular system. With no xylem to conduct water and minerals, and no phloem to transport photosynthetic products, the moss plant is necessarily restricted to short-distance internal transport. An unusually large moss plant would soon face water and food shortage, and would eventually starve to death. In addition, since xylem and phloem also function in supporting the plant body, the moss plant lacks sufficient support to enable it to attain a large size.

The reproductive process of the mosses also restricts their height. Fertilization must occur in a moist environment, since water must be present to provide a medium through which the sperm can swim. A low-growing plant body, being closer to the moist soil, is advantageous for the moss. A plant that is too tall would risk loss of its sperm by dispersing them into a non-fluid environment.

• PROBLEM 7-3

Why must mosses and liverworts (phylum Bryophyta) always live in close association with the water?

Solution: The bryophytes are generally considered to be primitive land plants because they are relatively ill-adapted to the terrestrial environment compared to the higher land plants. Mosses, for instance, are found frequently on stream banks or moist roadsides.

Liverworts, lacking a cuticle, are not as well protected against dessication as are the mosses, and are even more restricted in their distribution - the majority of them grow in moist, shady localities, and some are even true aquatic plants.

The fact that mosses and liverworts always live in close association with water can be explained by their anatomical structures and reproductive mechanism. They have rhizoids which are simple filaments of cells or cellular projections performing the function of water absorption. The rhizoids are, however, not efficient absorbers and in relatively dry areas cannot withdraw adequate materials from the ground. In addition, mosses and liverworts lack vascular tissues to conduct water, minerals, and organic substances to different parts of the plant. Therefore transport of materials in the mosses and liverworts depends to a great extent on simple diffusion and active transport through the leaves and plant axis. Furthermore, mosses and liverworts are small plants with a high surface to volume ratio. This means that evaporation can cause a rapid dehydration of inner as well as outer tissues. These disadvantages can be avoided in a moist habitat where evaporation by the atmosphere is slower and sufficient water is available through diffusion to compensate for water lost through the surface.

A moist habitat also favors the reproductive process of the bryophytes. The gametophyte plant produces flagellated sperm which can swim to the egg only when water is present. In order to reproduce successfully, mosses and liverworts must grow in close proximity to water.

• PROBLEM 7-4

In what ways do mosses and liverworts resemble and differ from each other?

Solution: Mosses and liverworts are two classes of bryophytes. Being in the same phylum, they resemble each other in several respects. First, they share an ability to live and reproduce on land. Second, they have evolved similar structures to help them survive on land: an epidermis to prevent excessive evaporation, pores on the surface to effect gaseous exchange, rootlike projections called rhizoids for anchorage, and small green leaflets to manufacture food. Third, the two classes have a similar life cycle in which the gametophyte is the dominant generation and the sporophyte is partially dependent upon the gametophyte for water, minerals, and anchorage. Fourth, both produce their sex cells in multicellular sex organs (called an antheridium in the male and an archegonium in the female) and have flagellated sperm requiring a moist medium for transport. Last, the embryos of both develop protected within the archegonia.

However, the mosses and liverworts have certain structural differences which separate them into these two distinct classes. First, the moss plants have an erect stem supporting spirally-arranged leaves. The more primitive liverworts are simply flat, sometimes branched, ribbonlike structures that lie on the ground, attached to the soil by numerous rhizoids. Second, the rhizoids of the liverworts are specialized, unicellular projections extending downward from the leaf-like plant body (thallus). The rhizoids of the mosses, on the other hand, are composed of filaments of cells extending from the base of the stem. Third, the liverworts' thallus, unlike the leaflets of the moss, is more than one cell layer thick and bear scales, frequently brown or red, on its lower, ground-facing surface. On the opposite, upper surface are small cups, known as gemmae cups. These form oval structures, the gemmae, which when detached from the parent plant give rise asexually to new gametophyte individuals. Finally, whereas in the mosses, the sex organs are borne at the tip of the stem, in the liverworts they are embedded in deep, lengthwise depressions or furrows in the dorsal surface of the thallus.

• PROBLEM 7-5

Describe sexual reproduction in the mosses.

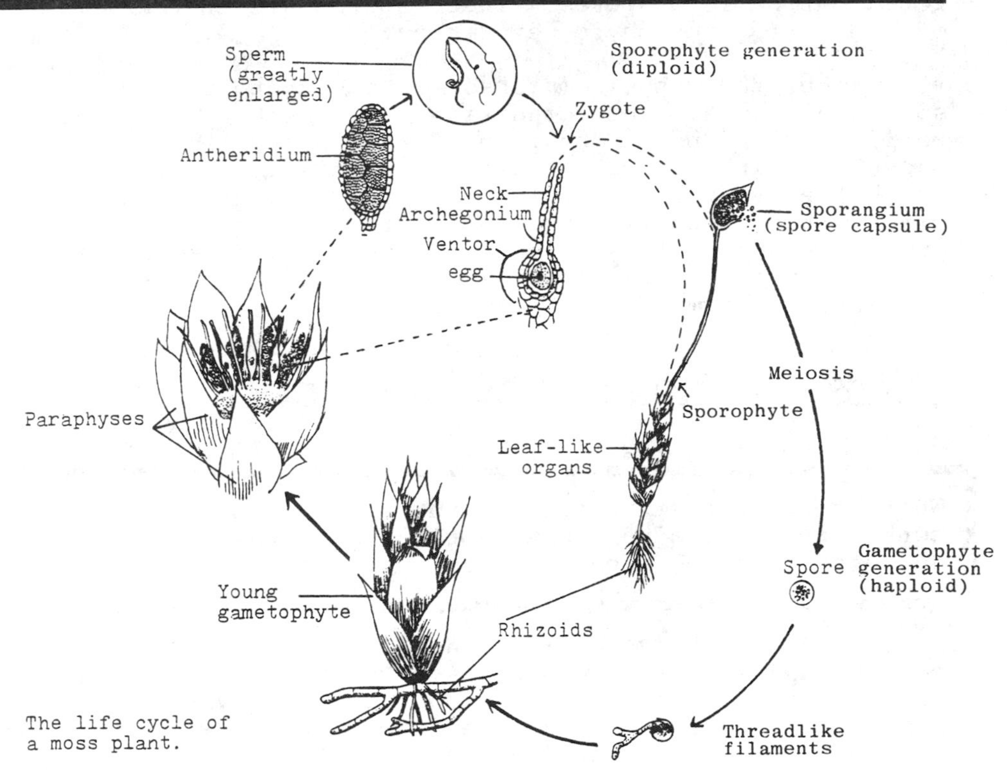

The life cycle of a moss plant.

Solution: The moss plant has a life cycle characterized by a marked alternation between the sexual and asexual generations. The sexual gametophyte generation is the familiar, small, green, leafy plant with an erect stem held to the ground by numerous rhizoids. When the gametophyte has attained full growth, sex organs develop at the tip of the stem, in the middle of a circle of leaves and sterile hairs called paraphyses. The male organs are sausage-shaped structures, called the antheridia. Each antheridium produces a large number of slender, spirally-coiled swimming sperm, each equipped with two flagellae. After a rain or in heavy dew, the sperm are released and swim through a film of moisture to a neighboring female organ, either on the same plant or on a different one. The female organ, the archegonium, is shaped like a flask and has one large egg at its broad base, the ventor. The archegonium releases a chemical substance that attracts the sperm and guides them in swimming down the archegonium to the ventor. Here one sperm fertilizes the egg. The resulting zygote is the beginning of the diploid, asexual sporophyte generation.

The mature sporophyte is composed of a foot embedded in the archegonium and a leafless, spindle-like stalk or seta which rises above the gametophyte. The sporophyte is nutritionally dependent on the gametophyte, absorbing water and nutrients from the archegonium via the tissues of the foot. A sporangium, called the capsule, forms at the upper end of the stalk. Within the capsule each diploid spore mother cell undergoes meiotic division to form four haploid spores. These spores are the beginning of the next gametophyte generation.

When the capsule matures, it opens and releases spores under favorable conditions in a mechanism specific to its kind. In the Sphagnum, for example, the mature capsule shrinks, bursts, pushes away the lid, and exposes the spores to the wind. If a spore drops in a suitable place, it germinates and develops into a protonema, a green, creeping, filamentous structure. The protonema buds and produces several leafy gametophytes, thereby completing the moss life cycle.

CLASSIFICATION OF LOWER VASCULAR PLANTS

• PROBLEM 7-6

What are the lower vascular plants? Why are they considered more advanced than the bryophytes but less advanced than the seed plants?

<u>Solution:</u> The vascular plants (phylum Tracheophyta) are subdivided into five groups. These are the Psilophyta (fork ferns), Lycophyta (club mosses, quill worts) Sphenophyta (horsetails or the "scouring rushes"), Pterophyte (ferns) and Spermophyta (the seed plants gymnosperms, and angiosperms). Vascular plants, in contrast to the algae, fungi, and bryophytes, posses a vascular system that serves for support and for the conduction of water, mineral salts, and foods. The lower vascular plants refer to all the

non-seed bearing classes, namely all classes except the gymnosperms and angiosperms.

Most of the lower vascular plants are land plants. As mentioned before, they contain xylem and phloem, a system missing in the bryophytes. They have also evolved a more complicated anatomy than the bryophytes - more efficient roots, more elaborate stem and leaves, more complicated embryonic structure and a larger body size. The sporophyte of the lower vascular plants has become the independent, dominant generation, and correspondingly, the gametophyte has become reduced. These characteristics of the lower vascular plants place them higher on the evolutionary ladder than the bryophytes.

Structurally, most lower vascular plants resemble the seed plants far more than the bryophytes in their roots, stems and leaves. However, they are not that much better adapted for a life on land than the bryophytes. Because their sperm retain the biflagellated structure necessitating a film of moisture to be active, and because the young sporophyte develops directly from the zygote without passing through any stage where it is protected by a seed, these plants are restricted to moderately moist habitats for active growth and reproduction. Most of the seed plants, on the other hand, have evolved nonflagellated sperm, a mechanism of pollination for sperm transport and gametic union, and a seed structure within which the embryonic sporophyte is protected and nourished. These three features make the seed plants extremely successful inhabitants of the land. In addition, while the lower vascular plants have a reduced gametophyte, the seed plants have reduced it even further in size and have simplified its structure to the point where it is completely dependent on the sporophyte. The seed plants have also evolved heterospory, that is, the production of two types of spores. The lower vascular plants are homosporous, and produce only one type of spore.

The seed plants are therefore most advanced in the evolutionary trend toward greater embryonic protection, flexible and efficient ways of fertilization, the production of two types of spores, and the reduction of the haploid generation. While the lower vascular plants, then, are more advanced than the bryophytes, they trail the seed plants in evolutionary development.

DIFFERENTIATION BETWEEN MOSSES AND FERNS

• PROBLEM 7-7

Describe the life cycle of Selaginella.

Solution: Selaginella is a member of the subphylum Lycophyta. Its dominant sporophyte generation generally consists of a branched, prostrate stem with short upright branches, normally only a few inches high. Both horizontal and upright stems are sheathed with small leaves in four longitudinal rows or ranks. At the ends of the upright branches, reproductive leaves, called sporophylls

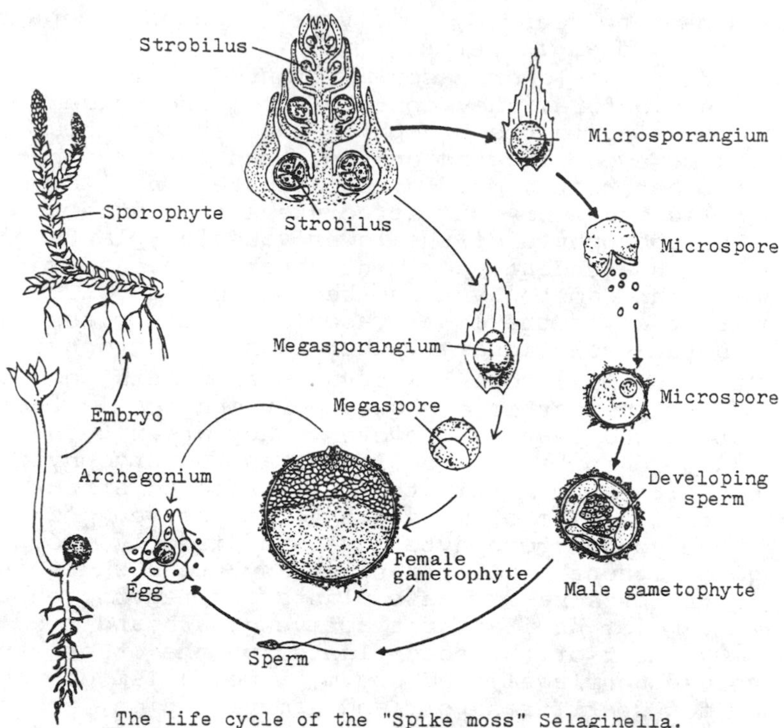

The life cycle of the "Spike moss" Selaginella.

group to form cones or strobili. Sporangia grow in or near the axils of the sporophylls. Two types of sporangia are present in Selaginella: the microsporangia, which produce small microspores; and the megasporangia, which produce large megaspores. A single strobilus usually contains both types of sporangia and thus produces both types of spores.

In a developing megasporangium, all but one of the spore mother cells degenerate. This remaining cell undergoes meiosis and forms four megaspores. Each megaspore is capable of giving rise to a female gametophyte. In the microsporangium, only a few spore mother cells degenerate. Each of the 250 or so cells that remain gives rise to four microspores by meiosis. All microspores are capable of developing into the male microgametophyte.

Repeated mitotic cell divisions of the megaspore result in the formation of the female gametophyte within the megaspore cell wall. The megaspore may be shed from the strobilus at any stage in the development of the megagametophyte, or may remain in the strobilus well after fertilization until the completion of early embryonic development. The material stored in the megaspore is the major source of food for the female gametophyte and the developing embryo. The gametophyte increases in size, eventually rupturing the megaspore wall, and the small, colorless megagametophyte tissue protrudes, exposing the archegonia.

The haploid microspore also undergoes mitosis and

forms two cells. The smaller cell, the prothallial cell, remains vegetative and does not divide further. The other cell, by repeated divisions, develops into a gametophyte, or antheridium, composed of a jacket of sterile cells enclosing either 128 or 256 biflagellated sperm, the gametophyte remains in the microspore cell, which is released from the microsporangia and may drop near a megaspore either on the ground or on a strobilus. When wet by dew or rain, the microspore wall splits, and the sperm within are free to swim to the megagametophyte and fertilize the haploid egg, thus initiating the diploid sporophyte generation.

Of the two cells formed by the first division of the zygote, only one develops into an embryo. The other grows into an elongated structure, the suspensor, which pushes the developing embryo into gametophytic tissue for food supply. The young Selaginella soon acquires its own food-making mechanism and becomes an independent sporophyte plant. In some species, as in the seed plants, the embryo is produced in the female gametophyte while it is still within the sporophyte.

• PROBLEM 7-8

How do the fork ferns differ from the true ferns? What unusual feature is present in Psilotum?

Solution: The fork ferns (Psilophyta) and true ferns (Pterophyta) are classified separately primarily because of the difference in their leaf structures. The leaves of the fork ferns are small, simple, scale-like, and are considered mere epidermal outgrowths rather then true leaves. The leaves of the true ferns are relatively large, elaborate, vein-containing, and usually compound, being finely divided into pinnae. The fork and true ferns also differ in that the well-developed roots present in the true ferns are lacking in the fork ferns. The latter have instead, numerous unicellular rhizoids which grow off the rhizome. The rhizome of the fork ferns is usually found in association with a fungus. Sporangia of the fork ferns are borne in axils of some of the scale-like leaves; sporangia of most true ferns are, however, carried on the undersurfaces of the leaves.

Psilotum, one of the two existing genera of fork ferns, is of particular interest to botanists because both its gametophyte and sporophyte have vascular tissues. Whereas the gametophyte of the bryophytes, seed plants, and almost all lower vascular plants lack a vascular system, the gametophyte of Psilotum has a stele complete with xylem and phloem surrounded by an endodermis.

• PROBLEM 7-9

Describe the life cycle of a typical fern plant.

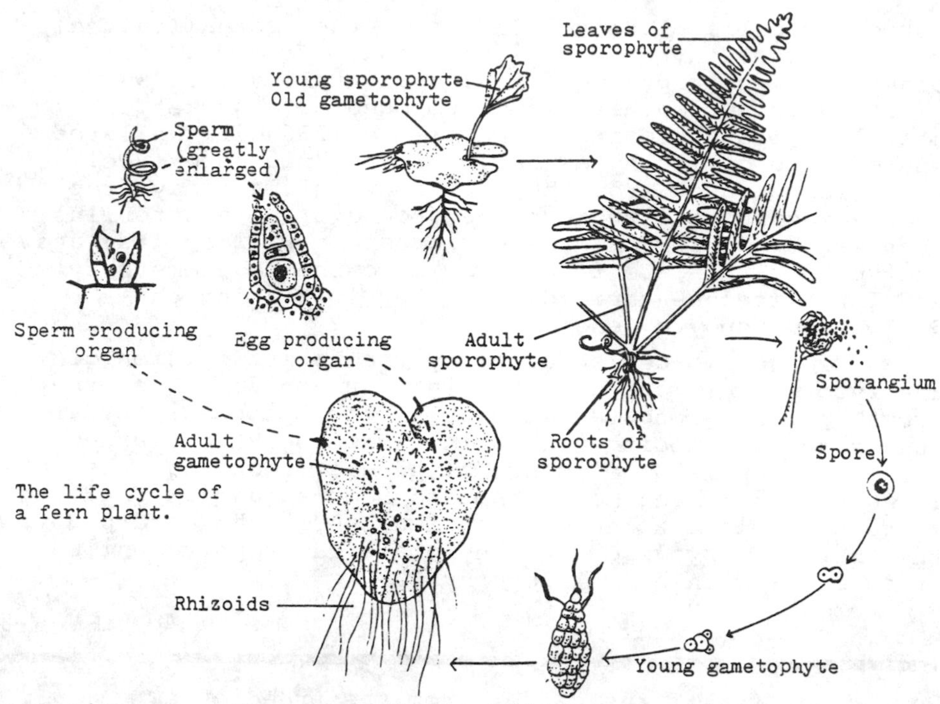

The life cycle of a fern plant.

Solution: The fern plant, like other vascular plants, has a life cycle in which the sporophyte generation is dominant. The sporophyte generation is the relatively large, leafy green plant we recognize as the fern plant growing in both the tropics and temperate regions. The sporophyte consists of a horizontal stem, or rhizome, lying at or just under the surface of the soil and bearing fibrous roots. From the rhizome grows several leaves or fronds, usually finely divided into pinnae. On the under surfaces of certain pinnae develop small, brown sori. A sorus is a cluster of sporangia or spore cases in which haploid spores are formed. Most modern ferns are homosporous, that is, all their spores are alike. When mature and proper conditions, the spores are shed. After germination, each spore develops into a gametophyte which is typically tiny, thin, and often heartshaped. Small and obscure as it is, the fern gametophyte, also called a prothallus, is an independent photosynthetic organism. It grows in moist, shady places, especially on decaying logs and on moist soil and rocks. A number of rhizoids grow from the gametophyte into the substrate, anchoring it and absorbing water and salts.

The male and female sex organs (antheridia and archegonia) develop on the under-surface of the gametophyte. Each archegonium, usually located near the notch of the heart-shaped prothallus, contains a single egg. The antheridia, located at the other end of the gametophyte, each develop a number of flagellated sperm, which are ovoid in shape and have many flagellae on a spiral band at their anterior end. The sperm, released after a rain and attracted by a chemical substance released by the archegonium, swim through a film of moisture to the egg. Although the fern plant is usually monoecious (that is, it possesses both the male and female sex organs), the sperm of one plant usually fertilizes the egg of another, thus accomplishing cross-fertilization.

The zygote begins to develop within the archegonium into a sporophyte embryo. At first the sporophyte develops as a parasite on the gametophyte, but it soon acquires its own roots, stems, and photosynthetic leaves and becomes an independent sporophyte, thus completing the life cycle.

• PROBLEM 7-10

What are spores? How are they formed in mosses and in ferns?

Solution: A spore is a haploid, usually unicellular structure formed by reduction division in the sporangium of a sporophyte. When released, it is capable of germinating and developing into an entire gametophyte individual, without fusion with another cell. This is in direct contrast to gametes, which, though also haploid, require fusion with another gamete in order to form a new individual, in this case, the sporophyte.

The spores of the moss are formed in the sporangium, or capsule, of the moss sporophyte. The capsule is surrounded by an epidermal layer composed of cells similar to those found in the epidermis of higher plants. The inner portion of the sporangium consists of a layer of sterile, supporting tissue covering a core of fertile sporogenous cells. Spore mother cells, each containing the diploid number of chromosomes, develop from the sporogenous cells. Meiosis follows, and each diploid spore mother cell gives rise to four haploid cells, known as the spores. They are the first cells of the gametophyte generation. When the capsule matures, the upper end forms a lid, which drops off, permitting the liberation of spores which are then likely to be dispersed by wind. If a spore falls on a good spot it will, under favorable conditions, germinate and develop into a new gametophyte individual.

The fern is a vascular plant, and, unlike the moss, which has a dominant gametophyte generation, the fern has a dominant sporophyte and an inconspicuous gametophyte generation. The relatively large, leafy green plant we call the fern is the sporophyte generation. Each leaf, or frond, of the fern is subdivided into a large number of leaflets. On the undersurfaces of certain leaflets develop clusters of small, brown sporangia. Depending on the genera and species, sporangia may be grouped in sori or grow on definite margins or edges of the leaflet. Within the sporangia, haploid spores are produced from spore mother cells by meiosis. The spores are released at the proper time, fall to the ground, and directly develop into small, flat, green, photosynthetic, heart-shaped gametophytes.

• PROBLEM 7-11

What features enable the ferns to grow to a much larger size than the mosses?

Solution: The ferns (Division Pterophyta) are placed higher up on the evolutionary ladder than the mosses. One obvious difference between the two lies in their sizes. While the mosses rarely grow beyond 15 cm in height, ferns have been found that reach a height of 16 meters, larger by a factor of 100. Ferns are able to attain these heights because of the presence of both more efficient roots and a vascular system. The root of a fern is much more advanced than the rhizoid of a moss. It is elaborated into tissues and varying zones of maturity. From the bottom up, it is composed of a protective cap, an apical region of rapid cell division, a zone of elongation, and above it, a zone of maturation. To increase the total surface area for absorption, the primary root branches and rebranches to form many smaller roots. This well-developed root system of the ferns, in contrast to the simple rhizoids of mosses which are merely filaments of cells, allows for firmer anchorage and more efficient absorption of materials as demanded by a larger plant. The absorbed water and minerals are then conducted up the stem and leaf petioles of the fern to the leaves by the xylem of the vascular system. In addition to xylem, there is the phloem, which transports organic products synthesized in the leaves down to the stem and roots. The xylem and phloem tissues, present in the ferns but not the mosses, make possible long-distance transport of the essential materials required by a big plant. Since the vascular system serves also as supportive tissue, it adds strength and rigidity to a fern plant, enabling it to grow to a large size.

In summary, because of the presence of an elaborate root system and a vascular system, the ferns are able to grow to a much larger size than are the mosses.

• PROBLEM 7-12

What evolutionary advanced features are present in Selaginella but not in the ferns?

Solution: The genus Selaginella has four advanced features which are not found in most of the other lower vascular plants, including the ferns. First, vessels which are missing in all but two genera of ferns are found in the xylem of many species of Selaginella. Second, the Selaginella gametophyte is greatly dependent on the sporophyte, it is highly reduced in size and complexity, and it developes entirely within the spore cell. The fern gametophyte, though also small and inconspicuous, is, however, a photosynthetic, independent organism.

Third, Selaginella has acquired heterospory, an evolutionary advanced feature of higher plants. In contrast to the ferns which are homosporous, Selaginella bears two types of sporangia and gives rise to two types of spores. As their names imply, the megasporangium forms larger spores called magaspores, and the microsporangium forms smaller spores called microspores.

Last, the Selaginella embryo, unlike that of the ferns, is equipped with a suspensor, a footlike structure that grows into the surrounding gametophytic tissue to absorb food. A similar, analogous structure by the same name is found in the gymnosperm embryo. These four advanced features of the Selaginella discussed above suggest that Selaginella may have evolved in a closer line with the higher plants than have the ferns.

COMPARISON BETWEEN VASCULAR AND NON - VASCULAR PLANTS

• PROBLEM 7-13

How does asexual reproduction take place in the bryophytes? In the lower vascular plants?

Solution: The mosses and liverworts carry out asexual as well as sexual reproduction. The young gametophyte of the moss, the protonema, is derived from a single spore but may give rise to many moss shoots simply by budding, a process of asexual reproduction. Some liverworts, such as Marchantia, form gemmae cups on the upper surface of the thallus. Small disks of green tissue, called gemmae, are produced within these cups. The mature gemmae are broken off and splashed out by the rain and scattered in the vicinity of the parent thallus where they grow into new plants.

Like the bryophytes, some lower vascular plants also propagate by vegetative reproduction. In some species of club mosses, small masses of tissue, called bulbils, are formed. These drop from the parent plant and grow directly into new young sporophyte plants. The ferns reproduce asexually either by death and decay of the older portions of the rhizome and the subsequent separation of the younger growing ends, or by the formation of deciduous leaf-borne buds, which detach and grow into new plants.

• PROBLEM 7-14

What evidence supports the theory that vascular plants evolved directly from green algae, and not from bryophytes?

Solution: At one time it was believed that the vascular plants evolved from a less advanced bryophyte, such as the liverworts or hornworts. But although there is fossil evidence of true vascular plants in the Silurian period some 360 million years ago, the first evidence of the bryophytes does not appear until the Pennsylvanian period, which began about 100 million years later. The fact that the bryophytes first appear in a later period than the vascular plants makes it unlikely that the vascular plants could have evolved from bryophytes. Since green algae contain pigments (chlorophylls), principal food reserves (starch) similar to those of the higher plants, and since fossil evidence shows they invaded the land some 140 million years earlier than the first group of vascular plants, botanists are now inclined to believe that ancient green algae were the common ancestors of both higher nonvascular and vascular plants. Hence bryophytes and vascular plants are now believed to have derived from green algae and to have diverged in their evolutionary processes long ago.

• PROBLEM 7-15

In what ways do the bryophytes resemble the tracheophytes, and in what ways are they different?

Solution: Botanists place the bryophytes and the tracheophytes (the vascular plants) in the same subkingdom Embryophyta. This is due to the fact that these two divisions of plants have evolved a life cycle in which the zygote is borne within the female sex organ of the gametophyte. Here it acquires protection, water and nutrients while it develops into a multicellular sporophyte embryo. Besides this common characteristic, the bryophytes and tracheophytes resemble each other in other ways. They have structures which are analogous and serve similar purposes. For example, the tracheophytes have roots and the bryophytes have rhizoids, both of which serve for anchorage and water absorption; the former have stomatae and the latter have pores for gaseous diffusion; the tracheophytes have well developed leaves to capture solar energy and manufacture food and the bryophytes have smaller, simpler leaf-like structures for the same purpose; both have an epidermis, which, in the tracheophytes and certain mosses is impregnated with cutin to prevent exce-sive water loss; and finally, both groups live on the land.

Despite these similarities, the tracheophytes and bryophytes are believed to have diverged in their evolutionary development long ago, and so also display significant differences between them. The tracheophytes are taller, larger, and more complicated than the bryophytes. They are organized into complex plant bodies in which many different kinds of tissues and structures can be recognized. They have developed a root system with an extensive surface area to extract large quantities of water and minerals from the soil and have evolved an efficient vascular system to conduct water and nutrients to different parts of the plant. The bryophytes, on the other hand, have only simple rhizoids and no true vascular tissues, and hence have to compensate for these disadvantages with a small body size and height. In addition, the higher tracheophytes (gymosperms and angiosperms) produce non-flagellated sperm, and are not dependent on moist conditions in order for fertilization to occur. Thus, while the bryophytes and lower vascular plants rigidly require a moist environment for sperm transport, the higher tracheophytes have evolved a greater variety of means by which the male sex cell can be carried to the female over long distances, such as by wind, water, insects, and other animals.

The tracheophytes also differ from the bryophytes in their reproductive cycles. The bryophytes have a pattern of alternation of generations in which the gametophyte is dominant; the sporophyte is greatly dependent upon it. On the contrary, the sporophyte dominates in the higher plants and, except for the earliest stages in the formation of an embryo, it is independent of the gametophyte. Correspondingly, the tracheo-gametophyte is highly reduced in size and complexity, and, in the conifers and flowering plants, it is completely dependent upon the sporophyte. The higher tracheophytes have also managed to enhance embryonic protection by the evolution of seeds. Seeds protect the embryonic sporophyte from the rigors of a life on land, nourish it, and allow it to travel over great distances for dispersal of the species.

The similarities and differences between the tracheophytes and bryophytes discussed above lead us to conclude that although both are land plants, the former are more advanced and better adapted to a terrestrial life than the latter.

• PROBLEM 7-16

In what ways do ferns resemble seed plants? In what respects do they differ from them?

Solution: The Pterophyte (ferns) and the Spermophyte (seed plants-gymnosperms and angiosperms) are alike in a number of respects. They are both terrestrial plants and as such have adapted certain similiar anatomical structures. The roots, of both plants are defferentiated into root cap, an apical meristem, a zone of elongation and a zone of maturation. Their stems have a protective epidermis, supporting, and vascular tissues: and their leaves have veins, chlorenchyma with chlorophylls, a protective epidermis and stomatae. The ferns also resemble the seed plants in that the sporophyte is the dominant generation.

The characteristics that distinguish ferns from the seed plants include the structure of the vascular system, the location of the sporangia, the absence of seeds, the structure and transport of sperm, and the patterns of reproduction and development. Unlike the seed plants, ferns have only tracheids in their xylem and no vessels. They bear their sporangia in clusters on their leaves (fronds), in contrast to the seed plants which carry their sporangia on specialized, non-photosynthetic organs, such as the cone scales of a gymnosperm. The ferns produce no seeds and the embryo develops directly into the new sporophyte without passing through any protected dormant stage as seen in the seed plants. The ferns retain flagellated sperm and require moisture for their transport and subsequent fertilization. The seed plants, on the other hand, have evolved a mechanism of gametic fusion by pollination, i.e., the growth of a pollen tube. The pollen tube eliminates the need for moisture and provides a means for the direct union of sex cells.

The ferns also differ from the seed plants in their life cycle. While in both, the sporophyte is the dominant generation, the fern gametophyte is an independent photosynthetic organism whereas the seed plant's gametophyte, bearing no chlorophyll, is parasitic upon the sporophyte. Also, the gametophyte of the seed plant is highly reduced in structure. The cycad (a gymnosperm) male gametophyte, for instance, consists of only three cells. The ferns, furthermore, are unlike the seed plants in being homosporous, that is, producing only one kind of spore. The seed plants, on the contrary, have two types of spores, the larger, female spores and the smaller, male spores.

SHORT ANSWER QUESTIONS FOR REVIEW

Answer

To be covered when testing yourself

Choose the correct answer.

1. The following are all characteristics of spores except: (a) they are haploid (b) they are usually unicellular (c) they are formed by mitosis (d) they germinate and develop into a gametophyte — c

2. Lower vascular plants are evolutionarily more advanced than the bryophytes. This is evident when considering all of the following traits except: (a) their supportive structures (b) their embryonic structure (c) the way in which fertilization occurs (d) their method of absorption — c

3. The young gametophyte of the moss which asexually reproduces many new moss shoots is called the: (a) gemmae (b) antheridia (c) strobilus (d) protenema — d

4. Lower vascular plants include which of the following sets? (a) Psilophyte, liverworts, Sphenophyta, Pterophyta (b) Bryophytes, liverworts, Selaginella, Psilophyta (c) Psilophyta, Lycophyta, Sphenophyta, Pterophyta (d) Psilo phyta, Bryophyta, Sphenophyta, Pterophyta — c

5. The characteristic that makes the Psilotum unique is that (a) it has an alternation of generations (b) both its gametophyte and sporophyte generations have vascular tissue (c) it lacks true roots (d) its leaves are replaced by pinnae. — b

6. Which of the following are common to both ferns and mosses? (a) a size greater than 15 cm (b) xylem tissue (c) chloroplasts (d) cambium — c

Question 7-14 pertain to characteristics of the bryophytes. Match column B to the characteristics in column A.

	A		B	Answer
7.	epidermis	(a)	reduces water loss	7-a,
8.	fertilization	(b)	n	8-e,
9.	gametophyte	(c)	absorbs water	9-b,
10.	rhizoids	(d)	reduced in size and can't exist independently	10-g,
11.	sporophyte			11-d,
12.	pores	(e)	requires moisture	12-f,

Answer

To be covered when testing yourself

				Answer
13.	plant axis	(f)	means by which gases are exchanged with the enviroment	13-c,
14.	height restriction	(g)	anchor the plant	14-h,
		(h)	lack of vascular tissue	

15. Although mosses and liverworts are both bryophytes, they differ in some respects. In the following list, if the trait is characteristic of mosses label it I, if it is characteristic of liverworts label it II, and if it is characteristic of both label it III. (a) gemmae cups (b) the gametophyte is the dominant generation (c) fertilization must take place in a moist environment (d) the sex organs are located at the tip of the stem (e) erect stem (f) the thallus bears scales and is more than one layer thick. (g) epidermis (h) the protenema is responsible for asexual reproduction.

Answer: a-II, b-III, c-III, d-I, e-I, f-II, g-III, h-I

16. Pictured below is a moss gametophyte. Label the indicated structures.

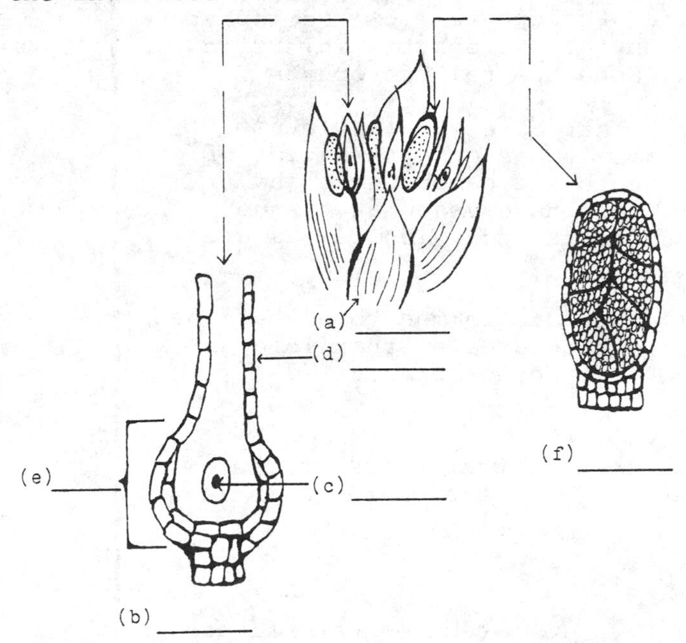

Answer: (a) paraphyses (b) archegonium (c) egg (d) neck (e) ventor (f) antheridium

In 17-23 are listed characteristics of the fern plant. Match column B to column A.

	A		B	Answer
17.	sporophyte generation	(a)	divided into pinnae	17-g,

Answer

To be covered when testing yourself

				Answer
18.	roots	(b)	n	18-f,
19.	antheridia	(c)	prothallus	19-d,
20.	fronds	(d)	male sex organ	20-a,
21.	spores	(e)	anchors the gametophyte generation to the substrate	21-b,
22.	rhizoids			22-e,
23.	gametophyte generation	(f)	anchors the sporophyte generation to the substrate	23-c
		(g)	dominant generation	

24. The following is a diagram of the life cycle of Selaginella. Label the structures indicated.

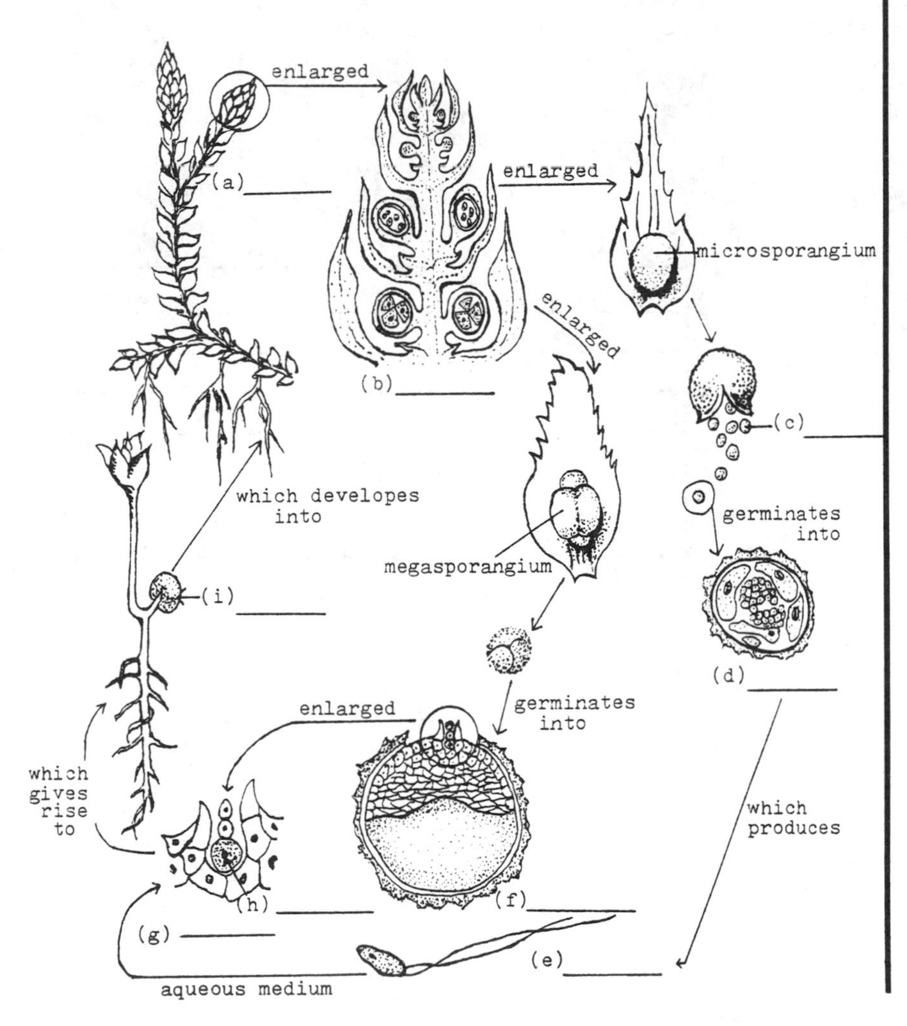

Answer: (a) sporophyte (b) strobilus (c) microspore (d) male gametophyte (e) sperm (f) femalc gametophyte (g) archegonium (h) egg (i) embryo

Answer

To be covered when testing yourself

Fill in the blanks.

25. The fern has a dominant _____ generation which produces _____ from the _____ cells within the _____ by the process of _____. At the proper time the _____ are released, fall to the ground and develop into _____.

sporophyte, spores, spore mother, sporangia, meiosis, spores, gametophytes

26. The following chart concerns three members of the subkingdom Embryophyta. Complete it with this set of words:

A) stomatae
B) rhizoids
C) vascular tissue
D) non-flagellated
E) leaves
F) pores
G) roots
H) dominant sporophyte
I) dominant gametophyte
J) flagellated
K) "leaf-like" structures
L) small body size to compensate for lack of vascular tissue

bryophytes: B,K,F,I,J,L
ferns: G,E,A,H,J,C
tracheophytes: G,E,A,H,D,C

	Bryophytes	Ferns	Tracheophyte
anchorage & water absorption			
photosynthetic organs			
gaseous diffusion			
alteration of generations			
sperm			
conduction			

CHAPTER 8

THE SEED PLANTS

CLASSIFICATION OF SEED PLANTS

• PROBLEM 8-1

What do the words 'gymnosperm' and 'angiosperm' mean? What characteristics distinguish conifers from flowering plants?

Solution: The term gymnosperm means "naked seeds" and the term angiosperm means "enclosed seeds." The gymnosperms and angiosperms are both seed-producing plants which differ in the degree of protective covering provided to the seeds by the structures producing them. The seeds of angiosperms are formed inside a fruit, and the seed covering is developed from the wall of the ovule of the flower. The seeds of gymnosperms are borne in various ways, usually on cones, but they are never really enclosed as are angiosperm seeds. Although embedded in the cone, the seeds lie open to the outside on the cone scales, and are not contained within any modified protective tissue.

Conifers are classified as members of the group gymnosperm, while flowering plants are classified as angiosperms. An easy way to distinguish conifers from flowering plants is to compare the gross anatomy of their leaves. Needle-like leaves having a heavily cutinized epidermis are peculiar to the conifers. Due to their small amount of surface area (as compared to the broad, flat leaves of angiosperms) and their waxy coat, these leaves enable the conifers to survive hot summers and cold winters and to withstand the mechanical abrasions of storms. Most angiosperms cannot. Both conifers and angiosperms utilize xylem for the transport of water. The two types of xylem cells through which water is conducted are tracheids and vessel elements (see 9-16). If we cut open the trunk of a conifer, we will find that its xylem consists almost entirely of tracheids with bordered pits. On the other hand, the xylem of many flowering plants are composed of both tracheids and vessel elements, although the vessel elements predominate. Conifers can also be distinguished from flowering plants by their reproductive structures. Whereas angiosperms produce flowers and fruits, conifers produce cones which are formed from spirally arranged scale-like leaves bearing either seeds or pollen on the inner surfaces. Seed-bearing cones are referred to as female or ovulate cones; pollen-bearing cones are called male or pollen cones. The method of fertilization differs in these two groups as well. The flower of the angiosperm has its pistil constructed in such a way that the germinating pollen tube must grow through both the stigma and the style in order to reach the egg. In the conifers, pollen lands on the surface of the ovule and its tube grows directly into the ovule. Moreover, there is the phenomenon of double fertilization in flowering plants, which gives rise to a diploid zygote and a triploid endosperm. Since fertilization is a single process in the conifers, the endosperm consists of the haploid tissue of the female gametophyte and is thus quite different from the triploid endosperm cells of the angiosperms.

GYMNOSPERMS

• **PROBLEM** 8-2

What is a cone? Describe its structure.

Solution: Cones are the typical reproductive structures of the gymnosperms. A cone or strobilus is a spiral aggregation of modified leaves called cone scales (see figure problem 8-3). These cone scales are the sporophylls, and each scale bears on its surface the sporangia. In most gymnosperms, the microsporangia and megasporangia are borne on separate cones. In the conifers, both male and female cones are borne on the same plants. It would thus appear that self-fertilization occurs. However, the male cones are borne on the lower branches of the conifers, while the female cones are borne on the higher branches. Thus, self-fertilization is essentially prevented. The ovulate, or female cones, are larger. In these cones, the megasporangia are enclosed by enveloping integuments to form structures called ovules. Each scale bears two ovules; each ovule is composed of the outer integuments and the inner nucellus (megasporangium) and has a small opening at one end known as the micropyle, through which the pollen grains will enter. It is from the tissue of the nucellus, that the megaspore mother cell will be produced. It is of interest to note that the megaspore, which gives rise to the megagametophyte, is never released from the megasporangium. It remains embedded in the sporangium, which is enclosed to form the ovule. It is the ovule, which when mature, constitutes the naked seed of the gymnosperms.

Male or staminate cones are smaller and bear the microsporangia on their scales. The number of microsporangia produced by each microsporophyll varies among gymnosperms; in conifers, two is the usual number, while cycads have many microsporangia scattered on the lower surface of each scale. Each microsporangium will produce the microspore mother cell, which will give rise to numerous microgametes or pollen grains.

Describe the life cycle of a pine tree.

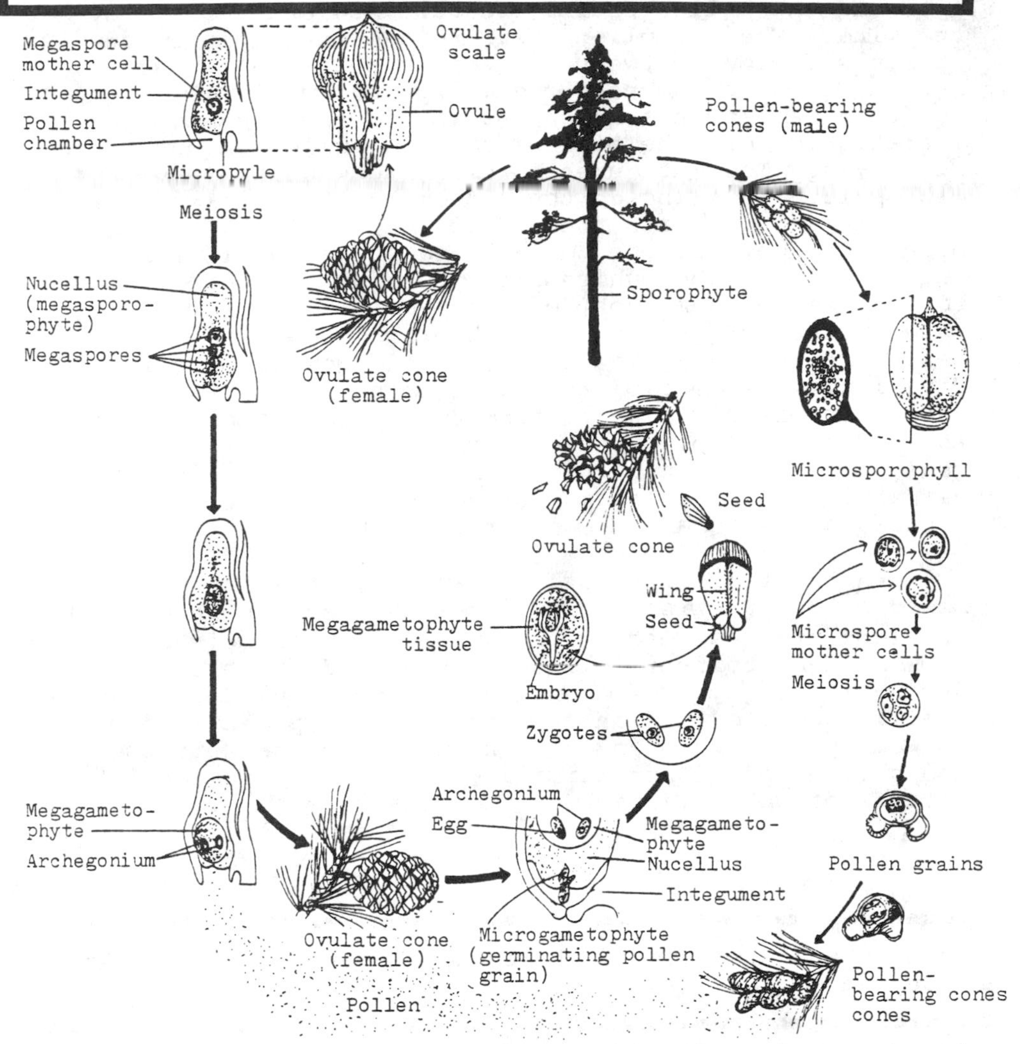

Life cycle of a pine. Note that more than one archegonia may be present and be fertilized within an ovule. However, only one embryo will survive and develop within the seed.

Solution: The pine tree, being a conifer, produces both male and female cones on the same plant. Each cone bears two sporangia on its surface. Within each ovule of the female cone, the nucellus, or megasporangia, contains a single megaspore mother cell. This cell divides by meiosis to form four haploid megaspores. Three of the megaspores disintegrate, leaving one functional megaspore which divides mitotically to form a multicellular haploid megagametophyte. The megagametophyte will form two to five archegonia (female sex organs) each containing a single large egg.

A similar process of gamete production occurs in the staminate (male) cone. Within each microsporangium are many microspore mother cells, each of which undergoes meiosis to form four haploid microspores. While still within the microsporangium, the microspores divide mitotically to form a four-celled microgametophyte or pollen grain. Upon maturity, the microsporangia burst open and the pollen grains are released and carried by the wind. When a pollen grain reaches an ovulate cone it may sift down between the scales and land on the region of the ovule near the micropyle, which is sticky due to a secretion from the nucellus. As the sticky secretion dries, the pollen grain is pulled through the micropyle. The integument swells and closes around the micropyle. Once inside the micropyle, the pollen grain comes in contact with the end of the nucellus. At this point, one cell of the four-celled pollen grain elongates into a pollen tube, which grows through the nucellus toward the megagametophyte. This cell is known as the tube nucleus. A second cell, known as the generative cell, enters the pollen tube and undergoes a mitotic division. Only one of the daughter nuclei is functional, and undergoes division to form two sperm nuclei. When the end of the pollen tube reaches the neck of an archegonia, it bursts open and discharges its sperm nuclei near the egg. One sperm fuses with the egg to form the diploid nucleus, and the other disintegrates.

After fertilization, the zygote, the surrounding megagametophyte tissue, the nucellus, and the integument develop into the seed. The haploid endosperm is derived from megagametophytic tissue. It will nourish the embryo during its early growth and development, in which several leaflike cotyledons, an epicotyl and a hypocotyl form. The embryo then remains dormant until the seed is shed and germinates. Upon germination, it will develop into a mature sporophyte, which is the pine tree.

• PROBLEM 8-4

What unusual feature of the cycads makes them a remarkable group among the seed plants?

Solution: Cycads are remarkable among the seed plants in that they form motile, swimming sperm within their pollen tubes. Of all the living seed plants, only the cycads and gingko, another gymnosperm, possess swimming sperm cells. The sperm formed within the pollen tubes of other seed plants are represented solely by nuclei.

The motile sperm swim by means of flagella, and those of the cycad have thousands of flagella per cell. Each of these flagella have the 9+2 pattern of axial filaments characteristic of eukaryotes, and are attached to a spiral band that encircles the anterior end of the sperm cell. The sperm cells, in addition, are remarkably large, measuring as much as 400 μm in one species, and may be seen with the naked eye. They are the largest motile male gametes known among the higher plants.

When the sperm become active and move about in the

pollen tube, the tube bursts and releases the sperm which then swim briefly in the fluid of the fertilization chamber. The sperm that will fertilize the egg enters the egg cytoplasm and loses its flagella.

The presence of swimming sperm is a primitive condition, characteristic of lower aquatic plants and early terrestrial plants, such as ferns and mosses, which require water for fertilization. Swimming sperm is of no advantage to a seed plant, and virtually all have been eliminated through natural selection as the plants have become more adapted to terrestrial life. For this reason, the cycad is considered to be a primitive gymnosperm which provides a connecting link between the ferns and lycopods and the more advanced seed plants.

ANGIOSPERMS

• PROBLEM 8-5

Describe the parts of a typical flower. What are their functions?

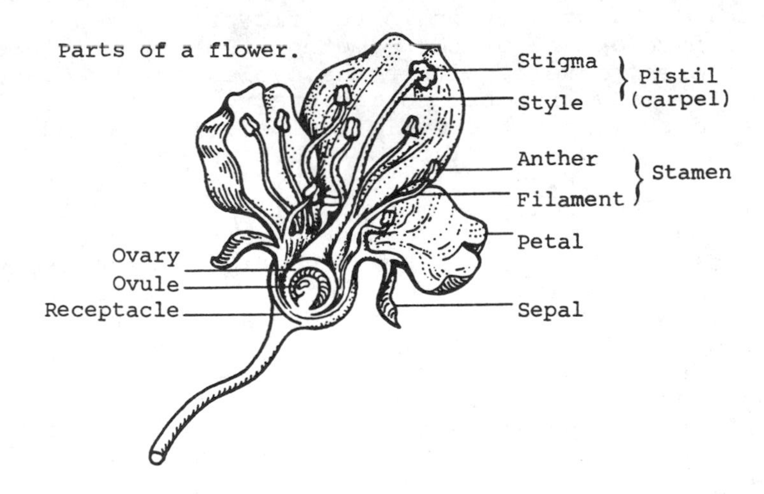

Solution: The flower of an angiosperm is a modified stem on which concentric rings of modified leaves are attached. A typical flower consists of four such rings of structures borne on the receptacle, the expanded end of the flower stalk. The stalk of the flower is also known as the pedicel. The outermost ring, usually green and most like ordinary leaves, is composed of sepals. These enclose and protect the flower bud until it is ready to open. Collectively, the sepals are known as the calyx. Internal to the calyx are the petals, which are often brilliantly colored in order to attract those insects or birds which promote pollination. Flowers which are pollinated by the wind need not be attractive to birds and insects, and so their petals tend to be less showy, and may even be absent.

Inside the corolla lie the stamens, the male reproductive organs of the flower. Each stamen consists of a slender stalk called the filament, which supports an expanded anther at the tip. The immature anther is composed primarily of pollen sacs. After reaching maturity, the anther contains numerous pollen grains, which produce haploid male gametes upon germination. In the center of the flower is a pistil (carpel) or pistils, the female reproductive organ. Each pistil is composed of a swollen portion, the ovary, at its base; a long, slender stalk which rises from the ovary, the style; and on top of this, an enlarged flattened crown called the stigma. The ovary contains one or more ovules, which are the future seeds of the angiosperm. The stigma's function is to secrete a moist, sticky substance to which the pollen grains will adhere. The style provides a lubricated pathway through which the pollen tube of the germinated pollen grain can grow on its way towards the egg cell.

• PROBLEM 8-6

Describe the process of gamete formation and fertilization in the angiosperms.

Solution: Reproduction in the flowering plants begins with the development of the gametes. The female gametes, or megaspores, develop within the ovules of the ovary, each ovule being attached to the ovary by a stalk. Embedded deep within the ovule is one cell, called a megasporangium, which enlarges to become the megaspore mother cell (see Figure 1), from which the gametes will be formed. The megaspore mother cell undergoes meiosis to form four megaspores. In most species, three of these disintegrate, leaving one functioning megaspore. This surviving cell then undergoes three mitotic divisions, producing eight nuclei which migrate so that three position themselves at the far end of the now enlarged megaspore and form the antipodal cells; three move towards the micropyle and form the egg and synergid cells. The synergid and antipodal cells are short-lived; their function is obscure. The remaining two central nuclei are the polar nuclei. The entire mature structure is termed the embryo sac, and becomes enclosed by the integument, layers of cells which develop from the megasporangia surrounding the megaspore mother cell.

The development of the sperm or microgametophytes begins within the tissues of the anther. Each anther typically contains four pollen sacs or microsporangia (see Figure 2). Early in its development, accelerated cell division in the microsporangia produces numerous microspore mother cells. Each of these microspore mother cells undergoes meiosis, resulting in a tetrad of four microspores. Each microspore then undergoes a mitotic division to form a tube nucleus and a generative nucleus. At this point the structure is termed a pollen grain. When the pollen is mature, the anthers split open and

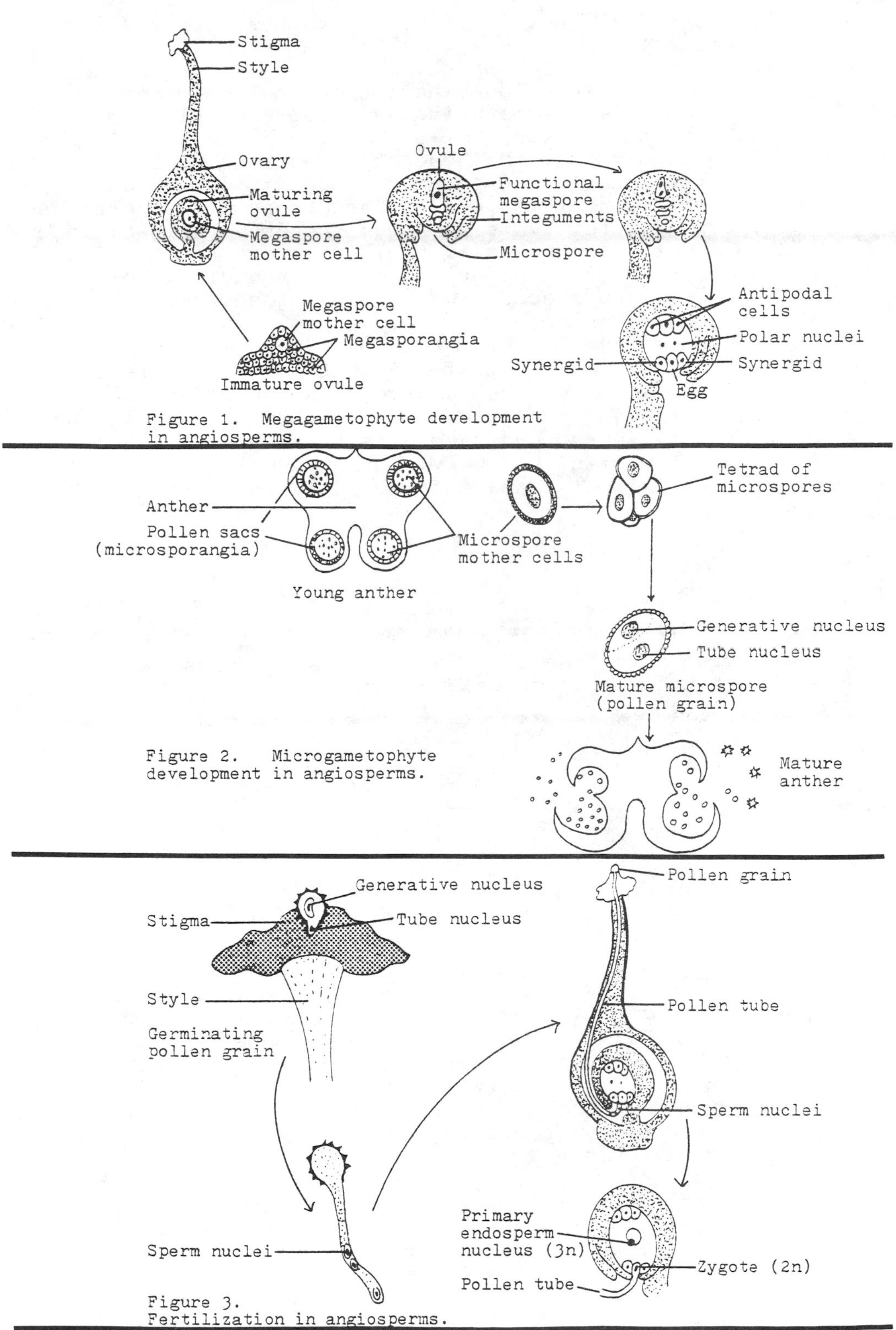

Figure 1. Megagametophyte development in angiosperms.

Figure 2. Microgametophyte development in angiosperms.

Figure 3. Fertilization in angiosperms.

shed the many pollen grains produced. Pollination may now occur by any one of various methods, resulting in the deposition of the pollen onto the stigma of either the same flower (self-pollination) or another flower (cross-pollination).

Upon successful pollination, the pollen produces a pollen tube which digests its way from the stigma down through the style (see Fig. 3) to the ovule. If it has not done so already, the generative nucleus will divide to form two functioning sperm during the journey down the pollen tube to the embryo sac, where they will be released. One sperm will fertilize the egg; the other will migrate and fuse with the two polar nuclei to form the triploid endosperm nucleus. The number of polar nuclei, however, may vary between species, with a consequent variation in the chromosome number of the endosperm. This process of the fusion of sperm nuclei with both the egg and polar nuclei is known as double fertilization. The fertilized egg will develop into the embryo, and the endosperm nucleus into the endosperm, or nutritive tissue for the embryo. The embryo undergoes its first stages of development within the ovary, but eventually becomes dormant and is shed with its surrounding tissues as a seed.

SEEDS

• PROBLEM 8-7

What exactly is a seed? What tissues are present and what are their respective functions?

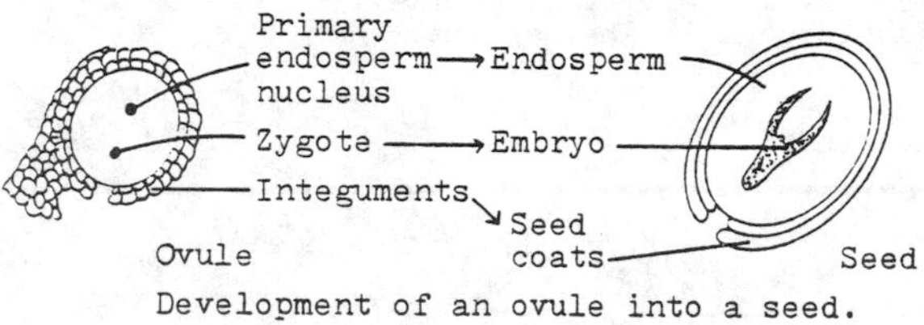

Development of an ovule into a seed.

Solution: A seed is actually a matured ovule, and consists of a seed coat surrounding a core of nutritive tissue in which the embryo is embedded. The seed is an interesting structure in that it is composed of tissues from three generations. The embryo consisting of 2n cells derived from the fusion of egg and sperm, is the new sporophyte generation and functions in the continuation of the species by developing into a new reproducing plant. The nutritive tissue, or endosperm, is derived from the female gametophyte. In gymnosperms it is haploid, but in angiosperms it is triploid, resulting from the fusion of both polar bodies with the sperm. Its high starch content provides a source of nourishment for the growing embryo. The seed coat differentiates from the outer layer of the ovule, known as the integument, and as such is 2n and belongs to the old sporophyte generation. The seed coat encloses the endosperm and embryo, and owing to its tough, resistant properties, protects the seed from heat, cold, desiccation, and parasites.

• PROBLEM 8-8

Describe the development of the seed.

Solution: After fertilization, the zygote undergoes a number of divisions and develops into a multicellular embryo. The triploid endosperm nucleus also undergoes a number of divisions and forms a mass of endosperm cells carrying a high content of nutrients. This endospermal mass fills the space around the embryo and provides it with nourishment. The sepals, petals, stamens, stigma and style usually wither and fall off after fertilization. The ovule with its contained embryo and endosperm becomes the seed; its wall, or integument, thickens to form the tough outer covering of the seed. The seed has an adaptive importance in dispersing the species to new locations and in enabling it to survive periods of unfavorable environmental conditions. This insures that germination will occur only when favorable growth is possible.

• PROBLEM 8-9

What is a fruit? What roles do fruits play in the dispersion of the seeds?

Solution: The ovary of an angiosperm is the basal part of the pistil. The ovary contains the ovules which will become the seeds after fertilization. Concomitant with the development of the zygote into an embryo, the ovary enlarges to form the fruit. A fruit therefore can be defined as a matured ovary, containing the matured ovules. A true fruit is one developed solely from the ovary. An accessory fruit is one developed from sepals, petals, or the receptacle as well as the ovary. The apple, for example, is mostly an enlarged receptacle; only the core is derived from the ovary. All angiosperms have fruits, either true or accessory. This characteristic makes them unique among living things.

Fruits represent an adaptation for the dispersal of the seeds by various means, and they may be classified according to this criterion as wind-borne fruits, water-borne fruits, or animal-borne fruits.

Wind-borne fruits are light and dry so that they can easily be carried by wind. In the tumbleweeds, the whole plant, or fruiting structure, is blown by the wind and scatters seeds as it goes. Other wind-borne fruits, such as the maple, have evolved wing - like structures. Still others, such as the dandelion, develop a plumelike pappus which keeps the light fruits aloft.

Water-borne fruits are adapted for floating, either because air is trapped in some part of the fruit, or because the fruit contains corky tissue. The coconut fruit has an outer coat especially adapted for carriage by ocean currents. Rain is another means of fruit dispersal by water, and is particularly important for plants living on hillsides or mountain slopes.

Animal-borne fruits are mostly fleshy. This makes them appetizing to vertebrates. When fleshy fruits ripen, they undergo a series of characteristic changes, mediated by the hormone ethylene. Among these are a rise in sugar content, a general softening of the fruit through the breakdown of pectic substances, and often a change in color to conspicuous bright red, yellow, blue, or black. When such fruits are eaten by birds or animals, they spread the seeds that lie within-them either passing them unharmed through their digestive tracts or carrying them as adherent passengers on their fur or feathers. Some fruits are further equipped with prickles, hooks, hairs, or sticky coverings, and so can be transported for long distances by animals. The modifications of seeds for dispersal by animals illustrate an evolutionary adaptation to the coexistence of plant and animal forms.

• PROBLEM 8-10

What is dormancy? What factors play a role in the termination of dormancy and the germination of a seed?

Solution: Dormancy is a special condition of rest which enables an embryo to survive long periods of unfavorable environmental conditions, such as water scarcity or low temperature. During this period of rest, the embryo ceases or limits its growth, and metabolizes at very low levels, although ordinary plant rest can be terminated and normal growth resumed by the onset of warmer temperatures or the availability of water. Germination or the resumption of normal growth by a dormant embryo requires certain, very precise combinations of environmental cues. This is of great survival importance to the plant in that it prevents the dormant seed from germinating in response to conditions such as a warm spell in winter, which, although apparently favorable, are only temporary. Should the seed germinate during such a period, the seedling would begin to grow, only to be killed by the ensuing cold and frost. Apparently the dormant seed contains certain endogenous inhibiting factors which must be overcome before germination can begin, and thus requires more complex triggers then just a favorable temperature or water supply.

Some seeds are required to pass through a period of cold before they are able to germinate. This is seen in the seeds of almost all plants growing in areas with marked seasonal variations. Many seeds require drying before they can germinate. Such a requirement has adaptive significance in that it prevents germination within the moist fruit of the parent plant. Some seeds such as those of the lettuce, require a light exposure, while the germination of others is inhibited by light. Certain seeds cannot germinate, even in highly favorable conditions of water, light, oxygen, and temperature, until they have been abraded by factors such as soil action. This abrasion wears away the seed coat, allowing water and oxygen to enter the seed and signal the initiation of germination. The seeds of some desert species germinate only when sufficient rainfall has drained inhibitory chemicals from their seed coats.

Most seeds "force" the embryo into dormancy by the fact that the seed coat is impermeable to undissolved oxygen and the seed itself has a low water content. The seed is protected from freezing and desiccation by the seed coat and provided with nourishment by the endosperm. During germination, the embryo and endosperm absorb water and swell; then the seed coat ruptures, freeing the embryo and enabling it to resume development. It is important to note that most seeds do not need soil nutrients in order to germinate-they germinate equally well on moist paper as in soil.

Dormancy plays a similar role in plant buds.

MONOCOTS AND DICOTS

• PROBLEM 8-11

Discuss the early development of the angiosperm embryo. What primary structures form and what does each become in the seedling?

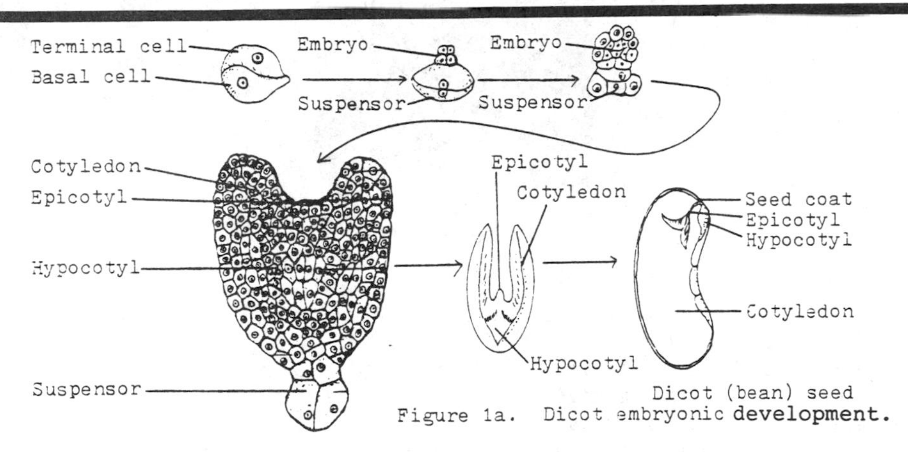

Figure 1a. Dicot embryonic development.

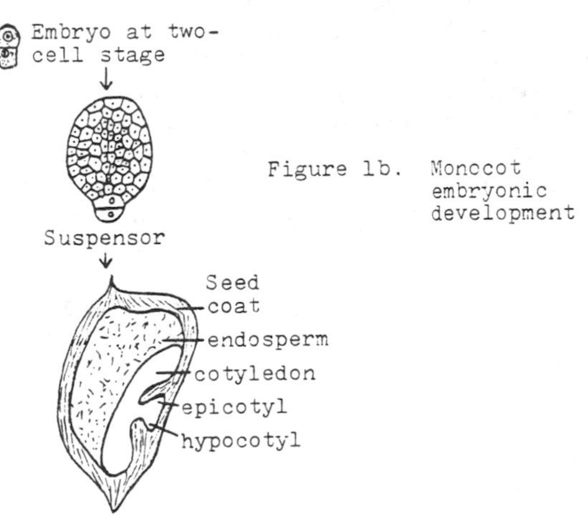

Figure 1b. Monocot embryonic development

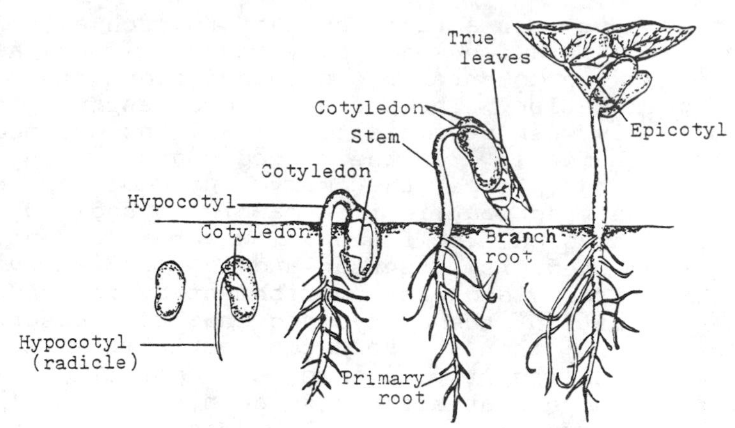

Figure 2a. Germination and early development of a dicot.

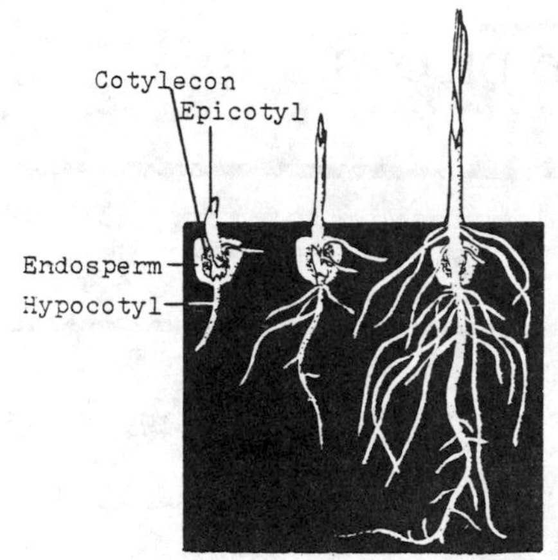

Figure 2b. Germination and early development of a monocot.

Solution: The first cell division that the zygote undergoes following fertilization produces a basal cell and a terminal cell. The basal cell develops into the suspensor, a filament of cells serving as a point of attachment for the embryo within the seed. The suspensor will eventually disintegrate as development of the embryo proceeds. The embryo will develop from the terminal cell, and as the cell undergoes successive divisions, the characteristic structures of the embryo begin to take shape (see Figure 1). The most obvious of these are the cotyledons, or primary leaves. Dicots develop two such leaves, monocots only one. In dicots, the cotyledons can serve to either absorb or to both store and absorb food from the endosperm while within the seed. In the monocots, the single cotyledon serves primarily to absorb, rather than store, the endosperm tissue. The portion of the embryo lying along the central axis below the point of attachment of the cotyledons is called the hypocotyl and the part above is called the epicotyl. At this point in its development, the embryo becomes dormant, and will remain so until conditions are favorable for its germination.

Upon germination, the hypocotyl elongates and emerges from the seed coat (see Fig. 2). It gives rise to the primitive root or radicle. Since the radicle is strongly and positively geotropic, it grows directly downward into the soil. The arching of the hypocotyl in the seed pulls the cotyledons and epicotyl out of the seed coat. The epicotyl, being negatively geotropic, grows upward out of the soil. It will develop into the stem and leaves.

In most dicots, by the time germination occurs, the cotyledons will have completely absorbed the endosperm. They now serve as reserves of food for the growing seedling until it has developed enough chlorophyll to become independent, at which point they shrivel and fall off. In some dicots, the cotyledons do not store nutrient material, but become photosynthetic foliage leaves upon germination. In monocots, the endosperm usually persists even after germination, and the cotyledon continues to absorb the nutrient material for the seedling until it can synthesize its own nutrients.

• PROBLEM 8-12

What are the differences between the growth patterns of dicots that use their cotyledons as an absorption organ and those that use them for both storage and absorption?

Solution: In the dicots, the cotyledons can either serve as only an absorption organ or as both an absorption and storage organ. When the cotyledons function only in absorption, the embryo remains relatively small and is surrounded by endosperm. The cotyledons of these plants absorb the stored food of the endosperm. After the seed germinates, and the endosperm has been depleted, these cotyledons develop into leaf-like photosynthetic organs. In other dicots, such as beans and peas, the cotyledons function in storage as well as absorption of nutrients. The embryo grows until all of the endosperm is absorbed by the cotyledons. Subsequently the cotyledons undertake the function of food reserve, resulting in their appearance as enlarged, thickened structures, which are not photosynthetic in function. In these dicots, the endosperm is usually completely absorbed before the seed germinates. The cotyledons remain as a food supply until the seedling is capable of photosynthesis, at which point they shrivel and fall off. off.

• PROBLEM 8-13

Differentiate between the monocots and dicots.

Solution: The two classes of angiosperms, the monocotyledons and dicotyledons, differ in eight respects. First, the monocot embryo has one cotyledon (seed leaf)

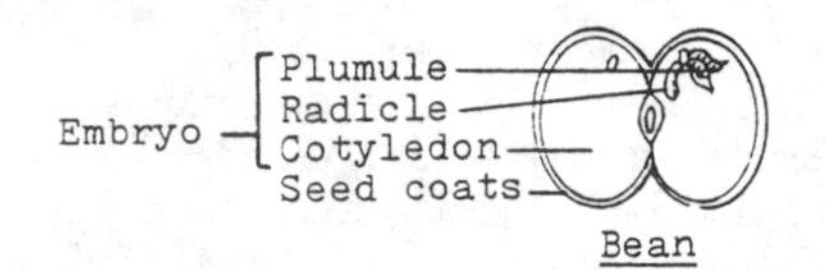

Bean

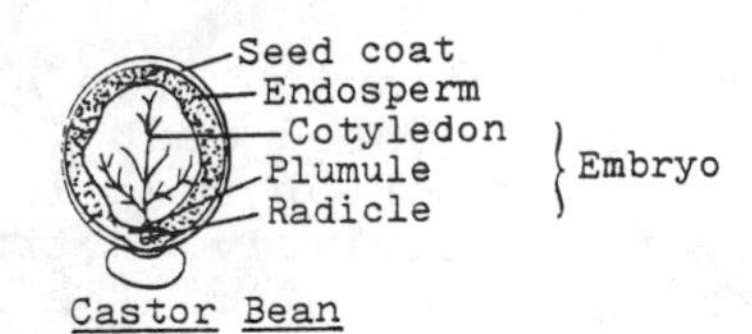

Castor Bean

Fruit
Endosperm
Cotyledon
Plumule
Hypocotyl
Embryo

Corn Grain

Sections through the seeds of three species with large seeds. All the endosperm of the bean used up before it matured, so only the embryo is found inside the seed coats. Note that the outer layer of a corn grain is fruit, not seed coats.

while the dicot has two. The cotyledon of the monocots functions generally in food absorption while that of the dicots can function both in absorption and storage.

Second, the endosperm typically persists in the mature seed of the monocots, while it is usually absent from the mature seed of the dicots.

Third, the leaves of the monocots have parallel veins and smooth edges; dicot leaves have net-like veins and lobed or indented edges.

Fourth, cambium or meristematic cork and vascular tissue is usually present in the stems of dicots but absent in the monocots.

Fifth, the vascular bundles of xylem and phloem are scattered throughout the stem of monocots. In the dicots they occur either as a single solid mass in the center of the stem or as a ring between the cortex and pith.

Sixth, the flowering parts of monocots - petals, sepals, stamens, and pistils - exist in threes or multiples of three , whereas dicot floral parts usually occur in fours, fives or multiples thereof.

Seventh, the monocots are mostly herbaceous plants while many dicots are woody plants.

Finally, the roots of monocots are typically fibrous and adventitious (outgrowths of the stem) whereas the root system of dicots usually consist of one or more primary tap roots and numerous secondary roots.

REPRODUCTION IN SEED PLANTS

• PROBLEM 8-14

Describe some of the methods which insure cross-pollination.

Solution: Cross-pollination involves pollen transfer between plants having different genetic constitutions. Since land plants are immobile, they must evolve specialized features to allow them to mate over distances. The seed plants have evolved structures which promote cross-pollination by wind, insects, bats, birds, and other animals.

Wind-pollination is common in plants having inconspicuous flowers, such as grasses and sedges. Flowers of grasses are dull-looking, and most of them lack both odor and nectar, making them unattractive to insects. Moreover, their pollen is light, dry and easily carried by the wind. Their stigmas are feathery and expose a large surface area to catch pollen. Their stamens are well exposed to the action of the wind. Wind-pollinated flowers, such as those of the grasses, probably represent a special adaptation to colder climates, where insects are less prevalent.

Insect-pollination is the method of pollen dispersal in plants with colorful, showy flowers. These flowers also produce a sweet nectar and volatile compounds having unique odors which attract insects. As an insect approaches a flower, floral architecture ensures transfer of pollen onto the insect's body. In this way, pollen of one flower is carried to the stigma of another as the insect goes from flower to flower.

Flower-visiting bats are also an agent of cross-pollination. Bats are attracted to the flowers largely through their sense of smell. Bat-pollinated flowers characteristically have very strong fermenting or fruit-like odors. Bats fly from plant to plant, eat floral parts, and carry pollen on their fur. Birds may also serve as pollinators. Birds have keen vision, and most bird-pollinated flowers are colorful. Some birds regularly visit these flowers to feed on nectar, floral parts, and flower-inhabiting insects. Pollen transfer is also aided by other animals with fur as they pass from plant to plant.

Certain species of plants are monoecious, meaning they have both sexes contained within the same plant. In order to effect cross-pollination and avoid inbreeding within the same plant, specific tactics have been devised. In some species, such as the goldenrod (Solidago), the male and female organs of the same plant mature at slightly different times, making self-pollination unlikely. In some other species, the pollen is unable to germinate on the same plant. In plants such as these, bisexuality has an advantage in that a pollinator can both pick up and deliver pollen at the same time.

• **PROBLEM** 8-15

"The ovule of a rose is analogous to the capsule of a moss." Explain what is wrong with this statement.

Solution: The ovule of a rose is analogous to the capsule of a moss only in that both are sporangia, meaning that they produce spores. The ovule is a megasporangium and produces female haploid spores by meiotic division of the megaspore mother cell. One of these spores will develop into the female gametophyte. Necessarily the rose plant must also produce male spores, and indeed the microsporangia or pollen sacs are the site of male haploid spore production. The male spores will give rise to male gametophytes. The rose plant, then, produces two types of spores which differ in size and function, and is said to be heterosporous.

The moss, like other lower plants, is homosporous. Its sporangium is the capsule, and it is the only type of sporangia that the moss possesses. The capsule gives rise to diploid mother spore cells which undergo meiotic divisions to form haploid spores. These spores are uniform in appearance, and all germinate alike into bisexual gametophytes bearing both male and female sex organs.

The ovule of a rose and the capsule of a moss differ then, in that the ovule produces only those spores which will develop into female gametophytes, while the capsule produces spores which will give rise to bisexual gametophytes.

• **PROBLEM** 8-16

Discuss the modes of asexual reproduction exhibited by higher plants.

Solution: A number of commercial plants - bananas, seedless grapes, navel oranges, and some varieties of potato, to name just a few - have lost the ability to produce functional seeds and must be propagated entirely by asexual means. Many cultivated trees and shrubs are asexually reproduced from the cuttings of stems, which will develop roots at their tips when placed in moist ground or in water containing a small amount of indoleacetic acid (auxin).

Many plants such as the strawberry develop long, horizontal stems called stolons or runners. These grow some distances along and above the ground in a single season and may develop new erect plants at every other mode. Other plants spread by comparable, but underground, stems called rhizomes. Such weeds as witch grass and crab grass spread very fast by means of either runners or rhizomes. Swollen underground stems, or tubers, also serve as an asexual means of reproduction in plants such as the potato. New plants grow out at the buds of the tuber. The stems of raspberries, currants, and wild roses, and the branches of several kinds of trees may droop to the ground. Adventitious roots and a new erect stem may develop at one of the nodes touching the ground. The stem or branch may lose its connection with the parent plant and grow into a new, independent plant.

• PROBLEM 8-17

What trends in the evolution of the life cycle are evident from the algae to flowering plants?

Solution: In the evolution from the algae to the flowering plants there is, first, a change from a dominating population that is mostly haploid to one that is almost entirely diploid. In multicellular filamentous algae such as Ulothrix, only one cell during the life cycle, namely the zygote, is diploid; all the rest are haploid. In mosses, the haploid phase is more conspicuous and long-lived than the diploid phase, but the latter has evolved a complex, multicellular plant body. The relative importance of the two phases is reversed in the ferns: the diploid phase is the obvious, larger plant, and the haploid gametophyte, though still an independent plant, is small and inconspicuous. The gymnosperms and angiosperms show progressive reductions of the haploid generation. In angiosperms, the male gametophyte consists only of three cells and the female gametophyte of eight cells. The evolutionary advantage of a long-lived, dominant sporophyte and a short-lived, inconspicuous gametophyte generation is that a diploid individual can survive despite the presence of a deleterious recessive gene. A haploid individual would be more susceptible to the effects of such genes, since there would be no way to counterbalance the lethality of such an allele.

Besides the reduction of the gametophyte, there is a gradual reduction in the dependence of fertilization on the presence of moisture. Aquatic plants such as algae have motile sperm that swim to the eggs. Bryophytes and lower vascular plants, which develop on the ground near water, also require a moist habitat. The seed plants have replaced sperm motility by pollination. Pollen may be very light and carried by the wind over large distances. The angiosperms have further evolved showy floral parts and nectar glands to attract insects, birds and other animals which serve to carry pollen on their feet or fur. Pollination has thus two evolutionary advantages: it allows sperm transport and fertilization to occur without moisture, and it enables long distance dispersal of sperm cells.

Greater embryonic protection is a third evolutionary trend in the life cycle of plants. The green algae Chlamydomonas has its zygote protected only within a thick wall. Bryophytes and lower vascular plants have their embryos developing within the multicellular archegonium which draws nutrients from the gametophyte. Gymnosperms and angiosperms have evolved a seed structure for maintaining and protecting the embryo independent of the gametophyte. The seed has contributed immensely to the success of the seed plants. The stored food nourishes the embryo and the tough outer coat protects it from heat, cold, desiccation, drying, and parasites. Seeds also provide an effective means for the dispersal of the species. They may be wind-borne, water-borne, or carried by animals. To secure even greater protection, the angiosperms have evolved fruit which encloses one or more seeds. Fruits may either decay or be eaten by animals, liberating the seeds.

SHORT ANSWER QUESTIONS FOR REVIEW

Answer

To be covered when testing yourself

Choose the correct answer.

1. The gymnosperms and angiosperms differ from the rest of the vascular plants in that (a) they have an independent gametophyte generation. (b) they have flagellated sperm. (c) they are heterosporous. (d) embryo development occurs within the female gametophyte. — c

2. In almost all natural populations, the seed plants greatly outnumber all other vascular plants. This is probably because (a) their vascular system is better adapted to a terrestrial existence. (b) they produce more spores than other vascular plants. (c) a greater number of their embryos survive than do those of other vascular plants. (d) a greater number of their gametophytes survive than do those of other plants. — c

3. Double fertilization refers to the fact that (a) two eggs are fertilized by a single sperm (b) two eggs are fertilized by two sperm (c) one sperm fuses with the egg; the other fuses with the polar nuclei (d) one sperm fuses with the egg; the other fuses with the vegetative nucleus. — c

4. In the gymnosperms, the megaspore mother cell is produced from the tissue of (a) the integument. (b) the nucellus, (c) the microsporangia, (d) the endosperm. — b

5. Which of the following characteristics of the angiosperms shows them to be more advanced than the gymnosperms? (a) covered seeds (b) broad, flat leaves (c) pollen tube (d) double fertilization. — a

6. The fact that the cycads form swimming sperm within their pollen tubes indicates that (a) they are aquatic plants (b) they are less advanced than the conifers. (c) they are more advanced than the conifers. (d) they require moisture in order for pollination to occur. — b

7. In monocots, the cotyledons, or embryonic leaves, function primarily (a) as photosynthetic organs (b) as protective coverings for the seed (c) in food storage (d) in nutrient absorption. — d

Answer

To be covered when testing yourself

8. In the seed plants, each megasporangium produces how many functioning megaspores? (a) two (b) four (c) one (d) many

c

9. The tough outer covering of the seed is derived from (a) the nucellus. (b) the integuments (c) the cuticle (d) the endoderm.

b

10. If a plant produces flowers that are the same color as its leaves, then it is likely that pollination of that plant occurs by (a) birds, (b) wind, (c) insects, (d) animals.

b

11. In the life cycle of the seed plants, the dominant generation is (a) haploid (1N) (b) diploid (2N) (c) triploid (3N) (d) haploid and diploid are equally dominant.

b

In 12-21, indicate whether the traits are characteristic of the monocots (M), dicots (D), both (B), or neither (N).

12. _____ single cotyledon
13. _____ leaves with cuticle
14. _____ food storage in cotyledon(s)
15. _____ petals in mutliples of threes
16. _____ woody plant tissues
17. _____ food absorption by cotyledon(s)
18. _____ cambium
19. _____ scattered vascular bundles
20. _____ both tracheids and vessel elements
21. _____ persistent endosperm

12-M, 13-N, 14-D, 15-M, 16-D, 17-B, 18-D, 19-M, 20-B, 21-M.

Fill in the blanks.

22. Upon germination, one cell of the pollen grain, called the _____, elongates to form the pollen tube. The _____ then enters the pollen tube and undergoes mitosis to form two sperm nuclei.

tube nucleus, generative nucleus

23. The _____ divides meiotically to produce haploid microspores, which divide mitotically to produce the _____.

microspore mother cell, microgametophyte (pollen grain)

24. A cone is a spiral aggregation of modified _____ called cone scales or _____.

leaves, sporophylls.

	Answer
	To be covered when testing yourself
25. The term gymnosperm means _____. The term angiosperm means _____.	naked seed, enclosed seed
26. A(n) _____ fruit is one derived from sepals, petals, or the receptacle as well as from the ovary.	acces-sory
27. Label the lettered items in the diagram below. 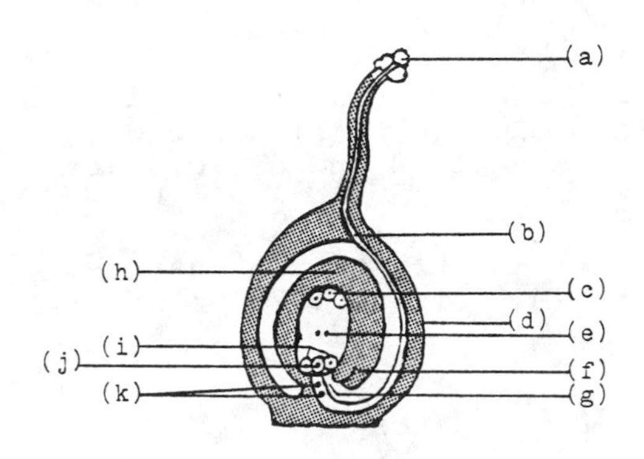	a-pollen grain, b-pollen tube, c-antipo-dal cells, d-recepta-cle, e-polar nuclei, f-integu-ments, g-micropyle h-ovule, i-synergid cells, j-egg, k-sperm nuclei
28. Label the following diagram of a seed. 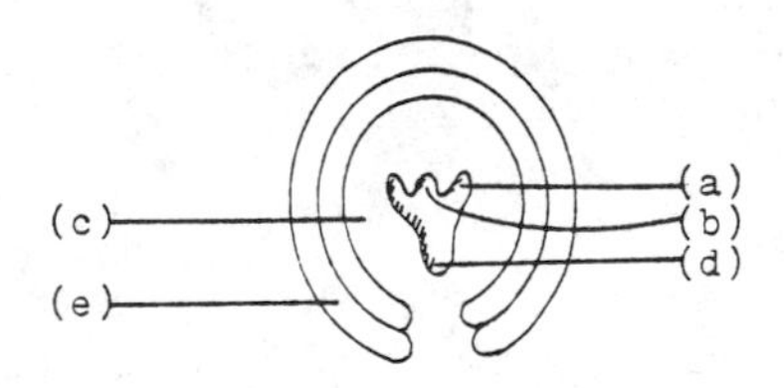	a-coty-ledon, b-epicotyl c-endo-sperm, d-hypocotyle e-integu-ment
29. The seed coat is derived from the _____ of the ovule.	integu-ment
30. The resumption of normal growth of a dormant embryo is termed _____.	germi-nation
31. The _____ gives rise to the primitive root or radicle.	hypocotyl
32. Indicate the chromosome number (N) of the following structures or tissues	
(a) _____ nucellus	a-2N,
(b) _____ endosperm	b-3N,
(c) _____ pollen grain	c-1N,
(d) _____ archegonia	d-1N,
(e) _____ strobilus	e-2N,
(f) _____ microspore mother cell	f-2N,

Answer

To be covered when testing yourself

(g) ______ megagametophyte
(h) ______ microspore
(i) ______ polar nuclei
(j) ______ seed coat
(k) ______ embryo
(l) ______ vegetative nucleus

Answer: g-1N, h-1N, i-1N, j-2N, k-2N, l-1N

33. Label the parts of the flower diagrammed below.

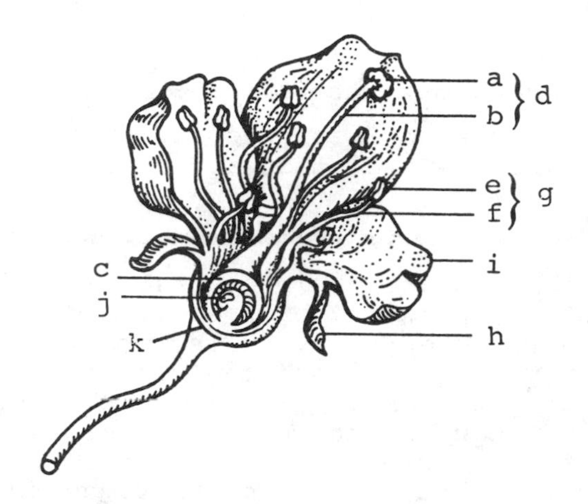

Answer: a-stigma, b-style, c-ovary, d-pistil, e-anther, f-filament, g-stamen, h-sepals, i-petals, j-ovule, k-receptacle

Determine whether the following statements are true or false.

34. During its period of dormancy, a seed limits its growth by ceasing to metabolize. — False

35. Warm spells in winter are deleterious because they cause seeds to germinate prematurely. — False

36. The seed coat of a gymnosperm is analogous to the fruit of an angiosperm. — False

37. The endosperm of dicots is usually completely absorbed by the time the seed germinates. — True

38. Natural asexual reproduction in seed plants most often occurs at the nodes. — True

39. A dominant haploid generation is a good indication that a plant is evolutionarily advanced. — False

CHAPTER 9

GENERAL CHARACTERISTICS OF GREEN PLANTS

REPRODUCTION

• PROBLEM 9-1

How does asexual reproduction in plants differ from sexual reproduction? What are the evolutionary advantages of the latter?

Solution: Asexual reproduction in plants takes place by the separation of any portion, or specialized portion, of a parent plant to form new individuals, or by the formation and germination of spores or single cells specialized in the replication of new individuals. Both these methods preserve the genetic makeup of the parent plants. Sexual reproduction involves the union of two gametes. The combination of two different sets of chromosomes from the gametes results in a new individual with a new genetic composition. Sexual reproduction therefore gives rise to new genotypes while asexual reproduction perpetuates the same ones.

Sexual reproduction is an important process in the evolution of a species, and is considered much more advantageous than asexual reproduction. The fusion of gametes from two genetically different parents makes possible new combinations of genetic material and thus increases the variation among individuals in the population. Genetic variation provides a pool of diversified phenotypes upon which natural selection can act. By this process, natural forces determine which individuals of the species will survive and which will become extinct, ultimately improving the species by sorting out the advantageous gene in the pool. An asexually reproducing species offers little genetic variation for selection and evolution is minimized.

In addition, the natural forces acting upon any species are always changing. The more a species is able to respond to these forces with evolutionary changes that maintain or increase its fitness, the more likely it is to survive and multiply. Such evolutionary changes can occur through genetic mutations, expressed in phenotypes which are then acted upon by natural selection. Because a mutated gene is usually recessive, sexual reproduction increases its chance of expression by bringing two mutated alleles together in the process of gametic fusion. Asexual reproduction will need a double mutation, which occurs in the order of 10^{-12} per generation, to result in an altered phenotype. Therefore, without sexual reproduction there could be little genotypic and thus phenotypic changes. Evolution could not progress, as there could be no response to inevitable environmental changes, and extinction of the species would soon follow.

PHOTOSYNTHETIC PIGMENTS

• PROBLEM 9-2

What pigments may be present in plant cells? What are the functions of these pigments?

Solution: Chlorophyll a occurs in all photosynthetic eucaryotic cells and is considered to be essential for photosynthesis of the type carried out by plants. It functions in the capture of light energy by either directly absorbing it or receiving it in the form of high energy electrons from the accessory pigments. These accessory pigments, such as chlorophyll b, are found in vascular plants, bryophytes, green algae and euglenoid algae. Chlorophyll b differs from chlorophyll a structurally and in its absorption spectra. Chlorophyll b shares with chlorophyll a the ability to absorb light energy and produce in the molecule some sort of excited state. The excited chlorophyll b molecule transfers light energy via high energy electrons to a chlorophyll a molecule, which then proceeds to transform it into chemical energy. Since chlorophyll b absorbs light of wavelengths that are different from chlorophyll a, it extends the range of light that the plant can use for photosynthesis. Chorophyll c or d takes the place of chlorophyll b in some algae and plant-like protists.

The carotenoids are another class of accessory pigments. Carotenoids are red, orange, or yellow fat-soluble pigments found in almost all chloroplasts. Carotenoids that do not contain oxygen are called carotenes, and are deep orange in color; those that contain oxygen are called xanthophylls, and are yellowish. Like the chlorophylls, the carotenoids are bound to proteins within the lamellae of the chloroplast. In the green leaf, the color of the carotenoids is masked by the much more abundant chlorophylls. In some tissues, such as those of a ripe red tomato or the petals of a yellow flower, the carotenoids predominate. As accessory pigments, the carotenoids function in absorbing light not usable by the chlorophylls and in transferring the absorbed energy to chlorophyll a.

Another pigment that may be found in plants is the light-sensitive, blue phytochrome. Phytochromes play a fundamental role in the circadian rhythms of plants; they allow the plant to detect whether it is in a light or dark environment.

• PROBLEM 9-3

What is meant by the statement that the atoms of the chlorophyll molecule constitute a "resonating system"? Of what importance is this in the process of photosynthesis?

Chlorophylls a and b contain a porphyrin ring composed of four nitrogen-containing rings and their connecting carbon atoms, with an atom of magnesium in the center and a **long lipid "tail" marked R. R' is the site of difference between Chlorophyll a and b. On chlorophyll a, R' is a methyl substituent (CH_3), while in chlorophyll b, it is a formyl group (CHO).**

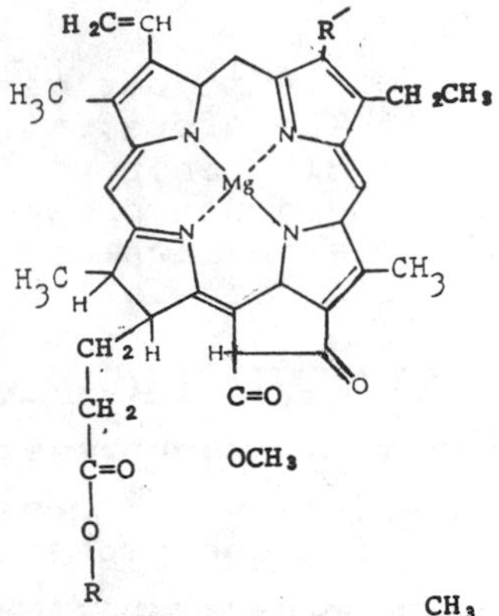

$$R = -CH_2 - CH = \overset{CH_3}{\overset{|}{C}} - CH_2 - (CH_2 - CH_2 - \overset{CH_3}{\overset{|}{CH}} - CH_2)_2 - CH_2 - CH_2 - CH(CH_3)_2$$

Solution: If we examine closely the structure of the chlorophyll molecule (see Figure) we will see that its atoms constitute a conjugated system, with double and single bonds alternating around the ring. Such a conjugated system provides for many possible resonance structures and is essentially a resonating system in which the pi electrons are spread, or delocalized, about the carbon atoms of the ring. In other words, a pi electron is no longer associated with a single atom or bond but with the conjugated system as a whole. This property, called resonance, gives the chlorophyll molecule considerable stability. As a consequence of resonance stabilization, only a small amount of energy is required to raise the pi electrons of the carbon atoms to a higher energy level.

The chlorophyll molecule absorbs energy by having its pi electrons excited and "pumped" to a new, more energetic orbital (energy level). The energy of excitation comes from the sun in the form of visible light energy, and is converted to chemical energy in the light reactions of the photosynthetic process. The absorbed energy is always in a definite, specific quantity, known as a photon. In order to raise an electron to a given new energy level, the energy of the photon must just equal the difference in energy content of the electron's old and new orbitals. Unless a photon can raise the energy of the electron by just the right amount for a defined energy level, that photon will not be absorbed. The adaptive importance of the chlorophyll molecule lies therefore in its unique capacity to function in capturing the energy of visible light with a high degree of efficiency.

• PROBLEM 9-4

Discuss the structure of the chloroplast and the possible relation of its structure to the process of photosynthesis. What is a quantosome?

Solution: Chloroplasts are small organelles found in plant cells that serve as the sites of photosynthesis. There are some 20 to 100 chloroplasts in a plant cell; these can grow and divide to form daughter chloroplasts. A chloroplast, like a mitochondrion, has a double - layered outer membrane. Lying within each chloroplast are many

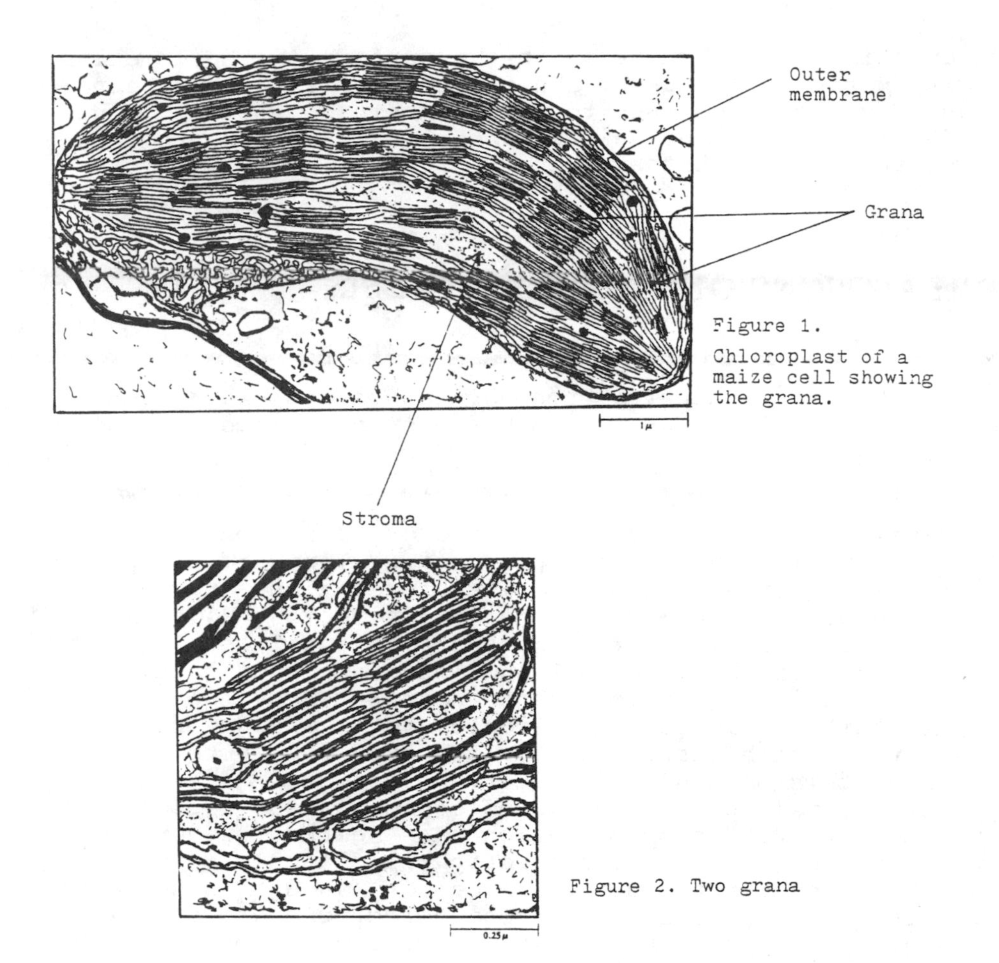

Figure 1. Chloroplast of a maize cell showing the grana.

Figure 2. Two grana

smaller stacks called grana (see figures). Under the electron microscope, the grana are seen as part of an elaborate system of membranes organized in parallel pairs. The pairs of membranes are joined at their ends to form a closed disk or thylakoid. The grana themselves are stacks of these thylakoids. Each granum is composed structurally of layers of protein molecules alternating with layers of chlorophyll, carotenes and other pigments, and special types of lipids containing galactose or sulfur but only one fatty acid. These surface-active lipids are believed to be absorbed between the layers and serve in stabilizing the lamellae composed of the alternate layers of protein and pigments. This lamellar structure of the grana is important in permitting the transfer of energy captured from the sun from one molecule to the adjacent one during the light phase of the photosynthetic process. This is evidenced by the fact that chlorophyll extracted from the chloroplast and isolated in a test tube can no longer carry out any photosynthetic reaction owing to the destruction of the lamellar structure.

The semifluid matrix within the chloroplast and surrounding the grana is called the stroma. The stroma contains the enzymes responsible for the dark reactions of photosynthesis.

Electron microscopy has revealed the presence of

repeating unit structures on the surface of the lamellae within the chloroplast. These structures, given the name quantosomes, are found to each contain some 230 molecules of chlorophyll. Botanists now believe that quantosomes are the functional photosynthetic units.

REACTIONS OF PHOTOSYNTHESIS

• PROBLEM 9-5

Compare photosynthetic phosphorylation and oxidative phosphorylation. What are the roles of ferredoxin and plastoquinone?

Solution: Photosynthetic and oxidative phosphorylation are both ATP-synthesizing processes. They differ, however, in their source of energy, site of reaction, and in the nature of their electron donor and acceptor. Photosynthetic phosphorylation, as its name suggests, utilizes light energy to synthesize ATP from ADP and inorganic phosphate, light energy striking a green plant is absorbed by the chlorophyll. The excited chlorophyll molecule ejects a high energy electron which passes down an electron transport chain. During this passage, the electron returns to its original energy level, producing ATP molecules in the process. Photosynthetic phosphorylation uses neither oxygen nor organic substrates. The chlorophyll molecule serves as both the electron donor and acceptor (though in non-cyclic photophosphorylation, additional electrons are supplied by water to reduce $NADP^+$). The excited chlorophyll molecule passes an energetic electron to ferrodoxin, and having lost an electron, is now able to accept another. The end result is that the energy from sunlight is converted to chemical energy stored in ATP. The entire process occurs in the chloroplasts.

Oxidative phosphorylation takes place in the mitochondria. The process is incapable of utilizing sunlight as its energy source. Rather, it uses the energy released when certain reduced substrates are oxidized. The electrons released are passed in a series of redox (oxidation/reduction) reactions down the electron transport chain, a system of electron acceptors of decreasing reduction potential. The energy of oxidation is captured in the form of an energy-rich phosphate-ester bond in ATP. In oxidative phosphorylation, the ultimate electron acceptor is oxygen and the primary electron donor is a sugar or some other organic substance.

Ferredoxin is a component of the electron carrier chain in the chloroplasts of plant cells. It is an iron-containing protein that functions in capturing energy-rich electrons from excited chlorophyll molecules and in passing them to the next electron acceptor of the chain. Plastoquinone is another electron carrier in the chloroplast; it can receive electrons from either ferredoxin or cytochrome and transfer them to the cytochromes.

Describe the light reactions of photosynthesis. What are the products of these reactions?

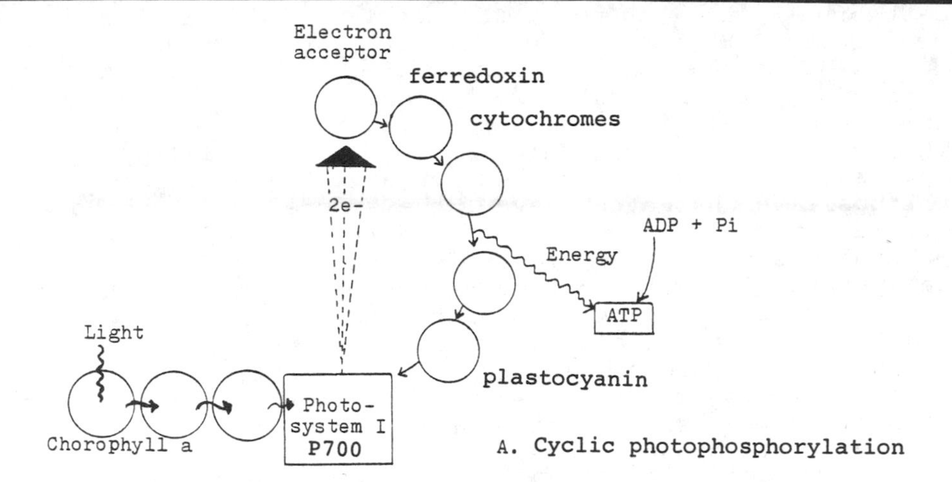

A. Cyclic photophosphorylation

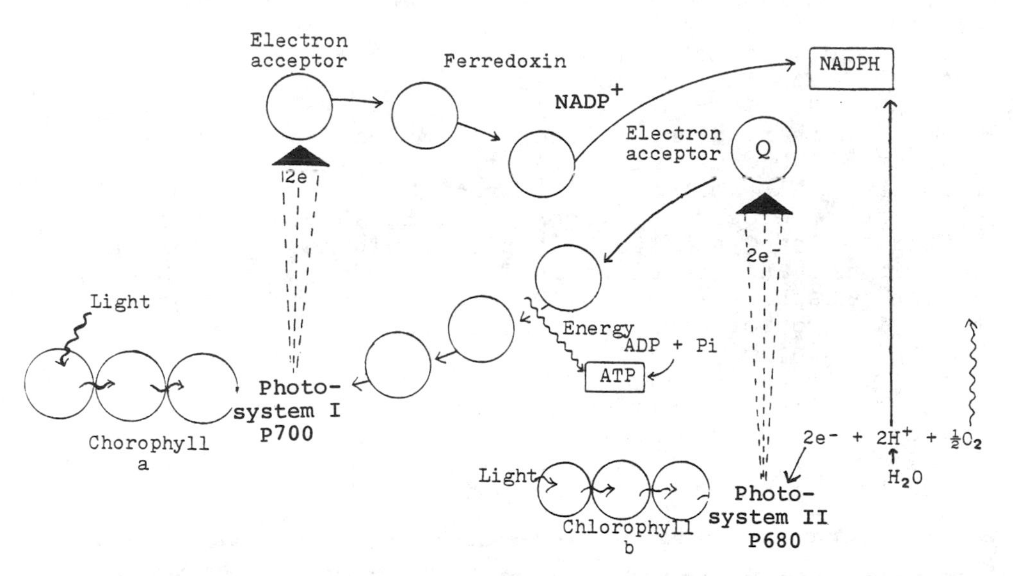

B. Noncyclic photophophorylation

Solution: The essential requirements for the energy-requiring dark reactions of photosynthesis are ATP and $NADPH_2$. These molecules are produced via the light reactions of photosynthesis, which occur entirely within the chloroplasts. This was proven by Daniel Arnon in 1954 who showed that intact chloroplasts carefully isolated from spinach leaves were able to carry out the complete photosynthetic reaction.

Scientists now agree that there are two reaction pathways occurring in the chloroplasts which generate ATP molecules from light energy, one of the pathways being designated cyclic photophosphorylation, the other noncyclic photophosphorylation. Both reactions require the presence of light. Each pathway is a photsystem. A photosystem is an energy-trapping center which receives packets of energy from excited chlorophyll molecules and converts them into a stream of high-energy electrons which can then travel down the electron transport chain. Photosystem I is called P700 due to the wavelength at which chlorophyll a absorbs light (700 nm). Photosystem II is similarly designated P680 due to the response by chlorophyll b.

In cyclic photophosphorylation, light striking a chlorophyll a molecule excites one of the electrons to an energy level high enough to allow it to leave the molecule. The chlorophyll a^+ molecule, having lost an electron, is now ready to serve as an electron acceptor because of its net positive charge. However, the ejected electron does not return ti its ground state and the chlorophyll a^+ molecule directly; instead it is taken up by ferredoxin and passed along an electron transport chain. As the electron passes from ferredoxin to the cytochromes, to plastocyanin and finally back to chlorophyll a^+, two ATP molecules are produced. In this way, light energy is converted into chemical bond energy, and not lost as heat as it would have been had the excited electron returned directly to its ground state.

In noncyclic photophosphorylation, oxygen is produced by the photolysis of water. NADPH and ATP molecules are formed. The excitation of chlorophyll a by light at Photosystem I ejects high energy electrons. These electrons pass to ferredoxin and then to $NADP^+$, reducing it to NADPH. To restore chlorophyll a+ to its original state, another set of chlorophyll molecules known as chlorophyll b come into play at Photosystem II. When light energy is absorbed by chlorophyll b, it also ejects high energy electrons. These pass to an electron acceptor called Q and then via plastoquinone, cytochromes and plastocyanin to chlorophyll a+. During these steps, ADP is phosphorylated to ATP. The electrons necessary to restore chlorophyll b+ to its ground level come from the splitting of water due to the high affinity that chlorophyll b+ has for electrons. The water molecule is thus split into component protons, electrons, and oxygen. The oxygen, after combining with another oxygen atom, is released in gaseous form as O_2. The two electrons are picked up by cytochrome b and the hydrogen ions are used in the conversion of $NADP^+$ to NADPH.

Compare the processes of cyclic and noncyclic photophosphorylation.

• PROBLEM 9-7

Solution: There are two types of photophosphorylation: cyclic and noncyclic. In cyclic photophosphorylation, the flow of electrons from light-excited chlorophyll molecules to electron acceptors proceeds in a cyclic fashion through electron carriers back to the original chlorophyll. Photosystem I, sensitive to longer wavelengths of light than photosystem II, functions in cyclic photophosphorylation. No oxygen is liberated, since water is not split; no $NADP^+$ is reduced, since it does not receive electrons. ATP is formed during the electron flow, and light energy is thus converted into chemical energy in the ATP molecules. However, since NADPH is not formed, cyclic photophosphorylation is not adequate to bring about CO_2 reduction and sugar formation,processes which require the energy of NADPH .

Noncyclic photophosphorylation, on the other

hand, produces both ATP and NADPH molecules necessary for the dark reactions of photosynthesis. In this process, electrons from excited chlorophyll a molecules are trapped by NADP in the formation of NADPH and do not cycle back to chlorophyll a. Electrons are ejected from chlorophyll b to be donated to chlorophyll a through a series of electron carriers. To restore chlorophyll b to its ground level, water is split into protons, electrons and oxygen. The electrons are picked up by chlorophyll b, the hydrogens are used to form NADPH, and oxygen escapes to the atmosphere in its molecular form. Both photosystems I and II are involved in the process of noncyclic photophosphorylation.

• PROBLEM 9-8

The overall reaction for photosynthesis is: $CO_2 + H_2O + h\nu$ (light energy) $(CH_2O) + O_2$. It was generally assumed in the past that the carbohydrate came from a combination of the carbon atoms and water molecules and that the oxygen was released from the carbon dioxide molecule. Do you agree with this assumption? Explain your answer.

Solution: Superficially, the assumption appears sound, and was believed for years to be true. It is however, quite wrong, and was refuted in the 1930's by C.B. van Niel of Stanford University. Van Niel was studying photosynthesis in different types of photosynthetic bacteria. He observed that in their photosynthetic reactions, bacteria reduce carbon dioxide to carbohydrate, but do not release oxygen. The purple sulfur bacteria, which require hydrogen sulfide for photosynthesis, were found to accumulate globules of sulfur within the cells. Based on these observations Van Niel then wrote the following equation for photosynthesis in purple sulfur bacteria:

$$CO_2 + 2H_2S \rightarrow CH_2O + H_2O + 2S$$

This equation shows that it is unlikely that carbon dioxide is split during photosynthesis; if it were, oxygen would be formed. Moreover, carbohydrate could not have come from the combination of carbon atoms and hydrogen sulfide molecules. It would be reasonable to propose that hydrogen sulfide was the molecule to be cleaved rather than carbon dioxide. This led Van Niel to propose the following general equation for photosynthesis,

$$CO_2 + 2H_2A \rightarrow (CH_2O) + H_2O + 2A$$

Here, H_2A represents an oxidizable species capable of donating its electrons upon being split. According to this hypothesis, in the photosynthesis of green plants, it is the water molecule, not carbon dioxide, that is split to release O_2. This brilliant speculation was not proven until many years later when advancement of radio-isotopic techniques made it possible to reveal the mechanism of the actual reaction. The use of isotopic oxygen, ^{18}O, allowed the successful tracing of the path of oxygen from water to free molecular oxygen:

$$CO_2 + 2H_2{}^{18}O - CH_2O + H_2O + {}^{18}O_2$$

Thus, it can readily be seen that molecular oxygen comes actually from water and not from carbon dioxide. It is now known that the splitting of water occurs during the light reactions of photosynthesis, releasing O_2 gas, electrons, which are energized by light to provide for ATP formation, and protons, which are used to form NADPH, the reducing power necessary for carbohydrate formation. Carbon dioxide is then reduced in the dark reactions of photosynthesis using both ATP and NADPH to form carbohydrates.

• PROBLEM 9-9

Discuss the possible sequence of events in the evolution of photosynthesis.

Solution: There are two types of photophosphorylation cyclic and noncyclic. It has been proposed that cyclic photophosphorylation is the more primitive of the two processes and perhaps was the earliest form of energy production. It might have developed at a time when the earth's atmosphere contained little or no oxygen gas. Cyclic photophosphorylation probably served as a means of synthesizing ATP for primitive organisms living in an anaerobic environment. The next evolutionary step was perhaps the development of noncyclic photophosphorylation, such as that found in certain bacteria today. In these bacteria, $NADP^+$ molecules are the electron acceptors but the electron donors that restore the excited chlorophyll molecules to ground state are molecules such as thiosulfate or succinate, not water. The third step, achieved by green plants, was probably the ability to use water molecules in noncyclic photophosphorylation. By evolving such an ability, green plants were able to live nearly everywhere and were not restricted to places where thiosulfate or succinate are found. Then, as plants spread and proliferated, they released into the atmosphere oxygen gas, a by-product of their noncyclic photophosphorylation. The accumulation of oxygen made possible the further biochemical evolution of animals and other organisms that use aerobic respiration (oxygen) for survival and active growth.

• PROBLEM 9-10

Discuss the sequence of reactions that constitute the "dark reactions" of photosynthesis. What are the products of these reactions?

Solution: The dark reactions of photosynthesis, in which carbohydrates are synthesized, occur in a cyclic

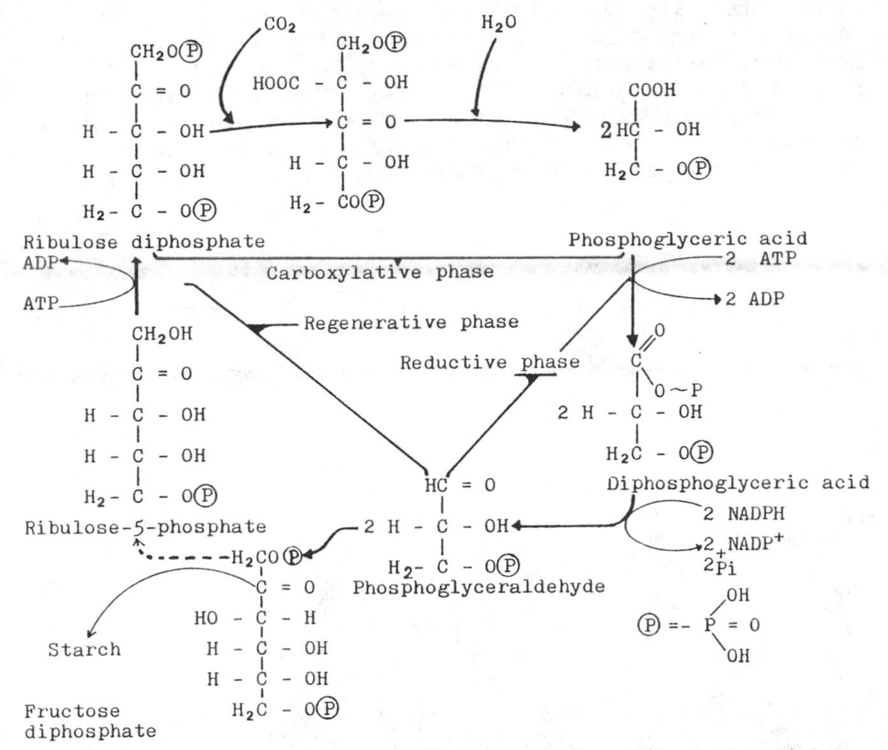

Diagram of the "dark" reactions of carbohydrate synthesis by which carbon dioxide is incorperated into sugars. The dotted line from fructose diphosphate to ribulose-5-phosphate indicates several reactions of the pentose phosphate pathway.

sequence of three phases - the carboxylative, reductive, and regenerative phases. These dark reactions do not require the presence of sunlight.

In the carboxylative phase, a five-carbon sugar, ribulose-5-phosphate, is phosphorylated by ATP to yield ribulose diphosphate (see Figure). Ribulose diphosphate is then carboxylated (that is, CO_2 is added), presumably yielding a six-carbon intermediate, which is split immediately by the addition of water to give two molecules of phosphoglyceric acid. The three-carbon phosphoglyceric acid is reduced by NADPH in an enzymatic reaction in the reductive phase, utilizing energy from ATP. The product of this reduction is phosphoglyceraldehyde, another triose. Two molecules of phosphoglyceraldehyde can then condense in the regenerative phase to form one hexose molecule, fructose diphosphate. This is subsequently converted to a glucose-6-phosphate molecule and finally transformed into starch.

Ribulose-5-phosphate, the cycle-initiating reactant, can be regenerated from fructose diphosphate via certain reactions of the pentose phosphate pathway. Some scientists tend to believe that the ribulose molecule is regenerated from phosphoglyceraldehyde by a complicated series of reactions. However, the cyclicity of the dark reactions is a well established and universally accepted fact.

The materials consumed in the production of one hexose molecule are one molecule each of CO_2 and H_2O, three ATP molecules, and four H atoms (from two molecules of $NADPH_2$). Ribulose-5-phosphate is not consumed but generated at some point in the cycle. The glucose-6-phosphate produced can either be polymerized into starch and stored, or broken down to yield energy for work.

• PROBLEM 9-11

How can you prove that oxygen is given off by green plants during photosynthesis?

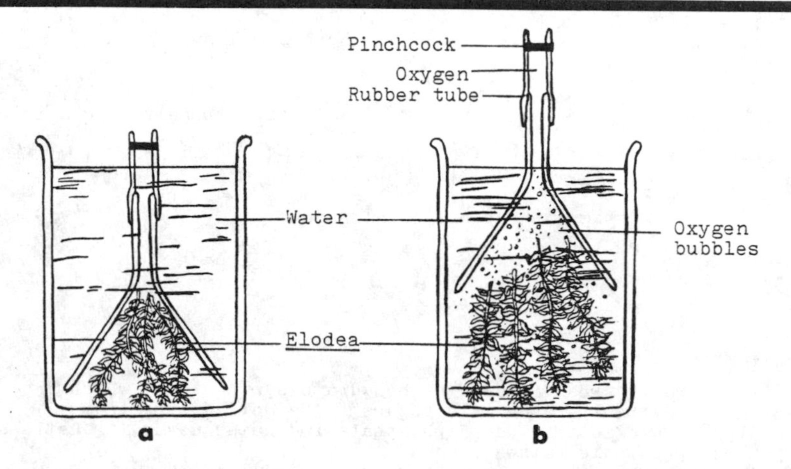

Solution: We can show that oxygen is one of the products of photosynthesis by performing a simple experiment. Elodea or some other photosynthetic aquatic plant is placed in a beaker of water and covered with an inverted glass funnel. A rubber tube is attached to the stem of the funnel. This tube is closed off with a pinch cock. The funnel is subsequently raised, creating a column of vacuum in its stem. (See accompanying diagram.) The plant is then exposed to light. The gas given off during the photosynthetic process bubbles upward through the water and fills the column of the stem. To identify the gas, a glowing splinter of wood is applied over the stem of the funnel. Upon removal of the pinch-cock, the glowing splinter bursts into flame. This means that the gas is oxygen, since it supports combustion.

PLANT RESPIRATION

• PROBLEM 9-12

Compare photosynthesis and cellular respiration in plants.

Solution: Photosynthesis in plants is restricted to those cells containing chlorophyll, such as the cells of the leaf. Respiration occurs in all living cells, whether they be the non-chlorophyll-containing root cells or the chlorophyll-containing leaf cells. Photosynthesis requires water and carbon dioxide as raw materials, and sunlight as an energy source. It produces oxygen and simple hexose molecules, such as glucose. Cellular respiration, on the contrary, consumes oxygen and organic molecules and releases water and carbon dioxide. Photosynthesis is an energy absorbing process, the energy ultimately coming from the sun; respiration releases energy: some is lost as heat and some is stored as chemical bond energy in ATP. Photosynthesis, as the name implies, occurs only when light is present, but respiration takes place continuously, during both night and day. Photosynthesis occurs in the chloroplasts, which contain the necessary enzymes and electron carriers; respiration occurs in the mitochondria, which contain the oxidative enzymes of the tricarboxylic acid cycle (also known as the Krebs cycle) and components of the electron transport system. Finally, photosynthesis results in an increase in the weight of the plant, as more organic substances are manufactured. Respiration results in a decrease in the weight of the plant, as organic materials are broken down for the synthesis and release of energy.

• PROBLEM 9-13

Based on your knowledge of plant nutrition, explain how a photosynthetic plant growing in iron - and magnesium - deficient soil would look.

Solution: Magnesium is an essential component in the synthesis of chlorophyll in green plants (see structures in Problem 9-3). A deficiency of magnesium results in an inability of the plant to manufacture enough chlorophyll, and the plant, deficient in green pigment, will look pale. Since chlorophyll is of vital importance in the light reactions of photsynthesis, very little if any radiant energy can be captured and hence little food can be manufactured. One obvious result is that the plant will cease to grow. It may survive on the limited food reserves for some time, but it will eventually die.

Iron is of particular importance in the synthesis of electron carriers of photophosphorylation and the electron transport chain. The cytochromes and ferredoxin, two important electron carriers, are both iron-containing proteins. An iron-deficient plant will suffer from a defective electron transport system. Thus, in the photosynthetic process, the stream of electrons activated by absorbed light energy will pass down an inefficient electron acceptor chain. Most of the energy of these electrons will be lost rather than stored in ATP and NADPH as would occur in a healthy plant. Similarly, because the iron-containing cytochromes are also involved in the process of oxidative phosphorylation, cellular respiration in an iron-deficient plant yields little usable energy stored in ATP and much energy would be lost as heat. The iron-deficient plant, therefore, not only fails to make enough ATP and NADPH in the light reactions, but also fails to generate ATP efficiently in the oxidation of food reserves.

In summary, a plant growing in soil deficient in both magnesium and iron contents will be pale, small, short, and weak, if it survies at all.

• PROBLEM 9-14

How do higher plants get rid of their waste products?

Solution: Since green plants undergo both photosynthesis and respiration, the products of one process may become the raw materials of the other and vice versa. Thus oxygen, the product of photosynthesis, is utilized in cellular respiration while the products of respiration, carbon dioxide and water, are used in photosynthesis. These products may also diffuse out through the pores of the leaves, depending on the dominance of either process at a particular time. For example, during the night, when respiration predominates, carbon dioxide and water vapor escape from the pores of the leaf surfaces.

The amount of nitrogenous wastes in plants is small compared to that in animals, and can be eliminated by diffusion either as ammonia gas through the pores of the leaves or as nitrogen-containing salts through the membrane of the root cells into the soil. Other wastes such as oxalic acid accumulate in the cells of the leaves and are eliminated from the plant when the leaves are shed.

TRANSPORT SYSTEMS IN PLANTS

• PROBLEM 9-15

"Since plants cannot move about as freely as animals, they have no need for a skeletal system." Do you agree or disagree with this statement? Justify your answer.

Solution: A skeletal system not only provides for locomotion but also for support. While it is true that most algae are small, aquatic, and have little need for a skeletal system, the land plants do need some skeletal structures to hold their leaves in position to receive sunlight and to help the stems stand erect on land. The land plants meet these requirements in one of two ways: either the cell wall can be very thick, as in the woody stems of trees and shrubs, and serve directly for the support of the plant body, or the cell wall can be rather thin yet provide support indirectly by way of turgor pressure.

Trees and shrubs have in their stems tough, woody cells called tracheids and vessels, whose structures are adapted for support in addition to conduction of water and minerals. These cells secrete a very thick wall made up largely of cellulose. Cellulose is a polysaccharide of repeating molecules of glucose attached end to end. The cellulose molecules are bound together in a complex matrix to form long, thin microfibrils. The microfibrils inter-

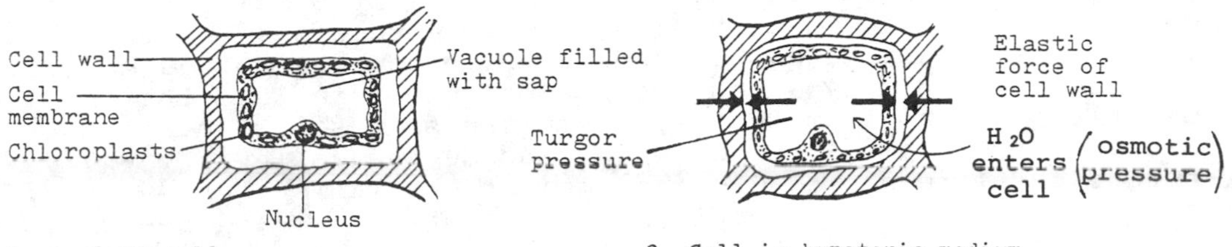

1. a plant cell

2. Cell in hypotonic medium (e.g., pure water). Increased turgor pressure.

3. Cell placed in hypertonic medium (strong salt solution). Cell shrinks. Loses turgidity.

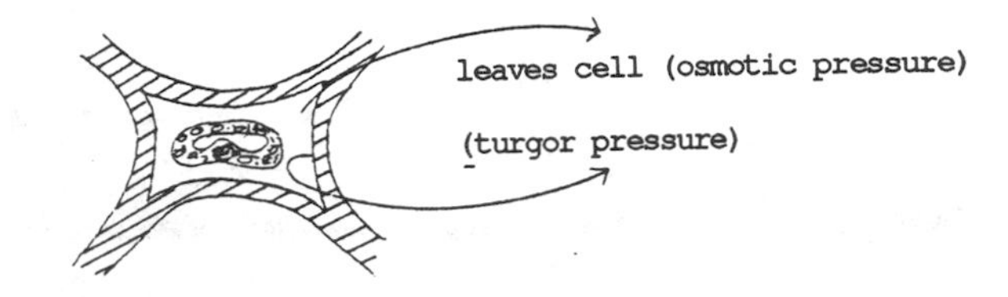

twine to form fine threads or fibrils, which may in turn coil around one another like strands in a cable. Cellulose molecules arranged in such an organized pattern form fibers as strong as steel cables of equivalent thickness. The cell walls of tracheids and vessels are also impregnated with lignin which adds to the strength of cellulose. Lignin is a highly branched, rigid organic molecule of repeating subunits. It is often deposited between the cellulose microfibrils in the walls of tracheids and vessels when the protoplast has ceased to function, and serves as a strengthening element because of its regular, rigid strucure and its resistance to decomposition. In addition, land plants have phloem cells which contain thick fibers in their walls to help support the trunk.

Non-woody plant cells are not equipped with thick walls but manage to support their shape and form with the aid of turgor pressure. Turgor pressure is the pressure exerted by the contents of the cell (such as organic materials and inorganic salts dissolved in water) against a cell wall which is slightly elastic and stretchable. Turgor can be maintained by a plant cell because it is naturally in a hypotonic medium. In such a medium, water tends to enter the cell, passing osmotically from a region of higher water concentration to a region of lower water concentration. The pressure resulting from a difference in the concentration of water molecules between two regions is known as osmotic pressure. The influx of water causes the cell to swell and the cell contents to exert a hydrostatic pressure, or turgor pressure, against the cell wall. Equilibrium is reached when the two opposing forces of hydrostatic pressure (built up in response to cell wall resistance) and osmotic pressure are equalized. Under such opposing pressures, the plant cell remains stiff or turgid.

The importance of turgor pressure in supporting the form of the plant body can be demonstrated by placing a crisp lettuce leaf in salt water. The lettuce leaf, after being in the salt water for some time, becomes limp and flaccid. In this hypertonic solution, water follows the osmotic gradient from the leaf into the salt solution. Without the accumulation of water inside the cell, no turgor pressure can be exerted against the walls of the cell. As a result, the cell loses its stiffness and becomes flaccid. For although the cells of the leaf still have their cellulose walls, these walls are alone unable to provide sufficient support. (See Figure.)

• PROBLEM 9-16

What are the functions of the xylem and the phloem?

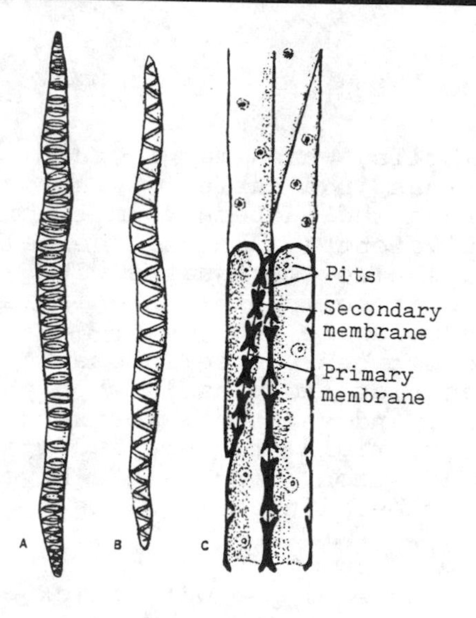

Figure 1.
Tracheids. (A) Primary tracheid with annular secondary walls. (B) Primary tracheid with spiral secondary walls. (C) Tracheids. Note the pits, which are **extremely abundant along the sides of the cells.**

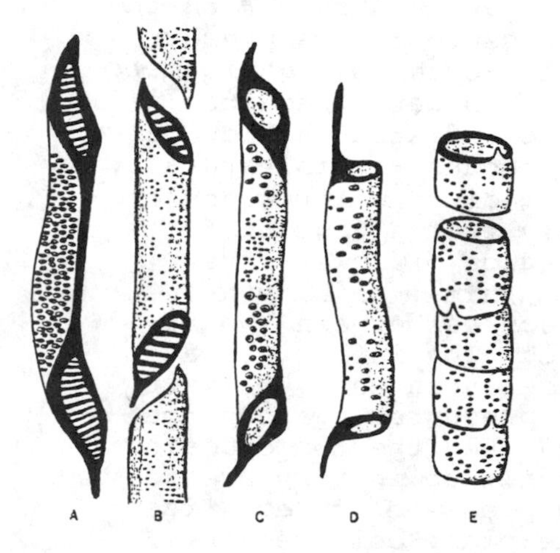

Figure 2.
Vessel cells. Five different types of vessel cells are shown - those thought to be the more primitive on the left, those thought to be the more advanced on the right. (E) shows a single vessel cell on top, four cells linked in sequence to form a vessel below. The evolutionary trend seems to have been toward shorter and wider cells, larger perforations in the end walls until no end walls remained, and less oblique, more nearly horizontal ends.

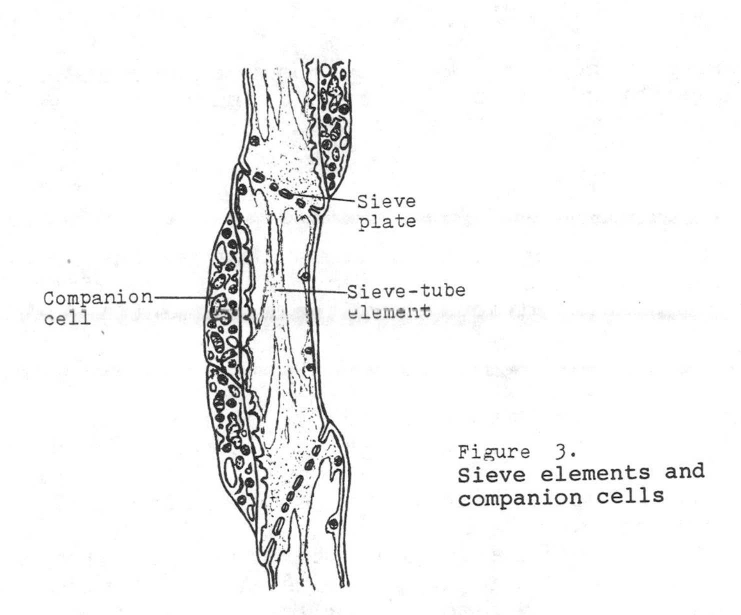

Figure 3.
Sieve elements and
companion cells

Solution: In vascular plants, the vital function of the transportation of food, water and minerals is performed by the vascular system, composed of the xylem and the phloem. The xylem is chiefly concerned with the conduction of water and mineral salts from the roots to the above-ground portion of the plant, where they are used for photosynthesis and other metabolic purposes. The xylem also serves as a means of support in larger vascular plants. It contains several specialized cell types, the two major kinds being the tracheids and vessel elements. Both are elongated cells with thickened cellulose walls heavily impregnated with lignin; both contain no living protoplasm, and hence are dead cells. The tracheids (see Figure 1) have tapering ends and lateral pores, or pits, which allow water to flow between adjacent cells. Although structurally similar, vessel elements are a more efficient system for conduction than tracheids. At maturity, the end of the vessels are broken down to form a perforation area, so that each individual cell is continuous with the cell above and the cell below. Vessel cells are thought to have evolved from tracheids, having increasing degrees of perforation with evolutionary advancement. (see Figure 2) Indeed, vessel elements are characteristic of the flowering plants and do not occur at all in most of the less advanced gymnosperms. Vessel elements have lateral pits but movement of materials is chiefly through their ends.

Carbohydrates are manufactured primarily in the leaves of a plant. They are transported to the other parts of the plant by the phloem, which runs parallel to the xylem throughout the plant body. The phloem consists of elongated, tube-like cells, called sieve-tube elements, with specialized pores at each end (See Figure 3). Unlike the tracheids and vessels, mature sieve-tube elements are living and have a protoplasm, but no nuclei. The sieve-tube-elements are associated with small, narrow cells having dense contents and prominent nuclei, knowns as the companion cells. These cells probably function as the 'nuclei' for the sieve-tube elements and make it possible for them to continue to function. Besides conduction, the phloem serves as a supporting tissue, due to the presence of strong fibers in the walls of its cells.

• PROBLEM 9-17

"Vascular plants are like animals because they both have circulatory systems." Using your knowledge of the vascular system of plants, discuss why the statement is basically incorrect.

Solution: The vascular system of a plant and the circulatory system of an animal are similar only in the sense that both systems have a common function of transportation of food and inorganic salts. The differences between these two systems are really more important than their similarity. Water moves through the xylem of a plant in a continuous stream and is constantly lost to the atmosphere as water vapor; this loss of water is called transpiration. Blood circulates inside an animal's body but, unless hemorrhage occurs, it is not continuously lost to the outside. It is true that under high temperature the body tends to lose water and salts to the outside through pores in the skin and through the urine, but this is a much slower process than transpiration. Compared to the 100 gallons of water lost by a large tree in a day, the sweat excreted by an animal is negligible. Moreover, the mechanism of flow in the vascular system is not comparable to the one found in animals. In plants, there are no pumps to drive the fluid along the vessels; in animals, a heart or analogous organs are always present to pump blood throughout the body. In addition, while the circulatory system of higher animals is designed to deliver oxygen to all parts of the body, only a very limited amount of oxygen moves through the vascular system of plants. Instead, oxygen enters the plant body by diffusion through the specialized pores or thin surfaces of the plant. Furthermore, the physical relationship between the xylem and phloem is fundamentally different from that between arteries and veins. Whereas arteries are connected to the veins by networks of capillaries, so that arterial blood is continuous with venous blood, the xylem and phloem are basically independent of each other, and very few substances pass from one to the other. Finally, while in the circulatory system of animals a large tube branches successively into smaller and smaller tubules, all xylem and phloem tubes are small and do not vary in their radii during their courses. They occur in bundles and it is the number of tubes per bundle which changes: in the lower part of the stem there are many per bundle; in the upper part there are fewer per bundle, some having entered the branches of the stems.

• PROBLEM 9-18

How might you argue that the transmission of impulses does occur in plants?

Solution: The reaction of the sensitive plant, Mimosa pudica, to touch serves as an excellent argument for the fact that transmission of impulses does indeed occur in plants. The leaf of Mimosa is subdivided into many leaflets. Normally the leaves are horizontal. However if one of the leaflets is lightly touched, that leaflet and all other leaflets of the stimulated leaf droop within two or three seconds leading to folding of the entire leaf. In addition, other neighboring leaves will soon fold and droop. The observation that touching one leaf leads other leaves to fold as well demonstrates that the excitation initiated by touching is transmitted from the site of stimulation to neighboring locations.

It is now believed that the folding of the leaves results from a decrease in the turgor pressure of the cells at the bases of the leaves. The excitation is transmitted along the sieve tubes of the leaves and stem in the form of an electrical impulse. This impulse involves a temporary physiological change in permeability of the plasma membrane, causing a change in intracellular and extracellular ionic distributions and consequently a change in turgor pressure. The nature of the impulse is fundamentally the same as that of a nerve impulse, although the latter travels about 2400 times faster.

TROPISMS

• PROBLEM 9-19

What are plant hormones? Discuss the important effects of the auxins.

Solution: Plant hormones, like animal hormones, are organic substances which can produce striking effects on cell metabolism and growth even though present in extremely small amounts. They are produced primarily in actively growing tissues, especially in the apical meristems of the plant. Like animal hormones, plant hormones usually exert their effects on parts of the plant body somewhat removed from the site of their production. Movement of plant hormones to target regions of the plant is made possible by the presence of phloem.

Auxins may be regarded as the most important of the plant hormones, since they have the most marked effects in correlating growth and differentiation to result in the normal pattern of development. The differential distribution of auxin in the stem of the plant as it moves down from the apex (where it is produced) causes the plant to elongate and bend toward light. Auxin from seeds induces the maturation of the fruit. Auxin from the tip of the stem passes down into the vascular cambium below and directs the tissue toward differentiating into secondary phloem and xylem. Auxin also stimulates the differentiation of roots-by placing a cut stem in a dilute solution of auxin, roots can be readily produced. The auxins, in addition, determine the growth correlations of the several parts of the plant. They inhibit development of the lateral buds and promote growth of the terminal bud. Finally, auxins control the shedding of leaves, flowers, fruits, and branches from the parent plant. By inhibiting the formation of the abscission layer between leaf petioles and the stem or branch, the auxins prevent the leaves from being shed.

• PROBLEM 9-20

What are tropisms? Where in a plant can tropistic responses occur?

Solution: Most animals perceive external stimuli via specialized sense organs and respond with the aid of an elaborate nervous system. Plants have neither sense organs nor nervous systems, but react to stimuli by means of tropisms. A tropism can be defined as a growth movement effected by an actively growing plant in response to a stimulus coming from a given direction. This results in the differential growth or elongation of the plant toward or away from the stimulus. Tropisms are named after the kind of stimulus eliciting them. Phototropism is a response to light; geotropism, a response to gravity; chemotropism, a response to a certain chemical; and thigmotropism, a response to contact or touch.

The mechanisms of tropisms are now under much investigation. At this point, they are believed to involve certain plant hormones which are produced primarily in meristemic regions (regions showing rapid cell division) of the plant. In appropriately low concentrations, these hormones stimulate growth and differentiation. These plant hormones are along the phloem and exert their effects on parts somewhat removed from the site of production. But because they can stimulate only those cells that are capable of rapid division, tropistic responses can occur only in those parts of the plant which are actively growing or elongating, such as the apical part of the stem and the tip of the root.

• PROBLEM 9-21

Suppose someone told you that plants bend toward the light so that their leaves will be better exposed for photosynthesis. Tell how you could demonstrate by experiments that this may not be a scientific explanation of the growth toward the light.

Solution: If plants bend toward the light in order to have their leaves exposed to sunlight, one would assume then, that the signal for stem-bending comes from the leaves. In other words, there will be a mechanism present in the leaves which allows them to receive the stimulus of sunlight and consequently direct the stem to elongate and bend in the direction of the light. We can test the validity of this hypothesis by defoliating the plant. If our hypothesis is correct, removing the leaves would also remove the receptors of the stimulus by light and the effectors of the signal to bend, and no bending of the stem should occur. We know, however, that as long as the apex is intact, a defoliated stem will bend toward the light. The coleoptile tip of the oat seedling, for instance, is a non-leaf-bearing structure which has been used in classical demonstrations of phototropism (the growth response to light).

Since a stem will bend toward the sun even if it has no leaves, one cannot really say that plants bend toward the sun purposefully in order to better expose their leaves for photosynthesis. Our experiment has shown that the leaves themselves do not hold the controlling key to bending. We can advance a step further and test the role of the tip of the stem in phototropism. If the apex is involved in the bending of the stem toward light, covering the tip from the light or removing it totally should prevent phototropism. It has, in fact, been observed in experiments that both operations do prevent phototropism as well as inhibit growth. Replacing the tip on the decapitated stem actually brings about resumed growth and bending of the stem. Thus it is clear that the apical meristem of the stem rather than the leaves of the plant is responsible for the phenomenon of phototropism.

• PROBLEM 9-22

Discuss the physiological basis of phototropism.

Solution: A physiological explanation of phototropism that has been proposed involves a group of plant hormones called the auxins. The bending of a plant shoot toward light in phototropism takes place as a result of unequal rates of growth between the side facing the light and the side shielded from the light. Since growth of an organism involves cell division and enlargement, the difference in the rate of growth in the two sides of the plant reflects a difference in the rate of cell division or enlargement or both. Auxins are known to accelerate cell growth in actively growing regions, such as stem tips and vascular cambium. These hormones have been demonstrated to be produced by the apical meristem, and shown to be transported down the stem from the apex in an active process. In response to unidirectional light, it is found that there is a differential distribution of auxin in the stem, the side of the stem facing light receiving a lower concentration of auxin than the side away from light.

Phototropism is proposed to be the effect of this uneven distribution of auxin as it moves down the stem from the apex. This mechanism suggests that as light strikes a plant unilaterally, the absorption of light energy by pigments such as carotenoids or flavins initiates the differential distribution of auxin. The exact mechanism by which this occurs is not known, but several hypotheses have been advanced. Some workers suggest that light causes greater destruction of auxin on the lighted side of the stem, or that light produces oxidation products of auxin which inhibit auxin transport down the lighted side. Others postulate that the plant redistributes the auxin that is being synthesized and transports it laterally to the shaded side.

However it is accomplished, it is certain that the shaded side has a higher auxin concentration and elongates faster than the illuminated side with a lower auxin level. The effect is that the tip and the top part of the stem curve toward the source of the light.

PLANT HORMONES

• PROBLEM 9-23

A bud on the side of a stem of an orange tree will only develop into a small scale if it is left on the stem. If this bud is removed and placed against the cambium layer of the main stem of a seedling and the terminal portion of the seedling is cut off, the bud may grow to form the entire crown of an orange tree. How can this be explained on the basis of auxin action?

Solution: One of the many effects of auxin is its action as a lateral bud inhibitor. The hormone, produced in the apical meristem, inhibits the growth of lateral buds, resulting in apical dominance or preferential growth at the apex.

In the case above, a bud on the side of a stem of an orange tree will produce only a small scale if it is left on the tree. This is due to the inhibitory action of auxin on the lateral bud, preventing the bud from actively growing and differentiating. If the bud is removed from the parent tree and grafted onto the side of a seedling, very likely the bud will not grow and develop if the dominant apical meristem of the seedling is intact. If, however, the apex is removed from the seedling, and the bud is grafted correctly so that the procambium layer of the bud is continuous with that of the seedling, the bud will grow to form the entire crown of an orange tree. This is because removal the apex of the seedling leads to the elimination of the apical dominance, and lateral buds are now freed from the inhibitory influence of auxin. Continuation of the procambium layers of the bud and of the seedling is important, for the vascular system is derived from the procambium. If the procambium layers are not continuous, the derived vascular systems will not be continuous, and the bud, soon deprived of water and minerals, will eventually die.

• PROBLEM 9-24

Describe the changes accompanying the ripening of a fruit.

Why do fruits kept in plastic bags or in the refrigerator ripen more slowly?

Solution: Ripening in fruit involves a number of changes that convert the fruit from a seed-manufacturing to a seed-dispersing organ. One transformation that frequently occurs is a color change from green to yellow, orange, red, or blue; this increases the fruit's chance of being spotted by potential dispersal agents such as animals and insects. Chlorophylls are broken down and other pigments such as the red and blue anthocyanins are synthesized. Another change is the conversion of starch, organic acids, and oils to sugar so that the fruit becomes sweet. Simultaneously, the fleshy part of the fruit softens as a result of enzymatic digestion of pectin, the principle component of the middle lamella of the cell wall, which gives the unripe fruit its toughness. When the middle lamella is weakened, cells are able to slip past one another. Volatile flavor components are synthesized, and, together with the conspicuous coloring of the fruit, they serve to attract fruit eaters.

During the ripening of a fruit, there is a large increase in cellular respiration, evidenced by a rapid uptake of oxygen. Cellular respiration can be suppressed or retarded by a decrease in available oxygen. Keeping fruits in plastic bags lowers the amount of oxygen available to the fruits and thus delays ripening. Cold also suppresses cellular respiration; this is why we put fruits in plastic bags and/or refrigerate them in order to slow down their rate of ripening.

• PROBLEM 9-25

In the early 1900's, many fruit growers made a practice of ripening fruits by keeping them in a room with a kerosene stove. They believed it was the heat that ripened the fruits. Was this belief based on correct facts? Explain your answer.

Solution: Fruit growers long believed that the heat generated from a kerosene stove had the effect of ripening fruit. However, as experiments showed, it was actually the incomplete combustion products of the kerosene that were responsible for the ripening effect. The most active component of the incomplete combustion products was identified as ethylene. As little as 1 part per million of ethylene in the air will speed the onset of fruit-ripening.

The onset of fruit-ripening occurs when there is a large increase in cellular respiration, as evidenced by a rapid uptake of oxygen. This increase in cellular respiration is triggered by the sudden production of ethylene, now thought to be a natural growth regulator in plants. The sudden production of ethylene, in turn, is induced by auxin. It is believed that some of the effects on fruits once attributed to auxin are related actually to auxin's effects on ethylene production. When a fruit ripens rapidly, its production of ethylene may increase many times over the rate initially necessary to reach the critical triggering level. Thus, "one bad apple in a barrel" producing ethylene can produce enough ethylene to trigger the ripening of the rest. Also, the ethylene produced by the mold on an orange may be enough to set off undesirable changes in the rest of the oranges. Unfortunately, the exact mechanism of ethylene in fruit-ripening is not known, but is a subject currently under much investigation.

• PROBLEM 9-26

Discuss the important effects of gibberellins, cytokinins, and abscisic acid on plant growth.

Solution: Gibberellins, cytokinins and abscisic acid are the three major types of plant hormones that have been identified in addition to auxins. These three hormones interact with auxins and with each other to regulate biological activities in plants. Gibberellins and cytokinins have dominant roles in controlling the early phases of growth and development; auxins become dominant later in controlling cell elongation, and abscisic acid opposes the functions of all these three growth-promoting factors to bring about a well-regulated, balanced pattern of plant growth.

Gibberellins function to lengthen stems, stimulate seed germination, induce flower formation, and increase the size of fruits in some species of plants. The former two are their most vital functions. Seed germination is examined here: Just before germination, the embryo of a seed secretes a gibberellin which induces the production of α-amylase, an enzyme which hydrolyzes the stored starch for energy. The hormone also activates other enzymes of the seed which breaks down the materials of the seed coat, weakening the coat to facilitate the breakthrough of the germinating embryo. It has been shown that actinomycin D inhibits the synthesis of α-amylase in response to gibberellin. Actinomycin-D is a known inhibitor of DNA-dependent RNA synthesis; this suggests that gibberellins cause the activation of certain enzymes by regulating, in some fashion, the expression of the genetic information contained in the DNA, perhaps by uncovering a portion of the DNA so that it can be transcribed to produce a specific RNA segment. This segment can then be translated to form an enzyme (which is a protein).

Cytokinins function in stimulating growth of cells and accelerating their rate of division. Cytokinins can also change the structure of plant tissue: at certain concentrations, cytokinins are shown to cause root and shoot formation in plant tissue cultures. Cytokinins are also found to oppose auxins by causing lateral buds to develop and thus modifying apical dominance. In addition, cytokinins have been demonstrated to prevent leaves from yellowing and hence play a role in delaying senescence. A possible explanation for the mechanism of this function is that cytokinins prevent the progressive "turning off" of segments of DNA (associated with aging) that are responsible for the synthesis of enzymes and protein and the production of other compounds such as chlorophyll.

Abscisic acid was first reported in the early 1960's, and is now known as a growth-inhibitor in plants. Abscisic acid is collected largely from the ovary bases of fruits, and its concentration here is highest at the time of fruit drop. The hormone acts antagonistically to gibberellins by inhibiting the production of α-amylase in seeds and triggering seed dormancy. It opposes cytokinins by inactivating vegetative buds, causing the yellowing of leaves, and reducing the growth rate of plants. Furthermore, it opposes auxins by accelerating abscission (hence its name) and promoting leaf, fruit, flower and branch shedding. It also promotes stomatal closure. In short, by checking growth and reproductive activities of the plant, abscisic acid interacts with the three growth-promoting hormones to bring about an optimal rate of growth and differentiation.

• PROBLEM 9-27

What are some practical applications of plant hormones?

Solution: Auxins have a variety of practical uses and are of tremendous economic importance. Because auxins stimulate root growth, it is possible to propagate a plant vegetatively by obtaining a cut stem containing an apical meristem which can produce auxin. Root formation will occur on the cut stem because of the presence of auxin. A new plant will grow from the cut stem. By treating the female flower part with auxin, it is possible to produce parthenocarpic fruits (fruits formed without pollination and hence without seed, such as seedless oranges and grapes). Auxin-treated fruits grow larger, ripen faster, and remain on the plant until they can be harvested. Plants sprayed with synthetic auxins retain their leaves, branches, and flowers, as well as their fruits, for longer periods of time. The synthetic auxin 2, 4-D (2,4-dichlorophenoxy acetic acid) is widely used as a weed killer. Most of the common weeds are dicotyledonous plants which are more susceptible to the auxin than are the monocots. The dicots take up the 2, 4-D and are stimulated to metabolize at such a high rate that they consume their own cellular constituents and eventually die. Many of the monocots are important food sources, such as corn, oats, rye, barley and wheat. Auxins are especially useful since they destroy the dicotyledonous weeds, but leave the monocotyledonous crop untouched.

Like auxins, gibberellins can cause the development of parthenocarpic fruits, including apples, currants, cucumbers, and eggplants. When treated with both auxin and gibberellin, these plants produce fruits more than twice as large as those obtainable by the application of either hormone alone. Application of gibberellin to some species of long-day (or short-night) plants induces flowering without appropriate long-day exposure. Gibberellins will also substitute for the dormancy-breaking requirement and induce germination by promoting the growth of the embryo and the emergence of the seedling.

Treatment of a plant with cytokinin causes the loss of apical dominance and the plant grows bushier. It also prevents leaves from yellowing, and thus has particular commercial value in indoor, decorative plants.

REGULATION OF PHOTOPERIODISM

• **PROBLEM** 9-28

What is meant by photoperiodism? How would you determine whether a particular flowering plant is a "long-night," a "short-night," or an "indeterminate" plant?

Solution: Photoperiodism is defined as the biological response to a change in the proportions of light and darkness in a 24-hour daily cycle. Photoperiodism has now been shown to control a wide range of biological activities, including the induction of flowering in flowering plants, the stimulation of germination in seeds of certain species, and the initiation of mating in certain insects, birds, fish and mammals.

Based on the phenomenon of photoperiodism, flowering plants can be classified as either long-night (short-day), short-night (long-day), or indeterminate (night-neutral). Experimentally, we can determine the photoperiodic classification of a given flowering plant by varying the length of darkness per day to which the plant is exposed and observing the effect of flowering - whether it occurs or whether it is inhibited. If the plant can produce flowers only when subject to a dark period of about nine hours or more per day, the plant is said to be a long-night plant. Long-night plants, such as asters, cosmos, chrysanthemums, dahlias, poinsettias and potatoes, normally flower in the early spring, late summer, or fall. If, on the other hand, the plant can be made to flower only when exposed to a period of darkness less than nine hours per day, it is said to be a short-night plant. Short-night plants, such as beets, clover, coreopsis, delphinium and gladiolus, normally flower in the late spring and early summer. If the flowering of a plant is unaffected by the amount of darkness per day, it is neither a long-night nor short-night plant, but rather an indeterminate plant. Examples of indeterminate plants are carnations, cotton, dandelions, sunflowers, tomatoes and corn.

It must be emphasized that the critical length of daily darkness for flowering depends on the individual species of plant, and the nine-hour period of darkness as a criterion for classifying flowering plants is at best an approximate figure. The cocklebur, for instance, is a long-night plant, at least 8½ hours of darkness per 24-hour cycle in order to flower.

• PROBLEM 9-29

> What are the functions of florigen and phytochrome in flowering plants?

Solution: In the 1930's, M. H. Chailakhyan of Russia was studying flowering in the long-night plant Chrysanthemum indicum. Long-night plants will flower only when exposed to darkness for nine or more hours. Chailakhyan showed that if the upper portion of the plant was defoliated and the leaves on the lower part exposed to a long-night induction period the plant would flower. If however, only the upper, defoliated part was kept on long-nights and only the lower, leafy part on short-nights, no flowering occurred. Based on this finding, Chailakhyan proposed the production by the leaves of a naturally-occurring flowering hormone, which he named florigen, the "flower maker." This hormone apparently moves from the leaves to the flower bud, stimulating it to mature and flower. Later experiments showed that florigen passes from the leaves to the bud by way of the phloem system, where most organic substances are transported. Unfortunately, nothing is known at present of the chemical composition of florigen nor how it may act to induce flowering.

In the late 1950's, a blue, light-sensitive protein pigment was isolated in plants. Borthwick and coworkers gave this pigment the name phytochrome. In the years that followed, it was discovered that phytochrome exists in two forms: P_r (P_{660}, responds to red light of day) and P_{fr} (P_{730}, responds to far-red light of night). P_{660}, the quiescent form, absorbs red light at a wavelength of 660 nanometers and is converted to P_{660}. In nature, the P_{660}-to-P_{730} conversion takes place in daylight, while the P_{730}-to-P_{660} conversion occurs in the dark. Thus during the day, a plant predominantly contains phytochrome in the form of P_{730}, and during the night it contains mostly the P_{660} form. The prevalence of either form is postulated to provide the plant with a means of detecting whether it is in a light or dark environment. The rate at which P_{730} is converted to P_{660} provides the plant with a "clock" for measuring the duration of darkness.

• PROBLEM 9-30

> Explain how the changes in the phytochrome system are related to blooming.

Solution: Phytochrome is a pigment found in certain plants in very small amounts. The pigment is thought to be responsible for detecting the light environment of the plant. Phytochrome exists in two different forms - P_{660} the inactive form, and P_{730}, the active form. P_{660} absorbs red light and is converted to P_{730}; P_{730} absorbs far-red light and is converted back to P_{660}. In nature, P_{660} is converted to P_{730} in the daytime (when red wavelengths predominate over far-red). The reverse occurs in the night time. Hence, P_{730} predominates in the plant during the day and P_{660} predominates during the night. This is summarized in the figure below. This differential response of phytochrome to light and dark is thought to be involved in the phenomenon of photoperiodism in plants.

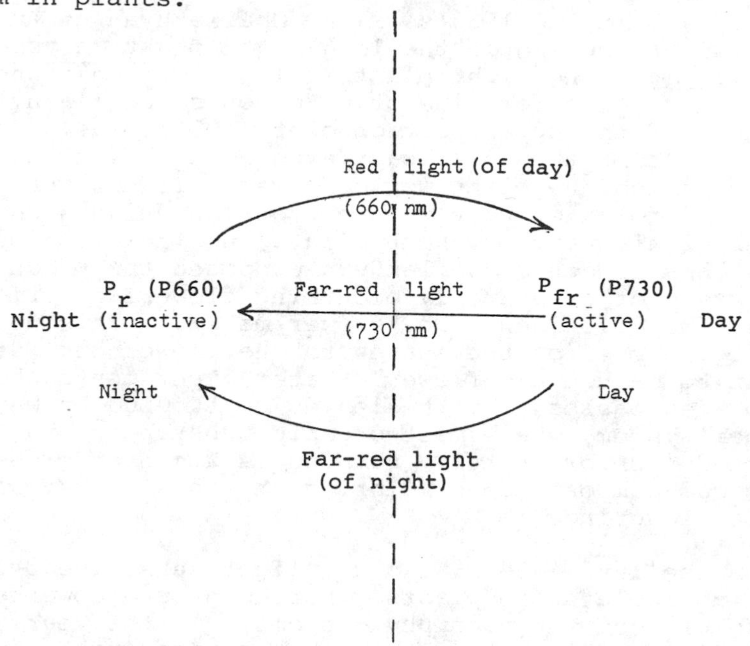

It has been found that red light inhibits flowering in short-night plants but promotes flowering in long-night plants, under conditions during which flowering normally takes place. This experimental observation led many investigators to hypothesize that the P_{730} - P_{660} interconversion might be the plant's time-regulator of flowering. According to this hypothesis, P_{730}, converted from P_{660} by the absorption of red light, would inhibit flowering in short-day plants but promote flowering in long-day plants. Because P_{730} accumulates in the day and diminishes at night, short-day plants could flower only if the nights were long enough, during which a great amount of P_{730} would be inactivated. Long-day plants, on the contrary, would require short nights, during which the P_{730} would not be completely inactivated, so that enough P_{730} would remain at the end of the night to promote flowering.

This hypothesis was highly regarded in the beginning. However, it was dealt a fatal blow by a later study showing that the reversion of all the P_{730} to P_{660} requires only three hours; this has been found to hold true in all plants, both long and short-day, that have been studied. What happens during the rest of the dark period? No satisfactory answer is as yet available. However, extended days provide P_{730} for longer periods.

It is generally agreed at present that the time-measuring phenomenon of flowering is not solely controlled by the interconversion of P_{660} and P_{730}. There seems to be two variables involved.

Experimentally it has been found that one is the presence or absence of light, the other is the length of the dark or light period. Phytochromes seem to be responsible for the detection of either light or darkness, but the actual measuring of the time between the moment the plant senses the onset of darkness and the moment it senses the next exposure to light depends on some mechanism, not yet known. Further investigation is needed before we can fully understand the "biological clock" in flowering.

• PROBLEM 9-31

Design an experiment to demonstrate that florigen, the flowering hormone, is indeed produced in the leaves and travels to the flower bud.

Figure 1. Graft-induce flowering.

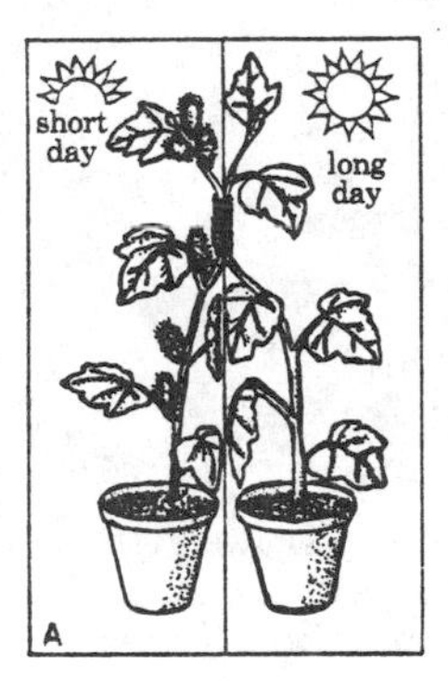

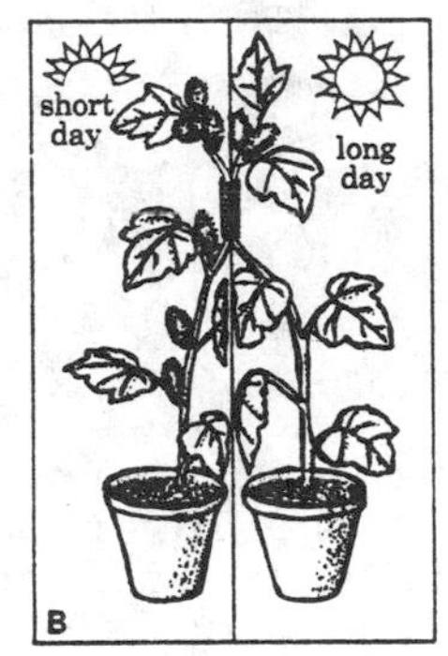

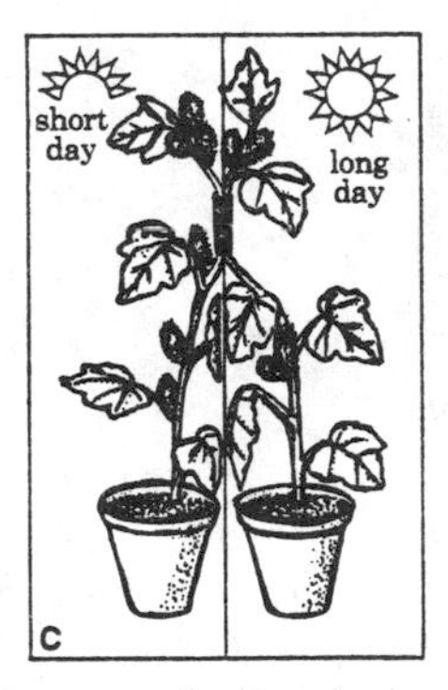

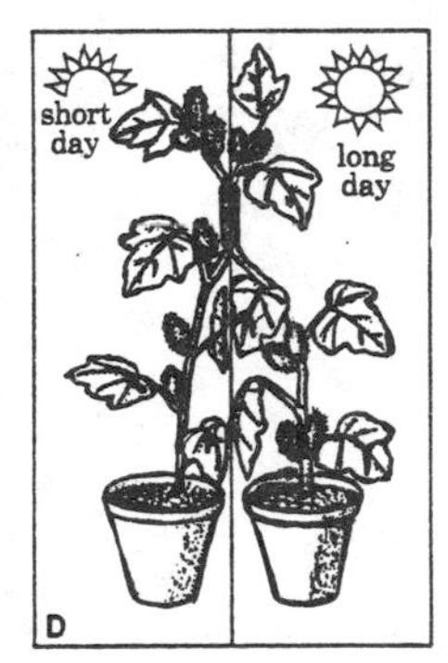

12 hrs per day - change to flower-inducing conditions

18 hrs per day - change to flower-inhibiting conditions

A → a
B → b
C → c
D → d

Solution: To demonstrate that florigen is indeed produced in the leaves, a flowering plant is defoliated and then exposed to appropriate conditions under which flowering normally occurs. The observation that the plant fails to flower indicates that florigen must have been produced in the leaves, and that florigen has been removed as a result of defoliation. A control experiment must of course be set up to show that defoliation of the plant produces no adverse effect on the plant.

It is postulated that in order to exert its effect on the buds, florigen must be transported from the leaves to the flower buds. This can be tested by defoliating the lower part of a plant and exposing it to flower inhibiting factors, while maintaning the upper, leafy part under

Figure 2. A short day plant grafted onto a long day plant.

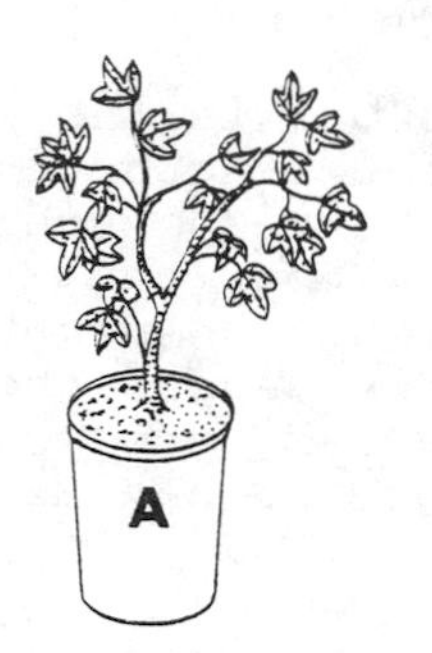

Plant A
(long-day plant)

Plant B
(short-day plant)

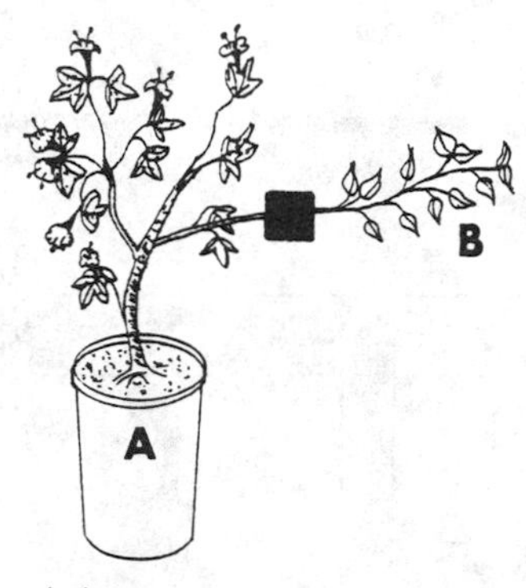

(a) Short-day exposure

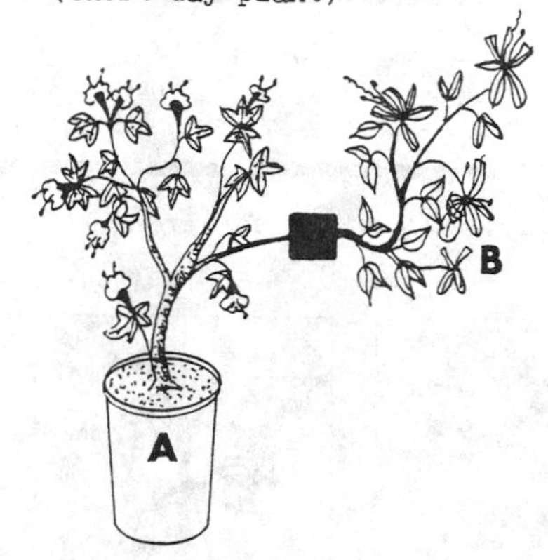

(b) Long-day exposure

flower-promoting conditions. The leafy part is found to flower as expected. Flowering in the lower part, however, can best be explained by the diffusion of florigen from the leaves of the upper part to the buds of the lower part of the plant. The defoliated lower part of the plant can never flower on its own because of the absence of florigen. Because flowers do appear on the defoliated part of the plant, it is concluded that florigen is a diffusible substance made in leaves.

Even more compelling evidence for this conclusion comes from grafting experiments. One can grow two plants, one plant exposed to flowering conditions, the other not. The plant exposed to flowering conditions will flower (fig. 1a). The flowering plant is now grafted onto the non-flowering plant, and a partition is used so that the grafted portion continues to be exposed to flowering conditions and the non-flowering plant to flower-inhibiting conditions (fig. 1b). Gradually, the non-flowering plant will begin to produce flowers, usually at the point nearest the graft (fig. 1c) and eventually throughout the entire plant (fig. 1d).

If one were to graft a short-day plant onto a long-day plant, and expose both to a short-day period, the short-day graft induces flowering in the long-day plant (fig. 2a). Likewise, one can expose both to a long-day period, and both portions will flower, the short-day portion having been induced by the long-day portions (fig. 2b). Similar results can be obtained using day-neutral plants. Thus florigen appears to be physiologically equivalent in the three groups of flowering plants.

Graft-induced flowering is successful only if there is a living tissue connection between the two plants. In addition, flower-induction is inhibited if one removes the phloem containing tissue in the graft. This supports the conclusion that florigen is transported via the phloem system of the plant. Some plants cannot be induced to flower by the above methods unless they are first defoliated. This suggests that, under flower-inhibiting conditions, some inhibiting substance may be produced by the leaves. Flowering may thus involve the interaction of both inducing and inhibiting factors.

SHORT ANSWER QUESTIONS FOR REVIEW

Answer

To be covered when testing yourself

Choose the correct answer.

1. Which process is responsible for the introduction of new genes into the gene pool of the population? (a) sexual reproduction (b) asexual reproduction (c) genetic mutation (d) genetic recombination (e) all of the above — c

2. Which of the following pigments absorb radiant energy and ultimately transfer it to chlorophyll a as high energy electrons? (a) chlorophyll b (b) carotene (c) xanthophyll (d) all of the above (e) none of the above — d

3. What do photosynthesis and oxidative phosphorylation have in common? (a) They both utilize the same source of energy, (b) they have the same site of reaction, (c) they have the same electron donor and acceptor. (d) they both result in the synthesis of ATP, (e) none of the above. — d

4. During the light reactions of photosynthesis, which is not produced? (a) ATP (b) NADPH (c) O_2 gas (d) NADP (e) all are produced. — d

5. What is the final electron acceptor during cyclic phosphorylation? (Light reaction) (a) chlorophyll a (b) ferredoxin (c) plastoquinone (d) cytochromes (e) oxygen — a

6. During non-cyclic phosphorylation how is chlorophyll a^+ reduced to its original form? (a) by the electron stored within the cytochrome (b) by the electrons released following irradiation of cytochrome b (c) by the electrons carried by $NADH_2$ (d) by the electrons released through the splitting of water molecules (e) all of the above — b

7. Which of the following occurs during the process of cyclic photophosphorylation? (a) Oxygen is liberated, (b) NADP is reduced (c) ATP is produced by phosphorylation of ADP, (d) CO_2 is reduced (e) sugar is formed as a final product of reaction. — c

8. Why are metallic ions vital to a growing photosynthetic plant? (a) magnesium is needed for absorption of water in the roots. (b) magnesium is needed for formation of a functional chlorophyll molecule. (c) magnesium is needed as a component of the cytochrome cycle. (d) magnesium is needed as a component of the ferredoxin molecule. (e) all of the above — b

Answer

To be covered when testing yourself

9. Which of the following is not a function of the hormone auxin? (a) it causes the plant to bend and elongate towards the light (b) it directs the tissue differentiation in the vascular cambium (c) it stimulates development of lateral buds and inhibits growth of the terminal bud (d) it controls the shedding of leaves, flowers, fruits, and branches (e) it determines growth correlations of plant parts. — c

10. Which of the following growth movements is responsible for deepening of the plant roots? (a) geotropism (b) chemotropism (c) thigmotropism (d) phototropism (e) all of the above. — a

11. Which statement is incorrect concerning the physiological basis of phototropism? (a) phototropism takes place as a result of unequal rates of growth (b) phototropism results from a difference in cell division and cell elongation (c) auxins are known to hinder cell growth in actively growing regions (d) all of the above (e) none of the above. — c

12. What is the most vital function of gibberellins in plants? (a) to lengthen stems and stimulate seed germination (b) accelerating the rate of division of cells (c) stimulating the elongation of the cells (d) acts as a growth inhibitor of the stems and leaves (e) promotes the shedding of branches and leaves — a

13. Which of the following hormones is produced by the leaves of the plant and is responsible for flowering? (a) phytochrome (b) auxin (c) gibberrelin (d) florigen (e) cytokinins — d

14. In which part(s) of the plant is the hormone florigen produced? (a) leaves (b) stems (c) roots (d) all of the above (e) none of the above. — a

15. In the overall reaction for photosynthesis;

$$CO_2 + H_2O \xrightarrow{hv} (CH_2O) + O_2$$

where does the carbohydrate oxygen come from? (a) CO_2 (b) H_2O (c) O_2 (d) All of the above can donate the oxygen atom. (e) none of the above can donate the oxygen atom. — a

In questions 16-25 match the appropriate terms to their corresponding descriptions.

16. Tricarboxylic acid cycle _____ (a) phloem — 16-b,

		Answer To be covered when testing yourself
17. Iron complex	_____ (b) Krebs cycle	17-e,
18. Nitrogenous excretion	_____ (c) tropism	18-i,
19. Microfibrils	_____ (d) phytochrome	19-f,
20. Water conduction	_____ (e) cytochromes	20-h,
21. Sieve tube	_____ (f) cellulose	21-a,
22. Actively growing region	_____ (g) ethylene	22-j,
23. Growth movement	_____ (h) xylem	23-c,
24. Growth regulator	_____ (i) diffusion	24-g,
25. P_{660}	_____ (j) apical meristem	25-d

Fill in the blanks.

Question	Answer
26. The excited _____ molecule transfers light energy via high energy electrons to a chlorophyll a molecule	chlorophyll b
27. Carotenoids that do not contain oxygen are called _____ and are deep orange in color, those that contain oxygen are called _____ and are yellowish.	carotenes, xanthophylls
28. _____ allows the plant to discern whether it is in a dark or light environment.	phytochrome
29. A series of alternating single and double bonds as is seen in chlorophyll is known as a(n) _____.	resonating system
30. The radiant energy absorbed during the light reactions by chlorophyll is always of a specific quantity known as a(n) _____.	photon
31. _____ are the organelles associated with photosynthesis in plants.	chloroplasts
32. The enzymes responsible for the dark reactions are contained within the _____.	stroma
33. _____ which are repeating unit structures found on the surface of the lamellae within the chloroplasts are now believed to be the functional photosynthetic units.	quantosomes
34. Oxidative phosphorylation takes place in the _____.	mitochondrion

	Answer
	To be covered when testing yourself
35. _____ is an iron containing protein that functions in capturing energy rich electrons from excited chlorophyll molecules.	ferre-doxin
36. The process of photosynthesis is restricted to those cells that contain _____.	chloro-phyll
37. The final products of cellular respiration are _____ and _____ .	CO_2, H_2O
38. The inorganic element _____ is an essential component of the chlorophyll molecule.	magne-sium
39. Most nitrogenous wastes are readily eliminated from plants by the process of _____.	dif-fusion
40. A polysaccharide which is vital to plants and composed of repeating glucose molecules is _____.	cellu-lose

CHAPTER 10

NUTRITION AND TRANSPORT IN SEED PLANTS

PROPERTIES OF ROOTS

• PROBLEM 10-1

If a ring is cut through the bark all the way around a tree, down to the wood, the tree will live for a while, then die. Explain why.

Solution: When a ring is cut through the bark down to the wood of a tree, the vascular system is being separated into two halves. Water moves continuously upward in the xylem in response to osmotic pressure build-up in the roots, transpiration and capillary action. Severing the xylem will prevent the upper region of the tree from obtaining water from the roots by the processes mentioned. The upper half of the tree, deprived of further supply of water could however continue to carry on photosynthesis using as much of the remaining water and minerals as it has stored in the upper half of the bisected xylem.

It will inevitably suffer rapid water loss as a result of evaporation at the surface of the leaves. As a consequence of the water loss at the leaf surface and the failure to receive water from the roots, the cells in the upper half of the tree gradually lose their turgor. This reduced turgidity causes the guard cells to collapse, closing the stomata in an attempt to prevent further water loss through the stomata.

What this indicates is that even though photosynthesis could theoretically be maintained in the upper region temporarily on stored nutrients and water, this usually does not occur. Photosynthesis is suspended as the stomata close to avoid desiccation.

Since the phloem is also discontinued at the cut, organic substances traveling down the phloem ooze out of the cut, and cannot be delivered to the lower portion of the tree. The lower half of the stem and the root system can nevertheless sustain a low level of respiration for a short while, using the food reserve, such as starch, stored in the cells. They will reduce their energy expenditure by reducing cell division and enlargement, which are energy requiring processes.

Thus, after the cut, the tree will strive to survive with minimal growth for a while. But once the water supply is exhausted in the upper half of the tree, and the food reserve is used up in the lower half, both halves cannot generate any more ATP to carry out the vital biological activities, and the tree will eventually die.

• PROBLEM 10-2

What are the functions of roots? What are the two types of root systems?

Solution: Roots serve two important functions: one is to anchor the plant in the soil and hold it in an upright position; the second and biologically more important function is to absorb water and minerals from the soil and conduct them to the stem. To perform these two functions, roots branch and rebranch extensively through the soil resulting in an enormous total surface area which usually exceeds that of the stem's. Roots can be classified as a taproot system (i.e. carrots, beets) in which theprimary (first) root increases in diameter and length and functions as a storage place for large quantities of food. A fibrous root system is composed of many thin main roots of equal size with smaller branches.

Additional roots that grow from the stem or leaf, or any structure other than the primary root or one of its branches are termed adventitious roots. Adventitious roots of climbing plants such as the ivy and other vines attach the plant body to a wall or a tree. Adventitious roots will arise from the stems of many plants when the main root system is removed. This accounts for the ease of vegetative propagation of plants that are able to produce adventitious roots.

• PROBLEM 10-3

Describe the growth zones of a typical root.

Solution: The tip of each root is covered by a protective root cap, a thimble-shaped group of cells. This root cap protects the rapidly growing meristemic region of the root tip. The outer part of the root cap is rough and uneven because its cells are constantly being worn away as the root pushes its way through the soil. The meristem, protected by the root cap, consists of actively dividing cells from which all the other tissues of the root are formed. It also gives rise to new root cap cells to replace the ones that are sloughed off. Immediately behind the meristem is the zone of elongation. The cells of this zone remain undifferentiated but grow rapidly in length by taking in large amounts of water. The meristematic region and the zone of elongation together account for the increase in length In the zone of maturation (or differentiation) of the root cells differentiate into the permanent tissues of

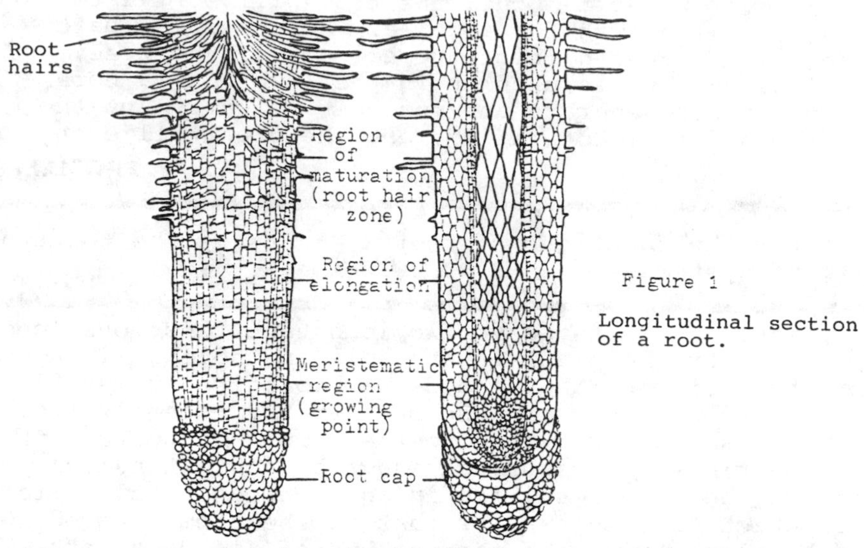

Figure 1

Longitudinal section of a root.

the root, for example, the xylem and the phloem tissues. Present in this region too, are slender, elongated, hairlike projections called root hairs that arise laterally. There is one root hair to each epidermal cell. Root hairs greatly increase the surface area of the root and enhance the absorption of the water and minerals into the root. As the root elongates, the delicate, short-lived root hairs wither and die, and are replaced by the ones newly formed by the meristematic region. Root hairs occur only in a short lower segment of the zone of maturation, usually only 1 to 6 cm long.

• PROBLEM 10-4

Describe the processes by which a root absorbs water and salts from the surrounding soil.

Solution: The movement of water from the soil into the root can be explained by purely physical principles. The water available to plants is present as a thin film loosely held to the soil particles and is called capillary water. The capillary water usually contains some dissolved inorganic salts and perhaps some organic compounds, but the concentration of these solutes in capillary water is lower than that inside the cells of the root. The cell sap in the root hair of the epidermal cells has a fairly high concentration of glucose and other organic compounds. Since the plasma membrane of this cell is semi-permeable (it is permeable to water but not to glucose and other organic molecules), water tends to diffuse through the membrane from a region of higher concentration (the capillary water of the soil) to a region of lower concentration (the cell sap of the root hair). This movement of water is controlled by a process called osmosis.

As an epidermal (root hair) cell takes in water, its

cell sap now has a lower solute concentration than that of the adjacent cortical cell. By the same process of osmosis, water passes from the root hair to the cortical cell. Because of the osmotic gradient, water will continue to diffuse inward toward the center of the root. In this way, water finally reaches the xylem and from there, water is transported upwards to the stem and leaves by a combination of root pressure and transpiration pull.

• PROBLEM 10-5

Explain why simple diffusion is not sufficient to account for the passage of all of the dissolved minerals into the cells of water plants.

Solution: Simple diffusion cannot account for the passage of all of the dissolved minerals into the cells of water plants. Plant cells frequently contain minerals in higher concentration than those available in the soil or water in which they are growing. Simple diffusion of materials into the cells cannot occur against a concentration gradient. Sometimes this gradient can be very steep. Studies of the vacuolar sap of *Nitella clavata*, a pond plant, for instance, have shown that the concentration of mineral anions in the sap exceeds that in the pond by 10^4 times! For this to happen, the plant must be able to transport minerals and other materials against this concentration gradient. Such a process is known as active transport. In contrast to simple diffusion which requires no energy expenditure, active transport can occur only with the expenditure of energy. In the living cell, the energy for active transport comes from the energy-rich phosphate bond of the ATP molecule.

• PROBLEM 10-6

What is wrong with the following statement? "In many desert plants the roots grow very deep because they are searching out for deep underground water to supply the plant." Explain your answer.

Solution: Although the observation of root length in desert plants is correct, the explanation given is oversimplified. It is a teleological interpretation of the observation, or an interpretation in terms of end results. Such a teleological account implies that the plant has the ability to perceive its needs and to respond in order to satisfy these needs. Yet we know that a plant has nothing like a brain or nerves so it would not be possible for it to make such perceptions nor to exhibit goal-directed behavior.

The true reason for desert plants having long roots may be the result of natural selection. We know that plant roots grow in response to gravity. This is an intrinsic response to their growth hormone auxin. Due to the fact that there is a low supply of water in the desert, those plants with short roots may not survive while plants with long roots may be capable of absorbing sufficient amounts of water from the deep underground, and thus survive. Therefore, natural elements favor plants which possess adaptive characteristics for desert survival and restrict those with unfavorable traits. This forms the basis of natural selection. After millions of years, desert plants would have evolved long roots which are essential for their survival. Thus, today we find desert plants with long roots.

DIFFERENTIATION BETWEEN ROOTS AND STEMS

• PROBLEM 10-7

What are the functions of the stem? How are stems and roots differentiated?

Solution: The stem is the connecting link between the roots, where water and minerals enter the plant, and the leaves, where organic foodstuffs are synthesized. The vascular tissues of the stem are continuous with those of the root and the leaves and provide a pathway for the transport of materials between these parts. The stem and its branches support the leaves so that each leaf is exposed to as much sunlight as is possible. The stem of a flowering plant also supports flowers in the proper orientation to enhance reproduction and later, seed dispersal. Along the stem are growing points where the primordia of leaves and flowers originate. Some stems have cells which carry out photosynthesis, others have cells specialized for the storage of starch and other nutrients.

Roots and stems are structurally quite different. Stems have an epidermis covered with a protective layer of cutin while the epidermis of roots has no cutin but gives rise to hairlike projections called the root hairs. Stems, but not roots, have nodes, which are junctions where leaves arise. Stems may have lenticels for "breathing" while roots do not. The tip of a root is always covered by a root cap whereas the tip of a stem is naked unless it terminates in a bud. The dicot stem typically contains separate rings of xylem and phloem, with the xylem central to the phloem, whereas in the roots, phloem tubes lie between the arms of the star-shaped xylem tissues.

• PROBLEM 10-8

What structures are present in both the root and the stem of a higher vascular plant? Describe the function of each structure.

Solution: Both the stem and the root of a plant are

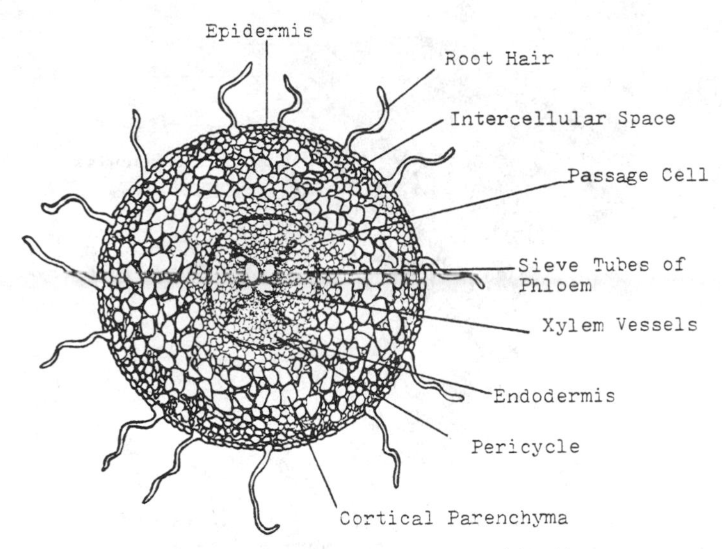

Figure 1 Cross section of a root near its tip.

covered on the outside by a layer of rectangular cells, known as the epidermis (see figures 1 and 2). The cells of the epidermis of the root are thin-walled and non-cutinized, in contrast to those epidermal cells of the stem which are heavily cutinized. If a waxy cuticle were present in the root epidermis, it would undoubtedly interfere with the absorption of water. The epidermis of the root gives rise to root hairs which increase the surface area for absorption. Just inside the epidermis is the cortex. The cortex of the root is a wide area composed of many layers of large, thin-walled, nearly spherical parenchymal cells. The cortex of the stem is narrower than that of the root. The cortical cells

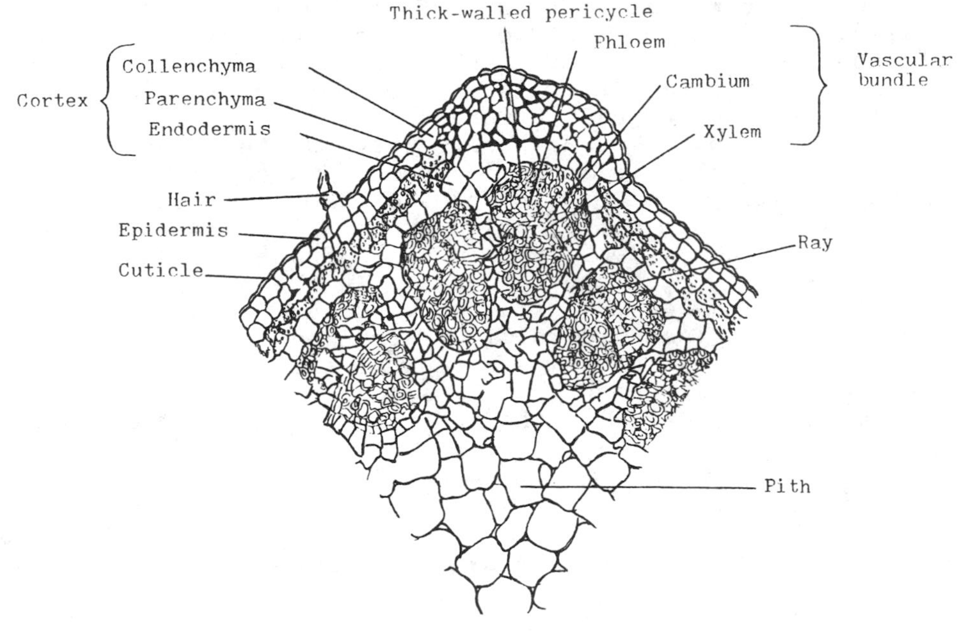

Figure 2 A sector of a cross section of a stem from a herbaceous dicot, alfalfa.

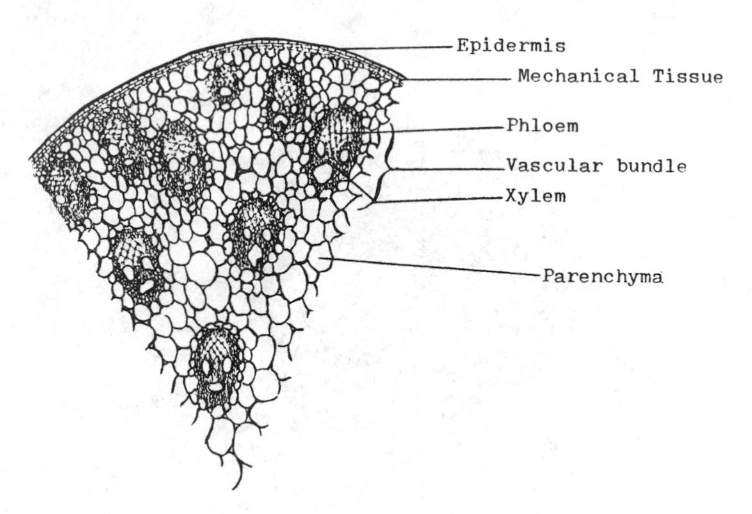

Figure 3 A sector of a cross section of a stem from a monocot, corn.

differ from those of the root in that the former is photosynthetic while the latter is not. In addition, there is an outer layer of thick-walled collenchymal cells in the stem, which serve as supportive tissue. Collenchymal cells are not present in the root. Inside the ring of cortex is a one-called thick layer called the endodermis. The endodermis is interrupted at some points by passage cells through which water and minerals can pass. Immediately adjacent to the endodermis is another layer knows as the pericycle. The pericycle of the root consists of thin-walled parenchymal cells which can be transformed into meristematic cells that later give rise to lateral roots. The pericycle of the stem, however, has thick-walled cells for the purpose of support. On the inner border of the pericycle is the vascular system, composed of xylem and phloem tissues. In the root, the vascular tissues occupy the central longitudinal portion. The cells of the root xylem, tracheids and vessels, are usually arranged in the shape of a star, or like the spokes of a wheel. They are thick-walled and elongated tubular cells which conduct water and dissolved mineral salts. Between the arms of the xylem star are bundles of phloem cells, which are smaller and thinner-walled than the xylem. Phloem cells function in the conduction of organic substances. In the stem, the xylem and phloem tissues are arranged in scattered bundles in the monocots (see figure 3) or arranged in a ring in the dicots. The stem contains a central axial core of tissue, called the pith, which is composed of colorless parenchymal cells that serve in the function of nutrient storage.

HERBACEOUS AND WOODY PLANTS

• PROBLEM 10-9

Differentiate between herbaceous and woody plants and between annual, biennial and perennial plants. How do the stem of woody plants increase in diameter?

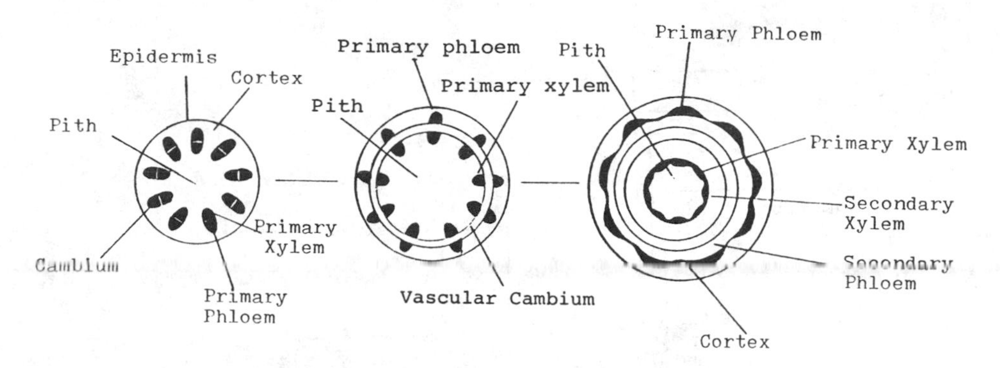

Stages of growth in a woody stem.

Solution: Plants are generally differentiated into two types based on the nature of their stems. Herbaceous plants have a supple, green, rather thin stem, woody plants have a thick, tough stem or trunk, covered with a layer of cork. This cork is derived from the cork cambium, a layer of meristematic cells in the outer cortex of woody stems. Herbaceous plants are typically annuals or biennials. Annual plants start from seed, develop flower, and produce seeds within a single growing season, and die before the following winter. Biennial plants characteristically have two-season growing cycles. During the first season, while the plant is growing, food is stored in the root. Then the top of the plant dies and is replaced in the second growing season by a second top which produces seeds. Carrots and beets are examples of biennials. Woody plants, exemplified by a wide variety of trees and shrubs, are usually perennial plants, which live longer than two years. Perennials have been found that live hundreds or even thousands of years. They produce seeds yearly.

The stems of woody plants resemble herbaceous ones during their first year of growth, with each containing an epidermis, a cortex, an endodermis, a pericycle, vascular bundles, and a pith. But by the end of the first growing season, a cambium layer has taken a form which extends as a continuous ring between the primary xylem and phloem, pushing the phloem to the periphery of the stem and keeping the xylem toward the pith (see figure). This cambium layer, called the vascular cambium, is a region of rapid cell division, that is, the cambium is meristemic. In each successive year, this ring of meristem divides to form two types of cells: those inside the ring differentiate into the secondary xylem elements, and those outside the ring become the secondary phloem cells. The yearly deposits of xylem form the annual rings. Addition of a new ring to the old ones each year causes the stem of a woody plant to increase in diameter. Monocotyledons and some dicotyledons have herbaceous stems, while all gymnosperms and some dicotyledons have woody stems.

• PROBLEM 10-10

Differentiate between sapwood and heartwood and between spring wood and summer wood. What are the vascular rays?

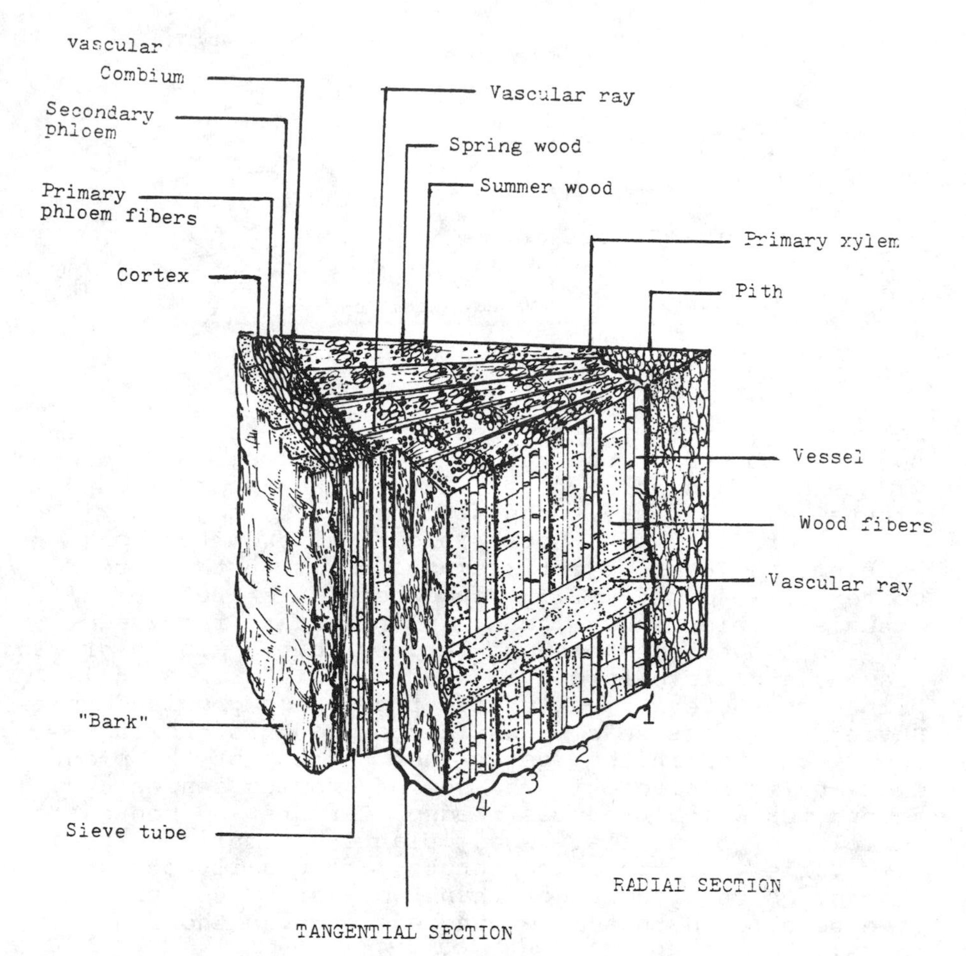

Diagram of a four year old woody stem.

<u>Solution:</u> The sapwood is the youngest, outermost portion of the xylem in a woody plant. It contains tracheids and vessels which function in carrying water and minerals from the root to the leaves. The heartwood is the older, inner portion of the xylem whose tracheids and vessels have lost the ability of conduction. The heartwood serves to increase the strength of the stem, and to support the increasing load of foliage as the tree grows.

The spring wood and summer wood together make up the secondary xylem ring (see figure) of each growing season. They account chiefly for the increase in diameter of a woody tree trunk. They are derived from the vascular cambium, which also gives rise to the secondary phloem. At the beginning of each growing season, the cambial cells divide to produce relatively large, thin-walled xylem cells, which make up the spring wood (early wood). Toward the end of each season, the xylem cells produced are smaller and have thicker walls. These smaller cells constitute the summer wood (late wood). Ordinarily, there is only one growth season per year and abrupt change in cell size between the spring wood and summer wood clearly marks off distinct alternate circular zones called annual rings. The number of annual rings found in the cross-section of a woody trunk is now used as an indicator of the age of the tree. In addition, because the width of the annual rings varies according to the climatic conditions prevailing when ring was formed, it is possible to infer what the climate was at a particular time by examining the rings of old trees.

The vascular rays are rows of small, parenchymal cells that lie at right angles to the long axis of the stem or root, reaching radially from the secondary xylem to the secondary phloem. They are, like the secondary xylem and phloem, derived from the vascular cambium, and extend in length as the plant increases in girth. The vascular rays conduct water and minerals from the secondary xylem to the vascular cambium, the secondary phloem, and the cortical cells. Nutrients form the secondary phloem are channeled to the vascular cambium, the living cells of the secondary xylem, and the cells of the pith (if living).

• PROBLEM 10-11

What is the function of the leaves? Describe the structures of a typical dicot leaf.

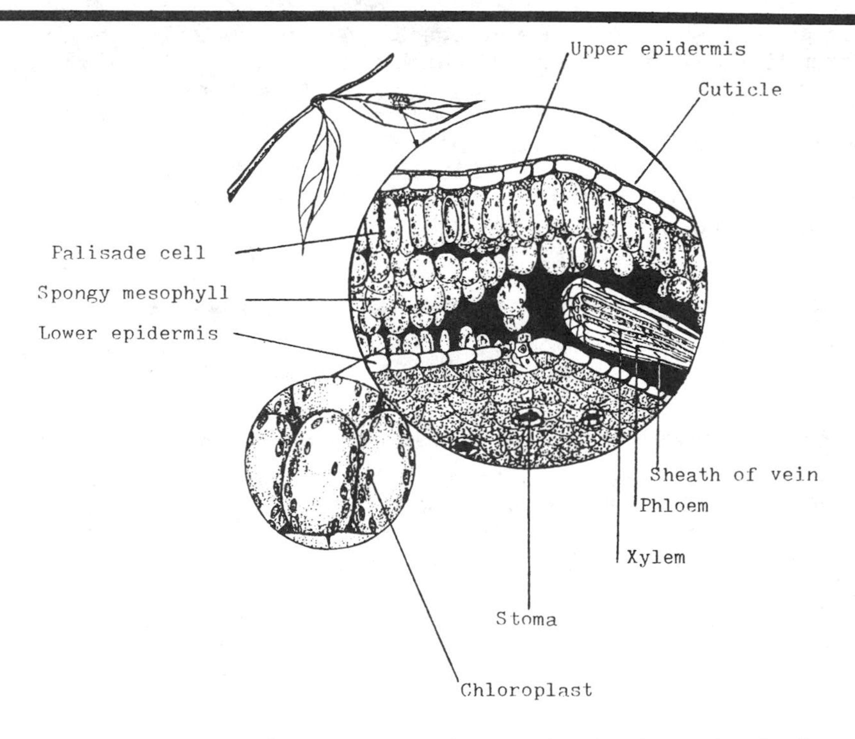

A diagram of the microscopic structure of a leaf.

Solution: The leaf is a specialized nutritive organ of the plant. It's function is to carry out photosynthesis, a process requiring a continuous supply of carbon dioxide, water, and radiant energy.

The leaf of a typical dicot consists of a stalk, called the petiole, attached to the stem, and a broad blade, which may be simple or compound. The petiole, like a stem in cross-section, contains vascular bundles. Within the blade, the vascular bundles fork repeatedly and form a network of veins. A cross-section of the leaf shows that it is composed of several types of cells. The outer cells, lining both the top and bottom surfaces of the leaf, make up the epidermis which secretes a protective, waterproof cutin covering. Distributed throughout the epidermal surface are many small, specialized pores, the stomata, each surrounded by two guard cells. There are many more stomata on the lower surface than the upper surface of the leaf in most species. The chloroplast-containing guard cells regulate the size of the stomata by changing their turgor pressure and thus their shape. The stomata allow oxygen to diffuse out and carbon dioxide to diffuse into the leaf. Most of the space between the upper and lower epidermal layers is filled with thin-walled cells, called mesophyll,
Most of the space between the upper and lower epidermal layers is filled with thin-walled cells, called mesophyll,
which are full of chloroplasts. The mesophyll layer near the upper epidermis is usually made of cylindrical palisade cells, closely packed together with their long axes perpendicular to the epidermal surface. The rest of the mesophyll cells are very loosely packed together, with large air spaces between them. These loosely-packed cells form the spongy layer of the leaf. The cells of the palisade layer, containing abundant chloroplasts, are chiefly responsible for the photosynthetic functioning of the leaf, while the air spaces between the mesophyll of the spongy layer hold moisture and gases. These air spaces are continuous with the stomata where exchange of gases takes place.

• PROBLEM 10-12

What are the functions of (a) the vascular cambium, (b) stomata, (c) heartwood, (d) lenticels (e) abscission layer and (f) cutin?

Solution: The vascular cambium is a meristemic region found in the stem. Sometimes it is found in the roots of higher plants. It exists as a continuous ring of cells extending between the cluster of phloem cells and xylem cells. The cambium produces these two types of cells during periods of active mitotic cell division. The ones facing the center of the stem differentiate into xylem cells; and the ones facing the periphery become phloem cells.

The stomata are small specialized pores scattered over the epidermal surfaces of leaves and some stems. These pores are found mainly on the lower surface of the leaves. The stomata are delineated by a pair of cells called the guard cells. The guard cells, by changing their shape (due to changes in their turgor pressure), regulate the size of the stomatal aperture. An open stoma permits the escape of water and the exchange of gases.

The heartwood is the inner layer of the xylem in a woody plant. They are composed of dead, thick-walled xylem cells which have lost the ability to conduct water and minerals. Instead it has become tough supporting fibers. The heartwood increases the strength of the stem, and accomodates the increasing load of foliage as the tree grows.

Lenticels are masses of cells which rupture the epidermis and form swellings. These masses of cells are formed from cells of the cork cambium which divide repeatedly and rupture the epidermis lying above. Lenticels represent a continuation of the inner plant tissues with the external environment, and permits a direct diffusion of gases into and out of the stem or twig. Such direct passages are necessary because the cork cambium forms a complete sheath around the vascular bundles and effectively obstructs the ventilation of the vascular bundles.

An abcission layer is a sheet of thin-walled cells, loosely joined together, which extends across the base of the petiole. The formation of the abscission layer separates and loosens the point of attachment of the petiole to the stem. After the abscission layer is formed, the petiole is held on only by the epidermis and the fragile vascular bundles, so that wind or other mechanical disturbances will cause the leaf to fall off. This is the mechanism that accounts for the fall of leaves during autumn.

Cutin is a waxy organic substance secreted by the epidermal cells of the stems and leaves, but not the roots. Its waterproof property retards evaporative water loss to the atmosphere.

GAS EXCHANGE

• **PROBLEM 10-13**

Discuss the mechanism by which the guard cells regulate the opening and closing of a stoma.

Solution: The opening and closing of a stoma is regulated by changes in the turgor pressure within the two guard cells that surround the stomatal opening. Each bean-shaped guard cell has a thicker wall on the side toward the stoma called the inner wall, and a thinner wall on the side away from the stomatal opening called the outer wall. Increased turgor pressure causes the cells to bulge. Since the outer walls are thinner than the inner walls, the former stretch more than the latter, causing the cells to bow in and the stoma to open. When

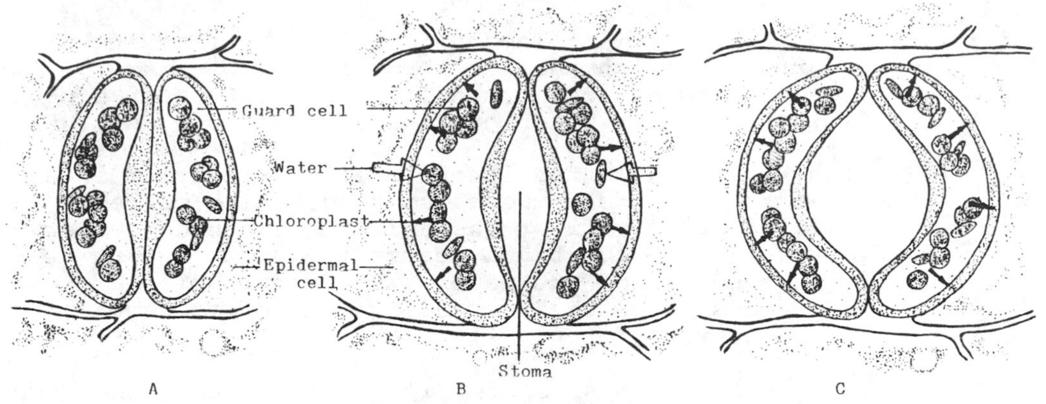

Figure Diagrams illustrating the regulation of the size of the stoma by the guard cells.

the turgor pressure in the guard cells decreases, the inner walls regain their original shape and the cells unbend, closing the stoma as a result.

Thus the opening and closing of a stoma depends on a mechanism which varies the turgor pressure within the guard cells. This variation of turgor involves in part the production of glucose and other osmotically active substances in the guard cells themselves. It is believed that light initiates a sequence of enzymatic reactions that lead to the conversion of starch stored in the guard cells into osmotically active glucose molecules, thus increasing the turgor pressure of the guard cells. As light increases the rate of photosynthesis, carbon dioxide is utilized and the decrease in carbon dioxide concentration, increases the pH of the guard cells (remember that guard cells contain chloroplasts and so they are able to undergo photosynthesis). This increased basicity stimulates the enzyme phosphorylase to convert starch to glucose-1-phosphate. The glucose-1-phosphate is subsequently converted to glucose, increasing the concentration of glucose in the guard cells and causing the influx of water into the cells by osmosis. (Another osmotically active solute may be K^+ -potassium ions- which are pumped into the guard cell from the epidermal cell as CO_2 levels decrease.) This phenomenon raises the turgor pressure in the guard cells and causes the stoma to open (see Fig.) In the dark the process is reversed. Thus under normal conditions, the stomata in many species open regularly in the morning and close in the evening.

• PROBLEM 10-14

Discuss the responses of stomata to environmental factors.

Solution: A number of environmental factors affect stomatal opening and closing. Water loss is the most important among them. When the turgor of a leaf drops below a certain critical point because of rapid evaporation to the atmosphere, the stomatal aperture responds to this loss by becoming smaller or closed. In this situation, the need for water conservation overrides the need for the intake of carbon dioxide for photosynthesis. With the stomata closed, the leaf cells can use the carbon dioxide produced by their respiration, and thus be able to sustain a very low level of photosynthesis.

Factors other than water loss that affect the stomata include light, carbon dioxide concentration, and temperature. It has been shown that when a turgid leaf which has been kept in darkness for a few hours is exposed to light, the stomata open. When the light is switched off, the stomata close. Light apparently plays a role in the opening and closing of the stomata. Light is now believed to initiate a series of enzymatic reactions which break down insoluble starch to many smaller, soluble glucose molecules in the guard cells of a stoma. As the solute concentration increases, water enters the guard cells by osmosis, increasing their turgor, thereby opening the stomata. The role of carbon dioxide is similar to that of light, in the opening and closing of the stomata. It is found experimentally that a decrease in carbon dioxide increases the pH in a cell and an increase in pH facilitates the conversion of starch to glucose. This is explained by the fact that the enzyme phosphorylase is stimulated by the increased pH to convert starch to glucose-1-phosphate, which is in turn converted to glucose. As discussed earlier, water now enters the leaf cells, turgor pressure increases and as the guard cells bend, the stomata open. Temperature also fits into this scheme because of its influence on carbon dioxide concentration.

An increase in temperature increases the rate of respiration, and so increases in turn the cellular carbon dioxide level. A decrease in temperature, on the other hand, retards respiration, and thus decreases the cellular carbon dioxide levels.

It is clear that the stomata are responsive to a complex interplay of factors and are able to respond efficiently to different combinations of these factors.

• **PROBLEM** 10-15

Explain why the stomata of a leaf tend to open up during the day and to close at night. What advantage is there in this?

Solution: During the day, the plant undergoes photosynthesis, a process which converts carbon dioxide and water into carbohydrates. This results in a drop of the carbon dioxide content of the cell as carbon dioxide is used up. When this happens, the acidity of the cell decreases, promoting the conversion of the insoluble starch to the soluble glucose. This can be explained by the fact that the cellular enzymes that catalyze the conversion of starch to glucose do not function in acidic surroundings. During the night hours, production of carbon dioxide occurs during respiration. In addition, no carbon dioxide is depleted by photosynthesis which occurs only during the day. This results in a high carbon dioxide level in the cells. When its levels are high, carbon dioxide will combine with water to form carbonic acid. This will inhibit the conversion of starch to sugar, due to a decrease in the pH. When the carbon dioxide level in the cell is as low as it is during the daytime when photosynthesis occurs, much of the carbonic acid is reconverted to carbon dioxide and water, increasing the alkalinity of the cell and thus raising the pH. The greater alkalinity of the plant cells during the day brings about a higher cellular glucose concentration with the reverse true at night.

During the day, the high cellular concentration of glucose causes water to enter the guard cells by osmosis, rendering them turgid. The turgor of the guard cells cause them to bend, opening the stoma as a result. At night, the acidity of the cell inhibits the breakdown of starch. Since starch is insoluble in the cell sap, it does not exert an osmotic pressure. In the absence of osmotic pressure, the guard cells lose their turgor, causing them to collapse and close the stoma.

What then is the purpose of having the stoma open in the day and closed at night? In the daytime, when the leaves are photosynthesizing, the carbon dioxide supply in the cell is constantly being depleted and its supply must be replenished in order to maintain photosynthesis. Thus it is highly advantageous to have the stomata open during the day, in order for carbon dioxide molecules to enter the leaf from the atmosphere through the stomata. Open stomata also result in continuous transpiration, hence the air spaces of the leaf are always kept moist for carbon dioxide to dissolve in. This is important since carbon dioxide must first dissolve in water before it can enter the cell. At night when photosynthesis ceases, an external supply of carbon dioxide is not required. The stomata close to retard water loss by transpiration which may occur rapidly if the night is dry or the temperature is high.

• PROBLEM 10-16

Explain the process of gaseous exchange in the leaf.

Solution: A leaf that is actively carrying on photosynthesis must have a high rate of gas exchange with the environment to remove the excess oxygen and to acquire an adequate supply of carbon dioxide. As carbon dioxide is utilized in photosynthesis, its concentration within the cell decreases and carbon dioxide diffuses into the cell from the film of water surrounding the cell. Other carbon dioxide molecules pass from the air spaces of the spongy layer of the leaf and dissolve in the water film. They will then diffuse into the cell as the cell rapidly uses up carbon dioxide. Still other

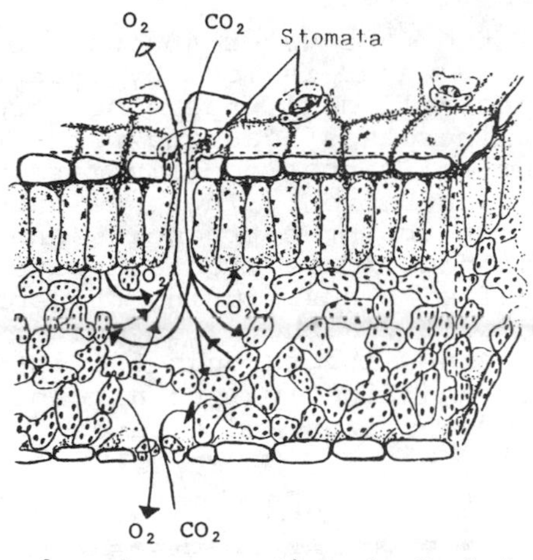

An enlarged cross section of a leaf showing its structure.

carbon dioxide molecules diffuse from the air outside the leaf through the stomata and into the air spaces within the leaf. In short, carbon dioxide moves into the cell along a concentration gradient.

Carbon dioxide molecules diffuse from a region of higher concentration such as the atmosphere, to a region of lower concentration, in this case, the cells of the leaf . In a similar manner but opposite direction, oxygen produced within the cells as a result of photosynthesis passes from cell to water film to air spaces and through the stomata to the exterior by diffusion along a concentration gradient.

TRANSPIRATION AND GUTTATION

• PROBLEM 10-17

What is the role of transpiration in plants? Describe the mechanism of transpiration.

Solution: The leaves of a plant exposed to the air will lose moisture by evaporation unless the air is saturated with water vapor. This loss of water by evaporation, mainly from the leaves (but also from the stems) is called transpiration. Transpiration plays a very important role in water transport in plants. It directs the upward movement of water and minerals from the soil to the leaves, keeps the air spaces of leaves moist for carbon dioxide to dissolve in, concentrates initially dilute leaf cell solutions of minerals that have been absorbed by the roots, and contributes to the cooling of the plant body by removing heat during the vaporization of water (540 calories of heat are needed to convert each gram of water to water vapor).

Transpiration is responsible for the great amount of water that passes through a plant body per day.

About 98 per cent of the water absorbed by the roots is lost as vapor. As water is lost by evaporation from the surface of a mesophyll cell, the concentration of solutes in the cell sap increases, causing water to pass into it from neighboring cells that contain more water, that is, lower solute concentration. These neighboring cells in turn receive water from the tracheids and vessels of the leaf veins, which ultimately obtain water from the soil via root hair cells. Thus, during transpiration, water passes from the soil via the xylem system of the roots, stem, and leaf veins, and through the intervening cells, to the mesophyll and finally into air spaces in the leaves, where most of it vaporizes and escapes.

• PROBLEM 10-18

What factors affect the rate of transpiration?

Solution: A number of different factors affect the rate of transpiration. The most important of these is temperature. Because temperature is a measurement of the average molecular kinetic energy, a high temperature heating up the plant body will provide the water molecules within the leaves with increased internal energy, and many of them may soon gain adequate kinetic energy to enter the gaseous phase and escape into the atmosphere. The higher the temperature rises, the greater will be the rate of transpiration, all other factors remaining constant. Plants of the desert, thus, must adapt some mechanisms to slow down water loss due to rapid evaporation by the hot desert air.

Humidity is a second important factor. Water is lost much more slowly into air already laden with water vapor. This is why plants growing in shady forests, where the humidity is generally high, typically spread large luxuriant leaf surfaces since their chief problem is getting enough sunlight, not losing water. In contrast, plants of grasslands or other exposed areas often have narrow leaves and relatively little total leaf surface, since they are able to get sufficient light but are constantly in danger of excessive water loss.

Air currents also affect the rate of transpiration. Wind blows away the water vapor from leaf surfaces, allowing more vapor molecules to escape from the leaf surface. If the air is humid, wind may actually decrease the transpiration rate by cooling the leaf, but if the breeze is dry, evaporation is increased and the transpiration rate is greatly enhanced. Leaves of plants growing in windy, exposed areas are often hairy. These hairs are believed to protect the leaf surface from wind action by trapping a layer of air saturated with water vapor over the surface of the leaf. As before, humid air reduces the rate of evaporation, thus preventing excessive loss of water.

• PROBLEM 10-19

What process in the plant is responsible for the phenomenon of guttation?

Solution: Droplets on a leaf commonly seen early in the morning are usually assumed to be dew, which is water that has condensed from the air. Actually, these droplets may have resulted from guttation, and come from the leaf itself through special openings on the leaf margins (and not through the stomata). Like transpiration, guttation is the loss of water through plant leaves, yet guttation occurs only in those plants with short stems. The phenomenon of guttation can be explained by the effect of root pressure.

Root pressure is a term given to the forces operating at the junction of root and stem in pushing a column of water up the xylem. The existence of root pressure has been experimentally demonstrated by removing the upper part of the stem of a well-watered plant, and attaching an air-tight piece of glass tubing to the remaining stump. Water actually is seen to rise in the tubing from the stump up to a height of one meter or more. If roots are killed or deprived of oxygen, root pressure falls to zero, indicating that the pushing force must be generated by the root. Root pressure has been found to equal 6 to 10 or more atmospheres of pressure even in plants that raise water to a height of less than 1 meter.

Guttation usually occurs in plants when little or no transpiration is taking place. At night, for instance, when the stomata are closed, transpiration is prevented or kept to a minimum. When the air is very moist or when the air currents are still, transpiration occurs only slowly. In these cases, guttation replaces transpiration, and is the chief force pushing water up the stem to the leaves. A small part of the water is actually utilized by the mesophyll in photosynthesis, and a large part of it passes through special pores of the leaf and appears as droplets on its surface.

• PROBLEM 10-20

How do transpiration and the theory of water cohesion explain the process of water transport in plants? When does root pressure become important in water transport?

Solution: Water transport in plants can be explained by transpiration and the theory of cohesion-tension. When a leaf transpires, water is lost from the leaf cells, molecule by molecule. As a consequence of the water loss, the osmolarity of the leaf cells increases and water molecules enter these cells by osmosis from adjacent cells with a lower osmolarity. These adjacent leaf cells in turn receive water from neighboring leaf cells and so on, which all ultimately get their water from the xylem cells in the vascular tissue.

Since water molecules are linked to each other by hydrogen bonds into continuous columns within the xylem vessels right down to the root tip, a molecule leaving a xylem cell to enter a mesophyll cell of the leaf will necessarily tug this entire column of water along behind it. Because water has sufficient tensile strength, the column does not snap when being pulled. The upward movement of this column causes a negative pressure to develop within the xylem. This negative pressure is similar to the suction produced while sucking a drink through a straw. This draws water molecules from the root tip cells into the vessels. This theory of water movement in the xylem is known as the cohesion-tension theory. In the root tip, water molecules from the soil pass down a gradient through the root hairs to the cells that have just lost water to the xylem. In this way, the upward stream of water is continuous. In short, while transpiration provides a pull at the top of the plant, the strong tendency of water molecules to stick together attributed to hydrogen bonds (cohesive force of water molecules) transmits this force through the length of the stem and roots and results in the movement of the entire water column up the xylem.

In the spring, before leaves are formed and transpiration becomes important, root pressure is probably the major force bringing about the rise of sap. In addition, under conditions of high humidity, still air currents, or extremely low temperatures, root pressure may become important in raising water to the leaves of some plants. Thus root pressure may contribute to the upward movement of water under some conditions, but it probably is not the principal force causing water to rise in the xylem of most plants most of the time.

NUTRIENT AND WATER TRANSPORT

• PROBLEM 10-21

What is meant by translocation? What theories have been advanced to explain translocation in plants? Discuss the value and weakness of each.

Solution: Translocation is the movement of nutrients from the leaves where they are synthesized to other parts of the plant body where they are needed for a wide variety of metabolic activities. Translocation takes place through the sieve tubes of the phloem.

Experimental data indicates that the high rate at which translocation occurs cannot be attributed to simple diffusion. Three theories have been offered to explain the mechanism of translocation. Each of them has its own value and weakness, but the first one presented (pressure-flow theory) holds the most support.

The pressure-flow theory, which had been widely accepted in the past, proposes that the nutrient sap moves as a result of differences in turgor pressure.

Sieve tubes of the phloem in the leaf contain a high concentration of sugars, which results in high osmotic pressure into the cells. This osmotic pressure causes an influx of water, and build up of turgor pressure against the walls of the sieve tube cells.

In the roots however, sugars are constantly being removed. Here, a lower concentration of solute is present resulting in a lowering of the osmotic pressure. There is consequently a lower turgor pressure exerted against the sieve tube walls of the root as compared to the leaf. This difference in turgor pressure along the different regions at the phloem is believed to bring about the mass flow of nutrients from a region of high turgor pressure - such as the leaves (where photosynthesis produces osmotically-active substances like glucose) - to regions of lower turgor pressure - such as the stem and roots. The fluid containing the nutrients is pushed by adjacent cells, along the gradient of decreasing turgor, from the leaves to the roots. This theory predicts that the sap should be under pressure as it moves down the phloem, and this is experimentally verified. But there are problems with this hypothesis. Under some conditions, sugar is clearly transported from cells of lesser turgor to cells of greater turgor. In addition, this theory fails to explain how two substances can flow along the phloem in different directions at the same time, a situation observed to occur by some investigators.

Another theory proposes that cyclosis, the streaming movement evident in many plant cells, is responsible for translocation. According to this theory, materials pass into one end of a sieve tube through the sieve plate and are picked up by the cytoplasm which streams up one side of the cell and down the other. At the other end of the sieve tube, the material passes across the sieve plate to the next adjacent sieve tube by diffusion or active transport. This theory is able to account for the simultaneous flowing of nutrient saps in different directions. However, it is attacked by some investigators on the basis that cyclosis has not been observed in mature sieve cells.

A third theory proposes that adjacent sieve-tube cells are connected by cytoplasmic tubules, in which sugars and other substances pass from cell to cell. The movement of these substances is powered by the ATP from the mitochondria-like particles that are believed to lie within these connecting tubules. This theory is, however, weak since it is supported by few experimental findings and much of its content is based on speculation.

• PROBLEM 10-22

In aquatic plants, which have protoplasms that are hypertonic in comparison to the water in which they live, the amount of water in the plant cells remains at a constant level. Explain how this is regulated.

Solution: Plants living in fresh water have a protoplasm more concentrated in solutes than their surroundings i.e. their surroundings are hypotonic. Accordingly by osmosis, water tends to enter the cells of these plants in order to equalize the concentration of solutes inside and outside the cells. Although plant cells do not have contractile vacuoles (the osmoregulatory organelle in amoeba) to pump out excess water, they do have a firm cellulose wall to prevent undue swelling. As water enters a plant cell, an internal pressure known as turgor pressure is built up within the cell which counterbalances the osmotic pressure and prevents the entrance of additional water molecules. Hence the turgor pressure is pressure that is generated against the cell wall to regulate the influx of water. Thus water plant cells are protected against excessive swelling and bursting.

ENVIRONMENTAL INFLUENCES ON PLANTS

• **PROBLEM** 10-23

What is an essential element in plants? Name all the macronutrients and micronutrients of plants. What are their respective functions in plant growth?

Solution: An element is considered essential (1) if the plant fails to grow normally and complete its life cycle when deprived of the element and no other element can replace the missing one or (2) if the element can be shown to be part of a molecule clearly essential in the plant's structure or function. Essential elements are usually separated into two categories - the macronutrients, each of which comprises at least .1% of the dry weight of the plant, and micronutrients, which are present in as little as a few parts per million of the dry weight of the plant.

The macronutrients include predominantly nitrogen and other mineral elements such as potassium, calcium, phosphorous, magnesium and sulfur. Nitrogen is an essential element in the synthesis of amino acids and thus proteins; nucleotides such as ATP, ADP, NAD, and NADP; chlorophyll (and other similar organic molecules with complex ring structures); the nucleic acids, DNA and RNA; and many vitamins, such as the vitamin B group. Potassium regulates the conformation of some proteins, and affects some enzymatic reactions in the synthesis of biomolecules. It also affects water and ion balance by its' osmotic effects. Calcium is a structural component of the cell wall - it combines with pectic acid in the middle lamella of the plant cell wall. It also plays a part in cell growth and division, and is a cofactor for some enzymes. Phosphorous occurs in the sugar phosphate backbone of DNA and RNA, in nucleotides such as ADP and ATP, and in the phospholipids of the cell membrane. Magnesium is a structural part of the chlorophyll molecule and an activator of many enzymes. It functions in formation of amino acids, vitamins, fats and sugars. Sulfer occurs in coenzyme A and in some amino acids and proteins.

The micronutrients include iron, chlorine, copper, manganese, zinc, molybdenum, and boron. Iron is a structural component of the electron carriers, the cytochromes and ferredoxin, and is thus extremely important in the processes of photosynthesis and cellular respiration. Chlorine regulates osmosis and ionic balance in the cell. It may play a role in root and shoot growth. Copper, manganese and zinc are currently found to be activators of some enzymes. Copper is used in carbohydrate and protein metabolism, and with manganese, it functions in chlorophyll synthesis. Zinc is used in the formation of auxin, chloroplasts and starch. Molybdenum is believed to be involved in nitrogen metabolism. Boron influences calcium ion uptake and utilization by the cell, and is suggested to be involved in carbohydrate transport as well. It affects cell division, flowering, pollen germination and nitrogen metabolism.

As can be readily seen, minerals fill a wide range of basic cell needs and are involved in a variety of fundamental biological and biochemical processes. The effects of mineral deficiencies are therefore widespread, affecting a number of structures and functions in the plant body.

• PROBLEM 10-24

What are the major constituents of a productive soil? What is the role of each of these constituents in plant growth? What measures can be taken to prevent the loss of topsoil?

Solution: A rich, fertile soil should contain sufficient quantities of the essential mineral nutrients. It should also contain an adequate amount of organic materials which is provided by humus. In addition, it should include innumerable bacteria, fungi, and other microorganisms to bring about decay of organic substances and aeration of the soil.

The essential minerals play important roles in the structure and functions of the plant. The broadest role of the minerals is their action in catalyzing enzymatic reactions in the cell. In some cases they act as an essential structural component of the enzyme, in others they act as regulators or activators of certain enzymes. Some minerals, absorbed in the ionic form by active transport, function in providing a hydrostatic pressure, or turgor within the cell. In other words, their concentrations inside the cell or in the surrounding medium regulate the movement of water in and out of the cell. Some minerals regulate the permeability and integrity of cell membranes, while others serve as structural components of cell parts. Other minerals constitute structural parts of the electron acceptors of the electron transport system, such as the cytochromes and ferredoxin. Still others act as buffering agents to prevent drastic changes in pH within the cell.

Humus, derived from the decaying remains of plants and animals, provides the soil with organic nutrients, among them mainly carbon- and nitrogen-containing compounds. In addition, humus increases the porosity of soil so that proper drainage and aeration can occur, and enhances the ability of the soil to absorb and hold water.

To bring about decomposition of the humus, a good soil should contain a large number of bacteria, fungi, and other small organisms. Such organisms, as the earthworms, benefit the soil further by constantly tilling the soil, aerating it, and mixing in additional organic substances.

A major conservation problem all over the world is to decrease the amount of valuable topsoil carried away each year by wind and water. Reforestation of mountain slopes, building check dams in gullies to decrease the spread of the run-off water, contour cultivation, terraces and the planting of windbreaks are some of the successful methods that are currently being used to protect the topsoil against erosion.

• **PROBLEM** 10-25

Differentiate between the hydrophytes, mesophytes and xerophytes. Give examples of each.

Solution: Plants have become adapted to grow in environments ranging from very wet to very arid. Botanists classify plants according to their water needs into three groups called hydrophytes, mesophytes, and xerophytes. Hydrophytes grow in a very wet environment, and are either completely aquatic or rooted in water or mud but with stems and leaves above the water for capturing sunlight. Water lilies, pondweeds and cattails are common hydrophytes. Mesophytes are land plants that prefer a habitat with a moderate amount of moisture. Common mesophytes are beech, maple, oak, dogwood and birch. Xerophytes are plants such as yuccas and cacti that have become adapted to areas where soil water is scarce. They manage to survive with a limited amount of water by minimizing its loss through a number of ways. For example, xerophytes may have a reduced number of stomata, a heavily cutinized epidermis to retard evaporation, or they may develop thick stems and leaves to store water.

SHORT ANSWER QUESTIONS FOR REVIEW

Answer

To be covered when testing yourself

Choose the correct answer.

1. The following are all characteristics of the meristemic region except: (a) it differentiates into the permanent tissues of the root (b) it is composed of actively dividing cells (c) it gives rise to new root cap cells as old ones are being sloughed off. (d) All other tissues of the root are formed from its cells.

a

2. The following traces the development of a young woody stem in cross section to one year of age.

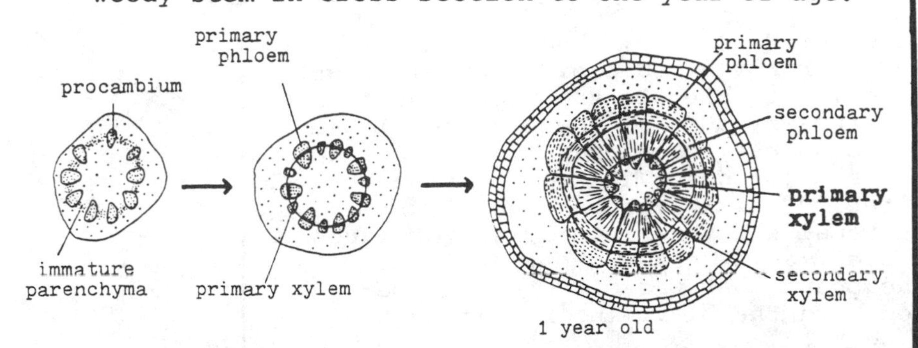

This woody stem at three years of age would be best illustrated as:

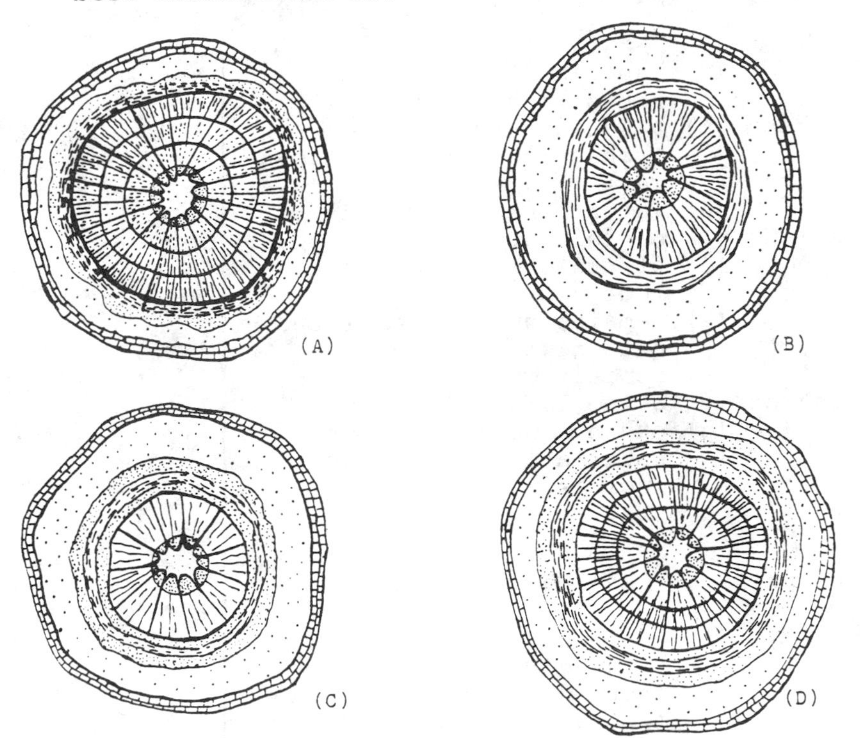

A

Answer

To be covered when testing yourself

3. Turgor pressure and thus the opening and closing of the stomata depends on all of the following except: (a) A change in the amount of carbon dioxide taken up (b) a change in the amount of oxygen taken up. (c) The conversion of starch to glucose by a light initiated reaction. (d) A change in the pH of the guard cells. — b

4. Transpiration (a) helps cool the plant body. (b) Keeps the air spaces of the leaves moist. (c) Directs the upward movement of water and minerals to the leaves. (d) all of the above. — d

5. A plant is watered heavily. Then the upper part of its stem is removed and a piece of air tight glass tubing is attached to the remaining stump. What will be observed? (a) The plant will die because it is unable to maintain the same internal pressures. (b) Water is seen to rise through the glass tubing. (c) The portion of the plant that remains is unaffected; it continues to maintain itself as if nothing happened. (d) The part of the plant that was cut off will be regenerated with the plant incorporating the glass tubing. — b

In 6-12 match the plant root part to its function or characteristic.

6. root hairs
7. zone of elongation
8. phloem
9. meristemic region
10. xylem
11. root cap
12. zone of maturation

(a) vascular tissue which moves water upward
(b) vascular tissue which transports nutrients downward
(c) is composed of rapidly dividing cells
(d) offers protection to the root
(e) remains undifferentiated but grows rapidly
(f) zone in which root hairs arise laterally
(g) enhance absorption of water and minerals by increasing the surface area.

Answer: 6-g, 7-e, 8-b, 9-c, 10-a, 11-d, 12-f.

In 13-21 match the parts of the leaf to their properties.

13. guard cells
14. veins

(a) attached to the stem and contains vascular bundles
(b) may be simple or compound

Answer

To be covered when testing yourself

		Answer
15. cells of palisade layer	(c) vascular bundles which have forked repeatedly	
16. spongy layer	(d) secretes protective, waterproof cutin	
17. epidermis	(e) responsible for gas exchange	
18. mesophyll cells	(f) changes turgor pressure to regulate the size of the stomata.	13-f,
19. petiole	(g) found between the lower and upper epidermal cells	14-c,
20. stomata	(h) contains a lot of chloroplasts and is chiefly responsible for photosynthesis	15-h.
21. blade	(i) holds moisture and gases	16-i,
		17-d,
		18-g,
		19-a,
		20-e,
		21-b

In 22-27 match the essential nutrients to their functions.

		Answer
22. calcium	(a) essential in the synthesis of chlorophyll and NADP	22-c,
23. sulfur	(b) regulates the conformation of some proteins	23-f,
24. magnesium	(c) combines with pectic acid in the middle lamella of the cell wall	24-e,
25. nitrogen	(d) occurs in the cell membrane as well as nucleic acids, ADP and ATP	25-a,
26. phosphorous	(e) structural part of the chlorophyll molecule	26-d,
27. potassium	(f) occurs in coenzyme A	27-b

Fill in the blanks.

28. Mary took a clipping from one of her plants and placed it in water. After some time passed, roots grew out of the stem allowing her to place this clipping into soil. It was now able to grow into a full, new plant. The roots that developed from the stem are called _____ roots. — *Answer:* adventitous

29. Label the regions of growth in the terminal portion of the root which are depicted:

Answer

To be covered when testing yourself

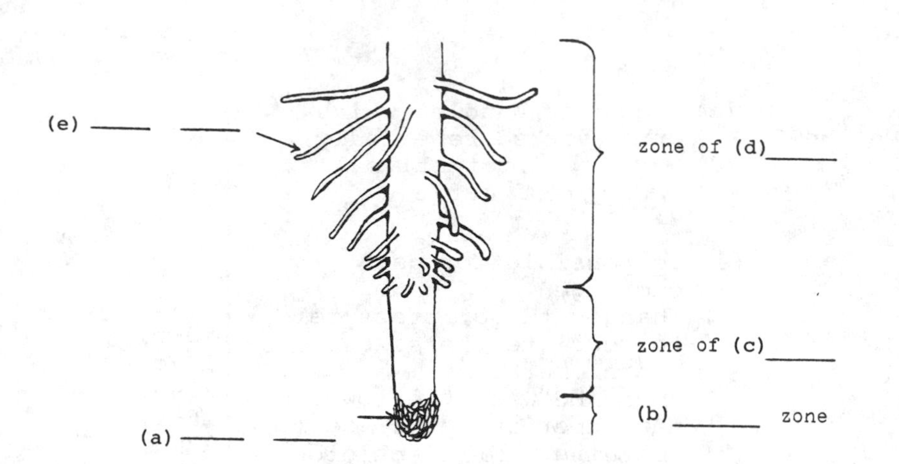

a-root cap
b-meristematic
c-elongation
d-maturation
e-root hair

30. The process by which root hairs first absorb water and then transport it to the cortical cells is _____. The water will continue to diffuse toward the root's center because of the _____. In this way, the water will reach the _____ which transports the water up through the stem to the leaves by a combination of _____ and _____.

osmosis, osmotic gradient, xylem, root pressure, transpiration pull

31. Pictured is a cross section of a young woody stem. Label it.

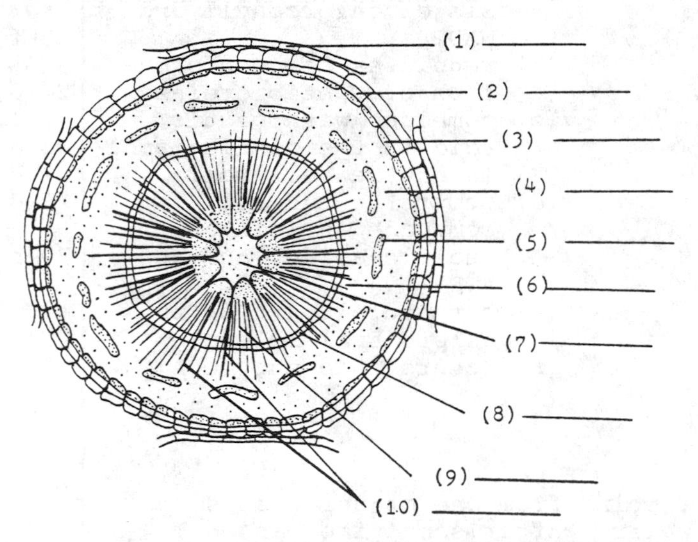

1-epidermis
2-cork
3-cork cambium
4-cortex
5-sclerenchyma fibers
6-phloem
7-pith
8-vascular cambium
9-xylem
10-vascular rays

32. There are various types of woody tissues. _____ is the youngest, outermost portion of the xylem in a woody plant, _____ is the older, inner portion of the xylem which has lost the ability to conduct. The secondary xylem is composed of _____ and _____.

sapwood, heartwood, spring wood, summer wood

	Answer
	To be covered when testing yourself
33. During the daytime, plants undergo the process of _____ and thus their stomata are _____. During the night, plants _____ and their stomata are _____.	photo-synthesis, open, respire, closed
34. The cell is in constant need of _____ when it is photosynthesizing. It is able to obtain it from the environment by_____ because these molecules move from a region of _____ concentration to a region of _____ concentration. In a similar manner, the _________ produced by this process must be removed from the plant through the _____ to the exterior by _____ down a _____.	carbon dioxide, diffusion, higher, lower, oxygen, stomata, diffusion, concent-ration gradient
35. Nutrients are moved from the _____ where they are synthesized to other parts of the plant where they are needed by _____. This takes place in the _____ cells of the _____.	leaves, translo-cation, sieve tube, phloem
36. There are three groupings of plants based upon their water needs. _____ grow in a very wet environment being either completely aquatic, or rooted in water or mud. _____ are land plants that live in an environment with a moderate amount of moisture. _____ are plants that grow in areas where water is scarce.	Hydro-phytes, Meso-phytes, Xero-phytes
Determine whether the following statements are true or false.	
37. Environmental factors have no effect on the rate of transpiration, which is constant depending only upon the plant species.	False
38. The droplets of water seen in a field early in the morning can be explained by both condensation from the air and guttation.	True
39. Guttation typically takes place in a plant when transpiration is taking place.	False
40. Water is moved up through the xylem by transpiration and cohesion tension. The upward movement of the column of water causes a positive pressure to develop in the xylem.	False
41. As sap moves down through the phloem, pressure is exerted on it.	True

CHAPTER 11

LOWER INVERTEBRATES

THE PROTOZOANS

Characteristics

• PROBLEM 11-1

What are the chief characteristics of the protozoans?

Solution: The protozoans are a heterogeneous assemblage of a large number of species, which are almost exclusively microscopic organisms.

The protozoans are grouped into the phylum Protozoa within the Animal Kingdom, although this classification remains controversial. Some biologists still believe that the protozoans have more in common with the Kingdom Protista. Protozoa live either singly or in colonies. These organisms are usually said to be unicellular. Therefore they contain no tissue or organs, which are defined as aggregations of differentiated cells. Instead of organs, they have functionally equivalent subcellular structures called organelles. These organelles do show a great deal of functional differentiation for the purposes of locomotion, food procurement, sensory reception, response, protection and water regulation. Certain protozoans have interesting plant-like characteristics in both structure and physiology.

Reproduction among the protozoans is variable. An individual may divide into two, usually equal halves, after which each grows to the original size and form. This form of reproduction is called binary fission and can be seen in the flagellates, among the ciliates, and in organisms such as the amoeba. Multiple fission, or sporulation, where the nucleus divides repeatedly and the cytoplasm becomes differentiated simultaneously around each nucleus resulting in the production of a number of offspring, is also seen among the protozoans. Other types of reproduction characteristics of the protozoans are plasmotomy, which is the cytoplasmic division of a multinucleate protozoan without nuclear division, resulting in smaller multinucleate products.

Budding is another reproductive process by which a new individual arises as an outgrowth from the parent organism differentiating before or after it becomes free. All the reproductory mechanisms thus far mentioned illustrate asexual means of reproduction. Sexual reproduction may also occur by the fusion of two cells, called gametes, to form a new individual, or by the temporary contact and nuclear exchange (conjugation) of two protozoans (for example, two paramecia). The result of conjugation may be "hybrid vigor", defined as the superior qualities of a hybrid organism over either of its parental lines. Some species have both sexual and asexual stages in their life cycles.

With regard to their ecology, protozoans are found in a great variety of habitats, including the sea, fresh water, soil, and the bodies of other organisms. Some protozoans are free-living; meaning that they are free-moving or free-floating, whereas others have sessile organisms. Some live in or upon other organisms in either a commensalistic, mutualistic, or parasitic relationship.

The mechanisms for the acquisition of nutrition is also variable among the protozoans. Some are holozoic, meaning that solid foods such as bacteria, yeasts, algae, protozoans, and small metazoans or multicellular organisms, are ingested. Others may be saprozoic, wherein dissolved nutrients are absorbed directly; holophytic, wherein manufacture of food takes place by photosynthesis; or mixotrophic, which use both the saprozoic and holophytic methods.

It should be pointed out that the unicellular level of organization is the only characteristic by which the phylum Protozoa can be described. In all other respects, such as symmetry and specialization of organelles, the phylum displays extreme diversity.

• PROBLEM 11-2

What problems are posed in referring to the many forms of protozoans as belonging to one single animal phylum?

Solution: The first problem lies in the fact that the unicellular level of organization is the only characteristic by which the phylum can be described; in all other respects the phylum displays extreme diversity. Protozoans exhibit all types of symmetry, a great range of structural complexity, and adaptations for all types of environmental conditions. Although all of them have remained at the unicellular level, they have evolved along numerous lines through the specialization of the protoplasm. Specialization has occurred through the evolution of an array of subcellular organelles. A second problem involves taxonomic organization.

As is true of any taxonomic category, a phlyum should contain members which are derived from a common ancestral form. In the classification of the protozoans, virtually all motile unicellular organisms have been grouped into a single phylum, with very little regard to evolutionary relationships. It is among the flagellate protozoans that the concept of a single subphylum produces the greatest aberration. Most of the unicellular free-living flagellates are organisms which when assembled constitute a collection of largely unrelated forms.

A third problem lies in the possession of certain plant-like features, such as autotrophism and the presence of chloroplasts. In fact, certain green flagellate "protozoans" appear to be rather closely related to unicellular green algae which belong to the Kingdom Plantae Thus, not only is the concept of a single phylum questionable, there is the additional problem of whether the Protozoa is a true animal phlyum.

Flagellates

• PROBLEM 11-3

Euglena is a common green flagellate protozoan found in fresh water ponds. Describe briefly the method of locomotion, nutrition, and asexual reproduction in this organism.

Solution: Normally, locomotion in Euglena is produced by undulating movements of the flagellum. These movements of the flagellum draw the organism after it in a characteristic spiral path. Actually, Euglena has two flagella, but only one extends from the body and is used for locomotion.

With regard to nutrition, Euglena carries on autotrophic nutrition, as do green plants, synthesizing food from inorganic substances in the presence of light. However, under condition of total darkness for long periods of time, in a medium containing necessary nutrients, Euglena will shift to a heterotrophic mode of nutrition. Under these conditions the chloroplasts, the organelles which carry on photosynthesis, disappear, and the organisms live by absorbing the necessary nutrients from the surrounding medium through their cell membranes. When returned to the light environment, the chloroplasts reappear, and the autotrophic mode of living is resumed. This behavior is one of the reasons why Euglena is said to be both plant-like and animal-like.

The life cycle of Euglena involves both an active phase, during which the organism moves about, and an encysted phase, during which the organism is rounded up with a protective cyst membrane surrounding it. Asexual reproduction can occur in both stages. The division is typically a longitudinal binary fission. The nucleus divides by mitosis, and then the cytoplasm divides, forming two cells each with a nucleus.

Sarcodines

• **PROBLEM** 11-4

The ameba has no mouth for ingestion of food, but can take food in at any part of the cell. Explain this method of ingestion. What occurs following ingestion?

Solution: In the amoeba, pseudopodia (false feet) are used for nutrient procurement. These are temporary projections of cytoplasm which extend around the prey in a cup-like fashion eventually enveloping it completely. These pseudopodia can form anywhere on the surface of the amoeba. The enclosing of the captured organism by cytoplasm results in the formation of a food vacuole within the amoeba. Usually, the pseudopodia are not in intimate contact with the prey during engulfment and a considerable amount of water is enclosed within the food vacuole along with the captured organism. Engulfment may also involve complete contact with the surface of the prey, and the resulting vacuole is then completely filled by food. Death of the prey takes from 3 to 60 minutes and results primarily from a lack of oxygen.

Digestion occurs within the food vacuole. In amoeba, as in man, digestion is controlled by enzymes, and different enzymes act at definite hydrogen-ion concentrations. The enzymes that function in the vacuoles of an amoeba enter by fusion of the vacuoles with lysosomes. The enzymes hydrolyze the proteins into amino acids, fats into fatty acids and glycerol, and carbohydrates into simple sugars. These end products of digestion are absorbed by the rest of the organism through the vacuolar membrane. The food vacuole is not stationary within the organism but circulates in the fluid cytoplasm. Undigested material is egested from the cytoplasm. Like that of ingestion, the point of egestion is not fixed; food vacuoles containing undigested material may break through the surface at any point.

• **PROBLEM** 11-5

Describe locomotion in the amoeba.

Solution: The locomotion of amoeba is considered to be the simplest type of animal locomotion. A moving amoeba sends out a projection, termed a pseudopodium. Following this, the organism advances as the inner, granular, gel-like endoplasm flows into the pseudopodium. Two or three pseudopodia may be formed simultaneously but ultimately one will become dominant for a time. As new pseudopodia are formed, the old ones withdraw into the general body region. In its locomotion the amoeba often changes its course in response to environmental stimulation, by forming a new dominant pseudopodium on the opposite side, thus moving in a very irregular fashion.

Currently there is no fully complete explanation for the changes, both physical and chemical, which are involved in amoeboid movement. The theory accepted by zoologists at the present time is based on changes in the texture of the cytoplasm. As a result of some initial stimulus, ectoplasm, the outer clear, thin layer of the organism, undergoes a liquefaction and becomes endoplasm, which is gel-like. As a result of this change, internal pressure builds up and causes the endoplasm to flow out at this point, forming a pseudopodium. In the interior of the pseudopodium, the endoplasm flows forward along the line of progression; around the periphery, endoplasm is converted to ectoplasm, thereby building up and extending the sides of the pseudopodium like a sleeve. At the posterior of the body, ectoplasm is assumed to be undergoing conversion to endoplasm. During this entire process, energy consumption is known to have taken place.

• **PROBLEM 11-6**

What is the function of the contractile vacuole?

Solution: Unicellular and simple multicellular organisms lack special excretory structures for the elimination of nitrogenous wastes. In these organisms, wastes are simply excreted across the general cell membranes. Some protozoans do, however, have a special excretory organelle called the contractile vacuole. It appears that this organelle eliminates water from the cell but not nitrogenous wastes. These organelles are more common in fresh water protozoans than in marine forms. This is because, in fresh water forms, the concentration of solutes is greater in the cytoplasm than in the surrounding water, and water passively flows into the cell through the cell membrane. The fresh water protozoan is said to live in hypotonic environment. This inflow of water would cause a fatal bloating if the water were not removed by the contractile vacuole. The vacuole swells and shrinks in a steady cycle, slowly ballooning as water collects in it, then rapidly contracting as it expells its contents, and then slowly ballooning again. The exact process of how the cell pumps water out of the vacuole is still unclear, but it is believed that the process is an energy consuming one.

Ciliates

• **PROBLEM 11-7**

Unlike Amoeba, Paramecium has a permanent structure, or organelle, that functions in feeding. Describe this organelle and the method of feeding and digestion in Paramecium.

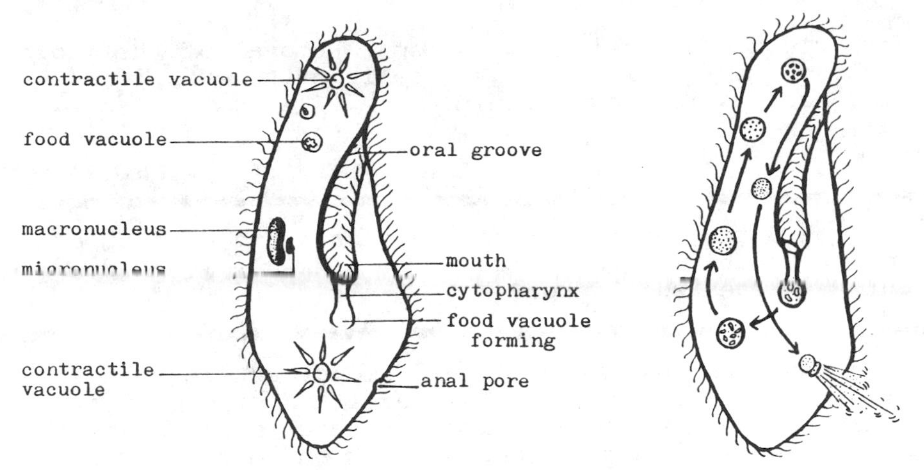

Paramecium. Left: Drawing showing major structures.
Right: Fate of ingested food particles.

Solution: The feeding organelle in Paramecium begins with an oral groove. This oral groove is a ciliated channel located on one side of the organism. Food particles are swept into the oral groove by water currents produced by the beating cilia, and are carried down the groove to a point called the cytostome, which can be thought of as the mouth. Food is carried through the cytostome into the cytopharynx, which is also lined with cilia. As food accumulates at the lower end of the cytopharynx, a food vacuole forms around it. The vacuole eventually breaks away and begins to move toward the anterior of the cell. Lysosomes fuse with the food vacuole and secrete digestive enzymes into the vacuole. As digestion proceeds, the end products, such as sugars and amino acids, diffuse across the membrane of the vacuole into the surrounding cytoplasm, and the vacuole begins to move toward the posterior end of the cell. When at the posterior end of the cell, the vacuole fuses with a structure there called the anal pore. Undigested material in the vacuole is expelled to the outside through the anal pore. In addition to serving as the digestive organ, the food vacuole also serves, by its movement, to distribute the products of digestion to all parts of the cell.

• PROBLEM 11-8

What results would you expect if the macronucleus were removed from a paramecium?

Solution: Ciliates differ from all other animals in possessing two distinct types of nuclei - a large macronucleus and one or more smaller micronuclei. The macronucleus is sometimes called the vegetative nucleus, since it is not critical in sexual reproduction, as are the micronuclei. The macronucleus is essential for normal metabolism, for mitotic division, and for the control of cellular differentiation. The macronucleus is considered to participate actively in the synthesis of RNA, which is used in cell metabolism. Removal of the macronucleus from a ciliate causes cell death, even if a micronucleus is present.

• PROBLEM 11-9

Why would you not expect conjugation among a group of paramecia that had descended from a single individual through repeated fission?

Solution: Conjugation occurs among certain protozoans, ciliates in particular. It is a method of exchanging genes between organisms. In paramecium conjugation may take place when parameciums of plus and minus strains are present in the same pond or culture. Conjugation will not however take place with two organsisms of similar mating types. During conjugation two animals come together and adhere to each other in the oral region. Next, the micronuclei divide by meiosis. Then all but two of the resulting haploid micronuclei in each cell disintegrate; the macronucleus usually also disintegrates. One of the two nuclei in each cell remains stationary and functions as the female nucleus. The other, the male nucleus, moves into the conjugant cell and fuses with the female nucleus in that cell in the process of fertilization. Therefore, each cell acts as both male and female, donating one nucleus and receiving another, and when the two cells part, each has a new diploid nucleus. This nucleus then undergoes more divisions, and some of the new nuclei become macronuclei. Following some cytoplasmic cleavages, the normal nuclear number of the cell both micro and macro, is restored.

As stated, conjugation involves the mixing of genes between organisms, and requires two different mating types. In a group of paramecia descended from a single individual through repeated fission,all individuals are identical genetically, and conjugation could not occur. The exchange of nuclear material through conjugation between these organisms would not result in greater genetic diversity and would, therefore, be pointless.

• PROBLEM 11-10

The ciliates are the fastest moving protozoans. Explain ciliary movement in a ciliate. What other function is served by the cilia?

Solution: Ciliary movement consists of an effective and a recovery stroke. As depicted in the accompanying figure, during the effective (power) stroke, the cilium is outstretched and moves from a forward to a backward position. During the recovery stroke, the cilium, bent over to the right against the body when viewed from above and looking anteriorly, is brought back to the forward position in a counterclockwise movement. The recovery position offers less water resistance. The beating of the cilia is synchronized and waves of ciliary beat progress down the length of the body from anterior to posterior. The direction of the waves is slightly oblique, which causes the ciliate to swim in a spiral course and at the same time to rotate on its longitudinal axis.

The beating of the cilia can be reversed in direction, and the animal can move backwards. The so called "avoidance reaction" is associated with the backward movement of the ciliate. When the organism comes in contact with some undesirable object, the ciliary beat is reversed. It moves backward a short distance, turns, and moves forward again. This avoidance reaction can be repeated.

Detection of external stimuli is another function of the cilia and though perhaps all of the cilia can act as sensory receptors in this respect, there are certain long, stiff cilia that play no role in movement and are probably exclusively sensory. The two functions, then, of the cilia are locomotion and sensory reception.

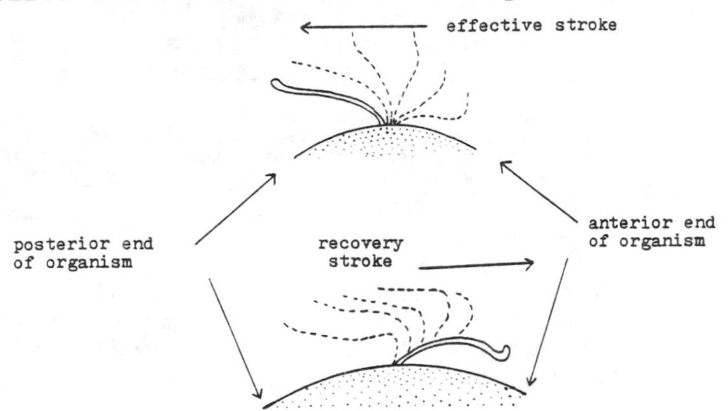

Figure. One cycle of ciliary movement. Only one cilium is shown. Result of cycle is that animal moves forward.

PORIFERA

• PROBLEM 11-11

What is the structure of a sponge? How do sponges obtain food and water?

Solution: The most primitive metazoans belong to the phylum Porifera and are commonly called the sponges. Sponges may be radially symmetrical, but more commonly they are asymmetrical animals, consisting of loose aggregations of cells which are poorly arranged into tissues.

The surface of a sponge is perforated by small openings, or incurrent pores, which open into the interior cavity, the atrium or spongocoel. The atrium opens to

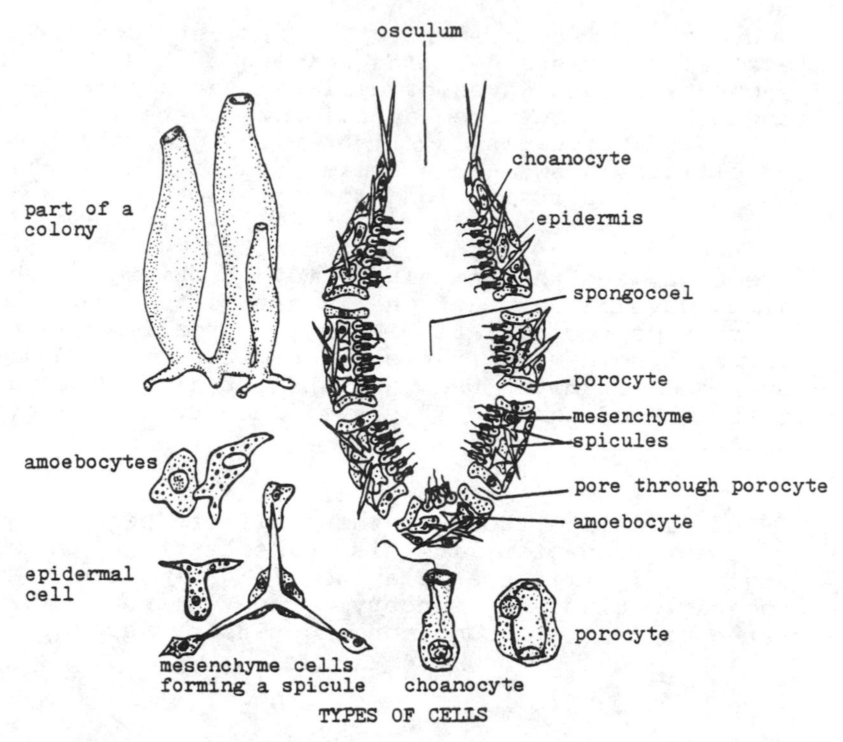

Sponges. Upper left, sketch of part of a colony of sponges. Upper right, diagram of a section through a simple sponge showing its cellular organization. Lower, sketches of the types of cells found in a sponge.

the outside by a large opening at the top of the tube-shaped sponge, called the osculum. There is a constant stream of water passing through the incurrent pores into the atrium, and out through the osculum.

The body wall is relatively simple. The outer surface is covered by flattened polygonal cells called pinacocytes. The pores are guarded by cells called porocytes, which are modified pinacocytes, and are shaped like tubes extending from the outside of the sponge to the spongocoel. The pore, or ostium, can be regulated by contractions in the outer end of the porocyte.

The mesohyl, a layer consisting of a gelatinous protein matrix, contains skeletal material and amoeboid cells, and lies directly beneath the pinacoderm. The skeletal material may be calcareous or siliceous spicules, or protein spongin fibers, or a combination of the last two. The composition, size and shape of these spicules form the basis for the classification of species of sponges. Though the spicules are located in the mesohyl, they frequently project through the pinacoderm. The skeleton, whether composed of spicules or spongin fibers, is secreted by amoebocytes called sclerocytes. Amoedboid cells in the mesohyl include the archaeocytes. These calls are capable of forming other types of cells that are needed by the sponge.

The inner side of the mesohyl, lining the atrium, is a layer of cells called choanocytes, or collar cells. The characteristic shape of a choanocyte is ovoid, with a flagellum surrounded by a collar. The choanocytes are responsible for the movement of water through the sponge and for obtaining food. The current is produced by the beating of the flagella of the choanocytes. The flagella are oriented toward the spongocoel, and beat in a spiral manner from their base to their tip. As a result of this, water is sucked into the spongocoel through the incurrent pores.

Sponges feed on very fine particulate matter. Only particles smaller than a certain size can pass through the pores. Food particles are brought inside the spongocoel by water currents. Larger food particles are phagocytized by amoebocytes lining the inhalant chamber. Particles of bacterial size or below are probably removed and engulfed by the choanocytes. It is thought that the amoebocytes act as storage centers for food reserves. Egested wastes leave the body in water currents.

There are three basic structural types of sponges. The simplest, called the asconoid type, has been described earlier when basic sponge structure was discussed. The syconoid and leuconoid types are more complex, having thicker mesohyls and a greater number of water channels, but their structures can be seen to be just complexes of asconoid sponges attached to one another.

COELENTERATA

• PROBLEM 11-12

Describe the body plan of a hydra.

Solution: Hydra belong to the animal phylum called the Coelenterata (or Cnidaria). This phylum includes also such animals as the jelly fish, sea anemones and corals. Most are marine except for Hydra which is freshwater. These animals display radial symmetry, a digestive cavity, which opens into a mouth, and a circle of tentacles surrounding the mouth. The body wall consists of three basic layers: an outer epidermis; an inner layer of cells lining the digestive cavity or gastrovascular cavity called the gastrodermis; and between these two, a layer called the mesoglea. This last layer is usually devoid of cells, containing rather a gelatinous matrix.

Though all coelenterates are basically tentaculate and radially symmetrical, two different structural types are encountered within the phylum. One type which is sessile, is known as the polyp. The other form is free-swimming and is called the medusa. The hydra exists only in the polypoid stage.

The body of Hydra is divided into several specialized regions (see Figure). The first region is called the

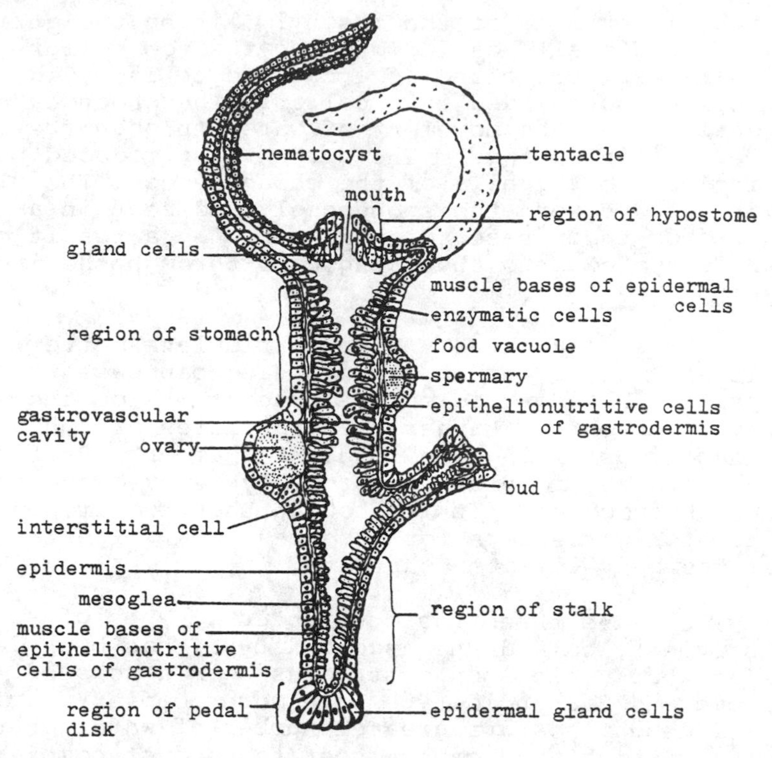

Diagram of a Hydra cut longitudinally to reveal its internal structure.

hypostome, or manubrium. This is a conical elevation located at the oral end of the body, the apex of which is the mouth. Surrounding the hypostome is a second region composed of tentacles. In Hydra, there are usually 5 or 6 hollow and highly extensile tentacles. Along the arms of the tentacles are located batteries of stinging cells called cnidocytes. These specialized cells, which are unique to and characteristic of all coelenterates, contain stinging structures called nematocysts. Nematocysts need not be restricted to the tentacles but may be found elsewhere in the hydra. A combination of physical and chemical stimuli causes the release of the nematocyst to take place. Undifferentiated cells in the epidermal layer, called interstitial cells, become modified to form the cnidocyte cells. These interstitial cells may also form generative cells, resulting in germ cell-producing bodies as well as buds.

The body of the hydra is in the shape of a cylindrical tube. The tube is more or less divided into a stomach or gastric region and a stalk region that terminates in the region of the pedal disk. The gastrodermis in the gastric region contains enzyme-secreting gland cells in addition to flagellated, amoeba-like cells. The cells in the stalk region do not produce enzymes, but are highly vacuolated, indicating intracellular absorption.

The pedal disk marks the attaching end of the body. An adhesive substance for attaching the body to some object is secreted by the epidermal cells. These cells can also produce a gas forming a bubble inside the adhesive secretion, which allows the organism to float to the surface of the water.

• PROBLEM 11-13

The tentacles of hydra are armed with stinging cells. What is the structure of these cells, how are the stingers fired and of what value are they to the animal?

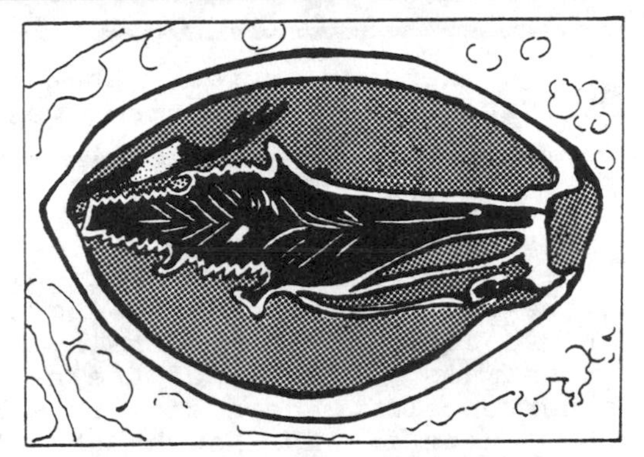

Sagittal section of an undischarged nematocyst of hydra, as seen through the electron microscope.

Solution: The stinging cells of hydra are called cnidocytes and are located throughout the epidermis. These cells contain stinging structures called nematocysts (see Figure). A cnidocyte is a rounded or ovoid cell with a short, stiff, bristle-like process, called a cnidocil, at one end. The cnidocil is exposed to the surface. The interior of the cell is filled by a capsule containing the nematocyst, which is a coiled tube, and the end of the capsule that is directed toward the outside is covered by a cap or lid. Supporting rods run the length of the cnidocyte.

The nematocysts are discharged from the cnidocyte and are used for anchorage, for defense, and for capture of prey. The discharge mechanism apparently involves a change in the permeability of the capsule wall. Under the combined influence of mechanical and chemical stimuli, which are initially received and conducted by the cnidocil, the lid of the nematocyst opens. Water pressure within the capsule everts the tube, and the entire nematocyst explodes to the outside. A discharged nematocyst consists of a bulb representing the old capsule, and a thread-like tube of varying length, which may be spiked.

From a functional standpoint, nematocysts can be divided into three major types. The first, called a volvent, is used to entangle prey. When discharged, the volvents wrap around the prey animal. The second type is called a penetrant. The tube of a penetrant is open at the tip and frequently armed with barbs and spines. At discharge, the tube penetrates into the tissues of the prey and injects a protein toxin that has a paralyzing action. The nematocysts of hydra do not have this effect on man; but the larger marine Cnidaria can produce a very severe burning sensation and irritation. The third type of nematocyst is a glutinant, in which the tube is open and sticky and is used in anchoring the animal under certain conditions.

The cnidocyte degenerates following discharge of its nematocyst, and new cnidocytes are produced from interstitial cells.

• PROBLEM 11-14

The hydra has a unique nervous system. What is this type of system called? Explain how the system operates.

Solution: The nervous system of the hydra is primitive. The nerve cells, arranged in an irregular fashion, are located beneath the epidermis and are particularly concentrated around the mouth. There is no aggregation or coordination of nerve cells to form a brain or spinal cord as in higher animals. Because of the netlike arrangement of the nerve cells, the system is called a nerve net. For some time it was thought that these nerve cells lacked synapses, but at the present time research indicates that synapses are indeed present. (Synapse is the junction between the axon of one nerve cell, or neuron, and the dendrite of the next.)

It is known that some synapses are symmetrical, that is, both the axon and the dendritic terminals secrete a transmitter substance and an impulse can be initiated in either direction across the synapse; while some are asymmetrical, permitting transmission only in one direction. Impulse in the nerve net can move in either direction along the fibers. The firing of the nerve net results primarily from the summation of impulses from the sensory cells involved, which pick up the external stimuli and the degree to which the response is local or general depends on the strength of the stimulus. When a sensory cell, or receptor, is stimulated, an impulse is relayed to a nerve cell. This in turn relays the impulse to other nerve cells, called effectors, which stimulate muscle fibers and nematocyst discharge.

The rate of transmission of nerve impulses in hydra is usually quite slow. In spite of the fact that the nerve net is very primitive in comparison to the vertebrate type of nervous system, it is apparently adequate for hydra.

• PROBLEM 11-15

Hydras, like most cnidarians, are carnivorous and feed mainly on small crustaceans. Describe the method of ingestion and digestion in these organisms.

Solution: Should a small crustacean make contact with the tentacles of the hydra, the nematocysts, which are lodged in the tentacle walls, would discharge. The nematocysts will entangle the prey and paralyze it. The captured organism is pulled toward the mouth by the tentacles. The mouth then opens to receive the prey. The retraction and bending inward of the tentacles and the opening of the mouth is a reflex response.

Mucus secretions in the mouth aid in swallowing. Eventually the prey arrives at the gastrovascular cavity. Secretory cells lining the cavity release enzymes that begin the digestion of protein. Eventually, the tissues of the prey animal are reduced to a soupy broth. Flagellated cells lining the cavity continually beat their flagella, thus aiding in mixing the digesting food with enzyme molecules.

Subsequent to this extracellular phase of digestion comes intracellular digestion. Amoeba-like cells lining the cavity extend pseudopodia which engulf small fragments of what is left of the prey animal. Continued digestion of proteins and the digestion of fats occur within food vacuoles in these amoeba-like cells lining the gastrodermis. Products of digestion are circulated by cellular diffusion. Glycogen and fat are the chief storage products of excess food. Undigested materials are ejected from the mouth upon contraction of the body.

• PROBLEM 11-16

Obelia, a marine hydrozoan, exists in a polypoid and medusoid form. Explain how these stages are formed in the life cycle of the organism.

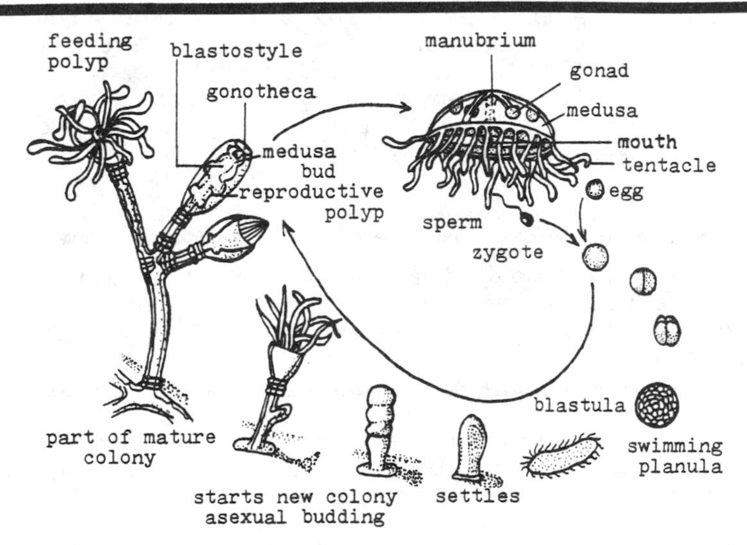

Life cycle of Obelia.

Solution: The coelenterate Obelia is a marine colonial organism, consisting of a group of polypoid individuals living together in close association. Within the colony are two types of polyps serving different functions. There are nutritive (or feeding) polyps and reproductive polyps. All the polyps of the colony are derived from an ancestral hydralike polyp by asexual budding. These buds fail to separate and the new polyps remain attached by hollow stemlike connections to the colony.

Each reproductive polyp has a central core, the blastostyle. When mature, the blastostyle buds off an entirely new body form, an umbrella-shaped medusa. In the process of budding, the medusa breaks free of the blastostyle and swims out through the opening of the blastostyle called the gonotheca.

The medusa of Obelia has a fringe of tentacles on its edge, and hanging down from the center, similar to the handle of an umbrella, is the manubrium containing the mouth. The gastrovascular cavity extends from the mouth in the manubrium into four radial canals, on which are located the reproductive organs, or gonads. Medusae also reproduce sexually. There are two types of medusae, each bearing either testes, the male gonads which produce sperm, or ovaries, the female gonads which produce eggs. Sperm and eggs are released into the surrounding sea water. Fertilization takes place in the sea water, and a zygote is formed. Following early embryonic cleavage, a free-swimming larva, called a planula, is formed which is ciliated and lacking a gastrovascular cavity or mouth. After a free-swimming existence lasting for several hours to several days, the planula attaches to an object and develops into a polypoid colony, thus completing the life cycle.

The alternation of a polypoid and medusoid stage is an example of "alternation of generations", or metagenesis. Asexually and sexually reproducing stages alternate with each other. It must be noted that it is different from the alternation of generations that occurs in plants in that both polyp and medusa are diploid, as in all multicellular animals, the only haploid stage in the life cycle are the gametes.

• PROBLEM 11-17

Mesoglea and mesoderm are two terms used to describe the middle tissue layer of an organism. What distinction can be made between the two and in what organisms can each be found?

Solution: Mesoglea is the term used to describe the layer between the outer epidermis and the inner layer of cells lining the gastrovascular cavity. It is found in adult Coelenterates (Cnidarians) such as hydra, Obelia, and Portuguese man-of-war (Physalia). The mesoglea consists of either a thin non-cellular membrane or thick, fibrous, jellylike, mucoid material with or without cells.

The term mesoderm is used to describe the middle

of the three embryonic tissue layers first delineated during an early developmental stage of the embryo. This layer gives rise to the skeleton, circulatory system, muscles, excretory system, and most of the reproductive system. It is found in higher invertebrates, the insects, and all vertebrate groups.

The mesoderm is always a cellular layer and always refers to an embryonic layer. Mesoglea, on the other hand, is usually acellular and merely denotes the middle layer of the body wall of an organism.

THE ACOELOMATES

Platyhelminthes

• PROBLEM 11-18

Describe feeding in planarians.

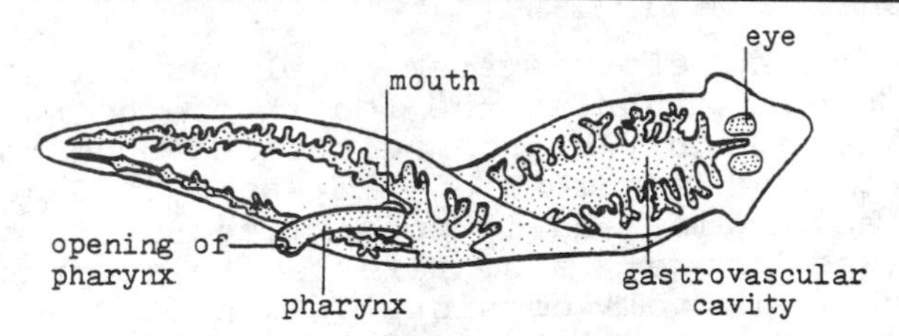

Planaria, showing much-branched gastrovascular cavity cavity and extruded pharynx.

Solution: Planarians belong to the group of fresh water flatworms and have a digestive system which lacks an anal opening. The mouth of the planarian opens into a cavity that contains the muscular pharynx, or proboscus, which can be protruded through the mouth directly onto the prey (see Figure). In feeding, the planarian moves over its food object, which may be a small worm, crustacean or insect larva, and traps it with its body. The pharynx is then extended and attached to the food material, and by sucking movements produced by muscles in the pharynx, the food is torn into bits and ingested. Digestive enzymes are released in order to assist the planarian in breaking down the food prior to ingestion. The pharynx delivers the food into the three-branched gastrovascular cavity. The branching of the planaria's gastrovascular cavity into one anterior and two posterior branches provides for the distribution of the end products of digestion to all parts of the body. Flatworms with three branched gut cavities are called triclads, in contrast to those with many branches, called polyclads.

Most of the digestion in planaria is intracellular, which means that it occurs in food vacuoles in cells lining the digestive cavity. The end products of digestion diffuse from these cells throughout the tissues of the body. Undigested materials are eliminated by the planaria through its mouth. The mouth, then serves as both the point of ingestion and the point of egestion.

It is interesting to note that planarians can survive without food for months, gradually digesting their own tissues, and growing smaller as time passes.

• PROBLEM 11-19

Discuss the excretory system of planaria.

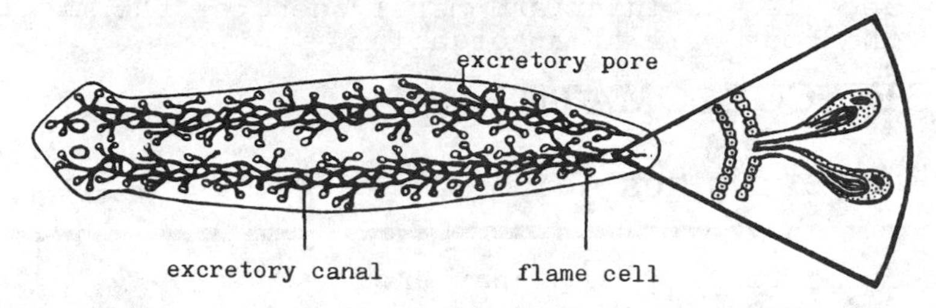

Excretory system of planaria

Solution: The excretory system of planaria involves a network of tubules running the length of the body on each side. These highly-branching tubules open to the body surface through a number of tiny pores. Side branches of the tubules end in specialized cells called flame cells (see Figure), also referred to as protonephridia. Each flame cell consists of a hollow, bulb-shaped cavity containing a tuft of long, beating cilia.

It is very probable that flame cells function primarily in the regulation of water balance. The presence of better developed flame cells in freshwater species lends support to the osmoregulatory function of these cells. Primarily, water, and some waste materials, move from the tissues into the flame cells. The constant undulating movement of the cilia creates a current that moves the collected liquid through the excretory tubules to the nephridiopores, through which it leaves the body. The motion of the cilia resembles a flickering flame, hence this type of exretory system is often called a flame-cell system.

Most metabolic wastes move from the body tissues into the gastrovascular cavity, and from there they are eliminated to the outside through the mouth. Nitrogenous wastes are excreted in the form of ammonia via diffusion across the general body surface to the external aquatic environment.

• PROBLEM 11-20

Explain how the planarian changes its body shape.

Solution: In the planarian, three kinds of muscle fibers can be differentiated: an outer, circular layer of muscle just beneath the epidermis; an inner, longitudinal layer; and dorsoventral muscles that occur in strands. Contraction of the longitudinal muscles causes constriction of the body, whereas contraction of the circular muscles causes an elongation of the body. Other alterations of body shape are produced by contraction of the dorsoventral muscle strands. Thus, by coordinated contraction and relaxation of these muscles, the body shape of a planarian can be varied.

• PROBLEM 11-21

What is meant by regeneration? Explain regeneration by using planaria as an example.

Solution: Regeneration is defined as the repair of lost or injured parts. Planaria, like hydra, has great powers of regeneration. Because of its larger size, the planaria has been studied the most. If a planarian is cut in two pieces, each part will regenerate into a whole worm; that is, a new head will form on the tail-piece and a new tail will form on the headpiece. If a slice is taken out from the animal immediately posterior to the head, this slice will be seen to regenerate two heads. A larger section cut out from the abdomen will regenerate a head from one end and a tail from the other. It is also possible to cut the head region down the middle, keeping the cut surfaces apart, and have two heads regenerated. The capacity for this regeneration is related to cells called neoblasts, which are part of the mesoderm, that retain the capability to proliferate and differentiate into the different tissues found in planaria.

The striking degree of regeneration, such as seen in hydra and planaria, becomes restricted in more complex animals due to greater cellular differentiation. A star-fish may regenerate its arms and an earthworm can re-generate a new head, but in man and other warm blooded vertebrates, the adult's capacity for regeneration is restricted to the healing of wounds.

• PROBLEM 11-22

The development of a parasitic mode of existence is a remarkable example of an adaptation that has evolved to permit one organism to exist at the expense of an-other. Among the flatworms the flukes are parasitic. Describe the body structure of a fluke, its modifica-tions for a parasitic existence, and its life cycle.

Solution: The adult fluke is roughly one inch long, and its body is quite flat. The worm is covered by a cuticle that is secreted by the underlying cells. The cuticle is one of its adaptation for parasitism, for it protects the worm from the enzymatic action of the host. With the development of the cuticle, the cilia and sense organs so characteristic of other flatworms have dis-

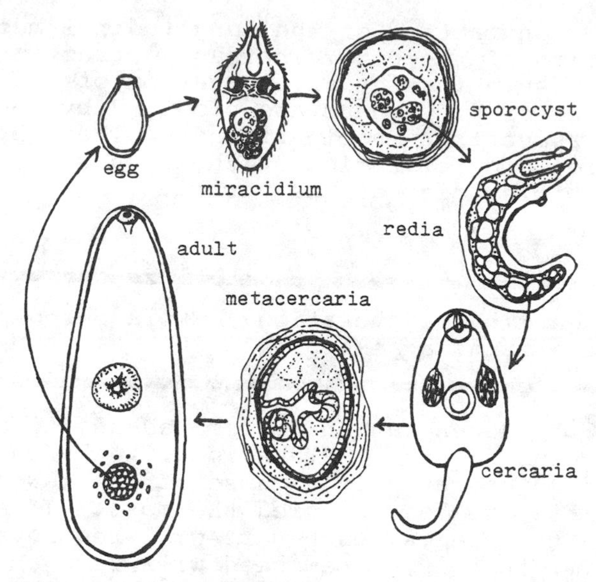

Life cycle of a fluke.

appeared. The mouth is located at the anterior end and is surrounded by an oral sucker. Another sucker, the ventral sucker, is located some distance posterior to the mouth. The ventral sucker is used to attach the fluke to the body of its victim, the host. The anterior sucker, with the aid of the muscular pharynx, is used to withdraw nutrients from the host. The mouth opens into a muscular pharynx, and just back of this, the alimentary tract branches into two long intestines that extend posteriorly almost the entire length of the body.

By far the most complicated aspect of these worms is their reproductive cycle. The fluke is hermaphroditic, that is, it contains both male and female sexual organs. There is no copulatory structure in the flukes and reproduction occurs through self fertilization. Following fertilization, the eggs are expelled from the fluke, which at this point is within the host's body; they then exit from the host's body in the feces. Hatching of the eggs in the oriental liver fluke occurs only when the eggs are eaten by certain species of freshwater snails, and takes place in the digestive system of the snail. Here, the egg hatches into a ciliated, free-swimming miracidium (see Figure). Within the digestive system of the snail, which is called the first intermediate host, the miracidium develops into a second larval stage, called a sporocyst. Inside the hollow sporocyst, germinal cells give rise to a number of embryonic masses. Each mass develops into another larval stage called a redia, or daughter sporocyst. Germinal cells within the redia again develop into a number of larvae called cercariae. The cercaria possesses a digestive tract, suckers and a tail. The cercaria is free swimming and leaves the snail. If it comes in contact with a second intermediate host, an invertebrate or a vertebrate, it penetrates the host and encysts. The encysted stage is called a metacercaria. If the host of the metacercaria is eaten by the final vertebrate host, the metacercaria is released, migrates, and develops into the adult form within a characteristic location in the host, usually the bile passages of the liver. Man, the final host, usually gets the fluke by eating a fish which contains the encysted cercaria.

• PROBLEM 11-23

What is a hermaphrodite? How and why is self-fertilization in hermaphrodites prevented?

Solution: Any organism that has both male and female gonads is said to be hermaphroditic. The majority of flowering plants are hermaphroditic in that they bear both the pistil (the female sex organ) and the stamen (the male sex organ). Hermaphroditism is rarer in animals than in plants. It occurs chiefly in sessile forms, notably the sponges and some mollusks. It is also found in the earthworms, the parasitic tapeworms and flukes. Hermaphroditism due to abnormal embryonic sexual differentiation resulting in individuals born with both a testis and an ovary has also been documented in man.

Some hermaphroditic animals, the parasitic tapeworms, for example, are capable of self-fertilization. Most hermaphrodites, however, do not reproduce by self-fertilization; instead, two animals, such as two earthworms, copulate and each inseminates the other. Self-fertilization is prevented by a variety of methods. In some animal species, self-fertilization is prevented by the development of the testes and ovaries at different times. In others, such as the oysters, the different gonads produce gametes at different times, so that self-fertilization cannot occur.

The advantage accompanying cross-fertilization is that genetic diversity is enhanced. Self-fertilization preserves the same genetic composition and with mutation occurring at a slow rate genetic variation is limited. Hermaphrodites undergoing cross-fertilization produce offspring showing the same genetic diversity as those resulted from individuals having a distinct sex. Thus, by ensuring gene exchange between individuals, cross-fertilization enhances genetic variation in the population. Only when a population carries a sufficient degree of genetic variation, can physical and environmental forces act to ultimately improve the characteristics of the individuals. And, as we shall see in a later chapter, the survival and propagative ability of a species depends to a large extent on its genetic diversity.

Nemertina

• PROBLEM 11-24

What are the evolutionary advances shown in the anatomy of the proboscis worms? How do these animals compare with the round worms?

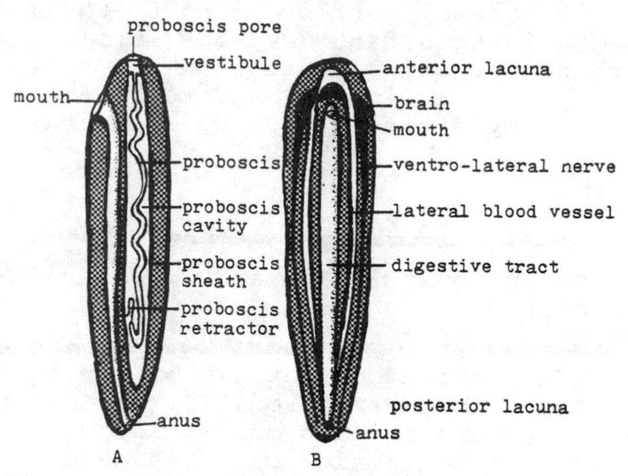

Diagrams of the structure of a typical proboscis worm or nemertean. A, Lateral view of the digestive tract and proboscis. B, Dorsal view of the digestive, circulatory and nervous system.

Solution: The proboscis worms (or nemerteans) are a relatively small group of animals. Almost all are marine forms, excepting a few which inhabit fresh water or damp soil. They have long, narrow bodies, either cylindrical or flattened, and range in length from 5 cm. to 20 meters. Their most remarkable organ is the proboscis, a long, hollow, muscular tube which they evert from the anterior end of the body to capture food. A mucus secreted by the proboscis helps the worm in catching and retaining the prey.

There are three important evolutionary advances achieved by the proboscis worms. First, these animals have a complete digestive tract, with a mouth at one end for ingesting food, an anus at the other for egesting feces, and an esophagus and intestine in between. This is in contrast to the coelenterates and planarians, in which food enters and wastes leave by the same opening. As in the flatworms, water and metabolic wastes are eliminated from the proboscis' body by flame cells. The second advance exhibited by the proboscis worms is the separation of digestive and circulatory functions. These organisms have a primitive circulatory system consisting of only three muscular blood vessels which extend the length of the body and are connected by transverse vessels. Hemoglobin, the oxygen-caryying protein present in higher animals, is found in the red blood cells of this rudimentary circulatory system. In the absence of a heart, the blood is circulated through the vessels by the movements of the body and the contractions of the muscular blood vessels. Capillaries have not developed in these animals. The third advance in the anatomy of the proboscis worms is seen in the structure of the nervous system. There is a primitive "brain" at the anterior end of the body, consisting of two groups of nerve cells, called ganglia, connected by a ring of nerves. Two nerve cords run posteriorly from the brain.

Compared to the proboscis worms, the round worms (or

nematodes) are a far more numerous and ecologically diversified group than the proboscis worms. There are about 8000 species of round worms, all with a similar basic body plan. They have elongate, cylindrical, threadlike bodies which are pointed at both ends. Their many habitats include the sea, fresh water, the soil, and other animals or plants which they parasitize. This last characteristic is in sharp contrast to the proboscis worms, none of which is parasitic and none is of economic importance. Common parasitic nematodes are the hookworm, trichina worm, ascaris worm, filaria worm and guinea worm, which all utilize man as the host. Because of their parasitic existence, these worms are covered with a protective cuticle and have only longitudinal muscles for simple bending movements. Contrary to the proboscis worms, which have cilia all over the epithelium and the lining of the digestive tract, none of the nematodes has any cilia at all. However, like the proboscis worms, the round worms have evolved a complete digestive system, a separate circulatory system, and a nervous system composed of a "brain" and nerve cords.

THE PSEUDOCOELOMATES

• **PROBLEM** 11-25

Parasitic organisms often have complex life cycles involving two or more host species. The nematode (round worm) causing elephantiasis has man as its primary host and the mosquito as its intermediate host. Describe the life cycle of this nematode.

Solution: Adult filaria worms of the genus-species Wucheria bancrofti, inhabit the lymph glands of man, particularly in the upper leg or hip. Large numbers of these nematodes block the lymph glands and prevent the return of fluid to the blood stream. This causes enormous swelling of the legs, a condition known as elephantiasis. Blocking of the lymph glands also predisposes the host to microbial infections.

These filara worms mature in the host's lymph glands where they later lay their eggs. The larval nematodes then migrate to the peripheral blood stream. The migration to the peripheral circulation has been found to correlate temporally with the activity of mosquitos. When a mosquito of a certain species bites the infected person, the larval filariae respond by entering the host's peripheral bloodstream. When other mosquitos subsequently bite the host and suck his blood, some larvae move into the bodies of these intermediate hosts. Inside a mosquito, the larvae migrate from the new host's digestive tract to its thoracic muscles, where they undergo further development. After a certain period, the larvae move to the proboscis of the mosquito. The proboscis consists of the upper lip or labium, the mandibles, maxillae, and the hypopharynx, which are long and sharp, modified for piercing man's skin and sucking his blood. The immature filariae are then introduced into another human whom the mosquito bites. Inside this primary host, they are carried by the circulatory system to the lymph glands, where they grow to adults and reach sexual maturity.

• PROBLEM 11-26

The animals having radial symmetry are considered to be at a lower evolutionary stage than those having bilateral symmetry. Why is this true? What is meant by the term secondarily radial?

Solution: The bodies of most animals are symmetrical, that is, the body can be cut into two equivalent halves. In radial symmetry any plane which runs through the central axis from top to bottom divides the body into two equal halves. In bilateral symmetry, only one plane passes through the central body axis, that can divide the body into equal halves. In a bilateral body plant, six sides are distinguished: front (ventral), back (dorsal), head-end (anterior), tailend (posterior), left, and right. (Most animals are not perfectly symmetrical; for example, in man, the heart is located more to the left, the right lung is larger than the left lung, and the liver is found on the right side of the body.)

Coelenterata (including the hydras, the true jellyfishes, the sea anemones, sea fans and corals) and Ctenophora (including the comb jellies and sea walnuts) are two radiate phyla. Their members have radially symmetrical bodies that are at a relatively simple level of construction. They have no distinct internal organs, no head, and no central nervous systems though they possess nerve nets.There is a digestive tract with only one opening serving as both mouth and anus, and there is no internal space or coelom between the wall of the digestive cavity and the outer body wall. These animals have two distinct tissue layers: an outer epidermis (derived from embryonic ectoderm) and an inner gastrodermis (derived from embryonic endoderm). A third tissue layer, mesoglea (mesoderm) is usually also present between the epidermis and gastrodermis. Often gelatinous in nature, it is not a well-developed layer, and has only a few scattered cells, which may be amoeboid or fibrous. Coelenterates and ctenophores are mostly sedentary organisms. Some are sessile at some stage of their life cycles, while others are completely sessile throughout their lives.

The higher animals are usually bilateral in symmetry. The flatworms and the proboscis worms are regarded as the most primitive bilaterally symmetrical animals. They are however far more advanced than the coelenterates and ctenophores. Both have bodies composed of three well-developed tissue layers - ectoderm, mesoderm, and endoderm. Their body structures show a greater degree of organization than the radially symmetrical animals. Although there are no respiratory and circulatory systems, there is a flame-cell excretory system, and well-developed reproductive organs (usually both male and female in each individual). Mesodermal muscles show an advance in construction; circular and longitudinal muscle layers are developed for purposes of locomotion and/or alteration of body shape. Several longitudinal nerve cords running the length of the body and a tiny "brain" ganglion located in the head are present which together constitute a central nervous system. In higher bilaterally symmetrical animals, there is observed a trend toward more complicated construction of the body. Separate organ systems are developed, specializing in different functions. There is also a separation of sexes in individuals so that each individual produces only one kind of gametes (male or female but not both).

The echinoderms are radially symmetrical animals. The phylum Echinodermata includes sea stars (starfish), sea urchins, sea cucumbers, sand dollars, brittle stars, and sea lillies. These animals are fairly complex; they have a digestive organ system, a nervous system, and a reproductive system. The echinoderms are believed to have evolved from a bilaterally symmetrical ancestor. Also, whereas the adults exhibit radial symmetry, echinoderm larvae are bilaterally symmetrical. For these reasons, the echinoderms are considered to be secondarily radially symmetrical. (The echinoderms will be discussed in greater detail in Chapter 12.)

SHORT ANSWER QUESTIONS FOR REVIEW

Answer

To be covered when testing yourself

Choose the correct answer.

1. Which of the following is incorrect concerning protozoans? (a) Protozoans are usually only found on land and fresh water. (b) Protozoans can take part in parasitic or mutualistic relationships. (c) Protozoans can feed holozoically, meaning they can ingest nutrients like bacteria directly, or holophytically, meaning they obtain nutrients via photosynthesis. (d) Because their shapes are so diverse, protozoans cannot be classified by symmetry. — a

2. In reference to the life cycle of the euglena, what is meant by the terms "active phase" and "encysted phase"? How and when does reproduction occur in relation to these phases? (a) The "active phase" refers to movement of the euglena without a protective coat, while the "encysted phase" is movement with a protective coat surrounding the euglena. Reproduction can only occur during the "active phase" by plasmotomy. (b) The "active phase" refers to movement of the euglena at times other than while digesting, while the "encysted phase" is movement during feeding, the protective coat necessary for proper digestion. Reproduction can take place during either phase sexually by conjugation. (c) The "active phase" refers to the portion of a euglena's life cycle in which it is free to move about, while the "encysted phase" is the portion of the cycle when the euglena is covered by a protective coat, not free to move about. Reproduction can take place in either phase and occurs asexually via binary fission. (d) None of the above. — c

3. Which of the following best explains the probable mechanism of amoeboid movement? (a) The simultaneous conversion of both ectoplasm and endoplasm to a gel-like mass causes the formation of a pseudopodia. This acts as a foot-like device, as the amoeba uses it to move around. (b) The simultaneous conversion of both ectoplasm and endoplasm to a liquid-like mass causes pressure to build up on the rim of the amoeba, forming a pseudopodia. The amoeba then uses the pseudopodia to move around. (c) neither ectoplasm nor endoplasm change their original consistency. A pseudopodia is formed due to a change in internal pressure inside the interior of the amoeba. The pseudopodia is then used by the amoeba to move around. (d) Ectoplasm moves

Answer

To be covered when testing yourself

from outside to inside along the periphery of the cell, causing pressure to build up inside the amoeba. This causes the endoplasm to move out to the periphery, causing the formation of a pseudopodia. The endoplasm then continues flowing outward, until its flow is fountain-like, moving back towards the cell interior. The process then begins again. — d

4. Which of the following is incorrect concerning the method of feeding of a Paramecium? (a) Food is taken into the organism via water currents by way of the oral groove. From there, it moves into the cytostome, the so-called mouth of the organism. (b) The nutriments are transported around the Paramecium by a food vacuole. (c) Undigested material leaves the paramecium when the food vacuole fuses with the cell membrane. (d) Digestive enzymes are secreted by lysosomes that fuse with the food vacuole. — c

5. Which of the following is incorrect concerning conjugation in Paramecium? (a) It can only take place between parameciums of different mating types. (b) Each Paramecium involved assumes both a male and a female role in the passing of genetic material. (c) Once the macronuclei disintegrate, they rarely appear again. (d) The micronuclei divide by meiosis, with each one then becoming haploid. — c

6. Which of the following best describes ciliary movement in a Paramecium? (a) simple back and forth movement (b) movement similar to a rowing motion (c) a highly coordinated motion consisting of a power stroke and a recovery stroke. (d) movement that begins at the tip of the cilia and moves downward in a wave-like motion. — c

7. Which of the following groups of cell types found in sponges are matched correctly with their function. (a) Pinacocytes - line the spongocoel. porocytes - guard the incurrent pores. Sclerocytes - specialized amoebocytes that secrete spongin fibers. Choanocytes - flat, polygonal cells that line the outer surface of the sponge. (b) Pinacocytes - flat, polygonal cells that cover the outer surface of the sponge. Porocytes - cells that guard the incurrent pores. Sclerocytes - specialized amoebocytes that secrete spongin fibers. Choanocytes - cells which line the spongocoel, responsible for movement of water and hence, nutriment procurement. (c) Porocytes - cells specialized for production of spicules.

Answer

To be covered when testing yourself

Choanocytes - guard the incurrent pores.
Sclerocytes - specialized archaeocytes.
Pinacocytes - Phagocytic cells. (d) Sclerocytes - cells specialized for spongin secretion.
Choanocytes - phagocytic cells. Pinacocytes - guard the incurrent pores
Porocytes - line the outer surface of the sponge. | b

8. A planarium has the characteristic of being a triclad because: (a) It posesses three different orifices for ingestion of food. (b) It posessed a three-branched muscular pharynx. (c) It posesses three different orifices for egestion of waste-products. (d) It posesses a three-branched gastrovascular cavity. | d

9. Which of the following is incorrect concerning reproduction during the medusoid stage of Obelia? (a) The medusa contains an extension from its center called the manubrium. The gastrovascular cavity begins in the manubrium and extends into radial canals. There, the reproductive organs are located. (b) A medusa can be male, posessing testes, or female, posessing ovaries. (c) Fusion of gametes occurs within a specialized area of the female manubrium.

(d) Once fusion of gametes occurs, embryonic cleavage soon begins, giving rise to a larval form called a planula. The planula will eventually form a polypoid colony. | c

10. Which of the following best describes the "nervous system" of a hydra? (a) A regularly arranged nerve net concentrated in no one particular area of the hydra. The system lacks synapses and nerve transmission is rapid. (b) An irregularly arranged nerve net, slightly concentrated around the mouth, but not aggregated enough to form a brain or spinal cord. Synapses are present, with some conducting in two directions and others in only one direction. Transmission is slow. (c) Nerve cells are arranged concentrically around the mouth forming a primitive brain. Axons radiate outward forming a "nerve net." The system lacks synapses, so transmission is fast. (d) None of the above. | b

11. Which of the following is not a type of a nematocyst? (a) Volvent type - used to entangle prey (b) Glutinant type - used for anchoring purposes (c) Buccinant type - used for digestive purposes (d) Penetrant type - used to paralyze prey. | c

Answer

To be covered when testing yourself

12. Which of the following is correct concerning the reproductive cycle of a fluke? (a) Flukes are hermaphroditic, so reproduction can occur via self-fertilization (b) The cycle consists of two intermediate hosts before it reaches its definitive host - man. (c) The order of the reproductive cycle, with appropriate names of each stage, is as follows: eggs released by fluke → eggs hatch into miracidium inside 1st intermediate host, the snail → 2nd larval stage, sporocyst, also takes place in snail. → sporocyst gives rise to redia which give rise to cercariae, which leaves the snail → cercariae ingested by 2nd intermediate host, such as a fish, encysts, and is then called a metacercariae → man ingests fish, and metacercariae is released into digestive system. (d) a, b, and c are all correct. — d

13. If an animal is hermaphroditic, does it automatically mean that it reproduces via self-fertilization? (a) Yes, because there is no differentiation of sex, so self-fertilization must take place. (b) No, because if the testes and ovaries develop at different times, self-fertilization cannot take place, but rather copulation can occur between two individuals containing different gonads. (c) No, because if the gonads release gametes at different times, self-fertilization is prevented. (d) Both b and c are correct. — d

14. Which statement is incorrect? (a) Round worms can be parasitic, while Proboscis worms are usually never parasitic. (b) Both round worms and Proboscis worms are considered to posess a primitive brain. (c) Both round worms and Proboscis worms contain an outer protective cuticle. (d) Round worms do not contain any cilia, while Probiscis worms have a ciliated epithelium. — c

15. Which of the following most accurately describes the life cycle of the nematode Wucheria bancrofti. (a) The adult worm lays its eggs inside a mosquito. The mosquito then bites a human, at which time the eggs are injected into the human. The eggs hatch and mature in the blood stream and are taken up by another mosquito when he bites that individual. (b) The adult worm matures and lays its eggs in the human small intestine. From there, the eggs move to the blood stream, where they mature, and eventually find their way back to the small intestine. (c) The adult worm matures and lays its eggs in human lymph

Answer

To be covered when testing yourself

glands. From there, they migrate to the blood stream, where they are taken up when a mosquito bites that individual. The larval form then further develops inside the mosquito. The mosquito then bites another individual, injecting the still immature worm. The worm then migrates to the individual's lymph glands and the cycle begins again. (d) None of the above.

c

Fill in the blanks.

16. Reproduction in protozoa occurs in a variety of ways. When the cell divides into two equal halves, this is known as _____. Organisms such as the _____ multiply by this process. Other multinucleate protozoa divide by a method called _____, where the cytoplasm divides with no concurrent _____ division.

binary fission, amoebas, plasmotomy, nuclear

17. A euglena can feed either _____ or _____. When feeding by the latter method, energy is captured via _____. During darkness these structures _____, and the euglena shifts to the other mode of feeding.

heterotrophically, autotrophically, chloroplasts, disappear

18. Once a pseudopod has surrounded its prey, the amoeba takes in the food via a _____. The nutriment is then broken down, as in man, by _____. Any unwanted material is egested from the cytoplasm via _____ also, at _____ point along the cell membrane.

food vacuole, digestive enzymes, food vacuoles, any

19. The presence of contractile vacuoles in some protozoans is usually limited to those that live in _____ water. The reason for this is the differences in the _____ effect due to the greater solute concentration in the _____.

fresh, osmotic, ocean

20. Conjugation will serve no purpose if the two Paramecia involved are of the _____ make-up.

same genetic

21. Sponges are members of the phylum _____. The surface of a sponge has small openings called _____ that open into a cavity called a _____. This cavity has a direct, larger opening to the top surface called an _____.

porifera, incurrent pores, spongocoel, osculum

Answer

To be covered when testing yourself

22. Regulation of water balance in planaria is a function of the _____, also known as _____. The method by which this occurs is caused by beating _____. Nitrogenous waste is usually eliminated from the planaria via its _____.

flame cells, protonephridia, cilia, mouth

23. In an organism such as the hydra, the acellular layer sandwiched between the outer epidermis and the cells lining the gastrovascular cavity is called the _____. In higher invertebrates and all vertebrates, a similarly-placed middle layer of embryonic tissue holds greater importance, as it gives rise to systems such as the _____ and _____ systems. This layer is called the _____.

mesoglea, circulatory, excretory mesoderm

24. The polyps of a colony of Obelia are derived by _____. The reproductive polyps have a central core, the _____, which eventually buds, forming a _____. The latter then breaks off from the former, and swims through an opening in the former called a _____.

asexual budding, blastostyle, medusa, gonotheca

25. In the hydra, after _____ aid swallowing in the mouth, the prey enters the _____. There, _____ are released that begin the digestive process. This phase of the digestive process is _____. The _____ phase involves amoeba-like cells that capture the broken down food via _____, continuing the digestive process inside _____.

mucus secretions, gastrovascular cavity, enzymes, extracellular, intracellular, pseudopodia, food vacuoles

26. Located in the _____ of the hydra are specialized stinging cells called _____, which contain a bristle-like process at one end called a _____. Inside the _____ is a capsule containing the actual stinging structure, called a _____.

epidermis, cnidocytes, cnidocil, cnidocyte, nematocyst

27. _____ cells, located in the epidermal layer of the hydra, begin their life _____, but can change into either _____ cells or _____ cells.

interstitial, undifferentiated, cnidocyte, generative

	Answer
	To be covered when testing yourself
28. Elongation of the body of planaria occurs via contraction of its _____, _____ layer of muscle, while constriction of its body occurs via contraction of its _____, _____ layer of muscle.	outer, circular, inner, longitudinal
29. When contrasting structural developement in proboscis worms versus coelenterates, the former posess an _____ for excretory purposes; the latter utilize the _____ opening for ingestion and excretion. Also, the latter posess nerves in a net-like arrangement, while the former have two aggregations of nerve cells, each known as a _____, which can be considered a _____.	anus, same, ganglia, primitive brain.
30. Severe swelling of the limbs due to accumulation of fluid is called _____. This can be caused by blockage of the _____ by a member of the worm phylum called _____. The genus-species of this worm is _____.	elephantiasis, lymph glands, nematodes Wucheria bancrofti
Determine whether the following statements are true or false.	
31. Although there are a number of methods by which protozoa reproduce asexually, sexual reproduction does not occur in protozoa.	False
32. The fact that protozoans have a very diverse evolutionary profile, and that some even resemble plants more than animals, has caused many to disregard their classification into a particular animal phylum.	True
33. Nutrient procurement in the amoeba is done via pseudopodia. These are projections of the cytoplasm at specific sites along the cell membrane of the organism.	False
34. The function of a contractile vacuole in protozoans is strictly limited to the elimination of nitrogenous waste materials.	False
35. In Ciliates, such as the Paramecium, both the micronucleus and the macronucleus are necessary for sexual reproduction.	False
36. Cilia serve no other function other than locomotion, and in that capacity, can only move in one direction.	False

	Answer
	To be covered when testing yourself
37. Although sponges are multicellular, there is little communication between the aggregation of cells, hence, there are no organ systems.	True
38. Planaria can go for long amounts of time without eating, but in the process, will get smaller due to digestion of parts of itself.	True
39. Concerning alternation of generations of the coelenterate Obelia, the polypoid stage is always diploid, while the medusoid stage is always haploid.	False
40. Concerning the food procurement of the hydra, nematocysts and tentacles work together to get the prey into the mouth of the hydra.	True
41. Although it was once thought that the primitive nervous system of a hydra contained synapses, it is now known that they do not occur.	False
42. A change in the air pressure inside the capsule of a cnidocyte is what eventually causes the release of the nematocyst.	False
43. The body wall of a hydra consists of three structural layers: an outer layer called the epidermis, an inner layer called the gastrodermis, and a middle layer called the mesoderm.	False
44. A fluke has two suckers: a ventral one, adapted for attaching the fluke to its host, and an anterior one, adapted for sucking nutrients from the host.	True
45. An animal that is secondarily radial refers to the fact that in its larval stage it is radially symmetrical, but in its adult form it becomes bilaterally symmetrical.	False

CHAPTER 12

HIGHER INVERTEBRATES

THE PROTOSTOMIA

MOLLUSCS

• PROBLEM 12-1

Discuss the basic features of the molluscan body. How do they differ in the several classes of molluscs?

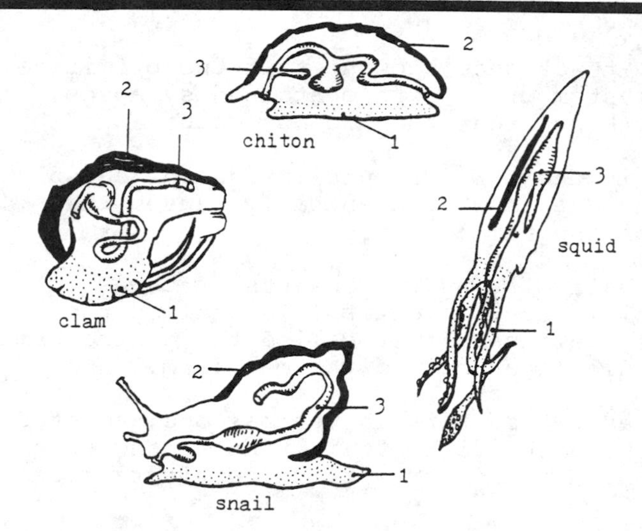

Variations in the basic molluscan body plan in chitons, snails, clams and squid. Note how the foot (1), shell (2) and alimentary tract (3) have changed their positions in the evolution.

Solution: The phylum Mollusca is the second largest in the animal kingdom. Snails, clams, oysters, slugs, squids, and octopuses are among the best known molluscs. The adult body plan is remarkably different from that of any other group of invertebrates. The soft body consists of three principal parts: (1) a large ventral muscular foot which can be extruded from the shell (if a shell is present) and functions in locomotion; (2) a visceral mass above the foot, containing most of the organs of the body; and (3) a mantle, a heavy fold of tissue that covers the visceral mass. In most species, the mantle contains glands that secrete a shell. The mantle often projects over the edges of the foot and overhangs the sides of the visceral mass, thus enclosing a mantle cavity, in which gills frequently lie.

The chitons, members of the class Amphineura, have

an ovoid bilaterally symmetrical body with an anterior mouth and posterior anus. They have paired excretory organs, nerves, gonads, and gills. The shell consists of eight dorsal plates.

Class Gastropoda is a large class, containing the snails, slugs and their relatives. Most gastropods have a coiled shell; however, in some species, coiling is minimal, and in others, the shell has been lost. The body plan of the adult gastropods is not symmetrical. During development, two rotations of the body occur, so that the anus comes to lie dorsal to the head in the anterior part of the body. The organs on one side of the body atrophy, so that the adult has one heart, one kidney, one gonad, and one gill. This embryonic twisting, called torsion, occurs in all gastropods, including those with a flat shell, slugs and other species without a shell.

Class Pelecypoda or Bivalvia contains the bivalves. These molluscs have two hinged shells. Scallops, clams, mussels, and oysters are well known bivalves. These animals have well-developed muscles for opening and closing the shells. These animals also have a muscular siphon for the intake and output of water.

The nautilus squid, and octopus are members of the class Cephalopoda. Cephalopod means head-foot, and in these molluscs the foot is fused with the head. In squids and octopuses, the foot is divided into ten or eight tentacles, and the shell is greatly reduced or absent. These molluscs have a well-developed nervous system with eyes similar to vertebrate eyes. The mantle is thick and muscular. Giant squids are the largest living invertebrates; they have attained lengths of 55 feet, and weights of two tons.

• PROBLEM 12-2

Snails can usually be seen crawling around during the night and on dark, damp days. What happens to them on bright, sunny days and why?

Solution: The soft body of the snail is protected by a hard calcareous shell secreted by the mantle. The mantle is a fold of tissue that covers the visceral mass and projects laterally over the edges of the foot. During the daytime, the snail usually draws up inside the shell. In order to move and to feed, the foot and head must emerge from the shell. This usually occurs at night. In bright sunlight, the soft body of the snail would dry out. Slugs are molluscs similar to snails, but without hard shells. Slugs are generally found in damp forest environments, and remain in the moist earth during bright daylight.

Most molluscs are aquatic animals, and hence are in no danger of drying out. Land snails are one of the few groups of fully terrestrial invertebrates. In most land snails, the gills have disappeared, but the mantle cavity has become highly vascularized and functions as a lung.

• PROBLEM 12-3

Discuss feeding mechanisms in different molluscs. Describe the differences in structural organization that have enabled the various classes to adapt to different lifestyles.

Solution: The chitons lead a sluggish, nearly sessile life. A horny-toothed organ, the radula, is contained within the pharynx, and is capable of being extended from the mouth. Chitons crawl slowly on rocks in shallow water, rasping off fragments of algae for food.

Gastropods occur in a wide variety of habitats. While most are marine forms, there are many fresh water and some land species. Gastropods have a well-developed head with simple eyes. Most feed on bits of plant or animal tissue that they grate or brush loose with a well-developed radula. Gastropods generally move about slowly.

Bivalves lack a radula, and are filter feeders. Sea water is brought to the gills by the siphon. Cilia on the surface of the gills keep the water in motion and food particles are trapped by mucus secreted by the gills. The cilia push the food particles towards the mouth. Most digestion is intracellular. Certain bivalves, such as oysters, are permanently attached to the sea or river floor; others, such as clams, burrow through sand or mud. Scallops are capable of rapid swimming by clapping their two shells together. Enormous amounts of water are filtered by bivalves - an average oyster filters 3 liters of sea water per hour. Bivalves do not have a well-developed head or nervous system.

The cephalopods are active, predatory molluscs. The tentacles of squids and octopuses enable them to capture and hold prey. Cephalopods have a radula as well as two horny beaks in their mouths. These structures enable the cephalopods to kill the prey and tear it to bits. The nervous system is very well-developed. There is a large and complex brain, and image-forming eyes. The muscular cephalopod mantle is fitted with a funnel. By filling the mantle cavity with water and then ejecting it through the funnel, cephalopods attain rapid speeds in swimming.

• PROBLEM 12-4

Compare the open and the closed type of circulatory systems. Describe molluscan circulation in a bivalve, such as a clam. What is a hemocoel?

Solution: A closed type of circulatory system is one in which the blood is always contained in well-defined vessels. In an open circulatory system, there are some sections where vessels are absent and the blood flows through large open spaces known as sinuses. All vertebrates have closed circulatory systems, as do the annelids (earthworms and their relatives). All arthropods (insects, spiders, crabs, crayfish, and others) as well as most molluscs have open circulatory systems. Movement of blood through an open system is not as fast, orderly or efficient as through a closed system.

Aquatic molluscs and most aquatic arthropods respire by means of gills. In bivalves, blood in large open sinuses bathes the tissues directly. The blood drains from the sinuses into vessels that go to the gills, where the blood is oxygenated. The blood then goes to the heart, which pumps it into vessels leading to the sinuses. A typical circuit is heart → sinuses → gills → heart. In arthropods the blood sinuses are termed hemocoels. The hemocoel is not derived from the coelom, but from the embryonic blastocoel.

ANNELIDS

• PROBLEM 12-5

In what ways are the bodies of earthworms (i.e. Lumbricus) and marine worms (i.e. Nereis) adapted to their habitats?

Solution: Earthworms and Nereis are both classified in the phylum Annelida, the segmented worms. At a finer taxonomic level, earthworms are in the class Oligochaeta (few bristles) and Nereis in the class Polychaeta (many bristles). These two classes of annelids are adapted to live different lifestyles.

Earthworms live in a moist terrestrial habitat. They are commonly seen burrowing through damp soil. Nereis, like many polychaetes are marine worms which swim freely in the sea. Often they are found burrowing in sand and mud near the shore, or living in tubes formed by secretions from their body wall. The obvious difference in lifestyles between the two worms, the earthworm being a soil burrower, Nereis being a free swimmer, will be important in analyzing their adaptive body structures.

The earthworm has four pairs of bristles or chaetae (also spelled setae) on each body segment except the first. These bristles enable the earthworm to grasp the ground or adhere to the walls of its burrow while in motion. Nereis also has chaetae which cover the paired paddles or parapodia, which appear on each of their body segments. The parapodia are the structures which give Nereis the ability to swim.

Adaptations are also apparent in the methods of nutrient procurement in the different habitats of the earthworm and Nereis. Earthworms feed on decomposing matter as they sweep along the ground. The anterior segment bears no special appendages. In contrast, Nereis is a predator and the anterior end of the body commonly bears tentacles, bristles, palps, and antennae.

The earthworm, living a subterranean life, has no well-developed sense organs. Nereis, an active swimmer, has two pairs of eyes, and organs sensitive to touch and to chemicals in the water.

Fertilization is external in both the earthworm and Nereis. However, Nereis, which relies on a watery medium for the union of egg and sperm, has no special protective device for the developing young. Earthworms produce special cocoons, and deposit eggs and sperm in them to facilitate union and so that the young can develop in a sheltered environment. Earthworms are hermaphroditic, while in Nereis, sexes are separate.

• PROBLEM 12-6

Explain excretion in the earthworm.

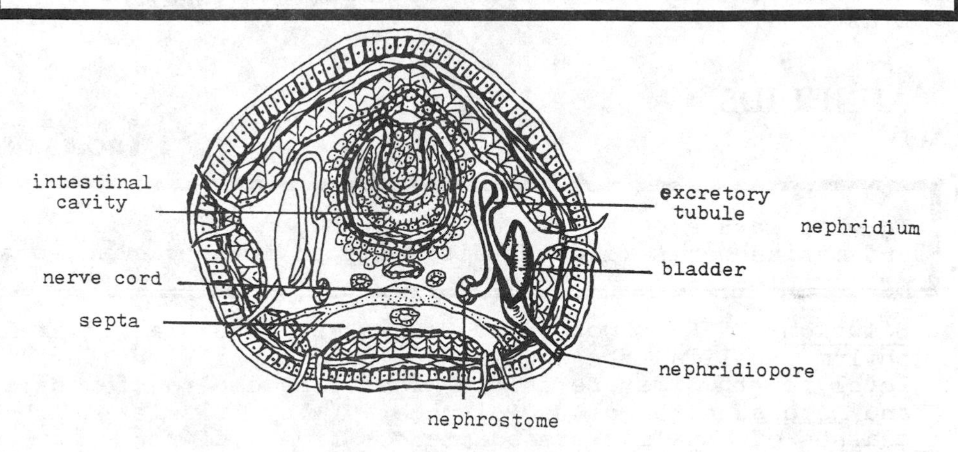

Excretory system of the earthworm (cross-section)

Solution: The earthworm's body is composed of a series of segments internally partitioned from each other by membranes. In each segment of its body, there are a pair of specialized excretory organs, called nephridia. They open independently from the body cavity to the outside. The various nephridia are not connected to each other. A nephridium consists of an open ciliated funnel or nephrostome (corresponding to the flame cell in planaria) which opens into the next anterior coelomic cavity. A coiled tubule running from the nephrostome empties into a large bladder, which in turn empties to the outside by way of the nephridiopore (see Figure). Around the coiled tube is a network of blood capillaries. Materials from the coelomic cavity move into the nephridium through the open nephrostome partly by the beating of the cilia of the nephrostome, and partly by currents created by the contraction of muscles in the body wall. Some materials are also picked up by the coiled tubule directly from the blood capillaries. Substances such as glucose and water are reabsorbed from the tubule into the blood capillaries, while the wastes are concentrated and passed out of the body through the nephridiopore. The earthworm daily excretes a very dilute, copious urine, which amounts to 60 percent of its total body weight.

The principal advantage of this type of excretory system over the flame cell is the association of blood vessels with the coiled tubule, where absorption and reabsorption of materials can occur.

• PROBLEM 12-7

Compare the functions of the seminal vesicles and the seminal receptacles in the earthworm.

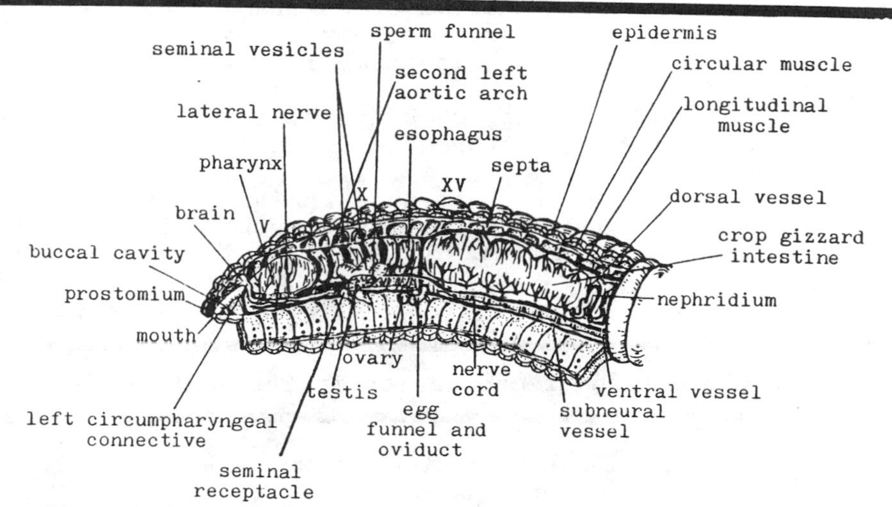

Figure 1. Internal structure of anterior portion of worm.

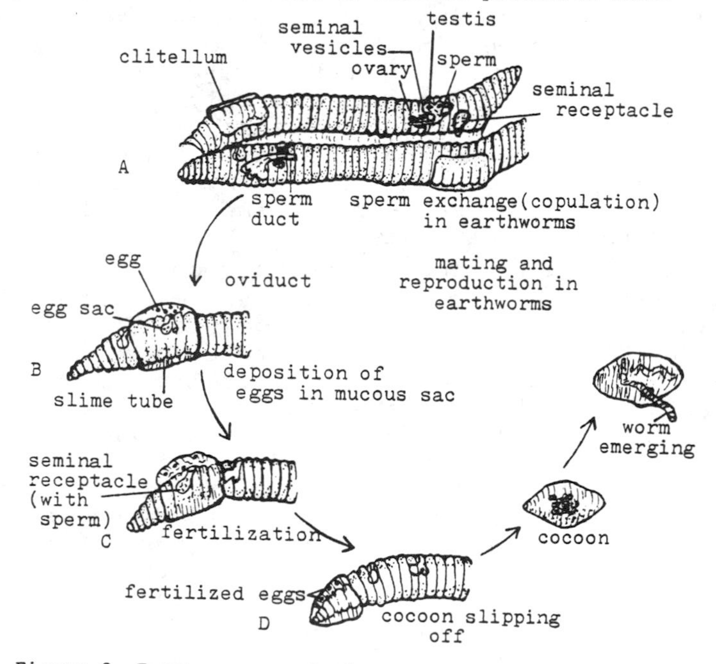

Figure 2. Earthworm copulation and formation of egg cocoons.

Solution: The seminal vesicles are glands that secrete a thick fluid which nourishes the sperm cells. The mixture consisting of sperm cells and fluid from the seminal vesicles is called semen. In the earthworm, there are three pairs of seminal vesicles. The testes, which produce the sperm, are housed within the seminal vesicles. There are two pairs of testes within the three pairs of seminal vesicles. Sperm are discharged from the testes into the vesicles, where they are stored and nourished. During copulation, the sperm enter the funnel-shaped openings of the vas deferens, and pass to the outside.

Forming a part of the female reproductive system, but completely separated from the female oviducts, are the seminal receptacles. These storage chambers are simple pairs of sacs, opening onto the ventral surface of the segment containing them. The number of seminal receptacles ranges from one to five pairs, each pair located in a separate segment. During copulation, sperm are deposited in the receptacles and stored for later use.

It should be noted that an earthworm contains both seminal vesicles and receptacles, therefore it is called a hermaphroditic organism. However, they do not self-fertilize.

• PROBLEM 12-8

Explain the method of copulation in the earthworms.

Solution: During copulation two worms are united, ventral surface to ventral surface, with the anterior ends in opposite directions and the anterior quarter of the length of the bodies overlapping (see figure in previous question). They are held together in this position in part by mucous secretions from a swollen glandular region called the clitellum. The mucus from the clitellum forms a sleeve around the animals. During copulation, each worm discharges sperm that pass from the vas deferens into the seminal receptacles of the other worm, through temporary longitudinal furrows that form in the skin. Following the exchange, the worms separate.

A few days after copulation, a cocoon is secreted for the deposition of the eggs and sperm. A mucous tube is secreted around the anterior segments, including the clitellum. The clitellum then secretes a tough chitin-like material that encircles the clitellum like a cigar band; this material forms the cocoon. When completely formed, the cocoon moves forward over the anterior end of the worm. The eggs are discharged from the female gonopores, and they enter the cocoon before it leaves the clitellum. Sperm are deposited in the cocoon as it passes over the seminal receptacles. As the cocoon slips over the head of the worm, and is freed from the body, the mucous tube quickly disintegrates, and the ends of the cocoon constrict and seal themselves. The cocoons are left in the damp soil, where development takes place. Development is direct, that is, there is no larval stage and the eggs develop into tiny worms within the cocoon.

Explain the locomotory pattern of the earthworm.

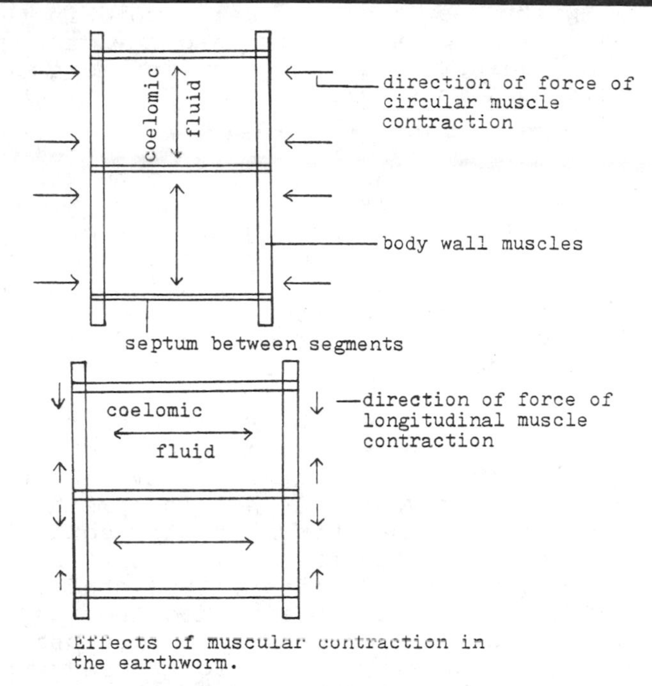

Effects of muscular contraction in the earthworm.

Solution: In the earthworm, two layers of muscle in the body wall and the four pairs of short bristles or chaetae of each body segment are the structures involved in locomotion. Circular muscle occupies the outer layer of the body wall, and just beneath it is the longitudinal muscle layer. Contractions of these two sets of muscles account for the changes in shape of the worm. Each segment of the body is filled with coelomic fluid that is incompressible. Thus the contraction of either layer of muscle results in a change of shape. Contractions of the longitudinal muscles make the worm shorter and thicker, while contractions of the circular muscles make it thinner and correspondingly longer (see Figure). Circular muscle contraction is most important in crawling, and always generates a pressure pulse in the coelomic fluid. Longitudinal muscle contraction is more important in burrowing. When the worm is crawling, longitudinal contraction may not be strong enough to generate a coelomic pressure pulse, but it does so in burrowing.

Chaetae are extended during longitudinal muscle contraction, functioning to anchor the worm in its burrow and to determine the direction of locomotion. They are retracted during circular muscle contraction. Because of the combined effects of the chaetae and longitudinal muscle contraction, each segment of the body moves in steps of 2 to 3 cm, at the rate of seven to ten steps per minute. The direction of contraction can be reversed, thus enabling the worm to reverse directions. It should be remembered that without the chaetae, no forward movement can take place, but only a change in shape.

• **PROBLEM 12-10**

The earthworm has a central nerve cord running along the entire length of the body. When an earthworm is cut into several parts, severing its nerve cord in the process, each part will go on crawling around for some time. How can each fragment carry out crawling movement even after the nerve cord is severed?

Solution: The arrangement of the nervous system in lower invertebrates is quite different from that in higher vertebrates like man. In higher vertebrates, there is an expanded and highly developed anterior end of the spinal cord, forming the brain. The brain coordinates and regulates the activities of the entire body. When the brain is separated from the rest of the body, the animal cannot perform any complicated functions.

However, in the earthworm, a higher invertebrate, coordinated activities such as crawling can still be observed when the body is cut into several transverse sections. This movement is possible because the earthworm has more than one neural center controlling and coordinating its activities. The nervous system of the earthworm consists of a large, two-lobed aggregation of nerve cells, called the brain, located just above the pharynx in the third segment, and a subpharyngeal ganglion just below the pharnyx in the fourth segment. A nerve cord connects the brain to the subpharyngeal ganglion and extends from the anterior to the posterior end of the body. In each segment of the body there is a swelling of the nerve cord, called a segmental ganglion. Sensory and motor nerves arise from each segmental ganglion to supply the muscles and organs of that segment. The segmental ganglia coordinate the contraction of the longitudinal and circular muscles of the body wall, so that the worm is able to crawl. When the earthworm is cut into several pieces, thus severing the connection to the brain (that is, the nerve cord), the resulting fragments still contain segmental ganglia which can fire impulses to the muscles of the body wall, resulting in crawling movement.

ARTHROPODS

Classification

• **PROBLEM 12-11**

What characteristics of arthropods have been of primary importance in their evolutionary success?

Solution: Arthropoda is an extraordinarily large and diverse phylum. There are about a million different arthropod species. Approximately 80 percent of all animal species belong to this phylum. The Class Insecta, containing more than 750,000 species, is the largest group of animals; in fact, it is larger than all other animal groups combined.

The development of a chitinous exoskeleton has been an important factor in the evolutionary success of the arthropods. An exoskeletal plate covers each body segment, and movement is possible because the plates of each segment are joined by a flexible, thin articular membrane. In most arthropods, there has been a reduction in the total number of body segments, fusion of many segments, and secondary subdivision of certain segments. This has resulted in body parts specialized for different functions, with movable joints between the parts. Infolding of the embryonic ectoderm has also resulted in an internal skeleton, providing sites for muscle insertions. Arthropod musculature is complex, and quite unlike that of most other invertebrates. The exoskeleton also lines both the anterior and posterior portions of the digestive tract.

The exoskeleton is termed a cuticle and is composed of protein and chitin. The cuticle consists of an epicuticle, an exocuticle, and an endocuticle. The outermost epicuticle often contains wax, as it does in insects, and serves to prevent water loss. The exocuticle is tougher than the innermost endocuticle. The epicuticle and exocuticle are absent at joints to provide for flexibility. The cuticle of the crustaceans is further strengthened by deposition of calcium salts. Where the epicuticle is absent, the cuticle is relatively permeable to gas and water. Fine pore canals in the cuticle permit elimination of secretions from ducts. The support and locomotion provided by the elaborate exoskeleton are major advantages that arthropods have over all other invertebrates, and the ability for modification has enabled the cuticle to serve many functions.

Another evolutionary advantage of the arthropods is the adaptation of well-developed organ systems. The digestive, respiratory, and blood-vascular systems are all complex, and often show improvements over other invertebrates' systems. The reproductive structures are well-developed in many arthropods. Internal fertilization occurs in all terrestrial forms, and in many arthropods, the eggs and the young are brooded and "nursed" by adults. The excretory system is well-developed in terrestrial arthropods.

The complex organization of the nervous tissues in the arthropods is yet another key factor in the evolutionary success. The brain in most arthropods is large, and many complex sense organs are present. Motor innervation to the muscles is precise and allows for many different movements, speeds, and strengths. The compound eyes of insects and crustaceans result in a wide visual field, and a great ability to detect movement. Other sensory receptors are sensitive to touch, chemicals, sound, position and movement. Arthropods often have complex behavioral patterns, many of which are under hormonal control.

• PROBLEM 12-12

What are the distinguishing features of the different classes of arthropods?

Solution: The arthropods are generally divided into the subphyla Chelicerates, Mandibulates (or Crustacea), Uniramia and the extinct Trilobita.

The chelicerate body is divided into two regions, a cephalothorax and an abdomen. Chelicerates have no antennae. The first pair of appendages are mouthparts termed chelicerae. The chelicerae are pincerlike or fanglike. The second pair serves various functions including capturing prey or serving as a sensory device, while the last four pairs of appendages on the cephalothorax function mainly as walking legs. Class Merostomata contains the horseshoe crabs; other members of this class are extinct. The horseshoe crabs have five or six pairs of abdominal appendages that functionas gills. The last abdominal segment or telson is long and spinelike. The exoskeleton contains a large amount of calcium salts. The horseshoe crabs are scavengers. Arachnids comprise the other living class of Chelicerates, including spiders, scorpions, mites and ticks. Arachnids are terrestrial. The first pair of legs is modified to function as feeding devices and are called pedipalps. In most arachnids, the pedipalps seize and tear apart prey. Most arachnids are carnivorous. In spiders, the pedipalps are poisonous fangs. Abdominal appendages are lost, or else function as book lungs, as in spiders. Arachnids usually have eight simple eyes, each eye having a lens, optic rods, and a retina. The eyes are chiefly for the perception of moving objects, and vision is poor.

The mandibulates have antennae and have mandibles as their first pair of mouthparts. Mandibles usually function in biting and chewing, but are never claw-like as are chelicerae. Maxillae are additional mouthparts found in most Mandibulates. The class Crustacea contains an enormous number of diverse aquatic animals. Crabs, shrimp, lobsters and crayfish are well known crustaceans. There are many small, lesser known crustacean species, such as water fleas, brine shrimp, barnacles, sowbugs, sandhoppers, and fairy shrimps. Crustaceans have two pairs of antennae, a pair of mandibles, and two pairs of maxillae. Compound eyes are often present in adult crustaceans. The larger crustaceans have a calcareous cuticle. Larger crustaceans have gills, but many smaller crustaceans rely on gas exchange across the body surface. There is enormous diversity among crustaceans. A great range of diet and feeding mechanisms are used. Appendages are modified in many different ways, and the body plan varies greatly.

Uniramia encompasses the next three classes. Class Chilopoda contains the centipedes. Their body consists of a head and trunk. There is a pair of mandibles, two pairs of maxillae, and the first pair of trunk appendages are poisonous claws that enable the centipede to capture prey. All other trunk segments bear a pair of walking legs. Centipedes are terrestrial. They have tracheae and Malpighian tubules.

Class Diplopoda contains the millipedes. The body is divided into head and trunk. The trunk segments bear two pairs of legs. There are mandibles and a single fused pair of maxillae, and no poisonous claws. Millipedes are not carnivorous. Respiration and excretion are similar to that of centipedes and insects.

Class Insecta is an enormous group. The insect body has 3 parts: a head, with completely fused segments; a thorax of three segments, each segment bearing a pair of legs; and an abdomen. Two pairs of wings, if present, are attached to the thorax. There is one pair of antennae, and usually a pair of compound eyes. The mouthparts of insects are mandibles, a pair of maxillae, and a lower lip which is formed from fused second maxillae. Respiration is by tracheae and excretion by Malpighian tubules. Insects are the only invertebrates which fly and are mainly terrestrial.

• PROBLEM 12-13

What are the distinguishing features of the several orders of insects?

Solution: Insects are divided into different orders based primarily upon wing structure, type of mouthparts, and the type of metamorphosis they undergo, i.e., complete or incomplete.

There are approximately twenty-five orders of insects. The class Insecta (Hexapoda) is divided into two subclasses. Subclass Apterygota contains the wingless insects, which are believed to be the most primitive living insects. Subclass Pterygota contains most of the insect orders. These are winged insects, or if wingless, the loss of wings is secondary. Some of the better known orders will be further discussed.

Order Thysanura contains primitive wingless insects with chewing mouthparts and long tail-like appendages. Silverfish and bristletails are members of this order. Some species are commonly found in houses and eat books and clothing.

Order Odonata contains dragonflies and damsel flies. The two pairs of long membranous wings are held permanently at right angles to the body. They have chewing mouthparts and large compound eyes. Immature forms of these insects (nymphs) are aquatic (fresh water).

Other winged insects can fold their wings back over the body when they are not flying. Order Orthoptera contains grasshoppers, crickets and cockroaches. The forewings are usually leather-like. They do not function in flying, but function as covers for the folded hindwings. The chewing mouthparts are strong.

Termites are social insects that belong to the order Isoptera. Both winged and wingless varieties comprise the termite colony.

Order Hemiptera contains the true bugs. They have

piercing-sucking mouthparts. The forewing has a distal membranous half and a basal, leathery, thick half.

Order Anoplura contains the sucking lice. These insects are wingless and have piercing-sucking mouthparts. The legs are adapted for attachment to the host. These lice are external parasites on birds and mammals- the headlouse and crablouse are parasites on man. They are often vectors of disease, such as typhus. All of the orders just discussed have incomplete metamorphosis.

Order Lepidoptera contains butterflies and moths. They undergo complete metamorphosis, as do the rest of the orders to be discussed. Lepidoptera have two large pairs of scale covered wings, and sucking mouthparts (in adults).

Order Diptera contains true flies, mosquitoes, gnats, and horseflies. They have one pair of flying wings, with the hindwings modified as balancing organs. Mouthparts are piercing-sucking, or licking. Adults are often disease vectors.

Order Coleoptera, containing beetles and weevils, is the largest order. They have hard forewings which cover membranous hindwings. They have chewing mouthparts and undergo complete metamorphosis.

Order Siphonaptera contains the fleas. They are small, wingless parasites. They have piercing-sucking mouthparts, and long legs adapted for jumping.

Order Hymenoptera contains ants, wasps, bees, and sawflies. There are winged and wingless species. Wings, when present, are two membranous pairs which interlock in flight. Mouthparts are for chewing, chewing-sucking or chewing-lapping.

External Morphology

• PROBLEM 12-14

What are the advantages and disadvantages of an exoskeleton? How have some of the disadvantages been overcome by the molluscs and by the arthropods?

Solution: An exoskeleton serves several functions: (1) it provides support for the soft body to counterbalance the force of gravity; (2) it serves as a point of attachment for muscles; (3) it protects the body against desiccation; and (4) it serves to protect the animal from predators. The disadvantages of an exoskeleton are that movement is somewhat restricted and that growth within the exoskeleton is limited.

Different species of molluscs have evolved different mechanisms for overcoming these limitations. Most molluscs lead very sluggish lives; many are sedentary or burrow in sand or wood. The fast moving squids have a greatly reduced exoskeleton enclosed within the mantle. Octopuses, also fast moving, have lost the exoskeleton entirely. Movement of bivalves is made possible by muscles attached to two hinged shells. Growth of most molluscs proceeds by additions to the shell, creating a larger exoskeleton. In gastropods, growth proceeds spirally, resulting in a spiral shell. Nautilus, a cephalopod, builds a new chamber as it grows, and lives within the latest and largest chamber of the shell. The animal secretes a gas into the other chambers, and is thus able to float.

The arthropods have a segmented body. The rigid exoskeleton is thinner in certain regions, such as the leg joints and between body segments. Muscle attachments at these points offer arthropods a fairly wide range of movement. Most arthropods walk, jump, or crawl. Some insects are capable of flying and many aquatic arthropods swim actively. Growth in arthropods occurs by a series of molts. When the growing animal reaches a certain size, the exoskeleton is shed and a new, larger exoskeleton is formed. During the process, the arthropod is left "naked", and is thus temporarily vulnerable.

• PROBLEM 12-15

What are the advantages of a segmented body plan? Which animal phyla have segmentation?

Solution: Segmentation offers the advantage of allowing specialization of different body segments for different functions. More primitive animals have a large number of segments, all very similar to one another. More complex animals have fewer segments, and the specialization of different segments may be so far advanced that the original segmentation of the body plan is obscured.

Annelida, the phylum containing earthworms, marine worms, and leeches, is the most primitive invertebrate group displaying segmentation. Earthworms and most marine worms, have both internal and external segmentation. The earthworm body is composed of one hundred or more segments. The segments are separated by partitions termed septa. The marine worms often have incomplete septa, and the leeches have lost the septa completely. Except for a few specialized head and tail segments, the appendages and musculature of all segments are exactly the same. Internally, each segment has a pair of excretory organs and its own ventral nerve ganglion. The earthworm has five pairs of muscular tube-like hearts, two pairs of testes, three pairs of seminal vesicles, and one to five pairs of seminal receptacles.

Molluscs exhibit embryonic development very similar to that of annelids, and the larval form, called the trochophore, is very similar in both phyla. This similarity suggests a possible evolutionary relationship between molluscs and annelids. Neopilina is the only living species of molluscs showing segmentation; it may, however, only be secondary.

Onycophora (Peripatus and relatives) is a very small phylum showing segmentation. The onycophorans have a combination of annelid and arthropod characteristics, and provide an evolutionary link between these two phyla. The onycophorans have many similar segments with 14 to 43 pairs of short, unjointed legs.

Arthropods are a large group of segmented animals that have evolved in many complex ways. More primitive forms have paired, similar appendages on each segment. More advanced forms have appendages modified for a variety of functions such as sensory antennae, feeding apparatus, walking legs, swimming legs, claws, reproductive apparatus, and respiratory structures, depending upon the species. The more advanced arthropods have a greatly reduced number of segments. These arthropods also have a fusion of body segments into distinct regions, such as a head and trunk, or head, thorax and abdomen. Internally, there is a great deal of specialization, and septa between segments have been lost. The first six segments always form a head, usually with a well-developed brain and sense organs. In primitive forms there are ganglia in every segment, but in many groups the ganglia have fused to form ganglionic masses. Aquatic arthropods may have one pair or several pairs of excretory organs, or coxal glands. In most groups of terrestrial arthropods, the excretory organs are Malpighian tubules. Generally, there is only one pair of reproductive organs; centipedes, however may have 24 testes.

Chordata is the other phylum characterized by segmentation. As with the arthropods, there also has been increasing specialization of segments in chordates. In man, the nervous system and the skeletal system are segmentally organized. In primitive fish, the gills and musculature are also segmented.

• PROBLEM 12-16

Of the terrestrial arthropods, the insects are among the best adapted for the prevention of water loss. Explain.

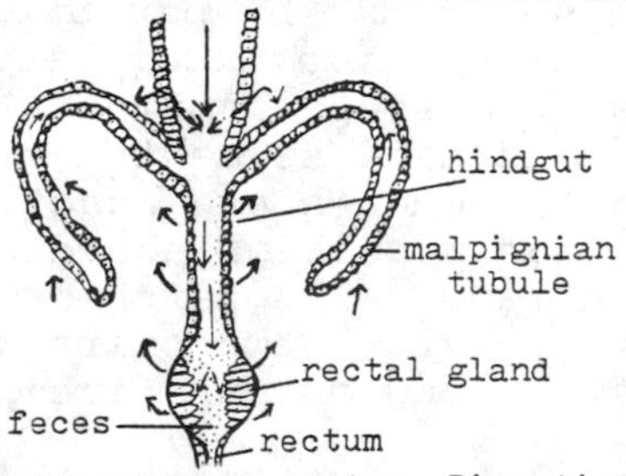

hindgut

malpighian tubule

rectal gland

feces

rectum

The insect excretory system. Directions of water movement indicated by arrows.

Solution: Insects are a large group of mostly terrestrial arthropods. Their success of survival on land is partly due to the adaptation of water-conserving mechanisms. Their epicuticle, the outermost exoskeletal layer, is impregnated with waxy compounds. This serves to prevent loss of water by evaporation. Their wastes are excreted as uric acid, a crystalline substance, thus minimizing water loss due to protein metabolism. Uric acid is excreted via Malpighian tubules. Malpighian tubules lie more or less free within the hemocoel (blood sinus), with the proximal end attached to the digestive tract at the junction of the midgut and hindgut (rectum). There may be 2 to 250 tubules. If there are only a few, the tubules are usually long, slender, and convoluted. If there are many, the tubules may be short and grouped in bunches. The cells lining the tubule lumen are cuboidal epithelial cells. The outer layer of the tubule wall is composed of elastic connective tissue and muscle fibers. The tubules are capable of peristalsis.

Uric acid, formed by the tissues and released into the blood, is selectively absorbed by the tubule cells from the blood within the hemocoel. Salt, water, and amino acids are also absorbed. Together, the substances are discharged from the Malpighian tubules into the hindgut. Nutritive substances and some of the salts are reabsorbed by the rectal epithelium and are returned to the blood. Much water is also reabsorbed to conserve water inside the body. The uric acid is excreted with the waste products of digestion.

There are several other methods of excretion in insects. Some excess salts and other substances are deposited in the cuticle and shed when the insect molts. Some particulate wastes are picked up and degraded intracellularly by nephrocytes, cells located on or near the heart.

Musculature

• PROBLEM 12-17

Why is the flexor muscle of the crayfish well developed?

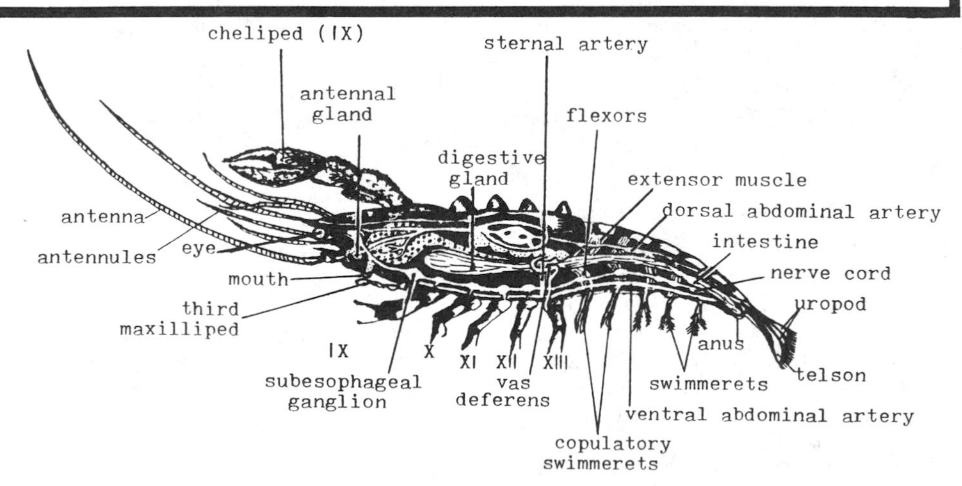

Internal structure of male crayfish.

Solution: In order to answer this question, it is necessary to understand the body plan of the crayfish. (Refer to figure in last solution.) Crustaceans have a head and trunk which is usually composed of a thorax and abdomen. The head and thoracic segments of the decapods are fused together dorsally by a carapace, which hangs over the sides of the animal and encloses the gills. Lobsters and crayfish have a large segmented abdomen, with appendages on each segment. The fourth thoracic appendages are chelipeds, walking legs that function as claws. The next four pairs of thoracic appendages are smaller, thinner walking legs. Lobsters, crayfish, and burrowing shrimps generally crawl with these walking legs. These animals are also capable of swimming backward rapidly. This is accomplished by flexing the abdomen ventrally. The flexor muscle of the crayfish is well-developed and powerful. It enables the animal to flex its abdomen rapidly in swimming. Many non-swimming decapods have a greatly reduced abdomen that is folded up under the thorax. This is often the case in crabs, as most crabs cannot swim. The blue crab, however, is a powerful and agile swimmer. Its last pair of legs are broad flat paddles that act as propellers. Shrimps are generally bottom-dwellers, i.e., they dwell at the bottom of a body of water. They swim intermittently.

The abdominal appendages, pleopods, are the principal swimming organs. The pleopods are large and often fringed. Shrimps utilize ventral flexion of the abdomen for quick backward darts and for vertical steering.

• PROBLEM 12-18

Insect flight differs from bird flight in that insects have no muscles attached to the wings. Describe the mechanism of insect flight.

Solution: Insect wings are attached directly to the body wall over a fulcrum. The fulcrum system of the wings and body wall function like a seesaw mechanism. As contractions of the muscles occur changing the shape of the thorax, the wings respond by up and down motions (Figure 1)

Figure 1. Fulcrum action of insect flight.

The dorsal plate of a segment in the insect body is termed the tergum, and the ventral plate is the sternum. Contraction of vertical muscles of a segment pulls the tergum to the sternum, causing the wings, on the opposite side of the fulcrum, to be raised. The contraction of the longitudinal muscles of a segment causes the tergum to bulge upward, and the wing moves downward. The movements of the thoracic body wall segments are barely perceptible, but the lengths of the levers on the two sides of the fulcrum are very different, and the movement of the wings is great in comparison (see Figure 2).

Figure 2.
Diagram of the arrangement of the flight muscles of an insect. The contraction of the longitudinal muscles forces the tergum up and the wings down. Contraction of the vertical muscles forces the tergum down and the wings up.

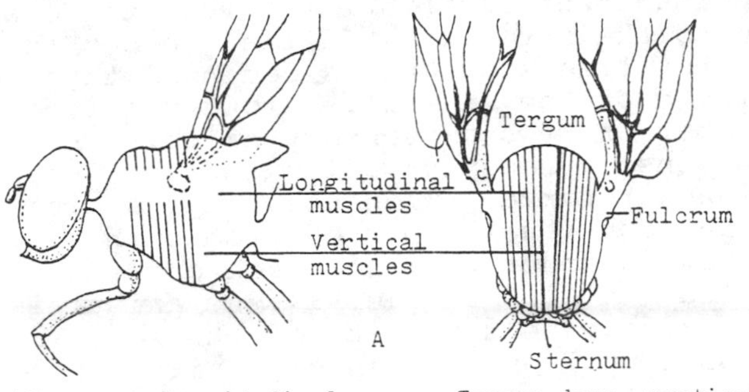

Wings up, longitudinal muscles relaxed

Tergum down, vertical muscles contracted

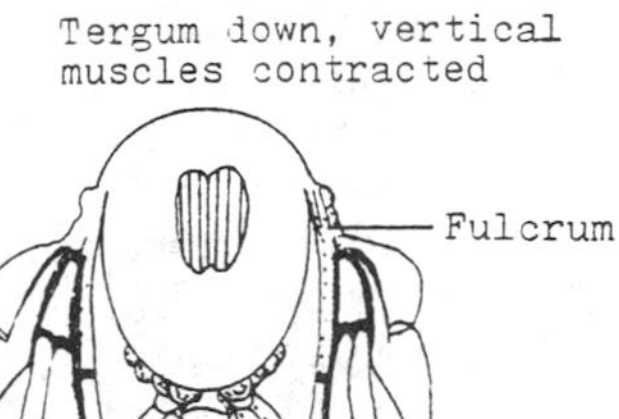

Fulcrum

B

Wings down, longitudinal muscles contracted

Tergum up, vertical muscles relaxed

The vertical muscles are referred to in some textbooks as the tergo-sternal muscles. Their name indicates their function, which is to move the tergum and sternum towards and away from each other.

In most insects, two pairs of wings are present, both usually function in flying. The anterior pair of wings functions in propulsion, while the posterior wings are balancing organs. They are termed halteres and function as gyroscopes for control of flight stability. This wing arrangement is seen in the mechanism of flight for mosquitoes and flies.

In the grasshoppers and beetles, only the posterior wings are flying organs. The anterior wings are protective devices for the flying pair.

Insect flight muscles are very powerful. The fibrils are relatively large and the mitochondria are huge - about half the size of a human red blood cell.

The Senses

• PROBLEM 12-19

Explain how the sense organ of equilibrium functions in the crayfish.

Solution: The sense organ of equilibrium in the crayfish is the antenna.

All crustaceans have two pairs of antennae. The second pair in lobsters and crayfish is extremely long; this pair functions in touch and taste. At each base of the first pair, a statocyst is present, which functions in the maintenance of equilibrium. The statocysts are sacs that open to the exterior. Inside the sac is a statolith, composed of fine sand grains cemented together by secretions from the statocyst wall. Along the floor of the sac are a number of rows of sensory hairs. These hairs arise from sensory receptor cells and are innervated by the antennular nerve. When the crayfish is in an upright position, continual impulses from both right and left statocysts counterbalance each other and the crayfish does not lean to either side. If the animal is rotated onto its right side, the statolith of the left side exerts a gravitational pull on the sensory hairs, which send impulses to the brain. The statolith on the right side does not exert a pull on the sensory hairs, and the crayfish as a result rotates to the left, thus balancing itself. If the animal is turned so that both statoliths are exerting forces, and impulses are sent from both statocysts, the crayfish rotates in the direction of whichever statocyst is most stimulated.

The statocyst is an invagination of the ectoderm, which forms the exoskeleton, and is shed with each molt. The animal must gather new sand grains to form the statolith. This is done by inserting sand into the statocyst sac or by burying the head in sand. Studies of the statocyst's function were done by inserting iron filings into sand when crayfish molted, and using a magnet to direct statolith forces on the sensory hairs.

The statocyst may also function to maintain position during movement. For this purpose there are free sensory hairs, which are not in contact with the statolith. These hairs are stimulated by the motion of fluid in the chamber when the animal moves.

• PROBLEM 12-20

How does the compound eye of crustaceans and insects differ from man's eyes? What advantages do insects' eyes have that man's do not?

Solution: Compound eyes consist of a varying number of units or ommatidia (see Figure). Each ommatidium has its own lens, cornea, and group of neurons. Some arthropods, such as ants, have only a few ommatidia while others, such as dragonflies, have as many as 10,000. Each ommatidium, in gathering light from a narrow sector of the visual field, projects a mean intensity of the light from that sector onto the total retinal field. All the points of light from the various ommatidia form an image, called a mosaic picture. The nature of a mosaic image can be understood by comparing it to the picture of a TV screen or a newspaper photo. The picture is a mosaic of many dots of different intensities. The clarity of the image depends upon the number of dots per unit area. The greater the number, the better the picture. The same is true of the compound eye. Each ommatidial receptor cell with its accompanying nerve-fiber projects a segment of the total picture. The

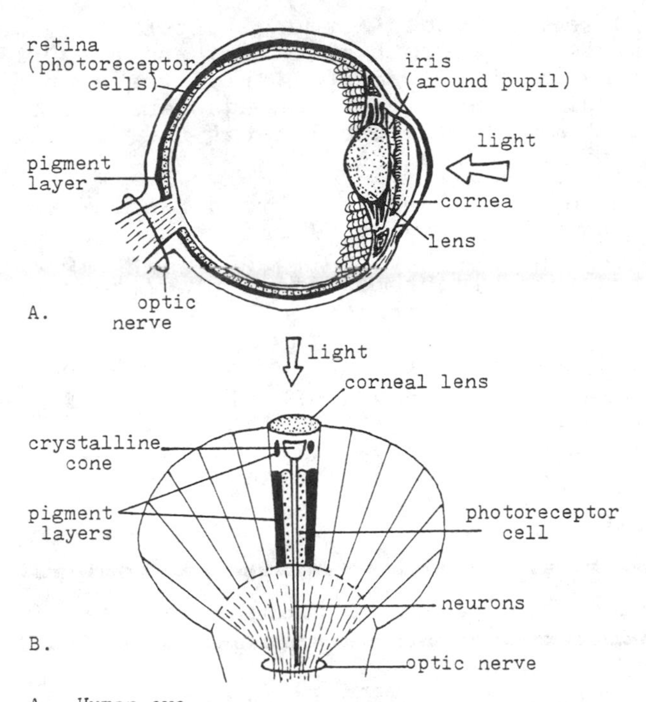

A. Human eye
B. Compound eye (one ommatidium shown in detail)

compound eye may contain hundreds of thousands of neurons, clustered in groups of seven or eight per ommatidium.

The human eye also forms a retinal picture composed of numerous points, each point corresponding to a rod or cone cell with its accompanying neuron. Considering the fact that the human eye contains approximately 125,000,000 rods and 6,500,000 cones, the concentration of corresponding points is much higher than in the compound eye, and the image formed is thus much finer.

In addition, while the compound eye of the arthropod is composed of numerous lenses, the human eye has only one lens for its entire corneal field. Hence, the insect has no structure strictly analogous to the human retina; their critical surface for vision is the outer surface of the compound eye itself, composed of the many closely packed individual lenses.

Many arthropods have very wide visual fields; in some crustaceans, the corneal surface covers an arc of 180°. However, the anthropod cornea and lens are developed from the skeleton, and the eyes are fixed in place. The human eye, has a much smaller visual field, but compensates for this to some extent with movability.

Compound eyes are particularly sensitive to motion. This is because the ommatidia recover very rapidly from a light impulse, making them receptive to a new impulse in a very short time. Compound eyes thus can detect flickers at extremely high frequency. Flies detect flickers of a frequency up to 265 per second, as compared to man's limit of about 53 per second. Because flickering light at higher frequency is seen as a continuous light by man, motion pictures are seen as smooth movement, and 60-cycle bulbs give off steady light. The ability of the compound eyes to detect high frequency light changes enables them to rapidly detect motion and facilitates the capture of prey and the avoidance of enemies.

Compound eyes are sensitive to a broader range of light wavelengths than human eyes. Insects can see well into the ultraviolet range. Colors that appear identical to man may reflect ultraviolet light to different degrees, and appear strikingly different to insects.

Compound eyes are also able to analyze the plane of polarization of light. A sky appearing evenly blue to man, reveals different patterns in different areas to insects, because the plane of polarization varies. Honeybees use this ability as an aid to navigation.

Organ Systems

• PROBLEM 12-21

Describe the entire digestive process in the crayfish.

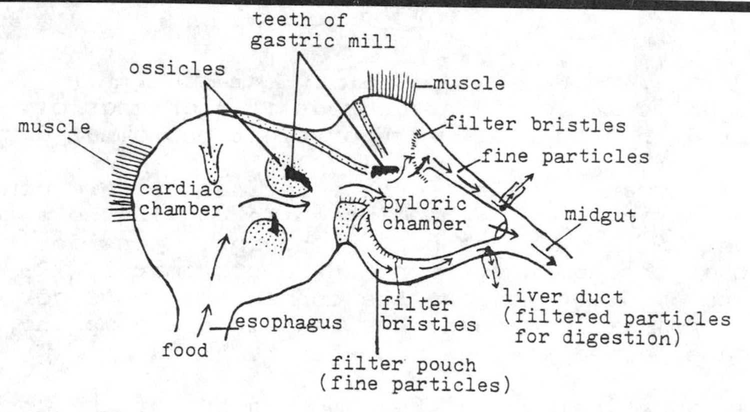

Digestion in the crustaceans.

Solution: The crayfish is a member of the class Crustacea, order Decapoda. Most crustaceans are marine animals. The crayfish are the most successful freshwater decapods. Most are about 10 cm long, but some are as large as lobsters, another decapod. Crayfish are mostly bottom-dwellers, while some may be semi-terrestrial. Crayfish are omnivorous. The feeding appendages of the decapods consist of a pair of mandibles, two pairs of maxillae, and three pairs of maxillipeds. Each pair of feeding appendages is a different modification of the leg-like processes of the primitive body segments. The mandibles are short and heavy with opposing grinding and biting surfaces. The accessory appendages serve to hold the food, tear it, or push it to the mandibles. The first pair of crayfish legs is larger and heavier than the other pairs, and function as claws or pincers. Food is caught or picked up by these pincers, called chelipeds, and then pushed to the third pair of maxillipeds which pushes the food between the other mouth parts. A piece is bitten off by the mandibles, while the remainder is pulled from the mouth by the maxillae and maxillipeds. The bitten piece is pushed into the pharynx and another bite is taken. When the crayfish is not eating, the third pair of maxillipeds covers the other mouth parts.

In the digestive tract, a short esophagus leads to a large cardiac stomach. The cardiac stomach is separated by a constriction from a smaller pyloric stomach. A long intestine extends from the pyloric stomach through the abdomen to the anus. The anus is located on the ventral surface of the last abdominal segment, the telson. Opening into the midgut (the posterior part of the pyloric stomach in crustaceans) is a single blind sac which functions as a digestive gland or hepatopancreas. It also functions in food absorption and storage. The hepatopancreas is composed of ducts and secretory tubules. Absorption of food occurs only in the tubules of the digestive gland and in the walls of the midgut.

Food is consumed in large pieces by crayfish and other decapods. The cardiac stomach serves to mix and break up these pieces. The walls of the cardiac stomach have a number of chitinous ridges and calcareous ossicles, that function as teeth to grind and tear the food.

Digestive enzymes, secreted by the hepatopancreas, are passed forward to the cardiac stomach. The cardiac stomach has a complex musculature which controls movement of the stomach walls. These factors enable the cardiac stomach to reduce food to a very fine consistency.

The pyloric stomach has setae, chitinous bristles, on complex folded walls. The setae act to filter the food and direct it to the hepatopancreas. Only fluid and the smallest particles are able to pass through the duct of the digestive gland. The musculature of the pyloric stomach serves to force food through the filters, and also directs it to the intestine. Food that is not absorbed is eliminated via the anus.

• PROBLEM 12-22

The circulatory system of insects does not function in gas exchange. What is its function? Describe the circulatory and respiratory systems in insects.

Solution: Insects, which have high metabolic rates, need oxygen in large amounts. However, insects do not rely on the blood to supply oxygen to their tissues. This function is fulfilled by the tracheal system. The blood serves only to deliver nutrients and remove wastes.

The insect heart is a muscular dorsal tube, usually located within the first nine abdominal segments. The heart lies within a pericardial sinus. The pericardial sinus is not derived from the coelom, but is instead a part of the hemocoel. It is separated by connective tissue from the perivisceral sinus which is the hemocoel surrounding the other internal structures. Usually, the only vessel besides the heart is an anterior aorta. Blood flow is normally posterior to anterior in the heart and anterior to posterior in the perivisceral sinus. Blood from the perivisceral sinus drains into the pericardial sinus. The heart is pierced by a series of openings or ostia, which are regulated by valves, so that blood only flows in one direction. When the heart contracts, the ostia close and blood is pumped forward. When the heart relaxes, the ostia open and blood from the pericardial sinus is drawn into the heart through the ostia. After leaving the heart and aorta, the blood fills the spaces between the internal organs, bathing them directly. The rate of blood flow is regulated by the motion of the muscles of the body wall or the gut.

A respiratory system delivers oxygen directly to the tissues in the insect. A pair of openings called spiracles is present on the first seven or eight abdominal segments and on the last one or two thoracic segments. Usually, the spiracle is provided with a valve for closing and with a filtering apparatus(composed of bristles) to prevent entrance of dust and parasites.

The organization of the internal tracheal system is quite variable, but usually a pair of longitudinal trunks with cross connections is found. Larger tracheae are supported by thickened rings of cuticle, called taenidia. The tracheae are widened in various places to form internal air sacs. The air sacs have no taenidia and are sensitive to ventilation pressures (see below). The tracheae branch to form smaller and smaller subdivisions, the smallest being the tracheoles. The smallest tracheoles are in direct contact with the tissues and are filled with fluid at their tips. This is where gas exchange takes place.

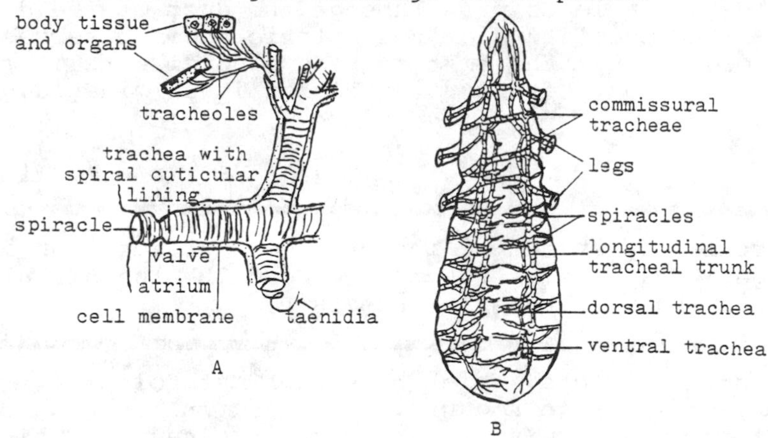

Figure. A, Relationship of spiracle, tracheae, taenidia (chitinous bands that strengthen the tracheae), and tracheoles (diagrammatic). B, Generalized arrangement of insect tracheal system (diagrammatic). Air sacs and tracheoles not shown.

Within the tracheae, gas transport is brought about by diffusion, ventilation pressures, or both. Ventilation pressure gradients result from body movements. Body movements causing compression of the air sacs and certain elastic tracheae force air out; those causing expansion of the body wall result in air rushing into the tracheal system. In some insects, the opening and closing of spiracles is coordinated with body movements. Grasshoppers, for example, draw air into the first four pairs of spiracles as the abdomen expands, and expel air through the last six pairs of spiracles as the abdomen contracts.

Reproduction and Development

• **PROBLEM** 12-23

Compare the function of the swimmerets in the male and female crayfish in sexual reproduction.

Solution: Swimmerets are the abdominal appendages of the crayfish. In many crustaceans these function as swimming legs. (The thoracic appendages generally function as feeding appendages, pincers, and walking legs.) Crayfish generally do not swim by means of the swimmerets. The swimmerets of both male and female crayfish are modified for reproduction and are called pleopods.

The male reproductive system consists of paired connected testes in the eighth thoracic segment. The testes lead to a sperm duct. At its terminal end, the sperm duct becomes a muscular ejaculatory duct, which opens to the outside on the ventral surface of the male on the base of the leg of the last thoracic segment. The first two pairs of pleopods are modified to aid in transfer of sperm to the female.

The ovaries are located in the sixth thoracic segment. The oviduct opens to the outside on the ventral base of the last thoracic appendage. Most female crayfish possess a median seminal receptacle. This is a pouch that opens to the outside on the ventral surface of the last, or second to last, thoracic segment.

In copulation, the tips of the male pleopods are inserted into the seminal receptacle of the female crayfish. Sperm flow along grooves in the pleopods and are deposited in the seminal receptacle. The sperm may be stored in the pouch until the female lays eggs. Fertilization occurs when eggs are laid. The female crayfish lies on its back and curls the abdomen forward. By beating the pleopods, a water current is created, and this drives the eggs into a chamber created by the curved ventral abdominal surface. The eggs remain attached to the abdomen while they are brooded. A cementing material is associated with the egg membrane, and the pleopods are modified to hold the eggs in place.

Male and female swimmerets are considerably different in decapods, making it easy to distinguish the two sexes. Almost all female decapods brood the eggs and have pleopods which are modified to hold the eggs. Female crayfish never use the pleopods for swimming.

• **PROBLEM** 12-24

How does metamorphosis differ from molting? What animals other than arthropods are known to undergo metamorphosis in their development?

Solution: In order to grow, arthropods must periodically shed their rigid exoskeletal shell and in its place grow a larger one. This process is termed molting. The crustacea molt many times during development. Each stage resembles the stage before it, and with each successive molt, the animal becomes larger and its body proportions become closer to the adult body proportions until its adult size has been attained.

Many insects pass through successive developmental stages that are quite unlike one another. For example, a fly passes from a worm-like larva to a sedentary pupa to an adult form in its maturation sequence. This striking change from the juvenile to the adult form is termed metamorphosis. Certain amphibians undergo metamorphosis. For instance, the larval stage of the frog is the tadpole. Aquatic animals adapted for sessile lives also often undergo metamorphosis. The tiny ciliated larvae either swim or are carried by sea currents to new locations, where they undergo metamorphosis. An example of this is the sea star, which changes from a bilaterally symmetrical larva to a radially symmetrical adult.

• PROBLEM 12-25

What controls the process of molting in crustacea?

Solution: Most crustaceans have successive molts. With each molt, they go through a gradual series of immature, larval stages, until they finally reach the body form characteristic of the adult. The lobster, for example, molts several times early in life, and at each molt it gets larger and resembles the adult more. After it reaches the adult stage, additional molts with only changes in size allow growth of the animal to continue.

Molting is under the control of a hormone released from nerve cells. Secretion by nerve cells rather than glands, is known as neurosecretion. An "antimolting" hormone is produced by a cluster of nerve cell bodies located near the eyestalk, called the X-organ. The axons of these cells are expanded at their tips, and it is in these expanded tips that the hormone is stored. The bundles of axons together constitute the sinus gland (note that the sinus gland is actually misnamed for it is not a glandular tissue). Proof that the sinus gland secretes a hormone inhibiting molting comes from surgical removal of this gland, which resulted in repeated molting by the crustacean.

The antimolting hormone prevents the secretion of a hormone by the Y-organ, a gland composed of ectodermal cells located at the base of the mandibular muscles. When the sinus gland hormone falls below a certain level in the blood, the Y-organ is stimulated to release its hormone. This hormone initiates the event of molting.

Just before molting, epidermal cells, stimulated by

the Y-organ hormone, secrete enzymes which digest the chitin and proteins of the inner layers of the old cuticle. A soft, flexible, new cuticle is then deposited under the old one, with folds to allow for growth. At this point, the crustacean seeks a protected retreat or remains in its burrow. In order to rid itself of its old shell, a crustacean will take in enough water to swell its body up to three to four times its normal size. The resulting pressure which builds up on the old, rigid shell causes it to burst. The underlying new shell, however, is pliable enough to accommodate the larger size. Eventually, the intaken water is replaced by the growth of new tissue. Water and air are taken in and the old skeleton swells up and bursts, as the new skeleton expands. The animal then extricates itself from the old skeleton. The epidermis secrete enzymes to harden the new cuticle by oxidizing some of the compounds and by adding calcium carbonate to the chitin. The new exoskeleton is completed after subsequent secretion of additional cuticule layers.

• PROBLEM 12-26

Discuss the structural and hormonal aspects of insect metamorphosis.

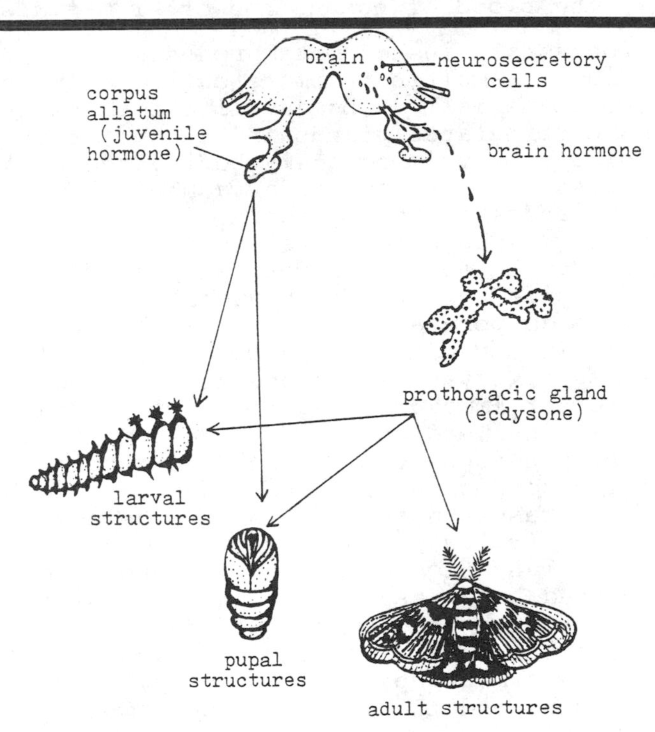

Interactions of juvenile hormone, brain hormone, and molting hormone (ecdysone) in cecropia silkworm.

Solution: Basically, there are two kinds of metamorphosis: incomplete metamorphosis which results in a similiar but larger form, and complete metamorphosis, which gives rise to strikingly different forms. Many insects, such as moths, butterflies and flies undergo metamorphosis.

Generally, a wormlike larva - called a caterpillar in moths, maggot in flies, and grub in bees - hatches from the egg. The larva is a relatively active form; it crawls about, eats voraciously, and molts several times, each time becoming larger in size. The last larval molt gives rise to a pupa, an inactive form which typically does not move or eat. Moth and butterfly larvae spin a cocoon around their bodies and molt within the cocoon to form the pupae. During this process all the structures of the larva are broken down and used as raw materials for the development of the adult. Each part of the adult body develops from a group of cells called a disc. The discs are embryonic cells derived directly from the egg and remain quiescent during larval stages. During the pupal stage they grow and differentiate into adult structures such as wings, legs, and eyes. These structures remain collapsed, folded and thus nonfunctional at first; when the pupa molts into the adult form, blood is pumped in to inflate them, and chitin is deposited to make them hard.

Metamorphosis in insects is characterized not only by sharp changes in appearance from the larval to the adult form, but also by striking changes in their modes of life. The butterfly larva eats leaves whereas the adult feeds on nectar from flowers. The mosquito larva lives in ponds and eats algae and protozoa, while the adult sucks the blood of humans and other mammals.

The process of insect metamorphosis involves a brain hormone, a molting hormone called ecdysone, and a regulatory hormone called the juvenile hormone. The brain hormone stimulates glands in the insect's prothorax (the part of the body immediately behind the head, to which the first pair of legs is attached). The prothoracic glands respond by secreting ecdysone which induces molting. Ecdysone is believed to be involved also in many growth and developmental processes in insects. Juvenile hormone is produced by a pair of glands, called corpora allata, which is located just behind the brain and closely associated with it. High concentrations of juvenile hormone at the time of molting result in another immature stage following the molt, while low concentrations lead to a more advanced stage after molting. This is supported by the demonstration that removal of the corpora allata from insects in a highly immature stage resulted in pupation at the next molt, followed by a molt that resulted in a midget adult. Conversely, implantation of an active corpora allata into an advanced pupa resulted in another immature stage after molting rather than an adult. In the larva, juvenile hormone is present in a high concentration, hence the larva molts to form a pupa. In the pupa, juvenile hormone is absent and thus when the pupa molts an adult results.

Notice that insect metamorphosis involves the nervous system as well as endocrine glands. Of the three hormones discussed, one is released directly by the brain (the brain hormone), one is secreted under the influence of the brain hormone (ecdysone),and the third is secreted by glands closely associated with the brain (juvenile hormone). The interactions between these three hormones are shown in the accompanying figure.

Social Orders

• PROBLEM 12-27

What are the differences between the queen bee, the worker bee, and the drone?

Solution: A honey bee colony consists of a single reproductive female, the queen, a few hundred males, called drones, and thousands of workers, which are sterile females. Drones hatch from unfertilized eggs, and are haploid, while females hatch from fertilized eggs, and are thus diploid. The type of food fed to the female larva determines whether it will develop into a worker or queen. A diet of "royal jelly" causes her to become a fertile queen. The work performed by a worker bee is determined by its age. For two weeks after metamorphosis, workers function as nurse bees, first incubating the brood and preparing brood cells, later feeding the larvae. Then the sterile females become house bees for one or two weeks, working as storekeepers, wax secreters, or guards and cleaning the hive, or ventilating it by fanning their wings. The worker finally becomes a food gatherer for four or five weeks, and collects nectar, pollen,and water.

A queen bee soon after hatching mates several times with male drone bees. The queen accumulates enough sperm to last for her lifetime, and stores the sperm internally in a spermatotheca. Thereafter she lays fertilized and unfertilized eggs, as many as a thousand a day. The drone dies after copulation; its reproductive organs literally explode into the female. When the number of bees in a hive becomes too large, about half the drones and workers, a new queen bee, and the old queen bee migrate to a new location and begin a new society. The old colony is left with developing queens.

• PROBLEM 12-28

What is meant by the term "social" and "solitary" as applied to insects?

Solution: "Solitary" insects are those capable of many functions including sexual reproduction, food gathering and other types of non-reproductive behavior. "Social" insects are insects living in aggregations in which there is extensive division of labor and internal societal organization. Colonial organization in insect societies is similar to that in a coelenterate colony, except that the insects are not anatomically joined, as are the members of an *Obelia* colony. The individual insects of a society are incapable of independent survival, and both their structure and function are specialized for their particular role in the society. Division of labor in insect societies is based upon major biological differences between individuals. Exactly how different castes are determined is not fully understood, and the process is not the same in different insect species. Ants, termites, certain wasp and bee species are social insects.

Insect behavior is based primarily on instinct, and is subject to relatively little modification, or learning. The individual has no choice of role in this system. Insect societies are not analogous to human societies. Most adult humans are capable of reproduction. Insect societies set apart several individuals for a solely reproductive role, while the other individuals are sterile and function in maintaining the colony. Human behavior is based primarily on learning, and division of labor is not chiefly based on biological differences. Humans can survive for extended periods apart from society, and are able to change their role within a society.

• PROBLEM 12-29

What mutually beneficial adaptations have arisen in the flowering plants and certain insects?

Solution: In the terrestrial environment, insects are of great ecological significance. Insects are crucial for the fertilization of many different species of angiosperms, the flowering plants. Approximately two-thirds of all angiosperms depend upon insects for pollination. Bees, wasps, butterflies, moths and flies are the principal insect pollinators. The three orders represented by these insects have an evolutionary history closely tied to that of flowering plants: both groups underwent an explosive period of development in the Cretaceous period, and both have evolved mechanisms or structures which promote insect pollination.

Many insects rely on the nectar produced by plants for food. As the insect sucks up the nectar, the sticky pollen grains adhere to its feet and body. When the insect flies to another plant, the pollen grains stick to the new flower and fertilize the egg.

A tremendous amount of adaptation between the flowering plants and the insects has occurred. The color, odor, and nectar of flowers all serve to attract insects. As an example, bees are attracted by bright blue and yellow colors, and by sweet, aromatic, or minty odors. The flowers pollinated by bees have these characteristics, and often have a protruding lip on which the bee can land. Flowers pollinated by the wind, on the other hand, usually lack conspicuous colors, odors, and nectar, and the pollen is light, not sticky. The mouthparts of different insects also show advanced specialization. For example, butterflies and moths suck liquid nectar, and have mouthparts modified to form a long tube. Bees, which gather both nectar and pollen for food, have mouthparts which are adapted for both sucking and chewing. A truly interesting adaptation for insect pollination is shown in some flowers with floral parts which evolved to resemble female insects of a species. The male insects are attracted to these flowers. The pollen grains are transferred to the bodies of the insects as they attempt to copulate with the insect-resembling flowers.

Thus insects help to insure cross-pollination of species of flowers. Some insects eat the fruits of angiosperms and eliminate the undigestible seeds in their waste. This is of significant advantage to the flowering plants in that their seeds can be dispersed over long distances by insects and new environments can be exploited.

THE DEUTEROSTOMIA

ECHINODERMS

• PROBLEM 12-30

Discuss the characteristics of the echinoderms and briefly describe several different classes.

Solution: The echinoderms have pentamerous radial symmetry, that is, the body can be divided into five similar parts along a central axis. Echinoderms have an endoskeleton composed of calcareous ossicles, these may articulate with each other as in sea stars, or may form interlocking plates, as in sand dollars. The skeleton usually bears projecting spines that give the skin a spiny or warty surface. The echinoderm coelom is well developed. The system of coelemic canals and projections, called the water vascular system, is found only in echinoderms.

Class Asteroidea contains the sea stars (starfish). Sea stars have a central disc and usually five arms, although some have more. The mouth is on the lower surface of the disc, and the anus on the upper surface. Locomotion is by means of the tube feet. The surface is studded with many short spines. They have no special respiratory system, but breathe by means of skin gills or dermal branchiae.

Class Ophiuroidea contains the brittle stars. They resemble sea stars, but their arms are longer, more slender, more flexible, and often branched. Locomotion occurs by rapidly lashing the arms, not through the use of tube feet. There is no anus or intestine. Gas exchange occurs through invaginated pouches at the periphery of the disc.

The sea urchins and sand dollars are members of Class Echinoidea. The endoskeletal plates are fused to form a rigid case. The body is spherical or oval, and is covered with long spines. Five bands of tube feet are present, and function in respiration in some species. There is a long, coiled intestine, and an anus.

Sea cucumbers, class Holothuroidea, have a much reduced endoskeleton and a leathery body. Many forms have five rows of tube feet. Unlike other echinoderms, they lie on their side. The mouth is surrounded by tentacles attached to the water vascular system. Gas exchange usually occurs through complexly branched respiratory trees attached to the anal opening.

Class Crinoidea contains the sea lilies. They are attached to the sea floor by a long stalk and are sessile. They have long, feathery, branched arms around the mouth, which is on the upper surface.

• PROBLEM 12-31

The same system which enables starfish to cling to rocky substrates is useless on sandy beaches. Explain.

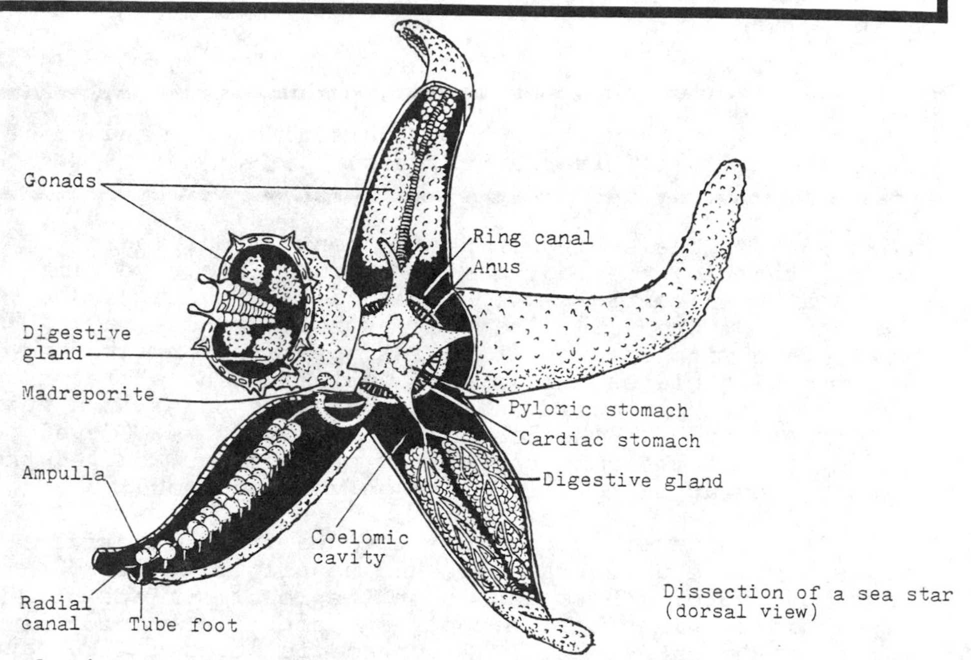

Dissection of a sea star (dorsal view)

Solution: Starfish (also called sea star) locomotion occurs by means of tube feet. The tube feet are part of a hydraulic system found only in echinoderms called the water vascular system. The under surface of each arm of a sea star has hundreds of pairs of tube feet, which are hollow, thin-walled, sucker-tipped, muscular cylinders. Internally, the tube foot is connected to a muscular sac, an ampulla, at its base. The tube feet are connected via short lateral canals to a radial canal which extends the length of the arm. The radial canals are joined at the central disc of the sea star by a ring canal. The ring canal leads via a short tube called the stone canal to a sieve-like plate, called a madreporite, on the surface of the animal. The entire system is filled with watery fluid. In order to extend the tube foot, the ampulla contracts, forcing fluid into the tube foot. A valve prevents the fluid from flowing into the radial canals. The tube foot attaches to the sea floor (or substratum) by a sucker at its tip, which acts as a vacuum. Longitudinal muscles cause the tube foot to shorten in length; this pulls the sea star forward, and water is forced back in to the ampulla. The tube foot is released from the substratum, and then again. The rapid repetition of this cycle of events enables sea stars to move slowly.

One can see that the use of suction in the tube feet would be useless to the sea star in moving on sandy beaches. On a soft surface, the tube feet are employed as legs. Locomotion becomes a stepping process, involving a backward swinging of the middle portion of the podia followed by a contraction, shoving the animal forward. Certain sea stars have tube feet without suckers at their tips; instead the tips are pointed. These sea stars live on soft bottoms, and can feed on buried animals.

Starfish are slow-moving carnivorous animals. Their prey are generally sedentary or slow moving animals, such as clams and oysters.

• PROBLEM 12-32

Describe the function of the sea star's stomach as the animal feeds upon a clam.

Solution: The sea star places its arms on the clam, and by applying suction with the tube feet, is able to pry the two shells slightly apart. The sea star then everts its cardiac stomach, as it is projected out of its mouth. The stomach then enters the crack between the bivalve shells. Digestive enzymes from the stomach cause the soft body of the clam to be degraded. The clam's adductor muscles are digested and the valves open. The partly digested food is taken into the pyloric stomach and then into the digestive glands, where digestion is completed and the products are absorbed. After the body of the clam is eaten, the sea star retracts the cardiac stomach.

The digestive tract of the sea star consists of a mouth on the lower surface, a short esophagus, an eversible cardiac stomach, a smaller pyloric stomach, a very small intestine, and an anus. Attached to the pyloric stomach are five pairs of large digestive glands. Each pair of digestive glands lies in the coelomic cavity of one of the arms. Both stomachs fill most of the interior of the central disc. The cardiac stomach is capable of entering a gap between the shells of a clam as small as 0.1 mm.

Certain species of sea stars spread the everted cardiac stomach over the ocean bottom, and digest all types of organic matter encountered. The everted stomach of sea stars is also capable of engulfing prey, enabling the prey to be swallowed whole.

Sea stars are carnivous and feed on many marine invertebrates, and even small fish. They also feed on dead matter. Some sea stars have extremely restricted diets, and feed only on certain species. Others have a wide range of prey, but exhibit preferences.

HEMICHORDATA

• PROBLEM 12-33

Phylum Hemichordata is believed to be the most closely related to our own phylum, Chordata. What is the evidence for this? How are the echinoderms linked to chordates?

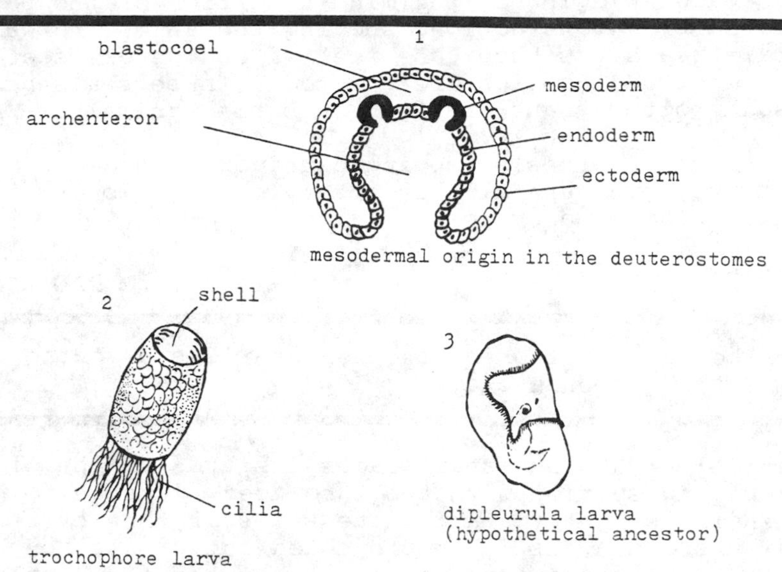

Solution: The hemichordates are marine animals, often called acorn worms. They live in U-shaped burrows in sand or mud. The body is bilaterally symmetrical and consists of a conical proboscis, a short collar, and a long trunk. The hemichordates' embryonic development is similar to that of the chordates. The echinoderms, hemichordates, and chordates are deuterostomes. The characteristics shared by these phyla are the formation of the anus from the embryonic blastopore. They also undergo radial and indeterminate cleavage, with the origin of mesoderm being from pouches of endoderm, and the formation of the coelom as the cavities in the mesodermal pouches. The other invertebrate phyla (mollusks, annelids and arthropods) are protosomes. In protosomes, the mouth forms from the blastopore, cleavage is determinate and spiral, mesoderm arises from ingrowth of a single cell near the blastopore, and the coelom developes from a split in an initially solid mass of mesoderm. The Deuterostomia and the Protosomia represent two important diverging lines of development from a primitive bilaterally symmetrical ancestor.

Adult hemichordates resemble chordates in two other ways. The hemichordates have pharyngeal gill slits. Water drawn into the mouth is forced out over these slits in the wall of the pharynx. Oxygen is removed from the water and carbon dioxide is released into it from the blood. Capillary beds are located in the septa between the gill slits. Only chordates and hemichordates have pharyngeal clefts. Hemichordates also have a thick, hollow, dorsal nerve cord in the collar. The anterior region contains a ventral nerve cord as well as a dorsal nerve cord. All chordates have a dorsal hollow nerve cord.

The echinoderms are related to the hemichordates. Hemichordates have a hydrostatic skeleton similar to the water vascular system of echinoderms. In addition, there are similarities between their larval stages. The dipleurula larvae of echinoderms and hemichordates are quite alike, and differ from the trochophore larvae of molluscs and annelids. Chordates have no ciliated larval stages similar to both of these. However, this ciliated larvae link between echinoderms and hemichordates is additional indirect evidence of the link between echinoderms and chordates.

SHORT ANSWER QUESTIONS FOR REVIEW

Answer

To be covered when testing yourself

Choose the correct answer.

1. In most species of Mollusca, the shell is secreted by (a) the excretory organ (b) the visceral mass (c) the mantle (d) the ventral foot. — c

2. Which of the following molluscs are filter feeders? (a) cephalopods (b) gastropods (c) bivalves (d) all of the above — c

3. The earthworm excretes (a) a very concentrated urine through the nephridiopore (b) a very concentrated urine through the nephrostome (c) a dilute, copious urine through the nephridiopore, (d) a dilute, copious urine through the nephrostome. — c

4. The function of the earthworm's seminal receptacles is to (a) secrete a thick fluid which nourishes the sperm cells in the male (b) store the sperm in the female after copulation until it is ready to be used. (c) store the sperm in the male (d) store the eggs in the female — b

5. The epicuticle layer of an arthropod's skeleton functions to (a) provide support at the joints (b) provide tensile strength (c) provide rigidity (d) prevent water loss — d

6. Segmented worms have the phylum name: (a) Platyhelminthes (b) Nemathelminthes (c) Annelida (d) Echinodermata — c

7. Which of the following characteristics does not apply to arthropods as a group? (a) have a notochord (b) includes more species than any other phyla (c) hard exoskeleton composed of chitin (d) jointed legs — a

8. In the class Insecta if wings are present they are attached to (a) the head (b) the thorax (c) the abdomen (d) it varies depending on the species — b

9. The sense organ of equilibrium in the crayfish is the (a) claws (b) ears (c) antenna (d) first mouth appendages. — c

10. The compound eye of an arthropod differs from the human eye in all of the following except: (a) the fact that the image formed is actually

Answer

To be covered when testing yourself

a mosaic. (b) the sensitivity to motion (c) the presence of a retina (d) the number of lenses. — a

11. The class of echinoderms which locomote by their long, slender, more flexible arms and do not possess either an anus or an intestine is the: (a) Holothuroidea (b) Asteroidea (c) Echinoidea (d) Ophiuroidea — d

Listed in 12-15 are four classes of molluscs. Match them to their members.

Classes:		Answer
12. Amphineura	(a) Nautilus, squid, octopi	12-d,
13. Gastropoda	(b) bivalves	13-c,
14. Pelecypoda	(c) snails	14-b,
15. Cephalopoda	(d) chitons	15-a.

Column A contains various parts of the earthworm. Match them to their descriptions in column B.

A	B	Answer
16. cocoon	(a) paired excretory organs	16-e,
17. seminal vesicles	(b) secretes a thick fluid which nourishes the sperm cells	17-b,
18. nerve cord	(c) sperm storage chambers found in the female	18-i,
19. segmental ganglion	(d) secretes mucus which holds together two earthworms during copulation	19-j,
20. brain	(e) eggs and sperm are deposited into it, fertilization and development occurs there	20-g,
21. subpharyngeal ganglion	(f) involved in locomotion	21-h,
22. nephridia	(g) large, 2-lobed aggregation of nerve cells	22-a,
23. clitellum	(h) located in the fourth segment	23-d,
24. chaetae	(i) connects the brain and the subpharyngeal ganglion.	24-f,
25. seminal receptacles	(j) give rise to the sensory and motor neurons.	25-c.

Answer

To be covered when testing yourself

Column A lists different structural and hormonal aspects of insect metamorphosis. Match them to their descriptions in column B.

A	B	Answer
26. juvenile hormone	(a) worm-like moth larva	26-g,
	(b) worm-like fly larva	
27. pupa	(c) worm-like bee larva	27-d,
28. ecdysone	(d) inactive form which arises from the last larval molt	28-f,
29. caterpillar	(e) group of cells from which each part of the adult body develops	29-a,
30. brain hormone	(f) molting hormone	30-h,
31. maggot	(g) regulatory hormone	31-b,
32. grub	(h) stimulates glands in the insect's prothorax.	32-c,
33. disc	(i) active form which hatches from an egg.	33-e,
34. larva		34-i.

Fill in the blanks.

35. Gastropods exhibit an embryonic twisting called _____ in which two rotations of the body occur.

    torsion

36. The bivalves are molluscs which have two hinged shells and a muscular _____ for the intake and output of water.

    siphon

37. Most molluscs and all arthropods have an _____ circulatory system which is characterized by some sections where vessels are _____ and the blood flows through _____.

    open, absent, sinuses

38. During copulation, two earthworms exchange sperm and store it. Fertilization does not occur until a few days later when a _____ is secreted by the _____ for the deposition of the _____ and _____ which are discharged from the _____ and _____, respectively. Then the development into _____ occurs within the _____.

    cocoon, clitellum, eggs, sperm, gonopores, seminal receptacles, eggs, worms, cocoon.

39. In most chelicerates the first pair of appendages are modified to serve as _____ parts.

    mouth.

Answer

To be covered when testing yourself

40. The following chart compares the earthworm to *Nereis*. Fill it in with the words supplied.

4 pairs of chaetae | marine | free swimmer | Oligochaeta
hermaphrodite | scavenger | Polychaeta | soil burrower
predator | moist and terrestrial | sexes are separate | one chaetae on each body segment

	earthworm	*Nereis*
class		
habitat		
locomotion		
bristles		
nutrient procurement		
sexes		

Answer: earthworm — oligochaeta, moist and terrestrial, soil burrower, 4 pairs of chaetae, scavenger, hermaphrodite; *Nereis* — polychaeta, marine, free swimmer, one chaetae on each body segment, predators, sexes are separate

41. Arthropods and annelids both share the property of having _____ bodies, but the arthropod's _____ differ from those of the annelids in that they are specialized for a variety of purposes.

Answer: segmented, somites

42. An arthropod sheds its exoskeleton in order to _____.

Answer: grow

43. The _____ cell of the compound eye is analogous to the retina of the human eye.

Answer: omatidial receptor

44. Insects excrete _____ which minimizes _____ loss due to protein metabolism. This is excreted through the _____.

Answer: uric acid, water, Malpighian tubules

45. The tissues of insects are supplied with oxygen by the _____.

Answer: tracheal system

46. Just prior to molting in a crustacea, the epidermal cells are stimulated by the _____ hormone to secret enzymes which digest the _____ and _____ of the inner layers of the old _____.

Answer: Y-organ, chitin, protein, cuticle

	Answer
	To be covered when testing yourself
47. Bees are an example of _____ insects.	social
48. _____ is a class of echinoderms which are sessile and are attached to the sea floor by a long stalk.	Crino-idea
Determine whether the following statements are true or false.	
49. The nephrostome of the earthworm is analagous in function to the flame cell of the planaria.	True
50. Since one earthworm contains both male and female sexual organs, the species is mainly propagated by self-fertilization.	False
51. Locomotion occurs in the earthworm with the contraction of the longitudinal and circular muscles. The contractions compress the coelomic fluid which fills the segments of the body, and thus moves the earthworm along the ground.	False
52. Spiders, scorpions, ticks and mites are all Arthropods.	True
53. The presence of calcium salts in the exoskeleton of crustaceans serve to maintain the sodium level.	False
54. The pores in the cuticle of an arthropod exoskeleton function solely in the uptake of oxygen.	False
55. The first pair of appendages of centipedes are poisonous.	True
56. There are many classes of invertebrates that can fly.	False
57. Insects are divided into different orders by taking into account their wing structures, mouth parts and the way in which they undergo metamorphosis.	True
58. Insects fly by using the radial muscles attached to their wings.	False
59. An insect has the ability to differentiate between more colors than a human.	True
60. In insects, blood can flow in two directions because of the many different valves and ostia that have control of the way the blood flows into and out of the heart.	False

Answer

To be covered when testing yourself

Question	Answer
61. Metamorphosis occurs with the animal maintaining a similar form but its body size becomes progressively larger.	False
62. The fact that both flowering plants and some insects underwent an extensive period of development during the Cretaceous implies that insects are important factors in a flowering plant's life cycle.	True
63. Echinoderms have a pentamerous radial symmetrical body plan.	True

CHAPTER 13

CHORDATES

CLASSIFICATION

• PROBLEM 13-1

What are the three chief characteristics of phylum Chordata?

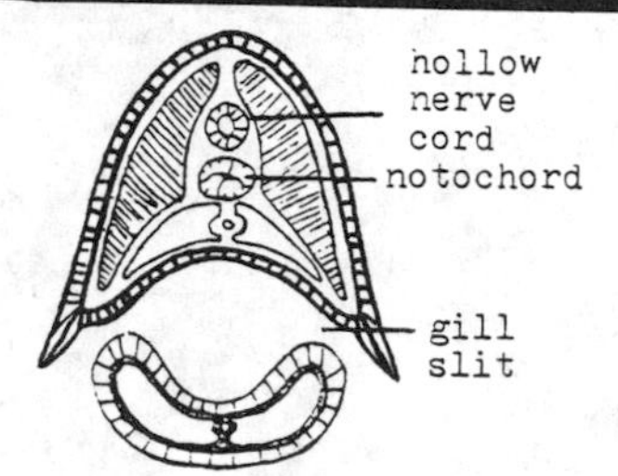

Schematic cross-section through throat region of a chordate, showing three characteristics of the group: hollow nerve cord, notochord beneath, gill slits connecting gut with exterior.

Solution: All chordates display the following three characteristics: first, the central nervous system is a hollow tube containing a single continuous cavity, and is situated on the dorsal side of the body. A second characteristic feature is the presence of clefts in the wall of the throat region, usually referred to as gill slits, originally utilized perhaps as a food-catching device. The third characteristic is the presence of a notochord, a rod lying dorsal to the intestine, extending from the anterior to posterior end, and serving as a skeletal support. In the Vertebrata, one of the three subphyla of the chordates, the notochord is partially or wholly replaced by the skull and vertebral column. It should be noted that the dorsal hollow nerve cord, the notochord and the gill slits need only be present at some time in the life of an organism for it to be considered a chordate. In a tunicate, for example, which belongs to the Chordata subphylum, Urochordata, the dorsal hollow nerve cord and the notochord are confined to the tail in the larval stage, and disappear in the adult stage. In the subphylum, Cephalochordata, all chordate characteristics are retained in the adult.

• PROBLEM 13-2

During low tide, tough, shapeless organisms can be seen attached to rocks near the low tide line. These organisms can be seen to eject water from their bodies when they contract. What are these organisms?

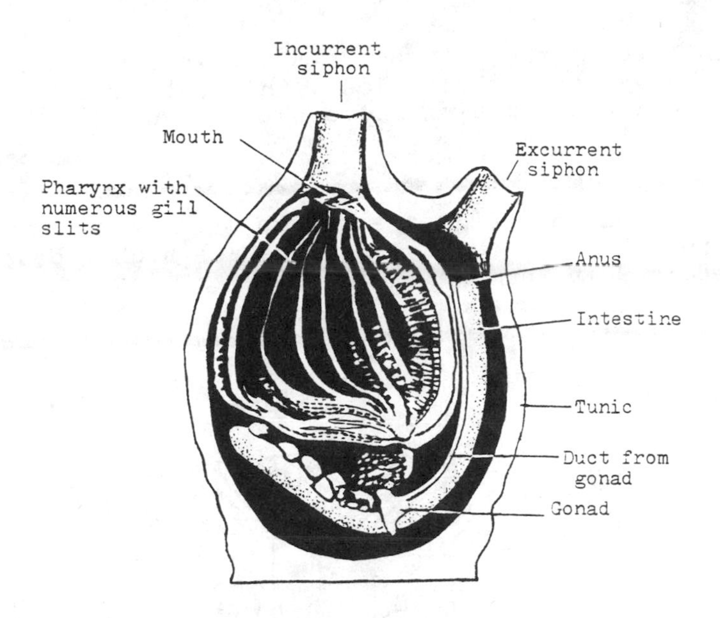

Figure 1. Cutaway model of adult tunicate.

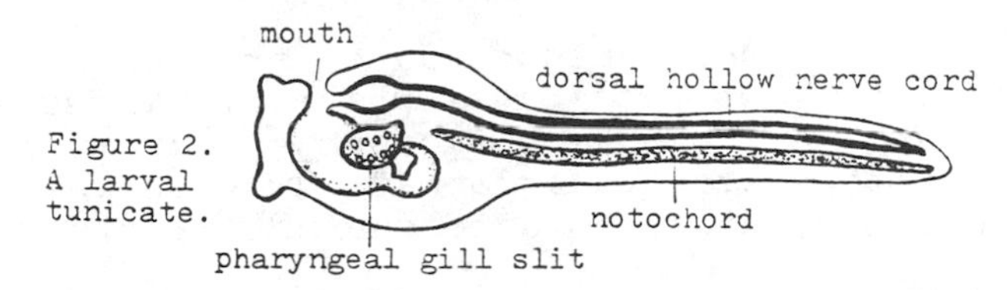

Figure 2. A larval tunicate.

Solution: These sessile marine organisms are known as the tunicates, or sea squirts, which belong to the subphylum Urochordata. There is nothing in the external appearance of the adult that suggests it is a member of the chordates. The adult body is covered by a tough, leathery tunic, composed principally of a kind of cellulose. Within the tunic, the greater portion of the body is comprised of a pharynx perforated by numerous gill slits. The tunic has two openings: the incurrent and excurrent siphens. Water and nutrients enter the body via the incurrent siphon. After entering the pharynx and having the food particles digested, the waste products are carried by the water to the excurrent siphon where they are released from the body. During reproduction, the gametes leave the animal via the excurrent siphon. Tunicates are filter-feeders, removing food particles from the stream of water passing through the pharynx. The food particles are trapped in mucus secreted by part of the pharynx called the endostyle, and are carried down by ciliated cells into the esophogus and digestive gland. The flow of water through the tunicate is maintained by the contraction of the body wall.

Whereas the adult tunicate is a sessile organism, the tunicate larva is motile, and shows a greater resemblance to the other chordates. They possess a well-developed dorsal hollow nerve cord, a notochord beneath it in the tail region, and pharyngeal gill slits. When the larvae settle down and undergo metamorphosis into the adult form, the notochord is completely lost. Most of the nerve cord is also lost, the rest remaining only as a ganglion or "brain" above the pharynx.

• PROBLEM 13-3

In the Cephalochordata, one of the three chordate subphyla, all three chordate characteristics are highly developed and retained in the adult. Outline the features of an adult Cephalochordate.

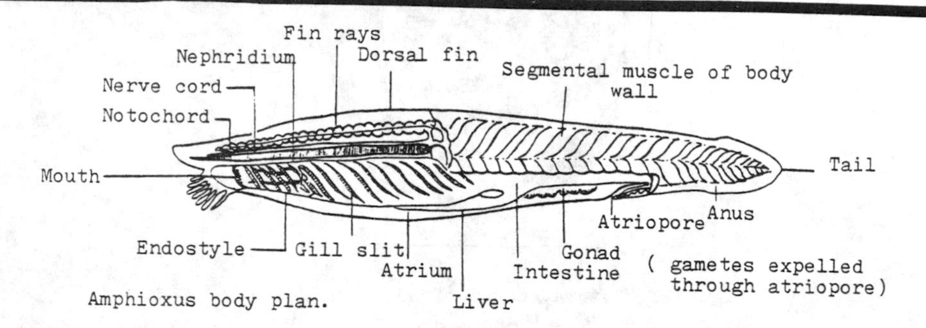

Amphioxus body plan.

Solution: The cephalochordates are represented by the lancelets, most notably Amphioxus. These small, marine chordates have been the subject of much study because of the almost diagrammatic way in which they illustrate the basic chordate characteristics.

In Amphioxus, the notochord extends from the tip of the head to the tip of the tail, and aids the animal in burrowing through the mud. There are numerous gill slits in the pharynx. The dorsal hollow nerve cord extends along the entire length of the body. Water enters the mouth (the opening into the pharynx), and passes through the gill slits into a chamber known as the atrium and leaves the body through the atriopore. Food is removed from the water in the pharynx with the aid of secretions from the endostyle in the ventral portion of the pharynx. The food is then sent posteriorly into the instestine; undigested food is eventually eliminated via the anus. Metabolic wastes are excreted by segmentally arranged, ciliated protonephridia that open into the atrium. The muscles of the body are V-shaped, segmentally arranged structures. A continuous dorsal fin, supported by fin rays, expands into a tail fin and then continues ventrally to the atriopore as the ventral fin. Although superficially similar to fishes, the lancelets are more primitive for they lack paired fins and jaws . Anterior to the atriopore, however, are two lateral folds that may have been the forerunners of the paired pectoral and pelvic fins of fishes. It hardly seems possible that any form living today is in the direct line of vertebrate ancestry, but the lancelet most closely represents the ancestral vertebrates. Its body plan, although relatively simple, resembles that of vertebrates, and the larva of the lamprey, one of the lowest vertebrates, is in many ways similar to the lancelets.

• PROBLEM 13-4

What are the chief characteristics of the subphylum Vertebrata?

Solusiton: In addition to the three basic chordate characteristics, the vertebrates have an endoskeleton of cartilage or bone that reinforces or replaces the notochord. The notochord is the only skeletal structure present in lower chordates, but in the vertebrates there are bony or cartilaginous segmental vertebrae that surround the notochord. In the higher vertebrates the notochord is visible only during embryonic development. Later the vertebrae replace the notochord completely. A part of each vertebra consists of an arch, and all of the vertebrae together form a tunnel-like protection for the dorsal nerve cord. A brain case, skull, or cranium, composed either of cartilage or bone, develops as a protective structure around the brain of all vertebrates.

The eyes in vertebrates are unique and differ both in structure and development from those of the invertebrates. The eyes of vertebrates develop as lateral outgrowths of the brain. Invertebrate eyes, such as those of insects, may be highly developed and quite efficient, but they develop from a folding of the skin. The formation of ears for detecting sounds is another vertebrate characteristic. The ears also function as organs of equilibrium, as is the major function of the ears in the lowest vertebrates.

The circulatory system of vertebrates is distinctive in that the blood is confined to blood vessels and is pumped by a ventral, muscular heart. The higher invertebrates such as arthropods and molluscs typically have hearts but they are located on the dorsal side of the body and pump blood into open spaces in the body, called hemocoels. Vertebrates are said to have a closed circulatory system. The invertebrate earthworm is an exception among the lower invertebrates in that it has a closed circulatory system. Arthropods and molluscs have an open circulatory system.

FISH

• PROBLEM 13-5

The earliest vertebrate fossils date back to the Ordovician period, 500 million years ago. What distinguishes these early vertebrates from later vertebrates?

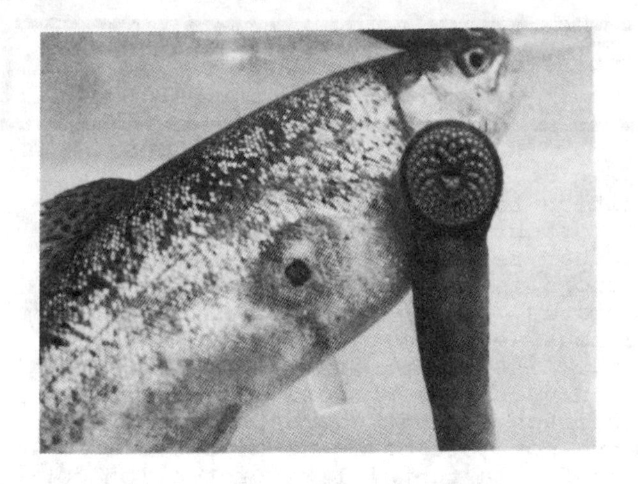

Lamprey detached from rainbow trout to show feeding wound that had penetrated body cavity and perforated gut.

Solution: These early vertebrates belong to the class Agnatha. The organisms of this class have cylindrical bodies up to a meter long, with smooth scaleless skin. Unlike other vertebrates, these organisms lack jaws ("agnatha" means jawless) and paired fins. In addition, they are also the only parasitic vertebrates, feeding primarily on the fishes. Members of this class include the lampreys and hagfishes, and the extinct ostracoderms (the earliest Agnaths). Lampreys and hagfishes constitute the cyclostomes, and have a circular sucking disc around the mouth, which is located on the ventral side of the anterior end. They attach themselves by this disc to other fish, and using the horny teeth on the disc and tongue, bore through the skin and feed on blood and soft tissues of the host. Some may bore completely through the skin and come to lie within the body of the host. The ostracoderms were the only Agnaths that were not parasitic. Since they were the first vertebrates, there were no fish as yet to exploit as food. It is assumed that the ostracoderms obtained food by filtering mud.

In the cyclostomes, the notochord persists in the adult as a functional supporting structure. The gill slits and dorsal hollow nerve cord are present in the larva called an ammocoetes and adult, and the vertebrae are rudimentary, consisting only of a series of cartilaginous arches that protect the nerve cord.

• PROBLEM 13-6

Which were the first vertebrates to exhibit a hinged jaw and paired fins? Describe the evolution of the hinged jaws of vertebrates.

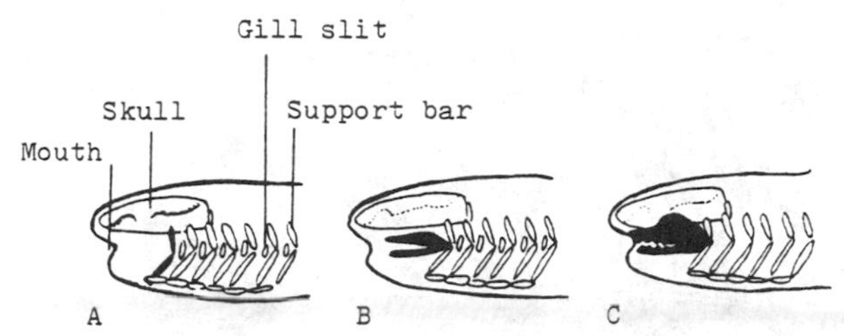

Evolution of the hinged jaws of vertebrates. (A) The earliest vertebrates had no jaws. The structures (heavy type) that in their descendants would become jaws were gill-support bars. (B) A pair of gill-support bars has been modified into weak jaws. (C) The jaws have become larger and stronger.

Solution: Roughly four hundred million years ago, there evolved a group of fishlike organisms with an armor of bony scales; usually, but not always, a cartilagenous endoskeleton; both pectoral and pelvic fins, and most notably, hinged jaws. These organisms were grouped in the class Placodermi.

The placoderms mark a notable advance in vertebrate evolution in that they possess hinged jaws. The acquisition of jaws was one of the most important events in the history of vertebrates, because it made possible a revolution in the method of feeding and hence in the entire mode of life of early fishes. Studies have indicated that the hinged jaws of the placoderms developed from a set of gill-support bars. Notice that hinged jaws arose independently in two important animal groups, the arthropods and the vertebrates, but that although they are functionally analogous structures, they arose in entirely different ways: from ancestral legs in the insects and from skeletal elements in the wall of the pharyngeal region in vertebrates.

Though the placoderms themselves have been extinct for at least 230 million years, two other classes that arose from them are still important elements of our fauna. These are the Chondrichthyes (cartilaginous fishes) and the Osteichthyes (bony fishes).

• PROBLEM 13-7

Contrast the major features of cartilaginous and bony fish.

Solution: The cartilaginous fish of the class Chondrichthyes are distinguished by, as their name implies, their cartilaginous skeletons. No bone is present in this group. The presence of a cartilaginous

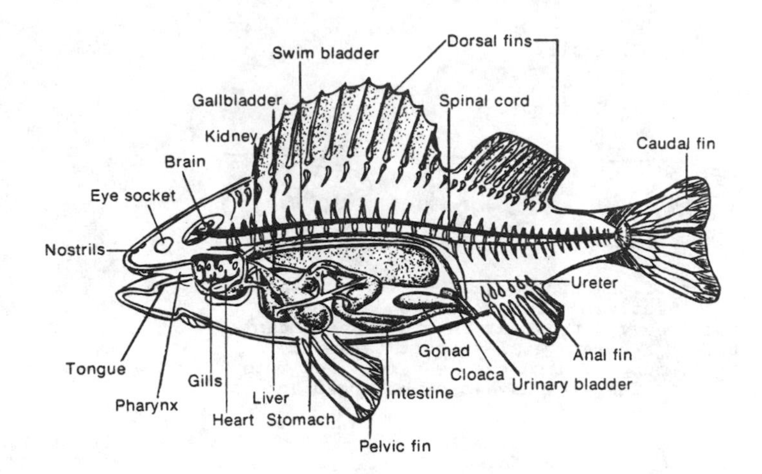

A diagram of the structure of a perch, a bony fish.

skeleton in this class must be regarded as a specialization because there is evidence that the ancestors of the group, the placoderms, had bony skeletons.

Included in the class Chondrichthyes are the sharks, skates, and rays. The sharks are streamlined, fast-swimming, voracious feeders. One kind, the whale shark, is the largest known fish. It may attain a length of 50 feet or more. Rays are sluggish, flattened bottom feeders.

Chondrichthyes have neither swim bladders nor lungs. Osmoregulation in some groups is unusual, involving retention of high concentrations of urea in the body fluids. Fertilization is internal, and the eggs have tough, leathery shells. Most species are predatory but a few are plankton feeders.

Like the cartilaginous fish, the bony fish, belonging to the class Osteichthyes, arose from the placoderms. In most of them, the adult skeleton is composed largely of bone. Unlike cartilaginous fish, fertilization is external in most bony fish. The gills of bony fish are covered by a hinged bony plate, the operculum, on each side of the pharynx. In most bony and cartilaginous fish the body is covered with scales. Unlike the cartilaginous fish, most bony fish have a swim bladder, which arose as a modification of the lungs of the earliest Osteichthyes. There are still some bony fish today that possess lungs instead of swim bladders. By modifying the gaseous content of the swim bladder, the fish can modify its buoyancy. Cartilaginous fish have no swim bladder and are denser than the sea water around them. They would sink to the bottom if it were not for the movements of the pelvic and tail fins which serve to maintain buoyancy in these fish.

The early Osteichthyes gave rise to three main lines

of bony fishes: the lobe-finned fish (Order Crossopterygii), the lungfish (Order Dipnoi), and the telecosts (Order Teleostei) which form the dominant group of marine and aquatic animals today). The lungfish and lobe-finned fish have functional lungs as well as gills. Teleosts posses a swim bladder, but no lungs.

• PROBLEM 13-8

Most aquatic animals have a special sensory system used for detecting water vibrations. Describe the anatomy of this system.

Solution: This system, peculiar to most aquatic vertebrates, consists of sensory cells arranged in a linear fashion. In cyclostomes, amphibians, and some other forms, the lines of sensory cells are exposed on the body surface. Usually they are embedded in canals, which may remain open as a groove (e.g., in primitive sharks), or more often are closed over, with pores at intervals. In most bony fish, the canals run through the scales. These canals are the canals of the lateral-line system and consist of the main lateral-line canal and four main branches on the head. The head canals are supplied by branches accompanying the facial nerve. The lateral-line system is found in cyclostomes, aquatic larvae of amphibians, and in most other fishes.

The sense organs of the lateral line system consist of neuromasts, which are clusters of sensory and supporting cells. The sensory cells have a terminal hair projecting into the lumen of the canal and are innervated by nerves of the system. The sensory cells functionally resemble those of the internal ear in terrestrial vertebrates.

• PROBLEM 13-9

Describe the function of the lateral-line system in fishes.

Solution: Just as the sensory hair cells in the semicircular canals of terrestrial vertebrates function in the detection of sound and acceleration, so does the lateral-line system in fishes. The lateral-line system consists of a series of grooves on the sides of a fish. There are sensory hair cells occurring at intervals along the grooves. These sensory cells are pressure sensitive, and enable the fish to detect localized as well as distant water disturbances. The lateral-line system bears evolutionary significance in that the sensory hair cells of terrestrial vertebrates is believed to have evolved from the sensory cells of the archaic lateral-line system in fishes. The lateral-line system of modern fishes functions primarily as an organ of equilibrium. Whether or not a fish can hear in the way that terrestrial vertebrates do is not known.

• PROBLEM 13-10

A marine animal had the following characteristics: a notochord, a dorsal hollow nerve cord, a backbone, hinged jaws, lungs and large, fleshy paired pelvic and pectoral fins. A marine biologist concluded that the animal very closely represented the direct ancestor of the land vertebrates. What kind of animal was this?

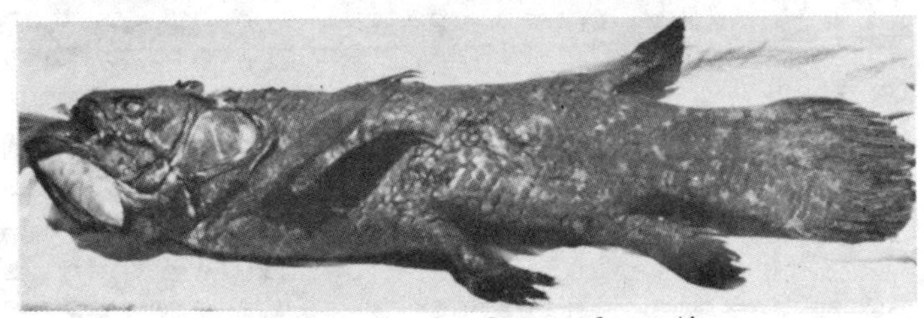

A photograph of a coelacanth

Solution: By assigning the observed characteristics to a class of organisms in a progressively specialized manner, we may deduce the identification of this marine animal. The presence of a notochord and a dorsal hollow nerve cord indicates that the animal belongs to the phylum Chordata. The presence of a backbone and hinged jaws leads us to reason that the animal is a vertebrate; more specifically, it is a bony fish (Osteichthyes). The existence of lungs and large, fleshy fins further classifies the animal as a lobe-finned fish. Since the animal was thought to represent the ancestors of the land vertebrates, our conclusion is confirmed. The animal is a lobe-finned fish, otherwise known as a coelacanth.

• PROBLEM 13-11

From which group of organisms did the amphibians develop? What two characteristics of the amphibian's ancestors were preadaptations for life on land?

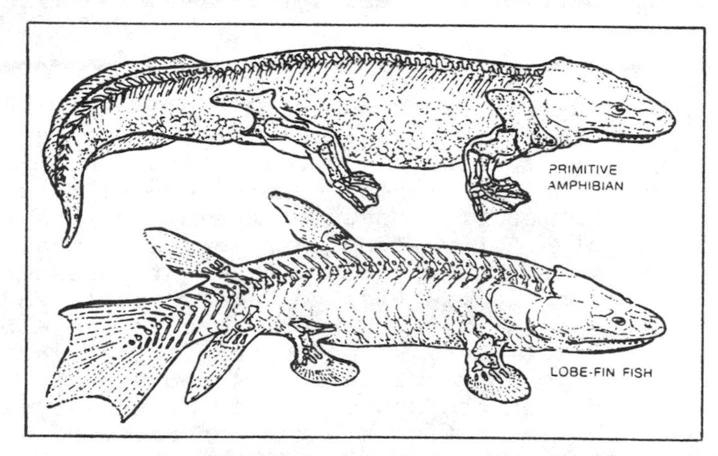

LEGS OF AMPHIBIANS evolved from the bony fins of lobe-fin fishes. The fishes were probably able to crawl from stream to stream or from pond to pond on their stubby fins.

Solution: It is believed that the lobe-finned fish were the ancestors of the land vertebrates. This group was thought to be extinct for about 75 million years. In 1939, however, a living lobe-finned fish of the superorder Crossopterygii, the coelacanth, was caught by a commercial fisherman from waters off the coast of South Africa. Since that time, many more of these fish have been caught. The coelacanths are not the particular lobe-finned fish thought to be the ancestors of land vertebrates, but they resemble those ancestral forms in many ways.

In addition to the presence of lungs used for breathing, these fish had another important preadaptation for life on land - the large fleshy bases of their paired pectoral and pelvic fins. These fleshy bases had a bony support through them which allowed for greater support. It is these supportive bones which later form the leg bones of the amphibians. At times, especially during droughts, lobe-finned fish probably used these leg-like fins to pull themselves onto sand bars and mudflats. They may even have managed to crawl to a new pond or stream when the one they previously occupied had dried up.

During ancestral times, before the vertebrates came onto land, the land was already filled with an abundance of plant life. Now, should any animal have the ability to survive on land, that animal would have a whole new range of habitats open to it without competition. Any lobe-finned fish that had appendages slightly better suited for land locomotion than those of its companions would have been able to exploit these habitats more fully. Through selection pressure exerted over millions of years, the fins of these first vertebrates to walk, or crawl, on land would slowly have evolved into legs. Thus, over a period of time, with a host of other adaptations for life on land evolving at the same time (e.g., atmospheric changes), one group of ancient lobe-finned fishes must have given rise to the first amphibians.

AMPHIBIA

• PROBLEM 13-12

Although some adult amphibia are quite successful as land animals and can live in comparatively dry places, they must return to water to reproduce. Describe the process of reproduction and development in the amphibians.

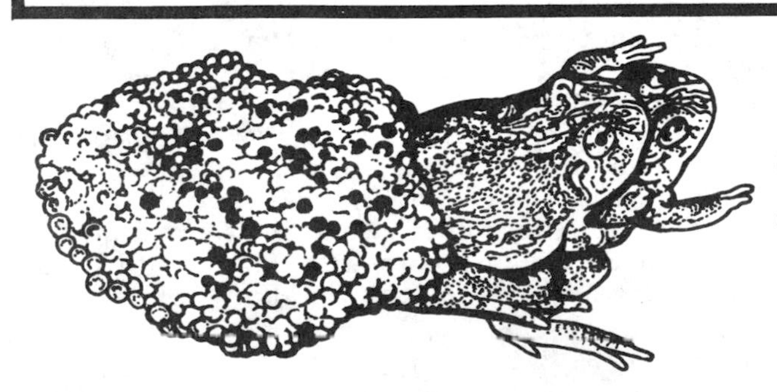

Amplexus in frogs. Male (on top) holds female and both discharge their gametes simultaneously into water. Mass of eggs is visible at left.

Solution: The frog (order Anura) will be used as a representative amphibian in this problem. Reproduction in the frog takes place in the water. The male siezes the female from above and both discharge their gametes simultaneously into the water. This process is known as amplexus. Fertilization then takes place externally, forming zygotes. A zygote develops into a larva, or tadpole. The tadpole breathes by means of gills and feeds on aquatic plants. After a time, the larva undergoes metamorphosis and becomes a young adult frog, with lungs and legs. The same type of system is seen in the salamanders (order Urodela).

Like the metamorphosis of insects and other arthropods, that of the amphibia is under hormonal control. Amphibia undergo a single change from larva to adult in contrast to the four or more molts involved in the development of arthropods to the adult form. Amphibian metamorphosis is regulated by thyroxin, the hormone secreted by the thyroid gland, and can be prevented by removing the thyroid, or the pituitary which secretes a thyroid-stimulating hormone modulating the secretion of thyroxin.

• PROBLEM 13-13

Describe the changes that take place in a larval amphibian which result in adaptation to life on land.

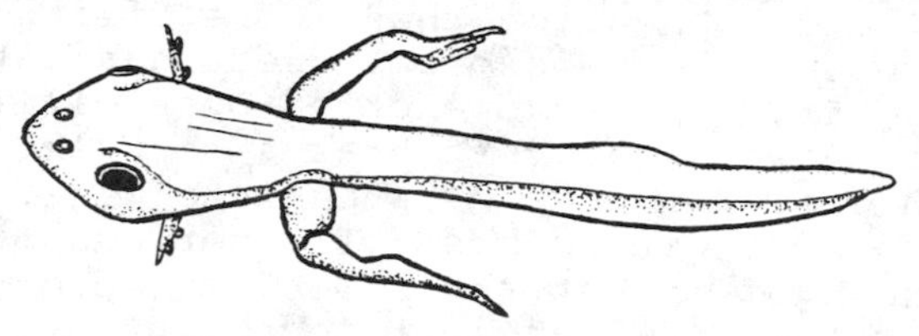

Metamorphosing tadpole

Solution: In this problem, the changes that occur in the frog larva which result in the adult frog will be considered. The frog larva is a tadpole which undergoes metamorphosis to become an adult frog. The metamorphosis is a single-step process as opposed to the many molts that an arthropod larva undergoes in reaching the adult form. As in arthropod metamorphosis, amphibian metamorphic processes are hormonally controlled. Thyroxin, secreted by the thyroid gland, is the regulating hormone in amphibian metamorphosis. The tadpole is a completely aquatic organism and breathes via gills. It feeds on microscopic plants in the water. The tadpole also has a tail. During metamorphosis, the gills and gill slits are lost, the lungs develop, forelegs grow out of skin folds, the tail is retracted, the tongue, the eyelids and the tympanic membrane are formed, and the shape of the lens changes. The adult form of the organism, the frog, is a semi-aquatic animal which feeds on insects. The frog breathes by means of its lungs as well as through its thin, moist skin. The skin of amphipians functions as a respiratory organ.

This is facilitated by the physical nature of the skin, which must remain moist in order to function as a respiratory organ. If terrestrial conditions become very dry, the frog's thin, moist skin will be in danger of dessication, and hence, the frog will retreat to the aquatic environment.

• PROBLEM 13-14

The four-legged land vertebrates (the amphibia, reptiles, birds and mammals) are placed together in the superclass Tetrapoda. Which animal was the first tetrapod and what did it give rise to?

Solution: The first successful land vertebrates were the labyrinthodonts, which became extinct in the first part of the Mesozoic Era. They were clumsy, salamander like, ancient amphibians with short necks and heavy muscular tails. These amphibians closely resembled the ancestral lobe-finned fish, but the labyrinthodonts had evolved pentadactyl limbs strong enough to support the weight of the body on land. The labyrinthodonts ranged in size from small, salamander-size animals up to ones as large as crocodiles. They gave rise to other primitive amphibians, to modern frogs and salamanders, and to the earliest reptiles, the cotylosaurs.

REPTILES

• PROBLEM 13-15

Describe the evolution of the reptilian excretory system to account for the transition from an aquatic to a terrestrial habitat.

Solution: The reptilian excretory system has evolved in a way that enables the animal to conserve most of its water. The conservation of body fluid is a necessary characteristic of a terrestrial animal. Without this feature, the body tissues would become dessicated in the dry environment.

Reptiles conserve water by having a coarse, dry, horny skin. In addition, the glomeruli of the reptilian kidney have diminished in size so that less water is filtered from the blood. Another modification is a greater degree of reabsorption of water from the glomerular filtrate by the kidney tubules. This occurs as the reptiles have evolved kidney tubules with two highly coiled regions and a long loop of Henle extending deep into the medulla of the kidney. These long portions of the tubule function in the reabsorption of water. Their ability to produce a concentrated hypertonic excretory product is an important adaptation to land life.

The conservation of water in the reptile has affected the nature of the excretory product. Nitrogenous wastes are excreted as uric acid in reptiles, as opposed to urea in amphibians (whose ancestors gave rise to primitive reptiles). Uric acid is excreted as a watery paste in reptiles. It is less toxic and less soluble in water than urea. Therefore, it is not necessary for reptiles to use water to dissolve their nitrogenous wastes. In fact, only a comparatively very small amount is required to simply flush the uric acid out of the excretory system. In this way, the reptile conserves most of its body water and prevents dehydration.

• **PROBLEM 13-16**

Which of the following is not a reptile: snake, lizard, salamander,turtle or alligator.

Solution: The class Reptilia is divided into four orders. Turtles are reptiles belonging to the order Chelonia; crocodiles and alligators belong to the Order Crocodilia; lizards and snakes are in the Order Squamata; and the tuatara belong to the Order Rynchocephalia, which has only one surviving species. Of the five animals mentioned in the question, only the salamander is not a reptile. The salamander is an amphibian in the Order Urodelia. By anatomical observation, it is easily deduced that the salamander is not of the same class as these other animals. The smooth and moist porous skin of the amphibious salamander contrasts sharply with the coarse dry skin of the reptilian animals. This epidermal variation is based upon the different environmental adaptations that these animals underwent. The skin of the amphibian is adapted for a semi-aquatic environment whereas that of the reptile is adapted for a strictly terrestrial environment.

BIRDS AND MAMMALS

• **PROBLEM 13-17**

Which of the following would maintain a constant body temperature in spite of changes in the enviromental temperature: frog, robin, fish, dog, or lizard?

Solution: Animals that can regulate their internal body temperature so that it remains constant in varying envirinmental temperatures are known as homeotherms. Of the phylum Chordata, only mammals and birds can therm-regulate- i.e., they possess temperature-regulating systems. They are considered to be warm-blooded animals. From the list of animals given, two of them are homeotherms. The robins, since it is a bird, belongs to the class Aves, and the dog is a member of the class Mammalia. Hence these two animals are homeothermic and will maintain a constant body temperature. The remaining animals are poikilotherms. They possess no temperature-regulating system and as a result, their internal body temperature varies directly with the environmental temparature. This poikilothermic characteristic places strict limitations on the environment exploited by the animal, which in turn affects its behavioral patterns. The frog is an amphibian whereas the lizard is a reptile. A fish belongs to the superclass pisces. These three animals are poikilotherms and hence do not maintain a constant body temperature independent of changes in environmental temperature.

• PROBLEM 13-18

How are whales adapted for their life in cold waters?

Solution: Whales are animals that belong to the order Cetacea of the class Mammalia. Only animals belonging to the classes Mammalia and Aves possess a temperature regulating system within their body. They are homeothermic. These animals maintain a nearly constant internal body temperature that is independent of the environmental temperature. Whales, sea cows and pigeons are animals that have this temperature maintenance ability. All animals not belonging to the class Mammalia or Aves are poikilothermic- their internal body temperature varies directly with the environmental temperature. Another feature of the whale that helps it survive in cold waters is the tremendous amount of subcutaneous fat serving an insulating function. All mammals have a layer of fat underlying the skin. An extreme case of this is seen in the whale.

• PROBLEM 13-19

What is meant by a cloaca and what is its relation to the mammals?

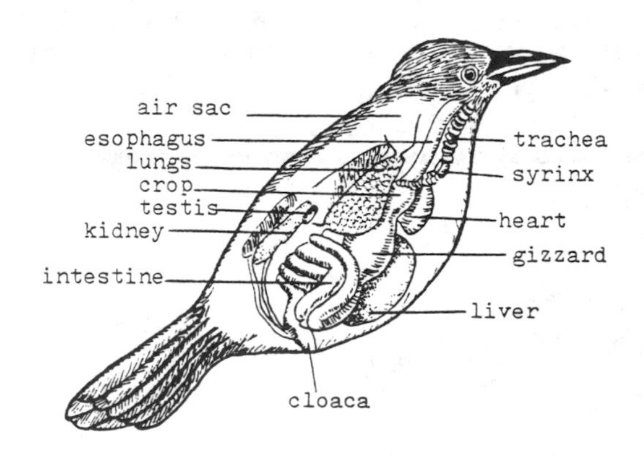

Figure 1. Internal anatomy of a male bird. The excretory and reproductive products exit through a common duct (cloaca).

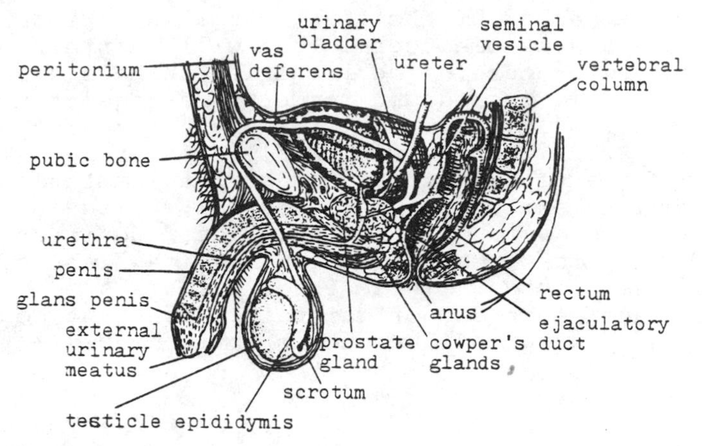

Figure 2. Human male reproductive system, median section of pelvis. Note that the urethra is a common duct for urine and sperm. Thus excretory and reproductive separation is incomplete.

Solution: Chordates, except for mammals and teleosts, have a cloaca. A cloaca is a common duct through which passes the reproductive, digestive and excretory matters. In the frogs, for example, sperm cells pass from the testes into the kidneys and down the excretory ducts to the cloaca. The reproductive system and the excretory system are very closely related.

The evolutionary trend in the vertebrates, with regard to the cloaca, has been one toward increasing separation between the reproductive and excretory systems. Mammals have, instead of a cloaca, ureters which release their excretory material (i.e., urine) into a urinary bladder. The urethra carries the urine from the bladder to the outside. In the mammals, however, sperms also pass through the urethra. Thus, although there is far more separation in mammalian males than in fish or frogs, the mammalian reproductive and excretory systems still share the urethra and hence do not have separate openings to the outside. Only in mammalian females has complete separation arisen.

• PROBLEM 13-20

Describe the characteristic differences between the monotreme, marsupial and placental mammals.

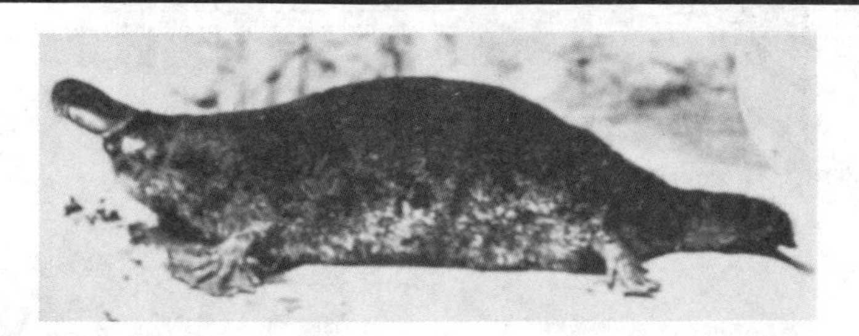

Duck billed platypus, an egg-laying mammal.

Euro kangaroo, a marsupial mammal, with young in pouch.

Ring-tailed lemur, a placental mammal.

Solution: The class Mammalia is divided into three subclasses: Prototheria (Monotremes), Metatheria (Marsupials), and Eutheria (placental mammals). Of these three, the monotremes can be considered the most primitive. They are an early offshoot of the main mammalian groups. The Marsupials and Placentials appeared at about the same time. Major characteristics of mammals include the possession of mammary glands and hair. Most mammals have sweat glands. All mammals are homeotherms.

Monotremes are peculiar in that they possess mammalian as well as reptilian traits. They are classified as mammals, however, because the number of mammalian traits far exceed the number of reptilian traits. Some of the more outstanding mammalian traits include: a layer of hair covering the body and the secretion of milk by mammary glands. There are only two surviving species of monotremes: the duck-billed platypus and the spiny anteater, both of which are found in Australia. The trait that most clearly distinguishes monotremes from other groups of mammals is their ability to lay eggs. In marsupials and placentals, embryonic development occurs in the uterus of the female.

The characteristic difference between the marsupials and the placentals is the time of embryonic development within the uterus. Marsupial embryos undergo a short developmental period before they leave the uterus. Embryonic development is completed in an abdominal pouch of the mother where the embryo is attached to a nipple. In contrast to this is the placentals.In this group, the embryo develops completely in the uterus of the mother. In both marsupial and placental mammals, the young are born alive.

SHORT ANSWER QUESTIONS FOR REVIEW

Answer

To be covered when testing yourself

Choose the correct answer.

1. Which of the following are chordates? (a) fish, (b) man (c) frogs (d) a, b, and c, (e) a and c — d

2. Tunicates are best described as (a) terrestrial animals very similar to chordates. (b) urochordates, with a chordate larval stage, and lacking a notochord as an adult. (c) always maintaining a notochord. (d) never having a notochord. — b

3. The cephalochordates such as Amphioxus (a) superficially resemble fishes but are more primitive in lacking paired fins, jaws, sense organs and brain. (b) are more advanced than fish with larger brains and more complex sense organs. (c) do not contain a notochord, gill slits and a nerve chord. (d) none of the above are correct. — a

4. Vertebrate characteristics include(a) an open circulatory system, internal skeleton of cartilage or bone, ears, and eyes developed from outgrowths of brain. (b) a closed circulatory system, external skeleton of cartilage or bone, ears, and eyes developed from a folding of skin. (c) a closed circulatory system, internal skeleton of cartilage or bone, ears, and eyes developed from outgrowths of brain. (d) none of the above. — c

5. Which statement is correct? (a) Agnatha, or jawless fishes include ostracoderms, and the living lamprey eels and hagfishes. (b) Lampreys and hagfishes are the only parasitic vertebrates. (c) Ostracoderms are the earliest known fossil chordates (d) all of the above are correct. — d

6. The chondrichthyes are characterized as (a) having a skeleton of cartilage which is always calcified. (b) having paired jaws and two pairs of fins. (c) the only fish that do not have highly vascular gills. (d) having swim bladders enabling them to float. — b

7. The bony fish or osteichthes (a) evolved independently of and at about the same time as, the cartilaginous fishes (b) evolved from the cartilaginous fishes (c) do not contain swim bladders (d) do not have their gills covered by a protective structure called the operculum. — a

	Answer
	To be covered when testing yourself
8. Choose the incorrect statement. (a) Amphibia undergo a single change from larva to adult. (b) Adult amphibia do not depend solely on their lungs for exchange of respiratory gasses. (c) Salamanders, but not frogs have the ability to regenerate lost parts of their bodies (d) Amphibian metamorphosis has nothing to do with hormonal control.	d
9. Choose the correct statement. (a) Reptiles like fish and amphibians have a body temperature regulating mechanism. (b) The bodies of reptiles are covered by moist,soft scales that help protect against dehydration (c) Due to the development of kidney tubules as water conserving structures, reptiles are able to survive in the desert (d) Reptiles secrete nitrogenous wastes as urea.	c
10. Which of the following are modified reptilian scales? (a) the gills of a fish (b) the feathers of a bird (c) both of the above (d) none of the above.	b
11. The correct characteristics of all birds include: (a) cold-blooded, all fly, internal fertilization (b) warm-blooded, some fly, internal fertilization (c) warm-blooded, all fly, internal fertilization (d) warm-blooded, some fly, external fertilization.	b
12. The distinguishing characteristics of mammals are: (a) hair and mammary glands (b) carnivorous (c) sweat glands and differentiation of teeth into incisors, canines, and molars (d) answers a and c are correct	d
Fill the blanks.	
13. _____ are present in all chordate embryos but are not found in higher adult vertebrates.	Gill slits
14. The subphylum Vertebrata is characterized by a _____ or _____ vertebral column.	cartilaginous, bony
15. Vertebrates have a _____ circulatory system.	closed
16. Unlike the bony fishes, cartilaginous fishes have no _____.	swim bladders
17. All chordates contain a hollow, tubular, dorsal _____.	nerve cord

	Answer
	To be covered when testing yourself
18. All chordates at some time in their life cycle have grooves in the _____ region.	pharyn-geal
19. The three different kinds of kidneys found in vertebrates are the _____, _____ and _____.	prone-phros, mesone-phros, metane-phros
20. Bird have a _____ chambered heart.	four
21. Placental mammals make up the subclass _____.	Eutheria
22. The organ within a female eutherian mammal where the young develop is called a _____.	uterus
23. Like the metamorphosis of insects and crustacea, that of amphibia is under _____ control.	hormonal
24. The operculum is the hinged bony plate which covers the _____ in bony fish.	gills
25. The agnatha are eel-shaped animals which are characterized by the complete lack of _____.	jaws
26. The marine lobe-finned fishes that are credited as being the ancestors of the land vertebrates are called _____.	coela-canths
Determine whether the following statements are true or false.	
27. The most advanced vertebrates contain a notochord and simpler vertebrates do not.	False
28. Higher forms of vertebrates lack the notochord as adults,and the vertebrae take over its functions.	True
29. Mammals tend to have a higher body temperature than birds.	False
30. The placenta of mammals allows an embryo to obtain nourishment from the mother before birth.	True
31. The bony fishes originated in fresh water and subsequently entered the oceans.	True
32. The swim bladder is used by bony fish in the sea to excrete excess salt.	False
33. Reptiles are true land dwellers and do not have to return to water to reproduce as amphibians do.	True

	Answer To be covered when testing yourself
34. Reptilian skin serves as the chief respiratory organ.	False
35. The amphibian heart consists of three chambers.	True
36. Birds have extensions of their lungs called air sacs to aid in bouyancy during flight.	True
37. Amphibian larvas breathe by means of gills.	True
38. Ninety-five percent of all living reptiles are snakes and lizards.	True
39. The skeleton of the ostheichthyes consists mostly of cartilage.	False
40. Penticles are tiny pointed structures that are found on the skin of cartilaginous fish.	True
41. The structure of the metanephros allows vertebrates to live in an aquatic environment.	False

CHAPTER 14

BLOOD AND IMMUNOLOGY

PROPERTIES OF BLOOD AND ITS COMPONENTS

• PROBLEM 14-1

Explain why blood is such an important tissue in many animals. Discuss the major functions of blood.

Solution: All cells, in order to survive, must obtain the necessary raw materials for metabolism, and have a means for the removal of waste products. In small plants and animals living in an aquatic environment, these needs are provided for by simple diffusion. The cells of such organisms are very near the external watery medium, and so nutrients and wastes do not have a large distance to travel. However, as the size of the organism increases, more and more cells become further removed from the media bathing the peripheral cells. Diffusion cannot provide sufficient means for transport. In the absence of a specialized transport system, the limit on the size of an aerobic organism would be about a millimeter, since the diffusion of oxygen and nutrients over great distances would be too slow to meet the metabolic needs of all the cells of the organism. In addition, without internal transport, organisms are restricted to watery environments, since the movement to land requires an efficient system for material exchange in non-aqueous surroundings. Therefore, larger animals have developed a system of internal transport, the circulatory system. This system, consisting of an extensive network of various vessels, provides each cell with an opportunity to exchange materials by diffusion.

Blood is the vital tissue in the circulatory system, transporting nutrients and oxygen to all the cells and removing carbon dioxide and other wastes from them. Blood also serves other important functions. It transports hormones, the secretions of the endocrine glands, which affect organs sensitive to them. Blood also acts to regulate the acidity and alkalinity of the cells via control of their salt and water content. In addition, the blood acts to regulate the body temperature by cooling certain organs and tissues when an excess of heat is produced (such as in exercising muscle) and warming tissues where heat loss is great (such as in the skin).

Some components of the blood act as a defense against

bacteria, viruses and other pathogenic (disease- causing) organisms. The blood also has a self-preservation system called a clotting mechanism so that loss of blood due to vessel rupture is reduced.

• PROBLEM 14-2

What components of the plasma can affect the blood pressure? Explain how they affect it. What other solutes are present in the plasma?

Solution: Although the blood appears to be a red-colored homogeneous fluid as it flows from a wound, it is actually composed of cells contained within a liquid matrix. The yellowish, extracellular liquid matrix of the blood is called plasma. Suspended in it are three major types of formed elements: the red blood cells, or erythrocytes, the white blood cells, or leukocytes, and the platelets, small disc-shaped cell fragments important in the clotting process. The cellular portion constitutes about 45% of the whole volume of blood, while plasma constitutes the other 55%. Since the cellular part of the blood has a higher specific gravity (the density of a substance relative to that of water) than plasma (1.09 vs. 1.03), the two may be separated by centrifugation. Although the plasma consists of 90% water, there are many substances dissolved in it. The concentrations of two of these substances, inorganic salts and plasma proteins, can affect the blood pressure.

The major inorganic cations (positively charged ions) in the plasma are sodium (Na^+), calcium (Ca^{2+}), potassium (K^+), and magnesium (Mg^{2+}). The major inorganic anions (negatively charged ions) in the plasma include chloride (Cl^-), bicarbonate (HCO_3^-), phosphate (PO_4^{3-}), and sulfate (SO_4^{2-}). The major salt is sodium chloride. In mammals, inorganic ions and salts make up .9% of the plasma by weight. Usually, the plasma concentration of these substances remain relatively stable due to a variety of regulatory mechanisms, such as the action of the kidneys and other excretory organs. But if the total concentration of dissolved substances is appreciably changed, serious disturbances can occur. For example, the concentration of sodium chloride ($NaCl$) and sodium bicarbonate ($NaHCO_3$) determines the osmotic pressure of the plasma relative to the extracellular medium bathing the cells of the body. Normally, the osmotic pressure is balanced; however, if the concentrations of these salts are changed, the osmotic pressure of the plasma is changed and an osmotic gradient established. Water will then enter or leave the plasma by diffusion from the extracellular fluid. The resulting water loss or uptake will change the total blood volume. As with any fluid in a closed system, the volume of the blood directly affects the pressure it exerts on the walls of the vessel containing it. If the concentration of sodium chloride in the plasma is increased, water will tend to diffuse into the blood by osmosis, increasing its volume, which in turn increases the blood pressure.

The proper balance of the individual ions is also essential to the normal functioning of nerves, muscles, and cell membranes. A deficiency of calcium ions, for instance, may cause severe muscular spasms, a condition called tetany. Certain ions, such as bicarbonate, affect the pH of the plasma, which is normally slightly alkaline at 7.4.

Other important substances in the plasma that can affect blood pressure are the plasma proteins. Plasma proteins, which constitute about 9% of the plasma by weight, are of three types: fibrinogen, albumins and globulins. These substances contribute to the osmotic pressure of the plasma since they are too large to readily pass through the blood vessels. Their concentration affects the amount of water which diffuses into the blood, thereby regulating the general water balance in the body. In addition, these proteins determine the viscosity of the plasma. Viscosity is a measure of the resistance to flow caused by the friction between molecules as they pass one another. The heart can maintain normal blood pressure only if the viscosity of the blood is normal. For example, if the viscosity were to decrease, there would be less resistance to flow. There is a physical law which states that an increased flow rate (due to decreased resistance) will exert more pressure on the walls of the vessels transporting the fluid. Hence, this decreased viscosity leads to increased blood pressure; the viscosity of the blood is therefore inversely proportional to the blood pressure.

In addition to inorganic ions and plasma proteins, plasma also contains organic nutrients such as glucose, fats, amino acids and phospholipids. It carries nitrogenous waste products, such as urea, to the kidneys. Hormones are also transported in the plasma. Dissolved gases are also transported, including carbon dioxide and oxygen, the gases involved in respiration. Most of these molecules are frequently bound to specific sites whithin the red blood cells, and not transported freely in the plasma.

• PROBLEM 14-3

How can one measure the volume of blood in a man's body (without draining it out)?

Solution: The volume of any fluid can be measured by placing a known quantity of a test substance in the fluid, allowing it to disperse evenly throughout the fluid and then measuring the extent to which it has become diluted (the final concentration of the substance). All one need know is the total quantity of the test substance (usually a dye) injected and its final concentration in the fluid after dispersion.

Dyes that are used for blood volume measurements must be able to stay in the plasma and not diffuse out into the tissues. Dyes that are usually used are those which bind to plasma proteins, these proteins being too large to diffuse out of the plasma. The most universally employed dye is called T-1824 or Evan's blue. A known amount of dye is injected into the blood, where it combines with the plasma proteins and disperses throughout the circulatory system within approximately ten minutes. A sample of blood is then taken and the red blood cells removed by centrifugation. The volume of the remaining plasma and the concentration of dispersed dye in this plasma sample are measured. (The concentration is determined using a spectrophotometer, an instrument which measures the absorbance of light by a solution. The higher the absorbance, the greater the concentration of the solution). Knowing the amount of dye in this plasma sample and the initial amount of dye injected, we can calculate the total plasma volume using a simple ratio:

$$\frac{\text{amount of dye in sample}}{\text{volume of plasma sample}} = \frac{\text{total amount of dye injected}}{\text{total plasma volume}} .$$

For example, if 2000 units of dye were injected, and we determined that there were 10 units of dye in a 15 milliliter sample of plasma after injection, then the total volume of plasma would be:

$$\frac{10}{15} = \frac{2000}{\text{total plasma volume.}}$$

$$\text{total plasma volume} = \frac{2000(15)}{10} = 3000 \text{ ml. or } 3 \text{ liters.}$$

Since this is only the plasma volume and not the whole blood volume, we must also determine the volume of the cellular blood component. The hematocrit of blood refers to the percentage of cellular components in the blood. Blood hematocrit is determined by centrifuging blood in a calibrated tube so that one can directly read the percentage of cells from the level of packed cells. The hematocrit of a normal man is about 42, whereas that of a normal woman is about 38. If the percentage of cells is 42%, the percentage of plasma is 58%.

The total blood volume of this normal man can then be calculated:

$$.58\,x = 3000 \text{ ml. where } x = \text{total blood volume (ml)}$$

$$x \simeq 5200 \text{ ml.} \simeq 5.2\ \ell .$$

The volume of blood in this man's body would be about 5200 millititers or about 5.2 liters.

There is a small degree of error ($\sim 1\%$) in this method arising from the loss of dye from the circulatory system during the dilution. Some dye is excreted into the urine and some is lost from the circulatory system by leakage of plasma proteins.

CLOTTING

• PROBLEM 14-4

If blood is carefully removed from a vessel and placed on a smooth plastic surface, will it clot? Explain why or why not.

Solution: Vertebrates have developed a mechanism for preventing the accidental loss of blood. Whenever a blood vessel is ruptured, one of the soluble plasma proteins, fibrinogen, is enzymatically converted into an insoluble protein, fibrin, which forms a semisolid clot.

Many people think that blood clots when it becomes exposed to the air or when it stops flowing. However, if one were to carefully remove blood from a vessel without allowing it to contact the damaged part of the vessel, and then place this blood on a smooth plastic dish or one lined with paraffin, it would not clot. However, if this blood were allowed to touch any damaged tissues or were placed on glass or some other relatively rough surface, the blood would clot. Either the damaged tissues or the blood itself must release some chemical which initiates the clotting mechanism. Actually, it is both damaged tissues and disintegrated platelets in the plasma that release substances responsible for the clotting reaction. Platelets are very small disc-shaped bodies found in mammalian blood and formed in the bone marrow from large cells called megakaryocytes. Platelets seem to disintegrate more readily upon contacting glass surfaces than on plastic surfaces. Platelets are also called thrombocytes.

When a blood vessel is cut, the damaged tissues release a lipoprotein, called thromboplastin, which initiates the clotting mechanism. Calcium ions and certain protein factors in the plasma must be present in order for thromboplastin to be effective. Thromboplastin interacts with Ca^{+2} and these proteins to produce prothrombinase, the enzyme that catalyzes the second step in the clotting mechanism. Prothrombinase can also be made by the interaction of a substance released from the disintegrated platelets (platelet factor #3) and other factors in the plasma, including Ca^{2+} and proteins. The prothrombinase made from either the tissues or the platelets catalyses the conversion of prothrombin, a plasma globulin, into thrombin. Finally, thrombin enzymatically converts fibrinogen into fibrin, an insoluble protein. Fibrin forms long fibers which mesh and trap red cells, white cells, and platelets, forming the clot. Usually the clot forms within 5 minutes of the rupturing of the vessel. The clot then begins to contract and squeeze out most of the plasma from itself within an hour. This process, called clot retraction, serves to increase the strength of the clot, and also pulls the vessel walls adhering to the clot closer together. The extruded plasma is now called serum, since all the fibrinogen and most other clotting factors have been removed. Because it lacks these constituents, serum cannot clot.

The entire process is summarized in the following diagram:

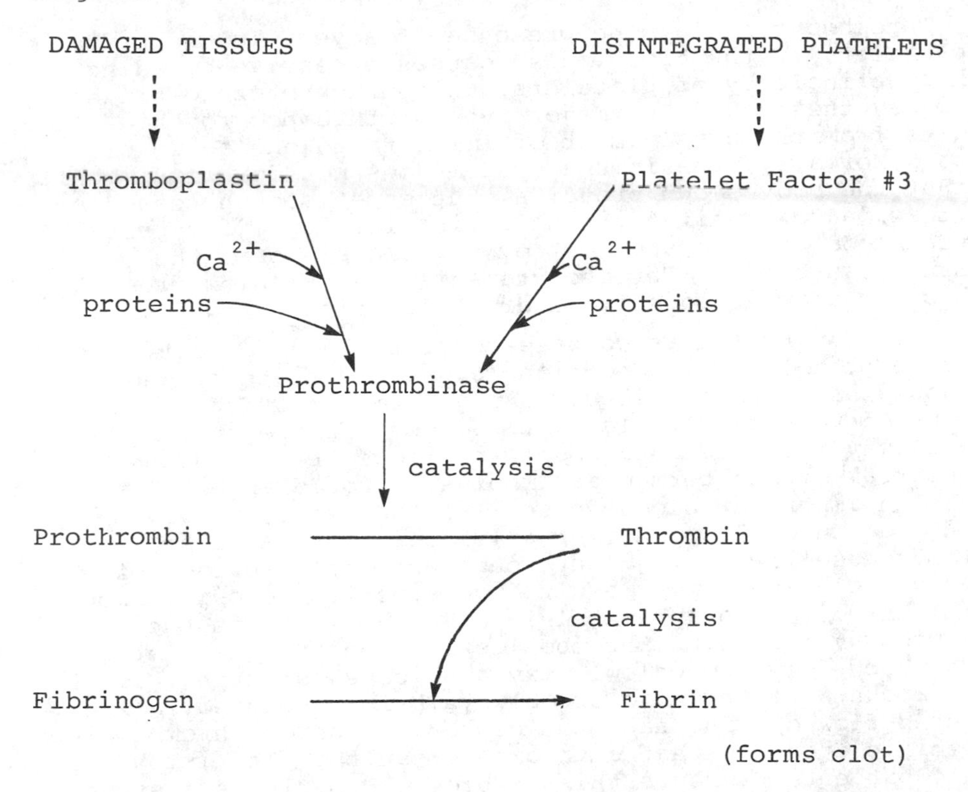

• PROBLEM 14-5

In an earlier question, we learned about the various factors which are involved in the clotting mechanism. Show how these factors are affected in conditions that cause excessive bleeding in humans, and in conditions that cause intravascular clotting.

Solution: Excessive bleeding can result from a deficiency of any one of the blood clotting factors. An insufficiency of prothrombin, one of the intermediates in the clotting mechanism, can cause a patient to develop a severe tendency to bleed. Both hepatitis and cirrhosis (diseases of the liver) can depress the formation of prothrombin in the liver. Vitamin K deficiency also depresses the levels of prothrombin. Vitamin K deficiency does not result from the absence of the vitamin from the diet, since it is continually synthesized in the gastrointestinal tract by bacteria. The deficiency results from poor absorption of fats (vitamin K is fat soluble) from the tract due to a lack of bile, which is secreted by the liver.

Hemophilia, or "bleeder's disease," is a term loosely applied to several different hereditary deficiencies in coagulation, resulting in bleeding tendencies. Most hemophilias result from a deficiency in one of the protein factors called Factor VIII or the antihemophilic factor necessary for production of prothrombinase by the platelets.

Thrombocytopenia, the presence of a very low quantity of platelets in the blood, also causes excessive bleeding. With a definciency of platelets, not enough prothrombinase can be synthesized. One major type of thrombocytopenia results from the development of immunity to one's own platelets. [Normally, the immune system does not develop immunity against the body's own proteins, but sometimes one can develop this "autoimmunity".] The antibodies (specific proteins in the blood) attack and destroy the platelets in the person's own blood. Thrombocytopenia also results from pernicious anemia and certain drug therapies.

These abnormalities cause excessive bleeding; but there are also pathological conditions caused by clotting when it should not normally occur. An abnormal clot that develops in a blood vessel is called a thrombus. If the thrombus breaks away from its attachment and flows freely in the bloodstream, it is termed an embolus. Should an embolus block an important blood vessel (to the heart, lungs, or brain), death could occur. What causes intravascular (within a blood vessel) clotting? Any roughened inner surface of a blood vessel, which may result from arteriosclerosis, bacterial infection, or physical injury, can initiate the clotting mechanism by releasing thromboplastin from the platelets. Blood which flows too slowly may also cause clotting. Since small amounts of thrombin are always being produced, a hampered flow can increase the concentration of thrombin in a specific area, so that a thrombus results. The immobility of bed patients presents this problem. Heparin is a strong anticoagulant normally produced in small amounts by special cells in the body. Heparin can be used as an intravenous anticoagulant in the prevention of thrombi formation.

• PROBLEM 14-6

The citrate ion forms a tight bond with calcium ions so that the calcium is unable to react with other substances. In view of this, why is sodium citrate added to whole blood immediately after it is drawn from a blood donor?

Solution: Sodium citrate dissociates into sodium and citrate ions, the latter of which binds with calcium ions to remove them from solution. Calcium ions must be present for thromboplastin, a substance released by damaged tissues, to be converted into the enzyme prothrombinase. Calcium must also be present for prothrombinase to convert prothrombin into thrombin, and for thrombin to catalyze the formation of fibrin. Fibrin ultumately forms the fibers which trap the cellular constituents of the blood, thus forming a clot. Without calcium ions, this process cannot occur.

Blood drawn from a donor is treated with sodium citrate to prevent coagulation. The blood can then be stored for weeks in a blood bank by keeping the blood at a temperature of about 4°C. After injection into a recipient, the citrate ion is removed from the blood within minutes by the liver. The liver polymerizes the citrate into glucose which is then metabolized for energy. If the liver is damaged or if too large a quantity of citrated blood is injected, the level of citrate will depress the level of calcium ions in the blood. If the level drops significantly, a disease called tetany can result. Tetany is characterized by muscle spasms, which can be severe enough to cause death by asphyxia (oxygen exhaustion).

GAS TRANSPORT

• PROBLEM 14-7

> One of the functions of blood is to carry oxygen to the cells of the body. Exactly how are the oxygen molecules carried, and why aren't the carriers found free in the plasma?

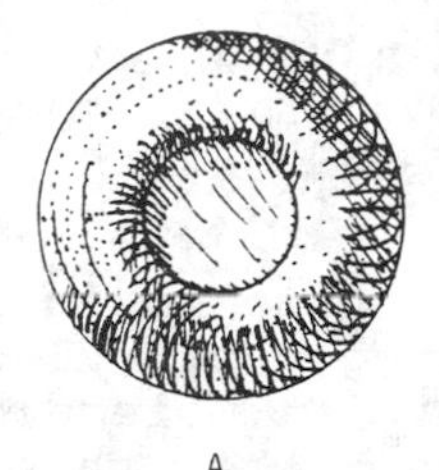

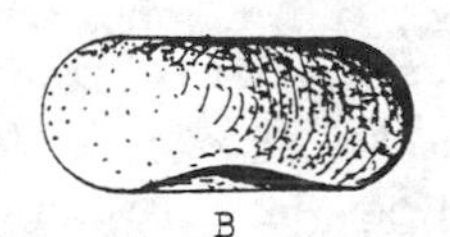

Figure 1. Erythrocytes. The cells have a biconcave disclike shape. (A) Top view. (B) Side view.

Figure 2. Structure of the heme group.

Solution: The respiratory and circulatory systems have evolved to provide the cells with a continuous and adequate supply of oxygen. There are limitations however: Because oxygen is highly insoluble in water, it has a very low solubility in plasma, since plasma is 90% water. A major adaptation is the evolution of oxygen-carrying molecules. The oxygen-carrying molecule found in the blood is hemoglobin. Hemoglobin is contained in the erythrocytes of vertebrates. Human erythrocytes, or red blood cells, are small, biconcave, disc-shaped cells (See Figure 1). During the maturation of red blood cells in mammals, they lose their nuclei, mitochondria, Golgi apparatus and other organelles. It is believed that the evolutionary loss of the nuclei in mammals has the adaptive advantage of leaving more space for hemoglobin. Similarly, since the red blood cell functions to transport oxygen, it would be counterproductive to have mitochondria which utilize oxygen. The mature erythrocyte does not need these organelles since its main function is to carry hemoglobin, which in turn transports oxygen from the lungs to the tissues. Each human erythrocyte contains about 280 million hemoglobin molecules.

A hemoglobin molecule consists of four polypeptide chains, two α chains and two β chains. The capacity of hemoglobin to bind oxygen depends on the presence of a non-protein unit called a heme group. The heme group consists of an organic portion, called protoporphyrin, and an inorganic iron atom in the center (See Figure 2). The iron atom is capable of binding reversibly to a molecule of oxygen. Since each polypeptide chain contains a heme group, each hemoglobin molecule can combine with a maximum of four oxygen molecules. When all four binding sites contain oxygen molecules ($Hb + 4O_2$), the resulting compound is called oxyhemoglobin, as opposed to deoxyhemoglobin. Oxy and deoxyhemoglobin differ in quaternary structure- that is, oxygenation changes the three-dimensional configuration of the four polypeptide chains. It is this change in structure which causes oxyhemoglobin to appear a brighter red than deoxyhemoglobin. This, in turn, accounts for the slight change in color between oxygen-rich blood and oxygen-poor blood.

While some invertebrates also utilize hemoglobin in oxygen transport, others have different oxygen-carrying molecules. Hemocyanin is a protein in which a molecule of oxygen is carried between two copper atoms. When this molecule is oxygenated, it is a pale blue color (deoxyhemocyanin is colorless). Hemocyanin is the oxygen-carrying pigment in molluscs, crustaceae and arthropods.

It was mentioned earlier that most of the hemoglobin molecules are contained within the red blood cells and not freely flowing in the plasma. By concentrating the hemoglobin molecules, the affinity for oxygen is increased, and the entry of oxygen from the outside to the lungs and from the lung tissues to the circulatory system is made faster and more efficient. Also, if the hemoglobin molecules were found in the plasma, the increased concentration of protein would thicken the blood. This increased viscosity would require that the heart perform more work in pumping the blood to peripheral organs and particularly to the brain. Erythrocytes thus act as hemoglobin containers for more efficient physiological functioning.

• PROBLEM 14-8

Both myoglobin and hemoglobin are oxygen-carrying molecules. Why is hemoglobin the molecule of choice to carry oxygen in the blood?

Solution: In vertebrates, both myoglobin and hemoglobin act as oxygen carriers. Hemoglobin, however, is the only one that acts in the blood, carrying oxygen from the lungs to the tissues where it is needed for metabolic processes.

Myoglobin is located in muscle, and serves as a reserve supply of oxygen. What accounts for the functional differences between these two proteins?

Myoglobin is a single-chain protein containing only one heme group, which serves to bind an oxygen molecule. Hemoglobin is composed of four polypeptide chains (each chain similar to the one of myoglobin) and thus has four binding sites for oxygen. It is the interaction of these four polypeptide chains that confers special properties to the hemoglobin molecule, making it a better oxygen-carrier in the blood.

One of these important properties is the ability to transport CO_2 and H^+ in addition to O_2 . Myoglobin does not have this ability. The binding of oxygen by hemoglobin is regulated by specific substances in its environment such as H^+, CO_2, and organic phosphate compounds. These regulatory substances bind to sites on hemoglobin that are far from the heme groups. The binding of these substances affects the binding of oxygen by producing conformational changes in the protein, that lower hemoglobin's affinity for oxygen. A change at the regulatory site (where these regulatory substances bind) is translated to the heme site by changes in the way the four polypeptide chains are spatially arranged. This results in a structural change in the protein. Such interactions between spatially distinct sites are termed allosteric interactions. Hemoglobin is an allosteric protein (whereas myoglobin is not) and therefore exhibits certain allosteric effects which account for the functional differences between the molecules.

This essential difference between myoglobin and hemoglobin is reflected in their respective oxygen dissociation curves. The curve is a plot of the amount of saturation of the oxygen-binding sites as a function of the partial pressure of oxygen:

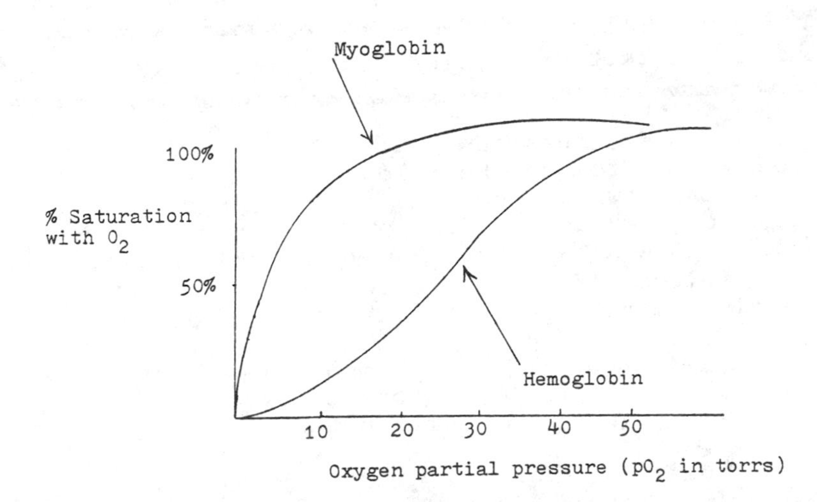

The shape of the oxygen dissociation curve reflects the crucial difference: the shape of the hemoglobin curve is sigmoidal (S-shaped) whereas that of myoglobin is hyperbolic. The sigmoidal shape is ideally suited to hemoglobin's role as an oxygen carrier in the blood. When the partial pressure of the oxygen is high, hemoglobin tends to bind oxygen; that is, many of the binding sites are filled with oxygen. There is a high partial pressure of oxygen (∿ 100 torrs) in the alveoli of the lungs, so that hemoglobin tends to pick up oxygen there. When the partial pressure of oxygen is low, hemoglobin tends to release oxygen. There is a low partial pressure of oxygen (∿ 20 torrs) in the capillaries in active muscle, where oxygen is being rapidly consumed by cellular respiration. CO_2 is being rapidly produced, and its concentration is correspondingly high. Hemoglobin tends to release oxygen.

Hemoglobin thus acts to pick up oxygen where it is available and release it where it is needed. Myoglobin, on the other hand, has a higher affinity for oxygen than does hemoglobin. Thus, even at relatively low partial pressures, myoglobin still tends to bind oxygen. Only when the pressure is very low does myoglobin release oxygen. This behavior makes it unsuitable to carry oxygen in the blood. An oxygen carrier in the blood must be able to readily release oxygen at places where oxygen is needed.

On a molecular level, the sigmoidal shape means that the binding of oxygen to hemoglobin is cooperative- that is, the binding of oxygen at one heme facilitates the binding of oxygen at another heme on the same hemoglobin molecule. The binding of the first oxygen molecule is thought to be most thermodynamically unfavorable. Subsequent oxygen molecules can bind more easily because the first molecule alters the structure (disrupt certain electostatic interactions) upon binding. The increasing affinity of the hemoglobin molecule for oxygen thus accounts for its sigmoidal dissociation curve.

• PROBLEM 14-9

How is the Bohr effect of physiological importance?

Solution: Both free protons (H^+) and carbon dioxide (CO_2) are known to promote the release of oxygen from hemoglobin. This is known as the Bohr effect. Myoglobin shows no change in oxygen binding over a range of pHs or carbon dioxide concentrations. But in hemoglobin, acidity enhances the release of oxygen. Lowering the pH shifts the oxygen dissociation curve to the right (see Figure).

A shift to the right enhances oxyhemoglobin dissociation. For example, at any partial pressure of oxygen, the curve at the lower pH is lower, signifying that less oxygen is

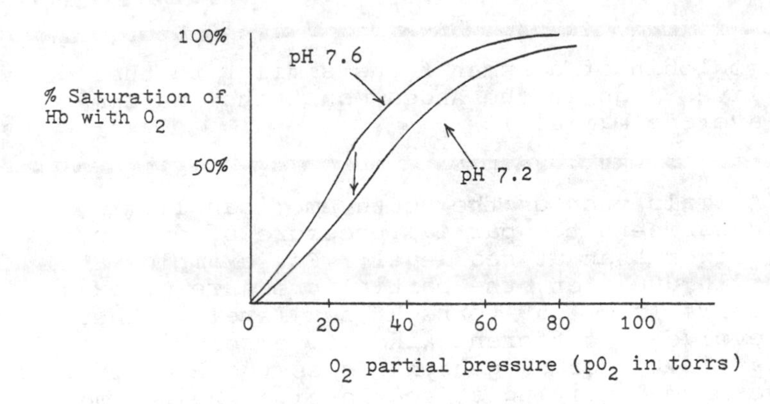

bound to the hemoglobin molecule (less % saturation) as a result of a decreased O_2 affinity.

An increase in the CO_2 concentration also lowers oxygen affinity. This could be due to, in part, a pH effect. CO_2 can react with the water in the erythrocyte to form carbonic acid:

$$CO_2 + H_2O \longrightarrow H_2CO_3$$

Carbonic acid then dissociates to form H^+ ions which lower the pH:

$$H_2CO_3 \longrightarrow H^+ + HCO_3^-$$

However, increasing the CO_2 concentration, while keeping pH constant, also lowers the oxygen affinity.

Both H^+ and CO_2 bind to the hemoglobin molecule at special regulatory sites spatially distinct from the heme site. The binding alters the structure of the molecule by changing the arrangement of its polypeptide chains. This altered conformation lowers hemoglobin's affinity for oxygen. Since H^+ and CO_2 affect hemoglobin's binding of oxygen by allosteric interaction between the hemoglobin subunits, they have no similar effect on the myoglobin molecule, since it has only one polypeptide chain.

Now that we understand the Bohr effect, what is its physiological significance? Rapid metabolism occurs in highly active tissue such as contracting muscle. Here much CO_2 is produced from cellular respiration. Acid production is also high, resulting from the increased CO_2 concentration and the reduction of pyruvate to lactic acid when the amount of oxygen is limiting. Thus, the presence of high levels of CO_2 and H^+ in actively metabolizing tissue enhances the release of oxygen from oxyhemoglobin. More oxygen is thus provided to the tissues where it is needed most. The opposite process occurs in the alveoli of the lungs. The increased concentration of oxygen drives off H^+ and CO_2, so that more oxygen may bind.

How is the hemoglobin of a human fetus similar to that of a llama that lives high in the Andes Mountains (as compared to the average mammal?)

Solution: Animals such as the South American llama live at high altitudes where the partial pressure of oxygen is significantly lower than at sea level. For example, at an altitude of 10,000 feet, the partial pressure of oxygen drops to 110 torrs (from 159 torrs at sea level). These animals have evolved a different kind of hemoglobin that has a higher affinity for oxygen than does the hemoglobin of the average mammal living at sea level. Their hemoglobin picks up oxygen more easily so that sufficient oxygen can be obtained at high altitudes.

Human fetal hemoglobin is similar to llama hemoglobin in that it is also adapted to pick up oxygen in a medium which has a partial pressure of oxygen lower than normal. Fetal hemoglobin thus has a higher affinity for oxygen than adult hemoglobin. This adaptation is significant since the fetus obtains its oxygen from the mother's blood. For the hemoglobin of the fetus to be able to take oxygen from the hemoglobin of the mother, it must have a higher affinity for the oxygen than does the mother's hemoglobin. Fetal hemoglobin is therefore oxygenated at the expense of adult hemoglobin in the maternal circulation. What accounts for the higher oxygen affinity?

The structure of fetal hemoglobin is slightly different from that of adult hemoglobin. This altered structure binds an organic phosphate called 2,3- diphosphoglycerate (DPG) less strongly than does adult hemoglobin. DPG, which is present in human erythrocytes, reduces the oxygen affinity of hemoglobin:

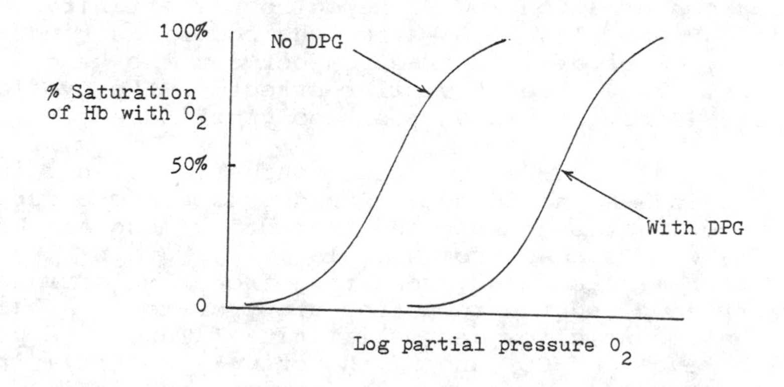

Figure 1.

DPG reduces the oxygen affinity by stabilizing the deoxyhemoglobin structure; the binding of oxygen is thus less favored. Since fetal hemoglobin binds DPG more weakly than does adult hemoglobin, its deoxygenated form does not undergo this degree of stabilization; its oxygen affinity is therefore enhanced. The oxygen dissociation curves of both fetal and maternal are shown in the following diagram:

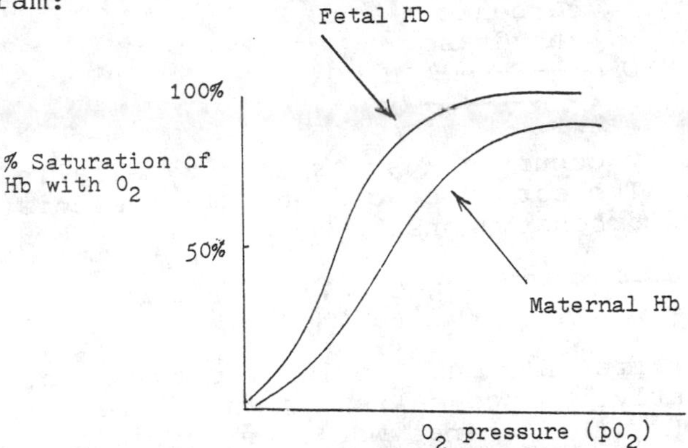

Figure 2. For fetal hemoglobin, the oxygen dissociation curve is shifted to the left, indicating higher oxygen affinity.

For fetal hemoglobin, the oxygen dissociation curve is shifted to the left, indicating a higher oxygen affinity.

When an adult human adapts to high altitude, there is an increase in DPG in the red blood cells which shifts the curve to the right, favoring unloading of oxygen from hemoglobin at any given partial pressure.

• PROBLEM 14-11

When carbon dioxide combines with water, carbonic acid is produced and rapidly dissociates to the bicarbonate ion and free hydrogen ions:

$$CO_2 + H_2O \longrightarrow H_2CO_3$$

$$H_2CO_3 \longrightarrow H^+ + HCO_3^-$$

When carbon dioxide is carried by the blood, why doesn't the blood become acidic owing to the presence of free H^+?

Solution: Not only does the blood transport oxygen from the lungs to the tissues (via hemoglobin), it also transports carbon dioxide in the reverse direction, from the tissues to the lungs. Some of the carbon dioxide ($\sim$ 7%) is carried as gas dissolved in the plasma. A slightly larger proportion ($\sim$ 23%) is carried by hemoglobin; the carbon dioxide-hemoglobin complex is called carbaminohemoglobin. The loosely bound CO_2 is combined with the terminal amino groups of all four polypeptide chains; each hemoglobin molecule having four polypeptide chains, can bind four CO_2 molecules. Since CO_2 does not combine with hemoglobin at the same site as oxygen, hemoglobin can transport both oxygen and carbon dioxide at the same time. However, most of the CO_2 (70%) is transported as the bicarbonate ion, HCO_3^-. Carbon dioxide enters red blood cells where it reacts with water to form carbonic acid:

$$CO_2 + H_2O \underset{}{\overset{\text{carbonic anhydrase}}{\rightleftarrows}} H_2CO_3$$

This reaction can only occur because of an enzyme called carbonic anhydrase. The carbonic acid quickly dissociates into hydrogen and bicarbonate ions:

$$H_2CO_3 \longrightarrow H^+ + HCO_3^-$$

The presence of free H^+ ions should drop the pH of the blood significantly. Since cells can only live within a narrow range of pHs, this change would be physiologically harmful. But the blood exhibits only very slight variations (±.04 pH units) around the normal pH of 7.4. What accounts for this phenomenon?

The pH would not drop if the H^+ ions could be combined with some substance, so as to remove them as free ions from solution. The H^+ ions combine with an ionized form of hemoglobin (Hb^-) to form acid hemoglobin (HHb) while the HCO_3^- leaves the red blood cell and enters the plasma.

$$HCO_3^- + H^+ + Hb^- \longrightarrow HCO_3^- + HHb$$

Acid hemoglobin dissociates to a much lesser degree than does carbonic acid, so that most H^+ ions remain bound, and therefore effectively removed from solution. Acid hemoglobin acts as a buffer in the blood. A buffer is a substance which prevents large changes in pH. When there is an excess of H^+ ions (when the pH drops), hemoglobin binds H^+ ions to remove them from solution, thus restoring the pH.

$$Hb^- + H^+ \longrightarrow HHb$$

When the blood becomes alkaline because of an excess of hydroxide ions (OH^-) or a decrease of H^+, the acid hemoglobin dissociates, releasing H^+ and restoring the pH:

$$OH^- + HHb \longrightarrow OH^- + H^+ + Hb^- \longrightarrow H_2O + Hb^-$$

Hemoglobin and some plasma proteins act to prevent the large deviations from the normal pH which would occur in carbon dioxide transport. A summary of CO_2 transport is provided in the following illustration:

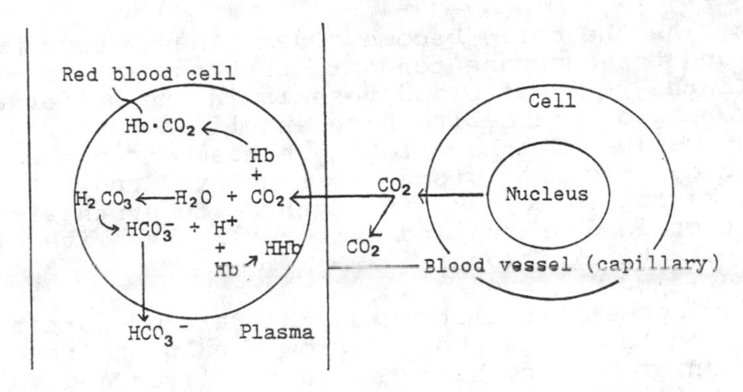

The reverse process occurs in the lungs. Carbamino-hemoglobin dissociates to release CO_2. Acid hemoglobin dissociates to produce H^+ ions which combine with bicarbonate ions to form carbonic acid. This acid then dissociates to form water and CO_2, which is expired from the lungs.

ERYTHROCYTE PRODUCTION AND MORPHOLOGY

• PROBLEM 14-12

A person exposed to gamma ray radiation from a nuclear bomb blast is likely to sustain complete destruction of his bone marrow. Why would this person become anemic?

Solution: The first step in the solution is to define anemia. Anemia is a deficiency of red blood cells, caused either by rapid loss or slow production of these cells. This results in decreased oxygen transport and thus impairment of the metabolism of all tissues. Gamma-ray radiation causes bone marrow aplasia, or a lack of functioning bone marrow. This disease is lethal if all the bone marrow is destroyed, because the bone marrow is the site of red blood cell production. During the middle part of development, the fetus produces erythrocytes in the liver,spleen and lymph nodes. But later in gestation and after birth, red blood cells are produced almost exclusively by the bone marrow. Until adolescence, erythrocytes are produced in almost all bone marrow. In the adult, however, only the marrow in the ends of the long bones and the shafts of the flat bones, like those of the skull and ribs, form erythrocytes.

Red blood cells are derived from cells, hemocytoblasts, which arise from unspecialized primordial stem cells in the bone marrow. The hemocytoblasts have a nucleus but no hemoglobin. They divide to form precursor cells which gradually transform into mature erythrocytes, having no nucleus and containing hemoglobin. Although mature erythrocytes have no nucleus, they already contain all the enzymes and machinery necessary for metabolism and energy production.

When red blood cells are released into the circulatory system from the bone marrow, they normally circulate about

120 days. As the cells become older, they become more fragile and the membrane tends to rupture, killing the cell. In the walls of blood vessels in the spleen and liver are phagocytic cells which engulf and destroy old red blood cells. The hemoglobin released from these cells is degraded: the iron atoms are recovered and returned to the bone marrow to be used in erythrocyte synthesis; the heme portion is degraded and excreted by the liver as bilirubin, a bile pigment.

Under normal circumstances, the rate of formation of new erythrocytes in the bone marrow equals the rate of destruction of old erythrocytes in the liver and spleen. About 2.5 million erythrocytes are made and destroyed each second. Thus, the total number of circulating erythrocytes remains constant.

• PROBLEM 14-13

Why does the number of red blood cells in the human body increase at high altitudes?

Solution: It has been observed that the loss of red cells by hemorrhage decreases the ability of the blood to deliver sufficient oxygen, and increased erythrocyte production results. Also, partial destruction of the bone marrow (by x-ray, for example) destroys the sites of erythrocyte production and diminishes the availability of oxygen to the tissues. Increased red cell production by the remaining healthy marrow follows. It seems that the initiator or stimulus for increased erythrocyte production could be a lack of oxygen. This is indeed found to be true. At high altitudes, the partial pressure of oxygen is decreased, and thus less oxygen is delivered to the tissues. Any condition that causes a decrease in the amount of oxygen transported to the tissues causes an increased rate of erythrocyte production.

Actually a decreased oxygen concentration in the bone marrow does not directly (by itself) expedite erythrocyte production. In response to decreased oxygen levels, the kidney, liver, and other tissues secrete erythropoietin, also called erythropoietic stimulating factor. This glycoprotein is transported via the blood to the bone marrow, where it stimulates erythrocyte production. When the number of red blood cells has increased to the point where the oxygen level in the tissues is again normal, erythropoietin secretion decreases and a normal amount of erythrocytes are again produced. This negative feedback mechanism prevents an overproduction of red cells, preventing the viscosity of the blood from increasing to a dangerous level.

It is interesting to note that erythrocyte production and hemoglobin synthesis are not necessarily correlated. An iron deficiency decreases hemoglobin synthesis (since the heme part of the hemoglobin molecule contains an iron atom). This low hemoglobin level leads to a reduction of oxygen brought to the tissues. As was cited before, a reduced oxygen level causes increased red blood cell production. In this case, the red blood cells have a lower hemoglobin content than normal (they are called hypochromic erythrocytes) and thus are less efficient oxygen carriers.

• PROBLEM 14-14

In sickle-cell anemia, what causes the erythrocytes to assume a sickle shape? Why is the incidence of sickle-cell anemia high in Africa, where the incidence of malaria is also high?

Solution: Sickle-cell anemia is caused by an abnormal type of hemoglobin, termed hemoglobin S. Hemoglobin S differs from normal adult hemoglobin in that it contains a different amino acid in one of its polypeptide chains (valine instead of glutamate in one of the β chains). This alteration greatly reduces the solubility of deoxygenated hemoglobin S. The amino acid substitution creates a site on the surface of the hemoglobin molecule, sometimes referred to as a "sticky patch." Only deoxyhemoglobin S has a complementary site to this sticky patch. Therefore, the deoxyhemoglobin molecules bind to each other, forming long aggregates which precipitate and distort the normal shape of the erythrocyte. Besides elongating the cell to the familiar sickle or crescent shape, the precipitated hemoglobin also damages the cell membrane, making the cell fragile. The sickled cells either become trapped in small blood vessels (because of their distorted shape), causing impaired circulation and tissue damage, or they readily lyse because of their fragility. Sometimes, a vicious cycle is initiated when sickled cells cause vascular blockage. The blockage of the vessels creates a region of tissues where the oxygen concentration is low. This deficiency causes more oxyhemoglobin S to change into the deoxy form, which in turn causes more sickling and further decreases the oxygen concentration. This cycle progresses quickly and the resulting anemia is usually severe, often resulting in shock and death.

Sickle-cell anemia is genetically transmitted. Sickle-cell anemics are homozygous for an abnormal gene which codes for synthesis of the abnormal hemoglobin S. Heterozygous individuals, who have one normal and one substitute allele, carry the sickle-cell trait. They do not usually show the symptoms of the disease except under conditions of low oxygen such as occurs in high altitude.

The incidence of malaria and the frequency of the sickle-cell gene in Africa are correlated. It has been discovered that individuals carrying the sickle-cell trait (heterozygotes) are protected against a lethal form of malaria. This protection acts as a strong selective pressure to maintain the high incidence of the sickle-cell gene in regions where the incidence of malaria is high. This is an example of balanced polymorphism: the heterozygote is protected against malaria and does not show the symptoms of sickle-cell disease, whereas the sickle-cell homozygote is eliminated by anemia and the normal homozygote is eliminated by malaria. Even though malaria is virtually eliminated in the U.S., there is still a significant incidence of sickle-cell anemia among American Negroes.

DEFENSE SYSTEMS

• PROBLEM 14-15

What are the basic structural and functional differences between white blood cells and red blood cells? Describe two different ways that white blood cells act to protect the body from foreign agents such as microorganisms.

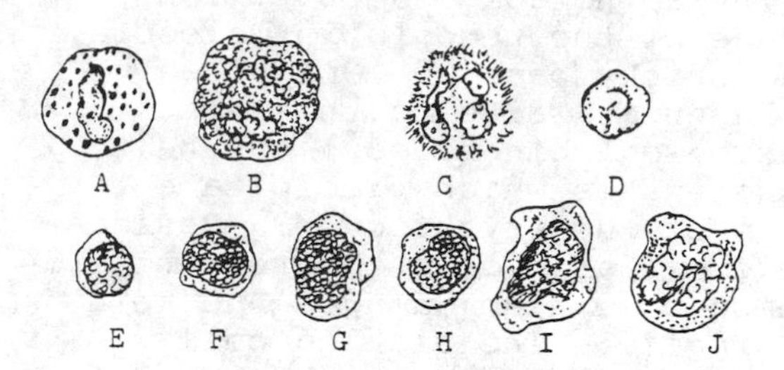

Types of white blood cells. (A) basophil; (B) eosinophil; (C) neutophil; (E-H) a variety of lymphocytes; (I and J) monocytes. (D) is a red blood cell drawn to the same scale.

Solution: In addition to red blood cells, human blood contains five types of white blood cells, or leukocytes. Unlike the red blood cells, all types of leukocytes contain nuclei and do not contain hemoglobin. Leukocytes are generally larger than erythrocytes and are far less numerous. There are approximately 5,400,000 red cells per cubic millimeter of blood in an adult male (4.6 million in females), while only about 7,000 leukocytes per cubic millimeter.

There are five different types of leukocytes (see Figure). Leukocytes are classified as granular (eosinophils, neutrophils and basophils) or angular (monocytes and lymphocytes). The granulocytes have granules in their cytoplasm, have lobed nuclei, and are produced in red bone marrow. The agranulocytes are produced in organs such as the lymph nodes, spleen and thymus.

Neutrophils make up over 60% of the leukocytes present in the body. Like other leukocytes, they can squeeze through the pores of blood vessels and enter the tissue spaces. They then move by amoeboid movement to sites of infected tissue. For example, when bacteria enter a certain tissue of the body, they can either attack cells or produce damaging toxins. The blood vessels in the infected region dilate and allow more blood flow to the site, causing the heat and redness characteristic of inflammation. Blood vessel permeability also increases, causing fluid to enter the tissue and swelling to result. Neutrophils and monocytes pass through the blood vessels and engulf, ingest and destroy the bacteria. Foreign particles and dead tissue can also be engulfed by this process of phagocytosis. These leukocytes are chemically attracted to the inflamed site by products released from the damaged tissues.

The monocytes (5.3%) usually appear after the neutrophils and are more important in fighting chronic infections. Monocytes can phagocytize bacteria but more often, they enlarge and become wandering macrophages, which can move more quickly and engulf more bacteria. Eosinophils (2.3%) are weakly phagocytic, and are involved in allergic reactions (by releasing antihistamines) and in fighting trichinosis (an infection caused by the parasitic worm, trichinella). Basophils (.5%) liberate the anticoagulant heparin which combats the coagulative processes that sometimes occurs in prolonged inflammation.

The second means by which leukocytes protect the body from invasion is carried out by the lymphocytes. Lymphocytes (30%) function in the process of immunity. Lymphocytes are involved in the production of antibodies, proteins made by the body in response to specific foreign substances. These foreign substances are subsequently attacked and destroyed by the anitbodies. Immunity and its agents will be discussed in further questions.

The main functions of the leukocytes are thus protective: phagocytosis and immunity. The major function of the erythrocytes is to carry hemoglobin for gaseous transport and exchange, which is essential to the continuation of cellular metabolic processes.

• PROBLEM 14-16

A man is exposed to a pathogenic bacterium. If inflammation occurs, has the body's first line of defense been penetrated?

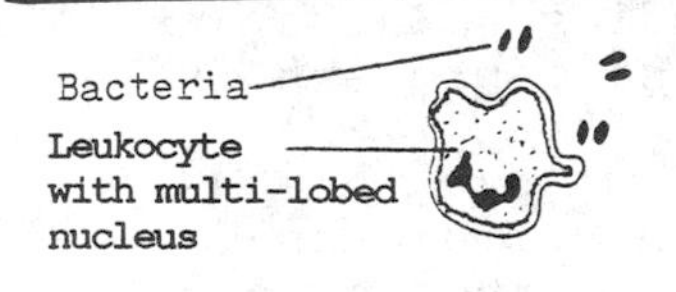

Bacteria englufment.

Bacteria engulfed may be destroyed.

Bacteria phagocytized by leukocytes. Foreign matter, such as bacteria, attracts the leukocytes to the invaded tissue and phagocytosis results.

Solution: This answer will describe the body's first two lines of defense against invading pathogenic microorganisms. The first line of defense is basically mechanical, involving physical barriers such as the skin and mucous membranes. The tough layer of skin forms a protective barrier as long as it is intact. The sticky mucous secretions of the respiratory and digestive tract, along with other tissues, collect and hold many pathogens until they can be disposed of. Cilia, small hairlike appendages lining the epithelial cells of the body cavities, help sweep bacteria way from susceptible areas. The mechanical processes of coughing, sneezing, tear-shedding, and salivating provide a clearing mechanism that flushes away bacteria. Both enzymes, such as lysozyme in tears, and the acidity or alkalinity of body secretions prevent bacteria from entering further into the body.

The second line of defense is cellular. If pathogens succeed in penetrating the first line of defense and enter deeper tissue, they may be engulfed and destroyed by cells in the body named phagocytes. This process is called phagocytosis. Phagocytes include macrophages and leukocytes (white blood cells). Inflammation results from damage to body tissue caused by the invading pathogen. There is an accumulation of serum and leukocytes in the area, and this produces the signs of inflammation: swelling, heat, pain and redness. Although the phagocytes destroy the bacteria by ameboid engulfment, some phagocytes are themselves also destroyed after the engulfment occurs. Dead phagocytes, bacteria and serum that result from an inflammation are together termed pus, which may be colored according to the pigment of the bacteria involved in the infection.

We thus see that inflammation occurs when the body's first line of defense is broken and the second line of defense, phagocytosis, takes over.

• PROBLEM 14-17

A man is exposed to a virus. The virus breaks through the body's first and second lines of defense. Does the body have a third line of defense? Outline the sequence of events leading to the destruction of the virus.

Solution: The first line of defense against invading pathogens, such as viruses, is the skin and mucous membranes. If this line is broken, the second defense mechanism, phagocytosis, takes over. If the virus breaks this line of defense, it enters the circulatory system. Its presence here stimulates specific white blood cells, the lymphocytes, to produce antibodies. These antibodies combine specifically with the virus to prevent it from further spreading.

Any substance, such as a virus, which stimulates antibody synthesis is called an antigen. Antibodies are not produced because the body realizes that the virus will produce a disease, but because the virus is a foreign substance. An individual is "immune" to a virus or any antigen as long as the specific antibody for that antigen is present in the circulatory system. The study of antigen-antibody interactions is called immunology.

The basic sequence of events leading to the destruction of the virus is similar to most antigen-antibody interactions. The virus makes contact with a lymphocyte which has a recognition site specific for the virus. The lymphocyte is then stimulated to reproduce rapidly, causing subsequent increased production of antibodies. The antibodies are released and form insoluble complexes with the virus. These insoluble aggregates are then engulfed and destroyed by macrophages.

• PROBLEM 14-18

Are all antibodies structurally and functionally similar? Explain.

Solution: All antibodies are proteins and are referred to as immunoglobulins. Five major classes of immunoglobulins can be distinguished based on their characteristic structures and separate functional roles: IgG, IgM, IgA, IgD and IgE.

IgG immunoglobulins constitute about 80% of all the antibodies in the blood serum. These are the major antibodies involved in the destruction of invading microorganisms. IgM immunoglobulins appear first in response to an antigen; they are later replaced by IgG. IgA immunoglobulins are found in secretions such as saliva and tears and are involved in the defense of body cavities such as the mouth and vagina. The function of IgD is still unknown, while IgE is thought to be involved in allergic reactions.

There are thus five basic types of immunoglobulins. But we are certainly exposed to more than five different antigens. What distinguishes one immunoglobulin molecule from the next? Let us look at the structure of an IgG molecule. It is composed of four polypeptide chains; two heavy chains (mol. wt. ∿ 53,000) and two light chains (mol. wt. ∿ 22,500). In a given IgG molecule, both heavy chains are identical and both light chains are identical. The polypeptide chains are held together by disulfide (-S-S-) linkages to form a Y-shaped molecule:

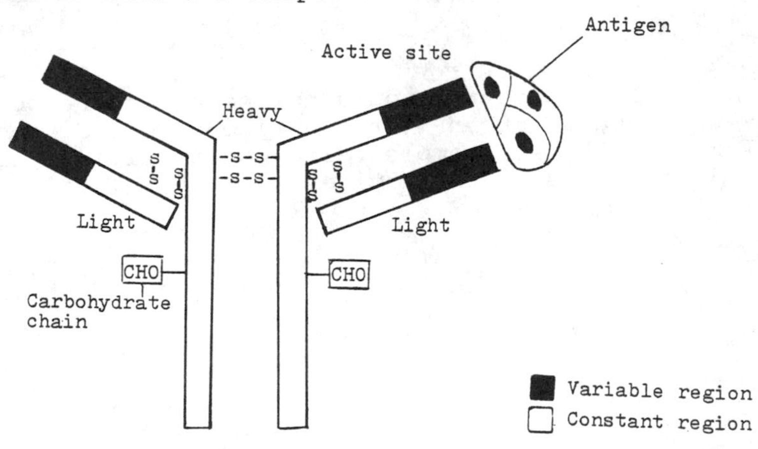

The ends of each arm of the IgG molecule contain an active site which can combine with a specific antigen. Thus, two identical binding sites permit each antibody molecule to form a complex with two antigens. Both light and heavy chains contain two regions: a "constant" region which has very similar amino acid sequences in all immunoglobulins, and a "variable" region which has different amino acid sequences for each type of antibody. This variability accounts for the specificity of antibodies-that is, one particular antibody molecule can bind only one particular kind of antigen. The specific sequence of amino acids in the variable region at the end of each arm of the antibody molecule determines the spatial configuration of the active site which will only bind one particular antigen.

• PROBLEM 14-19

Many antibodies react by causing agglutination and precipitation of the antigen-antibody complex. Explain why these antibodies and their respective antigens must each have more than one binding site per molecule.

Solution: The structure of the IgG molecule has already been visualized in the electron microscope. It is a Y-shaped molecule with one binding site, or valency, at the end of each arm. Each IgG molecule is thus divalent, having two binding sites capable of combining with two antigens. Each antigen molecule also has more than one binding site; that is, it is divalent or multivalent.

The reason that both antibodies and antigens are multivalent is to permit a high degree of agglutination (clumping). This allows precipitation of the antibody-antigen complex to occur. If only one binding site existed on each antibody and antigen molecule, then only one molecule of antibody could bind to one molecule of antigen:

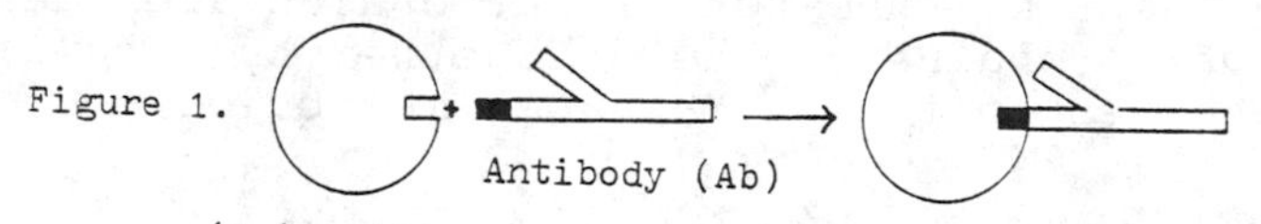

Figure 1.

Precipitation would not occur if the complex were this small. Even if the antibody were divalent, as we know to be so, complex size would still be limited to 3 molecules if the antigen were univalent:

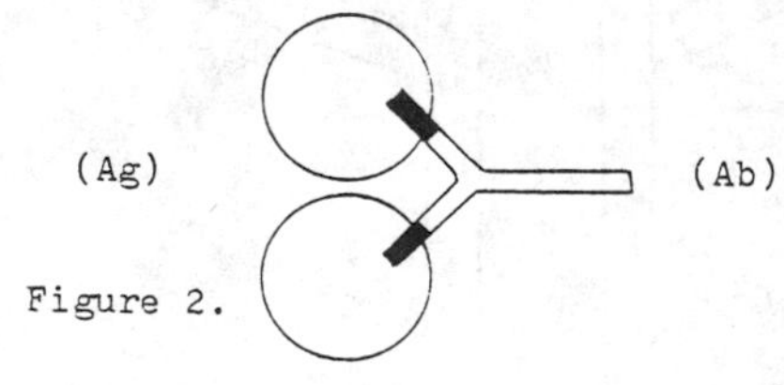

Figure 2.

Therefore, the antigen molecule must also be at least divalent for there to be extensive bridging of antigen-antibody complexes:

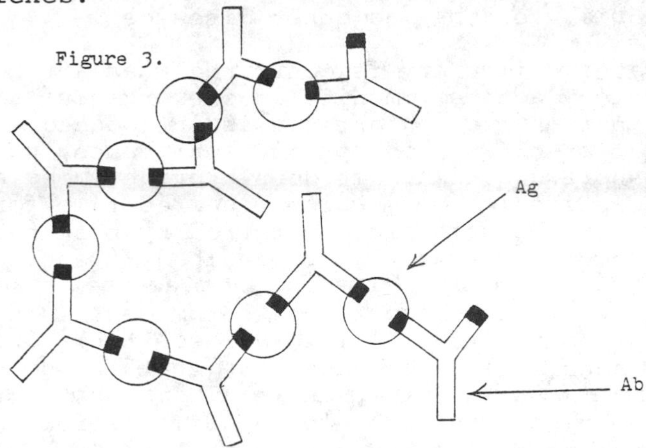

This large complex of antigens and antibodies, called a lattice, is now able to settle out of the body fluid. It then tends to be engulfed and destroyed by macrophages.

TYPES OF IMMUNITY

• PROBLEM 14-20

Distinguish between active and passive immunity. How are they produced? Are vaccines and toxoids used to induce active immunity or passive immunity?

Solution: Immunity is a natural or acquired resistance to a specific disease. A host is immune to a certain pathogen as long as the antibodies specific for the pathogen are present in his circulatory system. The crucial difference between active and passive immunity lies in the answer to the host's question, "Are these antibodies made by me or were they formed in some other organism?"

Active immunity results when antibodies are produced by the cells (lymphocytes) of the host as a result of contact with an antigen. The antigen may be a microorganism or its product. Active immunity usually develops slowly within a couple of weeks (as antibody production reaches a maximum level), yet it may last up to many years for some antigens. Active immunity may be induced naturally, while recovering from an infectious disease such as mumps, measles, or chicken pox; or artificially, by vaccines. Vaccines are inactivated or attenuated microorganisms which can still stimulate antibody production. Attenuated microorganisms are living, yet too weak to be virulent. Vaccines are available for typhoid fever, poliomyelitis, rabies, smallpox and various other diseases. Also used to induce active immunity are toxoids, made by destroying the poisonous parts of the toxins produced by some pathogens. The antigenic part of the toxin is not affected and can still induce antibody production for protection against diphtheria, tetanus and other diseases.

Passive immunity is conferred when antibodies produced by active immunity in one organism are transferred to another. Since no antigen is introduced, there is no stimulation of antibody production; hence, passive immunity is of short duration. It does provide immediate protection, unlike active immunity which requires time for development. Passive immunity is also conferred by both natural or artificial means: naturally, by antibody transfer to the fetus from an immune mother (that is, by placental transfer), or artificially, by injection of serum containing antibodies from an immune individual to a susceptible one. In the latter case, it is not necessary that the donor and the recipient be of the same species. It is, however, dangerous for people who are allergic to animal serum, and so must be used with caution.

If the answer to the host's question is "These are my own antibodies," then active immunity is involved; if the answer is "These are not my antibodies," then passive immunity is involved.

• PROBLEM 14-21

Distinguish between cell-mediated immunity and humoral immunity. What structures are responsible for each type of immunity?

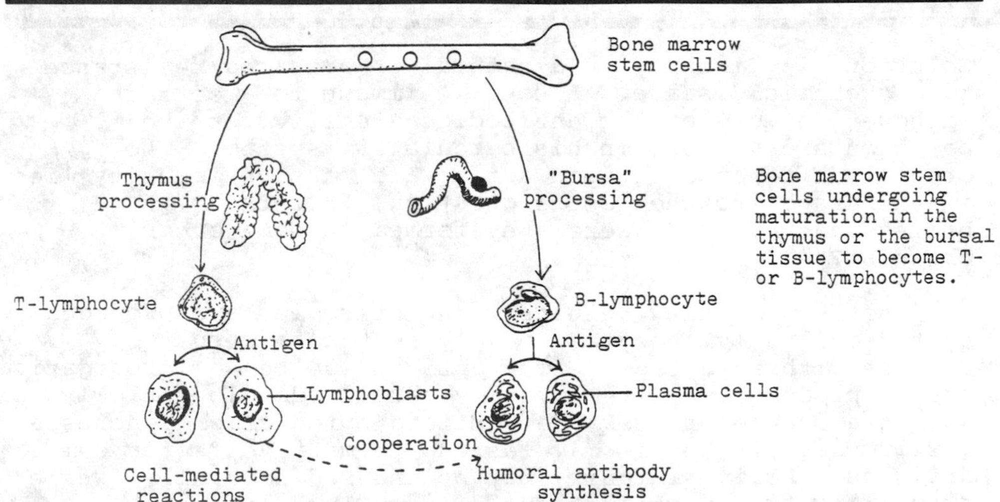

Solution: It was originally thought that all antibodies were secreted into the blood or other body fluids, such as saliva, mucous and tears. Such antibodies, which function only after they are secreted into body fluids, are called humoral antibodies, and the immunity mediated by them is called humoral immunity. It eventually became clear that other types of immunological responses were due to antibodies which remain bound to their parent lymphocyte. The immunity mediated by these special lymphocytes is called cell-mediated immunity. A variety of antigens evoke the cell-mediated immune response for example, the bacterium Mycobacterium tuberculosis (the causative agent of tuberculosis).

We can now distinguish two different types of lymphocytes involved in the immunological response. The T, or thymus lymphocytes are responsible for cell-mediated immunity, whereas the B, or bursal lymphocytes are responsible for humoral immunity. Both types of lymphocytes originate from bone marrow stem cells. Those stem cells which migrate to the thymus differentiate into T cells while those which migrate into the bursa differentiate into B cells. (The bursa is a gland in birds, although we do not yet know its mammalian counterpart.) Both B and T lymphocytes enter the circulation and colonize in lymphoid tissue such as the spleen and lymph nodes.

In the presence of antigens, T lymphocytes transform into lymphoblasts, cells which do not secrete their antibodies, but which are involved in cell-mediated immunity. When B lymphocytes are antigenically stimulated, they transform into plasma cells which produce and secrete the immunoglobulins involved in humoral immunity.

A given antigen predominately leads to either a cell-mediated or humoral immunological response. As a general rule, the humoral response is seen as a defense against bacteria and viruses in the body's fluids, while the cellular response is effective against bacteria and viruses in infected host cells as well as against fungal and protozoan infections.

ANTIGEN-ANTIBODY INTERACTIONS

• PROBLEM 14-22

How does a plasma cell make a specific antibody when stimulated by a specific antigen?

Solution: We still do not know the exact mechanism of antibody stimulation, although several theories have been proposed. One theory widely accepted some thirty to forty years ago was called the instructive theory of antibody formation. According to this theory, the antigen is needed to supply the information necessary for synthesis of the specific antibody. It was thought that the antigen combines with a newly synthesized antibody chain before it folds into a particular three-dimensional form. The developing antibody molecule would fold around the antigen, creating a complementary shape at the antibody's binding sites. The specificity of the antibody was then assumed to be due to the spatial configuration of its chains and not to the amino acid sequence of the antibody. This theory was questioned when it was discovered that different antibodies have different amino acid sequences. Moreover, if we simultaneously inject several different antigens into an animal, we find that each plasma cell produces only one specific antibody. According to the instructive theory, if several different antigens enter a particular plasma cell, we would expect that cell to produce different antibodies (since each antigen acts as a template for antibody formation). This, however, is not the case.

It has now been determined that the presence of an antigen simply increases the number of plasma cells which produce the corresponding antibody specific for that antigen. The presence of the antigen only stimulates division and differentiation of those lymphocytes which synthesize antibodies specific for that antigen. This theory thus assumes that there pre-exists a very large number of differentiated small lymphocytes already "programmed" genetically to produce one specific type of antibody. The sole function of the antigen is to stimulate the division of those lymphocytes which will produce the corresponding antibody. This generally accepted theory is called the theory of clonal selection. A clone is a group of identical cells all descended from a common ancestral cell.

How does a particular lymphocyte recognize the antigen that its antibodies are specific for? Both B and T lymphocytes have immunoglobulin molecules on the surface of their plasma membranes. These antibodies are positioned in such a way that the binding sites face outward and can thus bind antigens. Each particular lymphocyte contains only one type of surface antibody; when the antibody interacts with the corresponding antigen, the lymphocyte is stimulated to proliferate and produce more of this specific antibody. The exact mechanism of how the antigen stimulates division is still unknown.

The clonal selection theory of antibody formation can be outlined as follows:

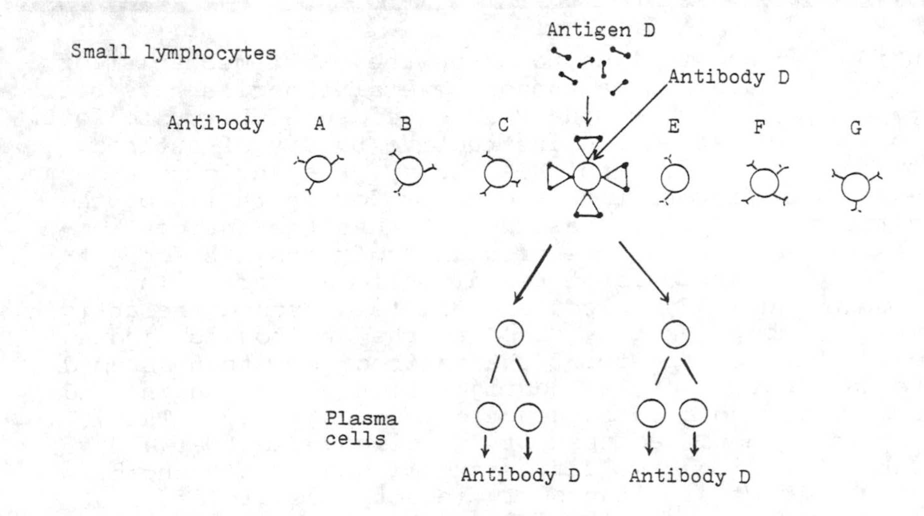

• PROBLEM 14-23

How would you explain the secondary response to antigens on the basis of the clonal selection theory?

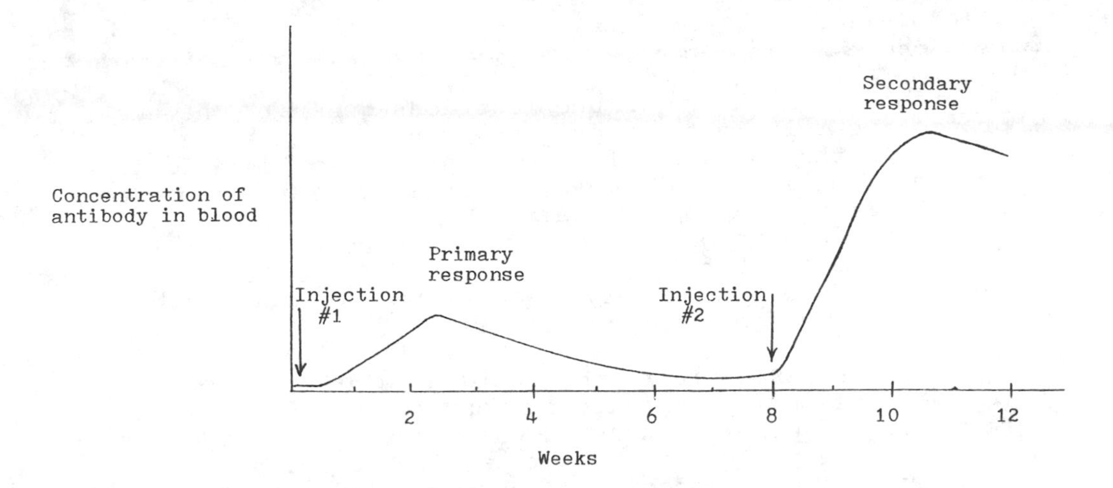

Solution: When an antigen is first injected into an animal, only a small number of lymphocytes become transformed into plasma cells. The amount of antibodies produced rises slowly to a low peak and then decreases. This is termed the primary response. If a second injection of the same antigen occurs some weeks later, a much more vigorous response is induced. Not only is the amount of antibodies much greater than in the primary response, but their production is much more rapid (see Fig.). Further injections will result in the maximum amount of antibody that can be produced. Primary and secondary responses of antibody formation can be explained on the basis of the clonal selection theory. When an antigen first comes in contact with its corresponding lymphocyte, it stimulates the lymphocyte to divide and differentiate into plasma cells (or lymphoblasts). The antibodies of these cells interact with the antigen to form complexes which re engulfed and destroyed by macrophages. Some days a ter the primary response, these plasma cells (or lymphol lasts) eventually die. However, the antigen caused production of many more copies of the originally stimul ted small lymphocytes. It is thought that these lymphocytes, called memory cells, arise from division of the originally stimulated cells without concomitant differentiation into plasma cells (lymphoblasts). This immunological memory explains the process in which an animal "remembers" previous exposure to an antigen. When the animal is exposed to the antigen the second time, many more lymphocytes (specific for the antigen) are stimulated, resulting in a greater and more rapid production of antibodies. This explains the characteristics of the secondary response.

• PROBLEM 14-24

What is the physiological basis of allergies? What measures can be taken to decrease the effect of allergies?

Solution: Although immune reactions generally have a protective role in the body, they can sometimes cause damage to the body itself. Such an adverse immunological response, called an allergy or hypersensitivity reaction, results from an acquired reactivity to an antigen. The reactivity can result in bodily damage upon subsequent re-exposure to that specific antigen. The allergic response can be elicited by a variety of antigens, including pollen, dust, certain foods and bee stings.

Initial exposure to the antigen leads to some antibody synthesis (the primary response), but also to the formation of more lymphocytes which act as memory cells. Subsequent exposure to the same antigen elicits a more vigorous response (secondary response); many memory cells are stimulated to differentiate into plasma cells which produce the corresponding antibody. These particular antigens stimulate production of IgE immunoglobins which circulate in the blood and bind to mast cells (found in connective tissue) and basophils (a type of white blood cell). When the antigen forms a complex with the IgE attached to the mast cell, release of histamine and other chemicals from the mast cell results. The symptoms of the allergy result from the effects of these chemicals. For example, when a previously sensitized person inhales pollen, the antigenic part of the pollen combines with the IgE-mast cell complex in the respiratory passages. The histamine released causes increased mucous secretion, increased blood flow and constriction of smooth muscle lining the passages. These physiological changes cause the familiar symptoms of running nose, congestion and difficulty of breathing. In some people, a very severe allergic reaction called anaphylaxis can occur which may have systemic (non-localized) symptoms. People have died from anaphylaxis resulting from a bee sting.

To decrease the effect of allergies, one can administer antihistamines. These only provide incomplete relief, however, since other chemicals are also released. A therapy called desensitization can help allergic people. In this process, the antigen is injected repeatedly into the person in small but increasing dosages. The theory behind desensitization is that this procedure induces IgG synthesis against the antigen. When the antigen is subsequently presented, it is more likely to be complexed by IgG antibodies than by IgE antibodies bound to mast cells.

Why are organ transplants generally unsuccessful?

Solution: Under normal circumstances, if a piece of skin, or some organ such as the heart or kidney, is transplanted from one person to another, it initially appears to be accepted. But within two weeks, the graft (transplanted tissue) is invaded by white blood cells and the tissue dies. This graft rejection is an immunological reaction, also known as a primary set reaction. Second grafts from the same donor are rejected within a few days. This accelerated rejection is termed the second set reaction and can be explained on the basis of the clonal selection theory.

Although humoral (circulating) antibodies may be involved in the rejection of transplanted tissues and organs, the active agents are intact small lymphocytes. Immunoglobulins on the surface of these lymphocytes form a complex with the surface antigens on the foreign tissue cells. These lymphocytes specifically destroy these cells by secreting toxic substances or by stimulating increased phagocytic activity of macrophages.

This immunological response is a reflection of the ability of the body's cells (self) to recognize and destroy anything that is foreign (unself). The rejection of tissue from another member of a different species is termed a heterograft reaction. Heterografts are usually unsuccessful. When the donor is a member of the same species, the transplantation, called a homograft, is also usually unsuccessful. But when the donor and recipient are either identical twins or members of the same purebred strain, the transplantation is usually successful. These transplants, called isografts, are successful because the donor and recipient are genetically very similar. The genes that determine the antigens responsible for the transplantation immunological response are termed histocompatibility genes. The greater the number of histocompatibility genes that the donor and recipient have in common, the greater the probability that the transplant will be successful. That is why autografts, or tissue transplants from one location to another on the same individual, are usually successful. The success of homografts depend on how well the donor's and recipient's tissues are matched, that is, how well the donor and recipient resemble each other genetically.

Sometimes certain tissues of an individual's own body may be rejected, resulting in disease. The immunological mechanism for distinguishing between self and non-self becomes disrupted and the body begins to destroy itself. These autoimmune diseases include rheumatoid arthritis, rheumatic fever and multiple sclerosis.

A person has two strains of mice which are genetically different. How can he successfully perform a skin graft on a mouse of one strain using skin from a mouse of the other strain?

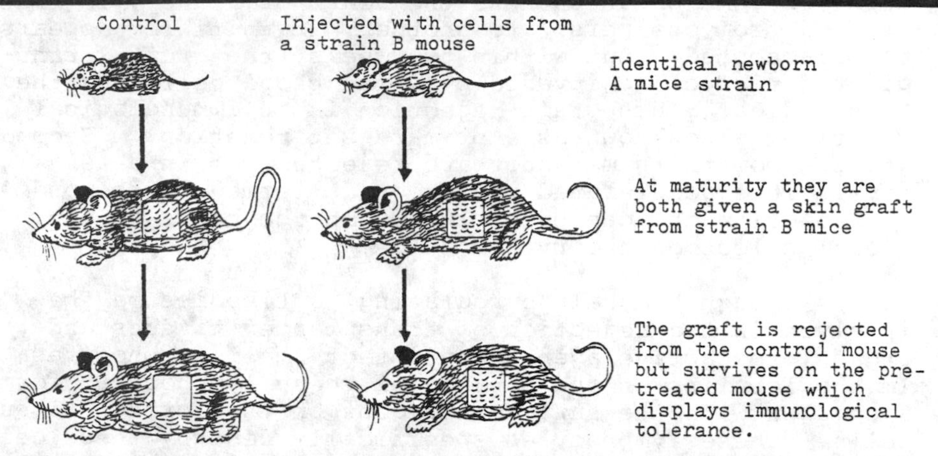

Experimental demonstration of immunological tolerance. Early exposure to foreign histocompatibility antigens enables a mouse to later accept skin grafts bearing these specific antigens.

Solution: The answer to this question will show how immunological capabilities develop. Since the two strains are genetically different, they should have very few histocompatibility genes in common. Therefore any grafting attempt should be unsuccessful, since the recipient mouse's cellular immunity would respond to the antigens on the donor's tissue cells. The solution to the problem is based on an observation made during the 1940's. It was discovered that non-identical twin calves each had the antigens normal to his own red blood cells and also the antigens of his twin's red blood cells. (The antigens on red blood cells determine the blood type.) Each calf, moreover, was not destroying the antigens of its twin. It was shown that the circulatory systems of the twins had been interconnected during their early embryonic development. Apparently, by being exposed to the other twin's antigens early in the calf's fetal development, it later recognizes these antigens as self. The immunological mechanisms had not developed at the time the antigens were first shared. This lack of immunological responsiveness to antigens present when the antibody-forming system is being developed is called immunological tolerance.

We can thus perform successful grafting experiments on the mice by exposing a mouse of one strain to the antigens of a mouse from the other strain. If an early embryo were injected with cells from a mouse of the other strain, the newborn mouse would accept grafts from that same mouse. Because the cells were injected before development of the antibody-forming system (which can last for several weeks after birth), the adult mouse will not form antibodies against the antigen of those cells.

According to one view, if developing lymphocytes

encounter their corresponding antigen at some critical stage in their development, they fail to mature and die. This not only explains why the mouse does not have antibodies against the antigens of the donor's tissue, but also why the mouse does not have antibodies complementary to its own antigens. Any lymphocyte which would recognize "self" antigens would be destroyed during fetal development. Lymphocytes would mature normally if their antigen were absent from the system. These mature lymphocytes are then able to recognize nonself, or foreign antigens.

Another method of increasing acceptance of tissue grafts is to remove the thymus from the newborn mice. The thymus is the organ where T lymphocytes, responsible for cellular immunity involved in transplants, become differentiated. These so-called "nude" mice, which lack thymuses, are incapable of producing T lymphocytes. Although they die within a few months, many of them can accept tissue grafts from different strains of mice, and even rats.

BLOOD TYPES

• PROBLEM 14-27

John is injured and badly needs a blood transfusion. He is type B. His girlfriend Jane volunteers, but she has type O blood. The physician, in view of the urgency of the situation, uses Jane's blood anyway. A few months later, Jane needs a transfusion. John volunteers but the physician turns him down. Explain.

Solution: Before we explain this particular problem, it would be worthwhile to discuss the A-B-O blood type system. There are four types of blood determined genetically: A, B, AB and O. The basis of these blood types is the presence or absence of certain antigens, called agglutinogens, on the surface of the erythrocytes. The erythrocytes of type A blood carry agglutinogen A; those of type B blood carry agglutinogen B. Type O blood has neither of these agglutinogens, while type AB has both A and B agglutinogens. These agglutinogens react with certain antibodies, called agglutinins, that may be present in the plasma. An agglutinin-agglutinogen reaction causes the cells to adhere to each other. This clumping, or agglutination, would then block the small blood vessels in the body, causing death.

Agglutination is simply the clumping of erythrocytes by agglutinins. The blood clotting mechanism, or coagulation is not involved. A Type A person has the A antigen (agglutinogen) and the anti-B antibody (agglutinin). A Type B person has the B antigen and anti-A antibody. A Type O person has no antigens but both antibodies. Conversely, a Type AB person has both anitgens but no antibodies. Basically, a person does not have the agglutinin in his plasma which would clump his own red blood cells. A summary of the antibodies and antigens that each blood type contains is outlined in the following table:

Blood Type	Agglutinogens (antigens)	Agglutinins (antibodies)
A	A	anti-B
B	B	anti-A
AB	A and B	none
O	none	anti-A and anti-B

Blood typing is a critical factor in blood transfusions. Since the agglutinins in the plasma of one type will react with the agglutinogens on the erythrocytes of the other types, transfusions are usually between people with the same blood type. However, one may use another blood type provided that the transfusion is small and the plasma of the recipient and the erythrocytes of the donor are compatible (does not cause a reaction). This type of transfusion was performed in the given case. Type O blood from Jane was given to John, a blood-type B patient. The anti-A agglutinin in John's plasma finds no agglutinogens to attack on Jane's erythrocytes. We can usually ignore the agglunitins, specifically anti-B, in Jane's plasma, which would normally react with the agglutinogen B on John's erythrocytes. This is because Jane's plasma is sufficiently diluted during transfusion so that the concentration of agglutinins is too low to cause appreciable agglutination. Jane's blood, type O, can be given to anyone since her erythrocytes contain no agglutinogens. Type O blood is therefore called the universal donor. Although type O blood can be donated to any type, it can only accept type O since it has the agglutinins, anti-A and anti-B, which react with the agglutinogens of every other blood type.

In contrast, type AB blood can receive from any type since it has no agglutinins in its plasma to react with the donor's erythrocytes. Type AB blood is therefore called the universal recipient. But it can only donate to other AB patients since its agglutinogens, A and B, react with the plasma of every other type.

The reason John's offer was turned down is that his blood, type B, would agglutinate with Jane's blood, type O. The agglutinins in Jane's plasma, specifically anti-B, would react with the B agglutinogen on John's erythrocytes and cause agglutination.

A woman who is Rh-negative marries a man who is Rh-positive. What possible problems could arise if the couple wants more than one child?

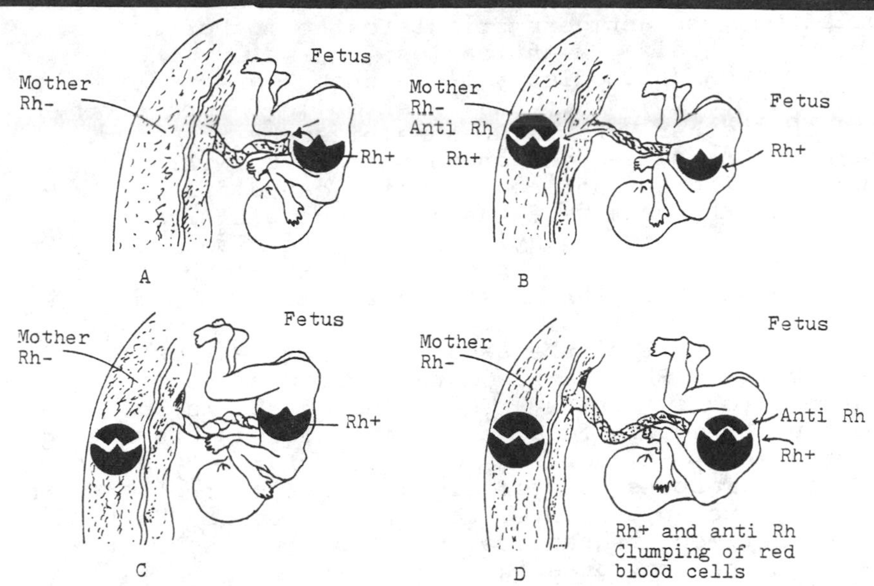

Diagram of the sequence of events leading to erythroblastosis fetalis, a condition in which the red cells of the fetus clump within the uterus.
(A) Red cells pass from an Rh+ fetus to its Rh- mother through some defect in the placenta.
(B) Maternal white cells produce anti Rh+ antibodies.
(C) During a subsequent pregnancy anti-Rh+ antibodies pass from the maternal blood to the fetal bloodstream.
(D) The reaction of Rh+ cells with anti-Rh+ causes clumping of the fetal red cells.

Solution: Besides the A and B agglutinogens, there is another type of blood antigen, designated the Rh factor. There are at least eight different types of Rh antigens which may be present on human erythrocytes. People are classified into two groups with respect to the Rh antigen: Rh-positive individuals have an Rh antigen, while Rh-negative individuals have such a weakly antigenic factor that it can be disregarded.

The A-B-O blood type system is an example of a rare case in which the body normally synthesizes antibodies in large amounts against an antigen to which it has not yet been exposed. The Rh system follows the rule rather than the exception. An individual's plasma will not contain any antibodies against the Rh antigen unless it has been sensitized to the antigen by previous exposure. Thus, while Rh-positive individuals have the Rh antigen, Rh-negative individuals have neither an effective antigen nor an antibody against the antigen. If Rh-positive blood is given to an Rh-negative individual, the Rh-negative individual will gradually synthesize antibodies against the Rh antigen introduced. Agglutination of the cells occurs, plugging up the small blood vessels, inactivating the erythrocytes as oxygen carriers, and eventually leading to death of the recipient.

In our particular problem, the couple's first child might inherit the Rh-negative characteristic. There is thus no particular problem since the mother is also Rh-negative. But if the fetus is Rh-positive, trouble may occur. If some blood from the Rh-positive fetus enters the maternal circulation, its antigens stimulate the mother's plasma cells to synthesize Rh antibodies. Normally, the mother cannot produce antibodies rapidly enough and in large enough amounts to affect the first child before birth. But if the mother, who is Rh-sensitized by the first pregnancy, has subsequent pregnancies, she will have produced sufficient antibodies which can diffuse into the blood of an Rh-positive fetus and react with its erythrocytes. This agglutination of the blood of the fetus with the antibodies produced in its mother is a disease known as erythroblastosis fetalis.

Only about 3% of the second or later Rh-positive fetuses born to Rh-negative mothers show signs of erythroblastosis fetalis. The blood of the mother may not diffuse into the fetal circulation, thus preventing agglutination. In addition, only one of the types of Rh antigens is responsible for erythroblastosis; other forms will not cause any trouble. Also, the longer the time interval between pregnancies, the less intense is the anti-Rh immunal response in the mother.

One method of preventing erythroblastosis is to inject Rh antibodies into a Rh-negative mother immediately after the birth of her first Rh-positive infant. The antibodies kill any fetal red blood cells that have entered her circulation and thus prevent stimulation of maternal antibody production. The injected antibodies eventually are destroyed and the mother is now unsensitized and able to bear a second Rh-positive infant.

SHORT ANSWER QUESTIONS FOR REVIEW

Answer

To be covered when testing yourself

Choose the correct answer.

1. Small animals and plants which live in an aquatic environment lack transport systems because (a) they evolved from ancestors that originated in dry habitats (b) they are so small, they have little need for the nutrients contained within water. (c) they can meet their needs of nutrition and waste disposal by diffusion. (d)all of the above (e) none of the above — c

2. Which is not a function of the blood? (a) transport of nutrients and oxygen to the cells (b) removal of carbon dioxide and other wastes (c) transport of hormones to target organs (d) regulation of acidity and alkalinity of the cells (e) all of the above are functions of the blood. — e

3. The blood plasma contains a number of inorganic cations and anions. If NaCl increases in concentration in the plasma, thus increasing the number of ions, what will be the result? (a) blood pressure will decrease (b) blood pressure will increase (c) blood pressure will remain unchanged. — b

4. If 3000 units of dye were injected into the bloodstream, and we determined that there were 20 units of dye in a 40 ml sample of plasma after injection, what is the total volume of plasma? (a) 2000 ml, (b) 4000 ml, (c) 6000 ml, (d) 8000 ml, (e) 10,000 ml. — c

5. Hematocrit refers to (a) the cellular component of blood (b) small blood sucking organisms (c) the leukocytes and erythrocytes found in the blood (d) a dye used to stain the blood plasma (e) none of the above. — a

6. During the clotting process, fibrinogen is converted into (a) thromboplastin (b) fibrin (c) platelets (d) prothrombin (e) thrombin — b

7. Which of the following proteins is required for the fibrinogen to fibrin conversion? (a) thromboplastin (b) fibrin (c) platelets (d) prothrombin (e) thrombin — e

8. Which of the following vitamin dificiencies depresses the levels of prothrombin? (a) vitamin A (b) vitamin B (c) vitamin C (d) vitamin K (e) vitamin D — d

Answer

To be covered when testing yourself

9. Which of the following plasma proteins are important in regulating the water content of the cells and body fluid? (a) albumins (b) globulins (c) gamma globulins (d) a + b (e) b + c — a

10. The hormone thyroxin is transported in the blood plasma (a) as a compound in simple solution (b) attached to the hemoglobin molecule (c) complexed globulin carriers (d) all of the above (e) none of the above. — c

11. Which correctly describes the binding of oxygen and hemoglobin? (a) the oxyhemoglobin molecule is formed when oxygen combines reversibly with the peptide portion of the hemoglobin molecule. (b) the oxyhemoglobin molecule is formed when oxygen combines irreversibly with the peptide portion of the hemoglobin molecule (c) the oxyhemoglobin molecule is formed when oxygen combines reversibly with the iron atoms of the hemoglobin molecule (d) the oxyhemoglobin molecule is formed when oxygen combines irreversibly with the iron atoms of the hemoglobin molecule (e) none of the above. — c

12. Hemoglobin is composed of (a) five peptide chains (2α, 3β). (b) six peptide chains (3α, 3β). (c) four peptide chains (3α, 1β). (d) four peptide chains (2α, 2β). (e) none of the above — d

13. The combination of oxygen with hemoglobin and its release from oxyhemoglobin are controlled by (a) carbon dioxide concentration (b) oxygen concentration (c) carbonic acid concentration (d) a + b (e) b + c — d

14. Under aqueous conditions carbon dioxide forms carbonic acid. What is the result of an increase of carbonic acid in the blood? (a) hemoglobin can bind less oxygen due to the increased acidity caused by carbonic acid ionization (b) hemoglobin can bind more oxygen due to the increased acidity caused by carbonic acid (c) hemoglobin binds less oxygen since carbonic acid competes with the former for points of attachment on the iron atoms (d) more than one of the above (e) none of the above. — a

15. Red blood cells originate in the (a) bone marrow (b) spleen (c) kidneys (d) liver (e) lungs — a

Fill in the blanks.

16. Severe loss of red blood cells by hemorrhage or disease will inevitably result in _____. — anemia

	Answer To be covered when testing yourself
17. A severe and formerly fatal disease is _____ characterized by red cells that are immature, very fragile and decreased in number	perni-cious anemia
18. An increase in the number of circulating red cells wherein the number may reach 11 to 15 million per cubic millimeter is called _____.	poly-cythemia
19. As a result of transfusing blood incorrectly, clumping may occur. This is referred to as _____.	agglu-tination
20. The universal blood donor is type _____, and the universal recipient is type _____.	O, AB
21. An Rh negative mother bearing her second Rh positive fetus may cause clumping of the fetal red blood cells. This disease is known as _____.	erythro-blastosis fetalis
22. The primary agent of protection against bacterial infection and antigen penetration is the _____.	skin
23. Some of the macrophages wander around in the tissue spaces, others are fixed in one place. Together the two constitute the _____ system of phagocytes.	reticulo-endothe-lial
24. The most severe reaction to an allergy, commonly associated with insect bites, is _____.	ana-phylaxis
25. Tissue transplanted from one part of the body to another in the same organism is termed a(n) _____.	autograph
Determine whether the following statements are true or false.	
26. Carbon monoxide has a greater affinity for hemoglobin than oxygen.	True
27. The amount of oxygen combining with hemo-globin is porportional to the partial pressure of O_2.	True
28. Cells containing fetal hemoglobin can take up oxygen at lower oxygen tensions than adult cells.	True
29. Birds, reptiles, amphibians and fishes have enucleated red cells containing hemoglobin.	False
30. The heme portion of the hemoglobin molecule undergoes chemical degradation and is excreted by the spleen as bile pigments.	False

	Answer
	To be covered when testing yourself
31. The constancy of the number of red cells provides us with an example of static equilibrium.	False
32. Under certain conditions the kidney and liver can provide the stimulus for red blood cell formation.	True
33. Monocytes and lymphocytes are capable of fighting against bacterial infection by phagocytosis.	True
34. The compound heparin can be classified as an anti-coagulant of the blood.	True
35. A detached thrombus which clots a blood vessel is termed an embolus.	True

CHAPTER 15

TRANSPORT SYSTEMS

NUTRIENT EXCHANGE

• PROBLEM 15-1

Explain why the capillaries, rather than the veins or arteries, are the only vessels where exchange of nutrients, gases, and wastes can take place.

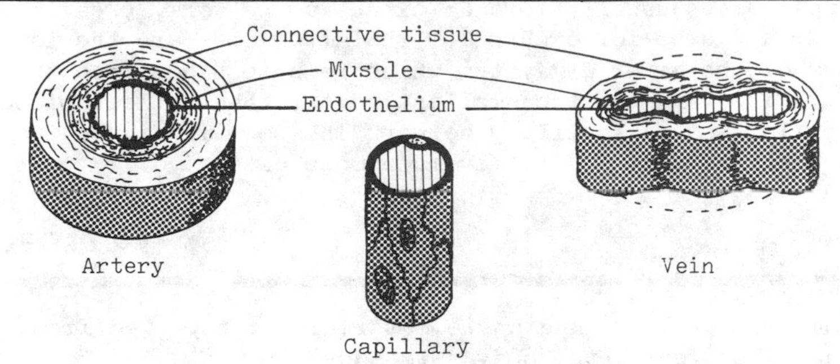

Figure 1. Walls of the artery, vein, and capillary compared. Arteries and veins have the same three layers in their walls, but the walls of the veins are much less rigid and they readily change shape when muscles press against them.

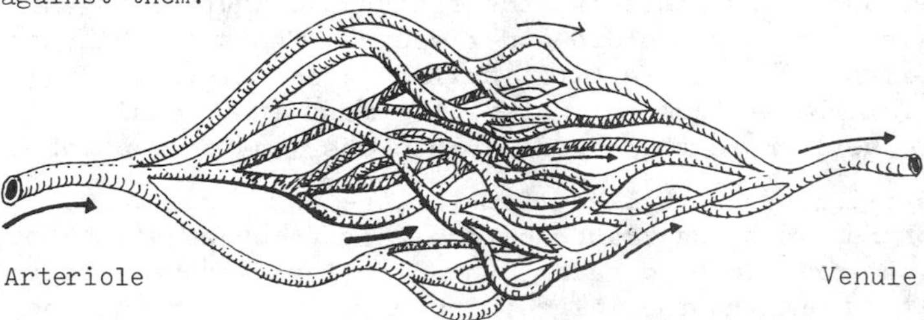

Figure 2. A capillary network.

Solution: The circulatory system of humans and other vertebrates consists basically of a heart and three types of blood vessels: arteries, veins and capillaries.

An artery is a blood vessel that carries blood away from the heart. The walls of the arteries are the thickest of all the blood vessels. They are composed of three layers: (1) an outer layer of connective tissue, containing elastic fibers which give the arteries their characteristic elasticity; (2) a middle layer of smooth muscle which can regulate the size of the arterial lumen (opening); and (3) an inner layer of connective tissue lined with endothelial cells. The largest artery, the aorta, has a wall about three millimeters thick.

A vein is a blood vessel that carries blood back toward the heart. The walls of veins are thinner, and the lumens larger than those of the arteries. Although the same three layers are present, the outer layer has fewer elastic fibers and the middle layer of smooth muscle is much thinner. The walls of veins are therefore much less rigid and easily change shape when muscles press against them. Unlike arteries, veins have valves which prevent the backflow of blood and thereby maintain the direction of flow to the heart.

Capillaries are the blood vessels that connect the arteries with the veins. The capillaries have walls composed of endothelium only one cell thick. It is the thinness of the capillary walls which allows for the diffusion of oxygen and nutrients (also hormones) from the blood into the tissues, and for carbon dioxide and nitrogenous wastes to be picked up from the tissues by the blood. The walls of arteries and veins are thick and impermeable, and are thus unsuitable for this exchange process. Much of the exchange, however, is not across the endothelial cells of the capillary but rather between the cells through pores. The exchanged substances do not enter the tissue cells directly but first enter the interstitial fluid which lies in the spaces between the cells. They then diffuse through the plasma membranes of the neighboring cells.

Capillaries usually form an extensive network, creating a large surface area for exchange. The arterioles are the last small branches of the arterial system which release blood into the capillaries . The venules, the smallest veins, then collect blood from the capillaries. A capillary network is shown in Figure 2. (The amount of branching is actually much more extensive).

• PROBLEM 15-2

How are nutrients and waste materials transported from the circulatory system to the body's tissues?

Solution: It is in the capillaries that the most important function of circulation occurs; that is, the exchange of nutrients and waste materials between the blood and tissues. There are billions of capillaries, providing a total surface area of over 100 square meters for the exchange of material. Most functional cells in the body are never more than about 25 microns away from a capillary.

Capillaries are well-suited for the exchange process. They are both numerous and branch extensively throughout the body. Unlike the arteries and veins, the capillaries have a very thin wall; it is one endothelial cell thick. This thin wall permits rapid diffusion of substances through the capillary membrane. The extensive branching increases the total cross-sectional area of the capillary system. This serves to slow down the flow of blood in the capillaries, allowing more time for the exchange process to occur. The very small diameter of each of the capillaries (which is not much larger than the diameter of the blood cells which pass through them) provides friction, which increases the resistance to flow. This causes a significant drop in blood pressure in the capillaries, which is important in the filtration of fluid between the capillaries and the interstitial fluid (lying in the space between cells).

The most important means by which nutrients and wastes are transferred between the plasma and interstitial fluid is by diffusion. Material must first diffuse into the interstitial fluid before it can enter the cells of the body's tissues. As the blood flows slowly through the capillary, large amounts of water molecules and dissolved substances diffuse across the capillary wall. The net diffusion of substances is proportional to the concentration difference between the two sides of the capillary membrane. For example, the concentration of oxygen in the blood is usually greater than that in the interstitial fluids. Therefore, oxygen diffuses into the interstitial fluid and then into the tissues. In the same way, if the concentration of nitrogenous wastes is higher in the tissues than in the blood, they will diffuse into the interstitial fluids and then into the blood.

Actual diffusion across the capillary wall occurs in three ways. Water-soluble substances (sodium ions, glucose) diffuse between the plasma and interstitial fluid only through the pores in the capillary membrane. The diameter of the pores (8 - 10 Å) allows small molecules such as water, urea, and glucose to pass through, but not larger molecules such as plasma proteins. Lipid-soluble substances diffuse directly through the cell membranes and do not go through the pores (which are filled with water). Oxygen and carbon dioxide permeate all areas of the capillary membranes. Another method by which substances can be transported through the membrane is pinocytosis. In pinocytosis, the cell membrane invaginates and sequesters the surrounding substances. Small vesicles are produced which migrate from one side of the endothelial cell to the other, where the contents are released. Pinocytosis accounts for the transport of larger substances, such as plasma proteins and glycoproteins.

PROPERTIES OF THE HEART

• PROBLEM 15-3

Trace the path of blood through the human heart.

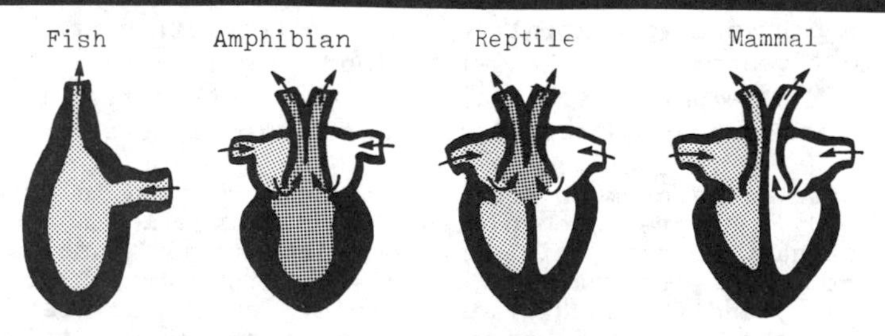

The hearts of four classes of vertebrates. From fish to mammal, there is increasing separation between the two sides of the heart, with consequent decrease in the amount of mixing between oxygenated and deoxygenated blood.

Solution: The heart is the muscular organ that causes the blood to circulate in the body. The heart of birds and mammals is a pulsatile four-chambered pump composed of an upper left and right atrium (pl., atria) and a lower left and right ventricle. The atria function mainly as entryways to the ventricles, whereas the ventricles supply the main force that propels blood to the lungs and throughout the body.

Depending on where the blood is flowing from, it would enter the heart via one of the two veins: the superior vena cava carries blood from the head, neck and arms; the inferior vena cava carries blood from the rest of the body. The blood from these two veins enters the right

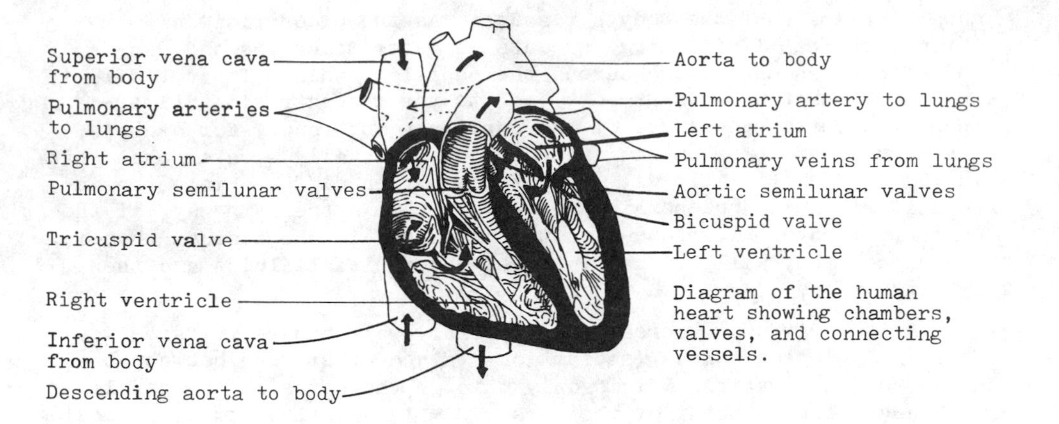

Diagram of the human heart showing chambers, valves, and connecting vessels.

atrium. When this chamber is filled with blood, the chamber contracts and forces the blood through a valve called the tricuspid valve and into the right ventricle. Since this blood has returned from its circulation in the body's tissues, it is deoxygenated and contains much carbon dioxide. It therefore must be transported to the lungs where gas exchange can take place. The right ventricle contracts, forces the blood through the pulmonary semilunar valve into the pulmonary artery. This artery is unlike most arteries in that it carries deoxygenated blood. The artery splits into two, with one branch leading to each lung. The pulmonary arteries further divide into many arterioles, which divide even further and connect with dense capillary networks surrounding the alveoli in the lungs. The alveoli are small sac-like cavities where gas exchange occurs. Carbon dioxide diffuses into the alveoli, where it is expelled, while oxygen is picked up by the hemoglobin of the erythrocytes. The capillaries join to form small venules which further combine to form the four pulmonary veins leading back to the heart. The pulmonary veins are unlike most veins in that they carry oxygenated blood. These veins empty into the left atrium, which contracts to force the blood through the bicuspid (or mitral) valve into the left ventricle. When the left ventricle, filled with blood, contracts, the blood is forced through the aortic semilunar valve into the aorta, the largest artery in the body (about 25 millimeters in diameter).

The aorta forms an arch and runs posteriorly and inferiorly along the body. Before it completes the arch, the aorta branches into the coronary artery, which carries blood to the muscular walls of the heart itself, the carotid arteries, which carry blood to the head and brain, and the subclavian arteries, which carries blood to the arms. As the aorta runs posteriorly, it branches into arteries which lead various organs such as the liver, kidney, intestines and spleen and also the legs.

The arteries divide into arterioles which further divide and become capillaries. It is here that the oxygen and nutrients diffuse into the tissues and carbon dioxide and nitrogenous wastes are picked up. The capillaries fuse to form venules which further fuse to become either the superior or inferior vena cava. The entire cycle starts once again.

The part of the circulatory system in which deoxygenated blood is pumped to the lungs and oxygenated blood returned to the heart is called the pulmonary circulation. The part in which oxygenated blood is pumped to all parts of the body by the arteries and deoxygenated blood is returned to the heart by the veins is called the systemic circulation.

• PROBLEM 15-4

Explain why a four-chambered heart is more efficient than a three-chambered heart. Why does an animal with a two-chambered heart not experience the problem that an animal with a three-chambered heart experiences?

Solution: A four-chambered heart is characteristic of "warm-blooded" animals such as birds and mammals. Since these animals maintain a relatively high constant body temperature, they must have a fairly high metabolic rate. To accomplish this, much oxygen must be continually provided to the body's tissues.

A four-chambered heart helps to maximize this oxygen transport by keeping the oxygenated blood completely separate from the deoxygenated blood. The right side of the heart, which carries the deoxygenated blood, is separated by a muscular wall, called the septum, from the left side of the heart, which carries the oxygenated blood. In amphibians, the atria are divided into two separate chambers, but a single ventricle exists. This three-chambered heart permits oxygen-rich blood returning from the pulmonary circulation to mix with oxygen-poor blood returning from the systemic circulation. It is less efficient than a four-chambered heart because the blood flowing to the tissues is not as oxygen-rich as it could be. Fortunately, amphibians, being cold-blooded, do not have to maintain a constant body temperature and hence do not need the efficiency of warm-blooded hearts. Reptiles also have three-chambered hearts but partial division of the ventricle in these animals has decreased the amount of mixing.

Fish, which possess a two-chambered heart, (one atrium and one ventricle) do not have this problem of mixing oxygenated and deoxygenated blood. The blood of fishes is oxygenated in the capillary beds of the gills. This oxygenated blood does not go back to the heart but goes directly into the body circulation. When the blood returns from the body, it drains into the heart, which simply pumps this deoxygenated blood to the gills. No mixing occurs in the two-chambered heart because only deoxygenated blood is passed through the heart.

• PROBLEM 15-5

The heart is removed from the body and placed in an isosmotic solution. Although it is completely separated from nerves, it continues to beat. Explain.

Solution: The initiation of the heartbeat and the beat itself are intrinsic properties of the heart and are not dependent upon stimulation from the central nervous system. The heart is stimulated by two sets of nerves (the sympathetic and vagus nerves), but these only partly regulate the rate of the beat, and are not responsible for the beat itself.

The initiation of the heartbeat originates from a small strip of specialized muscle in the wall of the right atrium called the sino-atrial (S-A) node, which is also called the pacemaker of the heart. It is the S-A node which generates the rhythmic self-excitatory impulse, causing a wave of contraction across the walls of the atria. (See figure). This

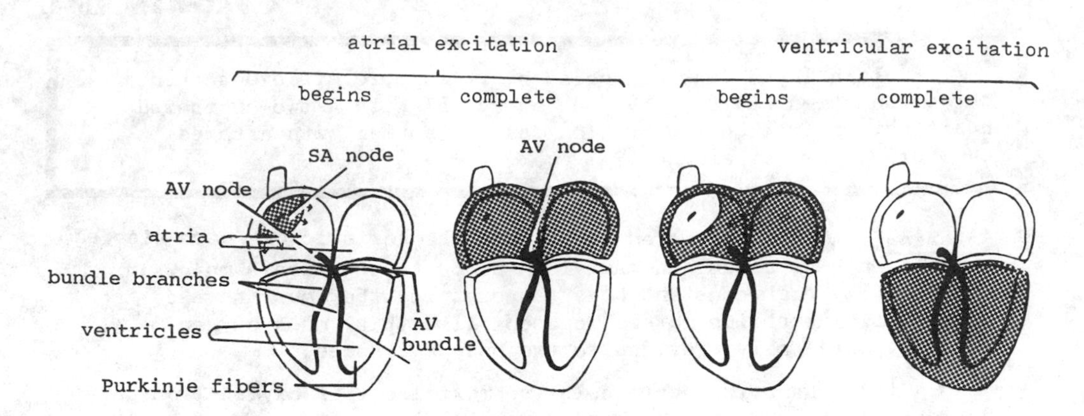

Sequence of cardiac excitation.

wave of contraction reaches a second mass of nodal muscle called the atrio-ventricular node, or A-V node. The atrio-ventricular node is found in the lower part of the interatrial septum. As the wave of contraction reaches the A-V node, the wave stimulates the node, which produces an excitatory impulse. This impulse is rapidly transmitted to a bundle of nodal fibers called the A-V bundle or bundle of His. This bundle divides into right and left bundle bunches which ramify to the ventricular myocarium (the muscular layer of the heart wall) via the Purkinje fibers. The impulse is then transmitted to all parts of the ventricles causing them to contract as a unit.

As the cardiac impulse travels from the S-A node through the atria, the wave of contraction forces blood through the valves and into the ventricles. (In the atria, the impulse is transmitted from cell to cell across the intercalated discs). It is a "wave" of contraction because parts of the atria farthest from the S-A node contract later than do the parts closest to the node. The parts of the atria that contract first are also the first to relax, thus causing a wavelike motion upon contraction. Once the impulse reaches the A-V node, it is delayed (by a fraction of a second) before it is passed on to the ventricles. This transmission delay allows time for the atria to pump the blood into the ventricles before ventricular contraction begins. Unlike the atria, the ventricles contract as a unit. This is due to the rapid transmission of the impulse by the Purkinje fibers. For example, in the atria, the impulse travels at a velocity of .3 meter per second while in the Purkinje fibers, the impulse travels at about 2 meters per second. This allows almost immediate transmission of the impulse throughout the ventricular system.

The adaptive significance of the contraction of the ventricles as a unit rather than as a wave is clear. The ventricles must exert as great a pressure as possible in order to force the blood throughout a long system of arteries, capillaries and veins. A greater pressure is more easily obtained if the ventricles contract as a unit, rather than in waves.

• PROBLEM 15-6

When the atria contract, only 30% of the ventricular space becomes filled. How does the additional 70% become filled? Describe the process of the cardiac cycle in general.

Solution: The cardiac cycle is the period from the end of one contraction to the end of the next contraction. Each heartbeat consists of a period of contraction, or systole, followed by a period of relaxation, or diastole. Since the normal heart rate is 70 beats per minute, each complete beat (systole and diastole) lasts .85 second. The atria and the ventricles do not contract simultaneously. The atria contract first, with atrial systole lasting .15 seconds. The ventricles then contract, with ventricular systole lasting .30 second. Diastole therefore lasts .40 second.

The cardiac cycle begins with atrial systole. The wavelike contraction of the atria is stimulated by the impulse of the sino-atrial or S-A node. The contraction forces into the ventricles only 30% of the blood that will fill them. How the other 70% enters the ventricles will be explained shortly. During contraction of the two atria, the tricuspid and bicuspid valves are opened. There is a very brief pause before ventricular systole begins due to the delay in transmission of the cardiac impulse at the atrio-ventricular or A-V node. The impulse is then transmitted to the A-V bundle and bundle branches and then very rapidly to the Purkinje fibers, which stimulate the ventricles to contract as a unit. The contraction of the ventricular muscles causes a rapid increase in pressure in the ventricles. This increased ventricular pressure immediately closes the tricuspid and bicuspid valves, preventing backflow of the blood into the atria. It is important to realize that the valves work passively. There is no nervous stimulation which directly regulates their opening and closing; it is only the difference in pressure due to the relative amount of blood in the atria and ventricles (or ventricles and arteries) which controls this. As ventricular systole progresses, the ventricular pressure further increases (due to contraction). At this point, no blood is entering the ventricles, since the tri- and bicuspid valves are closed, and no blood is leaving, since the semilunar valves are closed. When the ventricular pressure becomes greater than the pressure in the arteries (the pulmonary artery and the aorta), the semilunar valves open and blood is forced into these vessels. After the ventricles complete their contraction, ventricular diastole begins. As the ventricles relax, the pressure decreases. When the ventricular pressure is less than arterial pressure, the semilunar valves shut, preventing backflow. Since the pressure in the ventricles is still higher than in the atria, the tri- and bicuspid valves are still closed, and no blood enters or leaves the ventricles. But some blood is entering the now relaxed atria. During further relaxation of the ventricles, the ventricular pressure continues to decrease until it falls below the atrial pressure. At this point, the tri- and bicuspid valves open, allowing blood to rapidly flow into the ventricles. The atria do not contract to force this blood into the ventricles. It is a passive flow due to the fact that the atrial pressure is greater than the ventricular pressure. This flow allows for 70% of the ventricular filling before atrial systole. Thus, the major amount of ventricular filling occurs during diastole, not atrial systole, as one might expect.

• PROBLEM 15-7

What causes the characteristic sounds of a beating heart?

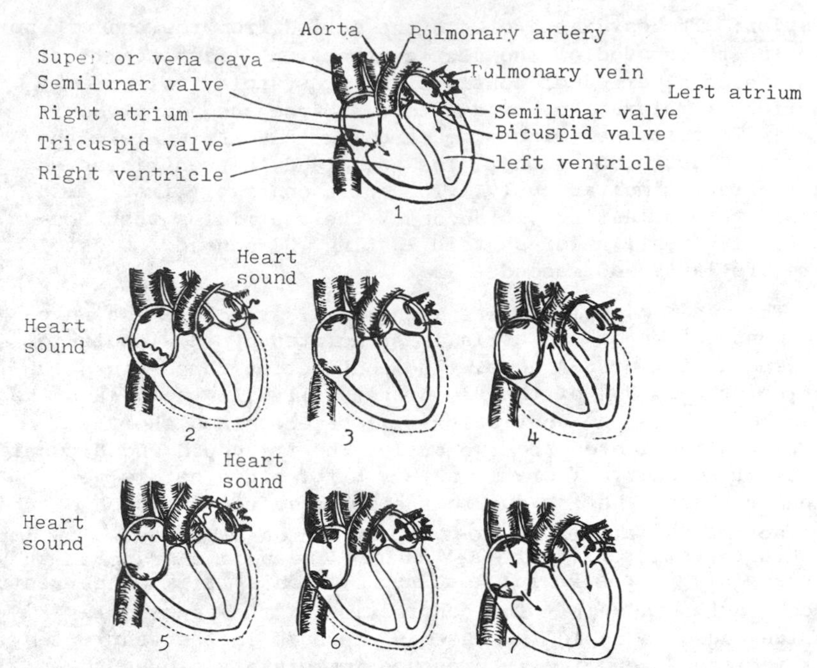

The heart cycle. Arrows indicate the direction of the flow of blood; dotted lines indicate the change in size as contraction occures.

Solution: The heart produces two sounds per heartbeat which can be heard by using a stethoscope, which is an instrument which magnifies sounds and transmits them to the ear. The sound is sometimes described as "lub, dub, lub, dub..." The first sound, lub, is low-pitched and of long duration. It is associated with the closure of the atrio-ventricular valves (the tricuspid and bicuspid) at the beginning of ventricular systole. The second sound, dub, is higher-pitched, louder and shorter in duration. It is associated with the closure of the semilunar valves at the beginning of diastole.

In the heart, the valves are cushioned by the blood, and it is difficult to understand why the valves create sound. It is not due to the slapping together of the "flaps" of the valves. The cause is believed to be the vibrations of the walls of the heart and the vessels around the heart. For example, contraction of the ventricles causes blood to flow against the A-V valves, closing them. The elasticity of the valves causes the blood to surge backwards and bounce back toward the ventricles. This sets the ventricular walls into vibrations, which are carried to the chest wall (where it contacts the heart), creating the sound waves audible in a stethoscope. A very small portion of the first sound is also caused by the contraction of the heart muscle itself. As any muscle contracts, a small amount of noise can be heard as the internal filaments rearrange.

The second heart sound is caused by the vibrations of the walls of the aorta, the pulmonary artery, and the ventricles. The closing of the semilunar valves causes the blood in the arteries to reverberate between the arterial walls and valves. When these vibrations of the walls come into contact with the chest wall, an audible sound is created.

When the heartbeat sounds normal, it is an indication that the valves are functioning properly. When the valves are damaged, a hissing or sloshing sound can be heard. This is caused by the blood leaking back through the damaged valve. This condition is known as a heart murmur. It is a common result of rheumatic fever, syphilis, and other diseases.

• PROBLEM 15-8

During exercise, certain changes occur within the body of the active animal. Explain in what ways these changes affect the activity of the heart.

Solution: During exercise, the tissues become increasingly metabolically active and require much more oxygen and nutrients than are needed when the tissues are at rest. The heart is stimulated by this increased activity and adapts to it by increasing cardiac output- the volume of blood pumped per minute. Cardiac output is the product of stroke volume (blood volume pumped per beat) and heart rate. Since stroke volume and heart rate increase with exercise, the cardiac output increases:

$$\text{Cardiac output} = \text{Stroke volume} \times \text{heart rate}$$
$$\text{ml/min} = \text{ml/beat} \times \text{beat/min}$$

The heart normally pumps about 75 milliliters of blood per beat, but can pump as much as 200 mls during strenuous exercise. The normal heart rate of 70 beats per minute can increase to as much as 200 beats per minute during exercise.

One change that occurs during exercise is an increase in the carbon dioxide content of the blood. As the tissues become more metabolically active, more ATP is made with a concomitant increase in carbon dioxide as a waste product. The CO_2 diffuses from the tissues into the bloodstream. This blood stimulates certain receptors in the vessels, which then stimulate the heart to increase its volume per beat.

Increased muscle activity also causes other changes. The movement of the muscle during exercise exerts a greater than normal pressure on the veins, causing more blood to flow into the heart. The increased blood volume stretches the heart muscle. Since the contractile power of any muscle is increased (up to a point) by the tension it experiences when it begins to contract, the increased blood volume causes a greater quantity of blood to be pumped per heartbeat.

The third change which occurs during exercise does not affect the blood volume per beat, but affects the heart rate. During exercise, excess heat is produced and raises the body temperature a few degrees. The extra heat causes the sino-atrial node to increase its rate of stimulation, thus increasing the number of beats per minute. It is thought that the increased temperature increases the permeability of the muscle membrane of the S-A node to various ions (Na^+ and K^+), thus accelerating the stimulatory process. (It is the change in membrane permeability to Na^+ ions which causes an impulse to arise). This temperature sensitivity of the S-A node explains the increased heart rate accompanying a fever.

The heart thus helps the tissues of the body acquire more oxygen and remove more waste by increasing both the volume of blood pumped per beat and the number of beats per minute. There are also autonomic nervous changes that occur during exercise that work to increase the cardiac output (see next question).

Summary of cardiovascular changes during exercies

(+ increase; - decrease)

skeletal muscle blood flow	+ 175%
systolic arterial pressure	+ 50%
diastolic arterial pressure	no change
stroke volume	+ 20%
heart rate	+ 100%
mean arterial pressure	+ 15%
cardiac output	+ 120%
blood flow to kidneys	- 30%
total peripheral resistance	- 50%

• PROBLEM 15-9

Simultaneous removal of all nerves to the heart causes the heart rate to increase to about 100 beats per minute. What does this tell us about the regulation of the heart rate? Hormones regulate many bodily functions. Do they influence heart rate?

Solution: The excitatory impulses of the sino-atrial node occur spontaneously without the presence of nervous stimulation or hormones. But the S-A node is always under the influence of both nerves and hormones.

Nervous control of the heart rate is found in the vasomotor center in the medulla oblongata of the brain. A large number of sympathetic nerve fibers run from this center to the sino-atrial node. Sympathetic nerve stimulation increases the heart rate. The parasympathetic nerve fibers which innervate the heart are contained in the vagus nerve, which also runs from the medulla to the S-A node. Stimulation of the vagus nerve has the opposite effect of decreasing the heart rate, and if strong enough, may stop the heart completely. Cutting the sympathetic nerves decreases the heart rate, while cutting the parasympathetic vagus nerve increases the heart rate. Cutting or removing all nerves from the S-A node causes the heart rate to increase from its normal 70 beats per minute to 100 b.p.m. This tells us two things: (1) that the heart rate is influenced by the nervous system; and (2) that the parasympathetic influence is stronger, since cutting both sympathetic and parasympathetic fibers increases the heart rate. This increased rate of 100 b.p.m. is the inherent autonomous discharge rate of the S-A node (i.e., without any external influencing factors).

There are sensory receptors in the walls of the right atrium and superior and inferior vena cava which are stimulated when these walls are stretched by an increased volume of blood. The stimulated receptors send impulses to the vasomotor center in the brain which stimulates the sympathetic nerves; this increases the heart rate and stroke volume to compensate for the increased volume of blood in these areas of the heart. In addition, there are sensory receptors in the walls of the aorta and carotid arteries. These receptors are also stimulated by distension of the vessel walls and send impulses to the vasomotor center. Here, however, the vagus nerve is stimulated and the heart rate decreases. The nervous system adjusts the heart rate quickly to the metabolic needs of the body (such as during exercise). It also provides a feedback control to prevent any excessive cardiac response. Sympathetic stimulation acts via the neurotransmitter norepinephrine, while parasympathetic stimulation acts via the neurotransmitter acetylcholine.

Besides nervous control, the heart rate is also under hormonal control. Epinephrine, a hormone secreted by the adrenal medulla, is produced in increased amounts during emergencies or "fight-or-flight" situations. Epinephrine causes vasoconstriction in some vascular beds of the body (smooth muscle of the digestive tract) and vasodilation in both skeletal and cardiac muscle. Overall, it causes an increase in blood pressure and an increase in heart rate. Thyroxine also increases the heart rate but it requires several hours to cause the acceleration. This hormone from the thyroid gland can only affect long-term responses of the heart, unlike the quick action of epinephrine.

• PROBLEM 15-10

The heart does not extract oxygen and nutrients from the blood within the atria and ventricles. How is the heart supplied with its metabolic needs?

Solution: The heart depends on its own blood supply for the extraction of necessary oxygen and nutrients. The blood vessels supplying the heart are known as the coronary vessels. The coronary artery originates from the aorta, just above the aortic valve, and leads to a branching network of small arteries, arterioles, capillaries, venules, and veins similar to those found in other organs. The rate of flow in the coronary artery depends primarily on the arterial blood pressure and the resistance offered by the coronary vessels. The arterioles in the heart can constrict or dilate, depending on the local metabolic requirements of the organ. There is little if any neural control. If the coronary vessels are blocked by fatty deposits, the heart muscle would become damaged because of decreased supply of nutrients and oxygen. If the block is very severe and presists for too long, death of heart muscle tissue may result; this condition is called heart attack. Low arterial pressure may also lead to a heart attack for the same reasons.

• PROBLEM 15-11

Compare cardiac muscle to skeletal and smooth muscle.

Solution: Cardiac muscle is the tissue of which the heart is composed. Cardiac muscle shows some characteristics of both skeletal and smooth muscle. Like skeletal muscle, it is striated; it has myofibrils composed of thick and thin myofilaments, and contains numerous nuclei per cell. The sliding filament mechanism of contraction is found in cardiac muscle. Cardiac muscle resembles smooth muscle in that it is innervated by the autonomic nervous system. The cells of cardiac muscle are very tightly compressed against each other and are so intricately interdigitated that previously no junctions were thought to exist between cells. They do, however, exist, and are visible under the light microscope as dark-colored discs, called intercalcalated discs. It is believed that these discs may help to transfer the electrical impulses generated by the S-A node between muscle cells due to their low resistance to the flow of current.

The metabolism of cardiac muscle is designed for endurance rather than speed or strength. A continuous supply of oxygen and ATP must be provided in order for the heart muscle to maintain its contractile machinery. Cardiac cells deprived of oxygen for as little as 30 seconds cease to contract, and heart failure ensues.

FACTORS AFFECTING BLOOD FLOW

• PROBLEM 15-12

What makes the blood flow throughout the body? In what part of the circulatory system does the blood flow most slowly? Why?

Solution: Blood flow is the quantity of blood that passes a given point in the circulation in a given period of time. It is usually expressed in milliliters or liters per minute. Blood flow through a blood vessel is determined by two factors: (1) the pressure difference that drives the blood through the vessel; and (2) the vascular resistance, or impedance to blood flow. These relations can be expressed mathematically:

$$\text{Flow} = \frac{\Delta \text{ Pressure}}{\text{Resistance}} .$$

As the pressure difference increases, flow increases; as the resistance increases, flow decreases. The pressure difference between the two ends of the vessel causes blood to flow from the high pressure end to the low pressure end. This pressure difference or gradient is the essential factor which causes blood to flow. Although the heart acts to pump the blood, it is the pressure difference (about 140 mm Hg in the aorta and near 0 in the vena cava at the right atrium) that is critical for blood flow. The overall blood flow in the circulation of an adult is about 5 liters per minute. This is called the cardiac output because it is the amount pumped by each ventricle of the heart per unit time. The cardiac output is determined by multiplying the heart rate and the volume of blood pumped by each ventricle during each beat (stroke volume): cardiac output = heart rate x stroke volume. Normally the heart rate is about 70 beats per minute and the stroke volume is about 70 milliliters per beat:

$$CO = 70 \frac{\text{beats}}{\text{min}} \times 70 \frac{\text{mls}}{\text{beat}} = 4900 \frac{\text{mls}}{\text{min}} .$$

The normal cardiac output is thus about 5 liters per minute.

The rate of flow throughout the circulation is not constant. The rate is rapid in the arteries (about 30 centimeters per second) but falls as it moves through the arterioles. The rate is slowest in the capillaries but then increases again in the venules and veins. There is a physical explanation for this phenomenon. The velocity of blood flow in each type of blood vessel is inversely proportional to its total cross-sectional area. If any fluid passes from one tube to another of larger radius, the rate of flow is less in the larger tube. The rate is fastest in the aorta since it has the smallest total cross-sectional area of any vessel type (2.5 cm^2) The rate decreases in the arteries (20 cm^2) and decreases further in the arterioles (40 cm^2). The rate of flow decreases drastically in the capillaries which have a total cross-sectional area of 2500 cm^2. Although the individual branches of the capillaries have a diameter much less than the aorta, their total cross-sectional area is much greater; therefore the rate in the capillaries is 1/1000 the rate in the aorta. The rate of flow increases in the venules (250 cm^2) and increases further in the veins (80 cm^2). Finally, the rate increases rapidly in the vena cava where the cross-sectional area is 8 cm^2 .

• **PROBLEM 15-13**

A person suffers a laceration on his arm. How does one know whether an artery or a vein has been severed?

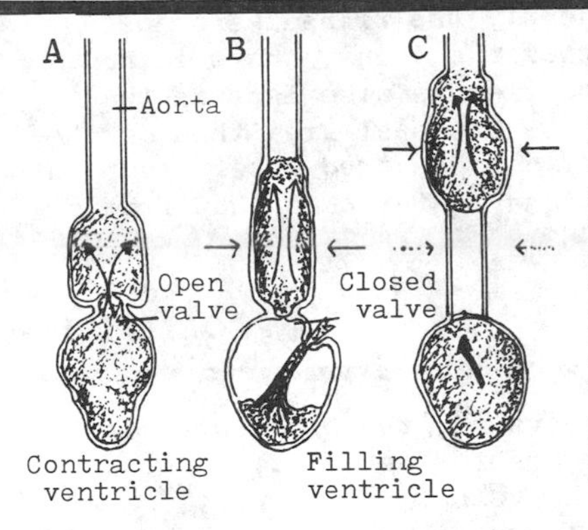

The movement of blood from the ventricle through the elastic arteries. For simplicity only one ventricle and artery are shown and the amount of stretching of the arterial wall is exaggerated. (A) as a ventricle contracts, blood is forced through semilunar valves and the adjacent wall of the aorta is stretched. (B) as the ventricle relaxes and begins to fill for the next stroke, the semilunar valve closes and the expanded part of the aorta contracts, causing the adjacent part of the aorta to expand as it is filled with blood. (C) the pulse wave of expansion and contraction is transmitted to the next adjoining section of the aorta.

Solution: One can distinguish whether an artery or a vein has been cut by observing the flow of blood as it leaves the wound. A severed artery will spurt blood intermittently, whereas in a severed vein, the blood will ooze out smoothly. The difference is due to the existence of pressure pulses in the arteries, which are absent in the veins.

The heart forces blood into the arteries only during ventricular systole. Therefore, blood in the arteries tends to flow rapidly during systole and more slowly during diastole. During systole, the force of the contracting left ventricle pushes the blood forward into the aorta. The extra volume of blood distends the walls of the arteries. These distended walls soon contract (during diastole) and squeeze the blood along the artery. The walls do not actively contract, but recoil passively as a stretched elastic band does upon release. The arterial pressure continues to propel the blood forward. During the next systole, another surge of blood causes the arterial walls to distend. This alternating distention and contraction along the arterial wall is called the pressure pulse. At the height of the pulse, the pressure (systolic) is usually 120 mm Hg, while at its lowest point, the pressure (diastolic) is usually 80 mm Hg. The difference between the two pressures, 40 mm Hg, is called the pulse pressure. These however, are only average values and depend on the interaction of many physical factors.

The pressure pulse wave of distention and contraction is transmitted along the arterial system. The velocity of transmission along the aorta is about 4 meters per second, and increases to about 8 meters per second along the large arterial branches and up to 35 meters per second in the smallest arteries. The velocity of transmission of the pulse is much greater than the velocity of the blood, which is about .3 meter/sec. in the arteries. As the pulse wave moves along the artery, the blood is sped up somewhat by the force of the wave. But the speed of the pulse can be 100 times as fast as the speed of the blood.

The pressure pulse becomes less intense as it passes through the smaller arteries and arterioles until it almost disappears in the capillaries. This damping effect is mainly caused by: (1) vascular distensibility, and (2) vascular resistance. The distensibility (expansion or stretching of the vascular walls) of the smaller vessels is greater, so that the small extra volume of blood within the distended section of vessel produces less of a pressure rise. The resistance in the smaller arteries and arterioles impedes the flow of blood and consequently the transmission of pressure. By the time the blood reaches the veins, there is no pressure pulse. The velocity of blood flow is also less in the veins because they have a greater cross-sectional area than do the arteries. Therefore, the blood flow from a severed artery is both spasmodic, due to the pressure pulse, and faster than that from a severed vein.

• PROBLEM 15-14

By the time the blood reaches the veins, the blood pressure is too low to effectively move blood back to the heart. What other mechanisms aid the flow of blood in the veins?

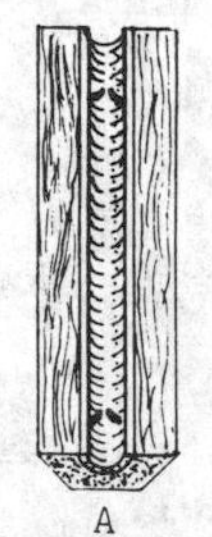

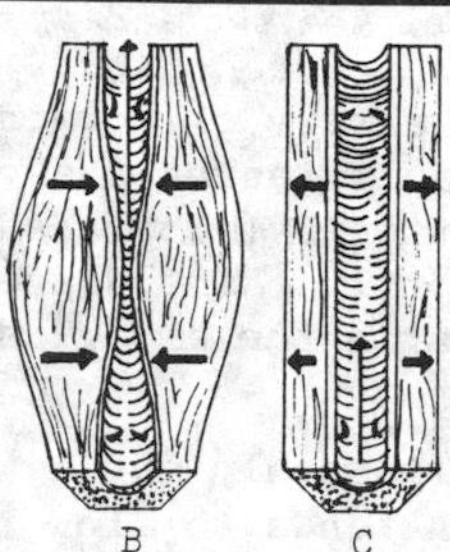

The action of skeletal muscles in milking blood through the veins. (A) Resting condition. (B) Muscles contract and bulge compressing veins and forcing blood toward the heart. The lower valve prevents backflow. (C) Muscles relax and the vein expands, filling with blood from below; the upper valve prevents a backflow.

Solution: The blood flow in the veins is partially aided by the movements of skeletal muscles near the veins. The middle layer of the venous walls has much less smooth muscle than does the middle layer of the arterial walls. Veins are therefore much (8 X) more distensible and easily collapsible. Most of the veins are surrounded by skeletal muscles, which contract as the body moves. This contraction exerts a pressure on the veins, compressing them. The compression of the veins forces the blood to move. As the muscle relaxes, the compressed section of vein fills with the blood, which is again pushed along by the next contraction. This "milking" action of skeletal muscle is shown in the accompanying figure.

Why does the blood only move in the direction of the heart? At numerous points along the interior of the vein, there are valves which allow blood to flow in only one direction - towards the heart. Thus when contracting muscle forces the blood to move in the veins, the valves act to ensure unidirectional movement of blood by preventing backflow.

When one stands perfectly still for a long period, the milking action of the skeletal muscles is decreased, venous pressure increases in the lower parts of the legs increasing capillary pressure. This forces fluid from the circulatory system into the tissue spaces, causing swelling (edema).

The movement of the chest during breathing also aids the flow of blood to the heart. As the diaphragm and chest muscles contract during inspiration, the chest cavity increases in size. The increased volume causes a decrease in pressure within the cavity to below atmospheric pressure. This pressure gradient causes air to flow into the lungs. But this decrease in pressure also lowers venous pressure in the chest cavity. During inhalation, blood thus tends to be drawn into the veins in the chest cavity. During exhalation, blood cannot be forced back because of the venous valves.

• PROBLEM 15-15

An inexperienced nurse measures the blood pressure in the artery of the upper arm of a man. She then measures the blood pressure in the artery of the man's leg. The nurse obtains a different value. Why?

Solution: During systole, the left ventricle contracts, forcing blood under pressure into the aorta and the arteries. Since the arterial walls are elastic (due to the elastic fibers in the outer layer of arteries), the rush of blood stretches the arteries and exerts a pressure on them. Blood pressure is the actual force exerted by the blood against any unit area of a vessel wall. The maximal pressure caused by the contraction of the heart is called the systolic pressure. In a normal adult at rest, the systolic pressure averages about 120 mm of mercury. This value means that the force exerted would be sufficient to push a column of mercury up to a level of 120 millimeters.

During diastole, the heart relaxes and does not exert a pressure on the arterial blood and the arterial pressure decreases. The stretched arterial walls recoil passively, maintaining some pressure which drives the blood. The pressure does not fall to zero during diastole because the next ventricular contraction occurs quickly enough to restore the pressure. The minimal pressure during the relaxation phase of the heart is called the diastolic pressure. The normal adult value is about 80 mm of mercury. The systolic and diastolic pressures are usually written as a fraction: 120/80.

These values of blood pressure are only specific for the arteries in the upper arms. As the nurse should know, the blood pressure values are different in other parts of the body. The blood pressure decreases as the blood moves further away from the heart, with the lowest value occurring in the vena cava.

The arterial pressure in the systemic circulation is inversely proportional to vascular resistance. That is, as resistance to blood flow increases, pressure decreases. Vascular resistance is directly proportional to the viscosity of the blood and the length of the vessel, and inversely proportional to the fourth power of the radius of the vessel. This is Poiseuille's Law:

$$R = \frac{8\,n\,\ell}{\pi r^4}$$

where n = viscosity
ℓ = length
r = radius

All these factors tend to produce friction between the blood flowing in the vessel and the vessel walls, decreasing the blood pressure. Note that the single most important determinant of resistance is the radius of the vessel.

In the aorta, the resistance is very low since the blood has not traveled far and the radius is large. Resistance in the large arteries is also low, so that the pressure is still about 100 mm Hg (Hg means mercury). The resistance begins to increase rapidly in the arterioles, as the radii of these vessels decrease. The pressure of the blood as it leaves the arterioles to enter the capillaries decreases to about 30 mm Hg. In the capillaries, the velocity of the blood flow is very low. The pressure at the beginning of the veins is about 10 mm Hg and decreases to almost 0 mm Hg at the right atrium. One would not expect the pressure to decrease so much in the veins since they are wider in width than the capillaries. This larger radius should cause a decrease in resistance. However, pressure in the veins does not increase because the walls of veins are easily distensible. Distensibility arises from the fact that veins are not rigid structures; they have only a thin muscle coat. We know that very little pressure is generated in filling an easily distensible object, hence the pressure of the veins is negligible.

• PROBLEM 15-16

Why is the drop in blood pressure important in retaining fluid within the capillaries?

Solution: As the blood passes through the capillaries, there is a pressure resulting from the beating of the heart, which tends to force the plasma through the capillary membrane and into the interstitial fluid. This pressure is called the capillary, or hydrostatic pressure. There are also two other factors determining whether fluid will move out of the blood or not: 1) the interstitial fluid pressure and 2) the plasma colloid osmotic pressure.

The interstitial fluid pressure is the hydrostatic pressure exerted by the interstitial fluid. It has been measured to be about -7 mm Hg, and so tends to move fluid outward through the capillary membrane. The plasma colloid osmotic pressure arises from the proteins found in the plasma. The plasma proteins do not diffuse readily through the capillary membrane. Therefore, there is a concentration gradient with four times as much protein in the plasma as in the interstitial fluid. An osmotic pressure develops since water tends to pass through the capillary membrane from the interstitial fluid (which has a higher concentration of water and a lower concentration of protein molecules) to the blood plasma. The plasma colloid ("colloid" because the protein solution resembles a colloidal solution) osmotic pressure is approximately 23.5 mm Hg. This pressure, unlike the interstitial fluid pressure, opposes the capillary pressure by tending to move fluid inward through the capillary.

An analysis of the forces at the arteriole end follows:

Forces moving plasma outward:	mm Hg
Capillary pressure	25
Interstitial fluid pressure	7
Total outward pressure	32
Forces moving plasma inward:	
Plasma colloid osmotic pressure	23.5
Total inward pressure	23.5
Net outward force	32 - 23.5 = 8.5

At the arteriole end, then, there is a net outward filtration pressure of about 8.5 mm Hg. If this force were constant throughout the length of the capillary, there would be a large net loss of fluid outward. There is, however, a drop in capillary pressure (to 9 mm Hg) at the venule end due to the filtration of fluid along the capillary. Therefore net reabsorption of the lost fluid occurs at the venule end as follows:

Forces moving plasma outward:

	mm Hg
Capillary pressure	9
(Negative)interstitial fluid pressure	7
Total outward pressure	16
Forces moving plasma inward:	
Plasma colloid osmotic pressure	23.5
Total inward pressure	23.5
Net inward force	23.5 - 16 = 7.5

At the venule end, there is a reabsorption pressure of 7.5 mm Hg This reabsorption causes about nine-tenths (7.5/8.5) of the fluid that had filtered out at the arteriole end to be recovered. The slight excess of filtration will be recovered by the lymphatic system (see next question).

THE LYMPHATIC SYSTEM • PROBLEM 15-17

The lymphatic system in man constitutes an extensive network of thin vessels resembling veins. What are the functions of the lymphatic system?

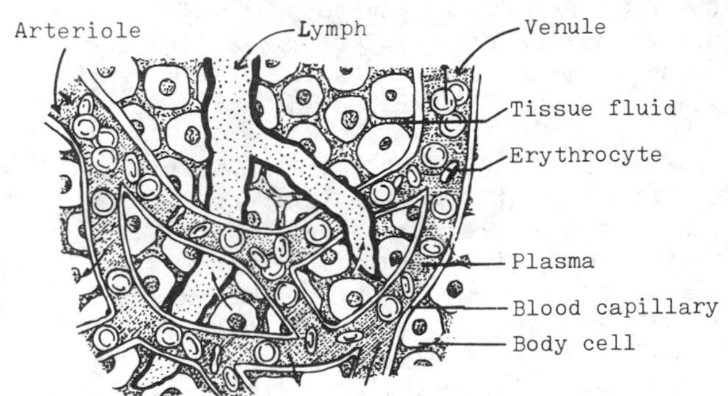

Figure 1.

Diagram of the relation of blood and lymph capillaries to tissue cells. Note that blood capillaries are connected at both ends whereas lymph capillaries, outlined in black, are "dead-end streets" and contain no erythrocytes. The arrows indicate direction of flow.

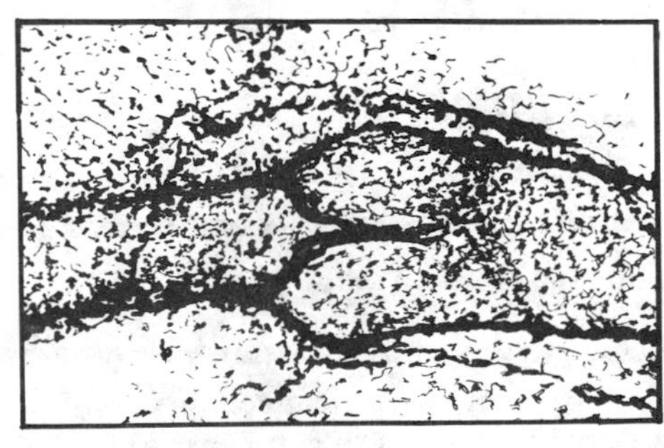

Figure 2. A lymph vessel valve.

Solution: The lymphatics are not part of the circulatory system per se, but constitute a one-way route from interstitial fluid to the blood. The lymphatic system in man constitutes an extensive network of thin-walled vessels resembling the veins. These vessels ultimately drain into veins in the lower neck. One of the functions of the lymphatic system is to transfer excess interstitial fluid back to the blood. There is a net movement of plasma out of the capillaries and into the tissues; the lymphatics restore the blood volume by returning this fluid that has filtered out. In addition, the lymphatic system serves to return proteins to the blood. Since the capillaries have a slight permeability to plasma proteins, and since the concentration of protein in the blood plasma is greater than that in the interstitial fluid, there is a small but steady loss of protein from the blood into the interstitial fluid. This protein returns to the blood via the lymphatics. Should there be a malfunction in the lymphatic system, the interstitial fluid protein concentration would increase to that of the plasma. This eliminates the protein concentration difference between the plasma and the interstitial fluid, thus eliminating the plasma colloid osmotic pressure. Only blood pressure and interstitial fluid pressure remain, and permit the net movement of fluid out of the capillary into the interstitial space (edema).

Another function of the lymphatic system is to provide the pathway by which fat and other substances absorbed from the gut reach the blood. It is also believed that certain high-molecular weight hormones reach the blood via the lymphatics.

In addition to its function in transport, the lymphatic system plays a critical role in the body's defense mechanism against disease. The lymph nodes, which are found at the junctions of lymph vessels, act as filters and are sites of formation of certain types of white blood cells. Lymph, the fluid in the lymph capillaries, flows slowly through the nodes where invading bacteria are phagocytosed by the cells of the lymph node. Indigestible particles such as dust and soot, which the phagocytic cells cannot destroy, are stored in the nodes. Since the nodes are particularly active during an infection, they often become swollen and sore, as the lymph nodes at the base of the jaw are apt to become during a throat infection.

It should be noted that since the lymphatic system is not connected to the arterial portion of the blood circulatory system, lymph is not moved by the hydrostatic pressure developed by the heart. Lymph flow, like that in the veins, depends primarily upon forces external to the vessels. These forces include the contractile action of the skeletal muscle (through which the lymphatics flow) and the effects of respiration on the pressures in the chest cavity. Since the lymphatics have valves similar to those in veins, external pressure would permit only unidirectional flow.

DISEASES OF THE CIRCULATION

• PROBLEM 15-18

Why does a physician advise reduced salt intake in cases of hypertension?

Solution: Hypertension or high blood pressure, is caused by abnormalities in arterial pressure regulation. The mean arterial pressure (M.A.P.) is the average pressure throughout the pressure pulse cycle. It is not equal to the average of the systolic and diastolic pressure, since the heart remains in diastole twice as long as in systole.

$$\text{M.A.P.} = \frac{\text{SBP} + 2\text{DBP}}{3} = \frac{120 + 2(80)}{3} = 93.3$$

Thus mean arterial pressure is closer to the diastolic than to the systolic blood pressure.

Arterial pressure is determined by two factors: cardiac output and total peripheral resistance. The relationship between them is:

Pressure = Cardiac Output x Total Peripheral Resistance

Cardiac output is the amount of blood pumped by each ventricle per unit time, and is determined by both the heart rate and the stroke volume. Total peripheral resistance is the resistance of the entire systemic circulation. Any factor that causes an increase in cardiac output or total peripheral resistance will cause an increase in mean arterial pressure.

One such factor is the secretion of the hormone aldosterone by the adrenal cortex. Aldosterone acts on the tubules of the kidney to increase the reabsorption of sodium. This increased retention of sodium causes more water to be retained, thereby increasing the volume of the blood. This increased blood volume increases the stroke volume and hence the cardiac output. Arterial blood pressure increases. When excessive aldosterone is secreted (such as by a tumor in the adrenal cortex), hypertension results. Salt, or sodium chloride, is a source of sodium. A decreased salt intake causes less water to be retained, decreasing blood volume, and in turn lowering blood pressure.

Blood pressure can also be increased by vasoconstriction (constriction of the blood vessels). Constriction causes an increased resistance to flow (since a decrease in vessel radius increases vascular resistance). A protein called angiotensin II causes both vasoconstriction and stimulation of aldosterone production and secretion. Angiotensin II is produced enzymatically from angiotensin I, which is converted from angiotensinogen, a glycoprotein made in the liver. The enzyme renin, formed in the kidney, catalyzes this conversion. Kidney damage can cause an oversecretion of renin, which causes an increase of angiotensin II, and thus hypertension.

Slight hypertension usually accompanies arteriosclerosis, the deposition of fatty material (such as cholesterol) in the arterial walls. The arteries become fibrous and calcified. This buildup of deposits inhibits the flow of blood, increasing resistance to flow and thus increasing the blood pressure needed to overcome the resistance.

Oversecretion of renin and aldosterone and arteriosclerosis can thus cause hypertension, but in most cases (90%), the cause is unknown. These cases, known as "essential" hypertension (because of unknown etiology) appear to be hereditary. Hypertension is harmful because it increases the work load of the heart and damages the arteries by subjecting them to excessive pressures. Limiting the amount of salt in foods may help to alleviate hypertension.

• **PROBLEM 15-19**

What is meant by arteriosclerosis and what is the cause of this condition?

Solution: Arteriosclerosis, commonly known as "hardening of the arteries," is a disease characterized by a thickening of the arterial wall with connective tissue and deposits of cholesterol. Though it is not really clear how this thickening occurs, it is known that smoking, obesity, high-fat diet, and nervous tension predispose one to this disease. The suspected relationship between arteriosclerosis and blood concentrations of cholesterol and saturated fatty acids has received widespread attention. Many studies are presently trying to evaluate the hypothesis that high blood concentration of these lipids increase the rate and the severity of the arteriosclerotic process. Physiologically, cholesterol is an important substance because it is the precursor of certain hormones and the bile acids. The major dietary source of cholesterol is animal fats. However, the liver is capable of producing large amounts of cholesterol, particularly from saturated fatty acids; even if the intake of cholesterol is markedly reduced, the blood cholesterol level will not be lowered considerably since the liver responds by producing more. It may be the high content of saturated fatty acids, rather than cholesterol, which causes the ingestion of animal fat to make one more susceptible to arteriosclerosis.

Arteriosclerosis in the coronary artery increases the risk of heart attack and death. Coronary artery arteriosclerosis is estimated to cause 500,000 deaths per year.

Answer

To be covered when testing yourself

SHORT ANSWER QUESTIONS FOR REVIEW

Choose the correct answer.

1. The only artery in the human body that carries deoxygenated blood is the (a) right coronary artery (b) left coronary artery (c) pulmonary artery (d) carotid artery. — c

2. All of the following can cause an increase in the amount of blood pumped by the heart except (a) a rise in carbon dioxide content of the blood (b) decreased temperature (c) hormones (d) nerves. — b

3. The characteristic sounds of the beating heart are caused by (a) the closure of the semilunar values (b) the closure of the tri- and bicuspid valves (c) the contraction of the muscle in the ventricle (d) all of the above. — d

4. What is responsible for the initiation of the heart beat and the regulation of the rate of contraction? (a) the sinoatrial node (b) the atrioventricular node (c) the pacemaker (d) a and c are correct. — d

5. Which of the following statements is incorrect? (a) the rate of blood flow throughout the circulation is not constant (b) the middle layer of a vein has much less smooth muscle than does the middle layer of the arterial wall (c) in humans, systolic pressure is about 120 mm Hg and diastolic pressure about 75 mm Hg. (d) angio tensin II is a protein that causes dilation of the arteries. — d

6. The connective tissue sac enclosing the heart is called the (a) endothelium, (b) myocardium (c) pericardium (d) vena cava. — c

7. Which of the following carries blood from the head, neck and arms to the heart? (a) superior vena cava (b) inferior vena cava (c) pulmonary artery (d) pulmonary vein. — a

8. Which of the following is the proper sequence for the cardiac impulse? (a) S-A node, A-V node, A-V bundle, Purkinje fibers (b) A-V node, A-V bundle, S-A node, Purkinje fibers (c) Purkinje fibers, A-V node, A-V bundle, S-A node (d) A-V bundle, A-V node, S-A node, Purkinje fibers. — a

	Answer
	To be covered when testing yourself
9. Which of the following is incorrect? (a) the atria and the ventricles contract simultaneously (b) the period of heart contraction is called systole (c) a heart murmur may be a symptom of a damaged valve (d) sympathetic nerve stimulation increases the heart rate.	a
10. Which of the following is not a function of the circulatory system? (a) transports nutrients to each cell (b) regulates body temperature (c) produces oxygen (d) protects body against invading microorganisms.	c
11. The pumping chambers of the heart are called (a) atria, (b) ventricles (c) pacemakers (d) cardiac muscle.	b
12. Sensory receptors in the walls of the vena cava and right atrium are stimulated when (a) vessels are distended with blood and the wall is stretched (b) blood temperature changes, (c) a rise in the carbon dioxide content of the blood occurs (d) none of the above.	a
13. Hypertension can be a result of (a) excess aldosterone secretion and increased sodium intake, (b) decreased aldosterone secretion and increased sodium intake (c) decreased sodium intake (d) none of the above	a
14. Choose the incorrect statement (a) the lymphatics are part of the circulatory system that constitute a two-way route between interstitial fluid and blood (b) lymph functions primarily in maintaining water, salt, pH, and osmotic equilibria between the interior and exterior of cells (c) lymph provides a medium for diffusion and transport of foodstuffs, respiratory gases, waste materials and some hormones (d) all of the above are correct.	a
Fill in the blanks.	
15. A heart rate center is located in the hindbrain in a region called the _____.	medulla oblongata
16. The large vein that drains the legs and the lower part of the body is the _____.	inferior vena cava
17. The valve between the right atrium and right ventricle is called the _____.	tricuspid valve

	Answer To be covered when testing yourself
18. The valve between the left atrium and left ventricle is called the _____.	bicuspid valve
19. The propogation of action potentials through the atrioventricular node is _____ for approximately 0.1 seconds.	delayed
20. Systole is the name of the period of _____ contraction and diastole is _____ relaxation.	ventricular, ventricular
21. The electrical currents recorded by an electrocardiograph corresponding to the P waves are from the depolarization of the ______.	atria
22. The Q.R.S. complex corresponds to the depolarization (electrical activation) of the _____.	ventricle
23. The _____ wave corresponds to the relaxation of the ventricle.	T
24. Increased temperature affects the sinoatrial node and the heart rate is _____ (increased or decreased).	increased
25. Epinephrine and thyroxine _____ the heart beat.	accelerate
26. During exercise there is a(an) _____ (increased or decreased) amount of blood flow to the stomach, spleen and intestine.	decreased
27. _____ = heart rate × stroke volume	cardiac output
28. The observation that there is a direct proportion between the diastolic volume of the heart and the force of contraction of the following systole is refered to as _____.	Starling's Law

Determine whether the following statements are true or false.

29. The rhythmic discharge of the SA node occurs spontaneously in the complete absence of any nervous or hormonal influences.	True
30. There is no muscular connection between the atria and ventricles.	True
31. The atrioventricular node initiates the heart beat.	False

Answer

To be covered when testing yourself

	Answer
32. The heart muscle is nourished by the blood within its chambers.	False
33. The total cross-sectional area of the capillaries is greater than that of the aorta. Therefore the rate of flow in the capillaries is correspondingly less.	True
34. Atrial excitation is complete before ventricular excitation begins because of the delay at the A.V. node.	True
35. Cutting the parasympathetic nerves to the heart causes the heart rate to decrease.	False
36. The atria and ventricles contract simultaneously to move blood rapidly out of the heart.	False
37. The small loss of fluid due to net filtration is compensated by the return of fluid via the lymphatic system.	True
38. In normal heart rates, the influence of the sympathetic branch is stronger than the parasympathetic branch.	False
39. Blood flow is determined by pressure difference and resistance.	True
40. Blood returning to the right atrium contains oxygenated blood.	False
41. The pressure pulse in the arteries is produced by alternating distension and recoil of the elastic fibers in the arterial wall.	True
42. The sounds heard during heart beats are produced by the closure of the semilunar valves and the atrio-ventricular valves.	True

CHAPTER 16

RESPIRATION

TYPES OF RESPIRATION

• PROBLEM 16-1

Differentiate clearly between "breathing" and "respiration".

Solution: Respiration has two distinct meanings. It refers to the oxidative degradation of nutrients such as glucose through metabolic reactions within the cell, resulting in the production of carbon dioxide, water and energy. Respiration also refers to the exchange of gases between the cells of an organism and the external environment. Many different methods for exchange are utilized by different organisms. In man, respiration can be categorized by three phases: ventilation (breathing), external respiration, and internal respiration.

Breathing may be defined as the mechanical process of taking air into the lungs (inspiration) and expelling it (expiration). It does not include the exchange of gases between the bloodstream and the alveoli. Breathing must occur in order for respiration to occur; that is, air must be brought to the alveolar cells before exchange can be effective. One distinction that can be made between respiration and breathing is that the former ultimately results in energy production in the cells. Breathing, on the other hand, is solely an energy consuming process because of the muscular activity required to move the diaphragm.

• PROBLEM 16-2

Differentiate between direct and indirect respiration, and between external and internal respiration.

Solution: The phenomenon in which the cells of an organism exchange oxygen and carbon dioxide directly with the surrounding environment is termed direct respiration. This form of respiration is a fairly simple process and occurs in small, aquatic animals such as paramecia or hydras. In these animals, dissolved oxygen from the surrounding water diffuses into the cells, while carbon dioxide within the cell diffuses out; no special respiratory system is needed.

With the evolution of animals into larger and more complex forms, it became impossible for each cell to exchange gases directly with the external environment. Consequently, it became necessary for these organisms to have a specialized organ system that would function in gas exchange with the environment. This structure must be thin-walled, and its membrane must be differentially permeable. In addition, the membrane must be kept moist so that oxygen and carbon dioxide could dissolve in it, and it must have a good blood supply. The process of respiration employing this organ system is called indirect respiration. For indirect respiration, the lower vertebrates developed gills and the higher vertebrates developed lungs.

During indirect respiration, gas exchange between the body cells and the environment may be categorized into two phases: an external and an internal phase. External respiration is the exchange of gases by diffusion that occurs between the lungs and the bloodstream. Oxygen passes from the lungs to the blood and carbon dioxide passes from the blood to the lungs. Internal respiration takes place throughout the body. In the latter, there is an exchange of gases between the blood and other tissues of the body, with oxygen passing from the blood to the tissue cells and carbon dioxide passing from the cells to the blood. This phase, along with the external phase, relies on the movement of gases from a region of higher concentration to one of lower concentration.

• PROBLEM 16-3

What function does oxygen serve in the body?

Solution: Oxygen is necessary for the process of cellular respiration. This process involves the catabolism of glucose to result in the production of energy, in the form of ATP, for utilization by the cell. Cellular respiration can be thought to have two parts. The first series of reactions can take place without the presence of oxygen, while the second series of reactions is dependent on oxygen. In the second reaction sequence, oxygen acts as the final hydrogen acceptor. This occurs in the electron transport system within the mitochondria. The oxygen combines with hydrogen, and is converted to a molecule of water. If anaerobic conditions existed (absence of oxygen), no ATP formation will occur in the mitochondria. It should be noted however that two ATP per glucose oxidized are still produced by the first series of reactions, collectively called glycolysis. Glycolysis will occur regardless of the availability of oxygen, and is restricted to the cytoplasm; it does not occur in the mitochondria. The amount of ATP produced via glycolysis is very small compared to that produced by the second reaction sequence. The second reaction sequence, which occurs only in the mitochondria, is collectively referred to as oxidative phosphorylation.

Since the mitochondria provide about 95% of all the ATP formed in the cell, the absence of oxygen results in less energy production. This energy is required to maintain cell structure and function, and the cell will ultimately die without it. Roughly 99% of all the molecular oxygen consumed by a cell is used for cellular respiration.

• PROBLEM 16-4

Discuss the role of hemoglobin in the transport of oxygen and of carbon dioxide.

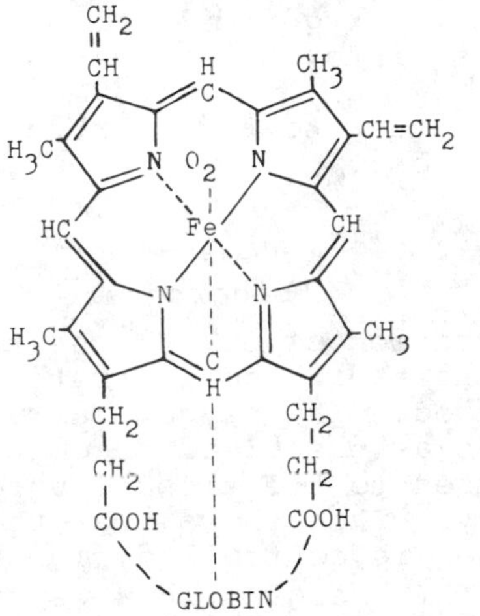

One of the subunits of the hemoglobin molecule. Four subunits, two α and two β chains, comprise a single hemoglobin molecule. Note how O_2 is bound to the iron atom.

Solution: Hemoglobin is the pigment in red blood cells that is responsible for transporting nearly all the oxygen and some of the carbon dioxide in the body. In arterial blood, oxygen is present in two forms. It is either physically dissolved in the blood plasma or chemically bound to hemoglobin. Because oxygen is relatively insoluble in water, and blood plasma is comprised primarily of water, only 3 ml. of oxygen can be dissolved in 1 liter of plasma at normal oxygen pressure (100 mm Hg on average). Hence, approximately 2% of the oxygen in the blood is dissolved in the plasma; the rest is transported via hemoglobin. In the lungs, oxygen enters the capillaries, and diffuses into the red blood cells where it binds to hemoglobin. Four oxygen molecules are attached to four iron atoms in a single hemoglobin molecule (see Figure). The chemical reaction between oxygen and hemoglobin is usually written as

$$O_2 + Hb \rightleftharpoons \underset{\text{oxyhemoglobin}}{HbO_2}$$

The reaction goes to the right in the lungs and to the left in the body tissues. Hemoglobin combined with oxygen (HbO_2) is called oxyhemoglobin; when not combined (Hb), it is called reduced hemoglobin or deoxyhemoglobin. Because there is a finite number (four) of binding sites for oxygen on the hemoglobin molecule, there is a maximum amount of oxygen which can combine with hemoglobin. When hemoglobin exists as both Hb and HbO_2, it is said to be partially saturated. When it exists as only HbO_2, it is said to be fully saturated.

The combination of oxygen with hemoglobin and its release from oxyhemoglobin are controlled by the concentration of oxygen present and, to a lesser extent, by the amount of carbon dioxide present. For example, the percentage of saturated hemoglobin undergoes little change between the lungs and the arteries, because the oxygen concentration in the lungs is very similar to that of the arteries. However, in the tissues, the concentration of oxygen is low, so by the law of mass action, the reaction will go to the left. As a result, oxygen will break off from the oxyhemoglobin to diffuse into the tissues.

Carbon dioxide reacts with water to form carbonic acid, H_2CO_3, which then dissociates:

$$CO_2 + H_2O \rightleftarrows H_2CO_3 \rightleftarrows H^+ + \underset{\text{bicarbonate}}{HCO_3^-}$$

An increase in the concentration of CO_2 drives the reaction to the right, increasing the H^+ concentration and hence the acidity of the blood. The oxygen carrying capacity of hemoglobin decreases as the blood becomes more acidic. Thus, more O_2 is released to the tissues under conditions of increasing acidity. This shows that the combination of hemoglobin and oxygen is controlled, in part, by the amount of CO_2 present. This results in a very efficient transport system. In the lung capillaries, oxygen is taken up by the hemoglobin due to the effects of high oxygen tension and low CO_2 tension. The situation **is reversed in the tissues. There,** CO_2 **pressure is** high and oxygen pressure is low, so oxygen dissociates from oxyhemoglobin and diffuses into the tissues.

It is important to realize that the principal factor that determines the mode of hemoglobin dissociation **and rate of diffusion is the partial pressure** (P_{gas}) **of each particular gas.**

As is true for oxygen, the quantity of carbon dioxide that can physically dissolve in blood is quite small. Carbon dioxide can undergo the following reaction:

$$CO_2 + H_2O \rightleftarrows \underset{\text{carbonic acid}}{H_2CO_3}$$

This reaction would go quite slowly if it were not catalyzed by the enzyme carbonic anhydrase. The quantities of both dissolved carbon dioxide and carbonic acid are directly proportional to the partial pressure of CO_2. In this case, as in the case with oxygen, we are only concerned with the pressure of the gas in question. This is because in a mixture of gases, each one acts independently of the others, and exerts the same pressure it would if it were present alone. The actual quantity of carbonic acid in the blood is small, because, as we saw previously, it dissociates into H^+ and HCO_3^- ions. These ions are quite soluble in the blood. Thus, the addition of carbon dioxide to blood plasma results ultimately in the production of hydrogen and bicarbonate ions. Carbon dioxide is transported primarily as bicarbonate ions to the lungs, where it is excreted as carbon dioxide. Carbon dioxide can also react with proteins, particularly hemoglobin, to form carbamino compounds.

$$CO_2 + Hb \rightleftarrows \underset{\text{carbamino hemoglobin}}{HbCO_2}$$

Carbon dioxide diffuses from the tissues into the blood. In the blood, some (8%) of the CO_2 stays dissolved, and some (25%) reacts with hemoglobin to form $HbCO_2$. The largest fraction (67%) is converted to H^+ and HCO_3^-. This occurs primarily within the red cells because they contain large quantities of carbonic anhydrase, whereas the plasma does not. This dissociation into H^+ and HCO_3^- explains why tissues and capillaries, where CO_2 concentration is high, have a hydrogen ion concentration higher than that of arterial blood. This also explains the increase in H^+ concentration as metabolic rate increases. The CO_2 itself passes from the tissues to the blood, and then to the lungs by diffusing from a region of high CO_2 tension to one of low CO_2 tension.

HUMAN RESPIRATION

• PROBLEM 16-5

What is meant by the "vital capacity" of a person? In what conditions is it increased or decreased?

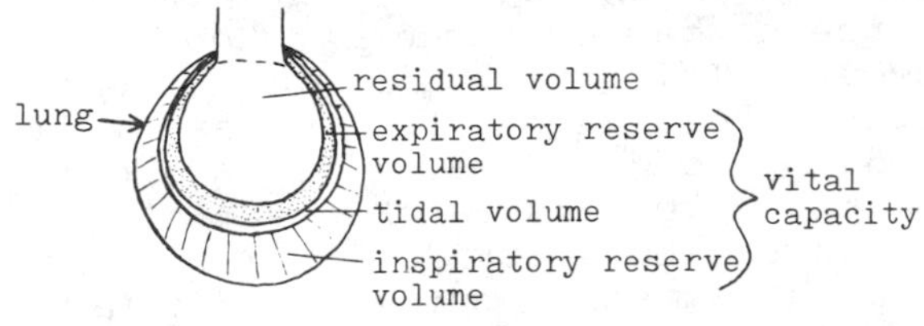

Lung volumes

Solution: During a single normal breath, the volume of air entering or leaving the lungs is called the tidal volume. Under conditions of rest, this volume is approximately 500 ml. on average. The volume of air that can be inspired over and above the resting tidal volume is called the inspiratory reserve volume, and amounts to about 3000 ml. of air. Similarly, the volume of air that can be expired below the resting tidal volume is called the expiratory reserve volume, and amounts to approximately 1000 ml. of air. Even after forced maximum expiration, some air (about 1000 ml.) still remains in the lungs, and is termed the residual volume. The vital capacity is the sum of the tidal volume and the inspiratory and expiratory reserve volumes. The vital capacity then represents the maximum amount of air that can be moved in and out during a single breath. The average vital capacity varies with sex, being 4.5 liters for the young adult male, and about 3.2 for the young adult female. During heavy work or exercise, a person uses part of both the inspiratory and expiratory reserves, but rarely uses more than 50% of his total vital capacity. This is because deeper breaths than this would require exhaustive activities of the inspiratory and expiratory muscles. Vital capacity is higher in an individual who is tall and thin than in one who is obese. A well developed athlete may have a vital capacity up to 55% above average. In some diseases of the heart and lungs, the vital capacity may be reduced considerably.

• PROBLEM 16-6

Why is it that alveolar air differs in its composition from atmospheric air? Of what significance is this fact?

Solution: The respiratory tract is composed of conducting airways and the alveoli. Gas exchange occurs only in the alveoli and not in the conducting airways. The maximum alveolar volume is about 500 ml. Let us consider then what takes place during expiration. Five hundred milliliters of air is forced out of the alveoli, and into the conducting airways of the respiratory tract. Of this air, 150 ml. remains in the respiratory airways following expiration, while 350 ml. of air is exhaled from the body. During the next inspiration 500 ml. of air are taken up by the alveoli, but 150 ml. of this air is not atmospheric, but rather is the air that remained in the tubes following the previous expiration. One can see, then, that only 350 ml. of fresh atmospheric air enters the alveoli during each inspiration. At the end of inspiration, 150 ml. of fresh air also fills the airways but does not reach the alveoli. Hence no gas exchange between this air and the blood can occur. This fresh air will be expelled from the body during the next expiration, and will be replaced again with 150 ml. of alveolar air, thus completing the cycle. From this cycle it can be seen that of the 500 ml. of air entering the body during inspiration, 150 ml. of it never reaches the alveoli, but instead remains in the conducting tubes of the respiratory system. The term "anatomical dead space" is given to the space within the conducting tubes because no gas exchange with the blood can take place there.

The question arises then, as to the significance of this dead space. Because the tubes are not completely emptied and filled with each breath, "new" air can mix with "old" air. Consequently, alveolar air contains less oxygen and more carbon dioxide than atmospheric air. In addition, the air that remains in the alveoli following expiration (the residual volume) can, to a limited extent, mix with the incoming air and thereby alter its composition.

List the parts of the human respiratory system. How is each adapted for its particular function?

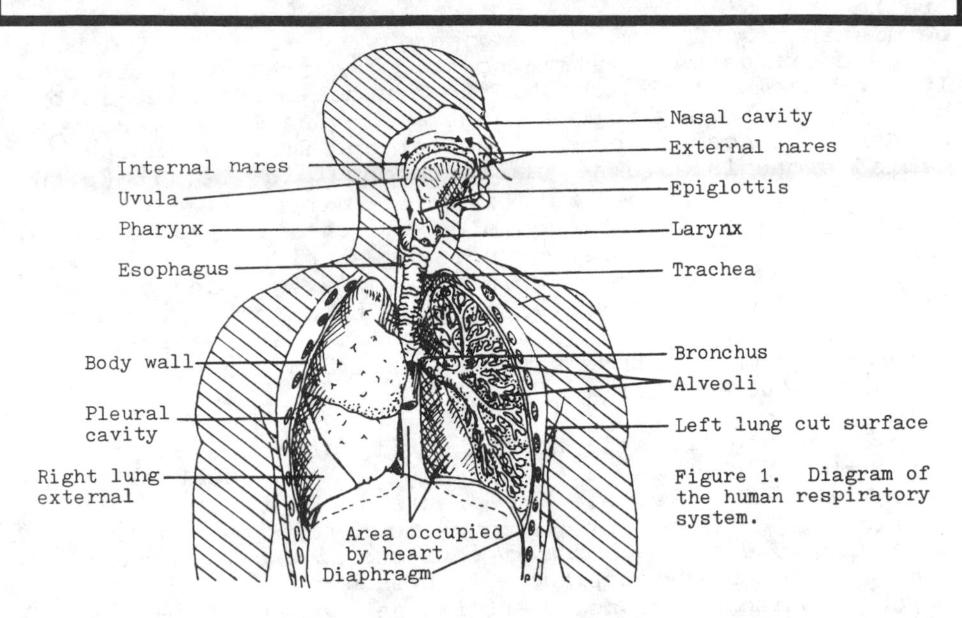

Figure 1. Diagram of the human respiratory system.

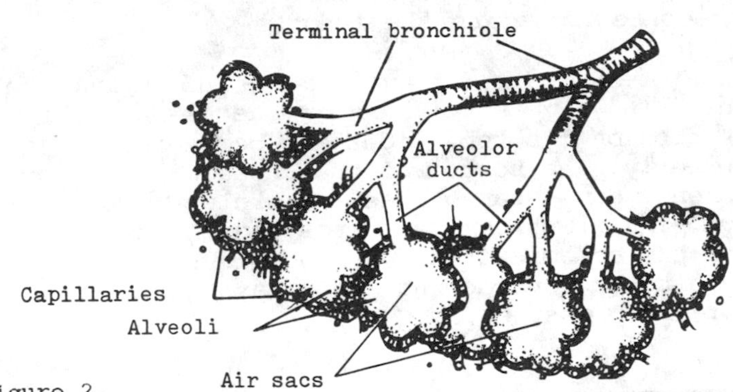

Figure 2.
Diagram of a small portion of the lung, highly magnified showing the air sacs at the end of the alveolar ducts, the alveoli in the walls of the air sacs, and the proximity of the alveoli and the pulmonary capillaries containing red blood cells.

Solution: The respiratory system in man and other air-breathing vertebrates includes the lungs and the tubes by which air reaches them. Normally air enters the human respiratory system by way of the external nares or nostrils, but it may also enter by way of the mouth. The nostrils, which contain small hairs to filter incoming air, lead into the nasal cavities, which are separated from the mouth below by the palate. The nasal cavities contain the sense organs of smell, and are lined with mucus-secreting epithelium which moistens the incoming air. Air passes from the nasal cavities via the internal nares into the pharynx, then through the glottis and into the larynx. The larynx is often called the "Adam's apple," and is more prominent in men than women. Stretched across the larynx are the vocal cords.

The opening to the larynx, called the glottis, is always open except in swallowing, when a flap-like structure (the epiglottis) covers it. Leading from the larynx to the chest region is a long cylindrical tube called the trachea, or windpipe. In a dissection, the trachea can be distinguished from the espophagus by its cartilaginous C-shaped rings which serve to hold the tracheal tube open. In the middle of the chest, the trachea bifurcates into bronchi which lead to the lungs. In the lungs, each bronchus branches, forming smaller and smaller tubes called bronchioles. The smaller bronchioles terminate in clusters of cup-shaped cavities, the air sacs. In the walls of the smaller bronchioles and the air sacs are the alveoli, which are moist structures supplied with a rich network of capillaries. Molecules of oxygen and carbon dioxide diffuse readily through the thin, moist walls of the alveoli. The total alveolar surface area across which gases may diffuse has been estimated to be greater than 100 square meters.

Each lung, as well as the cavity of the chest in which the lung rests, is covered by a thin sheet of smooth epithelium, the pleura. The pleura is kept moist, enabling the lungs to move without much friction during breathing. The pleura actually consists of two layers of membranes which are continuous with each other at the point at which the bronchus enters the lung, called the hilus (roof). Thus, the pleura is more correctly a sac than a single sheet covering the lungs.

The chest cavity is closed and has no communication with the outside. It is bounded by the chest wall, which contains the ribs on its top, sides and back, and the sternum anteriorly. The bottom of the chest wall is covered by a strong, dome-shaped sheet of skeletal muscle, the diaphragm. The diaphragm separates the chest region (thorax) from the abdominal region, and plays a crucial role in breathing by contracting and relaxing, changing the intrathoracic pressure.

• PROBLEM 16-8

Where is the respiratory center and what are its functions? Describe briefly the neural control of breathing.

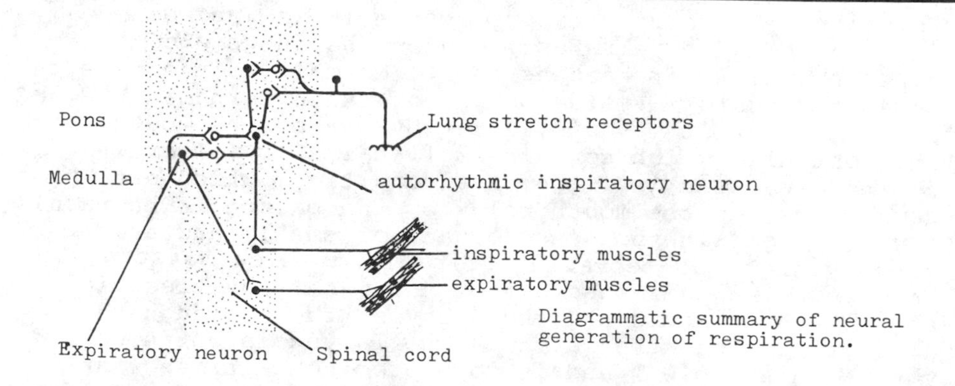

Diagrammatic summary of neural generation of respiration.

Solution: Breathing requires the coordinated contraction and relaxation of several muscles. This is achieved by the respiratory center, which is composed of special groups of cells in the medulla and pons of the brain. From this respiratory center, nerve impulses are rhythmically discharged to the intercostal (rib) muscles, resulting in their periodic contraction every 4 to 5 seconds. The breathing movements are automatic and occur without voluntary control under normal conditions. We can voluntarily hold our breath, but not indefinitely, since then the automatic center eventually takes over and forces us to exhale.

By experimentation, neurons have been found in the medulla which propagate action potentials in perfect synchrony with inspiration. A smaller number of other neurons have been discovered which discharge synchronously with expiration. These two types of neurons are called inspiratory and expiratory neurons, respectively. These may also be referred to as inspiratory and expiratory centers. Electrical stimulation of the inspiratory neurons can produce a maximal inspiration. Conversely, electrical stimulation of the expiratory neurons shuts off inspiration abruptly, and produces contraction of the expiratory muscles. The question arises as to what induces firing of the medullary inspiratory neurons. It appears that these neurons have inherent autorhythmicity - the capacity for periodic self-excitation. However, synaptic input from other neurons plays an essential role in regulating the rhythmicity of these inspiratory neurons. There are three vital inputs to the medullary inspiratory neurons, which play a role in modulating respiratory rhythm. They are (1) direct intracranial connections between the medulla and the pons (2) reciprocal connections with the medullary expiratory neurons, and (3) afferent input from stretch receptors in the lung (see Figure).

The connections between the inspiratory and expiratory centers are inhibitory in nature. Thus at the beginning of inspiration, when the inspiratory neurons are firing, the expiratory center is prevented from firing so that expiration is inhibited. When inspiration ceases, expiratory inhibition also stops, and expiration is able to occur which then, in turn, inhibits inspiration. These reciprocal connections serve to synchronize inspiration and expiration. As was previously mentioned, the medullary inspiratory center is connected to the pons. This area of the pons is often called the pneumotaxic center, and destruction of this center produces profound changes in respiration. The medullary neurons receive neural input from the pons, which exerts a tonic effect upon the inspiratory neurons. It is also likely that the pneumotaxic center serves as a central relay station for the respiratory inhibition initiated by the lung stretch receptors. As the lungs expand during inspiration, these receptors are stimulated, and impulses travel up the afferent nerves to the brainstem, where they aid in terminating inspiration.

To summarize, the medullary inspiratory neurons primarily control the cycle of ventilation. These inspiratory neurons innervate the inspiratory muscles. The spontaneous increase in the firing rate of these inspiratory neurons is a crucial factor for initiating ventilation. Due to the inhibitory impulses from medullary expiratory neurons and pulmonary stretch receptors which act through higher brain centers, these neurons will stop firing. The cessation of activity by the inspiratory neurons releases the inhibition on the expiratory neurons so that expiration can passively occur. Only in forced expiration are the expiratory muscles themselves used. This active expiratory movement is synchronized with the passive component of expiration. This is possible because of the reciprocal connections between the medullary inspiratory and expiratory centers.

• PROBLEM 16-9

Describe the homeostatic mechanisms related to breathing.

Solution: The control centers for breathing are located in the medulla of the brain. These centers function in maintaining respiratory rhythmicity, but it does not account for the variations in the rhythm. The question arises then, as to how the respiratory centers know what the body oxygen requirements are. The primary stimulus comes from the carbon dioxide which is released from the cells into the blood. This gas, which is generated in cellular respiration, is transported by the circulation in the form of carbonic acid (H_2CO_3), which freely dissociates to H^+ and HCO_3^- ions. The concentration of carbonic acid increases as a result of increased cellular respiration. The increased carbonic acid content of the blood stimulates the inspiratory centers of the medulla in two manners. The carbonic acid either acts directly upon the inspiratory neurons or upon nearby chemosensitive cells, from which there are neural connections to the inspiratory neurons. The inspiratory neurons increase the rate of stimulation of the inspiratory muscles, which results in a greater respiratory rate. The result of this increased rate of breathing is the increased rate of removal of carbon dioxide from the blood upon expiration. The loss of carbon dioxide favors the reduction of carbonic acid by the following formula, in accordance with the law of mass action.

$$CO_2 + H_2O \rightleftarrows H_2CO_3 \rightleftarrows H^+ + HCO_3^-$$

As the carbonic acid concentration decreases, there will be a corresponding reduction in the rate of breathing. Chemosensitive cells in the brain, called central chemoreceptors, play an integral role in the regulation of the respiratory rate. These cells are bathed by the cerebrospinal fluid, and they monitor the hydrogen ion concentration of the blood. An increased carbon dioxide content ultimately yields an increased hydrogen-ion concentration through the dissociation of carbonic acid. The brain is extremely sensitive to changes in hydrogen ion (H^+) concentration, and even small fluctuations can cause serious brain malfunction. The chemosensitive cells make synaptic connections with the inspiratory neurons, and stimulate them when the H^+ concentration becomes relatively high. Chemosensitive cells are also present at the bifurcation of the common carotid artery and the aortic arch. These receptors are termed peripheral chemoreceptors. Peripheral receptors respond to changes in the concentration of H^+, CO_2 and O_2. The rate of breathing is controlled more by fluctuations in H^+ and CO_2 levels than by fluctuations in O_2 levels. Hence, the respiratory stimulation due to low O_2 tension may be overridden by other simultaneously occurring changes to which a person is more sensitive, particularly changes in CO_2 and H^+ levels in the blood. It should be noted that only the peripheral chemoreceptors respond to decreased O_2 tension.

The concept that respiration is controlled at any instant by multiple factors is of great importance. Consider, for example, what happens when a person hyperventilates before holding his breath, and then does strenuous exercise. Hyperventilation causes CO_2 to be lost from the blood faster than it is produced by the cells of the body. The decreased CO_2 level inhibits the firing of chemoreceptors, and the respiratory rate tends to decrease. If the person then holds his breath and exercises, O_2 is lost, but CO_2 will not increase fast enough to immediately stimulate respiration. This is a very dangerous procedure. During exercise oxygen consumption is high. Because of our relative insensitivity to oxygen deficits, a rapidly decreasing oxygen level without a proportionately increasing CO_2 level may cause fainting, before ventilation is stimulated by the chemoreceptors.

Afferent nerves from the peripheral chemoreceptors enter the medulla and synapse with the medullary centers. A low oxygen content in the blood increases the rate of firing of the receptors. Normally, however, the much greater sensitivity to CO_2 and H^+ fluctuations in the blood prevent the oxygen level from dropping to such a low level.

Respiration is stimulated by other factors aside from variations in dissolved gas concentrations. Extremes of temperature, pain, initiation of exercise, and certain drugs all act to stimulate respiration. Respiration also changes with an individual's emotional state (probably via hormonal changes), and may also change by voluntary control.

• PROBLEM 16-10

Both food and air pass through the pharynx. Explain how these are channelled into the proper tubes so that the food goes into the stomach and the air goes to the lungs.

Solution: Under normal conditions the glottis (the opening to the larynx) is open and air passes freely into the larynx or voice box, and ultimately into the lungs. When we swallow, however, the larynx moves up so that the glottis is closed by the epiglottis (a flap of tissue) and the food or liquid passes into the esophagus behind the trachea. Swallowing is a complex reflex initiated when pressure receptors in the wall of the pharynx are stimulated. These receptors send impulses to the swallowing center in the medulla, which coordinates the swallowing process via nerves to the skeletal muscles in the pharynx, larynx, and upper esophagus, as well as the smooth muscles of the lower esophagus. Once swallowing has been initiated, it cannot be stopped. Following the swallowing reflex, the glottis once again opens, and remains open until the next swallowing reflex is initiated. Because of the placement of the epiglottis over the glottis, breathing ceases momentarily during swallowing. After the completion of the reflex, which takes about one second, breathing resumes.

• PROBLEM 16-11

Explain the physical changes which take place during inspiration.

Solution: Just prior to inspiration, at the conclusion of the previous expiration, the respiratory muscles are relaxed and no air is flowing into or out of the lungs. Inspiration is initiated by the contraction of the dome-shaped diaphragm and the intercostal muscles. When the diaphragm contracts, it moves downward into the abdomen. Simultaneously, the intercostal muscles which insert on the ribs contract, leading to an upward and outward movement of the ribs. As a result of these two physical changes, the volume of the chest cavity increases and hence the pressure within the chest decreases. Then, the atmospheric pressure, which is now greater than the intrathoracic pressure, forces air to enter the lungs, and causes them to inflate or expand. During exhalation, the intercostal muscles relax and the ribs move downward and inward. At the same time, the diaphragm relaxes and resumes its original dome shape. Consequently, the thoracic volume returns to its pre-inhalation state, and the pressure within the chest increases. This increase in pressure, together with the elastic recoil of the lungs, forces air out of the lungs causing them to deflate. The role of the diaphragm in breathing can be demonstrated by the figure below.

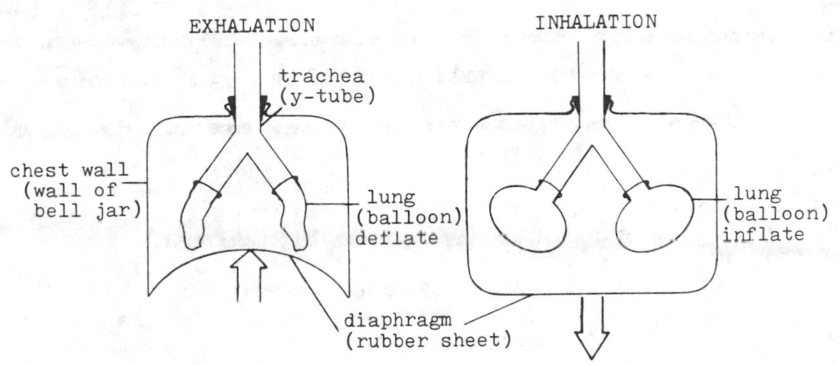

As the rubber sheet moves up, the volume in the bell jar decreases with a corresponding increase in pressure within the jar. This causes air to rush out of the y-tube, resulting in the collapse of the balloons.

When the rubber sheet moves down, the volume increases with a corresponding decrease in pressure. Thus the balloons inflate.

Model of how the diaphragm works.

• PROBLEM 16-12

Trace the path of an anesthetic, such as ether, from the ether cone over the nose to the cells in the brain.

<u>Solution</u>: The ether enters the body through the external nares, or nostrils, and would then enter the nasal cavities. The sense organs for olfaction are located here, and the individual would at this point smell the ether. The air containing the ether would be warmed, moistened and filtered here before passing via the internal nares into the pharynx. The pharynx is where the paths of the digestive and respiratory systems cross. From there the ether would pass through the glottis into the larynx, the trachea, the bronchi, the bronchioles and ultimately would reach the alveoli. At the alveoli, the ether will diffuse into the blood capillaries surrounding the alveoli, due to the concentration gradient of ether that exists across the alveolar membrane.

Diffusion is possible because of the extreme thinness and moistness of the membranes separating the pulmonary air from the blood in the capillaries. Electron micrographs have shown that these membranes, the alveolar epithelium and the endothelium of the capillary, are each one-cell thick. Once within the capillary, the ether will follow the circulatory routes to the heart. It is here that the blood containing the ether will be pumped to the brain, where the action of the gas will take effect.

• PROBLEM 16-13

Briefly discuss some abnormalities of the respiratory tract.

Solution: The respiratory tract is subject to various infections and damages which may interfere with normal functioning. Tuberculosis is caused by the bacteria Mycobacterium tuberculosis and results in the destruction of lung tissue. It was once a major cause of death. Infectious bacteria or viruses can bring about infections in the nasal cavities, the sinuses, the throat, the larynx (laryngitis), the bronchial tubes (bronchitis), and the lungs (pneumonia).

Pneumonia, which may result from infection by bacteria or viruses, results in the filling of the alveoli with fluid. As the infection spreads throughout the lungs, the surface area for gas exchange is reduced.

Allergic reactions may also occur in the respiratory tract. They occur when a person becomes sensitized to certain substances in the air, such as pollution, smoke or pollen. A person who is allergic to these substances has a condition known as hay fever. In such people, allergic reactions are initiated in the nasal cavities. Should these substances pass into the bronchioles and not be filtered in the nasal cavities, allergic asthma may result. The bronchiole tubes swell and become clogged with mucus, and the person has respiratory difficulty.

It has long been known that cigarette smoking can cause cancer in the lungs. It is not as widely known, however, that it can also result in emphysema. Emphysema results from a thickening of the lung tissues, which causes a loss of elasticity and a loss in the ability to absorb oxygen. In extreme cases the vital capacity is so reduced that breathing becomes a major effort. Since exercise requires increased oxygen consumption, physical activity is greatly reduced in people with emphysema.

Alveolar cells secrete a detergent-like chemical called surfactant, which intersperses with the water molecules in the alveoli. The air within each alveolus is separated from the alveolar membrane by an extremely thin layer of fluid. Surfactant markedly reduces the water molecules cohesive force, thereby decreasing the surface tension. If the surfactant were absent, this tension would be strong enough to cause the alveoli to collapse. In some newborn infants, predominantly those born prematurely, the lungs do not secrete an adequate amount of surfactant for their inspiratory forces to overcome the surface tension. Hence their alveoli are unable to expand when they inhale. Without treatment, these infants die of suffocation. The disease associated with insufficient surfactant production is referred to as respiratory distress syndrome. The recent discovery that the administration of hormones from the adrenal cortex enhances maturation of the surfactant-synthesizing cells may provide an important means of combatting this disease.

• PROBLEM 16-14

Why does one experience difficulty in breathing at high altitudes?

Solution: Barometric pressure progressively decreases as altitude increases. The air is still 21% oxygen, but the partial pressure of oxygen decreases along with barometric pressure. At sea level, the barometric pressure is 760 mm Hg, and thus, the partial pressure of O_2 is about 160 mm Hg. Arterial blood, under these conditions, contains nearly fully saturated hemoglobin. At an altitude of 9000 ft. the barometric pressure is about 525 mm Hg, and thus, the partial pressure of oxygen is 110 mm Hg. Under these conditions, there is only 90% saturation of the hemoglobin in arterial blood.

It must be understood that 90% saturation does not mean that some hemoglobin molecules are carrying less than four O_2 (recall that each hemoglobin molecule binds four O_2 molecules). The binding of oxygen to hemoglobin is cooperative; that is, the binding of the first oxygen molecule facilitates the binding of the second molecule, which in turn makes it easier to bind the third molecule, and so on. Therefore, through cooperative binding, a hemoglobin molecule binds four O_2 or none at all, and 90% saturation means that 10% of the hemoglobin molecules have no oxygen bound to them at all. The higher the altitude, the lower the percentage of hemoglobin saturated with oxygen. Thus at high altitudes, hypoxia, or the deficiency of oxygen at the tissue level, becomes a problem.

The effects of oxygen deprivation vary with individuals. Most persons who ascend rapidly to altitudes of about 10,000 ft. experience breathlessness, nausea, fatigue, decreased mental proficiency, and ultimately become comatose if the lack of oxygen is too severe. These symptoms will disappear if an individual remains at high altitudes for several days, but full physical capacity will be diminished.

The initial acute responses to high altitudes are increased rate and depth of breathing, increased cardiac output and increased heart rate. The latter increases the rate and volume of blood flow to the alveoli, enhancing CO_2 and O_2 exchange and insuring enough oxygen for every cell. As ventilation increases, thereby increasing the oxygen supply, the heart rate returns to normal. During this same time, erythrocyte and hemoglobin syntheses are stimulated by erythropoietin, a hormone produced by the kidney. Consequently, the total circulating red cell mass increases considerably. The production of more red cells, and the synthesis of more hemoglobin together compensate for the low partial pressure of oxygen. There are now more cells transporting oxygen, although each cell is carrying less oxygen than at atmospheric pressure.

Compared to the rapid increases in respiratory rate and heart rate, erythropoiesis is a slow process. A second adaptation of the body to an even greater altitude involves an increase in capillaries. The mechanism that stimulates the growth of new capillaries is unknown, but the result is a decrease in the distance which oxygen must diffuse to reach the tissues. Also, a compound 2,3-DPG increases in red blood cells. DPG decreases the affinity of hemoglobin for oxygen which eases the unloading of oxygen at the cells. In time, these three slow compensatory phenomena can once again allow a person to function effectively at higher altitudes.

• **PROBLEM** 16-15

Sea divers are aware of a danger known as the "bends". Explain the physiological mechanism of the bends.

Solution: In addition to hypoxia (lack of oxygen at the tissue level), decompression sickness may result from a rapid decrease in barometric pressure. In this event bubbles of nitrogen gas form in the blood and other tissue fluids, on the condition that the barometric pressure drops below the total pressure of all gases dissolved in the body fluids. This might cause dizziness, paralysis, and unconsciousness, and it is this set of symptoms that describes the condition known as the "bends".

Deep sea divers are greatly affected by the bends. Divers descend to depths where the pressure may be three times as high as atmospheric pressure. Under high pressure, the solubility of gases (particularly nitrogen) in the tissue fluids increases. As divers rise rapidly to the surface of the water, the accompanying sharp drop in barometric pressure causes nitrogen to diffuse out of the blood as bubbles, resulting in decompression sickness.

EVOLUTIONARY ADAPTATIONS

• **PROBLEM** 16-16

Describe the major trends in the evolution of the lungs.

Solution: When the vertebrates adopted a land habitat, the gill slits and the gills disappeared in the adults, and their physiological role was taken over by a pharyngeal outgrowth. This outgrowth already existed in some fish (i.e. the lung fish, the lobe-finned fish, and some bony fish), and is called the swim bladder. In these fish, this bladder is used primarily as a device to facilitate floating and maintain buoyancy, though it may also function as a respiratory organ. Swim bladders can occur in pairs, but they usually occur as a single structure. In some fish the swim bladder can be seen to have lost its pharnygeal connection. At the anterior end of the swim bladder are specialized cells which transfer oxygen from the blood into the bladder. This unique ability is unparalleled anywhere in the animal kingdom. In addition to these anterior cells, there are also specialized cells at the posterior end of the bladder, which function in the transference of oxygen from the bladder to the blood. Because of the pumping action of these two groups of cells, the fish can vary its buoyancy to maintain a certain depth in the water without muscular effort. While the swim bladder has been found to serve as an organ for the production of sound, this is relatively rare and is found in only a few fish.

Some stages in the evolution of the lungs.

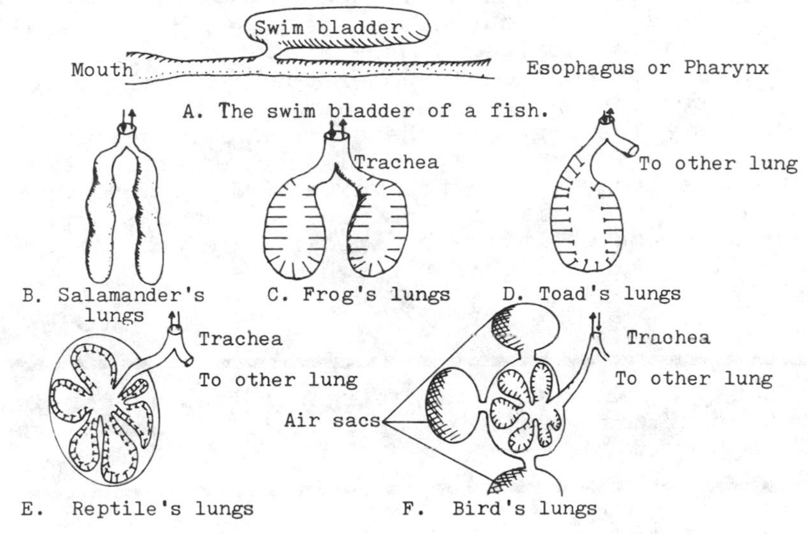

It is believed that the tetrapods evolved from the group of fish called the lobe-finned fish. These fish, which were once thought to be extinct but were rediscovered in 1939, have a swim bladder which can be used as a lung. In contrast to the bladder of other fish, the swim bladder of the lobe-finned fish is equipped with a pulmonary artery. These fish can force air into their swim bladder, which functions as a lung. In the event that their pond would dry up, the fish would survive by breathing air directly,and with the aid of their lobe fins they would be able to traverse land to other streams. The presence of a "lung" and lobe fins facilitated the evolution of these creatures into the first tetrapods, the amphibians.

The lungs of the tetrapods are homologous (arise embryologically in the same way) to the lungs of the lobe-finned and lung fish. The lungs of the mud puppies (the most primitive amphibians) are two long simple sacs, the outside of which is covered by capillaries. In frogs and toads, the surface area of the lung is increased by folds on the inside of the lung sacs. The method of breathing in frogs and toads is much different from that of humans, because of the absence of a diaphragm and rib muscles. Valves in the nostrils and muscles in the throat are incorporated into the respiratory anatomy of these amphibians.

Following the amphibian stage of lung development, the general evolutionary trend was toward a greater subdivision of the lung into smaller and smaller sacs. In this respect, the structure of the lung became increasingly complex in reptiles, birds and mammals. In some lizards, for example, the lungs have supplementary air sacs which, when inflated, are thought to be a protective device to frighten predators.

The respiratory system of birds shows some remarkable peculiarities. Birds possess large, thin-walled air sacs scattered throughout the body. These are situated amongst the muscles and within the interior of bones. Bronchial tubes connect these air sacs to the lung. The alveoli are spongy masses of minute air pockets where gas exchange takes place, and occur along the sides of the smaller bronchial tubes called bronchioles. The functioning of the system is not completely understood, but it appears that air passes through the lungs, and then into and out of the air sacs. The air sacs therefore act as bellows, so that the lungs are completely ventilated at each breath. This system is lacking in other vertebrates, where considerable amounts of used air remains behind in the lungs. The air sacs also act to lower the specific gravity of the bird, which is advantageous for flight.

• PROBLEM 16-17

In aquatic animals, external respiration is carried out by specialized structures called gills. How do these gills operate?

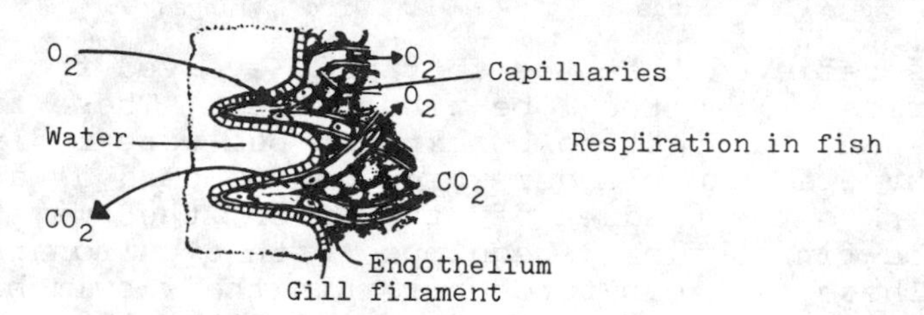

Solution: **Fish, molluscs and many arthropods (except insects) possess gills for respiratory purposes (see** figure). Organisms such as these require a mechanism to keep a fresh supply of water flowing about them. This is necessary because gas exchange occurs between the **blood vessels in the gills and the water which contains** dissolved oxygen. Continuous bathing of the gills ensures that there will always be enough oxygen available for the organism. It also ensures that the carbon dioxide diffusing from the gills into the water is removed from the organism. This is important in that the CO_2 concentration gradient is maintained, so diffusion of CO_2 will continue to occur from the gills to the water.

A fish opens its mouth and receives a quantity of water. It then closes its mouth and forces the water out past its gills by contracting its oral cavity. Gills have thin walls, are moist, and are well supplied with blood capillaries. Oxygen, which is dissolved in the water, diffuses though the gill epithelium and into the blood capillaries. Simultaneously, carbon dioxide diffuses in the opposite direction in accordance with the CO_2 concentration gradient. The amount of oxygen dissolved in sea water is relatively constant, but the amount in freshwater ponds may fluctuate greatly. A fish would suffocate in water that is lacking sufficient amounts of dissolved oxygen.

It is interesting to note that the direction of blood flow through a gill is opposite to the direction of water flow over the gill. This counter-current system maximizes the amount of gas exchange that can take place. If both the blood flow and water current were in the same direction, the CO_2 and O_2 concentration gradients across the gills would decrease. This would result in a slower rate of diffusion, and the amount of gas exchange would be reduced.

• PROBLEM 16-18

Gills are very important for the proper exchange of gases between an aquatic organism and its environment. Using the fish as an example, describe the structure of a gill.

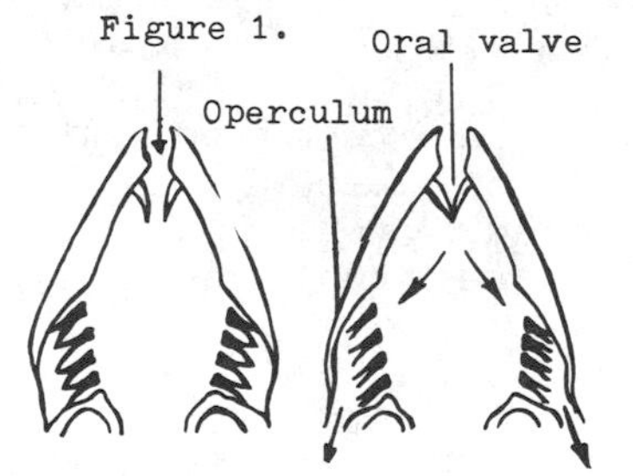

Figure 1. Water flow through the gills.

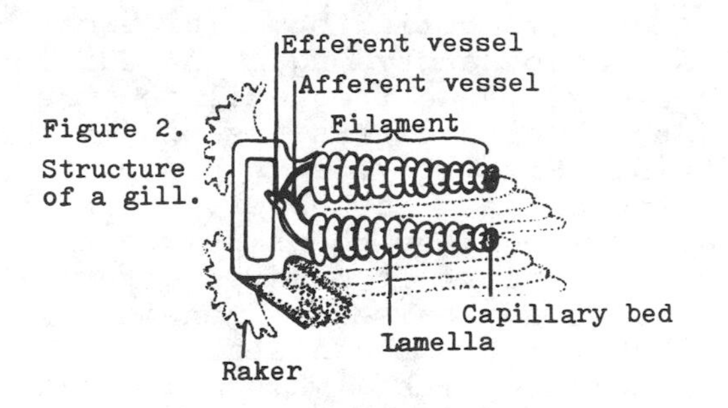

Figure 2. Structure of a gill.

Solution: The gills of a fish are located on both sides of the animal, slightly posterior to the eyes. A bony plate known as the operculum completely covers each gill except at the posterior margin (see figure 1).

This opening allows water, which has entered through the mouth and passed over the gill, to exit from the body. The gill is composed of many filaments, each subdivided into numerous lamellae which serve to increase the surface area for gas exchange (see Fig. 2). It is across these lamellae that gas exchange occurs. Running through the filaments are two blood vessels called the afferent and efferent blood vessels. The former carries deoxygenated blood to the gill, while the latter carries freshly oxygenated blood to the rest of the body. The gill filaments are supported at their base by bony extensions of the ventral skeletal column, called gill arches. The filaments themselves do not have any other means of support, because they receive sufficient support from the water around them.

The blood vessels in the filaments are separated from the water by only two layers of cells: the single cell layer of the wall of the blood vessel and the single layer of cells making up the filament surface. Oxygen moves by diffusion from the water, across the intervening cells and into the blood. This blood then leaves the gill through the efferent vessel and transports the oxygen to the rest of the body.

• PROBLEM 16-19

Compare the methods of obtaining oxygen in the earthworm, the frog, the fish and man.

Solution: The earthworm belongs to the group of invertebrates called the Annelids, or roundworms. The earthworm has no specialized respiratory system. Respiration takes place by diffusion of gases through the body integument or skin. The skin is kept moist by the secretions of mucous glands directly underneath the skin. The thin cuticle is quite permeable to both oxygen and carbon dioxide. The red blood cells contain dissolved hemoglobin, which aids in the transportation of oxygen to various parts of the body.

In most adult amphibians (ex. frogs) the moist skin, which is supplied by a large capillary network, functions as a respiratory organ in addition to the lungs. When submerged, the frog utilizes the dissolved oxygen in the water, which enters its body via diffusion through the skin. Similarly, carbon dioxide produced by the frog diffuses into the water. When on land, respiration is accomplished by the skin and the lungs. The lower part of the throat is covered with a flap of skin that can be lowered and raised. When lowered, air comes into the mouth through the nostrils; when raised, the nostrils are closed with a valve-like flap of skin and the air is forced down into the lungs. Unlike man, the frog does not have a diaphragm to facilitate breathing. The larval frog has a method of respiration entirely different from that of the adult; it is an aquatic animal and breathes via gills. Fish and other complex water animals have gills, as their main respiratory organ. The gills are made of finely divided structures that are very thin, and richly supplied with blood capillaries. Diffusion of gases occur readily across the thin surface of the gills. Gills are principally a respiratory adaptation to an aquatic environment.

Gas exchange in man takes place in the lungs. Lungs can be thought of as organs that serve the same purpose in terrestrial animals as gills do in aquatic animals. The lungs are located internally in order to prevent dessication and possible damage.

It should be noted that as an animal becomes larger and more complex, and more metabolic energy is needed, the surface area for gas exchange likewise increases.

This facilitates the uptake of oxygen in addition to the release of carbon dioxide. The lamellar, filamentous surface of the gills, and the extensive branching of the bronchioles into alveoli in the lungs are adaptations towards increased surface area for gas exchange.

• PROBLEM 16-20

How does respiration occur in insects?

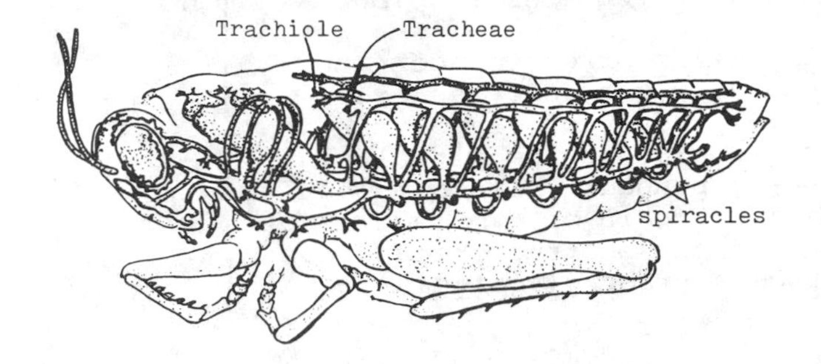

Tracheal system of the grasshopper.

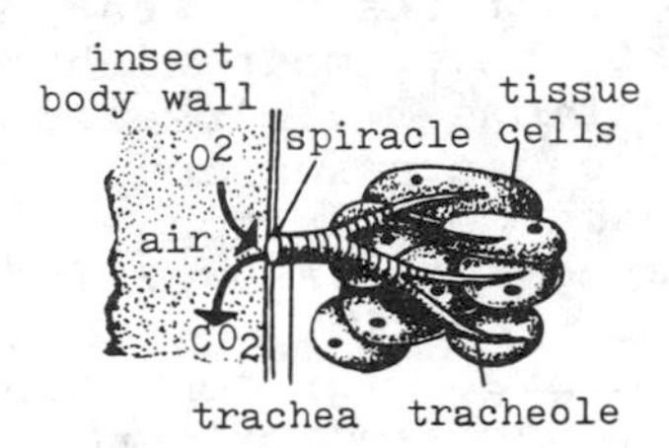

Terminal tracheole

Solution: The unique respiratory system of insects consists of a network of tubes called trachea (see figure 1). These trachea open to the outside by means of holes in the insects body. These holes are called spiracles and are conspicuous on the sides of the thorax and abdomen. Each spiracle is guarded by a valve which can be opened or closed to regulate air flow. The tracheal tubes extend to all the internal organs. The tubes terminate in microscopic fluid-filled tracheoles; oxygen and carbon dioxide diffuse through these into the adjacent cells. The body wall of the insect pulsates, drawing air into the trachea when the body expands, and forcing air out when the body contracts. Grasshoppers draw air into the body through the first four pairs of spiracles when the abdomen expands and expel it through the last six pairs when the body contracts. Therefore in contrast to a fish or a crab, which respire through gills, the tracheal system conducts air deep within the insect's body. The air is brought near enough to each cell so that gases can diffuse across the wall of the tracheal tube. For this reason, insects need not maintain a rapid blood flow, as vertebrates must, to supply their cells with oxygen.

SHORT ANSWER QUESTIONS FOR REVIEW

Answer

To be covered when testing yourself

Choose the correct answer.

1. The process by which gases are exchanged in the alveoli of our lungs is called: (a) external respiration (b) indirect respiration (c) internal respiration (d) direct respiration. — b

2. Which of the following is incorrect concerning glycolysis? (a) glycolysis can occur anaerobically. (b) the final product of glycolysis is 2 molecules of pyruvate, (c) during glycolysis, 2 molecules of ATP are expended, while 6 molecules are produced, making a net gain of 4 molecules of ATP, (d) hexokinase and phosphofructokinase are important enzymes necessary for glycolysis to occur. — c

3. Which of the following conditions will cause oxygen to be released from hemoglobin with the greatest ease? (a) conditions of increased CO_2 tension, increased pH, and decreased temperature, (b) conditions of decreased CO_2 tension, decreased pH, and increased temperature (c) conditions of increased CO_2 tension, decreased pH, and increased temperature, (d) conditions of increased CO_2 tension,increased pH, and increased temperature. — c

4. An individual's tidal volume can be represented by which of the following choices? (a) vital capacity - (expiratory reserve volume + residual volume) (b) vital capacity - (inspiratory reserve volume + residual volume) (c)(inspiratory reserve volume + expiratory reserve volume) - vital capacity. (d) vital capacity - (inspiratory reserve volume + expiratory reserve volume). — d

5. Which of the following is incorrect concerning the control of respiration? (a) the pneumotaxic center, located in the pons, can be said to exert an inhibitory effect over the medullary inspiratory neurons, (b) during normal expiration, impulses pass from the medullary expiratory center to all the muscles that aid in expiration, (c) the lungs themselves assist in the neural control of respiration, (d) in the medulla, there is direct inhibitory communication between the expiratory and inspiratory neurons; i.e., they inhibit each other. — b

6. Which of the following conditions would have the greatest effect on the rate of respiration? (a) low oxygen tension around vicinity of central chemoreceptors. (b) low oxygen tension around

Answer

To be covered when testing yourself

vicinity of carotid bifurcation (c) high carbon dioxide tension around vicinity of carotid bifurcation (d) high carbon dioxide with correspondingly low pH around central chemoreceptors. — d

7. Which of the following mechanisms cause air to rush into our lungs during inspiration? (a) the diaphragm moves up, while the external intercostal muscles contract, causing the chest cavity to decrease in size. That raises the intrathoracic pressure, and air rushes in, (b) the diaphragm moves down, while the internal inter - costal muscles relax, causing the chest cavity to increase in size. That raises the intrathoracic pressure above atmospheric pressure, and air rushes in, (c) the diaphragm contracts and moves downward, while the internal intercostal muscles contract, causing the ribs to move upward. The result is a larger chest cavity, causing intrathoracic pressure to fall below that of atmospheric pressure. Air will then rush in, (d) none of the above. — c

8. Choose the <u>incorrect</u> statement, (a) pneumonia, a condition characterized by fluid-filled alveoli, can be both of viral or bacterial origin, while the disease tuberculosis is specifically caused by bacteria, (b) emphysema patients must avoid strenuous activity, as their gas exchange ability is severly impaired due to thickened alveolar tissue, (c) respiratory distress syndrome is characterized by a lack of capillary flow around some alveoli, causing a decrease in gaseous diffusion, (d) since surfactant lowers the surface tension of an air-liquid interface, it can be stated that one of its functions is to equalize the size of alveoli. — c

9. Which of the following does <u>not</u> occur in response to high altitude breathing? (a) increased cardiac output and increased rate of respiration, (b) decreased urine production and excretion, (c) increased synthesis of red blood cells, (d) the formation of new capillaries. — b

10. Which of the following statements gives the basic physiological reason for the occurence of the "bends" in some deep sea divers? (a) the solubility of a gas in our blood increases with a decrease in barometric pressure (b) regardless of depth, nitrogen in our blood remains at the same solubility, (c) the solubility of a gas in our blood decreases with a decrease in barometric pressure (d) our lungs cannot be expanded at depths below 200 feet below sea level. — c

Answer

To be covered when testing yourself

11. Choose the incorrect statement concerning the "swim bladder." (a) this organ can be found in the bony fish group and is considered a pharyngeal outgrowth, (b) there is differentiation amongst the cells that comprise the organ, (c) it can be used for buoyancy purposes and in a few fishes, it is even used as a sound device, (d) this organ is never equipped with any vascularization. — d

12. What is meant by the "counter-current system" with regard to the respiratory mechanism of gilled-fish? (a) CO_2 and O_2 flow in opposite directions to each other, (b) the water inside the gills flows in the opposite direction to that of the water flow outside the gill, (c) there is no such thing as a "counter-current system" in fish, (d) blood inside the gills of a fish flows in the opposite direction to the water flow over the gill. — d

13. What function does the operculum of fish serve? (a) as an exit for waste products (b) as an entrance for water (c) as a protective device for the gills (d) as a protective device for the eyes. — c

14. Which of the following is not an advantageous evolutionary adaptation concerned with the process of respiration? (a) the proximity of the lungs in relation to the heart, (b) the increased surface area for gas exchange, (c) the shifting of the respiratory organs from outside the body to inside the body, (d) the appearance of a diaphragm. — a

15. Which of the following is incorrect concerning the respiratory system of a grasshopper? (a) the system consists of a network of small tubes called tracheal tubes that lead into smaller fluid-filled tracheoles. (b) air enters the tracheal tubes via openings called spiracles located on the thorax and abdomen of the grasshopper. (c) all spiracles conduct air in when the body expands, and conduct air out when the body contracts. (d) the tracheal system conducts the air close enough to the insect's cells, allowing for complete nourishment without the aid of a rapid blood flow. — c

Fill in the blanks.

16. Glycolysis occurs exclusively in the _____, — cytoplasm

	Answer To be covered when testing yourself
while the Krebs Cycle and the Electron Transport System occur only in the _____. In terms of ATP production, the mitochondrial reaction sequences yield a _____ (greater, lesser) amount of molecules per turn than does the cytoplasmic sequence, but only under _____ conditions.	mitochon-dria, greater, aerobic.
17. The majority of the oxygen in our blood is carried via _____, while the majority of the carbon dioxide in our blood is carried in the form of _____.	hemo-globin, bicar-bonate.
18. Inhaled air doesn't reach the alveoli for gas exchange, but rather remains in the conducting passageways called the ____ space. As a result of this phenomenom, alveolar air contains _____ (more, less) oxygen and _____ (more, less) carbon dioxide than _____ air.	anato-mical dead, less, more, atmos-pheric.
19. Air entering the nose passes through the internal nares, and from there, enters the _____. The larynx is the next landmark passed, air entering it via an opening called the ____, which is where the _____ are located. Air then passes through the trachea, which divides eventually into two _____.	pharynx, glottis, vocal cords, main bronchi.
20. Both peripheral and central chemoreceptors are sensitive to changes in the blood levels of _____ and _____, but just the peripheral chemo-receptors are sensitive to changes in the blood level of _____.	H^+, CO_2, O_2
21. When swallowing, the raising of the _____, with the concurrent closure of the glottis by the _____, prevents food from entering the _____.	larynx, epiglot-tis, trachea.
22. During exhalation,the intercostal muscles _____ (contract, relax) and the ribs move _____ (up, down) and _____ (out, in).	relax, down, in
23. Cigarette smoking can cause lung cancer as well as _____, which is a result of the thickening of the lung tissues.	emphy-sema
24. At high altitudes, there is a _____ (higher, lower) percentage of hemoglobin saturated with oxygen,therefore ____ occurs which is a deficiency of oxygen at the tissue level.	lower, hypoxia
25. Decompression sickness results in the formation of _____ gas bubbles.	nitrogen

	Answer
	To be covered when testing yourself
26. Tetrapods have a _____ which can be used as a lung	swim bladder
27. Insects need _____ (to, not) maintain a rapid blood flow due to air being brought very close to each cell.	not
28. As metabolic energy needs increase, the evolutionary trend is toward _____ (greater, less) surface area for gas exchange.	greater
29. Surfactant markedly _____ (increases, decreases), the water molecules' cohesive force, therefore _____ the surface tension.	decreases reducing
30. _____ temperature and _____ acidity increase the percent of hemoglobin saturated with oxygen.	low, low
Determine whether the following statements are true or false.	
31. Respiration is an energy consuming process and breathing is an energy producing process.	False
32. Glycolysis occurs only in the presence of oxygen.	False
33. In the following reaction, a decrease in the concentration of CO_2 drives the reaction to the left, $CO_2 + H_2O \rightleftarrows H_2CO_3 \rightleftarrows H^+ + HCO_3^-$	True
34. As the blood becomes more basic the capacity of hemoglobin to carry oxygen increases.	True
35. In the lung capillaries there is low oxygen tension and high CO_2 pressure. In the tissues the reverse is true.	False
36. The vital capacity includes the inspiratory reserve, tidal volume, expiratory reserve and residual volume.	False
37. Physiological dead space is the space within the conducting tubes where no gas exchange with the blood takes place.	True
38. The trachea can be distinguished from the esophagus, in a dissection, by its cartilaginous rings.	True
39. The firing of the medullary inspiratory neurons and the regulation of rhythmicity of these neurons is completely inherent within them.	False

	Answer To be covered when testing yourself
40. As the carbonic acid concentration increases, there will be a corresponding increase in the rate of breathing.	True
41. Respiratory distress syndrome is associated with insufficient surfactant production.	True
42. The swim bladder in some fish function in transferring carbon dioxide from the blood into the bladder.	False
43. The direction of blood flow through a gill is in the same direction as water flow over the gill.	False
44. The insect repiratory system includes spiracles which always remain open for air flow.	False
45. Frogs breathe without the use of a diaphram.	True

CHAPTER 17

NUTRITION

NUTRIENT METABOLISM

• **PROBLEM** 17-1

What roles does glucose play in cell metabolism?

Solution: Glucose, a six carbon monosaccharide, is the primary source of energy for all cells in the body. The complete oxidation of one molecule of glucose yields 36 ATP molecules in a cell. The energy stored in the form of ATP can then be used in a variety of ways including the contraction of a muscle or the secretion of an enzyme. The supply of glucose to all cells must be maintained at certain minimal levels so that every cell gets an adequate amount of glucose. Brain cells, unable to store glucose, are the first to suffer when the blood glucose concentration falls below a certain critical level. On the other hand, muscle cells are less affected by changes in the glucose level of the blood because of their local storage of glucose as glycogen.

In certain cells, glucose can be converted to glycogen and can be stored only in this form, because glucose will diffuse into and out of a cell readily. Glycogen is a highly branched polysaccharide of high molecular weight, composed of glucose units linked by α-glycosidic bonds. Once the conversion is complete, glycogen remains inside the cell as a storage substance. When glucose is needed, glycogen is reconverted to glucose.

In addition to being stored as glycogen or oxidized to generate energy, glucose can be converted into fat for storage. For example, after a heavy meal causes the supply of glucose to exceed the immediate need, glucose molecules can be transformed into fat which is stored in the liver and/or adipose (fat) tissue. This fat serves as an energy supply for later use.

• **PROBLEM** 17-2

What difference can be observed in carbohydrate metabolism between liver and skeletal muscle cells?

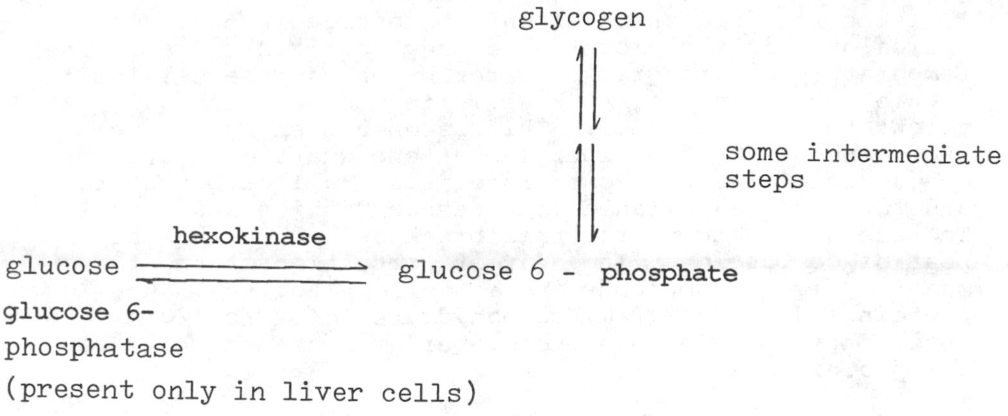

glycogen

⇅ some intermediate steps

glucose ⇄ glucose 6 - phosphate (hexokinase forward)

glucose 6-phosphatase (present only in liver cells)

Diagram outlining the interconversion of glucose and glycogen.

Solution: Both liver cells and muscle cells have the ability to store excess glucose as glycogen. Glucose is not transformed directly into glycogen. The first step in this conversion involves the change of glucose into glucose 6-phosphate. This ATP-dependent reaction is catalyzed by hexokinase. After several reversible intermediate steps, glucose 6-phosphate molecules are condensed to form the polysaccharide glycogen.

The major difference in carbohydrate metabolism between liver and muscle cells is the presence of the enzyme, glucose 6-phosphatase in liver cells. This enzyme converts glucose 6-phosphate to free glucose. The unphosphorylated glucose can then enter the circulatory system. The presence of glucose 6-phosphatase allows the liver cells to break down glycogen into glucose molecules and release the latter into the bloodstream. Thus, one of the functions of the liver is to regulate glucose level in the blood by releasing or storing the monosaccharide as the situation demands. Muscle cells are unable to release free glucose and thus cannot regulate the blood glucose level because it lacks glucose 6-phosphatase. However, muscle cells can still store glucose as glycogen due to the presence of hexokinase.

• PROBLEM 17-3

What are the major types of food and how much energy does each type yield?

Solution: The food consumed by a human is a combination of carbohydrates, fats, proteins, water, mineral salts, and vitamins. All of these are essential to life but only the carbohydrates, fats and proteins can provide energy. The vitamins and minerals act as coenzymes and chemical cofactors in many metabolic reactions.

The standard chemical unit of energy is the calorie. One calorie is the amount of energy required to raise the temperature of one gram of water by one degree Celsius. In biology, kilocalorie (kcal.) is used and is equivalent to 1000 calories. This is enough energy to raise the temperature of a liter (about one quart) of water by one degree Celsius. Most proteins and all carbohydrates and fats can be degraded into carbon dioxide and water. The energy released from this breakdown is called the heat of combustion. The heat of combustion of the three major types of compounds are as follows:fat - 9.5 kcal/gram, protein - 4.5 kcal/gram, carbohydrate - 4.0 kcal/gram. Thus, fats contain more energy per gram than carbohydrates and proteins.

Physiologically, more than one half of the energy of combustion is lost as heat while the rest (about 38%) is converted into cellular energy in the form of ATP. The actual yield from one mole of glucose is 36 ATP plus heat, and the yield from a mole of fat is 463 ATP plus heat. These are equivalent to .21 mole ATP/(gram glucose) and .55 moles ATP/(gram fat) respectively. In terms of the energy actually produced in the cell, fats produce twice that of carbohydrates. Proteins yield about as much energy as carbohydrates on a per gram basis. Proteins, however, are rarely fully metabolized. Their constituent amino acids are used for cellular protein production. It takes less energy for the cell to utilize existing amino acids than to degrade them and then synthesize new ones.

• PROBLEM 17-4

What is meant by the term basal metabolic rate? What are some conditions that cause the rate to change?

Solution: The "basal metabolic rate" refers to the amount of energy expended by ones body under resting conditions. For example, sitting at rest in a room at a comfortable temperature requires 100 kcal. each hour for an "average" 70 kg. man. If another man who weighs 70 kg. requires only 95 kcal. per resting hour, his metabolic rate is 5% below the norm.

The basal metabolic rate is the measure of the amount of energy utilized just to keep alive, when no food is being digested and no muscular work is being done. The average minimum energy requirement for young adult men is 1600 kcal. each day, and is about 5% less for women. To measure the basal rate, the subject is rested for 12 hours after the last meal, and reclined for 30 minutes at room temperature (18°-26°C). Pure oxygen is inhaled for an hour. It is known that the body absorbs only one fifth of the oxygen inhaled and that the body produces 4.8 kcal of energy for each liter of oxygen absorbed. Therefore, the basal metabolic rate can be calculated by measuring a person's oxygen consumption over a given time span. The amount of heat produced can be determined from the amount of oxygen consumed. By directly measuring the heat given off by a person in an insulated chamber, one can also measure the basal metabolic rate directly, but this is a more awkward process. The energy requirement per square meter of body surface can also be calculated, and is found to be directly proportional to one's basal metabolic rate (BMR). The average energy expenditure is 40 kcal. per square meter. The BMR is a useful indicator of the state of one's metabolism.

This is important for the diagnosis of certain hormonal disorders such as diabetes or goiter.

Several factors influence one's metabolic rate. Children have high metabolic rates and aging adults have low rates. The female metabolic rate is below that of a male of similar size but it increases during pregnancy and lactation. Digestion requires about one tenth of the metabolic energy. Any infection or disease raises the metabolic rate and causes fever. Fever, in turn, raises the rate further and results in a loss of weight. Hormones and environmental conditions also play a major role in determining one's metabolic rate. Mental tension increases adrenalin production, raising the metabolic rate, blood pressure and heart rate.

Adrenalin production can only continue for a few hours. The body will adjust to the presence of the stimulant and the sudden termination in its production drastically lowers the metabolic rate far below normal. There is a slowing of the heart rate, a drop in blood pressure,a shifting of blood to the intestines, and a decrease in blood sugar level. The largest influence on metabolism is muscular activity. During hard work, one's metabolic rate will double and may even reach fifteen times the BMR in brief bursts of activity.

• **PROBLEM** 17-5

A biochemist performed several experiments on rats. In one, he used two groups of young rats. Group A rats were fed on a diet of purified casein (cheese protein), starch, glucose, lard, minerals, and water. Group B rats were fed the same diet with the addition of 3 cm^3 of milk daily. It was observed that Group A rats stopped growing and lost weight, while Group B rats gained steadily in weight and size. After eighteen days the milk was given to Group A rats and removed from the diet of Group B. Group A rats now resumed growth and gained in weight, while Group B rats stopped growing and lost weight.

The investigator knew that 3 cm^3 of milk has an insignificant food value in terms of carbohydrates, fats, proteins and minerals. Consider yourself in his position, confronted with only the information given so far.

a) What conclusions could you reach about the presence in milk of substances other than these four foods, the possible functions of the hypothetical substances, and the quantities needed by the rats?

b) Why was it necessary to transfer the milk from Group B rats to Group A rats half-way through the experiment?

c) The rats received only one type of protein (casein). Why can protein starvation be ruled out as a possible cause of growth inhibition and weight loss?

Solution: One can conclude that some factors responsible for growth and weight increase are present in milk. Such factors are called vitamins. Vitamins are complex chemicals which are required in minute amounts in the diet of all heterotrophic organisms. Autotrophic organisms, such as green plants, also need to procure certain essential vitamins from the environment for their growth.

Vitamins have no energy value. They play a vital role in many chemical reactions in metabolism. Most vitamins function as coenzymes. Lack of any vitamin in the diet causes the reaction in which it takes part to slow down, and since most metabolic reactions are part of a long sequence of events, an alteration in the rate of any one of them can have widespread effects in the body. Since 3 cm^3 of milk is enough for normal growth in rats,it is shown that only small quantities of vitamins are needed.

b) By interchanging the diet of Group A and Group B a control experiment is performed, demonstrating that the lack of growth is due to the diet alone and not to any constitutional defects in the rats. I.e, when normally growing rats in Group B are fed with the vitamin-deficient diet,they stop growing.

c) Casein, a milk protein, is an example of an adequate protein. An adequate protein is one which contains all the essential amino acids necessary for growth. Since the essential amino acids are present, we can rule out protein starvation, a deficiency in essential amino acids, as a cause of the observed stunted growth.

• PROBLEM 17-6

A paramecium is introduced to a medium rich in yeast. How do the yeast cells get into the food vacuole of the paramecium? What is the fate of the yeast in the food vacuole?

Solution: Paramecia are ciliates (organisms covered with cilia over their entire surface). On one side of the paramecium there is a slight depression, the oral groove, which serves as a mouth. A paramecium feeds by creating a current (by the beating of the cilia around the oral groove) which sweeps the food particles into the oral groove. In this case, the food particles are the yeast cells. Under pressure from the current, the yeasts and fluids are pushed into the cytoplasm at the base of the groove, called the cytopharynx, and a food vacuole is formed. The vacuole breaks away from the groove and circulates around the cell, diminishing in size as it moves. As the vacuole circulates enzymes made in the cytoplasm are secreted into it. These enzymes break down the yeasts into simple biomolecules (sugars, amino acids, etc.) which diffuse out of the vacuole into the cytoplasm to be used later for the organism's metabolic processes.

When the vacuole reaches a tiny specialized region of the cell surface called the anal pore, it becomes attached to the pore and then ruptures, expelling to the outside any remaining bits of indigestible material. The sequence of events is reviewed in the figure below. Since the enzymatic hydrolysis of foodstuff occurs inside the cell of the paramecium, digestion in this organism is of the intracellular type, known as intracellular digestion.

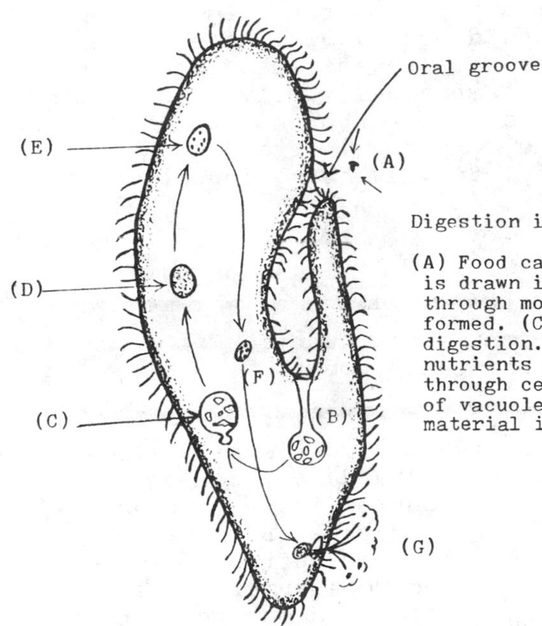

Digestion in Paramecium

(A) Food caught in current produced by cilia, is drawn into oral groove. (B) Food has passed through mouth into cytopharynx; vacuole being formed. (C) Lysosome fuses with vacuole, starting digestion. (D), (E), (F) Digestion continues, nutrients diffusing out of vacuole as it moves through cell; vacuole shrinks. (G) What remains of vacuole fuses with anal pore, where indigestible material is ejected.

• PROBLEM 17-7

Identify the minerals most essential to the body and give the function of each.

Solution: About fifteen minerals are known to be essential in the human diet. A few of these are required in minute quantities and are referred to as trace elements. The essential minerals include sodium, chlorine, potassium, magnesium, phosphorus, calcium, iron, copper, manganese and iodine.

Sodium chloride is a major salt in blood and other body fluids. Sodium and chloride ions play an important role in maintaining osmotic balance and acid-base balance in the body fluid. Secretions of the digestive tract such as the hydrochloric acid of the stomach and the pancreatic and intestinal juices, contain a large amount of sodium and chloride ions. Potassium and magnesium are required for muscle contraction and for the functioning of many enzymes. Calcium and phosphorus form the major components of bones and teeth, and are thus essential for maintenace of locomotion and support of the body. Phosphorus is also important in the synthesis of DNA, RNA, NAD^+ (NADH), $NADP^+$ (NADPH) and ATP. Moreover, phosphorylation (the addition of phosphate) of sugars and glycerol must take place before they can be catabolized to produce biologically useful energy.

Trace elements such as iodine, iron, copper and zinc generally function as metal or non-organic components of specific enzyme systems. Iodine is a constituent of thyroxine, the hormone secreted by the thyroid gland. Hemoglobin and the cytochromes contain iron. Small amounts of copper are necessary in the diet to facilitate proper utilization of iron for normal growth and as a component of certain enzymes. Traces of manganese, molybdenum, zinc and cobalt are required for normal growth and as activators of certain enzymes. Fluorine, present in trace amounts in drinking water,is effective in preventing dental decay.

• PROBLEM 17-8

Define the term vitamin. Describe the function of each essential vitamin needed by the human body.

Solution: Vitamins are relatively simple organic compounds required in minute quantities and not used as an energy source. Vitamins are required in very small quantities because they serve as coenzymes or as parts of coenzymes. A coenzyme is a substance required by an enzyme for catalytic activity. Since coenzymes, like enzymes, are not altered in a reaction and may be used repeatedly, and for more than one enzyme, only small amounts are needed.

There are two principal groups of vitamins: the fat-soluble vitamins (A, D, E, K) and the water-soluble vitamins (B complex and C). Vitamin A is necessary for the growth and maintenance of the epithelial cells of the skin, various tracts in the body and the eye. It is also a precursor to the visual pigment, retinaldehyde, in the eye. A deficiency of vitamin A results in lowered resistance of skin to infection and in night blindness.

Vitamin D can be synthesized by the skin in the presence of ultraviolet light. Vitamin D is required for normal growth of the skeleton and teeth because it controls the levels of calcium and phosphorus in the body. Malformation of the bones, known as rickets, is associated with a lack of vitamin D.

Some vitamins needed by man

Vitamin	Some deficiency symptoms	Important sources
FAT-SOLUBLE		
Vitamin A (retinol)	Dry, brittle epithelia of skin, respiratory system, and urogenital tract; night blindness and malformed rods	Green and yellow vegetables and fruit, dairy products, egg yolk, fish-liver oil
Vitamin D (calciferol)	Rickets or osteomalacia (very low blood calcium level, soft bones, distorted skeleton, poor muscular development)	Egg yolk, milk, fish oils
Vitamin E (tocopherol)	Male sterility in rats (and perhaps other animals); muscular dystrophy in some animals; abnormal red blood cells in infants; death of rat and chicken embryos	Widely distributed in both plant and animal food, e.g. meat, egg yolk, green vegetables, seed oils
Vitamin K	Slow blood clotting and hemorrhage	Green vegetables
WATER-SOLUBLE		
Thiamine (B_1)	Beriberi (muscle atrophy, paralysis, mental confusion, congestive heart failure)	Whole-grain cereals, yeast, nuts, liver, pork
Riboflavin (B_2)	Vascularization of the cornea, conjunctivitis, and disturbances of vision; sores on the lips and tongue; disorders of liver and nerves in experimental animals	Milk, cheese, eggs, yeast, liver, wheat germ, leafy vegetables
Pyridoxine (B_6)	Convulsions, dermatitis, impairment of antibody synthesis	Whole grains, fresh meat, eggs, liver, fresh vegetables
Pantothenic acid	Impairment of adrenal cortical function, numbness and pain in toes and feet, impairment of antibody synthesis	Present in almost all foods, especially fresh vegetables and meat, whole grains, eggs
Biotin	Clinical symptoms in man are extremely rare, but can be produced by great excess of raw egg white in diet; symptoms are dermatitis, conjunctivitis	Present in many foods, including liver, yeast, fresh vegetables
Nicotinamide	Pellagra (dermatitis, diarrhea, irritability, abdominal pain, numbness, mental disturbance)	Meat, yeast, whole wheat
Folic acid	Anemia, impairment of antibody synthesis, stunted growth in young animals	Leafy vegetables, liver
Cobalamin (B_{12})	Pernicious anemia	Liver and other meats
Ascorbic acid (C)	Scurvy (bleeding gums, loose teeth, anemia, painful and swollen joints, delayed healing of wounds, emaciation)	Citrus fruits, tomatoes

Vitamin E is important in some animals in maintaining good muscle condition and normal liver function. In addition, vitamin E has been shown to be necessary in preventing sterility in many test animals including rats, chickens and ducks. A deficiency of vitamin E produces progressive deterioration and paralysis of the muscles, abnormal red blood cells in infants, and male sterility in some animals. A deficiency of this vitamin rarely occurs in man.

Vitamin K plays an important role in the normal coagulation of blood by promoting synthesis in the liver of prothrombin and proconvertin. These are two factors of the blood-clotting system. Vitamin K-deficient patients are highly susceptible to hemorrhages. Bacteria living in the large intestine of man are capable of producing vitamin K.

Vitamin C, a water-soluble vitamin, is required in cellular metabolism, specifically in cellular reductions. It is also important in the formation of hydroxyproline, one of the constituents of collagen. The disease scurvy, resulting from a lack of this vitamin, is characterized by bleeding gums, bruised skin, painful swollen joints and general weakness.

Vitamin B complex is a group of water-soluble compounds, unrelated chemically but which tend to be associated together. Several of them act as parts of coenzymes functioning in cellular respiration. For example, thiamine (vitamin B_1) is a principal part of the coenzyme that catalyzes the oxidation of pyruvic acid. Riboflavin (vitamin B_2) is one of the hydrogen-carriers in the oxidative cytochrome system. Pyridoxine (vitamin B_6) is a component of a coenzyme involved in transamination-reactions transferring amino groups from one compound to another. A list of vitamins, their sources and deficiency symptoms appear in the table.

COMPARATIVE NUTRIENT INGESTION AND DIGESTION

• PROBLEM 17-9

Compare the processes of digestion in the hydra, flatworm and earthworm.

Solution: Hydra is a member of the coelenterate phylum. Hydra is a multicellular, sac-like organism whose body wall is two cells thick. Many of the cells are semi-specialized, differing in function from their immediate neighbors. The hydra's digestive system, the gastrovascular cavity, has only one opening-the mouth (refer to Figure 1). The tentacles of the hydra have special cells, called nematocysts, that spear tiny organisms swimming near the hydra. The tentacles then draw the prey into the mouth. The cells lining the gastrovascular cavity begin digestion by secreting enzymes. This mode of digestion, occuring outside cells, is extracellular digestion. Extracellular digestion is followed by intracellular digestion in the hydra. When small food

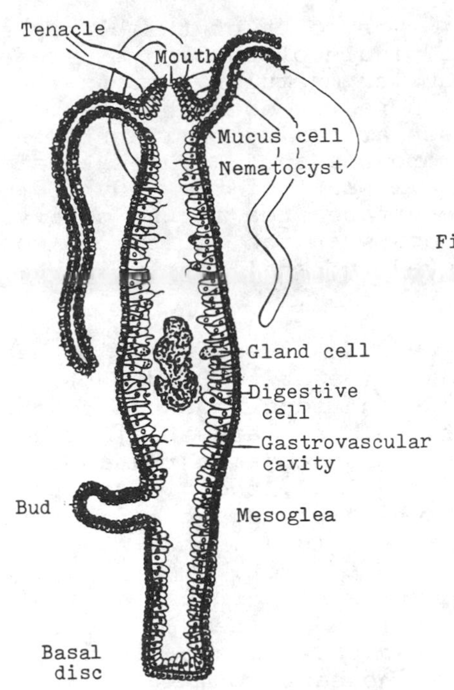

Figure 1. Hydra, showing gastrovascular cavity with food material in it.

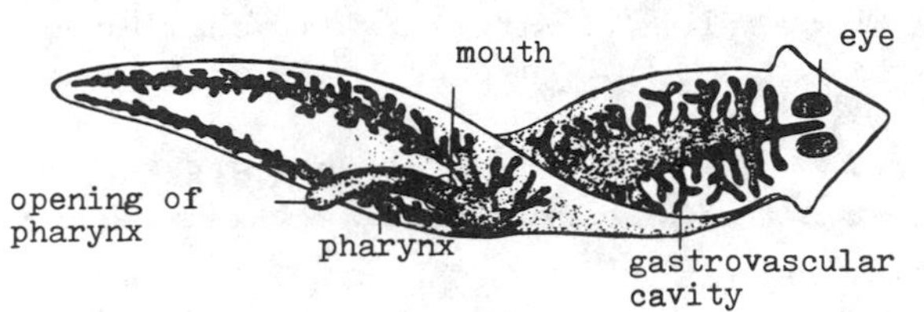

Figure 2. Planaria, showing much-branched gastrovascular cavity and extruded pharynx.

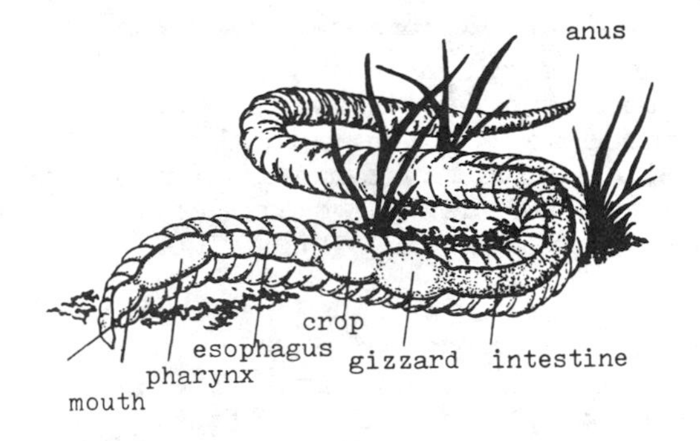

Figure 3. Digestive system of an earthworm.

fragments come into contact with the cells lining the cavities, these fragments are phagocytized and digestion is completed inside the cell. The indigestible material is expelled through the mouth. Thus, the hydra utilizes extracellular digestion to break down relatively large organisms into small pieces which are then digested intracellularly by individual cells.

Flatworms are small animals that are often parasitic. Parasitic flatworms have lost most of their ability to procure food and are provided with nutrients by their host. However, some flatworms such as planaria are non-parasitic and are free-living organisms. The mouth of the planaria is on the ventral side, in the middle of the body (refer to Figure 2). It opens into a tubular muscular pharynx that can protrude through the mouth directly onto the prey. The pharynx opens into a gastrovascular cavity similar to those other coelenterates but much more branched. The extensive branching brings nutrients to all parts of the body and helps increase the absorptive surface area of the cavity. Some extracellular digestion occurs in the cavity but most food is phagocytized and digested intracellularly in the cells lining the cavity.

An earthworm has a complete digestive tract. It has two openings, the mouth and anus, (refer to Figure 3), with wastes expelled through the anus and food ingested through the mouth. Various regions of the tract are specialized: the pharynx to suck in food, the esophagus to transport food to the crop, the crop to store the food, the gizzard to break the food up, and the intestine to digest and absorb the food. Most digestion is extracellular, utilizing the enzymes secreted by cells lining the tract. The digestive tract is not widely branched, but is rather like a simple gastrovascular cavity. However, large numbers of folds are found in the tract lining, which greatly increase the absorptive surface area. Since the earthworm is a land animal, it must also conserve its body fluids by reabsorbing water from the indigestible wastes. This occurs in the posterior region of the intestine, analogous to the human large intestine.

• PROBLEM 17-10

Compare the teeth of mammals with those of lower vertebrates.

Solution: The chief differences between mammalian teeth and those of lower vertebrates is their respective embryonic origins. Of the three main types of embryonic tissues (ectoderm, mesoderm and endoderm), mammalian teeth are made up of two (ectoderm and mesoderm) while teeth of the lower vertebrates are derived solely from the ectoderm.

Mammalian teeth result from outgrowths of the embryonic bone-producing tissue - the mesoderm. They are covered with a hard enamel derived from the embryonic ectoderm. Mammalian teeth are composed of living cells, supplied with blood vessels and nerves (see Figure 1). They are also highly evolved, consisting of different shapes adapted for different functions. The foremost teeth of most mammals are chisel-shaped incisors (see Figure 2). These are used for biting and cutting food. The incisors of rodents and horses grow continually and are worn down during eating. Meat eaters have sharp canine teeth that are used for tearing. All of a dog's teeth are sharp and pointed, and highly specialized for chewing raw meat. Herbivorous mammals have flat grinding teeth in the rear of their mouth that help crush and scrape the tough cellulose cell walls of plant cells.

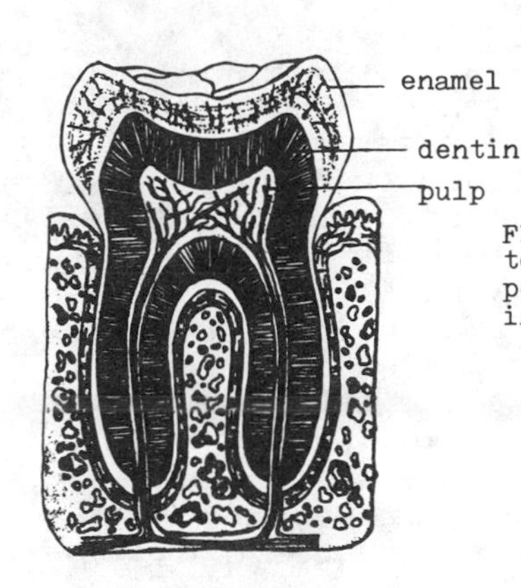

Figure 1. Internal structure of a tooth. Blood vessels and nerves penetrate into the pulp, but not into the outer harder layers.

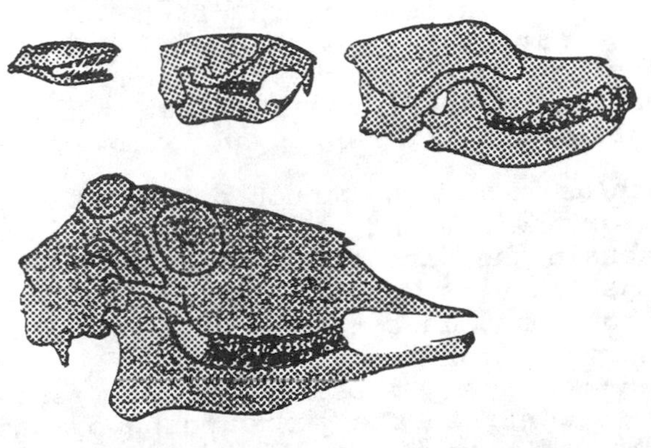

Figure 2. Structure and arrangement of teeth in different animals. (A) Snake: thin, sharp, backward-curved teeth that have no chewing function (the snake skull is here shown disproportionately large in relation to the other three). (B) Beaver (gnawing herbivore): few but very large incisors, no canines, premolars and molars with flat grinding surfaces. (C) Dog (carnivore): large canines, premolars and molars adapted for cutting and shearing. (D) Deer (grazing and browsing herbivore): six lower incisors (three on each side), but no upper incisors-these are functionally replaced by a horny gum; premolars and molars with very large grinding surfaces. Notice the large gap between the incisors and premolars.

All the teeth of the lower vertebrates are sharp and pointed. The teeth of fish, reptiles, and amphibians have the same general shape, whether they are meat or plant eaters. The triangular, conical shape of these teeth indicates their evolutionary origin. The teeth of these vertebrates have evolved from scales. They are formed from the ectoderm (exterior tissue) of the embryo and are replaced when lost. These teeth are not made up of living cells and do not contain blood vessels or nerve endings.

Thus mammalian teeth are highly specialized, living organs while the teeth of the lower vertebrates are non-living, less evolved extensions of their skin tissue.

• PROBLEM 17-11

How does nutrient procurement by green plants differ from that of animals, fungi, and protista?

Solution: All green plants are autotrophs, subsisting exclusively on inorganic substances taken from their environment. The necessary molecules, being small and soluble, are able to pass through cell membranes. Autotrophic organisms therefore do not need to digest their nutrients before taking them into their cells. Green plants are photosynthetic, taking in carbon dioxide and water from their environment, and using the energy from sunlight to synthesize carbohydrates from them.

93% of the dry weight of plant was originally carbon dioxide in the air. Plants also need to synthesize proteins. Protein synthesis requires other elements, such as nitrogen and sulfur. When rain water dissolves these nutrients in the soil, the solution is taken up by the root system and transported to various parts of the plant via the vascular system. Minerals required by plants are also transported up the plant body via the root and vascular systems. Some plants supplement their protein diet by trapping insects (the Venus flytrap is an example).

Many fungi such as mushroom, cannot produce their own carbohydrates from the air and water. They are heterotrophs: they must obtain the carbohydrates they require from the surroundings. This is usually supplied by dead plant matter, but plant matter is mainly cellulose and other polysaccharides which must be first digested. Specialized secretory cells in these organisms release the enzymes for digestion into the surrounding organic matter. The products of this extracellular digestion are then directly absorbed through the cells' surfaces or sometimes by specialized absorbing structures.

Almost all animals are dependent on plant matter, or other animals that feed on plant matter, for food. They too are heterotrophs. Without plants, there would be virtually no heterotrophs on earth. Like the mushroom, most animals digest the majority of their food extracellularly. The simple products are then taken into the cells where they are further degraded by intracellular enzymes. Digestion is unimportant in green plants but fungi and animals have digestive process of the extracellular type.

Protista are autotrophs or heterotrophs. Protozoa are heterotrophic, nonphotosynthetic protista. Protozoa includes the paremecium and amoeba, in which digestion is intracellular.

• PROBLEM 17-12

Name the major organs of the human digestive tract and explain their functions.

Solution: The human digestive system begins at the oral cavity. The teeth break up food by mechanical means, increasing the substrates' surface area available to the action of digestive enzymes. There are four types of teeth. The chisel-shaped incisors are used for biting, while the pointed canines function in tearing, and the flattened, ridged premolars and molars are used for grinding and

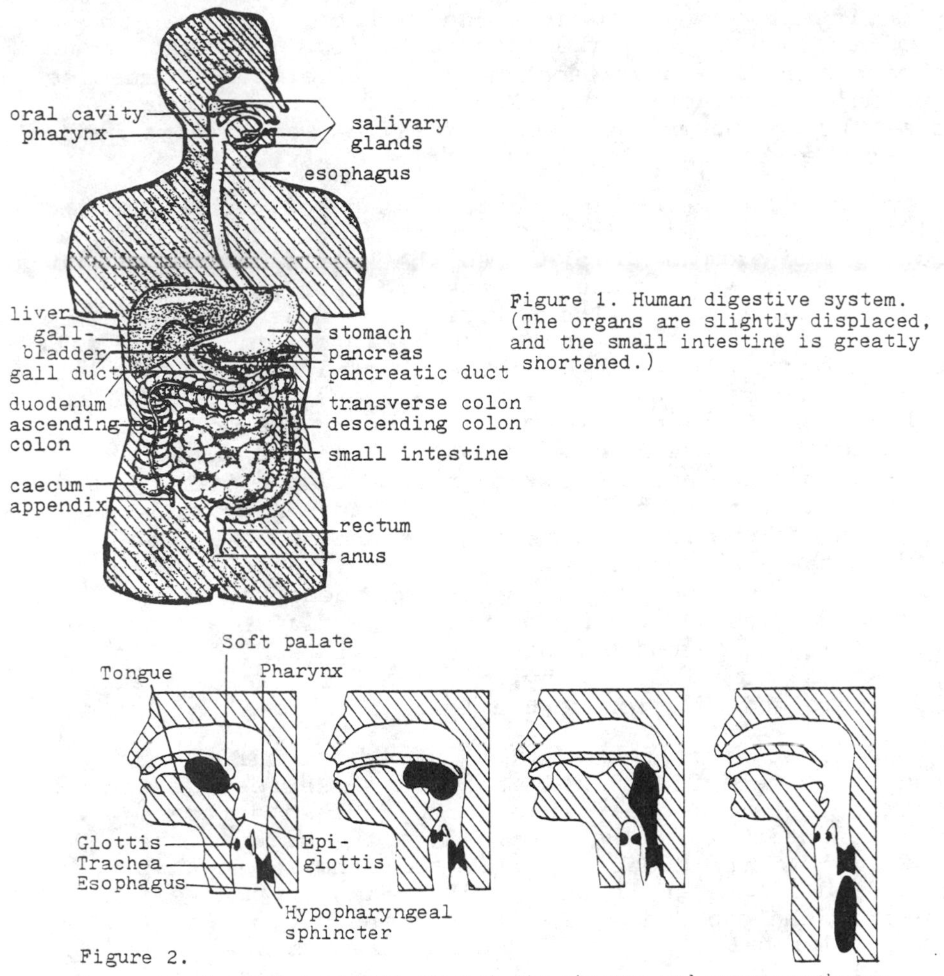

Figure 1. Human digestive system. (The organs are slightly displaced, and the small intestine is greatly shortened.)

Figure 2.
Movement of a bolus of food through the pharynx and upper esophagus during swallowing.

crushing food. In addition to tasting, the tongue manipulates food and forms it into a semi-spherical ball (bolus) with the aid of saliva.

The salivary glands consist of three pairs of glands. The parotid glands, located in the cheek in front of the ear, produce only watery saliva which dissolves dry foods. The submaxillary and sublingual glands, located at the base of the jaw and under the tongue, produce watery and mucous saliva which coagulates food particles and also lubricates the throat for the passage of the bolus. Saliva also contains amylases which break down starches.

The tongue pushes the bolus into the pharynx which is the cavity where the esophagus and trachea (windpipe) meet (see Fig. 2). The larynx is raised against the epiglottis and the glottis is closed, preventing food from passing into the trachea. The act of swallowing initiates the movement of food down through a tube connecting the mouth to the stomach called the esophagus. Once inside the esophagus, the food is moved by involuntary peristaltic waves towards the stomach. At the junction of the stomach and the esophagus, is a special ring of muscles called the lower esophageal sphincter. These muscles are normally contracted, but when food from the esophagus reaches the sphincter, it opens reflexively and allows food to enter the stomach.

The stomach is a thick muscular sac positioned on the left side of the body just beneath the ribs.The upper region of the stomach, closest to the heart, is called the cardiac region. Below that is the crescent part of the sac called the fundus. The pyloric region is tubular and connects the stomach to the small intestines. The wall of the stomach is made up of three thick layers of muscle. One layer is composed of longitudinal, one of circular, and the other of oblique (diagonal) fibers. The powerful contractions of these muscles break up the food, mix it with gastric juice and move it down the tract. Gastric juice is a mixture of hydrochloric acid and enzymes that further digest the food. Gastric juice and mucus are secreted by the small gastric glands in the lining of the stomach. The mucus helps protect the stomach from its own digestive enzymes and acid. The partially digested food, called chyme, is pushed through the pyloric sphincter into the small intestine. The pyloric sphincter is similar in structure and function to the lower esophageal sphincter.

The first part of the small intestine, called the duodenum, is held in a fixed position. The rest of the intestine is held loosely in place by a thin membrane called mesentery which is attached to the back of the body wall. In the duodenum, bile from the liver that has been stored in the gallbladder is mixed with pancreatic juice from the pancreas. The secretions of the pancreas and glandular cells of the intestinal tract contain enzymes that finish digesting the food. As digestion continues in the lower small intestine, muscular contractions mix the food and move it along. Small finger-like protrusions, called villi, line the small intestine facing the lumen. They greatly increase the intestinal surface area and it is through the villi that most of the nutrient absorption takes place.

The small intestine joins the large intestine (colon) at the cecum. The cecum is a blind sac that has the appendix protruding from one side. Neither the appendix nor the cecum are functional. At the junction of the cecum and the lower small intestine (ileum) is a sphincter called the ileocecal valve. Undigested food passes through this valve into the large intestine.

The large intestine has the function of removing water from the unabsorbed material. At times there is an excretion of certain calcium and iron salts when their concentrations in the blood are too high. Large numbers of bacteria exist in the colon. Their function in man is not fully understood, but some can synthesize vitamin K which is of great importance to man's blood clotting mechanism. The last section of the colon stores feces until it is excreted through the anal sphincter.

THE DIGESTIVE PATHWAY

• PROBLEM 17-13

What is the basic mechanism of digestion and what digestive processes take place in the mouth and stomach?

Solution: The process of digestion is the breakdown of large, ingested molecules into smaller, simple ones that can be absorbed and used by the body. The breakdown of these large molecules is called degradation. During degradation, some of the chemical bonds that hold the large molecules together are split. The digestive enzymes cleave molecular bonds by a process called hydrolysis. In hydrolysis a water molecule is added across the bond to cleave it.

$$\underset{\text{Dipeptide}}{H-\underset{H}{N}-\underset{H}{\overset{R}{C}}-\overset{O}{\overset{\|}{C}}-\underset{H}{N}-\underset{H}{\overset{R'}{C}}-\overset{O}{\overset{\|}{C}}-OH} + H_2O \xrightarrow{\text{Enzyme}} H_2N-\underset{H}{\overset{R}{C}}-\overset{O}{\overset{\|}{C}}-OH + H-\underset{H}{N}-\underset{H}{\overset{R'}{C}}-\overset{O}{\overset{\|}{C}}-OH$$

Hydrolysis of a dipeptide, R and R' represent different side chains.

Within living systems, chemical reactions require specific enzymes to act as catalysts. Enzymes are very specific, acting only on certain substrates. In addition, different enzymes work best under unlike conditions. Digestive enzymes work best outside of the cell, for their optimum pHs lie either on the acidic (e.g., gastric enzymes) or basic side (e.g., intestinal and pancreatic enzymes). The cell interior, however, requires an almost neutral (about 7.4) pH constantly. Digestive enzymes are secreted into the digestive tract by the cells that line or serve it.

Digestion begins in the mouth. Most foods contain polysaccharides, such as starch, which are long chains of glucose molecules. Saliva (and the intestinal secretions) contain enzymes that degrade such molecules. Salivary amylase, an enzyme that is also called ptyalin, hydrolyzes starch into maltose. (Compounds whose names end with -"ase" are enzymes, and those with the suffix "-ose" are sugars.) Glucose is eventually absorbed by the epithelial cells lining the small intestine.

The saliva has a pH of 6.5 - 6.8. This is the optimal range for salivary enzyme activity. Food spends a relatively short amount of time in the mouth, and eventually enters the stomach. The stomach is very acidic,with a pH of 1.5-2.5. The acid is secreted by special cells in the lining of the stomach called parietal cells. The low pH is required for the activity of the stomach enzyme pepsin. Rennin coagulates milk proteins in the infant's stomach,making them more susceptible to enzyme attack. Pepsin is a proteolytic enzyme (protease): it degrades proteins. Pepsin starts the protein digestion in the stomach by splitting the long proteins into shorter fragments, or peptides, that are further digested in the intestine. Pepsin will split any peptide bond involving the amino acids tyrosine or phenylala-

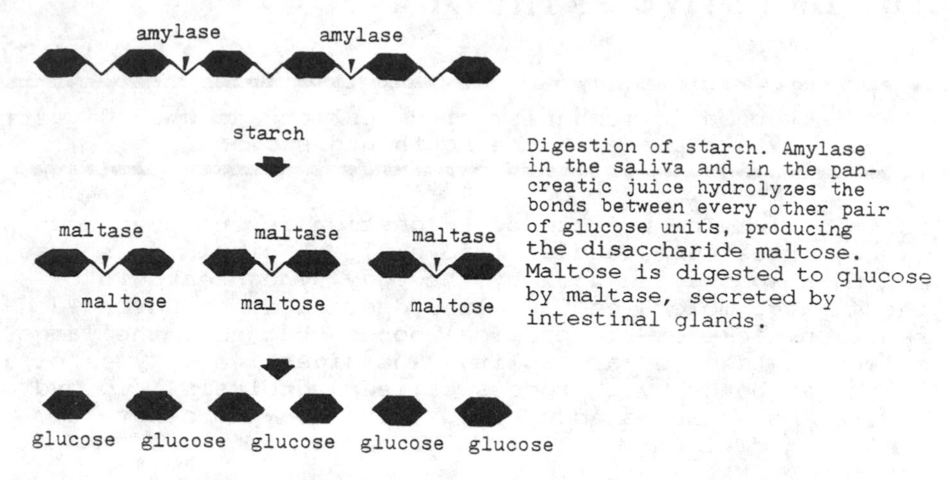

Digestion of starch. Amylase in the saliva and in the pancreatic juice hydrolyzes the bonds between every other pair of glucose units, producing the disaccharide maltose. Maltose is digested to glucose by maltase, secreted by intestinal glands.

nine. There are 20 different kinds of amino acids that can make up a protein and some proteins are thousands of amino acids long. The body needs the amino acids it obtains from digestion to synthesize its own proteins.

• PROBLEM 17-14

Describe the mechanism that prevents food from entering the trachea or nasal pasages during swallowing.

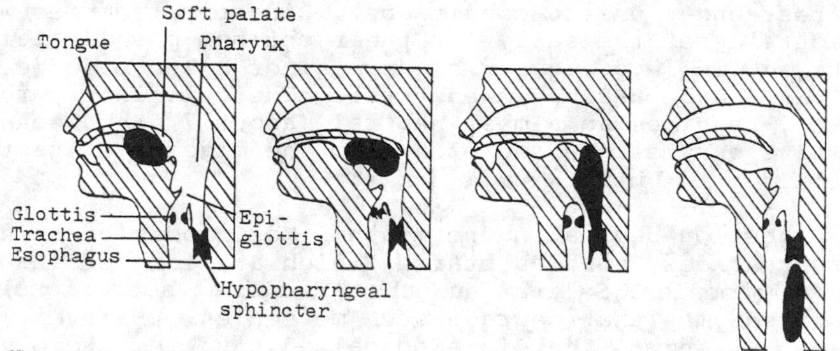

Movement of a bolus of food through the pharynx and upper esophagus during swallowing.

Solution: Swallowing is a complex reflex initiated when the tongue forces a bolus of food into the rear of the mouth. Here, pressure-sensitive cells are stimulated to send impulses to the swallowing control center in the medulla, which coordinates the remainder of the process. Swallowing is an all-or-none response, and once initiated, it cannot be stopped. The swallowing reflex is a sequential process in which several responses occur at timed intervals determined by nerve connections in the control center.

As the bolus is pushed into the pharynx, the soft palate rises and lodges against the back wall of the pharynx, preventing the entrance of food into the nasal cavity. Breathing is temporarily stopped and the glottis, the opening between the vocal cords, is closed as the larynx, the upper region of the trachea, rises. The rear portion of the tongue pushes the bolus down the throat over the epiglottis and the epiglottis is simultaneously bent backwards over the glottis. It is primarily the closure of the glottis, not the folding of the epiglottis, that prevents food from entering the trachea. At the top of the esophagus is the hypopharyngeal sphincter. This sphincter is normally closed but opens as food reaches it. The sphincter immediately closes after the food passes it, preventing the food from reversing its direction. Peristalsis, a series of rhythmic wave-like muscular contractions, pushes the food through the esophagus into the opening of the stomach.

• PROBLEM 17-15

Explain how peristalsis moves food through the digestive tract.

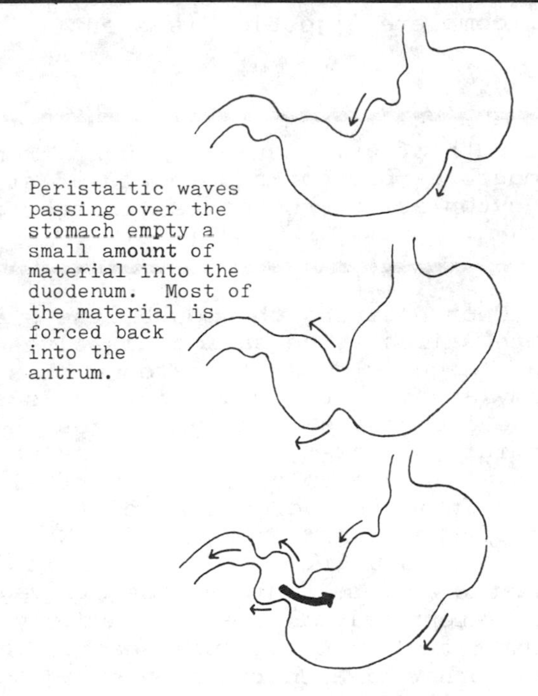

Solution: In each region of the digestive tract, rhythmic waves of constriction move food down the tract. This form of contractile activity is called peristalsis, and involves involuntary smooth muscles. There are two layers of smooth muscle throughout most of the digestive tract. Circular muscles run around the circumference of the tract while longitudinal muscles traverse its length.

Once a food bolus is moved into the lower esophagus, circular muscles in the esophageal wall just behind the bolus contract, squeezing and pushing the food downward. At the same time, longitudinal muscles in the esophageal wall in front of the bolus relax to facilitate movement of the food. As the bolus moves, the muscles it passes also contract, so that a wave of contraction follows the bolus and constantly pushes it forward. This wave of constriction alternates with a wave of relaxation.

Swallowing initiates peristalsis and once started, the waves of contraction cannot be stopped voluntarily. Like other involuntary responses, peristaltic waves are controlled by the autonomic nervous system. When a peristaltic wave reaches a sphincter, the sphincter opens slightly and a small amount of food is forced through. Immediately afterwards, the sphincter closes to prevent the food from moving back. In the stomach, the waves of peristalsis increase in speed and intensity as they approach the pyloric end. As this happens, the pyloric sphincter of the stomach opens slightly. Some chyme escapes into the duodenum but most of it is forced back into the stomach (see figure). This allows the food to be more efficiently digested. There is little peristalsis in the intestine, and more of a slower oscillating contraction. This is why most of the 12-24 hours that food requires for complete digestion is spent in the intestine.

• PROBLEM 17-16

Describe the path of a molecule of sugar from the time it enters the mouth as part of a molecule of starch, until it reaches the cytoplasm of the cells lining the small intestine.

Solution: Upon entering the mouth, the starch is chewed and mixed with saliva. The saliva dissolves small molecules and coats the larger clumps of starch with a solution of the enzyme amylase. As the mass of food is swallowed, the enzyme amylase from the saliva hydrolyzes the starch, degrading it into maltose units. Maltose is a disaccharide composed of two glucose molecules. Eventually, the acidity of the stomach inhibits the actions of the salivary amylase, leaving most of the starch undigested. No further digestion of the starch occurs until it reaches the small intestine. There, pancreatic amylase finishes the cleavage of starch into maltose. Intestinal maltase then hydrolyzes the maltose molecules into two glucose molecules each. Other intestinal enzymes cleave other disaccharides, such as sucrose, into monosaccharides.

When these free sugar molecules come into contact with the cells lining the small intestine, they are held by binding enzymes on the cell surface. Glucose is transported through the membrane into the cellular cytoplasm.

• PROBLEM 17-17

You have decided to build up your body, so you have a dinner that is heavy in protein content, followed by a quart of milk. Trace the movement of the food from the mouth to the duodenum. What happens to the food in the stomach and small intestine?

Solution: When food is introduced into the mouth, it is mixed with saliva and masticated by the teeth into small pieces to facilitate swallowing and digestion (by increasing the surface area exposed to digestive enzymes). With the help of the tongue, the chewed food is rolled into a ball, or bolus, which is then thrust into the back of the mouth for swallowing.

As soon as the food enters the esophagus, it is moved down the tract by rhythmic muscular contractions known as peristaltic waves. Peristalsis is the alternate contraction and relaxation of the smooth muscles lining the digestive tract, and helps to move the bolus along. Movement of the bolus is also assisted in this region by salivary mucin.

As the food nears the stomach, the sphincter muscle controlling the opening of the stomach relaxes and allows the food to enter the stomach. In the stomach, it is acted upon by the gastric juice, which works optimally in an acidic medium. Gastric juice contains:

(i) hydrochloric acid, which acidifies the medium, in order for the proteolytic enzymes to work optimally;

(ii) pepsin, a proteolytic enzyme that cleaves long protein molecules into shorter fragments called peptides; and

(iii) rennin, an enzyme that solidifies casein, a milk protein, so that it can be retained in the stomach long enough for pepsin to act upon it.

After about four hours in the stomach, the semi-digested food, or chyme, is released into the duodenum through the pyloric sphincter. The duodenum receives pancreatic juice from the pancreas and bile from the gallbladder. The proteases trypsin and chymotrypsin, found in the pancreatic juice, further degrade peptides into smaller fragments.

The digestion of protein is then completed in the ileum, the last part of the small intestine. Here, the remaining peptide fragments are further digested to free amino acids by carboxypeptidases and aminopeptidases, which split off amino acids from the carboxyl and amino ends of the peptide chains, respectively. The free amino acids are then actively transported across the intestinal wall into the bloodstream.

SECRETION AND ABSORPTION

• PROBLEM 17-18

What prevents the stomach from being digested by its own secretions?

Solution: The lining of the stomach is composed of cells that secrete hydrochloric acid, gastric juice, and mucus. Mucus is a polymer made up of repeating units of a protein-sugar complex. A coat of mucus, about 1 to 1½ millimeters in thickness, lines the inner surface of the stomach. Mucus is slightly basic. This alkalinity provides a barrier to acids, keeping the area next to the stomach lining nearly neutral. In addition, the membranes of the cells lining the stomach have a low permeability to hydrogen ions, preventing acid from entering the underlying cells.

The cells that make up the stomach (and duodenum) lining do not last long, even under this protection. Cell division and growth replace the entire stomach lining every 1 to 3 days. Thus the mucus layer, the permeability of the membranes and the continual replacement of the cells comprising the lining all help protect the underlying tissues from the action of proteolytic enzymes of the stomach and duodenum.

For many people, however, this does not provide enough protection. If too much acid is released, perhaps because of emotional strain or because the proteolytic enzymes have digested away the mucus, an ulcer will result. Ulcers are usually treated by eating many small, bland meals throughout the day. This helps to keep the acid level down.

• PROBLEM 17-19

In certain abnormal conditions, the stomach does not secrete hydrochloric acid. What effects might this have on the digestive process?

Solution: The high HCl concentration in the stomach serves several important functions. Without the acid, the gastric enzymes would show no activity. For example, pepsin, an important protein-digesting enzyme, would remain inactive. Pepsin exists in the form of a precursor compound, called pepsinogen, which has no proteolytic activity in the absence of acidity. The pepsinogen will remain in the stomach as is, until HCl is secreted and activates it. Thus the absence of HCl would lead to some loss of protein digestion. This loss, however, can be made up by the intestinal proteases.

HCl is also important for its direct effect on the conformation of proteins. The high acidity of HCl denatures proteins. Denaturation does not mean cleavage of a molecular bond, but rather an unwinding of the molecule. Proteins are normally tightly packed, woven chains, that are difficult for enzymes to attack. The acid alters the protein to facilitate enzyme action.

Another function of stomach acid is to kill ingested bacteria. Many of the bacteria ingested are harmless, but some, if allowed to grow, could have serious effects. HCl is also important to pancreatic secretion. Pancreatic juice secretion is controlled, in part, by the hormone secretin.

The HCl secreted by the stomach glands passes through the pyloric sphincter, along with the food, into the duodenum. Some of the acid stimulates special receptor cells in the intestinal lining. These cells, in turn, secrete secretin into the blood. When the secretin reaches the pancreas, a pancreatic juice rich in bicarbonate travels down the pancreatic duct into the intestine. If no HCl is present, there will be less than one quarter of the secretion, and less one quarter of the normal amount of digestion.

• PROBLEM 17-20

> The intestine, especially the small intestine, is a vital organ for absorption of nutrients required by the body. In what ways is it suitable for such a function?

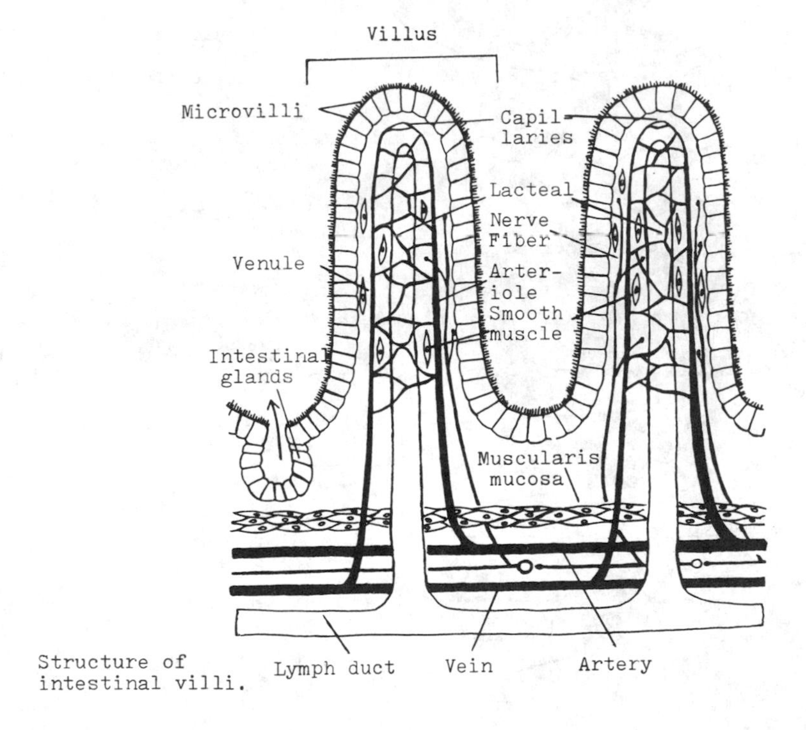

Structure of intestinal villi.

Solution: The small intestine is that region of the digestive tract between the stomach and the cecum. Its long, convoluted structure is an adaptation for absorption of nutrients. Structural modifications of the internal surfaces of the small intestine act to increase its surface area for absorption. First, the mucosa lining the intestine is thrown into numerous folds and ridges. Second, small fingerlike projections called villi cover the entire surface of the mucosa (see figure). These villi are richly supplied with blood capillaries and lacteals (for absorption of fats) in order to facilitate absorption of nutrients. Third, individual epithelial cells lining the folds and villi have a "brush-border" on the surface facing the lumen, consisting of countless, closely-packed, cylindrical processes known as microvilli. These microvilli add an enormous amount of surface area to that already present. The total internal surface area of the small intestine is thus incredibly large; this is advantageous for the purpose of absorption.

The large intestine also has villi to increase its surface area. However, the number of villi present in the small intestine far exceeds that found in the large intestine. The main function of the large intestine is to absorb water from the undigested food substances and reduce the remains to a semi-solid state before it is expelled through the anus.

• PROBLEM 17-21

Discuss the absorption of glucose and amino acids. How does it differ from the mechanism of absorption of fatty acids?

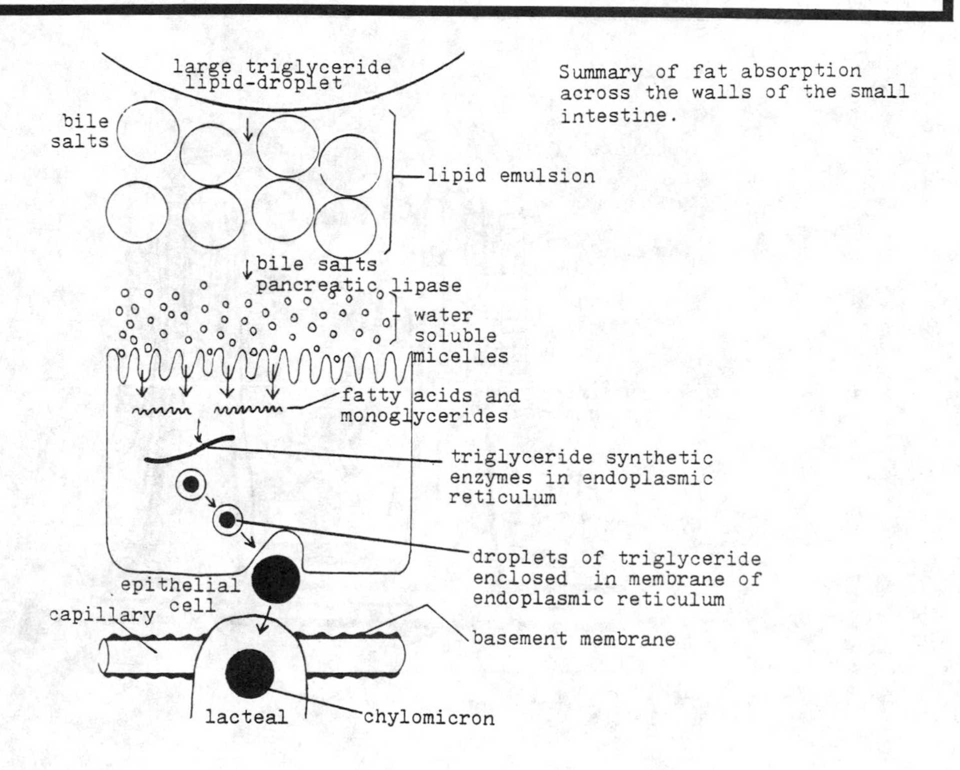

Summary of fat absorption across the walls of the small intestine.

Solution: Almost all the nutrient products obtained from digestion are absorbed through the walls of the small intestine. The total volume occupied by the small intestine is relatively small. However, the internal surface area of the small intestine is immense, allowing for very efficient absorption of digested nutrients. This large surface area is a result of the folds and ridges of the small intestine, compounded with the villi and microvilli that cover its internal surface. The villi are finger-like processes that extend into the lumen of the intestine. The microvilli are hair-like projections of the membrane of cells lining the villi. It is here that specific absorption takes place.

When glucose or free amino acids approach the cell membrane, they are bound by special groups of enzymes that are within the membrane. These enzymes quickly transport the molecules through the membrane into the cytoplasm. The nutrients are then transported across the intestinal epithelium into the blood. This occurs in part by simple diffusion, in part by facilitated diffusion, and in part by active transport. The presence of sodium in the lumen is required for monosaccharide transport. The various hexoses are absorbed by active transport. Amino acids are also absorbed by means of active transport. These nutrients are then circulated to the cells of the body, or transported to the liver for storage.

Glycerol and fatty acids are not selectively absorbed. After fats are digested in the intestinal lumen, the majority of their products, primarily fatty acids and monoglycerides, simply diffuse across the cell membrane because of their high solubility in membrane lipids. These are then brought into the endoplasmic reticulum (E.R.) of the epithelial cells, where they are resynthesized into triglycerides. These lipids, in turn, combine with specific proteins, forming lipoproteins. The lipoproteins aggregate into droplets called chylomicra within the E.R. The E.R. then fuses with the cell surface and empties the chylomicra into the extracellular space. Here, they are absorbed by the lacteals of the lymphatic system and eventually transferred into the bloodstream. After a high-fat meal, one's blood may become "milky-looking", due to the high concentration of chylomicra. Note that no energy is required for fat absorption.

Another important aspect of absorption is the uptake of partially digested material. When disaccharides and dipeptides are within the vicinity of the cell membrane, the membrane extends itself, forming a pocket around the partially digested molecules. This pocket invaginates into the cell in a process known as pinocytosis. The pinocytotic vesicle fuses with another vesicle called a lysosome. The lysosome contains the cells own digestive enzymes. As these enzymes digest the vesicle's contents, the degraded nutrients diffuse through the membrane into the cytoplasm and are used by the cell.

ENZYMATIC REGULATION OF DIGESTION

• PROBLEM 17-22

Discuss the action of enzymes in intestinal digestion.

Solution: Digestion and absorption of ingested food takes place mainly in the small intestine. The partially digested food that enters the upper part of the small intestine (the duodenum) from the stomach is very acidic, due to the high acid content of the stomach. This acidity stimulates special receptor cells in the intestinal lining. These cells, in turn, produce a hormone, cholecystokinin (CCK) that enhances the release of bile from the gall

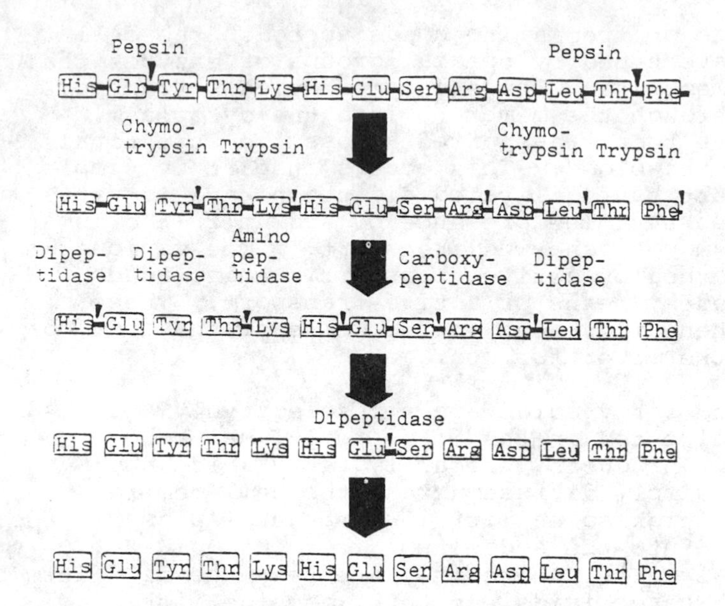

Digestion of protein. Pepsin in the stomach hydrolyzes peptide bonds at the amino end of tyrosine (Tyr) and phenylalanine (Phe). Then the food moves into the intestine, where trypsin and chymotrypsin from the pancreas hydrolyze bonds adjacent to lysine and arginine and to tyrosine, leucine, and phenylalanine, respectively (chymotrypsin also hydrolyzes bonds adjacent to tryptophan and methionine when they are present). Pepsin, trypsin, and chymotrypsin hydrolyze only internal bonds, not bonds attaching terminal amino acids to the chains. Terminal bonds at the amino end of chains may be split by aminopeptidase and those at the carboxyl end by carboxpeptidase. Bonds between pairs of amino acids are split by dipeptidases whereupon digestion is completed.

bladder. CCK also stimulates the release of a pancreatic juice rich in digestive enzymes (proteases, lipase, nucleases and amylase). Secretin, also produced by the intestinal mucosa, causes release of a pancreatic juice rich in sodium bicarbonate, which neutralizes the stomach acid.

Proteases are enzymes that hydrolyze proteins, which are long chains of amino acids joined together by peptide bonds. Generally, short chains of amino acids are called polypeptides. The intestinal proteases are usually called peptidases because they break the peptide bonds of peptide chains. There are many different kinds of peptidases (see figure). Pepsin, produced in the stomach, is an endopeptidase; it cleaves only those peptide bonds that are within the peptide chain, splitting the large chain into many shorter chains. Trypsin and chymotrypsin are the endopeptidases produced by the pancreas. Unlike pepsin, which requires the high acidity of the stomach in order to function, these intestinal enzymes prefer an alkaline environment. Like pepsin of the stomach, these enzymes will only break the peptide bonds between certain amino acids. Trypsin will split the bond linking the amino acids lysine or arginine. Chymotrypsin cleaves the peptide bonds next to phenylalanine, tyrosine, or tryptophan.

The second category of peptidases are the exopeptidases. Exopeptidases only attack the ends of the peptide. The many peptides produced as the result of endopeptidase activity provide numerous sites for attack by the exopeptidases.

The two ends of a peptide chain are chemically different. One end is called the amino end($-NH_2$) while the other is the carboxylic end ($-COOH$). Thus, there are two kinds of exopeptidases: aminopeptidase and carboxypeptidase. These enzymes are secreted by glands in the intestinal lining and the pancreas, respectively. When digestion of a protein is completed by these enzymes, all that remains are the free amino acids, which are taken up by the cells lining the villi.

Lipase, produced by the pancreas, digests the lipids in our food, with the help of the bile. Two other pancreatic enzymes, ribonuclease and deoxyribonuclease, break up any RNA and DNA from the nuclei of the cells we ingest. The products of their digestion are quickly absorbed through the villi. Pancreatic amylase breaks up any remaining starch into maltose.

In addition to aminopeptidase, the intestinal glands secrete enzymes which degrade sugars. Maltase splits maltose into two glucose molecules; sucrase splits sucrose into glucose and fructose; and lactase splits lactose into glucose and galactose molecules.

• PROBLEM 17-23

What controls the secretion of the digestive enzymes?

Activities of the gastrointestinal hormones

	Secretin	CCK	Gastrin
Secreted by:	Duodenum	Duodenum	Antrum of the stomach
Primary stimulus for hormone release	Acid in duodenum	Amino acids and fatty acids in duodenum	Peptides in stomach Parasympathetic nerves to stomach
Effect on:			
Gastric motility	Inhibits	Inhibits	Stimulates
Gastric HCl secretion	Inhibits	Inhibits	STIMULATES*
Pancreatic secretion			
Bicarbonate	STIMULATES	Stimulates	Stimulates
Enzymes	Stimulates	STIMULATES	Stimulates
Bile secretion	STIMULATES	Stimulates	Stimulates
Gallbladder contraction	Stimulates	STIMULATES	Stimulates
Intestinal juice secretion	stimulates	stimulates	

*STIMULATES denotes that this hormone is quantitatively more important than the other two.

<u>Solution:</u> Digestive enzyme secretion is regulated by two kinds of factors: neural and humoral (hormonal). Neural control is usually based upon the sensing of a physical mass of food. The stretching of the digestive tract (distension), is an example of this tyupe of stimulation. It is interesting to note that most of the salivary and some of the gastric secretions are stimulated by smelling, tasting and even thinking of food. Hormonal control, on the other hand, is much more specific. The presence of specific kinds of molecules in the ingested food stimulates receptor cells to pro-

duce their specific hormone. The hormone circulates in the blood until it reaches the secretory organs which it controls.

All secretory organs are under the influence of several factors. The presence of food in the mouth initiates gastric secretions by the stomach. When the food reaches the stomach, the distension of the stomach walls stimulates an increase in stomach motion and the partial production of gastrin. The hormone gastrin has several effects, the major of which is the stimulation of gastric juice secretion. It achieves this by stimulating gastric HCl secretion, which in turn induces protease activity. Pepsin, present in the gastric juice, cleaves proteins into many smaller chains. These small peptide chains stimulate receptor cells in the antrum of the stomach to produce more gastrin. Gastrin also has the ability to stimulate limited secretion in the intestine and pancreas. In additon, gastrin stimulates an increase in gastric motility and helps keep the sphincter of the esophagus tightly closed.

When partially digested food, or chyme, reaches the duodenum, its physical presence initiates peristalsis over the entire intestine. In addition, the acid in the chyme causes receptor cells to release the hormone secretin. Secretin has many functions. It slows down stomach motion and decreases gastric juice and gastrin production. This effectively keeps the food in the stomach for a longer period. At the same time, secretin prepares the intestine for neutralization of the acidic chyme by stimulating the release of alkaline bicarbonate ions in pancreatic juice.

Other receptor cells in the duodenum sense free amino acids and fats. These cells release a hormone called cholecystokinin or CCK. CCK stimulates the release of a pancreatic juice rich in digestive enzymes while slowing down the motion in the stomach. The more fat and protein there is in a meal, the longer will digestion occur. A review of the regulatory activities of gastrointestinal hormones is given in the accompanying table.

Note that CCK and secretin act synergistically. CCK stimulates secretion of a pancreatic juice rich in digestive enzymes, but it accentuates secretin's stimulation of a pancreatic juice rich in bicarbonate. Secretin stimulates secretion of a juice high in bicarbonate, but it accentuates CCK's secretion of a juice high in enzymes.

The control of secretions in digestion is a complex interaction of several factors. One's emotional state, as well as physical health, have a great influence on digestion. Age and diet are also critical factors. When a child is born, it does not have sufficient enzyme production to digest many foods other than milk. As we age, most adults stop producing rennin and lose the ability to fully digest milk. Humans have the most varied diet of all animals. This is reflected in the complex interactions of enzymes and hormones that control our digestion.

• **PROBLEM 17-24**

Summarize the roles of the various enzymes involved in digestion.

Solution: Enzymes in Digestion

Enzyme	Source	Optimum pH	Product(s)
Salivary amylase	Salivary glands	Neutral	Maltose
Pepsin	Stomach	Acid	Peptides
Rennin	Stomach	Acid	Coagulated casein (A milk protein)
Trypsin	Pancreas	Alkaline	Peptides
Chymotrypsin	Pancreas	Alkaline	Peptides
Lipase	Pancreas	Alkaline	Glycerol, Fatty acids, Mono- and Diglycerides
Amylase	Pancreas	Alkaline	Maltose
Ribonuclease	Pancreas	Alkaline	pentoses, nucleotides and nitrogen bases
Deoxyribonuclease	Pancreas	Alkaline	pentoses, nucleotides and nitrogen bases
Carboxypeptidase	pancreas	Alkaline	Free amino acids
Aminopeptidase	Intestinal Glands	Alkaline	Free amino acids
Maltase	Intestinal Glands	Alkaline	Glucose
Sucrase	Intestinal Glands	Alkaline	Glucose and fructose
Lactase	Intestinal Glands	Alkaline	Glucose and galactose

THE ROLE OF THE LIVER

• **PROBLEM 17-25**

In humans, the liver is one of the largest and most active organs in the body. Describe its functions.

Solution: One of the most important functions of the liver is to regulate the level of glucose in the blood stream. The maintenance of a constant level of blood glucose is essential because specialized cells, such as the brain cells, are easily damaged by slight fluctuations in glucose levels. Excess glucose is stored in the liver as glycogen while declining blood glucose can be restored by conversion of liver glycogen into glucose. In addition to the formation of glycogen, the liver can convert glucose into fat, which is then either stored in the liver itself or in special fat-deposit tissues known as adipose tissues.

The production of bile occurs in the liver. Bile is of major importance in aiding the digestion of fats. It is secreted by the liver cells into a number of small ducts which drain into a common bile duct. Bile is stored in the gallbladder, which releases its contents upon stimulation by the presence of fats in the duodenum.

When proteins are metabolized, ammonia, a toxic substance, is formed. The cells of the liver are able to detoxify ammonia rapidly by transforming it into the relatively inert substance urea. Urea can then be excreted from the body in the urine.

The liver also produces a lipid which has been frequently associated with heart diseases - cholesterol. In fact, a diet with reduced cholesterol content may not be effective in lowering cholesterol level in the body because the liver responds by producing more.

The liver plays some important roles in blood clotting. It is the site of production of prothrombin and fibrinogen, substances essential to the formation of a blood clot. In addition, bile secreted by the liver aids in the absorption of vitamin K by the intestine. Vitamin K is an essential cofactor in the synthesis of prothrombin and the plasma clotting cofactors.

Vitamin D, produced by the skin, is relatively inactive and requires biochemical transformation, first by the liver and then by the kidneys, before it is fully able to stimulate absorption of calcium by the gut. The mechanism of activating vitamin D is not yet fully understood. However, the presence of liver and kidney enzymes is essential to the process.

• PROBLEM 17-26

What is bile and what is its role in digestion? Where is bile manufactured and how does it reach the food undergoing digestion?

Solution: Bile is very important for proper digestion, although it contains no enzymes. It is highly alkaline and helps neutralize the acid in the chyme as it leaves the stomach and enters the small intestine. This is necessary in order for the intestinal enzymes to function properly. Bile is composed of bile salts, lecithin, cholesterol and bile pigments. The first three are involved in the emulsification of fat in the small intestine. The bile pigments give bile its color.

The major bile pigment is bilirubin. Bilirubin is actually a breakdown product of hemoglobin, the oxygen-carrying protein in red blood cells. In the large intestine, bilirubin and other bile pigments are further converted by bacteria into brown pigments, which give rise to the color of feces. If the bile duct is blocked so that the pigments cannot be excreted in the bile, they will be reabsorbed by the liver. Eventually, the pigments will accumulate in the blood and tissues, giving the skin a yellow color; this condition is called jaundice.

Cholesterol is a large fat-like molecule that has a very low solubility in body fluids. This may lead to deposits of cholesterol in the heart and arteries, which could result in heart disease or arteriosclerosis. The liver excretes excess cholesterol in the bile. Gallstones result from the accumulation of excess insoluble cholesterol in the gallbladder.

Bile salts are the most active part of bile. They are salts of glycocholic acid, which is made from cholesterol. Unlike cholesterol, bile salts are very soluble. These salts are essential for digestion of fats. Butter and oil are fats which constitute part of a group of molecules called lipids. Lipid molecules are insoluble in water and tend to coalesce to form globules. The enzymes that digest lipids, called lipases, can only work on the surface of these globules. Alone, it would take weeks for lipases to complete fat digestion in this manner. Bile salts solve this problem by having detergent-like properties - they coat the globules and break them up into millions of tiny droplets called micelles. This process, called emulsification, greatly increases the surface area exposed to attack by lipases, speeding up lipid digestion. Bile salts are conserved by the body, and are reabsorbed in the lower part of the intestine, carried back to the liver through the bloodstream, and secreted again.

The liver, one of the body's largest organs, constantly secretes bile. (600 - 800 ml. a day) A network of ducts collects the bile and passes it into the gallbladder, where it is stored until needed. The gallbladder is a small muscular sac that lies on the surface of the liver. When food enters the duodenum, certain receptor cells in the wall of the intestine sense the presence of fats in the chyme. Stimulated by the fats, the receptor cells in the duodenum secrete a hormone, called cholecystokinin (CCK) into the bloodstream. This hormone causes inhibiton of gastric motility and contraction of the gallbladder, forcing bile out through the bile duct and into the small intestine.

• PROBLEM 17-27

What are gallstones and how are they formed?

Solution: **Gallstones contain about 80% cholesterol. Cholesterol is a steroid molecule that is absorbed from** certain digested fats, such as butter, and is also produced in the liver. It is a precursor of many of the body's important steroid hormones (see fig.) Even though cholesterol is vital for proper body functioning, it is often not handled well by our bodies. This is because it is almost completely insoluble in body fluids. Excess cholesterol in the blood may be deposited along the arterial walls and in the heart. This can lead to arteriosclerosis and ultimately, to heart disease.

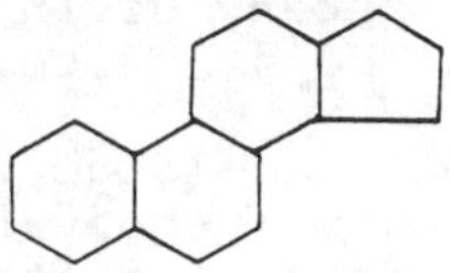

Steriod ring structure

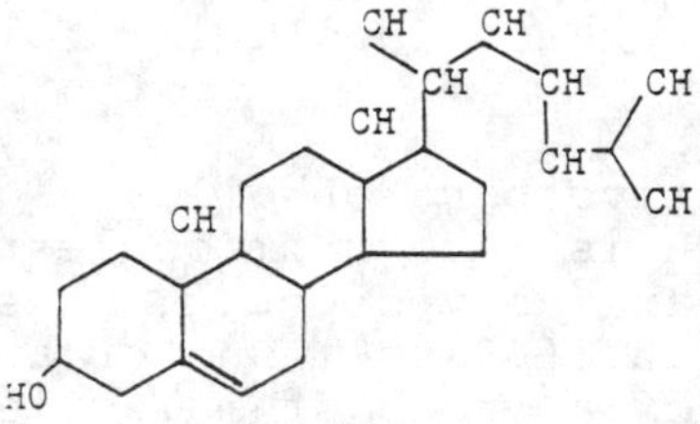

Cholesterol

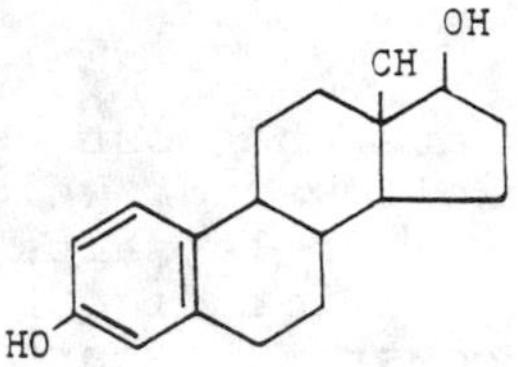

Estrogen

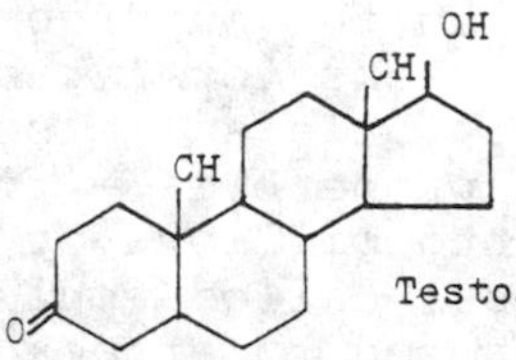

Testosterone

Carbon ring structure of various steroids, including the femal and male sex hormones: estrogen and testosterone.

The body's major mechanism for the removal of cholesterol from the blood takes place in the liver, which releases cholesterol into the intestine as part of the bile. Bile contains bile salts and lecithin that emulsify lipids such as cholesterol. In the gallbladder, bile is stored and slowly concentrated by the removal of water. As this happens, the concentration of cholesterol increases to a level near the saturation point. If there is an excess of cholesterol, or insufficient amounts of bile salts or lecithin to keep it soluble, the cholesterol will precipitate, forming a hard solid mass called a gallstone.

If a gallstone is small, it will pass through the bile duct and out into the intestine, causing no complications. A larger stone, however, may lodge in the neck of the gallbladder. This leads to pain and contractile spasms of the gallbladder wall. Since the gallbladder is off to the side of the bile duct, bile can still flow from the liver to the intestine for digestion. If the spasms of the gallbladder are powerful enough, the stone may be forced out of the gallbladder and become lodged in the bile duct, obstructing it. Complete blockage may lead to a condition known as jaundice, in which the skin yellows due to an accumulation of bile pigments in the body. Since no bile reaches the intestine, there is also decreased emulsification and absorption of fats. Gallstones may sometimes lodge at the point where the bile duct and pancreatic duct join before they enter the duodenum. When this happens, both bile and pancreatic juice are blocked in their passages, resulting in severe impairment of both digestion and absoprtion in the intestinal tract.

Gallstones may be removed surgically if they are not dissolved by the body. If necessary, the gallbladder may be removed without any great effect on digestion.

SHORT ANSWER QUESTIONS FOR REVIEW

Answer

To be covered when testing yourself

Choose the correct answer.

1. Hexokinase and glucose 6-phosphatase are important enzymes concerned with glucose usage and storage. Which of the following is incorrect concerning these enzymes? (a) Both liver cells and muscle cells contain the enzyme hexokinase, and can therefore store glucose as glycogen. (b) Liver cells have the ability to restore the glucose level of the blood to normal when it becomes too low. This is a function of the enzyme glucose 6-phosphatase. (c) Muscle cells, during times of extensive exercise, can utilize their stored glycogen via dephosphorylation by the enzyme glucose-6-phosphatase. (d) In a sense, these two enzymes are performing opposite functions: one puts a phosphate group on a glucose molecule in order to activate it for storage, while the other removes a phosphate from a glucose molecule so it may leave the cell. — c

2. A 70 kilogram man has been asleep for a normal 8 hours, and upon waking decides to remain in bed for a few more hours. He then suddenly remembers an errand he must attend to and he jumps out of bed, gets dressed, and runs 4 blocks to the nearest supermarket. Upon returning, he decides to sit down and relax again for an hour or two. Describe the man's metabolic rate (M.R.) from the time he wakes up to the end of his relaxation time after returning from his errand. (a) Upon awakening and for the subsequent hours before going to the supermarket, the man's B.M.R. (basal metabolic rate - the baseline or minimum energy requirement in a waking, post-absorptive state) is about 300 kcals per hour. During his trip to the supermarket, his M.R. goes up considerably and remains at that level for the next few hours. (b) Upon awakening and for the subsequent hours before going to the supermarket, the man's B.M.R. is about 100 kcals per hour. During his errand and for the subsequent few hours after that, his M.R. remains exactly the same as the original value. (c) Upon awakening and for the subsequent hours before going to the supermarket, the man's B.M.R. is about 200 kcals per hour. During his errand, his M.R. slightly decreases. While relaxing after his errand, his M.R. rises considerably. (d) Upon awakening and for the subsequent hours before going to the supermarket, the man's B.M.R. is about 100 kcals per hour. During his errand and for a short time after his return, his M.R. rises considerably. While relaxing after his errand, his M.R. will return to its original basal level. — d

Answer

To be covered when testing yourself

3. Which of the following is completely correct?
   (a) Vitamin A - associated with vision.
   Vitamin E - associated with coagulation.
   Vitamin K - associated with anti-sterility in some animals
   Vitamin D - associated with bone growth
   Vitamin C - associated with synthesis of collagen.
   (b) Vitamin D - associated with growth of teeth.
   Vitamin A - associated with skin epithelial maintenance.
   Vitamin K - associated with blood coagulation.
   Vitamin E - associated with liver function in some animals.
   Vitamin C - associated with the synthesis of collagen.
   (c) Vitamin E - associated with bone growth.
   Vitamin A - associated with oxidative decarboxylation.
   Vitamin K - associated with sterility.
   Vitamin C - associated with coagulation.
   Vitamin D - associated with collagen synthesis.
   (d) Vitamin D - associated with teeth and bone.
   Vitamin E - associated with liver function in some animals.
   Vitamin A - associated with our visual pigment.
   Vitamin C - associated with collagen synthesis.
   Vitamin K - associated with transamination reactions.

   b

4. Which of the following will result from deficiencies of the following minerals: Sodium, Potassium, Calcium, Iodine, and Iron? (a) Abnormal osmotic and acid-base balance, poor vision, and abnormal DNA synthesis. (b) Abnormal bone growth, poor vision, abnormal insulin, and many non-functioning enzymes. (c) Insufficient levels of the hormone thyroxine, abnormal and insufficient hemoglobin, poor bone growth, abnormal muscle contractions, and abnormal acid-base and osmotic balance. (d) Poor vision, abnormal coagulating properties, insufficient DNA synthesis, and abnormal acid-base balance.

   c

5. Which of the following is incorrect concerning the teeth of mammals and the teeth of lower vertebrates? (a) Nerves and blood vessels are an integral part of both mammalian and lower vertebrate teeth. (b) The shape of mammalian teeth are usually a function of what that particular mammal eats, while the shape of lower vertebrate teeth are fairly constant. (c) Embryonically, both mammalian and lower vertebrate teeth have ectodermal origin, but only mammals possess teeth that are

Answer

To be covered when testing yourself

partly of mesodermal origin. (d) Lower vertebrate teeth can be considered extensions of their outer covering. This cannot be said for mammalian teeth. a

6. Which of the following best contrasts nutrient procurement in a photosynthetic plant with a fungus such as a mushroom? (a) In photosynthetic plants, no actual digestion is necessary, as they can use what they take in from the environment directly to synthesize carbohydrates. This is not the case with fungi. (b) Fungi contain enzymes, located in secretory cells, for the purpose of polysaccharide breakdown. These enzymes are not necessary in photosynthetic plants. (c) Nutrients necessary for protein synthesis can be obtained from the ground via an extensive vascular system in photosynthetic plants. Fungi do not contain such a system, and nutrients are sometimes obtained via dead organic matter. (d) All three of the above contrast photosynthetic with fungal nutrient procurement. d

7. Which of the following statements is <u>incorrect</u> concerning digestion? (a) Although very little digestion actually occurs in the stomach, some cells lining the stomach do secrete hydrochloric acid, which aids in the breakdown of proteins. by activating a proteolytic enzyme. (b) Most diegstion takes place in the upper 1/3 of the small intestine, an area known as the duodenum. (c) Bile, in addition to pancreatic enzymes, is secreted into the cecum to put the "finishing touches" on digestion. (d) If food is not digested properly when it reaches the large intestine, it can cause severe problems, as the bacteria there can ferment the leftover carbohydrates, multiply rapidly, and cause infection. c

8. Which of the following best describes how a peristaltic wave is formed? (a) Circular smooth muscle immediately in front of a bolus of food contracts, while smooth muscle behind the bolus of food relaxes. (b) Skeletal muscle surrounding the entire alimentary tract relaxes and contracts intermittently, causing food to move in a forward direction. (c) Circular smooth muscle immediately behind the bolus of food contracts, while the smooth muscle in front of the bolus relaxes. (d) Smooth muscle both in front of and behind a bolus of food contract together, causing the bolus to move toward the anus. c

Answer

To be covered when testing yourself

9. Which of the following is <u>not</u> a proteolytic enzyme? (a) Trypsin (b) Lipase (c) Carboxypeptidase (d) Pepsin.

b

10. Which of the following is <u>not</u> a function of hydrochloric acid secreted into the stomach? (a) As a bacteriocidal agent, (b) as a "turn-on" factor for some proteolytic enzymes, (c) as an aid in the formation of chylomicrons for fat digestion, (d) as a denaturing agent.

c

11. Which of the following is <u>incorrect</u> concerning the pancreas? (a) Among its secretion products are the enzymes trypsin and chymotrysin. (b) The pancreas plays only a minor role in blood sugar level regulation. (c) The digestive secretions of the pancreas are controlled in part by the hormone secretin. (d) The pancreas can be considered an exocrine and an endocrine gland.

b

12. Which of the following is <u>incorrect</u> concerning the absorption of fats from our diet? (a) Fats need not be transported across the cell membrane via a carrier system due to their similarity in structure to the membrane itself. (b) Because fats are transported via chylomicra,a deficiency of protein in ones diet could lead to fat malabsorption. (c) To form a lipoprotein and the subsequent chylomicron, 3 molecules of ATP must be expended. (d) It is not abnormal for an individual's blood to appear "milky" after ingesting a meal high in fats.

c

13. Which of the following hormones is <u>incorrectly</u> paired with its stimulus? (a) Gastrin - distension of the stomach wall. (b) Secretin - low pH in doudenum, (c) Cholecystokinin - high pH in ileum. (d) Salivary amylase - the thought of a juicy steak.

c

14. Which of following is <u>incorrect</u> concerning bile? (a) It consists of bile acids, pigments, cholesterol, and lecithin. It is synthesized in the liver and stored in the gall bladder. (b) It is released when the gall bladder contracts. This is caused by stimulation from the hormone cholecystokinin. (c) Its main function is the emulcification of fats. It does this by breaking up fat globules into smaller units called micelles. This increases the surface area available for lipase activity. (d) All constituents of bile are excreted out of the body as part of the feces. Essentially, there is no recirculation of components.

d

	Answer
	To be covered when testing yourself
15. Which of the following is not a function of the liver? (a) Hydroxylation of vitamin D (b) Regulation of blood sugar levels via synthesis and secretion of insulin and glucagon. (c) Detoxification of ammonia via its conversion to urea. (d) Conversion of glucose into fat, which can then be stored in adipose tissue.	b
Fill in the blanks.	
16. The storage form of glucose is called _____. It is a polymer of glucose molecules linked by _____ bonds.	glycogen, α-glycosidic
17. Energy released from the breakdown of the three major foodstuffs is called the _____ _____, and can be measured in _____. These values for carbohydrates, proteins and fats respectively are _____, _____, _____. It can be seen that the most energy can be produced from the breakdown of _____.	heat of combustion, kcal/gram, 4.0, 4.5, 9.5, fats
18. Vitamin B_1 (Thiamine) and Vitamin B_2 (Riboflavin) act as part of _____ involved with electron accepting during cellular respiration. Vitamin B_6 (Pyridoxine) is necessary for _____ reactions, important in the synthesis of all the _____.	coenzymes, transamination, amino acids
19. A diet consisting of proteins, carbohydrates, fats, minerals, and water is not sufficient for normal, complete growth in a rat. The factors that are missing are the _____.	vitamins
20. In a paramecium, the structure analogous to our mouth is called the _____. The area of the aforementioned "mouth" where a food vacuole is formed is called the _____. After the vacuole has been enzymatically broken down and the nutrients used, any non-usable material leaves the paramecium via the _____.	oral groove, cytopharynx, anal pore
21. The _____ and _____ salivary glands secrete a fluid that is not only watery, but contains _____ to aid in the passage of food down the throat. The _____ salivary gland, on the other hand, secretes just a _____ saliva. All three salivary glands secrete a fluid that contains _____ for starch breakdown.	submaxillary, sublingual, mucus , parotid, watery, amylase
22. The action of the enzyme _____ takes place specifically in the stomach, where the pH lies in the range of _____ to _____. Its function is to break down _____ into smaller _____ chains. It does this by utilizing water in a _____ reaction.	pepsin, 1.5, 2.5, proteins, peptide, hydrolysis

	Answer
	To be covered when testing yourself
23. The sudden relaxing of the _____ sphincter allows food to pass from the pharynx into the esophagus. It then contracts and closes immediately thereafter. At the same time, the closure of the _____ prevents food from entering the trachea.	hypo-pharyngeal, glottis
24. When following the digestive path of a starch molecule, it is first exposed to the enzyme _____ in the _____. This breaks some of the starch into smaller disaccharide units called _____ units. Then, another _____ in the small intestine breaks down any left over starch, while another enzyme, _____, breaks down the maltose units into glucose molecules.	amylase, mouth, maltose, amylase, (from the pancreas) maltase
25. The fact that mucus protects the lining of the stomach from being digested by enzymes is due to the pH of mucus lying in the _____ range. This offsets the acidity caused mainly by _____ secreted in the stomach.	basic, hydrochloric acid
26. The surface area of the small intestine is increased at _____ different levels. First, the _____ lining the intestine is thrown into folds and ridges. Second, on the surface of the lining are fingerlike projections called _____. These contain numerous capillaries which aid in the actual _____ of nutrients. Finally, on each cell itself lining the intestine are microscopic projections called _____. These contribute most to the increased surface area.	three, mucosa, villi, absorption, microvilli
27. The absorption of monosaccharides at the lining of the small intestine is dependent upon the presence of _____ ions in the lumen of the intestine. This form of transport is called _____. Amino acids can also be transported by this method, in addition to _____ and _____.	sodium, facilitated diffusion, simple diffusion, active transport
28. An enzyme that attacks the end of a peptide chain is called an _____. There are two kinds of _____; an _____ and a _____. These enzymes function in the _____ and are necessary for digestion of _____.	exopeptidase, exopeptidases, amino-peptidase, carboxy-peptidase, intestine, proteins.
29. When bile pigments such as bilirubin are not able to be incorporated into bile, they can accumulate in the blood and tissues. This gives rise to a condition called _____.	jaundice

	Answer
	To be covered when testing yourself
30. The enzyme pepsin is produced in the _____; its digestive end product is small _____. The hormone lipase is produced by the _____; its digestive end-products are _____ and some free _____, _____ and _____. The hormone sucrase is produced in the _____; its digestive end-products are _____ and _____.	stomach, peptides, pancreas, glycerol, diglycerides, monoglycerides, fatty acids, intestine, glucose, fructose
Determine whether the following statements are true or false.	
31. Brain cells and muscle cells, both of which are dependent upon glucose for nutrient, do not suffer the immediate effects of glucose insufficiency due to their storage depots of glucose.	False
32. Muscle cells,like liver cells, have the ability to regulate the blood sugar level.	False
33. Although we derive all our energy from the breakdown of the three major foodstuffs, the energy used to form ATP is less than half of that available from the combustion of the foodstuffs, the rest being dissipated as heat.	True
34. A complete lack of phosphorus in one's diet would eventually be incompatible with life.	True
35. Digestion in the hydra occurs extracellularly via nematocysts and enzymes from the cells lining the gastrovascular cavity, and intracellularly via phagocytosis by the cells of the gastrovascular cavity.	True
36. When referring to the digestive systems of flatworms and earthworms, the one most similar to man's digestive system is that of the flatworm.	False
37. During the act of swallowing, food is prevented from entering the trachea by the elevation of the larynx at the same time that the glottis is closed. Once the bolus of food passes through the pharynx into the esophagus, peristaltic waves in the esophagus cause the bolus to pass into the stomach, with the concurrent relaxation of the cardiac (lower esophageal) sphincter.	True
38. If an individual so desires, swallowing can be voluntarily halted before completion.	False

Answer

To be covered when testing yourself

39. As chyme from the stomach approaches the pyloric sphincter, the sphincter opens wide enough and long enough to allow all of the chyme to enter the duodenum, with no further churning from the stomach necessary. — False

40. Although pancreatic juice and gastric juice are secreted at different locations, they contain exactly the same constituents. — False

41. The protection of the stomach lining by mucus is sufficient to allow the cells of the lining to live a long time, with cell turnover occuring only occasionally. — False

42. The large intestine has fewer villi than the small intestine, but plays an important role in concentrating feces. — True

43. Bile is produced in the liver and stored in the gallbladder. Its function is to aid in the digestion of the nucleic acids, DNA and RNA. — False

44. Gallstones are caused by an overload of cholesterol in the gallbladder, the excess precipitating out and forming a hard mass. The gallstone can be either passed out via the intestine or can lodge in the bile duct, making digestion of fats difficult. If necessary, the gallbladder itself can be removed with no subsequent harm to the body. — True

45. By reducing one's intake of cholesterol, one can effectively, with no exception, lower the amount of circulating cholesterol in the body. — False

CHAPTER 18

HOMEOSTASIS & EXCRETION

FLUID BALANCE

• PROBLEM 18-1

How is the fluid which is ingested into the body absorbed?

Solution: About 9,000 mls. of fluid normally enter the intestinal tract from the stomach. This fluid is composed of digested material (chyme) from the stomach, gastric secretions, and ingested fluids, either digested or undigested. The material is nearly isotonic, that is, having the same solute concentration as plasma. However, as a result of the extremely rapid absorption of monosaccharides, sodium, and amino acids in the small intestine, the total concentration of solutes in the lumenal fluid drops, and the fluid becomes relatively hypotonic. Because the intestinal walls are highly permeable to water, water rapidly diffuses from the lumen to the plasma. This movement is extremely rapid. The phenomenon is similar to that for sodium and water reabsorption by the proximal convoluted tubules of the nephrons in the kidneys. In addition, other ions, such as Cl^-, K^+, Mg^{2+} and Ca^{2+} are absorbed, some by active transport.

In the large intestine, additional water is absorbed. Here, active transport of sodium from the lumen to the blood causes the osmotic absorption of water. The longer fecal material remains in the colon, the more water is reabsorbed.

• PROBLEM 18-2

List the three primary nitrogenous excretory wastes; tell which animals excrete each, and why. What is uremia?

Solution:

The three primary nitrogenous waste products eliminated from animals are ammonia, urea, and uric acid. Nitrogenous wastes of aquatic animals, both marine and freshwater, are usually excreted as ammonia. Ammonia, a toxic substance, is very soluble in water. A considerable amount of water is needed, but since there is no danger of water loss in aquatic animals, the excretion of ammonia presents no difficulty.

Urea is a nitrogenous waste produced in the liver and excreted by the kidneys in mammals and amphibians. It is formed by combining ammonia with carbon dioxide, thus converting ammonia to a relatively non-toxic form. Urea is highly soluble, and requires water for its excretion. However, because it is non-toxic, urea, unlike ammonia, does not require extremely large amounts of water for excretion. Urea may be retained in the body for some time before being excreted.

Uric acid is also excreted by terrestrial animals, most notably, the egg-laying or oviparous animals such as birds, snakes, and lizards. Uric acid is a semi-solid, highly insoluble waste product. It is formed in the tissues and in insects is selectively absorbed by the malpighian tubule cells. In birds, it is absorbed by the kidney tubules. The rectum of uric acid-producing animals has powerful water-reabsorptive capacities; the uric acid thus leaves the rectum as a nearly dry powder or hard mass. In its excretion, nearly no water need be lost, and water is highly conserved.

Those animals excreting uric acid lay eggs enclosed within a relatively impermeable shell. Excretion of toxic ammonia is out of the question, and urea, if excreted, would build up to harmful concentrations over the period of embryonic development in the egg. Uric acid, on the other hand, being almost solid, can be safely stored in the egg until development is complete.

Uremia is a condition where there is a gradual accumulation of urea and other waste products in the blood. This condition can result from many different forms of kidney disease. The severity of uremia depends upon how well the impaired kidneys are able to preserve the constancy of the internal environment.

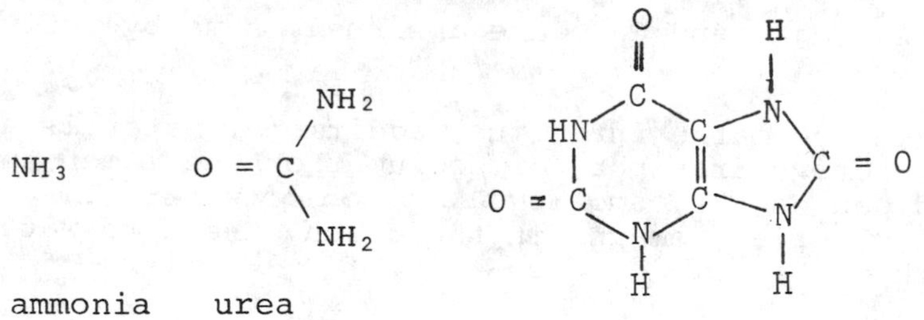

```
                               O        H
                               ‖        |
                  NH2          C        N
                 /        HN       C        
 NH3        O = C          |       ‖        C = O
                 \    O =  C       C        
                  NH2          N        N
                               |        |
                               H        H

 ammonia     urea
                                 uric acid
```

• **PROBLEM** 18-3

Most marine vertebrates have body fluids with osmotic pressure lower than that of their saline environment. How do these organisms osmoregulate in face of a perpetual threat of dehydration?

Solution: To survive their hostile saline environment, marine organisms show a variety of adaptations. The concentration of body fluids in marine fish is about one-third that of the sea water. Osmotic pressure, being higher in sea water, tends to drive water out of the body fluid of the fish. One site of severe water loss is the gill, which is exposed directly to the surrounding water for gaseous diffusion. To compensate for this loss, the fish must take in large amounts of water. However, the only source of water available is sea water; if consumed, the sea water would cause further water loss from the body cells because of its higher osmotic pressure. Marine fish, however, overcome this problem. In the gills, excess salt from the consumed sea water is actively transported out of the blood and passed back into the sea.

Marine birds and reptiles also face the same problem of losing water. Sea gulls and penguins take in much sea water along with the fish which they scoop from the sea. To remove the excess salt, there are glands in the bird's head which can secrete saltwater having double the osmolarity of sea water. Ducts from these glands lead into the nasal cavity and the saltwater drips out from the tip of the bill. The giant sea turtles, a marine reptile, has similar glands but the ducts open near the eyes. Tears from the eyes of turtles are not an emotional response, in contrast to popular belief. Rather, 'crying' is a physiological mechanism to get rid of excess salt.

• **PROBLEM** 18-4

In man, the kidney performs the bulk of the excretion of wastes from the body. Outline the structure of the human kidney and urinary system.

Solution: **Located on each posterior side of the human body just below the level of the stomach are the bean-shaped kidneys. Each kidney is about 10 cm long, and consists of three parts: an outer layer called the cortex, an inner layer called the medulla, and a sac-like chamber called the pelvis (see figure 1). The functional unit of a kidney is the nephron; there are about a million nephrons per kidney. A nephron consists of two components: a tubule for conducting cell-free fluid and a capillary network for carrying blood cells and plasma. The mechanisms by which the kidneys perform their functions depend on both the physical and physiological relationships between these two components of the nephron.**

Throughout its course, the kidney tubule is composed of a single layer of epithelial cells which differs in structure and function from one portion of the tubule to another. The blind end of the tubule is Bowman's capsule, a sac embedded in the cortex and lined with thin epithelial cells. The curved side of Bowman's capsule is in intimate contact with the glomerulus, a compact tuft of branching blood capillaries, while the other opens into the first portion of the tubular system called the proximal convoluted tubule. The proximal convoluted tubule leads to a portion of the tubule known as the loop of Henle. This hairpin loop consists of a descending and an ascending limb, both of which extend into the medulla. Following the loop, the tubule once more becomes coiled as the distal convoluted tubule. Finally, the tubule runs a straight course as the collecting duct. From the glomerulus to the beginning of the collecting duct, each of the million or so nephrons is completely separate from its neighbors. However, the collecting ducts from separate nephrons join to form common ducts, which in turn join to form even longer ducts, which finally empty into a large central cavity, the renal pelvis, at the base of each kidney. The renal pelvis is continuous with the ureter, which empties into the urinary bladder where urine is temporarily stored. The urine remains unchanged in the bladder, and when eventually excreted, has the same composition as when it left the collecting ducts.

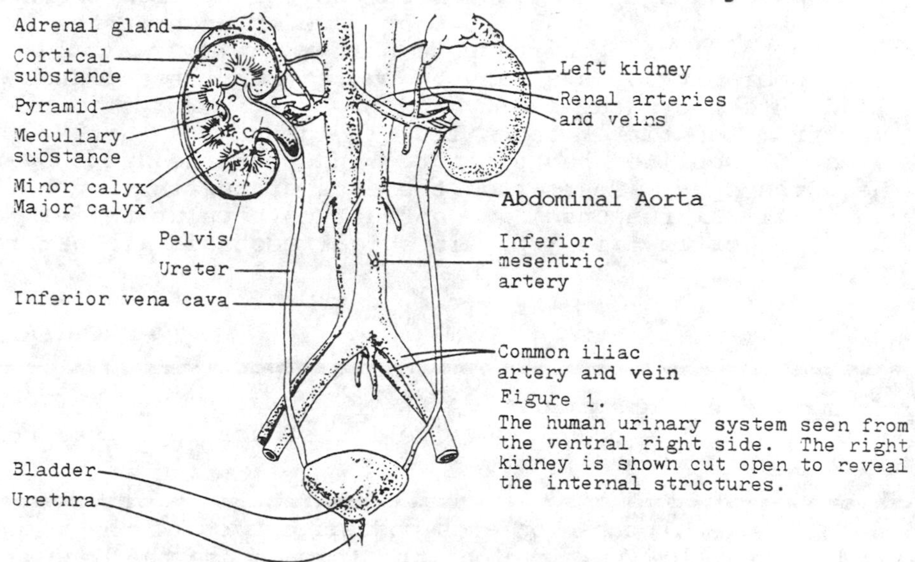

Figure 1.
The human urinary system seen from the ventral right side. The right kidney is shown cut open to reveal the internal structures.

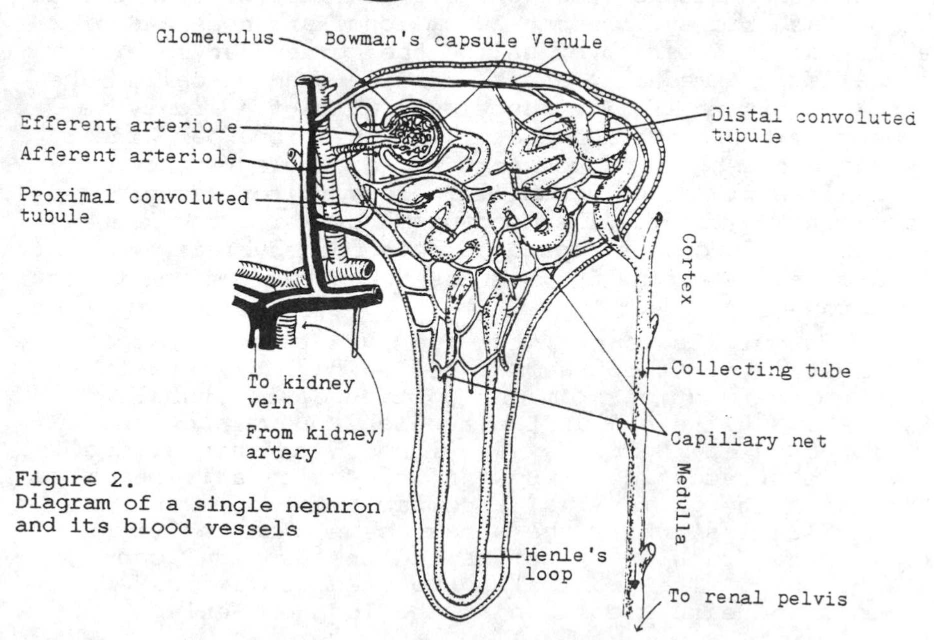

Figure 2.
Diagram of a single nephron and its blood vessels

Blood enters the kidney through the renal artery, which upon reaching the kidney divides into smaller and smaller branches. Each small artery gives off a series of arterioles, each of which leads to a glomerulus. The arterioles leading to the glomerulus are called afferent arterioles. The glomerulus protrudes into the cup of Bowman's capsule and is completely surrounded by the epithelial lining of the capsule. The functional significance of this anatomical arrangement is that blood in the capillaries of the glomerulus is separated from the space within Bowman's capsule only by two extremely thin layers: (1) the single-celled capillary wall, and (2) the one-celled lining of Bowman's capsule. This thin barrier permits the filtration of plasma (the non-cellular blood fraction) from the capillaries into Bowman's capsule.

Ordinarily, capillaries recombine to form the beginnings of the venous system. However, glomerular capillaries instead recombine to form another set of arterioles, called the efferent arterioles. Soon after leaving the region of the capsule, these arterioles branch again forming a capillary network surrounding the tubule. Each excretory tubule is thus well supplied with circulatory vessels. The capillaries eventually rejoin to form venous channels, through which the blood ultimately leaves the kidney.

• PROBLEM 18-5

When the higher vertebrates left the water to take on a terrestrial existence, they had to adopt mechanisms to conserve body water. One of these was the evolution of a concentrated urine. How is this achieved physiologically?

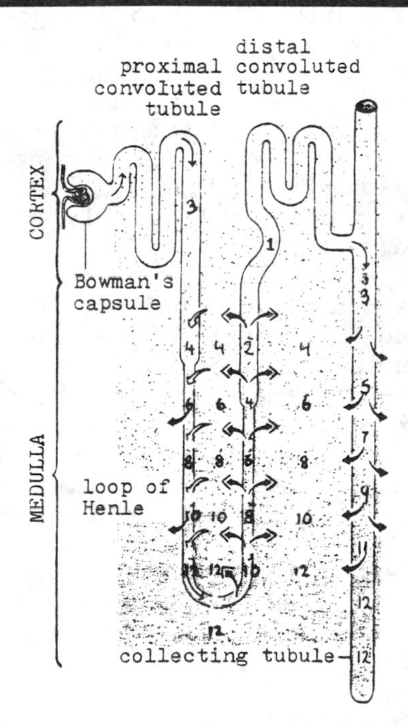

←flow of filtrate or urine
←active transport of Na^+
◇passive diffusion of Na^+
←passive diffusion of H_2O

Numbers indicate solute concentration in hundreds of milliosomoles per liter.

Diagram of countercurrent flow mechanism of the mammalian kidney.

Solution: The higher vertebrates have evolved a final urine having a greater osmolarity than blood plasma. This is advantageous in that water can be retained within the body making the organism less likely to suffer dehydration. In the absence of a special renal mechanism, the urine would be highly dilute and severe water loss from the organism could become fatal.

The loop of Henle is a hairpin segment of the renal tubule having an ascending limb and a descending limb. The connecting tubule lies in close proximity and parallel to the loop of Henle so that the three parts of the renal tubule can interact with one another (see figure).

The walls of these three portions differ in their permeabilities to substances such as sodium and water. Both the anatomical and physiological aspects of the renal tubule determine the concentration of the urine excreted.

The ascending limb of the loop of Henle actively pumps sodium ions out of its tubular space into the interstitial fluid, with chloride following passively due to electrostatic attraction. The ascending limb is impermeable to water and thus the osmolarity of fluid inside the ascending limb decreases as the loop ascends and a concentration gradient is established between the loop and the interstitial fluid. The descending limb is permeable both to water and sodium, so that some of the sodium and chloride ions diffuse passively into its tubular space. The cycling of sodium from ascending limb to interstitial fluid to descending limb results in the establishment of a concentration gradient of sodium and chloride in the tissue fluid surrounding the loop, with the lowest concentration near the cortex and the highest concentration deep in the medulla. (See accompanying figure). This mechanism of flow in the mammalian kidney, called the countercurrent flow, permits the concentration gradient in the interstitial fluid to be maintained; this gradient is essential in establishing the final concentration of the urine.

With this in mind, we can understand how the final urine becomes highly concentrated relative to the blood. The urine that leaves the ascending limb is not more concentrated than the glomerular filtrate since it has been diluted in the ascending loop. But as the urine flows through the collecting tubule, it is essentially flowing through a concentration gradient in the interstitial fluid from the cortex to the medulla. Moreover, the collecting tubule is permeable only to water. Since the osmolarity of the interstitial fluid is lowest near the cortex and increases toward the medulla, the urine flowing down the collecting tubule will lose water during it scourse to the medulla. The final urine that emerges from the collecting tubule is substantially hypertonic to blood and remains in this same concentration until it is excreted to the outside.

• PROBLEM 18-6

The kangaroo rat never drinks water, yet it survives in some of the hottest and driest regions of the world. Explain how it obtains and conserves water.

Solution: Animals which live in the desert have limited means of obtaining water. For these animals, water retention and conservation are very crucial. The kangarco rat can survive without ever drinking water, and in this respect is fairly unique. This animal is found in the Soutwestern deserts of the United States. It lives on dry seeds and obtains all of its water from carbohydrate metabolism.

The excretory organs of the kangaroo rat are modified to conserve as much water as possible. Its kidneys are remarkably efficient in concentrating the urine, allowing very little water loss. As was noted earlier, a concentration gradient is achieved in the interstitial medium surrounding the loop of Henle and the collecting duct in the mammalian kidney. The longer the loop, the greater the gradient can become, and therefore, the greater is the amount of water that can be reabsorbed. In the kangaroo rat, the loops are extremely long and thus a large amount of water is retained by the body and very little excreted.

• PROBLEM 18-7

There are three processes which together enable the kidney to remove wastes while conserving the useful components of the blood. What are these processes and where do they occur?

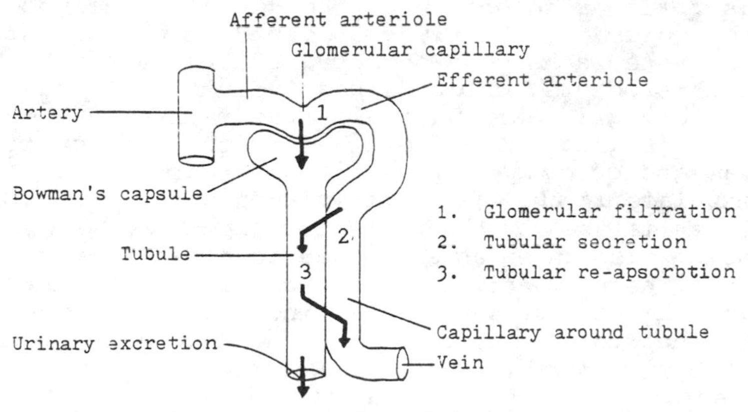

The three basic components of renal function.

Solution: Blood flowing to the kidneys first undergoes glomerular filtration. This occurs at the junction of the glomerular capillaries and the wall of Bowman's capsule. The blood plasma is filtered as it passes through the capillaries, which are freely permeable to water and solutes of small molecular dimension yet relatively impermeable to large molecules, especially the plasma proteins. Water, salts, glucose, urea and other small species pass from the blood into the cavity of Bowman's capsule to become the glomerular filtrate.

It has been demonstrated that the filtrate in Bowman's capsule contains virtually no protein and that all low weight crystalloids (glucose, protons, chloride ions, etc.) are present in the same concentrations as in plasma.

If it were not for the process of tubular reabsorption, the composition of the urine would be identical to that of the glomerular filtrate. This would be extremely wasteful, since a great deal of water, glucose, amino acids and other useful substances present in the filtrate would be lost. Tubular reabsorption is strictly defined as the transfer of material from the tubular lumen back to the blood through the walls of the capillary network in intimate contact with the tubule. The principal portion of the tubule involved in reabsorption is the proximal convoluted tubule. This tubule is lined with epithelial cells having many hair-like processes extending into its lumen. These processes are the chief sites of tubular reabsorption. As the filtrate passes through the tubule, the epithelial cells reabsorb much of the water and virtually all the glucose, amino acids and other substances useful to the body. The cells then secrete these back into the bloodstream. The secretion of these substances into the blood is accomplished against a concentration gradient, and is thus an energy consuming process - one utilizing ATP. The rates at which substances are reabsorbed, and therefore the rates at which wastes are excreted (because what is not reabsorbed is eliminated), are constantly subjected to physiological control. The ability to vary the excretion of water, sodium, potassium, hydrogen, calcium, and phosphate ions, and many other substances is the essence of the kidney's ability to regulate the internal environment. Reabsorption also occurs in the distal convoluted tubules, where sodium is actively reabsorbed under the influence of aldosterone, a hormone secreted by the adrenal cortex. When this occurs, chloride passively follows due to an electrical gradient; water is also reabsorbed because of an osmotic gradient established by the reabsorption of sodium and chloride. In addition, reabsorption of water takes place in the distal convoluted tubule and collecting duct, stimulated by the posterior pituitary hormone vasopressin, also known as antidiuretic hormone (ADH). ADH increases the permeability of the distal convoluted tubule and collecting duct to water, allowing water to leave the lumen of the nephron and render the urine more concentrated.

The kidney also removes wastes by means of tubular secretion. This process involves the movement of additional waste materials directly from the bloodstream into the lumen of the tubules, without passing through Bowman's capsule. Tubular secretion may be either active or passive, that is, it may or may not require energy. Of the large number of different substances transported into the tubules by tubular secretion, only a few are normally found in the body. The most important of these are potassium and hydrogen ions. Most other substances secreted are foreign substances, such as penicillin. In some animals, like the toadfish, whose kidneys lack glomeruli and Bowman's capsules, secretion by the tubules is the only method available for excretion.

GLOMERULAR FILTRATION

• PROBLEM 18-8

With reference to fluid pressures, explain the mechanism underlying glomerular filtration.

Solution: The mean blood pressure in the large arteries of the body is approximately 100 mm Hg. However, as the blood passes through the arterioles connecting the renal artery to the glomeruli, the blood pressure decreases, so that the pressure in the glomerular capillaries is usually only about half the mean arterial pressure or 50 mm Hg. This level is considerably higher than in other capillaries of the body, due to the fact that the arterioles leading to the glomeruli are wider than most other arterioles and therefore offer less resistance to flow. The force driving fluid out of the glomerulus and into Bowman's capsule is the pressure of the blood in the glomerular capillaries. But this hydrostatic pressure favoring filtration is not completely unopposed. There is fluid within Bowman's capsule, resulting in a capsular hydrostatic pressure of about 10 mm Hg, which resists the flow into the capsule. The presence of protein in the plasma and its absence in Bowman's capsule, results in a second force opposing filtration. This difference in protein concentration between the capsule and the blood induces an osmotic flow of water into the blood. This osmotic pressure, also known as colloidal osmotic pressure, equals roughly 30 mm Hg.

Summing the forces, there is a glomerular capillary blood pressure of 50 mm Hg favoring filtration, a 10 mm Hg opposition to filtration due to fluid in Bowman's capsule, and a 30 mm Hg. pressure working against filtration due to a protein concentration difference between the blood and the capsule fluid, resulting in a net glomerular filtration pressure of 10 mm Hg. This net pressure forces fluid from the blood into Bowman's capsule.

• PROBLEM 18-9

The glomerular filtration rate is controlled primarily by alteration of the glomerular capillary pressure. What are the two factors which produce these changes?

Solution: The glomerular filtration rate is defined as the volume of blood filtered by the glomerulus per minute. One of the factors expected to influence the glomerular filtration rate is the arterial blood pressure. A decrease in arterial pressure would decrease the filtration rate by lowering glomerular capillary pressure. An increase in arterial pressure would serve to increase the glomerular filtration rate. However, GFR stays relatively constant despite changes in mean arterial pressure. This is due to an intrinsic ability of the kidneys to autoregulate. The direct factor influencing filtration is the diameter of the afferent arterioles. A decrease in the diameter of the afferent arterioles lowers blood flow to the glomerulus, lowering glomerular capillary pressure and filtration rate. Conversely, an increase in the diameter of the afferent arterioles increases the volume of blood flow to the glomerulus, increasing glomerular capillary pressure and filtration rate. The main stimulus that controls afferent arteriolar size is the mean arterial pressure (MAP). When the MAP falls, the afferent arteriole dilates to maintain GFR. The converse occurs when the MAP increases.

• **PROBLEM 18-10**

What is meant by the term "renal threshold"? "renal clearance"?

Solution: Although glucose is present in the glomerular filtrate, there is normally little or no glucose in the urine, due to its reabsorption by the cells of the renal tubules. If the plasma level of glucose were increased drastically, the level of glucose in the glomerular filtrate would likewise increase. In this situation, not all the glucose could be reabsorbed as the filtrate passed through the kidney tubules, and some glucose would appear in the urine. The concentration in the plasma of a substance such as glucose at the point where it just begins to appear in the urine is referred to as the "renal threshold" of the substance. The renal threshold for glucose is about 150 mgs. of glucose per 100 mls. of blood. When the concentration of glucose in the blood exceeds this level, glucose will begin to appear in the urine. There are comparable thresholds for many other substances.

Physiologists have adopted the concept of "renal clearance" to express quantitatively the kidneys' ability to eliminate a given substance from the blood.

The elimination of a substance is dependent on its concentration in the blood, the filtration rate at the glomerulus (or volume of blood filtered per minute), and the rate of secretion or reabsorption of that substance by the renal tubules. The renal clearance relates the rate of appearance of a substance in the urine to its concentration in the blood, and is expressed as the volume of plasma cleared of that substance per minute. One can also think of renal clearance as the volume of plasma that would contain the amount of a substance excreted in the urine per minute.

The renal clearance of a substance x is defined as

$$C_x = \frac{\text{mgs. of substance x secreted in urine per minute}}{\text{mgs. of substance x per ml. of plasma}}$$

Thus, if x is present in the plasma in a concentration of .02 mg/ml, but secreted at a rate of .2 mg/min, its clearance would be

$$C_x = \frac{.2 \text{ mg x/min.}}{.02 \text{ mg x/ml plasma}}$$

$$= 10 \text{ mls plasma/min.}$$

In other words, 10 mls of plasma are completely cleared of substance x per minute, thus .2 mgs. of that substance appears in the urine per minute.

• PROBLEM 18-11

Certain substances are maintained in the blood at a high threshold while others are held at a very low threshold. What factor determines what the kidney threshold for a substance will be? Which substances characterize each level?

Solution: The amounts of the individual substances dissolved in the blood exhibit wide variation. Some substances are high threshold substances, which means that they are maintained in the blood at a high concentration relative to their concentration in the urine. Substances maintained at such levels include glucose, amino acids, fatty acids, vitamins, and hormones. These are filtered out of the blood at the glomerulus and almost fully reabsorbed by the tubules. Recall that it is through active transport by the cells of the proximal convoluted tubule that reabsorption takes place. The threshold level of reabsorption is thus a function of the rate of active transport across the tubule surface. Only if the concentration of a high threshold substance in the blood is very great will the amount filtered surpass the amount reabsorbed and cause the substance to spill into the urine. Substances with high threshold values are essential nutrients or other compounds needed in body metabolism.

Low threshold substances are those having a high concentration in the urine relative to their concentration in the blood. Only a small fraction of these substances is reabsorbed in the tubules. Low threshold substances are mostly waste products, such as urea; the concentration of urea in the urine may be up to 70 times that in the blood. Various drugs, such as aspirin and antibiotics, are also included in this group. This is not always advantageous, for the kidneys may remove a drug so fast that continued application is needed for maximal effectiveness. This problem is solved in some cases by mixing the medication with a binder so that it is released slowly over an extended period of time.

Substances having medium thresholds include inorganic mineral ions and certain organic substances. Such substances are roughly equally reabsorbed and excreted. Important ions in this group are the sodium and chloride ions. These ions are involved in the maintenance of the pH of the blood (7.3) by their association with hydrogen and hydroxyl ions. The cellular metabolism of food particles most often results in an excess of hydrogen ions; these are then excreted along with chloride ions, resulting in a slightly acidic urine.

THE INTERRELATIONSHIP BETWEEN THE KIDNEY AND THE CIRCULATION

• PROBLEM 18-12

The volume of plasma and other extracellular fluids are regulated by automatic feedback controls. What are these controls?

Solution: The mechanisms responsible for the regulation of body fluid volume depend on the fact that an increase in blood volume causes an increase in blood pressure. One such mechanism operates in the capillaries of all tissues. Here, increased blood volume, because it increases blood pressure, tends to increase the pressure within the capillaries. When this occurs, the pressure that tends to move fluid out of the capillaries into the interstitial space becomes greater than the colloid osmotic pressure, the force which tends to move fluid from the interstitium back into the capillaries (see previous solution). What results is a net movement of fluid out of the capillaries. Eventually, this movement causes a reduction of blood volume, which in turn reduces the blood pressure. Capillary pressure decreases, and there is a net movement of fluid from the extracellular space back into the capillaries. This is due both to the decreased capillary pressure and to a rise in the colloid osmotic pressure in the blood caused by the drop in plasma volume. Fluid thus enters the capillaries,restoring normal blood volume and normal blood pressure.

A second process stabilizing blood volume occurs at the glomerulus-capsule junction. As blood volume increases, the pressure within the glomeruli of the kidneys likewise increases. Filtration pressure, being proportional to the glomerular capillary pressure, rises also and more urine is subsequently formed. This reduces the fluid volume of the blood, thus decreasing the blood pressure.

A third fluid volume regulating process is based

on the presence of baroreceptors in the walls of the arteries in the chest and neck. Increased blood volume exerts a stronger pressure on the arteries, stretching them. The baroreceptors respond to this stretching of the arterial walls and send impulses to the vasomotor center in the brain. From here, impulses are sent to the smooth muscle in the walls of the afferent renal arterioles, causing them to relax. This relaxation allows the arterioles to dilate and increase the flow of blood into the glomeruli; more blood will be filtered and more water eliminated. By this autonomic feedback mechanism the blood volume can then be restored to normal.

• PROBLEM 18-13

Persons with kidney trouble usually also have high blood pressure. Explain why. What condition can result if the blood pressure is not restored to normal?

Solution: Any kidney disorder which interferes with normal filtration invariably affects blood pressure. Blood pressure is directly related to blood volume; an increase in blood volume will raise blood pressure, and a decrease will lower blood pressure. If for some reason filtration were drastically reduced, the amount of water excreted from the body would drop. This would result in more water retention by the body, increasing both blood volume and blood pressure. Conversely, should filtration be increased drastically, water loss from the body would increase, blood volume would decrease, and blood pressure would fall. A condition where filtration is increased is less common then a condition where it is decreased. For example, a common disorder in the vasomotor center causes constriction of the renal arterioles, leading to reduced filtration.

An increase in blood pressure causes an increase in capillary pressure, favoring movement of material out of the capillary into the surrounding tissue. If the blood pressure remains high, the lymph vessels which normally drain the tissues of excess fluid, become ineffective and fluid begins to accumulate in the surrounding tissue. This condition is known as edema.

Edema may also result from a dietary deficiency of protein. Under such a condition, the colloid osmotic pressure due to protein decreases, and the force holding fluid in the capillaries is reduced. Fluid from the blood diffuses out of the capillaries into the extracellular spaces in tissues, and swelling of the tissues occurs.

REGULATION OF SODIUM AND WATER EXCRETION

• PROBLEM 18-14

One of the functions of the kidney is to regulate the

amount of sodium in the body. What functions does sodium perform in the body?

Solution: Sodium is essential to the body for proper functioning. Sodium is responsible for changes in membrane potential, and is thus required in physiological processes involving excitation of cells. Recall that the plasma membrane is differentially permeable to potassium and sodium ions. The membrane also has a pumping mechanism which actively pumps sodium ions out of the cell. The net effect is a higher concentration of sodium outside the cell and a lower concentration inside the cell. Although potassium ions are also distributed unequally on the two sides of the membrane, the difference is not as great as that of the sodium ions. Thus there is a net negative charge inside the cell as compared to the outside. This electrical potential is the resting cell membrane potential. A change in the distribution of sodium ions will disturb the resting potential. In nerve cells, this change generates a nerve impulse. In muscle cells, a change in sodium ion distribution may lead to muscle contraction. Sodium is therefore important in transmission of nerve impulses and in muscle contraction. (Refer to the appropriate solutions for discussions of these mechanisms.)

Sodium is also important for the maintenance of the total solute concentration, or osmolarity, of the plasma and other body fluids. Any change in the sodium content of the plasma or extracellular fluid can have drastic effects on the body cells. For example, should the sodium concentration in the plasma increase, a gradient will be created whereby water would move osmotically out of the red blood cells, resulting in their crenation.

• PROBLEM 18-15

How is aldosterone related to proper kidney functioning? How is its secretion controlled?

Solution: Aldosterone is a hormone produced by the adrenal cortex. It is thought that this hormone stimulates sodium reabsorption in the distal convoluted tubules of the kidney by altering the permeability of the tubules to sodium. In the complete absence of this hormone, a patient may excrete close to 30 gms of sodium per day, whereas excretion may be virtually zero when aldosterone is present in large quantities. The major clue that some substance secreted by the adrenal cortex controls sodium reabsorption came from the observation of patients who had missing or diseased adrenal glands. These patients were seen to excrete large quantities of sodium in the urine, despite a decreased glomerular filtration rate (which would tend to reduce sodium loss). This indicated that decreased tubular reabsorption must be responsible for the sodium loss.

Aldosterone secretion is controlled by reflexes involving the kidneys themselves. Lining the arterioles in the kidney are specialized cells which synthesize and secrete an enzyme known as renin (not to be con - fused with rennin, a digestive enzyme secreted by the stomach) into the blood. Renin catalyzes a reaction in which angiotensin I, a small polypeptide, is split off from a large plasma protein called angiotensinogen. Angio-tensinogen is synthesized by the liver and is always present in the blood. It is still not certain how renin secretion is controlled by the kidney. There seems to be multiple nerve inputs to the renin-secreting cells, and their respective roles have not yet been determined.

Angiotensin I is converted to Angiotensin II via a different enzyme. Angiotensin II then acts to stimulate aldosterone secretion by the adrenal gland, and as such constitutes the primary input to the adrenal controlling the production and release of the hormone. Aldosterone then acts on the kidney tubules to stimulate sodium reabsorption. Angiotensin II has other effects on the body including stimulating thirst centers to ultimately increase blood volume by drinking, and causing vasoconstriction which increases blood pressure.

• PROBLEM 18-16

Sodium concentration in the plasma is largely determined by sodium excretion by the kidney. Describe how sodium excretion is regulated.

Solution: The control of sodium excretion depends upon two variables of renal function: the glomerular filtration rate and sodium reabsorption by the kidney tubules.

The plasma sodium level is intimately related to the plasma volume because of the principle of osmosis. If the sodium concentration in the blood increases, water enters the plasma from the surrounding tissue cells or interstitial fluid, raising the plasma volume. Conversely, if the plasma sodium concentration decreases, the osmolarity of the blood decreases. Water then leaves the plasma by osmosis, reducing the blood volume. Thus the sodium level in the body is able to be regulated in the same way that blood volume is regulated via the glomerular filtration mechanism. For instance, if the sodium level falls low in the blood, the blood volume drops due to osmosis of water out of the circulatory system. Decreased blood volume causes a fall in blood pressure, and consequently a drop in glomerular capillary pressure. The amount of glomerular filtrate declines, and more sodium is conserved in the body. Decreased blood volume also reduces filtration via an autonomic feedback mechanism involving baroreceptors in arterial walls and the central nervous system. The net effect is that water and sodium are retained in the body to maintain a normal physiological state.

The second renal mechanism regulating sodium excretion occurs in the distal convoluted tubule. It involves aldosterone, a hormone produced by the adrenal cortex, and a group of factors leading to the synthesis and release of aldosterone. (See previous question). Aldosterone acts on the epithelium of the distal convoluted tubule, increasing its permeability to sodium, and thus enhancing sodium reabsorption by the renal tubular cells. It is important to note that that as sodium is reabsorbed by the distal convoluted tubules under the influence of aldosterone, water is also reabsorbed. This occurs because as sodium is removed from the tubular space, an osmotic gradient is established. This gradient causes water to leave the tubular space and reenter the blood. Therefore, an increased sodium reabsorption results also in an increased water reabsorption. This is why aldosterone is stimulated when blood volume falls; the aldosterone released stimulates sodium and therefore water reabsorption. The water enters the bloodstream and helps restore the blood volume to normal.

• PROBLEM 18-17

The cells of humans and most animals have become adapted to surviving only within a relatively small range of hydrogen ion concentrations. What is the role of the kidneys in regulating hydrogen ion concentration?

Solution: Most metabolic reactions are sensitive to the hydrogen ion concentration of the fluid in which they occur, due primarily to the marked influence of protons on enzyme function. Accordingly, the hydrogen concentration of the extracellular fluid is one of the most critically and delicately regulated chemical levels in the entire body.

The kidneys regulate the hydrogen ion concentration of the extracellular fluids by excreting either acidic or basic constituents when the levels deviate from normal. This regulation is achieved by exchanging hydrogen ions for sodium ions in the tubular fluid. Sodium ions usually combine with bicarbonate ions in the tubular fluid to form sodium bicarbonate. However, when excess hydrogen ions are secreted into the urine, they also combine with the bicarbonate there, essentially replacing the sodium ions to form carbonic acid. The carbonic acid formed dissociates to produce CO_2 and water:

$$H^+ + HCO_3^- \rightleftarrows H_2CO_3 \rightleftarrows CO_2 + H_2O$$

bicarbonate carbonic acid bicarbonate

The CO_2 is reabsorbed into the blood from the tubules and transported to the lungs where it is eliminated during expiration. The decrease in the concentration of bicarbonate ions in the tubular fluid represents a net excretion of hydrogen ions. Usually the amount of hydrogen ions exchanged for sodium in the distal convoluted tubules is equivalent to the amount of bicarbonate ions in the tubular fluid. If the plasma becomes very acidic, the quantity of bicarbonate ions in the kidney filtrate is insufficient to react with the abnormally high quantity of hydrogen ions. When this happens, the hydrogen ions combine with other buffers, in the tubular fluid, such as phosphate, and are excreted.

If the extracellular fluids become too alkaline, the amount of bicarbonate in the glomerular filtrate becomes greater than the amount of hydrogen ions secreted by the tubules. The unreacted bicarbonate is simply excreted as sodium bicarbonate. The loss of sodium bicarbonate makes the body fluid more acidic, returning the acid-base balance to normal.

In addition to the bicarbonate technique, the kidney has another mechanism for coping with excessive acidity which involves the secretion of ammonia by the kidney tubules. Should the hydrogen ion concentration in the filtrate be too great for the bicarbonate to handle, ammonia would be secreted into the tubules. Here, the ammonia combines with hydrogen ions to produce ammonium ions (NH_4^+), which are then excreted. By combining with hydrogen ions, the ammonia effectively buffers the acidic tubular fluid by removing the hydrogen ions from the fluid. The more hydrogen ions there are in the tubule, the greater is the conversion of ammonia to ammonium ions.

• PROBLEM 18-18

How does the body control the amount of water excreted?

Solution: Factors such as blood volume and glomerular capillary pressure act to regulate the amount of fluid initially absorbed by the kidney. The volume of urine excreted, however, is ultimately controlled by the permeability of the walls of the distal convoluted tubules and collecting ducts to water. This permeability can be varied, and is regulated by a hormone known as vasopressin or antidiuretic hormone (ADH). In the absence of ADH, the water permeability of the distal convoluted tubule (DCT) and collecting tubule is very low, and the final urine volume is correspondingly high. In the presence of ADH, water permeability is high, and the final urine volume is small. ADH has no effect on sodium absorption, but regulates the ability of water to osmotically follow ionic absorption.

ADH is produced by a discrete group of hypothalamic neurons whose axons terminate in the posterior pituitary, from where ADH is released into the blood. The release of ADH is regulated by osmoreceptors in the hypothalamus. Increased plasma osmolarity causes increased secretion of ADH. Decreased osmolarity leads to decreased secretion of ADH. Blood volume also influences ADH secretion via stretch receptors in the left atrium. Increased blood volume stimulates this baroreceptor reflex to decrease secretion of ADH.

Let us look at the interaction of these regulatory mechanisms in a specific situation. If a man drinks an excess amount of water, but does not increase his sodium intake, the most logical way to maintain optimal osmotic concentrations in the body would be to excrete the excess water without altering normal salt excretion. Intake of the excess water results in an increase in extracellular and blood fluid volumes. This has a two-fold effect: First, the osmotic concentration of the blood decreases. This stimulates the osmoreceptors to cause decreased secretion of ADH. Second, the arterial baroreceptors are stimulated and send impulses to the hypothalamus, where ADH release is inhibited. The permeability of the collecting tubules to water is lowered, and consequently more water is excreted.

• PROBLEM 18-19

Large deficits of water can be only partly compensated for by renal conservation; drinking is the ultimate compensatory mechanism. What stimulates the subjective feeling of thirst which drives one to drink water?

Solution: The feeling of thirst is stimulated both by a low extracellular fluid volume and a high plasma osmolarity. The production of ADH by the hypothalamus is also stimulated by these factors. The centers which mediate thirst are located in the hypothalamus very close to those areas which produce ADH. Should this area of the hypothalamus be damaged, water intake would stop because the sensation of thirst would be impaired. Conversely, eletrical excitation of this area stimulates drinking. There has been much speculation that, because of the similarities between the stimuli for ADH secretion and thirst, the receptors which initiate the ADH controlling reflexes are identical to those for thirst. These are the osmoreceptors referred to in the previous solution.

In addition to hypothalamic pathways, there are also other pathways controlling thirst. For example, dryness of the mouth and throat causes profound thirst, which is relieved by moistening. There is also a learned control of thirst; the quantity of fluid drunk with each meal is, in large part, a learned response determined by past experience.

RELEASE OF SUBSTANCES FROM THE BODY

• PROBLEM 18-20

In the human, many organs in addition to the kidneys perform excretory functions. What are these organs and what do they excrete?

Solution: The organs of excretion include the lungs, liver, skin, and the digestive tract, in addition to the kidneys. Water and carbon dioxide, important metabolic wastes, are excreted by the lungs. Bile pigments, hemoglobin, red blood cells, some proteins, and some drugs are broken down by the liver for excretion. Certain metal ions, such as iron and calcium, are excreted by the colon. The sweat glands of the skin are primarily concerned with the regulation of body temperature, but they also serve in the excretion of 5 to 10 percent of the metabolic wastes formed in the body. Sweat and urine have similar composition (water, salts, urea and other organic compounds) but the former is much more dilute than the latter, having only about one eighth as much solute matter. The volume of perspiration varies from about 500 ml. on a cool day to as much as 2 to 3 liters on a hot day. While doing hard work at high temperatures, a man may excrete from 3 to 4 liters of sweat in an hour.

• PROBLEM 18-21

The terms defecation, excretion and secretion are sometimes confused. What are meant by these terms?

Solution: Defecation refers to the elimination of wastes and undigested food, collectively called feces, from the anus. Undigested food materials have never entered any of the body cells and have not taken part in cellular metabolism; hence they are not metabolic wastes. Excretion refers to the removal of metabolic wastes from the cells and bloodstream. The excretion of wastes by the kidneys involves an expenditure of energy by the cells of the kidney, but the act of defecation requires no such effort by the cells lining the large intestine. Secretion refers to the release from a cell of some substance which is utilized either locally or elsewhere in some body processes; for example, the salivary glands secrete saliva, which is used in the mouth in the first step of chemical digestion. Secretion also involves cellular activity and requires the expenditure of energy by the secreting cell.

• PROBLEM 18-22

Artificial kidneys have been devised for patients with kidney disease. How do these artificial kidneys work? What is the basic principle involved?

Solution: The artificial kidney is used to replace the diseased kidney in eliminating the excess ions and wastes which would accumulate in the blood as a result of failure of the latter. In an artificial kidney, the patient's blood is passed through a system of very fine tubes bounded by thin membranes. The other side of the membrane is bathed by a dialysis fluid into which waste products can pass from the blood. The fine tubes converge into a tubing which then conducts the blood back into the patient's body through a vein.

The fine tubes of the artificial kidney are made of cellophane. This material is used because it has much the same characteristics as the endothelium of blood capillaries; it is highly permeable to most small solutes but relatively impermeable to protein. The dialysis fluid which bathes the cellophane tubes is a salt solution with ionic concentrations similar to those of blood plasma.

The function of the artificial kidney can be explained by the principle of diffusion. Since the cellophane membrane is permeable to most small solutes, the concentrations of solutes in the blood, as it flows through the tubes, tend to equal those in the dialysis fluid. However, if there is an above normal level of a certain solute in the blood, this solute will diffuse out of the blood into the surrounding fluid, which has a lower concentration of the solute. In this way, waste products and other fluid substances in excess will leave the blood and pass into the dialysis fluid since their concentrations in the fluid are very low or non-existent. To prevent waste solutes from building up in the fluid and interfering with diffusion, the fluid is continually being replaced by fresh supply.

Great care must be taken to maintain the sterility of the artificial kidney. The total amount of blood in an artificial kidney at any one moment is roughly 400-500 ml. Usually heparin, an anticoagulant, is added to the blood as it enters the artificial kidney to prevent coagulation; an antiheparin is added to the blood before returning it to the patient's body in order to allow normal blood clotting to take place. An artificial kidney can clear urea from the blood at a rate of about 200 ml. of plasma per minute, - that is, 200 ml of plasma can be completely cleared of urea in one minute - this being almost three times as fast as the clearance rate of both normal kidneys working together. Thus we can see that the artificial kidney can be very effective, but it can be used for only 12 hours every 3 or 4 days, because of the danger to the blood's clotting mechanisms due to the addition of heparin.

SHORT ANSWER QUESTIONS FOR REVIEW

Answer

To be covered when testing yourself

Choose the correct answer.

1. The functional unit of the kidney is called a (a) tubule, (b) neuron, (c) urethra, (d) nephron. — d

2. If someone suffers from kidney failure, it might be traced to a breakdown in (a) secretion, (b) filtration, (c) reabsorption, (d) any of the above. — d

3. Nasal salt glands are found in (a) fresh water bony fish, (b) hag fish, (c) sea birds, (d) marine bony fish. — c

4. The removal of metabolic waste products that become toxic when they accumulate is known as (a) secretion, (b) excretion, (c) osmosis, (d) respiration. — b

5. Which of the following is released from specialized brain cells after a person eats a salty meal? (a) ADH (b) aldosterone (c) acetylcholine (d) ATP ase. — a

6. Which of the following is not a major nitrogenous waste product of protein and nucleic acid production? (a) ammonia, (b) sodium chloride, (c) uric acid, (d) urea. — b

7. In humans, urine is temporarily stored in the _____ before it is voided. (a) Intestine, (b) liver, (c) bladder, (d) nephron. — c

8. Choose the correct statement: (a) The ascending limb of the loop of Henle pumps sodium ions out. (b) Chloride ions pass into the ascending limb. (c) The ascending limb is permeable to water. (d) The descending limb is impermeable to water and sodium. — a

9. The thin barrier at Bowman's capsule allows for the filtration of (a) whole blood, (b) plasma, (c) oxygen, (d) electrolytes only. — b

10. The transfer of material from the collecting tubular lumen back to the blood is called (a) tubular secretion, (b) tubular excretion, (c) tubular reabsorption, (d) micturition. — c

11. Which of the following hormones influence the reabsorption of water in the kidney (a) ADH (b) aldosterone (c) insulin (d) a and b are correct. — d

Answer

To be covered when testing yourself

12. Which of the following are forces influencing the net glomerular filtration pressure? (a) Glomerular capillary blood pressure, (b) fluid pressure in Bowman's capsule, (c) colloidal osmotic pressure, (d) all of the above are correct.

d

13. All of the following are examples of actively transported solutes in the kidney except (a) urea, (b) glucose, (c) phosphate, (d) all of the above are correct.

a

14. Which of the following equations are physiologically correct

(a) $CO_2 + H_2O \rightarrow H_2CO_3 \rightleftarrows H^+ + HCO_3^-$

(b) $H^+ + HCO_3^- \rightleftarrows H_2CO_3 \rightleftarrows CO_2 + H_2O$

(c) $H^+ + HCO_3^- \rightarrow H_2O + CO_2 \rightleftarrows H_2CO_3$

(d) $H_2CO_3 + H^+ \rightleftarrows H_2O + H^+ \rightleftarrows HCO_3^-$

b

Fill in the blanks.

15. Sodium is actively reabsorbed under the influence of _____, a hormone secreted by the adrenal cortex.

aldo-sterone

16. _____ passively follows sodium across the cell membrane.

Chloride

17. Material that is _____ to plasma is considered to have the same solute concentration.

isotonic

18. _____ is a condition where there is a gradual accumulation of urea and other waste products in the blood.

Uremia

19. The capsular hydrostatic pressure present at Bowman's capsule _____ (favors, opposes) filtration.

opposes

20. Constriction of the afferent arteriole _____ (decreases, increases) the filtration rate.

decreases

21. If a substance is present in the plasma in a concentration of, .03 mg/ml and secreted at a rate of .1 mg/min its renal clearance would be _____ mls plasma/min.

3.3

22. The concentration of a substance in the plasma at the point where it just begins to appear in the urine is referred to as the _____ of the substance.

renal threshold

	Answer
	To be covered when testing yourself
23. The Kangaroo rat never drinks water but obtains water from _____ metabolism.	carbo-hydrate
24. Reabsorption of water in the collecting tubules is stimulated by the hormone _____.	vasopres-sin or antidiu-retic hormone (ADH)
25. Aspirin and antibiotics as well as urea are _____ threshold substances.	low
26. When blood pressure remains high, a condition of _____ occurs as the lymph vessels do not drain properly.	edema
27. Aldosterone is produced by the adrenal _____.	cortex
28. As blood volume decreases, aldosterone levels _____.	increase
29. The sensation of thirst would be lost if the _____ was destroyed.	hypo-thalamus
Determine whether the following statements are true or false.	
30. The net glomerular filtration pressure is about 10 mm Hg in a normal man.	True
31. Colloidal osmotic pressure and increased blood pressure both move fluid out of the capillaries and into the interstitial space.	False
32. Baroreceptors are found in the walls of the veins in the chest and neck.	False
33. The resting cell membrane potential is due to the fact that there is a net negative charge on the outside of a cell due to the distribution of sodium ions.	False
34. Aldosterone is a hormone produced by the adrenal cortex.	True
35. Aldosterone acts on the epithelium of the proximal convoluted tubule, increasing its permeability of sodium.	False
36. The release of ADH is regulated by specific osmoreceptors.	True

	Answer
	To be covered when testing yourself
37. If a substance is present in the plasma in a concentration of .04 mg/ml and is being secreted at a rate of .4 mg/min, its renal clearance would be at a rate of 10 mls of plasma per minute.	True
38. The longer the loop of Henle is in animals, the smaller is the amount of water that is retained by the body in all animals.	False
39. The final urine that emerges from the collecting tubule is hypotonic to blood.	False
40. ADH has no effect on sodium reabsorption.	True
41. The kidney is able to transport certain waste materials directly from the blood to the proximal convoluted tubule without passing through Bowman's capsule.	True
42. Aquatic animals usually excrete urea in a liquid form.	False
43. The control of urinary water loss is the <u>major</u> mechanism by which body water is regulated.	True
44. The arterioles leading to the glomerulus are called efferent arterioles.	False

CHAPTER 19

PROTECTION & LOCOMOTION

SKIN

• PROBLEM 19-1

The skin of all vertebrates consists of many layers of cells. It is the largest organ of the human body. Describe the major physical characteristics of skin. What structures are derived from human skin?

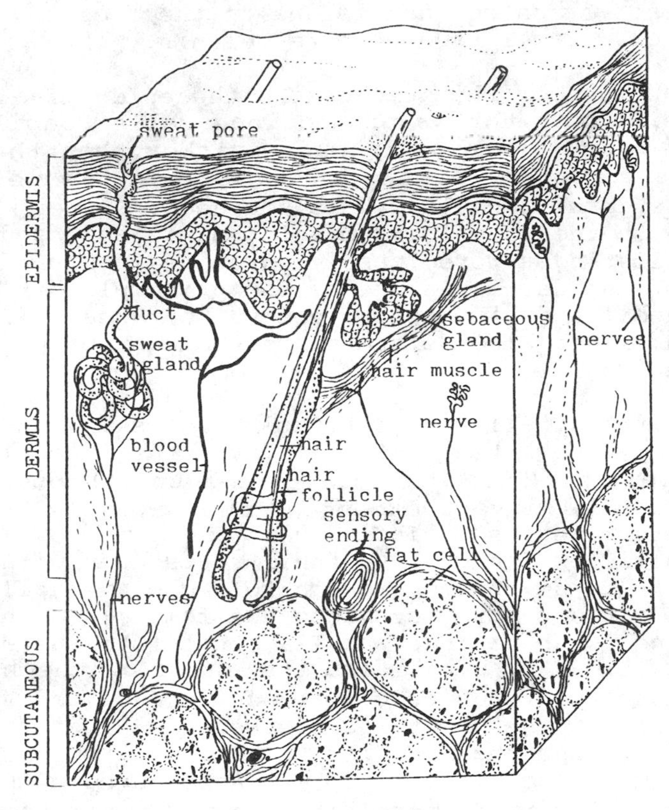

Figure Section of human skin.

Solution: Human skin is composed of a comparatively thin, outer layer, the epidermis, which is free of blood vessels, and an inner, thick layer, the dermis, which is packed with blood vessels and nerve endings. The epidermis is a stratified epithelium whose thickness varies in different parts of the body. It is thickest on the soles of the feet and the palms of the hands. The epidermis of the palms and fingers has numerous ridges, forming whorls and loops in very specific patterns. These unique fingerprints and palmprints are determined genetically, and result primarily from the orientation of the underlying fibers in the dermis. The outermost layers of the epidermis are composed of dead cells which are constantly being sloughed off and replaced by cells from beneath. As each cell is pushed outward by active cell division in the deeper layers of the epidermis, it is compressed into a flat (squamous), scalelike epithelial cell. Such cells synthesize large amounts of the fibrous protein, keratin, which serves to toughen the epidermis and make it more durable.

Scattered at the juncture between the deeper layers of the epidermis and the dermis are melanocytes, cells that produce the pigment melanin. Melanin serves as a protective device for the body by absorbing ultraviolet rays from the sun. Tanning results from an increase in melanin production as a result of exposure to ultraviolet radiation. All humans have about the same number of melanocytes in their skin. The difference between light and dark races is under genetic control and is due to the fact that melanocytes of dark races produce more melanin.

The juncture of the dermis with the epidermis is uneven. The dermis throws projections called papillae into the epidermis. The dermis is much thicker than the epidermis, and is composed largely of connective tissue. The lower level of the dermis, called the subcutaneous layer, is connected with the underlying muscle and is composed of many fat cells and a more loosely woven network of fibers. This part of the dermis is one of the principle sites of body fat deposits, which help preserve body heat. The subcutaneous layer also determines the amount of possible skin movement.

The hair and nails are derivatives of skin, and develop from inpocketings of cells from the inner layer of the epidermis. Hair follicles are found throughout the entire dermal layer, except on the palms, soles, and a few other regions. Individual hairs are formed in the hair follicles, which have their roots deep within the dermis. At the bottom of each follicle, a papilla of connective tissue projects into the follicle. The epithelial cells above this papilla constitute the hair root and, by cell division form the shaft of the hair, which ultimately extends beyond the surface of the skin. The hair cells of the shaft secrete keratin, then die and form a compact mass that becomes the hair. Growth occurs at the bottom of the follicle only. Associated with each hair follicle is one or more sebaceous glands, the secretions of which make the surface of the skin and hair more pliable. Like the sweat glands, the sebaceous glands are derived from the embryonic epidermis but are located in the dermis. To each hair follicle is attached smooth muscle called arrector pili, which pulls the hair erect upon contraction.

Nails grow in a manner similar to hair. Both hair follicles and nails develop from inpocketings of cells from the inner layer of the epidermis. The translucent, densely packed, dead cells of the nails allow the underlying capillaries to show through and give the nails their normal pink color.

Like the hair and nails found in man, the feathers, hoofs, claws, scales, and horns found in other vertebrates are also derivatives of the skin.

• PROBLEM 19-2

The skin is much more than merely an outer wrapping for an animal; it is an important organ system and performs many diverse functions. What are some of the primary functions of the skin in man?

Solution: Perhaps the most vital function ot the skin is to protect the body against a variety of external agents and to maintain a constant internal environment. The layers of the skin form a protective shield against blows, friction, and many injurious chemicals. These layers are essentially germproof, and as long as they are not broken, keep bacteria and other microorganisms from entering the body. The skin is water-repellent and therefore protects the body from excessive loss of moisture. In addition, the pigment in the outer layers protects the underlying layers from the ultraviolet rays of the sun.

In addition to its role in protection, the skin is involved in thermoregulation. Heat is constantly being produced by the metabolic processes of the body cells and distributed by the bloodstream. Heat may be lost from the body in expired breath, feces, and urine, but approximately 90 per cent of the total heat loss occurs through the skin. This is accomplished by changes in the blood supply to the capillaries in the skin. When the air temperature is high, the skin capillaries dilate, and the increased flow of blood results in increased heat loss. Due to the increased blood supply, the skin appears flushed. When the temperature is low, the arterioles of the skin are constricted, thereby decreasing the flow of blood through the skin and decreasing the rate of heat loss. Temperature-sensitive nerve endings in the skin reflexively control the arteriole diameters.

At high temperatures, the sweat glands are stimulated to secrete sweat. The evaporation of sweat from the surface of the skin lowers the body temperature by removing from the body the heat necessary to convert the liquid sweat into water vapor. In addition to their function in heat loss, the sweat glands also serve an excretory function. Five to ten per cent of all metabolic wastes are excreted by the sweat glands. Sweat contains similar substances as urine but is much more dilute.

MUSCLES: MORPHOLOGY AND PHYSIOLOGY

• **PROBLEM** 19-3

Skeletal muscle is the largest tissue in the body, accounting for 40 to 45 per cent of the total body weight. What are the structural characteristics of skeletal muscle?

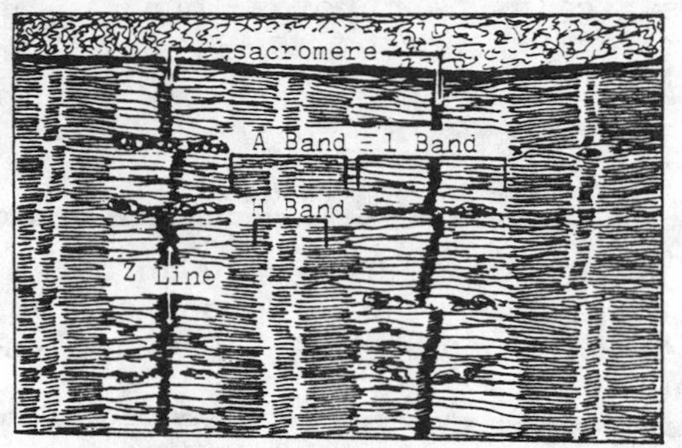

Figure 1. Photomicrograph of skeltal muscle myofibrils.

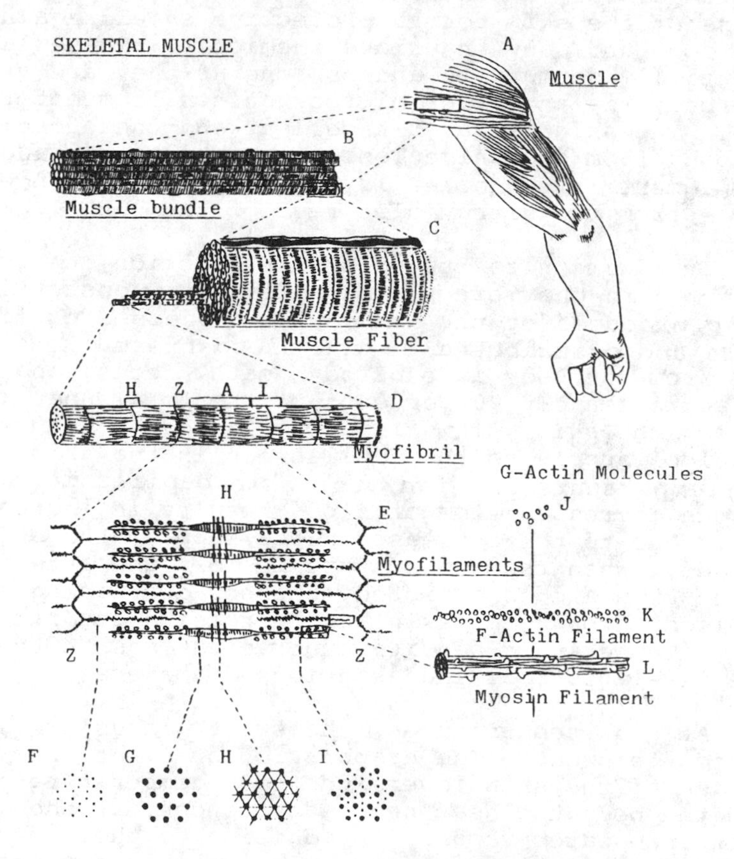

Figure 2. Diagram of the organization of skeletal muscle from the gross to the molecular level. F, G, H, and I are cross sections at the levels indicated.

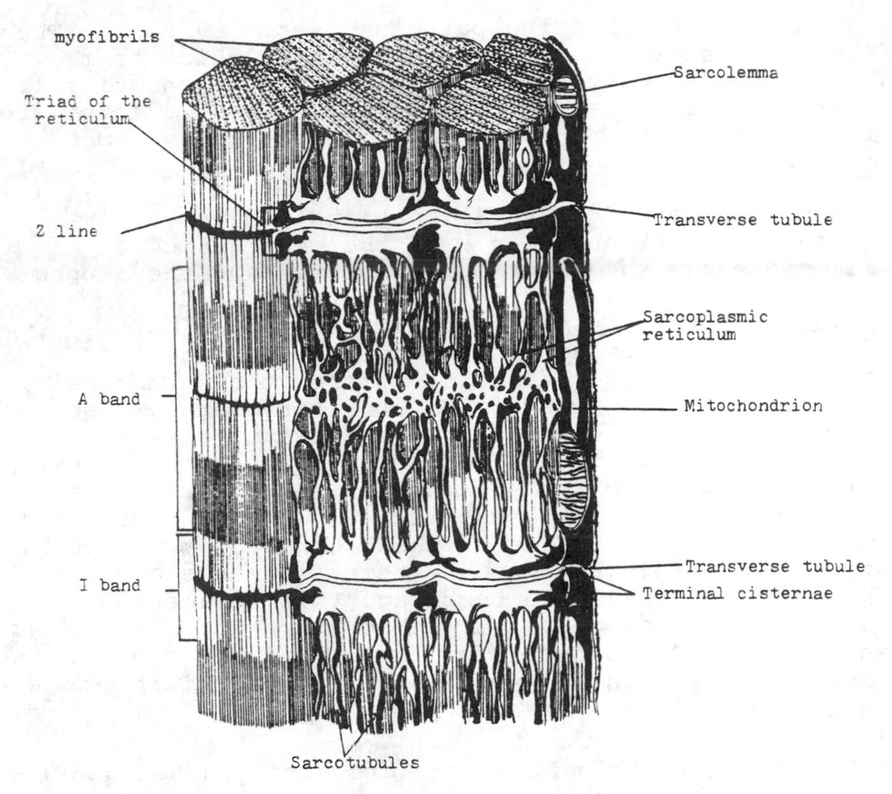

Figure 3 Schematic representation of the distribution of the sarcoplasmic reticulum around the myofibrils of amphibian skeletal muscle.

Solution: The term muscle, as it is commonly used, refers to a number of muscle fibers bound together by connective tissue. Skeletal muscle fibers are multinucleated cylindrical cells, 10 to 100μm in diameter, and may be up to 1 ft. long. Generally each end of an entire muscle is attached to a bone by bundles of collagen fibers known as tendons. Some tendons are very long, and the site of attachment of the tendon to the bone is far removed from the muscle. For example, some of the muscles which move the fingers are found in the lower portion of the arm, between the elbow and the wrist.

The most striking feature of muscle fibers is the series of transverse light and dark bands forming a regular pattern along the fiber. Both skeletal and cardiac muscle exhibit such banding patterns; smooth muscle does not. Though the pattern appears continuous across a fiber, the fiber is actually composed of a number of independent cylindrical elements in the cytoplasm of the fiber known as myofibrils. Bundles of these fibrils are enclosed by the muscle cell membrane or sarcolemma. Each myofibril is about a micron in diameter. Between the myofibrils are large numbers of mitochondria, which are to be expected in cells that have such a high energy requirement. The myofibrils show the same pattern of cross striations as the fibers of which they are a part.

When viewed with the electron microscope, the structures

responsible for the banding patterns become evident. The myofibrils consist of smaller myofilaments which form a regular repeating pattern along the length of the fibril. One unit of this repeating pattern is known as a sarcomere, and is the functional unit of the contractile system of the muscle.

Each sarcomere (Figure 1) contains two types of myofilaments: thick and thin. In the central region of the sarcomere are the thick myofilaments, and they appear as a dark band. This band is termed the A band. These **thick filaments contain the protein myosin. The** thin myofilaments contain the protein actin and are attached at either end of the sarcomere at a structure known as the Z line. The limits of a sarcomere are defined by two successive Z lines. The thin filaments extend from the Z lines toward the center of the sarcomere where they overlap with the thick filaments. Two more bands are distinguishable in the muscle. The I band represents the region between the ends of the A bands of two adjoining sarcomeres. In this region only thin filaments are present; there is no overlap of the thick and thin filaments. **Because it contains only thin filaments, the I band usually appears light. The H Zone corresponds to the space between the ends of the thin filaments; only thick filaments are present.**

The thick and thin filaments are arranged hexagonally with respect to one another. Each thick filament is surrounded by six thin filaments, and each thin filament is surrounded by three thick ones. **(See Figure 2)** An average muscle fiber contains about 15 billion thick and 65 billion thin filaments. Under extremely high magnification, the gap between thick and thin filaments in the region of the A band appears to be bridged by projections at intervals along the filaments. It appears that these projections, or cross bridges, are arranged in a spiral around the thick filament. These projections are thought to play an active role in the contraction of the muscle. (See following question.)

An important muscle cell organelle is the sarcoplasmic reticulum, a continuous system of tubules extending throughout the cytoplasm, forming a closely meshed canal network around each myofibril. This organelle corresponds to the endoplasmic reticulum of other cell types, but in muscle it is largely devoid of ribosomes and exhibits a highly specialized repeating pattern. The tubules of the reticulum overlying the A bands have a prevailing longitudinal orientation but anastomose freely in the region of the H band. At regular intervals along the length of the myofibrils the longitudinal tubules of the sarcoplasmic reticulum, called sarcotubules, come together with transversely oriented channels of larger caliber called terminal cisternae. **(See Figure 3)**

Pairs of parallel terminal cisternae run transversely across the myofibrils in close apposition to a slender intermediate element, the transverse tubule, commonly called the T tubule. These three transverse structures, namely the two parallel terminal cisternae and one T tubule, constitute the triads of skeletal muscle. In man, there are two triads to each sarcomere, situated at the junctions of each A band with the adjacent I bands.

The cavity of the T tubule does not open into the adjacent cisternae and, strictly speaking, is not part of the sarcoplasmic reticulum. It is merely a slender invagination of the sarcolemma. The lumen of the T tubule is thus continuous with the extracellular medium surrounding the muscle fiber. The T tubules are involved in transmitting electrical signals from neurons on the cell surface to deep within the muscle fiber.

• PROBLEM 19-4

The most widely accepted theory of muscle contraction is the sliding filament theory. What is the major point of this theory?

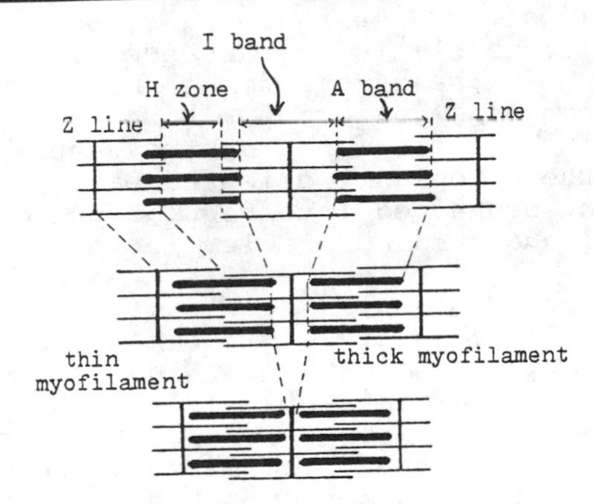

Figure. Changes in banding pattern resulting from the movements of thick and thin filaments past each other during contraction.

Solution: The major premise of the sliding filament theory is that muscle contraction occurs as the result of the sliding of the thick and thin filaments past one another; the lengths of the individual filaments remain unchanged. Thus the width of the A band remains constant, corresponding to the constant length of the thick filaments. The I band narrows as the thin filaments approach the center of the sarcomere. As the thin filaments move past the thick filaments, the width of the H zone between the ends of the thin filaments becomes smaller, and may disappear altogether when the thin filaments meet at the center of the sarcomere. With further shortening, new banding patterns appear as thin filaments from opposite ends of the sarcomere begin to overlap. The shortening of the sarcomeres in a myofibril is the direct cause of the shortening of the whole muscle.

The question arises as to which structures actually produce the sliding of the filaments. The answer is the myosin cross bridges. These cross bridges are actually part of the myosin molecules which compose the thick filaments. The bridges swivel in an arc around their fixed positions on the surface of the thick filaments, much like the oars of a boat. When bound to the actin filaments, the movement of the cross bridges causes the sliding of the thick and thin filaments past each other. Since one movement of a cross bridge will produce only a small displacement of the filaments relative to each other, the cross bridges must undergo many repeated cycles of movement during contraction.

• PROBLEM 19-5

What are the distinguishing characteristics of cardiac and smooth muscle.

Solution: Smooth muscle fibers are considerably smaller than skeletal muscle fibers. Each smooth muscle fiber has a single nucleus located in the central portion of the cell. By contrast, each skeletal muscle fiber is multinucleated with the nuclei located peripherally. The most noticeable morphological factor distingushing smooth from either skeletal or cardiac muscle is the absence of striated banding patterns in the cytoplasm of smooth muscle. Smooth muscle does not contain myofibrils. However, myosin thick filaments and actin thin filaments can be seen distributed throughout the cytoplasm, oriented parallel to the muscle fiber, but not organized into regular units of filaments as in skeletal and cardiac muscle.

It is believed that the molecular events of force generation in smooth muscle cells are similar to those in skeletal muscle. Smooth muscle exhibits wide variations in tonus; it may remain almost entirely relaxed or tightly contracted. Also, it apparently can maintain the contracted condition of tonus without the expenditure of energy, perhaps owing to a reorganization of the protein chains making up the fibers. Smooth muscle cells are not under voluntary control. Their activity is under the regulation of the autonomic nervous system and they have an ability to perform work for long periods of time, since their contraction is slow and sustained. Smooth muscle is often referred to as a visceral muscle since it regulates the internal environment of a great many systems and organs. It is found in the walls of hollow organs such as the intestinal tract, the bronchioles, the urinary bladder, and the uterus. Vascular smooth muscle lines the walls of blood vessels. It may also be found as single cells distributed throughout an organ such as the spleen or in small groups of cells attached to the hairs in the skin.

Cardiac muscle has properties similar to those of skeletal muscle. It is multinucleated and striated, with its thick and thin myofilaments organized into myofibrils. The sliding-filament type of contraction is also found in cardiac muscle. Unlike skeletal muscle but like smooth muscle, cardiac muscle is involuntary. Each beat of the heart represents a single twitch. Cardiac muscle has a long refractory period. Consequently, it is unable to contract tetanically, since one twitch cannot follow another quickly enough to maintain a contracted state. A unique feature of cardiac muscle is its inherent rhythmicity; it contracts at a rate of about 72 beats per minute. The muscle is innervated by nerves, but these nerves only serve to speed up or slow down the inherent cardiac rhythm. In addition, cardiac muscle is unique in having intercalated discs, or tight junctions, between cells; these aid in the transmission of electrical impulses throughout the heart.

• PROBLEM 19-6

What are the properties of actin and myosin which produce the cyclic activity of the cross bridges responsible for contraction? What causes rigor mortis?

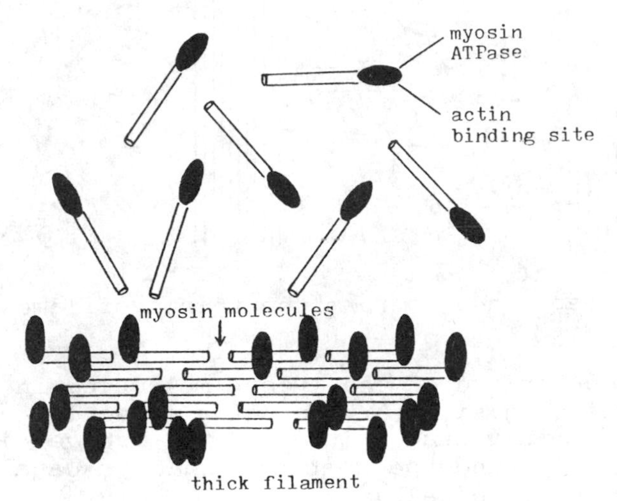

Figure 1. Aggregation of myosin molecules to form thick filaments, with the globular heads of the myosin molecules forming the cross bridges.

Figure 2.
Structure of thin myofilament composed of two helical chains of globular actin monomers.

Solution: Myosin, the larger of the two molecules, is shaped like a lollypop (see Figure 1). The myosin molecules are arranged within the thick filaments so that they are oriented tail-to-tail in the two halves of the filament; the globular ends extend to the sides, forming the cross briges which bind to the reactive site on the actin molecule. Actin is a globular-shaped molecule having a reactive site on its surface that is able to combine with myosin. These globular proteins are arranged in two chains which are

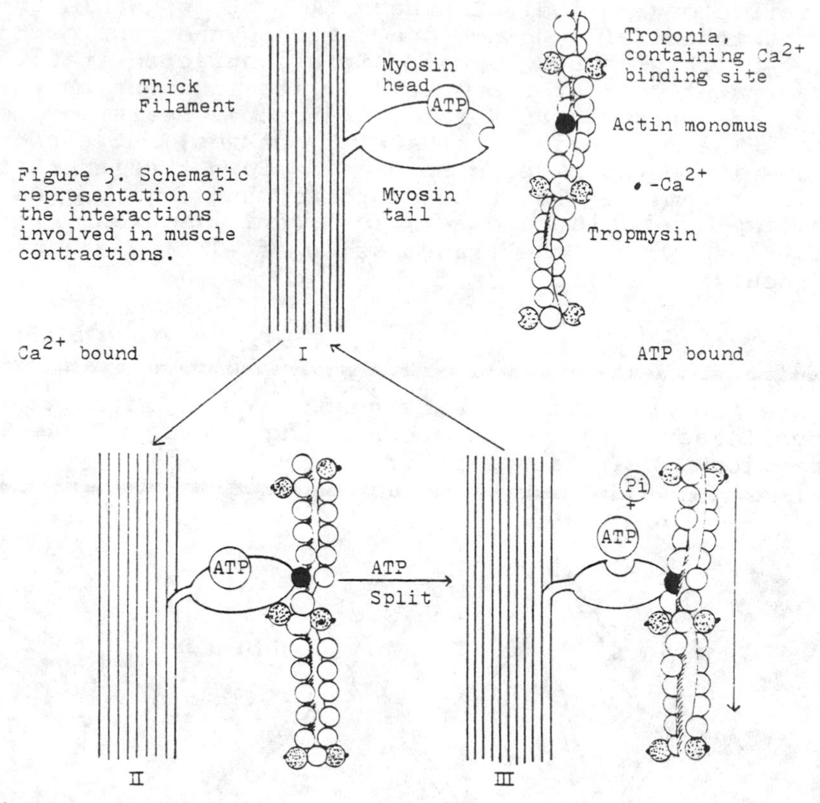

Figure 3. Schematic representation of the interactions involved in muscle contractions.

helically intertwined to form the thin myofilaments. (See Figure 2)

The globular end of the myosin molecule, in addition to being a binding site for the actin molecule, contains **a separate binding site for ATP. This active site has ATPase activity, and the reaction that is catalyzed is the hydrolysis of ATP:**

$$H_2O + ATP \longrightarrow ADP + Pi$$

However, myosin alone has a very low ATPase activity. It appears that an allosteric change occurs in the active site of myosin ATPase when the myosin cross bridge combines with actin in the thin filaments, considerably increasing the ATPase activity. The energy that is released from the splitting of ATP produces cross bridge movement by an as yet unknown mechanism. It is believed that the oscillatory movements of myosin cross bridges produce the relative movement of thick and thin filaments, resulting ultimately in the shortening of a muscle fiber (see Figure 3).

Since many cycles of activity are needed to produce the degree of shortening observed during muscle contraction, the myosin bridge must be able to detach from the actin and then rebind again. This is accomplished by the binding of ATP to the myosin in the cross bridge, forming what is known as a low-energy complex. The low-energy complex has only a weak affinity for actin; the actin-myosin bond is broken, allowing the cross bridges to dissociate from actin. Shortly after this event, a conformational change occurs in the myosin - ATP complex and a high energy complex is formed. The high energy complex has a very high affinity for actin, and the cross bridges are able to rebind to the actin. In this manner, the cross bridges are able to bind and dissociate from actin in a cycle of coordinated actions. This cycle may be summarized in the following sequence of events:

A = actin   M = myosin

M-ATP ——————> M^*-ATP
(low-energy complex)   (high-energy complex capable of binding actin)

A + M^*-ATP ——————> A-M^*-ATP
(with actin bound, myosin is able to split ATP)

A-M^*-ATP ——————> A-M+ADP+Pi
(as ATP is split, cross bridge movement occurs)

A-M+ATP ——————> A + M-ATP
(low-energy complex dissociates from actin)

Rigor mortis is a phenomenon in which the muscles of the body become very stiff and rigid after death. It results directly from the loss of ATP in the dead muscle cells; the myosin crossbridges are unable to combine with actin and those bonds already formed cannot be broken - thus the rigid condition.

At the molecular level, we can identify two specific roles for ATP: 1) to provide energy for movement of cross bridge, and 2) to dissociate actin from the myosin cross bridges during the contraction cycle of the bridges. ATP is also needed to restore Ca^+ in the sarcoplasmic reticulum following contraction (see next question).

Two regulatory proteins, troponin and tropomyosin, are associated with actin. During nervous stimulation of a muscle, there is an increase in free intracellular calcium ions: calcium diffuses in from the terminal cisternae and from the extracellular fluid in the T tubules. Calcium binds to troponin which causes tropomyosin to shift its position along the actin helix. This exposes the binding site on actin for myosin.

• **PROBLEM** 19-7

Trace the sequence of events between nerve action potential and contraction and relaxation of a muscle fiber.

Solution: The cell membranes of muscle fibers are excitable membranes capable of generating and propagating action potentials by mechanisms very similar to those found in nerve cells. An action potential in the muscle

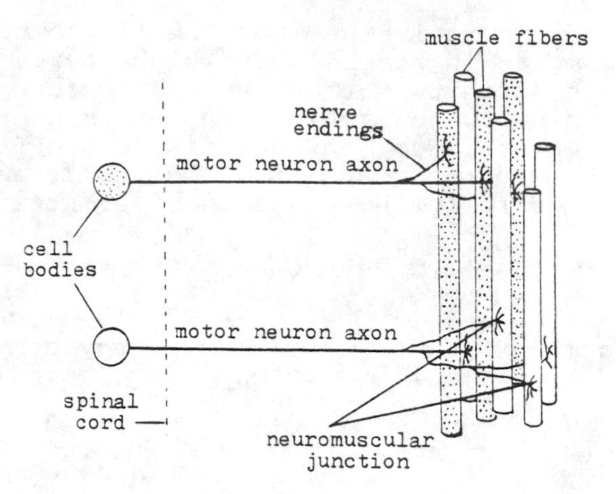

Figure 1. Two motor units, each of which consists of a motor neuron and the muscle fibers it immervates.

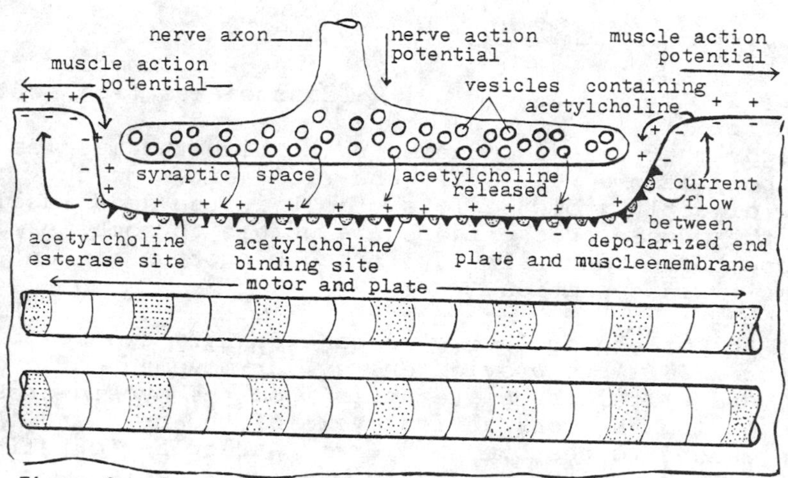

Figure 2. Events occuring at a neuromuscular junction which lead to an action potential in the muscle membrane.

cell membrane results from nervous stimulation and provides the signal for the initiation of contractile activity within the muscle cell by a mechanism known as excitation-contraction coupling. An action potential is initiated and propagated in a motor axon which innervates the muscle fiber. This potential is the result of synaptic events on the cell body and dendrites of the motor neuron in the central nervous system. (see Figure 1) The action potential in the motor axon causes the release of the neurotransmitter known as acetylcholine from the axon terminals into the synaptic space betweeen the nerve and muscle. (see Figure 2) This junction between nerve and skeletal muscles is known as the neuromuscular junction. The nerve cells which form junctions with skeletal muscles are known as motor neurons, and the cell bodies of these neurons are located in the brain and spinal cord.

Once released, acetylcholine binds to receptor sites on the muscle membrane which lie directly under the terminal portion of the axon. This region of the muscle membrane is known as the motor end plate. Acetylcholine increases the permeability of the motor end plate to sodium and potassium ions, causing a depolarization of the end plate called an end-plate potential (EPP). The EPP depolarizes the muscle membrane to its threshold potential, generating a muscle action potential which is propagated over the surface of the muscle membrane. The molecules of acetylcholine released from the motor-neuron have a lifetime of only about 5 milliseconds before they are destroyed by the acetylcholinesterase, an enzyme on the muscle cell membrane near the receptors for acetylcholine. Once acetylcholine is destroyed, the muscle membrane permeability to sodium and potassium ions returns to its initial state, and the depolarized end plate returns to its resting potential.

The muscle action potential depolarizes the T tubules at the A-I junction of the sarcomeres. This depolarization leads influx of calcium from the extracellular fluid and to the release of calcium ions from the terminal cisternae of the sarcoplasmic reticulum surrounding the myofibrils. Calcium ions bind to troponin in the thin actin myofilaments. This calcium-troponin complex causes tropomyosin to shift its position, releasing the inhibition that prevented actin from combining with myosin. Actin then combines with myosin. This binding activates the myosin ATPase, which splits ATP, releasing energy which is used to produce the movement of the cross bridge of the myosin molecule. ATP then binds to the myosin bridge, breaking the bond between the actin and myosin, thus allowing the cross bridge to dissociate from actin. As long as the concentration of calcium ions is high enough to counteract the inhibitory action of the troponin-tropomyosin system, cycles of cross-bridge contraction and relaxation will continue. The concentration of calcium ions falls as they are moved back into the sarcoplasmic reticulum by an energy-requiring process which splits ATP. Removal of calcium ions restores the inhibitory action of troponin-tropomyosin, and in the presence of ATP, actin and myosin remain in the dissociated, relaxed state.

• PROBLEM 19-8

What is meant by the term tonus, or tone?

Solution: The term tonus refers to the state of sustained partial contraction present in skeletal muscles as long as the nerves to the muscle are intact. Unlike skeletal muscle, cardiac and smooth muscle exhibit tonus even after their nerves are cut. Tonus is a mild state of tetanus. It is present at all times and involves only a small fraction of the fibers of a muscle at any one time. It is believed that the individual fibers contract in turn, working in relays, so that each fiber has a chance to recover completely while other fibers are contracting before it is called upon to contract again. A muscle under slight tension can react more rapidly and contract more strongly than one that is completely relaxed, because of changes in the elastic component in the latter.

What is meant by an antagonistic muscle? Give examples.

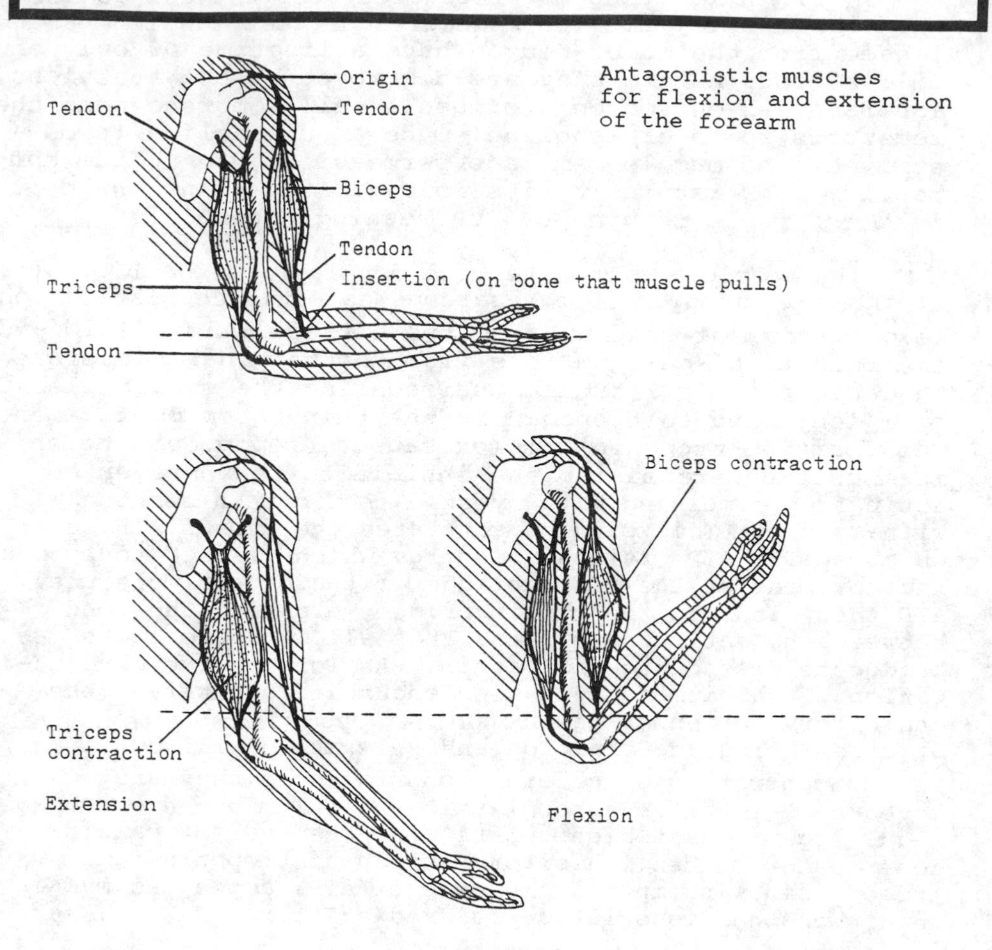

Solution: Muscles can exert a pull but not a push. For this reason, muscles are typically arranged in antagonistic pairs: one pulls a bone in one direction and the other pulls it in the opposite direction. The biceps, for example, bends or flexes the arm and is termed a flexor. Its antagonist, the triceps, straightens or extends the arm and is termed an extensor (see accompanying figure.) Such pairs of opposing extensors and flexors are found at the wrist, ankle and knee, as well as at other joints. When either the flexor or the extensor contracts, its antagonistic muscle must relax to permit the bone to move. The proper coordination of nerve impulses is necessary for antagonistic pairs to function properly.

Other antagonistic pairs of muscles are adductors and abductors, which move parts of the body toward or away from the central axis of the body, respectively; levators and depressors raise or lower parts of the body; and while pronators rotate parts of the body downward and backward, supinators rotate them upward and forward.

• **PROBLEM 19-10**

Differentiate between an isometric and an isotonic contraction.

Solution: Contraction refers to the active process of generating a force in a muscle. The force exerted by a contracting muscle on an object is known as the muscle tension, and the force exerted on a muscle by the weight of an object is known as the load. When a muscle shortens and lifts a load, the muscle contraction is said to be isotonic, since the load remains constant throughout the period of shortening.

When a load is greater than the muscle tension, shortening is prevented, and muscle length remains constant. Likewise, when a load is supported in a fixed position by the tension of the muscle, the muscle length remains constant. This development of muscle tension at a constant muscle length is said to be an isometric contraction. The internal physiochemical events are the same in both isotonic and isometric contraction. Movement of the limbs involves isotonic contractions, whereas maintaining one's posture requires isometric contractions.

• **PROBLEM 19-11**

The following is a graph of the mechanical response, the twitch, of a muscle cell to a single action potential. Describe the main features of the isotonic twitch shown.

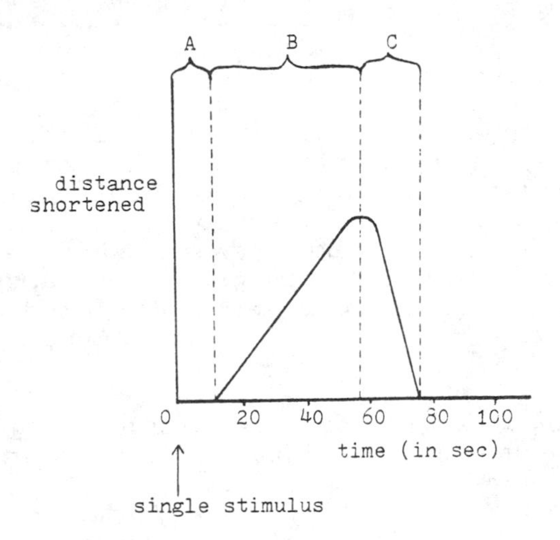

Solution: When a muscle is given a single stimulus, such as an electric shock, it responds with a single, quick twitch. The single twitch shown above can be separated into three phases. The first phase (A) is known as the latent period, and represents the interval between the application of the stimulus and the beginning of the visible shortening of the muscle. The next phase (B) is the contraction period which represents the time during which the muscle shortens. The last of the three phases (C) is the relaxation period during which the muscle returns to its original length.

Muscle fibers, like nerve fibers, have a refractory period, that is, a very short period of time immediately following one stimulus, during which they will not respond to a second stimulus. The refractory period in skeletal muscle is so short (about .002 second) that muscle can respond to a second stimulus while still contracting in response to the first. The result of this is the summation of contractions, which leads to a greater than normal shortening of the muscle fiber.

• PROBLEM 19-12

What is meant by the term muscle fatigue?

Solution: A muscle that has contracted several times, exhausting its stored supply of organic phosphates and glycogen, will accumulate lactic acid. (Review chapter 3, cellular metabolism.) It is unable to contract any longer and is said to be "fatigued". Fatigue is primarily induced by this accumulation of lactic acid, which correlates closely with the depletion of the muscle stores of glycogen. Fatigue, however, may actually be felt by the individual before the muscle reaches the exhausted condition.

The spot most susceptible to fatigue can be demonstrated experimentally. A muscle and its attached nerve can be dissected out and the nerve stimulated repeatedly by electric shock until the muscle no longer contracts. If the muscle is then stimulated directly by placing electrodes on the muscle tissue, it will contract. With the proper device for detecting the passage of nerve impulses, it can be shown that upon fatigue, the nerve leading to the muscle is not fatigued, but remains capable of conduction. Thus,since the nerve is still conducting impulses and the muscle is still capable of contracting, the point of fatigue must be at the junction between the nerve and the muscle, where nerve impulses initiate muscle contraction. Fatigue is then due in part to an accumulation of lactic acid, in part to depletion of stored energy reserves, and in part to breakdown in neuromuscular junction transmission.

In contrast to true muscle fatigue, psychological fatigue may cause an individual to stop exercising even though his muscles are not depleted of ATP and are still able to contract. An athlete's performance depends not only upon the physical state of his muscles but also upon his will to perform.

• PROBLEM 19-13

If a muscle had to rely on its stored supply of ATP for contraction, it would be completely fatigued within a few twitches. Therefore, if a muscle is to maintain its contractile activity, molecules of ATP must be made available as rapidly as they are broken down. What are the sources of ATP in a skeletal muscle? What is oxygen debt?

Solution: There are three sources of ATP in muscle cells: 1) creatine phosphate, 2) substrate level phosphorylation during glycolysis and the TCA cycle, and 3) oxidative phosphorylation in the mitochondria.

In an exercised muscle, the increased levels of ADP and inorganic phosphate resulting from the breakdown of ATP ultimately act as positive modulators for oxidative phosphorylation. However, a short time elapses before these multienzymatic pathways begin to deliver newly formed ATP at a high rate. It is the role of creatine phosphate to provide the energy for ATP formation during this interval. Creatine phosphate can transfer its phosphate group to a molecule of ADP, converting it to ATP and leaving a creatine molecule behind. A single enzyme catalyzes this reversible reaction.

During moderate exercise, the muscle cell is able to derive ATP from sources other than creatine phosphate. The source of ATP during moderate exercise comes from the complete oxidation of carbohydrate (e.g. glycogen) to carbon dioxide and water via glycolysis, the Krebs (TCA) cycle, and the electron transport chain. During the unavoidable delay before adjustments of the respiratory and circulatory systems increase the oxygen supply to the active muscles, some of the oxygen for aerobic respiration may come from oxymyoglobin. Myoglobin is a compound in muscle which is chemically similar to hemoglobin. It forms a loose combination with oxygen while the oxygen supply is plentiful and stores the oxygen until the demand for it increases. During heavy muscular activity, a number of factors begin to limit the cell's ability to replace ATP by oxidative phosphorylation: the delivery of oxygen to the muscle, the availability of substrates such as glucose, and the rates at which the enzymes in the metabolic pathways can process the substrates. Any of these may become rate-limiting under various conditions. Since oxidative phosphorylation depends upon oxygen as the final electron acceptor, the continued formation of ATP depends upon an adequate delivery of oxygen to the muscle. Should the oxygen delivery be insufficient, lactic acid (homolactic) fermentation occurs. The muscles obtain the extra energy they need from this anaerobic process and lactic acid, the end product, accumulates. The muscle thus incurs what physiologists call oxygen debt. When the violent activity is over, the muscle cells consume large quantities of oxygen as they convert the lactic acid into pyruvic acid. Pyruvic acid is oxidized via the TCA and the electron transport processes. The cells utilize the energy thus obtained to resynthesize glycogen from the lactic acid that remains. In this manner, the oxygen debt is paid off, and the lactic acid is removed. This is why one continues to breathe hard for some time after one has stopped high levels of activity.

• **PROBLEM 19-14**

The property of skeletal muscle contraction in which the mechanical response to one or more successive stimuli is added to the first is known as summation. What is the underlying explanation for this phenomenon?

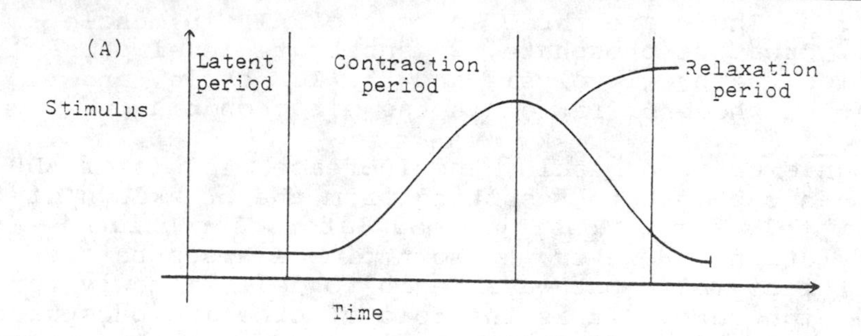

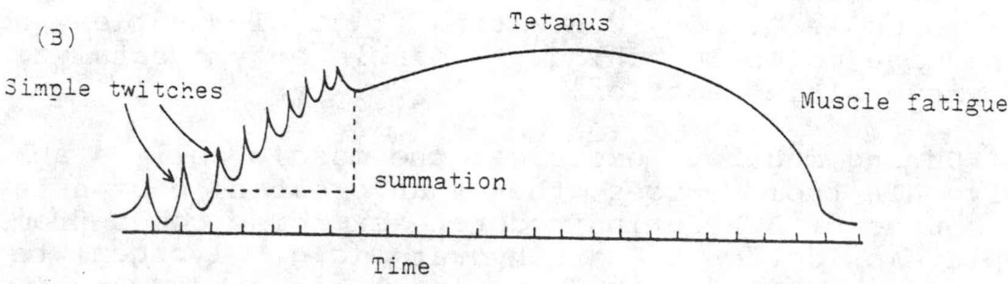

Diagrams showing kymograph (apparatus for studying muscle contraction) records of (A) a simple twitch, and (B) summation and tetanus. In (B) the time units are drawn closer in space so that a simple twitch shown as a curve in (A) appears as a sharp spike.

Solution: A possible explanation of this phenomenon, based on the role of calcium in excitation-contraction coupling, is that the amount of calcium released from the sarcoplasmic reticulum during a single action potential is sufficient to inhibit only some of the troponin-tropomyosin in the muscle. Multiple stimulation would then release more calcium so that more troponin-tropomyosin would be inhibited, allowing for further contraction. However, the truth is that more than enough calcium is released by the first action potential to inhibit all the troponin-tropomyosin, so this proposal must be discarded.

The explanation of summation involves the passive elastic properties of the muscle. Tension is transmitted from the cross bridges through the thick and thin filaments, across the Z lines, and eventually through the extracellular connective tissue and the tendons to the bone. All these structures have a certain amount of elasticity, analogous to a spring that is placed between the contractile components of the muscle and the external object. In the muscle, the contractile elements in their fully active state begin to stretch the passive elastic structures immediately following calcium release. Only when the elastic structures are all taut, can increasing contraction by the muscle occur. Summation occurs because a second stimulus is given, very close in time to the first, while the elastic structures are still a bit taut and not yet slack. Under this condition, the active state of the contractile proteins is maintained and the result is contractions that are stronger than any single simple twitch. Should sustained stimulation occur, the elastic elements would never have time to relax at all, and it is at this point that maximal force by the muscle fibers is attained; the individual contractions are indistinquishably fused into a single sustained contraction known as tetanus. (see accompanying figure). If stimulation of the muscle continues at this frequency, the ultimate result will be fatigue and possibly complete cessation of activity due to exhaustion of nutrients.

It is not surprising to note that cardiac muscle has an extremely long refractory period, allowing the elastic components to relax and thus avoiding tetanus, which would result in death due to loss of pumping action of the heart.

• PROBLEM 19-15

The muscle fibers innervated by a motor unit will contract if a stimulus of sufficient magnitude is propagated or it will not contract at all. This is known as the all-or-none law. If this is true, how can muscle force be graded? For example, we use the same muscles to lift a one ounce object that we use to lift a 20 lb. object.

Solution: The total tension that a muscle can develop depends upon two factors: 1) the number of muscle fibers in the muscle bundle that are contracting at any given time and 2) the amount of tension developed by each contracting fiber. The number of fibers in a muscle that are contractiong at any given time depends upon the number of motor neurons to the muscle that are being stimulated. Recall that each motor neuron innervates several muscle fibers, forming a motor unit. The number of motor units that are activated is determined by the activity of the synaptic inputs to the motor neurons in the brain and spinal cord. With proper stimulation by the brain and spinal cord, more motor units may be activated at any one time, thus increasing the number of contracting fibers, and therefore, the strength of the muscle bundle. The process of increasing the number of active motor neurons and thus the number of active motor units is known as recruitment.

The number of muscle fibers associated with a single motor neuron varies considerably in different types of muscle. In muscles of the hand, for example, which are able to produce very delicate movements, the size of the individual motor units is small. In the muscles of the back and legs, each motor unit contains hundreds of muscle fibers. The smaller the size of the motor units, the more precisely the tension of the muscle can be controlled by the recruitment of additional motor units.

In addition to the variability of the number of

active motor units, the tension produced by individual fibers can be varied. This can be accomplished by increasing the frequency of action potentials, resulting in stronger contractions, such as in summation or tetanus.

• PROBLEM 19-16

Describe the mechanism involved in the maintenance of an erect posture.

Solution: As long as an individual is conscious, all his muscles are contracted slightly; this phenomenon is known as tonus. It is by the partial contraction of the muscles of the back and neck and of the flexors and extensors of the legs that posture is maintained. When a person stands, both the flexors and extensors of the thigh must contract simultaneously so that the body sways neither forward nor backward. The simultaneous contractions of the flexors and extensors of the shank lock the knee in place and hold the leg rigid to support the body. When movement is added to posture, as in walking, a complex coordination of the contraction and relaxation of the leg muscles is required.

Receptors in the joints and associated ligaments and tendons, along with their pathways through the nervous system, play an important role in the unconscious control of posture and movement and also give rise to the conscious awareness of the position and movement of joints. Joint receptors are accurate indicators of movement and position. Skin receptors sensing contact of the body with other surfaces also play a role in the regulation of body posture.

• PROBLEM 19-17

What is the physiological purpose of shivering?

Solution: Shivering is a means by which the body maintains its normal temperature when the ambient temperature is cold. Recall that heat is produced by virtually all chemical reactions within the body. One way by which the basal level of heat production can be increased is to increase the rate of skeletal muscle contractions. The first muscle changes in response to cold are a gradual and general increase in skeletal muscle tone. This soon leads to shivering, which consists of oscillatory, rhythmic muscle tremors occurring at the rate of about 10 to 20 per second. This intensive muscle activity rapidly uses up ATP, thus stimulating more cellular respiration and more heat production. These contractions are so effective that body heat production may increase several fold within seconds. Because no external work is performed, all the energy liberated by the metabolic machinery becomes internal body heat. Thus shivering is an adaptation to cold.

Shivering tends to ruffle the body hair in most animals, creating dead air space that serves as insulation. In man, shivering causes the erection of body hair, but his sparse coat of hair is insufficient to act as an insulatory mechanism.

BONE

• PROBLEM 19-18

Bone, like other connective tissues, consists of cells and fibers, but unlike the others its extracellular components are calcified, making it a hard, unyielding substance ideally suited for its supportive and protective function in the skeleton. Describe the macroscopic and microscopic structure of bone.

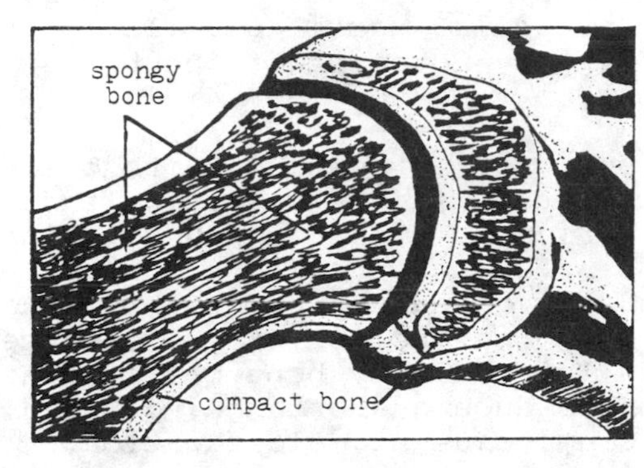

Figure 1. Longitudinal section of the end of a long bone.

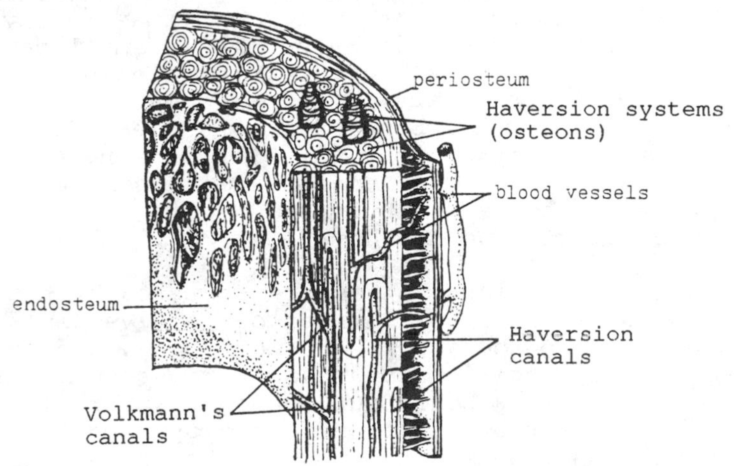

Figure 2. Cross-section of a long bone showing internal structures.

Solution: **Upon inspection of a long bone with the naked eye, two forms of bone are distinguishable:** cancellous (spongy) and compact. Spongy bone consists of a network of hardened bars having spaces between them filled with marrow. Compact bone appears as a solid, continuous mass, in which spaces can be seen only with the aid of a microscope. The two forms of bone grade into onc another without a sharp boundary. (See Fig. 1.)

In typical long bones, such as the femur or humerus, the shaft (diaphysis) consists of compact bone surrounding a large central marrow cavity composed of spongy bone. In adults, the marrow in the long bones is primarily of the yellow, fatty variety, while the marrow in the flat bones of the ribs and at the ends of long bones is primarily of the red variety and is active in the production of red blood cells. Even this red marrow contains about 70 percent fat.

The ends (epiphyses) of long bones consist mainly of spongy bone covered by a thin layer of compact bone. This region of the long bones contains a cartilaginous region known as an epiphyseal plate. The epiphyseal cartilage and the adjacent spongy bone constitute a growth zone, in which all growth in length of the bone occurs. The surfaces at the ends of long bones, where one bone articulates with another are covered by a layer of cartilage, called the articular cartilage. It is this cartilage which allows for easy movement of the bones over each other at a joint.

Compact bone is composed of structural units called Haversian systems. Each system is irregularly cylindrical and is composed of concentrically arranged layers of hard, inorganic matrix surrounding a microscopic central Haversian canal. Blood vessels and nerves pass through this canal, supplying and controlling the metabolism of the bone cells. The bone matrix itself is laid down by bone cells called osteoblasts. Osteoblasts produce a substance, osteoid, which is hardened by calcium, causing calcification. Some osteoblasts are trapped in the hardening osteoid and are converted into osteocytes which continue to live within the bone. These osteocytes lie in small cavities called lacunae, located along the interfaces between adjoining concentric layers of the hard matrix. Exchange of materials between the bone cells and the blood vessels in the Haversian canals is by way of radiating canals. Other canals, known as Volkmann's canals, penetrate and cross the layers of hard matrix, connecting the different Haversian canals to one another. (See Fig. 2.)

With few exceptions, bones are invested by the periosteum, a layer of specialized connective tissue. The periosteum has the ability to form bone, and contributes to the healing of fractures. Periosteum is lacking on those ends of long bones surrounded by articular cartilage. The marrow cavity of the diaphysis and the cavities of spongy bone are lined by the endosteum, a thin cellular layer which also has the ability to form bone (osteogenic potencies).

Haversian type systems are present in most compact bone. However, certain compact flat bones of the skull (the frontal, parietal, occipital, and temporal bones, and part of the mandible) do not have Haversian systems. These bones, termed membrane bones, have a different architecture and are formed differently than bones with Haversian systems.

• **PROBLEM 19-19**

Bone always develops by replacement of a preexisting connective tissue. When bone formation takes place in preexisting cartilage it is called endochondral ossification. Describe this method of bone formation.

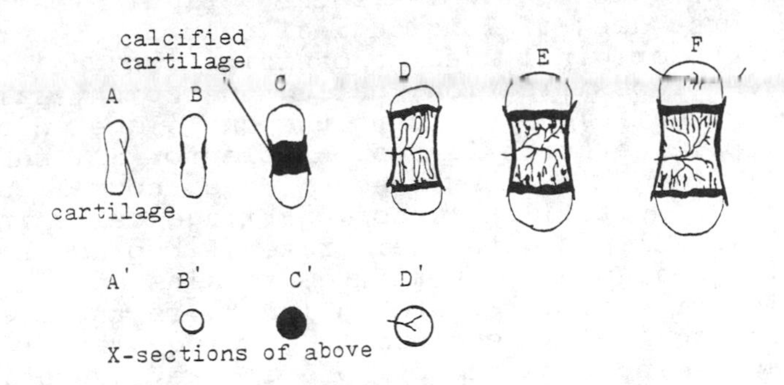

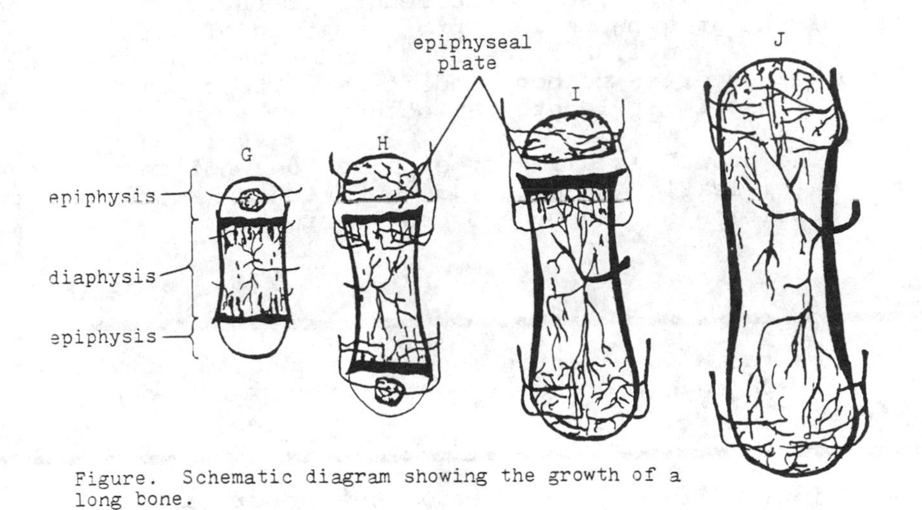

Figure. Schematic diagram showing the growth of a long bone.

Solution: Bones at the base of the skull, in the vertebral column, the pelvis, and the limbs are called cartilage bones because they originate from cartilage. This cartilage, present in the infant is replaced with bone in later years by means of a process called endochondral ossification. This can best be studied in one of the long bones of an extremity. We first start with a cartilaginous shaft. This shaft begins to ossify, or harden into bone, around its midportion, due to the deposition of calcium by the cartilage cells (chondrocytes). At the same time, blood vessels from the surrounding layer of connective tissue grow into the diaphysis. The calcified cartilage cells then die and are replaced by cells called osteoblasts, which form the bone matrix. The thinwalled blood vessels branch and grow toward either end of the cartilage model, forming capillary loops that extend into the blind ends of the cavities in the calcified cartilage. Cells are brought into the interior of the cartilage by these vessels. These cells later form bone marrow or bone matrix.

In the continuing growth in length, the cartilage cells in the epiphyses become arranged in longitudinal columns. The epiphyses are, like the diaphyses before them, invaded by blood vessels and begin undergoing ossification. The expansion of these centers of ossification gradually replaces all of the epiphyseal cartilage except that which persists as the articular cartilage, and a transverse disk of longitudinal columns of cartilage between the original area of ossification and the epiphyseal area of ossification, called the epiphyseal plate. The epiphyseal plate contains the cartilage columns whose zone of proliferation is responsible for all subsequent growth in length in long bones. Under normal conditions, the rate of multiplication of cartilage cells in this zone is in balance with their rate of replacement by bone. The epiphyseal plate, therefore, retains approximately the same thickness. Growth in length is the result of the cartilage cells continually growing away from the shaft and being replaced by bone as they recede. The net result is an increase in the length of the shaft. At the end of the growing period, proliferation of cartilage cells slows and finally ceases. The remaining cartilage becomes converted to bone and it is at this point that no further growth in length can occur.

The growth in diameter of bone does not depend upon the calcification of cartilage but rather is the result of deposition of new bone by the periosteum.

• PROBLEM 19-20

Besides their function in locomotion and support, bones also serve several other important functions. What are they?

Solution: **Bones are an important reservoir for** certain minerals. The mineral content of bones is constantly being renewed. Roughly all the mineral content of bone is removed and replaced every nine months. Calcium and phosphorus are especially abundant in the bones and these are minerals which must be maintained in the blood at a constant level. When the diet is low in these minerals, they can be withdrawn from the bones to maintain the proper concentration in the blood. Stress seems to be necessary for the maintenance of calcium and phosphate in the bones, for in the absence of stress these minerals pass from the bones into the blood faster than they are taken in. This elevates the blood concentration of these minerals to a very high level, which may ultimately lead to to the development of kidney stones. Before special stress exercise programs were developed, astronauts in space often became victims of this type of kidney trouble.

During pregnancy, when the demand for minerals to form bones of a growing fetus is great, a woman's own bones may become depleted unless her diet contains more of these minerals than is normally needed. During starvation, the blood can draw on the storehouse of minerals in the bones and maintain life much longer than would be possible without this means of storage.

Bones are also important in that they give rise to the fundamental elements of the circulatory system.

Bone marrow is the site of production of lymphocyte precursor cells, which play an integral role in the body's immune response system. Red blood cells, or erythrocytes, also originate in the bone marrow. As the erythrocytes mature, they accumulate hemoglobin, the oxygen carrier of blood. Mature erythrocytes, however, are incomplete cells lacking nuclei and the metabolic machinery to synthesize new protein. They are released into the blood-stream, where they circulate for approximately 120 days before being destroyed by the phagocytes. Thus, the bone marrow must perform the constant task of maintaining the level of erythrocytes for the packaging of hemoglobin.

TEETH

• PROBLEM 19-21

The teeth are usually considered as a part of the skull although they are formed from invaginations of the outer ectoderm in the embryo and hence should be part of the integumentary system. What are the different kinds of mammalian teeth? What is the structure of a typical human tooth?

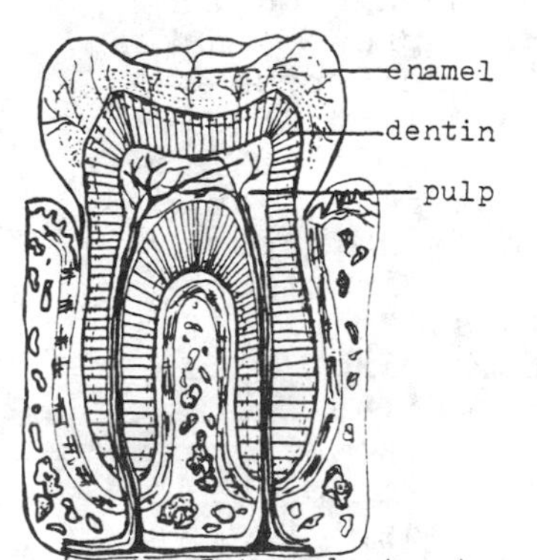

Figure 1. Internal structure of a human tooth. Blood vessels and nerves penetrate into the pulp, but not into the outer harder layers.

incisors
canine
premolars
molars
(A)

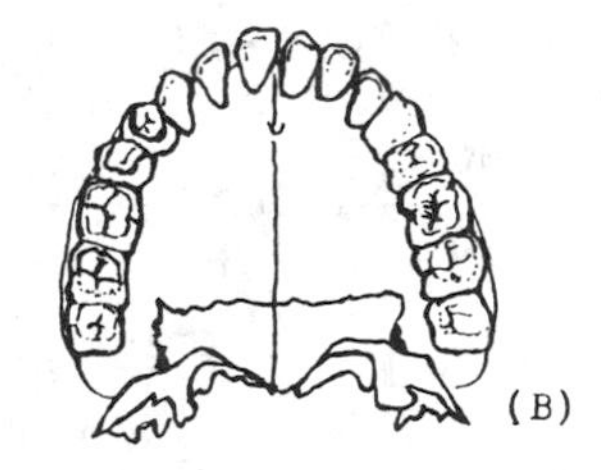

(C)

Figure 2. Human teeth.
(A) Lower jaw of adult.
(B) Upper jaw of adult.
(C) Lower jaw of child, showing permanent teeth in gums below milk teeth.

Solution: The teeth of mammals are differentiated into four types: incisors, canines, premolars, and molars. The incisors are usually simple teeth with a chisel-shaped cutting edge used for biting. In man, there are four incisors in the upper jaw and four in the lower. The canines are usually simple conical teeth which are often greatly enlarged, as in the tusks of the walrus, and are used for tearing food. Man has four canines, one on the side of each jaw. The premolars and molars are often spoken of together as cheek teeth. These teeth have flattened, ridged surfaces, and function in grinding, pounding, and crushing food. Behind the canines in man, there are two premolars and three molars on each side of each jaw. A child's first set of teeth does not include all those mentioned above; the first (or milk) teeth are lost as the child gets older, being replaced with permanent teeth that have been growing in the gums. The final permanent set of teeth totals 32 in man.

A typical tooth consists of a crown, neck, and root. The part projecting above the gum is the crown; that surrounded by the gum is the neck; below the neck is the root, embedded in the jawbone. The body of a tooth is composed of dentin, which resembles bone in its structure, chemical components, hardness, and development. The crown is covered with a layer of enamel the hardest substance in the body, and the root is fastened to the jawbone by a layer of cement. The central region of each tooth is the pulp cavity, which contains connective tissue with nerves and blood vessels.

• **PROBLEM** 19-22

The teeth of different species of vertebrates are specialized in a variety of ways and may be quite unlike those of man in number, structure, arrangement, and function. Give examples.

Solution: The teeth of snakes, unlike those of man, are very thin and sharp, and are usually curved backward. These teeth do not serve the function of mechanical breakdown of food; they serve in capturing prey. Snakes swallow their prey whole. The teeth of carnivorous mammals, such as cats and dogs, are more pointed than those of man; the canines are longer, and the premolars lack flat, grinding surfaces, being more adapted for cutting and shearing. On the other hand, herbivorous animals such as cows and horses may lack canines, but have very large, flat premolars and molars. Notice that sharp pointed teeth, poorly adapted for chewing, seem to characterize meat eaters, whereas broad flat teeth, well-adapted for chewing, seem to characterize vegetarians. In vegetarians, the flat teeth serve to break up the indigestible cell walls of the ingested plant, allowing the cellular contents to be exposed to the action of digestive enzymes. Animal cells do not have the indigestible armor of a cell wall and can be acted upon directly by digestive enzymes; hence, there is no need for flat, grinding teeth.

TYPES OF SKELETAL SYSTEMS

• **PROBLEM** 19-23

The vertebrate skeleton may be divided into two general

parts, the axial skeleton and the appendicular skeleton. What bones constitute these in man?

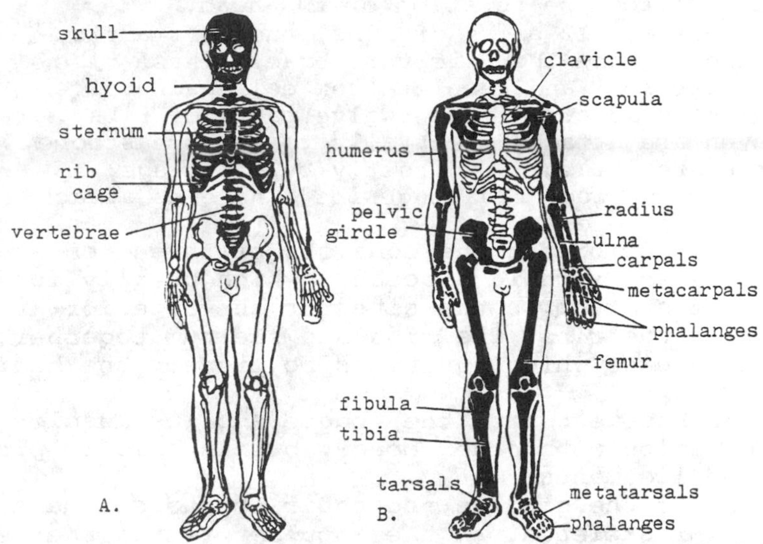

Diagrams of human body showing, A, the bones of the axial skeleton and B, the bones of the appendicular skeleton.

Solution: The axial skeleton consists of the skull, vertebral column, ribs, sternum and hyoid bone. The primary function of the vertebrate skull is the protection of the brain. The part of the skull that serves this function is the cranium. The rest of the skull is made up of the bones of the face. In all, the human skull is composed of twenty-eight bones, six of which are very small and located in the middle ear. At the time of birth, several of the bones of the cranium are not completely formed, leaving five membraneous regions called fontanelles. These regions are somewhat flexible and can undergo changes in shape as necessary for safe passage of the infant through the birth canal.

The human vertebral column, or spine, is made up of 33 separate bones known as vertebrae, which differ in size and shape in different regions of the spine. In the neck region there are 7 cervical vertebrae; in the thorax there are 12 thoracic vertebrae; in the lower back region there are 5 lumbar vertebrae, in the sacral or hip region, 5 fused vertebrae form the sacrum to which the pelvic girdle is attached; and at the end of the vertebral column is the coccyx or tailbone, which consists of four, or possibly five, small fused vertebrae. The vertebrae forming the sacrum and coccyx are separate in childhood, with fusion occurring by adulthood. The coccyx is man's vestige of a tail.

A typical vertebra consists of a basal portion, the centrum, and a dorsal ring of bone, the neural arch, which surrounds and protects the delicate spinal cord which runs through it. Each vertebra has projections for the attachment of ribs or muscles or both, and for articulating (joining) with neighboring vertebrae. The first vertebra, the atlas, has rounded depressions on its upper surface into which fit two projections from the base of the skull. This articulation allows for up and down movements of the head. The second vertebra, called the axis, has a pointed projection which fits into the atlas. This type of articulation allows for the rotation of the head.

In man there are 12 pairs of ribs, one pair articulating with each of the thoracic vertebrae. These ribs support the chest wall and keep it from collapsing as the diaphragm contracts. Of the twelve pairs of ribs, the first seven are attached ventrally to the breastbone, the next three are attached indirectly by cartilage, and the last two, called "floating ribs", have no attachments to the breastbone.

The sternum or breastbone consists of three bones - the manubrium, body and xiphoid process - which usually fuse by middle-age. The sternum is the site for the anterior attachment of most of the ribs. The ribs and sternum together make up the thoracic cage which functions to protect the heart and lungs.

The hyoid bone supports the tongue and its muscles. It has no articulation with other bones, but is held in place by muscles and ligaments.

The bones of the girdles and their appendages make up the appendicular skeleton. In the shoulder region the pectoral girdle, which is generally larger in males than in females, serves for the attachment of the forelimbs. The pectoral girdle consists of two collarbones, or clavicles, and two shoulder blades, or scapulas. In the hip region, the pelvic girdle serves for the attachment of the hindlimbs. The pelvic girdle, which is wider in females so as to allow room for fetal development, consists of three fused hipbones, called the ilium, ischum and pubis, which are attached to the sacrum. Collectively, the "hip bone" is called the oscoxae or innominate bone.

Articulating with the scapula is the single bone of the upper arm, called the humerus. Articulating with the other end of the humerus are the two bones of the forearm called the radius and the ulna. The radius and ulna permit rotation of the forearm. The ulna has on its end next to the humerus a process often referred to as the "funny bone." The wrist is composed of eight small bones called the carpals. The arrangement of these bones permits the rotating movements of the wrist. The palm of the hand consists of 5 bones, known as the metacarpals, each of which articulates with a bone of the finger, called a phalanx. Each finger has three phalanges, with the exception of the thumb, which has two.

The pattern of bones in the leg and foot is similar to that in the arm and hand. The upper leg bone, called the femur, articulates with the pelvic girdle. The two lower leg bones are the tibia (shinbone) and fibula, corresponding to the radius and ulna of the arm, respectively. These two bones are responsible for rotation of the lower leg. Ventral to the joint between the upper and lower leg bones is another bone, the patella or knee cap, which serves as a point of muscle attachment for upper and lower leg muscles. This bone has no counter part in the arm. The ankle contains seven irregularly shaped bones, the tarsals, corresponding to the carpals of the wrist. The foot proper contains five metatarsals, corresponding to the metacarpals of the hand, and the bones in the toes are the phalanges, two in the big toe and three in each of the others.

• PROBLEM 19-24

Differentiate between an exoskeleton and an endoskeleton.

Solution: In most of the large multicellular animals, some kind of skeleton is found that serves for protection, support and locomotion. The skeleton of an animal may be located outside or inside the body. If the former is the case, the skeleton is termed an exoskeleton, and is usually composed of a semiflexible substance known as chitin. If the latter is the case, the skeleton is termed an endoskeleton, and is composed of either cartilage bone, or both. The hard shells of lobsters, crabs, oysters, clams, and snails are examples of exoskeletons. In fact, the mollusc and arthropod phylums are characterized by the presence of an exoskeleton. The advantage of an exoskeleton as a protective device is obvious, but a serious disadvantage is the attendant difficulty of growth. Snails and clams solve this difficulty by secreting additions to their shells as they grow. Lobsters and crabs have evolved a complicated process, called molting, whereby the outer shell is first softened by the removal of some of its salts. A new similarly soft cuticle is secreted beneath the old shell. The old shell is then split down the back. Following this split, the animal crawls out of the old shell, grows rapidly for a short time, and then redeposits the removed mineral salts in the new shell, hardening it. During molting, the animal usually remains well-hidden, because this is a time when it is defenseless and easily killed by its enemies.

Human beings and all other chordates have an endoskeleton. The skeleton of sharks and rays is made of cartilage. The human skeleton consists of approximately 200 bones; the exact number varies at different periods of life as some bones, at first distinct, gradually become fused. The skeleton in man acts as a supporting framework for the other organs of the body. Since the skeleton is internal and grows with the organism, no problem such as that encountered in the growth of lobsters and crabs occurs.

• PROBLEM 19-25

Arthropods have an exoskeleton. How is movement accomplished?

Solution: Movement in the arthropods is possible in spite of the hard exoskeleton because the body is segmented and the segments are joined by a thin layer of flexible chitin. Jointed legs are especially characteristic of the arthropods; they consist of a series of cone-like sections with the small end of one fitting into the large end of the next. Only arthropods and vertebrates have jointed appendages; there are more joints, however, in the arthropod legs because each joint does not have as great a degree of movement as the joint of a vertebrates.

• **PROBLEM 19-26**

What is meant by a joint? What different types of joints are there? Why do older people usually complain of stiffness in their joints?

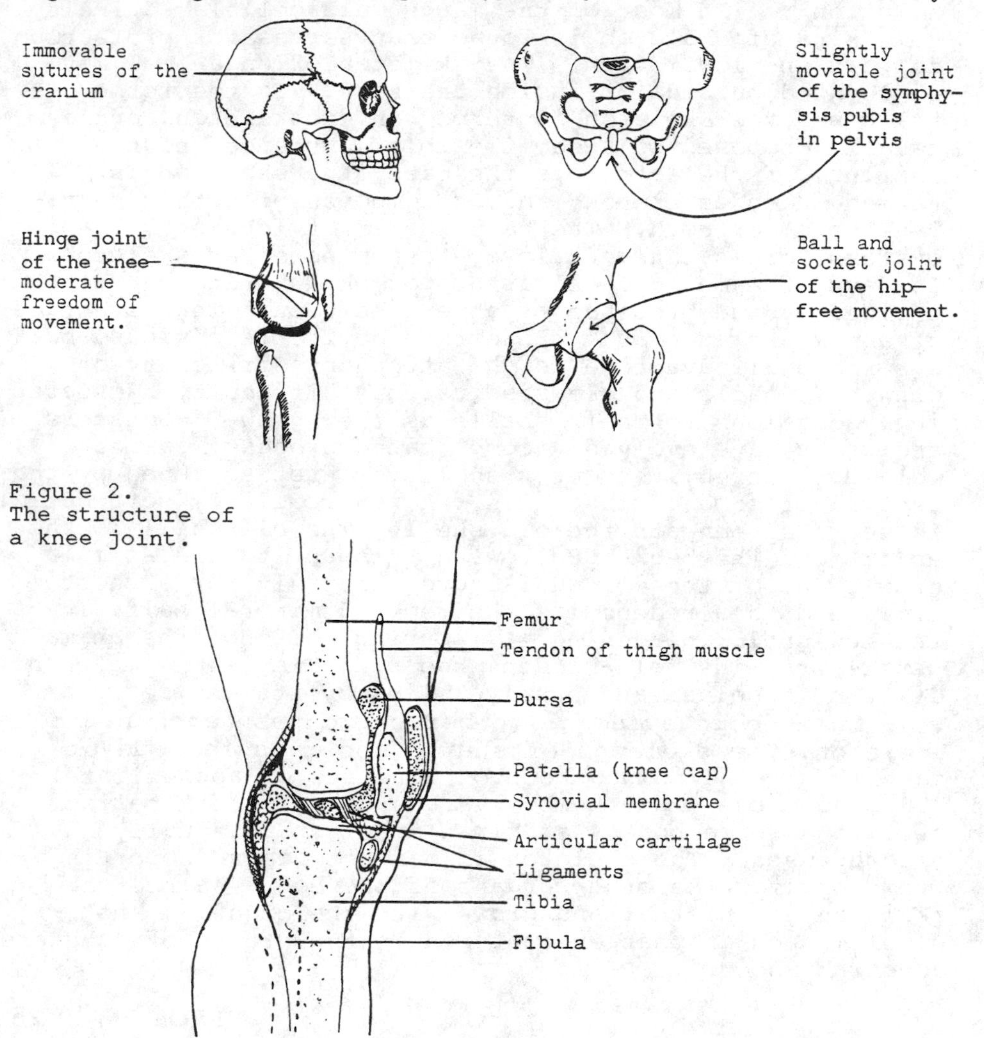

Solution: The point of junction between two bones is called a joint. Some joints, such as those between the bones of the skull, are immovable and extremely strong, owing to an intricate intermeshing of the edges of the bones. The truly movable joints of the skeleton are those that give the skeleton its importance in the total effector mechanism of locomotion. Some are ball and socket joints, such as the joint where the femur joins the pelvis, or where the humerus joins the pectoral girdle. These joints allow free movement in several directions. Both the pelvis and the pectoral girdle contain rounded, concave depressions to accomodate the rounded convex heads of the femur and humerus, respectively. Hinge joints, such as that of the human knee, permit movement in one direction only. The pivot joints at the wrists and ankles allow freedom of movement intermediate between that of the hinge and the ball and socket types. (Refer to Fig. 1.)

The different bones of a joint are held together by connective tissue strands called ligaments. Skeletal muscles, attached to the bones by means of another type of connective tissue strand known as a tendon, produce their effects by bending the skeleton at the movable joints. The ends of each bone at a movable joint are covered with a layer of smooth cartilage. These bearing surfaces are completely enclosed in a liquid-tight capsule, called the bursa.

The joint cavity is filled with a liquid lubricant, called the synovial fluid, which is secreted by the synovial membrane lining the cavity. (Refer to Fig. 2) During youth and early maturity the lubricant is replaced as needed, but in middle and old age the supply is often decreased, resulting in joint stiffness and difficulty of movement. A common disability known as bursitis is due to the inflammation of cells lining the bursa, and also results in restrained movement.

• PROBLEM 19-27

What are the basic differences in the mechanical arrangement of vertebrate and arthropod joints?

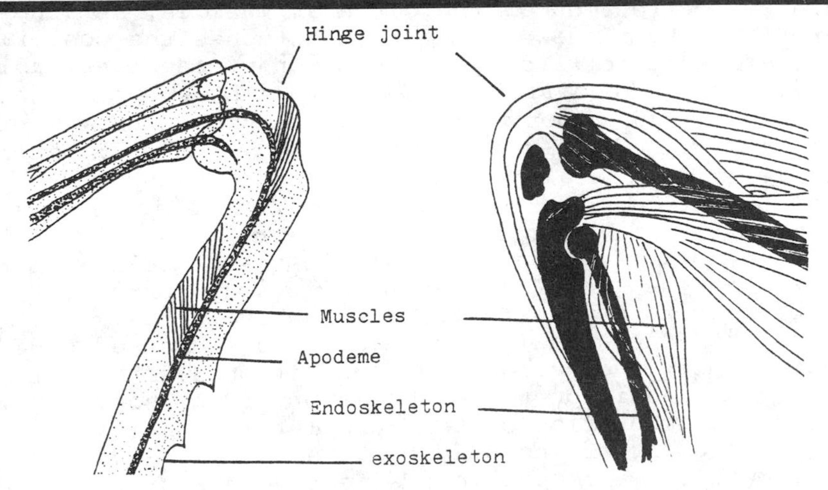

A comparison of the vertebrate endoskeleton and joint with the arthropod exoskeleton and joint.

Solution: The vertebrates are characterized by having an endoskeleton - a bony or cartilaginous framework lying within the body - surrounded by muscles. The arthropods, on the other hand, have an exoskeleton. The arthropod exoskeleton is a chitinous framework on the outside of the body surrounding the muscles. The contraction of muscles in the vertebrates move one bone with respect to another. One end of the muscle is attached to one bone and the other end is attached to another bone. When the muscle contracts, one end - called the origin - remains relatively fixed while the other - called the insertion - moves. The insertion pulls along the bone to which it is attached, and bone movement results. The muscles of the arthropod lie within the skeleton and are attached to the skeleton's inner surface. Regions in which the exoskeleton is thin and flexible allow bending to occur and serve as the joints in the arthropod exoskeleton. A muscle may stretch across the joint so that its contraction will move one part on the next; or a muscle may be located entirely within one section of the body and may be attached at one end to a tough apodeme - a long, thin, firm part of the exoskeleton extending into that section from the adjoining one. (See Figure)

• PROBLEM 19-28

What is meant by a hydrostatic skeleton? In which organisms is such a structure found?

Solution: Cnidarians (coelenterata) such as hydra, flatworms such as planaria, and annelids like the earth worms all move by the same basic principle of antagonistic muscles. They have no hard exo- or endo-skeleton to anchor the ends of their muscles. Instead the noncompressible fluid contents of the body cavity serve as a hydrostatic skeleton. Such animals typically have a set of circular muscles, the contraction of which decreases the diameter and increases the length of the animal, and a set of antagonistic longitudinal muscles, which, when contracted, decrease the length and increase the diameter of the animal. As an example, the leech is often used. The leech attaches its posterior end to the substrate by means of a sucker, extends its body forward by contraction of the circular muscles, attaches its front end by a sucker, and then detaches the posterior end and draws it forward by contraction of the longitudinal muscles. Note that circular and longitudinal muscle are found in man: the smooth muscle lining his hollow visceral organs, for example. By alternate contractions of each muscle layer, peristaltic waves propel the contents along the hollow organ. For instance, this occurs in the esophagus, intestines, oviducts, ureters, etc.

The relatively soft internal tissues of solid-bodied animals like the leech can function as a hydrostatic skeleton only to some extent. More active wormlike animals, like the annelids, have evolved segmented bodies in which fluid is contained in a partitioned series of cavities. In addition to segmentation of the body cavity, there is a segmentation of the musculature; the fact that each segment of the body has its circular and longitudinal muscles makes possible effective use of the compartmented hydrostatic skeleton. The arrangement of the hydrostatic skeleton allows for peristaltic movements down the length of the organism. This aids the organism in burrowing.

Some marine worms have additional diagonally arranged muscles that permit more complex movements of the body and the paddlelike parapodia that extend laterally from the body wall. Many marine worms live in tubes and the movements of the parapodia are important not only in locomotion, but in moving currents of water laden with oxygen and nutrients through these tubes.

STRUCTURAL ADAPTATIONS FOR VARIOUS MODES OF LOCOMOTION

• PROBLEM 19-29

How are the legs of a horse adapted for running? Compare the method of walking of a bear, a cat and a deer.

Solution: Animals differ with respect to which part of the foot they put on the ground when walking and running. Men and bears walk on the entire sole of the foot. This method of locomotion is known as platingrade. Animals such as dogs and cats, to increase their effective limb length and thus their running speed, have become adapted to running on their digits, or fingers. This type of locomotion is known as digitigrade.

Speed is increased still further in the horses, deer, and cattle, which walk on their hoofs, or nails. In these animals the lower limb bones are lengthened, raising the wrist and ankle farther off the ground. This type of locomotion is known as unguligrade. In the case of the horse, the leg is supported by one digit, terminating in a hoof, whereas in cattle, two digits are used for support, terminating in two hoofs. Those ungulates, such as the horse, which walk on one digit are known as the Perissodactyla. Those walking on two digits are known as the Artiodactyla.

• PROBLEM 19-30

What is the major difference between the method of swimming in a fish such as a tuna, and that of a dolphin?

Solution: The major difference between these two organisms with regard to swimming is in the movement of the tail fin. In the tuna, the tail fin is oriented dorso-ventrally. Swimming is accomplished by the side to side movement of the fin by muscle layers in the side of the body. In the dolphin, the tail fin is oriented laterally. Swimming here is accomplished by the movement **of the fin dorso-ventrally. In addition, the dolphin's** hind limbs, or pelvic fins, which are generally used for the maintenance of balance while swimming in other fish, are only vestigial. The forelimbs, which have been modified into swimming paddles, take over the role of balance. In the tuna, the pelvic fins are intact. In both organisms, steering is accomplished by contractions of muscles in the body wall.

• PROBLEM 19-31

Describe the structure of an insect wing. How are wings used in flight?

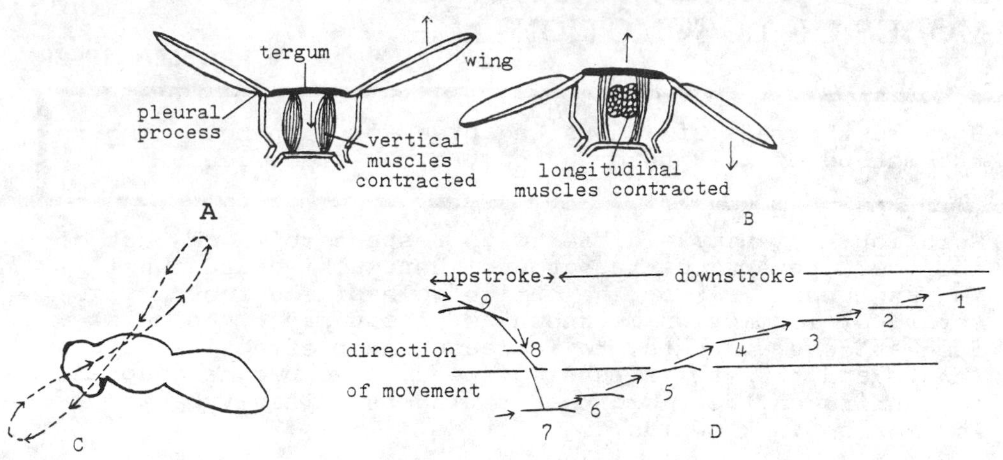

Diagrams showing relationship of wings to tergum and pleura, and the mechanism of the basic wing strokes in an insect. A. Upstroke resulting from the depression of the tergum through the contraction of vertical muscles. B. Downstroke resulting from the arching of the tergum through the contraction of longitudinal muscles. C. An insect in flight, showing the figure 8 described by the wing during an upstroke and a downstroke. D. Changes in the position of the forewing of a grasshopper during the course of a single beat. Short arrows indicate direction of wind flowing over wing and numbers indicate consecutive wing postions.

Solution: The wings of an insect are evaginations or folds of the integument, and are composed of two sheets of cuticle. A vein runs through the wing at a point where the two cuticular membranes are thickened and separated, forming an effective supporting skeletal rod for the wing. The wings of the more primitive insects are netlike, but there has been a general tendency in the evolution of wings toward reduction of this netlike appearance.

Each wing articulates with the edge of the dorsal cuticle, called the tergum, but its inner end rests on a dorsal pleural process (the pleura is the cuticle covering the side of the body), which acts as a fulcrum. The wing is thus somewhat analogous to an off-centered seesaw. Upward movement of the wings results indirectly from the contraction of vertical muscles within the thorax, depressing the tergum (see Figure A). Downward movement of the wings is produced either directly, by contraction of muscles attached to the wingbase, or indirectly, by the contraction of transverse horizontal muscles raising the tergum. Insects such as dragonflies and roaches exhibit direct contraction, while bees, wasps, and flies show indirect contraction. Downward movement can come about by both direct and indirect muscles in insects such as grasshoppers and beetles. The raising or lowering of the wings involves the alternate contraction of antagonistic muscles.

Up and down movement alone is not sufficient for flight; the wings must also move forward and backward. A complete cycle of a single wing beat describes an ellipse in grasshoppers, and a figure eight in bees and flies; the wings are held at different angles to provide both lift and forward thrust. The wing beat frequency varies from 4 beats per second in certain gnats. The fastest flying insects are the moths and horse flies, which can fly over 33 miles per hour. Gliding, an important form of flight in birds,occurs in only a few large insects.

Insect flight muscles are very powerful. The fibrils in the muscle cells are relatively large and the mitochondria are huge. Insects are the only poikilothermic (cold-blooded) fliers; their low body temperature and correspondingly low metabolic rate impose limitations on mobility. On a cold day some butterflies are known to literally "warm up" before flight. They remain stationary on a tree trunk or some other location, and move their wings up and down until sufficient internal heat is generated to permit the stroke rate necessary for flight.

• **PROBLEM** 19-32

How does the wing shape of a bird correlate with its distinctive flight?

Solution: The shape of a wing is correlated with both the power and type of flight for which it is used. Long, slender, pointed wings, sometimes reaching beyond the tail, are seen in birds having great flying powers and soaring habits, such as gulls, eagles, hawks, and vultures. Birds which do not soar, but which fly by continuous wing strokes, have shorter, more rounded wings. Very short, broad wings occur in the fowls, pheasants, grouses, and quails. These are habitual ground dwellers with feet adapted for running. They occasionally make short powerful flights by rapid wing strokes.

Degeneration of the wings to a flightless condition has occurred in a number of birds, such as the penguins and the ostriches. In the ostrich group, many other changes accompany the loss of flight, such as the disappearance of the keel of the breast bone, which supports the flight muscles, and the development of strong running legs and feet.

SHORT ANSWER QUESTIONS FOR REVIEW

Answer

To be covered when testing yourself

Choose the correct answer.

1. Which statement concerning the skin is true? (a) The epidermis is a comparatively thin outer layer of skin containing numerous blood vessels. (b) The dermis is a thicker layer of skin (than the epidermis) containing blood vessels and nerve endings. (c) Most of the fat cells of the skin are contained in the epidermis. (d) Humans have varying numbers of melanocytes in their skin. — b

2. Transverse light and dark bands form a regular pattern along (a) all muscle fibers, (b) only smooth muscle fibers, (c) skeletal and cardiac muscle fibers, (d) skeletal and smooth muscle fibers. — c

3. The A band is (a) the thick dark band of myofilaments contained within a sarcomere, (b) composed of the protein actin, (c) the thin myofilament containing the protein myosin, (d) none of the above. — a

4. The region between the ends of the A bands of two adjoining sarcomeres is called (a) the Z band, (b) the H zone, (c) the T tubule, (d) the I band. — d

5. Myosin bridges can be prevented from combining with actin by (a) troponin, (b) tropomyosin, (c) calcium, (d) only a and b are correct. — d

6. According to the sliding filament theory of muscle contraction (a) the thick and thin filaments change their length as they slide past each other, (b) the lengths of the individual thick and thin filaments remain the same during muscle contraction, (c) groups of muscles slide over each other during stress, (d) none of the above. — b

7. Choose the correct statement. (a) Skeletal muscle fibers are multinucleated, (b) smooth muscle shows a striated banding pattern, (c) cardiac muscle is striated and has a single nucleus, (d) all of the above are correct. — a

8. The phenomenon of rigor mortis is a direct result of (A) the breaking of ATP to myosin bonds, (b) the loss of ATP in the dead muscle cells, (c) the myosin crossbridges being unable to combine with actin, (d) none of the above. — b

	Answer
	To be covered when testing yourself
9. Which of the following statements is incorrect? (a) Cardiac and smooth muscle exhibit tonus even after their nerves are cut. (b) A fatigued muscle will accumulate lactic acid. (c) The mechanical response of a muscle cell to a single action potential is known as a twitch. (d) The number of muscle fibers in a muscle bundle is not important for muscle tension.	d
10. Bone matrix is laid down by cells called (a) osteocytes, (b) chondrocytes, (c) osteoblasts, (d) chondroblasts.	c
11. All of the following contribute to bone formation except (a) periosteum, (b) endosteum, (c) sarcoplasmic reticulum, (d) osteoid.	c
12. When bone formation takes place in pre-existing cartilage it is called (a) intramembranous bone formation, (b) primary ossification, (c) endochondral ossification, (d) osteolysis.	c
13. Which of the following is not a function of bone? (a) Replacement of certain minerals, (b) production of lymphocyte precursor cells, (c) production of erythrocytes, (d) all of the above are functions of bone.	d
14. The region of the tooth containing nerves and blood vessels is called the (a) pulp, (b) enamel (c) dentin, (d) none of the above.	a
15. Which of the following is not a part of the axial skeleton? (a) Skull, (b) humerus, (c) ribs, (d) sternum.	b
Fill in the blanks.	
16. A serious disadvantage of an exoskeleton is the difficulty of _____.	growth
17. Connective tissue strands called _____ hold the bones of a joint together.	ligaments
18. Connective tissue strands called _____ hold muscle to bone.	tendons
19. An exoskeleton is usually composed of a semi-flexible substance known as _____.	chitin
20. The fibrous protein synthesized by epithelial cells is called _____.	keratin
21. The functional unit of the contractile system of the muscle is called a ____.	sarcomere

	Answer
	To be covered when testing yourself
22. Two successive _____ lines define the limit of a sarcomere.	Z
23. Between the ends of the A bands of two adjoining sarcomeres is a region represented by the _____ band.	I
24. The triad of skeletal muscle is composed of two parallel terminal _____ and one _____.	cisternae T tubule
25. During muscle contraction, the length of the individual thick and thin filaments are _____ (changed or unchanged).	un- changed
26. The globular-shaped molecule called _____ has a surface reactive site able to combine with _____.	actin, myosin
27. The _____ is a specialized connective tissue that has the ability to form bone and contributes to fracture healing.	peri- osteum
28. Haversian systems are characteristic of _____ bone.	compact
29. Endochondral ossification is the formation of bone in pre-existing _____.	cartilage
30. The three predominant materials that a tooth consists of are _____, _____ and _____.	enamel, dentin, cementum
Determine whether the following statements are true or false.	
31. Digitigrade is the type of locomotion where an animal such as a bear walks on the soles of its feet.	False
32. The enamel of a tooth contains connective tissue with nerves and blood vessels.	False
33. Possible explanations for depletion of minerals in bones are pregnancy and starvation.	True
34. Endochondral ossification can be the formation of bone from cartilage or cartilage from bone.	False
35. Tonus is the phenomenon of muscle tissue resting in a slightly contracted state.	True
36. The number of muscle fibers in a contracting muscle bundle is not important for muscle tension.	False

	Answer
	To be covered when testing yourself
37. The amount of calcium released during a single action potential is not the deciding factor in excitation-contraction coupling of a muscle.	True
38. During moderate exercise, the source of ATP for muscle cells comes from the complete oxidation of carbohydrates.	True
39. Summation of contractions in skeletal muscle is possible due to the very short refractory period.	True
40. A muscle under slight tension reacts more slowly than one that is more relaxed.	False
41. Smooth muscle does not contain sarcomeres.	True
42. Smooth muscle cells are not under the influence of the autonomic nervous system.	False
43. When a muscle cell exhausts its supply of organic phosphates and glycogen, it will use the available lactic acid.	False
44. The total tension a muscle can develop depends partly on the amount of tension developed by each contracting fiber.	True
45. The number of fibers in a muscle that are contracting is independent of the number of motor neurons connected to the muscle.	False

CHAPTER 20

COORDINATION

REGULATORY SYSTEMS

• PROBLEM 20-1

Describe the primary functions of the nervous system. What other systems serve similar functions?

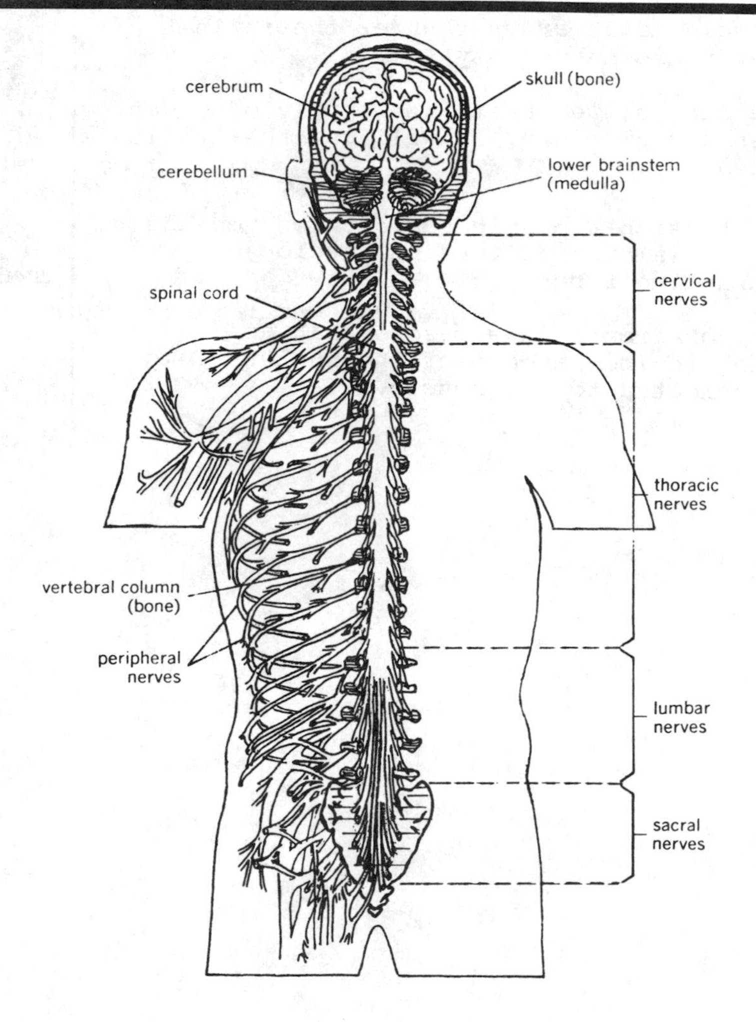

Nervous system viewed from behind.

Solution: The human nervous system, composed of the brain, spinal cord, and peripheral nerves, connects the eyes, ears, skin, and other sense organs (the receptors) with the muscles, organs, and glands (the effectors). The nervous system functions in such a way that when a given receptor is stimulated, the proper effector responds appropriately.

The chief functions of the nervous system are the conduction of impulses and the integration of the activities of various parts of the body. Integration means a putting together of generally dissimilar things to achieve unity.

Other systems involved in similar functions are the endocrine system and the regulatory controls intrinsic in the enzyme systems within each cell. Examples of the latter are inhibition and stimulation of enzymatic activities. The endocrine system utilizes substances, knows as hormones, to regulate metabolic activities within the body.

VISION

• PROBLEM 20-2

Draw a diagram of the human eye, labeling all parts. Briefly describe the function of each part.

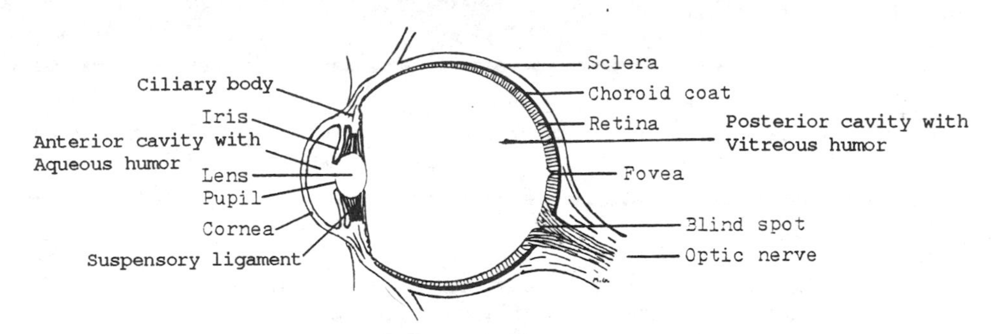

Diagrammatic section of the human eye.

Solution: The human eye consists of three layers or tunics. The fibrous tunic is the outermost layer of the eye: it consists of the sclera posteriorly and the cornea anteriorly. The sclera is a protective coat and is the "white of the eye." The cornea is the transparent covering which functions in the refraction (bending) of light.

The middle coat is the vascular layer. Posteriorly, it consists of the choroid coat. This pigmented layer absorbs excess light and nourishes the retina through its rich vasculature. The choroid continues anteriorly as the ciliary body. The ciliary body consists of the ciliary muscle, a smooth muscle which functions in the accommodation reflex, and the ciliary process which secretes aqueous humor (fluid) into the anterior part of the eye. Finally, the ciliary body continues anteriorly as the iris. The iris is the colored part of the eye and also contains smooth muscle which functions in controlling the size of the pupil - the small opening through which light passes.

The innermost layer is the nervous tunic. It consists solely of the retina in the back of the eye. The retina consists of photoreceptor cells called rods and cones. In the center of the retina is a small depressed area called the fovea centralis, the region of keenest vision. Medial to this is the optic disc. The optic nerve exits the back of the eye at this point. Because there are no photoreceptors at the disc, there is a blind spot in the peripheral field of vision. There is no anterior continuation of the retina.

There are two fluid filled cavities inside the eye. The anterior cavity contains an aqueous humor (a watery fluid), and is located between the cornea and the lens. The posterior cavity is filled with vitreous humor (a gel-like fluid), and is located between the lens and the retina. The transparent lens which separates these two cavities is responsible for focusing incoming light rays on the retina.

• **PROBLEM 20-3**

Our eyes are the principal organs involved in seeing. Yet we are able to perform manual acts such as dressing or tying knots when our eyes are closed. Why?

Solution: There are many sense organs (receptors) in the body. Our pair of eyes is just one example. Although our eyes are extremely important to our perception of this world, we will not be totally and helplessly lost when we can no longer use our eyes. For instance, we are still aware of the relation of our body to the environment even if our eyes are closed. We know whether we are standing or sitting, we know where our limbs are, and we know where one part of our body is in relationship to another. Such perception without the use of our eyes in achieved with a different set of sense organs known as proprioceptors.

Proprioceptors are receptors found in muscles, tendons and joints, and are sensitive to muscle tension and stretch. They pick up impulses from the movements and positions of muscles and limbs relative to each other and relay them to the cerebellum for coordination. Impulses from the proprioceptors are extremely important in ensuring the coordinated and harmonious contraction of different muscles involved in a single movement. Without them, complicated skillful acts would not be possible.

Proprioceptors have other important functions. They help maintain the sense of balance and give the body general awareness of its environment.

• PROBLEM 20-4

How is the human eye regulated for far and near vision? In what respects does the frog eye differ in these regulatory mechanisms?

Solution: The human eye can focus near or distant images by changing the curvature of the lens. The lens is bound to ciliary muscles via the suspensory ligaments. When inverted to focus on a distant object, the ciliary muscles contract, stretching the suspensory ligaments as well as the flexible lens. A flat lens correctly focuses the image of a distant object on the retina. Ciliary muscles relax when one focuses on a near object. This allows the lens to contract into a shape that will correctly focus the closer object.

Lens regulation in frogs differs from man in one important aspect. The frog focuses objects by moving the eye lens forward or backward, whereas in man accommodation is achieved by changing the shape of the lens, without any change in its position.

• PROBLEM 20-5

What are the defects that produce myopia, hypermetropia and astigmatism? What corrective measure can be taken for each defect?

Solution: The most common defects of the human eye are nearsightedness (myopia), which is the inability to see distant objects clearly; farsightedness (hypermetropia), which is the inability to see nearby objects distinctly; and astigmatism, a defect owing to an improperly shaped eyeball or irregularities in the cornea. In the normal eye (Figure A) the shape of the eyeball is such that the retina is located at a proper distance behind the lens for light rays to converge in the fovea. In a nearsighted eye (Figure B), the eyeball is too long, and the retina is too far from the lens. The light rays converge at a point in front of the retina, and are again diverging when they reach it,

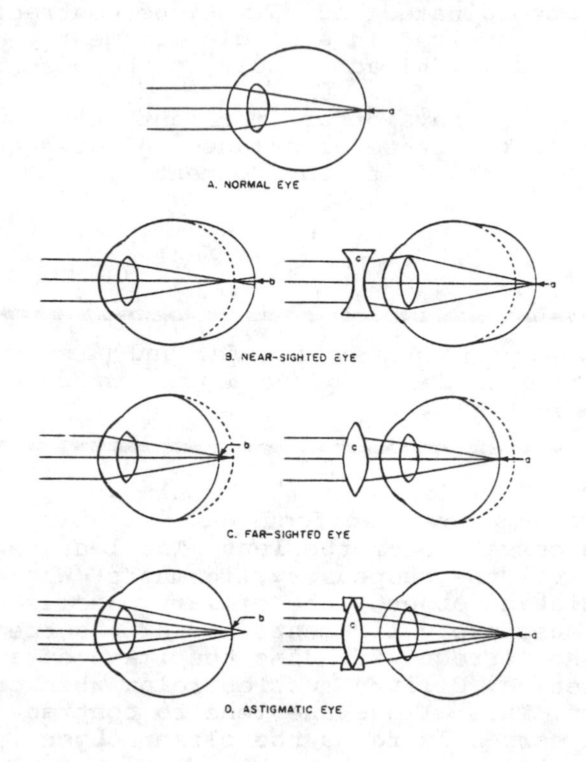

Diagram illustrating common abnormalities of the eye.
a. Sharp image b. Blurred image c. Correction

resulting in a blurred image. This defect can be corrected by placing a concave lens in front of the eye, which diverges the light rays before they reach the lens, making it possible for the eye to focus these rays properly on the retina. In a farsighted eye (Figure C), the eyeball is too short and the retina is thus too close to the lens. Light rays strike the retina before they have converged, again resulting in a blurred image. Convex lenses, when placed in front of the defective eyes, correct for the farsighted condition by causing the light rays to converge farther forward, so that they can come to a focal point on the retina.

In contrast to length of the eyeball as the cause for these visual disorders, a myopic eye may have a lens which is too strong - it bends light rays too much. A hypermetropic eye has a lens which is too weak. Whether the cause be from eyeball length or lens strength, a myopic eye will focus the image in front of the retina, while a hypermetropic eye will focus the image behind the retina.

In astigmatism (Figure D), the cornea is curved unequally in different planes, so that light rays in one plane are focused at a different point from those in another plane. To correct for astigmatism, a cylindrical lens is used which bends light rays going through certain irregular parts of the cornea.

• PROBLEM 20-6

Discuss the mechanism by which the photoreceptors are stimulated by light. How are rods and cones distributed in the retina?

Solution: Photoreceptors are sensory cells that are sensitive to light. In the human retina, they are called rods and cones according to their shapes. Both types of cells contain light-sensitive molecules called visual pigments, whose primary function is to absorb light. The rods contain rhodopsin (visual purple) which is composed of a chromophore (a variant of vitamin A) and a protein (opsin). The cones contain iodopsin which is made up of the same chromophore as in rhodopsin but with a different protein.

Light from the outside enters the eye and stimulates the rods or cones, thus triggering the emission of nerve impulses by the receptor cells. Light does not directly provide the necessary energy to set off the impulse. The energy comes from the chemical bonds in the rhodopsin or iodopsin molecule. In this respect, the phenomenon of vision is basically different from the phenomenon of photosynthesis, in which light supplies the energy to drive the series of chemical reactions.

When light strikes the visual pigments, it acts upon the chromophore, which then splits away from the opsin. This splitting occurs because light changes the molecular configuration of the chromophore in such a way that it no longer can bind to the opsin. Simultaneosly, impulses are triggered and these travel to the brain (via the optic nerve) where they are interpreted.

Rhodopsin is sensitive to a very small amount of light. Rod cells, which contain this kind of visual pigment, are used to detect objects in poor illumination such as in night vision. They are not responsible for color vision, but are important in the perception of shades of gray, and brightness. Their acuity - ability to distinguish one point in space from another nearby point - is very poor. Rods are most numerous in the peripheral retina, that is, that part closest to the lens, and are absent from the very center of the retina (the fovea). On the other hand, cones operate only at high levels of illumination and are used for day vision. The primary function of the cones is to perceive colors. Their visual acuity is very high, and because they are concentrated in the center of the retina, it is that part which we use for fine, detailed vision.

• PROBLEM 20-7

Describe the chemical reactions that occur in the human retina bringing about visual sensation.

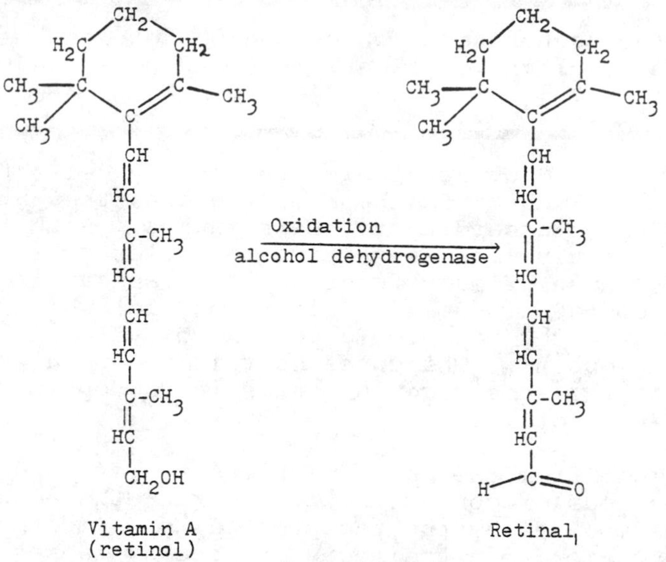

Oxidation
alcohol dehydrogenase

Vitamin A (retinol) — Retinal$_1$

Figure 1. Structures of retinol and retinal$_1$.

Solution: Rhodopsin is a visual pigment in the human eye that is responsible for the perception of an enormous range of light intensities. Located in the rod cell of the retina, rhodopsin has a chromophore (retinal$_1$) bound to a protein (opsin). Retinal is the aldehyde (molecule containing the $-\mathrm{C}{\overset{\mathrm{O}}{\underset{\mathrm{H}}{\scriptstyle{\nearrow\!\searrow}}}}$ functional group) of vitamin A, also called retinol (see Figure 1). Retinal can be synthesized from vitamin A via an oxidation reaction, catalyzed by the enzyme alcohol dehydrogenase. Retinal exists in the cis-configuration before light hits the rhodopsin molecule.

When light strikes the rhodopsin molecule, cis-

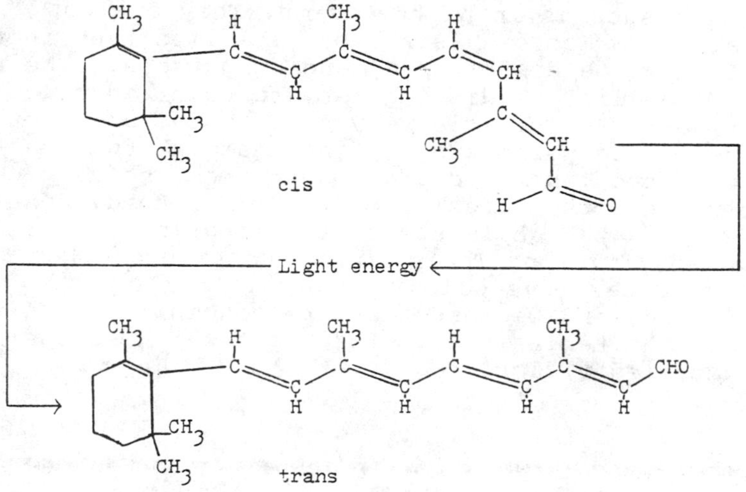

cis

Light energy

trans

Figure 2. Light-induced conversion of cis-retinal to trans-retinal.

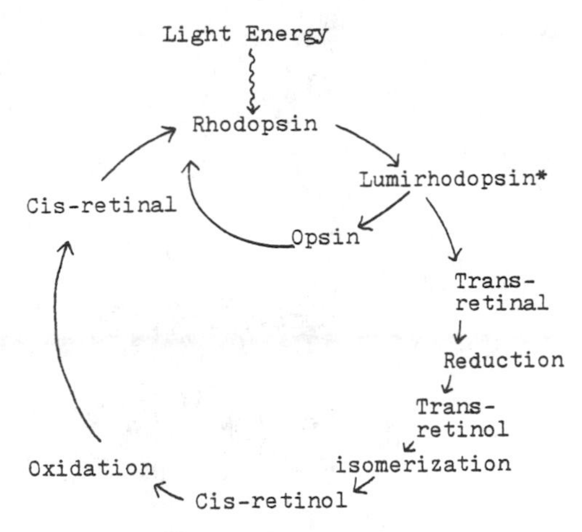

* indicates high energy state

Figure 3. The visual cycle

retinal absorbs the light energy and is converted to the trans form (Figure 2). Simultaneously, the sensation of vision, the stimulation of nerve impulses by the rod cells, occurs. The conversion causes a rearrangement of the molecular structure, resulting in an unstable compound called lumirhodopsin. Lumirhodopsin rapidly decomposes into free trans-retinal and opsin. This is due to the inability of trans-retinal to fit onto the opsin molecule as does the cis form.

Rhodopsin has to be resynthesized from cis-retinal and opsin to provide for further excitation of the rods. Trans-retinal is first reduced to form trans-retinol, which is subsequently enzymatically converted into cis-retinol. The cis-retinol then undergoes oxidation to regenerate cis-retinal, which then combines with opsin to yield rhodopsin. The series of reactions from the light-induced breakdown of rhodopsin to its resynthesis, collectively termed the visual cycle, is outlined in figure 3.

Because vitamin A is the precursor of retinal, synthesis of rhodopsin is dependent on its presence. In vitamin A deficiency, vision is most conspicuously affected. This deficiency is responsible for a condition known as xerophthalmia ("dry eyes") and may cause blindness. An early sign of vitamin A deficiency is night blindness, which is a maladaptation of the eyes to the dark.

TASTE

• PROBLEM 20-8

Certain connoisseurs can recognize hundreds of varieties of wine by tasting small samples. How is this possible when there are only four types of taste receptors?

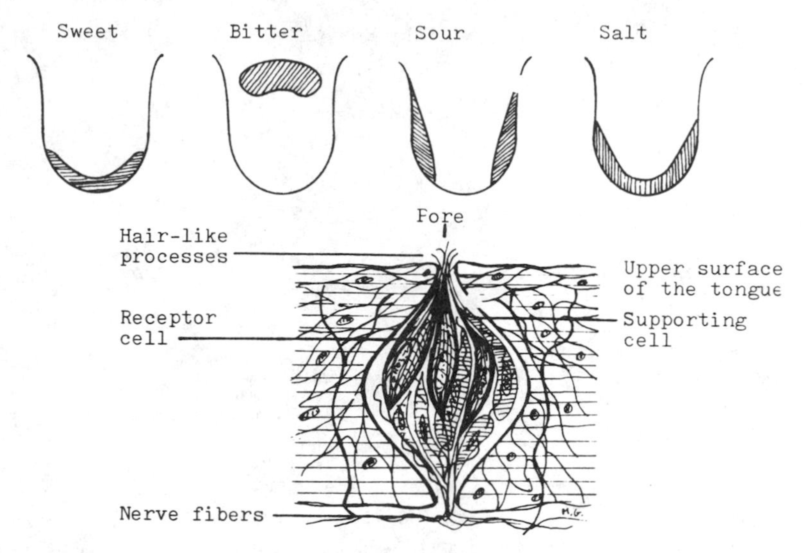

The distribution on the surface of the tongue of taste buds sensitive to sweet, bitter, sour, and salt. Below, cells of a taste bud in the epithelium of the tongue.

Solution: Taste buds on the tongue and the soft palate are the organs of taste in human beings. Each taste bud contains supportive cells as well as epithelial cells which function as receptors. These epithelial cells have numerous microvilli that are exposed on the tongue surface. Each receptor is innervated by one or more neurons, and when a receptor is excited, it generates impulses in the neurons. There are four basic taste senses: sweet, sour, bitter, and salty. The receptors for each of these four basic tastes are concentrated in different regions of the tongue-sweet and saltly on the front, bitter on the back, and sour on the sides (see Figure). The sensitivity of these four regions on the tongue to the four different tastes can be demonstrated by placing solutions with various tastes on each region. A dry tongue is insensitive to taste.

Few substances stimulate only one of the four kinds of receptors, but most stimulate two or more types in varying degrees. The common taste sensations we experience daily are created by combinations of the four basic tastes in different relative intensities. Moreover, taste does not depend on the perception of the receptors in the taste buds alone. Olfaction plays an important role in the sense of taste. Together they help us distinguish an enormous number of different tastes. We can now understand how a connoisseur, using a combination of his taste buds and his sense of smell, can recognize hundreds of varieties of wine.

THE AUDITORY SENSE

• PROBLEM 20-9

With the aid of a diagram, describe the structures found in the human ear.

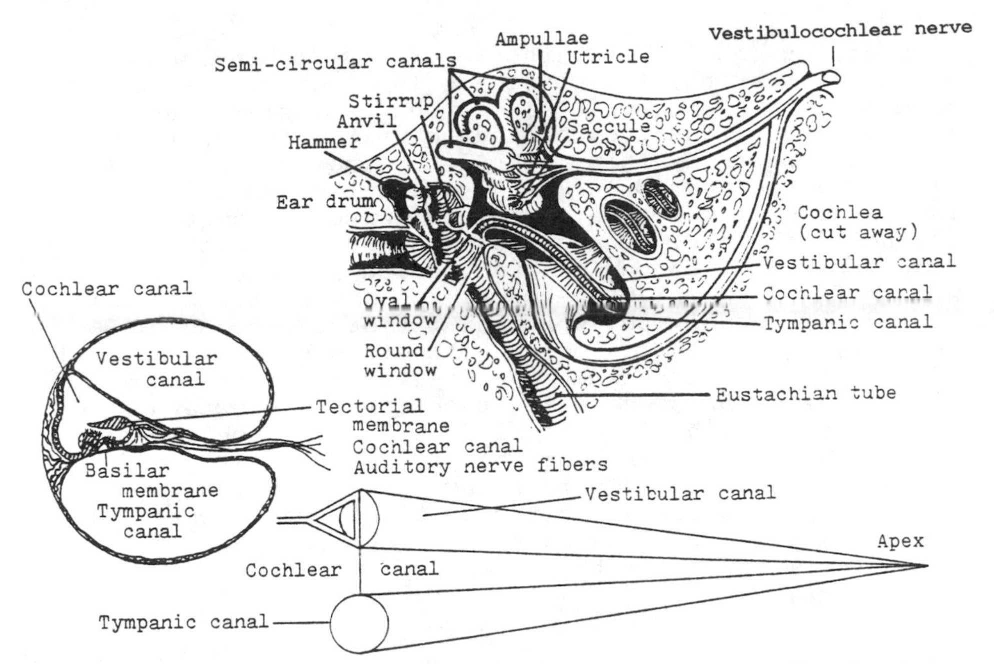

Upper right. The coiled cochlea shown dissected out of the skull and cut open to reveal the vestibular canals. Lower right. A diagram of the cochlea as though it were uncoiled and drawn out in a straight line. Lower left. A cross section through the cochlea to show the organ of Corti resting on the basilar membrane and covered by the tectorial membrane.

Solution: Three parts of the human ear can be distinguished: the outer ear, the middle ear, and the inner ear. The outer ear consists of the skin-covered cartilaginous flap, or pinna, and a channel known as the external auditory meatus. At the inner end of this canal is a membrane called the tympanic membrane (eardrum).

On the other side of the tympanic membrane is a small chamber, the middle ear, which contains three tiny bones. These three bones are the malleus (hammer), incus (anvil) and stapes (stirrup) and are arranged in sequence across the middle ear from the tympanic membrane to another membrane, the oval window, which separates the middle ear from the inner ear. The middle ear is connected to the pharynx via the Eustachian tube, which serves to equalize the air pressure between the outer and middle ear.

The inner ear consists of a complicated labyrinth of interconnected fluid-filled chambers and canals. One group of chambers and canals is involved with the sense of equilibrium. There are two small chambers - the sacculus and utriculus - and three semicircular canals arranged perpendicularly to one another. The utricle and saccule are small, hollow sacs lined with sensitive hair cells and containing small ear stones, or otoliths, made of calcium carbonate. The semicircular canals contain hair cells and are filled with a fluid known as the endolymph. Movement of the endolymph over the hair cells stimulates the latter to send impulses to the brain. Besides the structures involved in maintaining equilibrium, the inner ear also houses the organ for hearing, the cochlea. The cochlea is a coiled tube made up of three canals - the vestibular canal, the tympanic canal and the cochlear canal, each separated from each other by thin membranes. The oval window is linked to the vestibular canal while the tympanic canal is sealed by a membrane, the round window, which leads to the middle ear. These two canals are connected with each other at the apex of the cochlea and are filled with a fluid known as the perilymph. The cochlear canal, filled with endolymph, contains the actual organ of hearing, the organ of Corti. The organ of Corti consists of a layer of epithelium, the basilar membrane, on which lie five rows of specialized receptor cells extending the entire length of the coiled cochlea. Each receptor cell is equipped with hairlike projections extending into the cochlear canal. Overhanging the hair cells is a gelatinous structure, the tectorial membrane, into which the hairs project. The hair cells of the organ of Corti initiate impulses via the fibers of the auditory nerve to the brain.

• PROBLEM 20-10

A person seated on a swivel chair is rotated rapidly. The rotation is suddenly stopped and the person is told to stand up. He complains that he feels dizzy. How can you account for his dizziness?

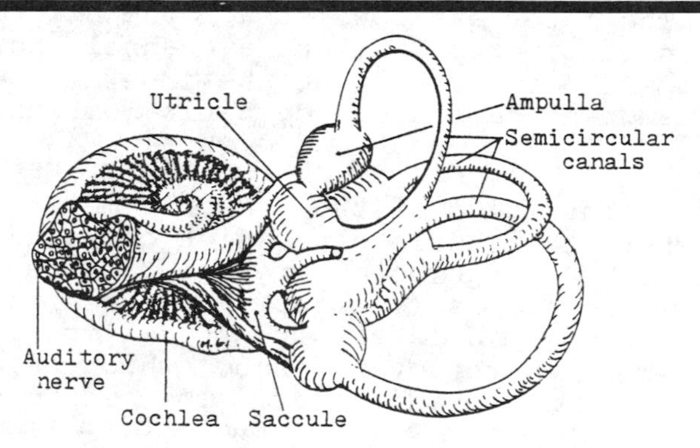

The right semicircular canals and cochlea of an adult man, shown dissected free of surrounding bone and seen from the inner and posterior side.

Solution: The labyrinth of the inner ear has three semicircular canals, each consisting of a semicircular tube connected at both ends to the utricle. Each canal lies in a plane perpendicular to the other two. At the base of each canal, where it leads into the utriculus, is a bulb-like enlargement (the ampulla) containing tufts of hair cells similar to those in the utriculus and sacculus, but lacking otoliths. These cells are stimulated by movements of the fluid (endolymph) in the canals. When a person's head is rotated, there is a lag in the movement of the endolymph in the canals. Thus, the hair cells in the ampulla attached to the head rotate, in effect, in relation to the fluid. This movement of the hair cells with respect to the endolymph stimulates the former to send impulses to the cerebellum of the brain. There, these impulses are interpreted and a sensation of dizziness is felt.

• PROBLEM 20-11

When we hold our head upright, there is very little problem keeping our balance. Yet when our head is turned upside down, as in a head stand, we still manage to maintain our equilibrium. How is balance achieved in the latter case? Discuss your answer in relation to the ear.

Solution: In the inner ear, there are two small, hollow sacs known as the utriculus and sacculus. Each sac is lined with sensitive hair cells, upon which rest small crystals of calcium carbonate (otoliths). Normally when the head is upright, the pull of gravity causes the otoliths to press against particular hair cells, stimulating them to initiate impulses to the brain via sensory nerve fibers at their bases. Changes in the position of the head cause these crystals to fall on and stimulate some other hair cells, thereby giving the brain a different set of signals. The relative strength of the signals to the brain indicates the position of the head at any given moment.

When a head stand is being done, the otoliths in the utriculus and sacculus are reoriented by gravity and come to rest on other hair cells. These hair cells are stimulated and send impulses to our brain, informing us that now our head is in a new position. The brain will then transmit the appropriate response to the effectors so that we can maintain our balance even when our head is upside down.

It is important to realize that equilibrium in humans depends upon the sense of vision, stimuli from the proprioceptors, and stimuli from cells sensitive to pressure in the soles of the feet in addition to the stimuli from the organs in the inner ear. This is why in certain types of deafness, in which the cochlea as well as the equilibrium organs of the inner ear are impaired, the sense of equilibrium is still present.

• PROBLEM 20-12

A student hears the ringing of a bell at the end of the day and begins to pack his books to go home. Trace the path of the stimulus from the source (bell) to its appropriate center in the brain.

Solution: When the bell is struck, it sets off vibrations in the air in the form of sound waves. These sound waves travel through the air and some of them are collected by the pinna. They are then channeled along the auditory canal of the outer ear towards the middle

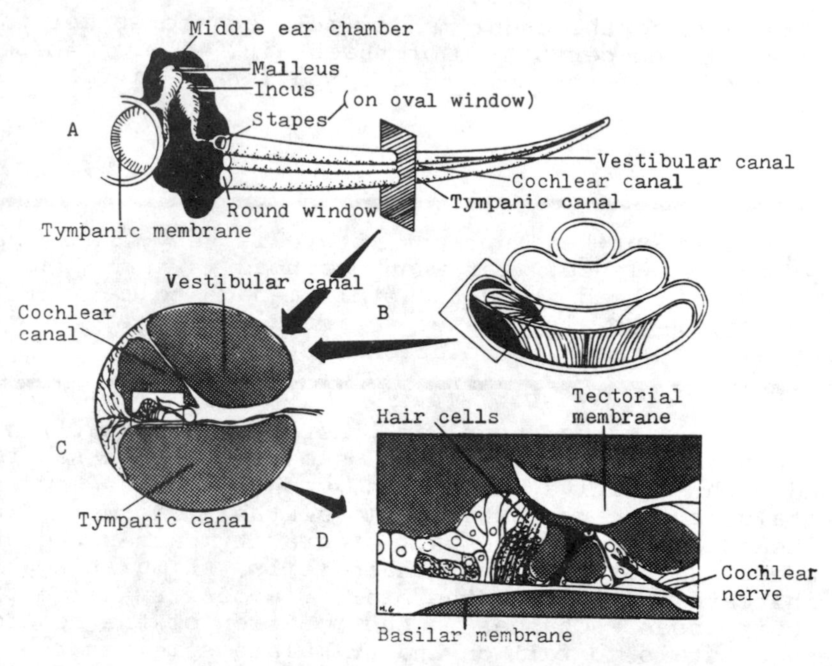

(A) Diagram of the relationship between the middle ear and the cochlea, which has here been uncoiled to show its canal system more clearly. (B) A section through the cochlea in its normal coiled state. (C) Enlarged cross section through one unit of the coil, showing the relationship between the vestibular, cochlear, and tympanic canals and the location of the organ of Corti. (D) Enlarged diagram of the organ of Corti, which rests on the basilar membrane separating the cochlear and tympanic

ear. At the inner end of the auditory canal, the sound waves impinge on the tympanic membrane and set the membrane vibrating. These vibrations are first transmitted to the malleus which is in direct contact with the tympanic membrane. Then the vibrations are relayed across the middle ear by the incus and stapes, which are so arranged that they decrease the amplitude but increase the force of vibrations. The stapes transmits the vibrations, via the oval window, to the perilymph in the vestibular canal. This occurs as the vibrating stapes strikes the membrane of the oval window, causing it to oscillate. Since fluids are incompressible, movement of the membrane causes displacement of the perilymph from the vestibular canal to the tympanic canal. (Recall that these two canals are connected at the apex of the cochlea.) The perilymph in the tympanic canal hits the membrane of the round window, resulting in movement of the membrane. Thus vibrations from the bones of the middle ear are transformed into oscillations of the perilymph in the cochlea.

The pressure waves of the perilypmh set the thin membranes separating the three canals into vibration, most important of which is the basilar membrane in the cochlea canal, since on it lie the receptor cells for hearing.Vibrations of the basilar membrane cause the sensory hairs of the receptor cells to move up and down against the less movable tectorial membrane. This movement results in the deformation of the hairs, which in some unknown fashion initiates and sends nerve impulses along the sensory neurons, lying at the base of the receptor cells. Axons of the sensory neurons come together to form the auditory nerve which leads the impulses from the organ of Corti to the auditory center in the temporal lobe of the brain. There the impulses are interpreted as the ringing of the bell.

ANESTHETICS

• PROBLEM 20-13

How do anesthetics act to reduce or eliminate the sensation of pain?

Solution: In human beings, the interpretation of sensations (vision, hearing, taste, smell, pain, etc.) takes place in the brain. For example, the rods and cones (receptor cells) of the retina do not "see." Only the combination of rods, cones, optic nerve, and the visual center of the brain lead to the sensation of vision.

Since only those nerve impulses that reach the brain can result in sensations, any blockage of the impulses along the nerve fibers by an anesthetic has the same effect as removing the original stimulus entirely. The sense organs, when stimulated, will continue to initiate the impulses, which can be detected by the proper electrical apparatus, but the anesthetic prevents the impulses from reaching their destination in the brain. The tranquilizing effect of anesthetics results from the diminished amounts of neurotransmitter in the synapse. As a consequence of this, a nerve impulse generated by the receptor cells will not be transmitted to the next neuron. This is because the amount of neurotransmitter released will be insufficient to stimulate the neuron to propagate the nerve impulse. Since the impulses going to the pain center of the brain are blocked, the brain is unable to receive and interpret the original stimulus. Therefore there is no sensation of pain.

THE BRAIN

• PROBLEM 20-14

Make a diagram of the human brain. Label the principal parts and list the function(s) carried out by each.

Solution: The human brain is the enlarged, anterior end of the spinal cord. This enlargement is so great that the resemblance to the spinal cord is obscured. The adult

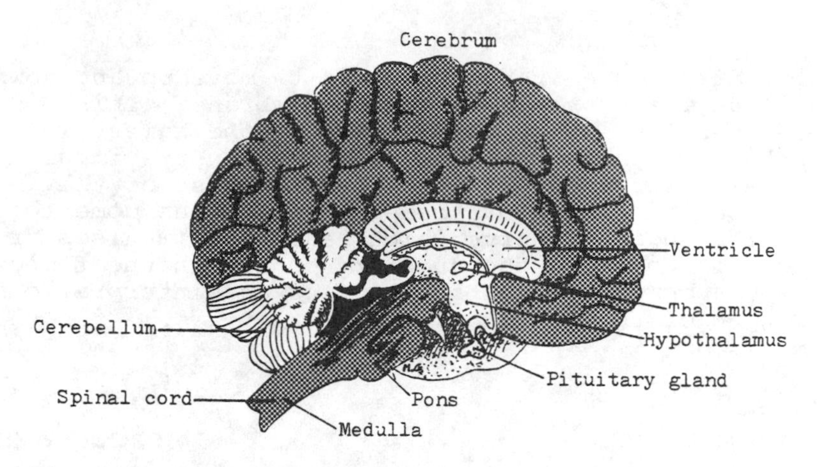

Figure 1. Interior portion of one side of the human brain.

human brain has six major regions: the medulla, pons and cerebellum which constitute the hindbrain; the thalamus and cerebrum, both of which are in the forebrain; and the midbrain.

The most inferior part of the brain, connected immediately to the spinal cord, is the medulla. Here the central canal of the spinal cord (spinal lumen) enlarges to form a fluid-filled cavity called the fourth ventricle. The medulla has numerous nerve tracts (bundles of nerves) which bring impulses to and from the brain. The medulla also contains a number of clusters of nerve cell bodies, known as nerve centers. These reflex centers control respiration, heart rate, the dilation and constriction of blood vessels, swallowing and vomiting.

Above the medulla is the cerebellum, which is made up of a central part and two hemispheres extending sideways. The size of the cerebellum in different animals is roughly correlated with the amount of their muscular activity. It regulates and coordinates muscle contraction and is relatively large in active animals such as birds. Removal or injury of the cerebellum is accompanied not by paralysis of the muscles but by impairment of muscle coordination. A bird, with its cerebellum surgically removed, is unable to fly and its wings are seen to thrash about without coordination.

The pons is an area of the hindbrain containing a large number of nerve fibers which pass through it and make connections between the two hemispheres of the cerebellum, thus coordinating muscle movements on the two sides of the body. The pons also contains the nerve centers that aid in the regulation of breathing.

In front of the cerebellum and pons lies the thick-walled midbrain. The midbrain is an important integrating region and contains the centers for certain visual and auditory reflexes. A cluster of nerve cells regulating muscle tone and posture is also present in the midbrain. A small canal runs through the midbrain and connects the fourth ventricle behind it to the third ventricle in front of it.

The midbrain, pons and medulla are collectively called the brainstem. Ten of the twelve cranial nerves originate in the brainstem.

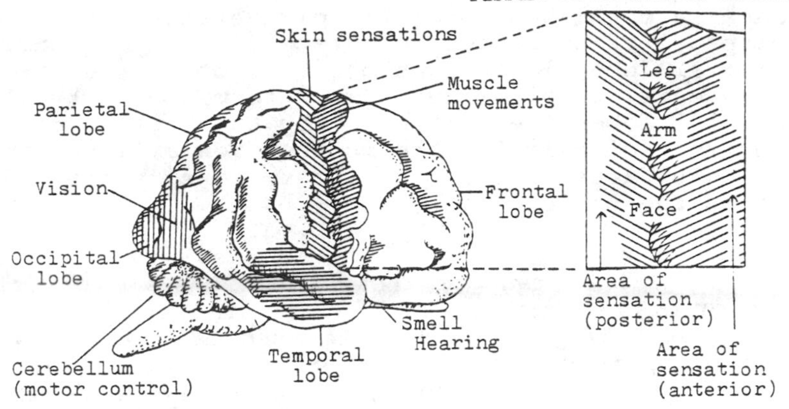

Figure 2. The right cerebral hemisphere of the human brain, seen from the side. The striped areas are regions of special function; the light areas are "association" areas. Inset: Enlarged view of the sensory and motor areas adjacent to the fissure of Rolando, showing the location of the nerve cells supplying the various parts of the body.

The thalamus of the forebrain serves as a relay center for sensory impulses. Fibers from the spinal cord and parts of the brain synapse here with other neurons going to the various sensory areas of the cerebrum. The thalamus seems to regulate and coordinate the external signs of emotions. By stimulating the thalamus with an electrode, a sham rage can be elicited in a cat - the hair stands on end, the claws protrude, and the back becomes humped. However, as soon as the stimulation ceases, the rage responses disappear.

The hypothalamus, located under the thalamus, is a collection of nuclei (a cluster of nerve cell bodies located within the central nervous system), concerned with many important homeostatic regulations. Electrical stimulation of certain cells in the hypothalamus produces sensations of hunger, thirst, pain, pleasure, or sexual drive. The hypothalamus is also important for its influence on the pituitary gland, which is functionally under its control. Cells of the hypothalamus synthesize chemical factors that modulate the release of hormones produced and stored in the anterior pituitary. The hypothalamus has neural connections with the posterior pitiutary. The hypothalamus and thalamus are collectively called the diencehpalon.

The cerebrum, consisting of two hemispheres, is the largest and most anterior part of the human brain. In human beings, the cerebral hemispheres grow back over the rest of the brain, hiding it from view. Each hemisphere contains one cavity (one contains the first and the other the second ventricle, collectively known as the lateral ventricles), which is connected to the third ventricle. The outer portion of the cerebrum, the cortex, is made up of gray matter which comprises the nerve cell bodies. The grey matter folds greatly, producing many convolutions of the cerebral surface. These convolutions increase the surface area of the gray matter. The inner part of the brain is the white matter which is composed of masses of nerve fibers.

Each cerebral hemisphere is divided into four lobes. The occipital or posterior lobe receives visual information. The temporal lobe, located on the side of the cerebrum just above the ears, receives auditory information. The frontal and parietal lobes are demarcated by the fissure of Rolando. Just in front of this fissure in the frontal lobe is the primary motor area, while the primary sensory area lies behind it in the parietal lobe.

When all the areas of known functions are plotted, they cover only a small part of the total area of the human cortex. The rest, known as association areas, are regions responsible for the higher intellectual faculties of memory, reasoning, learning and imagination, all of which help to make up one's personality. In some unknown way, the association regions integrate into a meaningful unit, all the diverse impulses constantly reaching the brain, so that proper response is made.

• PROBLEM 20-15

What general trends are observed in the evolution of the brain?

Solution: Generally, the higher an animal is on the ladder of evolution, the more complex and heavy is its brain (see Figure 1). In the earthworm, an invertebrate, the 'brain' is merely a ganglion (a clump of nerve cell bodies) of about the same size as the other ganglia present in each segment of the body. Even among the mammals, the size and complexity of the brain vary to a great extent. The weight of the brain of a full grown man is approximately three pounds whereas a gorilla, about four times heavier than man, has a brain that weighs only one and a half pounds. This discrepancy in brain weight can account partly for the difference in intellectual capabilities of the two mammals.

Increasing size is not the only feature seen in the evolution of the brain. The brains of higher mammals display increasing foldings of their surface. These foldings, or convolutions, are particularly pronounced in man. By stimulating various regions of the brain with an electrode, and observing the corresponding results in behavior, functional mapping of the brains of various mammals has been achieved. The mapping shows that different regions of the brain are associated with different functions. There are the areas for motor functions, somatic sensory functions, vision, hearing, olfaction, and so forth. The mapping also shows that the proportion of the total area of the cerebrum devoted to sensory and motor functions differs greatly among species. It has been found that, in general, the more convoluted the surface of the cerebrum (called the cerebral cortex), the smaller the proportion devoted exclusively to sensory and motor activities. Since, as mentioned, there is a trend toward increasing convolutions in the evolution of the brain, we can anticipate that

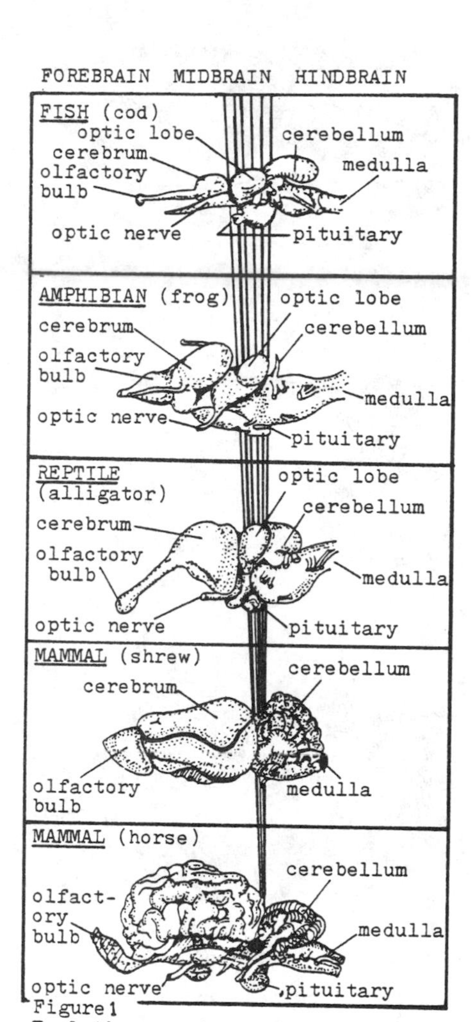

Figure 1
Evolutionary change in relative size of midbrain and forebrain in vertebrates.

higher animals have a more convoluted cerebral cortex, and that at the same time, a smaller proportion of their brain is involved in sensory and motor functions. This is supported by the finding that in the cat the sensory and motor functions occupy a major portion of the cortex, whereas in man, the area of the cortex devoted to those functions is relatively small (see Figure 2).

The rest of the brain not associated with motor and sensory activities contain the so-called associative areas. These areas are made up of neurons that are not directly connected to sense organs or muscles but supply interconnections between the other areas. They are believed to be responsible for the higher intellectual faculties of memory, reasoning, learning, imagination, and personality. They are, in short, responsible for the behavioral consequences that most clearly distinguish man from other animals.

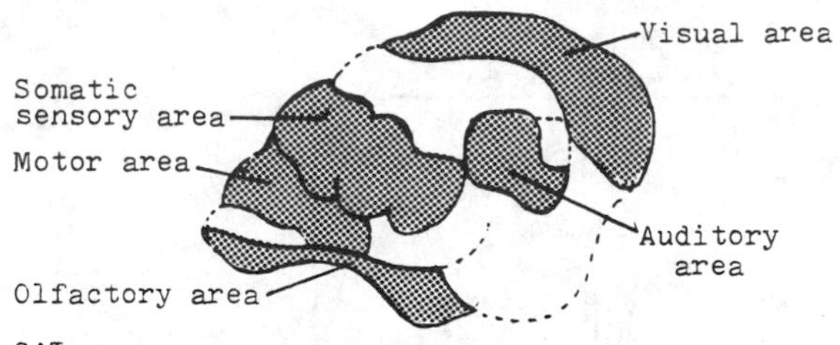

(A) CAT

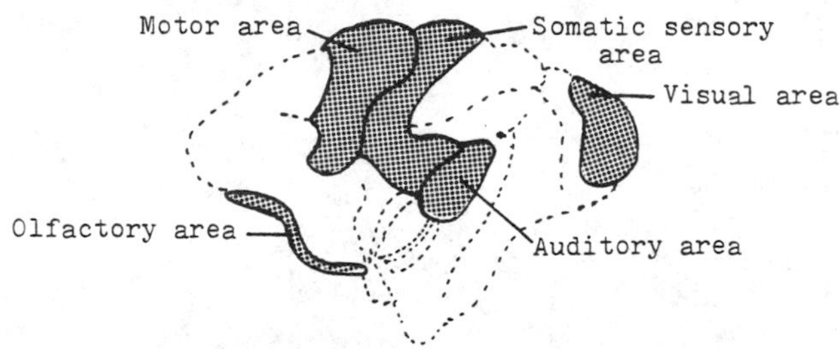

(B) MONKEY

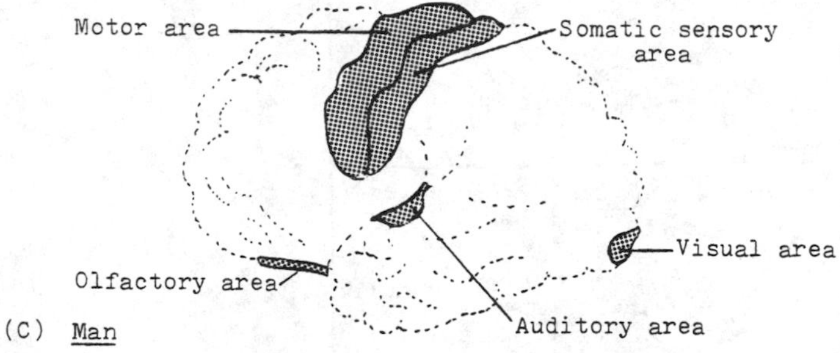

(C) Man

Figure 2. Proportion of cerebral cortex devoted to sensory and motor functions in three mammals. (A) In the cat sensory and motor areas, (color) constitute a major portion of the cortex. (B) In the monkey the proportion of cortex devoted to association areas (white) is much greater than in the cat. (C) In man the sensory and motor areas occupy a relatively small percentage of the cortex, most of the cortical area being devoted to association.

THE SPINAL CORD

• PROBLEM 20-16

Describe the structures found in a cross section of the mammalian spinal cord.

Solution: The mammalian spinal cord extends from the base of the brain and is tubular in shape. Along with the brain, the spinal cord makes up the central nervous system of all vertebrates. It has two very important functions: to transmit impulses to and from the brain, and to act as a reflex center.

The spinal cord is protected by the vertebral

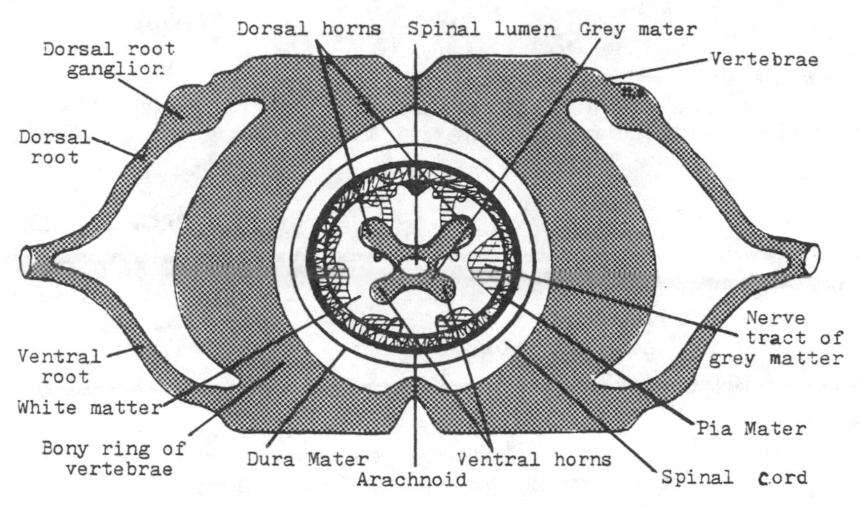

Cross section of the mammalian spinal cord surrounded by the bony vertebrae.

column. The vertebral column is composed of segments of bone, called the vertebrae, connected to each other by cartilage. The cartilage gives it flexibility, while the bone gives it strength. It is possible to anatomically divide the spinal cord into five different regions using the vertebrae as a guide. These are (1) cervical region, (2) thoracic region, (3) lumbar region, (4) sacral region, and (5) coccygeal region. The regions of the spinal cord send out nerves which innervate different parts of the body (see Figure 1).

The spinal cord is surrounded by three membranes: the dura mater, arachnoid, and pia mater. It is a hollow, tubular structure. The hollow portion, called the spinal lumen, runs through the entire length of the spinal cord and is filled with spinal fluid. A cross-section shows 2 regions, an inner mass of gray matter composed of nerve cell bodies, and an outer mass of white matter made up of bundles of axons and dendrites (see Figure 2). The coloration of the white matter region is due to the presence of white myelinated fibers. The nerve cells of the gray matter lack myelin, which is white, and hence the "natural" gray color of nerve cells is seen. Four protuberances of the gray matter are noted. The two anterior processes are called the ventral horns, and the two posterior ones are called the dorsal horns. The axons and dendrites of the white matter carry impulses between lower levels and higher levels of the spinal cord, or between various levels of the spinal cord and the brain.

The spinal cord receives sensory fibers from peripheral receptors at the dorsal root. The cell bodies from which the sensory fibers arise are located in a cluster, called the dorsal root ganglion, in the dorsal root. These sensory fibers pass into the dorsal horns of the gray matter, where they synapse with interneurons in the dorsal horns and/or motor neurons in the ventral horns. Axons from the motor neurons leave the spinal cord at the ventral root and soon join the sensory fibers - together they constitute the spinal nerve. Spinal nerves arising from the spinal cord branch to supply various parts of the body except the head, part of the neck, the thorax, abdomen, and the upper and lower extremities.

• PROBLEM 20-17

The dorsal root of a nerve leading to the leg of a rat is severed. How will this affect the rat? How would the effect vary if the ventral root were severed instead?

Solution: A cross-section of the spinal cord (see problem 26-16) reveals that the paired dorsal roots are the junctions where sensory fibers from the peripheral receptor areas enter, while the paired ventral roots are where axons of the motor neurons leave the spinal cord to effectors, such as muscles. When the dorsal root of a spinal nerve is cut, the sensory fibers are severed and afferent impulses (that is impulses to the nervous system) can no longer reach the spinal cord and thence the brain.

If a hot iron is brought to touch the leg of a rat, in which the dorsal root of the nerve leading to the leg is severed, the rat will not feel any burn sensation, as evidenced by its lack of attempt to avoid the stimulus. Since the receptor cells on the skin of the leg are not impaired, they do respond to the stimulus by firing impulses along the sensory fibers to the spinal cord. However, the impulses cannot reach the central nervous system and therefore cannot be translated and interpreted in the brain. Also, simple reflex action such as the rapid withdrawal of the leg cannot occur in the rat, as impulses from the sensory neurons cannot be transmitted to the motor neurons in the reflex arc. As a consequence, the rat will not feel the stimulation of the hot iron, nor would it perform reflexive withdrawal of the leg from the iron. In higher animals such as man in which memory and experience play an important role in behavior, severing the dorsal root may cause loss of sensation and loss of reflex action, but the somatic response to the same stimulus may be present. For instance, by past experience one has learned that the hot iron causes an unpleasant sensation. Although by lesion of the dorsal root he does not sense the burning sensation, he may remove his hand or leg from the hot iron based on this information stored in the memory center of his brain.

If instead of the dorsal root, the ventral root were severed, the injury would be inflicted on the motor neurons. In this instance, the rat would sense the stimulation of the hot iron, but would find itself unable to withdraw the leg since the impulses generated by the motor neurons cannot reach the effector muscles of the leg. However, the rat may demonstrate other responses, for example, it may squeak in pain.

• PROBLEM 20-18

Describe and give an example of a reflex arc.

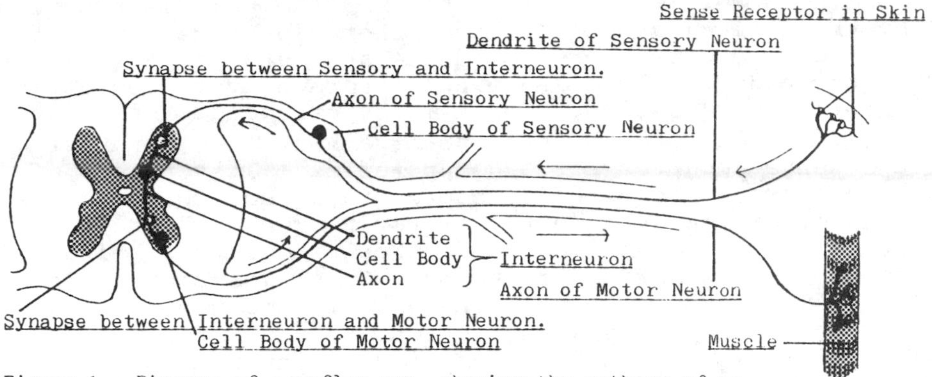

Figure 1. Diagram of a reflex arc, showing the pathway of an impulse, indicated by arrows.

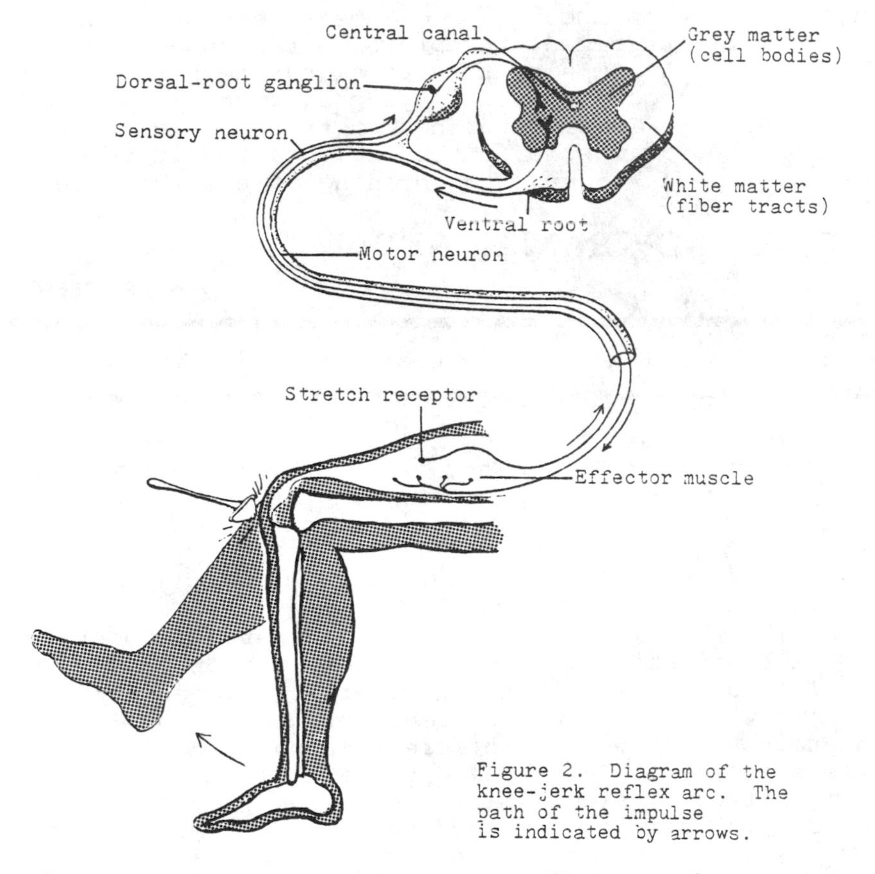

Figure 2. Diagram of the knee-jerk reflex arc. The path of the impulse is indicated by arrows.

Solution: To understand what a reflex arc is, we must know something about reflexes. A reflex is an innate, stereotyped, automatic response to a given stimulus. A popular example of a reflex is the knee jerk. No matter how many times we rap on the tendon of a person's knee cap, his leg will invariably straighten out. This experiment demonstrates one of the chief characteristics of a reflex: fidelity of repetition.

Reflexes are important because responses to certain stimuli have to be made instantaneously. For example, when we step on something sharp or come into contact with something hot, we do not wait until the pain is experienced by the brain and then after deliberation decide what to do. Our responses are immediate and automatic. The part of the body involved is being withdrawn by reflex action before the sensation of pain is experienced.

A reflex arc is the neural pathway that conducts the nerve impulses for a given reflex. It consists of a sensory neuron with a receptor to detect the stimulus, connected by a synapse to a motor neuron, which is attached to a muscle or some other tissue that brings about the appropriate response. Thus, the simplest type of reflex arc is termed monosynaptic because there is only one synapse between the sensory and motor neurons. Most reflex arcs include one or more interneurons between the sensory and motor neurons (see Figure 1).

An example of a monosynaptic reflex arc is the knee jerk. When the tendon of the knee cap is tapped, and thereby stretched, receptors in the tendon are stimulated. An impulse travels along the sensory neuron to the spinal cord where it synapses directly with a motor neuron. This latter neuron transmits an impulse to the effector muscle in the leg, causing it to contract, resulting in a sudden straightening of the leg (see Figure 2).

SPINAL AND CRANIAL NERVES

• PROBLEM 20-19

What are spinal nerves and what are their functions?

Solution: Spinal nerves belong to the peripheral nervous system (PNS). They arise as pairs at regular intervals from the spinal cord, branch, and run to various parts of the body to innervate them. In human beings, there are 31 symmetrical pairs of spinal nerves. The size of each spinal nerve is related to the size of the body area it innervates.

All the spinal nerves are mixed nerves, in the sense that all have both motor and sensory components in roughly equal amounts. They contain fibers of the somatic nervous system, which are both afferent (conducting impulses towards the nervous system) and efferent (conducting impulses to effector organs). They also contain fibers of the autonomic nervous system, which are uniquely efferent and are separable into parasympathetic and sympathetic nervous systems. Because of this nature, the spinal nerves function to convey messages from the external environment to the central nervous system, and from the central nervous system to various effectors of the body. In other words, the spinal nerves, as part of the peripheral nervous system, serve as a link between the central nervous system and the effector organs.

Somatic fibers of the spinal nerves innervate skeletal muscles of the body, and are under voluntary

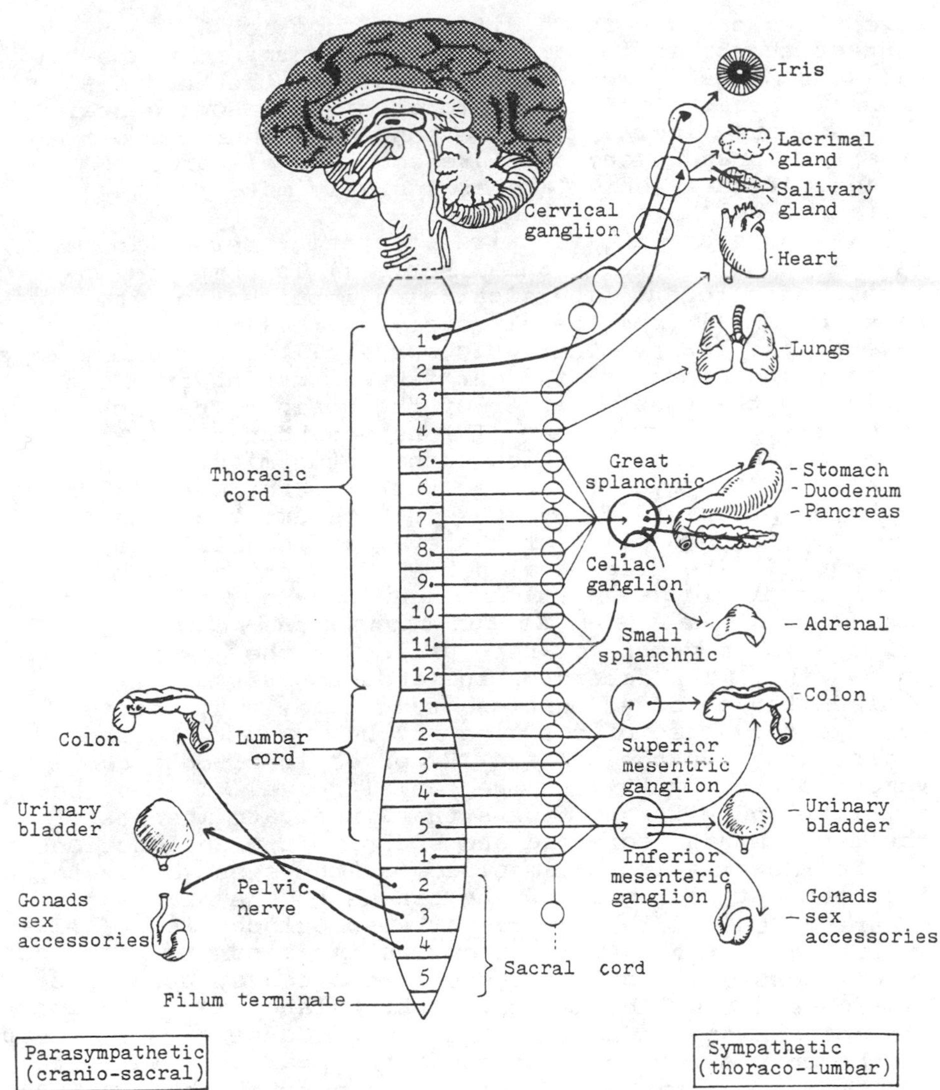

The spinal nerves and their autonomic fibers. The somatic fibers leave the spinal cord together with the autonomic fibers and then seperate.

regulation. We can bend our arm or leg at will. The autonomic nerves innervate the smooth and cardiac muscles and glands of the body, and cannot be voluntarily controlled. We cannot speed up our stomach contractions or heart beat at will. The autonomic nerve fibers leave the spinal cord and run for a certain distance with the somatic fibers in the spinal nerves. Then the two types of fibers diverge and run to their respective body areas which they innervate.

• PROBLEM 20-20

There are twelve pairs of cranial nerves in man. Give the type(s) of fibers found in each and briefly discuss the function(s) of each of them.

Solution: Cranial nerves leave the brain at different regions and serve different functions. They connect the brain to various effector organs, primarily the sense organs, muscles and glands of the head. Some cranial nerves contain only motor fibers, some contain only sensory fibers, and some contain both types of fibers. In addition, some are composed of parasympathetic fibers (which are exclusively motor), the action of which is involuntary.

There are twelve pairs of cranial nerves in man. The olfactory nerve (cranial nerve I) is composed entirely of afferent fibers carrying impulses for the sense of smell from the olfactory epithelium to the base of the brain. The optic nerve (II) is also entirely sensory and contains afferent fibers running from the retina to the visual center of the brain. The occulomotor (III) and trochlear (IV) nerves both have afferent and efferent branches which connect the midbrain to the muscles of the eye. Together they are responsible for proprioception of the eye muscles, movement of the eyeball, accomodation of the eye, and constriction of the pupil. The trigeminal nerve (V) contains both afferent and efferent fibers running between the pons and the face and jaws. It functions mainly in stimulating movement of the muscles of the jaws involved in chewing. The fibers of the abducens nerve (VI) run between the pons and muscles of the eye. This nerve conveys the sense of position of the eyeball to the brain, and aids the III and IV nerves in effecting movement of the eyeball. The facial nerve (VII) has both afferent and efferent fibers that innervate muscles of the face, mouth, forehead and scalp. It also functions in the transmission of impulses for the sense of taste from the anterior part of the tongue to the brain. The fibers of the auditory nerve or vestibulocochlear (VIII), exclusively afferent, run from the inner ear to the junction of the pons and medulla. These fibers convey the senses of hearing and equilibrium (movement, balance and rotation) to the appropriate centers in the brain. Connecting the medulla to the epithelium and muscles of the pharynx, and to the salivary gland and tongue are the fibers of the glossopharyngeal nerve (IX). This nerve is responsible for the sense of taste from the posterior part of the tongue and from the lining of the larynx. It is also responsible for the reflexive act of swallowing. The tenth cranial nerve, the vagus, has both sensory and motor branches, but the motor fibers are parasympathetic autonomic fibers and thus their action is involuntary. Its sensory fibers originate in many of the internal organs - lungs, stomach, aorta, larynx, to name a few - and its motor (parasympathetic) fibers run to the heart, stomach, small intestine, larynx and esophagus. Besides the vagus, the III, VII, and IX nerves also contain parasympathetic fibers, but in smaller amounts. The efferent and afferent branches of the accessory nerve (XI) connect the medulla to the pharynx and larynx and muscles of the shoulder, which they innervate. Afferent fibers of the last cranial nerve, the hypoglossal, convey the sense of proprioception from the tongue to the medulla, and the efferent fibers stimulate movement of the tongue.

The cranial nerves are numbered anteriorly to posteriorly. I and II originate in the olfactory epithelium and retina respectively. III and IV originate in the midbrain; V through VIII (vestibular branch) originate in the pons; and VIII (cochlear branch) through XII originate in the medulla.

THE AUTONOMIC NERVOUS SYSTEM

• PROBLEM 20-21

What is meant by the autonomic nervous system? What are its subdivisions and to which of the large divisions of the entire nervous system does it belong?

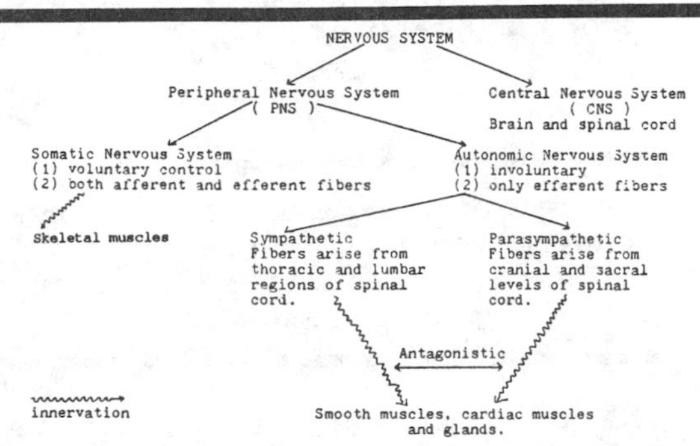

Summary of the divisions and subdivisions of the mammalian nervous system.

Solution: Nerves from the central nervous system (CNS) that innervate the cardiac muscles, smooth muscles and secretory glands form the autonomic nervous system. Structural and physiological differences within the autonomic nervous system are the basis for its further subdivision into sympathetic and parasympathetic systems. Nerves of these two divisions leave the central nervous system at different regions. The sympathetic nerves emerge from the thoracic and lumbar regions of the spinal cord, and the parasympathetic nerves arise from the brainstem and sacral portions of the spinal cord. Thus the sympathetic system is known as the thoracolumbar outflow (originating in T_1-T_{12} and L_1-L_2), while the parasympathetic system is referred to as the craniosacral outflow (originating in cranial nerves III, VII, IX, and sacral nerves 2-4).

The autonomic nervous system contains only motor nerves and is distinguished from the rest of the nervous system by several characteristics. There is no voluntary control by the cerebrum over these nerves. We cannot control our heart beat or the action of the muscles of the stomach or intestines. Another important characteristic of the autonomic nervous system is that most internal organs receive a double set of fibers, one set belonging to the sympathetic system and the other to the parasympathetic system. Impulses from the sympathetic and parasympathetic nerves always have antagonistic effects on the organs innervated. Thus if one functions to increase a certain activity, the other functions to decrease it. However, many tissues/cells do not receive parasympathetic innervation (i.e. fat cells, most arterioles, spleen).

The autonomic nervous system is part of a larger unit called the peripheral nervous system. The peripheral nervous system contains two types of nerve fibers: the afferent fibers which convey information from receptors in the periphery to the CNS, and the efferent fibers which carry information from the CNS to the effectors. Effectors are tissues or organs which bring about appropriate responses to certain stimuli, both internal (such as from the brain) and external (such as from the environment). Some examples of effectors are the skeletal muscles, cardiac and smooth muscles, and secretory glands. The peripheral nervous system can be divided functionally into two parts. That part which innervates the skeletal muscles is known as the somatic nervous system, and that which innervates smooth muscle, cardiac muscle and glands is the autonomic nervous system.

• PROBLEM 20-22

Homeostatic regulation of many body functions is not achieved by the central nervous system but by the autonomic nervous system. Explain how the autonomic nervous system can bring about such regulation.

Solution: The autonomic nervous system is divided in two parts, both structurally and functionally. One part is called the sympathetic nervous system and the other is known as the parasympathetic nervous system. These two branches act antagonistically to each other. If one system stimulates an effector, the other would inhibit its action. The basis for homeostatic regulation by the autonomic nervous system lies in the fact that the sympathetic and parasympathetic systems each sends a branch to the same organ, causing the phenomenon of double innervation.

Double innervation by the autonomic nervous system, together with the action of the endocrine system, is the basis for maintaining homeostasis inside the body. If through the stimulation by one system of an organ, a substance or action is produced excessively, then the other system will operate to inhibit the same organ, thus reducing the production of that substance or inhibiting that action. This is basically how the internal condition of the body is kept constant.

An example will illustrate the above. Both the sympathetic and parasympathetic systems innervate the heart. The action of the former strengthens and accelerates the heart beat while the latter weakens and slows the same. When a person is in fright, his heart beat involuntarily quickens owing, in part, to stimulation by the sympathetic system. To regain the normal state, the sympathetic system is overridden by the parasympathetic system which decelerates the heart beat. In the normal condition, the heart receives impulses from the two antagonizing systems, and it is through a balance of the two that the proper rate of heart beat is maintained.

• PROBLEM 20-23

List the antagonistic activities of the sympathetic and parasympathetic systems.

Solution: Below is tabulated the antagonistic activities of the sympathetic and parasympathetic systems.

Organs innervated	Sympathetic action	Parasympathetic action
Heart	Strengthens and speeds up heart beat	Weakens and slows down heart beat
Arteries	constricts lumen and raises blood pressure	No innervation
Digestive tract	slows peristalsis	speeds peristalsis
Digestive glands	decreases secretion	increases secretion
Urinary bladder	relaxes bladder	constricts bladder
Bronchial muscles	dilates passages, facilitating breathing	constricts passages
Muscles of iris	dilates pupil	constricts pupil
Muscles attached to hair	causes erection of hair	No innervation
Sweat glands	increases secretion	No innervation

In general, the sympathetic system produces the effects which prepare an animal for emergency situations, such as quickening the heart and breathing rates and dilating the pupil. These alert responses are together termed the fight-or-flight reactions. The parasympathetic systems reverse the fight-or-flight responses to restore an animal to a calm state.

• PROBLEM 20-24

Besides their actions, the sympathetic and parasympathetic systems also differ in the neurotransmitter they release. Explain.

Solution: Nerves of the parasympathetic system secrete a neurotransmitter called acetylcholine. For this reason they are usually referred to as cholinergic neurons. Acetylcholine is also the transmitter at the neuromuscular junction. Nerves of the sympathetic system release noradrenaline, also called norepinephrine, and are thus noradrenergic.

Acetylcholine is a strong base, containing a choline moiety. $[-CH_2CH_2-\overset{+}{N}-(CH_3)_3]$

In exists as a cation (positive ion) at physiological pH (about 7.4). Because of its ability to attach to a membrane and create a reversible change in the membrane's permeability to different ions, acetylcholine released by the presynaptic neuron acts to bring about depolarization and generate an impulse in the postsynaptic neuron. The molecular structure of acetylcholine is:

$$CH_3-\overset{O}{\overset{\|}{C}}-CH_2-CH_2-\overset{+}{N}-(CH_3)_3$$

Noradrenaline has a molecular structure containing a ring moiety:

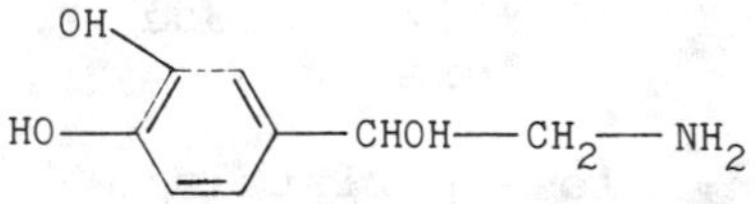

$$HO-C_6H_3(OH)-CHOH-CH_2-NH_2$$

The closely related compound adrenaline (epinephrine) -

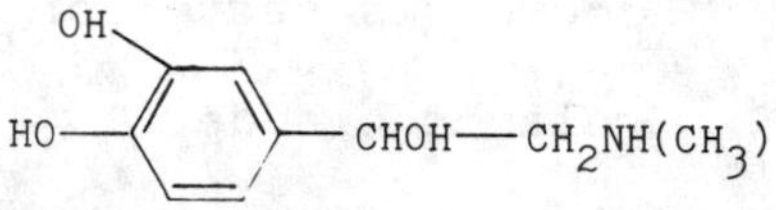

$$HO-C_6H_3(OH)-CHOH-CH_2NH(CH_3)$$

is a hormone released by the adrenal glands, and although its action is similar to that of noradrenaline, it is not a neurotransmitter. The adrenal medulla (inner part of the adrenal gland) acts like a modified post-ganglionic sympathetic "neuron": it is derived embryologically from neural tissue; it is innervated by a sympathetic preganglionic fiber; and it has become specialized to secrete noradrenaline and adrenaline and thus prolongs sympathetic activation.

• PROBLEM 20-25

Adrenalin is a hormone which stimulates the sympathetic system. What effects would you expect adrenaline to have on (a) the digestive tract, (b) the iris of the eye, and (c) the heart beat?

Solution: The sympathetic nervous system is comprised of nerves coming out of the spinal cord at the thoracic and lumbar regions. After each neuron (the preganglionic neuron) leaves the spinal cord, it synapses with another neuron (the postganglionic neuron). It is the post-ganglionic neuron that innervates the effectors supplied by the sympathetic system.

(a) The action of the sympathetic nerves on the digestive tract is to slow peristalsis and reduce secretion to decrease the rate of digestion. Adrenaline, which stimulates the sympathetic system will hence cause peristalsis and digestion to slow down.

(b) The iris of the eye consists of two sets of smooth muscle - circular and radial. Sympathetic stimulation results in dilation by causing the radial muscles to contract. Adrenaline, which excites the sympathetic system, produces exactly this effect.

(c) Sympathetic stimulation to the heart is twofold: it stimulates the pacemaker to increase heart rate and it stimulates the myocardium (muscular layer of heart wall) to beat more forcefully. Thus adrenaline has the effect of accelerating and strengthening the heartbeat.

• PROBLEM 20-26

Pilocarpine is a drug that stimulates the nerve endings of parasympathetic nerves. What effects would you expect this drug to have on (a) the digestive tract, (b) the iris of the eye, and (c) the heart rate?

Solution: The parasympathetic system consists exclusively of motor fibers originating from the brain and emerging via the third, seventh, ninth and tenth cranial nerves, and of fibers originating from the sacral region of the spinal cord and emerging by way of the spinal nerves in that region. The parasympathetic system therefore has nerves from two separate regions - the brain and the sacral region of the spinal cord.

(a) The ninth cranial nerve, the glossopharyngeal, innervates the salivary gland. It stimulates the salivary gland to release digestive enzymes. A more important cranial nerve, the vagus, sends branches to the stomach, duodenum and the pancreas. This nerve enchances motility and peristalsis of the digestive tract. (Motility and peristalsis of the digestive tract should not be confused with each other. Motility of the tract causes food to mix and churn but no movement of it. Peristalsis moves the food down the digestive tract). It also stimulates the secretion of digestive enzymes by digestive glands. For example, stimulation of the pancreas by the vagus nerve causes secretion of the pancreatic juice. Thus the glossopharyngeal and vagus together act to promote digestion. When pilocarpine is present, the parasympathetic fibers are stimulated. The result is that digestion is facilitated or enhanced.

(b) The third cranial nerve supplies the muscles of the iris which controls the size of the pupil of the eye. Excitation of this nerve causes constriction of the pupil, by contraction of the circular smooth muscle. Since pilocarpine excites the parasympathetic system, its presence would cause the pupil to reduce in size.

(c) The parasympathetic branch supplying the heart is part of the vagus nerve. Its action is to weaken and slow down the heart beat. Pilocarpine introduced into the blood stream would activate the parasympathetic nerve endings, causing the heart beat to slow and weaken.

NEURONAL MORPHOLOGY

• PROBLEM 20-27

What is the structure of a typical neuron?

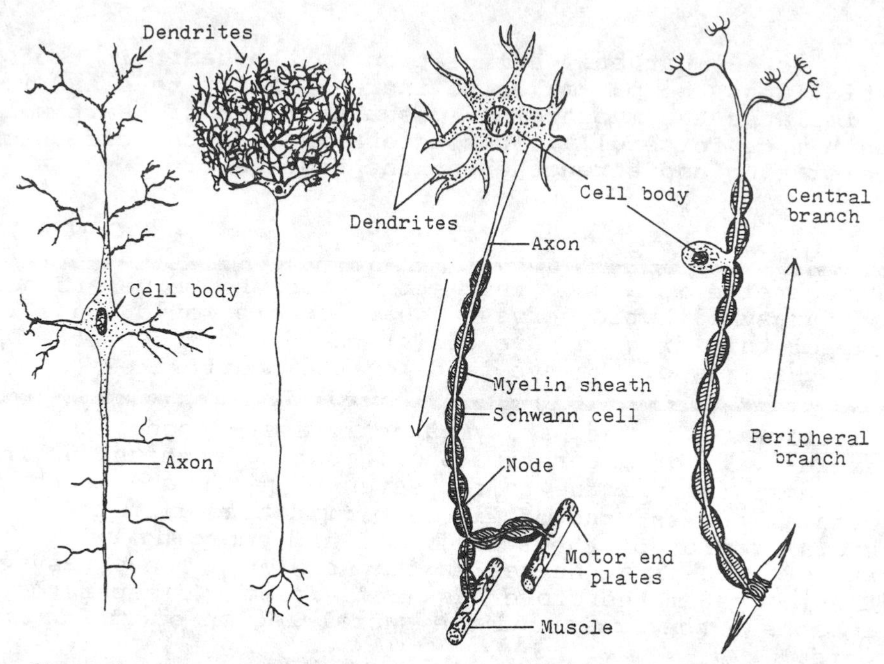

Figure 1. A variety of neuron types in human beings.

Solution: Before we get to the answer, it is important to recognize the fact that the term 'typical neuron' is rather vague. Neurons, which are the basic structural and functional units of the nervous systems of multicellular animals, show a great diversity in types. In humans, many types of neurons are present (Figure 1). Nevertheless, often three parts of a neuron can be distinguished: a cell body, an axon, and a group of processes called dendrites.

Dendrites are usually rather short and numerous extensions from the cell body. They frequently branch profusely, and their many short terminals may give them a spiny appearance. When stained, they ordinarily show many dark granules. There is usually only one axon per neuron (very rarely two), and it is frequently longer than the dendrites. It may branch extensively, but unlike dendrites it does not have a spiny appearance and does not show dark granules when stained. The most fundamental distinction between dendrites and axons is that dendrites receive excitation from other cells whereas axons generally do not, and that axons can stimulate other cells whereas dendrites cannot. Thus, dendrites carry information to the cell body while axons carry information away from the cell body.

Axons may be several feet long in some neurons. A bundle of many axons wrapped together by a sheath of connective tissue is what we commonly call a nerve. Each vertebrate axon is usually enveloped in a myelin sheath formed by special cells, the Schwann cells, that almost completely encircle the axon. Schwann cells play a role in the nutrition of the axons, and provide a conduit within which damaged axons can grow from the cell body back to their original position. The myelin sheath is interrupted at regular intervals; the interruptions are called nodes of Ranvier. At these nodes, the myelin sheath disappears. The

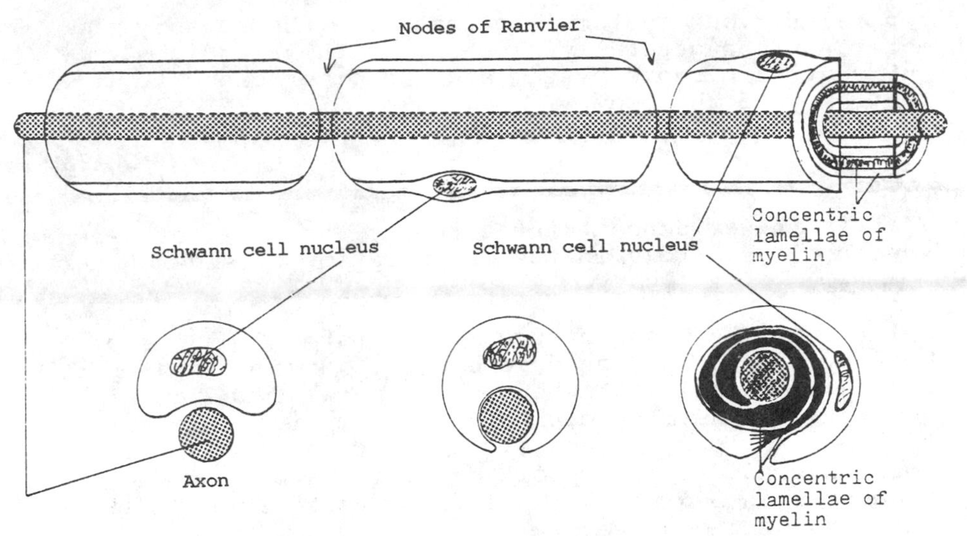

Figure 2 Sheath cells on a neuron. (upper) Dissection of a myelinated nerve fiber. (lower) Envelopement of axis cylinder by a sheath cell.

myelin functions in speeding up the transmission of impulses in the axon it envelopes. It is crucial to note that the myelin sheath is not a separate layer by itself. Through electron microscopy, it has been proved that the sheath is not a secretion product of the axon or Schwann cells, as was once believed, but a tightly packed spiral of the cell membrane of the Schwann cells. Thus the sheath is composed of the lipid from the membrane's bilayer. The nucleus and cytoplasm of the Schwann cell are pushed aside to form the neurolemma. The nodes in the myelin sheath are simply the points at which one Schwann cell ends and another begins (see Figure 2).

The neuronal cell body, or perikaryon, contains the nucleus and cytoplasmic organelles distributed around the nucleus. The perikaryon has well-developed endoplasmic reticulum and Golgi apparatus for manufacturing all the substances needed for the maintenance and functioning of the axon and dendrites.

• PROBLEM 20-28

How many kinds of neurons are there? What functions are performed by each type?

Solution: Neurons are classified on the basis of their functions into sensory, motor or interneurons. Sensory neurons, or afferent neurons, are either receptors that receive impulses or connectors of receptors that conduct information to the central nervous system (brain and spinal cord). Motor neurons, or efferent neurons, conduct information away from the central nervous system to the effectors, for example, muscles and glands. Interneurons, which connect two or more neurons, usually lie wholly within the central nervous system. Therefore, they have both axonal and dendritic ends in the central nervous system. In contrast, the sensory and motor neurons generally have one of their endings in the central nervous system and the other close to the periphery of the body.

• PROBLEM 20-29

What is the evidence for the theory that the neurilemma sheath plays a role in the regeneration of severed nerves?

Solution: When an axon is separated from its cell body by a cut, it soon degenerates. However, the part of the axon still attached to the perikaryon can regenerate. A healthy perikaryon is important in regeneration. As long as the cell body of the neuron has not been injured, it is capable of making a new axon. Regeneration begins within a few days following the severing of the nerve.

Axon regeneration is believed to involve the neurilemma, a cellular sheath composed of Schwann cells which envelopes the axon. The role the neurilemma sheath plays in regeneration is to provide a channel for the axon to grow back to its former position. What happens is that the growing axon enters the old sheath tube and proceeds along it to its final destination in the central nervous system or periphery.

In some experiments, the neurilemma sheath is removed and replaced by a conduit, for example, a section of blood vessel or extremely fine plastic tube. The severed axon is able to regenerate normally within the substituted conduit. This result shows that the sheath is not an absolute requirement for the regeneration of an axon in vitro, since a plastic sheath serves the same function as well.

• PROBLEM 20-30

A myelinated nerve can conduct impulses of velocities of up to 100 meters per second, whereas unmyelinated nerve conducts at velocities of 20 to 50 meters per second. Explain this difference.

Solution: The big discrepancy between the rates of conduction of a myelinated and an unmyelinated nerve lies in the morphology of the nerves. Myelinated nerves are bundles of axons, each enveloped in a sheath of fatty substance, known as myelin. Axons of unmyelinated nerves have no myelin sheath, and are for this reason sometimes referred to as bare axons.

A myelin sheath around a neuron is a highly insulating covering which prevents the flow of ions between the cell interior and the exterior of the membrane. Hence no action potential will be generated at the myelinated regions of an axon. More importantly, the insulating sheath prevents the loss of charges when Na^+ ions flow from one region to another.

At successive nodes of Ranvier there are gaps in the sheath where there is no insulation. At these points, free ionic communication between the inside and outside of the membrane is possible. Thus, when Na^+ ions arrive at the gaps, the membrane is depolarized and an action potential is set off. Therefore, impulses are generated only at the nodes, and nerve impulses leap from one node to the next. This kind of conduction, by "jumping" from node to node, is called saltatory transmission. The "jumping" is responsible for the more rapid impulse conduction by myelinated nerves, as compared with unmyelinated nerves. The latter lack the extensive insulation that makes saltatory conduction possible.

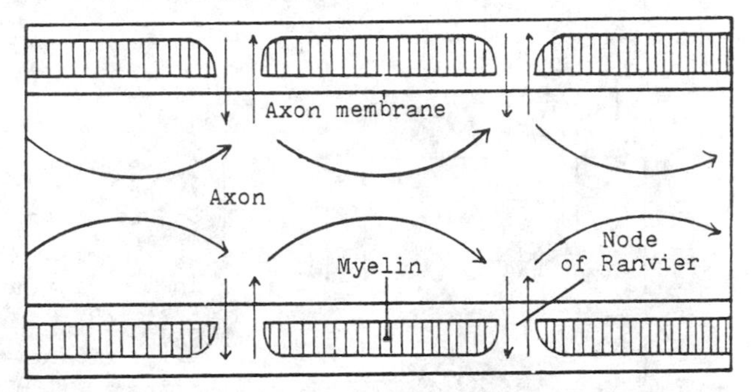

In myelinated fibers the key permeability changes take place at the nodes where the insulationg myelin sheath is interrupted. Between the nodes, the flow of Na^+ ions carries the impulse which "jumps" from node to node along the axon.

In myelinated fibers the key permeability changes take place at the nodes where the insulating myelin sheath is interrupted. Between the nodes, the flow of Na^+ ions carries the impulse which 'jumps' from node to node along the axon.

THE NERVE IMPULSE

• PROBLEM 20-31

What is a resting potential? Describe the chemical mechanism responsible for the resting potential. How can a resting potential be detected?

Solution: There is a difference in electrical potential between the inside and the outside of all living cells. For example, the potential difference across the plasma membrane of the neuron is measured to be about 60 millivolts, the inside being negative with respect to the outside. This potential difference is called the resting potential.

The chemical basis for the resting potential is as follows (refer to Figure). By active transport, an energy-

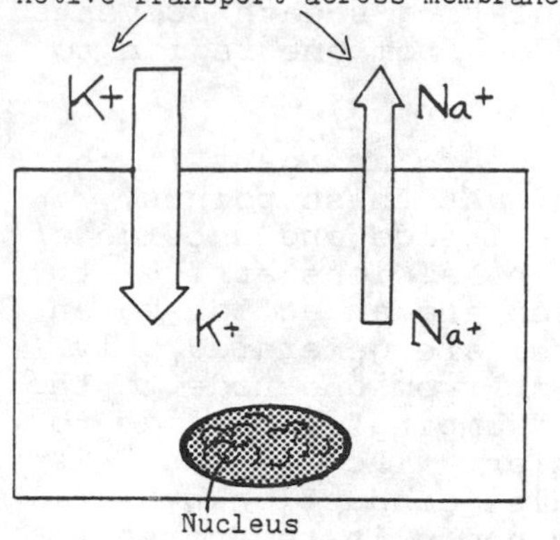

1. K^+ ions are actively transported into the cell while Na^+ ions are actively transported out of the cell.

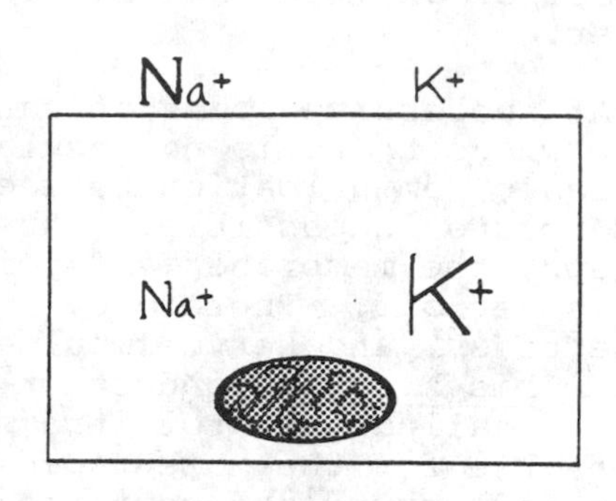

2. This leads to an accumalation of K^+ ions inside and Na^+ ions outside the cell.

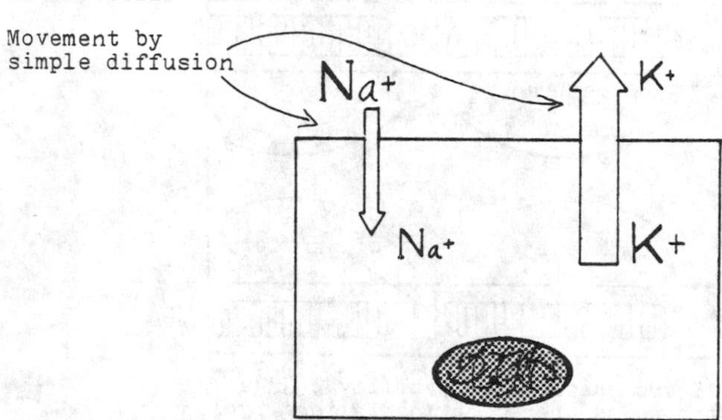

3. Permeability differences of K^+ ions and Na^+ ions across the membrane result in the net negative charge inside and the net positive charge outside the cell. This gives rise to the resting potential.

Diagrams showing the chemical mechanism responsible for the resting potential. Sizes of letters and arrows represent relative amounts present.

requiring process that transfers substances across the plasma membrane against their concentration gradients, the concentration of potassium (K^+) ions is kept higher inside the cell than outside. At the same time there is a lower concentration of sodium (Na^+) ions in the cell interior than the exterior. Moreover, in the resting state the permeability of the plasma membrane is different to K^+ and Na^+ ions. The membrane is more permeable to K^+ ions than to Na^+ ions. Hence, K^+ ions can move across the membrane by simple diffusion to the outside more easily than Na^+ ions can move in to replace them. Because more positive charges (K^+) leave the inside of the cell than are replenished (by Na^+), there is a net negative charge on the inside and a net positive charge on the outside. **An electrical potential is established across the membrane. This potential is the resting membrane potential.**

We can measure the resting potential by placing one electrode, insulated except at the tip, inside the cell and a second electrode on the outside surface and connecting the two with a suitable recording device such as a sensitive galvanometer. The reading on the galvanometer should be approximately 60 millivolts if the cell tested is a neuron. (Different types of cells, such as skeletal and cardiac muscle cells, vary in their values of resting potential.) Note, however, that if both electrodes are placed on the outside surface of the cell, no potential difference between them is registered because all points on the outside are at equal potential. The same is true if both electrodes are placed on the inside surface of the plasma membrane.

• PROBLEM 20-32

What is an action potential? Discuss the physical and electrochemical changes during an action potential.

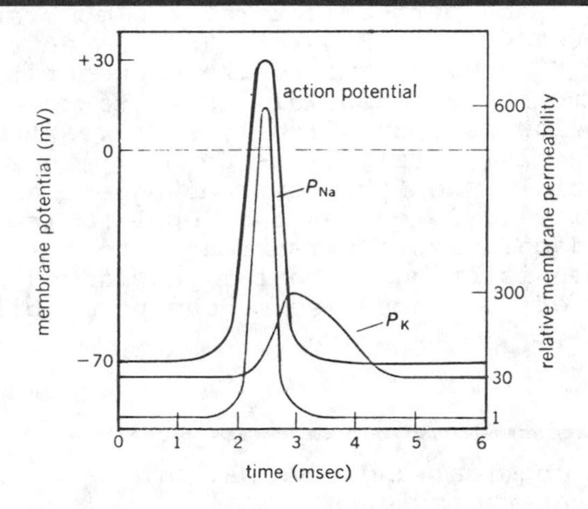

Changes in membrane permeability to sodium and potassium ions during an action potential.

Solution: We know that an unstimulated nerve cell exhibits a resting potential of about 60 millivolts across its membrane. An active pumping mechanism within the membrane causes the cell interior to accumulate a high concentration of K^+ ions and the exterior a high concentration of Na^+ ions. Since the resting membrane is 50 to 75 times more permeable to K^+ than to Na^+, more K^+ moves by simple diffusion out of the cell than Na^+ moves into the cell in the resting state, and hence a potential difference exists across the membrane. Diffusion moves ions down their concentration gradients, which were established by the active transport pump.

When a region of the axonal membrane is stimulated by a neurotransmitter, that region undergoes a set of electrochemical changes that constitutes an action potential. An action potential can be detected as a spike on a recording device, with a rising phase, a peak, and a falling phase, each corresponding to the characteristic flow of ions across the membrane of the axon. The duration of an action potential in a neuron is about 1 msec (.001 sec).

During an action potential, the permeability of the membrane to Na^+ and K^+ ions is markedly altered. Initially, the membrane permeability to Na^+ undergoes a thousand-fold increase, whereas that to K^+ remains relatively unchanged. Consequently Na^+ ions rush into the cell. If one inserts an electrode into this region of the axon, one will find that the potential at the inside of the membrane starts to rise, as the influx of positive (Na^+) ions reduces the negativity of the cell interior. Soon, the inside of the membrane registers a net positive charge. The positivity rises until a peak is reached at about + 40 millivolts. This phase of the action potential, when the membrane potential approaches or even rises above zero, is known as the depolarizing (rising) phase. At the peak of the action potential the increased sodium permeability is rapidly turned off, and immediately following this the permeability of the membrane to K^+ suddenly increases (see Figure). Sodium entry stops and an efflux of K^+ results due to the concentration gradient of K^+. The membrane potential starts to move toward zero, then drops below zero and finally restores the pre-excitation, or resting, state (at - 60 mV). This phase of the action potential, when the membrane potential moves toward its resting level, is called the repolarizing (declining) phase. How depolarization is brought to a stop and repolarization is brought about is explained by a rapid shutting off of the increased sodium permeability (sodium inactivation).

• PROBLEM 20-33

What are the chemical and physical processes involved in transmission at the synapse?

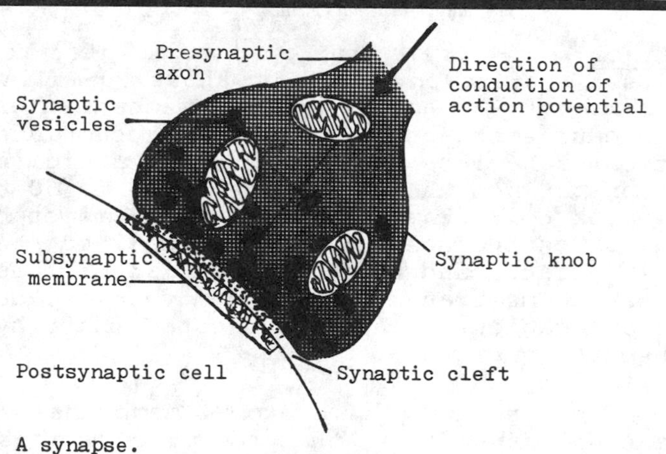

A synapse.

Solution: The nervous system is composed of discrete units, the neurons, yet it behaves like a continuous system of transmission of impulses. For this to occur, there have to be functional connections between neurons. These connections are known as synapses. A synapse is an anatomically specialized junction between two neurons, lying adjacent to each other, where the electric activity in one neuron (the presynaptic neuron) influences the excitability of the second (the postsynaptic neuron). At the synapse, the electric impulse is transformed into a chemical form of transmission.

Chemical transmission at the synapse involves the processes of neurosecretion and chemoreception. The arrival of a nerve impulse at the axon terminal stimulates the release of a specific chemical substance, which has been synthesized in the cell body and stored in the tip of the axon, into the narrow synaptic space between the adjacent neurons. This process constitutes neurosecretion. The chemical secreted, known as neurotransmitter, can cause local depolarization of the membrane of the postsynaptic region and thus transmit the excitation to the adjacent neuron. Chemoreception is the process in which the neurotransmitter becomes attached to specific molecular sites on the membrane of the dendrite (postsynaptic region), producing a change in the properties of the cell membrane so that a new impulse is established.

The chemical transmitter, say acetylcholine, passes from the presynaptic axon to the postsynaptic dendrite by simple diffusion across the narrow space, called the synaptic cleft, separating the two neurons involved in the synapse. The synaptic clefts have been measured under the electron microscope to be about 200 Å in width. Diffusion is rapid enough to account for the speed of transmission observed at the synapse. After the neurotransmitter has exerted its effect on the postsynaptic membrane, it is promptly destroyed by an enzyme called cholinesterase. This destruction is of critical importance. If the acetylcholine were not destroyed, it would continue its stimulatory action indefinitely and all control would be lost.

• PROBLEM 20-34

Certain nerve gases are known to cause a breakdown of cholinesterase. How would these gases affect the human body and why?

Solution: Acetylcholine is a chemical responsible for transmitting impulses across synaptic junctions. If the supply of acetylcholine is continuous, repeated stimulation of the postsynaptic neuron will result. Hence, this neurotransmitter has to be destroyed after performing its function. The substance acetylcholinesterase specifically breaks down acetylcholine by splitting off the choline moiety from the molecule, thus inactivating it.

Certain nerve gases, when inhaled, act to destroy acetylcholinesterase. This action leads to serious consequences. Following the breakdown of acetylcholinesterase, acetylcholine would not be cleaved but would remain in the synaptic cleft and continue to stimulate the postsynaptic neuron. If this neural pathway leads to a muscle, continual excitation of the neuron would produce sustained contraction of that muscle. The person involved may, as a result, enter a state of tremors and spasms, or may even die depending on the quantity of gas inhaled.

• PROBLEM 20-35

It is observed that the chemical transmission occurring at a synapse is unidirectional that is, the neurotransmitter always travels from the presynaptic axon to the postsynaptic dendrite. Explain why.

Solution: Because there are physical separations (synaptic clefts) between neurons, an impulse traveling down the axon of one neuron must first be converted from an electrical form into a chemical form at the synapse, which can diffuse across the physical separation. This chemical form of transmission utilizes a chemical substance secreted by the presynaptic axon terminal, known as a neurotransmitter. This neurotransmitter has the ability of changing the permeability properties of the postsynaptic cell membrane, causing it to depolarize and an action potential to be generated. The action potential then propagates down the axon of the postsynaptic neuron, forming a wave of depolarization. This new impulse, when it reaches the axon terminal, again can induce a neurotransmitter to be released at the new synapse, and subsequently another neuron can be excited. Hence, by this mechanism, there can be continuous conveyance of neural passages by neurons, despite the presence of physical disconnections between them.

The chemical transmission that occurs at a synapse is unidirectional since the neurotransmitter is present in the axonal end of a neuron only. This chemical is contained within small, sac-like structures called synaptic vesicles. The synaptic vesicles fuse with the membrane of the axon, induced by an electric impulse that reaches the axon terminal. They then break open, discharging the contents into the synaptic cleft. Because of the fact that no synaptic vesicle is present inside the dendrites of a neuron, chemical transmission is always from the axonal end to the dendritic terminal, and not the reverse.

• PROBLEM 20-36

An axon can conduct impulses in both directions simultaneously. However, an impulse normally propagates down an axon in one direction only, either axon to dendrite or dendrite to axon. Explain how this unidirectional conduction of impulses along the neuron is possible.

Solution: Normally, an impulse travels from the dendrites to the cell body, then along the axon to the synaptic junction. However, impulses can be made to travel along an axon to the dendrites. But these impulses are not relayed to the next neuron since dendrites do not contain the chemical substance (neurotransmitter) required in the transmission of an impulse across a synapse.

Although two directions of transmission along an axon are potentially possible, once an impulse is started

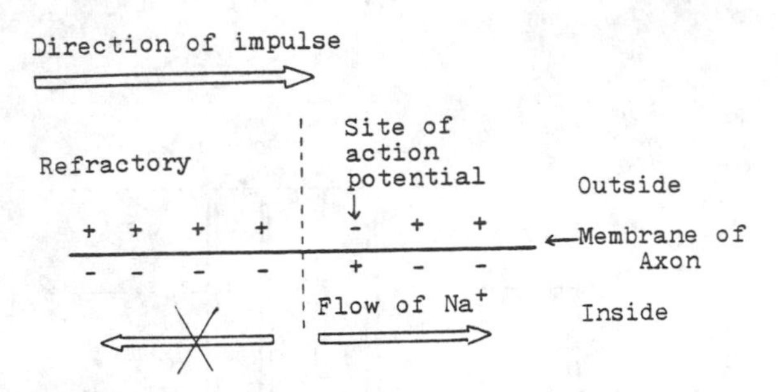

Figure. The impulse is traveling from left to right. Na^+ ions can potentially flow in both directions from the point of origin. However, the region that has just undergone an action potential cannot undergo another since it is in the refractory period. Hence the depolarization of a membrane can only occur to the right, indicating the unidirectionality of the flow of impulse.

at either end of a neuron, it will pass on through the neuron toward the opposite end. This phenomenon is due to the existence of a recovery period, known as the refractory period in the neuronal membrane. After a region of the axon is stimulated, a refractory period must elapse before it can be excited again. To understand the relation between the refractory period and the one-way conduction of impulses, one must recall that a region undergoing an action potential is positive with respect to other points inside the axon. Since that region is of higher potential than the surrounding region, a flow of Na^+ ions spreads from that local region to the adjacent regions on both sides of it. This is confirmed by observations which indicate that within a neuron, an impulse can travel in either direction from its point of origin. However, because the region that has just been stimulated cannot be made to depolarize until it has recovered, the impulse moves in one direction only (see Figure). By the time that region has regained the ability to be depolarized, the impulse has traveled too far away from it, and hence cannot depolarize it. It should be remembered that the unidirectional propagation of impulses is determined by the stimulus location (i.e., at which end the stimulus is given) rather than an intrinsic inability of the axon to conduct in the opposite direction. If action potentials are initiated near the middle of a neuron, impulses propagate from this region toward the two ends. In most nerves, however, action potentials are initiated at one end of the cell and impulses therefore propagate in only one direction toward the other end of the cell.

• PROBLEM 20-37

Explain the following statements.

(1) An impulse will be initiated only when the stimulus is of a certain intensity.

(2) Although a neuron exhibits all-or-none response, an animal is able to detect different degrees of intensity of a stimulus.

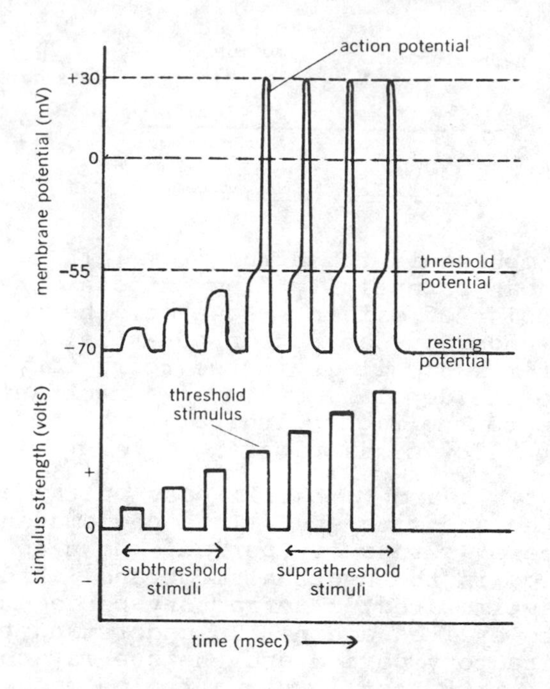

Decreasing membrane potential with increasing strength of depolarizing stimulus. When the membrane potential reaches the threshold potential, action potentials are generated. Increasing the stimulus strength above the threshold does not alter the action-potential response.

Solution: The two statements above come from experiments done on an isolated nerve. Two electrodes are placed at points several centimeters apart on the surface of a nerve. The electrodes are connected to recording devices, which are able to detect any electrical changes that may occur at the points on the nerve the electrodes are in contact with. An extremely mild electrical stimulus is applied to the nerve initially. The recording device shows no change. Next, a slightly more intense stimulus is applied, and still there is no change. The intensity of the stimulus is increased further and this time the device records an electrical change occurring at the point of the membrane in contact with the first electrode and a fraction of a second later, a similar electric change at the point in contact with the second electrode. The nerve is now stimulated, as evidenced by the wave of electrical changes that moves down the fiber, passing first one electrode and then the other. This shows that an impulse will be generated in a nerve fiber only if the stimulus reaches a certain intensity. If a more intense stimulus is applied next, an impulse is again triggered and a wave of electrical changes moves down the nerve. However, the intensity and speed of this impulse are the same as those of the one recorded from the previous stimulation.

Based on these experiments, we are ready to account for the two statements.

(1) A potential stimulus must be above a critical intensity if it is to generate an action potential. This critical intensity is known as the threshold value, and it varies in different neurons. Physiologically, what happens is that below the threshold value, the membrane permeability to sodium ions, though increased, is still less than that of potassium ions. This means that the inside of the membrane is still too negative, and no action potential is generated. Only when the intensity of the stimulus reaches the threshold value does the membrane permeability of Na^+ ions increase so much that its inflow exceeds the potassium outflow. This inflow of sodium ions is now sufficient to depolarize the membrane potential to a critical level, known as the threshold potential. Consequently, an action potential is triggered. Once threshold depolarization is reached, the action potential is no longer dependent upon stimulus strength. Thus, action potentials generated by threshold stimuli are identical to those generated by above-threshold (or suprathreshold) stimuli.

(2) Increasing the intensity of the stimulus above the threshold value does not alter the intensity or speed of the nervous impulse produced, that is, the nerve fires maximally or not at all, a type of reaction commonly called an all-or-none response.

An animal can detect the different degrees of intensity of a stimulus in two ways. The more intense the stimulus, the more frequent are the impulses moving along the fiber. This is called the frequency code. Second, a more intense stimulus ordinarily stimulates a greater number of nerve fibers than does a weak stimulus. This is called recruitment or the population code. Hence, the frequency of firing of impulses and the number of axons stimulated can be interpreted by the brain, and the strength of the stimulus can be detected.

SHORT ANSWER QUESTIONS FOR REVIEW

Answer

To be covered when testing yourself

Choose the correct answer.

1. In accomodation, relaxation and contraction of which of the following determines the thickness of the lens? (a) Bruch's membrane (b) Ciliary muscle (c) Iris (d) Choroid plexus.

   b

2. Proprioception involves (a) coordination of the eyes and ears, (b) the ability to repeat something that is spoken, (c) the awareness of body position involving muscles, tendons, and the cerebellum, (d) the ability to develop short and long-term memory.

   c

3. Which of the following is incorrect? (a) Myopia (nearsightedness) is a condition caused by light rays converging in front of the retina, and can be corrected by using a concave lens in front of the eye, (b) Hypermetropia (farsightedness) is a condition caused by light rays converging behind the retina, and can be corrected by use of a convex lens in front of the eye. (c) Astigmatism is a condition caused by an irregularly shaped cornea. (d) Myopia and hypermetropia are both caused by a combination of irregular focusing properties and an irregularly shaped cornea.

   d

4. The fovea consists of (a) equal amounts of rods and cones, (b) more rods than cones, (c) more cones than rods, (d) no rods or cones. (e) only cones

   e

5. Which of the following gives the approximate sequence of the visual cycle? (a) Rhodopsin (cis-retinal+opsin) $\xrightarrow{\text{light}}$ lumirhodopsin (trans-retinal+opsin) → trans-retinol → cis-retinol cis-retinal+opsin → rhodopsin (b) Rhodopsin (cis-retinal+opsin) $\xrightarrow{\text{light}}$ trans-retinol → cis-retinol → lumirhodopsin (trans-retinal+opsin) cis-retinal+opsin → rhodopsin (c) Lumirhodopsin (trans-retinal+opsin) $\xrightarrow{\text{light}}$ rhodopsin (cis-retinal +opsin) → trans-retinal → cis-retinal+opsin → rhodopsin (d) trans-retinol → cis-retinol → cis-retinal+opsin → lumirhodopsin $\xrightarrow{\text{light}}$ rhodopsin (trans-retinal+opsin).

   a

6. Which of the following is <u>not</u> involved with the maintenance of equilibrium? (a) Sacculus and utriculus (b) semicircular canals (c) cochlea (d) endolymph

   c

Answer

To be covered when testing yourself

7. Which of the following is a correct explanation for the feeling of dizziness after being rotated in a swivel chair? (a) The movement of hair cells with respect to the movement of perilymph in the cochlea of the inner ear causes the hair cells to send impulses to the cerebellum. (b) The movement of hair cells with respect to the movement of endolymph in the ampulla of the inner ear causes the hair cells to send impulses to the cerebellum. (c) The movement of perilymph with respect to the movement of hair cells in the ampulla of the inner ear causes the hair cells to send impulses to the cerebrum. (d) The movement of endolymph with respect to the movement of perilymph in the ampulla of the inner ear causes the former to send impulses to the cerebellum. — b

8. If both the dorsal root and the ventral root of a peripheral nerve to the foot of a dog were severed, what would be observed upon touching the foot of the dog with a hot iron? (a) The dog would show signs of experiencing pain, but would be unable to move his foot from the iron. (b) The dog would show signs of experiencing pain, and accordingly would move his foot from the iron. (c) The dog would show no signs of experiencing pain nor would he be able to move his foot away from the iron. (d) The dog would show no signs of experiencing pain, but would be able to remove his foot away from the iron. — c

9. Which of the following statements is incorrect? (a) Fibers from the parasympathetic division of the autonomic nervous system (ANS) arise from the brainstem and sacral portions of the spinal cord, and, as an example of one of its actions, decelerate the heart rate. (b) Fibers from the sympathetic divison of the A.N.S. arise from the thoracic and lumbar regions of the spinal cord, and, as an example of one of its actions, accelerate the heart rate. (c) Both the parasympathetic and sympathetic divisions of the A.N.S. arise from all five regions of the spinal cord, and act cooperatively to accelerate the heart rate. (d) The parasympathetic division of the A.N.S. is responsible for the so-called "rest and digest" reflex, while the sympathetic division of the A.N.S. is responsible for the so-called "fight or flight" reflex. — c

10. Which of the following statements is <u>completely</u> correct? (a) Actions of the parasympathetic division of the A.N.S. include slowing of the heart rate, dilation of blood vessels, dilation

Answer

To be covered when testing yourself

of the pupils, and bladder constriction. (b) Actions of the sympathetic division of the A.N.S. include increased non-digestive glandular secretion, dilation of the pupils, decreased digestive gland secretions, and dilation of airway passages. (c) Actions of the sympathetic division of the A.N.S. include constriction of blood vessels, inhibition of peristalsis for digestion, constriction of the bladder and acceleration of the heart rate. (d) Actions of the parasympathetic division of the A.N.S. include decreased non-digestive glandular secretion, constriction of airway passages, constriction of the pupils, and acceleration of peristalsis for digestion. — d

11. Which of the following would occur after administration of an anti-cholinesterase? (a) Acetylcholine would be broken down at a faster rate and would lead to a cessation of nerve transmission. (b) Acetylcholine would not be broken down as it normally is, and nerve transmission would continue for an abnormally long period of time. (c) Acetylcholine would not be effected, but norepinephrine would be broken down at a faster rate. (d) Acetylcholine would not be broken down as it normally is, but this would have no effect on nerve transmission. — b

12. Which of the following is the correct explanation of the composition and function of the myelin sheath? (a) The myelin sheath is a secretion product of Schwann cells composed of protein. Its function is that of synthesis of neurotransmitter substance. (b) The myelin sheath is a secretion product of Schwann cells composed of carbohydrates. It functions to slow the speed of nerve conduction. (c) The myelin sheath is the coiled membrane of a Schwann cell, and is mainly composed of phospholipids. It functions to increase the speed with which a nerve impulse is transmitted. (d) The myelin sheath is composed of connective tissue and has no relationship to Schwann cells. Its function is to act as a protective covering for nerves. — c

13. Which of the following is not important for the generation of a resting potential? (a) The so-called "Na^+-K^+" pump" causes a high concentration of K^+ inside the cell and a high concentration of Na^+ outside the cell. (b) The cell membrane is more permeable to K^+ than it is to Na^+. (c) Due to the difference in permeability of the membrane, more K^+ ions leave the cell interior than Na^+ ions that enter the cell interior. (d) Cl^- ions are

	Answer
	To be covered when testing yourself
directly responsible for the negative charge inside the cell relative to the outside of the cell.	d
14. Which of the following is not involved in the process of synaptic transmission? (a) The release of a neurotransmitter from synaptic vesicles at the pre-synaptic neuron. (b) The destruction of the post-synaptic membrane after the neurotransmitter has come in contact with it. (c) Diffusion of the neurotransmitter across the synaptic cleft. (d) Destruction of the neuro-transmitter after transmission of the impulse has taken place.	b
15. Which of the following statements is incorrect? (a) A peripheral nerve, even though composed of many nerve fibers, will elicit an all-or-none response with only one threshold value. (b) The all-or-none response means that a nerve fiber will "fire" only if the threshold value of stimulation is reached. (c) Once threshold value is reached, the fiber will fire to the same level regardless of how strong the stimulus is. The frequency of firing will change though, as the stimulus gets stronger. (d) As a stimulation gets more intense, more individual nerve fibers in a peripheral nerve will fire, exhibiting their own all-or-none responses.	a

In questions 16-24 match the numbers from column A with the letters in column B.

Column A	Column B	Answer
16. medulla, pons, cerebellum	(a) constitute the fore-brain	16-e,
17. contains centers controlling res-piration and heartrate	(b) medulla	17-b,
18. removal of it will most effect coor-dination	(c) pons	18-i,
19. center for some auditory and visual reflexes	(d) thalamus	19-h,
20. coordinating center for emotions	(e) constitute the hind-brain	20-d,

Continued on following page.

Answer

To be covered when testing yourself

		Answer
21. has direct control over function of pituitary gland	(f) cerebral cortex	21-g,
22. mostly consists of "association areas"	(g) hypothalamus	22-f,
23. plays important role in regulation of respiration	(h) midbrain	23-c,
24. thalamus and cerebrum	(i) cerebellum	24-a.

Match the numbers from column A with the letters in column B.

Column A	Column B	Answer
25. this nerve is responsible for smiling	(a) oculomotor nerve (III)	25-e,
26. strictly afferent, this nerve runs from the retina to the visual center of the brain	(b) trochlear nerve (IV)	26-d,
	(c) trigeminal nerve (V)	
	(d) optic nerve (II)	
27. this nerve is responsible for tongue movements	(e) facial nerve (VII)	27-i,
	(f) auditory nerve (VIII)	
28. this nerve transmits taste from the posterior 1/3 of the tongue	(g) olfactory nerve (I)	28-k,
	(h) vagus nerve (X)	
29. this nerve is both efferent and afferent and is responsible for movements of the muscles of mastication (chewing)	(i) hypoglossal nerve (XII)	29-c,
	(j) accessory nerve (XI)	
	(k) glossopharyngeal nerve (IX)	
30. this nerve is responsible for the movement of the lateral rectus muscle of the eye	(l) abducens nerve (VI)	30-l,
31. strictly afferent, this nerve carries impulses for smelling		31-g,

Continued on following page.

	Answer
	To be covered when testing yourself
32. this nerve is responsible for the digestive movements of the small intestine	32-h,
33. this nerve is responsible for the movements of the superior oblique of the eye	33-b,
34. strictly afferent, the nerve is involved with hearing and equilibrium	34-f,
35. this muscle is responsible for movement of the trapezius muscle of the shoulder	35-j
36. this nerve is involved with movements of the eyeball and constriction of the pupils.	36-a
Fill in the blanks.	
37. The clear, curved, anterior portion of the eye is known as the _____.	cornea
38. Light enters the eye via an opening called the _____.	pupil
39. The eye contains two separate types of fluids. They are the _____ and the _____.	aqueous humor, vitreous humor.
40. For distance vision, the smooth muscle of the ciliary body _____, causing the ciliary ligaments to become _____. This causes the lens to become _____, making it suitable for focusing on distant objects.	relaxes, taut, thinner
41. Rods contain the visual pigment, _____, which is made up of a _____ and the protein _____. Cones contain the visual pigment _____, which consists of a different protein than that found in rods.	Rhodopsin Chromophore Opsin, Iodopsin.
42. Rods, which are most abundant at the _____ of the retina, are most effective for _____ vision, since they react to very small amounts of light.	periphery, night
43. Vitamin _____ is necessary for proper visual function.	A

Answer

To be covered when testing yourself

44. The _____ separates the outer ear and middle ear, while the ____ separates the middle ear and inner ear.

Tympanic membrane oval window

45. The organ for hearing located in the inner ear is called the _____. It consists of three canals, two of which are filled with _____, while the third is filled with _____. This third canal contains the organ of corti. It contains the special receptors, the _____ _____, necessary for normal hearing. The _____ membrane overhangs these receptors, and plays a role in the transmission of impulses to the brain.

Cochlea, Perilymph, Endolymph, Hair cells, Tectorial

46. The five functional regions of the vertebral column and spinal cord are the _____, _____, _____, _____, and _____ regions.

Cervical, Thoracic, Lumbar, Sacral, Coccygeal

47. The autonomic nervous system, one part of the _____ nervous system, contains only _____ fibers, while the _____ part of the _____ nervous system contains both _____ and _____ fibers that innervate _____ _____.

Periphe-ral, Efferent, Somatic, Peripheral Efferent, Afferent, Skeletal Muscle

48. The neurotransmitter of the parasympathetic division of the autonomic nervous system is _____.

Acetyl-choline

49. Two or more neurons can be connected by an _____. This type of neuron will usually lie completely, from end to end, in the _____.

Inter-neuron, central nervous system.

50. The flow of sodium ions necessary to cause an action potential is prevented by the presence of the _____. This structure is not present at certain areas called _____. There, an action potential can be propagated. This phenomenon causes an action potential to be propagated _____, as it "jumps" from _____ to _____. This is called _____.

myelin sheath, nodes of Ranvier, faster node to node, saltatory conduction

51. An action potential is caused by the sudden influx of _____ ions at the cell membrane.

Na^+

Answer

To be covered when testing yourself

52. The period of time in which a previously stimulated axon cannot be restimulated is known as the _____ _____. Due to this phenomenon, flow of an impulse will occur _____, as the portion of the axon that was depolarized first, cannot be depolarized again, until it is out of its _____ _____. So, the direction of propagation of the impulse will be _____ from the point of original stimulation.

refractory period, unidirectionally, refractory period, opposite

Determine whether the following statements are true or false.

53. In general terms, the nervous system can be explained using the following sequence: receptor → connector → effector.

True

54. Nourishment to the retina is obtained via the layer of the eye called the sclera.

False

55. The portion of the retina that gives the greatest visual acuity is called the optic disc.

False

56. Rod cells are used for color vision, while cone cells are adapted for night vision.

False

57. Accomodation in man and frog differs with respect to the position of the lens within the eye.

True

58. In photosynthesis, the energy needed for the chemical reactions is supplied directly by light. In vision though, this energy is supplied by the breaking of chemical bonds within the rods and cones.

True

59. The four different sensations of taste - sweet, sour, bitter, and salty - are located in different regions of the tongue.

True

60. When a sound is heard, the correct sequence the sound waves follow from ear to brain is: auditory canal → tympanic membrane → malleus → incus → stapes → oval window → vestibular canal → tympanic canal → round window → basilar membrane of cochlea canal → movement of receptor cells against the tectorial membrane → sensory neurons forming the auditory nerve → temporal lobe of brain.

True

61. Sensory neurons enter the spinal cord via the ventral horn, while motor neurons leave the spinal cord via the dorsal horn.

False

Answer

To be covered when testing yourself

62. A spinal nerve can contain both afferent and efferent fibers and can contain fibers associated with the autonomic nervous system. True

63. Homeostasis is effectively maintained via the antagonistic action of the parasympathetic and sympathetic divisions of the autonomic nervous system. True

64. Nerve impulses are conducted to the cell body of a neuron via an axon and away from the cell body via a dendrite. False

65. The presence of the neurilemma of the original axon is absolutely essential for nerve regeneration. False

66. Synaptic transmission is unidirectional because the synaptic vesicles are stored in the dendritic branch of the neuron, not the axonal branch of the neuron. False

67. The threshold value is the minimum amount of stimulation needed to cause an action potential. That is, it is the level of stimulation needed to cause the influx of Na^+ ions to exceed the efflux of K^+ ions, and hence, cause depolarization. True

CHAPTER 21

HORMONAL CONTROL

DISTINGUISHING CHARACTERISTICS OF HORMONES

• PROBLEM 21-1

Compare the modes of action of the nervous and endocrine systems.

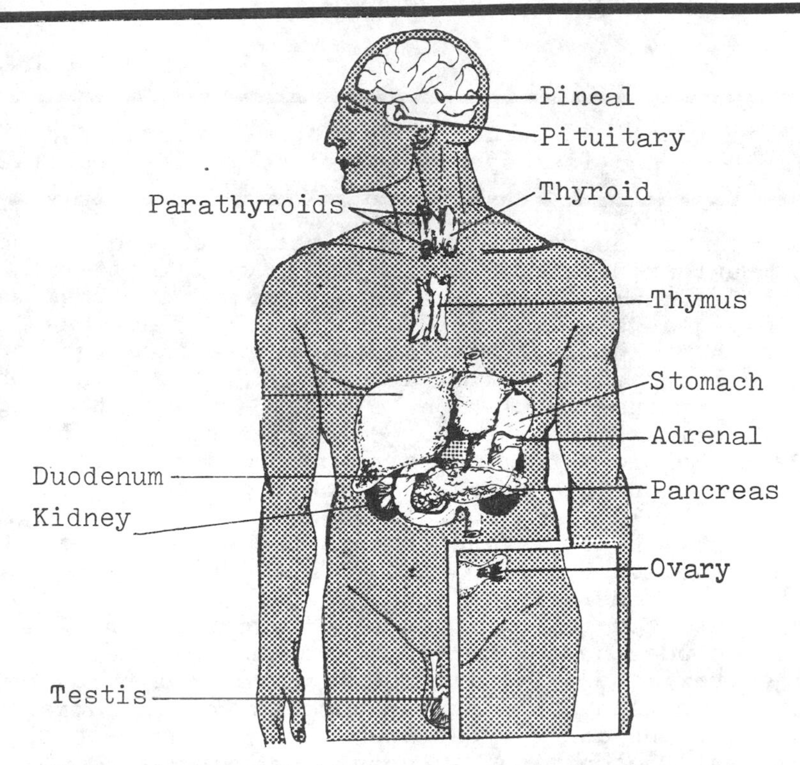

The major endocrine organs in man.

Solution: The activities of the various part of the body of higher animals are integrated by the nervous and endocrine systems. The endocrine system consists of a number of ductless glands which secrete hormones. The swift responses of muscles and glands, measured in milliseconds, are typically under nervous control. Nerve impulses are transmitted along pathways consisting of neurons. The hormones secreted by the endocrine glands are transported by the bloodstream to other cells of the body in order to control and regulate their activities.

Nervous stimulation is required by some endocrine glands to release their hormones, particularly the posterior pituitary gland. The responses controlled by hormones are in general somewhat slower (measured in minutes, hours, or even weeks), but of longer duration than those under nervous control. The long term adjustments of metabolism, growth and reproduction are typically under endocrine regulation.

We already mentioned that hormones travel in the blood and are therefore able to reach all tissues. This is very different from the nervous system, which can send messages selectively to specific organs. However, the body's response to hormones is highly specific. Despite the ubiquitous distribution of a particular hormone via the blood, only certain types of cells may respond to that hormone. These cells are known as target-organ cells.

The central nervous system, particularly the hypothalamus plays a critical role in controlling hormone secretion; conversely hormones may markedly influence neural function and behavior as well.

• PROBLEM 21-2

Define a hormone. How would you go about proving that a particular gland is responsible for a specific function?

Solution: The endocrine system constitutes the second great communicating system of the body, with the first being the nervous system. The endocrine system consists of ductless glands which secrete hormones. A hormone is a chemical substance synthesized by a specific organ or tissue and secreted directly into the blood. The hormone is carried via the circulation to other sites of the body where its actions are exerted. Hormones are typically carried in the blood from the site of production to the site(s) of action, but certain hormones produced by neurosecretory cells in the hypothalamus act directly on their target areas without passing through the blood. The distance travelled by hormones before reaching their target area varies considerably. In terms of chemical structure, hormones generally fall into three categories: The steriod hormones include the sex hormones and the hormones of the adrenal cortex; the amino acid derivatives (of tyrosine) include the thyroid hormones and hormones of the adrenal medulla; proteins and polypeptides make up the majority of the hormones. The chemical structure determines the mechanism of action of the hormone. Hormones serve to control and integrate many body functions such as reproduction, organic metabolism and energy balance, and mineral metabolism. Hormones also regulate a variety of behaviors, including sexual behaviors.

To determine whether a gland is responsible for a particular function or behavior, an investigator usually begins by surgically removing the gland and observing the effect upon the animal. The investigator would then replace the gland with one transplanted from a closely related animal, and determine whether the changes induced by removing the gland can be reversed by replacing it. When replacing the gland, the experimenter must ensure that the new gland becomes connected with the vascular system of the recipient so that secretions from the transplanted gland can enter the blood of the recipient. The experimenter may then try feeding dried glands to an animal from which the gland was previously removed. This is done to see if the hormone can be replaced in the body in this manner. The substance in the glands will enter the blood stream via the digestive system and be carried to the target organ by the circulatory system. Finally, the experimenter may make an extract of the gland and purify it to determine its chemical struture. Very often the chemical structure of a substance is very much related to its function. Studying the chemical structure may enable the investigator to deduce a mechanism by which the gland-extract functions on a molecular level. The investigator may also inject the purified gland-extract into an experimental animal devoid of such a gland, and see whether the injection effected replacement of the missing function or behavior. Some hormonal chemicals have additive effects. The investigator may inject a dosage of the purified gland-extract to an intact animal to observe if there was any augmentation of the particular function or behavior under study.

• PROBLEM 21-3

Distinguish between a hormone and a vitamin, and a hormone and an enzyme.

Solution: A hormone is an organic substance synthesized by a specific organ or tissue and secreted into the blood, which carries it to other sites in the body where the hormonal actions are exerted. Vitamins are chemical compounds, a small quantity of which is essential for the survival of certain organisms. Unlike hormones, most vitamins can not be synthesized by the body but must be obtained from external sources (i.e. diet). Like hormones, vitamins are necessary in only small quantities to ensure proper body functioning. Vitamins ordinarily function as coenzymes, which are non-protein portions of enzymes that help in binding the enzyme to the substrate.

Enzymes are organic compounds that function as biochemical catalysts produced by living organisms, and are primarily constituted of protein. Unlike hormones which may travel considerable distances to target sites, enzymes generally are used at, or very near to, their site of production. For example, pancreatic enzymes exert their effect in the duodenum, which is nearby the pancreas. Protein hormones that act through the adenyl cyclase system generally elicit their responses by activating certain enzymes in their target cells. Whereas enzymes are always made up of protein, hormones may consist of amino acids or steroids. Also, enzymes are not released into the blood as are hormones.

• **PROBLEM 21-4**

What is a pheromone, and how does it differ from a hormone?

Solution: The behavior of animals may be influenced by hormones - organic chemicals that are released into the internal environment by endocrine glands which regulate the activites of other tissues located some distances away. Animal behavior is also controlled by pheromones- substances that are secreted by exocrine glands into the external environment. Pheromones influence the behavior of other members of the same species. Pheromones represent a means of communication and of transferring information by smell or taste. Pheromones evoke specific behavioral, developmental or reproductive responses in the recipient; these responses may be of great significance for the survival of the species.

Pheromones act in a specific manner upon the recipient's central nervous system, and produce either a temporary or a long-term effect on its development or behavior. Pheromones are of two classes: releaser pheromones and primer pheromones. Among the releaser pheromones are the sex attractants of moths and the trail pheromones secreted by ants, which may cause an immediate behavioral change in conspecific individuals. Primer pheromones act more slowly and play a role in the organism's growth and differentiation. For example, the growth of locusts and the number of reproductive members and soldiers in termite colonies are all controlled by primer pheromones.

THE PITUITARY GLAND

• **PROBLEM 21-5**

The pituitary gland has been called the master gland. Is this term justified? Where is the gland located and what does it secrete?

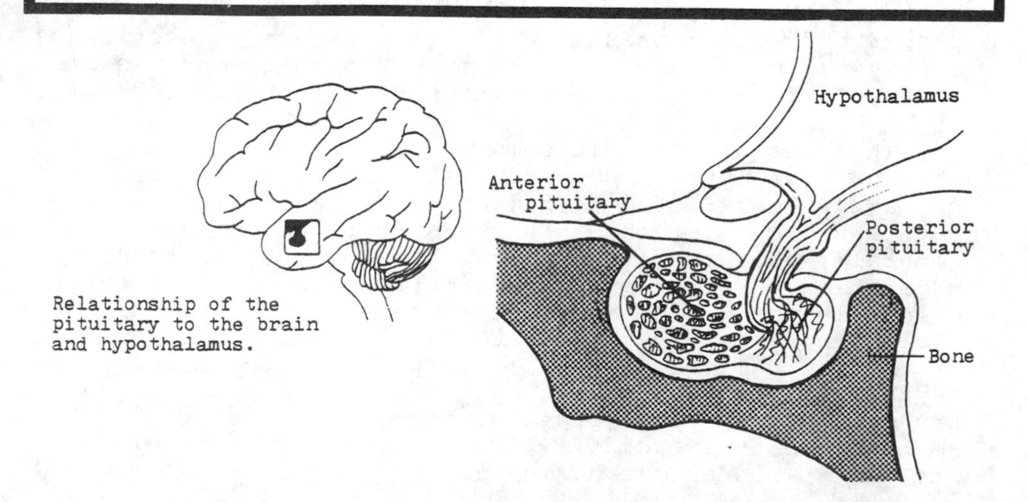

Relationship of the pituitary to the brain and hypothalamus.

Solution: The pituitary gland, also known as the hypophysis, lies in a pocket of bone just below the hypothalamus. The pituitary gland is composed of three lobes, each of which is a functionally distinct gland. They are the anterior, intermediate and posterior lobes. The anterior and posterior lobes are also known as the adenohypophysis and neurohypophysis, respectively.

In man, the intermediate lobe is only rudimentary and its function remains unclear. It secretes melanocyte stimulating hormones (MSH) which is known to cause skin darkening in the lower vertebrates.

The anterior lobe is made up of glandular tissue which produces at least six different protein hormones. Evidence suggests that each hormone is secreted by a different cell type. Secretion of each of the six hormones occurs independently of the others.

One of the hormones secreted by the anterior pituitary is known as TSH (thyroid stimulating hormone), which induces secretion of the thyroid hormones from the thyroid gland. Thyroid hormones refers to two closely related hormones, thyroxine (T_4) and triiodothyronine (T_3). Another hormone secreted by the anterior pituitary is ACTH (adrenocorticotrophic hormone), which stimulates the adrenal cortex to secrete cortisol. The anterior pituitary is also responsible for the release of the gonadotropic hormones, FSH (follicle stimulating hormone) and LH (luteinizing hormone). These hormones primarily control the secretion of the sex hormones estrogen, progesterone and testosterone by the gonads. FSH and LH also regulate the growth and development of the reproductive cells (sperm and ovum). TSH, ACTH, FSH and LH are all trophic hormones in that they stimulate other endocrine glands to secrete hormones. There are two hormones secreted by the anterior pituitary that do not affect other hormonal secretions, but rather act directly on target tissues. One of these is called prolactin, which stimulates milk production by the mammary glands of the female shortly after giving birth. The other hormone is called growth hormone, which plays a critical role in the normal processes of growth.

The posterior lobe of the pituitary gland is actually an outgrowth of the hypothalamus and is true neural tissue. The posterior pituitary differs from the anterior pituitary with respect to embryological origin as well as types of hormones secreted. It releases two hormones called oxytocin and vasopressin. Oxytocin principally acts to stimulate contraction of the uterine muscles as an aid to parturition. Emotional stress may also cause the release of this hormone, and is frequently the cause of a miscarriage. Oxytocin is also responsible for the milk let-down reflex. It stimulates smooth muscle cells of the mammary glands which causes milk ejection.

Antidiuretic hormone (ADH) stimulates the kidney tubules to reabsorb water and thus plays an important role in the control of plasma volume. In addition, ADH can increase blood pressure by causing arteriolar constriction. Thus ADH is also called vasopressin.

It should now be clear why the pituitary is called the master gland; it secretes at least nine hormones, some of which directly regulate life processes while others control the secretion of other glands important in development, behavior and reproduction.

• PROBLEM 21-6

Removal of the pituitary in young animals arrests growth owing to termination of supply of growth hormone. What are the effects of growth hormone in the body? What is acromegaly?

Solution: The pituitary, under the influence of the hypothalamus, produces a growth-promoting hormone. One of the major effects of the growth hormone is to promote protein synthesis. It does this by increasing membrane transport of amino acids into cells, and also by stimulating RNA synthesis. These two events are essential for protein synthesis. Growth hormone also causes large increases in mitotic activity and cell division.

Growth hormone has its most profound effect on bone. It promotes the lengthening of bones by stimulating protein synthesis in the growth centers. The cartilaginous center and bony edge of the epiphyseal plates constitute growth centers in bone. Growth hormone also lengthens bones by increasing the rate of osteoblast (young bone cells) mitosis.

Should excess growth hormone be secreted by young animals, perhaps due to a tumor in the pituitary, their growth would be excessive and would result in the production of a giant. Undersecretion of growth hormone in young animals results in stunted growth. Should a tumor arise in an adult animal after the actively growing cartilaginous areas of the long bones have disappeared, further growth in length is impossible. Instead, excessive secretion of growth hormone produces bone thickening in the face, fingers, and toes, and can cause an overgrowth of other organs. Such a condition is known as acromegaly.

GASTROINTESTINAL ENDOCRINOLOGY

• PROBLEM 21-7

Gland secretions and smooth muscle contraction in the gastrointestinal region are regulated by both nerves and hormones. Describe these hormones and their effects.

Solution: The gastrointestinal region consists of the stomach, the small and large intestines and parts of the liver and pancreas. It is the site where complex molecules are catabolized into simpler forms which are absorbed into the circulation. These simpler molecules are transported to different synthetic sites of the body.

The deposition of proteins into the stomach provides

the major chemical stimulus for the release of the hormone gastrin. This hormone may also be released as a result of stimulation of the parasympathetic fibers of the vagus nerves to the stomach. Gastrin is secreted by cells in the mucosa of the pyloric region of the stomach. It has mutiple sites of action throughout the gastrointestinal system. Gastrin stimulates secretion of hydrochloric acid into the stomach as well as the contraction of the gastroesophageal sphincter, a smooth muscle. Gastrin also increases gastric motility.

Stimulation of the wall of the duodenum causes release of the hormone enterogastrone. This hormone inhibits the secretion of gastrin. Under normal conditions, the amount of gastric juices secreted by the stomach is determined by the balance between levels of enterogastrone and gastrin.

The pancreas can be divided into two portions. The endocrine portion of the pancreas secretes the hormones insulin and glucagon. The exocrine portion of the pancreas secretes solutions containing high concentrations of sodium bicarbonate and large numbers of digestive enzymes.

Secretin is a hormone that is released by the mucosal cells of the duodenum. The stimulus for secretin release is the presence of acid in the duodenum. Secretin will elicit an increase in the amount of bicarbonate and the volume of fluid released by the pancreas, but stimulates little enzymatic secretion. The action of secretin is **predominantly upon the acinar cells of the pancreas, which** secrete bicarbonate. This bicarbonate will neutralize the acid secretions entering the small intestine from the stomach.

Another hormone secreted by the duodenum is cholecystokinin (CCK). The presence of amino acids and fatty acids in the duodenum will stimulate its release. CCK will increase pancreatic enzyme secretion, which leads to additional fat and protein digestion. It may also cause contraction of the gallbladder, especially after a fatty meal. This contraction will release bile into the intestinal tract. Bile is essential for digestion of fat in the small intestine. It is produced in the liver and stored in the gallbladder until the stimulatory action of CCK causes its release.

As chyme leaves the stomach to enter the duodenum, its presence stimulates the release of CCK and secretin from the wall of the duodenum. These hormones then stimulate the secretion of enzymes and bicarbonate by the pancreas. These pancreatic enzymes will act upon large molecules in the chyme to produce simple sugars, amino acids and fatty acids which can readily be absorbed by the intestinal cells. These enzymes also serve as a chemical stimulus for hormonal release until the nutrients are absorbed. Ulcerative damage to the intestinal walls by acids released from the stomach is prevented by pancreatic secretion of bicarbonate. In addition, bicarbonate creates an alkaline environment, which favors the action of pancreatic enzymes. The release of CCK and secretin also inhibits gastric secretion and motility, thus slowing the movement of digested nutrients and allowing neutralization, digestion and absorption to occur.

• PROBLEM 21-8

The pancreas is a mixed gland having both endocrine and exocrine functions. The exocrine portion secretes digestive enzymes into the duodenum via the pancreatic duct. The endocrine portion secretes two hormones (insulin and glucagon) into the blood. What are the effects of these two hormones?

Solution: Insulin is a hormone which acts directly or indirectly on most tissues of the body, with the exception of the brain. The most important action of insulin is the stimulation of the uptake of glucose by many tissues, particularly liver muscle and fat. The uptake of glucose by the cells decreases blood glucose and increases the availability of glucose for those cellular reactions in which glucose participates. Thus, glucose oxidation, fat synthesis, and glycogen synthesis are all accentuated by an uptake of glucose. It is important to note that insulin does not alter glucose uptake by the brain, nor does it influence the active transport of glucose across the renal tubules and gastrointestinal epithelium.

As stated, insulin stimulates glycogen synthesis. In addition, it also increases the activity of the enzyme which catalyzes the rate-limiting step in glycogen synthesis. Insulin also increases triglyceride levels by inhibiting triglyceride breakdown, and by stimulating production of triglyceride through fatty acid and glycerophosphate synthesis. The net protein synthesis is also increased by insulin, which stimulates the active membrane transport of amino acids, particularly into muscle cells. Insulin also has effects on other liver enzymes, but the precise mechanisms by which insulin induces these changes are poorly understood.

Insulin secretion is directly controlled by the glucose concentration of the blood flowing through the pancreas. This is a simple system which requires no participation of nerves or other hormones.

Insulin is secreted by beta cells, which are located in the part of the pancreas known as the Islets of Langerhans. These groups of cells, which are located randomly throughout the pancreas, also consist of other secretory cells called alpha cells. It is these alpha cells which secrete glucagon. Glucagon is a hormone which has the following major effects: it increases glycogen breakdown thereby raising the plasma glucose level; it increases hepatic synthesis of glucose from pyruvate, lactate, glycerol, and amino acids, (a process called gluconeogenesis, which also raises the plasma glucose level); it increases the breakdown of adipose-tissue triglyceride, thereby raising the plasma levels of fatty acids and glycerol. The glucagon-secreting alpha cells in the pancreas, like the beta cells, respond to changes in the concentration of glucose in the blood flowing through the pancreas; no other nerves or hormones are involved.

It should be noted that glucagon has the opposite effects as insulin. Glucagon elevates the plasma glucose whereas insulin stimulates its uptake and thereby reduces plasma glucose levels; glucagon elevates fatty acid concentrations whereas insulin converts fatty acids (and glycerol) into triglycerides, thereby inhibiting triglyceride breakdown.

Thus the alpha and beta cells of the pancreas constitute a "push-pull" system for regulating the plasma glucose level.

• PROBLEM 21-9

A man with diabetes mellitus has a sudden fit of anger over a minor traffic accident, and lapses into a coma. Explain the physiological events leading to the coma and what should be done to bring him out of the coma.

Solution: Diabetes mellitus is a disease which results from a hyposecretion of insulin. Insulin stimulates glucose uptake by all cells with the exception of brain cells. Hyposecretion of insulin reduces the cellular uptake of glucose, leading to an accumulation of glucose in the blood. The blood glucose may rise to such a level that a good deal of it cannot be reabsorbed by the kidney tubules. When this happens the level of glucose excreted into the urine increases tremendously. The osmotic force exerted by glucose in the urine retains water and prevents its reabsorption. Thus, the volume of water excreted is also abnormally high in people with diabetes. This loss of water, if sufficiently great, can ultimately reduce blood pressure to such an extent that there is a marked decrease in blood flow to the brain, resulting in coma and possibly death.

In our specific example, coma is induced by the aforementioned means. Emotional stress stimulates the release of glucose from the liver through the action of epinephrine, thereby resulting in a rapid increase in blood glucose. In the case of a person with diabetes, this increased blood glucose results in a large amount of water loss, which in turn leads to a decreased blood pressure. Blood flow to the brain is reduced owing to the drop in blood pressure. Coma usually results as blood supply to the brain cells becomes insufficient. Insulin should be given as treatment for this comatose person. Insulin will stimulate glucose uptake, thereby reducing its level in the blood. Glucose concentration in the urine subsequently declines, and water loss is diminished. The pressure of the blood is thereby restored to normal.

It is important to note that the dosage of insulin injected is critical. If an overdose of insulin is administered, the situation is equally fatal had no insulin been injected. This is because insulin can not enter the brain, and hence the brain cells will not alter their rate of glucose uptake. Thus they are greatly dependent on the blood supply for glucose to maintain their energy metabolism. When too much insulin is given to the diabetic, his blood stream is essentially depleted of glucose units which have entered most of the body cells excepting the brain cells. Hence the brain will starve from an insufficient supply of glucose, and its functions will break down. A coma is usually what follows.

• PROBLEM 21-10

The disease diabetes mellitus is caused by an insulin deficiency. Can this disease be inherited? What happens if a person with diabetes goes untreated? What is the treatment given for diabetes mellitus?

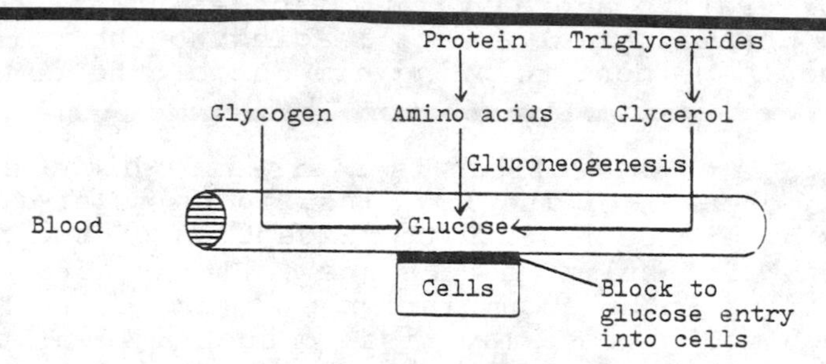

Factors which elevate blood glucose concentration in insulin deficiency.

Solution: It is believed that diabetes mellitus probably results from an insufficiency in insulin production. This may happen either as a genetic abnormality in which insulin synthesis is impaired, or as an acquired disease in which insulin-secreting cells of the pancreas are hypofunctional or destroyed. In fact, the latter may also have a genetic origin. Thus in many cases of diabetes mellitus, the disease can be inherited. The children of the person with diabetes may tend to exhibit the symptoms of diabetes. The disease, however, may develop slowly or to a varying degree, for diabetes is not an "all-or-none" disease. Moreover the overt symptoms of diabetes mellitus may all but disappear with appropriate measures such as a change in diet. Thus it would be more valid to say that a tendency toward diabetes mellitus can be inherited.

The untreated person with diabetes, suffering from a deficiency of insulin, will have an excess of glucose in the blood. A large amount of glucose will be excreted and lost from the body in the urine. The person with diabetes will therefore fail to acquire enough energy by glucose metabolism alone. He will have to obtain energy by way of triglyceride catabolism. The breakdown of triglyceride results in an elevation of fatty acid and glycerol levels in the blood. If further energy is needed, the diabetic body will convert his protein reserve into glucose, which would be accompanied by an accumulation of ketones. The products of these two metabolic pathways (i.e. fatty acids and ketones) will upset the delicately balanced pH system of the body, and thus may be considered toxic. Because insulin also has the effect on hindering glycogen formation from glucose, the diabetic individual has very little carbohydrate reserves. Most of the glucose consumed in food, unavailable to the cells for use, simply passes out of the body. In short, the untreated person with diabetes will face the paradox of cell starvation in the presence of an elevated plasma glucose concentration. Only the brain cells are spared the glucose deprivation since their uptake of glucose is not hormonally regulated. The consequence of these catabolic processes is progressive loss of weight despite any increase in food intake, and a lowered pH of the body fluids.

Another well known consequence of untreated diabetes is dehydration of the body tissues. The large amount of glucose excreted into the kidney tubules induces water to passively follow suit (osmosis), and prevents it from being reabsorbed into the body cells. For some complicated reason, sodium reabsorption is also reduced. The net effect is the marked excretion of water and sodium, resulting in a decreased plasma volume and hence blood pressure. The lowered blood pressure may reduce blood flow to the brain, leading to brain damage, coma and possibly death.

Associated with diabetes are arteriosclerosis, small-vessel and nerve disease, susceptibility to infection, and a variety of other complications. At one time it was thought that these problems resulted only from insulin deficiency. However, this view is now being questioned in light of the evidence which has shown that these problems can be delayed but not abolished by administration of insulin. Many researchers now believe that excessive hyperglycemia or possibly even hypoglycemia due to inappropriate insulin administration contribute to the occurrence of complications.

In treating persons with diabetes, the aim is to maintain the plasma glucose at a near normal value. The administration of insulin is the major therapy, and must be given through injection. This is because insulin, being a protein hormone, would be broken down by gastrointestinal enzymes if given orally. The dosage must be carefully determined since an overdose would drastically lower the plasma glucose concentration, thereby depriving brain cells of needed glucose and ultimately causing death. Non-insulin drugs which stimulate the islets of Langerhans to produce more insulin have been used in some cases, but are not effective therapeutic measures.

There are actually two types of diabetes. Juvenile diabetes (Type I) is the insulin-deficient diabests just described. These patients require insulin and show some of the more severe effects of the disease. Adult-onset diabetes (Type II) is much more common and less threatening. This type of diabetes is seen more frequently in older often obese patients. Their bodies produce and release insulin yet the receptors for insulin are less sensitive. The receptors are "down-regulated." Type II patients are not insulin-dependent. Improvement may occur with proper diet and exercise.

THE THYROID GLAND

• PROBLEM 21-11

The thyroid gland is located in the neck and secretes several hormones, the principal one being thyroxine. Trace the formation of thyroxine. What functions does it serve in the body? What happens when there is a decreased or increased amount of thyroxine in the body?

Solution: The thyroid gland is a two-lobed gland which manifests a remarkably powerful active transport mechanism for uptaking iodide ions from the blood. As blood flows through the gland, iodide is actively transported into the cells. Once within the cell, the iodide is converted to an active form of iodine. This iodine combines with an amino acid called tyrosine. Two molecules of iodinated tyrosine then combine to form thyroxine. Following its formation, the thyroxine becomes bound to a polysaccharide-protein material called thryroglobulin. The normal thyroid gland may store several weeks supply of thyroxine in this 'bound' form. An enzymatic splitting of the thyroxine from the thyroglobulin occurs when a specific hormone is released into the blood. This hormone, produced by the pituitary gland, is known as thyroid-stimulating hormone (TSH). TSH stimulates certain major rate-limiting steps in thyroxine secretion, and thereby alters its rate of release. A variety of bodily defects, either dietary, hereditary, or disease-induced, may decrease the amount of thyroxine released into the blood. The most popular of these defects is one which results from dietary iodine deficiency. The thyroid gland enlarges, in the continued presence of TSH from the pituitary, to form a goiter. This is a futile attempt to synthesize thyroid hormones, for iodine levels are low. Normally, thyroid hormones act via a negative feedback loop on the pituitary to decrease stimulation of the thyroid. In goiter, the feedback loop cannot be in operation - hence continual stimulation of the thyroid and the inevitable protuberance on the neck. Formerly, the principal source of iodine came from seafood. As a result, goiter was prevalent amongst inland areas far removed from the sea. Today, the incidence of goiter has been drastically reduced by adding iodine to table salt.

Thyroxine serves to stimulate oxidative metabolism in cells; it increases the oxygen consumption and heat production of most body tissues, a notable exception being the brain. Thyroxine is also necessary for normal growth, the most likely explanation being that thyroxine promotes the effects of growth hormone on protein synthesis. The absence of thyroxine significantly reduces the ability of growth hormone to stimulate amino acid uptake and RNA synthesis. Thyroxine also plays a crucial role in the closely related area of organ development, particularly that of the central nervous system.

If there is an insufficient amount of thyroxine, a condition referred to as hypothyroidism results. Symptoms of hypothyroidism stem from the fact that there is a reduction in the rate of oxidative energy-releasing reactions within the body cells. Usually the patient shows puffy skin, sluggishness, and lowered vitality. Hypothyroidism in children, a condition known as cretinism, can result in mental retardation, dwarfism, and permanent sexual immaturity. Sometimes the thyroid gland produces too much thyroxine, a condition known as hyperthyroidism. This condition produces symptoms such as an abnormally high body temperature, profuse sweating, high blood pressure, loss of weight, irritability, and muscular weakness. It also produces one very characteristic symptom that may not be predicted because it lacks an obvious connection to a high metabolic rate. This symptom is exophthalmia, a condition where the eyeballs protrude in a startling manner. Hyperthyroidism has been treated by partial removal or by partial radiation destruction of the gland. More recently, several drugs that inhibit thyroid activity have been discovered, and their use is supplanting the former surgical procedures.

• PROBLEM 21-12

Explain why the feeding of thyroid glands may cure myxedema, and why the feeding of pancreas does not cure diabetes.

Solution: Myxedema is a disease caused by a deficiency of thyroxine, a hormone secreted by the thyroid gland. Thyroxine has the following structural formula:

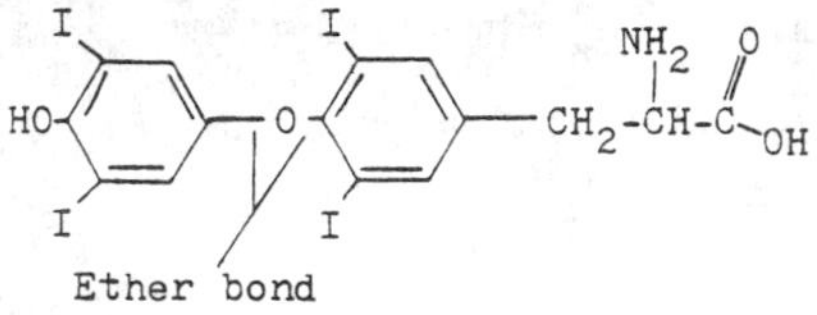

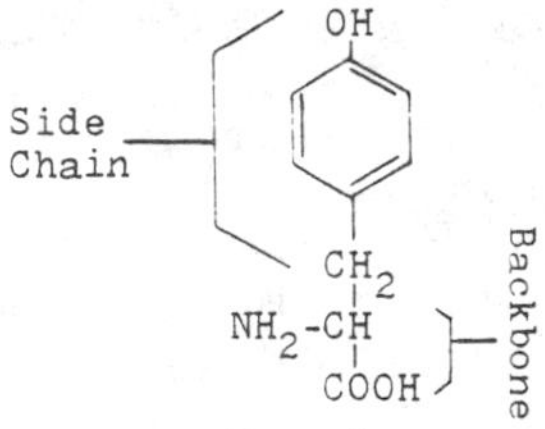

I I NH2 O
HO- O -CH2-CH-C-OH
I I
Ether bond

Side Chain
OH
CH2
NH2-CH
COOH
Backbone
Tyrosine

The molecule basically consists of two iodinated amino acid molecules (tyrosine) linked together by an ether bond. One of the tyrosine molecules loses its backbone and part of its sidechain during the synthesis of thyroxine. Although thyroxine is made up of two amino acids, they are not linked together by a peptide bond, which is characteristic of amino acid linkages. It is this fact which allows thyroxine to remain intact when subjected to proteolytic enzymes in the stomach following ingestion. The thyroxine ingested can be used to treat myxedema because it is absorbed by the gut in an unaltered form. Thus the feeding of thyroid glands is one possible way to cure myxedema.

The hormone produced by the Islets of Langerhans in the pancreas is insulin. Insulin is one of the smallest proteins known, and like other proteins it consists of a sequence of amino acids linked by peptide bonds. Amino acids join together to form peptide bonds in the following way:

$$\underset{H}{\overset{H}{>}}N-\underset{H}{\overset{R}{C}}-C\underset{OH}{\overset{O}{\lessgtr}} + \underset{H}{\overset{H}{>}}N-\underset{H}{\overset{R}{C}}-\underset{OH}{\overset{O}{\overset{\|}{C}}} \underset{\text{Hydrolysis}}{\overset{\text{Dehydration synthesis}}{\rightleftharpoons}} NH_2-\overset{R}{CH}-\overset{O}{\overset{\|}{C}}-\underset{H}{N}-\overset{R}{CH}-\overset{O}{\overset{\|}{C}}-OH + H_2O$$

Peptide bond

Should insulin be ingested and enter the stomach, proteases there would hydrolyze the hormone into individual amino acids, thus destroying the hormone. This is why diabetes cannot be cured by oral administration of insulin. Instead, insulin is usually directly introduced into the blood stream.

REGULATION OF METAMORPHOSIS AND DEVELOPMENT

• PROBLEM 21-13

What hormone is responsible for metamorphosis in amphibians?

Solution: In amphibians, the thyroid hormones control metamorphosis. This is seen by the fact that very young larvae that have had their thyroids removed did not undergo metamorphosis unless they were immersed in or injected with thyroxine solutions. In addition, normal larvae with thyroids metamorphosize precociously when they have thyroxine added to their water or administered by injection.

The matemorphosis includes regressive changes (ex: resorption of the tadpole tail, gills, and teeth) as well as constructive changes (ex: the development of limbs, tongue, and middle ear). The pituitary gland initiates metamorphosis by releasing thyroid stimulating hormone (TSH), which is carried by the bloodstream to the thyroid. The thyroid is induced to release thyroxine into the circulation.

Metamorphosis involves extensive changes at the biochemical level. A new set of enzymes appears in the liver, a new type of hemoglobin protein is synthesized; and a novel protein is formed in the retina. Structures that are resorbed contain high levels of proteases, nucleases and other digestive enzymes. These changes are all inducible with a single substance - thyroxine.

• PROBLEM 21-14

The transformation of insect larvae into mature adults is known to be under hormonal control. What hormones are involved and what are their effects?

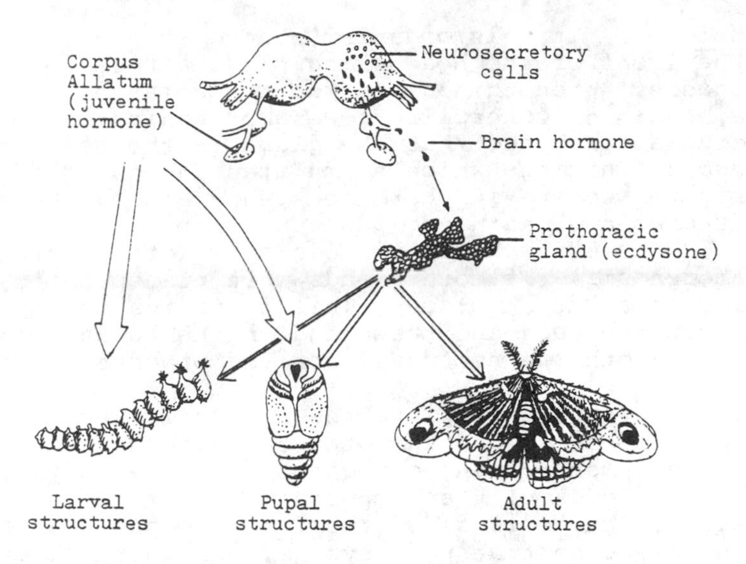

Interactions of juvenile hormon, brain hormone, and molting hormone (ecdysone) in Cecropia silkworm.

Solution: A hormonal secretion from the brain stimulates the release of the hormone ecdysone by the prothoracic gland. Ecdysone is the hormone that stimulates growth and molting in insects. Another hormone, present in insects prior to adulthood, is secreted by a portion of the brain known as the corpora allata. This hormone, referred to as juvenile hormone, is responsible for the maintenance of larval structures and inhibits metamorphosis. The juvenile hormone can exert its effect only after the molting process has been initiated. It thus must act in conjunction with the prothoracis hormone. When a relatively high level of juvenile hormone is present in the blood, the result is a larva-to-larva molt. When the level of the juvenile hormone is low, the molt is larva-to-pupa. Finally, in the absence of the juvenile hormone, molting gives rise to adults.

Ecdysone has been chemically identified as a steroid and it has been obtained in pure form. The pure extract of ecdysone has been used to investigate the functional mechanism of the hormone. In one recent experiment, small amounts of ecdysone were injected into a gnat. Within fifteen minutes, chromosome puffs (enlargements of certain regions of the chromosome) were observed to form at certain sites on the salivary gland. Chromosomes puff formation is taken as an indication of the activation of a certain gene or set of genes. This is one piece of evidence that certain genes may be activated by hormones.

• **PROBLEM 21-15**

The hormone which brings about metamorphosis of the cecropia moth is secreted by a gland in the thorax. However, no metamorphosis takes place if the head is cut off, even though the larva continues to live. Explain.

Solution: In this organism, there is a two-lobed gland at the front of the head which plays a role in metamorphosis. When this gland was transplanted into the body of a headless caterpillar, metamorphosis took place. A theory was proposed that the gland in the head of the moth produces a hormone which stimulates the gland in the thorax to secrete its hormone. When either the head or the thoracic gland alone was transplanted into the excised posterior half of a cecropia moth, there was no metamorphosis. If both glands were transplanted, the metamorphosis took place. This indicates that both the gland in the head and the gland in the thorax of the cecropia moth are necessary for the process of metamorphosis.

An investigation into the nature of the action of these hormones showed that the thoracic hormone stimulates the development of the cytochromes in the body cells. The cytochromes play an important role in the energy release of the cell since they are acceptors of hydrogen in the electron transport system. An increase in the cytochromes results in an increase in the metabolic rate. It is this increase in metabolism which supplies the additional energy required for metamorphosis.

THE PARATHYROID GLAND

• PROBLEM 21-16

The parathyroid glands in man are small, pealike organs. They are ususally four in number and are located on the surface of the thyroid. What is the function of these glands?

Solution: The parathyroids were long thought to be part of the thyroid or to be functionally associated with it. Now, however, we know that their close proximity to the thyroid is misleading; both developmentally and functionally, they are totally distinct from the thyroid.

The parathyroid hormone, called parathormone, regulates the calcium-phosphate balance between the blood and other tissues. Production of this hormone is directly controlled by the calcium concentration of the extracellular fluid bathing the cells of these glands. Parathormone exerts at least the following four effects: (1) it increases gastrointestinal absorption of calcium by stimulating the active-transport system which moves calcium from the gut lumen into the blood; (2) it increases the movement of calcium and phosphate from bone into extracellular fluid. This is accomplished by stimulating osteoclasts to break down bone structure, thus liberating calcium phosphate into the blood. In this way, the store of calcium contained in bone is tapped; (3) it increases reabsorption of calcium by the renal tubules, thereby decreasing urinary calcium excretion; (4) it reduces the reabsorption of phosphate by the renal tubules.

The first three effects result in a higher extracellular calcium concentration. The adaptive value of the fourth is to prevent the formation of kidney stones.

Should the parathyroids be removed accidentally during surgery on the thyroid, there would be a rise in the phosphate concentration in the blood. There would also be a drop in the calcium concentration as more calcium is excreted by the kidneys and intestines, and more is incorporated into bone. This can produce serious disturbances, particularly in the muscles and nerve, which utilize calcium ions for normal functioning. Overactivity of the parathyroids, which can result from a tumor on the glands, produces a weakening of the bones. This is a condition which makes them much more vulnerable to fracturing, because of excessive withdrawal of calcium from the bones.

• PROBLEM 21-17

A deficiency in which endocrine gland would result in tetany?

Solution: Tetany, a disease characterized by skeletal muscular tremors, cramps, and convulsions is caused by an insufficiency of calcium available to cells. It is of great physiological importance that an adequate quantity of calcium ions be present in body cells in order to maintain a normal membrance excitability. In tetany, due to the deficiency of calcium, irritability of nerve and muscle cell membranes is increased. Thus, muscle contractions become spastic and poorly controlled.

It is the primary function of parathormone to increase the availability of calcium to the body cells. Parathormone promotes the transport of calcium from the intestinal lumen into the blood, the release of calcium from the bones, and the reabsorption of calcium by the kidney tubules. Should the parathyroid gland be removed, or hypoactivity of the gland occur, the person would suffer from tetany.

This would happen because the calcium levels in his blood would decrease. Severe cases of tetany can cause oxygen exhaustion leading to death by asphyxia. If a solution of a calcium salt is injected into the vein of a person in tetanic convulsions, the convulsions will cease.

• PROBLEM 21-18

What hormone, if any, acts antagonistically to the parathormone?

Solution: In 1961 a hormone called calcitonin was

discovered, and was found to be directly antagonistic to parathormone, the hormone secreted by the parathyroid glands. Calcitonin, also called thyrocalcitonin, is secreted by cells within the thyroid gland which surround but are completely distinct from the thyroxine secreting cells. It is the function of calcitonin to lower the plasma calcium concentration. This is achieved primarily by the deposition of calcium-phosphate in the bones. Parathormone has an exactly opposite effect in calcium-phosphate metabolism.

The secretion of calcitonin is regulated by the calcium content of the blood supplying the thyroid gland. When blood calcium rises above a certain level, there is an increase in calcitonin secretion in order to restore the normal concentration of calcium. In lower invertebrates calcitonin is produced by separate glands, which in mammals become incorporated into the thyroid during embryonic development.

THE PINEAL GLAND

• PROBLEM 21-19

The pineal gland is a lobe on the upper portion of the forebrain, and has long intrigued investigators by its glandular appearance. Is this lobe an endocrine gland? If yes, what is its function(s)?

Solution: It was only recently demonstrated that the pineal gland has an endocrine function. In some primitive vertebrates, the pineal responds to light by generating nerve impulses and by secreting a hormone called melatonin. Melatonin lightens the skin by concentrating the pigment granules in melanophores, which are specialized pigment-containing cells. In frogs, the action of melatonin is opposed by melanocyte-stimulating hormone (MSH). MSH is secreted by the pituitary gland, and serves to disperse the pigment granules, causing a darkening of the skin. In mammals, the pineal secretes melatonin but does not generate impulses because light-sensitive cells are not present. Although the pineal does secrete melatonin in mammals, a problem arises when we realize that mammals have no melanophores (the sites of the hormonal action).

Thus, the function of the pineal and its hormone in mammals has been a source of puzzlement for a long time.

Recently, a theory has been proposed that the pineal gland functions as a neuroendocrine transducer. It was theorized that the pineal receives neural input from the visual receptors, and transduces this data into chemical messages (hormones). Apparently, impulses from the eyes reach the pineal via sympathetic nervous pathways from the cervical ganglia. The pineal responds by secreting more or less melatonin. In mammals, melatonin no longer affects skin pigmentation, but appears to have become a regulator of the anterior pituitary. Melatonin seems to inhibit reproductive activities of rodents and birds by inhibiting the release of gonadotrophic hormones (LH and FSH). The diurnal rhythm of the light/dark cycles seems to mediate this function of melatonin. Thus, in these animals, reproduction coincides with the optimal season, as light of day is interpreted by the brain via stimulation of the photoreceptors. The function of the pineal gland in man remains highly speculative, though it may play a role in circadian rhythms.

THE THYMUS GLAND

• PROBLEM 21-20

The thymus gland is a two-lobed, glandular-appearing structure located in the upper region of the chest just behind the sternum. What are the two principal functions that have been attributed to this gland?

Solution: It is thought that one of the functions of the thymus is to provide the initial supply of lymphocytes for other lymphoid areas, such as the lymph nodes and spleen. These primary cells then give rise to descendent lines of lymphocytes, making further release from the thymus unnecessary. This first function of the thymus is non-endocrine in nature.

The second function attributed to the thymus is the release of the hormone thymosin which stimulates the differentiation of incipient plasma cells in the lymphoid tissues. The cells then develop into functional plasma cells, capable of producing antibodies when stimulated by the appropriate antigens. To summarize, full development of plasma cells requires two types of inducible stimuli. They are: (1) stimulation by thymosin which initiates differentiation of all types of incipient plasma cells, and (2) stimulation by a specific antigen, which affects the functional maturation of only those cells with a potential for making antibodies against that particular antigen.

THE ADRENAL GLANDS

• PROBLEM 21-21

The two adrenal glands lie very close to the kidneys. Each adrenal gland in mammals is actually a double gland, composed of an inner corelike medulla and an outer cortex. Each of these is functionally unrelated. Outline the function of the adrenal medulla.

Solution: The adrenal medulla secretes two hormones, adrenalin (epinephrine) and noradrenalin (norepinephrine, NE), whose functions are very similar but not identical. The adrenal medulla is derived embryologically from neural tissue. It has been likened to an overgrown sympathetic ganglion whose cell bodies do not send out nerve fibers, but release their active substances directly into the blood, thereby fulfilling the criteria for an endocrine gland. In controlling epinephrine secretion, the adrenal medulla behaves just like any sympathetic ganglion, and is dependent upon stimulation by sympathetic preganglionic fibers, as shown below:

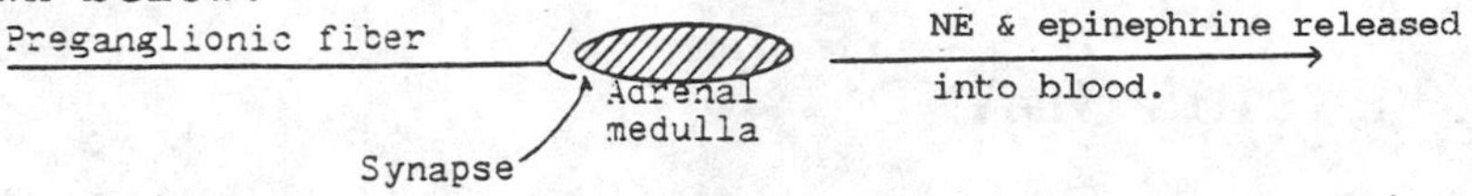

Preganglionic fiber — Adrenal medulla → NE & epinephrine released into blood.
Synapse

Epinephrine promotes several responses, all of which are helpful in coping with emergencies: the blood pressure rises, the heart rate increases, the glucose content of the blood rises because of glycogen breakdown, the spleen contracts and squeezes out a reserve supply of blood, the clotting time of blood is decreased, the pupils dilate, the blood flow to skeletal muscle increases, the blood supply to intestinal smooth muscle decreases, and hairs become erect. These adrenal functions, which mobilize the resources of the body in emergencies, have been called the fight-or-flight response. Norepinephrine stimulates reactions similar to those produced by epinephrine, but is less effective in the conversion of glycogen into glucose.

The significance of the adrenal medulla may seem questionable since the complete removal of the gland causes few noticeable changes in the animal; the animal can still exhibit the fight-or-flight response. This occurs because the sympathetic nervous system compliments the adrenal medulla in stimulating the fight-or-flight response, and the absence of the hormonal control will be compensated for by the nervous system.

• PROBLEM 21-22

The cortex of the adrenal gland is known to produce over 20 hormones, but their study can be simplified by classifying them into three categories: glucocorticoids, mineralocorticoids, and sex hormones. Explain the function of and give examples of each of these three groups of hormones. What will result from a hypofunctional adrenal cortex?

Solution: The glucocorticoids include corticosterone, cortisone, and hydrocortisone (cortisol). These hormones serve to stimulate the conversion of amino acids into carbohydrates (a process known as gluconeogenesis), and the formation of glycogen by the liver. They also stimulate formation of reserve glycogen in the tissues, such as in muscles. The glucocorticoids also participate in lipid and protein tabolism. Glucocorticoids are also essential for coping with stress and acting as anti-inflammatory agents. Glucocorticoid secretion is controlled by the anterior pituitary hormone ACTH (adrenocorticotrophic hormone).

Aldosterone, the major mineralocorticoid, stimulates the cells of the distal convoluted tubules of the kidneys to decrease reabsorption of potassium and increase reabsorption of sodium. This in turn leads to an increased reabsorption of chloride and water. These hormones, together with such hormones, together with such hormones as insulin and glucagon, are important regulators of the ionic environment of the internal fluid.

The adrenal sex hormones consist mainly of male sex hormones (androgens) and lesser amounts of female sex hormones (estrogens and progesterone). Normally, the sex hormones released from the adrenal cortex are insignificant due to the low concentration of secretion. However, in cases of excess secretion, masculinizing or feminizing effects appear. The most common syndrome of this sort is virilism of the female.

Should there be an insufficient supply of cortical hormones, a condition known as Addison's disease would result. This disease is characterized by an excessive excretion of sodium ions, and hence water, due to a lack of mineralocorticoids. Accompanying this is a decreased blood glucose level, due to a deficient supply of glucocorticoids. The effect of a decreased androgen supply can not be observed immediately. Injections of adrenal cortical hormones promptly relieves these symptoms.

Hormonal production in the adrenal cortex is directly controlled by the anterior pituitary hormone called adrenocorticotrophic hormone (ACTH).

THE MECHANISMS OF HORMONAL ACTION

• PROBLEM 21-23

The specific cell processes accelerated or decelerated by hormones are numerous and varied, but most of them fit into one of two general categories. One is the alteration of the rate of membrane transport of substances, and the other is the alteration of the rate of enzyme activity. Explain.

Solution: The effect of many hormones is to facilitate or inhibit the transport of substances across the membrane into the cell. Insulin, for example, somehow affects cell membranes so as to increase the rate of glucose transport into the cells. Other hormones, such as glucagon (an antagonist of insulin), inhibit glucose transport. A similar pattern involving hormone-mediated inhibition and stimulation also operates for the membrane transport of amino acids and other sugars. The transport of ions and water into and out of kidney cells is also under the influence of hormones.

The hormonal regulation of enzymatic activity usually involves catalysis for which there are different enzymes for the forward and reverse reactions. An example of this is the relationship that exists between glucose and glycogen in the liver:

$$\text{Glycogen} \underset{\text{enzyme 2}}{\overset{\text{enzyme 1}}{\rightleftharpoons}} \text{Glucose}$$

In this particular example, the activity of enzyme 2 is increased by the hormone insulin, and that of enzyme 1 is increased by the hormones glucagon (a pancreatic hormone) and epinephrine (an adrenal medullary hormone). The rate of either the forward or reverse reaction can be increased by production of more of the appropriate enzyme. But there is another way which does not change the total number of enzyme molecules; it does not have any effects on enzyme synthesis. Most enzymes exist in both an active and inactive state. The active state is the functional form of the enzyme. It is believed that the amount of active enzyme can be increased by converting inactive enzyme molecules into active ones. Thus, a hormone may function by converting a certain specific inactive enzyme into its active form, thereby augmenting the enzyme activity and thus the reaction rate. The advantage of this mechanism over synthesis of new enzyme molecules is that protein synthesis requires some time, whereas the activation of enzymes within the cell requires only a few minutes. A hormone may also function by inactivating certain enzymes to bring about a specific physiological change.

Note again that in this functional mechanism no new enzymes are being formed, although there is a change in enzyme activity and hence protein production.

• PROBLEM 21-24

It has become apparent that a large number of hormones actually have identical initial biochemical actions. This involves the activation of a system containing the enzyme adenyl cyclase, which is found in cell membranes throughout the body. This system has become known as the second-messenger or cyclic AMP system. Using this system; outline the sequence of events involved in producing glucose from glycogen via the action of epinephrine.

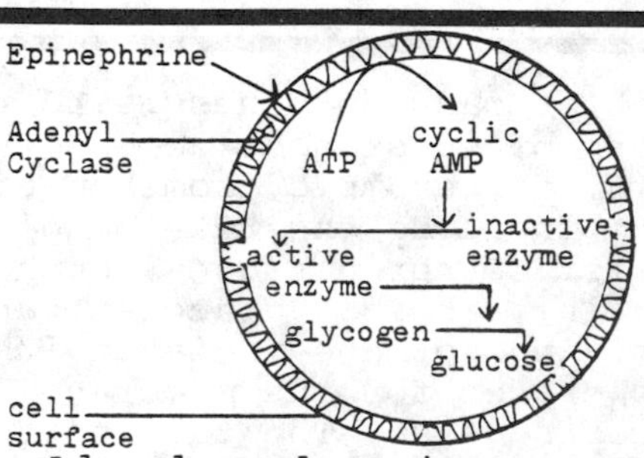

Solution: Adenyl cyclase is an enzyme found in membranes which, when activated, catalyzes the modification of ATP into cyclic AMP. It is the sole function of the hormone to interact with receptors on the cell membranes and activate the adenyl cyclase. The cyclic AMP is called the second messenger, and it serves to initiate a complex sequence of reactions, leading to the ultimate generation of the final active enzyme. For the case of epinephrine, the following sequence of events occurs: epinephrine, released from the adrenal medulla by stimulation of sympathetic nerves, activates adenyl cyclase on the surface of the liver cell (note that the hormone itself does not gain entry into the cell); adenyl cyclase in turn catalyzes the conversion of ATP into cyclic AMP; the cyclic AMP then stimulates conversion of the inactive enzyme in the cell into an active form; this active enzyme, in turn, catalyzes the critical reaction leading to the breakdown of glycogen.

There may appear to be a small inconsistency in this adenylcyclase system. If the generation of cyclic AMP is the common biochemical action of many hormones, why don't all these hormones produce identical effects on different cell types of the body? The reason for this is that the adenyl cyclase systems in different cell types differ in their abilities to be activated by different hormones. There are qualitative differences in membrane receptor sites in different tissues. A second inconsistency which arises is how several hormones, all of which influence cyclic AMP, have different effects on the same cell. One possible explanation is that there are qualitatively different receptor sites in the cell that respond to different hormones. Within the cell there are separate "compartments" of cyclic AMP, each of which when stimulated gains access to a different intracellular site and catalyzes a reaction there. At least twelve hormones act through intracellular synthesis of cyclic AMP, the most notable exception being the steroid hormones. It is interesting to note that cyclic AMP also plays important roles in non-endocrine mechanisms, such as the control of antibody production and the regulation of vision.

• **PROBLEM 21-25**

Steroid hormones do not act through the adenyl cyclase system. What is the mechanism of steroid hormone action?

Solution: Steroid hormones serve to increase the synthesis of proteins in target cells. The mechanism by which steroid hormones do this has been called the mobile-receptor model. The first step in this mechanism is the entry of the hormone into the cytoplasm. Steroid hormones are lipid soluble and can readily cross cell membranes. Once within the cell, the hormone binds to a soluble hormone-specific cytoplasmic protein, known as the receptor. The hormone-receptor complex then moves into the nucleus and combines with specific proteins associated with DNA in the chromosome. The molecular interaction triggers off the specific RNA synthesis which results in increased protein synthesis by the cell.

The thyroid hormones (amino acid derivatives) also act via direct diffusion into target cells. The distinction to be made is that thyroid hormones enter the nucleus and bind to a nuclear receptor.

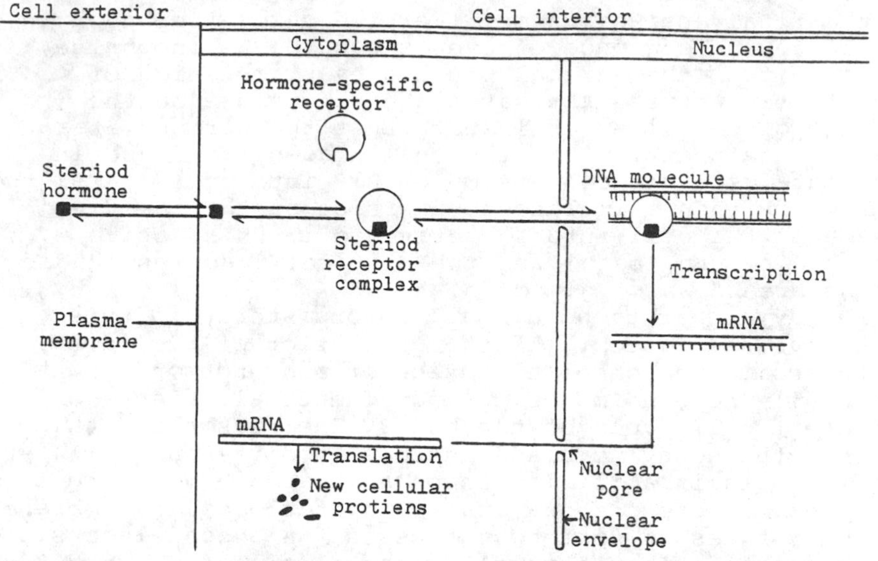

Diagram showing mobile-receptor model of a steriod hormone.

THE GONADOTROPHIC HORMONES

Sexual Development

• PROBLEM 21-26

What hormones are involved with the changes that occur in a pubescent female and male?

Solution: Puberty begins in the female when the hypothalamus stimulates the anterior pituitary to release increased amounts of FSH (follicle stimulating hormone) and LH (luteinizing hormone). These hormones cause the ovaries to mature, which then begin secreting estrogen and progesterone, the female sex hormones. These hormones, particularly estrogen, are responsible for the development of the female secondary sexual characteristics. These characteristics include the growth of pubic hair, an increase in the size of the uterus and vagina, a broadening of the hips and development of the breasts, a change in voice quality, and the onset of the menstrual cycles.

Before the onset of puberty in the male, no sperm and very little male sex hormone are produced by the testes. The onset of puberty begins, as in the female, when the hypothalamus stimulates the anterior pituitary to release increased amounts of FSH and LH. In the male, FSH stimulates maturation of the seminiferous tubules which produce the sperm. LH is responsible for the maturation of the interstitial cells of the testes. It also induces them to begin secretion of testosterone, the male sex hormone. When enough testosterone accumulates, it brings about the whole spectrum of secondary sexual characteristics normally associated with puberty. These include growth of facial and pubic hair, deepening of the voice, maturation of the seminal vesicles and the prostate gland, broadening of the shoulders, and the development of the muscles.

If the testes were removed before puberty, the secondary sexual characteristics would fail to develop. If they were removed after puberty, there would be some retrogression of the adult sexual characteristics, but they would not disappear entirely.

• PROBLEM 21-27

Describe the hormonal interactions that control the development and functioning of the breasts.

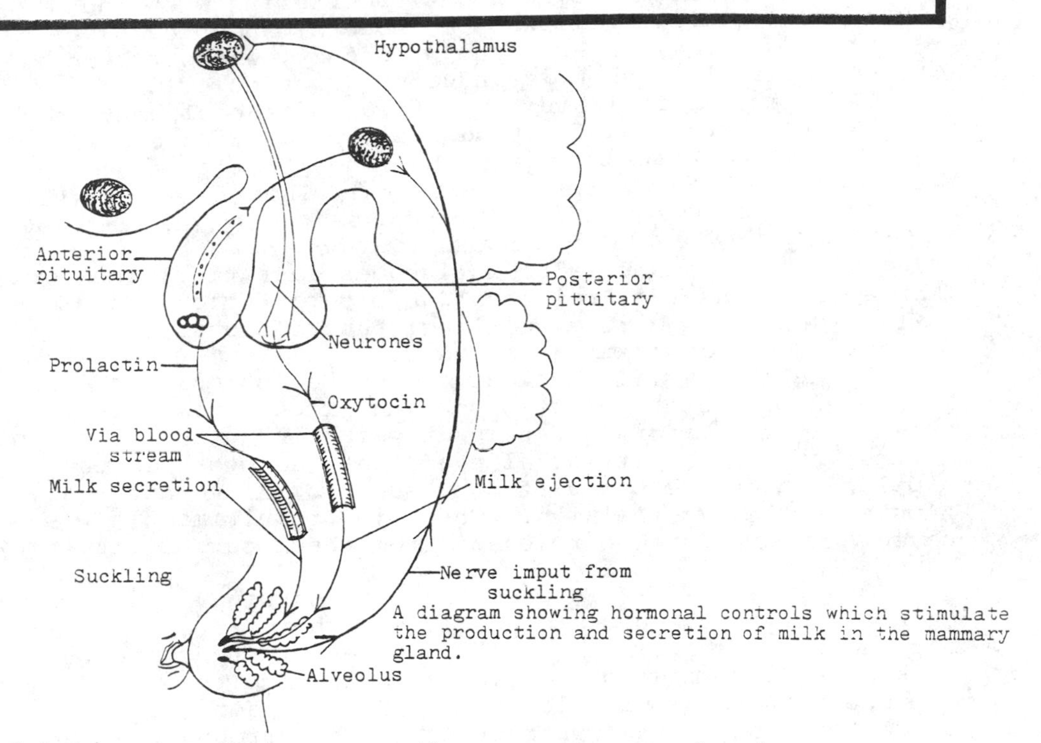

A diagram showing hormonal controls which stimulate the production and secretion of milk in the mammary gland.

Solution: Perhaps no other process so clearly demonstrates the intricate interplay of various hormonal control mechanisms as milk production. The breasts are composed of numerous ducts which branch through the breast tissue and converge at the nipples. These ducts terminate at their other end within the breast, in saclike structures (alveoli) typical of exocrine glands. It should be noted that an exocrine gland, by definition, secretes material into ducts leading to a specific compartment or surface. The alveoli are the actual milk secreting structures. Both the alveoli and the ducts are lined by contractile cells whose role will be eludicated later. Before puberty, the breasts are small with little internal structure. With the onset of puberty, estrogen and progesterone act upon the ductile tissue and alveoli to produce the basic architecture of the adult breast. In addition to these two hormones, normal breast development at puberty requires prolactin and growth hormone, both secreted by the anterior pituitary.

During each menstrual cycle, the morphology of the

breast changes in response to fluctuating blood concentrations of estrogen and progesterone. The greatest change in breast morphology occurs, however, during pregnancy.

The most important hormone which promotes milk production is prolactin. But the onset of lactation requires more than just high prolactin levels. This is seen in that the level of prolactin is elevated and the breasts are enlarged as pregnancy progresses, and yet there is no secretion of milk. The question arises as to what occurs during delivery to permit the onset of lactation. There have been numerous experiments performed with different mammals, which has shown that if the fetus is removed during pregnancy without interfering with the placenta, lactation will not be induced; when the placenta is removed at any stage of the pregnancy, without removal of the fetus, lactation is induced. Present evidence indicates that this inhibitory effect of the placenta is due to its secretion of estrogen and progesterone, which in large concentration appears to inhibit milk release by some direct action on the breasts. Therefore, delivery removes the source of the two hormones, and thereby removes the inhibition to allow lactation to occur.

The hypothalamus also influences lactation. The hypothalamus releases a Prolactin Inhibitory Factor. This chemical is decreased during suckling; thus increased prolactin allows secretion of milk to occur.

While it is still unclear as to what stimulates increased prolactin secretion during pregnancy, it is known what the major factor is in maintaining the secretion during lactation. It is believed that receptors in the nipples, which are stimulated by suckling, have nervous inputs to the hypothalamus. These inputs ultimately cause prolactin to be released from the anterior pituitary.

One final reflex is essential for nursing. The infant cannot suckle milk out of the alveoli where it is produced, but can only remove milk that is present in the ducts. Milk must first move out of the alveoli into the ducts, a process known as milk let-down. This is accomplished by contraction of the contractile cells surrounding the alveoli. The contraction of these cells is under the direct control of the posterior pituitary hormone oxytocin.This hormone is reflexively released by suckling just like prolactin. Oxytocin is also responsible for the contraction of the uterine muscles during delivery, and it is for this reason that many women experience uterine contractions during nursing.

Another important neuroendocrine reflex triggered by suckling is the inhibition of follicle stimulating hormone and luteinizing hormone released by the pituitary. Low levels of FSH and LH subsequently prevent ovulation. This inhibition is relatively short-lived in many women, and approximately 50 per cent begin to ovulate despite continued nursing.

The end product of all these processes is the milk which contains four major constituents: water, protein, fat, and the carbohydrate lactose. All of these components are very important for the proper development of the infant.

The Menstrual Cycle

• PROBLEM 21-28

The female menstrual cycle last roughly 28 days. Trace the ovarian and hormonal changes which occur during a normal menstrual cycle.

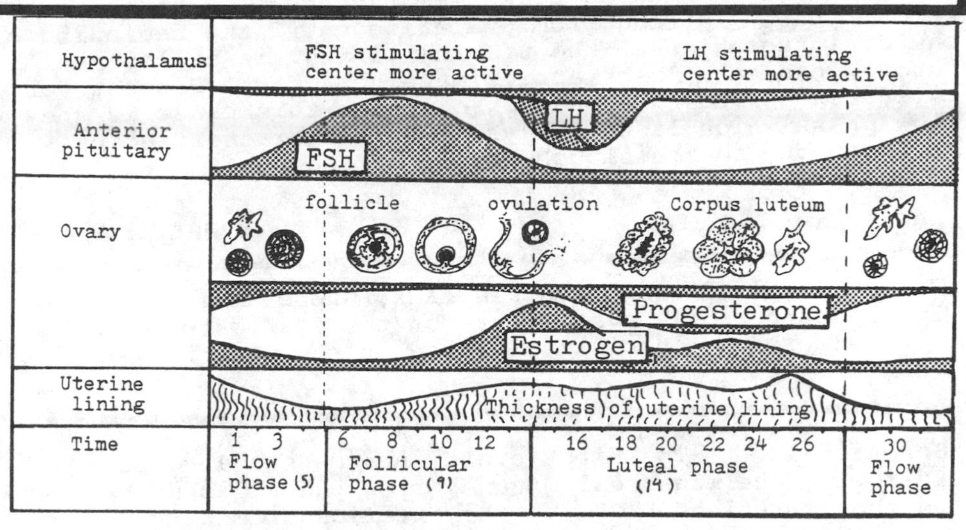

The sequence of events in the 28-day human menstrual cycle.

Solution: Under the influence of primarily FSH from the pituitary, a single ovarian follicle containing an ovum reaches maturity in about two weeks. During the second week the follicle cells and other ovarian cell types are stimulated by FSH to increase their secretion of estrogen. Near the middle of the cycle, or about the fourteenth day, the heightened level of estrogen in the blood reaches a critical value which stimulates the hypothalamus to signal the pituitary to release a surge of LH. This surge of LH induces rupture of the ovarian follicle. This releases the mature ovum into the fallopian tube of the uterus, a process called ovulation. Estrogen secretion decreases for several days following ovulation, perhaps because of the negative feedback effect of excess estrogen on FSH, which stimulates its production. The ruptured follicle is rapidly transformed into the corpus luteum, which secretes progesterone and, in lesser amounts, estrogen. The estrogen secreted by the corpus luteum raises its level in the blood, which inhibits further FSH secretion. Progesterone suppresses additional LH production, and thus prevents additional ovulation from occurring. The combined influence of progesterone and estrogen thickens and prepares the uterus for implantation of the embryo (in the event of fertilization). Failure of oocyte fertilization leads to the degeneration of the corpus luteum during the last few days of the cycle. The disintegrating corpus luteum is unable to maintain its secretion of estrogen and progesterone, and their blood concentrations drop rapidly. The marked decrease of estrogen and progesterone lead to degeneration of the uterine wall followed by sloughing off of the uterine surface tissue. It also results in the removal of inhibition of FSH secretion. The blood concentration of FSH begins to rise, follicle and ovum development are stimulated, and the cycle begins anew.

The cycle is usually broken up into three general phases. The first phase, encompassing follicle development up to ovulation, is called the follicular phase and lasts about 9 days FSH and estrogen dominate during the phase. The second phase, lasting from ovulation until the beginning of the disintegration of the uterine lining, is called the luteal phase and lasts approximately 14 days. Progesterone dominates during this phase. Following this phase is the flow phase which lasts 5 days, and is characterized by bleeding due to the sloughing off of the uterine lining.

• PROBLEM 21-29

The principle of negative feedback, where an increase in the output of a system acts on the system to halt its further production, is manifested in the menstrual cycle. Where?

Solution: The first event in a menstrual cycle following the menstrual phase (flow phase) is an increase in the secretion of FSH by the anterior pituitary as a result of stimulation by the hypothalamus. This hormone stimulates the development of the follicles in the ovaries, which contain the eggs. After some time, one of the follicle gains predominance. At this point, only this one follicle continues to grow. Under the combined influence of both FSH and LH, the latter of which is maintained, at a low level in the blood by the follicular cells, the cells surrounding the egg within the follicle begin to secrete estrogen. As more and more estrogen is produced, the increased level of this hormone in the blood exerts an inhibitory effect on the FSH-stimulating center in the hypothalamus. This results in a decrease in the level of FSH produced by the anterior pituitary. This is the negative feedback mechanism between estrogen and FSH. High concentration of a secreted substance (estrogen) turns off the original stimulus for its production (FSH), so that an excessive amount of the secreted substance (estrogen) would not be made. This occurs roughly 8 to 10 days following the menstrual phase, or 13 to 15 days after the start of the cycle. While the high level of estrogen inhibits FSH production, it activates the LH-stimulating center in the hypothalamus, ultimately inducing a surge of LH into the circulatory system, which initiates ovulation.

After ovulation the ovarian follicle with the egg discharged develops into the progesterone-secreting corpus luteum. The corpus luteum promotes the plasma progesterone level to such a state that progesterone begins to exert an inhibitory effect on the LH-stimulating center in the hypothalamus. This is the negative feed-back mechanism between progesterone and LH. The high progesterone concentration suppresses further release of LH so that another ovulation is prevented; hence no other corupus luteum is formed to contribute more progesterone to the circulatory system.

• PROBLEM 21-30

What would happen if, following ovulation, no corpus luteum was formed?

Solution: After rupture of the follicle and discharge of the ovum during ovulation, a transformation occurs within the remaining follicle in the ovary, giving rise to a yellowish glandlike structure called the corpus luteum. If pregnancy does not occur, the corpus luteum degenerates in 10 days. If pregnancy does occur, the corpus luteum grows and persists until near the end of pregnancy. The corpus luteum, which is composed of follicular cells, continues to secrete estrogen as the follicular cells did prior to ovulation. In addition, the corpus luteum cells also secrete progesterone.

Progesterone functions in preparing the uterus to receive the embryo. This hormone acts on the uterine lining, causing maturation of the complex system of glands in the lining. Estrogen also acts on the uterus to cause a thickening of the lining. This, however, occurs primarily during the stage of follicle formation prior to ovulation, called the follicular phase. The luteal phase follows ovulation, and during this phase progesterone has the primary influence on the development of the uterine lining. This hormone also stimulates breast growth, particularly of glandular tissue.

If a female failed to develop a corpus luteum following ovulation, a hormonal imbalance would result. There would be a lower than normal concentration of estrogen and progesterone in the circulation. As the levels of these two hormones become low, the feedback inhibition of FSH and LH production is removed. The FSH concentration begins to rise in the plasma, and when it reaches a sufficiently high level, follicle and oocyte development are stimulated. Thus a new menstrual cycle would begin shortly after ovulation if the corpus luteum did not develop. Even worse would be the consequence if fertilization occurred and the corpus luteum did not form. The uterus would then fail to mature and implantation of the fertilized egg would be prevented. Even if implantation did occur, subsequent development would be abnormal because of incomplete placental development.

• PROBLEM 21-31

Profound changes in uterine morphology occur during the menstrual cycle and are completely attributable to the effects of estrogen and progesterone. Describe these changes.

Solution: In the uterus estrogen stimulates growth

of the smooth muscle layer called the myometrium, and the glandular epithelium called the endometrium, lining its inner surface. Progesterone acts on the endometrium and converts it to an actively secreting tissue. The glands become coiled and filled with glycogen; the blood vessels become more numerous; various enzymes accumulate in the glands and connective tissue of the lining. These changes are ideally suited to provide a favorable environment for implantation of a fertilized ovum. Progesterone also causes the mucus secreted by the cervix (a muscular ring of tissue at the mouth of the uterus which projects into the vagina) to become thick and sticky. This forms a "plug" which may constitute an important blockage against the entry of bacteria from the vagina. This is a protective measure for the fetus should conception occur.

As a result of the regression of the corpus luteum when no fertilization occurs, there is a fall in blood progesterone and estrogen levels. Consequently, the highly developed endometrial lining becomes deprived of its hormonal support. The most immediate result of this deprivation is the constriction of the uterine blood vessels, resulting in a decreased flow of oxygen and nutrients to the tissue. Subsequently, the lining disintegrates and gets sloughed off as the mentrual flow begins. This phase is called the menstrual phase and usually lasts for about 5 days. The flow ceases as the endometrium repairs itself, and then grows under the influence of the rising blood estrogen concentration. This phase is called the proliferative phase, and lasts for 9 days. This phase covers the time from menstruation up until ovulation. After ovulation and formation of the corpus luteum, progesterone acts synergistically with estrogen to induce formation of the glandular endometrium. This period, called the secretory phase (lasting about 14 days) is terminated by disintegration of the corpus luteum. Thus the cycle is completed and a new one begins.

The phases of the menstrual cycle can be named either in terms of the ovarian or uterine events. The cycle is divided into follicular and luteal phases with respect to the ovary; proliferative and secretory phases are the terms referring to the uterine state. It is important to realize that the uterine changes simply reflect the effects of varying blood concentrations of estrogen and progesterone throughout the cycle.

Contraception

• PROBLEM 21-32

The"pill"is a birth control device available to women, and is one of the most effective and widely used contraceptives in today's society. How does it function? Are there any side effects when taking the pill?

Solution: The pill contains a small dose of synthetically produced female sex hormones, primarily estrogen and progesterone. These two substances act together to suppress ovulation, thus disrupting the normal reproductive process.

The progesterone-estrogen pill acts on the cervix of the uterus to increase mucus secretion which impairs the motility of sperms. It acts upon the uterine lining, preventing implantation because the fertilization of the egg is not in synchrony with the development of the uterus. But most importantly, the pill prevents FSH and LH secretions by direct effect on the pituitary, causing the uterus to be sustained in its post-ovulatory state. It achieves this by acting as a feedback inhibitor on the anterior pituitary, so that FSH and LH secretions are suppressed. With minimal amounts of FSH and LH, follicular maturation and ovulation cannot occur.

The pill must be taken daily for 20 or 21 days of the 28-day menstrual cycle, starting with the fifth day after the beginning of menstruation. When taken each day without fail, it is virtually 100% effective. There are, however, some side effects when using the pill. Up to 25% of the women taking the pill develop some of the outward symptoms of pregnancy: swelling and tenderness of the breasts, nausea, irritability, etc. Some users of the pill also develop increased susceptibility to inflammation of the veins combined with problems of blood clotting.

The principle concern over the use of the pill is that the levels of the synthetic estrogens in certain pills are dangerous, since these synthetic estrogens had been shown to be carcinogenic in experimental animals. There are two major points that should be considered in the use of the pill. In the first place, do oral contraceptives significantly threaten the users with blood clot problems? Secondly, is prolonged use of the pill by humans going to result in a marked rise in breast cancer? It is the view of many endocrinologists that the evidence in experimental animals suggests that such a threat may exist among women presently planning to use oral contraceptives.

• PROBLEM 21-33

One group of compounds that may become quite important in birth control is the prostaglandins. What are the advantages of prostaglandins in birth control?

Solution: The compounds known as prostaglandins have been identified as cyclic, oxygenated, 20-carbon fatty acids. There are now 16 known natural prostaglandins. The prostaglandins, which are secreted by the seminal vesicles and other tissues, appear to mediate hormonal action by influencing the formation of cyclic AMP. The end result of their actions is varied. Though they are expected to be valuable in treating many ailments, from nasal congestion to hypertension, their most significant use may be in birth control. Prostaglandins induce strong contractions in the uterus, and hence they may be powerful agents for abortion. As abortive agents, the prostaglandins are efficient in almost all of the cases where they are used. When more is known about the side effects of prostaglandins, it is believed that their application in abortion would be safer than mechanical means. The prostaglandins are also being considered as a once-a-month replacement for birth control pills. The powerful uterine contractions induced by prostaglandins prevent the fertilized ovum from being implanted. Since they can be introduced directly into the uterus through the vagina, the prostaglandins will probably be safer contraceptives than the oral chemical agents, which must travel through the blood and may affect other parts of the body.

Pregnancy and Parturition

• PROBLEM 21-34

High levels of progesterone are essential for the maintenance of the uterine lining during pregnancy. But high levels of progesterone exerts a negative feedback action on LH secretion, with a consequent decrease in LH secretion and eventual atrophy of the corpus luteum. Atrophy of the corpus luteum cuts the progesterone supply resulting in menstruation. Clearly, these events can not take place after conception, or the embryo would get sloughed off. When conception occurs, the corpus luteum is seen to last through most of pregnancy. How can this be explained?

Solution: It has been shown that the portion of the placenta called the chorion secretes a hormone, called chorionic gonadotrophin, that is functionally very similar to LH. The function of this hormone is to take the place of LH in preserving the corpus luteum, LH production being inhibited by a high progesterone level. The corpus luteum is then able to secrete progesterone at high levels without being shut off.

The amount of this chorionic hormone produced in a pregnant woman is very great. A pregnant woman produces so much of this hormone that most of it is excreted in the urine. Many commonly used tests for pregnancy are based on this fact. One test involves the injection of the patient's urine into a test animal such as a rabbit. If chorionic hormone is in the urine, development of a corpus luteum within 24 hours will take place in the rabbit, and pregnancy can be confirmed. Although the corpus luteum is essential during early pregnancy and is present during most of pregnancy, it has been shown that it is no longer necessary after about the first two months. It seems that the placenta begins to secrete progesterone (and estrogen) early in pregnancy, and once this secretion has reached a sufficiently high level the placenta itself can maintain the pregnancy in the absence of the corpus luteum.

• PROBLEM 21-35

What hormonal changes occur during the delivery of an infant (parturition)?

Solution: Present theory holds that a shift in the balance of estrogen and progesterone is one important factor in parturition. Estrogen is known to stimulate contractions of the uterine muscles and progesterone is known to inhibit muscular contraction. It is known that just prior to parturition, estrogen secretion by the placenta rises sharply, and this increase may play a role in the contraction of the uterus. It is therefore believed that during pregnancy progesterone suppresses contraction of the uterine muscles, but the rise of estrogen late in pregnancy overcomes the effects of the progesterone and initiates the contractions necessary for birth. Oxytocin, one of the hormones released from the posterior pituitary, is an extremely potent uterine muscle stimulant and is released as a result of stimulation of the hypothalamus by receptors in the cervix. Relaxin a hormone secreted by the ovaries and placenta during pregnancy, is another hormone which may be important in parturition. Relaxin loosens the connections between the bones of the pelvis, thereby enlarging the birth canal to provide easy exit for the newborn. Prostaglandins are fatty acid derivatives secreted by animal tissues. Prostaglandins stimulate the smooth muscle of many organs, including the smooth muscle of the wall of the uterus.

Menopause

• PROBLEM 21-36

The process of ovulation continues until a period of the female's life called the menopause. What are the changes that occur during this period?

Solution: The menopause usually comes between the ages of 45 and 50. It is apparently attributable to two reasons: (1) the cessation of secretion of LH by the pituitary as a result of declining activity of the LH-stimulating center in the hypothalamus, and (2) the declining sensitivity of the ovaries to the stimulatory activity of gonadotropins. In the absence of LH, secretory activity in the ovaries falls off. Neither estrogen nor progesterone can be produced in significant quantities. Ovulation and menstruation become irregular and ultimately cease completely. Some secretion of estrogen generally continues beyond these events but gradually diminishes until it is inadequate to maintain the estrogen-dependent tissues, most notably the breasts and genital organs, which become atrophied. Sexual drive, however, is frequently not diminished and may even be heightened. Severe emotional disturbances are not uncommon during menopause and are generally ascribed not to the direct effect of estrogen deficiency, but to the disturbing nature of the whole period-the awareness that reproductive potential has ended. Woman during menopause often experience hot flashes. These result from dilation of the skin arterioles, causing a feeling of warmth and marked sweating. The reason for this is unknown. Many of the symptoms of the menopause can be reduced by the administration of small amounts of estrogen.

Just recently, another aspect of menopause was discovered, that being the relationship between plasma estrogen and plasma cholesterol. Estrogen significantly lowers the level of plasma cholesterol. This explains why women have a significantly lower incidence of arteriosclerosis (a disease positively correlated to the level of cholesterol in the plasma) than men until the menopause, when the incidence becomes similar in both sexes.

Furthermore, decreased levels of estrogens are implicated in osteoporosis since estrogen normally stimulates osteoblasts to form new bone.

• PROBLEM 21-37

Are there comparable changes, in the male, to female menopause?

Solution: The changes that occur in the male due to aging are much less drastic than those which occur in the female. The testosterone and pituitary secretions which are initiated when the male reaches puberty continue throughout his adult life. Should the testosterone level decrease, this would result, through the switching-off of the negative-feedback inhibition on the pituitary, in a rise in the pituitary secretions of FSH and LH. Increased levels of FSH and LH stimulate testosterone production. Despite this, there is a steady decrease in testosterone secretion in later life, but this is probably caused by the slow deterioration of testicular function. Despite this decrease, testosterone secretion remains high enough in most men to maintain sexual vigor throughout life, and fertility has been documented in men in their seventies and eighties. Thus, there is usually no complete cessation of reproductive function in males analogous to female menopause.

SHORT ANSWER QUESTIONS FOR REVIEW

Answer

To be covered when testing yourself

Choose the correct answer.

1. In which of the following cases will gastric secretion by the stomach wall fail to occur? (a) Feeding a dog whose esophagus has been surgically bisected and both ends led to the outside through an incision in the neck. (b) Feeding a dog whose nerves leading to the stomach have been severed. (c) Inserting partially digested food in the end of the esophagus, and allowing it to pass to the stomach without the dog having tasted it. (d) Stimulation of the stomach wall of the dog with a glass rod. (e) Inserting partially digested meat directly into the stomach. — b

2. Which of the following is (are) correct? (a) Gastrin is released by stimulation of the pyloric region of the stomach. (b) Enterogastrone is released by stimulation of the duodenal wall of the small intestines. (c) Gastrin secretion results in the inhibition of the gastric glands. (d) Enterogastrone secretion results in the inhibition of the gastric glands. (e) Secretin is released by the mucosal cells of the small intestine when stimulated by the basicity of food coming from the stomach. — a,b,d

3. Which of the following hormones stimulates release of bile from the gall bladder? (a) Glucagon, (b) histamine, (c) gastrin and Secretin, (d) gastrin and Cholecystokinin, (e) cholecystokin. — e

4. Which substance may constrict the bronchioles, resulting in asthma? (a) Calcitonin, (b) thymosin, (c) melatonin, (d) histamine, (e) vasopressin. — d

5. Which are the following statements are true concerning diabetes? (a) Diabetics excrete an unusually small amount of urine. (b) Diabetes results from the over-production of insulin. (c) Diabetes results from the inability of the liver to convert glucose into glycogen due to a lack of insulin. (d) All of the above. (e) None of the above. — c

6. The hormone glucagon: (a) has the opposite effect on the liver that insulin does, (b) converts glycogen into glucose, (c) is produced in the beta cells of the pancreas, (d) a + b, (e) b + c. — d

7. It has long been known that kidney malfunction is commonly associated with high blood pressure (hypertension). What is a possible explanation

Answer

To be covered when testing yourself

of this? (a) Underproduction of thyroxin, (b) overproduction of the hormone called renin, (C) relaxation of the smooth muscles in the walls of the blood vessels, (d) decrease in the level of secretion of aldosterone by the adrenal cortex, (e) none of the above.

b

8. Goiter is a disease associated with a malfunctioning of the thyroid. Which is not a symptom of this disease? (a) Swollen neck, (b) dry and puffy skin, (c) loss of hair, (d) rapid heartbeat (e) mental dullness.

d

9. Which of the following hormones is produced by the thyroid? (a) Thyroxin, (b) triiodothyronine, (c) calcitonin, (d) all of the above, (e) none of the above.

d

10. The level of calcium concentration in the blood is controlled by two hormones. In which manner do these hormones operate? (a) Calcitonin results in a decrease of blood calcium levels, while parathormone increases the calcium level in the blood. (b) Calcitonin results in an increase of blood calcium levels, while parathormone decreases the calcium level in the blood. (c) Calcitonin and parathormone both increase the calcium level in the blood. (d) Calcitonin and parathormone both decrease the calcium level in the blood. (e) Calcitonin and parathormone have no effect on calcium levels in the blood.

a

11. A person is afflicted with a disease resulting in weak bones which are easily fractured. This problem resulted from an excessive depletion of calcium reserves in the bones. What is a possible cause? (a) A tumor of the parathyroid resulting in oversecretion of parathormone, (b) a disease in which parathyroid tissue is gradually destroyed, (c) a tumor of the thyroid resulting in oversecretion of calcitonin, (d) none of the above, (e) all of the above.

a

12. Which of the following hormones are secreted by the adrenal medulla? (a) Cortical sex hormones and aldosterone, (b) aldosterone and adrenaline, (c) noradrenalin and adrenalin, (d) cortisone and noradrenalin (e) none of the above.

c

13. The "Fight-or-Flight" reaction is characterized by the increase of glucose levels and oxygen content to the skeleton and muscles, with a decrease in glucose level and oxygen content to the digestive system. What hormone(s) are responsible for this reaction? (a) Adrenalin, (b) noradrenalin, (c) cortisone, (d) a + b, (e) b + c.

d

Answer

To be covered when testing yourself

14. Which of the following hormones should be associated with the bearded lady in the circus? (a) Glucocorticoids, (b) cortical sex hormones, (c) testosterone, (d) progesterone, (e) ecdysone. — b

15. An animal sees the approach of a potential mate. Which is the correct pathway that leads to the production of the proper sex hormones in response to this situation? (a) Stimulation of hypothalamus; secretion of gonadotrophic releasing factors; secretion of gonadotrophic hormones; stimulation of gonads; sex hormone secretion, (b) secretion of gonadotrophic releasing factors; stimulation of hypothalamus; secretion of gonadotrophic hormones; stimulation of gonads; sex hormone secretion, (c) secretion of gonadotrophic hormones; stimulation of hypothalamus; secretion of gonadotrophic releasing factors; stimulation of gonads; sex hormone secretion, (d) all of the above are reasonable pathways, (e) none of the above are reasonable pathways. — a

16. Vasopressin and oxytocin are two hormones associated with the (a) posterior pituitary, (b) anterior pituitary, (c) pineal gland, (d) thyroid, (e) thymus. — a

In questions 17-36 match each function to its appropriate hormone.

17. stimulates follicle formation in ovaries and sperm formation in testes
18. stimulates formation of corpus luteum in ovaries and secretion in testes
19. stimulates secretion of thyroxin from the thyroid
20. stimulates secretion of cortisone from the adrenal cortex
21. stimulates secretion of milk in mammary glands
22. controls rate of metabolism and physical and mental development
23. controls calcium metabolism

(a) LH
(b) ACTH
(c) FSH
(d) TSH
(e) prolactin
(f) vasopressin
(g) oxytocin
(h) thyroxin
(i) calcitonin
(j) parathormone
(k) glucagon
(l) aldosterone
(m) insulin

17-c, 18-a, 19-d, 20-b, 21-e, 22-h, 23-i,

Answer

To be covered when testing yourself

24. controls narrowing of arteries and rate of water absorption in kidney tubules
25. stimulates contractions of smooth muscles of uterus
26. regulates calcium and phosphate level of blood
27. stimulates formation of antibody system
28. promotes storage and oxidation of glucose
29. releases glucose into the blood stream
30. promotes glucose formation from amino acids and fatty acids
31. controls water and salt balance
32. influences sexual development
33. releases glucose into bloodstream, increase rate of heartbeat, increases rate of respiration, reduces clotting time, relaxes smooth muscle in air passage
34. controls female secondary sex characteristics
35. helps maintain attachment of embryo to mother
36. controls male secondary sex characteristics

(n) thymus hormone

(o) cortisones

(p) estrogen

(q) testosterone

(r) sex hormone

(s) progesterone

(t) epinephrine or norepinephrine

24-f, 25-g, 26-j, 27-n, 28-m, 29-k, 30-o, 31-l, 32-r, 33-t, 34-p, 35-s, 36-q.

Question 37-40, match answer to definition.

37. Non-protein portions of enzymes which help bind an enzyme to its substrate

(a) hormone

(b) pheromone

37-c,

Answer

To be covered when testing yourself

38. An organic substance synthesized by a specific organ or tissue and secreted into the blood	(c) vitamin (d) enzyme	38-a,
39. Organic catalysts produced by living organisms, and primarily composed of protein		39-d,
40. Chemicals involved in the sexual attraction mechanism of moths and ants.		40-b.

Fill in the blanks.

41. The hormone secreted by the posterior pituitary and responsible for contraction of the uterine muscles is _____.	oxytocin
42. Vasopressin reabsorbs water at the kidney. An alternate name for this hormone is _____.	anti-diuretic hormone
43. The entrance of proteins into the stomach provides the major chemical stimulus for the release of the hormone _____.	gastrin
44. The hormone which is responsible for decreasing the level of gastric juices secreted by the stomach is _____.	entero-gastrone
45. The two hormones secreted by the endocrine portion of the pancreas are _____ and _____.	insulin, glucogon
46. TSH produced by the pituitary gland stimulates the secretion of the hormone _____.	thyroxin
47. Follicles in the ovaries produce the hormone _____.	estrogen
48. The endocrine gland located on top of the kidney is the ______.	adrenal gland
49. The hormone responsible for maturation of the seminiferous tubules is _____.	testos-terone
50. The mammary gland secretes milk when the pituitary hormone _____ is present.	prolactin

CHAPTER 22

REPRODUCTION

ASEXUAL VS. SEXUAL REPRODUCTION

• PROBLEM 22-1

Define and compare asexual reproduction with sexual reproduction in animals. Give examples.

Solution: The fundamental difference between asexual reproduction and sexual reproduction is the number of parents involved in the production of offspring. In asexual reproduction only a single parent is needed. Offspring identical to the parent are produced when the parent splits, buds, or fragments. Asexual reproduction typically occurs in animals such as Euglena, Paramecium, Amoeba, Hydra, flatworms, and starfish.

Two parents are usually involved in sexual reproduction. A fertilized egg is produced through the union of specialized sex cells (egg and sperm) from each parent. Instead of having traits identical to a parent, the offspring possess a variety of recombined traits inherited from both parents. In higher animals such as man, sexual reproduction is the most common form.

It must be noted that in some cases, sexual reproduction can take place with a single parent. In the hermaphroditic worms, such as the earthworm and the fluke, both male and female sex organs occur in the same individual. Whereas self-fertilization is avoided in the earthworm by certain adaptive mechanisms, it does take place in the parasitic fluke. This is an exception to the general observation that sexual reproduction involves two parents.

• PROBLEM 22-2

Differentiate between fission, budding, and fragmentation as means of asexual reproduction.

Solution: Most protozoans reproduce asexually by fission, which is the simplest form of asexual reproduction. Fission involves the splitting of the body of the parent into two approximately equal parts, each of which becomes an entire, new, independent organism. In this case, the cell division involved is mitotic.

Hydras and yeasts reproduce by budding, a process in which a small part of the parent's body separates from the rest and develops into a new individual. It may split away from the parent and take up an independent existence or it may remain attached and maintain an independent yet colonial existence.

Lizards, starfish, and crabs can grow a new tail, leg, or arm if the original one is lost. In some cases, this ability to regenerate a missing part occurs to such an extent that it becomes a method of reproduction. The body of the parent may break into several pieces, after which each piece regenerates its respective missing parts and develops into a whole animal. Such reproduction by fragmentation is common among the flatworms, such as the planaria.

• PROBLEM 22-3

While encystment is not, in most cases, a means of reproduction, still it is a means of continuation of the species. Explain how encystment is used, using a cyst-forming organism as illustration.

Solution: Encystment is used by a number of protozoans which live in the water. Should their pond become desiccated or uninhabitable, these organisms would develop hard, protective coatings and the organisms would go into a state of dormancy. In this encysted state they can withstand extremely unfavorable conditions. When favorable condirions return (such as clean water), the wall of the cyst softens and the organisms escape.

Some pathogenic protozoans develop cysts which play a role in the spread of infection. Entamoeba histolytica is a parasitic protozoan which lives in the lining of the human large intestine, causing the disease amoebic dysentery. This parasite secretes a tissue-dissolving enzyme which causes ulceration of the large intestine. Some of these protozoans die when they pass from the body into the feces, but some of these organisms develop cysts and continue to live in dry, unfavorable conditions outside the body for a long time. Should such cysts enter the body of another person, the wall of the cysts would dissolve and the parasites would escape into the new host's body. A new infection would thus begin.

• PROBLEM 22-4

Name an organism which produces spores in its life cycle and explain how this aids in its proliferation.

Solution: While spores are primarily used as a means of

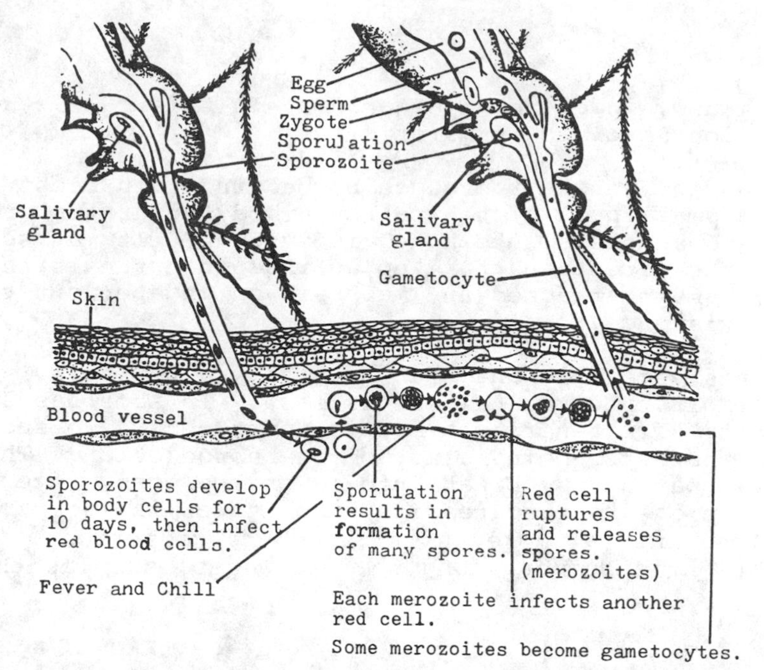

A diagram of the life cycle of the malaria parasite, Plasmodium.

reproduction in plants, there is one class of protozoans (the Sporozoa) which produces spores. These spores are highly resistant forms of the organism which can withstand extreme environmental conditions and are specialized for asexual reproduction.

The sporozoan Plasmodium is a spore-forming protozoan that causes the disease malaria. Plasmodium spores are produced initially by sexual reproduction in the body of a mosquito (see figure 1) which has sucked the blood of an individual infected with the organism. These spores (sporozoites) may now be injected into another individual's bloodstream when the mosquito bites the new host. After invading the red blood cells, the spores reproduce asexually, producing many spores (merozoites) within each invaded cell. This causes the cell to burst, releasing the spores. Each is able to invade another red blood cell. It is thus through repeated sporulation that the organism proliferates in the bloodstream of the infected individual. Some merozoites develop into male and female gametocytes. It is these gametocytes which a mosquito sucks up when she bites. In the mosquito the gametocytes develop into gametes. By sexual reproduction, the gametes fuse into zygotes which become sporozoites and the cycle repeats.

• PROBLEM 22-5

Differentiate between hermaphroditism and parthenogenesis.

Solution: Parthenogenesis is defined as the development of an egg without fertilization. A classic example of parthenogenesis is found among the Hymenoptera (ants, bees, and wasps). In the honeybees, males (or drones) develop from unfertilized eggs. The queen bee, though having been inseminated, has the ability to lay some eggs that have not been fertilized. The sperm she receives from a male bee (males are all haploid and produce gametes by mitosis) are stored in a pouch connected to her genital tract but closed off by a muscular valve. As the queen lays eggs, she may either open this valve, permitting the sperm to fertilize the eggs,or keep the valve closed, so that the eggs are laid without being fertilized. Fertilization usually occurs in the fall, and the fertilized eggs are quiescent during the winter. The fertilized eggs become females (queens and workers) and the unfertilized eggs become males (drones).

Eggs from species that normally do not exhibit parthenogenetic development may be stimulated artificially to develop without fertilization. This can be done by changing the temperature, pH or salinity of the surrounding media, or by chemical or mechanical stimulation of the egg. Eggs of the sea urchin (an echinoderm) will develop if they are treated with strong salt water, pricked with a needle, or exposed to organic acids. Frog eggs can also be stimulated. In addition, it has been possible to stimulate rabbit eggs to successfully develop parthenogenetically provided the cleaving egg that results is placed in the uterus of a hormonally primed female. However only weak and sterile female rabbits have been produced parthenogenetically.

Hermaphroditism refers to the presence within a given individual of both male and female sex organs. Many of the lower animals are hermaphroditic. Some, such as the parasitic flukes, are self-fertilizing, but more often two hermaphroditic animals copulate and each inseminates the other. This latter method is used by earthworms. In other species, self-fertilization is prevented by the development of the testes and ovaries at different times.

• PROBLEM 22-6

Distinguish between viviparous, ovoviviparous, and oviparous reproduction.

Solution: The females of all birds, amphibians and bony fish, most insects, and many aquatic invertebrates lay eggs from which the young eventually hatch; such animals are said to be oviparous (egg-bearing). In such cases, the major part of embryonic development takes place outside the female body, even though fertilization may be internal. The eggs of oviparous animals therefore contain relatively large amounts of yolk, which serves as a nutrient source for the developing embryo. Mammals, on the other hand, have small eggs with comparatively little yolk. The mammalian embryo develops within the female's body, deriving nutrients via the maternal bloodstream, until its development has proceeded to the stage where it can survive independently. Such animals are termed viviparous (live-bearing).

The third type of reproduction is intermediate in character. This is ovoviviparous reproduction and it involves the production of large, yolk-filled eggs which remain in the female reproductive tract for considerable periods following fertilization. The yolk of the egg is the nutrient source for the developing embryo, which in this case, usually forms no close connection with the wall of the oviduct or uterus and does not receive nourishment from the maternal blood. A diverse group of animals, including snails, trichina worms, flesh flies, sharks, and rattlesnakes are in this category.

GAMETOGENESIS

• PROBLEM 22-7

Describe the events of meiosis and account for its importance.

Solution: Meiosis is the process of gamete formation in which two cell divisions occur in succession to give rise to four haploid daughter cells with each bearing only one copy of an individual chromosome, unlike the diploid parent cell which carried two copies.

The sequence of events in meiosis begins as in mitosis. Chromosomes that have already duplicated in interphase begin to shorten in the first meiotic prophase to form condensed, easily visible structures; each consists of two identical chromatids joined together at the centromere. It is during this condensation of the chromosomes that the first unique event of meiosis occurs: the synapsis (pairing) of homologous chromosomes, each made up of two chromatids. Each joined pair of homologous chromosomes is known as a tetrad. As the chromatids become more and more condensed, it is possible to see points at which they connect to one another, forming cross-bridges called chiasmata. These may be considered to be the visible manifestations of crossing over between a pair of chromatids. Only two of the four chromatids (non-sister chromatids) present in the tetrad can be involved in the formation of a chiasma. The end of the first meiotic prophase is marked by the disappearance of the nuclear membrane and the nucleolus.

At the first meiotic metaphase, the homologous pairs (tetrads), having reached maximum condensation, line up along the equator of the meiotic spindle. This is followed by the first meiotic anaphase, during which the chromosomes separate from their homologs without separation of the sister chromatids, which remain attached at the centromere. Gradually, the duplicated chromosomes, now known as dyads, move to the two poles of the meiotic spindle.

At telophase, when the chromosomes have reached the two poles, cytoplasmic division occurs. At this point, the chromosomes may become less condensed and enter into a brief, second meiotic interphase, during which the nuclear membrane forms again. The chromosomes grow diffuse but do not replicate. A second prophase follows rapidly, and the chromosomes again condense. There is, however, no synapsis, because the homologous chromosomes had separated at the first meiotic anaphase. The nuclear membrane and the nucleolus now disappear as during the first meiotic division. Those species which do not exhibit a second interphase go directly from the first telophase into a brief second prophase,

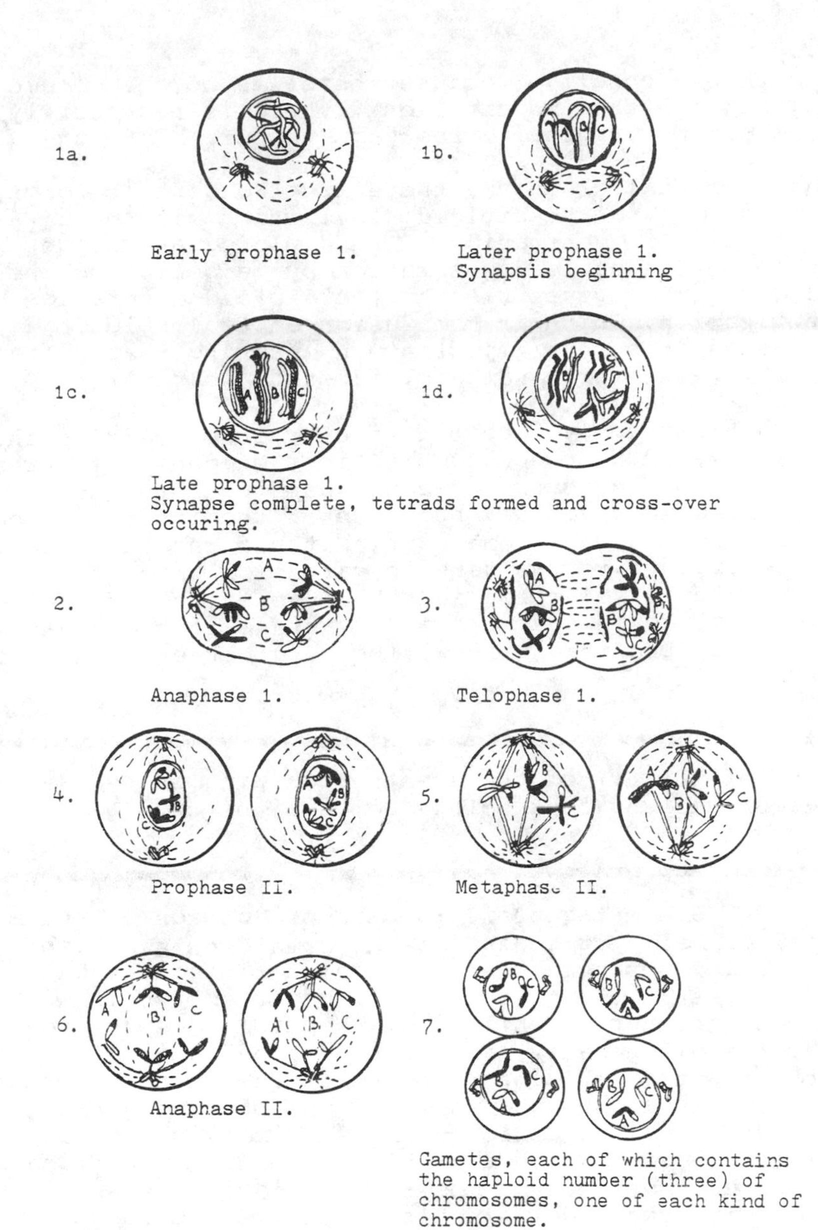

Early prophase 1.

Later prophase 1.
Synapsis beginning

Late prophase 1.
Synapse complete, tetrads formed and cross-over occuring.

Anaphase 1.

Telophase 1.

Prophase II.

Metaphase II.

Anaphase II.

Gametes, each of which contains the haploid number (three) of chromosomes, one of each kind of chromosome.

during which spindle formation occurs. During metaphase, the two members of each dyad line up at the equator.

At the second meiotic anaphase, the sister chromatids separate at the centromere, one going to each pole. When they have reached the poles, the nuclear membrane and nucleolus reappear. Cytoplasmic cleavage follows, giving rise to distinct daughter cells. After the second meiotic division, the number of chromosomes in each cell is half the original number, and reduction is complete.

The process of meiosis began with a single cell (2n) which doubled its chromosomes, resulting in four copies of each type of chromosome (4n).

Two meiotic cell divisions then produced four cells

(1n) from this one cell. Because there was no chromosome doubling between the two divisions, there is necessarily a halving of the original chromosome number.

In any sexual organism, there are two genetic forms, the haploid (1n) and the diploid (2n), which alternate from generation to generation. The diploid generation gives rise to the haploid generation by meiosis, and the haploid generation gives rise to the diploid generation by fertilization, which is the fusion of two haploid cells. Meiosis is thus important in ensuring that, after fertilization, the original diploid state is retained.

Another extremely important factor in meiosis is the possibility of getting new combinations of genes as a result not only of crossing over between synapsing chromosomes in the first meiotic division, but also of the random segregation of the homologous chromosomes during the first anaphase. These factors contribute to the genetic variation which forms the basis for natural selection and thus evolution. If the new combination of genes confer some advantage to the organism possessing it, that organism will have a better chance of survival.

• PROBLEM 22-8

Give a brief summary of the events occurring in spermatogenesis and oogenesis. Compare the two processes.

Solution: Spermatogenesis is the production of sperm (male sex cells). It begins with spermatogonia, which are primitive, unspecialized germ cells lining the walls of the tubules in the testes. Throughout embryonic development and during childhood, the diploid spermatogonia divide mitotically, giving rise to additional spermatogonia to provide for growth of the testes. After sexual maturity is reached, some of the spermatogonia grow and enlarge into cells known as primary spermatocytes (see figure), which are still diploid. The primary spermatocytes then undergo the first meiotic division, each producing two equal-sized secondary spermatocytes. Each of these in turn undergoes the second meiotic division, forming two spermatids of equal size. The haploid spermatids then undergo a complicated process of maturation, including development of a tail, to become functional sperm.

Oogenesis is the process of ovum (female gamete) formation. The immature germ cells of the female are called oogonia and are located in the ovary. Early in development, they undergo numerous mitotic divisions. Then some oogonia develop and enlarge into primary oocytes. At maturation, the primary oocyte undergoes the first meiotic division. The nuclear events are the same as in spermatogenesis, but the cytoplasm divides unequally to produce two cells of different sizes. The small cell is the first polar body and the large cell is the secondary oocyte. This haploid secondary oocyte undergoes the second meiotic division, which also involves unequal, cytoplasmic

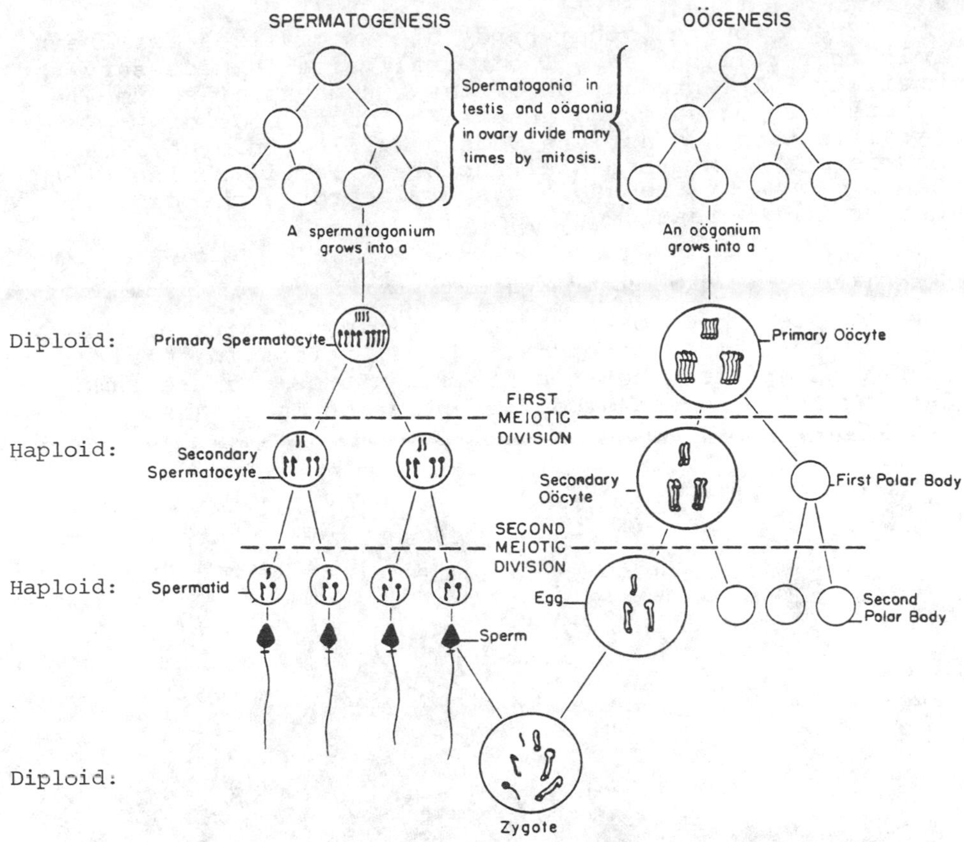

Comparison of the formation of sperm and eggs.

distribution, and forms a second polar body and a large ootid. The first polar body divides to form two additional second polar bodies. The ootid matures into the egg, or ovum. The polar bodies disintegrate and do not become functional gametes.

Both spermatogenesis and oogenesis involve the production and maturation of the gametes. However, there are a few notable differences intrinsic in each process with the most important being the differences in the amount of cytoplasm and the number of viable gametes produced.

The egg cell accumulates much more nutrient material during its development than does the sperm. The sperm actually loses cytoplasm during maturation. In addition, the unequal meiotic division of the oocyte results in a much larger gamete than would normal division. As a result, one viable gamete with a considerable amount of cytoplasm is produced in oogenesis. In spermatogenesis, four viable gametes with very little cytoplasm are produced. These differences are a reflection of the different roles of the egg and sperm in reproduction. It is the fertilized egg which will ultimately develop into the new individual. Large amounts of cytoplasm are necessary to provide adequate nutrition for the developing embryo. The unequal cytoplasmic division solves the problem of reducing chromosome number without losing cytoplasm.

Sperm, on the other hand, are specialized for their role in fertilization. Their small size is necessary for motility. The production of large numbers of sperm, in contrast to eggs, is to ensure the occurrence of fertilization against the odds resulting from physical barriers and environmental factors. Fertilization of the haploid ovum by the haploid sperm restores the diploid number in the resultant zygote.

• PROBLEM 22-9

The transformation of a spermatid into a functioning sperm is a process of differentiation. It varies in detail in different species, but the basic features are the same. Outline this process of differentiation in humans.

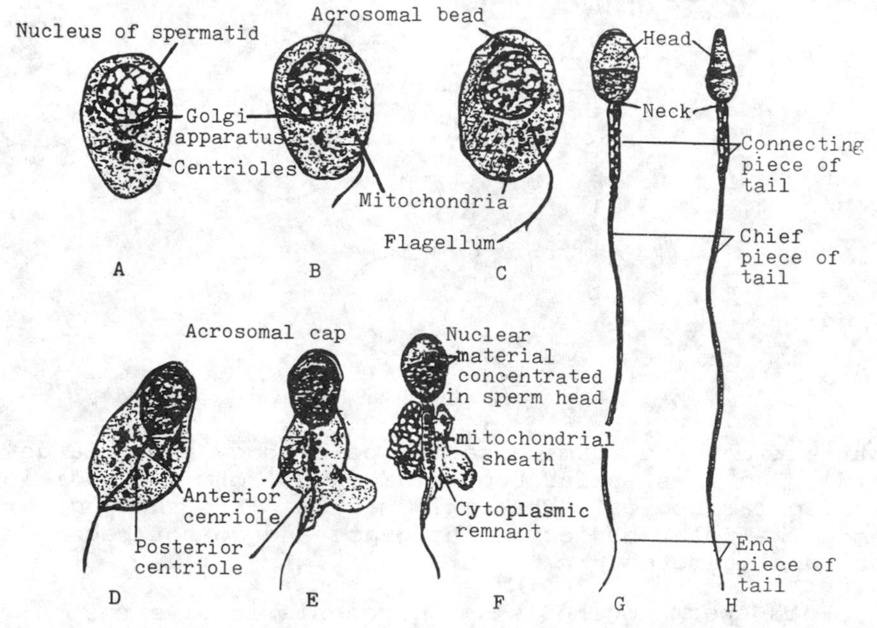

The development of the human sperm cell.

Solution: Within the cytoplasm of the human spermatid, one can find proacrosomic granules, centrioles, and mitochondria. During differentiation of the spermatid, a process called spermiogenesis, the proacrosomic granules, which were formed from the Golgi complex, gather toward one pole of the cell and condense into the form of a bead. This acrosomal bead contains appreciable amounts of carbohydrates and several enzymes of lysosomal nature. This region of the differentiating spermatid eventually becomes the acrosome, and is chiefly concerned with enzymatic penetration of the egg. The two centrioles of the spermatid move toward the opposite side of the nucleus. Here the centriole nearest the nucleus spins a filament that grows outward and becomes the axis of the sperm's tail or flagellum. This flagellum shows the usual 9 + 2 tubule pattern, and is often accompanied by additional fibrous structures that run alongside the tubules. (It is interesting to note that whereas basal bodies usually form flagella, here the centriole forms the flagellum.) The centrioles then move apart. The proximal one remains close to the posterior surface of the nucleus. The distal one forms a ring around the axial filament and recedes a short distance. Some of the mitochondria now gather around the axial filament in the space between the two centrioles and form a spiral in an area called the middle piece. Most of the cytoplasm pinches away and is discarded, but a little remains as a delicate sheath over the head area. Thus the resultant sperm consists of 3 pieces. The head contains an anterior acrosome, a posterior nucleus, and little cytoplasmic material, and by virtue of its lytic enzymes, functions in penetration of the egg. The midpiece contains the mitochondria necessary for energy-requiring motility. Finally, the tail with a 9 + 2 microtubule arrangement, contains the contractile machinery for whiplike movements.

It was shown that before fertilization, sperm in the vicinity of an ovum in the oviduct undergo a sequence of structural changes called the acrosome reaction. The outer membrane of the acrosome fuses at multiple points with the overlying plasma membrane of the sperm head to create numerous openings, through which the enzyme-rich contents of the acrosome are liberated in a process not unlike the release of secretory products from a glandular cell. The released enzymes help the sperm to digest its way through the protective coating of the egg. The result is the entry of the sperm into the egg cytoplasm.

• PROBLEM 22-10

In what ways are eggs and sperm adapted structurally for their respective functions in reproduction?

Solution: The human sperm consists of a head, a middle piece and a tail. About two-thirds of the head is comprised of the nucleus. The apical one-third of the head is the acrosome. The acrosome has a conglomerate of enzymes which aid the sperm in penetrating the egg surface in fertilization. Before a sperm can fertilize an egg, it must undergo a process called capacitation followed by the acrosome reaction. Capacitation is the breakdown or removal of a protective coating of the sperm. Following this, small perforations are produced in the acrosome wall, allowing for the escape of enzymes as part of the acrosome reaction.

The middle piece contains many mitochondria that provide energy for the movement of the tail. The mitochondria oxidize nutrients which diffuse into the middle piece from the seminal fluid and thus generate energy for the sperm's motion. The middle piece also contains a centriole which, after fertilization, aids in the division of the zygote. The tail lashes back and forth and propels the sperm.

The ovum is massive compared to the sperm, with an abundance of cytoplasm (recall the unequal cytoplasmic division during oogenesis) containing yolk granules. The yolk provides nutrition for the embryo during the early stages of its development. The human ovum is truly large and is just visible to the unaided eye as a tiny speck. Because of its large size and lack of flagella, the egg is nonmotile in contrast to the sperm. The ovum is surrounded by two extracellular layers which can be broken down by the enzymatic action of the sperm's acrosome. The inner layer is called the zona pellucida and the outer layer consists of follicular cells which become radially arranged as the corona radiata. The zona pellucida, upon enzymatic degradation by the acrosome, allows the sperm to pass through the egg cell membrane into the cytoplasm. It is also thought to be responsible for the inhibition of entry of more sperm, so that only one sperm will fertilize one egg. The mechanism by which this occurs is not well understood, but it appears to involve some structural changes of the zona pellucida controlled by granules released from the ovum's cytoplasm.

• PROBLEM 22-11

In addition to its function in secreting female sex hormones, the human ovary serves as the organ of egg production. Explain the process of egg formation in the mammalian ovary. What happens to the follicle after ovulation?

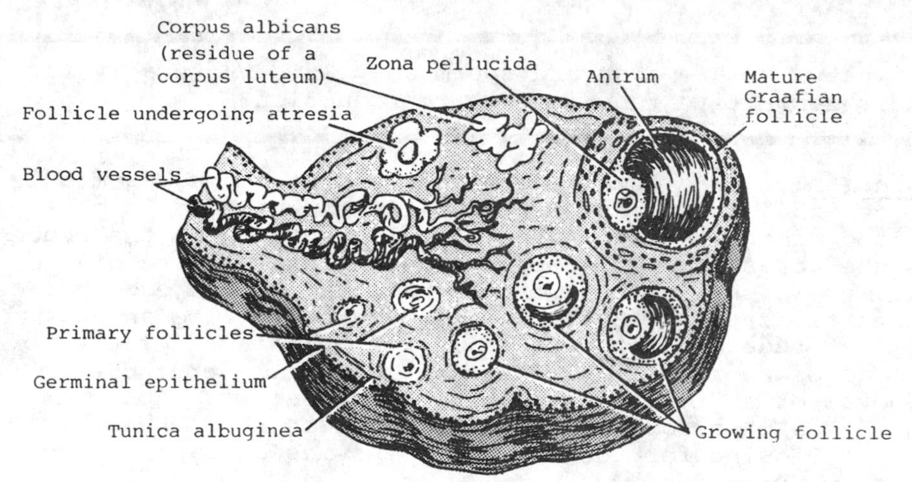

Diagram of the ovary, illustrating development and fate of ovarian follicles. (1) Oogonium, surrounded by follicular cells, the latter arising from germinal epithelium. (2) Primary follicle. (3-5) Maturing follicles. (6) Graafian follicle. (7) Ruptured follicle. (8) Corpus luteum. (9) Corpus luteum of menstruation.

Solution: The mammalian ovary (see figure) consists of two types of cells: the germ cells and the germinal epithelial cells. The germ cells are the oogonia, which will eventually become the eggs. The germinal epithelial cells line the walls of the ovary and will eventually form part of the follicle.

At birth, the human ovaries contain an estimated

400,000 immature eggs or oogonia. Thus, in marked contrast to the male, the newborn female already has all the germ cells she will ever have: only a few, perhaps 400, are destined to reach full maturity during her reproductive life. The process of maturation of an oogonium is called oogeneis. Its beginning is marked by the enlargement of the oogonium to form a primary oocyte. Following the formation of the primary oocyte, the prospective egg undergoes the first meiotic division. Although the nuclear division proceeds normally, there is unequal division of the cytoplasm, resulting in one large secondary oocyte and a smaller cell made up almost entirely of nuclear material, called the first polar body. The secondary oocyte also divides unequally in the second meiotic division, producing a large ootid and a small second polar body. The first polar body may or may not divide; in any event, the polar bodies all soon disintegrate. The ootid becomes the mature egg or ovum. The single haploid ovum that results from the meiotic divisions of the primary oocyte contains most of the original cytoplasm and also some stored food.

During oogenesis, the developing egg becomes increasingly surrounded by a layer of germinal epithelial cells, and by the secondary oocyte stage, it is well enveloped by these cells. The oocyte and its surrounding cells are together termed the follicle. After secondary oocyte formation is completed, about two weeks after the onset of oogenesis, the oocyte alone is released from the follicle in a process called ovulation. This involves the rupture of part of the ovarian wall and the subsequent expulsion of the oocyte from the ovary. The secondary oocyte then enters the Fallopian tube, and remains at the secondary oocyte stage until after fertilization. It is important to realize that the ovary is not directly connected to the Fallopian tube; during ovulation the secondary oocyte is expelled into the body cavity. However, because of the close proximity of the ovary and Fallopian tube and the presence of finger-like projections from the mouth of the Fallopian tube which help direct the oocyte into it, almost all oocytes that are ovulated do end up in the Fallopian tube. The follicular cells which stay behind in the ovary following ovulation enlarge greatly and become filled with a lipid substance; the entire gland-like yellowish structure is known as the corpus luteum (meaning "yellow body").

Under hormonal influence, the corpus luteum secretes progesterone and estrogen, two female sex hormones. These hormones stimulate the proliferation and vascularization of the uterine lining, preparing it for implantation of the embryo. If the ovum is fertilized, the corpus luteum enlarges to form a corpus luteum of pregnancy, which increases its hormone production to maintain pregnancy. If the ovum is not fertilized, the corpus luteum begins to degenerate about nine days after ovulation and is called a corpus luteum of menstruation.

FERTILIZATION

• PROBLEM 22-12

What are the results of fertilization?

Solution: There are four results of fertilization. First, fusion of the two haploid sex cells produces a zygote with a diploid number of chromosomes. For higher organisms in which the prominent generation is the diploid one, fertilization allows the diploid state to be restored. Second, fertilization results in the initiation of cleavage of the zygote. It sets off cleavage by stimulating the zygote to undergo a series of rapid cell divisions, leading to the formation of the embryo. Third, fertilization results in sex determination. It is at fertilization that the genetic sex of a zygote is determined. Hormonal factors, which are dependent upon the genotypic sex, regulate the development of the reproductive system during the embryonic period and of the secondary sex characteristics after birth, completing the sex differentiation of an individual. In humans, the male genetic sex is determined by the presence of the XY chromosomal pair, whereas the female is determined by the XX chromosomal pair. The fourth result of fertilization is the rendering of species variation. Because half of its chromosomes have a maternal source and the other half a paternal source, the zygote contains a new, unique combination of chromosomes, and thus a new set of genetic information. Hence fertilization provides for the genetic diversity of a species.

• PROBLEM 22-13

Trace the path of the sperm from the testes to its union with the egg in the human.

Solution: In order to trace this path, we must first outline some aspects of the human male and female reproductive systems, illustrated in Figures 1 and 2, respectively. In normal males, the testes lie in the scrotum, a sac attached to the lower anterior wall of the abdomen. The testes reside in the body cavity during early embryonic development, but before birth they descend into the cavities of the scrotum. The inguinal canals are connections between the scrotum and body cavity; these canals are blocked off by connective tissue after the testes descend. The testes are located in the scrotal sacs because the sperm within them require cooler temperatures than the internal body temperature in order to survive and mature.

Each testis consists of roughly 1,000 coiled tubules called seminiferous tubules. It is in these tubules that the germ cells produce the sperm. Sertoli cells are also present and nourish the developing sperm. Connected to the seminiferous tubules via fine tubes called vasa efferentia is the epididymis, which is a single,

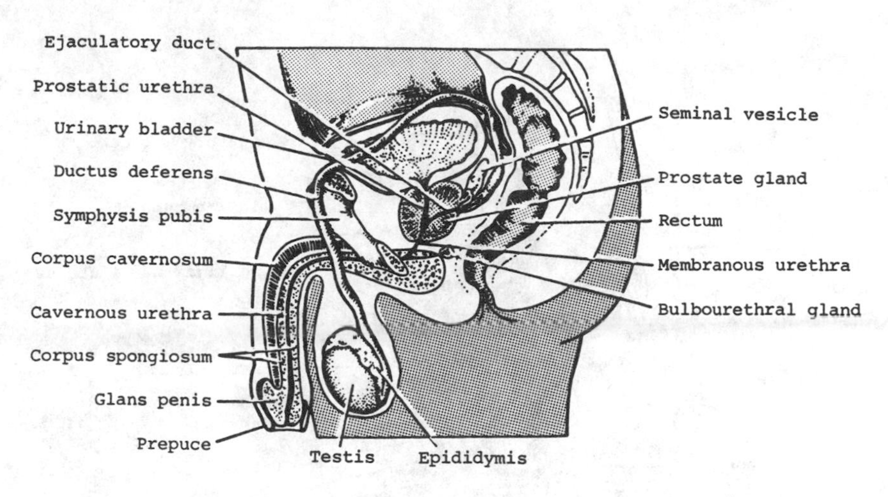

Reproductive system of the human male, lateral view.

complexly coiled tube in which sperm are stored and mature. The epididymis, which is derived from the embryonic kidney, empties into the vas deferens. This duct passes from the scrotum through the inguinal canal into the abdominal cavity, over the urainary bladder to a point where it opens into the ejaculatory duct, which empties into the urethra, the duct that leads from the urinary bladder to the exterior. Thus in the male, the urethra serves as a common passageway for both reproductive and excretory functions. The urethra passes through the penis and is flanked by three columns of erectile tissue. This tissue becomes enlarged with blood during peroids of sexual excitement, causing the penis to become erect. The engorgement of the penis is caused by arterial dilation and increased blood flow at unchanged arterial pressure. Prior to ejaculation, sperm from the testes pass through the vasa efferentia into the epididymis. During ejaculation, sperm from the epididymis are moved through the vas deferens by peristaltic contractions of its walls. Fluids are added to the sperm at the time of ejaculation. These fluids come from three pairs of glands; the seminal vesicles, the prostate glands (which in the human are fused into a single gland), and Cowper's bulbourethral glands, The mixture of secretions from these three sets of glands is termed seminal fluid. The semen consists of the sperm and seminal fluid. The seminal fluid consists of mucus (from seminal and bulbourethral secretions) and nutrients (seminal secretions), in a milky alkaline fluid (prostatic secretions).

During copulation, the male's penis is inserted into the female vagina and the semen is released there. Surrounding the vagina are the labia majora, two folds of fatty tissue covered by skin richly endowed with hair and sebaceous glands. These folds extend dorsally and down, enclosing the openings of the urethra and the vagina and merging behind them. The labia minora - thin folds of tissue devoid of hair - lie within the folds of the labia majora and are usually concealed by them. At the ventral junction of these two is the clitoris, a sensitive, erectile organ, which is the major site of female sexual excitement. The external female sex organs are collectively known as the vulva.

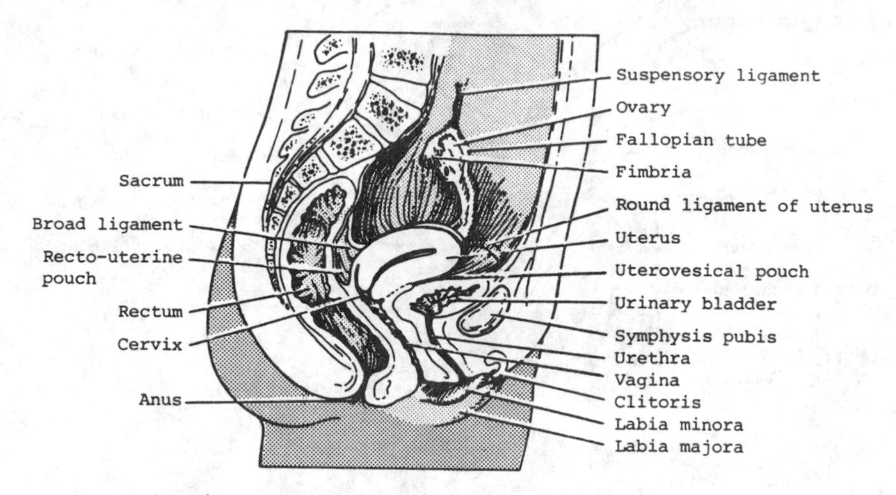

Reproductive system of the human female, lateral view.

The vagina is a single, muscular tube which extends from the exterior to the uterus. From the vagina, the sperm swim, by motion of their flagella, through the cervix, the muscular ring of the uterus which projects into the vagina and pass into the uterus. From there, they enter the Fallopian tubes (also known as the oviducts) where one may fertilize the secondary oocyte, if it is present. If fertilization does occur, the oocyte completes the second meiotic division and the zygote (fertilized egg) is formed. Cleavage of the zygote begins in the Fallopian tube and will have proceeded to a multicellular state by the time the egg enters into the uterus and is implanted.

• PROBLEM 22-14

How does mating behavior serve reproduction? Give examples. Define and give examples of courtship behavior.

Solution: Mating behavior is part of the synchronization of sexual activity. For most animals there exists a particular period or season of the year when it is most advantageous to mate. Therefore the production of sperm and release of eggs must be synchronized. Often environmental cues will initiate mating behavior. Changes in length of daylight, temperature, rainfall, tidal and lunar cycles are some stimuli that can elicit mating behavior. Examples of mating behavior are the migration of the gray whale to Baja California, and the migration of salmon upstream to breed.

Courtship behavior is a specific form of reproductive synchronization. It has two main purposes: to decrease aggression between two potential mates and to aid them in identifying one another. The song patterns of birds and the enlargement of colorful pouches in some fowls are two examples of courtship behavior. Courtship brings two potential mating partners together so that copulation, insemination and fertilization can follow for the production of offspring.

PARTURITION AND EMBRYONIC FORMATION AND DEVELOPMENT

• PROBLEM 22-15

Distinguish between copulation, insemination and fertilization.

Solution: Copulation is the act of physical joining of two sex partners. In copulation, accessory sex organs play an important role and synchronized hormonal control is often extensively involved.

Insemination is the actual process of depositing semen into the vagina. This occurs during copulation or sexual intercourse, but could also be accomplished by means of artificial injection.

When copulation and insemination are complete, fertilization can occur. Fertilization is the union of the male and female gametes to produce the zygote. It should be noted that successful copulation and insemination do not guarantee that fertilization will occur. This may be due simply to the fact that the female has not ovulated and thus provided a mature ovum at the time of insemination. However, it is possible that an ovum may be present, and yet not be fertilized. For fertilization to occur, sperm must not only survive and reach the ovum, they must also succeed in penetrating it. In addition, if one or both partners are sterile, fertilization cannot take place. Sterility refers to any number of specific conditions which result in the failure to produce viable gametes. For example, sterile males may produce viable sperm, but in such low amounts that sufficient numbers do not survive in the female to insure fertilization. Sterile females may produce viable eggs but have a blockage preventing their passage from the ovaries to the oviducts or uterus. Some may produce sperm antibodies in their cervical secretions which resist sperm passage.

• PROBLEM 22-16

Describe the structure and function of the human placenta.

Solution: The placenta is an organ composed of both maternal and fetal tissues that supplies the developing embryo with nutrients and oxygen and enables the embryo to excrete carbon dioxide and other metabolic wastes. It also functions as an endrocrine gland, secreting estrogens, progesterone, human chorionic gonadotropin (HCG), and human placental lactogen (HPL). The former two are steroids, while

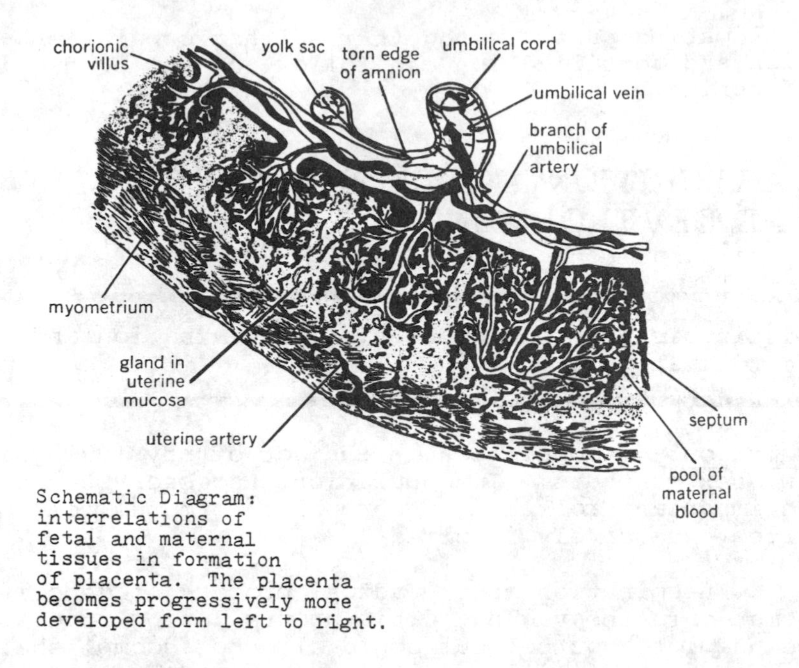

Schematic Diagram: interrelations of fetal and maternal tissues in formation of placenta. The placenta becomes progressively more developed form left to right.

the latter two are protein hormones. HCG functions to preven regression of the corpus luteum, while HPL stimulates secretion of the steroid hormones from the corpus luteum.

The placenta forms following the implantation of the embryo within the uterine wall. Lytic enzymes secreted by the trophoblasts break down the capillaries of the uterine endometrium causing the maternal blood to pool in that area. Subsequently, the trophoblast layer surrounds and sends out villi into this region (see Figure). These villi become integrated by the umbilical arteries and veins which are extensions of the fetal circulatory system. The fetal blood in the capillaries of the villi come in close contact with the maternal blood in the tissue sinuses between the villi. Nutrients diffuse from the maternal blood through the trophoblast layer into the fetal capillaries, and waste products from the fetus is eliminated into the maternal blood. Note, however, that at no time or place does the blood of the fetus mix with that of the mother. Blood from the mother enters the placenta via the uterine artery and leaves the region via the uterine veins. Similarly, the fetal blood never leaves the fetal vessels, which consist of umbilical arteries, the capillary network in the placental villi, and the umbilical vein.

• PROBLEM 22-17

Describe the early development and implantation of the human embryo.

Solution: The ovum must be fertilized while in the upper regions of the Fallopian tube or oviduct to develop successfully. Upon fertilization, it begins the series of cell divisions and differentiation which will eventually lead to development of the embryo (see Figure 1). The fertilized ovum continues its passage through the oviduct to the uterus, which requires three to four days. Once in the uterus, the egg floats free in the intrauterine fluid, from which it receives nutrients for several more days. During this time, it continues dividing and begins to differentiate into the blastocyst. The blastocyst stage (a fluid-filled sphere of cells) is attained by the time or shortly after the egg reaches the uterus. The egg would have slowly disintegrated during its passage to the uterus if fertilization had not occurred.

While the egg had been traveling to the uterus, the uterus was being prepared under the combined stimulation of estrogen and progesterone to receive the egg. If the fertilized egg were to pass too quickly through the oviduct, it would reach the uterus prematurely, when both it and the uterus had not yet reached a suitable state for implantation.

For the egg, the necessary event for implantation is the disintegration of the zona pellucida, a remnant of the

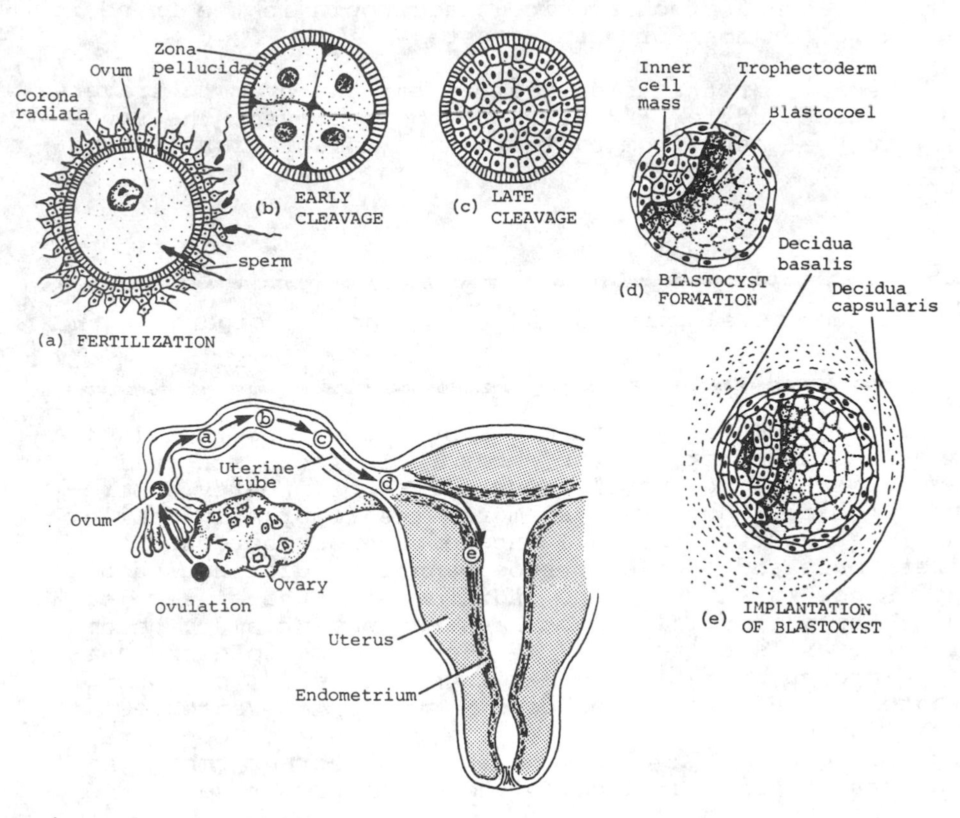

Figure 1 Schematic diagram of the maturation of an egg in a follicle in the ovary following its release (ovulation). a) fertilization in the upper part of the oviduct b) early cleavage as it desends c) late cleavage continuing desent d) blastocyst development in the uterus before implantation e) implantation of blastocyst in the wall of the uterus.

follicle. This disintegration is completed at the blastocyst stage. The blastocyst is composed of an outer layer of specialized cells called trophoblasts, which are responsible for implantation, and an inner cell mass destined to become the fetus itself. The trophoblast layer, upon disintegration of the zona pellucida, enlarges and adheres to the uterine wall. These trophoblasts secrete lytic enzymes which will digest the uterine endometrium, allowing the blastocyst to completely embed itself in the uterine lining. This digestion of tissue continues for a period, with the breakdown products of the endometrium serving as a nutrient source for the developing embryo until both the fetal circulation and the placenta have developed. The placenta will form from an intermeshing of the trophoblast layer with the endometrium, and provides for contact and nutrient exchange between the mother and the fetus.

In addition to enzymes, the trophoblast secrets hormones, including chorionic gonadotropin. This important hormone is necessary for retention of the corpus luteum, which would have involuted if its ovum had not been fertilized. The hormonal secretions of the corpus luteum and later the placenta, maintain the uterus in the proper condition for pregnancy and prevent menstruation. Detection of the presence of chorionic gonadotropin in the urine is the basis for most pregnancy tests.

Because of the time required for the ovum to traverse the oviduct and for the uterus to be prepared, the total time that elapses from ovulation to implantation is from seven to eleven days.

• PROBLEM 22-18

Outline the development of the reproductive organs in the human embryo.

Solution: The reproductive system develops relatively late in the history of the embryo. In fact, its full development is not complete untill after puberty. Moreover, the reproductive system remains remarkably labile and is subject to the influence of the sex hormones. At first, male and female embryos cannot be distinguished. This stage is known as the indifferent stage of development; the primordial gonads have the potential to become either ovaries or testes. If the gonad is to differentiate into an ovary, the outer layer of cells, or cortex, progresses, and the inner core, or medulla, regresses (see Figure 1). If it is to be a testis, the opposite process occurs. In addition, the indifferent embryo has two sets of primordial cells that will give rise to the reproductive ducts.

The indifferent germ glands, or gonads, first appear as thickenings of the coelomic epithelium on the side of the embryonic kidney, termed the mesonephros (see Figure 2) and contain the germ cells. There has been much dispute as to the origin of the germ cells themselves.

The bulk of the evidence indicates that they are not derived from the initial germinal epithelium but migrate into it via the blood from some other part of the embryo. Soon each gonad protrudes into the body cavity as a longitudinal ridge, called the germinal ridge. Cord-like clumps of cells, called the medullary cords, from the germinal epithelium push into the gonad. In the male, the medullary cords push into the core of the gonad and become the seminiferous tubules. Their wallş consist of immature germ cells (the spermatogonia) and accompanying nurse or Sertoli cells which serve a nourishing function. Ultimately, the spermatogonia will transform into mature sperm. The seminiferous tubules become connected to the rete testes, a system of thin tubules that develop from the dorsal part of the gonad, and form connections with the adjoining tubules of the mesonephros, (which later become the epididymis). The epididymis is attached to the

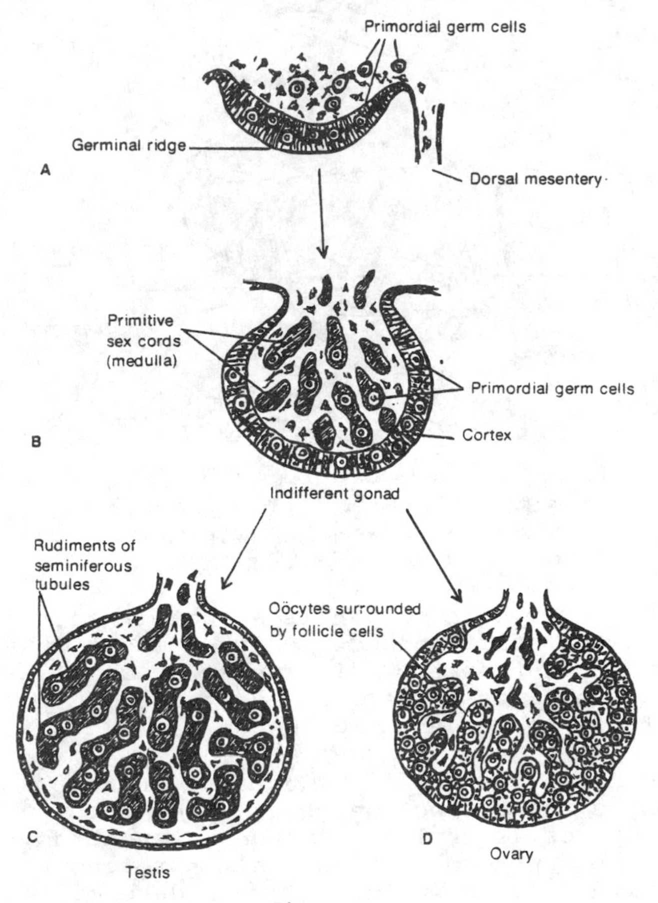

Figure 1

Diagram showing development of gonads in higher vertebrates.
(A) Germinal ridge stage; primordial germ cells partly embedded in epithelium of the ridge and located partly in the adjacent mesenchyme.
(B) Indifferent gonad, germ cells in the cortex and in primary sex cords.
(C) Gonad differentiating as testis; cortex reduced; germ cells in sex cords (future seminiferous tubules).
(D) Gonad differentiating as ovary; primary sex cords reduced; proliferating cortex contains germ cells.

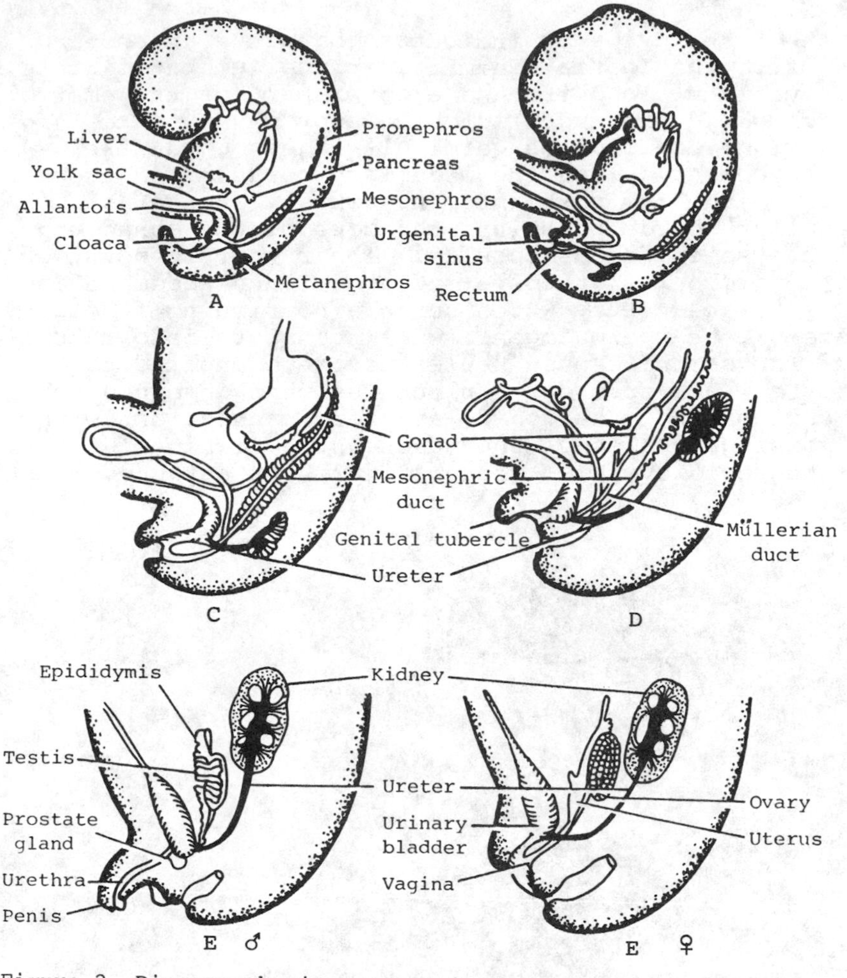

Figure 2 Diagram showing stages in development of the urinary and reproductive sydtems in the human body. A) Early in the fifth week of development B) sixth week C) seventh week D) eighth week E) three months ♂ - male, ♀ - female

vas deferens, whose embryonic origin is a duct in the mesonephros called the Wolffian duct, which opens to the outside. The development of the male system proceeds under the stimulation of testosterone secreted by the interstitial cells of the developing testes.

In female embryos, the Wolffian ducts degenerate when the mesonephros stops functioning as an excretory organ. The mesonephros is replaced by the metanephros, which remains as the functioning kidney all throughout life. New formations called Müllerian ducts develop lateral to the Wolffian ducts. These open at the anterior end by a funnel-shaped ostium into the coelomic cavity, and at the posterior end to the outside. The anterior portions become the Fallopian tubes. Farther back, the Müllerian ducts fuse and give rise to the uterus. The posterior parts of the Müllerian ducts unite to produce a single vagina. In females, the mesonephric and Wolffian ducts degenerate while the Müllerian ducts proceed to develop. In the male, the testis appears to secrete, in addition to testosterone, a second hormone, that inhibits the development of the Müllerian ducts so that no oviduct or uterus appears.

The external genitals of the male and female are derived from the same embryonic source, and at the indifferent stage they cannot be distinguished (see Figure 3).

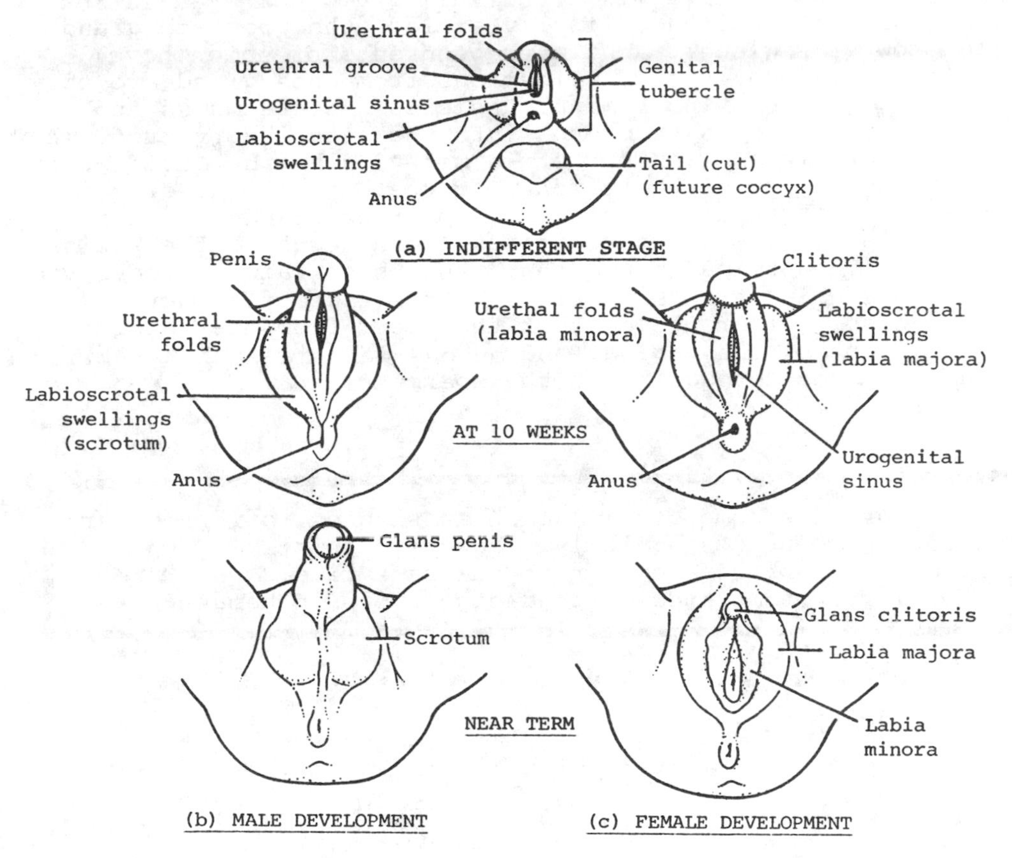

Fig. 3 Embryonic development of male and female external reproductive structures.

In mammals the genital opening of the embryo is flanked on both sides by elongated thickenings, the genital folds. A genital tubercle (phallus) is located on the ventral end where the genital folds meet. From this sexually indifferent condition, under the stimulation of the sex hormones, development proceeds toward the male or female condition. In the male, the phallus grows large and elongates to form the penis. A groove on its ventral surface is bounded by the genital folds. These close in beneath and form a tube, extending from the urethra to the tip of the penis.

In the female, the phallus is smaller and is known as the clitoris. The genital folds do not fuse but remain as the labia minora. An outer fold bordering each labia minor develops into the labia majora, which is comparable to the skin of the scrotum.

• PROBLEM 22-19

How and where is seminal fluid produced in man?

Solution: Semen containing seminal fluid and sperm is transferred to the vagina during sexual intercourse. There are three pairs of glands that together contribute to produce the two to five ml of seminal fluid per ejaculation. These are the paired seminal vesicles, the prostate gland, and Cowper's glands. While the contents of semen have been determined, the contributions from each gland are only suggested below. The seminal vesicles secrete mucus and nutrients (such as fructose) into the ejaculatory duct. The single prostate gland secrets a milky, alkaline fluid which increase the pH of the semen. At the base of the erectile tissue of the penis lies a third pair of glands, called Cowper's glands, which contribute more mucus to the seminal fluid. The seminal fluid that results contains glucose and fructose which are metabolized by the sperm for energy, acid-base buffers to protect the sperm from the acidic secretions of the vagina, and mucous materials that lubricate the passages through which the sperm travel.

• PROBLEM 22-20

In humans and some other mammalian species, offspring are usually produced singularly. However, sometimes two or more offspring are born at the same time. What are the different methods of producing twins in humans?

Twinning in man. (A) Human uterus with a single embyro. (B) Fraternal "twins"(from two eggs) with seperate chorions and placentas. (C) Identical twins (from one egg) with one placenta but with seperate amnions and yolk sacs (the latter are not shown). This results when two inner cell masses arise from one blastocyst. (D) Identical twins with one set of fetal membranes. This presumably is the result of two embryonic axes forming from one inner cell mass.

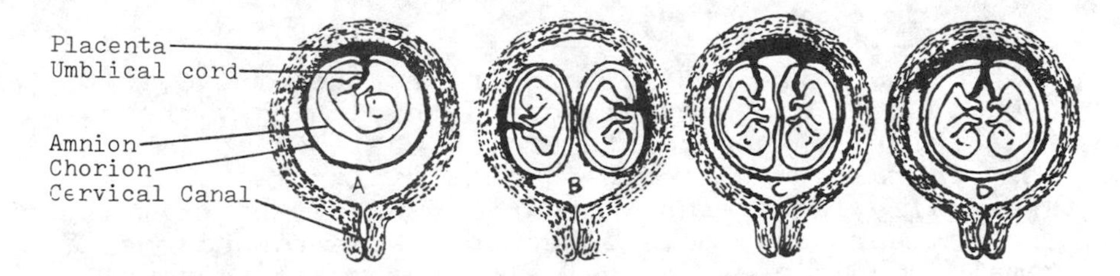

Solution: It is not unusual for human mothers to give birth to more than one offspring at the same time. Basically, there are two kinds of twinning that can occur. One results from the simultaneous release of two eggs (one from each ovary). Both of these eggs can be fertilized and develop. Such fraternal twins or heterozygous twins may be of the same or different sex and have the same degree of resemblance that normal brothers and sisters have. They are entirely independent individuals who have about 25% of their genetic information in common, as do normal siblings. Although they may be situated close together in the uterus, fraternal twins have separate fetal membranes.

True or identical twins, also termed homozygous twins, are the products of a single egg fertilized by a single sperm. At some early stage of development, the egg divides into two (or more) independent parts, with each developing into a separate fetus. Since they come from the same fertilized egg, they are genetically identical and, therefore, are of the same sex. Such twins usually arise from the same blastocyst. Two separate inner cell masses may arise from a single blastocyst or a single cell mass may divide into two. In these cases, the twins will have separate amnions and umbilical cords, but share the chorion and placenta. If, however, a single cell mass develops into two embryos, the twins will share amnion, chorion, and placenta (see Figure).

Occasionally, identical twins develop without separating completely and are born joined together, and are termed Siamese twins. All grades of union have been known to occur, from almost complete separation to fusion throughout most of the body so that only the head or legs are double. Sometimes the two twins are of different sizes and degrees of development. One might be quite normal, while the other might be an incompletely formed parasite of the first. Such errors of development usually cause death during or shortly after birth.

• PROBLEM 22-21

Explain what occurs during the human "birth process" and when these phenomena take place.

Solution: The actual process of birth is preceded by the gestation period. The gestation period or duration of pregnancy varies in different species. It normally lasts about 280 days in humans (the time of the last menstrual period to the birth of the baby). Little is known about the factors that initiate the birth process, or parturition, after gestation is complete, but it is believed that hormonal factors play an important role. It has been found experimentally in mice that the level of progesterone in the blood declines sharply, while that of estrogen rises shortly before parturition.

The start of parturition is marked by a long series of involuntary uterine contractions called "labor pains." Labor is divided into three periods. The first lasts about twelve hours. The fetus is moved down toward the cervix which becomes dilated to enable the fetus to pass through. The amnion ruptures, releasing the amniotic fluid through the vagina. The second period is the birth, where the fetus passes through the cervix and vagina and is "delivered." This stage takes about twenty minutes to an hour, and consists of combined involuntary uterine contractions and voluntary contractions of the abdomen by the mother. The last stage of labor begins after birth and lasts for ten to fifteen minutes. The placenta and the fetal membranes are loosened from the uterine lining by another series of contractions and expelled as the afterbirth.

Sometimes, a woman will require more help than usual during childbrirth. Drugs such as oxytocin and prostaglandins will help increase the uterine contractions. Special forceps can be used by an obstetrician to pull the infant through the birth canal. When the canal through which the baby passes is too small, a caesarean delivery is performed. This is an operation in which the abdominal wall and uterus are cut open from the front and the baby is removed.

HUMAN REPRODUCTION AND CONTRACEPTION

• PROBLEM 22-22

Describe and compare the different forms of currently used methods of contraception. Discuss the safety and efficiency of each.

Solution: Reproduction in man is an extremely complex process involving many different events. Contraception or birth control acts on the principle that these events are sequential; preventing any given event from occurring will prevent the end result, which is pregnancy. The currently used methods of birth control interfere with the normal reproductive process by (1) suppressing the formation and/or release of gametes; (2) preventing the union of gametes in fertilization; or (3) preventing implantation of the fertilized egg.

The oral contraceptive or "pill" acts to prevent ovulation in the female. The pill contains combinations of estrogen and progesterone, which inhibit the release of FSH (follicle stimulating hormone) and LH (luteinizing hormone) by the pituitary via a feedback mechanism; these hormones respectively stimulate follicular growth and trigger ovulation. A woman who is taking the pill inhibits their release and does not ovulate. The pill is usually taken daily from the fifth to the twenty-fifth day of the female cycle. Withdrawal of the first five days triggers menstruation. Some synthetic hormones used in oral contraception have additional antifertility effects in that they cause a thickening of the cervical mucus and as a result, sperm have more difficulty entering the uterus.

The effects of long-term use of the pill are not yet known. There does exist a relationship between increased amounts of estrogen and a higher risk of thromboembolism (the formation of blood clots). Presently, the efficiency of the oral contraceptive is close to 100%. Women using the pill have between zero and one pregnancy per hundred woman-years (a group of one hundred women using that method of contraception for one year). Fifty to eighty pregnancies per hungred woman-years will occur if no attempt is made to restrict conception.

Condoms and diaphragms interpose a barrier between the male and female gametes. A condom prevents semen from entering the vagina. A diaphragm blocks the cervix and prevents sperm entry into the uterus. Spermicidal jellies and foams are used with the diaphragm to kill sperm. Condoms and diaphragms are relatively safe, and their efficiency results in about 12 to 14 pregnancies per hundred woman-years.

The intrauterine device or IUD prevents the implantation of the fertilized egg. The IUD is a small object, such as a coil, placed within the uterus, which causes a slight local inflammatory response in the endometrium. This interferes with the endometrial preparation for proper implantation of the blastocyst. In some women, the IUD causes severe cramping and is sometimes spontaneously expelled. Its efficiency is about two pregnancies per hundred woman-years.

Restricting intercourse to a "safe" period is known as the rhythm method. Normally, a woman ovulates once a month and abstaining from intercourse during this time can be used as a form of contraception. However, it is extremely difficult to precisely determine when ovulation occurs, even when taking body temperature daily (the body temperature rises slighlty following ovulation). Therefore this method is not a very reliable method. About twenty-four pregnancies per hundred woman-years occur using the rhythm method.

A more permanent form of birth control is sterilization. In females, this is accomplished by tubal ligation, an operation in which the Fallopian tubes are cut and tied, preventing passage of the ovum from the ovaries to the uterus. This method is 100% effective if performed properly, but involves obvious surgical risks. Male sterilization involves a vasectomy, the cutting and tying off of the vas deferens. Again, if done properly, it is 100% effective and does not involve complicated surgery or anesthesia. The development of the vasectomy marks the beginning of the search for effective male contraceptive techniques.

• PROBLEM 22-23

Describe the human sex act.

Solution: The human sex act is generally divided into four phases. These are the excitement, plateau, orgasm and resolution phases.

The excitement phase in the female is characterized by a rise in heart rate and blood pressure. Muscle tension also increases as the breasts swell and the genitalia become filled with blood. A lubricating fluid is secreted during this phase by the vaginal wall.

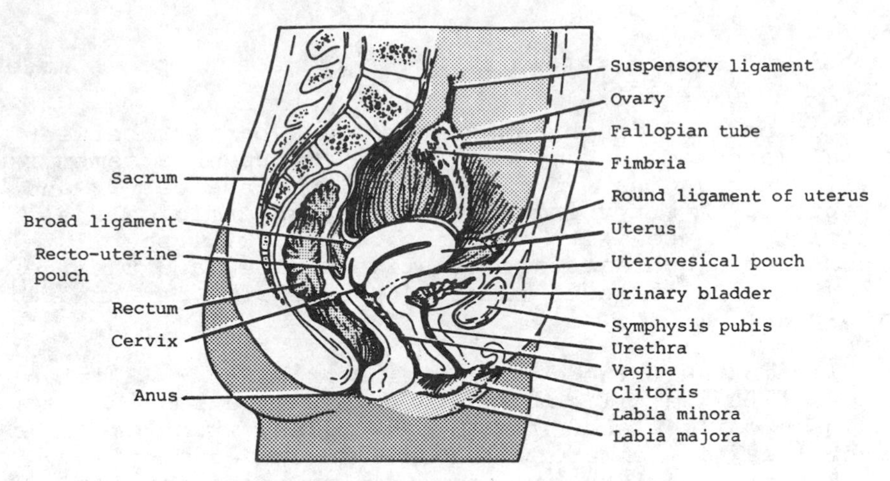

Female cross-section

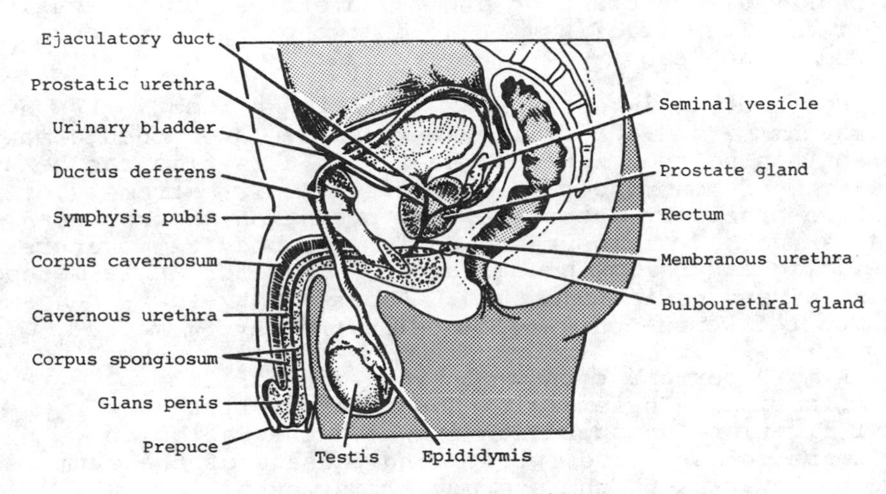

Male cross-section

In the male, the excitement phase is also characterized by increased blood pressure, heart rate and muscle tension. The penis becomes rigid as the erectile tissue becomes engorged with blood.

The plateau phase is characterized by increased breathing rate, pulse rate and muscle tension in male and female.

The orgasm phase is characterized by muscular contractions of the vagina in the female and by the muscles in the genital area of the male. The orgasm phase is initiated by frictional force against the vaginal wall and clitoris in the female. In the male, orgasm is caused by pressure and friction against the nerve endings in the glans penis and in the skin of the shaft of the penis.

The resolution phase is the final stage in the sex act. During this period body functions return to normal levels. Heart beat, pulse rate and breathing levels return to the precoital state.

Answer

To be covered when testing yourself

SHORT ANSWER QUESTIONS FOR REVIEW

Choose the correct answer.

1. Which statement is true concerning asexual and sexual reproduction? (a) Asexual reproduction involves one or more parents. (b) Asexual reproduction rarely results in identical offspring. (c) Sexual reproduction is predominant amongst the flatworms and starfish. (d) Two parents are usually involved in sexual reproduction. (e) Sexual reproduction never involves only one parent. — d

2. In the life cycle of the sporozoan which causes malaria, which is the animal that is the vector of the disease? (a) man (b) Anopheles mosquito (c) Plasmodium (d) all of the above (e) none of the above. — b

3. Parthenogenesis is most strictly defined as (a) the development of an egg without fertilization, (b) the mode of reproduction in insects, (c) a mode of asexual reproduction which produces strictly male offspring, (d) the development of sperm into a viable organism without the presence of an egg, (e) none of the above. — a

4. A queen bee which has a yellow body color (Yy) was fertilized by a drone which also has a yellow body color. Assuming the yellow body color is dominant and knowing that the drone, being haploid, possesses only the gene for yellow body color (Y), how is it possible that a brown bodied female (yy) is produced? Assume the two bees are isolated in a vial. (a) A mutation has occurred during development. (b) Body color has been environmentally influenced, following fertilization. (c) The queen bee has used stored sperm from previous inseminations, resulting in this offspring. (d) Male bees are not haploid so the above experiment is based on incorrect assumptions. (e) None of the above is a reasonable explanation. — c

5. Taking into account the needs of the developing organism, which type of reproduction would have the least amount of yolk? (a) oviparous (b) ovoviviparous (c) viviparous (d) all have the same amount (e) the needs of the developing organism are unrelated to the type of reproduction used. — c

6. Which mode of reproduction is most closely defined by the following statement? "In this type of reproduction, large yolk - filled

Answer

To be covered when testing yourself

eggs are produced which remain for a considerable time in the female reproductive tract. Following fertilization, the embryo makes no contact with the uterus or oviduct, and obtains no nourishment from the maternal blood". (a) oviparous (b) ovoviviparous (c) viviparous (d) the description is applicable to all of the above (e) the description is applicable to none of the above. — b

7. Which is the correct sequence of events leading to the formation of mature sperm? (a) primary spermatocytes → secondary spermatocytes → spermatids → spermatagonia → sperm, (b) spermatids → spermatagonia → primary spermatocytes → secondary spermatocytes → sperm, (c) spermatogonia → spermatids → secondary spermatocytes → primary spermatocytes → sperm, (d) spermatogonia → primary spermatocytes → secondary spermatocytes → spermatids → sperm, (e) none of the above. — d

8. Which of the following structures is not found in the cytoplasm of the human spermatid? (a) proacrosomic granules (b) centrioles (c) mitochondria (d) ribosomes (e) none of the above. — d

9. The lysozymes necessary for penetration of the egg are contained within the (a) acrosome (b) midpiece (c) tail (d) a and b (e) b and c. — a

10. The human ovaries contain an estimated _____ immature eggs or oogonia. Of these, perhaps _____ will reach full maturity during the female reproductive life, (a) forty thousand; four thousand (b) four thousand; four hundred (c) four hundred; forty (d) forty thousand; forty thousand (e) four hundred thousand; four hundred. — e

11. At which stage of development is the egg ovulated? (a) oogonium (b) primary oocyte (c) secondary oocyte (d) ootid (e) none of the above. — c

12. Which is not a result of fertilization? (a) restoration of the diploid state in the zygote (b) stimulation of zygotic cleavage (c) genetic determination of sex (d) increase in species variability (e) none of the above. — e

13. The coiled tube in which mature sperm are stored is known as the (a) vasa efferentia (b) epididymis (c) vas deferens (d) scrotal sac (e) Sertoli cells. — b

	Answer
	To be covered when testing yourself
14. Which is not true concerning fraternal twins? (a) They result from the simultaneous release of two eggs. (b) They are usually siblings of the same sex. (c) They have approximately 25% of their genetic information in common. (d) They have separate fetal membranes during their development. (e) None of the above.	b
15. Which of the following birth control methods is closest to being 100% effective? (a) rhythm (b) IUD (c) diaphragm (d) pill (e) spermicidal jelly.	d

In questions 16-26 match the letter to the appropriate number.

Number		Letter	Answer
16. regeneration	_____	(a) mesonephros	16-f,
17. <u>Plasmodium</u>	_____	(b) parthenogenesis	17-i,
18. Hymenoptera	_____	(c) childbirth	18-b,
19. oviparous	_____	(d) egg bearing	19-d,
20. chiasmata	_____	(e) Cowper's gland	20-j,
21. embryonic kidney	_____	(f) fragmentation	21-a,
22. uterus	_____	(g) chorionic gonado-tropin	22-h,
23. seminal fluid	_____	(h) Müllerian ducts	23-e,
24. pregnancy hormone	_____	(i) sporozoan	24-g,
25. parturition	_____	(j) crossing over	25-c

In questions 26-35 match the letter to the appropriate number.

Number		Letter	Answer
26. <u>Paramecium</u>	_____	(a) ootid	26-i,
27. earthworm	_____	(b) middle piece	27-c,
28. shark	_____	(c) hermaphroditic	28-j,
29. bird	_____	(d) sterilization	29-e,
30. diploid	_____	(e) oviparous	30-h,
31. haploid	_____	(f) fertilization membrane	31-a,
32. acrosome	_____	(g) head piece	32-g,

Answer

To be covered when testing yourself

			Answer
33. mitochondrion	_____	(h) spermatogonia	33-b,
34. zona pellucida	_____	(i) asexual reproduction	34-f,
35. vasectomy	_____	(j) ovoviviparous	35-d

Fill in the blanks.

36. A process during which a part of the parent organism separates from the rest and develops into a new individual is known as _____.

budding

37. The organism which may undergo encystment during unfavorable conditions, and later cause amoebic dysentery following invasion of the large intestine is _____.

Entamoeba histolytica

38. The protozoan Plasmodium is the organism responsible for the disease malaria. It is one of the rare cases in which an animal reproduces by means of _____.

spores

39. If an organism has functional male and female sex organs, it is known as a(n) _____.

hermaphrodite

40. _____ is the event during meiosis when homologous chromosomes pair together to form a complex known as a tetrad.

Synapsis

41. Condensation of chromosomes occurs during the _____ stage of meiosis.

prophase

42. The ovum is surrounded by two extracellular layers. The inner layer is called the _____ and the outer layer, consisting of follicular cells, is known as the _____.

zona pellucida, corona radiata

43. Following the formation of the primary oocyte from mitotic enlargement of the oogonia, meiosis takes place. Unequal meiotic division of the primary oocyte results in the formation of the _____ and a(n) _____.

secondary oocyte, polar body

44. Cleavage of the egg is stimulated by the process of _____.

fertilization

45. Production of sperm from germ cells occurs in the _____.

seminiferous tubules

46. Specialized cells present in the testes which nourish the sperm during their production are known as _____ cells.

Sertoli

	Answer To be covered when testing yourself
47. Semen is a mixture of the sperm and seminal fluid. The seminal fluid is composed of the secretions from the _____ gland, _____ glands, and the _____ vesicles.	prostate, Cowper's, seminal
48. If conditions are favorable, fertilization of the secondary oocyte by sperm will occur in the _____, and implantation will occur when the cleaving multicellular egg reaches the _____.	Fallopian tube, uterus
49. If the indifferent gonad is to develop into a testis, the cortex (progresses/regresses) and the medulla (progresses/regresses).	regresses, progresses
50. The development of the system of male sex ducts proceeds under the stimulus of the hormone _____ , secreted by the _____ of the developing testes.	testos-terone, intersti-tial cells
51. During female embryonic development, the _____ ducts undergo gradual degeneration and the _____ ducts undergo continuous modification and specialization.	Wolffian, Müllerian
52. Under the control of sex hormones, the _____ will either form the male or female erectile tissue in the adult.	genital tubercle
53. A male organism incapable of producing sufficient numbers of viable sperm to insure fertilization is _____.	sterile
54. The _____ is an organ composed of both maternal and fetal tissue that supplies the developing embryo with nutrients and oxygen, and enables the embryo to excrete carbon dioxide and other metabolic wastes.	placenta
55. The implantation of the egg in the endometrium of the female human occurs in the _____ stage.	blasto-cyst

CHAPTER 23

EMBRYONIC DEVELOPMENT

CLEAVAGE

• PROBLEM 23-1

What is meant by embryonic development? Describe the various stages of embryonic development.

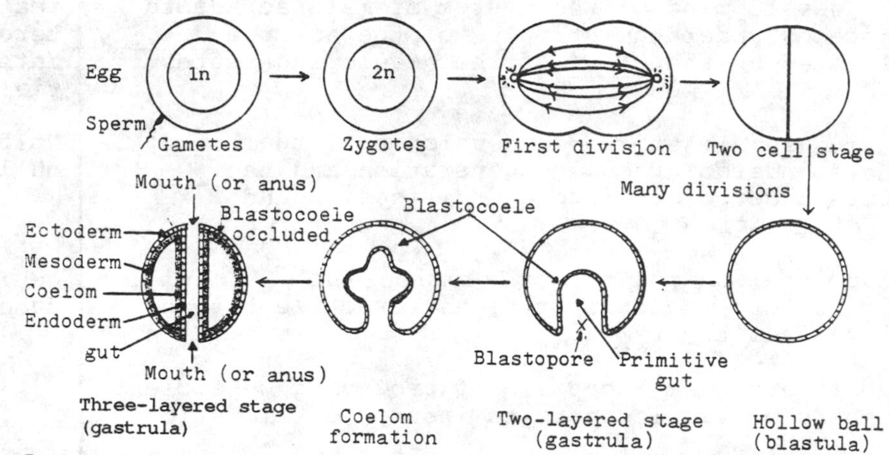

Early embryonic development in animals.

Solution: Embryonic development begins when an ovum is fertilized by a sperm and ends at parturition (birth). It is a process of change and growth which transforms a single cell zygote into a multicellular organism.

The earliest stage of embryonic development is the one-celled, diploid zygote which results from the fertilization of an ovum by a sperm. Next is a period called cleavage, in which mitotic division of the zygote results in the formation of daughter cells called blastomeres. At each succeeding division, the blastomeres become smaller and smaller. When 16 or so blastomeres have formed, the solid ball of cells is called a morula. As the morula divides further, a fluid-filled cavity is formed in the center of the sphere, converting the morula into a hollow ball of cells called a blastula. The fluid filled cavity is called the blastocoel. When cells of the blastula differentiate into two, and later three, embryonic germ layers, the blastula is called a gastrula. The gastrula period generally extends until the early forms of all major structures (for example, the heart) are laid down. After this period, the developing organism is called a fetus. During the fetal period (the duration of which varies with different species), the various systems develop further. Though developmental changes in the fetal period are not as dramatic as those occurring during the earlier embryonic periods, they are extremely important.

Congenital defects may result from abnormal development during this period.

• PROBLEM 23-2

Differentiate between isolecithal, telolecithal, and centrolecithal eggs. Which organisms are characteristic of each?

Solution: The eggs of different animals vary greatly in the amount and distribution of the yolk they contain. Some, such as those of many annelids, mollusks, and echinoderms, are small and contain only a little yolk. If the yolk is evenly distributed they are called isolecithal eggs. Other eggs are larger and have a moderate supply of yolk. Of this sort are amphibian eggs and those of primitive fish. Still others, such as those of reptiles and birds, have a tremendous supply of yolk. Since, in these types, the yolk is concentrated toward one pole of the egg (known as the vegetal pole), these eggs are termed telolecithal. Arthropod eggs usually have the yolk massed toward the center, and hence are called centrolecithal.

• PROBLEM 23-3

Following fertilization, the zygote begins to cleave. Describe the different cleavage patterns found in the animal kingdom.

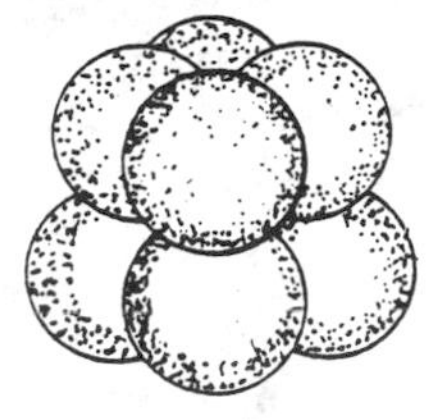

Radial and spiral cleavage patterns. Left: Radial cleavage, characteristic of deuterostomes. The cells of the two layers are arranged directly above each other. Right: Spiral cleavage, characteristic of protostomes. The cells in the upper layer are located in the angles between the cells of the lower layer.

Solution: Cleavage cells are known as blastomeres. They vary in size and content, principally by reason of differences in the amount and distribution of the yolk and other cytoplasmic inclusions which they contain. In isolecithal eggs, the cleavage cells are of approximately the same size. In telolecithal eggs, such as those of birds, the yolk is more abundant and is concentrated toward that area of the egg known as the vegetal pole. The opposite side of the egg, an area where the nuclear material is located, is called the animal pole. In this case, the blastomeres nearer the animal pole tend to be smaller and are called micromeres. The cells near the vegetal pole are usually larger and are termed macromeres. As long as the entire egg divides into cells, the cleavage is said to be complete, or holoblastic. In the case of sharks, reptiles, and birds, however, yolk is abundant and fills the egg, except for a thin disc of cytoplasm at the animal pole and an even thinner layer of cytoplasm around the periphery of the egg. In such eggs, cleavage is incomplete and is confined to a small area which surrounds the animal pole. The rest of the egg remains uncleaved. Such incomplete cleavage is termed meroblastic. In the centrolecithal eggs of arthropods, the yolk is concentrated toward the center. After the nucleus divides several times, the offspring nuclei migrate to the periphery of the egg where meroblastic cleavage of the peripheral cytoplasm takes place. There are three basic patterns of cleavage in the animal kingdom: radial cleavage, spiral cleavage, and superficial cleavage. In radial cleavage, the first mitotic spindle elongates in a direction at right angles to the egg axis. The second spindle also elongates transversely to the egg axis and at right angles to the first spindle. The third cleavage spindle elongates in a direction parallel to the egg axis. Consequently, a ball-like stage of eight cleavage cells is formed. This type of cleavage is found in the echinoderms and chordates. In spiral cleavage, the mitotic spindles are oblique to the polar axis of the embryo, and give rise to a spiral arrangement of newly formed cells. This type of cleavage is found in the annelids, mollusks and in some flatworms. A third type of cleavage, associated with arthropods, is known as superficial cleavage. This involves mitotic division of the nucleus without any cleavage, resulting in an uncleaved egg containing a large number of nuclei within the center. Later these nuclei migrate to the periphery where meroblastic cleavage of the egg takes place. The rate of cleavage in all three cases is rapid at first and then slows a bit, and can be regulated by temperature changes.

• PROBLEM 23-4

Contrast the development of an amphibian, bird, and starfish egg up to the point of gastrulation.

Solution: In the isolecithal starfish egg, cleavage is radial (Figure 1). The first cleavage division passes through the long axis of the egg, splitting the egg into two equal daughter cells. The second cleavage plane is also vertical but at right angles to the first, separating the two cells into four. The third division is horizontal, at right angles to the other two, and splits the four

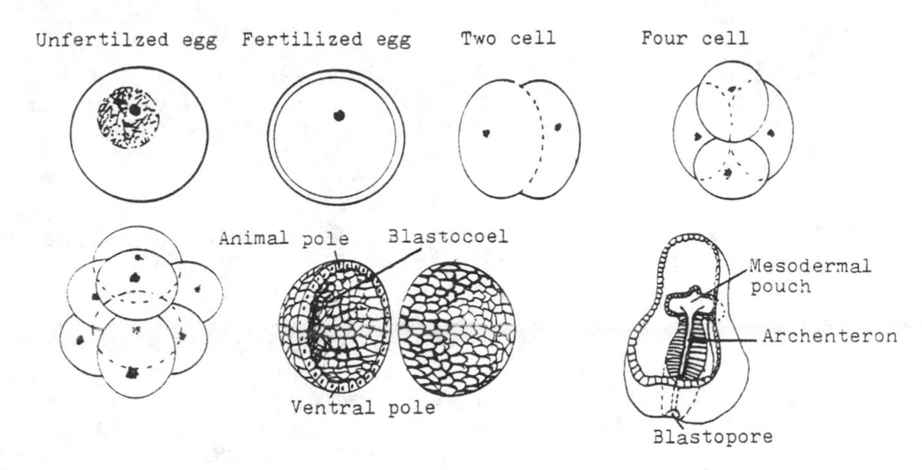

Figure 1. Some developmental stages in the sea star (starfish).

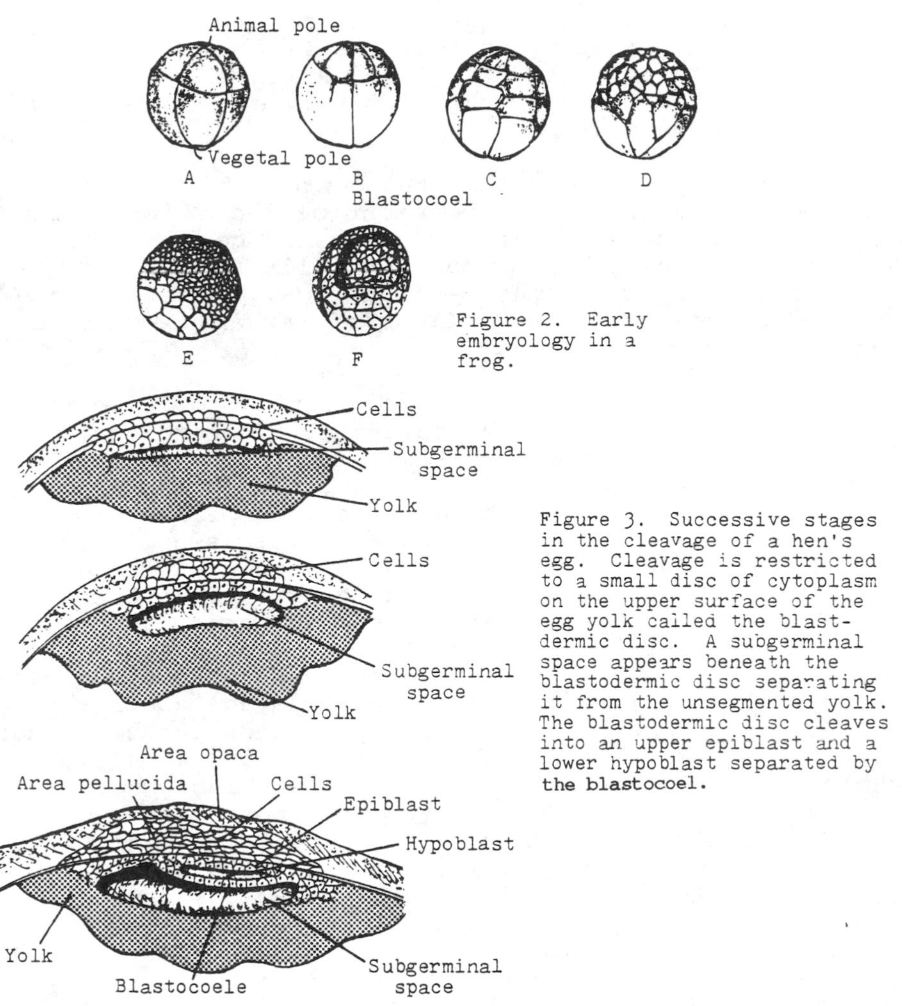

Figure 2. Early embryology in a frog.

Figure 3. Successive stages in the cleavage of a hen's egg. Cleavage is restricted to a small disc of cytoplasm on the upper surface of the egg yolk called the blastdermic disc. A subgerminal space appears beneath the blastodermic disc separating it from the unsegmented yolk. The blastodermic disc cleaves into an upper epiblast and a lower hypoblast separated by the blastocoel.

cells into eight. Further divisions result in embryos containing 16, 32, 64, 128 cells, and so on, until a hollow ball of cells called the blastula, is formed.

The wall of the blastula is a single layer of cells, the blastoderm, surrounding the blastocoel, the cavity in the center. It is this single-layered blastula which is converted later into a double-layered sphere, the gastrula.

Amphibians also display radial cleavage but because of the unequal distribution of yolk in the telolecithal egg, the resulting blastomeres are not of equal size (Figure 2). The first cleavage begins as a shallow furrow at the animal pole and progresses gradually through to the vegetal pole. This first cleavage usually results in two equal blastomeres. The second cleavage is at right angles to the first. It begins at the animal pole even before the first cleavage has been completed at the vegetal pole. The result is four blastomeres of nearly equal size. The third cleavage is typically at right angles to the first two cleavages, and hence cuts horizontally across the egg's vertical axis. It passes well above the equator, so that the eight-celled stage commonly consists of four smaller animal blastomeres (micromeres), and four larger, yolk-laden vegetal blastomeres (macromeres).

A cavity is present at the center of the group of blastomeres beginning at the eight-celled or 16-celled stage, which increases in size as cleavage progresses. The egg thus becomes a hollow ball of cells, namely, a blastula. Its blastocoel is roughly hemispherical. It has a dome of smaller animal cells and a floor of larger yolk-laden vegetal cells.

Telolecithal bird eggs also display radial cleavage, but the cleavage is meroblastic. In bird eggs and other eggs which contain a large amount of yolk, cleavage occurs only in a small disk of cytoplasm in the animal pole (Figure 3). At first, all cleavage planes are vertical and all the blastomeres lie in a single plane. The cleavage furrows separate the blastomeres from each other but not from the yolk. The central blastomeres are continuous with the yolk at their lower ends, and the blastomeres at the circumference of the disk are continuous both with the yolk beneath them and with the uncleaved cytoplasm at their outer edge. As cleavage continues, more cells become cut off to join the ones in the center, but the new blastomeres are also continuous with the uncleaved underlying yolk. The central blastomeres eventually become separated from the underlying yolk either by cell divisions with horizontal cleavage planes or by the appearance of slits beneath the nucleated portions of the cells. Horizontal cleavages separate upper blastomeres and lower blastomeres. The upper blastomere is a cell with a complete plasma membrane, which is separated from its neighbors and from the yolk. The lower blastomere is a cell which remains connected with the yolk. The blastomeres at the margin of the disk, and the lower cells in contact with yolk, eventually lose the furrows that partially separated them and fuse into a continuous syncytium. This syncytium contains many nuclei and is termed the periblast. The periblast is believed to break down the yolk, thereby making its nutrients available for the growing embryo. The free blastomeres (with complete plasma membranes) become incorporated into two layers, an upper epiblast, and a lower hypoblast. Between these two layers is a cavity, the blastocoel. Below the hypoblast and above the yolk is the shallow subgerminal cavity, which appears only under the central portion of the blastoderm. The area of the blastoderm over the subgerminal space is relatively transparent (due to lack of yolk) and is called the area pellucida, whereas the more opaque part of the blastoderm (which contains some yolk) that rests directly on the yolk is called the area opaca.

• PROBLEM 23-5

The result of cleavage is a blastula, typically a ball of cells surrounding a central cavity, or blastocoel. Cell divisions by mitosis continue after the blastula is formed, but generally speaking, the divisions alternate with intervals of cell growth. Hence no further decrease in size of cells takes place. Instead, mass movements of cells take place by which a gastrula is formed. What is a gastrula? Using echinoderms or amphibians as an example, describe the process of gastrulation.

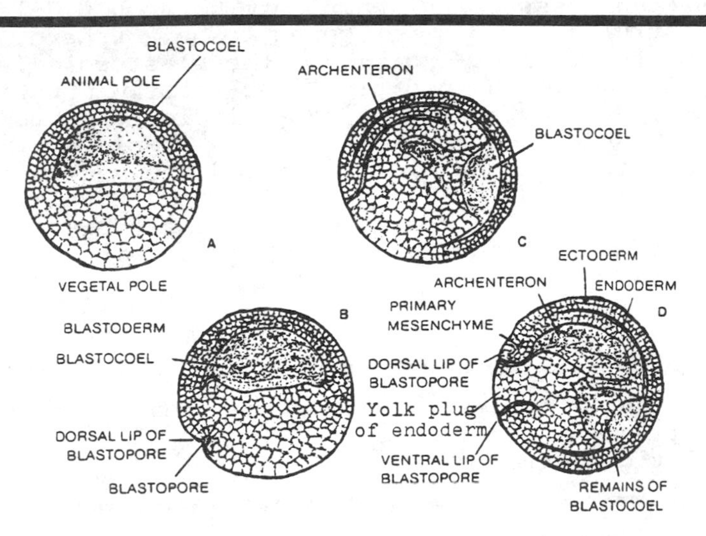

Gastrulation in frog. (A) is late blastula stage. In gastrula stages (B through D) cells migrate from animal pole, moving downward and inward to form the three primary germ layers and the primitive gut (archenteron). Blastocoel diminishes and a yolk plug forms from the vegetal cells.

Solution: A gastrula is a two or three-layered ball of cells, these layers being known as the germ layers. In amphibians and echinoderms, the gastrula is formed by the inward folding of one side of the blastula, partially obliterating the original cavity, or blastocoel. A new cavity is formed which is open to the outside, and is called the archenteron, or primitive gut. The blastopore is the opening from the archenteron to the outside. It lies on one side of the vegetal pole and is the site where infolding began. The cells which border the blastopore constitute the blastoporal lips. When gastrulation is finally complete, the outer germ layer is the ectoderm, and its inner layer, the endoderm. A third germ layer is present, the mesoderm. It occupies a position intermediate between the ectoderm and endoderm. Loose cells which are commonly present between the germ layers are termed mesenchyme. They are usually thought of as mesoderm, although they may arise from any of the three germ layers.

Gastrulation in amphibians and echinoderms is essentially an inward folding and is accomplished by three easily defined processes: invagination, involution, and expansion. During invagination, cells of the vegetal area pocket inward to form the archenteron. The invaginating layer of cells is concave when viewed from outside. Involution is the process by which the cells that border the vegetal area move toward the blastopore. They progressively roll around beneath the lips of the blastopore, eventually coming to lie within the mass of cells. Expansion is the process by which the cells of the animal hemisphere extend out and converge toward the blastopore. As a result of the coordination between these three processes, an echinoderm or amphibian egg during gastrulation does not cease to be a sphere. Throughout these three processes, the blastopore is changing its shape and position. It first widens from its crescent-shape, to form a semi-circle, then to the shape of a horseshoe, and finally to a complete circle. As it does so, its lips migrate toward the vegetal pole, an area of endodermal cells which is rich in yolk. Some of the yolk cells protrude through the blastopore lips as the yolk plug. This yolk plug is withrawn into the gastrula following gastrulation.

• PROBLEM 23-6

Are the daughter nuclei formed during cleavage exactly equivalent, or is there some sort of parceling out of potentialities during cleavage?

Solution: There is evidence that there is no segregation of genetic potentialities during cleavage. Experiments have shown that isolated blastomeres (the cleavage cells) are capable of developing into an entire organism. In a classical experiment, the nucleus is removed from a ripe frog's egg with a micropipette. A cell from an advanced stage of embryonic development is then separated from its neighbors. The plasma membrane of the cell is ruptured in the process and the nucleus is injected into the enucleated frog egg. The egg is found to develop normally. When nuclei obtained from late blastula or early gastrula stages are transplanted into enucleated eggs, the eggs can develop normally. Even nuclei from later stages of development, such as from the neural plate (the precursor of the spinal cord and brain) or from cells lining the gut may lead to the development of normal embryos.

In humans, there is no segregation of genetic potentialities during early cleavage. Identical twins are thus possible. More experimentation has to be done to determine just how far into development the "no segregation" rule will apply.

GASTRULATION

• PROBLEM 23-7

Describe the development of a mammalian egg from cleavage up to the development of the fetal membranes.

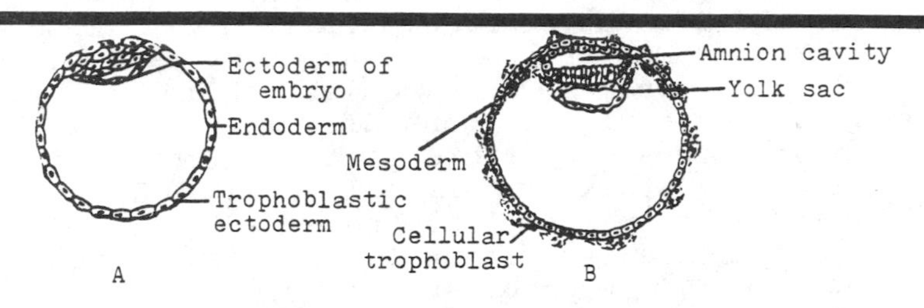

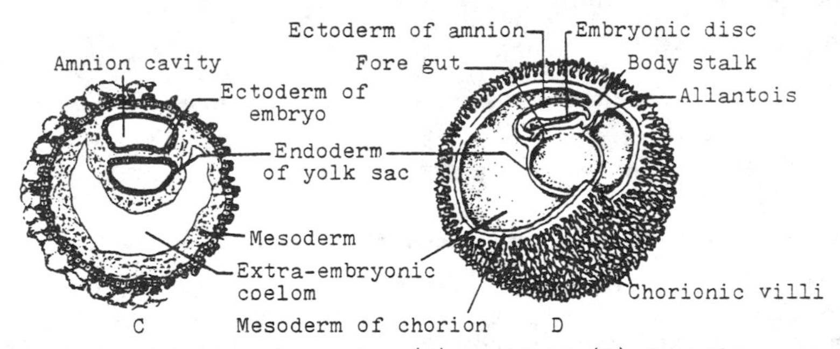

Diagrams of human embryos ten (A) to twenty (D) days old showing the formation of the amniotic and yolk sac cavities and the origin of the embryonic disc.

Solution: Cleavage in a mammalian egg takes place as the egg moves slowly down the oviduct, driven along by cilia and muscular contraction of the wall of the duct. The journey takes about four days in the case of cells, the morula, which divides and rearranges into a hollow ball of cells, the blastocyst. The blastocyst subdivides into an inner cell mass from which the embryo develops, and an enveloping layer of cells, the trophoblast. The cavity of the blastocyst may be compared to the blastocoel, but the embryo is not a blastula, for its cells are differentiated into two types. The cells of the inner cell mass differentiate further into a thin layer of flat cells, the hypoblast, which is located on the interior surface of the mass adjacent to the blastocoel, and which represents the endoderm. The remaining cells of the inner cell mass become the epiblast. The cells of the hypoblast, spreading along the inner surface of the trophoblast, eventually surround the cavity of the blastocyst, forming a "yolk sac," which is filled with fluid, not yolk.

As the hypoblast spreads out, the inner cell mass also spreads and becomes a disc shaped plate of cells similar to the blastodisc of bird and reptilian eggs. The blastodisc becomes delimitated from the rest of the embryo. Gastrulation begins with the formation of a primitive streak and Hensen's node in which cells migrate downward, laterally and anteriorly between the epiblast and hypoblast. Those cells which remain between these two layers comprise the mesoderm. Those cells which join the hypoblast become endoderm. A crevice appears between the cells of the inner cell mass, which then enlarges to become the amniotic cavity. The cavity of the blastocyst becomes filled with mesodermal cells, and is comparable to the subgerminal cavity in the bird. The embryonic disc comes to lie as a plate between the two cavities, connected to the trophoblast at the posterior end by a group of extraembryonic mesoderm cells, the body stalk or allantoic stalk. The non-functional endodermal part of the allantois, which develops as a tube from the yolk sac, is rudimentary and never reaches the trophoblast. Thus, after two weeks of development the human embryo is a flat, two-layered disc of cells about 250 microns across, connected by a stalk to the trophoblast.

• PROBLEM 23-8

What are the four extraembryonic membranes found in animals which lay their eggs on land, and what are their functions?

Solution: The development of extraembryonic membranes freed early reptiles from an obligatory embryonic development in water and set the scene for the flourishing of the land-dwelling vertebrates. The mammals have capitalized on their inheritance of extraembryonic membranes from the reptiles by utilizing them in intra-uterine development.

The chorion, or serosa, is an outer covering which, in reptiles, rests in contact with the inner surface of the shell, and which in mammals is established next to the cells of the uterine wall. This layer is the site where exchange of substances, largely gases, foods and wastes takes place between embryonic tissue and the maternal environment. To facilitate the exchange, various adaptations have been acquired. In the pig, the chorion elongates tremendously. Its great surface area affords abundant contact with the uterine wall. In most mammals, the chorion develops complex outgrowths known as chorionic villi. These penetrate into the tissue of the uterine wall and greatly increase the area and intimacy of contact between fetal and maternal tissues. Usually there is a specialized region of the chorion where exchange between the mother and embryo is most efficient. Such an area is called the placenta. The human chorion also secretes human chorionic gonadotropin (HCG), a hormone which functions to maintain the corpus luteum.

The amnion is a thin fluid-filled, membranous sac which surrounds the embryo and provides it with protection from pressure, abrasion, irritation, and loss of water. The fluid within the amnion is the amniotic fluid. During labor of humans and other mammals, the pressure of the contracting uterus is transmitted via the amniotic fluid and helps to dilate the neck of the uterus. Later, shortly before the fetus is born, the amnion normally ruptures, releasing about a liter of amniotic fluid. This is referred to as the breaking of the water bag. Sometimes it fails to burst and the child is born with the amnion still enveloping its head. The amnion is essentially an outfolding of the body wall of the embryo consisting of ectodermal and mesodermal tissues. The inner part of this fold forms the amnion and the outer part forms the chorion.

A third extraembryonic layer found in reptiles, birds, and mammals is the yolk sac, which is an outgrowth of the digestive tract of the embryo. This sac is richly endowed with blood vessels and functions to store and supply food. The human egg cell contains minimal yolk, but the sac has other functions. In addition, the yolk sac is the first source of blood cells and blood vessels of the embryo, and also performs biochemical functions which are later taken over by the liver.

The allantois, like the yolk sac, is an outgrowth of the digestive tract. It grows between the amnion and chorion and it fills almost all the space between the two. In reptiles and birds, the allantois serves as a depot for nitrogenous wastes. The products of nitrogen metabolism are excreted as uric acid by the kidney of the developing embryo. The poorly soluble uric acid is deposited as crystals in the cavity of the allantois and is discarded along with the allantois when the young hatch out of the egg shells. In birds and reptiles, the allantois fuses with the chorion to form a compound chorioallantoic membrane, across which gas exchange takes place. Recall that in mammals both food and gases are exchanged across the chorion. In reptiles and birds, only gas is exchanged across this membrane which is here associated with the allantois. The egg shell is porous and oxygen coming in through it is picked up by blood vessels in the chorioallantois. In the human, wastes are transported through the placenta into the mother. The allantois is thus less important as an exchange site. But the allantoic stalk becomes the umbilical cord, linking the chorion to the embryo.

• PROBLEM 23-9

Gastrulation in the bird is accomplished by a form of cell migration that differs from that of the frog. Explain.

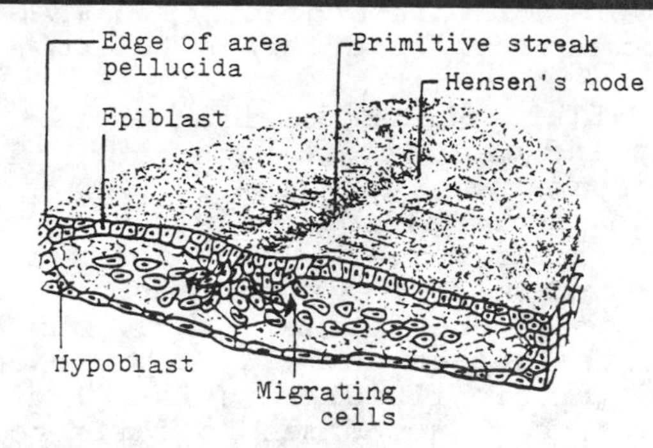

Gastrulation in the bird. The anterior half of the area pellucida of a chick embryo is cut transversely to show the migration of mesodermal and endodermal cells from the primitive streak.

Solution: In birds, the cells from the epiblast (the upper of two embryonic cell layers) migrate downward. Initially, a strip of epiblast extending forward in the midline of the embryo from the posterior edge of the area pellucida becomes thickened as the primitive streak, with a narrow furrow, the primitive groove, in its center. Hensen's node, a thickened knot of cells, is present at the anterior end of the primitive streak. It is by the migration of cells from the lateral portion of the epiblast that the primitive streak is formed. The cells of the epiblast invaginate at the primitive streak, move into the space between the epiblast and hypoblast and reach the latter, forming a mass of moving cells. The cells continue to migrate, moving laterally and anteriorly from the primitive streak area. Though cells are migrating, the shape of the primitive streak remains intact. The primitive streak, where invagination occurs, is considered to be homologous to the amphibian blastopore.

The cells which form the notochord arise from Henson's node. The notochord is the rod-shaped structure that forms a skeletal axis in chordates which is replaced in the vertebrates by vertebrae (in frogs, a special area of mesoderm develops into the notochord). Cells which grow laterally and anteriorly from the primitive streak between the epiblast and the hypoblast become the mesoderm. The original hypoblast cells plus other cells that migrate into the lower layer from the primitive streak form the endoderm, which gives rise to the digestive tract and yolk sac. The epiblast forms the ectoderm.

• PROBLEM 23-10

What is an organizer and how does it influence development?

Solution: An organizer is a chemical secreted by

certain cells in the embryo, which diffuses into and influences surrounding cells. The presence of organizers is the whole basis for the concept of embryonic induction.

In the frog and salamander, it was found that the dorsal lip of the blastopore was very important in organizing the development of the body parts. When a portion of this dorsal lip was transplanted to the ventral region of another embryo, it acted as an organizer. The cells surrounding it formed into a separate head, and part of the body. The result was a Siamese type of twin salamander. This demonstrated that the cells in one part of the body of the embryo had the power to form body parts characteristic of other regions, but they needed the stimulation of the organizer from the dorsal lip in order to function. Hans Spemann did pioneering work on organizers. The optic vesicle in the embryo, he noted, stimulates the adjacent ectoderm to form the lens of the eye. When the optic vesicle is removed from the head and inserted at any point under the skin of the embryo's body, it will organize the overlying ectoderm into a lens. This principle of embryonic induction also takes place in many other areas of the embryo and plays a role in the development of such tissues as the nervous system and the limbs.

• PROBLEM 23-11

What do each of the embryonic germ layers give rise to?

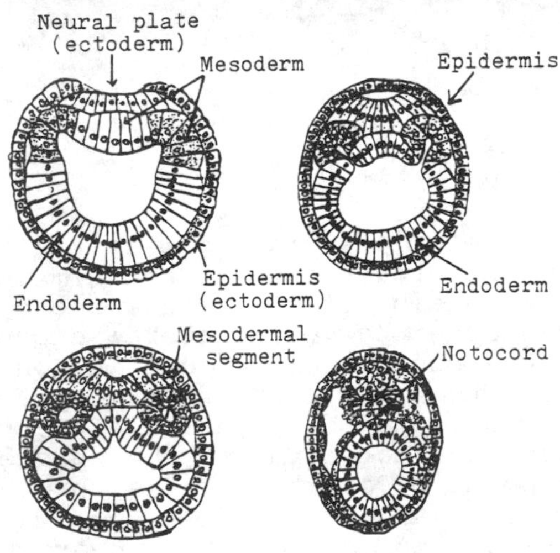

Early germ layer development.

Solution: Ectoderm gives rise to the epidermis of the skin, including the skin glands, hair, nails, and enamel of teeth. In addition, the epithelial lining of the mouth, nasal cavity, sinuses, sense organs, and the anal canal are ectodermal in origin. Nervous tissue, including the brain, spinal cord, and nerves, are all derived from embryonic ectoderm.

Mesoderm gives rise to muscle tissue, cartilage, bone, and the notochord which in man is replaced in the embryo by vertebrae. It also provides the foundation for the organs of circulation (bone marrow, blood, lymphoid tissue, blood vessels), excretion (kidneys, ureters) and reproduction (gonads, genital ducts). The mesoderm produces by far the greatest amount of tissue in the vertebrate body.

Endoderm gives rise to the epithelium of the digestive tract, the tonsils, the parathyroid and thymus glands, the larynx, trachea, lungs, the bladder and the urethra with its associated glands.

DIFFERENTIATION OF THE PRIMARY ORGAN RUDIMENTS

• PROBLEM 23-12

When changes are occurring in the ectoderm during the process of neurulation, differentiation of the pharynx is also occurring. What are the steps involved?

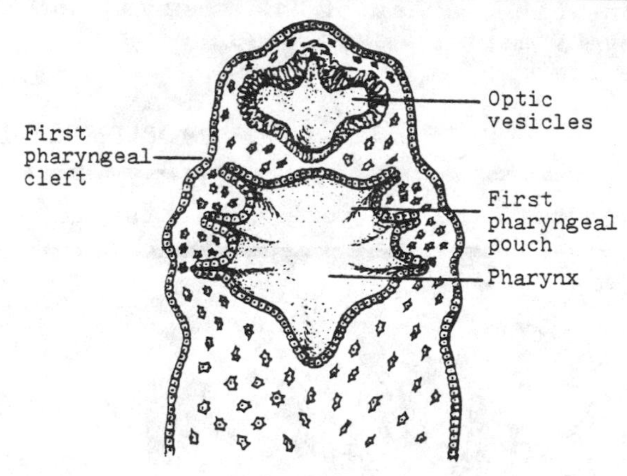

Head and pharyngeal region of amphibian embryo in frontal section. The pharynx is the expanded, endodermally lined, anterior end of the archenteron. Endodermal pouches extend to the surface and meet phayngeal clefts of ectodermal origin.

Soluion: One of the principal events occurring in the region of the archenteron at the time of neurulation is the rapid expansion of the anterior end to form the region known as the pharynx. Lateral expansion of the pharynx produces pouches which come in contact with the ectoderm. The two tissues form, respectively, pharyngeal pouches and pharyngeal clefts. Shortly after contact between the ectoderm and the endoderm of the pharynx, the point of contact breaks through to create pharyngeal clefts. The tissue now remaining between the two pharyngeal clefts is known as a pharyngeal arch. This structure is covered on the exterior by ectoderm and on the interior by endoderm. The arch is composed of loosely organized mesodermal cells called mesenchyme. In amphibians, the gills develop from the pharyngeal arches. In the human and other higher vertebrates, the pouches give rise to other structures or disappear. The first pair of pouches becomes the cavities of the middle ear and the connection with the pharynx, the Eustachian tube. The second pair of pouches becomes a pair of tonsils, while parts of the third and fourth pairs of pouches become the thymus gland, and other parts of them become the parathyroids. The fifth pair of pouches forms part of the thyroid. The bulk of the thyroid develops from a separate outgrowth on the floor of the pharynx. The cavity of the mouth arises as a shallow pocket of ectoderm that grows in to meet the anterior end of the pharynx. The membrane between the two ruptures and disappears during the fifth week of development.

• PROBLEM 23-13

How are eyes formed in embryos of the birds and mammals?

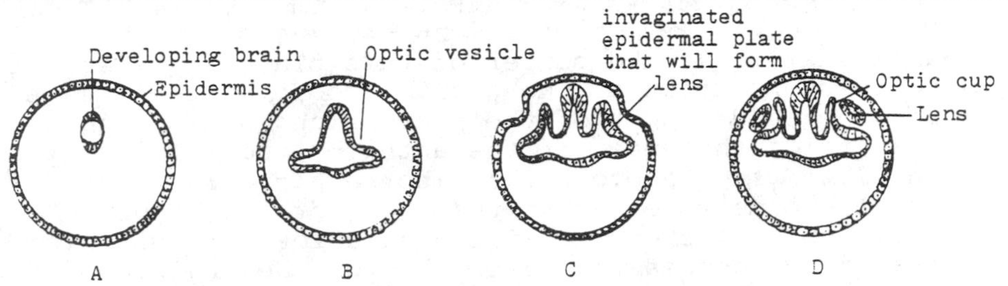

Development of optic vesicles and their induction of lenses in a frog.

Solution: In the area of the forebrain, there appears a lateral growth on each side of the neural tube. These saclike protrusions are termed the optic vesicles. These vesicles grow outward and come into contact with the inner surface of the overlying epidermis. When contact is made, the optic vesicle flattens out and invaginates to form a double walled optic cup. The inner, much thicker layer of the cup ultimately will become the sensory retina of the eye, and the outer layer becomes the pigment layer of the retina. When the optic cup touches the overlying epidermis it stimulates the latter to develop into a lens rudiment. In birds and mammals, the epidermis thickens and folds in to produce a pocket. This pinches off and forms a lens vesicle that lies in the opening of the optic cup and is surrounded by the iris, formed from the rim of the optic cup. The cells on the inner side of the lens vesicle become columnar and then are transformed into long fibers. The nuclei degenerate and the cytoplasm becomes hard, transparent and able to refract light.

• PROBLEM 23-14

Which are the first two organs to appear in the human embryo, and how are they formed?

Solution: The first two organs to appear in the embryo are the brain and spinal cord, which form by a process termed neurulation. During the third week of development, the ectoderm in front of the primitive streak develops a thickened plate of cells called the

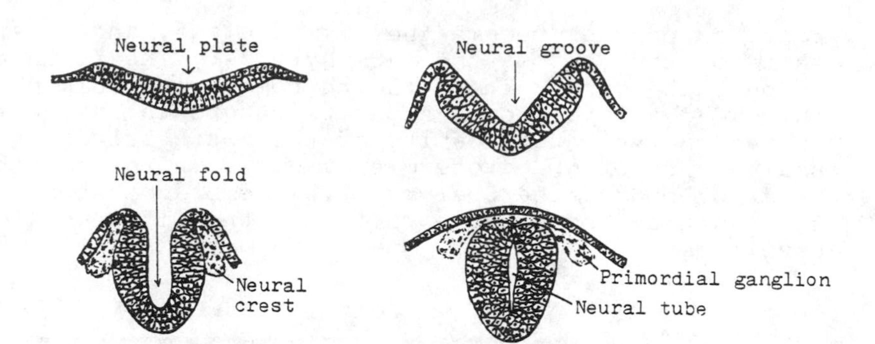

Cross section of the ectoderm of human embryos at successively later stages illustrating the origin of the neural tube and neural crest, which forms the dorsal root ganglia and the sympathetic nerve ganglia.

neural plate. At the center of this plate there appears a depression, known as the neural groove. At the same time, the outer edges of the plate rise in two longitudinal neural folds that meet at the anterior end. These appear, when viewed from above, like a horseshoe. These folds gradually come together at the top, forming a hollow neural tube. The cavity of the anterior part of this neural tube becomes the ventricles of the brain. At the same time, the cavity of the posterior part becomes the neural canal, extending the length of the spinal cord. The brain region is the first to form, and the long spinal cord develops slightly later. The anterior part of the neural tube, which gives rise to the brain, is much larger than the posterior part and continues to grow so rapidly that the head region comes to bend down at the anterior end of the embryo. By the fifth week of development,all the regions of the brain, i.e., the forebrain, midbrain, and hindbrain are established. A short time later, the outgrowths that will form the large cerebral hemispheres begin to grow. While the various motor nerves grow out of the brain and spinal cord, the sensory nerves do not, having a separate origin. During the formation of the neural tube, bits of nervous tissue, known as neural crest cells, are left over on each side of the tube. These cells migrate downward from their original position and form the dorsal root ganglia of the spinal nerves and the postganglionic sympathetic neurons. From sensory cells in the dorsal root ganglia, dendrites grow out to the sense organs and axons grow in to the spinal cord. Other neural crest cells migrate and form the medullary cells of the adrenal glands, the neurilemma sheath cells of the peripheral neurons, and certain other structures.

• PROBLEM 23-15

Many organs develop in the embryo without having to function at the same time, but the heart must function while undergoing development. Describe the steps in the formation of the heart.

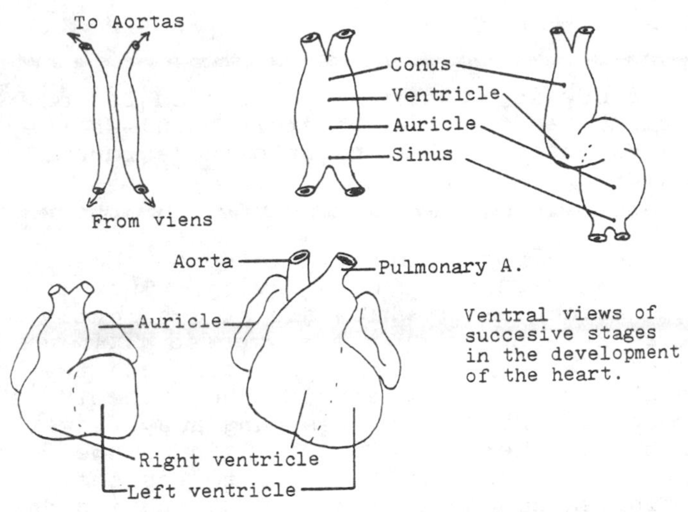

Ventral views of succesive stages in the development of the heart.

Solution: In its early stage of development, the heart consists only of a simple tube, formed from the fusion of two thin-walled blood vessels located below the head. In this early condition, the human heart resembles a frog heart. A frog heart consists of four "chambers" arranged in series: the sinus venosus, which receives blood from the veins, an atrium, a ventricle, and the arterial cone, which leads to the aortic arches. Since the tube grows faster than the points to which its front and rear ends are attached, it bulges to one side. The ventricle then twists in an S-shaped curve down and in front of the atrium, coming to lie ventral to it as it exists in the adult. Gradually, the sinus venosus becomes incorporated into the atrium as the atrium grows around it, and most of the arterial cone is merged with the wall of the ventricle.

The heart becomes partitioned into a right side and a left side in all air-breathing vertebrates. This takes place rather early in development, long before the lungs have begun to function at the time of birth. No such division occurs in the hearts of vertebrates which breathe in water. Their hearts pump just one kind of blood, namely, "venous blood" (blood low in oxygen). Aeration takes place in the gills, and only "arterial blood" (blood rich in oxygen) is distributed to the body. The advent of lungs necessitated a radical change in the arrangement of the circulation. The heart now receives two sorts of blood: venous blood from the body, and arterial blood from the lungs. It is important that the two streams be kept separate so that the venous blood is sent to the lungs for aeration and only the arterial blood is distributed to the body.

The heart begins separating into four chambers at an early stage. By the end of the second month of development, the two ventricles are completely separated. The septum separating the atria, the foramen ovale, however, is not completely fused until the fetus is born. If this septum remains open at birth, the baby is termed a "blue baby" due to the flow of deoxygenated blood through the body. Surgical correction is performed for this condition.

• PROBLEM 23-16

Occasionally a boy is born with undescended testes (cryptorchidism), and he will be sterile unless they are surgically caused to descend into the scrotum. Explain the reason for this.

Solution: Human sperm cannot develop at the high temperature found within the body cavity. They can only develop in the testes which are at a slightly lower temperature since they are suspended from the body in the scrotum. When the weather is warm, and the body is radiating considerable heat, the skin of the scrotum loosens so that the testes hang away from the body. When the weather is cold, the skin of the testes contracts and draws the testes closer to the body, thus preventing excess loss of heat. During its development in the mother, the testes remain inside the body cavity of the male fetus. At birth, the testes descend into the scrotum. If there is a deficiency of male hormone or if the canal is too narrow, the testes will not descend. The result of this would be sterility, the sperm being unable to develop at the higher temperature in the body cavity.

PARTURITION

• PROBLEM 23-17

Differentiate between a placenta and an umbilical cord.

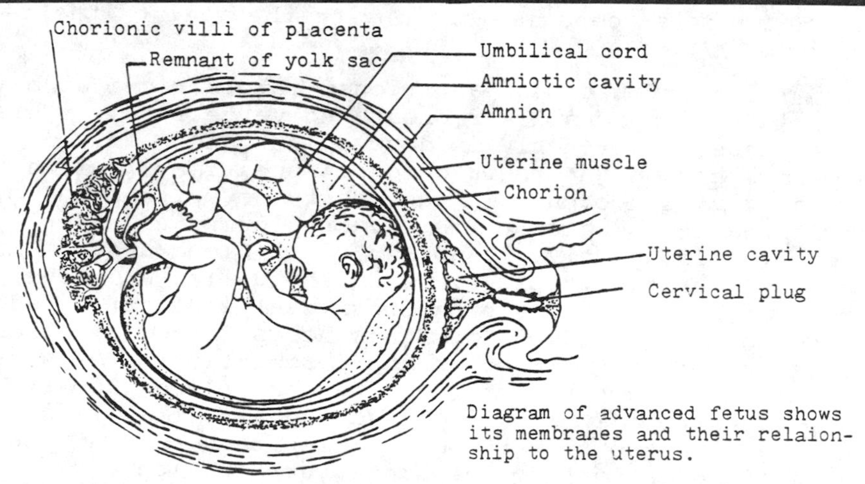

Diagram of advanced fetus shows its membranes and their relaionship to the uterus.

Solution: As the human embryo grows, the region on the ventral side from which the folds of the amnion, the yolk sac, and the allantois grow, becomes relatively smaller, and the edges of the amniotic folds come together to form a tube which encloses the other membranes. This tube is called the umbilical cord and contains, in addition to the yolk sac and allantois, the large umbilical blood vessels through which the fetus obtains nourishment from the wall of the uterus. The umbilical cord, about 1 cm in diameter and about 70 cm long at birth, is composed chiefly of a peculiar jelly-like material found nowhere else.

The placenta is the area where a portion of the chorion and a portion of the uterine wall join. The placenta is thus a structure of double origin, partly embryonic (chorion) and partly maternal (uterus). It functions to exchange nutrients, wastes, and gases between the mother and the fetus. In mammals such as the human, the placenta is a disk about 7 inches in diameter and about 1 to 1½ inches thick. Not all mammals have such disklike placentas. For example, cows have numerous isolated clusters of fetal-maternal associations, other animals have fingerlike projections of the chorion that provide a very diffuse type of association. In cats and dogs the placenta forms a ring around the fetus.

• PROBLEM 23-18

Birth is associated with a number of rather drastic but important changes that occur in a short period of time. Discuss these changes.

Solution: Several important changes must occur in the newborn baby in a very short period of time. During the time of its development in the mother, the fetus obtained all its oxygen and eliminated its CO_2 through the placenta. Its lungs were collapsed. Shortly after birth, when the placenta is no longer present, the carbon dioxide level in the blood of the baby rises to the point at which it stimulates the respiratory center in the medulla of the brain to initiate breathing. The lungs fill with air for the first time and normal breathing follows.

A second important change occurs in the heart and blood vessels. In the fetal heart there exists a hole between the right and left atria called the foramen ovale, allowing blood from both sides of the heart to mix. In addition, in the fetus there exists a duct between the pulmonary artery and the aorta. This duct, termed the ductus arteriosus, effectively cuts down the blood supply to the lungs by diverting it into the aorta. At the time of birth, the ductus arteriosus becomes constricted and eventually is completely blocked. This allows more blood to enter the lungs, the flow increasing by approximately seven times. Also at this time the flaps around the opening between the atria fuse together, thus effectively separating the two atria and allowing the heart to function as two separate pumps. While in the mother's womb the heart serves to pump oxygenated blood, obtained from the mother, to the rest of the fetal body. When the fetus is born, however, the heart must serve the function of pumping deoxygenated blood to the lungs as well as oxygenated blood to the rest of the body. Should the foramen ovale fail to close, deoxygenated and oxygenated blood would be able to mix and the infant would be deprived of its full oxygen supply. When this happens, the newborn is termed a "blue baby," because that is the color of the infant due to the deficiency of oxygen in blood. The foramen ovale must be surgically closed.

• PROBLEM 23-19

Parturition is the separation of the fetus and its membranes from the mother's body at the time of birth. Describe this process.

Solution: As the period of gestation ends, a hormonal substance termed relaxin appears in the bloodstream. This substance seems to be a product of the corpus luteum of pregnancy and of the placenta and causes the cervix of the uterus to become readily dilatable. Oxytocin, one of the hormones released from the posterior pituitary, is an extremely potent uterine-muscle stimulant. Oxytocin is reflexly released into the bloodstream as a result of afferent input to the hypothalamus from receptors in the uterus, particularly in the cervix. Oxytocin stimulates the uterine muscles to contract with greater force and frequency during labor. Relaxin and oxytocin are two important hormones involved in parturition.

At the time of birth, rhythmic contractions of the smooth muscle of the uterus begin. Soon they are accompanied by reflex and voluntary activity of skeletal muscle. The result is that internal pressure forces the fetus, normally head first, into the cervix of the uterus. The fetal membranes then bulge through the cervix and burst. Amniotic fluid is discharged. The child then enters the vagina and is born, a process lasting from twenty minutes to one hour. Some fifteen minutes after birth, the uterus again goes into rhythmic contraction, and an "afterbirth" consisting of the fetal membranes is expelled. Surprisingly enough, there is only a minor amount of bleeding. Within the next few days, the lining of the uterus is restored and the uterus returns to nearly its original dimensions.

SHORT ANSWER QUESTIONS FOR REVIEW

Answer

To be covered when testing yourself

Choose the correct answer.

1. Following fertilization, the zygote undergoes a series of rapid mitotic divisions. The stage at which a solid ball of cells is formed is called: (a) the morula (b) the blastula (c) the gastrula (d) the fetus (e) none of the above. — a

2. The first stage of embryonic development in which three distinct germ layers are seen is: (a) the morula (b) the blastula (c) the gastrula (d) the fetus (e) none of the above. — c

3. The structure responsible for prevention of multiple fertilizations is the: (a) hyaline layer (b) acrosomal layer (c) corona radiata (d) allantoic membrane (e) tympanic membrane. — a

4. An egg in which the yolk is concentrated towards one pole is called: (a) telolecithal (b) centrolecithal (c) isolecithal (d) mesolecithal (e) perilecithal. — a

5. The smaller sized blastomeres are called _____, and they re found at the _____ pole. (a) Macromeres, animal (b) macromeres, vegetal (c) micromeres, animal (d) micromeres, vegetal (e) none of the above. — c

6. Which is true concerning the process of cleavage? (a) Meroblastic cleavage is characteristic of human embryology. (b) Meroblastic cleavage results in complete cell cleavage. (c) Radial cleavage is characteristic of chordates. (d) Spiral cleavage occurs only in vertebrates. (e) Temperature has no effect on the rate of cleavage. — c

7. The primitive gut formed during gastrulation is called the (a) blastopore (b) gastrocoel (c) blastocoel (d) archenteron (e) ventral pore. — d

8. Which of the following structures are not ectodermal in origin? (a) Kidney, ureter, gonads (b) sense organs, anal canal, nasal cavity (c) epidermis, skin glands, tooth enamel (d) brain, spinal cord, nerves (e) all are ectodermal in origin. — a

9. Which of the following are not mesodermal in origin? (a) Muscle tissue, cartilage, bone (b) bone marrow, blood, lymphoid tissue (c) blood vessels, genital ducts, kidneys (d) thymus gland, trachea, bladder (e) all are mesodermal in origin. — d

Answer

To be covered when testing yourself

10. Which of the following embryonic precursors eventually forms the extraembryonic membranes? (a) vitelline membrane (b) hypoblast (c) epiblast (d) trophoblast (e) none of the above.

d

11. The development of an extraembryonic membrane was a key adaptation between which two groups of organisms? (a) Reptiles and mammals (b) birds and mammals (c) amphibians and birds (d) reptiles and birds (e) amphibians and reptiles.

e

12-17 Match the description to the word that best defines it:

			Answer
12. chorion	_____	(a) liquid medium surrounding the embryo which prevents shock, abrasion and desiccation,	12-f,
13. chorionic villi	_____	(b) richly vascularized sac which stores and supplies food for the embryo,	13-e,
14. placenta	_____	(c) most efficient area of exchange between the mother and the embryo,	14-c,
15. amnion	_____	(d) layer between amnion and chorion; depot of nitrogenous wastes,	15-a,
16. yolk sac	_____	(e) tissue outgrowths which penetrate the uterine wall thus increasing the absorptive area,	16-b,
17. allantois	_____	(f) functions in the exchange of gases and wastes.	17-d.

Fill in the blanks.

Questions 18-22 deal with the process of gastrulation concerning the bird:

18. In birds the upper layer of embryonic cells is known as the _____.

epiblast

19. A thickened knot of cells present at the anterior end of the primitive streak is known as _____.

Hensen's node

20. The primitive streak contains a narrow furrow referred to as the _____.

primitive groove

21. The cells of the epiblast during gastrulation invaginate at the _____.

primitive streak

	Answer
	To be covered when testing yourself
22. As the lateral cells continue to invaginate, they move into the space between the _____ and the _____.	hypoblast, epiblast
23. An _____ is a chemical secreted by embryonic cells which influences adjacent cells.	organizer
24. In the frog and salamander it was determined that the _____ was important in organizing the development of body parts.	dorsal lip of the blasto-pore
25. The _____ stimulates the adjacent ectoderm to form the lens of the eye.	optic vesicle
26. When the optic vesicle is transplanted from the head region to the flank region, the overlying ectoderm will be organized into a lens. This illustrates the principle of _____.	induc-tion
27. Loosely organized mesodermal cells are referred to as _____.	mesen-chyme
28. The region where the chorion and uterine wall join is called the _____.	placenta
29. _____ is the separation of the fetus and its membranes at the time of birth.	parturition
In questions 30-39 fill in the blanks using the following words:	
parturition umbilical cord morula	
yolk sac chorion inner cell mass	
amnion trophoblast allantois	
gestation period pseudopregnancy	
30. The stage of development when the ovum is a solid ball of cells is called the _____ stage.	morula
31. In human embryology, when the developing ovum reaches the uterine cavity it begins to differentiate into a blastocyst. The two layers of the blastocyst are the _____ and _____.	tropho-blast, inner cell mass
32. When a uterus develops due to a mechanical stimulation, the condition which results is called a _____.	pseudo-pregnancy

	Answer To be covered when testing yourself
33. The embryos of reptiles and birds have a digestive tract outgrowth containing nutrients which is called the _____.	yolk sac
34. The liquid containing sac which acts as a shock absorber for the developing embryo is the _____.	amnion
35. The _____ acts as a storage sac for nitrogenous wastes in reptilian and avian eggs.	allantois
36. The membrane which is in intimate contact with the shell in reptilian and avian eggs is the _____.	chorion
37. The structure through which the embryo receives nourishment from the uterine wall is the _____.	umbilical cord
38. The duration of pregnancy which is approximately 280 days is termed the _____ period.	gestation
39. The birth process which occurs at the end of pregnancy is called _____.	parturition
Determine whether the following statements are true or false.	
40. In animals, the sperm and ovaries unite to form a zygote.	False
41. After numerous divisions, the embryo develops into a hollow ball called a morula.	False
42. The blastula indents on one side to form a two-layered gastrula.	True
43. After continued development, the embryo contains three layers: epiderm, mesoderm, endoderm.	False
44. The inner layer of endoderm forms the digestive tract.	True

CHAPTER 24

STRUCTURE & FUNCTION OF GENES

DNA: THE GENETIC MATERIAL

• **PROBLEM** 24-1

What evidence regarding the nature of the genetic material has been obtained from experiments with bacteriophage?

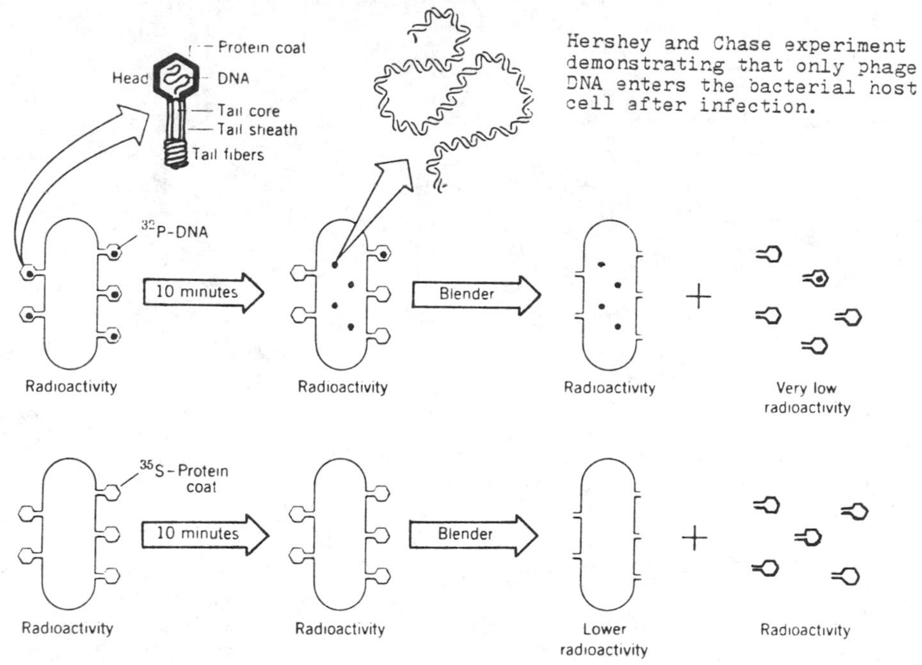

Hershey and Chase experiment demonstrating that only phage DNA enters the bacterial host cell after infection.

Solution: By using radioactive tracers, A.D. Hershey and M. Chase furnished proof that DNA is the genetic material. This evidence was obtained by use of the bacterium Escherichia coli; and the bacteriophage which attack it. The bacteriophage were grown on a medium containing limited amounts of radioactive sulfur and phosphorous. The radioactive labels used were phosphorous 32 (^{32}P) and sulfur 35 (^{35}S). Phage proteins do not contain appreciable amounts of phosphorus. Therefore, only the DNA was labeled with ^{32}P. Similarly, only the protein envelope around the phage was labeled with ^{35}S. By this method, the DNA phage and proteins were differentially labeled.

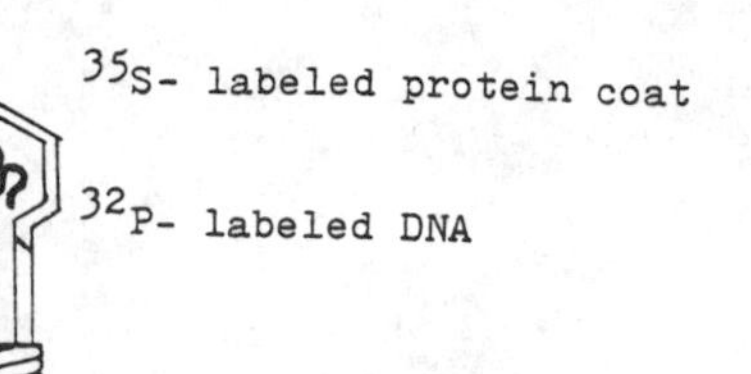

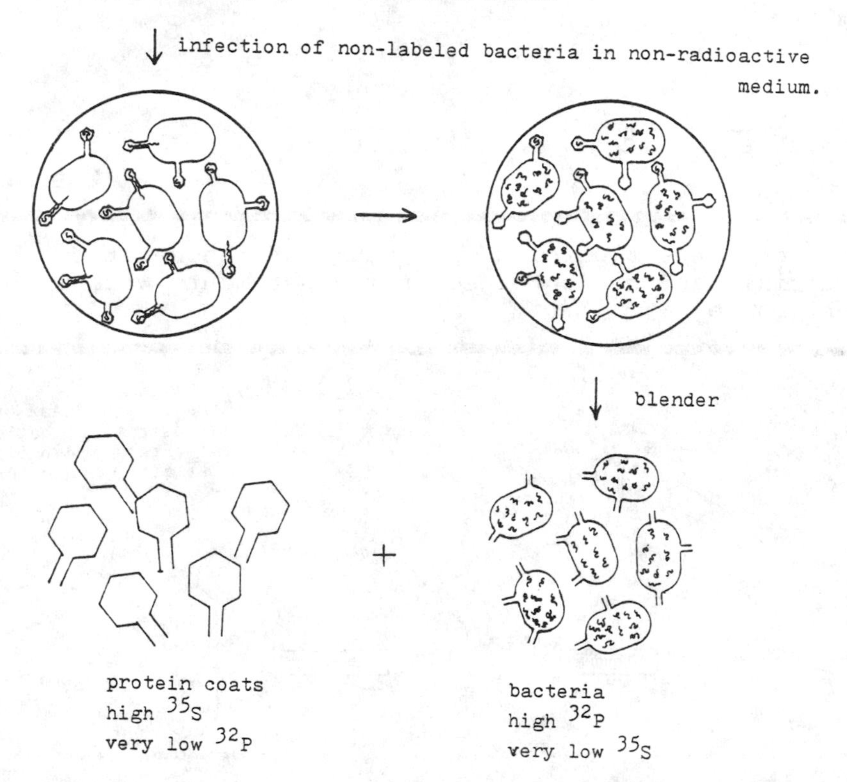

Following growth of the phage and incorporation of the radioactive labels, the radioactive viruses were exposed to non-radioactive bacteria. After absorption and infection of the bacteria, the phage were separated from the bacterial cells by centrifugation. The location of the ^{32}P labeled DNA and ^{35}S labeled proteins of the virus were then determined. The phosphorus label was found to be associated with the bacterial cells and the sulfur label was in the protein coats left in the medium. This indicated that the DNA had penetrated the cells but that the protein coats of the phage was left outside the walls of the bacteria and had been separated from the bacterial cells during centrifugation. Only the labeled DNA was passed on to the bacteria.

The significant conclusion of this experiment is that only the DNA of the virus entered the host bacteria. All the protein of the phage remained outside the bacteria cells wall. No protein was in the bacteria during viral replication. This experiment illustrates that it is DNA and not protein that is the genetic material. (See diagram)

After the DNA part of the virus was reproduced within the bacterial cell, new protein was synthesized and became associated with the DNA units. New infective virus particles were thus formed. When the host (bacterium) lysed, numerous infectious virus particles emerged, ready to enter other bacterial cells and repeat the cycle.

• PROBLEM 24-2

What is the nature of a "transforming agent"? What importance may this phenomenon have on our understanding of the chemical basis of inheritance?

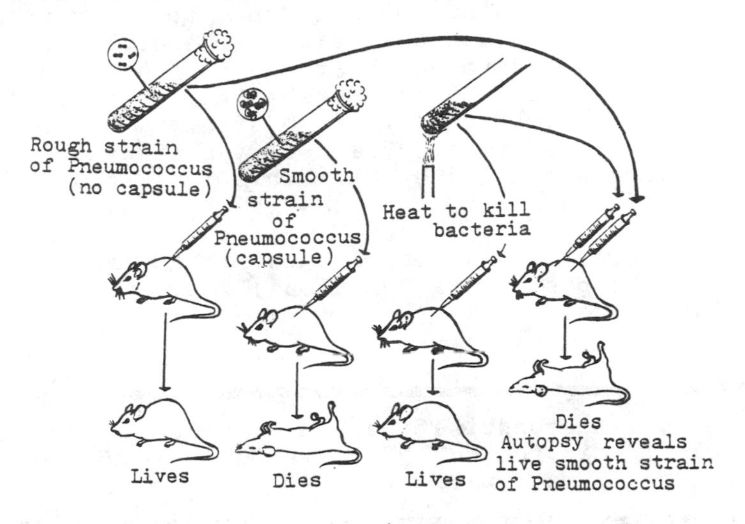

The experiments of Fred Griffiths, which demonstrated the transfer of genetic information from dead, heat-killed bacteria to living bacteria of a different strain. Although neither the rough strain of Pneumococcus nor heat-killed smooth strain pneumococci would kill a mouse, a combination of the two did. Autopsy of the dead mouse showed the presence of living, smooth strain pneumococci.

Solution: The concept of transformation arises from the experiments performed by Griffith in 1928. It was observed that when injected into mice, some strains of pneumococcus bacteria caused pneumonia and usually death. Other strains of the bacteria, were relatively harmless. The infective form always had a capsule (a complicated polysaccharide coating). The non-infective form did not have a virulent capsule. The encapsulated strain was called the "smooth strain" because the colonies looked smooth on a culture plate, and the harmless, unencapsulated strain was called the "rough strain" because of the rough appearance of its colonies.

In his famous experiment, Griffith injected one group of mice with the virulent smooth strain and another with the harmless rough strain. As expected, mice from the former group died while the latter group survived. Griffith then injected a third group of mice with the heat-killed smooth strain. This group lived, showing that bacteria killed by heat were no longer virulent. However, when a fourth group of mice was injected simultaneously with both the harmless rough strain and the heat-killed smooth bacteria, the mice died. The disease-causing organisms were of the smooth type bacteria. This process, by which something from the heat killed bacteria converted rough bacteria into smooth bacteria, is known as <u>transformation</u>.

Griffith felt that protein from the dead bacteria might be the active transforming agent. But Griffith's interpretation of his experiment was later shown to be incorrect by Avery, MacLeod, and McCarthy in 1944. They made an extract from heat-killed smooth cells, and purified the extract by removing any substance that did not cause transformation of rough bacteria into smooth bacteria. Eventually they determined that DNA was the essential transforming agent. Moreover, when this extracted DNA was added to other types of rough strains, the bacteria formed from transformation were always identical to the bacteria that donated the DNA. This indicated that hereditary information must be carried by DNA.

Transformation suggests that in bacteria, genetic traits can be passed via DNA alone, from one bacterium to another.

• PROBLEM 24-3

Discuss how the quantitative measurements of the dioxyribonucleic acid content of cells is evidence that DNA is the genetic material.

<u>Solution:</u> In the 1940's, A.E. Mirsky and Hans Ris working at the Rockefeller Institute, and Andre Boivin and Roger Vendrely, working at the University at Strasbourg, independently showed that the amount of DNA per nucleus is constant in all of the body cells of a given organism. By making cell counts and chemical analyses, Mirsky and Vendrely showed that there is about 6×10^{-9} milligrams of DNA per nucleus in somatic cells, but only 3×10^{-9} milligrams of DNA per nucleus in egg cells or sperm cells. In tissues that are polyploid, (having more than two sets of chromosomes per nucleus), the amount of DNA was found to be a corresponding multiple of the usual amount. From the amount of DNA per cell, one can estimate the number of nucleotide pairs per cell, and thus the amount of genetic information in each kind of cell.

Only the amount of DNA and the amount of certain basic positively charged proteins called histones are relatively constant from one cell to the next. The amounts of other types of proteins and RNA vary considerably from cell to cell. Thus the fact that the amount of DNA, like the number of genes, is constant in all the cells of the body, and the fact that the amount of DNA in germ cells is only half the amount in somatic cells, is strong evidence that DNA is an essential part of the genetic material.

STRUCTURE AND PROPERTIES OF DNA

• PROBLEM 24-4

Discuss the chemical composition of DNA.

Solution: Deoxyribonucleic acid or DNA is one type of nucleic acid present in the nucleus of all cells. Nucleic acids are rich in phosphorus, and contain carbon, oxygen, hydrogen, and nitrogen. DNA is composed of nitrogenous bases, a five-carbon sugar called deoxyribose, and phosphate groups. There are four kinds of nitrogenous bases: adenine, guanine, cytosine, and thymine. Adenine,and guanine belong to the class of organic compounds called purines, and cytosine and thymine belong to the pyrimidines. Each nitrogenous base is attached to deoxyribose molecule via a glycosidic linkage, and the sugar is attached to the phosphate by an ester bond. This combination of base-sugar-phosphate comprises the fundamental unit, termed a nucleotide, of nucleic acid. Four kinds of deoxyribonucleotides are found in DNA, each containing one of the four types of nitrogenous bases. The nucleotides are joined by phosphate ester bonds into a chain. Two complementary chains of nucleotides form a DNA molecule.

*= site of attachment to deoxyribose

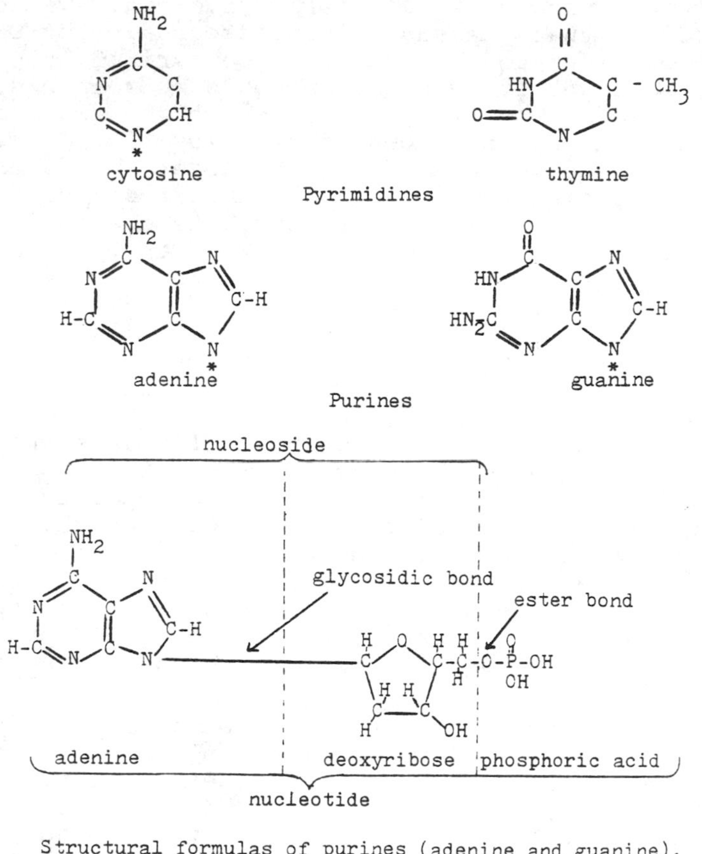

cytosine thymine

Pyrimidines

adenine guanine

Purines

nucleoside

glycosidic bond

ester bond

adenine deoxyribose phosphoric acid

nucleotide

Structural formulas of purines (adenine and guanine), pyrimidines (thymine and cytosine), and a nucleotide.

• PROBLEM 24-5

How does the Watson-Crick model of DNA account for its observed properties?

Solution: By 1950, several properties of DNA were well established. Chargaff showed that the four nitrogenous bases of DNA did not occur in equal proportions, however, the total amount of purines equaled the total amount of pyrimidines (A+G = T + C). In addition, the amount of adenine equaled the amount of thymine (A = T), and likewise for guanine and cytosine (G = C). Pauling had suggested that, the structure of DNA might be some sort of an α-helix held together by hydrogen bonds. The final observation was made by Franklin and Wilkins. They inferred from x-ray diffraction studies that the nucleotide bases (which are planar molecules) were stacked one on top of the other like a pile of saucers.

On the basis of these observed properties of DNA, Watson and Crick in 1953 proposed a model of the DNA molecule. The Watson and Crick model consisted of a double helix of nucleotides in which the two nucleotide helices were wound around each other. Each full turn of the helix measured 34 Å and contained 10 nucteotides equally spaced from each other. The radius of the helix was 10 Å. These measurements were in agreement with those obtained from x-ray diffraction patterns of DNA.

To account for Pauling's observation, Watson and Crick proposed that the sugar-phosphate chains of DNA should be on the outside, and the purines and pyrimidines on the inside, held together by hydrogen bonds between bases on opposite chains. When they tried to put the two chains of the helix together, they found the chains fitted best when they ran in opposite directions. Moveover, because x-ray diffraction studies specified the diameter of the helix to be 20 Å, the space could only accommodate one purine and one pyrimidine. If two purine bases paired, they would be too large to fit into the helix without destroying the regular shape of the double stranded structure. Similarly, two pyrimidines would be too small. Moreover, for one purine to be hydrogen-bonded with one pyrimidine properly, adenine must pair with thymine and guanine with cytosine. An A-T base pair is almost exactly the same in width as a C-G base pair, accounting for the regularity of the helix as seen in x-ray diffraction pictures. This concept of specific base pairing explained Chargaff's observation that the amounts of adenine and thymine in any DNA molecule are always equal, and the amounts of guanine and cytosine are always equal. Two hydrogen bonds can form between adenine and thymine, and three hydrogen bonds between guanine and cytosine (Fig. 2). The specificity of the kind of hydrogen bond that can be formed provides for correct base pairing during replication and transcription.

The two chains are thus complementary to each other, that is, the sequence of nucleotides in one chain dictates the sequence of nucleotides in the other. The strands are also antiparallel, i.e., they extend in opposite directions and have their terminal phosphate groups at opposite ends of the double helix.

• PROBLEM 24-6

Describe the three major structural distinctions between DNA and RNA.

Solution: DNA or deoxyribonucleic acid is in the form of a double helix having a deoxyribose sugar and phosphate backbone. The two helices are linked together by hydrogen bonds between nitrogenous bases bound to the sugar moiety of the backbone. In DNA, the bases are adenine, guanine, cytosine and thymine.

RNA or ribonucleic acid differs from DNA in three important respects. First, the sugar in the sugar-phosphate backbone of RNA is ribose rather than deoxyribose. Ribose has hydroxyl groups on the number 2 and 3 carbons, whereas deoxyribose has a hydroxyl on the number 3 carbon only (see figure 1).

Figure 1.

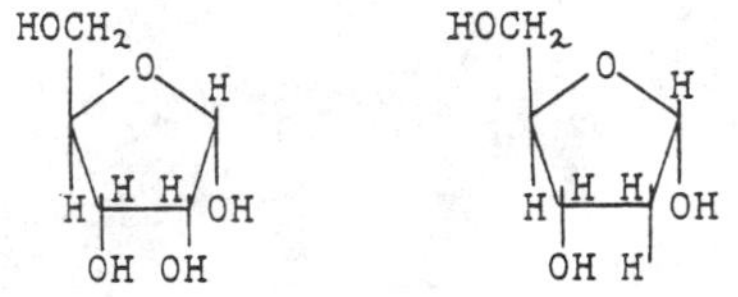

D-ribose (RNA) 2-deoxy-D-ribose (DNA)

Secondly, RNA has only a single sugar-phosphate backbone with attached single bases. Although hydrogen bonding may occur between the bases of a single RNA strand causing it to fold back on itself, it is not regular like that of DNA, where each base has its complement on the other strand. Thus, RNA is similar to a single strand of DNA. Finally, the pyrimidine base uracil (see figure 2) is found in RNA instead of the pyrimidine base thymine found in DNA.

Therefore, the nitrogenous bases present in RNA are adenine, guanine, cytosine and uracil.

Figure 2.

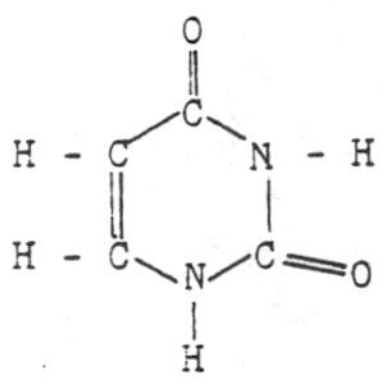

Uracil

• PROBLEM 24-7

How was it proven that DNA replication is semi-conservative?

Solution: Three possible models were considered to account for DNA replication, the process by which the two complementary strands of the double helical DNA replicate

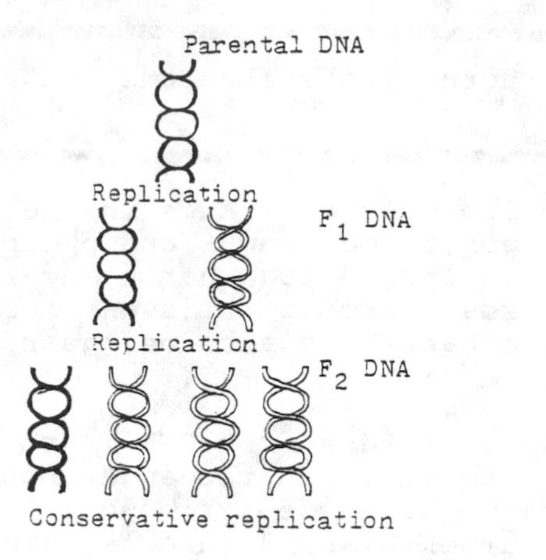

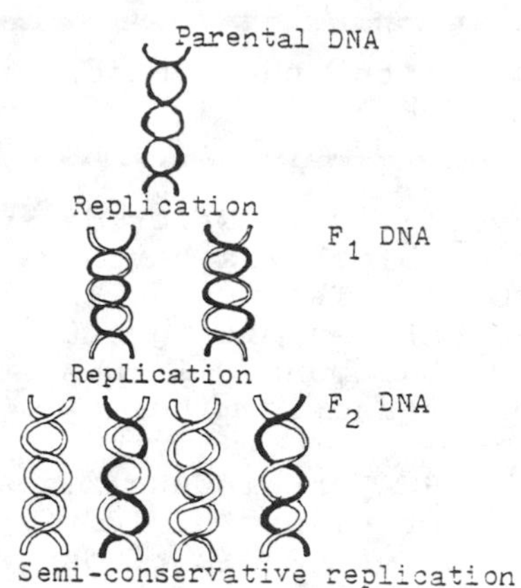

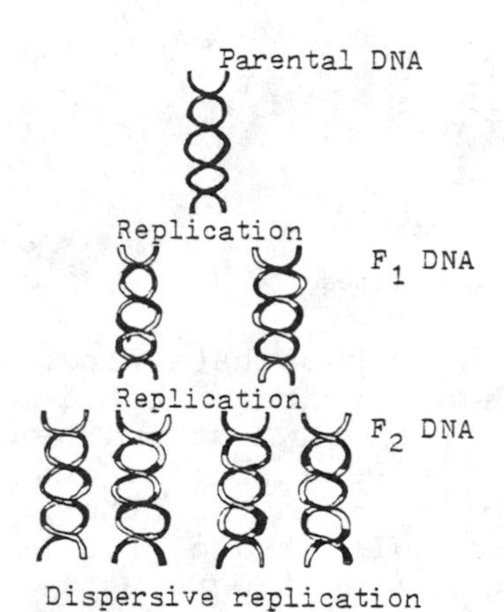

Figure 1. Models proposed for DNA replication, newly replicated strands are white while parental segments are black.

to form new complementary strands. In the first model, called conservative replication, the two strands of parental DNA replicate to yield the unchanged parental DNA and a newly synthesized DNA. The second model, semi-conservative replication, claims that the replication of one DNA molecule yields two hybrids, each composed of one parental strand and one newly synthesized strand. Finally, the third model, dispersive replication, has the parental strand breaking at intervals. These parental segments then combine with new segments to form the daughter strands (see Figure 1).

Meselson and Stahl's classic experiments proved that DNA replicates by a semi-conservative mechanism. They grew E. coli cells for several generations on a medium which contained "heavy" nitrogen, ^{15}N. Thus, all the nitrogen bases of the E. coli DNA were labeled with ^{15}N instead of the normal or "light" nitrogen, ^{15}N. The ^{14}N - DNA of these cells could be separated from the ^{14}N-DNA by equilibrium centrifugation in a cesium-chloride gradient because the ^{15}N-DNA has a significantly greater density than the ^{14}N-DNA.

Meselson and Stahl took the "heavy" (^{15}N containing)

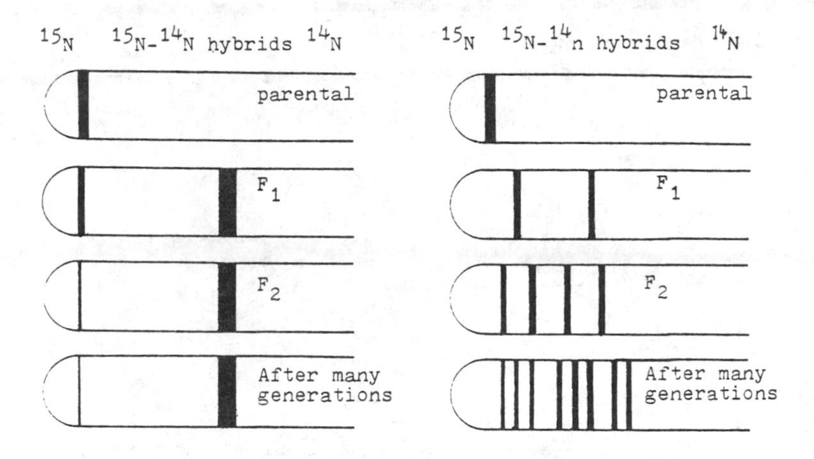

Conservative model of replication dispersive replication

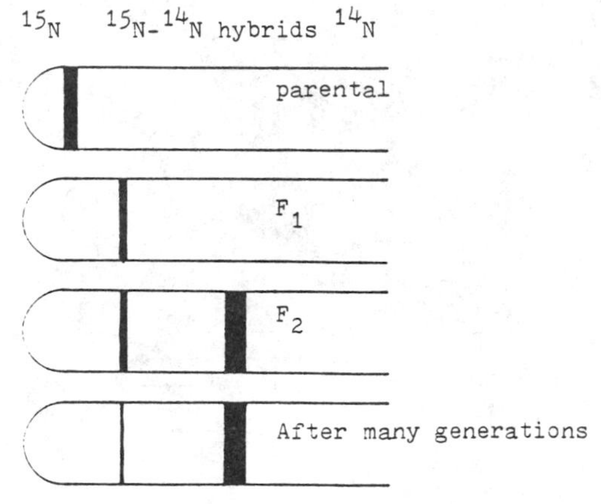

Semi-conservative model of DNA replication

⟵ Direction of sedimentation

Figure 2 The cesium-choride density gradients expected from the three models of DNA replication. The heavier fractions lie to the left.

cells and placed them in a "light" (^{14}N containing) medium. The cells were allowed to grow for several generations and were analyzed at various generation times. From the samples collected, the DNA was examined by determining its buoyant density with cesium-chloride density gradients. It was found that after one generation, the isolated DNA showed only a single band falling midway between the places where the heavy and light bands lie in a cesium-chloride density gradient. This is to be expected if the first generation DNA is a hybrid of one parental strand and one newly synthesized strand. After two generations, the isolated DNA exhibited two bands, one equal to the hybrid density of the first generation and one having a density equal to that of normal "light" DNA. After many generations passed, a "light" density band of DNA predominated. These observations are consistent with those expected from the hypothesis of semi-conservative replication but not with the other two models (see Figure 2).

• PROBLEM 24-8

How is DNA replicated?

Replication of DNA

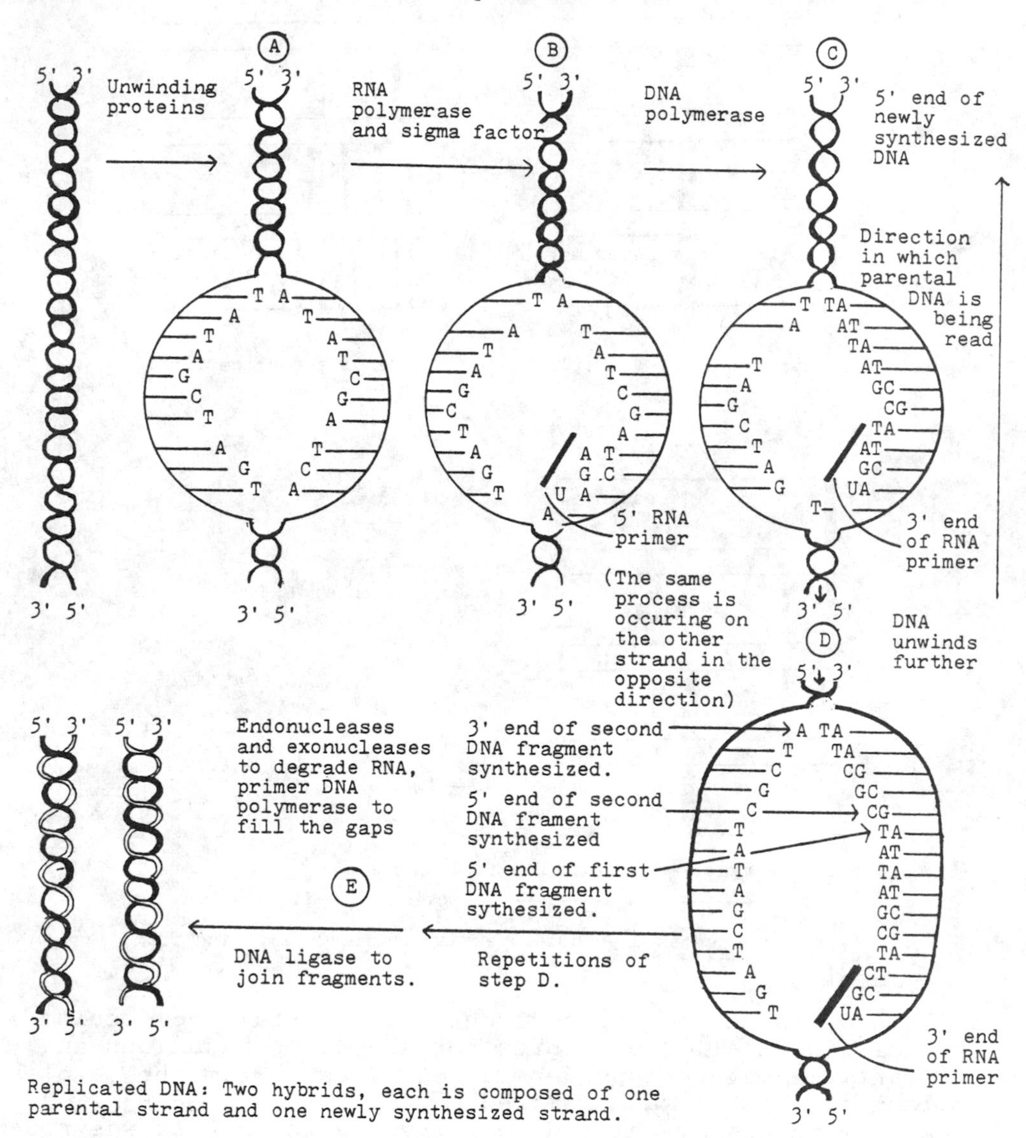

Replicated DNA: Two hybrids, each is composed of one parental strand and one newly synthesized strand.

<u>Solution:</u> By studying the Watson-Crick model of DNA, Kornberg and his colleagues determined the mechanism of DNA replication. Since DNA is the genetic material of the cell, it must have the information to replicate itself built into it. The Watson-Crick model of DNA seems to offer a good explanation of how DNA is replicated. There are only four different nucleotides found in DNA. Chargaff's rule of specific base pairing states that these nucleotides (adenine, guanine, thymine and cytosine) are ordered on both strands of the DNA helix so that adenine is always paired to thymine and cytosine to guanine. Thus, in order for DNA to be replicated, the helix need only be unwound to form a template. The nucleotide building blocks can line up in

a sequence complementary to the order presented. In this way, two double stranded helices identical to the original molecule are synthesized in the nucleus. This is a general overview of the mechanism of DNA replication.

Actually, the first step of DNA replication occurs when unwinding proteins uncoil a part of the DNA helix (see figure - step A). At this time, RNA polymerase combines with the sigma factor to initiate RNA synthesis. This is an essential preliminary step because the RNA that is synthesized serves as a primer which is necessary before DNA replication can occur (Figure-step B). The RNA that is synthesized is not a final product because before DNA replication is completed, it is destroyed by endonucleases and exonucleases. After the primer is made, DNA polymerase initiates DNA replication. There are free nucleotides in the nucleus which complementarily join to form a new strand along the 5' to 3' direction of the unwound portion, however, the DNA strand itself is synthesized in a 3' to 5' manner (Figure-step C). DNA replication is thought to occur in short, uncontinuous bursts allowing the DNA helix to unwind farther (Figure-step D). DNA polymerase then fills in the gaps between the fragments synthesized. DNA ligase joins the 5' end of one fragment to the 3' end of the adjacent fragment. In this way, a strand of DNA is replicated according to the semi-conservative model, a parental DNA replicates to give two first generation hybrids containing one strand of the parental DNA and one newly synthesized strand. (Figure-step E).

THE GENETIC CODE

• PROBLEM 24-9

Discuss the problem of the 'coding' of information in the gene.

The Genetic Code

First position (5' end)	Second position	Third position (3' end) U	C	A	G
U	U	Phe	Phe	Leu	leu
	C	Ser	Ser	Ser	Ser
	A	Tyr	Tyr	Terminator	Terminator
	G	Cys	Cys	Terminator	Trp
C	U	Leu	Leu	Leu	Leu
	C	Pro	Pro	Pro	Pro
	A	His	His	Glu NH_2	Glu NH_2
	G	Arg	Arg	Arg	Arg
A	U	lleu	lleu	lleu	Met
	C	Thr	Thr	Thr	Thr
	A	Asp NH_2	Asp NH_2	Lys	Lys
	G	Ser	Ser	Arg	Arg
G	U	Val	Val	Val	Val
	C	Ala	Ala	Ala	Ala
	A	Asp	Asp	Asp	Asp
	G	Gly	Gly	Gly	Gly

Solution: Although the knowledge that genes consist of nucleotide sequences was a major breakthrough in the study of genetics, some new problems arose out of this knowledge. One of the problems was to find out how many nucleotides are needed to code for one amino acid. We know that there are four different types of nucleotides because there are four different nitrogenous bases. We also know that there are 20 kinds of amino acids. Four nucleotides taken two at a time provide only 16 different combinations ($4^2 = 16$), which are insufficient to code for the 20 different amino acids. Four nucleotides taken three at a time provide 64 different combinations ($4^3 = 64$). At first glance, this would seem to provide many more terms than are needed, since there are only 20 different amino acids. It was believed at one time that the excess combinations were not used to specify any amino acids. However, there is now strong evidence that all but three of the 64 combinations do, in fact, code for amino acids, and that as many as six different nucleotide triplets may specify the same amino acid. The term degeneracy is used to describe the fact that a given amino acid may be specified by more than one codon. The fundamental characteristics of the genetic code are now well established: it is a triplet code, with three adjacent nucleotides, termed a codon, specifying each amino acid.

A second problem with the genetic code is whether there is overlapping or not. For example, is the sequence CAGAUAGAC read only as CAG, AUA, GAC or can it also be read CAG, AGA, GAU, AUA, UAG, AGA, GAC? Is each nucleotide part of one codon or three? The amino acid sequences of several mutant forms of the hemoglobin molecule have been analyzed. In each, only a single amino acid in the molecule is substituted. In contrast, if the code were overlapping and a given nucleotide were part of three adjacent codons, we would expect three adjacent amino acids to be changed. For example, a single substitution (circled) in the sequence CAG A UA GAC, resulting in CAGCUAGAC, would affect only the amino acid specified by AUA if no overlapping occurred. If there were overlapping, such a change would affect 3 amino acids, namely those coded for by AGA, GAU, and AUA. Experiments with synthetic polynucleotides having known base sequences have shown conclusively that the code is not overlapping. One can synthesize a messenger RNA strand containing only uracil, UUUUU... (polyuridylic acid). For every 3 bases added to the mRNA, only one additional phenylalanine will be incorporated into its peptide chain (the UUU codon codes for phenylalanine).

Finally, the code is commaless, i.e., no punctuation is necessary, since the code is read beginning at a fixed point, three nucleotides at a time, until the reading mechanism comes to a specific 'termination' codon, which signals the end of the message. The fact that there is punctuation between codons becomes important in deletion or insertion mutations. Adding or subtracting a nucleotide within a sequence of codons throws the entire reading frame and can change out of line every codon past the point of insertion or deletion. For example, the gene sequence AGAUCUUGG would normally be read as AGA, UCU, UGG and would code for the amino acids arginine, serine, and tryptophan. The insertion of an extra nucleotide (circled), would result in the sequence AGAU G CUUGG, which would read AGA, UGC, UUG and code for arginine, cysteine, and leucine. The deletion of a nucleotide would have a similar effect. If C were eliminated, the code would read AGA, UUU, and would code for arginine and phenylalanine.

• PROBLEM 24-10

Suppose the codons for amino acids consisted of only two bases rather than three. Would there be a sufficient number of codons for all twenty amino acids? Show how you obtain your answer.

Solution: There are two ways to approach this question. One way is to use a mathematical principle known as permutation and the other is to do it by common sense. The latter method will be discussed first.

In this question, we are told that a codon consists of two bases only. We know there are four different kinds of bases in DNA, namely adenine, guanine, cytosine and thymine. To get the total number of possible two-base codons, we will have to pick from the four bases and put them into two positions on the codon. For the first position, we can have any one of the four bases. That means there are four possible ways of filling the first position. For each of these four possible first positions, we again can put any one of the four bases into the second position. The only restriction we face is the number of different bases we have, which is four. So the total number of codons consisting of two bases is 4×4 or 16 codons.

The other way to solve this problem is to use permutation. The general formula used in permutation is $nPr = n^r$ where n is the total number of objects and r is the number of times permuted. nPr is read as "n permuted r times," and n^r as "n to the power of r." Applying this formula:

n is the number of different bases (that is, 4) and r is the number of bases in a codon (which is 2). Substituting in the numerical data, we have the total possible number of codons made up of 2 bases.

$$(n^r) = 4^2 = 16$$

In either case, we arrive at the same answer. There will be 16 different codons if each codon contains two bases. However, there are 20 different kinds of amino acids, thus 16 codons will be insufficient to code for all 20 amino acids.

• PROBLEM 24-11

Explain how the codon for phenylalanine was discovered.

Solution: It has been known for some time now that the genetic code is a triplet code - three bases in a messenger RNA molecule code for one amino acid. Each triplet is known as a codon and the entire set of codons comprising the RNA molecule constitute the genetic code for a polypeptide chain.

The translation of the genetic code resulted from recent research the aim of which was to determine the exact sequence of the bases on the mRNA molecule which specifies a particular amino acid. An important breakthrough in this research came when a method of artificially synthesizing mRNA was developed. Ribose sugar, phosphate ions, and various bases, when combined under the proper conditions, result in the formation of a synthetic RNA molecule. By varying the kinds of bases used in producing this RNA, it has been possible to determine which base sequences result in which amino acids. The synthetic mRNA is placed with ribosomes, amino acids, ATP, and tRNA. The resulting polypeptide chain is analyzed for its component amino acids.

The codon for the amino acid phenylalanine was the first to be discovered using the above procedure. The only base used in the synthesis of the mRNA, was uracil, and therefore the molecule had a series of UUU codons. This mRNA was added to ribosomes, amino acids, ATP and tRNA, and the resulting polypeptide consisted only of phenylalanine. Thus, it was deduced that the codon UUU is the sequence of bases which causes phenylalanine to be placed in the polypeptide chain.

• PROBLEM 24-12

What is the possible function of nonsense codons in the genes?

Solution: Nonsense codons are those which do not code. There are three such nonsense triplets for an amino acid: UAA, UAG, and UGA. They are found at the end of nucleotide sequences coding for given polypeptides, and code specifically for the termination of the chain. They signify that peptide elongation at the ribosome should stop. These codons are read by specific proteins known as release factors. These release factors are necessary to release the chain from the ribosome. Two release factors have so far been identified, each of which recognizes two codons. One is specific for UAA and UAG and the other for UAA and UGA.

Nonsense codons may be employed in the following manner: the nucleotide sequence etc.- CUU AGG UAU AUA UAG GCC ACG - etc.on the mRNA might represent two genes, with the UAG triplet acting as a terminating codon between them:

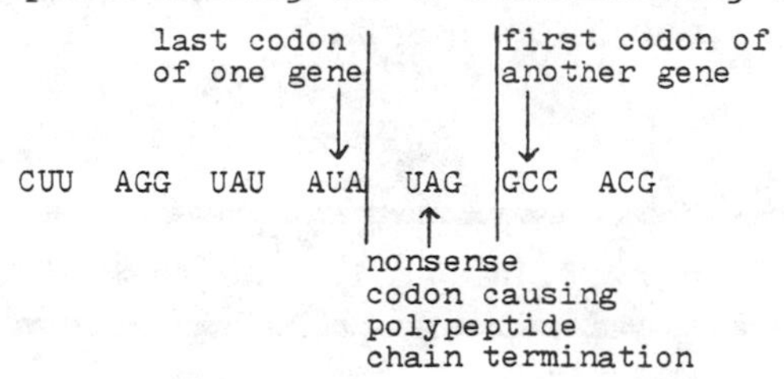

• **PROBLEM 24-13**

How can one estimate the total number of genes for a given organism?

Solution: Since genetic information is stored as DNA base sequences, the amount of DNA in a cell provides an initial basis for estimating the genetic information available to the cell. It is a well-established fact that chromosomes carry the genetic information for the ordering of amino acid sequences of the proteins that are made in the cell. Proteins consist of one or more polypeptide chains, which are composed of numerous amino acids. A given polypeptide chain is coded for by a specific gene, one gene carrying the information for only one polypeptide.

Genes are arranged in linear order on the chromosomes and are composed of a specific number and sequence of DNA base pairs. There is a direct correlation between the size of the gene (or the number of base pairs it contains) and the size of the protein it codes for (or the number of amino acids in that protein). A protein can consist of a few dozen to a few hundred amino acids, with 300-400 amino acids often used as the size for an "average" protein. Three mRNA bases are required to code for a single amino acid, and since mRNA is transcribed from only one of the two complementary DNA strands, this means that three DNA base pairs are needed to code for an amino acid. Thus, a very rough measure of the number of different proteins for which a cell might carry information is obtained by dividing the number of base pairs in its DNA by three, giving a theoretical number of amino acids in DNA. Dividing that number by the average number of amino acids per polypeptide gives a value for the possible number of polypeptides. If an estimate is made of the number of polypeptides the chromosomes can code for, then the number of genes found on the chromosomes can be estimated from the simple fact that one gene codes for one polypeptide.

In order for us to determine the total number of genes in an organism by this method, we must get a rather good estimate of the total number of DNA base pairs contained in the chromosomes. This is rather difficult to do with diploid organisms. In addition to the difficulty caused by the sheer number of chromosomes, large amounts of associated proteins known as histones, as well as other proteins, are found associated with the chromosomes. Moreover, the general structure of a diploid chromosome is for the most part not stretched out, but folded and convoluted into supercoils, further complicating the examination of the DNA. The chromosome structure of bacteria and viruses is simpler and better understood. Their chromosome is a pure DNA molecule, having no associated proteins, and is often circular in structure. Bacteria usually contain one DNA molecule per cell and the chromosome can be seen in rather good detail using an electron microscope. Owing to its simple chromosome structure then, let us estimate the total number of genes in the bacterium, Escherichia coli.

In order to proceed, we must first know the number of base pairs in the chromosome of E. coli. This figure can be estimated either from the length of the DNA molecule, which has been found to be approximately 1000 μm or 1 mm, or from the molecular weight, which is about 2×10^9 daltons. Each base pair has a molecular weight of approximately 660 daltons, and there are about 3000 base pairs per μm of DNA double helix molecule. Either way of determination (3000 base pairs/μm × 1000 μm or 2×10^9 daltons/660 daltons/base pair) indicates that there are approximately three million base pairs in the DNA molecule of E. coli. Dividing this by three, we can find out how many amino acids there are (recall that three DNA base pairs code for an amino acid). This gives us a value of about 1 million amino acids. Since the average protein contains about 400 amino acids, the chromosome of E. coli could possibly code for a maximum of 2500 proteins. (1,000,000/400)

This figure of 2500 genes may seem like an extremely large amount for one cell to contain. However, E. coli is a relatively complex organism. It is considered complex because it is relatively self-sufficient, living on a minimal medium of a glucose solution, from which it produces all other vitamins and nutrients necessary for life. This requires a great deal of enzyme and protein machinery, and thus a relatively large number of genes.

The mammalian cell, being more complex, should contain a larger number of genes than E. coli. It can be determined how much DNA is present per mammalian cell. The amount is approximately 800 times that in E. coli. This number gives us an upper limit of the number of different genes. Thus, a mammalian cell would be capable of synthesizing 800 × 2500, or over two million different proteins.

However, measurements of the amount of DNA present can be misleading, and we must thus be very cautious about relating DNA content directly to the number of different proteins that may be synthesized by a given cell and therefore, the number of genes in that cell. There is a definite lack of correspondence between DNA content and the number of genes. While in procaryotes and lower eucaryotes, most DNA codes for amino acid sequences, the vast majority of DNA from multicellular organisms has no apparent genetic function. In Drosophila only about 5% of the total DNA codes for amino acid gene products, while in humans even less DNA is so employed. There are a number of explanations for the presence of this "excess" DNA. There are multiple repetitive sequences of DNA, some of which are never transcribed, although their function is not clear. In Drosophila, about 25% of the DNA shows this kind of repetitive sequence, and in the higher Eucaryotes, 10% to 20% of the DNA is repetitive. Some DNA might play structural roles, such as involvement in chromosome folding or pairing during meiosis. Other regions may have regulatory functions, acting as binding sites for repressor molecules or polymerases. Most genes in higher plants and animals are present in only one copy per haploid genome. Exceptions, however, do exist and there are often multiple copies of some genes. For example, in sea urchins the various genes are present in some 100 to 200 copies. Such repetition of genetic information enables an organism to produce large amounts of a protein if necessary.

The existence of DNA base sequences which do not code for proteins, as well as the presence of multiple copies of genes, would make any estimate of the number of genes in a diploid organism highly imprecise.

• PROBLEM 24-14

Distinguish between a cistron, a muton, and a recon. Which of these is the largest and which is the smallest?

Solution: A cistron is defined as the genetic unit of biochemical function. One cistron corresponds to one functional unit (one gene coding for a polypeptide chain, protein, or enzyme) on a chromosome. A cistron can be defined by means of complementation tests. Complementation refers to the production of a normal phenotype through the combined activities of two chromosomes, each of which alone is incapable, due a recessive mutation, of producing a normal phenotype. If two homologous chromosomes contain mutations in different cistrons, the normal gene of one can compensate for the mutated gene of the other, and vice-versa, when the two chromosomes are in combination with each other in the organism (see figure 1). If each chromosome of the pair contains a mutation within the same cistron, neither can compensate for the other, and a mutant phenotype will

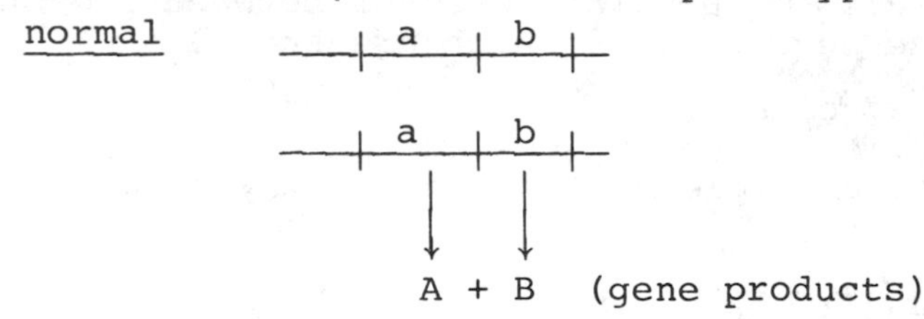

Complementation

A only <——— | a | b* | ———> A + B

B only <——— | a* | b |

(mutant phenotypes) (normal phenotype)

Non-complementation

B only <——— | a* | b | ———> B only

B only <——— | a* | b |

(mutant phenotype) (mutant phenotype)

a and b are cistrons
* denotes a recessive mutation

FIGURE 1

result. A cistron is thus the unit within which two chromosomes can have mutations and not complement each other when they occur in combination. It follows then that a cistron is the smallest unit capable of producing a gene product, and is equivalent to a gene.

A muton is defined as the smallest unit which, if altered, gives rise to a mutant phenotype. A muton may thus be as small as a single nucleotide, since a change in a single nucleotide can cause a mutant phenotype (think, for example, of sickle-cell anemia). Of course, some single-point mutations may not result in a mutant phenotype. If a given codon were changed via a single base substitution, for example, from CCA to CCG, that codon would still specify proline, due to the degeneracy of the genetic code.

A recon is defined as the smallest unit that is interchangeable, but not divisible, by recombination. In other words, recons are the smallest units between which crossing over can occur. This is known to be equivalent to a single nucleotide. This is because the breakage point in crossing over is between any two given nucleotides. Given two crossovers, it is thus possible for a single nucleotide to be exchanged (see figure 2).

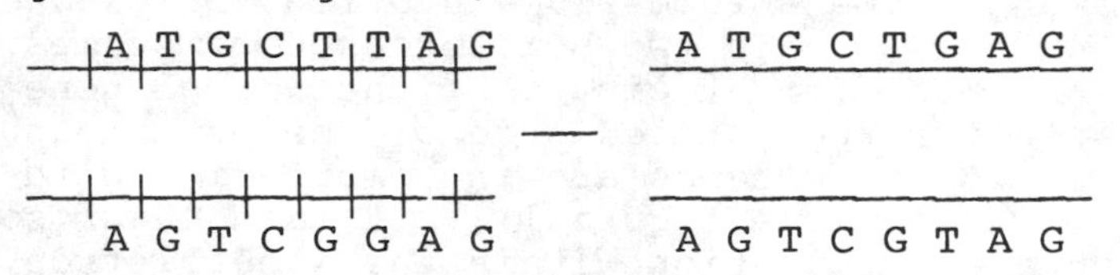

FIGURE 2. Double crossover between two strands of DNA, the base sequences of which are shown.

Since a cistron consists of a sequence of mutons or recons, it is the largest of the three. A recon may be as small as a single nucleotide, as may be a muton.

RNA AND PROTEIN SYNTHESIS

• PROBLEM 24-15

How can autoradiography be used to show that cells without nuclei do not synthesize RNA?

Solution: A valuable technique for tracing events in cells is autoradiography. This method relies on the fact that radioactive precursors (metabolic forerunners) taken up by the cells are incorporated into macromolecules when these cells are grown in a radioactive medium. The most widely used radioactive precursor is tritiated thymine. This is thymine (used by the cell to synthesize DNA) made radioactive by the substitution of tritium [^{3}H], a radioactive form of hydrogen, for some of its hydrogen atoms. Typically, radioactive precursors are presented to cells. The cells are then sliced into thin strips called sections. The sections are coated with photographic emulsion similar to ordinary camera film. Exposure of camera film to light followed by photographic development leads to reduction of exposed silver salts in the film and to production of metallic silver grains. These grains form the image in the negative. Similarly, exposure to radioactivity and subsequent development produces grains in the autoradiographic emulsion. By examining the sections with an electron microscope, both the underlying structure and the small grains in the emulsion are seen.

Autoradiography can be used to show that cells without nuclei do not synthesize RNA. This can be shown by growing enucleated cells in a medium containing a radioactive precursor for RNA. One such precursor is tritiated uracil, because uracil is the pyrimidine base found only in RNA. Tritiated uracil can be made by substituting the hydrogen atoms of uracil with tritium atoms. When we make sections of enucleated cells grown in tritiated uracil and examine them under the electron microscope, we observe no grains in the sections. This is because no RNA is produced and therefore no radioactive precursors are incorporated into the cell.

However, cells with nuclei, when treated in a similar manner, exhibit grains in their sections. This shows that normal cells do make RNA, whereas cells without nuclei do not synthesize RNA.

• PROBLEM 24-16

What is a polypeptide chain and how is it related to the proteins in a cell?

Solution: A polypeptide chain consists of linked units called amino acids. Each amino acid consists of a carboxyl group (C terminal), an amino group (N-terminal), and a side chain which varies with each amino acid. Structurally an amino acid is represented as follows:

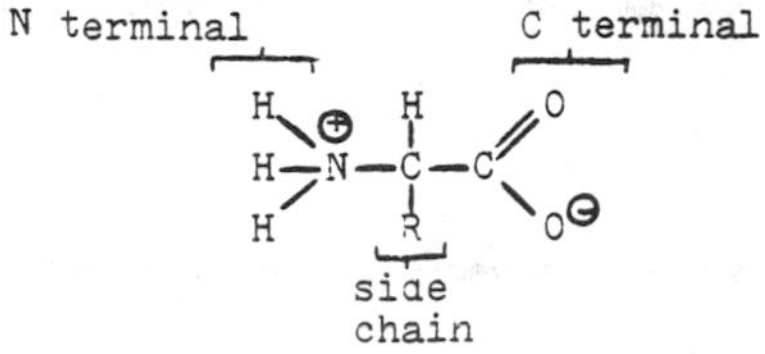

N terminal — C terminal

$$H_3N^{\oplus}-\underset{R}{\overset{H}{C}}-C(=O)O^{\ominus}$$

side chain

Two examples of amino acids are:

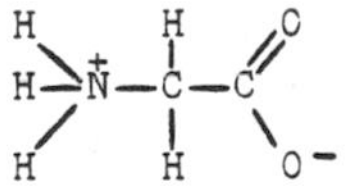

$$H_3\overset{+}{N}-\underset{H}{\overset{H}{C}}-C(=O)O^{-}$$

glycine (R=H)

$$H_3\overset{+}{N}-\underset{H-C(OH)-H}{\overset{H}{C}}-C(=O)O^{-}$$

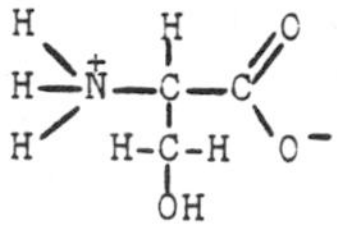

serine ($R=CH_2OH$)

Amino acids can be linked together by peptide bonds, which form between the N terminal of an amino acid on a C terminal. An example of a peptide bond is illustrated below:

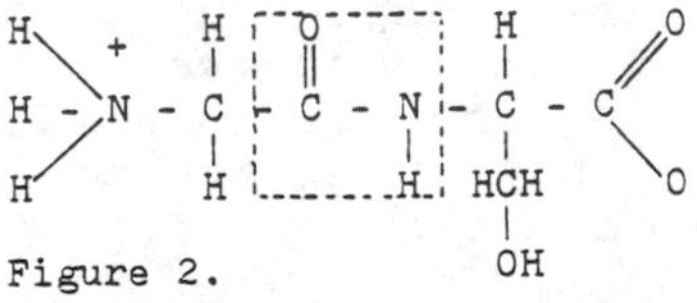

```
H     +    H  ┌─O─────┐  H         O
 \         |  │ ||    │  |       //
H - N  -  C  ├ C  -  N ┤- C  -  C
 /         |  │       |│  |       \
H          H  └──────H─┘ HCH       O
                          |
                          OH
```

Figure 2.

A peptide bond between glycerine and serine.

When many amino acids are linked together by peptide bonds, we have a chain of amino acids known as a polypeptide chain.

Proteins are actually-polypeptide chains. A protein molecule may be composed of one such chain, usually folded and cross-linked within itself to form a more compact and stable structure than a very long expanded chain. Some protein molecules are made of combinations of two or more polypeptide chains. The human protein hemoglobin, for instance, is composed of four polypeptides, two identical alpha chains and two identical beta chains. Each alpha chain has 141 amino acids, and each beta chain has 146.

An important distinction must be made between polypeptides and proteins at the genetic level. A given gene can code for only one polypeptide. Many proteins, however, are composed of a number of polypeptides. When a protein is composed of more than one polypeptide chain, each chain may be coded for and made separately. These chains aggregate only after their individual syntheses to form the final protein. In addition, some proteins exist in their functional form only after certain changes have taken place in the original polypeptide, such as the cleavage or loss of certain portions of the chain. Thus a given gene does not necessarily code directly for an active, functioning protein.

• PROBLEM 24-17

Distinguish between messenger RNA (mRNA), ribosomal RNA (rRNA), and transfer RNA (tRNA). What is the role of each in protein synthesis?

Solution: These three types of RNA are all single-stranded and are transcribed from a DNA template by RNA polymerase in the nucleus. However, the function of each type after they leave the nucleus is what determines their differences.

Messenger RNA carries the genetic information coded for in the DNA and is responsible for the translation of that information into a polypeptide chain. Each set of 3 bases of the mRNA comprises a codon which directs the incorporation of a specific amino acid into the polypeptide chain. Messenger RNA binds reversibly to the smaller subunit of the ribosome, where protein synthesis is initiated. It can dissociate from the ribosome without jeopardizing the

Figure 1. Schematic folding of an RNA chain showing several double-helical regions held together by hydrogen bonds.

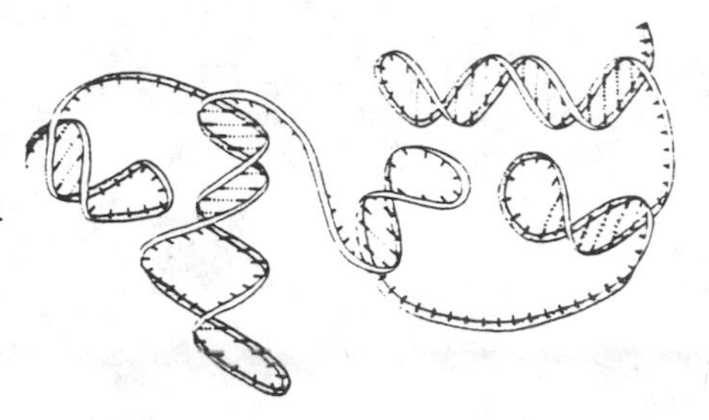

integrity of the ribosome. Degradation of mRNA by ribonucleases allows the cell to control its metabolism. Messenger RNA in E. coli has a lifetime of about 2 minutes. Degradation of mRNA is in the 5' ⟶ 3' direction and may proceed even as ribosomes are translating the mRNA. Ribosomes initiate protein synthesis by binding to the 5' end of the mRNA and reading the codons in the 5' ⟶3' direction. If degradation were 3' ⟶5', there would be synthesis of incomplete protein chains. Messenger RNA is heterogeneous in size due to the different lengths of polypeptide chains which a cell needs to synthesize, therefore, it is the most variable type of RNA.

While mRNA can dissociate from the ribosomes, rRNA is an integral part of the ribosome and its removal results in the destruction of the ribosome. The 30 S subunit of a bacterial ribosome contains 21 proteins and a 16 S fragment of rRNA. The 50 S subunit is composed of 35 proteins and 2 fragments of rRNA with Svedberg constants of 23S and 5 S, when placed together, the ribosomal proteins and rRNA undergo spontaneous self-assembly into a fully functional ribosome. This suggests that rRNA interacts with the ribosomal proteins and helps maintain the characteristic 3-D shape of the ribosome. While the function of rRNA is not completely understood, regions of unpaired bases may be involved in the binding of tRNA and mRNA to the ribosome. Ribosomal RNA has an average molecular weight of 5×10^5 to 5×10^6 daltons and is the least variable type of RNA, having only the three types mentioned above (5S, 16S, 23S). The single strand of both mRNA and rRNA is able to fold back on itself with subsequent formation of hydrogen bonds between complementary regions, giving rise to apparent "double-stranded" helical regions. (see figure 1)

Transfer RNA is the smallest type of RNA. With approximately 80 nucleotides, its molecular weight is 2.5×10^4 daltons. While there are at least 20 different types of tRNA's (one specific for each amino acid), the sequence of bases in the different tRNA molecules is highly conserved. This leads to the sharing of the same basic shape among the different types (see figure 2 a). The 3' end always ends in the sequence CCA-3' and the 5' end is a guanine. The amino acid residue is bound to the 3' adenine. Like mRNA and rRNA, tRNA folds back upon itself. Unlike the other two RNA's, tRNA, forms loops and double-stranded sections (see figure 2 b). In one of these loops is located the anticodon, which is what distinguishes the

Figure 2a. The complete neucleotide sequence of alanine tRNA showing the unusual bases and codon/anticodon position. Structure shown is two-demensional.

⊛ Unusual bases
Ψ Pseudouridine
$\xrightarrow{GCC}$ Codon
$\xrightarrow{IGC}$ Anticodon
I = Inosine
UH_2 = Dihydrouridine
T = Ribothymidine
GMe = Methyl guanosine
GMe_2 = Dimethyl guanosine
IMe = Methyl inosine

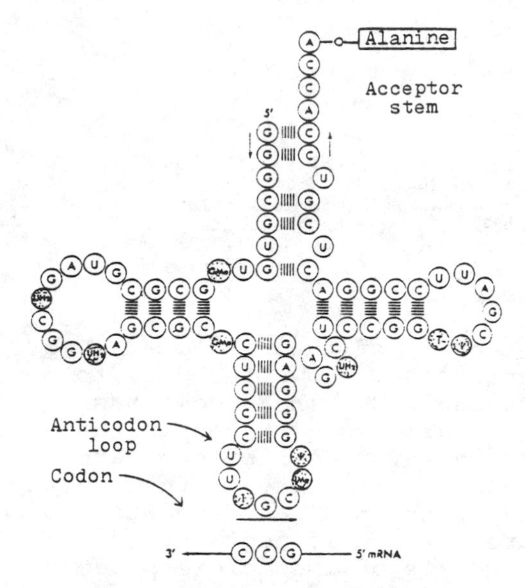

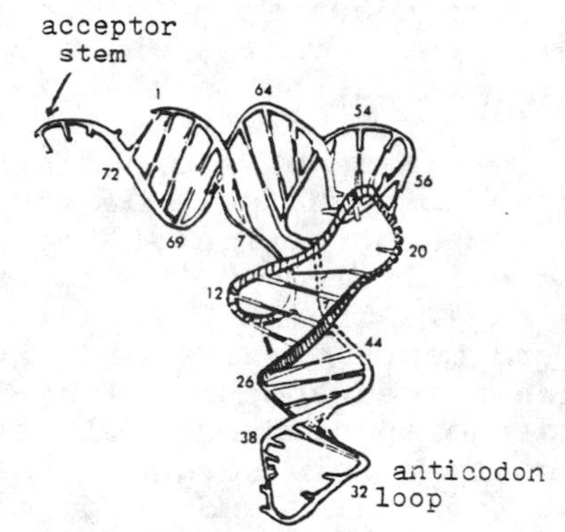

Figure 2b. A schematic three-demensional diagram illustrating the folding of the yeast tRNA molecule. The ribose-phosphate backbone is drawn as a continuous cylinder, and internal hydrogen bonding is indicated by crossbars. Positions of single bases are indicated by rods which are intentionally shortened. The variable regions, in terms of the number of nucleotides in different tRNA molecules, are shown in dotted outline, and two of the variable regions in the D loop, α and β, are labeled. The anticodon is at the bottom of the figure, while the amino acid acceptor is at the upper left.

different types of tRNA. The function of tRNA is to insert the amino acid specified by the codon on mRNA into the polypeptide chain, and it is through the complementation of anticodon and codon that the appropriate amino acid is incorporated. Hydrogen bonds form between the complementary codons and anticodons, thereby orienting the amino acid for peptide bond formation. A distinguishing feature of tRNA is the presence of unusual bases such as methyl inosine, dimethylguanosine, and ribothymidine. It is the presence of these alkylated bases which protects tRNA from digestion by ribonucleases.

• PROBLEM 24-18

A specific enzyme is found in the endoplasmic reticulum of a cell. The code for this enzyme is contained in a molecule of DNA in the nucleus. Start with this DNA and trace all steps involved in the production of the enzyme and its movement to the endoplasmic reticulum.

Solution: This question deals with a series of processes beginning with transcription (production of mRNA), followed by translation (production of a polypeptide chain) and ending with the passage of the polypeptide into the endoplasmic reticulum. Inside the nucleus, transcription of the DNA is mediated by RNA polymerase, an enzyme composed of five polypeptide subunits comprising the core, plus a loosely bound protein known as the sigma factor. The DNA undergoes a localized unfolding in the vicinity of the gene to be transcribed. The weak hydrogen bonds are broken and the two strands separate. Only one of the strands within a particular genome serves as a template for transcription by RNA polymerase. The sigma factor is responsible for recognizing the promoter region and attaches the enzyme core, to the DNA where transcription is initiated. Without the sigma factor RNA polymerase would bind at any place along the template and transcription would proceed on both strands. Once the core has been properly attached, the sigma factor can be released to become associated with another core polymerase molecule. Several core molecules may be transcribing a certain genome simultaneously.

RNA polymerase selects precursor ribonucleotides complementary to the DNA template. That is, adenine on the template is paired with uracil, cytosine is paired with guanine, and thymine with adenine. The precurser ribonucleotides consist of a base attached to a ribose triphosphate molecule. Other enzymes catalyze phosphate bond formation between adjacent ribonucleotides, and the new strand of mRNA is held together by these bonds in the backbone. The polymerization of the mRNA is driven by the energy of hydrolysis of the triphosphate of the precursor ribonucleotide.

The direction of transcription is specific. RNA polymerase reads from the 5' end of the DNA, with subsequent synthesis of mRNA in the 3' to 5' direction. The newly synthesized single-stranded mRNA peels away from the template allowing a new RNA polymerase molecule to attach or the DNA strand to reunite.

Now the synthesized mRNA has to move into the cytoplasm where protein synthesis occurs. The mRNA is able to diffuse into the cytoplasm because the nuclear membrane surrounding the nucleus has numerous rather large openings (nuclear pores) which allow the mRNA to pass through.

Once the mRNA gets into the cytoplasm, a ribosome attaches to the 5' end of the mRNA and translation begins. Several ribosomes may bind sequentially one mRNA strand, forming a polyribosome or polysome. The sequence of genetic information present in the mRNA directs the order of incorporation of amino acids into the protein chain. But before the amino acids can line up in the correct sequence, several things must happen.

Amino acids have to be oriented in the correct sequence on the ribosome. This is achieved by transfer RNA, or tRNA, whose function is to transport amino acids to the ribosome and translate the genetic code of the mRNA. To be transported, an amino acid must first be activated.

This occurs when the amino acid, which is bound to a loading enzyme, joins with a molecule of ATP, forming the activated amino acid. The activated amino acid is then transferred to the correct tRNA, which carries it to the ribosome. (see fig. 1)

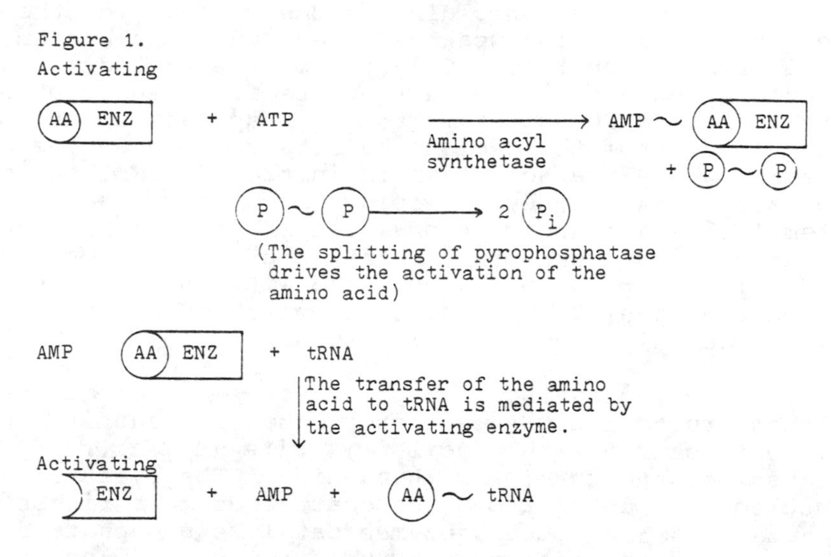

The high energy bond originally derived from the activation by ATP is retained in the amino acid ∿ tRNA complex. The activating enzyme is now free to bind another amino acid. The amino acid must be transferred to a tRNA molecule which is specific for the amino acid carried by the activating enzyme. Therefore, an activating enzyme must be able to recognize both a specific amino acid and a tRNA specific for that amino acid. There are at least twenty different activating enzymes and each one is specific for one amino acid and tRNA. There are also at least 20 different tRNA's, each one is specific for one amino acid or the DNA strands to reunite.

The correct sequencing of the amino acids is specified by the codon, a sequence of three bases on the mRNA. Each group of 3 bases codes for an amino acid. The tRNA has an anticodon, also a nitrogenous base triplet, which is complementary code to the codon specific for the amino acid carried by that tRNA. It is through the complementation of the codon and anticodon that the right order of amino acids is obtained.

As each amino acid gets into the right position in the sequence, an enzyme, peptidyl transferase, forms the peptide bond between two adjacent amino acid residues, (see fig.2).

The reaction is driven by the splitting of the high energy bond between the amino acid and the tRNA. As more and more amino acids are linked together, a polypeptide chain is formed. When translation is complete, this chain forms the enzyme that the DNA coded for.

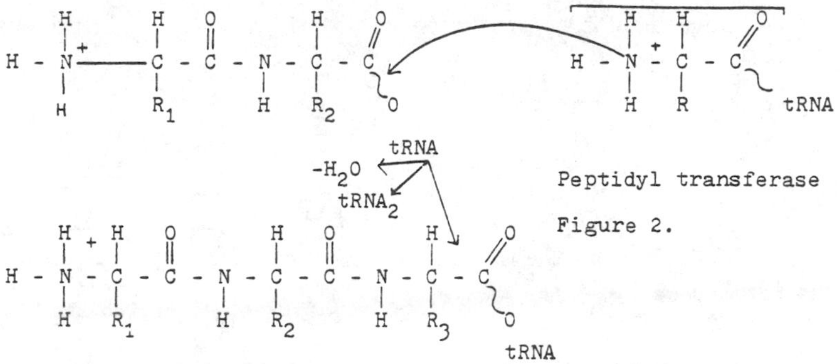

```
  H       H   O       H   O                         H    H    O
  |+      |   ||      |   //                        |  + |   //
H - N———— C - C - N - C - C     <———————————   H - N  - C - C~
  |       |       |   |     ~O                      |    |      tRNA
  H       R1      H   R2                            H    R

                      tRNA
               -H2O <
                    tRNA2                Peptidyl transferase

  H + H   O       H   O       H     O    Figure 2.
  |   |   ||      |   ||      |    //
H - N - C - C - N - C - C - N - C - C
  |   |       |   |       |   |      ~O
  H   R1      H   R2      H   R3
                                       tRNA
```

($tRNA_2$ is free to pick up another amino acid.)

Ribosomes are composed of two subunits, the 30S and 50S subunits. These two subunits will dissociate upon reaching the stop codon at the 3' end of the mRNA and the newly synthesized polypeptide chain is released. The subunits can reassociate at the 5' end of the mRNA, forming a functional 70S complex, and translation of the mRNA may proceed again.

The formation of the polypeptide begins with the amino terminal end. In bacterial polypeptides, the amino terminal residue is always N-formyl-methionine,

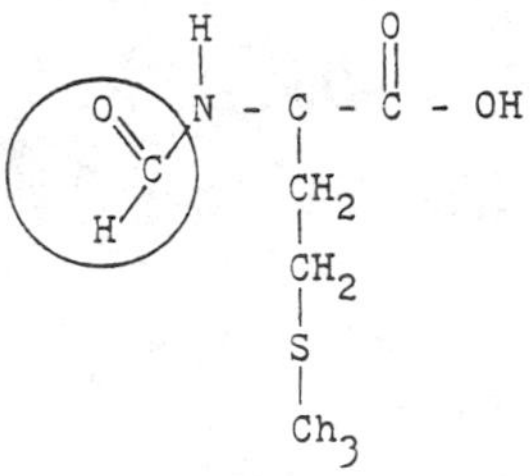

```
          H       O
          |       ||
   O\\    N - C - C - OH
      C/      |
    H/        CH2
              |
              CH2
              |
              S
              |
              Ch3
```

in which a hydrogen atom on the amino group of methionine has been replaced by a formyl group. The function of this modified residue is not entirely clear, but it may be involved in the protection of the growing polypeptide from digestion by various peptidases.

To get the enzyme into the endoplasmic reticulum, transport across the membrane of the endoplasmic reticulum has to occur. The actual mechanism of this transport is not known but there is a fairly sound hypothesis to account for it. This hypothesis states that when the enzyme is first formed on the ribosome, it has not assumed its three-dimensional shape, but remains an unfolded chain. In this conformation, the unfolded chain is transported across the membrane into the interior of the endoplasmic reticulum. We have to recall the fact that polyribosomes are associated with the endoplasmic reticulum. So the polypeptide chain made on the ribosome is next to the membrane of the endoplasmic reticulum. Once it gets across the membrane and reaches the interior, the polypeptide chain becomes folded into its normal three-dimensional configuration. Because it is now folded, it can no longer squeeze out through the membrane the way it came in.

A summary of the transfer of genetic information is presented below:

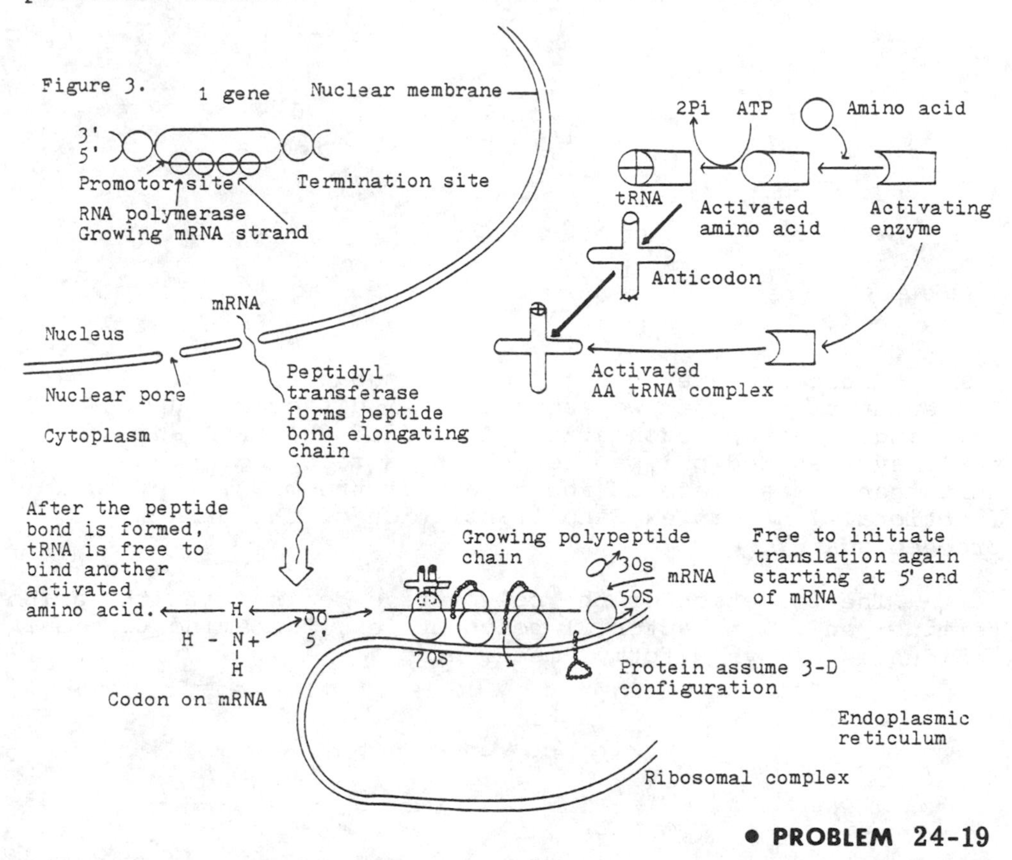

• **PROBLEM 24-19**

Mitosis is only one of four stages of the complete cell life cycle. List the other three and indicate the amount of RNA synthesis in each of the four stages. Explain why RNA output is greater in some stages than in others.

Solution: The length of the cell cycle under optimal conditions varies from 18 - 24 hours for most cells. Cell cycles are usually divided into four phases: m (mitosis), G_1 (period prior to DNA synthesis), S (period of DNA synthesis) and G_2 (period between DNA synthesis and mitosis). M is the shortest stage, usually lasting about one hour. There is no synthesis of either DNA or RNA in this stage. The absence of RNA synthesis may be the result of the tightly coiled and condensed nature of the DNA at this point, making it impossible to transcribe. Recall that for transcription to occur, DNA must first unfold.

During G_1, there is production of mRNA and the cell increases in size. Next, in the S phase, the DNA of the cell becomes doubled because the genes are duplicating. Duplicating genes cannot produce RNA, but since all the genes do not duplicate at once, there is still some output of RNA, although in reduced quantity. Finally, full output of RNA is resumed in G_2.

RNA production is greater in G_1 and G_2 because during these two stages the DNA is relaxed (uncondensed) and is not involved in duplication. The relaxed nature of the DNA favors RNA synthesis, which can only occur when the double stranded DNA separates into two individual strands.

• PROBLEM 24-20

Describe experimental evidence which indicates that it is not the kind of ribosomes but the kind of mRNA which determines the type of proteins which will be produced.

Solution: Most DNA is found in the nucleus. Transcription of the DNA gives rise to mRNA. So when the nucleus is removed from a cell, the source of mRNA is removed. Such an enucleated cell can live for a time, but growth and enzyme production will soon stop. For example, an enucleated ameoba (an ameoba with its nucleus removed) will crawl around for a time, it may engulf food but it will eventually be unable to digest it, growth stops and in time, the organism dies.

The above experimental observation can be explained in the following manner. The removal of the nucleus of the organism prevents further transcription of mRNA. The supply of digestive enzymes normally found in the organism becomes depleted because the soruce of mRNA is cut off and the translation of mRNA into enzyme cannot occur. Therefore, the food engulfed cannot be digested. This explanation shows that mRNA manufactured in the cell nucleus is essential for protein production of the cell.

In order to test what effect the type of mRNA has on the type of proteins produced, one can perform experiments where the nucleus of one species, producing certain types of mRNA, is introduced into an enucleated cell of a different species. For example, the nucleus of a mammalian reticulocyte, a cell that synthesizes hemoglobin, can be extracted and then introduced into an enucleated cell, such as an amphibian oocyte. This egg cell with a transplanted nucleus survives and produces mammalian hemoglobin instead of amphibian hemoglobin. This is good evidence showing that mRNA controls the type of proteins produced because the transplanted nucleus is the nucleus of a mammalian cell involved in the production of hemoglobin, the mRNA transcribed in the nucleus, when translated on the ribosomes, produces mammalian hemoglobin.

This experiment also implies that it is not the kind of ribosomes that determines the type of proteins produced. If this were true, then the hemoglobin produced would be amphibian hemoglobin, but this does not happen.

• PROBLEM 24-21

Would you expect the transfer RNA molecules and the messenger RNA molecules to be the same in the cells of a horse as in similar human cells? Give reasons for your answer in each case.

Solution: We have to realize that all tRNA and mRNA molecules are composed of the same chemical constituents, regardless of what species they come from. That is, they all contain the bases adenine, thymine, uracil and cytosine, (tRNA may have other bases in addition) ribose, and phosphate.

Observations from biochemical experiments aimed at elucidating the sequence and chemical composition of tRNA from different species have demonstrated that between species, the different types of tRNA molecules have basically the same nucleotide sequence and the same three-dimensional configuration (cloverleaf-shaped). The reason for this lies in the fact that the function of tRNA is to transfer the amino acids to their correct positions specified by the mRNA. Since all organisms make use of the same 20 amino acids to make their proteins, the tRNA used to transfer these amino acids are basically the same for different species.

The different mRNA molecules made by the cells of the same animal differ considerably in length. The great variety in lengths can be explained when we recall that mRNA carries the information for synthesizing different proteins. Different proteins are of different lengths, and therefore, there is a corresponding difference in lengths of the mRNA.

When we compare the mRNA from different species, such as man and horse, we have to bear in mind that there exist both equivalent and contrasting systems in the two species. Equivalent systems such as the Krebs cycle and electron transport system, which are almost universal among higher organisms utilize similar enzymes. The digestive system is an example of a contrasting system. The horse is a herbivore, whereas man is an omnivore. Because of the different modes of nutrition,some very different enzymes are involved.

We know enzymes, which are protein molecules, are the products of translation of mRNA. When translated, different mRNA molecules will give rise to different enzymes. Because horses and men rely on different enzymes, at least for digestion, we would infer that the mRNA from the two animals are different. However, we must not forget that both animals also use similar enzymes, as in Krebs cycle and electron transport system. Therefore they would also possess some similar if not identical mRNA.

In summary, we have determined that tRNA from cells of horse and man are basically the same, whereas the mRNA from the cells of these two animals would show much more difference.

• PROBLEM 24-22

Explain what difficulties would arise if messenger RNA molecules were not destroyed after they had produced some polypeptide chains.

Solution: Upon translation, mRNA molecules give rise

to polypeptide chains. Normally, mRNA molecules are short-lived and are broken down by RNAase after a few translations, thereby allowing a cell to control its metabolic activity. If an mRNA molecule were not destroyed, it would continue to synthesize its protein, and soon there would be an excess of this protein in the cell. This condition leads to some important implications for the cell.

The continual translation of mRNA into proteins would entail a serious depletion of the energy store in the cell. For example, before an amino acid can attach to a tRNA molecule, it has to be activated. Activation is brought about by the hydrolysis of one molecule of ATP. So for each amino acid in the polypeptide chain, one molecule of ATP must be used. If translation were to proceed ceaselessly because the mRNA is not destroyed, large amounts of ATP molecules would be consumed and the energy supply in the cell depleted.

In addition, the cell would accumulate proteins that it may not need. This use of large amounts of energy to produce unneeded protein would be a wasteful process. Indeed, excessive accumulation of a protein may even be harmful to the cell or organism. For example, given the right environment, a protease such as pepsin, if present in more than sufficient amounts, will eat away the wall of the stomach, forming an ulcer.

• PROBLEM 24-23

Distinguish between the two types of proteins produced by ribosomes.

Solution: Ribosomes are the cellular organelles that provide the sites for protein synthesis. The two types of proteins that are synthesized on the ribosomes are structural proteins and functional proteins.

Structural proteins are actual structual components of the cell. They constitute part of the cell protoplasm, and as such, are generally insoluble. They are an important component of cell membranes. Some structural proteins, such as collagen, or keratin are specialized for strength and support. Structural proteins help to form the cell, and the organism of which the cell is a part, therefore, structural proteins contribute to both cell and body growth.

Functional proteins are responsible for the control of cell activity, such as hormone production or nutrient metablism. Some of these proteins are known as enzymes - their function is to catalyze chemical reactions in the cell. Others function as carrier molecules. Still others function as regulatory molecules which control the synthesis of other proteins. Most functional proteins are soluble in the cell, since their action depends on their interaction with various molecules throughout the cell. Some are soluble in cell membranes.

The distinction between structural and functional proteins, then, is based on what they do in the cell. The former are involved in the structure of the cell or organism, whereas the latter deal with its functions.

GENETIC REGULATORY SYSTEMS

• PROBLEM 24-24

What is an operon? With the aid of diagrams, show how a known operon works.

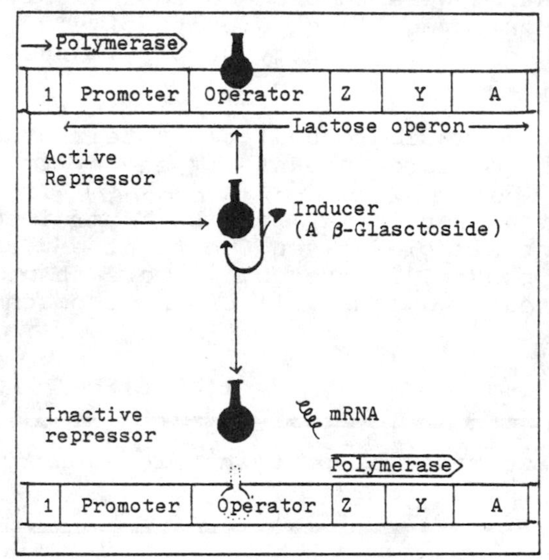

Figure a. Model of lactose operon depicts the induction of enzyme synthesis. Repressor molecule, shown here as a flask-shaped object, binds to the operator (top) and prevents transcription by RNA polymerase. The small inducer molecule can change the shape of the repressor so that it can no longer bind to the operator. Removal of repressor allows transcription to proceed (bottom).

Z = Galactosidase
Y = Galactoside permease
A = Galactoside acetylase

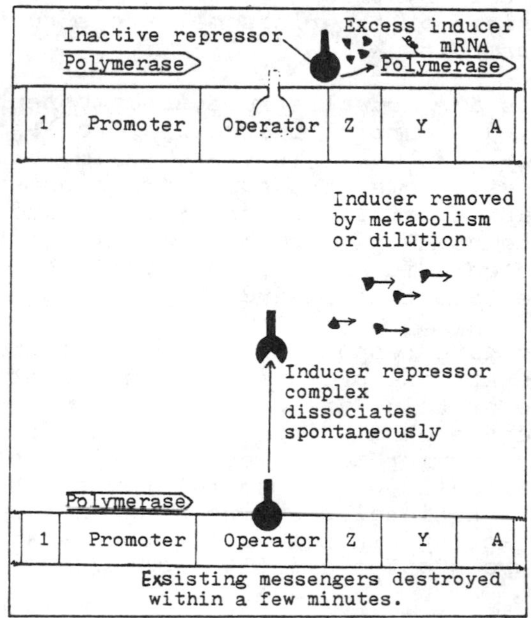

Figure b.

Translation stops when inducer is removed. Symbols are the same as in figure a. Inducer repressor complex has a tendency to dissociate. If inducer is used up or removed from cell, repressor molecules then bind to the operator, preventing both transcription and translation.

Solution: A controllable unit of transcription is called an operon. An operon consists of a binding site for RNA polymerase (a promoter), a binding site for a specific repressor (an operator), and one or more structural genes (see figure). These structural genes are usually associated in their functions, and their coordinated synthesis is desirable. Thus they are linked together in the operon, and when they are transcribed, a single mRNA molecule is produced. The promotor and operator act as controls for this transcription, and the entire system is controlled by an associated regulatory gene, which codes for the repressor molecule of the operon.

A known example of an operon is the lactose (lac) operon, present in E. coli. This operon contains the structural genes for β-galactosidase, an enzyme which cleaves lactose into glucose and galactose; galactoside permease, which allows for the increased rate of entry of lactose into the bacteria; and galactoside acetylase, whose function is not known. When lactose is absent from the medium, these enzymes are not needed, and so their levels are low. But when lactose is added to the medium containing these bacteria, the production of the enzymes increases tremendously, almost a thousand fold. How is the manufacture of this prodigious quantity of enzyme initiated and controlled? The regulatory gene (l) for the lac operon is transcribed constitutively, that is, its product, the repressor, is constantly being synthesized. This repressor binds to the operator(o) and blocks the transcription of the structural genes. That is why the levels of the three enzymes are normally so low. The situation is altered when lactose is introduced into the medium. Lactose molecules move into the nucleus of the bacterium, and bind to any repressor molecules in such a way that the repressors are inactivated. An inactivated repressor cannot bind to the operator. It therefore cannot block the transcription of the structural genes by RNA polymerase which binds to the promotor. Messenger RNA now transcribed from the genes of the operon is translated on the ribosomes into the three enzymes of lactose metabolism. As long as the inducer (lactose) is present, the synthesis of the enzymes can occur. But when the inducer is depleted by metabolism, repressor proteins (remember that synthesis of the lac repressor is constitutive) become free to again bind to the operator and effectively turn off the synthesis of the enzymes.

This model implies that the usual condition of the lactose operon is to be turned off. This is accomplished through the control of the regulator gene, which constitutively produces the repressor. In other words, the system is under negative control. Because the operon can be turned on by the presence of lactose (the inducer), it is said to be inducible. An important implication is that some genes, such as (l), produce proteins whose function is to regulate the expression of other genes, while certain other genes, namely operators and promoters, very probably do not code for any proteins at all.

• PROBLEM 24-25

Distinguish between structural genes, regulatory genes, promotor genes, and operator genes.

Solution: Structural genes are those genes which code directly for the synthesis of proteins required either as enzymes for specific metabolic processes or as structural units of the cell or organism. These genes specify the primary amino acid sequence of polypeptides. Oftentimes, related structural genes, such as those whose proteins catalyze sequential reactions in a metabolic pathway, will be positioned in a linear sequence along a region of the chromosome. When these genes are transcribed, a single continuous mRNA molecule is formed.

Regulatory genes code for inhibitory protein molecules known as repressors. These repressors act to prevent the activity of one or more structural genes by blocking the synthesis of their gene products. They do this by binding to operator genes. An operator gene is a specific region of the DNA molecule that exerts control over a specific group of structural genes by serving as a binding site for a given repressor molecule. When not bound to the repressor, the operator is transcribed along with the structural genes with which it is associated, though it is probably not translated.

Promotor genes are also associated with a given gene or group of genes. It serves as a binding site for RNA polymerase, the enzyme responsible for transcription of DNA. The operator gene is located between the promotor and structural genes on the chromosome. When the repressor binds to the operator, it prevents the movement of RNA polymerase along the DNA molecule, thereby inhibiting transcription of the structural genes. When the repressor is not bound to the operator, transcription is free to occur.

The term operon is used to designate a given system of structural genes, along with their associated promotor, operator, and regulatory genes. Unlike the promotor and operator, a regulatory gene is not necessarily located in the proximity of the operon with which it is associated. In addition, there is evidence that a given repressor molecule may control more than one group of structural genes.

Although the specific mechanistic and physical relationships of repressor, operator, promotor, and structural genes are relatively clear for microorganisms, they have not as yet been clearly defined in higher organisms. This is because crossing over, translocations, and inversions that occur in diploid organisms can act to disrupt clusters of linked genes. The overall regulation model, however, probably does operate in higher organisms.

• PROBLEM 24-26

What is meant by a gene "repressor"? A "corepressor"? An "inducer"? How are they used to regulate protein synthesis?

Solution: It is an established fact that all the cells of a given organism contain the same kind and number of chromosomes. However, different cells perform different functions. For example, muscle cells are involved in locomotion, whereas nerve cells form the basis of perception and response. The variation in function of different cell types implies that each is using a different portion of its total store of genetic information. In addition, in both simple and complex organisms, different cell stages, environmental conditions, and available nutrients require the functioning of different enzyme systems at different times. It would be wasteful indeed for a cell to be manufacturing certain proteins at a time when they are not needed. Thus, both cell differentiation and function require the presence of a control system to regulate the expression of the genes in the cells. The agents for these control systems are repressors, corepressors, and inducers.

A repressor is a regulatory protein molecule that "turns off" or represses the transcription of a gene and hence inhibits the synthesis of the protein (enzyme) which that gene specifies. Remember that enzymes play a direct role in regulating reactions at the metabolic level. However, the production of that enzyme depends on the transcription of the corresponding gene. In the absence of transcription of that gene, the cell cannot perform the certain reaction which requires that corresponding enzyme, hence the activity of the cell is requlated.

Repressors prevent transcription by binding to the operator of an operon, thereby preventing the transcription of its genes by RNA polymerase. There are two distinct systems by which repressors operate to regulate transcription. In an inducible system, the repressor protein is produced in an active form and as such is able to bind to the operator. However, in the presence of an inducer molecule, the repressor is no longer able to bind to the operator, and the gene is "turned on". The inducer molecule may be a protein, a given substrate or a metabolite whose presence indicates a need for the enzymes that the inducible genes code for. Consider an inducible system such as the lac operon. When E. coli is grown on a glucose medium, there is no need to produce the enzymes which metabolize lactose and the corresponding genes are not transcribed. However, if E.coli is grown on a lactose medium, the lac operon must be "turned on" in order for the lactose to be metabolized. Here, the lactose molecule functions as an inducer. The lac repressor protein has two active sites: one that interacts with the operator, and one that can bind lactose. The repressor can interact with lactose while it is bound to the operator. However, the interaction with lactose produces a conformational change in the repressor protein (the repressor is an allosteric enzyme) and this change prevents the binding of the repressor with the operator. Transcription by RNA polymerase can now occur.

The second mode of regulation is found in a repressible system. An example of a repressible system is the histidine biosynthetic pathway. Here, the repressor protein is produced in an inactive form which is unable to bind to the operator. This allows production of the enzymes which synthesize histidine. The repressor protein has a binding site for histidine, the corepressor molecule. As the level of histidine in the cell increases, histidine binds to the repressor protein, forming an active repressor - corepressor complex which can bind to the operator and prevent transcription. Histidine is called the corepressor, since in its absence there is no repression. Repressible systems are widely employed in biosynthetic pathways with the end products acting as corepressors to repress their own synthesis when they reach sufficient levels in the cell.

As seen in E. coli, the synthesis of a repressor protein may be constitutive. In some systems, however, the synthesis of the repressor is itself under control. In these systems, the repressor protein, in addition to binding to the operator of a given operon, can bind to the operator of its own gene or genes. It thus, acts as a repressor of its own synthesis when its level in the cell reaches a given quantity. Such a system is depicted in the accompanying figure.

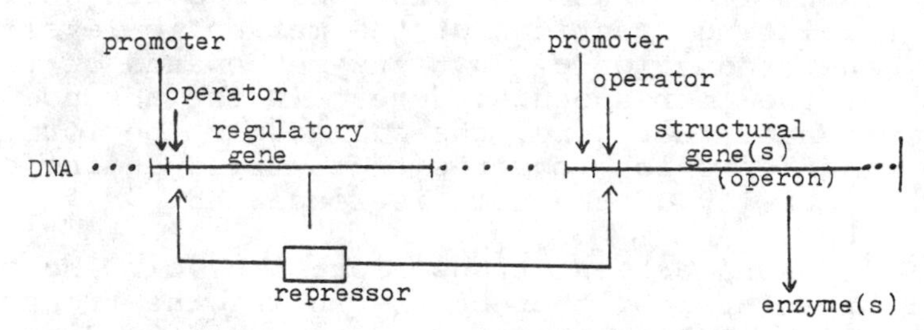

Repressor coded for by regulatory gene can bind to both operators shown.

It is important to differentiate between repression and feedback (end product) inhibition. Repression operates at the level of transcription to prevent enzyme production. Feedback inhibition acts at the enzyme level to prevent enzyme activity. In feedback inhibition, an endproduct acts to inactivate the enzymes responsible for the early steps of its synthesis. Regulation of cellular functions is much quicker by feedback inhibition than by repression, since repression merely prevents further enzyme production, leaving the enzymes already present in a cell completely active. It is through feedback inhibition that a cell can "turn off" the enzymes already present.

One can see that by using a combination of systems, a cell can execute rather fine control over its metabolic activities.

MUTATION

• PROBLEM 24-27

Define the term mutation. What types of mutations can be distinguished and how may each be produced?

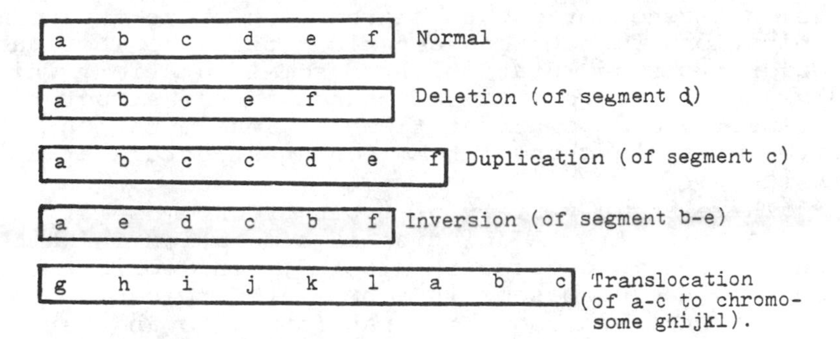

Figure 1
Diagram illustrating the type of mutations that involve changes in the structure of the chromosome.

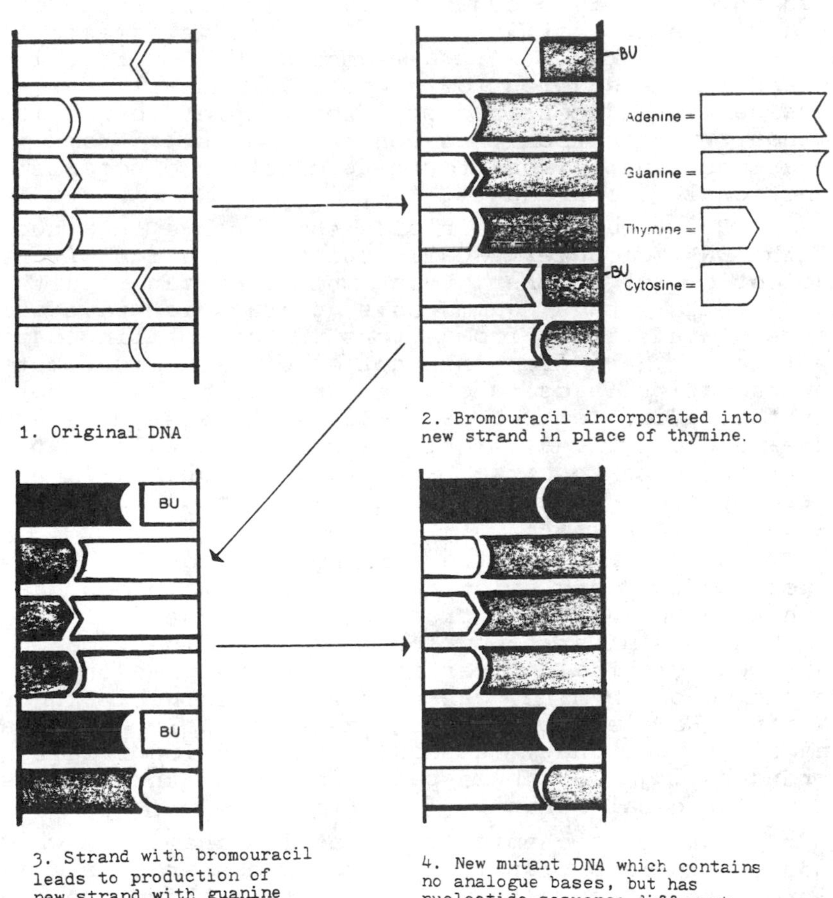

1. Original DNA

2. Bromouracil incorporated into new strand in place of thymine.

3. Strand with bromouracil leads to production of new strand with guanine paired to the bromouracil.

4. New mutant DNA which contains no analogue bases, but has nucleotide sequence different from original, with G-C pairs in place of A-T.

Figure 2
Diagrammatic scheme of how an analogue of a purine or pyrimidine might interfere with the replication process and cause a mutation, an altered sequence of nucleotides in the DNA(indicated in black). The nucleotides of the new chain at each replication are indicated by the gray blocks. In this instance, two new G-C pairs are indicated.

Solution: A mutation can be defined as any inheritable change in a gene not due to segregation or to the normal recombination of unchanged genetic material. There are two major types of mutation:chromosomal mutation, which can involve an extensive chemical change in the structure of a chromosome, and point mutation (or gene mutation) which involves a single change in molecular structure at a given locus.

There are a variety of types of chromosomal mutations (see fig. 1). A deletion is a mutation in which a segment of the chromosome is missing. In duplication, a portion of the chromosome is represented twice. When a segment of one chromosome is transferred to another nonhomologous chromosome, the mutation is known as a translocation. An inversion results when a segment is removed and reinserted in the same location, but in the opposite direction. (See Figure 1). Chromosomal mutations can also involve only one nucleotide. The insertion or deletion of a single nucleotide can have extensive effects. Such an occurrence results in a frame-shift by throwing the entire message out of register. (Recall that the gene message is read continuously from triplet to triplet, without "punctuation" between the codons.) Frame-shift mutations usually lead to the production of completely nonfunctional gene products.

Point mutations involve some change in a nucleotide of the DNA molecule, usually the substitution of one nucleotide for another. Point mutations can result from exposure to x-rays, gamma rays, ultraviolet rays and other types of radiation, from errors in base pairing during replication, and from interaction with chemical mutagens. How radiation leads to changes in base pairs is not clear, but the radiant energy may react with water molecules to release short-lived, highly reactive radicals that attack and react with specific bases to cause chemical changes. These changes can block normal replication or cause errors is base pairing.

Errors arising from DNA replication are better understood. There exist analogs of nitrogenous bases, such as azoguanine and bromouracil. These analogs possess structures similar to purines or pyrimidines, but have different chemical properties. For example, bromouracil, an analog of thymine, can pair with adenine during replication. However, bromouracil can also pair with guanine and if it does that, subsequent replication results in a strand having a G-C base pair, replacing the A-T base pair of the original strand. (See Fig. 2)

Mutagenic chemicals - that is, chemicals which induce mutations - include nitrogen mustards, epoxides, nitrous acids and alkylating agents. These are chemicals that react with specific nucleotide bases in DNA and can convert one base to another or cause mispairing of bases during replication.

If a mutated gene is transcribed, and results in changes at or near the active site of an enzyme, the altered enzyme may have markedly decreased or altered enzymatic properties. However, if the mutation occurs elsewhere in the enzyme molecule, it may have little or no effect on the catalytic activities of the enzyme and may go undetected. Therefore, the true number of mutations may indeed be much greater than the number observed.

• PROBLEM 24-28

Live bacteria, all of which were red in color, were placed under an ultra-violet lamp. After several days, groups of white bacteria began to appear among the red. What conclusions, if any, can be made at this stage? What further experiments should be performed?

Solution: At this stage, we can guess that the white bacteria probably arose as the result of a mutation. We know that exposure to ultraviolet radiation usually causes an increase in the mutation rate, i.e., it is said to induce mutations. We also know that the white bacteria arose from among the red, and in the same culture medium. If the original culture was indeed genetically pure (all the individuals having come from the same cell and thus having identical genotypes), there is no reason for any of the bacteria to exhibit different behavior or characteristics while remaining in the same environment. A change in phenotype could only be caused by a change in the genetic structure, in this case, a genetic mutation.

In order to be absolutely sure that the change in color is indeed caused by a genetic mutation, we can test the permanence of the change. If there has been a change in genetic structure, it must necessarily be passed on to offspring of the cell with the original mutation. We can remove the white cells from the culture and grow them in a separate, identical medium. If they reproduce true-to-type, and produce a colony of only white bacteria, we can be sure that a mutation has indeed occurred. Of course there is always a possibility that back mutation will occur, with a reversion to red color, but this is very rare.

It is important to note that while we know that UV radiation is a potential mutagen (agent inducing mutation), we cannot be entirely sure that all of the mutations resulted from UV rays. Some could have arisen as a result of a spontaneous change. Spontaneous mutations are chance occurrences, which are not caused by a specific external agent. The facts of the experiment, however, do fit well with what we know of the mechanisms of UV-induced mutations. Ultraviolet rays are theorized to affect the pyrimidines in DNA. They do this by (1) hydrolysing cytosine to a product that may cause mispairing, and (2) joining thymine molecules together to form dimers which block normal transcription from occurring. If such changes did occur, they would affect the production of substances that is coded for genetically. In this case, the gene product could have been a specific enzyme necessary for the production of red color. Its absence would account for the reversion to white color in the mutants. Of course an equally probable cause would be the production of an enzyme responsible for the white color morphology in the mutants.

• PROBLEM 24-29

Explain how a change of one base on the DNA can result in sickle-cell hemoglobin rather than normal hemoglobin.

Solution: If one base within a portion of DNA comprising a gene is altered, and that gene is transcribed, the codon for one amino acid on the mRNA will be different. The polypeptide chain which is translated from that mRNA may then have a different amino acid at this position. Although there are hundreds of amino acids in a protein, a change of only one can have far-reaching effects.

The human hemoglobin molecule is composed of two halves of protein, each half formed by two kinds of polypeptide chains, namely an alpha chain and a beta chain. In persons suffering from sickle-cell anemia, a mutation has occurred in the gene which forms the beta chain. As a result, a single amino acid substitution occurs, where glutamic acid is replaced by valine. One codon for glutamic acid is AUG, while one for valine is UUG. (A = adenine, U = uracil, G = guanine). If U is substituted for A in the codon for glutamic acid, the codon will specify valine. Thus, it is easy to see how a change in a single base has resulted in this amino acid substitution. This change is so important, however, that the entire hemoglobin molecule behaves differently. When the oxygen level in the blood drops, the altered molecules tend to form end to end associations and the entire red blood cell is forced out of shape, forming a sickle-shaped body. These may aggregate to block the capillaries. More importantly, these sickled cells cannot carry oxygen properly and a person having them suffers from severe hemolytic anemia, which usually leads to death early in life. The condition is known as sickle-cell anemia.

In a normal hemoglobin molecule, the glutamic acid residue in question, being charged, is located on the outside of the hemoglobin molecule where it is in contact with the aqueous medium. When this residue is replaced by valine, a non-polar residue, the solubility of the hemoglobin molecule decreases. In fact, the hydrophobic side chain of valine tends to form weak van der Waals bonds with other hydrophobic side chains leading to the aggregation of hemoglobin molecules.

• PROBLEM 24-30

What is meant by an "inborn error of metabolism"? What are some examples of inborn errors of metabolism in man?

Solution: The expression of a given structural or functional trait is often the result of a number, perhaps large, of chemical reactions in series, with the product of each serving as the substrate for the next: $A \rightarrow B \rightarrow C \rightarrow D$. For example, the color of most mammalian skin is due to the pigment melanin (D). Melanin is produced from dihyd-

Fig.1

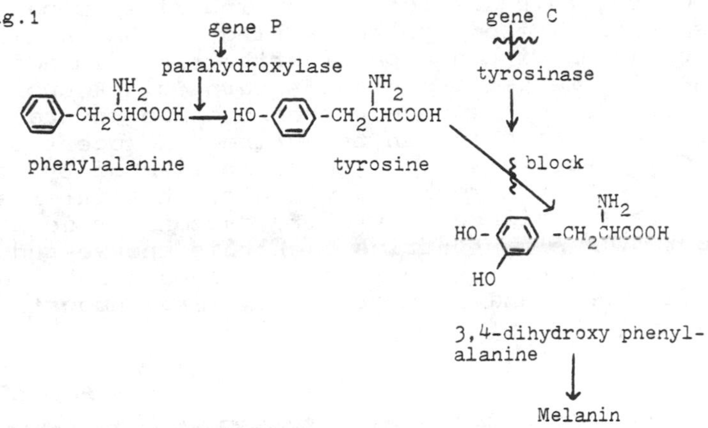

gene P
parahydroxylase
phenylalanine → tyrosine
gene C
tyrosinase
block
3,4-dihydroxy phenylalanine
Melanin

roxyphenylalanine (C), which is derived from tyrosine (B), which is in turn made from phenylalanine (A), (see figure 1). Each reaction is controlled by an enzyme. The conversion of tyrosine to dihydroxyphenylalanine is mediated by the enzyme tyrosinase. When this enzyme is lacking in an individual, melanin production is blocked, and albanism results.

We know that the synthesis of proteins is specified by DNA. It follows that part of a DNA molecule, a gene, codes for the production of an enzyme, which is a protein molecule. In albinos, however, this process is inhibited for the protein product tyrosinase. While the normal gene produces the enzyme tyrosinase, its recessive allele does not. When the recessive allele is in the homozygous state, no tyrosinase is produced in the individual. The absence of a normal functioning enzyme, due to either the absence of or a mutation in the gene for that enzyme, is known as an inborn error of metabolism.

Alcaptonuria is another example of an inborn error

Fig.2

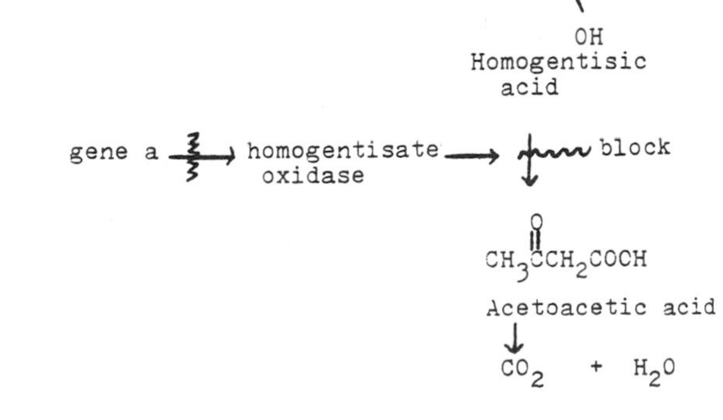

tyrosine → hydroxyphenyl pyruvic acid → Homogentisic acid
gene a → homogentisate oxidase → block
$CH_3\overset{O}{\overset{\|}{C}}CH_2COOH$
Acetoacetic acid
CO_2 + H_2O

of metabolism. This disorder is an inherited condition which is characterized by a darkening of the cartilage often associated with a type of arthritis. In addition, the urine of affected individuals turns black upon exposure to air. This is due to the presence of an unusual component, homogentisic acid, in the urine. Homogentisic acid is a normal intermediate in the metabolism of tyrosine and phenylalanine. In normal individuals, this intermediate is oxidized via an oxidase enzyme, to carbondioxide and water (see figure 2). Alcaptonurics lack this enzyme and consequently the acid builds up in the tissues and blood of these individuals and is excreted in large amounts in the urine.

• PROBLEM 24-31

Why does a child with a certain type of heredity have PKU? Explain in detail.

Solution: To understand PKU, or phenylketonuria, we have to trace metabolic pathways. When we digest protein foods, they are broken down into their component amino acids, one of which is phenylalanine. Eventually, these amino acids are taken up by the cells. Within the cells, some enzymes cause the phenylalanine to be combined with amino acids to form structural proteins that become part of the cell protoplasm. If phenylalanine is not used this way by the body cells, it remains in the blood and as it passes through the liver, it may be acted on by an enzyme which converts it into tyrosine, another amino acid. Still another enzyme will convert some of the phenylalanine into phenylpyruvic acid.

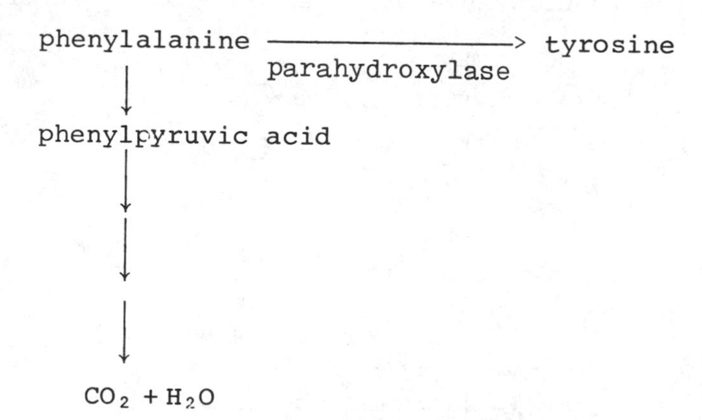

Each arrow in the diagram indicates a metabolic pathway mediated by an enzyme. Each enzyme is produced by a gene and a mutation in any one of these genes can break the chains of conversions.

In PKU, the gene which codes for —— phenylalanine H-monooxygenase the enzyme responsible for conversion of phenylalanine to tyrosine, is altered so that it no longer codes for that protein. Individuals homozygous for this recessive gene cannot produce the mono-oxygenase. Since the conversion of phenylalanine to tyrosine is blocked, there is an increase in the levels of the other by products of phenylalanine metabolism, the important one here being phenylpyruvate. Phenylpyruvate accumulates in blood, at levels from 50 to 100 times the normal values, and is excreted in large amounts in the urine. In childhood, excess circulating phenylpyruvate impairs normal brain development, causing severe mental retardation.

Since the disease results from a genetic defect, it can be inherited and passed from generation to generation.

• PROBLEM 24-32

What is reverse transcriptase and what is its function? How might it be used in cancer research?

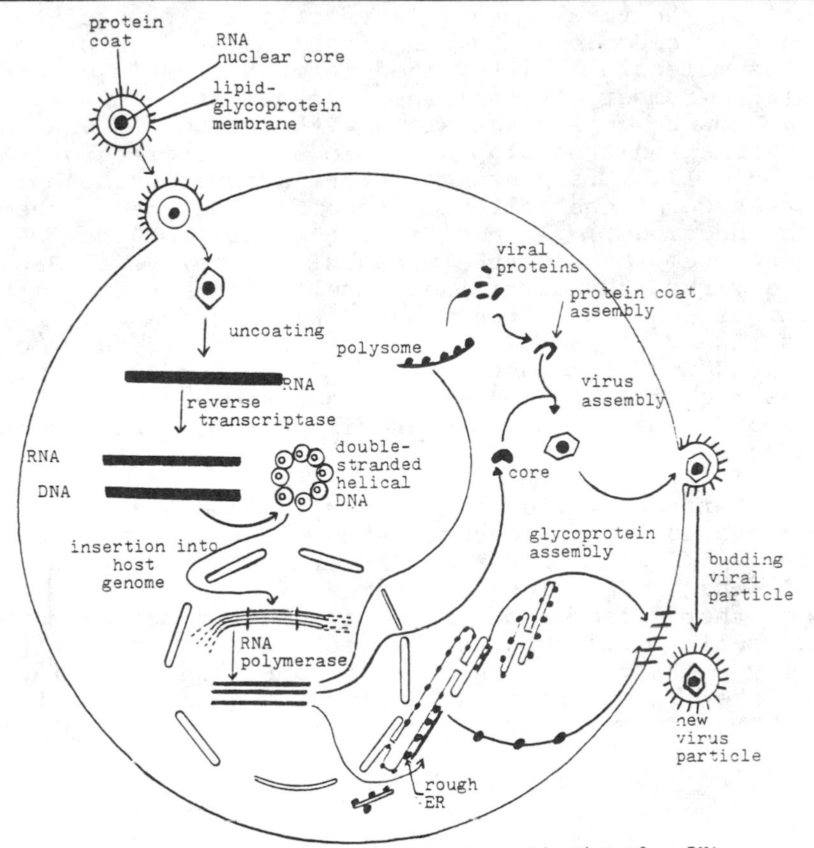

Schematic diagram of the major steps in the replication of an RNA tumor virus. Note that entire RNA tumor virus is enclosed in a lipid-glycoprotein membrane envelope. Unlike bacteriophages, the entire virus (RNA and protein coat) enters the cell.

<u>Solution:</u> We have seen in protein synthesis that biological information flows from DNA to RNA to protein. This flow had become the rule until 1964, when Temin

found that infection with certain RNA tumor viruses (cancer-causing substances) is blocked by inhibitors of DNA synthesis and by inhibitors of DNA transcription. This suggested that DNA synthesis and transcription are required for the multiplication of RNA tumor viruses. This would mean that the information carried by the RNA of the virus is first transferred to DNA, whereupon it is transcribed and translated, and consequently, that information flows in the reverse direction, that is, from RNA to DNA. Temin proposes that the RNA of these tumor viruses, in their replication, are able to form DNA. His hypothesis requires a new kind of enzyme - one that would catalyze the synthesis of DNA using RNA as a template. Such an enzyme was discovered by Temin and by Baltimore in 1970. This RNA - directed DNA polymerase, also known as reverse transcriptase, has been found to be present in all RNA tumor viruses.

An infecting RNA virus binds to and enters the host cell (the cell which the virus attacks). Once in the cytoplasm, the RNA genome is separated from its protein coat (see figure). Then, through an as yet unknown mechanism, the viral reverse transcriptase is used to form a DNA molecule using the viral RNA as a template. This DNA molecule is integrated into the host's chromosome, with a number of possible consequences. The viral DNA may now be duplicated along with the host's DNA, and its information thus propogated to all offspring of the infected cell. Its presence in the genome of the host may "transform" the cell and its offspring (cause them to become cancerous). In addition, the viral DNA may be used as a template for the synthesis of new viral RNA, and the virus thus multiplies itself and continues its infectious process, often without killing the cell. Because the viral genome is RNA, it cannot as such be integrated into the host's genome. Reverse transcriptase enables the virus to convert its genetic material to DNA, whereupon it is capable of inserting itself into the host chromosome.

There are three reasons why reverse transcriptase is so important in cancer research. First, human leukemia sarcomas (different forms of cancer) have been shown to contain large RNA molecules similar to those of the tumor viruses that cause cancer in mice. Second, these human cancer cells contain particles with reverse transcriptase activity. Third, the DNA of some human cancer cells have virus-like sequences of nucleotides, sequences not found in the DNA of comparable normal cells.

Since the action of RNA from tumor viruses depends on the presence of the enzyme reverse transcriptase, reseach into its chemical composition (amino acid sequence) may bring better understanding of its function. The ultimate objective of such reseach would of course be the prevention of cancer, perhaps through the inhibition of reverse transcriptase activity.

SHORT ANSWER QUESTIONS FOR REVIEW

Answer

To be covered when testing yourself

Choose the correct answer.

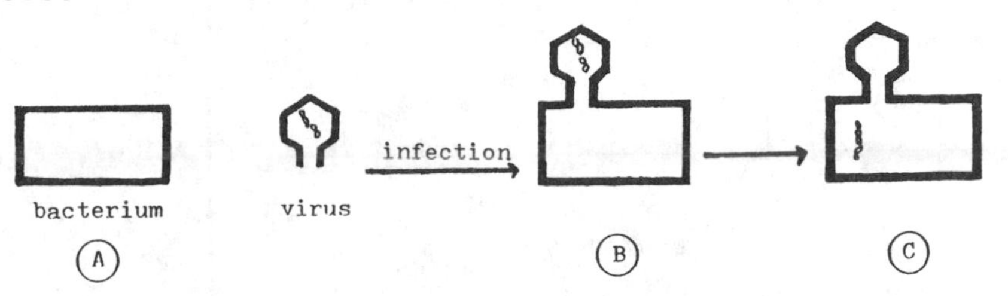

1. In the infection process illustrated above, the material that enters the bacterium from the virus in C is (a) the sulfur fraction, (b) nucleic acid, (c) protein, (d) a mutagen. — b

2. Which is not a part of a DNA nucleotide? (a) guanine, (b) deoxyribose (c) phosphate (d) adrenalin. — d

3. The nitrogenous base not found in RNA is (a) adenine, (b) guanine, (c) thymine, (d) cytosine. — c

4. The sequence of bases found in a strand of DNA which served as a template for the synthesis of a m-RNA is adenine-guanine-cytosine-thymine. Thus the sequence of bases found in the newly synthesized m-RNA is (a) guanine-cytosine-thymine-adenine, (b) thymine-cytosine-guanine-adenine, (c) uracil-adenine-cytosine-guanine, (d) uracil-cytosine-guanine-adenine. — d

5. The ribosomal - RNA of bacteria is composed of (a) a 30S and a 50S subunit, (b) a 23S and a 53S subunit, (c) a 16S and a 46S subunit, (d) a 40S and a 60S subunit. — a

6. Which of the following is not part of the cell's complete life cycle? (a) mitosis (b) G_1 phase (c) S phase (d) P phase. — d

7. The genetic code is (a) commaless, (b) degenerate, (c) non-overlapping, (d) all of the above. — d

8. In DNA replication the fragments of newly synthesized DNA are joined together, so that the 5^1 end of one fragment is joined to the 3^1 end of another fragment by the action of (a) DNA polymerase, (b) RNA primer, (c) DNA ligase, (d) DNA endonuclease. — c

9. A bacterium contains 3960 amino acids. Given the information that 4 is the average number of amino acids per polynucleotide for this bacterium, what is a good estimate for the total number of

Answer

To be covered when testing yourself

genes of this organism? (a) 160 (b) 330 (c) 990 (d) 1320

b

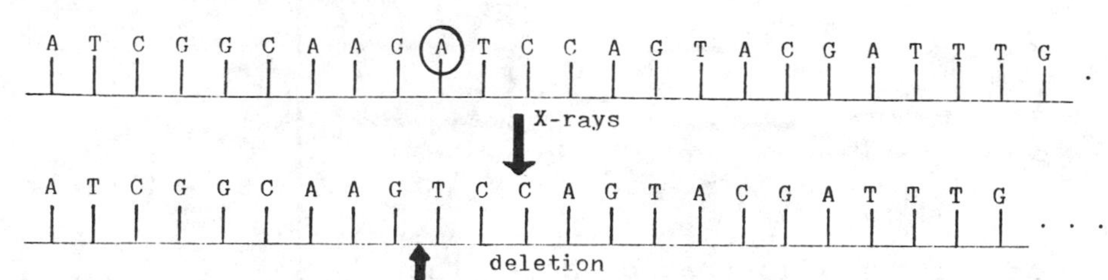

10. If the above mutation occured it would cause the mutated genetic code to be read as: (a) ATC GGC AAG TC CAG TAC GAT TTG ... (b) ATC GGC AAG TCC AGT ACG ATT TG ... (c) ATC GGC AAG CAG TAC GAT TTG ... (d) ATC GGC AAG TC C AG TAC GAT TTG ...

b

11. The genes which serve as a binding site for RNA polymerase are called: (a) operator genes, (b) structural genes, (c) promoter genes, (d) regulatory genes.

c

12. The fact that different cells of the same organisms perform different functions is proof of the fact that: (a) all cells contain the same genetic material but different cells use different portions of it, (b) something other than genetic information dictates the functioning of a cell, (c) a corepressor turns off the transcription of a gene, (d) all cells contain different genetic material, specific for its functions.

a

phenylalanine —(enzyme coded for by gene P)→ tyrosine —(enzyme coded for by gene C)→ dihydroxy phenylalanine → melanin

13. In the above series, if gene C was blocked and could not furnish the enzyme it codes for, then (a) tyrosine would be converted to dihydroxyphenylalanine by the same enzyme coded for by another gene, (b) melanin would be produced by another mechanism, (c) melanin would still be produced but to a lesser extent, thus the skin coloration would be lighter, (d) albinism would result.

d

14. The triplet code for phenylalanine was shown to be UUU by (a) analyzing the m-RNA and finding the codon to be UUU, (b) using synthetic RNA containing only uracil (c) labeling the cell with radioactive uracil (d) breaking up the DNA that coded for it and finding AAA to be the template.

b

Answer

To be covered when testing yourself

15. Twenty thousand plants in a given population are expected to have fuzzy leaves. But only 14,000 of these plants actually show this phenotype. Which statement best explains this? (a) Mutation has occurred, (b) the gene has 70% penetrance, (c) the environment probably has changed from dry to moist, eliminating the need for the hair on the leaves, (d) the expressivity of this population like all others must vary to some extent.

b

Fill in the blanks.

16. In 1928, Griffith demonstrated the phenomenon of _____ by infecting mice with smooth and rough strains of pneumococcus.

trans-
formation

17. The method of DNA replication is called _____.

semi-
conser-
vative

18. In order for DNA synthesis to begin, the double helix must be unzipped. This is done by breaking the _____ between the purine and pyrimidine bases.

hydrogen
bonds

19. In a protein, amino acids are linked together by _____ bonds.

peptide

20. In nucleic acid synthesis, the template is read in a _____ to _____ direction.

5^1 to
3^1

21. RNA polymerase must combine with the _____ before it can initiate DNA synthesis.

sigma
factor

22. Chargaff observed that in a DNA molecule the amount of cytosine is always equal to the amount of _____.

guanine

23. A chromosomal aberration in which a strand obtains genetic material from two different homologous pairs usually as a result of crossing over between these two different homologous pairs is a _____.

trans-
location

24. In the lac-operon model the presence of lactose turns on the operon. Thus the system is said to be _____.

inducible

25. _____ are regulatory proteins that can inhibit the synthesis of the protein (enzyme) by stopping the transcription of a gene.

repressors

Sickle cell hemoglobin differs from normal hemoglobin in _____ base pair(s).

one

26. _____ RNA carries the hereditary code from the nucleus to the cytoplasm.

messenger

	Answer
	To be covered when testing yourself
27. The smallest unit which when altered gives rise to a mutant phenotype is a _____.	muton

```
        | a   | b |
normal  | a   | b |      →    A + B  (gene products)

              | a*| b |
mutant₁       | a*| b |      →    B  only

              | a  | b*|
mutant₂       | a  | b*|      →    A  only
```

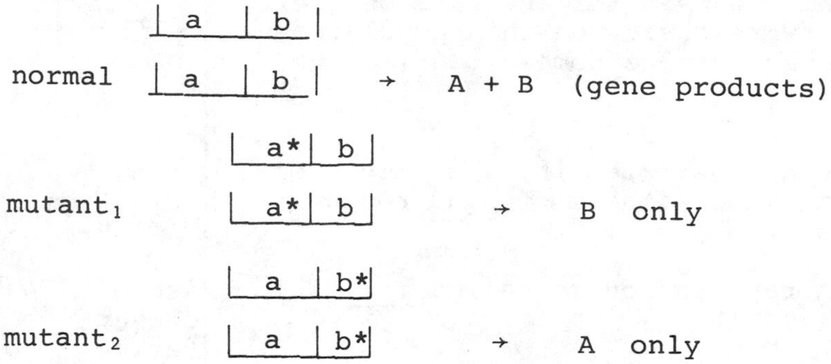

28. According to complementation, if the following gene combination was made: | a* | b | / | a | b* |, the gene products obtained would be _____ (refer to above illustration).	both A and B
29. _____ genes are inhibited by repressor substances produced by repressor genes.	operator
Determine whether the following statements are true or false.	
30. The amount of DNA is constant in all the cells of the body while the amount of protein and RNAs varies.	True
31. In the lac-operon model, lactose inactivates the repressor by destroying it.	False
32. Phenylketonuria results when a person produces an enzyme which changes phenylalanine into tyrosine rather than phenylpyruvic acid.	False
33. The only way in which biological information can flow is: DNA → RNA → protein.	False
34. Cretinism results when a person fails to produce the amino acid which converts tyrosine into thyroxine.	True
35. Before replication of DNA can take place, the bonds which must break lie between a sugar and a phosphate.	False

Answer

To be covered when testing yourself

36. If ... dines in a se... T, then the co... trand would be... — True

37. Pr... y in the nu... — False

38. T... ald be recons a... — False

39. E... dividual level w... n level. — True

Preguntas para Naomie Ranatunge

① ¿Cuántos años has jugado al tenis y por qué este deporte?

② ¿Por qué es tan importante la familia para ti?

Preguntas para mi (Nikita Bakhu)

① where have you traveled? A dónde te viajaste?

② where do you want to go in future? Adonde quieres viajar en el futuro.

CHAPTER 25

PRINCIPLES & THEORIES OF GENETICS

GENETIC INVESTIGATIONS

• PROBLEM 25-1

Why did Mendel succeed in discovering the principles of transmissions of hereditary traits, whereas others who had conducted such investigations had failed?

Solution: Mendels success was a combination of good experimental technique and luck. He chose the garden pea for his studies because it existed in a number of clearly defined varieties; the different phenotypes were thus evident and distinguishable. In addition, he only studied the transmission of one trait at a time, and crossed plants differing with respect to just that one trait. Previous investigators had studied entire organisms at a time, and had crossed individuals differing in many traits. Their results were a conglomeration of mixed traits and interacting factors, making it impossible to figure out what was happening.

Mendel was also careful to begin with pure line plants developed through many generations of natural self-fertilization. However, luck was involved in that the traits which Mendel randomly chose were coded for by genes that were located on different chromosomes, and thus assorted independently. If he had chosen traits whose genes were linked, his ratios would not have been understandable in terms of what he knew then.

• PROBLEM 25-2

What advantages do microorganisms have over larger forms of life as subjects for genetic investigation?

Solution: If we consider the methods by which genetic studies are carried out, the advantages of using microorganisms such as bacteria become obvious. Most genetic studies involve breeding and reproduction, and subsequent frequencies and ratios in the offspring often involve several generations. Here, microorganisms are prime experimental subjects, because of their extremely rapid rate of reproduction. Generations can be obtained in a matter of hours, and so genetic experiments can be done in relatively short periods of time. Large populations can be quickly grown, enabling rare events, such as mutations, to occur in observable quantities. Also, determination of genetic frequencies becomes extremely accurate with large numbers.

Microorganisms are easily obtained and stored. They live on inexpensive nutrients. This makes them perfectly suited to laboratory studies. In laboratories, the environment can be carefully controlled, and by selectively varying the culture media, one can easily study specific factors of metabolism. Single mutant cells can be isolated from among millions by simply varying the nutrient media.

Most of what we know about genes is learned from observations of their phenotypic expressions, and in this respect microorganisms display their biggest advantage. Many microorganism, particularly fungi, have a conspicuous and relatively long-lived haploid generation. In the haploid stage (n), there is only one set of genetic information; that set is expressed phenotypically, whether it be dominant or recessive. In the diploid stage, dominant alleles may mask recessive ones, and the genotype cannot be accurately determined solely from the phenotype. Lengthy procedures such as back-crosses are required to deduce the genotype, but can be eliminated if the test subjects arc haploid individuals.

Much of recent genetics has involved both microscopic chromosomal studies and gene effects at the cellular level. Here, the simplicity of the genetic material and metabolism of microorganisms makes them much better candidates for such studies than the complex multicellular organisms.

MITOSIS AND MEIOSIS

• PROBLEM 25-3

Outline briefly the events occuring in each stage of mitosis. Illustrate your discussion with diagrams if necessary.

Solution: Mitosis refers to the process by which a cell divides to form two daughter cells, each with exactly the same number and kind of chromosomes as the parent cell. In a strict sence, mitosis refers to the division of nuclear material (karyokinesis). Cytokinesis is the term used to refer to the division of the cytoplasm. Although each cell division is a continuous process, in order for it to be studied, can be artificially divided up into a number of stages. We will describe each stage separately, beginning with interphase.

1) Interphase: This phase is called the resting stage.

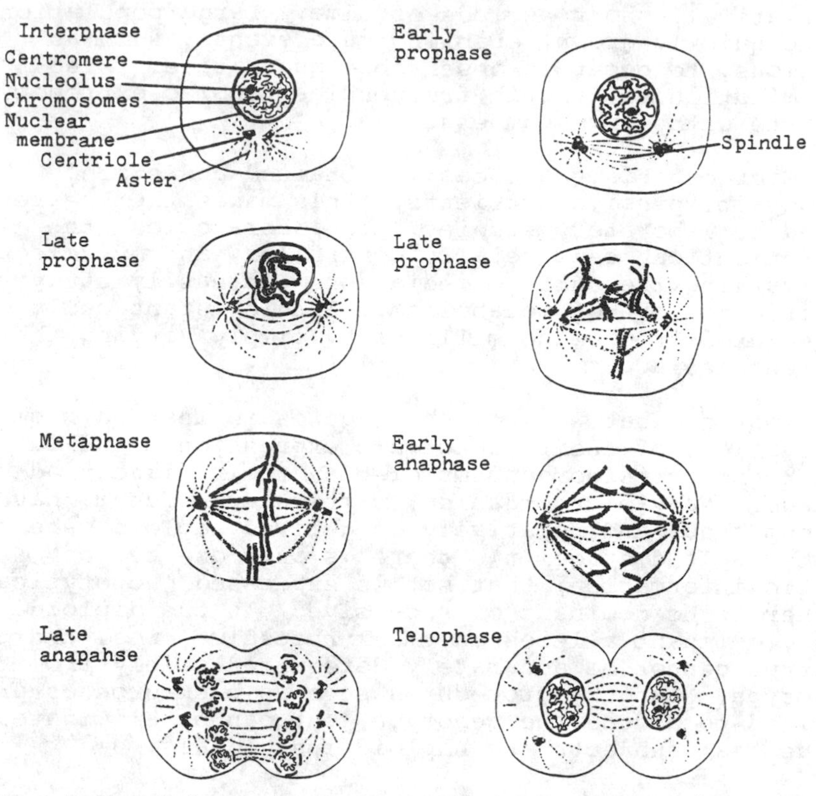

The stages of mitosis in a cell with a diploid number of 4.

However, the cell is "resting" only with respect to the visible events of division in later phases. During this phase, the nucleus is metabolically very active and chromosomal duplication is occurring. During interphase, the chromosomes appear as vague, dispersed thread-like structures, and are referred to as chromatin material.

2) Prophase: Prophase begins when the chromatin threads begin to condense and appear as a tangled mass of threads within the nucleus. Each prophase chromosome is composed of two identical members resulting from duplication in interphase. Each member of the pair is called a chromatid. The two chromatids are held together at a dark, constricted area called the centromere. At this point the centromere is a single structure.

The above events occur in the nucleus of the cell. In the cytoplasm, the centriole (a cytoplasmic structure involved in division) divides and the two daughter centrioles migrate to opposite sides of the cell. From each centriole there extends a cluster of raylike filaments called an aster. Between the separating centrioles, a mitotic spindle forms, composed of protein fibrils with contractile properties. In late prophase the chromosomes are fully contracted and appear as short, rod-like bodies. At this point individual chromosomes can be distingushed by their characteristic shapes and sizes. They then begin to migrate and line up along the equatorial plane of the spindle. Each doubled chromosome appears to be attached to the spindle at its centromere. The nucleolus (spherical body within the nucleus while RNA synthesis is believed to occur) has been undergoing dissolution during prophase. In addition, the nuclear envelope breaks down, and its disintegration marks the end of prophase.

3) Metaphase: When the chromosomes have all lined up along the equatorial plane, the dividing cell is in metaphase. At this time, the centromere divides and the chromatids become completelty separate daughter chromosomes. The division of the centromeres occurs simultaneously in all the chromosomes.

4) Anaphase: The beginning of anaphase is marked by the movement of the separated chromatids (or daughter chromosomes) to opposite poles of the cell. It is thought that the chromosomes are pulled as a result of contraction of the spindle fibers in the presence of ATP. The chromosomes moving toward the poles usually assume a V shape, with the centromere at the apex pointing toward the pole.

5) Telophase: When the chromosomes reach the poles, telophase begins. The chromosomes relax, elongate, and return to the resting condition in which only chromatin threads are visible. A nuclear membrane forms around each new daughter nucleus. This completes karyokinesis, and cytokinesis follows.

The cytoplasmic division of animal cells is accomplished by the formation of furrow in the equatorial plane. The furrow gradually deepens and separates the cytoplasm into daughter cells, each with a nucleus. In plants, this division occurs by the formation of a cell plate, a partition which forms in the center of the spindle and grows laterally outwards to the cell wall. After the cell plate is completed, a cellulose cell wall is laid down on either side of the plate, and two complete plant cells form.

• PROBLEM 25-4

Outline briefly the events occurring in each stage of meiosis. Illustrate your discussion with diagrams if necessary.

Solution: Meiosis is the process by which diploid organisms (having two sets of chromosomes) produce haploid gametes (having only one set of chromosomes). When two gametes fuse in fertilization, the zygote formed will thus have the full diploid chromosomal complement.

Meiosis consists of two cell divisions, the first (Meiosis I) called a reduction division, and the second (Meiosis II) a mitotic type division.

Interphase I: This phase is similar to mitotic inter-

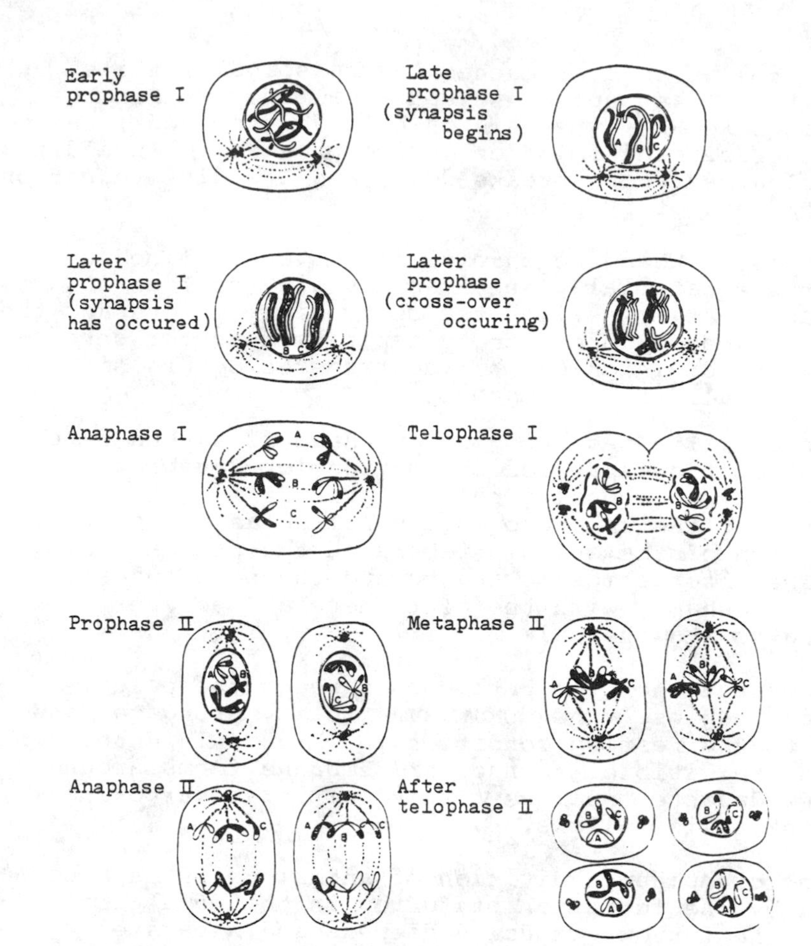

The stages of meiosis in a cell with a diploid number of 6.

phase. The cell appears inactive in reference to cell division; but it is during interphase that chromosome duplication occurs.

Prophase I: The chromosomes become thicker and more visible. While they are still long thin threads, an attractive force (as yet not identified) causes homologous chromosomes to come together in pairs, a process known as synapsis. This is the stage during which cross-over between homologous chromosomes will occur.

After synapsis, the chromosomes continue to shorten and thicken; their double nature becomes visible, so that each homologous pair appears as a bundle of four chromatids called a tetrad. Each tetrad is composed of two doubled homologous chromosomes. The number of tetrads is thus equal to the haploid number. The centromeres of homologous chromosomes are connected, and there are thus two centromeres for the four chromatids.

While these events are occurring, the centrioles migrate to opposite poles, the spindle begins to form between them, and the nucleolus and nuclear membrane dissolve. The tetrads move to the equatorial plane of the spindle.

Metaphase I: Migration to the equatorial plane is complete, and the nuclear membrane and nucleolus have dissolved.

Anaphase I: At this point the homologoues chromosomes that had paired in prophase separate and move to opposite poles of the cell. Each is still composed of two identical daughter chromatids joined at the centromere. Thus the number of chromosome types in each resultant cell is reduced to the haploid number.

Telophase I: Cytoplasmic division occurs as in mitosis. Meiosis I concludes and meiosis II begins. There is no definable interphase between the two series of divisions. The chromosomes do not separate or duplicate, nor do they form chromatin threads.

Prophase II: The centrioles that had migrated to each pole of the parental cell, now incorporated in each haploid daughter cell, divide, and a new spindle forms in each cell. The chromosomes move to the equator.

Metaphase II: The chromosomes are lined up at the equator of the new spindle, which is at right angles to the old spindle.

Anaphase II: The centromeres divide and the daughter chromatids, now chromosomes, separate and move to opposite poles.

Telophase II: Cytoplasmic division occurs. The chromosomes gradually return to the dispersed form and a nuclear membrane forms.

The two meiotic divisions yield four cells, each carrying only one member of each homologous pair of chromosomes. These cells are for this reason called haploid cells.

• PROBLEM 25-5

Compare the events of mitosis with the events of meiosis, consider chromosome duplication, centromere duplication, cytoplasmic division and homologous chromosomes in making the comparisons.

Solution: In mitosis, the chromosomes are duplicated once, and the cytoplasm divides once. In this way, two identical daughter cells are formed, each with the same chromosome number as the mother cell. In meiosis, however, the chromosomes are duplicated once, but the cytoplasm divides two times, resulting in four daughter cells having only half the diploid chromosomal complement. This difference arises in the fact that there is no real interphase, and thus no duplication of chromosomal material, between the two meiotic divisions.

In mitosis, there is no pairing of homologous chromosomes in prophase as there is in meiosis. Identical chromatids joined by their centromere are separated when the centromere divides. In meiosis, duplicated homologous chromosomes pair, forming tetrads. The daughter chromatids of each homolog are joined by a centromereas in mitosis, but it does not split in the first meiotic division. The centromeres of each duplicated member of the homologous pair are joined in the tetrad, and it is these centromeres which separate from one another in anaphase of meiosis I. Thus the first meiotic division results in two haploid daughter cells, each having chromosomes composed of two identical chromatids. Only in meiosis II, after the reduction division has already occurred, does the centromere joining daughter chromatids split as in mitosis, thus separating identical chromosomes.

• PROBLEM 25-6

Show by diagrams how genes located on different pairs of chromosomes segregate independently in meiosis in an organism with a diploid number of four.

Solution: Because there are two possible ways in which the chromosomes can line up and segregate in meiosis I, there are two possible sets of gametes that could be formed by meiosis. Each set has 2 pairs of identical gametes. There are thus 4 types of gametes possible. We know this from looking at the genotypes of the parental cell (Aa Bb). Four combinations of these genes are possible: AB, Ab, aB, and ab. All these gametes have been produced in the two possible meiotic divisions.

It is important to note, however, that each separate meiosis produced only two types of gametes - AB and ab or Ab and aB - and only in these combinations. Gametic genotype combinations such as AB and Ab or aB and ab could not be produced in the same meiotic division. Although there are eight chromosomes in total, the separation of homologs occurs as a function of duplicated chromosomes. Thus, if in meiosis I, AA is paired with BB, aa must be paired with bb. If AA is paired with bb, then aa must be paired with BB. The genotypes of the gametes produced in a meiotic division are determined in meiosis I, and meiosis II is merely a mitotic type division resulting in the separation of identical daughter chromosomes.

In meiotic gamete formation, the randomness of the way the chromosomes line up is one method that ensures genetic variation. Diverse gamates, differing in their combinations of chromosomes, are produced by the independent assortment of chromosomes when they first segregate in meiosis. In a cell with many chromosomes, such as a human cell with 23 pairs, the number of possible gametes that could be produced is very large indeed.

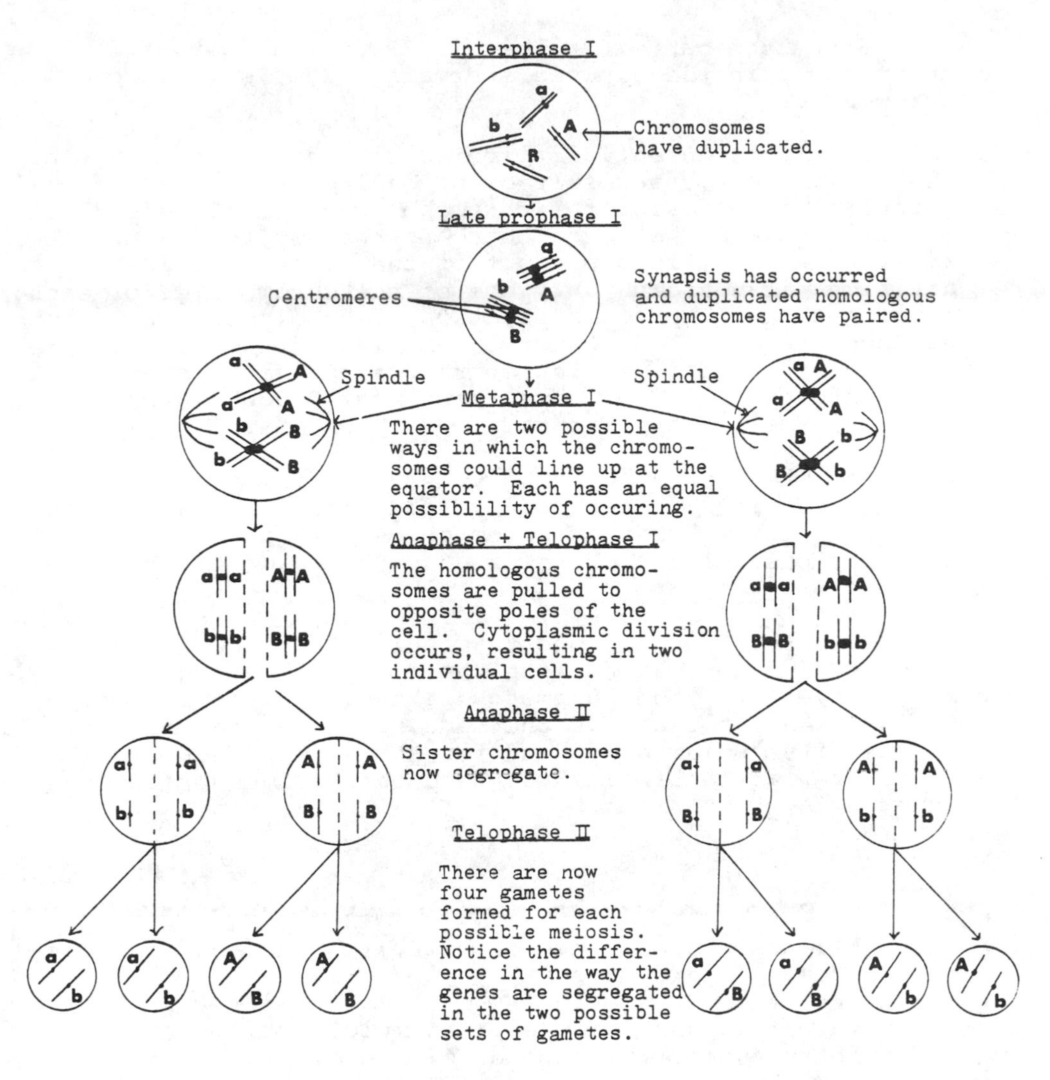

• PROBLEM 25-7

In an animal with a haploid number of 10, how many chromosomes are present in
a) a spermatogonium?
b) In the first polar body?
c) In the second polar body?
d) In the secondary oocyte?
Assume, that the animal is diploid.

Solution: In solving this problem, one must keep in mind how meiosis is coordinated with spermatogenesis and oogenesis.

a) Spermatogonia are the male primordial, germ cells. These are the cells that may undergo spermatogenesis to produce haploid gametes. But until spermatogenesis occurs, a spermatogonium is diploid just like any other body cell.

Since the haploid number is 10, the number of chromosomes in the diploid spermatogonium is 2×10 or 20 chromosomes.

b) It is essential to remember that while the polar body is formed as a result of unequal distribution of cytoplasm in meiosis, the chromosomes are still distributed equally between the polar body and the oocyte. Since the first polar body is a product of the first meiotic division, it contains only one of the chromosomes of each homologous pair, since separation of homologous chromosomes has occurred. But daughter chromatids of each chromosome have not separated, so there are 2 identical members in each chromosome. Therefore there are 10 doubled chromosomes in the first polar body, or 20 chromatids.

c) The second polar body results from the second meiotic division. In this division, the duplicate copies of the haploid number of chromosomes separate, forming true haploid cells. Therefore the chromosome number is 10.

d) The secondary oocyte results from the first meiotic division, along with the first polar body. Since, as we have said, the chromosomes have segregated equally, the secondary oocyte has the same number of chromosomes as the first polar body, and for the same reasons. Therefore, it contains 10 doubled chromosomes or 20 chromatids.

• PROBLEM 25-8

An animal has a dipoid number of 8. During meiosis, how many chromatids are present
a) in the tetrad stage?
b) In late telophase of the first meiotic division?
c) In metaphase of the second meiotic division?

Solution: In doing this problem, one must remember that meiosis involves both the duplication of chromosomes and the separation of homologous pairs.

a) In the tetrad stage of meiosis, homologous chromosomes synapse, or pair. But prior to this, every chromosome had been duplicated. Synapsis therefore results in a tetrad, a bundle of 4 chromatids (2 copies of each one of the homologous chromosomes.) The number of tetrads equals the number of haploid chromosomes. Therefore, there are $\frac{1}{2} \times 8$ or 4 tetrads. Since each tetrad has 4 chromatids, there are a total of 4×4 or 16 chromatids in the tetrad stage.

b) In late telophase of the first meiotic division, the homologous chromosomes of each pair have separated. But each chromosome is still double and composed of two daughter chromatids. So there are the haploid number (4) of doubled chromosomes, or a total of 8 chromatids.

c) In metaphase of the second meiotic division, the doubled chromosomes have lined up along the equator of the cell, but daughter chromatids have not yet separated. So the number of chromatids is still 4×2 or 8.

MENDELIAN GENETICS

• PROBLEM 25-9

Suppose pure line lima bean plants having green pods were crossed with pure line plants having yellow pods. If all the F_1 plants had green pods and were allowed to interbreed, 580 F_2 plants, 435 with green pods and 145 with yellow pods were obtained. Which characteristic is dominant and which is recessive? Of the F_2 plants, how many are homozygous recessive, homozygous dominant and heterozygous? Using G to represent the dominant gene and g to represent the recessive gene, write out a plan showing the segregation of genes from the parents to the F_2 plants.

Solution: This example is used to illustrate the basic concepts of genetics and the methods of solving genetic problems. First one must understand the common nomenclature used in the study of inheritance:

a) chromosomes - filamentous or rod-shaped bodies in the cell nucleus which contain the hereditary units, the genes.

b) gene - the part of a chromosome which codes for a certain hereditary trait.

c) genotype - the genetic makeup of an organism, or the set of genes which it possesses.

d) phenotype - the outward, visible expression of the hereditary constitution of an organism.

e) homologous chromosomes - chromosomes bearing genes for the same characters.

f) homozygote - an organism possessing an identical pair of alleles on homologous chromosomes for a given character or for all given characters.

g) heterozygote - an organism possessing different alleles on homologous chromosomes for a given character or for all given characters.

In solving genetic problems, one uses letters to represent the genotype of the organism. For example, "a" represents the gene for blue color and "A" represents the gene for red color. A capital letter is used for a dominant gene; that is, the phenotype of that gene will be expressed in a heterozygous state. For example, if the genotype Aa is expressed as red, then A is the dominant gene. Small letter a represents the recessive gene; that is, one whose phenotype will be expressed only in the homozygous state. Therefore, aa would be expressed as blue.

To solve a genetic problem, one writes down the genotypes of the two parents in the cross. Mendel's First Law tells us what to do. This law, also known as the Law of Independent Segregation, states that genes, the units of heredity, exist in individuals in pairs. In the formation of sex cells or gametes, the two genes separate and pass independent of one another into different gametes,so that each gamete has one, and only one, member of each pair. During fertilization, the gamete of one parent fuses with that of the other parent. Fusion brings the genes from each parent together, giving rise to offspring with paired genes. Now we will illustrate the Law of Segregation as it applies to the problem given. Let G represent the gene for green pod color and g represent the gene for yellow pod color. Since the parents come from pure lines, meaning that the two members of each gene pair are identical, we write the genotype of the parent plant with green pods as GG, and that of the parent plant with yellow pods as gg. Each gamete from the first parent will have one G and each gamete from the second will have one g. (Recall, the Law of Independent Segregation). Writing out a schematic cross, we have:

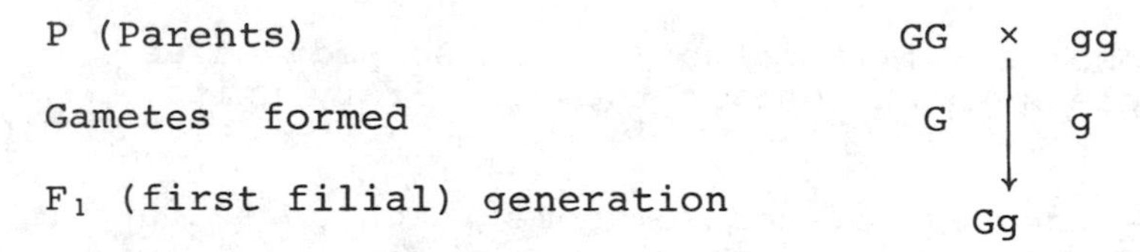

The genotype of the F_1 generation is written as Gg because it results from the fusion of the two gametes.

We are told that all the F_1 generation from the above cross are green, and we observe that they are all heterozygous. Therefore G, which stands for the gene for green pods, must be dominant (by definition), and g, which stands for the gene for yellow pods, must be recessive.

To determine the possible genotypes of the F_1 plants, or the second generation offspring, we mate two F_2 plants. The possible gametes derived from the parents are again obtained using Mendel's First Law. But now we obtain two types of gametes, G and g, from each parent, because either gene can come from the parental genotype of Gg.

Schematically, then:

P Gg × Gg

↓

gametes G_1 g G_1 g

It is easier to determine the F_2 generation using the Punnet square. The square is constructed as follows:

	G	g
G	GG	Gg
g	Gg	gg

possible gametes from female parent → (rows); ← possible gametes from male parent (columns)

It gives all possible combinations of the parental gametes, so in F_2 we have genotypically:

1 GG : 2 Gg : 1 gg,

Phenotypically:

GG is homozygous dominant and green because G is dominant;

Gg is hetorozygous and green because G is dominant; and

gg is homozygous recessive and yellow because g is recessive.

It is important to note that we cannot observe the genotype itself because it lies in the constitution of the gene. Our observations come only from what we actually see, that is, the phenotypic differences.

We know from the Punnet square that in the F_2 generation, the ratio of homozygous dominant to the entire progeny is ¼; the ratio of heterozygous is 2/4 or ½, and the ratio of homozygous recessive is ¼. Therefore, the number of homozygous dominant (GG) plants is

$$1/4 \times 580 \quad \text{or} \quad 145,$$

the number of heterozygote (Gg) plants is

$$1/2 \times 580 \quad \text{or} \quad 290$$

and the number of homozygous recessive (gg) plants is

$$1/4 \times 580 \quad \text{or} \quad 145.$$

When added together, they total 580, the number of F_2 plants.

• PROBLEM 25-10

In peas, yellow color is dominant to green. What will be the colors of the offspring of the following crosses?

a) homozygous yellow × green
b) heterozygous yellow × green
c) heterozygous yellow × homozygous yellow
d) heteroygous yellow × heterozygous yellow

Solution: This problem offers practice in genetic crosses. First we must determine the alphabetic representation of the genes involved. Let Y be the gene for yellow color in peas and y be the gene for green color in peas. For each cross, we write down the genotypes of the parents, the possible types of gametes, and determine the F_1 offspring. If it helps we may use the Punnett Square to obtain the first generation.

In the crosses then:

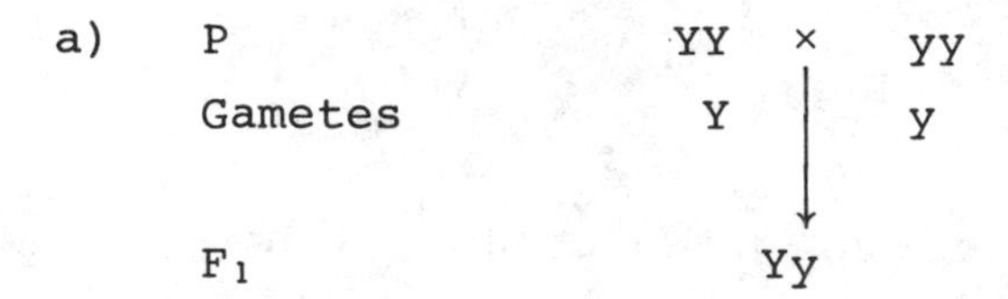

All the progeny of F_1 are yellow because Y is dominant.

b) P Yy × yy

Gametes Y; y y

F_1 1 Yy : 1 yy

Half the progeny is yellow (Yy) and half is green (yy).

c) P Yy × yy

Gametes Y; y Y

F_1 1 YY : 1 Yy

All the progeny will be phenotypically alike (yellow). Genotypically, half will be homozygous and half will be heterozygous.

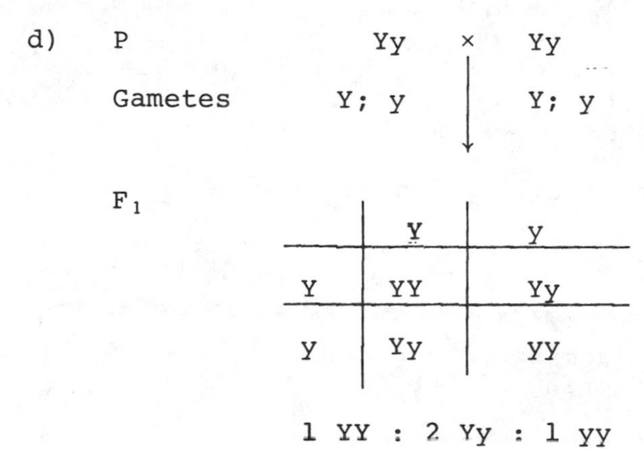

	Y	y
Y	YY	Yy
y	Yy	yy

1 YY : 2 Yy : 1 yy

A ratio is obtained of 3 yellow progeny (1 YY + 2 Yy) to 1 green (yy).

• **PROBLEM 25-11**

What are the expected types of offspring produced by a cross between a heterozygous black, short-haired guinea pig and a homozygous white, long-haired guinea pig?
Assume black color and short hair are dominant characteristics.

Solution: This is an example of a dihybrid cross. A dihybrid cross is one involving parents who differ with respect to two different traits. The principles of the cross are the same as those of monohybrid crosses.

Let B be the gene for black color, b the gene for white color, S the gene for short hair and s the gene for long hair. The parental genotypes are Bb Ss (heterozygous black, short-haired) and bbss (homozygous white, long-haired).

The gametes produced are obtained, as they are in monohybrid crosses, by Mendel's Second Law, such that the genes segregate independently. Note that each gamete formed can contain only one of the allele from each allelic pair. Thus there are only four possible gametes from the heterozygous parent, as shown below:

B b S s

BS Bs bS bs

The homozygous parent can produce only one gamete type, bbss. Doing the cross:

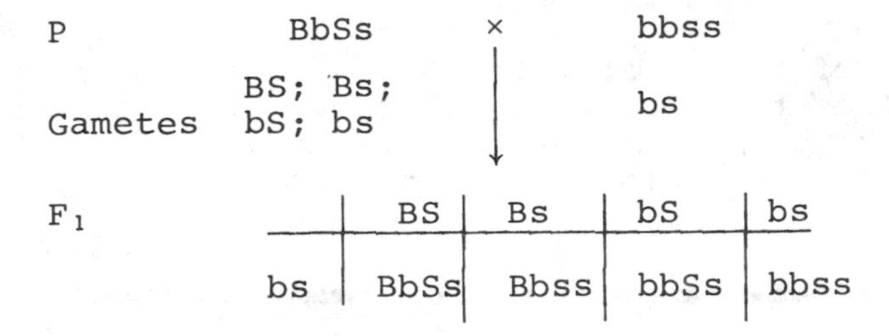

Thus the offspring are:

1/4 black, short-haired (BbSs),
1/4 black, long-haired (Bbss),
1/4 white, short-haired (bbSs), and
1/4 white, long-haired (bbss).

• **PROBLEM 25-12**

Two long-winged flies were mated. The offspring consisted of 77 with long wings and 24 with short wings. Is the short-winged condition dominant or recessive? What are the genotypes of the parents?

Solution: When we are not told which of the characteristics is dominant and which is recessive, we can deduce it from the ratio of phenotypes in the progeny. We know that 77 flies have long wings and 24 have short wings. This gives us an approximate ratio of 3 long-winged flies to every 1 short-winged fly

$$\frac{77}{24} \sim \frac{3}{1}$$

As previously noted, the three-to-one ratio signifies that dominant and recessive characteristics are most likely involved. Moreover, because there are three long-winged flies to every short-winged one, it suggests that short-winged is the recessive characteristic, and long-winged is dominant.

We cannot immediately conclude that both the long-winged parents are homozygous. In fact they are not, because if they were, no short-winged offspring could have resulted in the cross. So the presence of short-winged flies (homozygous recessive) in the progeny suggests that both parents carry the recessive gene and are thus heterozygotes.

Let L be the gene for long wings in flies and ℓ be the gene for short wings in flies. In the cross between two long-winged heterozygous parents:

P	$L\ell$	×	$L\ell$
Gametes	L; ℓ	↓	L; ℓ
F_1	1 LL: 2 $L\ell$:		1 $\ell\ell$
	long wing		short wing

The phenotypes of the F_1 show the three-to-one ratio of long-winged flies to short-winged flies, which concurs with the data given. Therefore the genotypes of the parents are the same, both being heterozygous ($L\ell$).

• PROBLEM 25-13

The ability to roll the tongue into almost a complete circle is conferred by a dominant gene, while its recessive allele fails to confer this ability. A man and his wife can both roll their tongues and are surprised to find that their son cannot. Explain this by showing the genotypes of all three persons.

Solution: Let us represent the dominant allele for the trait of tongue-rolling by R, and the recessive allele by r. We know that if a gene is dominant for a trait, it will always be expressed if it is present. Since the son cannot roll his tongue, he cannot possess the dominant gene for this trait, and his genotype must be rr. This means that, each parent must have at least one recessive allele to donate to their son. Also, each parent must have a dominant allele for this trait because they have the ability to roll their tongues. Thus the genotype of parents for this trait must be Rr.

We can illustrate this by looking at the mating of two such parents,

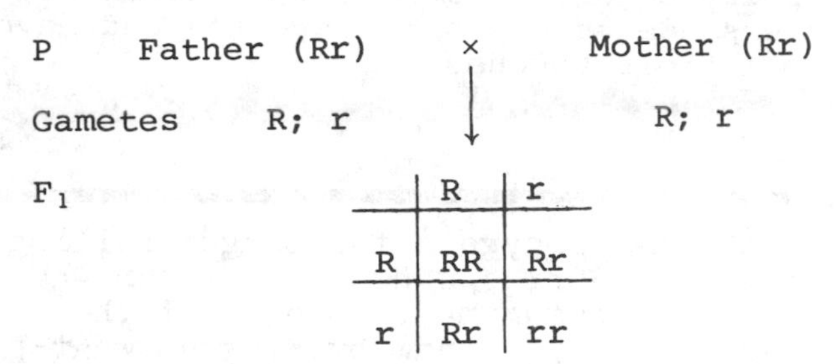

The offspring will be obtained in the following ratio:

F 1/4 RR, Homozygous dominant; tongue roller.
1/2 Rr, Heterozygous; tongue roller.
1/4 rr, Homozygous recessive; non-tongue roller.

We see then that there is a one-in-four chance that two parents who are heterozygous for a dominant trait will produce offspring without that trait. It is not unlikely then, that the parents in this problem could have had a son who does not have the ability to roll his tongue.

• PROBLEM 25-14

The ability to roll the tongue is conferred by a dominant gene, while its recessive allele fails to confer this ability. If 16% of the students in a school cannot roll their tongues, what percentage of students are heterozygous.

Solution: The 16 percent of the students that cannot roll their tongues must have a homozygous recessive genotype because the trait is determined by a dominant allele. The homozygous dominant or heterozygous condition will give one the ability to roll his tongue. If we represent the dominant allele by R and the recessive allele by r, then the three possible genotypes will be RR, Rr, and rr.

We are told that the non-tongue-rolling phenotype and thus the homozygous recessive genotype (rr) occurs with a frequency of 16% or .16. To find the frequency of the recessive allele r we take the square root of .16 $\sqrt{.16}$, = .40
Thus, .4 is the frequency of the non-roller allele. Knowing that the frequencies of the two alleles must add to 1 (i.e.,: R+r=1) we can calculate the frequency of the roller allele (R) to be .60 (i.e., R + .40 = 1; R = 1 - .40; R = .60). Squaring .60 will give us the frequency of the homozygous dominant rollers (RR). It is .36 ($.60^2$ = .36).

The total homozygous proportion of the student population can be obtained by adding the homozygous recessive's frequency (rr) with the homozygous dominant's frequency (RR). 0.36 + 0.16 = 0.52. To obtain the heterozygous proportion of the population we can use the fact that total population = homozygous population + heterozygous population. Thus, 1 = .52 + heterozygous population or, heterozygous population = 1 - .52 = .48. Therefore, 48 per cent of the students are heterozygotes and have the ability to roll their tongues.

• PROBLEM 25-15

A man and a woman are heterozygous for tongue rolling, and have three sons. The three sons marry women who are not tongue rollers. Assuming that each of the three sons has a different genotype, show by diagram what proportion of their children might have the ability to roll their tongues.

Solution: Let us again use the letter R to represent the dominant tongue rolling allele, and r to represent the recessive allele. Only three possible genotypes could result from the cross of the heterozygous parents: These alleles are: RR, Rr and rr. Let one of the sons be RR and therefore a tongue roller, the second son Rr and also a tongue roller, and the third son rr (homozygous recessive) and so unable to roll his tongue. All three sons have married woman who are not tongue rollers and who are therefore rr.

Let us now determine the proportion of the offspring who will be able to roll their tongues.

Case 1

		Son 1 × Woman 1	
P_1	RR	×	rr
Gametes	R	↓	r
F_1		Rr	100% Rr: tongue rollers

Case 2

		Son 2 × Woman 2	
P_1	Rr		rr
Gametes	R; r	↓	r
F_1	Rr;	rr	50% Rr: tongue rollers 50% rr: non-tongue rollers

Case 3

		Son 3 × Woman 3	
P	rr	×	rr
Ganetes	r	↓	r
F_1		rr	100% rr: non-tongue rollers

As a result of the three crosses, we can see that of the total offspring from the three marriages, half will be tongue rollers and half will be non-tongue rollers. The tongue rollers consist of all the children of the first son plus half the children of the second son. The non-tongue rolling children consist of all the offspring of the third son's marriage plus half the offspring from the second son's marriage.

• PROBLEM 25-16

If two fruit flies, heterozygous for genes of one allelic pair, were bred together and had 200 offspring
a) about how many would have the dominant phenotype?
b) Of these offspring, some will be homozygous dominant and some heterozygous. How is it possible to establish which is which?

Solution: Let A be the dominant gene and a be the recessive gene. Since the flies are heterozygous, each must therefore be Aa.

a) In the cross, we have:

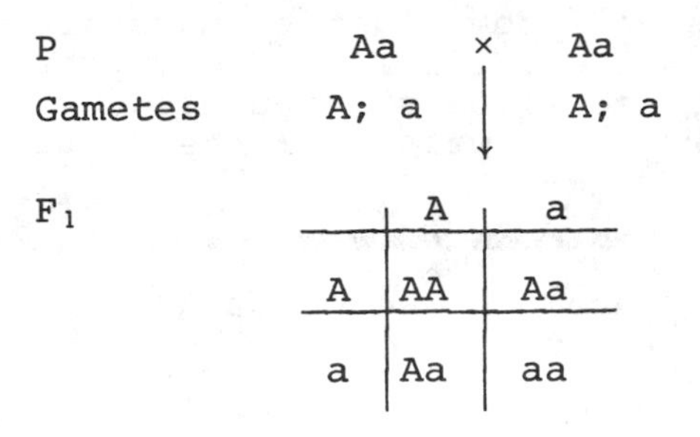

P Aa × Aa

Gametes A; a A; a

F_1

	A	a
A	AA	Aa
a	Aa	aa

1 AA : 2 Aa : 1 aa

homozygous dominant, hetero-zygous, homozygous recessive

The proportion of offspring expressing the dominant phenotype is $1/4 + 2/4 = 3/4$. Therefore, the expected number with the dominant phenotype is $3/4 \times 200$ or 150.

b) To establish which of the dominant phenotypes are homozygous and which are heterozygous, we have to perform a test cross; i.e., crossing one fly whose genotype is to be determined with a homozygous recessive fly. If the test fly is homozygous dominant:

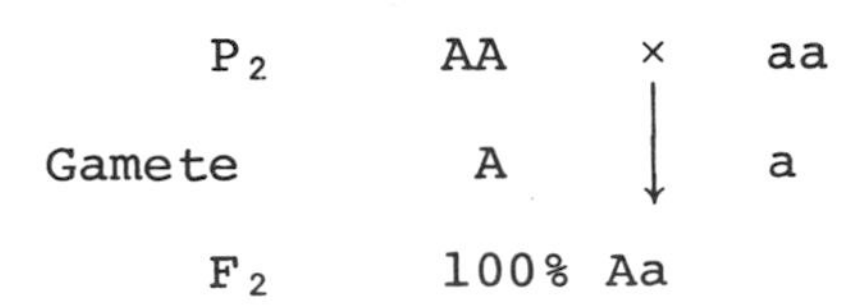

All the progeny will show the dominant phenotype. This implies that the test fly is homozygous dominant.

If the test fly is heterozygous, then in the back cross:

P_2	Aa	×	aa
Gamete	A, a	↓	a
F_2	Aa ;		aa

So half the progeny will show dominant phenotype and the other half will show recessive phenotype. This is indicative that the test fly was heterozygous.

Therefore, by performing a test cross between a fly whose genotype we do not know and a homozygous recessive fly, we can determine if the fly is homozygous or heterozygous by looking at the phenotypes of the progeny.

CODOMINANCE

• PROBLEM 25-17

Mendel believed that hereditary factors were always either dominant or recessive. How might he have altered this view had he performed the following cross? When pure line sweet peas with red flowers are crossed with pure line plants having white flowers, all the F_1 plants have pink flowers.

Solution: Let R be the gene for red color and W be the gene for white color. In the cross:

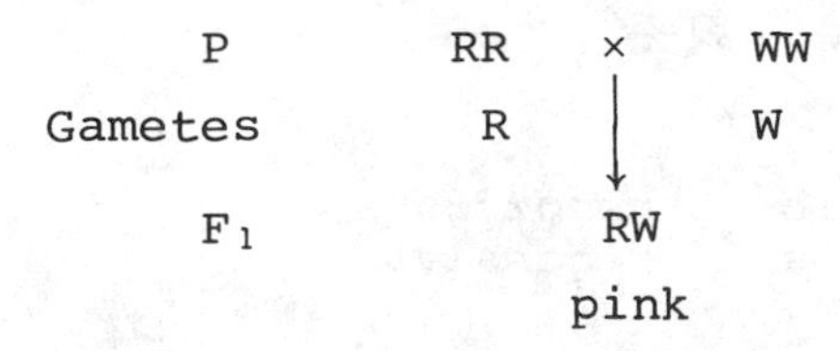

After observing such a cross, Mendel could not have proposed the concepts of dominance or recessiveness, because there is evidence for neither in the results. It is possible that he might have proposed the idea of "blending"; saying that the heterozygous genotype is the result of a genotypic blending of the two alleles. This would be erroneous however, because the two alleles still act and separate independently. This could be evidenced if a cross between two F_1 plants were done:

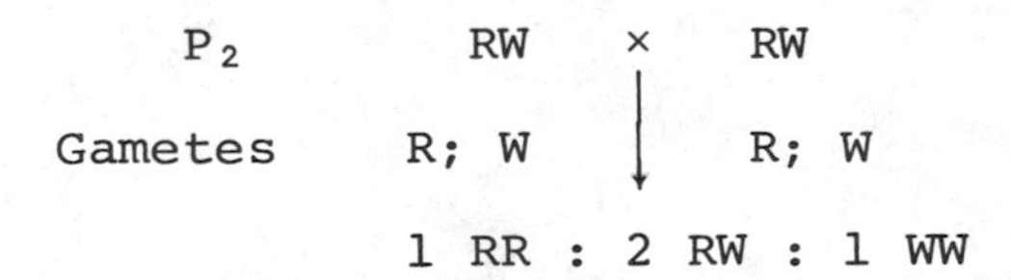

Two pink flowered plants have produced not only pink offspring, but also offspring having the red and white homozygous traits. Therefore the genes are still separating independently in the heterozygotes.

Mendel could have proposed that the two genes products interracted to form some sort of phenotypically blended product, but this is not exactly what occurs. If one looked closely at the pink flowers of the heterozygote, the independent action of each of the alleles is obvious. The pink color is not the result of some sort of blending to produce pink pigment, but results from the independent expressions of the red and the white pigments in the flower. The flower appears pink because of the interspersion of the red and white pigment granules in the petal.

The two alleles are said to be codominant. In such a case, neither allele is expressed in preference to the other. Each allele is capable of some degree of expression of its own trait in the heterozygous state.

• PROBLEM 25-18

Outline a breeding procedure whereby a true breeding strain of red cattle could be established from a roan bull and a white cow.

Solution: This problem deals with establishing a pure line having a trait which is codominant. A pure line is a strain of organisms that when inbred, always produce offspring having the same phenotype. They are homozygous and can never be heteroyzgous because when heterozygotes are inbred, they produce offspring of more than one phenotype, (which is not true breeding).

Color in cattle is determined by two alleles which are codominant. Neither allele is dominant or recessive to the other. In the heterozygous state, each allele acts to produce its own gene product independent of the other, and the total product is a sum of the contributions from each allele. Phenotypically, this usually results in a trait which is intermediate between the two dominant alleles. In this case, red and white color are codominant characteristics, and roan color is the intermediate phenotype of a heterozygote.

Let R respresent the gene for red color in cattle and W the gene for white color. By definition then, red cattle have the genotype RR, and white cattle have the genotype WW, and roan cattle are RW. (Note that we no longer use capital and lower case forms of the same letter to represent the alleles, since neither is dominant or recessive). Now since red cattle by definition must be homozygous, any red cattle produced will breed true when inbred. Therefore, in order to establish a pure line of red cattle from a roan bull and a white cow, all we need do is obtain male and female red cattle among the offspring. Then these red cattle should only be bred among themselves.

In the cross between the roan bull and the white cow:

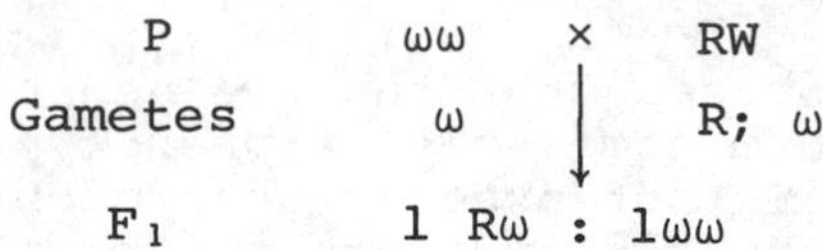

No red cattle were obtained in the F_1. The only way to obtain red cattle would be to cross two of the roan cattle from the F_1.
In the mating then:

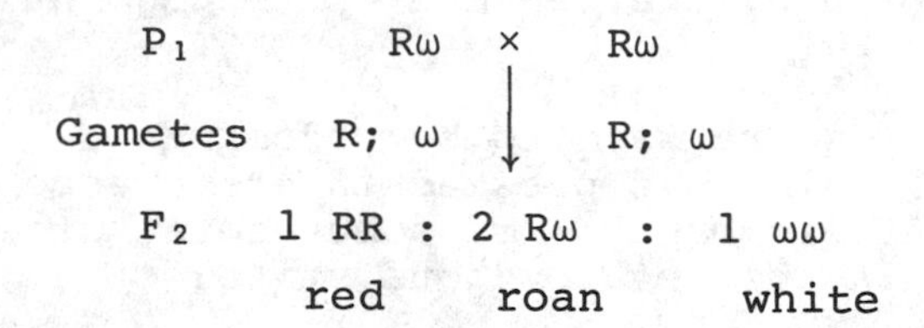

In the F_2, 1/4 of the offspring will be red, and thus homozygous for red color. They will always breed true.

• PROBLEM 25-19

A cattle breeder wants to establish a pure-breeding herd of roan short-horned cattle. What could you tell him about his chances for success in such a venture?

Solution: Using information obtained in the previous problem we can determine if it is possible to establish a pure-breeding line of roan cattle. Roan short-horned cattle are heterozygotes. Since they have both codominant genes (the gene for red color and the gene for white color) it is impossible for them to ever breed pure. To illustrate this, let us look at the results of a mating between a roan cow and bull.

Let R represent the gene for red color and W represent the gene for white color. Red cattle have the genotype RR. White cattle have the genotype WW. Roan cattle, because their color results from equal contribution of both red and white, have the genotype RW. If we mate two roan cattle: P Rω × Rω

Gametes R; ω R; ω

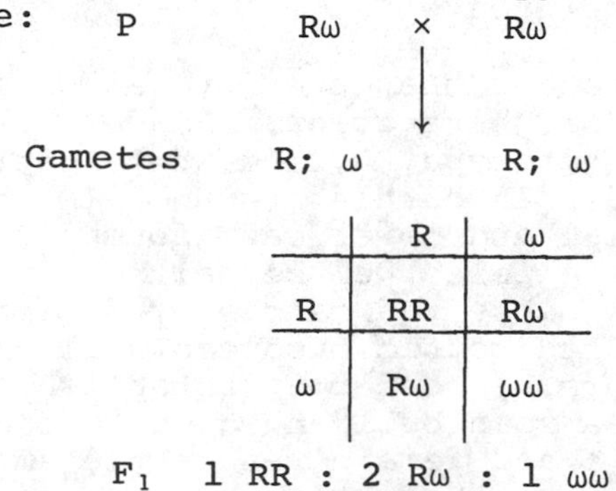

	R	ω
R	RR	Rω
ω	Rω	ωω

F_1 1 RR : 2 Rω : 1 ωω

1 red : 2 roan : 1 white

We see that one red and one white individual will be produced for every two roan individuals. Therefore, a cattle breeder might get twice as many roan cattle than red or white, but he can never be assured of having a pure-breeding herd of roan cattle.

If the breeder would settle for having large numbers of roan cattle within his herd, he could continue to mate roan cattle together, but now, whenever red or white cattle are produced, mate them to each other. Such a cross would result only in roan individuals.

P_1	RR (red)	×	ωω (white)
Gametes	R	↓	ω
F_1	100% Rω		or roan individuals.

• PROBLEM 25-20

Suppose you learned that "shmoos" may have long, oval or round bodies and that matings of shmoos resulted in the following:

(a) long × oval gave 52 long : 48 oval;

(b) long × round gave 99 oval; and

(c) oval × oval gave 24 long : 53 oval : 27 round.

What hypothesis about the inheritance of shmoo shape would be consistent with these results? Assume that shmoos are diploid. (Note that the shmoo is a hypothetical genetic organism.)

Solution: In this problem we have to decide what type of inheritance is used to determine a shmoo's body shape. From the data given in (b), we observe that the offspring of that cross possess an intermediate characteristic rather than a parental one; that is, oval instead of long or round. (Note that an oval, geometrically, is somewhere in between round and long.) This would immediately suggest the concept of codominance. Codominance is defined as the expression in the heterozygote of neither the dominant nor recessive phenotype, but something intermediate between the two.

Let us test our hypothesis. Let A be the gene for long body in shmoos, and B be the gene for round body in shmoos. According to our hypothesis of codominance, AB would then be the genotype of an oval-bodied shmoo. Outlining the given crosses in the usual way:

(a)

P	AA long	×	AB oval
Gametes	A	↓	A; B
F_1	AA long	;	AB oval.

The ratio of long to oval offspring is 1:1, which agrees

with the data in (a).

(b) P AA × BB

long round

Gametes A B

F_1 AB

oval

All the offspring in F_1 are oval-shaped; this is in agreement with the data in (b).

(c) P AB × AB

oval oval

Gametes A; B A; B

F_1 1 AA : 2 AB : 1 BB

long oval round

the ratio predicted in F_1 tallies with the data in (c).

Thus, body shape in shmoos is determined by means of the genes for long body and short body, and these two genes are codominant alleles.

DI- AND TRIHYBRID CROSSES

• PROBLEM 25-21

In rabbits, spotted coat (S) is dominant to solid color (s), and black (B) is dominant to brown (b). In a large population, brown spotted rabbits are mated to solid black ones and all the offspring are black spotted. What are the genotypes of the parents? What would be the appearance of the F_2 if two of these F_1 black spotted rabbits were mated? Illustrate your answer with a diagram.

Solution: The first thing to note about this problem is that the cross is dihybrid, and so deals with two separate traits. We can determine the genotypes of the parents by comparing their phenotypes with those of the F_1 offspring.

The brown spotted parents have two possible genotypes, bbSS and bbSs. The solid black parents have two possible genotypes, BBss and Bbss. There are four possible matings then:

1) bbSS × BBss,
2) bbSS × Bbss,
3) bbSs × BBss, and
4) bbSs × Bbss.

We can see that crosses 2, 3, and 4 could produce offspring with either or both of the homozygous recessive traits, brown and solid. But we are told that we are dealing with a large sample size and that all the offspring were black spotted. Therefore the genotypes in 1) must be the parental genotypes. Looking at the cross:

P_1 bbSS × BBss

Gametes bS | Bs

↓

F_1 BbSs

100% black spotted

When we cross two of these F_1 offspring:

P_2 BbSs × BbSs

Gametes BS; BS; bS; bs | BS; Bs: bS; bs

↓

	BS	Bs	bS	bs
BS	BBSS	BBSs	BbSS	BbSs
Bs	BBSs	BBss	BbSs	Bbss
bS	BbSS	BbSs	bbSS	bbSs
bs	BbSs	Bbss	bbSs	bbss

Phenotypically:

BBSS
BBSs
BbSS
BbSs
BBSs — 9 black, spotted
BbSs
BbSS
BbSs
BbSs

BBss
Bbss — 3 black, solid
Bbss

bbSS
bbSs — 3 brown, spotted
bbSs

bbss — 1 brown; solid

That is, a 9 : 3 : 3 : 1 ratio

9	3	3	1
black spotted	black solid	brown spotted	brown solid

Note that whenever we come across a 9 : 3 : 3 : 1 ratio in a dihybrid cross, it indicates that the cross involved dominant-recessive characteristics, and was between two heterozygotes.

• PROBLEM 25-22

A walnut-combed rooster is mated to three hens. Hen A, which is walnut-combed, has offspring in ratio of 3 walnut : 1 rose. Hen B, which is pea-combed, has offspring in the ratio of 3 walnut : 3 pea : 1 rose : 1 single. Hen C, which is walnut-combed, has only walnut-combed offspring. What are the genotypes of the rooster and the three hens?

Solution: This problem is an illustration of crosses in which there is gene product interaction. Interaction is usually indicated when the ratios observed in the offspring cannot be explained by simple dominant-recessive or codominant relationships, or linkage. The usual method of interaction is one in which one gene or gene pair masks the expression of another non-allelic gene or gene pair. This is called epistasis. Epistasis can occur in conjunction with dominant-recessive and/or codominant relationships, as it does in this problem. This can best be illustrated by looking at the different genotypes possible.

Let R be the dominant allele for rose comb, and r its recessive allele, P be the dominant allele for pea comb and p its recessive allele. These four alleles interact in the following manner. The alleles R and P act codominantly to produce walnut comb. (They are called coepistatic alleles). Each is dominant over both recessive alleles. When neither dominant allele is present, the single comb trait is expressed by the recessive alleles. In summary:

R-pp (RRpp; Rrpp) = rose comb
rrP- (rrPP; rrPp) = pea comb
R-P-(RRPP; RRPp; RrPP; RrPp) = walnut comb
rrpp = single comb

Let us look at each cross and see what they tell us about the possible genotypes of the rooster and hens involved. Then we can coordinate what we have learned from each cross to determine the exact phenotypes.

In the cross between the rooster and Hen A (each walnut and R-P-):

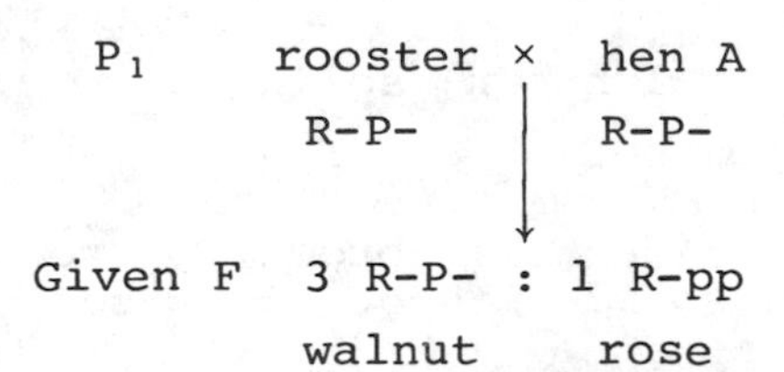

Since rose-combed offspring (R-pp) were produced, each parent must have had one recessive P gene to donate. This tells us that the rooster is R-Pp and the Hen R-Pp.

In the cross between the rooster (now known to be R-Pp) and Hen B (pea-combed and therefore rrP-):

P rooster × hen B

R-Pp × rrP-

↓

Given F_1 3 R-P- : 3 rrP- : 1 R-pp : 1 rrpp

walnut pea rose single

The fact that single-combed progeny resulted means that each parent must have both an r and a p to donate. Therefore the R-Pp rooster has to be RrPp and the rrP- hen has to be rrPp.

In the cross between the rooster (RrPp) and hen C (walnut-combed and R-P-):

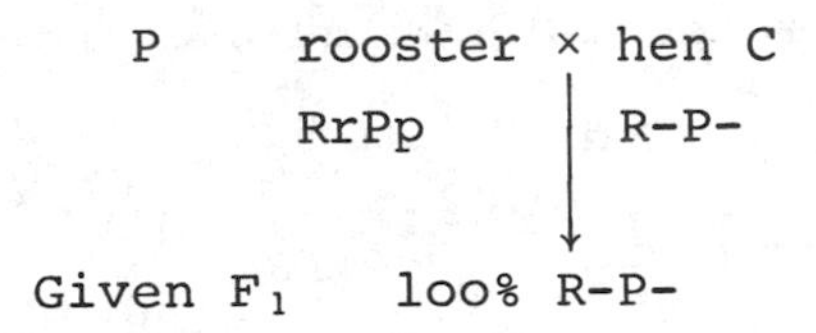

Here, since only walnut-combed progeny resulted, hen c can carry no reccessive alleles. Otherwise pea (R-pp) or rose (rrP-) would have been produced Therefore then c is RRPP.

We can now go back and solve the first cross.

If the rooster is RrPp:

P rooster × hen A

RrPp R-Pp

↓

3 R-P- : 1 R-pp

walnut rose

If Hen A had an r gene at (-), then the progeny of the rrP- or pea-combed type would have been produced; but none were. So Hen A must be RRPp. Note that we also know that no single-combed progeny were produced. Therefore all 4 recessive alleles cannot be present in the total genotypes of both parents. Since we already have 3 recessive known alleles (-) must be R, and again, Hen A is RRPp.

Summarizing:

rooster is RrPp,
hen A is RRPp,
hen B is rrPp, and
hen C is RRPP.

Doing the crosses verifies the results with the ratios obtained.

1) P rooster × hen A

RrPp RRPp

Gametes RP; Rp; rP, rp | RP; Rp

F_1

	RP	Rp	rP	rp
RP	RRPP	RRPp	RrPP	RrPp
Rp	RRPp	RRpp	RrPp	Rrpp

1 RRPP
2 RRPp
1 RrPP
2 RrPp — 6 walnut
1 RRpp
1 Rrpp — 2 rose

Since 6/2 = 3/1, the ratio is 3 walnut: 1 rose.

2) P rooster hen B

RrPp × rrPp

Gametes RP; Rp; rP; rp | RP; rp

F_1

	RP	Rp	rP	rp
rp	RrPP	RrPp	rrPP	rrPp
rp	RrPp	Rrpp	rrPp	rrpp

1 RrPP
2 RrPp — 3 walnut
1 Rrpp — 1 rose
1 rrPp
2 rrPp — 3 pea
1 rrpp — 1 single

The ratio is 3 walnut: 3 pea: 1 rose: 1 single.

3) P rooster × hen C

RrPp RRPP

Gametes RP; Rp; rP; rp | RP

	RP	Rp	rP	rp
RP	RRPP	RRPp	RrPP	RrPp

All the offspring are walnut.

• PROBLEM 25-23

In pea plants, tall plants (T) are dominant to dwarf (t), yellow color (Y) is dominant to green (y), and smooth seeds (s) are dominant to wrinkled seeds (s). What would be the phenotypes of the following matings?

a) Tt Yy Ss × ttyyss
b) Tt yy Ss × ttYySs

Solution: This is an example of a trihybrid cross, which means a cross involving three different traits. The method of solving follows the same principles as that for any other cross.

a) Before we can do the cross, we must determine the gametes involved. We can obtain the total number of different possible gametes from each parent by using a simple mathematical rule. The total number of possible combinations is 2^n, where n is the number of heterozygous traits involved in the cross. For the first parent (Tt Yy Ss), n equals 3 and the total combinations is 2^3 or 8. The genotypes of these eight gametes can be determined by using a dichotimous branching system (or forkedline method). By following any set of lines from T (or t) to S (or s) one can determine all possible combinations of factors in the following fashion:

			Gametes				Gametes
T	Y	S	TYS	t	Y	S	tyS
		s	TYs			s	tYs
	y	S	TyS		y	S	tyS
		s	Tys			s	tys

Since the second parent is homozygous (ttyyss) there is only 1 possible gamete type, tys, because n = 0 and $2^0 = 1$. In the cross:

P_1	TtYySs	× ttyyss
Gametes	TYS; TYs; TyS; Tys; tyS; tYs; tyS; tys	tys

F_1

	tys
TYS	TtYySs
TYs	TtYyss
TyS	TtyySs
Tys	Ttyyss
tYS	ttYySs
tYs	ttYyss
tyS	ttyySs
tys	ttyyss

Phenotypically,	TtYySs	is	tall,	yellow,	smooth;
	TtYyss	is	tall,	yellow,	wrinkled;
	TtyySs	is	tall,	green,	smooth;
	Ttyyss	is	tall,	green,	wrinkled;
	ttYySs	is	short,	yellow,	smooth;
	ttYyss	is	short,	yellow,	wrinkled;
	ttyySs	is	short,	green,	smooth; and
	ttyyss	is	short,	green,	wrinkled.

b) There are two heterozygous traits in the parent with genotype TtyySs, so n equals 2. The number of different gametes is 2^2 or 4.

Determining their genotypes:

		Gametes
T-y	S	TyS
	s	Tys
t-y	S	tyS
	s	tys

There are four different gametes from the second parent with genotype ttYySs ($2^2 = 4$).

			Gametes
t	Y	S	tYS
		s	tYs
	y	S	tyS
		s	tys

In the cross:

P_1 TtyySs × ttYySs

TyS; Tys, tyS, tys | tYS; tYs; tyS; tys

F_1

	TyS	tyS	Tys	tys
tYS	TtYySS	ttYySS	TtYySs	ttYySs
tyS	TtyySS	ttyySS	TtyySs	ttyySs
tYs	TtYySs	ttYySs	TtYyss	ttYyss
tys	TtyySs	ttyySs	Ttyyss	ttyyss

This cross could also be done using the forked line method. View the trihybrid cross as three separate monohybrid crosses. (Since genes segregate independently, we can legitimately do this). Then the genotypic pairs resulting from each cross can occur in any possible combinations with the gene pairs produced in any other cross. The ratios from each cross remain the same as they would if done separately. (Note that this method assumes that you can do the fairly simple monohybrid crosses by inspection, having had sufficient exposure and experience with them.)

1 2 3	×	1 2 3
TtyySs		ttYySs

1 × 1 results	2 × 2 results	3 × 3 results	Genotyses
		1 SS	1 TtYySS
		1 Yy-2 Ss	2 TtYySs
		1 ss	1 TtYyss
	1 Tt	1 SS	1 TtyySS
		1 yy-2 Ss	2 TtyySs
		1 ss	1 Ttyyss
		1 SS	1 ttYySS
		1 Yy-2 Ss	2 ttYySs
		1 ss	1 ttYyss
	2 tt	1 SS	1 ttyySS
		1 yy-2 Ss	2 ttyySs
		1 ss	1 ttyyss

Summarizing our results:

Phenotypes	Genotypes	Genotypic Frequency	Phenotypic Frequency
Tall, yellow, smooth	TtYySS	1	3
	TtYySs	2	
Tall, yellow, wrinkled	TtYyss	1	1
Tall, green, smooth	TtyySS	1	3
	TtyySs	2	
Tall, green, wrinkled	Ttyyss	1	1
Short, yellow, smooth	ttYySS	1	
	ttYySs	2	3
Short, yellow, wrinkled	ttYyss	1	1
Short, green, smooth	ttyySS	1	3
	ttyySs	2	
Short, green, wrinkled	ttyyss	1	1

• PROBLEM 25-24

The weight of the fruit in one variety of squash is determined by three pairs of genes. The homozygous dominant condition, AABBCC, results in 6-pound squashes, and the homozygous recessive condition, aabbcc, results in 3-pound squashes. Each dominant gene adds 1/2 pound to the minimum 3-pound weight. When a plant having 6-pound squashes is crossed with one having 3-pound squashes, all the offspring have 4 1/2-pound fruit.
What would be the weights of the F_2 fruit, if two of these F_1 plants were crossed?

Solution: This problem deals with polygenic inheritance; that is, the situation in which two or more independent pairs of genes have similar and additive effects on the same characteristic. Examples of such inheritance are height and skin color in man, and commercially important characteristics in animals and plants, such as the amount of eggs and milk produced, the size of fruit, and so on.

In the cross between a 6-pound squash plant and a 3-pound squash plant:

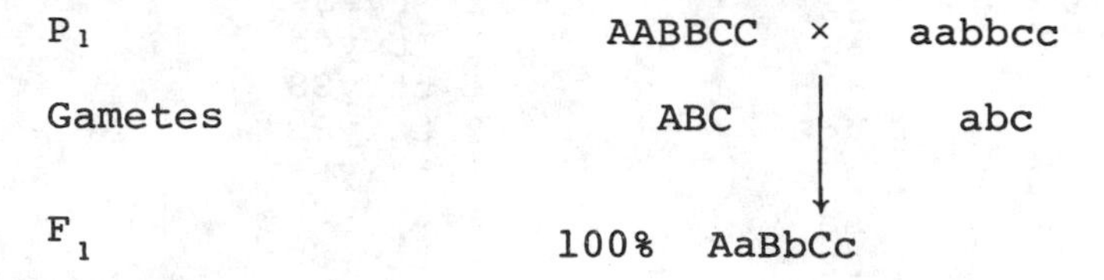

We are told that each dominant gene adds 1/2 pound to the weight. The presence of 3 dominant genes in F_1 (A, B, and C) would increase the weight by 1/2 + 1/2 + 1/2 or 1 1/2 lbs. Therefore each squash from F_1 weighs

3 + 1 1/2 or 4 1/2 pounds.

In the cross between two F_1 plants, we must first determine the possible gametes from each plant. The number of possible gametes is obtained using the 2^n rule, where n is the number of heterozygous traits and 2^n is the total number of different gametes formed. For each parent, n is equal to 3. Therefore the number of gametes is 2^3 or 8.

The genotypes of these gametes can be obtained by dichotomous branching, which ensures that all possible combinations are considered. Each possible gene from each allelic pair is matched to every possible gene combination of the other two pairs as follows:

			Gametes
A	B	C	ABC
		c	ABc
	b	C	AbC
		c	Abc
a	B	C	aBC
		c	aBc
	b	C	abC
		c	abc

Looking at the cross:

	1 2 3		1 2 3
P_1	Aa Bb Cc	×	Aa Bb Cc
Gametes	ABC;ABc;AbC;Abc; aBC;aBc'abc;abc;		ABC;ABC;AbC;Abc, aBC;aBc;abC;abc

Separating the trihybrid cross into three monohybrid crosses, and then using dichotomous branching to determine all possible combinations of the results of these crosses, one obtains:

1 × 1 results	2 × 2 results	3 × 3 results	Genotypes	Phenotypes (weight in pounds)
		1CC	1 AABBCC	6
	1BB —	2Cc	2 AABBCc	5 1/2
		1cc	1 AABBcc	5
		1CC	2 AABbCC	5 1/2
AA —	2Bb —	2cc	4(2×2) AABbCc	5
		1cc		
			2 AABbcc	4 1/2
		1C C		
	1bb —	2Cc	1 AAbbCC	5
		1cc		
			2 AAbbCc	4 1/2
		1cc		
			1 AAbbcc	4
	1BB —	2Cc		
		1cc	2 AaBBCC	5 1/2
		1CC	4(2×2) AaBBCc	5
2Aa —	2Bb —	1Cc		
			2 AaBBcc	4 1/2
		1cc		
		1CC	4(2×2) AaBbCC	5
	1bb			
		2Cc	8(2×2 2) AaBbCc	4 1/2
		1cc		
			4(2×2) AaBbcc	4
			2 AabbCC	4 1/2
			4(2×2) AabbCc	4
		1CC		
	1BB —	2Cc	2 Aabbcc	3 1/2
		1cc		
			1 aaBBCC	5
		1cc		
			2 aaBBCc	4 1/2
1aa —	2Bb —	2Cc		
		1cc	1 aaBBcc	4
		1cc	2 aaBbCC	4 1/2
	1bb —	2Cc		
			4(2×2) aaBbCc	4
		1cc	2 aaBbcc	3 1/2
			1 aabbCC	4
			2 aabbCc	3 1/2
			1 aabbcc	3

Summarizing:

1/64 weighs 6 pounds
6/64 weighs 5 1/2 pounds
15/64 weighs 5 pounds
20/64 weighs 4 1/2 pounds
15/64 weigh 4 pounds
6/64 weigh 3 1/2 pounds
1/64 weighs 3 pounds

It is important to be able to associate quantitative characteristics with polygenic inheritance. The cross itself is not difficult but may at times be tedious (Note: this problem could also have been done using the Punnett square, but dichotomous branching is more frequently used in crosses involving three or more traits).

MULTIPLE ALLELES

• PROBLEM 25-25

Mrs. Doe and Mrs. Roe had babies at the same hospital at the same time. Mrs. Roe brought home a baby girl and named her Nancy. Mrs. Doe received a baby boy and named him Richard. However, she was sure she had had a girl and brought suit against the hospital. Blood tests showed that Mr. Doe was type 0, Mrs. Doe was type AB, and Mr. and Mrs. Roe were both type B. Nancy was type A and Richard type 0. Had an exchange occurred?

Solution: Inheritance of blood groups is an example of a trait controlled by multiple alleles. The term multiple alleles is applied to three or more genes that can control a single trait; different combinations of any two genes may be present in a gene pair determining that trait in an individual. Any individual in the population may have any two of the possible alleles, but never more than two, because only two genes for a particular trait can be carried by an individual. Any gamete may have only one of the possible alleles. However, in the population as a whole, three or more different alleles will occur.

The blood types of man, 0, A, B and AB, are coded for by multiple alleles. Gene I^A provides the code for the synthesis of a specific protein, agglutinogen A, in the red cells. Gene I^B leads to the production of a different protein, agglutinogen B. Gene i produces no agglutinogen. Gene i is recessive to the other two genes, but neither gene I^A nor I^B is dominant to the other. The symbols I^A, I^B and i are used to emphasize that all three alleles are alleles at the same locus. Individuals with genotypes I^A I^A and I^A i make up blood group A, and produce agglutinogen A. Those with genotypes I^B I^B and I^B i compose blood group B, and produce agglutinogen B. Blood group 0 individuals have genotype ii, and produce no agglutinogens. When an individual has the genetic make-up of I^A I^B, he has both agglutinogens A and B and he belongs to blood group AB. These blood types are genetically determined and do not change during an individual's lifetime.

In reference to the problem given, it is possible to determine if an exchange occurred by comparing the bloodtypes of the babies with the possible bloodtypes that could be found in any offspring of each set of parents. In other words, we can cross the genotypes of each set of parents and determine what genotypes are possible for their offspring. Then we can see if the bloodtypes of the babies that each family brought home is compatable with the possibilities, and if the parents could then have indeed produced a child with that given bloodtype.

We are told that Mrs. Doe is type AB. Therefore, her genotype must be $I^A I^B$. Mr. Doe is type 0, so his genotype is ii. The possible bloodtypes of any Doe family offspring are obtained as follows:

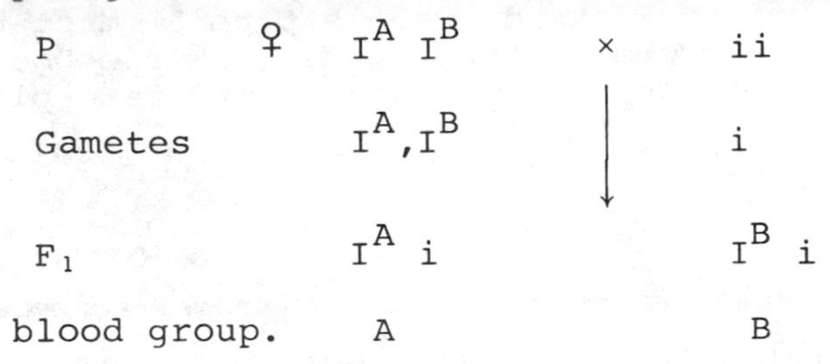

Thus, Mr. and Mrs. Doe can only produce offspring having bloodtypes A or B.

In the Roe family, we are told that Mr. and Mrs. Roe are both type B. Therefore, each parent has one of two possible genotypes, $I^B I^B$ or I^B i. Taking each possiblility:

1) P ♀ $I^B I^B$ × $I^B I^B$

gametes I^B ↓ I^B

F_1 $I^B I^B$

blood group = B

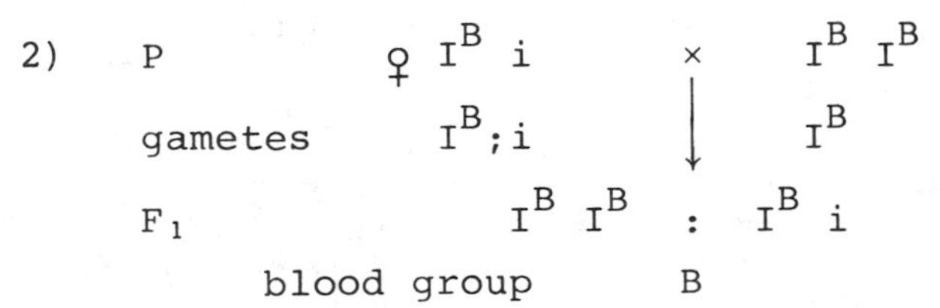

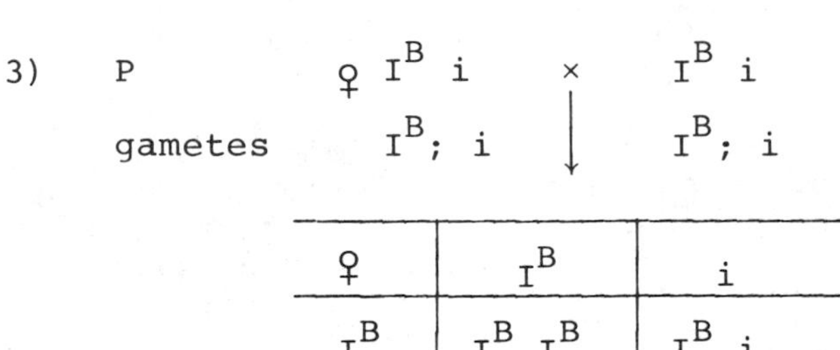

♀	I^B	i
I^B	$I^B I^B$	I^B i
i	I^B i	ii

F_1 1 $I^B I^B$: 2 I^B i : 1 ii

blood groups = B 0

Thus, the Roes can produce only offspring having bloodtype B or O.

Now let us look at the bloodtypes of the babies that each family brought home. The Does brought home Richard, who is type O. But, as we have seen, the Does can produce only offspring of type A or B. The Roes brought home Nancy, who is type A. But it would be impossible for the Roes to have a child with bloodtype A, for the only possible bloodtypes for their offspring are B and O. Therefore Richard, who is type O, must be their child, and Nancy who is type A, must be the Doe's daughter. We see that an exchange did indeed take place.

SEX LINKED TRAITS

• PROBLEM 25-26

Explain the mechanism of the genetic determination of sex in man.

Solution: The sex chromosomes are an exception to the general rule that the members of a pair of chromosomes are identical in size and shape and carry allelic pairs. The sex chromosomes are not homologous chromosomes. In man, the cells of females contain two identical sex chromosomes or X chromosomes. In males there is only one X chromosome and a smaller Y chromosome with which the X pairs during meiotic synapsis. Men have 22 pairs of ordinary chromosomes (autosomes), plus one X and one Y chromosome, and women have 22 pairs of autosomes plus two X chromosomes.

Thus it is the presence of the Y chromosome which determines that an individual will be male. Although the mechanism is quite complex, we know that the presence of the Y chromosome stimulates the gonadal medulla, or sex-organ forming portion of the egg; to develop into male gonads, or sex-organs. In the abscence of the Y chromosome, and in the presence of two X chromosomes, the medulla develops into female gametes. [Note, that a full complement of two X chromosomes are needed for normal female development.]

In man, since the male has one X and one Y chromosome, two types of sperm, or male gametes, are produced during spermatogenesis (the process of sperm formation, which includes meiosis) One half of the sperm population contains an X chromosome and the other half contains a Y chromosome. Each egg, or female gamete, contains a single X chromosome. This is because a female has only X chromosomes, and meiosis produces only gametes with X chromosomes. Fertilization of the X-bearing egg by an X-bearing sperm results in an XX, or female offspring. The fertilization of an X-bearing egg by a Y-bearing sperm results in an XY, or male offspring. Since there are approximately equal numbers of X- and Y- bearing sperm, the numbers of boys and girls born in a population are nearly equal.

• **PROBLEM 25-27**

Some sex-linked traits are expressed more often in girls than in boys, while others are expressed more often in boys than in girls.

Solution: The X chromosome carries not only the genes for sex, but also many other genes not related to sex, such as the gene for color blindness or the gene for hemophilia. Such genes are called sex-linked, because they are located on the sex chromosome. The Y chromosome does not carry any genes that we know of other than those which are related to the expression of the male sex.

Girls have two X chromosomes. Boys have one X chromosome and one Y chromosome. Each individual inherits one sex chromosome from each parent, so that a girl gets one X chromosome from her mother and one X chromosome from her father. A boy receives an X chromosome from his mother and a Y chromosome from his father. (His mother cannot give him a Y chromosome, as she has only X chromosomes to give).

Since the genes for sex-linked traits are carried only on the X chromosome, it follows that a girl would have twice the chance of receiving a sex-linked gene than would a boy, because she receives two X chromosomes, while a boy receive only one. Thus a girl has a chance to receive a sex-linked gene if either of her parents carry that gene. A boy, however, could only receive such a gene from his mother, even if his father carried that gene on his X chromosome.

Although a girl has a greater chance than a boy of receiving a sex-linked gene, the chance that either of them will express the trait coded for by that gene varies according to the dominant or recessive nature of the gene.

For example, if the gene for a sex-linked trait is dominant, that trait would be more commonly expressed in girls than in boys. Because it is dominant, only one copy of the gene is necessary for its expression. Since girls have a greater chance of receiving a copy of a sex-linked gene, they have a greater chance of expressing its trait.

If a sex-linked trait is recessive, a boy would have a greater chance of expressing the trait. In order for a girl to express that trait, she would have to have two copies of the recessive gene, because its expression would be masked by the presence of a normal X chromosome or one with a dominant allele. Thus both her parents would have to carry the gene. A boy however, need only have one copy of the gene in order to express its trait, because his Y chromosome does not carry any genes that would mask the recessive gene. So only his mother need carry the gene. The chances of this happening are much greater than the chance that two people carrying the gene will mate and have a girl, so the trait is more commonly expressed in boys.

It is important to note that girls still have a greater chance of receiving a single copy of a recessive gene, although they may not express it. Such individuals are called carriers, and their frequency in the population is greater than that of individuals expressing the trait, be they male or female.

• PROBLEM 25-28

One pair of genes for coat color in cats is sex-linked. The gene B produces yellow coat, b produces black coat, and the heterozygous Bb produces tortoise-shell coat. What kind of offspring will result from the mating of a black male and a tortoise-shell female?

Solution: Sex determination and sex linkage in cats is similar to that found in man, and indeed, in most animals and plants that have been investigated. So for cats, we can assume that if a gene is sex-linked, it is carried on the X chromosome. We can also assume that the Y chromosome carries few genes, and none that will mask the expression of a sex-linked gene or an X chromosome. Female cats, like female humans, are XX, and male cats are XY.

Let X^B represnt the chromosome carrying the gene for yellow coat, and X^b represent the chromosome carrying the gene for black coat. The male parent in this problem is black, so his genotype must be X^bY. The female is tortoise-shell, which means that she is carrying both the gene for yellow color and the gene for black color. Her genotype is X^BX^b. In the cross:

P X^bY × X^BX^b ♀

Gametes X^b; Y ↓ X^B; X^b

F₁

♀	X^b	Y
X^B	X^BX^b	X^BY
X^b	X^bX^b	X^bY

Phenotypically, the offspring consist of:

1/4 tortoise-shell females (X^BX^b),

1/4 black females (X^bX^b),

1/4 yellow males (X^BY), and

1/4 black males (X^bY).

Note that there can never be a tortoise-shell male, because a male can carry only one of the two possible alleles at a time.

• PROBLEM 25-29

The barred pattern of chicken feathers is inherited by a pair of sex-linked genes, B for barred and b for no bars. If a barred female is mated to a non-barred male, what will be the appearance of the progeny?

Solution: This is an example of a sex-linked cross. Using the notations for the sex-linked traits:

let X^B be the chromosome carrying the gene for the barred pattern, and

X^b be the chromosome carrying the gene for the non-barred pattern.

Since B is a dominant gene, a barred female could have one of the 2 genotypes: X^BX^B or X^BX^b. The only possible genotype for a non-barred male is X^bY. Therefore there are two possible crosses between a barred female and a non-barred male:

i) P ♀ $X^B X^B$ × $X^b Y$

Gametes X^B ↓ X^b; Y

F_1 $X^B X^b$: $X^B Y$

barred female barred male.

Phenotypically, all the progeny of this cross have the barred feather pattern.

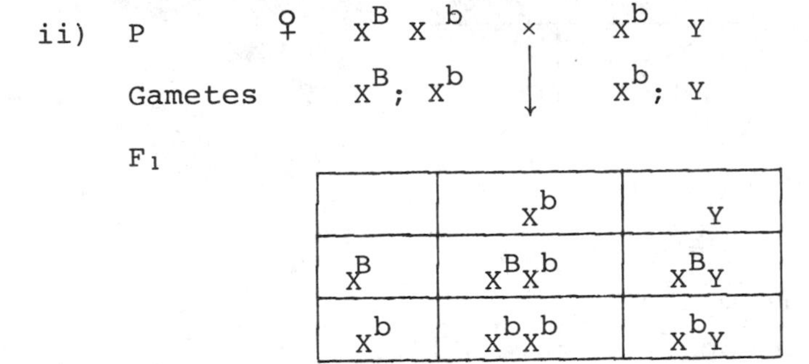

ii) P ♀ $X^B X^b$ × $X^b Y$

Gametes X^B; X^b ↓ X^b; Y

F_1

	X^b	Y
X^B	X^BX^b	X^BY
X^b	X^bX^b	X^bY

Phenotypically: $X^B X^b$ is a barred female,
$X^b X^b$ is a non-barred female,
$X^B Y$ is a barred male, and
$X^b Y$ is a non-barred male.

• PROBLEM 25-30

A girl is a hemophiliac. a) What are the possible genotypes and phenotypes of her parents? b) Assuming that her mother is normal, what were this girl's chances of being born with the disease? c) Several cases of hemophilia in girls have been reported within a small region in England where there is much close intermarriage. Explain this high frequency of hemophilia in girls.

Solution: a) Hemophilia is a sex-linked trait and its gene is carried on the X chromosome. It is also a recessive trait, and so will not be expressed in the presence of a normal X chromosome. It will be expressed, however, in the presence of a normal Y chromosome, because that chromosome does not carry any genes that would mask its expression. This means that in order for a boy to be a hemophiliac, he need only have one parent who carries the gene. A girl, however, would have to have both parents carry the gene, since she receives one X chromosome from each parent.

Let ^{+}X represent the chromosome carrying the gene for hemophilia, and X and Y represent the normal sex chromosomes. The father of a hemophiliac girl, in order to be able to give his daughter the gene for hemophilia, must be $^{+}X\ Y$, and therefore a hemophiliac himself. The mother could be either $^{+}X^{+}X$, (a hemophiliac), or ^{+}XX, (a carrier), in order to pass the hemophilia gene to her daughter.

b) If the mother is normal, her genotype must be ^{+}XX. Let us look at the cross that produced the hemophiliac girl.

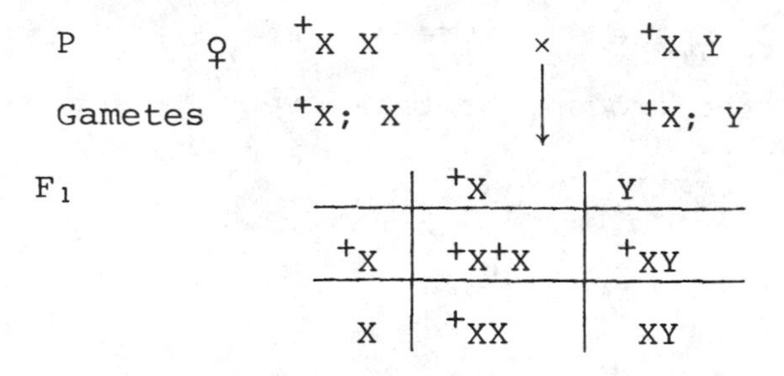

The offspring will be as follows:

1/2 normal	1/4 female carrier ($X^{+}X$)
	1/4 normal male ($^{+}X^{+}X$)
1/2 hemophiliac	1/4 hemophiliac male ($^{+}X\ Y$)
	1/4 hemophiliac female ($^{+}X^{+}X$)

Since half the possible offspring are hemophiliac, the girl had a 50% chance of being born with the disease.

c) While within the family above, a female has the same chance as a male of being a hemophiliac, this is not the case in the population as a whole. Because of the fact that it requires that a father be hemophiliac and a mother be at least a carrier in order to produce a hemophiliac girl, there is usually only a rare case now and then of a girl being born with hemophilia. The hemophilia gene has a lower frequency than the normal gene in the population, and the chances that both parents will have this gene are very low.

However, within the particular region in question, there is much intermarriage. If a gene is carried by a family member, its frequency among other family members will be much higher than in the population as a whole. If an individual carries a particular gene, in this case the gene for hemophilia, and marries within the family, there is a good chance that their spouse will also carry the gene. There is thus an increased probability that it will be carried by offspring in the homozygous state, in this case resulting in hemophiliac girls.

EXTRACHROMOSOMAL INHERITANCE

• PROBLEM 25-31

If a particular character in a certain species of animal were always transmitted from the mother to the offspring, but never from the father to the offspring, what could you conclude about its mode of inheritance?

Solution: The inheritance pattern outlined in this question cannot be explained by a mutation. Such a mutation would have to have occurred somewhere in the genes of the normal chromosomal complement of the egg during meiosis, and so coded for a trait not found in the unmutated genes of the sperm. If the trait was indeed caused by such a mutation, we would have to propose a hypothesis that the same mutation happens to all mothers of that species, and that every mother, but not the fathers, are affected. Such a hypothesis is extremely unlikely since mutation rates are by definition very low.

This inheritance also cannot be accounted for by the absence of certain genes in the father. If this were true, the first generation offspring would inherit that from their mothers but not their fathers. However in the matings of the offspring of the first generation, the special mode of inheritance would no longer be observed. This is because both the male and female offspring of the first generation will have equal chance of possessing the gene responsible for that trait from their mother. Thus, second generation offsprings will not necessarily inherit the characteristic from their mothers as opposed to the fathers.

Let us apply our knowledge of gamete formation to the problem. We know that spermatogenesis gives rise to sperm which contain little or no cytoplasm. Oogenesis produces large eggs with enormous amounts of cytoplasm. A small but significant amount of DNA may be present in the cytoplasm. This DNA may carry genes that are not carried on the DNA of the chromosomes. Such genes are called extrachromosomal genes. A male is unable to transmit the traits of such genes, even if he possesses them in his body cells, because his gametes, the sperm, are not formed with any substantial amount of cytoplasm. The female, because of the large amounts of cytoplasm incorporated in the formation of her gamete, the egg, transmit the gene.

Thus, we see that an extrachromosomal trait could only be transmitted by the mother. It can be expressed in both male and female offspring, but only the females will be able to transmit the trait to the next generation.

THE LAW OF INDEPENDENT SEGREGATION

• PROBLEM 25-32

Soon after the Mendelian laws became firmly established, numerous exceptions to Mendel's second law, the Law of Independent Segregation, were demonstrated by experiments. Parental non-allelic gene combinations were found to occur with much greater frequencies in offspring than were the non-parental combinations. How can this be explained?

Solution: The Law of Independent Segregation states that when two or more pairs of genes are involved in a cross, the members of one pair segregate independently of the members of all other pairs. This law was substantiated by Mendel's experiments with sea plants. However, in Mendel's time, the physical nature of genes was not known, nor was it known how they are carried. When it was learned that chromosomes are the bearers of genes, the reasons why Mendel's law was both supported by some experiments and negated by others became obvious.

It was seen that the chromosomes are relatively few in number. For example, Orosophila have only four pairs of chromosomes. Man has twenty-three pairs. In comparison, however, the number of genes possessed by each species is very large, often in the thousands. Since there are so many more genes than chromosomes, and the chromosomes carry the genes, it follows that there must be many genes on each chromosome. And since it is now known that it is whole chromosomes which segregate independently during meiosis, gene separation can only be independent if the genes in question are on different chromosomes. Genes located on the same chromosomes are physically forced to move together during meiosis. Such genes are said to show linkage.

When genetic experiments are performed using genes that are linked, very different ratios from the expected Mendelian ratios are obtained. Genes that were linked together on parental chromosomes tend to remain together in the gametes, and so occur in conjuction with one another more frequently in the offspring than they would if they had segregated independently.

Exactly how linkage produces these variant ratios will be dealt with in later problems. At the present we can see that linkage explains how exceptions to the Law of Independent Segregation could possibly occur.

• PROBLEM 25-33

In Drosophilia, black body and vestigial wings are both recessive traits, while gray body and normal wings are dominant. How would you determine if the genes for black body and vestigial wings are on the same chromosome? Show the expected results if they are on the same chromosome, and the results if they are on different chromosomes.

Solution: To determine whether or not the genes for body color and wing type are on the same chromosome, we can do a testcross between an individual heterozygous for both traits and a double recessive individual. A testcross with a double recessive allows the genotype of the other parent to be expressed in the offspring in the same proportions as the gametes produced by that parent. This is because the double recessive parent has no genes that will obscure the genotype of the other parent. We use a heterozygote to determine if linkage is present because such a parent carries all the genes involved. If the offspring of the testcross give a 1:1:1:1 phenotypic ratio, then the genes have assorted independently and are not linked. If they give some other ratio, then linkage is indicated, and they are probably located on the same chromosome. This can best be illustrated by doing the crosses.

Let G be the gene for gray body, g be the gene for black body, N be the gene for normal wings, and n be the gene for vestigial wings. We can obtain the heterozygote by crossing a known homozygous dominant with a homozygous recessive. Then we do the test cross. If the genes are not linked we have:

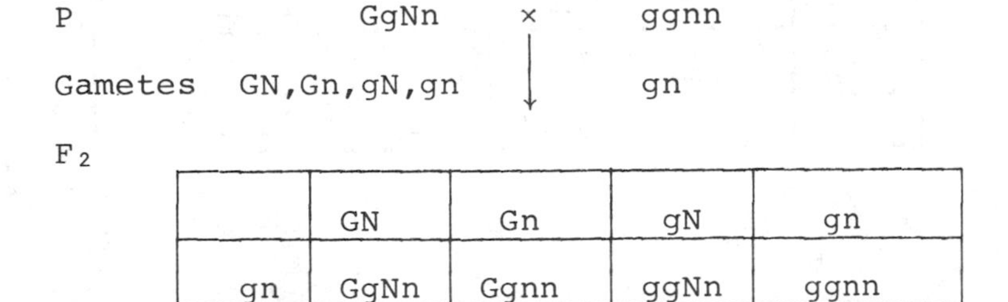

P GgNn × ggnn

Gametes GN, Gn, gN, gn ↓ gn

F_2

	GN	Gn	gN	gn
gn	GgNn	Ggnn	ggNn	ggnn

Phenotypically:

GgNn is gray-normal,
Ggnn is gray-vestigial,
ggNn is black-normal, and
ggnn is black-vestigial.

According to the Law of Independent Segregation, the gametes formed from the GgNn parent (GN, Gn, gN, gn) are in equal numbers. Thus the proportion of different F_2 offspring are equal; that is, 1:1:1:1.

If the genes are located on the same chromosome, the heterozygous parent will have the genotype $\frac{GN}{gn}$.

(Note: the genes are written on a bar $\underline{GN}$ to show linkage). Since the genes are linked, there is a chance that crossover will occur between them during meiosis. The heterozygous parent can thus form four kind of gametes in the following manner:

$\underline{G\ N}$		$\underline{G\ N}$	- parental type
$\underline{G\ N}$		$\underline{G\ n}$	- recombinant
$\underline{g \times n}$	————>	$\underline{g\ N}$	- recombinant
$\underline{g\ n}$		$\underline{g\ n}$	- parental type

Note that since their presence in the gametic population depends on the occurence of a crossover event, the frequency of the recombinants (those arising from cross-over)will be lower than that of the parental types (those having the same gene combination as the parent).

In the test cross we obtain:

P $\quad \frac{G\quad N}{g\quad n} \quad \times \quad \frac{g\quad n}{g\quad n}$ ♀

Gametes $\quad \underline{G\quad N}$, $\underline{G\quad n}$, $\underline{g\quad N}$, $\underline{g\quad n}$ $\qquad \underline{g\quad n}$

F_2

	$\underline{G\ N}$	$\underline{G\ n}$	$\underline{g\ N}$	$\underline{g\ n}$
$\underline{g\ n}$	$\frac{G\ N}{g\ n}$	$\frac{G\ n}{g\ n}$	$\frac{g\ N}{g\ n}$	$\frac{g\ n}{g\ n}$

Phenotypically:

$\frac{G\ N}{g\ n}$ is gray-normal,

$\frac{G\ n}{g\ n}$ is gray-vestigial,

$\frac{g\ N}{g\ n}$ is black-normal, and

$\frac{g\ n}{g\ n}$ is black-vestigial.

Although the same phenotypes are produced in this cross as in the cross between genes on separate chromosomes, the ratios will not be the same. Because their frequency as gametes is much higher than the recombinant-types, the parental types (gray-normal and black-vestigial) will occur with much higher frequency in the offspring.

The frequency of the recombinant types will depend on the distance between the two genes. If the genes are far apart, there is more chance of a crossover event between them, and the occurence of recombinants will be higher than if the genes were very close together on the chromosome. Also, since the formation of one recombinant type necessitates that the other type also be formed (refer to the diagram of the crossover), their frequencies will be about equal, as will the frequencies of the two parental types.

Thus whenever the observed ratio in a test cross varies from the predicted Mendelian ratio, and resembles the ratios outlined above, linkage is usually indicated.

GENETIC LINKAGE AND MAPPING

• PROBLEM 25-34

In Drosophilia, the dominant gene G codes for gray body color and the dominant gene N codes for normal wings. The recessive alleles of these two genes result in black body color (g) and vestigial wings (n) respectively. Flies homozygous for gray body and normal wings were crossed with flies that had black bodies and vestigial wings. The F_1 progeny were then test-crossed, with the following results:

Gray body, normal wings	236
Black body, vestigial wings	253
Gray body, vestigial wings	50
Black body, normal wing	61

Would you say that these two genes are linked? If so, how many map units apart are they on the chromosomes?

Solution: To determine if the two genes are linked, we look at the F_2 progeny. We notice that there are, two large groups (236, 253) which are approximately equal, and two small groups (50, 61) which are also nearly equal. We can reduce these numbers to small whole number ratios:

gray body, normal wings	236/50	roughly equals	5
black body, vestigial wings	253/50	roughly equals	5
gray body, vestigial wings	61/50	roughly equals	1
black body, normal wings	50/50	roughly equals	1

Since the heterozygous F_2 parent (the result of a cross from two homozygous parents) is crossed with a double recessive, we would expect a 1:1:1:1 phenotypic ratio among the offspring (see previous problem). However, the ratio here is 5:5:1:1. This significant departure from the 1:1:1:1 ratio indicates that linkage is indeed likely. The fact that the ratio divides the progeny into four groups, two large and of equal size and two small and of equal size, is also a typical result of a cross involving linked genes.

To calculate the distance between the genes, we will have to rely on the fact that the frequency of crossing over depends on this distance. It is logical to assume that the farther apart two genes are on a chromosome, the more likely it is that a crossover will occur, because there is a larger possible region in which it can occur. The process of locating genes on chromosomes is called mapping.

The percentage of crossing over or recombination gives us no information about the absolute distances between the genes but it does give us relative distances between them. Suppose, for example, that genes A and B have a recombination frequency of 20% and genes A and C have a frequency of 40%. From this data, we cannot tell the exact distance of B from A or C from A but we can say that C must be twice as far from A as B is from A, since twice as much crossing over occurred in the distance between C and A. Instead of having to compare two distances every time we want to refer to the separation of two genes, we have, by convention, established a standard measurement known as a map unit. A map unit is defined as the distance on the chromosome within which a crossover occurs one percent of the time. When genes on a chromosome are allocated their respective positions on that chromosome, we can obtain what we call a genetic map.

To obtain the map distance between two genes, we have to know the frequency of crossing over between them. Recombination frequency is defined as the ratio of the number of recombinants to the whole progeny, that is,

$$\text{recombination frequency} \overset{(RF)}{=} \frac{\text{number of recombinants}}{\text{total number of offspring in that generation containing the recombinants}}$$

Because recombinants can only result from a crossover event, their numbers in the population will be small for closely linked genes. The recombinants in this problem are found in the two smaller groups, and their total is 50 + 61 or 111. The total number of progeny is 236 + 253 + 50 + 61 or 600. Therefore:

$$RF = \frac{\text{number of recombinants}}{\text{total number of progeny}} = \frac{111}{600} = 0.185.$$

Since the map units are expressed as a percentage of recombination, a RF of 0.185 is the same as 0.185×100 or 18.5 map units.

Hence a map distance of 18.5 map units separates the gene for body color from the gene for wing size (See Fig.)

G N

| <—18.5 map units—> |

The recessive gene s produces shrunken endosperm in corn kernels and its dominant allele S produces full, plump kernels. The recessive gene c produces colorless endosperm and its dominant allele C produces colored endosperm. Two homozygous plants are crossed, producing an F_1 which are all phenotypically plump and colored. The F_1 plants are test-crossed with homozygous recessive plants and produce the following progeny:

shrunken, colorless	4035
plump, colored	4032
shrunken, colored	149
plump, colorless	152

a) What were the phenotypes and genotypes of the original parents? b) How are the genes linked in the F_1? c) Calculate the map distance between the two gene loci.

Solution: In solving this problem we will apply principles that have been demonstrated in earlier problems.

a) The genotypes and phenotypes of the parents can be determined by looking at the F_1 and F_2. We are told that all the F_1 offspring are plump and colored. This indicates that the genotypes of these offspring are either SSCC or SsCC, because they show the dominant trait. Now we must decide which one of the two alternatives is correct. This is where the F_2 comes in. If the F_1 plants were SSCC, only offspring expressing the dominant trait could have been produced in the F_2, since the other parent in the test cross was a homozygous recessive. All offspring would have been heterozygous, and in such a state would have expressed only the plump and colored traits. However, four phenotypes were actually produced in the F_1 , including the homozygous recessive (shrunken, colorless). Therefore, the F_1 parent had to have carried a copy of both types of recessive genes. The genotype of the F_1 is then SsCc.

Since we are told that both parents of the F_1 offspring were homozygous, their genotypes must be either SSCC and sscc, with phenotypes plump, colored and shrunken, colorless respectively or SScc and ssCC with phenotypes plump, colorless and shrunken, colored respectively. We illustrate this with a diagramatic cross:

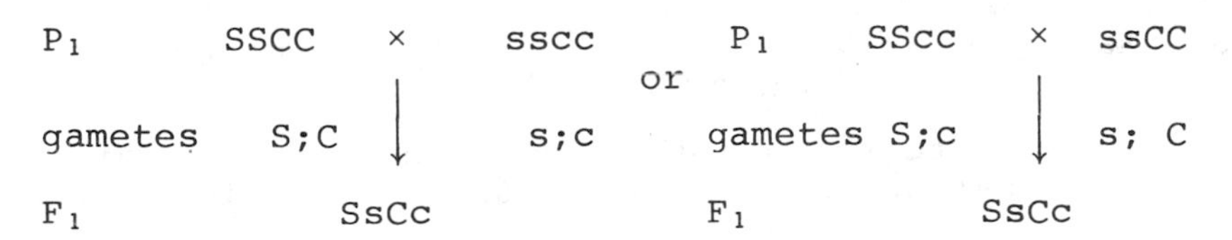

b) Now, looking at the ratios of offspring in the F_2 generation, we see that there is a significant departure from the expected 1:1:1:1 ratio when a test cross of a heterozygote with a recessive homozygote is performed.

This is indicative of crossover. (See previous problems for more detailed explanation of this and other concepts of linkage).

Looking at the ratios in the F_2 generation, we see that the larger groups, which represent the parental linkage combinations, are plump, colored and shrunken, colorless. Each of these separate combinations represents the combination of genes on the parental chromosomes. Therefore, the parental genes must have been linked $\underline{S\ C}$ and $\underline{s\ c}$. We have determined the F_1 genotype to be SsCc. Therefore, in the F_1, the genes are linked $\frac{S\ C}{s\ c}$.

In order to prove this, let us look at how this configuration could indeed give rise to the progency in F_2.

In meiosis, $\frac{S\ C}{s\ c}$ can produce four types of gametes if crossover occurs:

$$\begin{array}{ccl} \underline{S\ C} & & \underline{S\ C} \text{ - parental type} \\ S\ C & & \underline{S\ c} \\ \times & \longrightarrow & \qquad \text{recombitants} \\ s\ c & & \underline{s\ C} \\ s\ c & & \underline{s\ c} \text{ - parental type} \end{array}$$

The homozygous recessive $\left(\frac{s\ c}{s\ c}\right)$ can only produce gametes of the $\underline{s\ c}$ type.

In the cross, then:

$$P_2 \quad \frac{S\ C}{s\ c} \times \frac{s\ c}{s c}$$

$$\text{gametes} \quad \underline{S\ C},\ \underline{S\ c},\ \underline{s\ c} \qquad \underline{s\ c}$$

$$F_2 \quad \frac{S\ C}{s\ c} \qquad \frac{S\ c}{s\ c} \qquad \frac{s\ C}{s\ c} \qquad \frac{s\ c}{s\ c}$$

plump, colored (parental type) — plump, colorless; shrunken, colored (recombinants) — shrunken, colorless (parental type)

c) To calculate the map distances, we first determine the recombination frequency (RF) between the two genes:

$$RF = \frac{\text{number of recombinants}}{\text{total number of progeny}} = \frac{149 + 152}{149+152+4035+4032}$$

$$= \frac{301}{8368} = 0.036$$

An RF of 0.036 would give a map distance of 0.036 × 100 or 3.6 map units. Hence the map distance between the locus for shape and the locus for color is 3.6 map units apart.

```
S                                    C
|____________________________________|

      <——3.6 units ——>
```

• **PROBLEM 25-36**

In corn, the gene R for red color is dominant over the gene r for green color. The gene N for normal seed is dominant over the gene n for tassel seed.

A fully heterozygous red plant with normal seeds was crossed with a green plant with tassel seeds, and the following ratios were obtained in the offspring: 124 red, normal: 77 red, tassel: 126 green, tassel: 72 green, normal. Does this indicate linkage? If so, what is the map distance between the two loci?

Solution: Let us look at the cross.

```
P             RrNn          ×        rrnn
         (red, normal)      |    (green, tassel)
                            |
                            ↓
Gametes       RN, Rn,                r n
             rN, rn
```

F

	RN	RN	rN	rn
rn	RrNn	Rrnn	rrNn	rrnn

124 red, normal (RrNn) - parental
77 red, tassel (Rrnn) - recombinant
73 green, normal (rrNn) - recombinant
126 green, tassel (rrnn) - parental

The phenotypic ratio is indeed indicative of linkage. The two parental types (RrNn and rrnn) are found in much greater numbers than the recombinant types (Rrnn and rrNn). The recombination frequency (RF) is:

$$RF = \frac{\text{total recombitants}}{\text{total offspring}}$$

$$= \frac{73 + 77}{124+77+126+73} = \frac{150}{400}$$

$$= .375$$

An RF of .375 would give a map distance of 0.375 × 100 or 37.5 map units. The two loci are thus located on the same chromosome and are 37.5 map units apart.

• PROBLEM 25-37

Suppose genes A and B are 14.5 map units apart. Another gene, C, linked to these, is found to cross with gene B, 7 percent of the time. Is this data sufficient to determine the exact order of the three genes? If not, what other information is needed to order the genes?

Solution: A good way to order genes is to do it visually. We are told that genes A and B are separated by a map distance of 14.5 units. Translating this into a genetic map, we get:

A B

← 14.5 units →

We also know that C recombines with B, 7 percent of the time, which means that the map distance between B and C is 7 map units. On the genetic map, this fact becomes:

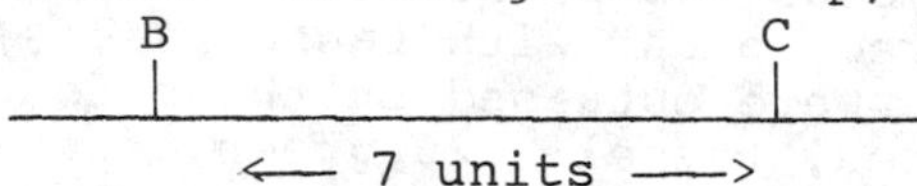

Combining the three genes on a genetic map, we can have:

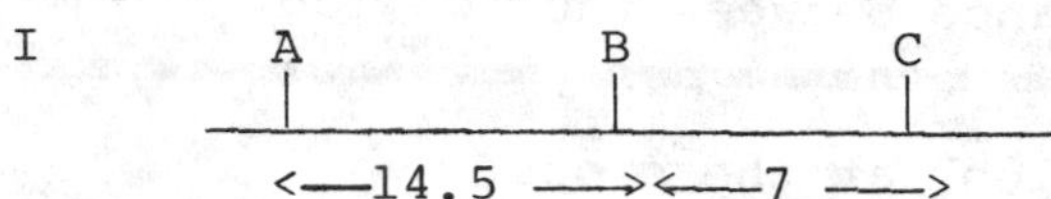

Or we can have:

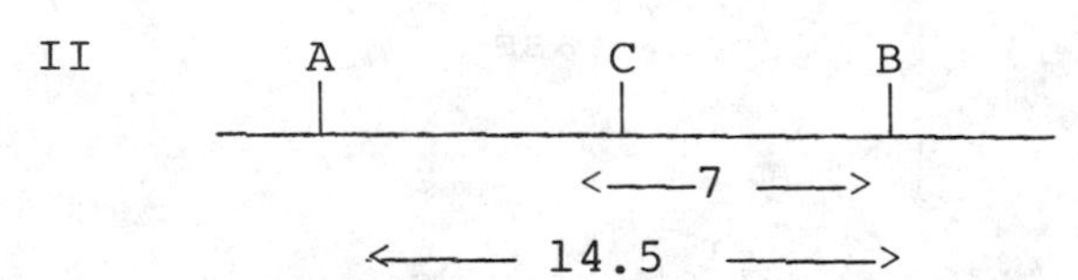

It is important to realize that the genes are probably not ordered CAB because the distance from B to C is less than the distance from B to A.

Thus both I and II are equally probable if we depend only on the data given. To decide which one of the two is correct, we would have to know the map distance between A and C. If this distance is greater than 14.5 map units, (the distance between A and B) then we know that I is correct.

If the distance between A and C is less than 14.5 map units (i.e., 7.5 m.u.), then we know that II is correct. In this way, the exact sequence of the three genes can be determined.

• PROBLEM 25-38

The cross-over percentage between linked genes A and B is 40%, between B and C, 20%, between C and D, 10%; between A and C, 20%; between B and D, 10%. What is the sequence of genes on the chromosome?

Solution: This question allows us further practice in ordering genes. Again we rely on the visual method because it is the most convenient. From the data, we know that A and B are 40 map units apart (recall that map units are directly proportional to cross-over percentage). B and C are 20 units apart. In order to determine whether C is to the right or left of B, we look at the map distance between A and C. This distance is 20 units. With this information, we now know that C is to the left of B because this is the only way that A and C can be 20 units apart.

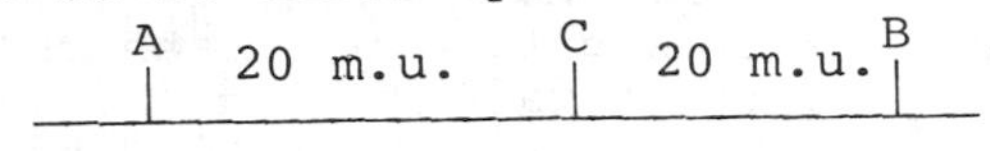

<——— 40 m.u. ———>

If C were to the right of B, then the distance between C and A would be 60 units, which is not the case.

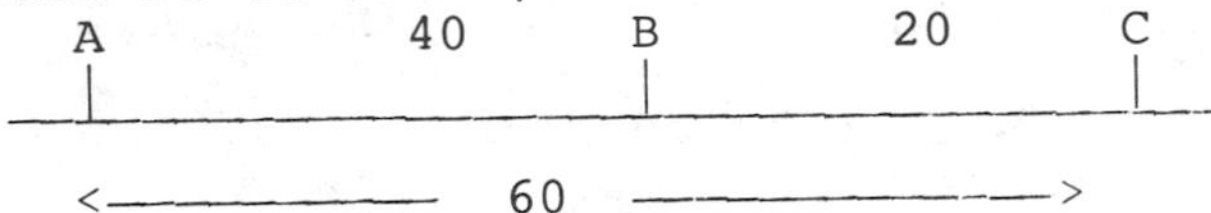

C and D are 10 units apart. We must determine whether D is to the right or left of C. We know that B and D are also separated by 10 units. Since B and C are 20 units apart, D must be between B and C.

A 20 m.u. C 10 m.u. D 10 m.u. B

<——— 40 m.u. ———>

Note: when ordering genes, we have first to establish two points (genes) on the chromosome. Using these as reference points, we make comparisons with other genes. In the case above, our reference points were A and B and we then proceeded to compare these points with C and D. Although it is helpful to pick the two most widely separated points as reference points, the choice is actually completely arbitrary. We could just as well have started with B and D as our reference points. The answer obtained would still be the same.

• PROBLEM 25-39

Two true-breeding varieties of unicorns differ with respect to three traits. The first has a straight horn, a green coat, and a straight mane. The other variety has a twisted horn, a blue coat, and a curly mane. When a male of either variety is crossed with a female of the other, all the progeny have straight horns, green coats, and straight manes. These hybrid offspring are then crossed to unicorns with twisted horns, blue coats, and curly manes. This cross produces offspring in the following numbers:

straight horn, green coat, straight mane	855
twisted horn, blue coat, curly mane	855
twisted horn, green coat, straight mane	95
straight horn, blue coat, curly mane	95
straight horn, blue coat, straight mane	50
twisted horn, green coat, curly mane	50
	2000

Construct a linkage map for the three genes involved in this cross.

Solution: This is a complicated problem on linkage. Go over the previous problems on recombination. Some new concepts will be introduced here.

This cross involves three traits (that is, a three-point cross) each governed by a gene. To begin, we have to determine the dominance or recessiveness of the phenotypes. This can be obtained from the F_1 since all the F_1 descendants have straight horn, green coat, and straight mane, the gene for straight horn (H) is dominant over twisted horn (h); green coat (C) is dominant over blue coat (c); straight mane (M) over curly mane (m). Since the parents are true-breeding, one of them is $\frac{H\ C\ M}{H\ C\ M}$ and the other $\frac{h\ c\ m}{h\ c\ m}$; The actual sequence of the genes has yet to be determined. In the cross between two homozygous parents,

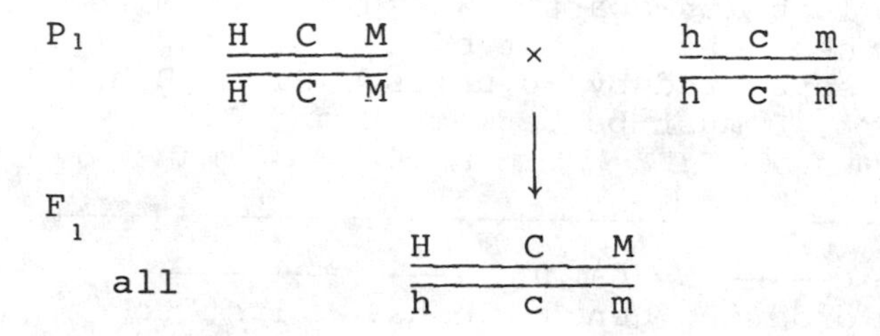

Note that all the dominant genes are on one chromosome while all the recessive ones are on the other. These are the parental types HCM and hcm. All other combinations of genes are recombinants.

To calculate the recombination frequency we have to know the number of recombinants. There are several ways to do this; one of them begins by translating the F_2 phenotypes to genotypes keeping the F_2 progeny in the same order we get:

1	H	C	M	855
2	h	c	m	855
3	h	C	M	95
4	H	c	m	95
5	H	c	M	50
6	h	C	m	50

Consider a cross-over between H and C. The number of recombinants is 95 3 + 95 4 + 50 5 + 50 6 = 290.

The RF between H and C = $\frac{290}{2000} = 0.145$.

An RF of 0.145 gives $0.145 \times 100 = 14.5$ map units, meaning H and C are 14.5 map units apart.

Consider a recombination between C and M . By similar reasoning, the number of recombinants

= 50 5 + 50 6 = 100

$$RF = \frac{100}{2000} = 0.05$$

That is, a map distance of 5 units separates C and M.

Finally, consider a cross-over between H and M . The number of recombinants

$$= 95\ 3 + 95\ 4 = 190$$

$$RF = \frac{190}{2000} = 0.095$$

That is, H and M are 9.5 map units apart.

Now that we have calculated a map distance for any two of the three genes, we are ready to plot the linkage map. As our reference points we use H and M , which are 9.5 units apart.

H M

<—— 9.5 m.u.——>

We know that H and C are 14.5 units apart. But C can be to the right or left of H. To determine which side of H,C is on, we look at the distance between C and M , which is 5 units. This means C must be on the right side of H.

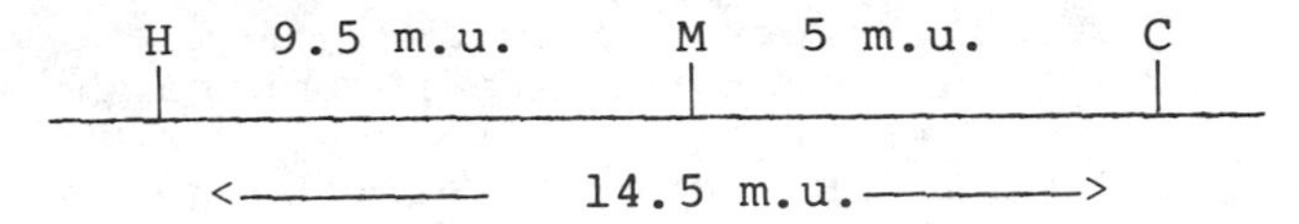

If C were 14.5 units to the left of H, then C and M would be separated by 24 units: This is not the case.

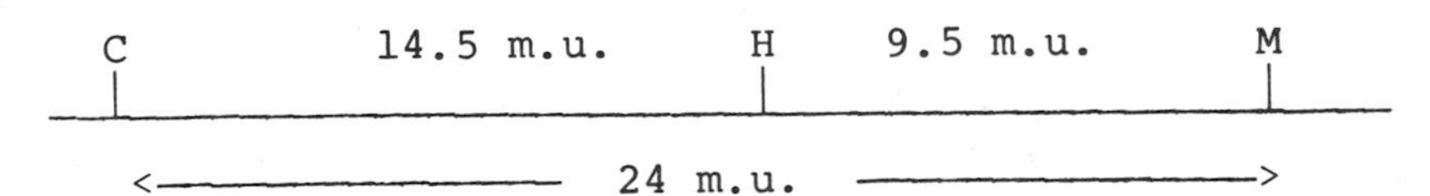

Therefore, the correct linkage map for the three genes is

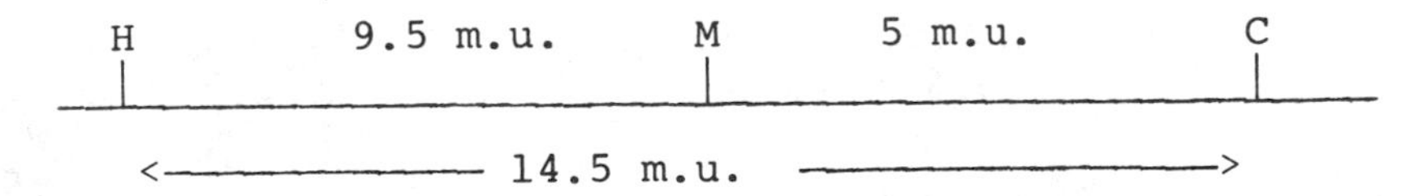

• **PROBLEM 25-40**

An experimenter crossed a stock of corn homozygous for the linked recessive genes colorless (c), shrunken (s) and waxy (w) to a stock homozygous for the dominant alleles of these genes (C, S, W) and then backcrossed the F_1 plants to the homozygous recessive stock. The progeny were as follows:

	Phenotype (all are -c-s-w)	Number
1	C S W	17,959
2	c s w	17,699
3	C s w	2,009
4	c S W	2,024
5	C S w	4,455
6	c s W	4,654
7	C s W	288
8	c S w	312
		49,400

Draw a linkage map for the three genes.

Solution: In order to draw a linkage map for the three genes, we must determine the distances between the genes in map units and the order of the genes on the chromosome.

Let us first determine the parental type. We know that since the F_1 plants resulted from a cross between a homozygous dominant (CC,SS,WW) and a homozygous recessive (cc,ss,ww) parent, they are heterozygous (CcSsWw). They were crossed with homozygous recessive plants (ccssww) Therefore the parental types are CcSsWs (1) and ccssww (2). The other six phenotypes are recombinants.

To determine the distances between the genes, let us determine the RF for two genes at a time. Consider a cross-over anywhere between c and S. The number or recombinants (from F_2) = 2009 3 + 2024 4 + 288 7 + 312 8 = 4633.

$$RF = \frac{4633}{49400}$$

$$= 0.094$$

An RF of 0.094 gives a separation of 9.4 units between C and S.

Consider a cross-over between S and W. The number of recombinants = 4455 5 + 4654 6 + 288 7 + 312 8 = 9709.

$$RF = \frac{9709}{49400}$$

$$= 0.196$$

That is, S and W are 19.6 units apart.

Consider a cross-over between C and Wx. The number of recombinants = 2009 3 + 2024 4 + 4455 5 + 4654 6 = 13142.

$$RF = \frac{13142}{49400}$$

$$= 0.266$$

That is, C and W are 26.6 map units apart.

To construct a linkage map, we choose randomly C and Wx as our reference points. S is 19.6 units away from w. The only way s can be 9.4 units away from C is for s to lie between C and W. On a genetic map, we have:

C 9.4 m.u. S 19.6 m.u. W

<———— 26.6 m.u. ————>

However, the sum of the map distances between C and S, and between S and W = 9.4 + 19.6 or 29.0 unit. This sum is greater than the distance between C and W (26.6 units). How do we explain this discrepancy?

The explanation lies in the fact that more than one cross-over can occur between two genes, expecially if they are far apart. In this case, there is a cross-over between C and S and between S and W making two cross-overs between C and Wx. Diagrammatically, this can be represented as follows:

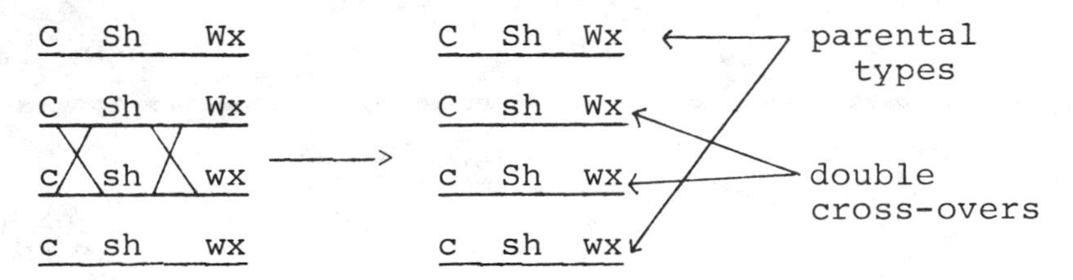

The result of two cross-overs between two genes is called a double recombinant. Since single cross-overs are quite infrequent, the frequency of double recombinants is even lower. The double recombinants are usually identified by their lowest frequency in the F_2.

From F_2, the number of double recombinants = 288 7 + 312 8 or 600.

$$\text{Percentage of} = \frac{\text{number of double cross-over}}{\text{total number of progeny}} \times 100$$

$$= \frac{600}{49400} \times 100$$

$$= 1.2\%$$

The cause of the discrepancy between the map distances can be traced to the fact that when we calculate the RF between C and S and between S and W, we include the number of double recombinants into the number of recombinants each time. (Refer to previous calculations). However,when we determine the RF between C and W, we neglect the double cross-overs because they do not affect the parental types CW and cw. To compensate for the double cross-overs, we have to add to the map distance between C and W twice the percentage of double cross-overs. This is because each double cross-over is the result of two separate events, a cross between c and S and a cross between S and W. In essence, four more crosses have occured between c and w that had not been considered before, having a total frequency of:

$$\frac{288}{49400} \text{ (between c and s)} \qquad \frac{288}{49400} \text{ (between s and w)}$$

$$+\frac{312}{49400} \text{ (between c and s)} \qquad +\frac{312}{49400} \text{ (between c and w)}$$

or 2 x 1. 2.

Thus the distance between c and w is 26.6 + (2 × 1.2) or 29.0 units. After the correction is made, everything falls in place.

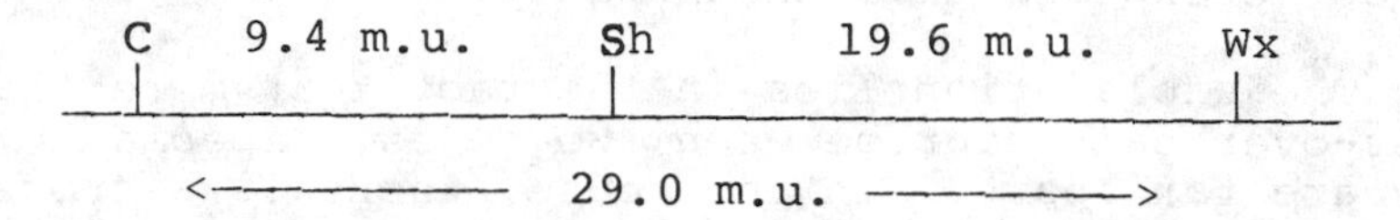

Note that the recombination frequency between each pair of genes is increased by the frequency of the double cross-overs.

• PROBLEM 25-41

In corn, a strain homozygous for the recessive genes a (green), d (dwarf) and r (normal leaves) was crossed to a strain homozygous for their dominant alleles A (red), D (tall), and R (ragged leaves). Offspring of this cross were then backcrossed to homozygous recessive plants. Listed below are the phenotypes produced from the backcross.

	Phenotype (all are -a-d-r)	Number
1	A D R	265
2	a d r	275
3	A D r	24
4	a d R	16
5	A d r	90
6	a D R	70
7	A d Rg	120
8	a D r	140
		1000

Propose a linkage map with distances between the three genes.

Solution: This problem is an application of the concepts discussed before. The F_1 parents of the cross arising from the mating of homozygous parents must be heterozygous $\left(\frac{ADR}{adr}\right)$. The parental types in the progeny are therefore 1 ADR and 2 adr. The rest are recombitants.

Consider a cross-over between A and D. From the F_2 the number of recombinants = 90 5 + 70 6 + 120 7 + 140 8 = 420.

$$RF = \frac{420}{1000}$$

$$= 0.42$$

This means that A and D are separated by 42 units.

Consider a cross-over between D and R. The number of recombinants = 24 3 + 16 4 + 120 7 + 140 8 = 300

$$RF = \frac{300}{1000}$$

$$= 0.30$$

That is, D and R are 30 map units apart.

Consider a cross-over between A and R. The number of recombinants = 24 3 + 16 4 + 90 5 + 70 6 = 200.

$$RF = \frac{200}{1000}$$

$$= 0.20$$

That is, there are 20 units between A and R.

A genetic map can be constructed by arbitarily making A and D our reference points. R is 30 units away from D. At the same time, R is also 20 units from A. These conditions allow only one arrangement of the genes; that is, with R between A and D.

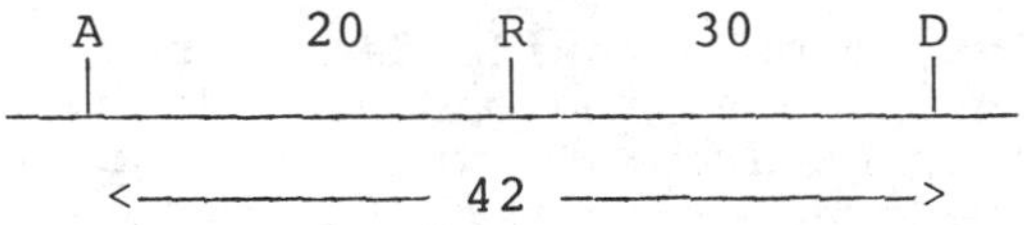

```
A        20        R        30        D
|                  |                  |
---------------------------------------
<---------------- 42 ---------------->
```

However, the sum of the distances between A and R and between R and D is greater than the distance between A and D. The reason why these distances on the map do not agree is because of the presence of double cross-overs. Double cross-overs can be identifiea by their lowest frequencies, namely 24 3 and 16 4, in the F_2. The percentage of double cross-overs

$$= \frac{24 + 16}{1000} = 4\%.$$

The map distance between A and D should be longer because we have ignored the double cross-overs in our calculation of it. The corrected distance between A and D should be $= 42 + (2 \times 4) = 50$ units. Using this distance, the new genetic map does not show any discrepancy.

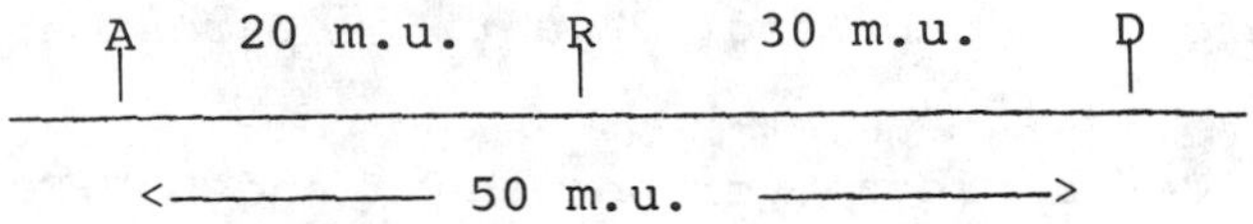

```
    A    20 m.u.    R     30 m.u.     D
    |               |                 |
 ________________________________________

      <---------  50 m.u.  --------->
```

• PROBLEM 25-42

The actual physical distances between linked genes bear no direct relationship to the map distances calculated on the basis of crossover percentages. Explain.

Solution: In certain organisms, such as Drosophilia, the actual physical locations of genes can be observed. The chromosomes of the salivary gland cells in these insects have been found to duplicate themselves repeatedly without separating, giving rise to giant bundled chromosomes, called polytene chromosomes. Such chromosomes show extreme magnification of any differences in density along their length, producing light and dark regions known as banding patterns. Each band on the chromosome has been shown by experiment to correspond to a single gene on the same chromosome. The physical location of genes determined by banding patterns gives rise to a physical map, giving absolute distances between genes on a chromosome.

Since crossover percentage is theoretically directly proportional to the physical distance separating linked genes, we would expect a direct correspondence between physical distance and map distance. This, however, is not necessarily so. An important reason for this is the fact that the frequency of crossing over is not the same for all regions of the chromosome. Chromosome sections near the centromere regions and elsewhere have been found to cross over with less frequency than other parts near the free end of the chromosome.

In addition, mapping units determined from crossover percentages can be deceiving. Due to double crossing over (which results in a parental type), the actual amount of crossover may be greater than that indicated by recombinant type percentages. However, crossover percentages are nevertheless invaluable because the linear order of the genes obtained is identical to that determined by physical mapping.

• PROBLEM 25-43

Can you distinguish between two gene loci located on the same chromosome that have 50 percent crossing over and two gene loci each located on different chromosomes?

<u>Solution:</u> When two linked genes have 50 percent crossing over between them, it means that 50 percent of the progeny will be recombinants. For example, a heterozygous parent genotypically $\frac{A\ \ B}{a\ \ b}$ will form four types of gametes if a cross-occurs between A and B:

A B	A B	- parental type
A B	A b	- recombinant
a b	a B	- recombinant
a b	a b	- parental type

Now if the recombinant types occurred 50% of the time, this means that they are produced in numbers equal to the parental types. This could happen only if crossover occurred between the genes during every meiosis in every individual. In other words, if the genes are so far apart that it is certain that crossover will occur between them, then in any meiotic event, recombinant types will always be formed, and in equal numbers with parental types. Therefore, the four types of gametes occur in a 1:1:1:1 ratio. When testcrossed, the progeny will also occur in a 1:1:1:1 phenotypic ratio.

This means that the ratio of the progeny from parents having linked genes showing 50% crossing over is indistinguishable from that of parents with genes that are not linked. Thus when genes are separated by 50 map units, it is impossible to differentiate between linkage and non-linkage.

Note that no more than 50% recombination is ever expected between any two loci because only two of the four chromatids are involved in crossing over. In fact, the ratio is usually less than 50% because crossover may not always occur and because double or multiple crossovers can reduce the apparent number of crossover events, as shown below.

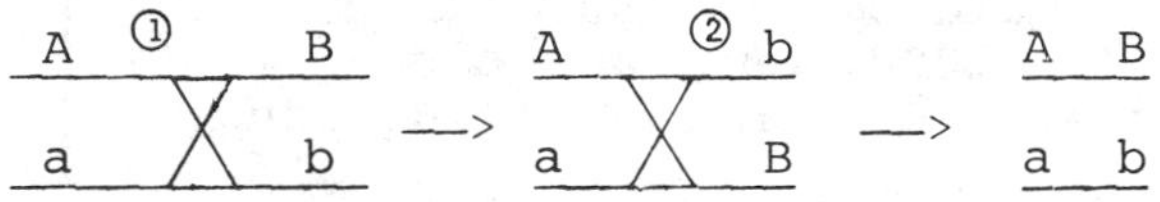

$$\frac{A \quad ① \quad B}{a \quad \times \quad b} \longrightarrow \frac{A \quad ② \quad b}{a \quad \times \quad B} \longrightarrow \frac{A\ \ B}{a\ \ b}$$

SHORT ANSWER QUESTIONS FOR REVIEW

Answer

To be covered when testing yourself

Choose the correct answer.

1. Gene duplication takes place during (a)telophase (b) anaphase (c) interphase (d) metaphase. — c

2. If the diploid number of an organism is 38, then (a) the first polar body will have 38 chromosomes, (b) a primary oocyte will have 19 chromosomes, (c) in mitosis, there will be 38 chromosomes present at metaphase, (d) a spermatid will have 19 chromosomes. — d

3. In an organism, pink spots is a sex-linked, recessive trait and black hair is dominant to white. If a pink spotted, black heterozygote female is mated to a white male which is not spotted, the phenotypic ratio of the male offspring would be (a) 1/4 pink spotted black: 1/4 pink spotted white: 1/4 unspotted black: 1/4 unspotted white, (b) 1/2 pink spotted black: 1/2 pink spotted white, (c) 3/4 pink spotted black: 1/4 pink spotted white, (d) 3/4 unspotted black: 1/4 pink spotted white. — b

4. If the parents of five children are both carriers of phenylketonuria the probability that they have 3 normal children and 2 affected is appoximately: (a) 0.26 (b) 0.51 (c) 0.76 (d) 0.04. — a

5. The shape and color of radishes are controlled by two independent pairs of alleles that show no dominance. The color may be red (RR), purple (RR') or white (R'R'), and the shape may be long (LL), oval (LL') or round (L'L'). Red, long radishes are crossed with white, round radishes and then the F_1's are allowed to interbreed. If 1600 F_2's are obtained, then the expected ratio of white offspring would be (a) 400 long: 800 oval: 400 round. (b) 1200 long: 400 round. (c) 50 long: 300 oval: 50 round. (d) 100 long: 200 oval: 100 round. — d

6. A black color coat in cocker spaniels is governed by a dominant allele, and red coat color by its recessive allele. A solid pattern is governed by a dominant independently assorting allele and a spotted pattern by a recessive allele. Two solid black puppies, two solid red pupples, one black and white puppy, and one red and white puppy are born when a solid black male is mated to a solid red female. It can be concluded that the genotype of (a) the father is BBSs, and the mother is bbSs.

Answer

To be covered when testing yourself

(b) the father is bbSs, and the mother is BbSs.
(c) the father is BbSs, and the mother is bbSs.
(d) the father is BbSS, and the mother is bbss. — c

7. If height is controlled by polygenetic inheritance in man, which of the following people has a genetic potential which would allow them to be taller than a person with the genotype Aa BB Cc Dd EE?
(a) Aa BB Cc Dd Ee (b) Aa bb cc DD EE (c) AA BB Cc DD Ee (d) aa BB cc DD ee — c

8. The cross-over percentage between linked genes J and M is 20%, J and L 35%, J and N 70%, L and K 15%, M and N 50%, and M and L 15%. Thus, the sequence of genes on this chromosome is:
(a) M, J, L, K, N (b) J, M, L, K, N (c) J, M, L, N, K (d) J, N, M, L, K — b

9. Which of the following is not a part of a prophase chromosome? (a) centromere (b) centrosome (c) chromatid (d) DNA — b

10. Human body cells usually have 46 chromosomes. During the anaphase stage of mitosis, a cell will have (a) 92 chromosomes, (b) 46 chromosomes, (c) 23 chromosomes, (d) 44 chromosomes. — a

11. Gametes are produced by (a) crossing-over, (b) mitosis (c) meiosis (d) conjugation. — c

12. In peas, yellow color is dominant to green. If a homozygous yellow plant is crossed with a green plant, then the expected phenotypic ratios of the offspring would be (a) 3/4 yellow: 1/4 green (b) all green (c) 3/4 green: 1/4 yellow (d) all yellow. — d

13. In certain types of dogs the allele for a spotted coat is recessive to that of a solid pattern. If 4% of the dogs in a given population of 10,000 are spotted, how many are expected to be heterozygous? (a) 4,800 (b) 9,600 (c) 3,200 (d) 800 — c

14. Usually a recessive sex-linked trait (a) will be expressed more often in males than in females (b) will be expressed more often in females than in males (c) will be expressed to the same extent in both males and females (d) will only be expressed in males. — a

15. In cats, a pair of genes for coat color is sex linked; gene B produces a yellow coat,gene b produces a black coat, and the heterozygous form produces a tortoise shell coat. If a yellow female is mated to a black male, then (a) all

Answer

To be covered when testing yourself

their offspring will have a tortoise shell coat (b) all their sons will have a tortoise shell coat, While 1/2 their daughters will be yellow and 1/2 will be tortoise shell. (c) 3/4 of their offspring will be black and 1/4 will be yellow (d) all their sons will be yellow and all their daughters will be tortoise shell.

d

Fill in the blanks.

16. The _____ is part of cell duplication in animal cells but not in plant cells.

cleavage furrow

17. In _____, the chromosomes are duplicated once, the cytoplasm divides once and two identical daughter cells are formed.

mitosis

18. The organism with the genotype, Aa BB Cc is expected to have _____ different types of gametes.

four

19. An animal which has a diploid number of twenty, will have _____ chromatids in its tetrad stage.

forty

20. In Drosophila, a dominant gene for a phenotype called "dichaete" alters the bristles, causes the wings to remain extended from the body while the fly is at rest and is lethal in its homozygote state. If two dichaete flies are crossed and 99 viable offspring are obtained, _____ of them are expected to have the dichaete trait.

66

21. In a dihybrid cross, the first loci's allelic relationship is dominant - recessive, the second loci's allelic relationship is codominant. Then the expected phenotypic ratio of the progeny is ___.

3:6:3:1:2:1

22. The spermatogonia of an organism with a haploid number of 21 has _____ chromosomes.

42

23. In Drosophila, a kidney bean shaped eye is governed by a recessive gene, k; cardinal eye color is produced by the recessive gene cd on the same chromosome; and another recessive allele, e, on the same chromosome produces ebony body color. Homozygous kidney, cardinal females were mated to homozygous ebony males and the trihybrid F_1 females were test crossed to produce the following F_2 progeny:

1761 kidney, cardinal	97 kidney
1773 ebony	89 ebony, cardinal
128 kidney, ebony	6 kidney, ebony, cardinal
138 cardinal	8 wild type

Draw a genetic map of this chromosome.

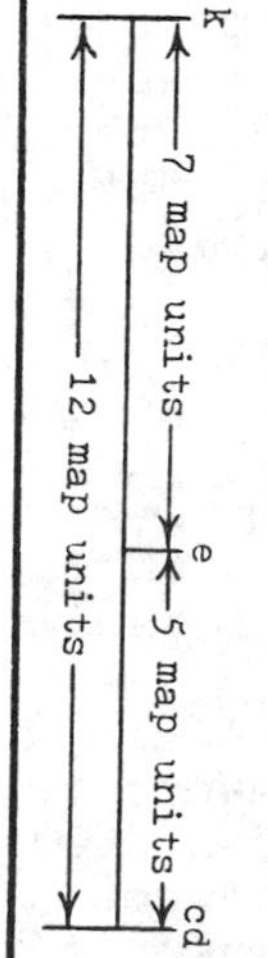

	Answer
	To be covered when testing yourself
24. In meiosis, the chromosomes line up in a random manner. This is called _____ and ensures genetic variation.	inde-pendent assortment
25. If 16% of a population of pea plants is green then the frequency of the yellow allele is _____ (green is recessive to yellow).	.6
26. A _____ is a strain of organisms that when inbred always produce offspring having the same phenotype.	pure line
27. The _____ chromosomes are non-homologous chromosomes which are not identical in size and shape.	sex
28. Chromosomes line up along the equatorial plate during the _____ stage of mitosis.	metaphase
29. To determine whether an organism is homozygous or heterozygous for a particular trait a test cross is performed in which the organism is crossed with a _____ organism for that trait.	homo-zygous recessive
30. The number of sex chromosomes in a human brain cell is _____.	two
Determine whether the following statements are true or false.	
31. Centromere duplication takes place during anaphase.	False
32. Oogenesis gives rise to one egg and two polar bodies.	False
33. If an allele is not dominant then it must be recessive.	False
34. In a family of six, the probability of one boy and five girls is 3/32,	True
35. In rabbits, color results from a dominant gene and albinism from its recessive allele. Black is the result of a dominant gene, brown of its recessive allele. Fully homozygous brown rabbits were crossed with albinos carrying the gene for black in the homozygous state. F_1 rabbits were crossed to double recessives. From many such crosses the offspring obtained were: 34 black, 66 brown, and 100 albino. From this it can be inferred that these traits are linked.	True
36. A woman whose blood type is A gives birth to a son with blood type O. She can rightfully claim that her husband whose blood type is AB is the father of her child.	False

Answer

To be covered when testing yourself

37. In turkeys, the gene for bronze coloration is dominant over the gene for red coloration. The gene for normal wings is dominant over the gene for hairy wings. In crosses between homozygous bronze, hairy birds and red, homozygous normal-feathered birds, the expected proportion of the F_2 bronze, normal-feathered progeny is 9/16. — True

38. An animal that has a diploid number of 24 has 12 chromatids in its secondary oocyte. — False

39. Karyokinesis defines mitosis in its strictest sense referring only to the division of nuclear material. — True

40. The diploid number of man is 46, there will therefore be 46 chromosomes in the anaphase of the second meiosis. — True

41. In humans, black hair is dominant over red, and having six fingers is dominant over having five fingers. Thus there is only one possible genotype for a red-haired, five fingered individual. — False

42. In the grasshopper, an egg fertilized by a sperm cell that lacks an X chromosome develops into a female. — False

43. If in a test cross, two of ten organisms showed the recessive trait it would mean that the organism being tested was a heterozygote. — True

44. One thing that proves that hereditary factors do not have to be either dominant or recessive are the A and B alleles of blood types. — True

45. The actual physical distances between linked genes bear a direct relationship to the map distances calculated on the basis of crossover. — True

CHAPTER 26

HUMAN INHERITANCE & POPULATION GENETICS

EXPRESSION OF GENES

• PROBLEM 26-1

In what ways are human beings not favorable for studies of inheritance?

Solution: In order to study inheritance, it is necessary that the phenotypic manifestations of the genes be clear cut and distinct. In this way, correlation of phenotypes and genotypes is direct. This is difficult to do with humans. Many human traits, such as height, are a result of the cooperative action of a number of a genes (polygenetic traits) which are difficult to identify or localize individually. Many broadly defined traits, such as hereditary deafness, can result from the manifestation of any one of a number of genes. Also, any one gene can often have a variety of phenotypic manifestations (pleiotropism), as can be seen in the widespread bodily effects that may result from the lack of a specific enzyme.

Ideally, genetic studies are conducted in controlled environmental situations. This is virtually impossible to do with human beings. Man's environment is far more complex and diversified than that of any other organism and in addition, he himself largely controls it. Unlike most other organisms, not only physiological and natural factors, but psychological and social factors, must be taken into account in any valid investigation.

Man also has a much longer life span and generation time than is feasible for genetic studies. His family sizes are far too small for the establishment of good genetic ratios and reliable statistics. Other organisms, such as bacteria and fruit flies, produce large numbers of offspring in relatively short periods of time, making them much more desirable subjects for genetic studies. Furthermore, ethical standards understandably prevent the use of humans for the establishment of standard genetic stocks and the controlled breeding that is a large part of the study of genetics.

These obstacles nonetheless do not totally prevent valid and useful human genetic studies. Modern statistical analysis has increased the reliability of human statistics. Pedigree analysis is still a valuable tool for geneticists. In addition, studies using identical twins are employed widely to control for the effect of the environment on the expression of genes. While organisms such as bacteria and fruit flies are by far the best subjects for genetic investigation, studies using human beings will always be essential for the understanding of inheritance in man.

• PROBLEM 26-2

In what ways are studies of twins useful in supplying information about the relative importance of inheritance and environment in the determination of a given trait?

Solution: Although both environment and heredity are involved in the development of any trait, a change in some aspect of the environment may alter one character relatively little as compared to its effect on another character. An individual's blood type, for example, seems fairly impervious to practically all environmental effects, while the phenotype of a diabetic can be radically changed by a mere alteration in diet. Moreover, not all characters are determined by simple genetic effects which have easily observable relationships with simple environmental changes. Some traits, such as intelligence in animals, are probably determined by complex genetic-environmental interactions, the results of which are then viewed as a single trait.

Performing studies using twins is one way of determining the relative contributions of genetic composition and environment to the expression of a given trait.

There are two kinds of twins: identical or monozygotic twins, which arise from a single fertilized egg, and fraternal or dizygotic twins, which result from the fertilization of two separate eggs. From this definition, we can see that monozygotic twins are genetically the same in every respect, while dizygotic twins may differ genetically for any character. By comparing both kinds of twins with regard to a particular trait, we can evaluate the roles of environment and heredity on the development of that character. On the one hand, we have genetically identical individuals of the same age and sex, raised in a single uterine environment; and on the other, we have genetically dissimilar individuals of the same age, though not necessarily of the same sex, raised in a common uterine environment. Presumably the difference between these two kinds of twins is only in the extent of their genetic similarity.

Using twins for genetic studies, then, provides built-in controls for both the effect of environment and the effect of heredity on the expression of a given trait. For example, if phenotypic similarities for a particular character are greater among identical twins than among fraternal twins, we can ascribe this to the genetic similarity of the identical pair and the genetic dissimilarity of the fraternal pair. On the other hand, if both identical and fraternal pairs show the same extent of phenotypic differences for a particular character, we can assume that genetic similarity or dissimilarity plays less of a role than the differences which may occur in the post-uterine environment of the twins. As a concrete illustration of this, consider a pair of monozygotic twins reared in different postnatal environments and a pair of dizygotic twins reared in a common postnatal environment. If the monozygotic twins were found to exhibit a wide range of dissimilarities and the dizygotic twins a significant range of similarities, the importance of the environment as a phenotypic modifier of the genetic composition of an individual can readily be appreciated.

• **PROBLEM** 26-3

Distinguish between penetrance and expressivity.

Solution: A recessive gene produces a given trait when it is present in the homozygous state. A dominant gene produces its effect in both the homozygous and heterozygous states. Geneticists, however, have found that many genes do not always produce their phenotypes when they should. Genes that always produce the expected phenotype in individuals who carry the gene in an expressible combination are said to have complete penetrance. If only 70 percent of such individuals express the character phenotypically, then the gene is said to have 70 percent penetrance. Penetrance is thus defined as the percentage of individuals in a population who carry a gene in the correct combination for its expression (homozygous for recessive, homozygous or heterozygous for dominant) and who express the gene phenotypically.

Some genes that are expressed may show wide variations in the appearance of the character. Fruit flies homozygous for the recessive gene producing shortening of the wings exhibit variations in the degree of shortening. Expressivity is defined as the degree of effect or the extent to which a gene expresses itself in different individuals. If it exhibits the expected trait fully, then the gene is said to be completely expressed. If the expected trait is not expressed fully, the gene shows incomplete expressivity.

The difference between the two terms-penetrance and expressivity, lies in the fact that the former is a function of the gene at the population level, while the latter varies on an individual level. Thus a given gene having a certain penetrance within a population may have varying expressivity in individuals of that population who express it.

Both penetrance and expressivity are functions of the interraction of a given genotype with the environment.

Changes in environmental conditions can change both the penetrance and expressivity of a gene. For example, a given gene may code for an enzyme required for the synthesis of a given metabolite in bacteria. If that metabolite is provided in the organisms nutrient environment, the organism might not produce the enzymes needed for its synthesis, and then the gene will not be expressed. If however, the nutrient is depleted from the media, the organism will begin to manufacture the enzyme, and thus express the gene. In humans, it is thought that allergy is caused by a single dominant gene; the different types of allergies are due to the varying expressivity of the gene, as a result of the interaction of the gene with both the environment and a given individual's genetic and physical make up.

PEDIGREES

• PROBLEM 26-4

Using the pedigree below, fill in the genotype of each individual. Assume the trait is recessive and an individual who marries into the family and does not exhibit a trait does not carry the recessive gene for it.

Left- handedness

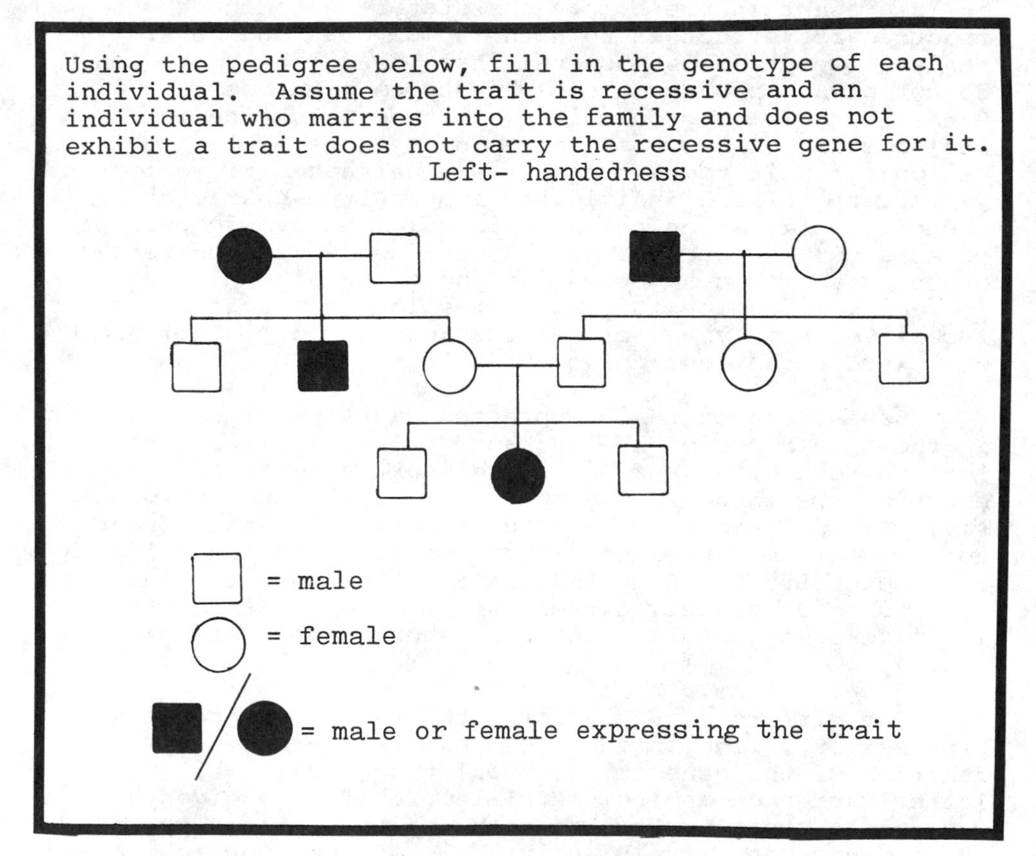

Solution: For simplicity and conciseness, we shall use the following system to refer to any one of the members in the pedigree:

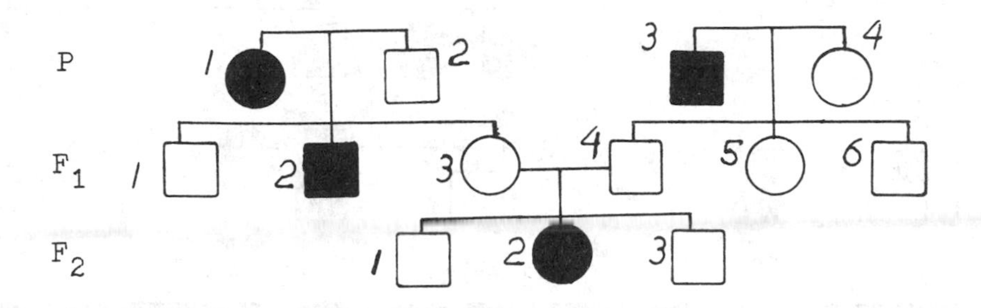

Let the allele for left-handedness be a and the allele for right-handedness be A. Consider P_1 (that is, the first member of the Parental generation). Since she expresses the trait, her genotype is aa. To determine the genotype of P_2 , we look at the progeny, F_1 (first generation). From the diagram, we see that $F_{1,2}$ is left-handed; hence his genetic make-up is aa. This means that he must have received a copy of a from each of his parents. Therefore we know that P_2 carries a copy of the recessive gene. But since P_2 is phenotypically normal, his genotype must be Aa. $F_{1,1}$ and $F_{1,3}$ are right-handed. Yet they must have received an a allele from P , since she has only a alleles to transmit. Therefore their genotype of $F_{1,1}$ and $F_{1,3}$ is Aa.

Consider the cross between P_3 and P_4. We know P_3 is aa (because he is left-handed) and P_4 can be AA or Aa (since she is left-handed). Since all the offspring are right-handed, it seems probable that P_4 is AA; however, we cannot be entirely certain of this, since both genotypes AA and A_2 are compatible with the phenotypes of the offspring. We know, however, that $F_{1,4}$, $F_{1,5}$, and $F_{1,6}$ must all have the genotype Aa since they carry an a allele donated by P_3.

In the F_2 generation, we see that $F_{2,2}$ is left-handed, and therefore her genotype is aa. This is compatible with what we have determined to be the genotypes of the parents ($F_{1,3}$ and $F_{1,4}$) namely Aa. $F_{2,1}$ and $F_{2,3}$ are both right-handed, and are either AA or Aa. Since both parents are Aa, either genotype is possible, though Aa is more probable.

• PROBLEM 26-5

Using the pedigree below, determine the method of inheritance of the trait and, as far as possible, fill in the genotypes of each individual. Assume that the trait is recessive.

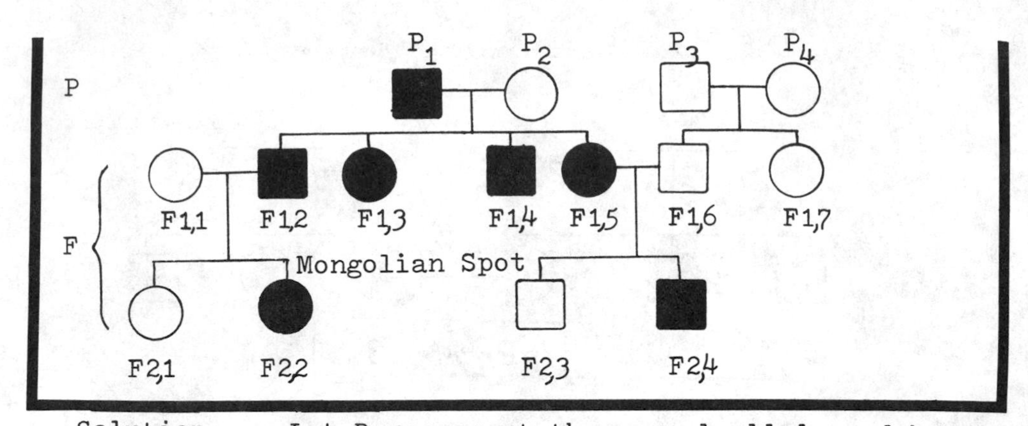

Solution: Let B represent the normal allele and b represent the allele for Mongolian spot. Using the system established in the previous question, we can say that P_1, $F_{1,2}$, $F_{1,3}$, $F_{1,4}$, $F_{1,5}$, $F_{2,2}$ and $F_{1,4}$ (see pedigree) are genotypically bb, because they all express the recessive Mongolian spot trait. Since all of the progeny of P_1 and P_2 carry the trait, the genetic make-up of P_2 who does not express the trait but must carry the recessive gene, is Bb. In the cross between $F_{1,1}$ and $F_{1,2}$, we see that one of their children ($F_{2,3}$) shows the trait. Since $F_{1,2}$ must have contributed one recessive gene, the other recessive gene must have come from $F_{1,1}$. But $F_{1,1}$ is normal, so her genotype must be Bb. $F_{2,1}$ is also Bb.

To determine the genotypes of F_3 and P_4, we have to examine the cross between $F_{1,5}$ and $F_{1,6}$. One of the offspring in this cross ($F_{2,4}$) shows Mongolian spots. This implies that both parents must have a copy of b. We already know that $F_{1,5}$ is bb. Therefore $F_{1,6}$ must be Bb, since he is normal $F_{2,3}$ is also Bb because he can receive only the recessive allele from $F_{1,5}$. Since $F_{1,6}$ is heterozygous, one of his parents must have allele b. We cannot assume which parent carries the b allele, because the information is not available for us to be certain. Conceivably, both parents can be Bb, the number of their progeny being too small to necessarily include a homozygous recessive. Or one parent may be BB, in which case all their offspring would show the normal phenotype. The genotype of $F_{1,7}$ can thus be either BB or Bb.

We can summarize the genotypes of the individuals in the following diagram:

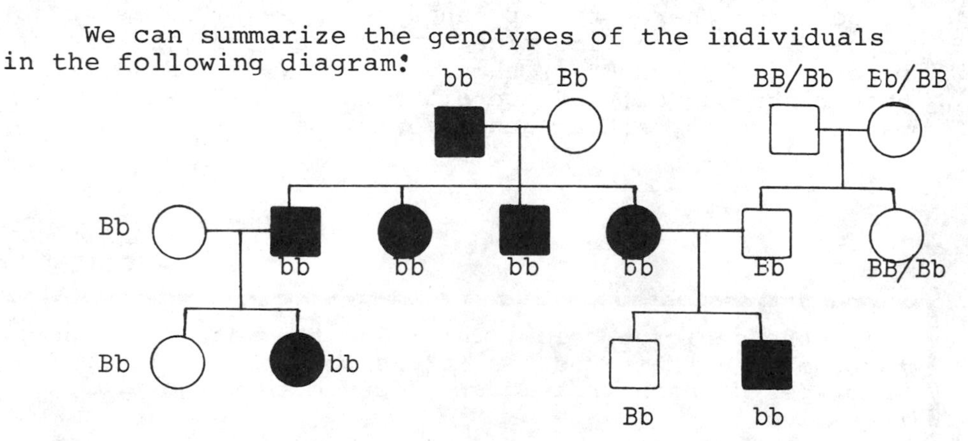

• PROBLEM 26-6

Using the pedigree below, determine the method of inheritance of the trait, and as far as possible, fill in the genotypes of each individual.

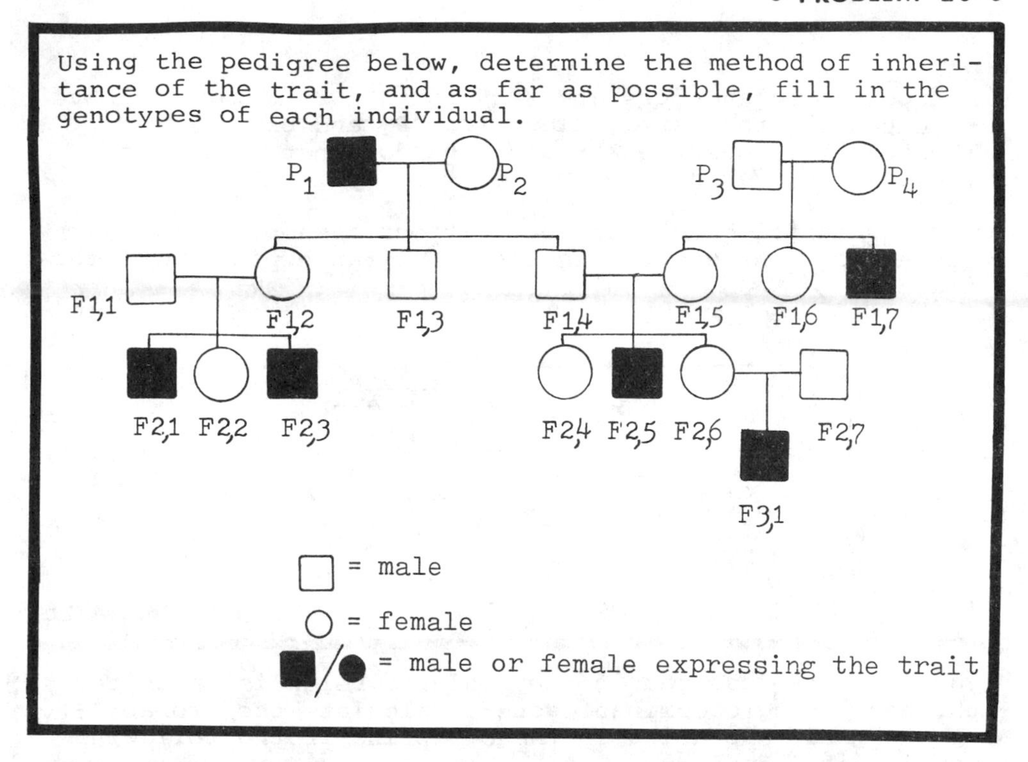

Solution: One should immediately note that the trait is only exhibited by male members of the family. Although the number of individuals is small enough that this could have been a a chance occurrence, it strongly suggests that the trait is sex-linked. Proceeding on this assumption, we can deduce the possible genotype of each member.

Letting r represent the recessive, sex-linked allele, we can assign the genotype X^rY to P_1, $F_{1,7}$, $F_{2,1}$, $F_{2,3}$, $F_{2,5}$, and $F_{3,1}$, since all are males expressing the trait. Since $F_{2,7}$, a male, does not express the trait, and therefore does not carry the r allele, his genotype is X^RY, where R represents the normal allele. We know that $F_{3,1}$ must have received the r allele from his mother, $F_{2,6}$ since to a male offspring, the mother contributes the X while the father contributes the Y chromosome. The genotype of $F_{2,6}$ is then X^RX^r, since she does not express the trait. $F_{1,5}$ must also be a carrier (X^RX^r) since $F_{2,6}$ is a carrier, $F_{2,5}$ expresses the trait, and $F_{1,4}$, who must be X^RY, could not have donated the recessive allele to his offspring. Since we have no information concerning the offspring of $F_{2,4}$, her genotype could be either X^RX^R or X^RX^r, depending on whether she received a dominant or recessive allele from $F_{1,5}$.

Since $F_{1,5}$ is a carrier, and her father P_3, who does not express the trait, is X^RY, we can assign P_4 the genotype X^RXr, which shows that she is also a carrier. $F_{1,6}$ can be either X^RX^R or X^RX^r, one dominant allele in either case being necessarily donated by P_3.

$F_{1,3}$ a male not expressing the trait, is X^RY. Since no male offspring of P_1 and P_2 express the trait, it is likely though not necessary, that P_2 is not a carrier.

Her possible genotypes are X^RX^R and X^RX^r. $F_{1,2}$ must have received the recessive allele from P_1 and be a carrier (X^RX^r). $F_{1,1}$ must be X^RY since he does not express the trait. $F_{2,2}$ could be either X^RX^Ror X^RX^r.

Summarizing, we can see that our assumption of sex linkage is consistent with the inheritance pattern of the trait:

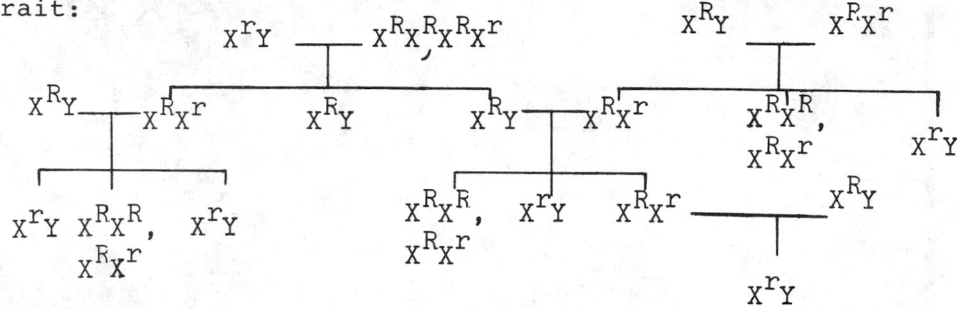

• PROBLEM 26-7

The trait represented in the pedigree below is inherited through a single dominant gene. Calculate the probability of the trait appearing in the offspring if the following cousins should marry.

(a) $F_{2,2} \times F_{2,4}$; (b) $F_{2,1} \times F_{2,3}$.

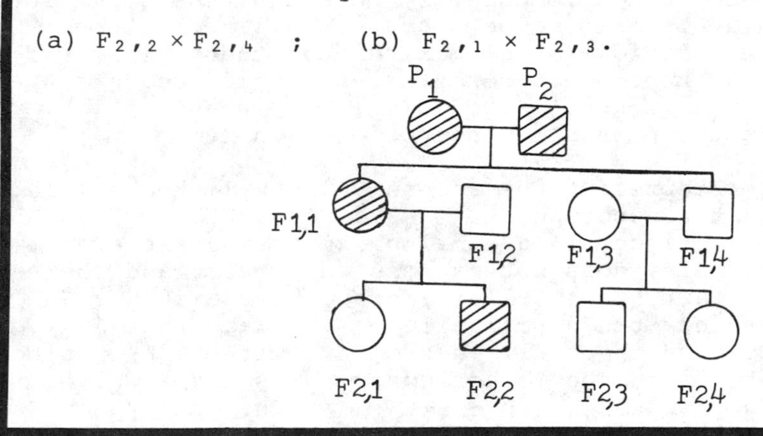

Solution: Let R represent the dominant allele for the trait, and r the recessive allele. First, we can determine the possible genotypes of the individuals involved. We know that $F_{1,2}$, $F_{1,3}$, $F_{1,4}$, $F_{2,1}$, $F_{2,3}$, and $F_{2,4}$ are homozygous recessive (rr), since they do not express the trait. P_1 and P_2 must be heterozygous dominant (Rr) since they produced a child ($F_{1,4}$) who is homoyzygous recessive. F_1, could be either RR or Rr. However, F_2, is homozygous recessive. Therefore $F_{1,1}$ must carry the recessive allele and be Rr.

$F_{2,2}$ is also a heterozygote. We know this because one of his parents ($F_{1,2}$) is a homozygous recessive, and the cross (Rr × rr) is incapable of producing a homozygous dominant.

Summarizing, we have

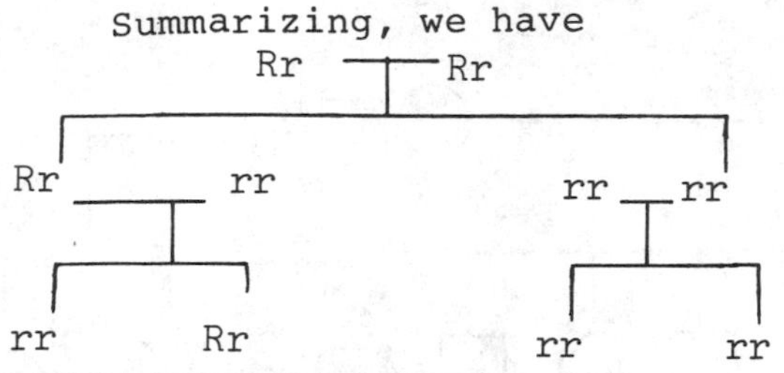

We can now do the crosses in question.

a)
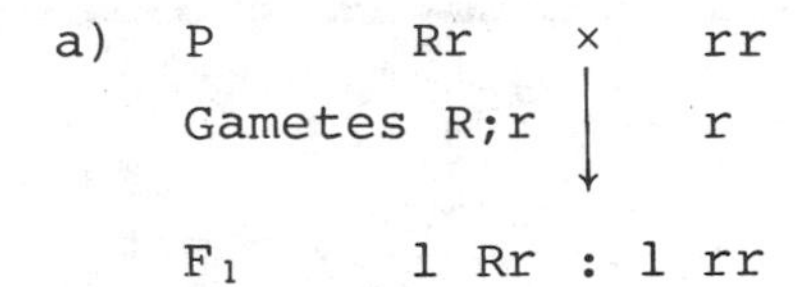

In this cross, half the offspring are heterozygous dominant and half are monozygous recessive. Thus we can say that there is a .5 probability that any offspring of the mating will express the dominant trait.

b) P rr × rr

gametes r r

F_1 rr

Since neither parent carries the dominant allele, it should be obvious even without doing the cross that no offspring expressing the dominant trait will be produced.

Thus we can say that there is 0 probability that any offspring will express the trait.

• PROBLEM 26-8

A trait has the inheritance pattern shown below for four different families. a) Determine whether the trait is dominant or recessive b) Based on your answer to a), determine the possible genotypes for each individual in the four families. (Solid blocks or circles indicate individuals who express the trait).

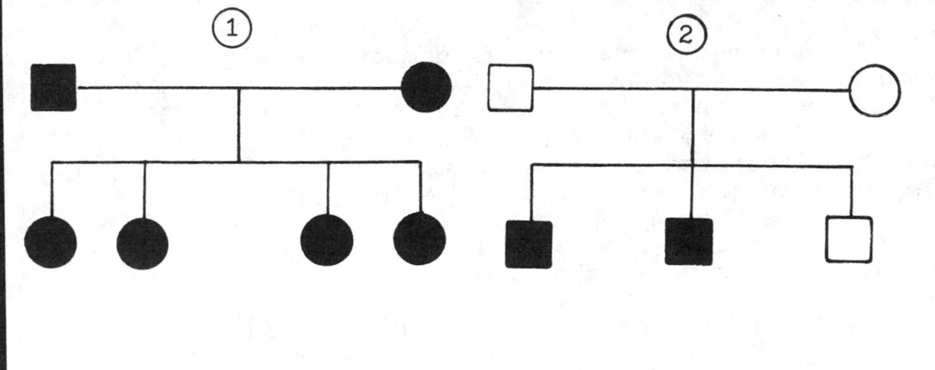

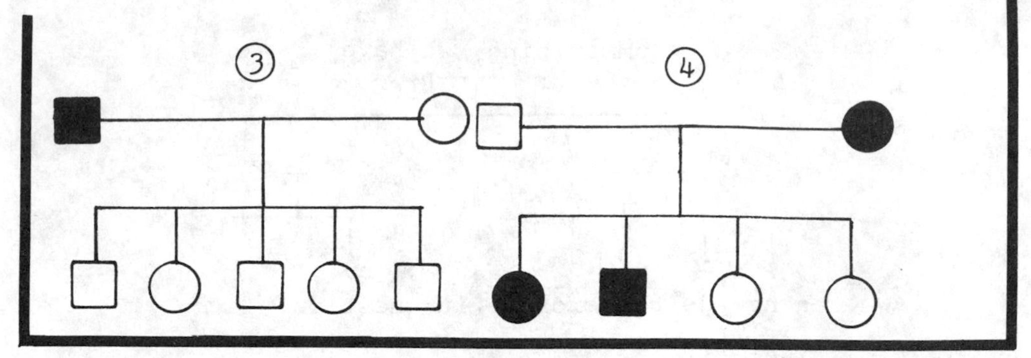

<u>Solution:</u> a) Let us look at what the pattern within each individual family tells us about the trait.

[1] Both parents express the trait, and all the offspring express it. This does not really tell us anything about the trait, for such a pattern could result from the inheritance of either a dominant or a recessive trait.

[2] Neither parent expresses the trait, yet it is expressed among the offspring. This is definite evidence that the trait must be recessive. We know that since the trait is present in the offspring, then one or both of the parents must carry the gene for that trait. Yet neither parent expresses the trait. Therefore, the gene must be present, but masked, and so it must be recessive.

We need go no further since we have established that the trait is recessive. However, let us see what we might have been able to learn about the trait from the remaining pedigrees.

[3] In this family, one parent expresses the trait, yet none of the offspring express it. This is inconclusive, because such a pattern could result from either dominant or recessive inheritance. If the trait were dominant, the parent expressing it could be heterozygous, and so carrying some recessive allele. Although it is not highly probable, it is possible that only the recessive allele was given to the offspring, and none received the dominant allele. If the trait were recessive, the parent not expressing the trait would carry at least one dominant gene which could have been inherited by all the offspring. This gene would mask any recessive alleles the offspring may have.

[4] Again this pattern is compatible with both dominant and recessive inheritance, assuming that one of the parents is homozygous and the other heterozygous.

b) Since we know from family two's pedigree chart that the inherited trait is recessive, we can determine the possible genotypes of all the individuals in all the families. Let A represent the dominant trait (whose inheritance is not shown: ,O) and a the recessive trait (whose inheritance is shown: ,) the genotypes are:

[1] P^1( ) = aa [2] P^1( ) = Aa

$P^2(\ ) = aa$ $\quad$ $P^3(\ ) = Aa$

$F_1(\ ) = aa$ $\quad$ $F_1^3(\ ,\ ) = AA$ or Aa

$F_1^{1,2}(\ ) = aa$

3 $\quad P^1(\ ) = aa$ $\quad$ 4 $\quad P^1(\ ) = Aa$

$P^2(\ ) = AA$ or Aa $\quad$ $P^2(\ ) = aa$

$F^{1-5}(\ ,\ ) = Aa$ $\quad$ $F^{1,2}(\ ,\ ) = aa$

$F^{3,4}(o,o) = Aa$

GENETIC PROBABILITIES

• PROBLEM 26-9

In man, two abnormal conditions, cataracts in the eye and excessive fragility of the bones, seem to depend on separate dominant genes located on different chromosomes. A man with cataracts and normal bones, whose father had normal eyes, married a woman free from cataracts but with fragile bones. Her father had normal bones. Their daughter marries a man with normal eyes and bones. If these two people have a child, what is the probability that it will have both cataracts and fragile bones?

Solution: First, let us determine the possible genotypes of the individuals involved. Let C represent the dominant allele for cataracts, c the recessive allele for normal eyes, B the dominant allele for fragile bones, and b the recessive allele for normal bones. The man has cataracts, but he must be heterozygous for that trait (Cc) since his father had normal eyes (cc). Since the man has normal bones, his genotype for that trait is bb. The woman whom he married has normal eyes (cc). Since she has fragile bones, and her father had normal bones (bb), she must be heterozygous for that trait (Bb).

Thus the following cross takes place:

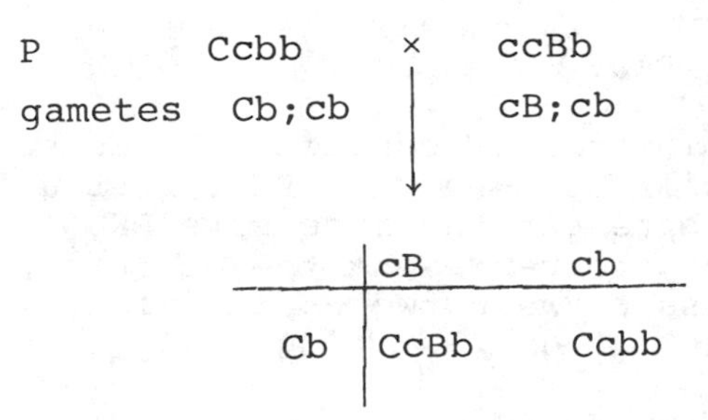

The daughter has one of four possible genotypes: CcBb, Ccbb, ccBb, and ccbb. Her husband, who is normal for both traits involved, is ccbb. In order for their child to express both abnormalities, its genotype must be CcBb (since the daughter's husband can transmit only recessive alleles). This means that the daughter must donate both dominant alleles. Looking at the cross outlined above, we see that there is only one genotype possible in which the daughter carries both dominant alleles, namely CcBb. The probability of this occurring is 1/4 (one out of four possible genotypes). Given this, there is a 1/2 probability that the daughter will donate the B, rather than the b allele, to her child, and a 1/2 probability that she will donate the C, rather than the c allele. The probability that she will both possess and subsequently donate both dominant alleles is equal to the product of the separate probabilities of each of the three events, namely $1/4 \times 1/2 \times 1/2$ or 1/16 probability that a child having both cataracts and fragile bones will be produced.

• PROBLEM 26-10

The gene for PKU is found in the heterozygous state in about one person in fifty.
About how frequently would you expect babies to be born with PKU?
Assume here that the homozygous state for PKU is either lethal or severely restricts the affected person's ability to produce offspring.

Solution: The trait for PKU disease is recessive, and thus the genotype of babies born with the disease must be homozygous recessive. This genotype could be produced through the mating of two phenotypically normal people who are heterozygous for the PKU gene. The chance that two such heterozygotes will marry can be determined by using simple probability procedures. We know that the probability of two separate events occurring together is the product of their separate probabilities. If the heterozygous genotype occurs in one person out of fifty, then the chance that any two people who marry will both be carriers is

$$\frac{1}{50} \times \frac{1}{50} \text{ or } \frac{1}{2,500},$$

and only one marriage out of 2,500 will be capable of producing children with PKU.

We must now determine the probability within such a marriage that a child will be born with PKU. Let us use the small letter p to represent the recessive PKU allele, and the capital letter P, to represent the dominant, normal allele. The cross between two Pp individuals is illustrated below: P Pp × Pp

Gametes P;p ↓ P;p

	P	p
P	PP	Pp
p	Pp	pp

The offspring are obtained in the ratio:

1/4 PP : 1/2 Pp : 1/4 pp

We see that only one child in four will be homozygous recessive for the PKU allele, and consequently have the disease. Now, if only one of every four offspring produced has the PKU disease, and only one of every 2,500 marriages has the genetic makeup to produce a PKU child, then the frequency of children born with PKU would be

$$\frac{1}{2,000} \times \frac{1}{4} \quad \text{or} \quad \frac{1}{10,000} .$$

Therefore, one out of every 10,000 children born will have PKU disease.

• PROBLEM 26-11

Why are genetic ratios more reliable when there are large numbers of offspring? Discuss the above with a reference to one inherited character in human beings.

Solution: Genetic ratios are often misunderstood. When we say that a certain type of cross yields a three-to-one ratio, this does not mean that for every four offspring, there will always be three of one type and one of another. A ratio is worked out on the basis of mathematical probability and will be approached when large numbers are considered.

In tossing a coin there is an equal probability of obtaining either a head or a tail, and the ratio of heads to tails would be 1:1. This is because each toss or event is independent of, and therefore not influenced by, the results of any preceding or subsequent tosses. The 50% chance of obtaining either side of the coin is the probability within which each individual event operates, and does not change, regardless of the number of time the event occurs. However, when the event takes place a large number of times, the results do tend to average out to the expected probability, and the actual ratio approaches the anticipated ratio of 1/2.

The same is true for certain genetic events. For instance, the cross between a heterozygous brown-eyed man (Bb) and a blue-eyed woman (bb) would be depicted as the following:

P	bb	×	Bb
gametes	b	↓	B;b

F_1

	B	b
b	Bb	bb
b	Bb	bb

Each offspring produced has a 1/2 chance of being brown eyed and a 1/2 chance of being blue eyed. Thus, the expected ratios are; 1 Bb : 1 bb; or phenotypically,

1 brown : 1 blue.

We might expect then, a one-to-one ratio in eye color among their offspring. But suppose their first child had blue eyes. This does not automatically mean that their second child must have brown eyes. That child has, like the first child, an equal probability of being either brown or blue-eyed; either outcome is equally probable regardless of what eye color the first child has. This is because the separation of alleles in gamate formation is, like a coin toss, a purely random event, not effected by preceding events. So, while we know that in this case the egg can only be carrying the gene for blue eyes, (because the female in this case is homozygous for blue eyes). The gene carried by the sperm that will fertilize the egg has an equal chance of being brown or blue, and we cannot predict which it will be because of the randomness of gene segregation. Therefore. though we might expect a one-to-one ratio for brown-eyed and blue-eyed offspring, the actual ratio may deviate from expected entirely by chance.

However - and this is the important point - if we were to tabulate the eye color of thousands of children from many families of parents having the same genotypic combinations as this couple, we would indeed find that close to one half of them will have blue eyes and the other half will have brown eyes. As was the case for the coin toss, the larger our sampling population, the closer our ratio approaches one-to-one, the expected ratio. Our actual ratio approaches the probable ratio, and our results become more reliable.

THE HARDY-WEINBERG LAW

• PROBLEM 26-12

What are the implications of the Hardy-Weinberg Law?

Solution: The Hardy-Weinberg Law states that in a population at equilibrium both gene and genotype frequencies remain constant from generation to generation. An equilibrium population refers to a large interbreeding population in which mating is random and no selection or other factor which tends to change gene frequencies occurs.

The Hardy-Weinberg Law is a mathematical formulation which resolves the puzzle of why recessive genes do not disappear in a population over time. To illustrate the principle, let us look at the distribution in a population of a single gene pair, A and a. Any member of the population will have the genotype AA, Aa, or aa. If these genotypes are present in the population in the ratio of 1/4 AA : 1/2 Aa : 1/4 aa, we can show that, given random mating and comparable viability of progeny in each cross, the genotypes and gene frequencies should remain the same in the next generation. Table 1 below shows how the genotypic frequencies of AA, Aa, and aa compare in the population and among the offspring.

Table 1

The Offspring of the Random Mating of a Population Composed of 1/4 AA, 1/2 Aa and 1/4 aa Individuals

Mating Male	×	Mating Female	Frequency	Offspring
AA	×	AA	1/4 × 1/4	1/16 AA
AA	×	Aa	1/4 × 1/2	1/16 AA + 1/16 Aa
AA	×	aa	1/4 × 1/4	1/16 Aa
Aa	×	AA	1/2 × 1/4	1/16 AA + 1/16 Aa
Aa	×	Aa	1/2 × 1/2	1/16 AA + 1/8 Aa + 1/16aa
Aa	×	aa	1/2 × 1/4	1/16 Aa + 1/16aa
aa	×	AA	1/4 × 1/4	1/16 Aa
aa	×	Aa	1/4 × 1/2	1/16 Aa + 1/16aa
aa	×	aa	1/4 × 1/4	1/16aa
			Sum:	4/16 AA + 8/16 Aa + 4/16aa

Since the genotype frequencies are identical, it follows that the gene frequencies are also the same.

It is very important to realize that the Hardy-Weinberg law is theoretical in nature and holds true only when factors which tend to change gene frequencies are absent. Examples of such factors are natural selection, mutation, migration, and genetic drift.

• PROBLEM 26-13

Contrast the meanings of the terms "gene pool" and "genotype".

Solution: A gene pool is the total genetic information possessed by all the reproductive members of a population of sexually reproducing organisms. As such, it comprises every gene that any organism in that population could possibly carry. The genotype is the genetic constitution of a given individual in a population. It includes only those alleles which that individual actually carries. In a normal diploid organism, there is a maximum of two alleles for any one given locus. In the gene pool, however, there can be any number of alleles for a given locus. For example, human blood type is determined by three alleles, I^A, I^B, and i. The gene pool contains copies of all three alleles, since all these are found throughout the entire population. Any given individual in the population, however, can have at the most two of the three alleles, the combination of which will determine his blood type.

• PROBLEM 26-14

Can complete equilibrium in a gene pool exist in real situations?

Solution: Of the four conditions necessary for the genetic equilibrium described by the Hardy-Weinberg Law, the first, large population size is met reasonably often; the second, absence of mutations, is never met; the third, no migration, is met sometimes; and the fourth, random reproduction, is rarely met in real situations. Therefore it follows that complete equilibrium in a gene pool is not expected.

With regard to the first condition, many natural populations are large enough so that chance alone is not likely to cause any appreciable alteration in gene frequencies in their gene pools. Any breeding population with more than 10,000 members of breeding age is probably not significantly affected by random changes.

The second condition for genetic equlibrium the absence of mutations, is never met in any population because spontaneous mutations are always occurring. Most genes probably undergo mutation once in every 50,000 to 1,000,000 duplications, with the rate of mutation for different genes varying greatly. However, since the rate of spontaneous mutation is usually low, it is usually insignificant in altering the gene frequencies in a large population.

The third condition for genetic equilibrium, implies that a gene pool cannot exchange its genes with the outside. Immigration or emigration of individuals would change the gene frequencies in the gene pool. A high percentage of natural populations, however, experience some amount of migration. This factor, which enhances variation, tends to upset Hardy-Weinberg equilibrium.

The fourth condition, random reproduction, refers not only to the indiscriminate selection of a mate but also to a host of other requirements that contribute to success in propagating the viable offspring. Such factors include the fertility of the mating pair, and the survival of the young to reproductive age. An organism's genotype actually influences its selection of a mate, the physical efficiency and frequency of its mating, its fertility, and so on. Thus entirely random reproduction in reality is not possible.

We can conclude, therefore, that if any of the conditions of the Hardy-Weinberg Law are not met, then the gene pool of a population will not be in equilibrium and there will be an accompanying change in gene frequency for that population. Since it is virtually impossible to have a population existing in genetic equilibrium, even with animals under laboratory conditions, there must then be a continuous process of changing genetic constitutions in all populations. This is ultimately related to evolution in that evolutionary change is not usually automatic, but occurs only when something disturbs the genetic equilibrium.

• PROBLEM 26-15

How may the gene pool be altered?

Solution: The total genetic material of a given population is termed the gene pool. The Hardy-Weinberg principle tells us that the relative gene frequencies within the gene pool of a population will remain constant from generation to generation unless certain factors alter the equilibrium of gene frequencies in the gene pool. Such factors include mutation, natural selection, migration, random genetic drift, and meiotic drive.

Mutations can change one allele to another, such as A to a, and if they are recurrent or "one-way", the relative frequencies of the two alleles will be changed by an increase in the proportion of a at the expense of A. Owing to the low mutation rate of most genes, such direct change in allelic frequency alone probably does not cause significant change in the gene pool. More important is the fact that mutations are a source of new genes, and thus traits, within a population. Together with sexual reproduction, which creates new combinations of existing genes, mutations provide the variation which is the basis for the operation of natural selection.

Selection is the nonrandom differential retaining of favored genotypes. Unlike mutation, which operates directly on a gene to alter its frequency, selection indirectly alters a gene's frequency by acting on its carriers as a function of their ability to reproduce viable offspring. For example, if individuals carrying gene A are more successful in reproduction than individuals carrying gene B, the frequency of the former gene will tend to increase generation after generation at the expense of the latter. Any trait which gives an organism a better chance at survival in a given environment, will increase that organism's ability to grow and reproduce, and increase the proportion of the gene for that trait in the gene pool. By the same token, any gene which confers a disadvantage to its carrier within its environment will decrease that organism's chance of survival to reproductive age, and thus decrease the frequency of that gene in the gene pool. This is the principle underlying natural selection.

Migration acts both directly and indirectly to alter gene frequencies. Directly, a population may receive alleles through immigration of individuals from a nearby population. The effectiveness of immigration in changing allelic frequencies is dependent on two factors: the difference in gene frequencies between the two populations and the proportion of migrant genes that are incorporated. Alternatively, there can be emigration of members from a population, which, depending on the size of the emigration and whether or not it is selective, can result in a change in allelic frequencies in the gene pool.

Indirectly, migration acts as a source of variation, similar to mutation, upon which the forces of natural selection can operate. Migration can also enhance natural selection within a population by upsetting the equilibrium that may exist among the given genotypes in a population, concerning an advantage by sheer numbers to a given group. It can blur the effects of natural selection by replacing genes removed by selection.

Allelic frequencies may fluctuate purely by chance about their mean from generation to generation. This is termed random genetic drift. Its effect on the gene pool of a large population is negligible, but in a small effectively interbreeding population, chance alteration in Mendelian ratios can have a significant effect on gene frequencies, and may lead to the fixation of one allele and loss of another. For example, isolated communities within a given population have been found to have different frequencies for blood group alleles than the population as a whole. Chance fluctuations in allelic frequency presumably caused these changes.

Another factor that may alter allelic frequencies is meiotic drive. This is the term for preferential segregation of genes that may occur in meiosis. For example, if a particular chromosome is continually segregated to the polar body in female gametogenesis, its genes would tend to be excluded from the gene pool since the polar bodies are nonfunctional and will disintegrate. There is significant evidence that, due to physical differences between certain homologous chromosomes, preferential selection of one over the other often occurs at other than random proportions.

GENE FREQUENCIES

• PROBLEM 26-16

In a given population of 610 individuals, the gene frequencies of the L^M and L^N alleles were found to be .62 and .38, respectively. Calculate the number of individuals with M, MN, and N type blood.

Solution: The Hardy-Weinberg principle tells us that the proportions of the genotypes in a population are described by the expansion of the binomial equation:

$$(PA + qa)^2 = p^2AA + 2\ pqAa + q^2aa,$$

where p represents the frequency of a given allele A and q represents the frequency of its homologous allele, a. In other words, if we were to consider all the matings in a given generation, a p number of A-containing eggs and a q number of a-containing eggs are fertilized by a q number of A-containing sperm and a g-number of a-containing sperm; and the total possible allelic combinations produced are

$$(pA + qa) \times (pA + qa) \text{ or } (pA + qa)^2.$$

In this instance, p and q are respectively the frequencies of the L^M and L^N alleles. The phenotypic classes for these alleles are

$M(_V L^M L^M)$ genotype, $MN(_V L^M L^N)$ genotype, and $N(_V L^N L^N)$ genotype.

This allelic system is ideal for studying genotypic ratios since the genes are codominant and the heterozygotes therefore constitute a distinct phenotypic class, rather then being obscured by the dominant trait.

The frequencies of the phenotypic classes in the population can be determined using the binomial expansion:

$$\begin{aligned} & (pL^M + qL^N)^2 \\ = & (.62L^M + .38L^N)^2 \\ = & (.62)^2 L^M L^M + 2(.62)(.38)\ L^M L^N + (.38)^2 L^N \\ = & .384\ M + .471\ MN + .144\ N \end{aligned}$$

Converting the frequencies to actual numbers of individuals, we have

$$\begin{aligned} (.384)(610)M &= 235\ M \\ (.471)(610)MN &= 287\ MN \\ \underline{(.144)(610)N} &= \underline{\ 88\ N\ } \\ (1)\ (610) &= 610\ \text{TOTAL} \end{aligned}$$

In this calculation, we multiply each frequency by the total population to obtain the actual numbers of those having each genotype. Note that the frequencies sum up to 1, and the total number of individuals for the three phenotypes equals the total population.

• PROBLEM 26-17

How can you determine whether or not a given population is in genetic equilibrium?

Solution: In an equilibrium population, both gene frequencies and genotype frequencies remain constant from generation to generation. The best way to determine if a given population is in equilibrium is to compare these frequencies of two generations to see if they are indeed constant. One way to do this is to determine the gene frequencies for a given adult population representative (or a group of that population) and for their offspring. Two generations are thus represented in the sample. For more reliable results, however, more generations should be compared with respect to their gene frequencies.

Suppose one wanted to know if the population of Australian aborigines were in genetic equilibrium. Data from that population concerning the distribution of MN blood groups are gathered, and are shown below:

	Phenotype			
Generation (G)	M	MN	N	Total
1	241	604	195	1040
2	183	460	154	797

MN blood groups are determined by two alleles, L^M and L^N, having intermediate or codominant inheritance. Thus M type is coded for by L^ML^M, MN type by L^ML^N, and N type by L^NL^N. We can calculate the genotype frequencies using the data given:

Genotype	Frequency G_1	Frequency G_2
L^ML^M	241/1040 = .231	183/797 = .230
L^ML^N	604/1040 = .581	460/797 = .577
L^NL^N	195/1040 = .188	154/797 = .193
	1.0	1.0

The genotype frequencies are thus found to be essentially the same for both generations.

We could also have determined the allelic frequencies from the data, the number of L^M alleles in the population is twice the number of M individuals (L^ML^M) plus the number of MN individuals (L^ML^N). The number of L^N alleles is twice the number of N individuals plus the number of MN individuals. Since each individual carries two alleles, the total number of alleles in the population is twice the number of individuals, or the sum of the L^M and L^N alleles. For our population, then, we have the following:

	$L^M = 2(M) + MN$	$L^N = 2(N) + MN$
G_1	2(241)+604 = 1086	2(195)+604 = 994
G_2	2(183)+460 = 826	2(154)+460 = 768

	Total Alleles (L^M+L^N)	Frequency L^M	Frequency L^N
G_1	1086+994=2(1040)=2080	1086/2080=.522	994/2080= .478
G_2	826+768=2(797)=1594	826/1594=.518	768/1594= .482

As with the genotype frequencies, the allelic frequencies remain nearly the same in both generations. Thus we can say that, with respect to the MN blood group alleles, the population is in equilibrium. Note that we can only determine equilibrium in a population in reference to a given gene or a small number of genes. Though our population may be in equilibrium for the MN alleles, other allelic systems may not be. For example, migration may be occuring from nearby populations having similar MN allelic frequencies but very different allelic frequencies for color blindness, thus affecting the equlibrium of these genes in the population.

• PROBLEM 26-18

The following MN blood types were determined from the entire population of a small isolated mountain village.

M	N	MN	Total
53	4	29	86

What is the frequency of the L^M and L^N alleles in the population?

Solution: According to the Hardy-Weinberg principle, we know that we can express the distribution of the MN blood type phenotypes in the population as

$$p^2 + 2\,pq + q^2 = 1,$$

where p equals the frequency of the L^M allele and q equals the frequency of the L^N allele. (Remember that $L^M L^M$ is phenotypically M, $L^M L^N$ is MN, and $L^N L^N$, N.) p^2 is the frequency, or proportion, of M individuals in the population, 2 pq the proportion of MN individuals, and q^2 the proportion of N individuals. Therefore, for this case:

$$\begin{aligned} q^2 &= 53/86 = .616 \\ g^2 &= 4/86 = .047 \\ 2\,pq &= 29/86 = \underline{.337} \\ & \qquad\qquad 1.0 \end{aligned}$$

From this we can determine the frequencies of the alleles themselves.

$$\begin{aligned} \text{frequency } L^M &= p = \sqrt{.616} = .785 \\ \text{frequency } L^N &= q = \sqrt{.047} = \underline{.215} \\ & \qquad\qquad\qquad 1.0 \end{aligned}$$

As a check:

$$2\,pq = 2(.785)(.215) = .337$$

• PROBLEM 26-19

A group of students were invited to taste phenylthiocarbamide (PTC). The ability to taste PTC is inherited by a single pair of genes and tasting (T) is dominant to nontasting (t). Among 798 students, 60.4 percent were tasters. a) Calculate the allelic frequency of T and t. b) How many of the students were TT? Tt? tt?

Solution: Putting this problem in terms of the Hardy-Weinberg principle, let p represent the frequency of the T allele and q the frequency of the t allele. We know that

$$p^2 + 2\,pq + q^2 = 1$$

describes the distribution of phenotypes in the population, where p^2 equals the frequency of TT, 2 pq equals the frequency of Tt, and q^2 equals the frequency of tt. Note that the sum of all proportions, or frequencies, equals 1.

a) We are told that 60.4% of our given population are tasters. Since tasting is dominant to nontasting, tasters can have one of two genotypes, TT and Tt. This means that the sum of the frequencies of the TT and Tt genotypes is .604. Since all frequencies total 1, we can infer that the frequency of nontasters (tt) in the sample of students is

1- .604 or .396.

Thus $q^2 = .396$, and

$q = \sqrt{.396}$ or .629. Since $p+q = 1$, $p = 1-.629$ or .371.

The frequency of T is therefore .371, and that of t, .629.

b) frequency TT = $p^2 = (.371)^2 = .137$

frequency Tt = $2p = 2(.371)(.629) = .467$

frequency tt = $q^2 = (.629)^2 = .396$

1.0

To convert the actual number of students carrying each genotype, we multiply the corresponding genotypic frequencies (or proportions) by the total sample size (i.e., 798):

TT = (.137)(798) = 109

Tt = (.467)(798) = 373

tt = (.396)(798) = 316

798

• PROBLEM 26-20

In an isolated mountain village, the gene frequencies of A, B and O blood alleles are 0.95, 0.04, and 0.01, respectively. If the total population is 424 calculate the number of individuals with O, A, B, and AB type blood.

Solution: Multiple allelic systems establish equilibrium in the same way as single-pair alleles, and the same equilibrium principles can be applied, though naturally the system is more complex. In the inheritance of blood groups, three genes, A, A^B, and a are involved. Letting p, q, and r, respectively, represent their frequencies, we can say that the total of these frequencies equals 1 ($p + q + r = 1$). We can represent the total possible gene combinations (genotypes) in the population by means of the same type of expansion used with single-pair allelic systems; in this case, however, three alleles are involved. Thus:

$$(p+q+r)^2 = p^2 + 2pq + q^2 + 2pr + r^2 + 2qr$$

where:

p = frequency A

q = frequency A^B

r = frequency a

The following table summarizes what we know of the genotypes of each blood group, (see previous chapter for an explanation of the inheritance of blood groups) and the corresponding allelic and genotypic frequencies.

blood type	genotypes	genotypic frequency
O	aa	r^2
A	AA, Aa	p^2, $2pr$
B	A^BA^B, A^Ba	q^2, $2qr$
AB	AA^B	$2pq$

In the given population, $p = 0.95$, $q = 0.04$, and $r = 0.01$. Knowing this we can calculate, the blood type frequencies using the results of the expansion. The frequency of type O blood in the population is:

$$r^2 = (0.01)^2 = 0.0001$$

The frequency of type A blood is the sum of the frequencies of the two genotypes comprising this phenotypic blood group:

$$\begin{aligned} p^2 + 2pr &= (0.95)^2 + 2(0.95)(0.01) \\ &= .9025 + .019 \\ &= 0.9215 \end{aligned}$$

The frequency of type B blood is the sum of the frequencies of the genotypes comprising the group:

$$q^2 + 2\,qr = (0.04)^2 + 2(0.04)(0.01)$$
$$= (.0016) + .0008$$
$$= .0024$$

The frequency of type AB blood is simply:

$$2\,pq = 2(0.95)(0.04) = .076$$

We can see that the sum of the frequencies is 1:

$$\begin{aligned} \text{frequency O} &= r^2 &= .0001 \\ A &= p^2+2\,pr &= .9215 \\ B &= q^2 + 2\,qr &= .0024 \\ AB &= 2\,pq &= \underline{.076} \\ & & 1.000 \end{aligned}$$

Converting the frequencies to actual numbers of individuals in the population:

$$\begin{aligned} O &= (r^2)(424) = (.0001)(424) \sim 0 \quad (.0424) \\ A &= (p^2+2pr)(424) = (.9215)(424) = 391 \\ B &= (q^2+2qr)(424) = (.0024)(424) = 1 \\ AB &= (2pq)(424) = (.076)(424) = \underline{32} \\ & \qquad\qquad\qquad\qquad\qquad\qquad 424 \end{aligned}$$

• PROBLEM 26-21

The frequency of the gene for sickle-cell anemia in American Blacks is less than that found in the people living in their ancestral home in Africa. What factors might account for this difference?

Solution: The sickle-cell disease is a homozygous recessive trait that usually results in death before reproductive age. In the heterozygous form, the sickle-cell gene is not harmful enough to cause death, but is instead beneficial in certain environments because it gives the carrier an immunity to malaria. In Africa, malaria is a severe problem and the heterozygous individuals have a survival advantage over their fellow Africans. Therefore, the frequency of the sickle-cell gene has been kept fairly constant in the gene pool in Africa.

In America, where the incidence of malaria is insignificant, an individual carrying the sickle-cell gene has no survival advantage, and the sickle-cell allele is slowly being lost and diluted in the population. Those with the homozygous sickle-cell genotype usually die, hence the frequency of the sickle-cell allele declines.

In addition, with more interracial marriages in America, the sickle-cell genes from blacks are being diluted by the normal genes from the non-Black population. Thus, in the American Black population, a trend is observed in which the frequency of the sickle-cell gene decreases gradually over generations.

• PROBLEM 26-22

A certain recessive gene (r) in a population has a frequency of 0.5. As a result of movement of the population to a new environment, homozygous recessive individuals (rr) are now selected against, with a loss of 80% of the homozygotes before maturity. Homozygous dominant (RR) and heterozygotes (Rr) are not affected. What is the frequency of the gene in the population after one generation? Has equilibrium in the new environment been reestablished?

Solution: Since there are only two alleles involved, and we know that the sum of their frequencies must be one, we can calculate the frequency of the dominant gene (R) in the population. Letting p equal the frequency of R, and q equal the frequency of r (which is known to be 0.5):

$$p + q = 1$$
$$p = 1-q$$
$$= 1-0.5$$
$$p = 0.5$$

We can now use the Hardy-Weinberg binomial expansion to determine the corresponding genotype frequencies after one generation.

$$(p+q)^2 = p^2 + 2\,pq + r^2$$
$$= (.5)^2 + 2(.5)(.5) + (.5)^2$$
$$= .25 + .50 + .25 \quad (=1),$$

where p^2 = frequency RR

$2\,pq$ = frequency Rr

q^2 = frequency rr

The frequency q^2, however, is only the initial frequency of recessive homozygotes in the offspring. 80% of these will be lost to the population before reproductive age. Thus, the number of rr individuals in the population will in effect be only 20% of what it originally was. The effect that this will have on actual gene frequencies can best be seen by first converting the frequencies to actual individual numbers. For convenience sake, let us assume an original population of 100. The original number of rr individuals is (.25)(100) or 25. Loss of 80% of these would give a final effective number of

$$25-(.8)(25) = 25 - 20 = 5.$$

The new frequencies now are calculated for an effective total population of 100 - 20 or 80. The number of RR and Rr individuals, 25 and 50 respectively, remains unchanged. The new genotype frequencies are:

$$\begin{aligned}\text{frequency RR} &= 25/80 = .313\\ \text{frequency Rr} &= 50/80 = .625\\ \text{frequency rr} &= 5/80 = \underline{.062}\\ &\qquad\quad 1.0\end{aligned}$$

The allelic frequencies can be calculated using a new effective allelic population of 2(80) or 160 (two alleles/individual).

$$\begin{aligned}\text{number R} &= 2\ \text{RR} + \text{Rr} = 2(25) + 50 = 100\\ \text{number r} &= 2\ \text{rr} + \text{Rr} = 2(\ 5) + 50 = \underline{\ 60}\\ &\qquad\qquad\qquad\qquad\qquad\quad 160\end{aligned}$$

$$\begin{aligned}\text{frequency R} &= p = 100/160 = .625\\ \text{frequency r} &= q = 60/160 = \underline{.375}\\ &\qquad\qquad\qquad 1.0\end{aligned}$$

Notice that the actual distribution of genotypes does not correspond to the binomial expansion of the gene frequencies, $(p+q)^2$ (i.e., $.313 \neq p$, $.625 \neq 2\ pq$, $.062 \neq q^2$). This is because our population is not in equilibrium, and the Hardy-Weinberg principle applies only to populations in equilibrium. Selection has upset the equilibrium in the population. Until the selection factor, whatever it is, is removed, the frequency of the recessive gene will continue to decrease as it is being removed from the population with each successive generation. We can see, however, that it will never be entirely removed, due to the presence of heterozygotes in the population. Nonetheless, its frequency can be reduced to almost zero if the selection is strong enough, and if enough generations have gone by.

Answer

To be covered when testing yourself

SHORT ANSWER QUESTIONS FOR REVIEW

Choose the correct answer.

1. Which one of the following is not a reason why human beings are unfavorable for studies of inheritance? (a) Some human traits are determined by more than one gene. (b) It is difficult to perform human genetics studies in controlled environment. (c) It is difficult to find human beings with traits researchers want to investigate. (d) The life span and generation time of humans are too long. — c

2. Studies on twins are important because (a) single human traits are difficult to compare. (b) Twins are more common as a result of increasing use of fertility drugs. (c) They provide more reliable results since there are two of a kind. (d) They provide information about environmental or genetic influence on certain human traits. (e) None of the above. — d

3. Hemophilia is a recessive sex-linked trait. What would be the phenotype of the progeny from a mating between a female hemophilia carrier and a normal male? (a) 1/2 of the sons will be hemophilic, (b) 1/2 of the daughters will be hemophilic, (c) all the daughters will be carriers (d) all the sons will be hemophilic, (e) 1/2 the daughters will be hemophilic, the other half carriers. — a

4. "Mongolian spots" is a recessive sex-linked trait. Only one son out of 6 offspring from two "normal-looking" parents shows the trait. Which one of the following is correct? (a) the mother is homozygous normal, (b) the mother is heterozygous carrier, (c) the father's genotype cannot be determined, (d) none of the above. — b

5. In a certain European community, the frequency of phenylketonuria carriers in the population is 1/20. Suppose 800 marriages are consummated in a year. How many of these marriages will have the potential of producing offspring showing the trait? (a) 8 (b) 20 (c) 2 (d) 40 (e) cannot be determined from the information given. — c

6. For the Hardy-Weinberg Law to hold in a population, which one of the following is not necessary? (a) large interbreeding population, (b) high mutation rates, (c) random mating (d) no natural selection. — b

Answer

To be covered when testing yourself

7. The term genotype refers to (a) the genetic constitution of an individual, (b) the type of genes of an individual, (c) the interaction between closely linked genes, (d) the number of copies of genes in an individual, (e) the physical appearance of an individual. — a

8. Meiotic drive is (a) preferential segregation of genes during meiosis in a heterozygous individual, (b) preferential segregation of genes during meiosis in a homozygous individual, (c) force that prevents meiosis from occurring, (d) force that disrupts the normal segregation processes in meiosis. — d

9. In a small isolated community, the following frequencies of MN blood types are found in the population.

M	MN	N	Total
64	32	4	100.

What is the frequency of L^M allele in the population? (a) .64 (b) .32 (c) .80 (d) .02 (e) none of the above. — c

10. Albinism is a recessive trait. In a certain community of 200 people, 18 persons are albinos. How many are normal homozygotes? (a) 182 (b) 164 (c) 100 (d) 98 (e) 84. — d

11. The inability to taste phenylthiocarbamide is a recessive trait. In a sample of 1000 students, 840 were tasters. How many of the students were heterozygous tasters? (a) 420 (b) 160 (c) 480 (d) 240 (e) cannot be determined from the data given. — c

12. Which one of the following is a reason for the lower frequency of the gene for sickle-cell anemia observed among American blacks than that found in Africa? (a) Difference in dietary intake (b) barriers imposed by immigration rules (c) sickle-cell anemia can be easily treated in America (d) increasing incidences of interracial marriages (e) difference in culture. — d

The following questions refer to the family pedigree shown below, in which an abnormal trait is inherited as simple recessive. Colored squares and circles represent expressions of the trait. Unless there is evidence to the contrary, individuals who have married into the family do not carry the recessive gene.

Answer

To be covered when testing yourself

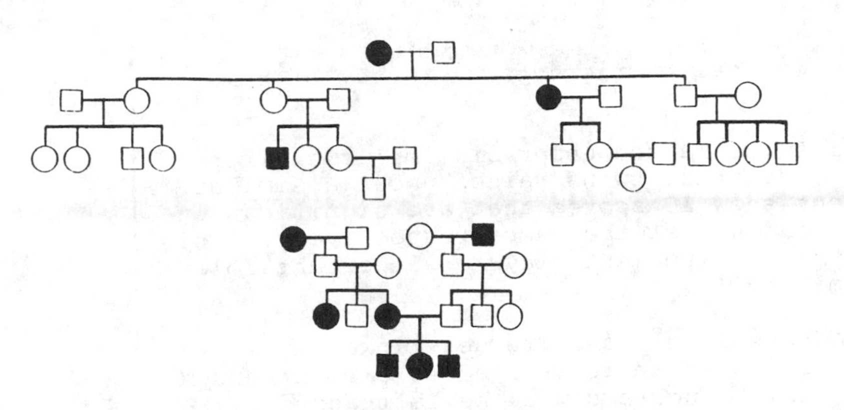

13. The best explanation for the offspring of the cross between F_1-3 and F_1-4 is that (a) both parents are carriers, (b) a mutation has occurred (c) F_1-4 has brought a dominant allele for the trait into the family (d) F_1-3 has a mutant phenotype which can be attributed to her parents. — a

14. If F_2-1 and F_2-9 should marry, the probability of the trait appearing in a given offspring is: (a) 1/16 (b) 1/4 (c) 1/2 (d) 1/8 — d

15. If F_3-1 and F_3-2 marry, the probability of any given offspring expressing the trait is (a) 1/16 (b) 1/4 (c) 1/32 (d) 1/64 — e

Fill in the blanks.

16. Two individuals arising out of a single fertilized egg are known as _____ twins. — mono-zygotic/identical.

17. The Hardy-Weinberg Law only applies to a population in _____. — equilib-rium

18. In real situations, Hardy-Weinberg equilibrium _____ exist in a gene pool. — does not

19. In a small population, genetic drift will tend to produce _____ for a certain gene pair. — fixation/homo-zygosity

20. In a dominant-recessive relationship, if the frequency of the dominant gene is .724 the frequency of the recessive gene is _____. — .276

21. In America, there is no selective advantage for sickle-cell genes among the blacks because _____ is almost eradicated. — malaria

Determine whether the following statements are true or false.

	Answer
	To be covered when testing yourself
22. There are ethical barriers to the study of human genetics.	True
23. Identical twins raised under dissimilar environments and fraternal twins raised under a similar environment show IQ scores that are significantly different and almost the same, respectively. This implies that environment plays an important role in determining IQ.	True
24. Lefthandedness is a recessive sex-linked trait. If a lefthanded woman is married to a righthanded man, all their daughters will be lefthanded.	False
25. Phenylketonurics are unable to convert phenylalanine to tyrosine.	True
26. It only takes one generation to bring a population initially in disequilibrium to equilibrium if the conditions for Hardy-Weinberg Law exist.	True
27. Immigration, but not emigration, changes the existing gene frequencies in a population.	False
28. Mutations provide a source of new genes which are acted on by natural selection.	True
29. The effect of genetic drift is significant in altering gene frequencies in a large population.	False
30. Populations in Hardy-Weinberg equilibrium show similar genotype frequencies from generation to generation.	True
31. There is selective advantage for sickle-cell genes in Africa.	True
32. The genotype frequency of a particular dominant-recessive trait such that $p^2 = .4$, $2pq = .4$ $q^2 = .2$ is in Hardy-Weinberg equilibrium.	False
33. The above pedigree is best described by a sex-linked recessive trait.	False

CHAPTER 27

PRINCIPLES & THEORIES OF EVOLUTION

DEFINITIONS

• PROBLEM 27-1

What is meant by a gene pool, balanced polymorphism, and genetic load?

Solution: Population geneticists visualize a population of organisms as a collection of all the genes possessed by the individuals of the population. Such a sum total of genes is termed a gene pool. Each allele in the gene pool has a certain probability of being expressed in the population. The frequency of occurrence of a given genotype in the population is given by the product of the frequencies of its alleles in the gene pool. A gene pool is said to be in equilibrium if the frequency of each individual allele remains constant from generation to generation; that is to say, when there is no change in the genetic composition of the population over a period of time.

The term polymorphism refers to the existence in the same interbreeding population of two or more distinct phenotypic forms of a genetically determined trait. The human blood groups 0, A, B, and AB are a classic example of polymorphism. Balanced polymorphism is a state in which the different forms of the polymorphic genotype are maintained together in equilibrium in the population over a period of time. Such a balance can be achieved and maintained through a variety of means, involving, what are often complex genetic-environmental interactions. One condition which produces balanced polymorphism is heterozygote superiority or heterosis. Here the heterozygote (Aa) has a survival advantage over both the dominant (AA) and recessive (aa) homozygotes. Thus both alleles are maintained in the population, and neither can eliminate the other. A classic example of this is sickle-cell anemia in a malarial environment. Contrary to what one might expect, the sickle-cell trait is maintained at a relatively high frequency in the population despite its obvious harmful effects.

Individuals carrying the recessive sickle-cell gene in the heterozygous state have resistance to malarial infection. When both malaria and sickle-cell anemia are strong selection factors, the heterozygote has superiority, since normal homozygotes are subject to malaria and those homozygous for the sickle-cell gene have sickle-cell anemia. Thus the recessive gene is maintained along with the normal gene, and their traits are kept in balance.

Often, a given species maintains different forms in the population because one or the other is favored at varying times in the year. The polymorphism gives the species an overall survival ability throughout the year.

In geographic polymorphism, the species population may maintain a variety of forms, each of which is advantageous under specific environment conditions; one form may do better in one area, another in a different area. On the whole, the species is thus able to maintain itself in a variety of areas and environmental conditions, allowing it to become more widespread.

Balanced polymorphism is important in evolution because it maintains a certain amount of variability in the population to permit further adaptive changes. However, although a certain amount of variability is advantageous, too great a genetic variety in the population can be harmful. Geneticists reason that different forms have differing effects upon the overall fitness of a population. Given that there is one optimal phenotype, the greater the number of other suboptimal types present, the less the average fitness of the population will be. The presence of these suboptimal forms imposes a condition of genetic load in terms of a reduction in average population fitness. It is possible that the population could be driven to extinction as a result of too heavy a genetic load.

• **PROBLEM 27-2**

What is meant by "genetic drift"? What role may this play in evolution?

Solution: The random fluctuation of gene frequencies due to chance processes in a finite population is known as genetic drift. Genetic drift may occur by what is commonly referred to as the Founder Principle. This principle operates by chance factors alone. A group of individuals may leave their dwelling and establish a new feeding or breeding ground in an unexploited area. The individuals that find the new area may not represent the same genetic makeup as the original population, since the individuals that left were a random group. This exhibits how one population may establish itself with a new genetic composition by chance alone. Genetic drift may also affect a population without the migration of individuals. Catastrophes in one area of a certain population, such as a flood, may randomly kill individuals of a population. Those individuals which survive, do so by chance alone and not by possessing any special selective advantage. Thus genetic drift may result from natural catastrophes where a random group of individuals may survive by chance alone and change the original population's allelic frequencies. The alteration of allelic frequencies by chance due to genetic drift is unlike the directional movements caused by the systematic pressures of mutation, selection, and differential migration. The situation is quite different in a large population.

In large populations, the relatively small numbers of chance variations in gene frequencies are absorbed into the population in succeeding generations and the overall effect is negligable. In small interbreeding populations, however, where there are limited numbers of progeny in a given generation, random variation has a significant effect and genetic drift can become an important factor. Random fluctuation can most often lead to fixation of a given allele in a small population.

That is, one allele is slowly replacing the other as a result of chance, and the former becomes fixed in the population as the latter is lost. This process of reduction of heterozygosity through loss and fixation at various loci is also known as the decay of variability.

The fixation of certain genes by means of genetic drift may explain the appearance of seemingly unimportant and useless structures in some organisms. These structures may be the expression of homozygous gene pairs which have accumulated in the population by chance alone. Such a loss of variability can also inactivate natural selection, which can act only when a certain degree of phenotypic variation is present. If there is a lack of variation, any adverse situation, such as a period of extreme cold or the arrival of a group of predators, could terminate an entire population. In such a homogeneous population, only genetic mutation could hopefully preserve the population. This happens as some mutants arise which are resistant to the unfavorable force. However, since the frequency of an allelic mutation is in the area of 10^{-6}, a homogeneous population could find itself in great danger of extinction.

• PROBLEM 27-3

Define a genetic mutation. After a mutation has occurred in a population, what events must occur if the mutant trait is to become established in the population?

Solution: A genetic mutation occurs as a change in a specific point (allele) or segment of a chromosome. This gives rise to an altered genotype, which often leads to the expression of an altered phenotype. Genetic mutations occur constantly, bringing about a variety of phenotypes in the population, upon which natural selection can act to choose the most fit. Because genetic mutations occur in the chromosomes and are therefore inherited, they are also referred to as the ultimate raw materials of evolution.

When a mutation first appears, only one or very few organisms in the population will bear the mutant gene. The mutation will establish itself in the population only if the mutants survive and breed with other members of the population. In other words, the mutants must be able to reproduce. Not only must they be able to reproduce, their zygotes must be viable and grow to become fertile adults of the next generation. By breeding of the mutants or mutation carriers either with each other or with normal individuals, the mutant gene can be transmitted to successive generations. Given the above conditions for establishment, over many generations, the mutant gene will appear with greater and greater frequency and eventually become an established constituant of the population's gene pool.

• PROBLEM 27-4

Differentiate between adaptive radiation and convergent evolution.

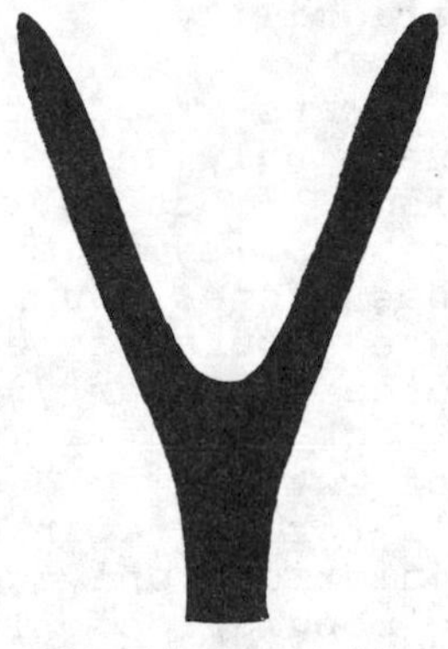

Adaptive radiation
divergent evolution

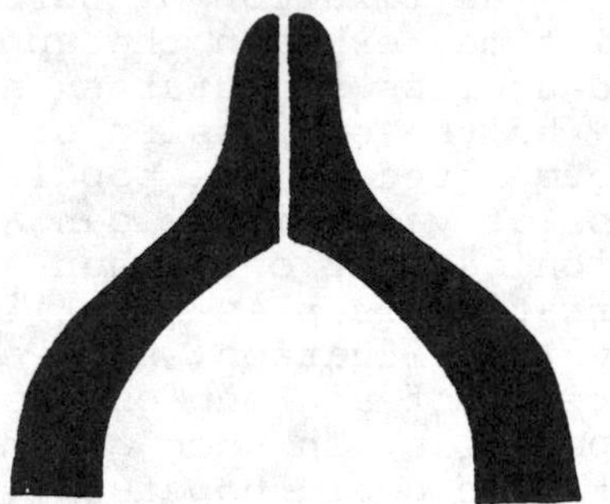

Convergent evolution

Diagram illustrating the difference between divergent and convergent evolution. In adaptive radiation, a single stock may branch to give many diverging stocks (in the given diagram, 1 stock <u>diverges</u> > 2 stocks). In convergent evolution, many stocks that are originally quite different can come to resemble each other more and more; as time passes, they converge. In the given diagram, 2 stocks are shown converging.

<u>Solution:</u> Because of the constant competition for food and living space, a group of organisms will tend to spread out and occupy as many different habitats as possible. This process, by which a single ancestral species evolves to a variety of forms that occupy somewhat different habitats, is termed adaptive radiation. Adaptive radiation is clearly an advantageous process in evolution in that it enables the organisms to tap new sources of food or to escape from predators. A classical illustration of adaptive radiation is the great variety of finches found on the Galapagos Islands west of Ecuador.

These finches, derived from a single common ancestor, exhibit diversity in beak size and structures, as well as in feeding habits, all of which are mutually related. Some of these birds feed on seeds, others feed mainly on cacti, and still others live in trees and eat insects. The diversity of food sources on the island has allowed each of the many forms of finches to survive in its particular habitat, and thus, prevent intra specific competition for food and space. Such adaptive radiation is often called divergent evolution, since its result is a diversity of adaptive forms evolving from a common ancestor.

An opposite process is convergent evolution, which also occurs quite frequently. Convergent evolution is the process by which two or more unrelated groups become adapted to a similar environment, and in doing so, develop characteristics that are more or less similar. For example, wings have evolved not only in birds but also in mammals (bats), reptiles (pterosaurs), and insects (grasshoppers). Likewise, the shark (a cartilagenous fish), the dolphin, and the porpoise (both mammals) have developed marked superficial similarity because of their adaptation to a similar environment.

CLASSICAL THEORIES OF EVOLUTION

• PROBLEM 27-5

What is meant by evolution? Explain briefly the concepts of inorganic and organic evolutions.

Solution: Evolution, in the broadest sense, means a gradual, orderly change from one condition to another. Evolution may be either inorganic or organic.

The planets and stars, the earth's topography, the chemical compounds of the universe, and even the chemical elements and their subatomic particles have undergone gradual, orderly changes. This is referred to as inorganic evolution since these changes do not involve living matter. Evolution that involves changes in the substance of living organisms is known as organic (meaning "living") evolution. The principle of organic evolution states that all the various plants and animals existing at the present time have descended from other, usually simpler, organisms by gradual modifications accumulated in successive generations. These modifications may be anatomical, physiological, biochemical, reproductive, or any other aspects of a species, that could have been instrumental in the emergence, proliferation, or complete extinction of a species.

• PROBLEM 27-6

What is implied by the theory of uniformitarianism?

Solution: The concept of uniformitarianism, first developed by James Hutton in 1785, states that the geologic forces at work in the past were the same as those of the present. After carefully studying the erosion of valleys by rivers and the formation of sedimentary deposits at the river mouths, Hutton concluded that the processes of erosion, sedimentation, disruption and uplift, over long periods of time, could account for the formation of the fossil-bearing rock strata which had interested many scientists.

Since the geologic factors operating on earth remain uniform throughout the ages, the rock strata bearing fossils of varying life forms reflect the gradual changes in the structures and living habits of organisms through time. The theory of uniformitarianism therefore implies that animals and plants continually undergo a process of organic evolution; their remains being fossilized in the earth's crust in layers by gradual deposition of sediment. Uniformitarianism also implies that the earth can not be only a few thousand years in age, as was widely accepted to be true two centuries ago. It must be much older, at least old enough for the process of organic evolution and the geological evolution of the earth's crust, both very slow, and gradual processes, to have occurred.

Charles Lyell expanded on Hutton's work. Lyell posited the idea that slight geological changes can have large-scale effects over long time intervals. He linked the concept of uniformitarianism with geological time. This association had a profound effect on the models of evolution developed by Darwin.

• PROBLEM 27-7

Describe the steps by which simple inorganic substances may have undergone chemical evolution to yield the complex system of organic chemicals we recognize as a living thing. Which of these steps have been duplicated experimentally?

Solution: Life did not appear on earth until about three billion years ago. This was some two billion years after the formation of the earth, either from a portion broken off from the sun or by the gradual condensation of interstellar dust. The primitive atmosphere before the appearance of any form of life is believed to have contained essentially no free oxygen; all the oxygen atoms present were combined as water or as oxides. Deprived of free oxygen, it was thus a strongly reducing environment composed of methane, ammonia, and water which originated from the earth's interior. At that time there were obviously no organic compounds on earth.

Reactions by which organic substances can be synthesized from inorganic ones are now well known. Originally, the carbon atoms were present mainly as metallic carbides. These could have reacted with water to form acetylene, which could subsequently have polymerized to form larger organic compounds. That such reactions occurred was suggested by Melvin Calvin's experiment in which solutions of carbon dioxide and water were energetically irradiated and formic, oxalic, and succinic acids were produced.

These organic acids are important because they are intermediates in certain metabolic pathways of living organisms.

After the appearance of organic compounds, it is believed, simple amino acids evolved. How this came about was demonstrated by Urey and Miller, who in 1953 exposed a mixture of water vapor, methane, ammonia and hydrogen gases to electric charges for a week. Amino acids such as glycine and alanine resulted. The earth's crust in prebiotic times probably contained carbides, water vapor, methane, ammonia and hydrogen gases. Ultraviolet radiation or lightning discharges could have provided energies analogous to the Urey-Miller apparatus, and in this manner, simple organic compounds could have been produced.

Most, if not all, of the reactions by which the more complex organic substances were formed probably occurred in the sea, in which the inorganic precursors and organic products of the reaction were dissolved and mixed. These molecules collided, reacted, and aggregated in the sea to form new molecules of increasing size and complexity. Intermolecular attraction provided the means by which large, complex, specific molecules could have formed spontaneously. Once protein molecules had been formed, they acted as enzymes to catalyze other organic reactions, speeding up the rate of formation of additional molecules. As evolution progressed, proteins catalyzed the polymerization of nucleic acids, giving rise to complex DNA molecules, the hereditary materials and regulators of important functions in living organisms. Enzymes also probably catalyzed the structural combination of proteins and lipids to form membranes, permitting the accumulation of some molecules and the exclusion of others. With DNA and a membrane structure, the stage was set for life to begin some three billion years ago.

More recent work by Clifford Matthews led to the discovery that ammonia and methane form polymers of hydrogen cyanide. It reacts with water to form amino acids. During the origin of life on the Earth these reactions occurred in clay and tide pools. Also, recent experiments on polymerization have shown RNA formations where these nucleic acids are enzymatic and self-replicating.

• PROBLEM 27-8

Discuss the current theory of the evolution of autotrophs from primitive heterotrophs.

Solution: The first organisms on earth were almost certainly heterotrophs, which, in an atmosphere lacking free oxygen, presumably obtained energy by the fermentation of certain organic substances in the sea. Because of their dependence upon the external environment for food, they could survive only as long as the supply of organic molecules in the sea lasted. Because food in the sea was necessarily limited, before the supply was exhausted, some of the then existing heterotrophs must have evolved the mechanisms to become autotrophs. They would then be able to make their own organic molecules either by chemosynthesis or photosynthesis.

An explanation of how an autotroph may have evolved from one of the primitive heterotrophs was offered by N.H. Horowitz in 1945. Horowitz postulated that the evolution was a result of successive gene mutations in certain heterotrophs which resulted in the enzymes needed to synthesize complex substances from simple ones. According to this hypothesis, the primitive heterotrophs were able to survive only if a certain specific organic substance x, limited in supply, was present in the surrounding sea. If a mutation occurred for a new enzyme enabling the heterotroph to synthesize substance x from some other more abundantly available organic form, the strain of heterotrophs with this mutation would be able to survive when the supply of x in the sea was exhausted.A second mutation that established an enzyme catalyzing another reaction by which x could be made from inorganic substances would again have survival value. Finally, when all the organic molecules in the sea were exhausted, a mutation for the synthesis of x from some inorganic substances, present in large quantities, marked an important evolutionary advantage.

In this way, successive mutations over time led to the production of a strain of organisms able to synthesize all of its metabolic requirements from simple inorganic compounds. Such organisms, called the autotrophs, are exemplified by the many photosynthetic algae and chemosynthetic bacteria that are believed to be among the earliest inhabitants of the earth.

• PROBLEM 27-9

In what way does Lamarck's theory of acquired characteristics not agree with present evidence?

Solution: Lamarck, like most biologists of his time, believed that organisms are guided through their lives by innate and mysterious forces that enable them to adapt to and overcome adverse evironmental forces. He believed that, once made, these adaptations are transmitted from generation to generation - that is, that acquired characteristics are inherited. In Lamarck's belief, then, new organs can arise in response to demands of the evironment, their size being proportional to their use. Changes in the size of such organs are inherited by succeeding generations. An example of this, according to Lamarck was the giraffe's long neck. He claimed that it resulted when an ancestor of the present giraffe took to eating the leaves of trees, instead of grass. Since the giraffe had to continually reach up to eat, its neck became stretched. The longer neck that developed was then inherited by the giraffe's descendants.

Lamarck's theory of the inheritance of acquired traits, though attractive, is unacceptable. Overwhelming genetic evidence indicates that acquired characteristics cannot be inherited. Many experiments have been performed in attempts to demonstrate the inheritance of acquired traits, all ending in failure. Geneticists now firmly believe that acquired traits cannot be inherited because they are not translated into a genetic message and incorporated into the genetic apparatus - the chromosomes - of cells. In particular, the genes of the sex cells are not changed. In other words, acquired traits involve changes only in the phenotype, not in the genotype; for this reason they are not inherited, and thus, are of no evolutionary significance.

• **PROBLEM 27-10**

What contributions did Darwin make to the theory of evolution?

Solution: Darwin made a twofold contribution to the study of evolution. He presented a mass of detailed evidence and convincing argument to prove that organic evolution actually occurred, and he devised the theory of natural selection, to explain how organic evolution works.

Darwin spent a major part of his life studying the animals, plants, and geologic formations of coasts and islands, at the same time making extensive collections and notes. When he was studying the native inhabitants of the Galapagos Islands, he was fascinated by the diversity of the giant tortoises and the finches that lived on each of the islands. The diversity was gradual and continual, and could not be explained by the theory of special creation, which stated that all living things are periodically destroyed and recreated anew by special, unknown acts of creation. As Darwin mused over his observations, he was led to reject this commonly held theory and seek an alternative explanation for his observations.

Then Darwin came up with the idea of natural selection. He believed that the process of evolution occurred largely as a result of the selection of traits by constantly operating natural forces, such as wind, flood, heat, cold and so forth. Since a larger number of individuals are born than can survive, there is a struggle for survival, necessitating a competition for food and space. Those individuals with characteristics that better equip them to survive in a given environment will be favored over others that are not as well adapted. The surviving individuals will give rise to the next generation, transmitting the environmentally favored traits to the descendents. Since the environment is continually changing, the traits to be selected also change. Therefore, the operation of natural selection over many years could lead ultimately to the development of descendants that are quite different from their ancestors - different enough to emerge as a new species. The formation, extinction, and modification of a species, therefore, are regulated by the process of natural selection. This principle, according to Darwin, is the major governing force of evolution.

• **PROBLEM 27-11**

What is differential reproduction? What factors bring

about differential reproduction?

Solution: When the percentage of viable offspring carrying a certain genotype cannot be accounted for by purely random mating, differential reproduction is said to have occurred. Differential reproduction can result from nonrandom mating, differential fecundity, or differences in either zygote viability or offspring fertility.

Nonrandom mating may occur due to genotypic variaton, genetic mutations, or both. There are well established behavior patterns of courtship and mating in many species, which are primarily genetically controlled. For example, in many species of birds and fish, a brightly colored spot on the male serves to stimulate the female for copulation. Males having this characteristic will succeed in finding a mate easily and more frequently than males without this trait, hence producing in their lifetime more offspring carrying their genes. Males lacking this trait will find it difficult to pass their genes on to succeeding generations. Mutations that lead to formation of bigger or brighter spots will benefit the males with the mutated gene by rendering them more attractive to females. Conversely, mutations that lead to disappearance of these spots or formation of smaller, duller ones, will have a negative selective value. These mutations are inherited by the progeny, perpetuating differential reproduction through nonrandom mating.

Success in mating, however, can neither guarantee successful fertilization nor assure large numbers of offspring. Such factors can also be important in differential reproduction. Differences in the number of gametes produced by different individuals and the proportion of their gametes that will successfully unite with others to form zygotes can be collectively termed differential fecundity. Differential fecundity may be determined genetically. Within a species, the individuals producing large numbers of gametes or having a large percentage of successful matings (resulting in fertilization) will necessarily contribute the greatest percentage of genes to the next generation. This depends, of course, on the viability of the zygotes formed. In environments which result in low zygote viability, individuals having high fecundity will tend to reproduce in larger numbers. In fish, for example, zygotes must be fertilized and also develop in an external environment; those species producing large numbers of both gametes and zygotes will tend to be preserved. Mammals, on the other hand, whose zygotes develop internally, cannot and need not produce such large numbers of gametes or zygotes.
In fact, large numbers of zygotes may actually be a disadvantage, since this limits the amount of care and feeding available to each offspring.

Another factor affecting the reproducing ability of a given individual or individuals is the fertility of the offspring. The offspring produced as a result of a given mating may be viable, but unless they are themselves fertile, the result is the same as if they were not viable.

Thus, it is the interaction of a variety of factors which is responsible for the reproductive capacity of a given individual or species. Any one of these factors could be responsible for differential reproduction.

APPLICATIONS OF CLASSICAL THEORY

• PROBLEM 27-12

Describe briefly the Darwin-Wallace theory of natural selection. What is meant by "survival of the fittest"? Do you think this applies to human populations today?

Solution: The Darwin-Wallace theory of natural selection states that a significant part of evolution is dictated by natural forces, which select for survival those organisms that can respond best to them. Since more organisms are born than can be accomodated by the environment, there must be chosen among the large numbers born the limited number to live and reproduce. Variation is characteristic of all animals and plants, and it is this variety which provides the means for this choice. Those individuals who are chosen for survival will be the ones with the most and best adaptive traits. These include the ability to compete successfully for food, water, shelter, and other essential elements, also the ability to reproduce and perpetuate the species, and the ability to resist adverse natural forces, which are the agents of selection.

Essential to the theory of natural selection are the ideas of the "struggle for existance" and "survival of the fittest." Because the resources of the environment are naturally limited, individuals must struggle among themselves for food and space. Those individuals with the traits that are best suited for the given environment will survive and multiply, while others will slowly decline in number or disappear completely. Since the natural forces are constantly operating and changing, struggle for survival goes on forever and the fittest of the competitors survive.

Survival of the fittest holds true for human populations, but it is less obvious among human beings than lower organisms. Whereas it is common to see wild dogs fighting each other for a piece of food, it is less common to see two human beings fighting for the same reason. Among human beings, the problems of food and shelter have been conquered successfully, such as by farming or by raising cattle. Actual physical struggle between human beings for food has become less important. The traits of strong muscles, great agility and alertness, and other fighting abilities that were once important to the cavemen have more or less lost their significance today.

• **PROBLEM 27-13**

Some scientists object to the Darwin-Wallace theory of natural selection on the basis that it cannot explain the presence of many apparently useless structures in some organisms. These vestigial structures seem to defy the processes of natural selection which tend to provide successive generations with better adaptations to their environment. How can you defend the theory of natural selection?

Solution: While certain structures may be useless to the organism in which they occur, they may be present because they are controlled by genes that are closely linked to other genes controlling characteristics important for survival. Such closely linked genes will be inherited together in most cases, and so the vestigial characteristic will tend to be inherited along with the more important trait. Other vestigial structures may have physiological effects of survival value which are not visible to the observer. They may represent intermediate stages in the evolution of some important adaptive structure, or be an incidental by-product of an important but invisible physiological trait. They may also be present because they have become fixed in a population by chance or genetic drift.

Genetic drift is important in the evolution of small, isolated populations. Within a small breeding population, heterozygous gene pairs tend to become homozygous for one allele or the other by chance rather than selection. This may lead to the expression and accumulation of certain phenotypes of no particular adaptive value. If the phenotypes are not lethal, they will remain in the population. Because natural selection can act only when there is a certain level of variation among the individuals, it loses its significance in a small breeding population. Under this condition, chance becomes the important factor in evolution.

Thus, the presence of apparently useless traits does not necessarily invalidate the theory of natural selection. It should be noted that natural selection and the presence of vestigial structures are mutually exclusive phenomena, and therefore, the inheritance of vestigial traits should not be governed by selection processes. Natural selection is defined as the process by which a beneficial trait is selected for, such that more organisms possessing the gene for this trait will be present in the population. These are the organisms producing the most offspring, and thus contributing the most genes to the population's gene pool. Vestigial structures are defined as those structures which no longer have any biological significance, neither advantageous, nor deleterious to the organism possessing them. Since natural selection operates to select for advantageous traits by eliminating deleterious traits, and vestigial structures are neither beneficial nor harmful, these structures will neither be selected for nor eliminated. However, it should be noted that many vestigial traits are remnants of what were once important structures. While the genes for the vestigial traits were not directly selected for, their precursor traits were, and so more individuals than not will tend to possess the vestigial trait. Its inheritance in future generations, however, will not be governed by natural selection.

• PROBLEM 27-14

Why is the Hardy-Weinberg population only an idealistic model and not representative of every natural population?

Solution: A Hardy-Weinberg population, like all statistical samples, exists only when the number of individuals involved is significantly large. In addition to a large population size, there must be no selection for or against any specific characteristic, no occurrence of genetic mutations, no movement of individuals into or out of the population, and complete random mating.

The Hardy-Weinberg Principle fails to explain the genetic composition of the population if any of the above conditions is not met. Natural populations rarely meet more than one, if any, of the conditions. While it is true that many populations in the natural environment are large, many are small and relatively isolated on islands, mountain slopes and other physical or geographical barriers. These isolated populations may contain homogeneous individuals as a result of continued inbreeding over long periods of time or through the fixation of genes caused by genetic drift. The condition of no selection for or against any specific trait is also readily violated in most natural populations. Selection by natural forces including wind, cold, heat, disease, and predators is clearly evident, and the individuals that survive tend to have certain common adaptive traits.

Mating of individuals in a population or species is rarely random. The more colorful male birds of some species, for example, are more effective in attracting and stimulating female birds. It is also not unusual to observe that in some species, bigger individiuals tend to mate with each other, and the same is true of smaller individuals. Mutations in the natural population have been shown to occur constantly, contributing a great number of changes to a species in its evolutionary development. Finally, whereas some populations are indeed isolated and have no free migration of individuals, many populations do allow the "come and go" of their members. Those migrating into the population may be able to mate with the native members, adding new genes to the population as a result. Those leaving the population are in effect, removing genes from the population.

• PROBLEM 27-15

Which is of greater significance in evolution - the individual or the population? Explain your answer.

Solution: Evolution is not the change in an individual during its life time, but the change in the characteristics of a population over many generations. In other words, on individual cannot evolve, but a population can. The genetic makeup of an individual is set from the moment of conception, and most of the changes during its lifetime are simply changes in the expression of the developmental potential inherent in its genes. But in a population, both the genetic makeup (the gene pool) and the expression of the developmental potential can change. The change in allelic frequencies of a given population over time is evolution.

EVOLUTIONARY FACTORS

• PROBLEM 27-16

What are the major forces of evolution?

Solution: Evolution is the result of the interaction of four major forces. These are 1) mutation, 2) genetic drift, 3) migration, and 4) natural selection. These four forces have one thing in common - each can bring about evolution by changing the allele frequencies in the gene pool of a population over time.

Mutations are random events that occur at a very low rate of approximately one out of every 10^6 genes. If evolution depended on the occurrence of mutations alone, life would not exist as we know it today. However, mutation is important because it introduces variety into a population. Natural selection can act upon these variations to increase the frequencies of the few adaptive mutations and decrease the frequencies of the many more maladaptive ones over time.

Genetic drift acts in the evolutionary process by causing chance fluctuations in gene pool frequencies. Although this has essentially no effect in large populations, it is probably the major cause of evolution in small populations. In these populations, chance - including accidents, mating preferences and other unpredictable events - can cause an allele's frequency to fluctuate widely from one generation to the next. This can result in either the fixation of that allele or its elimination from the gene pool.

Genetic drift can also interact with the processes by which small populations become established to produce major evolutionary changes. Two processes of population establishment are the 1) Founder Effect and 2) isolation.

In the Founder Effect, a small group of individuals may leave a large population and re-establish in another location. The main population has certain gene frequencies - for example, the frequency of gene A may be .6, and that of gene a may be .4. It is very possible that the small group will, entirely by chance, have allele frequencies different than those of the main population - perhaps A= .2 and a=.8. Thus, the new population being founded will begin with a different assortment and proportion of alleles than the parent population. If the new population survives, genetic drift over time may result in its becoming drastically different from the original population. This has been shown to be especially true on small isolated islands that have been settled by a few animals. These same principles may be applied to small, naturally occurring populations that have been isolated by events in nature. For example, after a severe dry season, marine organisms formerly inhabiting one large lake may become isolated from one another as the lake dries up into many small pools. This could result in a variety of small populations having varying allele frequencies and gene pool composition.

Migration occurs when individuals from one breeding population leave to join another. Migration can be either inward (immigration) or outward (emigration). The effect of migration on allele frequency in a population is a function of the size of that population in relation to the size of the migration. In a small population, the frequency of many genes can be significantly altered by a small migration, while large populations can usually absorb large amounts of migration before a noticeable change in frequencies occurs. Migration may lead to either more variation in a population, due to the introduction of new genes through immigration, or less variation due to the loss of genes through emigration.

Natural selection ensures that the changes in allele frequency caused by mutation, migration, and genetic drift are adaptive. As such, evolution will more likely give rise to a species better adapted to live in its environment. Although the previously discussed forces are non-directional, natural selection is a directional process through which populations adapt to their particular environments by changing their allele frequencies in response to environmental or selection pressure. More specifically, natural selection operates through differential reproduction, which is said to have occurred when certain individuals are able, by surviving and/or reproducing at a higher rate, to preferentially propogate and transmit their respective genes over those of other individuals. Natural selection is a very gradual process, with the changes accumulating and altering the physical appearance of a population over very long periods of time.

• PROBLEM 27-17

What are the five principles of evolution?

Solution: There are now five principles of evolution on which all biologists agree. First, evolution occurs more rapidly at some times than at others. At the present geological time it is occurring rapidly, with many new forms appearing and many old ones becoming extinct. Secondly, evolution does not proceed at the same rate in all types of organisms. For example, the lampshells have remained unchanged for at least five hundred million years, while several species of man have appeared and become extinct in the past few hundred thousand years. Thirdly, most new species do not evolve from the most advanced and specialized forms already living, but from relatively simple, unspecialized forms. Thus, mammals did not evolve from the large,specialized dinosaurs but from a group of rather small and unspecialized reptiles. Likewise, the vascular plants did not derive from the bryophytes but from lower and more primitive plants, like green algae. Fourthly, evolution is not always from the simple to the complex. There are many examples of "regressive" evolution, in which a complex form has given rise to simpler ones. For instance, many wingless birds have evolved from winged ones, and legless snakes from reptiles with appendages. This is because mutations occur at random, and if there is some advantage for a species in having a simpler structure, any mutations that happen to occur for such conditions will tend to accumulate by natural selection. Those mutations that result in a complicated delerious structure will be selected against. Fifthly, evolution occurs in populations (not in individuals) by the processes of mutation, migration, natural selection, and genetic drift.

SPECIATION

• PROBLEM 27-18

From what you have learned of the processes of heredity and evolution, how do you think new species arise with respect to mutation? Which is more important, the accumulation of small mutations or a few mutations with large phenotypic effects? Give the reasons for your answer.

Solution: Most adaptive mutations produce barely distinguishable changes in the structure and function of the organism in which they occur. On the other hand, mutations with the potential to produce large phenotypic effects, necessary to elicit major changes in one jump, are practically always lethal or at least detrimental to survival. Major mutations are deleterious because they constitute a great disturbance to the delicately balanced genetic systems in which they arise. Therefore, major genotypic changes through mutation are not likely to be important in evolution.

In contrast, the accumulation of small mutations over the course of many generations has significant evolutionary value. This has been supported by fossil evidence, laboratory experiments and field studies. Evolutionary changes appear to arise as a result of many small changes that have accumulated over time, ultimately producing a change in the composition of a given population's gene pool. An example of this can be seen in the evolution of the orchid. The modern orchid flower, in both shape and color, resembles the female wasp. Male wasps, thinking it is a female, are attracted to the flower and stimulated to attempt copulation. In doing so, they become coated with the sticky pollen grains of the flower and carry them along to the next orchid. Any mutation that gave the orchid any similarity, however slight, to the female wasp would increase its chances of being pollinated. Thus, such mutations would give those orchids bearing them a selective advantage over orchids without them. The increased rate of pollination of the mutated forms, would promote the propogation of the mutant gene. It is easy to see how, over time, such mutations would tend to be propogated and accumulate to the point where the orchid flower is today.

• PROBLEM 27-19

Discuss the establishment of a new species of plant or animal by the theory of isolation.

Solution: It is currently believed that new species of animals and plants may arise because some physical separation discontinues the distribution of a species for long periods of time. Such separation may be caused by geographic changes, such as an emerging piece of land dividing a marine habitat, or climatic changes, such as a heavy drought causing large lakes to divide into numerous smaller lakes or rivers to be reduced to isolated series of pools. In some species, behavioral patterns may prevent dispersal across areas that could easily be traversed if the attempt was made. For example, rivers may serve to isolate bird populations on opposite banks, or a narrow strip of woods may effectively separate two meadow populations of butterflies.

Such factors, then, may separate a single species population into two or more isolated groups. Each group will respond to the selection pressure of its own environment. Since mutation is a random process, it is not to be expected that identical mutations will show up in the different populations. A single evolving unit has thus become two independently evolving units. As long as such evolutionary units remain isolated, they continue to respond independently to evolutionary forces. Gradually, the two isolated groups will find their gene pools diverging, and if isolation and independent evolution continue for a long enough period of time, the genetic composition of the two groups may become so different that they become unable to breed with each other, even when brought very close together. Since two populations must be able to interbreed in nature if they belong to the same species, genetic isolation, brought about in this case by a long duration of physical isolation, causes two new distinct species to arise.

• PROBLEM 27-20

Besides geographic isolation, what mechanisms maintain isolation and differentiation?

Solution: Geographic isolation is an extrinsic mechanism of speciation. There are, in addition, a variety of intrinsic isolating mechanisms which establish and maintain the differentiation of species. In ecogeographic isolation, two populations initially separated by some extrinsic barrier may in time become so specialized to their respective environmental conditions that even if the original extrinsic barrier was removed, they would not be able to interbreed with each other. In effect, they have evolved genetic differences great enough to maintain them as two distinct species. In habitat isolation, two populations occupying different habitats which are within reach of each other find their individuals mating only with members of their own population simply due to a greater frequency of encounter. In time they may evolve into two different species. In behavioral isolation, two populations diverge genetically as a result of some behavioral differences such as courtship patterns and mate preference, that prevent them from interbreeding. In mechanical isolation, structural differences between two closely related groups make it physically impossible for matings between males of one and females of the other to occur. In this way the two populations are prevented from interbreeding. In seasonal isolation, two groups cannot interbreed even when brought together because they breed during different seasons of the year. In gametic isolation, even if individuals of two groups succeed in mating with each other, actual fertilization fails to take place. In developmental isolation, even when fertilization occurs, the development of the embryo is abnormal and may cease before birth. In hybrid inviability, hybrids are often weak and malformed, and frequently die before they reach their reproductive age. In hybrid sterility, the hybrids are normal structurally but are sterile. In selective hybrid elimination, the hybrids are normal and fertile, but their offspring are often less well adapted, and may soon be eliminated. Finally, in polyploidy speciation (which occurs in plants,) polyploids carrying an even-numbered multiple of chromosomal complements (4N, 6N, and so on) are unable to cross with the normal diploid individuals, but can breed among themselves.

The polyploids are genetically distinctive and reproductively isolated, they are usually more vigorous than the parental diploid form and are able to maintain themselves as a separate species.

• **PROBLEM 27-21**

> Explain how speciation by hybridization occurs. Contrast it with speciation by geographic isolation.

Solution: Occasionally members of two different but closely related species may cross and successfully produce viable hybrid offspring. Such hybrids of two different species, however, are usually not fertile. Their unlike chromosomes cannot pair in meiosis, and the resulting eggs and sperm do not receive the proper assortment of chromosomes. The hybrids, therefore, cannot continue their genetic line unless their chromosomes undergo some change that will allow them to synapse. One such change, which has been frequently observed in some species of plants, is the doubling of chromosomes in the gametes. This results in the production of tetraploids from the normal diploids. In the tetraploid progeny synapsis can occur between duplicated chromosomes, meiosis can take place in a normal fashion, and normal fertile eggs and sperm will be produced. Thereafter, the tetraploid hybrids will breed true, but will be unable to cross with either of the parental types. This is because a cross between a tetraploid hybrid and a diploid parent will produce triploid offspring, which, because of the highly irregular distribution of their chromosomes at meiosis, are sterile. Thus, the tetraploid hybrids (or any other polyploids with an even chromosome number) become a new, distinct species.

Through hybridization, the best characters of each of the original species may be combined in descendents better able to survive than the original species, and the worst combined in descendents that are doomed to be eliminated by natural selection. Hence, hybridization tends to produce new species that are better adapted than the parental species.

As was previously seen, hybridization with doubling of chromosomes produces new species with characteristics bridging the two parental species. In such cases, there is a convergence of characteristics. Geographic isolation, in contrast tends to produce new species with diverging characteristics. Where hybridization results in formation of one species from two parental forms, geographic isolation results in two or more distinct species from one common parental form.

• **PROBLEM 27-22**

> If mating is observed to occur freely between organisms of two different populations in the laboratory, and the cross produces viable offspring, can you assume that the two populations belong to the same species?

Solution: Successful interbreeding in the laboratory does not offer enough evidence that two populations must be of the same species. It only indicates that certain types of intrinsic isolation does not exist between the populations. It does not say whether effective interbreeding will occur in nature, nor does it clarify whether or not other kinds of intrinsic isolation are present.

By definition, two individuals are from the same species if no intrinsic reproductive barrier exists between them to prevent a genetic exchange in nature. Extrinsic isolation, commonly exemplified by geographic separation, will not cause speciation if intrinsic isolation is not established. In other words, two groups of organisms separated by, say, a mountain range, cannot be regarded as two species unless they fail to successfully interbreed when the separating factor is removed. Whereas interbreeding of two populations under a laboratory setting demonstrates that certain intrinsic isolating mechanisms, such as mechanical and gametic isolations, does not exist between them, it does not preclude the presence of other intrinsic isolating factors that are very much dependent on the natural environment. For example, under natural conditions, ecogeographic, habitat isolation, or seasonal isolation may exist, yet be inoperative under laboratory conditions. Observation of breeding in the laboratory cannot ascertain whether fertile mating would occur, should individuals of the populations be released from the laboratory and returned to their own natural habitats, which we assume here to be continuous with each other. Under this assumption, even though no physical barrier is present, the existence of some type of intrinsic isolation which is nature-dependent would necessarily inhibit gene flow from one population to the other. In short, since experimental finding cannot predict accurately what is to happen in nature, and since speciation is a phenomenon of the natural world rather than the artificial laboratory, we cannot assume that two populations that interbreed in the laboratory necessarily would do so in nature.

SHORT ANSWER QUESTIONS FOR REVIEW

Answer

To be covered when testing yourself

Choose the correct answer.

1. The evolutionary concept that variation was due to use and disuse of parts and the concept of inheritance of acquired characteristics is associated with (a) Darwin. (b) Aristotle. (c) Lamarck. (d) Wallace.

   c

2. The percentage of occurrence of a particular gene in a population is known as the gene (a) frequency, (b) pool, (c) number, (d) total.

   a

3. All of the following are part of the theory of evolution originally proposed by Darwin and Wallace except (a) natural selection, (b) inheritance and variation, (c) overproduction of offspring, (d) use and disuse of organs.

   d

4. The random establishment of nonadaptive, bizarre types in small populations is known as (a) genetic drift, (b) migration, (c) selection pressure, (d) the Hardy-Weinberg law.

   a

5. The most recent theories of the origin of life include all of the following elements in the primitive atmosphere except (a) oxygen, (b) hydrogen, (c) ammonia, (d) methane, (e) carbon dioxide.

   a

6. Evolutionary convergence is the development of (a) dissimilar structures in organisms of dissimilar ancestry, (b) similar structures in organisms of dissimilar ancestry, (c) similar structures in organisms of similar ancestry, (d) dissimilar structures in organisms of similar ancestry.

   b

7. According to recent views on the origin of life, the earliest organisms (a) were autotrophs, (b) respired aerobically, (c) were heterotrophs, (d) none of the above.

   c

8. The expression $p^2 + 2pq + q^2 = 1$ is an algebraic statement illustrating (a) the distribution of genotypes for a particular trait in a large population, (b) the rate of mutation, (c) the relationship between isolation and variation, (d) the constant change of gene frequencies in a randomly mating population.

   a

9. A definition of a species is that it is a group of populations all sharing the same (a) habitat, (b) mutations, (c) gene pool, (d) acquired character.

   c

	Answer
	To be covered when testing yourself
10. Differential reproduction can result from all of the following except (a) random mating, (b) non-random mating, (c) differences in zygote viability, (d) differences in offspring fertility.	a
11. Choose the correct statement. (a) Evolution always proceeds from the simplest to the more complex. (b) Evolution occurs at a steady rate. (c) Evolution occurs at the same rate in all types of organisms. (d) Most new species evolve from relatively simpler forms rather than from the more advanced form of an established species.	d
12. According to population genetics, evolution is best defined as (a) any change in gene frequency, (b) variation due to the environment, (c) a tendency toward geographic isolation, (d) the result of interbreeding.	a
13. The frequency of a gene in a population will not be largely affected by (a) mutation, (b) small population size, (c) selective migration, (d) random mating.	d
Fill in the blanks.	
14. To the extent that genetic variations originate at _____, evolutionary innovations similarly appear at _____.	random, random
15. Natural selection operates basically through _____, not through struggle for survival.	reproduction
16. The role of chance in the evolution of small breeding populations has been described as _____.	genetic drift
17. Evolution occurs in populations by the processes of _____, _____, _____ and _____.	mutation, differential reproduction, natural selection, genetic drift
18. One _____ is isolated from the next by reproductive barriers.	species
19. If the Hardy-Weinberg equilibrium exists, the relative frequencies of the genes in a population will remain _____ from one generation to another.	constant
20. _____ suggested that traits which an organism acquired during its lifetime could be inherited.	Lamarck

Answer

To be covered when testing yourself

Question	Answer
21. Darwin and Wallace observed that although a species tends to multiply _____, (arithmetically/ geometrically), populations tend to remain constant.	geometri- cally
22. Evolution has produced not what is theorectically best, but what is practically possible. There is no predetermined plan or goal. This characteristic of evolution is called _____.	random oppor- tunism
23. Stanley Miller, in a classic experiment, mixed water vapor, methane, ammonia and hydrogen continuously for a week over an electric spark, to obtain certain _____ and other products.	amino acids
24. In the past ages, sudden changes in environment not followed by readaptation of local species led to _____ of those species.	extinction
25. _____ occurs when, after one group of organisms has become extinct, another group evolves that adopts the vacated environment and way of life.	repla- cement
26. Organisms undergo evolutionary changes which allow them to fit into all available ecological niches. This is an example of _____.	adaptive radiation
27. _____ acting on genetic variability is the prime source of evolutionary change.	natural selection
Determine whether the following statements are true of false.	
28. In nature, populations may be small, and consequently the Hardy-Weinberg law will be strictly followed.	False
29. Evolution has been less rapid in terrestrial types than in aquatic types.	False
30. The usual pattern for major evolutionary changes usually is one of gradual accumulation of many small changes.	True
31. A species can be defined as the largest unit of population within which effective gene flow can occur.	True
32. The term natural selection is synonymous with differential reproduction, a phrase which indicates simply that some members of a population have more offpsring than others.	True
33. Within a large population, natural selection helps to maintain a constant gene pool.	True

Answer

To be covered when testing yourself

34. A great variety of mammals evolve from a single common ancestor exhibiting many types of modifications. Such diversification from a single ancestor is called convergent evolution. — False

35. Polymorphism has adaptive value in that it adapts the species to different environmental conditions. — True

36. In 1952 Stanley Miller used an apparatus to mix water, methane, ammonia and hydrogen and subject this to an electrical change. The outcome was the presence of a very primitive form of life. — False

37. In the equation representing the Hardy-Weinberg principle, $p^2 + 2pq + q^2 = 1$, the expression 2pq represents the frequency of both homozygous dominants and homozygous recessives. — False

38. In any population of organisms, the genes responsible for traits having survival values usually increase in frequency. — True

39. Life can never redevelop as it did because the O_2 in the air has produced an ozone layer which screens out ultraviolet radiation. — True

40. Darwin was successful in identifying the genetic causes of evolutionary change. — False

41. Appearance of genetic variation can only occur by mutation. — False

CHAPTER 28

EVIDENCE FOR EVOLUTION

DEFINITIONS

• **PROBLEM** 28-1

What exactly do the terms "primitive" and "advanced" mean in evolution? Are "generalized" organisms necessarily more advanced than "specialized" ones?

Solution: "Primitive" and "advanced" are two contrasting terms. "Primitive" means older, more like the ancestral condition. "Advanced" means newer, less like the ancestral form. "Primitive" and "advanced" are just two terms referring to relative sequence in evolutionary time, and do not imply that one is necessarily superior to or more complex than the other.

The terms "specialized" and "generalized" refer to the relationships of organisms, or particular characteristics of organisms, to their environment. "Specialized" organisms or structures are adapted to a special, usually rather narrow, mode of life. "Generalized" organisms or structures are broadly adapted to a greater variety of environments and ways of life. The terms "specialized" and "generalized" are sometimes confused with "advanced" and "primative." Generalized characteristics are likely to be more primitive, and specialized ones more advanced, but this is not an absolute rule.

• **PROBLEM** 28-2

Define the terms "range" and "center of origin."

Solution: The range of a given species is defined as that portion of earth in which it is found. The range may be small or large. The kangeroos, for example, are found only in Australia, while man is found almost all over the world. The range of a species does not necessarily include its center of origin. The center of origin of a species is that particular place where it originated, since each species of plant or animal originated only once. It is now known that a new species arises in a particular geographic area. If the species is favored for survival, it spreads out until it is stopped by a barrier of some kind - physical, such as an ocean; environmental, such as an adverse climate; or biological, such as the absence of food, danger from predators, the failure to compete with native members, or the inability to produce fertile young.

• PROBLEM 28-3

Differentiate between homologous and analogous structures. Give examples of each.

Solution: Homologous structures are structures derived from a similar evolutionary origin. They may have diverged in their functions and phenotypic appearance but their relationships to adjacent structures and embryonic development are basically the same. Homologous structures, such as a seal's front flipper, a bat's wing, a cat's paw, a horse's front leg, and the human hand and arm, all have a single evolutionary origin, but have diverged in order to adapt to the different methods of locomotion required by different lifestyles.

Analogous structures are similar in function and often in superficial appearance but, in direct contrast to homologous structures, they are of different evolutionary origins. The wings of robins and the wings of butterflies are examples of analogous structures. Although both are for the same purpose, namely flying, they are not inherited from a common ancestor. Instead, they evolved independently from different ancestral structures.

FOSSILS AND DATING

• PROBLEM 28-4

What is paleontology? List the various kinds of paleontologic evidence. Why is it that fossils are frequently difficult to find?

Solution: Paleontology is the science of the finding, cataloguing and interpretation of the abundant and diverse evidence of life in ancient times. Fossils are the most important tools of the paleontologist in studying the past. Fossils can be bones, shells, teeth, and other hard parts of an animal or plant body which have been preserved, and indeed, include any impression or trace left by some previous organism. As such they constitute the paleontologic evidence of evolution.

Most of the vertebrate fossils found are skeletal parts. From these parts paleontologists can infer the contours of the body, the posture, and the style of walking of the ancient animal. Using this information, reconstructions can be made showing how the animal may have looked in life. Footprints or trails are another type of fossil. These are made by the animal in soft mud, which subsequently hardened. They enable one to infer the size, body proportion, structures, and style of walking of the animal that made them. Tools made by early man are also important artifacts. They aid in the determination of how and when bipedalism evolved (hands had to be free in order to use tools). The discovery of tools also serves as invaluable information, providing clues of early man's culture, and can even be used to infer the origin of language.

Petrified fossils result from the replacement by minerals of the original animal or plant material. The minerals that replace the once living tissues may be iron pyrites: silica, calcium carbonate or a variety of other substances. Petrifaction has been shown to be an effective type of preservation for tissues as old as 300,000,000 years. Molds are still another type of fossil, and are formed by the hardening of the material surrounding the buried organism, followed by the decay and removal of the organism by ground water. Sometimes the molds became accidentally filled with minerals, which then hardened to form casts or replicas of the original organism.

In the far north, frozen fossils of whole animal bodies have been found in the ground or ice. Old forms of plants, insects and spiders have been found preserved in amber, a fossil resin from pine trees. The resin was at first a soft sap engulfing the fragile insect; then it slowly hardened, preserving the insect intact.

Usually, then, the formation and preservation of a fossil follows the burying of some structure or entire organism. This may occur at the bottom of a deep body of water, or on land hidden deep in the earth's crust. Generally, fossils are preserved in areas that are well hidden or hard to reach. This is because fossils under shallow water or near the earth's surface are usually destroyed due to constant disturbances from natural forces, such as wind and rain, and from living organisms. Therefore, fossils are generally well concealed and difficult to find.

• PROBLEM 28-5

What are some factors that interfere with our obtaining a complete and unbiased picture of life in the past from a study of the fossil strata?

Solution: There are five major rock strata in the earth's crust, each of which is subdivided into lesser strata. Theoretically, because the strata were formed by the chronological deposition of sediment, the characteristics of each ancient era or period should become evident upon examining the fossil composition of each stratum and substratum. Also, one should be able to estimate the duration of each era or period from the thickness of each layer and the known rate of sediment deposition.

Knowledge of the fossil strata has been useful in the past and is still being used today in those cases when radioactive dating is not applicable. In relative dating, a time or era can be assigned to a fossil or artifact relative to other fossils or artifacts of known age in other stata. But this view is oversimplified. The relative length of each ancient epoch can not, in reality, be determined in great accuracy since the rate of sedimentary deposition varied at different times and in different places. While the layers of sedimentary rock should occur in the sequence of their deposition, with the newer, later strata on top of the older, earlier ones, geologic disturbances at some point of history may have changed the order and relationship of the layers. It is known that certain sections of the earth's crust have undergone tremendous foldings and splittings, and old layers may now cover new ones. In addition, in some regions, the strata formed previously have emerged from the sea due to the continuous addition of new layers; the top layers were then very easily carried away by running water or other natural disturbances, so that relatively recent strata were deposited directly on top of very ancient ones. Such factors make it difficult for us to obtain a complete, accurate picture of the ancient ages in chronological order.

• PROBLEM 28-6

Explain how an estimate of the age of a rock is made on the basis of the radioactive elements present.

Solution: Certain radioactive elements are spontaneously transformed into other elements at rates which are slow and essentially unaffected by external factors, such as the temperatures and pressures to which the elements are subjected. The transformation, or decay, of each individual element takes place at a rate which can be measured. For example, half of a given sample of the element uranium will be converted into lead in 4.5 billion years. Thus, 4.5 billion years is the half-life of uranium. By measuring the proportion of uranium and lead in a given rock, we can estimate with a high degree of accuracy the absolute age of the rock. For instance, assume we determine that the ratio of uranium to lead in a given sample is 1:1. We know, that the rock has existed long enough for half of the original amount of uranium to be converted into lead, or a total of 4.5 billion years, which is equal to one half-life of uranium. If the ratio is 1:3, three-fourths of the original amount of uranium is now lead, and one-fourth remains. Therefore, two half-lives must have passed, and half of the uranium remaining after one half-life has turned into lead ($\frac{1}{2} \times \frac{1}{4}x = \frac{1}{4}x$, where x equals the original amount of uranium in the sample).

Two half-lives are equivalent to a time period of 9 (= 2 × 4.5) billion years, and the age of the rock, then, is 9 billion years.

Other methods are used to determine the age of rocks which are not old enough to use the uranium - lead metod of dating. One of these is the potassium 40 - argon (K40 - Ar) method. Potassium - 40 decays into argon and has a half-life of 1.3 billion years. Because these two elements are commonly found in volcanic rock, this method is particularly useful in areas where volcanic activity is known to have occurred in the past. Other absolute dating techniques include fission - track and paleomagnetic reversals. These, along with others being developed, promise to provide a firm basis in understanding the chronology of the events in the evolutionary record.

• PROBLEM 28-7

An ancient Egyptian mummy is found. A count of the radiation emitted from the C^{14} in the body is only one eighth that found in persons who have died recently. About how old is the mummy? (The half-life of C^{14} is 5568 years.)

Solution: Certain radioactive elements are transformed spontaneously into other elements at rates which are gradual and essentially independent of the temperatures and pressures to which the elements have been subjected. In addition, unstable isotopes of certain elements also undergo gradual transformation to more stable forms. An example of this is the decay of radioactive isotope carbon-14 into the element nitrogen. This process of transformation is termed radioactive decay.

It is established that carbon-14 has a half-life of 5568 years. A half-life is defined as the period of time (in units of seconds, minutes, hours, years and so forth) that it takes for one half of a sample to undergo radioactive decay. Since one half of the original sample remains after one half-life, one fourth will remain after two half lives, and one eight after three half lives.

Therefore, the fact that the mummy contains one-eighth the amount of C^{14} of that in persons who have died recently indicates to us that three half lives have passed since the mummification. During this period, 7/8 of the total C-14 has decayed into the more stable element, nitrogen. The mummy is hence 3×5568 or 16,704 years old.

THE PALEOZOIC ERA

• PROBLEM 28-8

What was the Appalachian Revolution? When did it occur and what were its effects?

Solution: The Appalachian Revolution refers to the general folding of the earth's crust which raised the great Appalachian mountain chain running from Nova Scotia to Alabama. It occurred at the end of the Permian Period, the final period of Paleozoic Era, some 280 million years ago. The Appalachian Revolution also brought into existence some of the mountain ranges of Europe. Because of a great glaciation which affected most of the lands existing at that time, the climate became colder and drier. Many of the Paleozoic forms of life were unable to adjust to the climatic changes and became extinct. Many marine forms were lost during the Appalachian Revolution, due to the cooling of the water and the decrease in the amount of available space caused by the drying up of the shallow seas. Other new forms of organisms arose, and the Paleozoic Era gave way to the Mesozoic Era.

• PROBLEM 28-9

Explain the four important evolutionary events that occurred in the Paleozoic Era.

Solution: The Paleozoic Era was marked by four important evolutionary events. First, in the Ordovician Period, the ancestral backbone-bearing animals or vertebrates appeared. These early vertebrates were the ostracoderms, small, jawless, armored, fresh-water, bottom-dwelling fishes without fins. The ostracoderms then gave rise in the later periods to a great variety of finned fish. In the period following the Ordovician, the Silurian, two events of great biologic importance marked the second and third major evolutionary advancement of the Paleozoic Era. The first land plants appeared, signifying the beginning of the invasion of the land by plants. These plants most closely resembled the ferns. At about the same time, the first air-breathing animals arose. These were the arachnids - land dwellers resembling to some extent modern scorpions. Finally, in the Carboniferous Period, the first reptiles, called the stem reptiles, appeared. This was an important evolutionary event because the stem reptiles were the common primal ancestors of a great variety of animals that still exist today. The modern reptiles, birds, and mammals are all derivatives of the stem reptiles.

• PROBLEM 28-10

Why are the stem reptiles especially important in evolution?

Solution: The stem reptiles (cotylosaurs) were the first reptiles to arise. They are an important group of organisms because they were the ancestral form of a variety of important animals, including the mammals. The stem reptiles originated in the Paleozoic Era and gave rise directly to the many forms of reptiles that flourished in the Mesozoic Era - the turtles, tuatera, lizards, and snakes. These forms have survived a long evolutionary history to become the modern reptiles. The stem reptiles also gave rise to several other lineages, including the therapsids that ultimately led to the mammals, and the thecodonts that in turn gave rise to crocodiles, birds, flying reptiles called pterosaurs, and the great assemblage of reptiles called dinosaurs.

THE MESOZOIC ERA

• PROBLEM 28-11

Which animal groups dominated the Mesozoic Era? What factors may have contributed to the extinction of the dinosaurs?

Solution: The Mesozoic Era began some 230 million years ago and was characterized by a wide variety of reptiles. In fact, the Mesozoic Era is commonly referred to as the "Age of Reptiles"; common reptiles of this era were the primitive lizards: snakes, turtles, alligators, crocodiles, pterosaurs (flying reptiles) and later in the era, the dinosaurs. All of these, and also the mammals which came later, evolved from an important paleozoic group called the stem or root reptiles (cotylosaurs).

By the end of the Mesozoic Era the vast majority of the reptiles, of which the dinosaurs were the prominent group, dramatically disappeared. No totally satisfactory explanation has been established for their disappearance. One possible reason may be the climatic changes brought about by the Rocky Mountain Revolution. As the climate became increasingly cold and dry, many of the plants died out. The dinosaurs, many of which were herbivorous, may have disappeared because of the lack of food. Some of the dinosaurs did not favor a dry environment, and they subsequently declined in number as the swamps dried up. Another reason may have been the emergance of the mammals. The smaller, warm-blooded mammals were probably better able to compete for food, escape from enemies, and adapt to a colder environment than the larger, cold-blooded reptiles. The early mammals fed on the dinosaurs' eggs, and the late dinosaurs, it is believed, resorted to using reptilian eggs for food as the supply of food from the environment became scarce. The fatality of the reptilian eggs may be another factor contributing to the extinction of the dinosaurs. It is generally agreed that the demise of the dinosaurs was most probably the result of a combination of factors, rather than any single one.

• PROBLEM 28-12

When did the first mammals appear? Compare the monotremes and marsupials.

Solution: The first mammals arose in the earliest period (the Triassic) of the Mesozoic Era which dates to some 230 million years ago. They were warm-blooded animals which managed to adapt to the colder climate because of the ability to maintain a constant body temperature. Mammals then proliferated and have become the dominant animal form of modern ages.

The monotremes are the earliest known mammals. Today, the survivors of the monotremes are the duck-billed platypus and the spiny anteater of Australia. Both have fur and suckle their young, but lay eggs like turtles. The marsupials are another group of mammals, whose ancestors evolved into existence during the Jurassic Period, some 181 million years ago. The marsupials are more advanced than the monotremes because they bring forth their young alive rather than laying eggs. They are, however, considered among the most primitive of modern mammals because their young are underdeveloped and must remain for several months in a pouch of the mother's abdomen, which contains the nipples.

• **PROBLEM** 28-13

Why are marsupials widespread in Australia and almost nonexistant elsewhere?

Solution: The marsupials are a relatively primitive type of mammal in that their young continue their development in the abdominal pouch for a period of time even after birth. For a considerable length of time after parturition, the young are extremely dependent upon the mother for protection and nutrition. If they are removed from the mother's pouch, they will soon encounter starvation, cold, predation, and/or death. For this reason the marsupials, compared to placental mammals, are at a selective disadvantage. In most parts of the world, the marsupials cannot compete with the other mammals for survival. But during the Mesozoic Era, the ancestral marsupials enjoyed a period of competition-free existence in Australia, which was then isolated from the rest of the world. The ancient marsupials in the isolated continent never had any intense competition from the better-adapted placental mammals. Such competition had succeeded in eliminating most of the marsupials outside Australia. But in Australia, the marsupials grew, multiplied, radiated into many niches and evolved into a variety of better-adapted forms found exclusively there today.

BIOGEOGRAPHIC REALMS

• **PROBLEM** 28-14

Why is it that animals and plants of England and Japan are very similar despite the fact that they lie on nearly the opposite sides of the world?

Solution: The observation that animals and plants

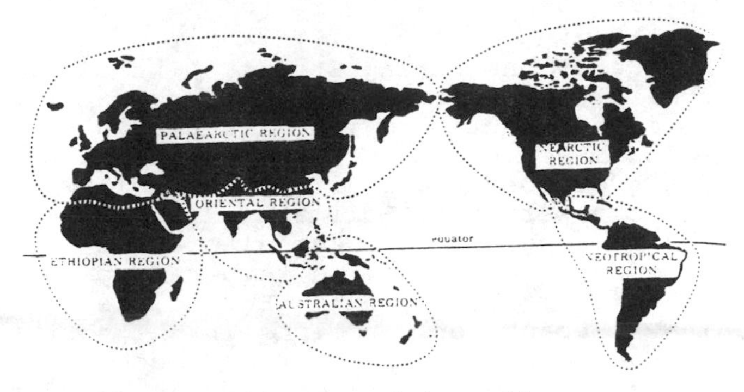

Biogeographic realms of the world

of England and Japan closely resemble each other is explained by the fact that England and Japan belong to the same biogeographic realm. A realm is a geographic division separated from other realms by major physical barriers, and characterized by the existence of certain unique organisms. Inside a realm there may be variations of climate and topography, but organisms are nevertheless able to pass more or less freely from one region of the realm to another.

It must be understood that some species have more ready physical access to a region suitable for habitation than others. What may be an effective barrier for one species may be a possible but difficult route for another or an easily crossed part for a third. Thus, keeping in mind the fact that different species have varying access to widely separated habitats, like England and Japan, and realizing that historically, geogrophic and environmental conditions were different than they are today, we can begin to understand the underlying concept ot this question.

England and Japan both belong to the Palearctic realm (see Figure). By definition then, the animals of these two countries are free to migrate from one country to the other, either by way of the sea or the air. Though plants themselves cannot migrate, their seeds can be carried by wind, water or migratory animals to distant locations. In other words, there are no absolutely impassable physical barriers between England and Japan, and genes can flow from one country to the other. Therefore England and Japan share many common groups of animals and plants.

• PROBLEM 28-15

Why are magnolias found both in the eastern United States and eastern China, which belong to different realms, and hardly anywhere else?

Solution: Early in the Cenozoic Era, the northern hemisphere was much flatter than it is now and North America was geographically continuous with Asia by means of a land bridge at the Bering Strait. The climate then was much warmer than at present, and magnolias spread over this entire connected region. In the later Cenozoic, the Rockies increased in height, and the western part of North America grew colder and drier. Magnolias, which were adapted to a warm moist climate, disappeared from western North America. Subsequent glaciation moving from the north wiped out any surviving temperate plants in western North America, and many in Europe, and the western part of China. Because only regions in the southeastern United States and eastern China were untouched by the glaciation, magnolias were left growing in these two regions. North America, however, soon broke away from the land mass of Asia, and two separate continents were formed. Because of the resulting geographic barrier (ocean) between the magnolias of the two countries, the plants for several million years followed separate evolutionary pathways, and have consequently become slightly different from each other.

TYPES OF EVOLUTIONARY EVIDENCE

• PROBLEM 28-16

Describe the various types of evidence from living organisms which support the theory of evolution.

Solution: There are five lines of evidence from living organisms that support the theory of evolution. First, there is the evidence from taxonomy. The characteristics of living things differ in so orderly a pattern that they can be fitted into a hierarchical scheme of categories. Our present, well-established classification scheme of living organisms, developed by Carolus Linnaeus in the 1750's, groups organisms into the Kingdom, Phylum or Division, Class, Order, Family, Genus, and Species. The relationships between organisms evident in this scheme indicates evolutionary development. If the kinds of plants and animals were not related by evolutionary descent, their characteristics would most probably be distributed in a confused, random fashion, and a well-organized classification scheme would be impossible. Secondly, there is the evidence from morphology. Comparisons of the structures of groups of organisms show that their organ systems have a fundamentally similar pattern that is varied to some extent among the members of a given phylum. This is readily exemplified by the structures of the skeletal, circulatory, and excretory systems of the vertebrates. The observation of homologous organs - organs that are basically similar in their structures, site of occurrence in the body, and embryonic development, but are adapted for quite different functions, provides a strong argument for a common ancestral origin. In addition, the presence of vestigial organs, which are useless or degenerate structures found in the body, points to the existence of some ancestral forms in which these organs were once functional. Thirdly, there is the evidence from comparative biochemistry. For example, the degree of similarity between the plasma proteins of various animal groups, tested by an antigen-antibody technique, indicates an evolutionary relationship between these groups. Fourthly, embryological structures and development further support the occurrence of evolution. Different animal groups have been shown to have a similar embryological form. It is now clear that at certain stages of development, the embryos of the higher animals resemble the embryos of lower forms. The similarity in the early developmental stage of all vertebrate embryos indicates that the various vertebrate groups must have evolved from a common ancestral form. Finally, there is the evidence from genetics. Breeding experiments and results demonstrate that species are not unchangeable biologic entities which were created separately, but groups of organisms that have arisen from other species and that can give rise to still others.

• PROBLEM 28-17

If, in tracing evolutionary relationships, anatomic evidence pointed one way and biochemical evidence the other, which do you think would be the more reliable? Why?

Solution: It has been stressed again and again that evolution can not occur without a change in the genotype. Biochemical properties are controlled to a greater extent by genes than are anatomic ones. Anatomic characteristics are more susceptible to modification by the external environment in which the organism possessing them lives. Plants of the same species growing in two quite different habitats may demonstrate strikingly different characteristics; this is because the two groups of plants are exposed to different environmental forces and achieve a different developmental potential even though the genes present in them are very similar.

When we compare two organisms or groups of organisms in seeking an evolutionary relatonship, we are essentially looking for a genetic relationship between them. Since biochemical characteristics are more greatly controlled by genes than are anatomic characteristics, a biochemical similarity between the two organisms should be more reliable than an anatomic similarity in indicating a genetic linkage and thus an evolutionary relationship. Therefore, biochemical evidence is more useful in the tracing of evolutionary relationships between organisms.

• PROBLEM 28-18

How can tissue enzymes and plasma proteins be used in the determination of evolutionary relationships between different species?

Solution: By this time it should be clearly understood that genetic change is the basis of evolution. It then follows that evolutionary relationships between species essentially result from genetic relationships. In general, we know that the more closely two species are related, the more they resemble each other in their genetic makeup. Since the biochemical functions in an organism are genetically controlled, the degree of biochemical simarilarity between two species should reflect the degree of similarity in their genetic composition, and how closely they are related evolutionarily.

Since enzymes are proteins whose synthesis is dictated by genes in the chromosomes, the properties of enzymes, such as their comformation and rate of catalysis, are dependent upon the genetic makeup of the organism in which they occur. Enzymes from various species, therefore, can be compared in their properties, and the degree of genetic similarity between the species can be inferred. It is found that enzymes from animals deemed to be closely related on the basis of anatomic or other evidence reveal very similar patterns of their rates of reaction. Enzymes from remotely related species show very different patterns, as can be expected.

Studies of the plasma proteins of various groups of mammals can also be used to infer the evolutionary relationships between species. The degree of similarity of the various plasma proteins can be tested using the antigen-antibody method. In this technique, an animal, such as a rabbit, is given repeated injections of the plasma from a different animal, such as a human (see Figure). The plasma cells of the rabbit produce antibodies specific for human blood-plasma antigens. The antibodies are then isolated from the blood of the rabbit in the serum. A diluted sample of this serum, when mixed with human blood,results in a visible precipitation caused by the combination of antigens and antibodies. By using a number of rabbits, each injected with the blood of different species of mammal, it is possible to obtain a series of antibodies, each specific for the blood proteins of a particular species.

The relationship between any two species can be determined by the strength of recognition when the antibody specific for one species' protein is reacted with the corresponding protein of the other species. The stronger the recognition, the more closely the two are related. For example, suppose one has obtained, by the method discussed, a rabbit-made antibody to human serum albumin. Bovine (sheep) serum albumin can be reacted with this antibody, and the strength of recognition of the antibody determined by the dilution at which precipitation (or recognition) no longer occurs. Such a determination could also be made between human serum albumin and that of any other mammal. The relative strengths of recognition enable one to place the different species in order of the closeness of their relationship to man. The evolutionary relationships determined by this method parallel those determined in other ways.

Evolutionary relationships between species can also be characterized by comparing the amino acid sequences of their respective plasma proteins and enzymes. Cytochrome c, a respiratory chain enzyme, has been studied extensively in this regard, and many species have been compared. The fact that certain crucial amino acids are consistently present in the same sites in all species suggests that the different cytochrome sequences arose through mutations from an ancestral molecule. As expected, the further apart any two species are evolutionarily, the greater the number of differences in their amino acid sequences. In fact, estimates of the chronological distance between the emergence of one species and another can be made using knowledge of mutation rates and the assumption that any given amino acid difference resulted from a single mutation. Such clear cut relationships however cannot always be found for all proteins studied.

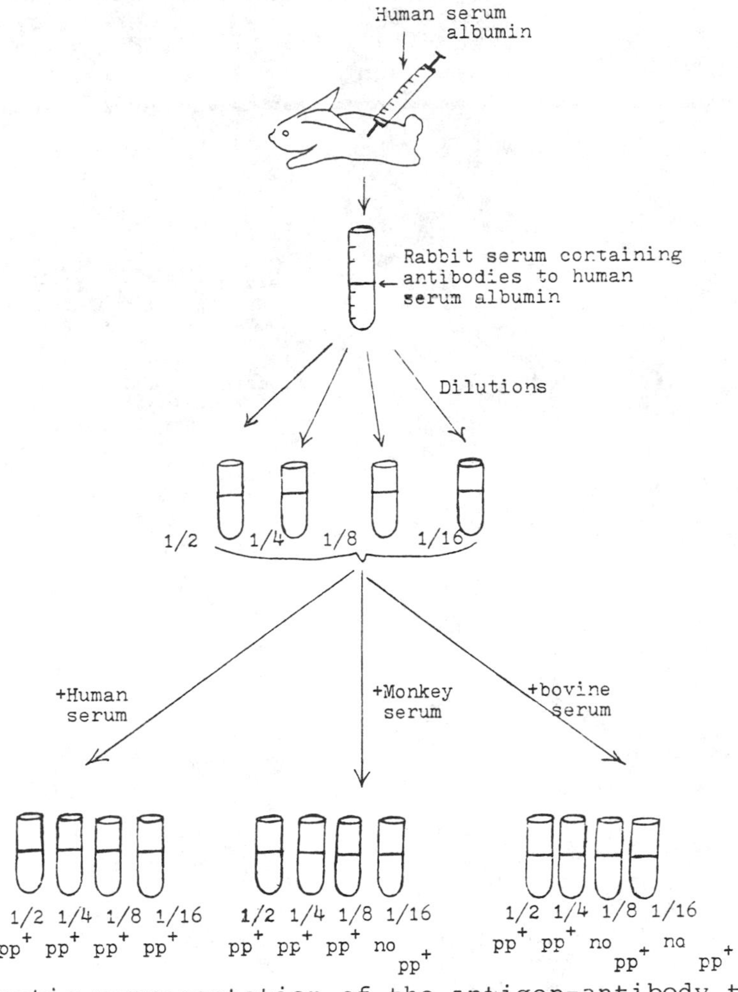

Schematic representation of the antigen-antibody technique. The strength of recognition between human serum albumin antibodies (made by the rabbit) and human, monkey, and-bovine serum is indicated by the dilution at which precipitation no longer occurs. According to this, monkey serum albumin is closer to human albumin than is bovine albumin.

Studies are also being done which use a measure of the extent of DNA hybridization as an indication of evolutionary relationships. Hybridization is said to occur when a single strand of a species' DNA reforms a double-stranded helix by pairing with a single strand of another species' DNA. The extent to which the pairing occurs is related to the similarities in the two species' DNA molecules. This is direct evidence of genetic similarity or dissimilarity between species.

ONTOGENY

• PROBLEM 28-19

What is meant by the statement "ontogeny recapitulates phylogeny"? With the advent of knowledge in genetics, how has this theory of recapitulation been modified and what significance does its present form have?

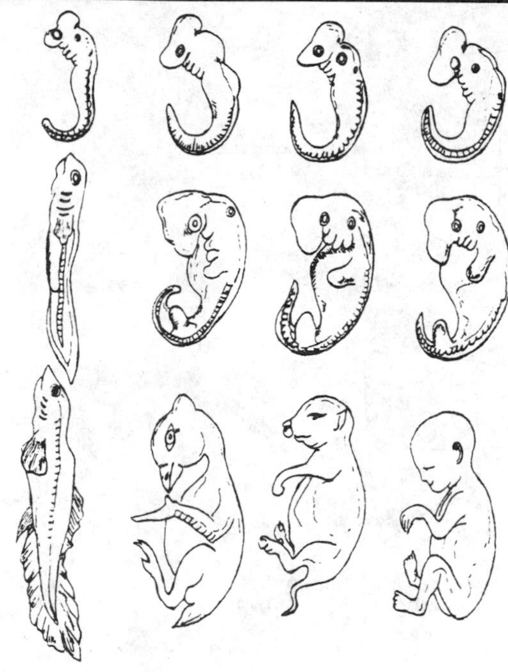

Successive stages (top to bottom) in the embryonic development of the fish, chick, pig and human. Note that the earlier stages of development (top row) are remarkably similar and that differences become more marked as development proceeds (bottom row).

Solution: The idea of recapitulation was first proposed by Ernst Haeckel in 1866, in the principle that "ontogeny recapitulates phylogeny." Haeckle claimed that embryos, in the course of development, recapitulate, or repeat, the evolutionary history of their ancestors in some abbreviated form. Thus, if mammals evolved from reptiles, which evolved from amphibians, which in turn evolved from fish, then a mammalian embryo should first resemble a fish, then an amphibian, then a reptile, and finally a mammal.

Through research and experimental findings, later workers modified Haeckle's idea and developed the modern theory of recapitulation. This theory states that the embryos of the higher animals resemble the embryos of lower forms, not the adults, as Haeckle had believed. An embryo passes through developmental stages similar to some of those which lower animal forms pass through. Ontogeny does not repeat the adult stages of phylogeny, but it does repeat, in an altered form, some of the ontogenic characteristics of ancestral forms.

Inherent in the theory of recapitulation is the existence of an evolutionary relationship between animal groups. There is also the implication that a species arises from some previous species and may give rise to new ones. Because embryonic development must primarily be controlled by genes, ontogeny can be rightfully thought of as an expression of the inherited genetic composition of an organism. Differences in ontogeny can be related, then, to differences in genetic composition. The modern theory of recapitulation thus also holds that a given species expresses a slightly altered version of the ontogenic program that it inherited genetically from the previous species, which in turn acquired its developmental program genetically from its ancestor. According to this, a species will closely resemble, both genetically and ontogenically, its recent ancestors and less closely its more distant ones. In other words, the more remote the ancestry, the fewer the similarities. It is also observed that in remote relationships, the similarities are restricted to the very early stages of embryonic development. Beyond those early stages evolutionary changes have grossly altered the direction of development. Thus, the ontogenies of fish and mammals resemble each other only in their earliest stages, while the ontogenies of reptiles and mammals show a close resemblance through later stages, the reptiles being a more recent ancestor of the mammals than the fish (see Figure).

The theory of recapitulation, as we can see, is significant for three reasons. First, it confirms that evolution does exist; second, it shows that evolution is a genetic occurrence; and third, taken with due caution, it may indicate the evolutionary hierarchy of some animals.

• PROBLEM 28-20

Describe briefly the characteristics of the ontogeny of a human being from a zygote to a mature fetus. What is the significance of this?

Solution: A human being, like any other organism, starts out as a fertilized egg, or zygote. The fertilized egg may be compared to the single-celled flagellate zygote of some ancient organism. The zygote then undergoes cellular divisions and gives rise to the blastula. The blastula may be compared to some spherical multicellular embryonic form of a primitive species. The blastula next develops into a gastrula with two layers of cells, then an embryo with three layers of cells. The early human embryo resembles a fish embryo, with gill slits, pairs of aortic arches, a fishlike heart with a single atrium and ventricle, a primitive fish kidney, and a tail that can wag. Later in its development, the human embryo resembles a reptilian embryo, with closed gill slits, fused vertebra, a new kidney, and an atrium partitioned into left and right chambers. Still later, the human embryo develops a mammalian, four-chambered heart, and a newer kidney. The seven-month old human embryo, covered with hair, resembles a baby ape more than an adult human.

The ontogeny of the human embryo reveals to some extent the evolutionary history of man. The one-celled, zygotic stage may reflect the early period of life on earth. Some unicellular flagellate organism is believed to be the ultimate origin of all living things. The fish-like embryonic stage may parallel the age of the fish, the retipilian stage the age of the reptiles, the mammalian stage the age of the mammals, and the apelike stage the period of the early primates. The sequence of all these embryonic stages parallels the chronological development of the ages as well as the evolution of animal forms. Thus, one can trace man's evolution back from some more primitive, apelike primate, through the forms of some ancient mammal reptile, amphibian, fish, primitive vertebrate, invertebrate, and multicellular structure, to an ultimate one-celled ancestor. The significance of this is that ontogeny recapitulates phylogeny.

SHORT ANSWER QUESTIONS FOR REVIEW

Answer

To be covered when testing yourself

Choose the correct answer.

1. Which of the following can be used as a criterion in determining whether two structures are homologous? (a) similar phenotypic appearance (b) similar functions (c) relationship to adjacent structures (d) all of the above (e) none of the above. — c

2. The wings of birds and the wings of butterflies are examples of (a) analogous structures (b) homologous structures (c) paralogous structures (d) orthologous structures (e) de novo structures. — a

3. The half life of the radioactive decay of uranium to lead is 4.5 billion years. If the ratio of uranium to lead is 1:7 in a given rock, what would be its approximate age? (a) 9.0 billion years (b) 11.2 billion years (c) 13.5 billion years (d) 18.0 billion years (e) 26.2 billion years. — c

4. A fossil animal is found frozen to death in a region of the North Pole. Scientists estimate the age of the animal at 28,000 years. Assuming that the half life of Carbon-14 is 5,600 years, what percentage of the preserved carcass would be expected to be in the form of nitrogen? (a) 3/4 (b) 7/8 (c) 15/16 (d) 31/32 (e) 63/64. — a

5. Due to the dominant fauna in the period, the Mesozoic is often referred to as the (a) "Age of Reptiles" (b) "Age of Fishes" (c) "Age of Amphibians" (d) "Age of Birds" (e) "Age of Mammals" — a

6. Which may have been a cause of dinosaur extinction? (a) decreasing floral supply on which they subsisted (b) predation upon dinosaur eggs by mammalian forms (c) changes in climactic conditions in regions of dominant dinosaur population, (d) all of the above (e) none of the above. — d

7. How can we account for the widespread distribution of marsupials in Australia? (a) Marsupial mammals are more adapted to the Australian climate than placentals. (b) Marsupials enjoyed an extensive adaptive radiation of available niches due to their isolation from placental competitors. (c) Marsupial mammals are more prolific than placental mammals, and this led to their eventual domination in the fauna. (d) Marsupials are

Answer

To be covered when testing yourself

resistant to many Australian diseases to which placentals were susceptible. (e) Marsupials became established in greater numbers due to their adaptations of greater offspring protection by the mother. — b

8. According to our most recent investigations, the lithosphere is composed of (a) 4 plates, (b) 5 plates, (c) 6 plates, (d) more than 6 plates, (e) none of the above. — b

9. The aging of fossil strata is often inconclusive because (a) erosion is constantly occuring (b) mountain building, earthquakes and other geologic phenomenon disrupt ordered strata formation (c) the rate of sedimentation varies at different times in the geologic record (d) the amount of sediment deposited may be different in various areas, depending on geographic barriers and means of transport by rivers and streams (e) all of the above. — e

10. To which forms did the stem reptiles not give rise? (a) reptiles: turtles, lizards, snakes (b) thecadonts: ancestral birds (c) therapsids: ancestral mammals (d) dinosaurs and pterosaurs (e) none of the above. — e

Fill in the blanks.

11. The _____ of a given species is defined as that portion of the earth in which it is found. — range

12. Those elements which can be traced in structure and function to a common ancestor of two different organisms are termed _____ structures. — homologous

13. The vertebrate eye and the cephalopod eye are _____ structures. This is further evidenced by dissimilar optic nerve formation in embryonic development. — analogous

14. Any evidence which indicates that an organism once lived can be classified as a _____. — fossil

15. When fossilization occurs by mineral infiltration of living tissues, it is known as _____. — petrifaction

16. Half of the uranium present in a rock will be converted to lead in 4.5 billion years. This time period is referred to as the _____ of uranium. — half-life

	Answer
	To be covered when testing yourself
17. Potassium is used as a radioactive marker in dating fossils. Following its decay, potassium forms the element _____.	argon
18. The rates of radioactive decay used in dating fossils are temperature and pressure _____ (dependent/independent).	independent
19. If the ratio of uranium to lead in a sample of rock is 1:3, the rock is _____ half lives old.	two
20. The radioactive element used to date organic remains is _____.	carbon-14
21. The mammals arose from a group of stem reptiles known as the _____.	cotylosaurs
22. The _____ are the most primitive mammals. This is indicated by their oviparaus lifestyle.	monotremes
23. The earliest vertebrates were small, jawless, armored, fresh water fish lacking fins known as the _____.	ostracoderms
24. The stem reptiles gave rise to many lineages including the _____, which in turn gave rise to crocodiles, birds, pterosaurs and the dinosaurs.	thecadonts
25. A _____ is a geographic division separated from other realms by major physical barriers, and characterized by unique organisms.	biogeographic realm
26. Ernst Haeckel claimed that embryos in the course of development repeat the evolutionary history of their ancestors in abbreviated form. This was propounded by the statement: "_____".	ontogeny recapitulates phylogeny
27. The _____ stage of embryonic development may be compared in an evolutionary scheme to some spherical multicellular embryonic form of a primitive species.	blastula
Determine whether the following statements are true or false.	
28. The range of an organism invariably overlaps with its center of origin.	False
29. Advanced structures are more complex than primitive structures.	False
30. Analogous structures are derived from a similar evolutionary origin.	False

Answer

To be covered when testing yourself

31. The wing of a bird and the wing of a fly are homologous. — False

32. The marsupials are the earliest known mammals. — False

33. The first vertebrates made their appearance in the Ordovician era. — True

34. The first plants to invade the land did so during the Silurian period. — True

35. Anatomic characteristics are more susceptible to modification by the external environment than are biochemical characteristics. — True

36. Genetic similarities can be inferred between species by comparison of enzyme properties and various other proteins. — True

CHAPTER 29

HUMAN EVOLUTION

FOSSILS

• PROBLEM 29-1

> Define a fossil and distinguish between the two major kinds of fossils. Discuss their preservation, dating, and usefulness in tracing human evolution.

Solution: Fossils are remains or traces of a plant or animal found in the earth. There are basically two types of fossils: physical (anatomical) fossils and cultural fossils (archaeological artifacts).

Physical fossils are those which reveal the form and structure of the animal or plant. Examples of physical fossils are bones, teeth and footprints. Cultural fossils differ in that they reveal information concerning the activities of an organism, or a group of organisms. Cultural fossils include tools and ceremonial artifacts.

Successful preservation of any type of fossil depends on the material of which it is composed and the environment in which it was formed. Whereas soft tissues of the body deteriorate rapidly in the air and leave little trace, hard bones may remain preserved for millions of years. The most frequently preserved anatomical fossils are teeth, which are coated with dentin. This is one of the hardest substances found in the body. Also commonly found are the larger and thicker bones of the cranium, femur, jaw, and pelvis.

Archaeological artifacts found are made of materials that are difficult to decompose. This explains why the obsidian tools (obsidian is a dark volcanic rock resembling glass) made by early man are found in large quantity, whereas no wooden tools have been uncovered from the same period.

Preservation of the body of an organism is most effective when death occurs rapidly and an air supply is immediately eliminated. This may occur when an organism falls into a quicksand pit, or is trapped in an avalanche or in the lava flow of a recently erupted volcano. Once the animal has died and the soft tissues begin to decay, minerals from soil or rocks may start to infiltrate these tissues resulting in a well preserved, petrified specimen.

However, this preservation of an entire organism trapped by some natural phenomenon is relatively rare. More often only fragments of fossilized teeth and bones are discovered, from which the nature of the unpreserved body parts must be deduced.

Digging for fossils is usually a costly enterprise requiring a great deal of time and precision. It is therefore important that a logical site be chosen before the project has begun. Historic water sites, caves and natural shelters are suspicious sites for a digging. When fossils are found,precise maps are drawn relating their positions relative to one another and to the surroundings. Digging is done in levels since the lower a fossil is found the older it usually is. This may not be true in areas of extreme erosion or regions where mountain formation or other natural phenomena such as earthquakes have upset the ordered layering of sediment.

The dating of fossils is essential in fitting together the pieces of the puzzle of man's history. The scheme of the evolutionary process can only be determined by knowing which features appeared first in the evolution leading to modern man. There are two methods of dating fossils; (1) relative dating, and (2) radioactive dating. A fossil is dated relatively when the position in which it is found is used as the reference point in time. In other words, what stratum the fossil was found in, along with the depth of the find and the other fossils occurring with it can be used to determine its relative age.

Radioactive dating is more exact than relative dating. There are several different types of radioactive dating, for example, carbon-14, potassium-argon, and uranium-lead. They all work on the principle of half-lives. Fossils are dated by measuring the decrease in the level of radioactivity from the calculated original value. C-14 is the only method that directly dates the fossils themselves. The others determine the approximate year the fossil was deposited by calculating the age of the rock the fossil is found in.

After enough fossils have been obtained and dated, and structural differences determined, the story of man's evolution can be better understood. The story will never be complete, but some trends in structural changes leading to modern man can be observed. Some forms of pre-historic man are believed to have become extinct due to changing climate, competition with other animals (including hominoid forms), and by disease. In the end, the hominoids who were our ancestors, must have possessed distinctive selection factors that permitted their survival, perpetuation and evolution.

• **PROBLEM 29-2**

How can we deduce information about prehistoric man from the fossil teeth, jaws, and craniums found?

Solution: Of all the fossils discovered relevant to the evolution of man, the craniums, jaws, and teeth have been the most numerous and useful. Teeth have been found in large numbers at archaeological sites because their enamel coating gives them great durability.

Teeth are complex structures with many distinctive features. The size and shape of the cusps, and the root size of the fossil teeth can be compared with modern man's dental anatomy and to that of modern apes. In this manner, a fossil man can be placed in a chronological and evolutionary relationship to other hominiod-like forms, whether they be extinct or extant. Much can also be said about our ancestor's diet by studying the structure of a fossil's teeth. For example, there is an evolutionary trend towards reduction in size of the canine teeth in man's evolution (see Figure). Man and apes are characterized by larger canines. Since the canines are used in tearing raw meat, the reduction in their size would seem to parallel the evolution of man's mode of feeding and his diet.

A major problem encountered by anthropologists is the absence of teeth from many of the jaws that are found. It becomes difficult to determine the exact number and size of the teeth when the jaw is not complete. As a general rule, the size of the jaw is directly related to the size of the teeth. For instance, man has characteristically both smaller teeth and smaller jaws than his predecessors. The shape of the jaw alone is also an important evolutionary indicator. The dental arch, or portion of the jaw where the teeth are attached, is rectangular in the apes and parabolic in man. This indicates that in the apes, the back teeth on either side of the jaw are parallel to each other, with the incisors in a straight row. In man and some of his fossil ancestors, parabolic dental arches are present wherein the distance between the canines on either side of the jaw is smaller than that between the molars.

The structure and size of the teeth, the ridges on the jaw, and the dimensions and shape of the dental arch indicate how the face may have looked, and the nature of the diet. If the fossil man ate hard, rough objects that required teeth with effective chewing and grinding surfaces, we would most likely observe huge molars that have undergone much wear, and jaws with large ridges for attachment of strong chewing muscles. Enlarged chewing muscles are also accompanied by large ridges on the sides and top of the skull, since the temporalis muscles used for chewing originate in the skull region in hominoids, although not always in the same place.

Huge neck muscles would also be accompanied by large ridges on the skull. This would indicate quadrupedalism. If the animal was quadruped, huge neck muscles would be needed to support the neck, and obvious muscle ridges would be present in the skull. If the animal was bipedal, the neck muscles and ridges would be reduced since the head would now be balanced on the spinal column

The centered position of the head indicating bipedalism can be further determined by the position of the foramen magnum. Bidepal organisms like man have the foramen magnum located underneath the center of the brain, while in apes, the foramen magnum is located toward the rear of the skull.

The overall dimensions of the braincase are important in determining an animal's cranial capacity. If the skulls of the various fossil men are put into a chronological sequence, a trend toward increasing brain size is noticed as well as a general thinning of the skull bones.

A great deal of the information indicated above is still to a large degree speculative, since skulls, jaws, or teeth are rarely found in their original form without great amounts of weathering and fragmentation.

DISTINGUISHING FEATURES

• PROBLEM 29-3

Define the terms Anthropoidea, Hominidae and Hominoidea?

Solution: Man belongs to the order of mammals called the primates. This order is commonly divided into two suborders - Prosimii and Anthropoidea. The Prosimii are the primitive primates (lemurs, tarsiers, etc.) and the Anthropoidea are the higher primates.

The suborder Anthropoidea is divided into the Platyrrhini or New World monkeys, and Catarrhini or Old World monkeys. The infraorder of Catarrhine monkeys is further divided into two superfamilies = Cercopithecoidea (tailed anthropoids) and Hominoidea (great apes and man).

The superfamily Hominoidea (hominoids) is represented by three families; Hylobatidae (gibbons), Pongidae (gorilla, chimpanzee, orangutan) and the Hominidae (fossil men and modern man).

Kingdom : Animalia

Phylum : Chordata (sub: Vertebrata)

Class : Mammalia (sub: Theria)

Order : Primates

Suborder : Anthropoidea
Infraorder : Catyrrhinii
Superfamily : Hominoidea

Family : Hominidae

Genus : Homo

Species : Sapiens (sub: Sapiens)

Distinguish between Cebidae (New World monkeys) and Cercopithecidae (Old World monkeys).

Solution: The first members of the suborder Anthropoidea diverged from the prosimians during the Oligocene era. The prosimians are the earliest known primates in evolutionary history. Two lines of anthropoid monkeys (Catarrhines and Platyrrhines), probably arose at about the same time from closely related prosimian ancestors. One of the lines (Platyrrhines) gave rise to the super family Ceboidea and the other line (Catarrhine) gave rise to the two superfamilies Cercopithecoidea and Hominoidea. The Cebidae is a family grouping in the super family Ceboidea, and is therefore a New World monkey. In addition to the Cebidae, there is another family in the superfamily Ceboidea, the Callithricidae or marmosets. The family Cercopithecidae is in the superfamily Cercopithecoidea and are thus Old World monkeys. The family from which man descended is the Hominidae which is in the superfamily Hominoidea, the second lineage of catarrhine monkeys.

The members of the family Cebidae are New World monkeys, with the adjective "New World" referring to their presence in South America and Central America. The ceboids, isolated in South America, underwent evolution independent of the members of the Cercopithecoidea and Hominoidea. Among the living ceboids are the howler monkey, the marmoset (squirrel monkey), the capuchin monkey (organ grinder's monkey) and the spider monkey. All these monkeys are found in tropical forest habitats and show a wide variety of adaptations to arboreal life, some of which parallel those of the Old World monkeys. The cercopithecoids are Old World Monkeys, "Old World" referring to their widespread distribution in Africa and Southeast Asia. The large group of Old World monkeys includes the macaque, mandrill, mangabey, baboon, langur, proboscis monkeys and others. They live in a variety of environments, both arboreal and terrestrial, from grassland savannas to jungle forests.

The New World and Old World monkeys differ in many ways and most differences are readily observable. Most New World monkeys, have a prehensile tail that they use almost like another hand for grasping objects and hanging from trees. The prehensile tail has a tactile pad at the tip with ridge patterns similar to fingerprints. The tail of the Old World monkeys is not prehensile, but is used instead for balance. The nostrils of the Ceboidea (New World monkeys) are separated by a wide partition, and are flat and facing in a lateral (outward) position. This condition is known as platyrrhini, hence the name platyrrhini monkey. The ceboids are all platyrrhini monkeys. The nostrils of the cercopithecoids and hominoids are closer together and are directed downward and forward. This condition is known as catarrhini, and thus primates with nostrils of this type are called catarrhine monkeys. The cercopithecoids and hominoids are all classified as catarrhine monkeys. New World monkeys have no adaptations for terrestrial life styles. Old World monkeys tend to sit upright, and have buttocks with a hardened, insensitive sitting pad called the ischial callosities. The ischial callosities is usually very brightly colored, being either red or blue. Old World monkeys possess distinctive molars (cheek teeth), which can be used to distinguish them from the New World monkeys. These are known as the bilophodent molars. Furthermore, Old World monkeys have a bony auditory tube which New World monkeys do not have. This may aid New World monkeys in distinguishing the direction of sounds.

• PROBLEM 29-5

Why is it incorrect to say that man evolved from monkeys? What did he possibly evolve from?

Solution: The oldest fossils of Old World monkeys are of Parapithecus and several related genera that have been found in lower Oligocene rocks. Parapithecus was smaller than any of the modern monkeys or apes and was near the stem leading to humans (it had the same dental formula). Parapithecus may therefore represent a common ancestor of today's Old World monkeys, anthropoid apes and man. From that point, the evolution of the modern Old World monkeys diverged from that of the hominoids. The ancestry shared by New World monkeys, Old World monkeys, and the hominoids was separated even earlier in time. The New World monkeys diverged from the latter two at the beginning of the Oligocene; while Parapithecus fossils are dated to the middle Oligocene.

By the Miocene (20 million years ago), the divergence of the Old World monkeys and the hominoid lineage is clearly evidenced by the fossil records. This can be verified by the examination of dental patterns. The hominoid fossils have 5-cusped molars rather than the 4-cusped molars of the monkeys. This indicates that man probably did not evolve directly from the monkeys.

It is more correct to say that monkeys, man, and apes all evolved from a common ancestor independently. This ancestor had characteristics of the higher primates which evolved along different lines of specialization over millions of years. The Oligocene hominoid that gave rise to the lineage leading to man was a small, arboreal, monkey-like anthropoid equipped with binocular vision, living in a society of small troops which used visual and vocal communication.

The oldest fossil found so far that shows features considered to be those of an ancestral hominoid is the skull of Dryopithecus. Dryopithecus is now known to have been a wide-ranging genus extending from western Europe to China that had some definite hominoid features. Dryopithecus is the most likely ancestor of the chimpanzee, gorilla, orangutan and man. The lineage of the gibbon is believed to have diverged into a distinct species before any of the other apes.

What characteristics distinguish humans from the great apes?

Solution: The living great apes fall into four groups: gibbons, orangutans, gorillas, and chimpanzees. All are fairly large animals that have no tail, a relatively large skull and brain, and very long arms. The earliest members of the family of man (Hominidae) probably shared a common ancestor with the great apes, in particular the chimp and gorilla. Both fossil evidence and biochemical data indicate that gorillas, chimpanzees, and man are more closely related in terms of common ancestry, than any one of them is to orangutans or gibbons.

There are distinctive anatomical features that separate man into a class of his own. These anatomical features are directly related to the differences in appearance of men and apes, as well as the differences in behavior, intelligence and physical capabilities. Some of the major differences may be found in the skull and jaws. Though the skulls vary greatly in size in the great apes, they are generally more bony and thicker than man's. The ridges on the skulls of apes are more pronounced in order to support heavier musculature. For example, the brow ridges or supraorbital ridges in apes protrude to a great extent, while in humans this ridge is hardly noticeable. Man has a high vertical forehead while apes do not have this feature. Further, man has a pronounced nose with a distinct bridge and tip. The facial features of apes are more outward, such as the upper and lower jaws, while man's face is relatively flat with the jaws placed further back into the head. The jaws of apes are more massive than man's, showing many ridges for muscle attachment. This indicates a rougher diet and a greater dependence on the chewing mechanism. In comparing the teeth, it can be seen that man's are somewhat smaller, but they are of the same general structure. Apes' canines project more than man's beyond the line of the other teeth. One last point to make concerning the facial differences between man and the apes is that the former has a jutting chin whereas the latter do not.

The brain of man is two to three times larger than that of most apes and it is more convoluted, which allows for greater surface area. The human brain, also has a more highly developed cerebral cortex region, where the processes of thought, memory, reasoning and speech are coordinated.

The attachment of the skull to the vertebral column in man is located at the center of the base of the skull, while the skull of apes is attached toward the rear. Man balances his head so that he can hold it upright with relative ease while apes require massive neck muscles for this purpose.

Keeping the head upright is one of the major requirements for upright posture and walking on two limbs, known as bipedalism. Bipedalism enabled man to free his hands for tool making and other tasks. This ability permitted the biological and cultural evolution of man to take its present course. Orangutans and gibbons live mostly in the trees and accomplish locomotion by swinging from branch to branch. This is called brachiation. They occasionally walk on the ground using all four limbs. Gorillas and chimpanzees spend time on the ground and also inhabit the trees. They run and walk on all four limbs. They are classified as quadrupeds and are sometimes referred to as knuckle walkers because they place the knuckles of their fingers on the ground when they walk. Although gorillas and chimpanzees do use bipedal locomotion for short distances, their anatomy does not allow for very efficient bipedalism, and their posture is only semi-erect when they stand.

Man's big toe is moved back in line with his other toes making it non-opposable, unlike his thumb. Apes' feet have very long toes and an opposable big toe. The digits of apes are relatively long compared to their thumbs and big toes, while the digits of man are closer in size and shorter in overall length. The feet of man are flat and have short toes. His feet have developed two arches, with one lengthwise and the other crosswise to better support the weight of his body for bipedalism.

The basic skeleton of man, gorilla and chimpanzee are generally the same. They differ mainly in the proportion of certain bones to others, such as that of the four limbs. In apes, the arms are longer than the legs, while in man the arms are much shorter than the legs. The shape of the pelvis, and the general position of the attachment of the legs to it, constitute another reason as to why man is bipedal.

Finally, man and apes differ in their diet. The diet of the apes is basically herbivorous with occasional supplements of insects, birds, eggs and other small animals. Man, on the other hand, is omnivorous.

• PROBLEM 29-7

What social and behavioral characteristics are paralleled in the primates, in particular chimpanzees, gorillas and man?

Solution: Chimpanzees, gorillas and man, as well as some other primates, have evolved grouping lifestyles that are composed of members of their family and relatives. These social groups, as they are referred to by anthropologists, are highly organized and structured. These primates grow up with similar life cycles: (1) a prenatal stage, (2) birth, (3) infancy, (4) childhood, (5) adolescence, and (6) adulthood. All higher primates usually space their births from 2 to 5 years apart for adequate nutrition and proper maternal care. As one ascends the evolutionary ladder of primates, there is observed a lengthening of period in which the offspring remains dependent on the mother. The lengthened prenatal and infancy periods might be necessary in the higher primates for proper development of the nervous system, which is characteristically highly complex, expecially in certain regions of the brain. After birth the young stay with their mothers for a relatively long time compared to other animals. During infancy and childhood, chimpanzees, gorillas and man are relatively devoid of responsibilities. They spend their days in a play situation with their siblings and/or other young, where they learn how to conduct themselves in a social atmosphere. These peer play groups are very important in that discrete behaviors learned in this environment will help to ensure the groups continued existence.

As these primates enter adulthood, hunting for food becomes a primary reponsibility. Food procurement by the primates highlights the great degree of sexual dimorphism in the primates, as in many other animals as well. For example, male chimpanzees possess greater strength and longer canine teeth than the females. Some sexually dimorphic characteristics have been documented to correlate to a division of labor between the males and females. Male chimpanzees hunt and are the protectors of the group, while female chimpanzees bear and care for the young and gather vegetation.

Although gorillas are herbivores, both chimpanzees and humans are omnivores. The chimpanzees' diet consists of approximately 25% meat and 75% gathered vegetables. The human's diet is more variable. Hunting for meat by primates used to be regarded only as a human feature, but it has also been observed in chimpanzees and baboons. The social organization involved in hunting requires a form of group communication for trapping the prey, even when only small animals are being hunted. Chimpanzees and man bring the hunted food back to the social group where it is shared. The organization for sharing food between the young, adult males and females is a rather complex feature when compared to the lifestyles of other carnivorous animals. The social interaction arising from the hunting and sharing of food may have been the step that bound communication and group unity in the beginnings of culture. All primates communicate with each other either by vocal or nonvocal signals. Finally, primate groups occupy a home range: a specific geographic area in which they remain and protect.

THE RISE OF EARLY MAN

• PROBLEM 29-8

What led to the demise of the western Neanderthal population? How and in which area or areas of the world did modern man, represented first by Cro-Magnon man, arise?

Solution: The Neanderthal race can be differentiated into two forms: the classical Neanderthals in western Europe, and progressive Neanderthals, or transitionals, in central and eastern Europe and the Middle East. There are conflicting points of view as to the taxonomy of the western group. Some would designate it as a separate species with the name Homo neanderthalensis. Others include it in the species sapiens and simply recognize a subspecies designation: Homo sapiens neaderthalensis. One of the major reasons for this taxonomic discrepancy is the recognition of transitional forms that cannot be assigned to either the Neanderthal or modern sapiens group. An important point is that the western European group was most likely genetically isolated during the Würm I glacial period (40,000 to 75,000 years ago), while groups to the east and south presumably had contact with each other. The populations at this point in time were geographically widespread. A great deal of morphological variability existed among them and the gene pool as a whole evolved from either Homo erectus or some undiscovered group very similar to it.

Cro-Magnon man, the early representative of Homo sapiens (or modern man) appeared around 40,000 years ago (the date is disputed). There are various viewpoints concerning the origin of Homo sapiens. Some believe that the western European neanderthal populations died out and lent little of their genes to modern man, while the transitional forms in eastern and central Europe and in the Middle East evolved further into modern man. Two other points of view - the Presapien and the Preneanderthal - have much in common. Both agree with the first in considering Neanderthals of western Europe as dead ends of hominid evolution, becoming extinct as the climatic conditions to which they were adapted changed. The Presapien theory suggests that Homo sapiens originated as a distinct lineage, completely separate from that line leading to the Neaderthals. The Preneanderthal theory explains that there was a broad and varying population, one segment of which was an isolated, European, cold-adapted group (that is, the western Neanderthals): the other segment inhabited the Near East and became modern man.

Another general theory views that modern man arose from a stock that existed about 250,000 years ago during an interglacial period, through a number of intermediate populations. While the western European Neanderthal group was slowly evolving in isolation in a cold climate, more favorable climatic conditions farther to the east and south and continual migration of hunting groups produced modern sapien populations. This does not exclude the possibility of some western European Neanderthal genes being incorporated into modern populations. It is very likely that some of their genes did find their way into modern populations.

The western Neanderthals became extinct, but how this occurred is not known for sure. Some anthropologists believe that they did not become extinct in the strict sense of the term. Rather, they were absorbed by interbreeding with emigrating, anatomically and culturally more modern populations, probably Cro-Magnon man. They reasoned that climatic amelioration probably allowed greater and greater population contact. If this was the case, then neanderthalensis and sapiens should not be considered as separate species (due to their ability to interbreed). For this and the reason mentioned earlier, some scientists classify neanderthalensis as a sub-species of sapiens. Some scientists on the other hand, believe that these two hominids belong to separate species, and Neanderthal man was eliminated by Homo sapiens by either combat or competition. Some even postulated a war, in which Cro-Magnon populations flooded western Europe and killed off the Neanderthals. We have yet to wait for further archaeological evidence to settle this conflict. But we can be certain at this stage that the process of cultural exploitation did occur. Old tools and methods gave way to newer, finer and better adapted methods. Thus, less technologically advanced cultures were slowly forced out of existence.

Where then did modern man arise? The center of origin of modern man appears to have been in Asia, specifically in the general region of the Caspian Sea, but this is being disputed. It is important to point out as a concluding comment that there is at present a great deal of differing opinions concerning the origin and evolution of Homo sapiens, as seen in the many disagreements among various textbooks on this issue.

• PROBLEM 29-9

Discuss why one would place Cro-Magnon man higher on the evolution ladder than Neanderthal man.

Solution: During the period ranging from 40,000 to 10,000 year ago, members of our species - Homo sapiens - continued to live as small-band hunters and gatherers. They spread to various parts of the world where they encountered and adapted to a variety of environmental conditions. Due to their burial of the dead, several nearly complete Cro-Magnon skeletons have been preserved.

The bones of Cro-Magnon remains indicate a robust build, but not as robust as the Neanderthal's. Restorations reveal that Cro-Magnon man had smaller eye ridges, a less receding forehead, smaller and shorter face, narrower and higher-bridged nose, and a more prominent chin than Neanderthal man. His brain size resembled that of modern man more than that of the Neanderthals. Generally, therefore, Cro-Magnon man had a more modern appearance.

Aside from having more modern physical features, Cro-Magnon man had a culture that was more refined and diverse than the one of the Neanderthals. Due to the great variety of environments entered, there was a great variety of habitations. Besides rock shelters and caves which were widely inhabited, Cro-Magnon populations also lived in oval huts constructed out of large bones, tree branches and animal skins. The Cro-Magnons inherited and continued many of the Neanderthal tool technology. They also manufactured finer, more complex stone tools and other delicate implements. For example, knifelike blades of high quality were produced, and fishing hooks and needles with an eye were invented.

The Cro-Magnons were painters as well as skilled craftsmen. They also domesticated the dog and developed fur clothing.

The most important achievement of the Cro-Magnon culture is probably the development of a fully articulate language with speech. A refined communication system can lead to a refined cultural assemblage, which in turn leads to greater adaptability and powers of exploiting the environment. Another great evolutionary advantage of Cro-Magnon man over Neanderthal man is the ability to make notational counts enabling him to record events for the future. Hence, the culture of the Cro-Magnon man is seen to involve a greater degree of higher level functioning, which may reflect an increase in intelligence in this hominid group.

• PROBLEM 29-10

In what ways do the Upper Paleolithic (Cro-Magnon) and the Neolithic cultures differ?

Solution: The Neolithic (or New Stone Age) culture originated somewhere between Egypt and India about 10,000 years ago. It is characterized by implements bearing the marks of careful grinding and polishing and by the beginnings of agriculture and animal husbandry. The earliest animals to be domesticated after the dog (first by Cro-Magnon man) were the pig, sheep, goat, and cow. The horse was not domesticated until much later. Our human ancestors gradually changed from wandering hunters and food gatherers of the Cro-Magnon stage to more settled and efficient food producers, raising grain, making pottery and cloth and living in villages. At this time, more diversified tools were made, the hunting men stayed closer to the home and migrated less, people became more self-sufficient and made or obtained what they needed from nearby sources. They no longer lived in small breeding groups but expanded their communities. The increase in food supply enabled an increase in the size of the population. Neighboring populations interbred and the tendency toward genetic drift was thus greatly reduced. Neolithic people also invented the dugout canoe and the wheel. With the Neolithic age we enter the historical times, for the oldest cultures of Egypt and Mesopotamia were Neolithic. The use of metals began with copper and was followed with bronze. They were used for making tools, vessels and weapons beginning about 4000 B.C.

The culture of Cro-Magnon man in comparison to the Neolithic culture may at first seem primitive. Actually, both peoples had about the same intelligence. But culture is an additive phenomenon i.e., one generation will usually build on what is inherited from the previous generations. Consequently, culture gets more and more complex and sophisticated. As we can see in the transition from Upper Paleolithic to Neolithic ages, people are living in larger social groups, language is progressing and is an expanding communication device for expressing ideas and experiences, both past and future. The interaction of language and a large social order speed up cultural advancements and the attainment of new knowledge.

MODERN MAN

• PROBLEM 29-11

Why is genetic drift less important in human evolution at present than it was twenty thousand or more years ago?

Solution: The communities that man lived in about twenty thousand years ago all had something in common - they all were very small villages. The total world population then was low and people living in these small social groups were relatively distant from people in other places. With no means of transportation and the danger of traveling alone, people tended to stay close to their shelters. As a result, most breeding took place within individual social groups with each probably ranging up to about 100 members. In other words, these people were probably genetically similar for in inbreeding, no genes could be exchanged with other populations that existed. As an illustration, we can consider the western European Neanderthal people who were cut off from the rest of the world population by geographic factors. They were no longer able to exchange genes and consequently their further evolution was independent. Thus, they were sheltered from different genes present in the world's gene pool. These people became more unlike the rest of the world population and more like each other. Because of their reduced variability, when faced with an environmental crisis such as sudden cold or disease, they were not able to adapt successfully and faced extinction.

Let us take a look at the genetic effects involved with small inbreeding populations. Evolutionary change is not usually automatic, it occurs only when something disturbs the genetic equilibrium. A condition necessary for a population to be in equilibrium is that it must be large enough to make it highly unlikely that chance alone could significantly alter gene frequencies. Any breeding population with more than 10,000 members of reproductive age is probably not significantly affected by random change. This is the condition that exists in the world today. But gene frequencies in small isolated populations of less than 100 reproductive members are highly susceptible to random fluctuations, which can easily lead to loss of an allele from the gene pool and the fixation of another even if the former allele is an adaptively superior one. This is the phenomenon called genetic drift. Genetic drift explains why small populations tend to have a high degree of homozygosity, while large populations tend to be more variable and diverse.

• PROBLEM 29-12

Do you believe the various races of man living today constitute one species or several? Why? What characteristics distinguish the present races of man?

Solution: The populations of man are now more diverse than ever in history. From all of the stereotyping that we are accustomed to, we all know each race's distinct physical and cultural features. But because the populations of large cities consist of samplings from almost every race, and interracial matings, the human races are slowly losing their characteristic distinctions. It would be improper to consider each race a separate species since individuals of any race are capable of mating with any other individual of another race and of producing fertile offspring. This is the major characteristic in deciding whether organisms belong to the same species. Thus, the various races of man living today constitute only one species.

Genetically, a race is a subdivision of a species that consists of a population with a characteristic combination of gene frequencies. Races are populations that differ from one another not in any single feature, but in having different frequencies of many genes that affect body proportions, skull shape, degree of skin pigmentation, texture of hair, abundance of body hair, form of eyelids, thickness of lips, frequencies of various blood groups, tasting abilities and many other anatomical and physiological traits. While each race has its fundamental characteristics, no race is "pure" because the evolutionary history of human beings has been one of continuous intermixture of races as peoples migrated, invaded and conquered their neighbors or were conquered by them. The worldwide interbreeding typical of mankind is a characteristic which is absent in other higher primates.

The daily or lifetime range of other individual primates or primate social groups is relatively small. For example, most apes need to be close to a source of water and to the same food supply that they are accustomed to, thus their diet is not very diverse. They also need a form of shelter for protection when they sleep at night. In addition, most apes remain in the small close-knit social group that they were born into during their entire lifetime. Because of the non-migrating feature of ages, there is only little genetic diversity within the various species.

Widespread migration and interbreeding in humans are thus primarily responsible for the genetic diversity of races. Almost every combination of genes has the possibility of occurring in mankind and this might be related to the evolutionary advances that have taken place in our history as well as those that will take place in the future. Mankind is evolving even today, but the rate is extremely slow. Usually an organism will tend to evolve or adapt in response to a change or stress in its environment. Living in a large city, with all of its pollution, technology, tension, and stress, amy prove to be a selection pressure for individuals who can better function in this type of situation. It is hard to imagine what the races of human beings will be like in the future.

OVERVIEW

• **PROBLEM 29-13**

Trace the evolution of man using the existing classifications and characteristics of fossil men.

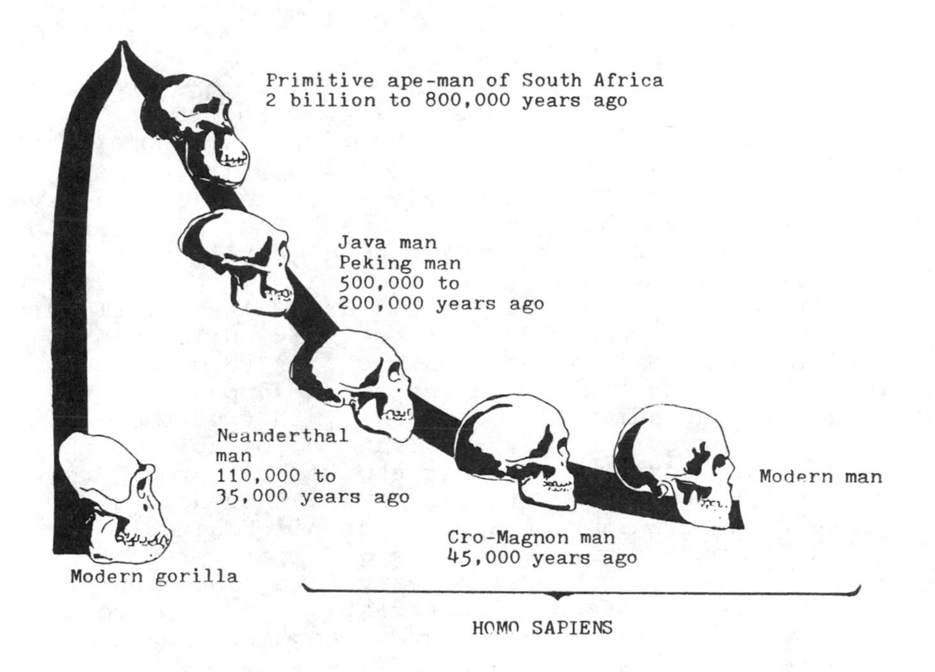

Figure 1. The evolution of the human skull.

Solution: The fossils of the Dryopithecine date back to about 30 million years ago during the oligocene era. It is believed that there once existed several species of Dryopithecus: Dryopithecus africanus which was thought to be ancestral to the chimpanzee, Dryopithecus major - thought to be ancestral to the gorilla, and Dryopithecus sivolensis - thought to be ancestral to the present day orangutan. Although not enough information has been collected to tell exactly when and where the human lineage first started, it is believed that it may have had its roots in the Dryopithecine line. The apelike Dryopithecus was probably the ancestor of man.

By the middle of Miocene (about 20 million years ago), the evolutionary lines leading to the various types of modern anthropoids were distinct. The first set of fossils found to be considered in the hominid line, representing Ramapithecus, dates back to the upper Miocene epoch (about 10 to 14 million years ago). It must be pointed out, however, that this is supported only by vague evidence since only the jaws and teeth from Ramapithecus have been found. The canine and incisor teeth are relatively small in man, and this may be an important distinction from the apes. If Ramapithecus did not use his teeth for defense and his diet changed to one that required more chewing, then this change may have been correlated with the increased use of hands. Since no fossil skull of Ramapithecus has been found, little can be learned about his intelligence. If he did rely more on his hands, there is the implication that he might stand at least semi-erect and use bipedalism at times.

A gap of about five million years in man's evolutionary history is found between Ramapithecus and the next distinct homonid form, called Australapithecus, who appeared about 5.5 to 1.2 million years ago. Relatively large portions and numbers of the Australopithecine skeleton have been found scattered throughout South and East Africa. An important advance toward the modern human condition shown by the Australopithecines is the acquisition of some degree of bipedalism. They might have stood and walked completely erect or perhaps they were fully erect only while running. From fossil evidence it is deduced that Australopithecus was relatively small, an adult weighing about 75 pounds. His dental anatomy was humanlike, especially that of the canines. His cranial capacity was very small and comparable to that of a present day gorilla. It must however be noted here that although the Australopithecine cranial capacity is comparable to a gorilla's, the ratio of his brain size to body weight is much higher that the gorilla's proportion. Australopithecus probably was not very intelligent but was obviously developing human characteristics and some form of culture. He probably lived in a group and had a daily hunting-gathering mode of living. This is supported by the finding of bones of small animals and evidence of gathered fruits, nuts, and shoots at or around excavated sites. Stone tools are found also at these sites and are of a style either known as pebble tools or as the Oldowan industry. Therefore, Australopithecus was a toolmaker as well as a tool user, a feature much less common among animals other than man himself.

A later stage in human evolution is represented by a group of fossil men, including Java and Peking man, that can be classified as Homo erectus. This species first appeared about 1.2 million years ago and existed till about 300,000 years ago. As compared to Australopithecus, Homo erectus had greater body size, increased cranial capacity, reduced facial shape and dental structure, and the beginning of a human type culture. A major feature found in Homo erectus is the elaboration of cultural adaptations that originated in the earlier Australopithecine stage. The cultural associations of Homo erectus, in the making of tools (called Acheulian tools) and in social activities, are more complex than those associated with the Australopithecines.

By the Homo erectus stage of human evolution, adaptations based on culture are seemingly well established. If culture is the basis of human adaptation, then natural selection would favor a more efficient culture. In these fossil men, an enlarged brain is related to an increased capacity for more complex cultural skills and behavior patterns. Reduction in facial and dental structure also may be linked to more efficient use of tools and cooking of food. Homo erectus appear to have led the same kind of life, including the hunting and gathering that characterizes some groups of modern man. Because of their similarities to modern men, these fossil men are included within the genus Homo.

The ancestral line, or lines, leading to modern man becomes hazy approximately 300,000 years ago. From around 300,000 to 150,000 years ago, the fossil record is rather scanty. However, a collection of fossils found in Terra Amata, France, which is about 200,000 to 35,000 years old, gives us a decent picture of man's transition from Homo erectus to Homo sapiens. The structures of these skeletal remains clearly show an increase in brain size, as well as the increasing likeness of facial features to those of modern man. The Homo sapiens neanderthalensis, known as the cavemen, is a hominid group with skulls that were large and massive with a thick, bony ridge over the eyes and a receeding forehead. Their nose was broad and short and there was almost no chin at all. Their cranial capacity was as large as, or larger than that of modern man. In general, these characteristics are intermediate between those of Homo erectus and those of Homo sapiens.

The population of Neanderthal men that inhabited regions that are now Europe, North Africa and the Middle East some 75,000 to 35,000 years ago seemed to be both culturally and physically modern. They were a widely spreading species - their remains have been found over a wide area including parts of western Europe, Asia and Africa, indicating that they were able to adapt to various climates. They were a society of small family groups who lived primarily in caves, used fire, made flint weapons of the Mousterian tool tradition, hunted a variety of game ranging in size up to the mammoth and rhinoceros, and buried their dead reverently with food, flowers and ornaments. This last mentioned fact indicates that they were capable of abstract thinking including the concept of a life after death.

Reconstruction of one Neanderthal skeleton caused these men to be depicted as standing stooped, with their knees bent forward. However, it has been determined later that this portrayal of Neanderthal man is incorrect. The skeleton used to form this illustration actually come from a man who died at a very old age with rickets, which accounts for the skeleton's deformation. It is now known through examination of many other skeletons that Neanderthal man stood as erect as modern man and was probably just as efficient a bipedal walker. Presently, Neanderthal man is considered to belong to a sub-species of Homo sapiens.

Answer

To be covered when testing yourself

SHORT ANSWER QUESTIONS FOR REVIEW

Choose the correct answer.

1. The bones which are the most frequently found anatomical fossils are (a) femurs (b) cranial bones (c) teeth (d) radii. — c

2. The shape of man's dental arch is: (a) rectangular (b) parabolic (c) v-shaped (d) cuboidal. — b

3. Which group is not a member of the superfamily Hominoidea? (a) Hylobotids, (b) Pongids (c) Hominids (d) Cercopithecoids. — d

4. The oldest found fossils of ancestral Old World monkeys are: (a) Ramapithecus (b) Parapithecus (c) Dryopithecus (d) Oreopithecus. — b

5. Gibbons and orangutans swing from branch to branch. This form of locomotion is called (a) arboreal locomotion (b) brachiation (c) hand-over-hand swing (d) bipedalism. — b

6. Sexual dimorphism is a term which refers to (a) traits in females which are superior to males (b) traits which are the same in both males and females (c) difference in traits between males and females (d) traits in males which are superior to females. — c

7. The correct evolutionary sequence is: (a) Parapithecus → Dryopithecus → Ramapithecus → Australopithecus → Homo erectus → Homo sapiens → Homo sapiens neanderthalensis → Homo sapiens sapiens (b) Ramapithecus → Parapithecus → Australopithecus → Dryopithecus → Homo erectus → Homo sapiens → Homo sapiens neanderthalensis → Homo sapiens sapiens. (c) Parapithecus → Dryopithecus → Ramapithecus → Australopithecus → Homo erectus →Homo sapiens sapiens → Homo sapiens neanderthalensis → Homo sapiens. (d) Dryopithecus → Australopithecus → Ramapithecus → Parapithecus → Homo erectus → Homo sapiens → Homo sapiens neanderthalensis → Homo sapiens sapiens. — a

8. The acquisition of some degree of bipedalism is first documented in (a) Ramapithecus (b) Dryopithecus (c) Australopithecus (d) Homo erectus. — a

9. The first hominid to make use of fire was: (a) Homo sapiens (b) Homo sapiens neanderthalensis (c) Australopithecus (d) Homo erectus. — d

Answer

To be covered when testing yourself

10. Which is not a trait of the Cercopithecoids: (a) bilophodent molars (b) ischial callosities (c) prehensile tail (d) catarrhini.

c

11. The presapien theory states that: (a) Homo sapiens originated as a distinct linage, completely separate from the line leading to Neanderthals, (b) there was a broad and varying Neanderthal population, one segment of which was an isolated European, cold-adapted group, the other inhabited the Near East and became modern man. (c) Neanderthals gave rise to Oligopithecus which diversified and evolved to give rise to pangids as well as modern man, (d) the European, cold adapted segment of the Neanderthals gave rise to modern man.

a

Fill in the blanks.

12. The superfamily Hominoidea is represented by three families: _____, _____, and _____.

Hylobatids Pongids, Hominids.

13. The oldest known fossils of Old World monkeys are _____.

Parapithecus

14. _____ enabled man to free his hands for tool making and other tasks.

Bipedalism

15. Dryopithecine fossils date back to the _____ era.

Oligocene

16. In the Dryopithecine line _____ is thought to be ancestral to the chimpanzee, _____ is thought to be ancestral to the gorilla; _____ is thought to be ancestral to the orangutan.

Dryopithecus africanus, Dryopithecus major, Dryopithecus sivalensis

17. Australopithecines had _____ tools or _____ industry.

pebble; Oldowan

18. Java and Peking man are classified as _____.

Homo erectus

19. During the _____, agriculture and animal husbandry began.

Neolithic

20. _____ are populations that have different frequencies of many genes and thus different phenotypic trends.

Races

21. _____ showed that they were capable of abstract thinking by being the first hominid to leave evidence of burying their dead.

Homo sapien neanderthalensis.

Answer

To be covered when testing yourself

Determine whether the following statements are true or false.

level C

level B

level A

Statement	Answer
22. An archaeologist excavated the above site. The fossils that were found in level B were older than those of level A.	False
23. The infraorder, catarrhine is divided into two superfamilies, cercopithecoids and Hominoids.	True
24. The most likely common ancestor of the chimpanzee, gorilla, orangutan and man is Dryopithecus.	True
25. The only great ape which has a tail is the gibbon.	False
26. Chimpanzees, gorillas and man are all omnivorous.	False
27. The Dryopithecines emerged during the Oligocene.	True
28. Ramapithecus dates back to the upper Paleocene.	False
29. The order that hominid evolution occurred in is Ramapithecus → Australopithecus → Homo erectus → Homo sapiens → Homo sapiens neanderthalensis → Homo sapiens sapiens.	True
30. Genetic drift is less important in human evolution at the present time than it was 20 thousand or more years ago.	True
31. Today there is more than one species of man. This can be observed by the different races of man.	False

CHAPTER 30

PRINCIPLES OF ECOLOGY

DEFINITIONS

• PROBLEM 30-1

How would you define ecology? Differentiate between autecology and synecology.

Solution: Ecology can be defined as the study of the interactions between groups of organisms and their environment. The term autecology refers to studies of individual organisms or populations, of single species and their interactions with the environment. Synecology refers to studies of various groups of organisms which are associated to form a functional unit of the environment.

Groups of organisms are characterized by three levels of organization - populations, communitites and ecocystems. A population is a group of organisms belonging to the same species which occupy a given area. A community is a unit composed of a group of populations living in a given area. The community and the physical environment considered together is an ecosystem. Each of these designations may be applied to a small local entity or to a large widespread one. Thus a small group of sycamore trees in a park may be regarded as a population, as could be the sycamore trees in the eastern United States. Similarly, a small pond and its inhabitants or the forest in which the pond is located may be treated as an ecosystem. From these examples, we see that the limit of an ecosystem depends on how we define our ecosystem. However each ecosystem must consist of at least some living organisms inhabiting a physical environment.

Various ecosystems are linked to one another by biological, chemical and physical processes. The entire earth is itself a true ecosystem because no part is fully isolated from the rest. This global ecosystem is usually referred to as the biosphere.

COMPETITION

• PROBLEM 30-2

Explain what is meant by an animal's ecological niche,

and define competitive exclusion.

Solution: Ecologically, niche is defined as the functional role and position of an organism within its ecosystem. The term niche should not be confused with habitat, which is the physical area where the organism lives. The characteristics of the habitat help define the niche but do not specify it completely. Each local population of a particular species has a niche that is defined with many variables. Each has a temperature range along with other climatic factors. There are required nutrients and specific activities that also help characterize the niche.

The principal of competitive exclusion states that unless the niches of two species differ, they cannot coexist in the same habitat. Two species of organisms that occupy the same or similar ecologic niches in different geographical locations are termed ecological equivalents. The array of species present in a given type of community in different biogeographic regions may differ widely. However, similar ecosystems tend to develop wherever there are similar physical habitats. The equivalent functional niches are occupied by whatever biological groups happen to be present in the region. Thus, a savannah-type vegetation tends to develop wherever the climate permits the development of extensive grass-lands, but the species of grass and the species of animals feeding on the grass may differ significantly in various parts of the world.

• PROBLEM 30-3

Explain the types and effects of competition.

Solution: Competition is the active demand of two or more organisms for a common vital resource. There are many ways for organisms to compete but basically competition can be categorized in two ways. Contest competition is the active physical confrontation between two organisms which allows one to win the resource. Extreme contest competition involves direct aggression. Scramble competition is the exploitation of a common vital resource by both species. In scramble competition, one organism is able to find and utilize vital resources more efficiently than another.

The aggressive behavior of ant colonies is an example of contest competition. Actual colony warfare can establish territorial limits within and between species. Often, contest competition takes the form of threatening gestures which involves safer means to achieve the same goal.

Scramble competition can best be demonstrated with observations of the fruit fly Drosophila. Combat and aggressive behavior have nothing to do with the type of competition these flies demonstrate. It has been observed that when the food supply was limited, there were four conditions for survival for the fly larvae. Those that survived were the quickest feeders, best adapted for the particular medium, heaviest at the beginning of competition, and most resistant to changes due to population increase.

Every species occupies a particular niche that to a large extent helps avoid competition. But niches do overlap in many areas and competition can never be clearly eliminated.

• PROBLEM 30-4

What are the consequences of intense interspecific competition?

Solution: The more similar two niches are, the more likely it is that both species will utilize at least one common limited resource (food, shelter, nesting sites, etc). They will therefore be competing for that limited resource.

The competitive superiority of one of the rival species may be such that the other is driven to extinction. Two species may fluctuate in competitive superiority according to the habitats they share, with the result that one is eliminated in some places and the other is eliminated in other places. One or both of the competing species may evolve in divergent directions under the strong selection pressure resulting from their intense competition. Natural selection would favor individuals with characteristics differing from those of the other species, because such characteristics would tend to minimize competition. In other words, the two species would evolve greater differences in their niches.

Whether any one or a combination of these effects will be the outcome in any given case of intense competition is determined by many complex factors. If a pair of species occupies identical ecological niches, one of them will indeed become extinct. This generalization is usually referred to as Gause's principle, or the principle of competitive exclusion, which states that two different species cannot simultaneously occupy the same niche in the same place. It is implicit in the definition of niche that no two species could ever occupy the same niche. To do so, they would have to be identical in every respect and hence they would be one species, not two. Thus, the competitive exclusion principle helps us see why the different species living together in a stable community occupy quite distinct niches. Only if organisms living together avoid competition can they successfully coexist. A fit organism is one that avoids competition, and is free from the necessity to struggle with others in its own niche.

INTERSPECIFIC RELATIONSHIPS

• PROBLEM 30-5

What is mutualism? Give three examples of this type of relationship.

Solution: Mutualism, (or symbiosis) like parasitism and commensalism, occurs widely through most of the principal plant and animal groups and includes an astonishing diversity of physiological and behavioral adaptations. In the mutualistic type of relationship, both species benefit from each other. Some of the most advanced and ecologically important examples occur among the plants. Nitrogen - fixing bacteria of the genus Rhizobium live in special nodules in the roots of legumes. In exchange for protection and shelter, the bacteria provide the legumes with substantial amounts of nitrates which aid in their growth.

Another example of plant mutualism is the lichens. They are actually "compound" organisms consisting of highly modified fungi that harbor blue-green and green algae among their hyphae (filaments). Together the two components form a compact and highly efficient unit. In general, the fungus absorbs water and nutrients and forms most of the supporting structure, while the algae provides nutrients for both components via photosynthesis.

In a different form of mutualism, many kinds of ants depend partly or wholly upon aphids and scale insects for their food supply. They milk aphids by stroking them with their fore legs and antennae. The aphids respond by excreting droplets of honeydew, which is simply partly digested plant sap. In return for this sugar-rich food, the ants protect their symbionts from parasitic wasps, predatory beetles and other natural enemies.

In man, certain bacteria that synthesize vitamin K live mutualistically in the human intestine which provides them with nutrients and a favorable environment.

• PROBLEM 30-6

In a commensalistic relationship between two species living symbiotically what are the consequences of their interactions?

Solution: Commensalism is a relationship between two species in which one species benefits while the other receives neither benefit nor harm. The advantage derived by the commensal species from its association with the host frequently involves shelter, support, transport, food or a combination of these. For example, in tropical forests, numerous small plants called epiphytes usually grow on the branches of the larger trees or in the forks of their trunks. They use the host trees only as a base of attachment and do not obtain nourishment from them. A similar type of commensalism is the use of trees as nesting places by birds. Such relationships do not produce any apparent harm to the hosts.

Commensalism is widespread throughout the animal kingdom, and is especially common among the marine invertebrates. The host organisms are typically slow-moving or sessile and live in shells or burrows, which can be readily shared by the smaller commensal species. One example is a certain species of fish that regularly lives in association with sea anemones, deriving protection and shelter from them and sometimes sharing some of their food. These fish swim freely among the tentacles of the anemones even though these tentacles quickly paralyze other species of fish that touch them. The anemones feed on fish, yet the particular species that live as commensals with them can actually enter the gastrovascular cavity of their host, emerging later with no ill effects. This implies a certain amount of physiological and behavioral adaptation between members making up a commensal relationship.

Still another example is a small tropical fish (Fierasfer) that lives in the rectum of a particular species of sea cucumber. The fish periodically emerges to feed and then returns by first poking its host's rectal opening with its snout and then quickly turning so that it is drawn tail first into the rectal chamber.

• PROBLEM 30-7

Explain and exemplify internal and external parasitism.

Solution: Parasites can be classified as external or internal parasites. The former live on the outer surface of their host, usually feeding on the hair, feathers, scales or skin of the host or sucking its blood. Internal parasites may live in the various tubes and ducts of the host's body, such as the digestive tract, respiratory passages or urinary ducts. They may also bore into and live embedded in tissues such as muscle or liver. However, pathogens like viruses and some bacteria are believed to actually live inside the individual cells of their host.

External parasitism may have evolved from what was once a commensal relationship. In some cases, internal parasitism may have evolved from external parasitism. External parasites might have wandered into one of the body openings of their host, such as the mouth nasal openings, or anus. Most of them cannot survive in the different environment they encounter there. But it is possible that in the course of thousands or millions of years some wanderers might have had the genetic constitutions enabling them to survive in the new habitat they found inside the host's body. From such a beginning, the specializations that ordinarily characterize the more advanced internal parasites could have evolved. Other internal parasites may have evolved from free living forms that were swallowed or inhaled.

Internal parasitism is usually marked by much more extreme specializations than external parasitism because the habitats available inside the body of a living organism are completely unlike those outside it. The unusual internal conditions have resulted in adaptations different from those seen in free-living forms. For example, internal parasites have frequently lost individual organs or whole organ systems that would be essential to a free-living species. Tapeworms, for instance, have no digestive system. They live in their host's intestine, where they are bathed by products of the host's digestion, which they can absorb directly across their body wall without having to digest anything themselves.

Loss of certain structures is not the only sort of special adaptation commonly seen in internal parasites. They often have body walls highly resistant to the destructive enzymes and antibodies of the host. Moreover, they frequently possess specialized structures such as the specialized heads of tapeworms with hooks and suckers that enable them to anchor themselves to the intestinal wall.

One of the most striking of all adaptations of internal parasites concerns their life cycles and reproductive capability. Individual hosts do not live forever. If the parasitic species is to be perpetuated, a mechanism is needed for leaving one host and penetrating another. Thus, at some point in their life cycle, all internal parasites must try to move from one host individual to another. But this is seldom simple. Rarely can a parasite move directly from one host individual to another. It is not unusual for a life cycle to include two or three intermediate hosts and a free living larval stage. But such a complex development makes the chances that any one larval parasite will encounter the right hosts in the right sequence exceedingly poor. The vast majority die without completing their life cycle. As a result, internal parasites characteristically produce huge numbers of eggs. Although they may be deficient in some organs through specialization they usually have extremely well-developed reproductive structures. In fact, some internal parasites seem to be little more than a sac of highly efficient reproductive organs.

• PROBLEM 30-8

Why would it be disadvantageous for a host-specific parasite to kill its host?

Solution: In the course of their evolution, parasites usually develop special features of behavior and physiology that make them better adjusted to the particular characteristics of their host. This means that they often tend to become more and more specific for their host. Where an ancestral organism might have parasitized all species in a particular family, its various descendants may parasitize only one species of host at each stage in its development. This means that many parasites are capable of living only in one specific host species and that if they should cause the extinction of their host, then they themselves would also become extinct. A dynamic balance exists where the host usually survives without being seriously damaged and at the same time allowing the parasites to moderately prosper. Probably most long-established host-parasite relationships are balanced ones. The ideally adapted parasite is one that can flourish without reducing its host's ability to grow and reproduce. For every one of the parasitic species that cause serious disease in man and other organisms, there are many others that give their hosts little or no trouble. It is generally true that the deadliest of the parasites are the ones that are the most poorly adapted to the species affected. Relationships that result in serious disease in the host are usually relatively new ones, or ones in which a new and more virulent form of the parasite has recently arisen, or where the host showing the serious disease symptoms is not the primary host of the parasite. Many examples are known where man is only an occasional host for a particular parasite and suffers severe disease symptoms, although the wild animal that is the primary host shows few ill effects from its relationship with the same parasite.

CHARACTERISTICS OF POPULATION DENSITIES

• PROBLEM 30-9

Why does population growth follow a logistic (S-shaped) form until it reaches the carrying capacity of the environment?

Solution:

Since nearly all mature individuals in a population can produce offspring, the rate at which an unrestricted population increases is a function of the generation time, which is the period required for an organism to reproduce. When a population initially grows, the size of the population will double every generation time. This kind of growth rate is referred to as exponential growth (Fig. I).

A population of animals has a potential through reproduction to increase at a given rate which is termed the intrinsic rate of increase. The intrinsic rate of increase is equal to the difference between the average birth rate and average death rate of a population. It varies enormously from species to species.

In a typical experiment, 40 individual paramecia were placed in a small tube of water and each day thereafter fresh food was added. Therefore, a constant but limited daily input of food to the system was provided. Under these conditions, the paramecia reproduced quickly at first, with their number increasing exponentially until there were about 4000 animals in the tube. Then the rate of growth leveled off, and the population remained at a steady state. The rate of replacement was finally about equal to the death rate, and the population became balanced. It seems likely that the balance reached by experimental populations is equivalent to the general balance existing in nature.

If animals breed prolifically when there is plenty of space and food, but cease to do so when crowded, then it is reasonable to suggest that cessation may be a result of increasing intraspecific (between the members of one species) competition for the resources of the habitat. Breeding may also become suppressed by some effect of the environment which inhibits the population more strongly as crowding increases. The growth of a population in a confined space with a limited input of energy is described by an "S"-shaped graph (see Fig. II).

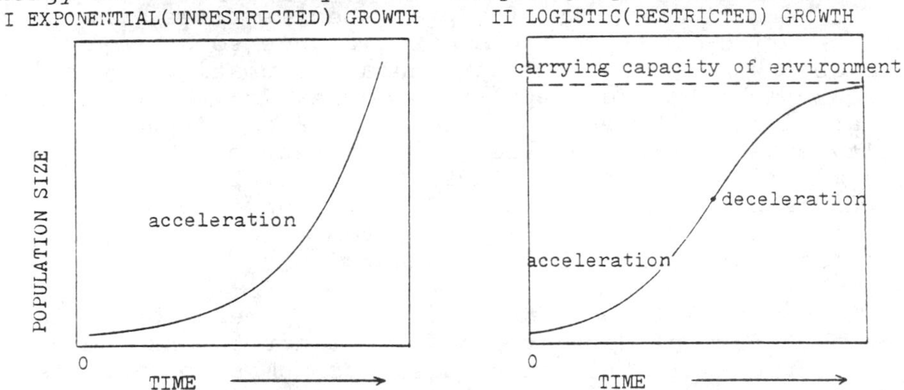

The number of individuals increases rapidly, but in time there is a leveling off of the rate of growth of the population so that the growth rate eventually becomes zero. This suppressing effect of competition gets stronger as the animals get more numerous, until it is strong enough to bring the rate of increase down to zero and an upper limit is reached where the environment in which they live cannot support any more of them.

The limiting size is called the carrying capacity of the environment. At this population size, the death rate equals the birth rate, and therefore the rate of increase in the number of individuals is zero. Temporary deviations from zero growth will probably occur causing the population to grow for a short time, or to decline, but the average value over long periods of time will be zero.

The actual value of the carrying capacity of the environment for a specific species is determined by the interaction of several factors, including the total energy flow in the ecosystem, the trophic level to which the species belongs, and the size and metabolic rate of the individuals. Eventually, energy from food will be a limiting factor for any population.

• PROBLEM 30-10

Explain how density dependent controls of population occur.

Solution: An important characteristic of a population is its density, which is the number of individuals per unit area. A more useful term to ecologists is ecologic density, which is the number of individuals per habitable unit area. As the ecologic density of a given population begins to increase, there are regulatory factors that tend to oppose the growth. These regulating mechanisms operate to maintain a population at its optimal size within a given environment. This overall process of regulation is known as the density dependent effect.

Predation is an example of a density dependent regulator. As the density of a prey species rises, the hunting patterns of predators often change so as to increase predation on that particular population of prey. Consequently, the prey population decreases; the predators then are left with less of a food resource and their density subsequently declines. The effect of this is often a series of density fluctuations until an equilibrium is reached between the predator and prey populations. Thus, in a stable predator-prey system, the two populations are actually regulated by each other.

Emigration of individuals from the parent population is another form of density dependent control. As the population density increases, a larger number of animals tend to move outward in search of new sources of food. Emigration is a distinctive behavior pattern acting to disperse part of the overcrowded population.

Competition is also a density dependent control. As the population density increases, the competition for limited resources becomes more intense. Consequently, the deleterious results of unsuccessful competition such as starvation and injury become more and more effective in limiting the population size.

Physiological as well as behavioral mechanisms have evolved that help to regulate population growth. It has been observed that an increase in population density is accompanied by a marked depression in inflammatory response and antibody formation. This form of inhibited immune response allows for an increase in susceptibility to infection and parasitism. Observation of laboratory mice has shown that as population density increases, aggressive behavior increases, reproduction rate falls, sexual maturity is impaired, and the growth rate becomes suppressed. These effects are attributable to changes in the endocrine system. It appears that the endocrine system can help regulate and limit population size through control of both reproductive and aggressive behavior. Although these regulatory mechanisms have been demonstrated with laboratory mice, it is not clear to what extent they operate in other species.

• PROBLEM 30-11

In contrast to density dependent effects, there are also density independent effects which operate without reference to population size. Elaborate on the major density independent effects and their relationship to population size.

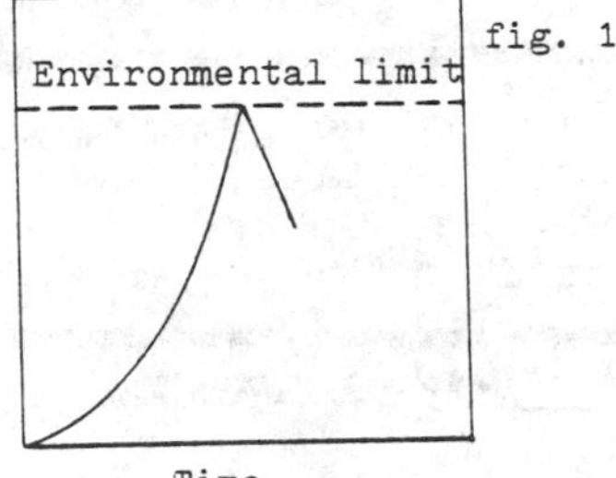

fig. 1

A growth curve in which the environmental limiting factors did not become effective until late, and then produced a sudden sharp decline.

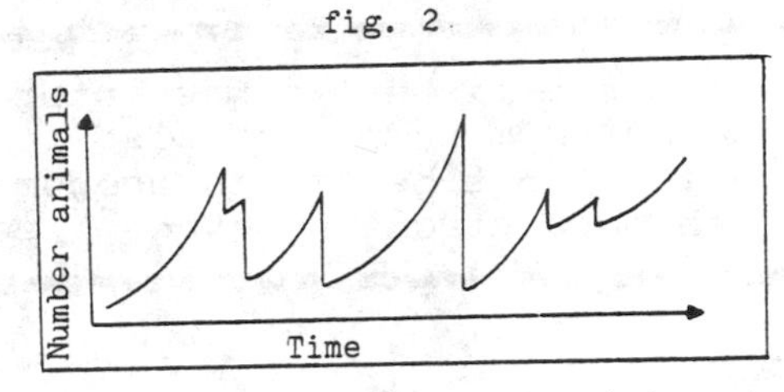

fig. 2

Growth curve under density independant effect.

Solution: There is no scientific principle which states that populations must be controlled as a function of density. Density effects will always be working in crowded populations, but it is also possible for actual control of population size to be exercised before crowding occurs. Control agents that are not dependent on density can take the form of sudden events that are catastrophic to animal populations. Catastrophic events can solve the problem of overcrowding just as efficiently as a density dependent device.

Hurricanes and volcanic explosions can destroy entire populations but these events are too scattered and local to be used as examples for general population controls. More applicable are the common catastrophes which we know as changes in the weather. Nearly all places on the earth suffer seasonal changes from summer to winter, from warm to cold, and from wet to dry. Each of these cyclic changes represents hazards to the animals of each area. Thus, growing populations are frequently cut back, making their normal lives a race to reproduce so quickly that there shall be at least some survivors following the next catastrophe. Unless something happens to prevent the catastrophe from occurring such a population may never grow large enough to suffer the effects of crowding.

Density independent factors may play an important role in limiting some organisms, particularly those with very short life cycles characterized by a growth curve in which the environmental limiting factors do not become effective until after many generations (see fig. 1). A sudden strong density independent limitation then brings growth to an end before density dependent factors can become operative.

The density independent effects are definite factors in moving population growth upward or downward, but they cannot hold the population size at a constant level. As a result, populations effected by density independent factors are under control but their numbers fluctuate within wide limits (see fig. 2) and can hardly be described as "in balance."

• PROBLEM 30-12

Suppose a fisherman was interested in trying to extract the maximum number of fish from a pond or lake. Why would it not be to his economic advantage to reduce the fish populations by excessive harvesting?

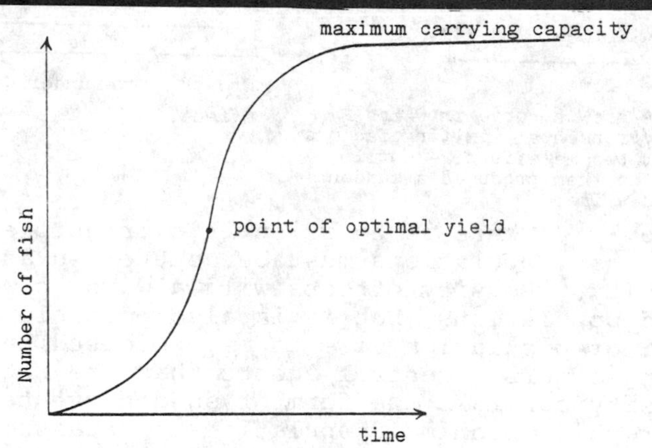

An S- shaped growth curve showing the point of optimal yield.

Solution: The ideal economic strategy for the fisherman to follow would be to catch only enough fish so as to keep the population at a level of optimal yield. If the population is allowed to reach the maximum that the environment can support, or if it is exploited to the point where the organisms become scarce, the yield will decline. This phenomenon exists because the greatest amount of increase in population size does not occur either when the population is very low or when the population has reached the carrying capacity of the environment. The point of greatest growth and hence the maximum replacement capacity of a population occurs at a point midway on the exponential part of an S-shaped growth curve, that is, half way between the baseline and the maximum carrying capacity (see figure).

This point corresponds to the point of steepest rise on the S-chaped curve. At this point, the greatest number of new individuals are added to the population in a given amount of time. This maximum rate of population growth is referred to as the optim·l yield.

Fish populations, as well as other animal populations that man harvests, can be obtained with much less effort at a point of optimal yield than populations reduced by overfishing. Moreover, the greater the exploitation of populations, the more the population will come to consist of younger and smaller individuals, which are usually the least valuable commercially. The proportion of large fish caught is higher when the population is at a point of optimal yield. Consequently, as the fish populations are reduced, more expensive and sophisticated ships and trawling techniques are required to produce the same amount of yield.

• PROBLEM 30-13

How is a survivorship curve used in determining the age distribution of a population?

Solution:

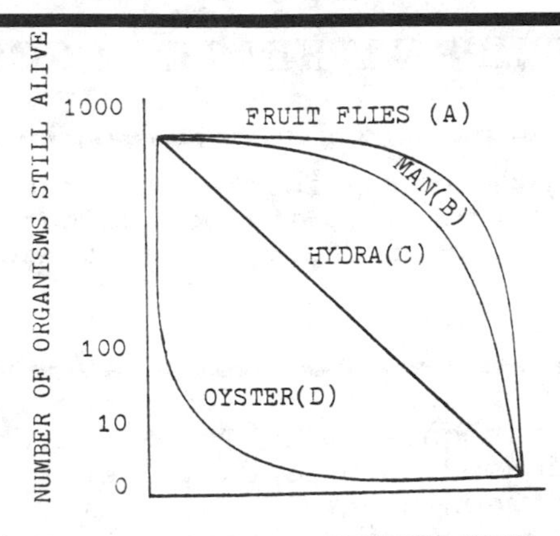

Survivorship curve for four species of animals, starting with equal population size. The curves show the number of living individuals at different ages.

The age distribution of a population is the proportion of individuals belonging to each age group. When a population is allowed to exist in a stable environment for several generations, so that its birth and death rates become balanced, its age distribution becomes stable. The stable age distribution differs greatly from species to species, and is expressed by a survivorship curve.

The oyster is an example of a species in which large numbers of young are produced, but the majority die in a short time. Only a tiny fraction succeeds in attaching themselves to a rock or to some other support which is the necessary step for continuing the life cycle. The survival rate among oysters after reaching this point is much higher. The hydra exemplifies species with constant mortality rates. In this case, an individual is just as likely to die when it is one year old as when it is one day old. Man and fruit flies, in contrast to the hydra and oyster, are species with a definite period of senescence. If provided with a good environment, most individuals live to a certain age in reasonably sound health. Then diseases and infirmities of old age begin to set in, and death becomes increasingly probable with each passing year.

Changes in environmental conditions may radically alter the shape of the survivorship curve for any given population, and the altered mortality rates in turn have profound effects on the age distributions and on its future size. For example, the chief cause for the enormous increase in the population size of human beings has been a great reduction in mortality during the early life stages as a result of improvements in sanitation, nutrition and medical care. These improvements have caused a shift in the human survivorship curve from one intermediate between the curves C and D in primitive societies to one approaching A in most advanced societies.

INTERRELATIONSHIPS WITHIN THE ECOSYSTEM

• PROBLEM 30-14

Sunlight is the ultimate energy source on earth. Energy from sunlight is not returned to its source but is transformed to other forms of energy which are closely tied together in an energy cycle. Describe the energy cycle.

Solution: The energy cycle starts with sunlight being utilized by green plants on earth. The Kinetic energy of sunlight is transformed into potential energy stored in chemical bonds in green plants. The chemical bonds are synthesized by the process of photosynthesis. The potential energy is released in cell respiration and is used in various ways. Thus, fundamental to the energy cycle is the ability of energy to be transformed.

Inorganic (nonliving) matter in the ecosystem is closely tied to organic (living) matter in the energy cycle. For example, photosynthesis requires carbon dioxide from the air and water and minerals from the soil to occur besides sunlight. These are the nonliving components of photosynthesis. Chlorophyll in green plants captures the sunlight, and organic substances (i.e., glucose) are generated from inorganic ingredients via a series of enzymatic reactions in the plant cell.

Chlorophyll, enzymes, and other cellular components form the living part of photosynthesis.

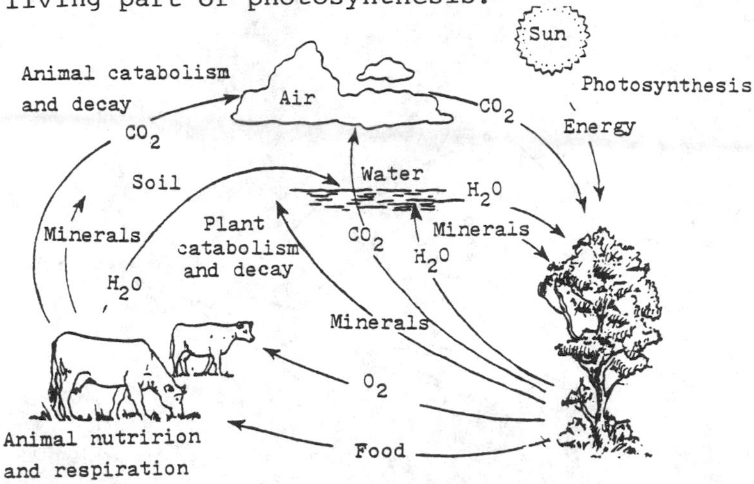

The energy cycle. This diagram shows the relationships between plants and animals and the nonliving materials of the earth. The energy of the sunlight is the only thing that is not returned to its source.

In the energy cycle (Figure), some of the food synthesized by green plants are broken down by the plants for energy, and consequently carbon dioxide and water are released. These again become available for green plants in capturing more energy of the sunlight. Some of the synthesized compounds are used in building the bodies of the plants and are hence stored as potential energy until the plants die. The bacteria and fungi of decay break down the bodies of the dead plants, using the liberated energy for their own metabolism. In these processes, carbon dioxide and water are released, and the minerals go back into the soil. These substances are thus recycled. Animals which feed on plants utilize a part of the energy from the food in cell respiration, with a release of carbon dioxide and water which are again recycled. Some of the minerals in the plant food are excreted by the animals and are thus available to be reused. Animals which feed on other animals utilize some energy from the food in building their own bodies. They break down some of their stored food to yield energy for daily activities such as locomotion. Food degradation is accompanied by release of carbon dioxide and water, which are returned to the ecosystem. When animals die, their bodies decay and all of the materials that were used in the construction are restored to a state which can be reused by the action of the bacteria and fungi of decay.

It must be remembered that at no point in the cycle is energy destroyed. The energy from sunlight is not destroyed but is transformed into heat, chemical or mechanical energy.

In contrasting a food chain and a food web, which one do you think actually operates in a real community?

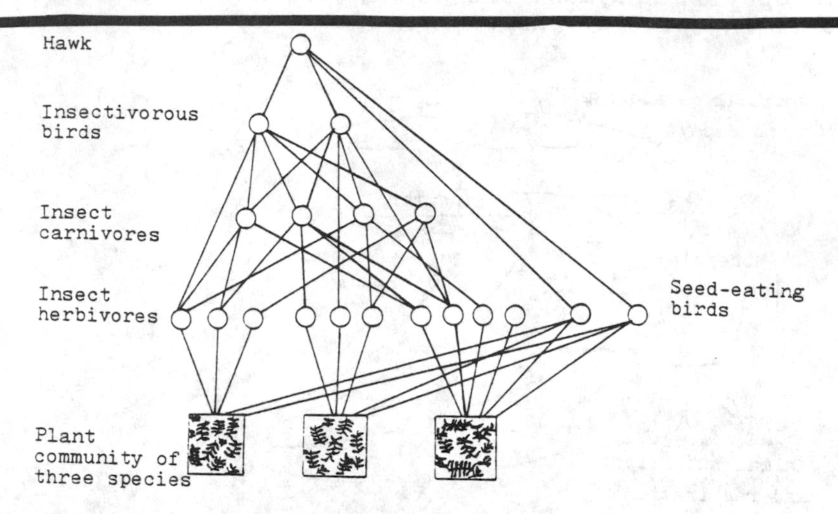

Hypothetical food web. It is assumed that there are three hundred species of plants, ten species of insect herbivores, two bird herbivores, two bird insectivores, and one hawk. In a real community, there would not only be many more species at each trophic level but also many animals that feed at more than one level, or that change level as they grow older. Some general conclusions emerge from even an oversimplified model like this however. There is an initial diversity introduced by the number of plants. This diversity is multiplied at the plant-eating level. At each subsequent level the diversity is reduced as the food chains converge.

Solution: Food is the common word used to describe the various nutrients that all living heterotrophic organisms must ingest in order to obtain energy to sustain their life processes. Autotrophic photosynthetic organisms, such as the green plants, can manufacture their own food from simple inorganic molecules with the energy from sunlight. Life on earth ultimately depends on food energy which in turn is dependent on the sun.

The radiant energy from the sun that reaches the photosynthetic green plants is responsible for the transformation of basic raw materials such as water, carbon dioxide, nitrogenous compounds and minerals into the development of the plants themselves. Plants store nutrients (starches and sugars) within them and in turn are eaten by heterotrophic organisms such as animals.

A food chain is most commonly a sequence of organisms that are related to each other as prey and predators. One species is eaten by another, which is eaten in turn by a third, and so on. Each species forms a step or link in one or more food chains.

Every food chain begins with the autotrophic organisms (mainly green plants) that serve as the producers of the community. Any organism that does not produce its own food and must therefore depend on another for nutrients is a consumer. Every food chain ends with decomposers, the organisms of decay, which are usually bacteria and fungi that degrade complex organic materials to simple substances which are reusable by the producers. The links between the producers and the decomposers are variable because the producers may die and be acted upon directly by the decomposers, or the producers may be eaten by primary consumers, called the herbivores. The herbivores consume the green plants and in turn may be either acted upon directly by the decomposers or fed upon by the secondary consumers, the carnivores (animal eaters). Some food chains consist of tertiary consumers, or secondary carnivores, which eat the secondary consumers and sometimes also the primary consumers. In almost all ecosystems there are top carnivores: one or more large, specialized animal species that prey on organisms on the lower steps, but are not ordinarily consumed by predators themselves. The larger whales enjoy this status, as do lions, wolves, and man. The decomposers can feed on any dead organism in the food chain. In addition, there are parasite chains in which small organisms live on or within larger ones from where they obtain food.

The successive levels in the food chains of a community are referred to as trophic levels. Thus all the producers together constitute the first (or lowest) trophic level and the primary consumers (herbivores) constitute the second trophic level. The herbivore-eating carnivores constitute the third trophic level, and so on.

Energy-flow in an ecosystem is actually more complicated than the flow of nutrients that simplified food chains imply. In most real communities, there are many different possible food chains that are tied together by cross linkages. Any one animal usually eats a variety of food and thus may be part of several food chains that intersect. Also, food chains starting from a common plant source may radiate outward as the plant food is eaten by different herbivores, and as the latter are eaten by different carnivores, and so on. Thus, there is actually a complex food web formed as the food chains first radiate outward from the plants and then come together at the top carnivore level of the web (see Figure). For example, the grasshoppers of the grasslands of Canada eat grass and are eaten by robins. Robins eat many kinds of insects as well as grasshoppers. The great horned owl feeds upon the robin as well as other mammals such as the prairie vole. But the prairie vole is also eaten by the coyote. Hence, even though organisms in a community can be classified into the various trophic levels according to the nature of their prey, it is more characteristic to assign organisms to a food web than to a food chain in a real community.

• PROBLEM 30-16

The biomass of each trophic level is usually much less than that of the preceding lower trophic level. Define the term biomass and explain the factors that determine the biomass of trophic levels in ecosystems.

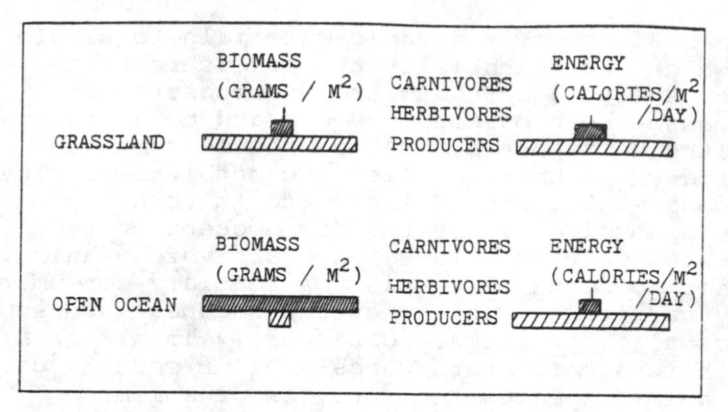

Biomass and energy pyramids for two very different ecosystems. The form of energy pyramids is similar from one system to another, but form of biomass pyramids varies considerably.

Solution: Biomass is the total mass of the living material present in a certain category, whether it is a trophic level or an ecosystem. The distribution of biomass within some ecosystems can be represented by a pyramid, with the first trophic level (producers) at the base and the last consumer level at the apex. In general, the decrease of energy at each successive trophic level means that less of the biomass can be supported at each level.

If the animals of each trophic level were of the same size, they would have to be rarer and rarer toward the top of the pyramid where energy is in shortest supply. Since the animals at the top are usually larger, they are sharing an even smaller supply of energy. The animals high in food chains therefore, must be few, and the pyramid of biomass is a direct consequence of this. Thus the total mass of carnivores in a given community is almost always less than the total mass of herbivores. The size, growth rate and longevity of the species at the various trophic levels of a community are important in determining whether or not the pyramidal model will hold for the biomass of the community. In fact, biomass pyramids of different communities vary greatly.

The variability of biomass pyramids exist because the plant producer organisms exhibit extreme variability in their ability to undergo photosynthesis. The small algae of some ocean communities can greatly outproduce most land plants on a per gram basis. This is also because of the high metabolic and reproductive rates of algae. Consequently, they are able to support a proportionately much larger biomass of herbivores. The production of ocean herbivores is still only about ten percent (due to ecological efficiency) that of algae. But the biomass of the herbivores is increased because the turnover rate (the rate at which they are consumed and replaced) of the algae is very high. Biomass normally tends to decrease with each successive trophic level, forming the shape of a pyramid. Some communities show an inverse pyramid biomass relationship, such as the case of open ocean algae primary producers.

• PROBLEM 30-17

What factors limit the number of trophic levels in a food chain?

Solution: As energy flows through the various food chains, it is being constantly channeled into three areas. Some of it goes into production, which is the creation of new tissues by growth and reproduction. Energy is used also for the manufacture of storage products in the form of fats and carbohydrates. Some of the energy is lost from the ecosystem by way of decomposing dead organic materials. The rest of the energy is lost permanently to the ecosystem through respiration. The loss of energy due to respiration is very high and only a small fraction of energy is transferred successfully from one trophic level to the next. Each trophic level depends on the preceding level for its energy source. The number of organisms supportable by any given trophic level depends on the efficiency of transforming the energy available in that level to useful energy of the subsequent level. Ecological efficiencies may vary widely from one kind of animal to another, even at the same trophic level. It has been shown that the average ecological efficiency of any one trophic level is about 10%.

The rate at which energy can be transformed by plants is controlled by the carbon dioxide concentration of the atmosphere, by the amount of photosynthetic surfaces available to light and by various limiting factors of the environment such as water and nutrients.

Animals have their energy supply already in the usable form of high-energy compounds. The efficiency of an animal trophic level is primarily a function of the food-getting and digesting process of that trophic level.

The actual energy flow into primary consumers or herbivores is a small fraction of the total energy converted by plants because a substantial part of what they eat cannot be digested and must be returned back to the environment. Moreover, much of the energy converted by vegetation is used by the vegetation itself for respiration.

Carnivores face similar but even more rigorous restrictions on their energy supplies, because their energy depends on how many of the herbivores they can manage to catch, eat and digest. For secondary carnivores the restrictions are even more rigorous and increase through the successive trophic levels.

Since energy is so important to animals, and is in restricted supply, the number of animals that can be supported is determined by the efficiency with which they utilize energy. From this can be calculated the frequency of a species and the total number that a given area can support. The number of trophic levels or steps in a food chain is normally limited to perhaps four or five because of the great decrease (90%) in available energy at each level; i.e., only 10% of the energy is transferable to the next trophic level.

• PROBLEM 30-18

In a simple ecosystem there exists a field of clover. Mice eat the clover and skunks eat the mice. The skunks are in turn eaten by wolves. According to the ten percent rule of ecological efficiency, how much energy would a wolf receive from the original clover plants?

Solution: The average ecological efficiency per trophic level is 10 percent. Therefore we expect that for every 10,000 calories available from clover plants, 1,000 calories will be obtained for use by a mouse. When a skunk eats this mouse, only 100 of the 1,000 calories will be available to the skunk. In the last trophic level of this food chain, the wolf will obtain 10 percent of the 100 calories transferred to the skunk. Thus, a mere 10 calories of the original 10,000 calories can be used for the metabolic processes of the wolf. A top trophic level carnivore that is receiving only one thousandth of the original calories of the plants must be sparsely distributed and far ranging in its activities because of its high food consumption requirements. Wolves must travel as much as twenty miles a day to acquire enough food. The territories of individual tigers and other great cats often cover hundreds of square miles. If there were predators of wolves, these predators would only be able to make use of 10 percent of the 10 calories that the wolf obtained or, one calorie. Hence, it is hardly worthwhile preying on animals in the upper trophic levels.

ECOLOGICAL SUCCESSION

• PROBLEM 30-19

What are the characteristics of ecological succession?

Solution: Succession is a fairly orderly process of changes of communities in a region. It involves replacement, in the course of time, of the dominant species within a given area by other species. Communities succeed each other in an orderly sequence in which each successive stage is thought to be dependent on the one which preceded it. Each community in the succession is called a seral stage.

The first stage of succession is usually the colonization of barren space such as a sandy beach or a rock, by simple pioneer species such as grasses. The pioneer species are able to grow and breed rapidly. They have adopted the strategy of finding and utilizing empty space. Then, gradually the pioneers are replaced by more complex and bulky species until finally the community is characterized by climax species which represent the final stage of succession. Climax communities are the most stable and will only if some outside agent such as new species or a geographic change displaces them. Most succession can be classified into either primary or secondary successions. Primary successions proceed by pioneering new uninhabited sites. Secondary successsions occur on disturbed land where a climax community that existed before had been destroyed by such occurrences as fire, severe wind storms, flooding and landslides.

Although succession in different places and at different times is not identical (the species involved are often completely different) some ecologists have nevertheless formulated generalizations that hold true for most cases where both autotrophs and heterotrophs are involved:

1) The species composition changes continuously during the succession, but the change is usually more rapid in the earlier stages than in the later ones. 2) The total number of species represented increases initially and then becomes fairly stabilized in the older stages. This is particularly true of the heterotrophs, whose variety is usually much greater in the later stages of the succession. 3) Both the total biomass in the ecosystem and the amount of nonliving organic matter increase during the succession until a more stable stage is reached. 4) The food webs become more complex, and the relations between species in them become better defined. 5) Although the amount of new organic matter synthesized by the producers remains approximately the same, except at the beginning of succession, the percentage utilized at the various trophic levels rises.

In summary, the trend of most successions is toward a more complex ecosystem in which less energy is wasted and hence a greater biomass can be supported.

• PROBLEM 30-20

Why does succession occur? Why can the first colonists not simply seize an area and hold it against subsequent intruders?

Solution: Succession occurs independently of the climate. Climate may be a major factor in determining what types of species will follow one another, but succession itself results from other changes. The most important of these are the modifications of the physical environment produced by the community itself. Most successional communities tend to alter the area in which they occur in such a way as to make it less favorable for themselves and more favorable for other communities.

In effect, each community in the succession sows the seeds of its own destruction.

Consider the alterations initiated by pioneer communities on land. Usually these communities will produce a layer of decay matter or litter on the surface of the soil. The accumulation affects the runoff of rainwater, the soil temperature, and the formation of humus (organic decomposition products). The humus, in turn contributes to the soil development and thus alters the availability of nutrients, the water, the pH and aeration of the soil, and the types of soil organisms that will be present. But the organisms that are characteristic of the pioneer communities that produced these changes may not prosper under the new conditions, and they may be replaced by invading competitors that do better in an area with the new type of soil. In another manner, plants sometimes foreclose future reproduction by the process of their own growth. In Australia, for example, the Eucalyptus trees of the open, sunny savannas provide shade in which young tress from the nearby rain forests can sprout and grow. When these intruders reach their full size, they cast too much shade for the Eucalyptus to reproduce themselves. The Eucalyptus finally die out as the rain forest takes over the land completely.

Not all successional stages prepare the way for their own decline and fall. In the eastern United States, mixed forests of pine and oak modify the soil in a way that makes it more favorable for the growth of their own seedlings than for those of their competitors. Succession in such cases occurs simply because slower growing trees rise to dominance at a later time and alteration of the environment therefore may have nothing to do with replacement.

An orderly sequence of communities in a succession usually occurs because as the species specialized for each stage persists and changes the environment, an excellent habitat for the next set of species is provided. In fact, ecologists recognize two broad classes of species involved in this ordered kind of successsion. <u>Opportunistic species</u> are able to disperse widely, and they grow and breed rapidly. The opportunistic species include the pioneers and the species involved in the early succession stages. They have adopted the strategy of finding and utilizing empty space before other species preempt it. <u>Stable species,</u> on the other hand, specialize in competitive superiority. Forest trees of most kinds belong to this category. They grow and disperse more slowly, but in their encounters with opportunistic species, they are able to grow successfully at the expense of the other species and remain for longer periods of time. Forest trees commonly make up a large percentage of the land climax communities.

• PROBLEM 30-21

Why is it that in most ecological successions the last stage is long lasting?

Solution: If no disruptive factors interfere, most successions eventually reach a stage that is much more stable than those that preceded it. The community of this stage is called the climax community. It has much less tendency than earlier successional communities to alter its environment in a manner injurious to itself. In fact, its more complex organization, larger organic structure, and more balanced metabolism enable it to control its own physical environment to such an extent that it can be self-perpetuating. Consequently, it may persist for centuries without being replaced by another stage so long as climate, physiography and other major environmental factors remain essentially the same. However, a climax community is not static. It does slowly change and will change rapidly if there are major shifts in the environment. For example, fifty years ago, chestnut trees were among the dominant plants in the climax forests of much of eastern North America, but they have been almost completely eliminated by a fungus. The present-day climax forests of this region are dominated by other species.

• PROBLEM 30-22

Describe the sequence of succession leading to a climax community from a bare rock surface.

Solution: Ecological successions are common in most regions, including those which develop on bare rocks. The first pioneer plants may be lichens (algae and fungi living symbiotically) which grow during the brief periods when the rock surface is wet and lies dormant during periods when the surface is dry. The lichens release acids and other substances that corrode the rock on which they grow. Dust particles and bits of dead lichen may collect in the tiny crevices formed and consequently pave the way for other pioneer communities. Mosses can gain anchorage in these crevices and grow in clumps. These clumps allow the formation of a thickening layer of soil composed of humus rotted from dead mosses, lichen parts, and also grains of sand and silt caught from the wind or from the runoff waters of a rainstorm. A few fern spores or seeds of grasses and annual herbs may land on the mat of soil and germinate. These may be followed by perennial herbs. As more and more such plants survive and grow, they catch and hold still more mineral and organic materials, and the new soil layer becomes thicker. Later, shrubs and even trees may start to grow in the soil that now covers what once was a bare rock surface. This type of succession is referred to as primary succession in that the region involved has never supported life forms before.

Secondary succession is best understood by considering what happens to a farm field when it is abandoned. In the very first year, the abandoned field is covered with annual weeds. By the second summer the soil is completely covered with short plants, and the perennial herbs have seized the available space. Grasses start to form a turf. In the next few years, the turf of herbs thickens, but also woody plants, such as brambles and thorny shrubs, become established. The shrubbery grows until the field is blocked by an almost impenetrable tangle of thorny plants. Trees grow up through them, changing the field in a decade or two into a rough wood forest. Other communities of trees replace these trees in the succession. The stages from ploughed fields to forests have been witnessed many times and the successions of vegetation before the climax community is reached are better known as seral stages.

ENVIRONMENTAL CHARACTERISTICS OF THE ECOSYSTEM

• **PROBLEM** 30-23

What factors determine the characteristics of particular organisms that exist in an ecosystem.

Solution: Temperature and precipitation are among the most crucial physical factors in determining what type of vegetation will exist in a certain area. It is often possible to predict the type of biome at a given locality from the characteristics of the climate. However, other physical factors are also important. The structure and chemistry of the soil can be equally vital. There are, for example, certain trace elements known to be essential for the full development of plants. These include boron, chlorine, cobalt, copper, iron, manganese, molybdenum, sodium, vanadium, and zinc. When one or more of these substances is not present in sufficient concentration, land which should carry a forest according to the temperature - precipitation features, will instead be covered with shrubby vegetation, or be grassland bearing a few scattered trees. Where grassland was indicated, there may exist sparce vegetation typical of a desert. Any single factor such as the amount of sunlight, temperature, humidity, or the concentration of trace elements is capable of determining the presence or absence of species and thus the characteristics of the entire biome. The Law of Minimum is a generalization that is sometimes used to explain this phenomenon. It states that the factor that is most deficient is the one that determines the presence and absence of species. It does not matter, for example, how favorable temperature and sunlight are in a given locality, or how rich the nutrients and trace elements of the soil are if the precipitation is very low, because the result will still be a desert. Only a consideration of all the climatic and soil characteristics of a region can reveal the reasons for the existence of certain plant formations.

In turn, animals are indirectly influenced in their distributions by soil types because of their dependence upon plants as the source of high-energy organic nutrients.

• PROBLEM 30-24

In attempting to map out the climatic conditions of the biosphere, would a climatologist find it helpful to correspond with an ecologist who has mapped the vegetation formations or biomes of the biosphere?

Solution: If you traveled on the earth several hundreds of miles at a time, you would frequently encounter vegetation types that were very different from one another. You would also get the impression of a world set out into different blocks of vegetation. These blocks are termed formations. The succession of formations between the equator and the arctic is roughly mirrored in both the Northern and Southern Hemispheres.

But why should the vegetations of different parts of the world look so different? Why should enormous stretches of land have vegetation of a special form? Also, why should the plants of different formations respect each other's territory? Some aspects of the climate seem to be the controlling factors, with the most important two being temperature and the availability of water.

Mapping a thing such as climate is a very hard task, but mapping vegetation types is easy by comparison. So climatologists could use vegetation maps as a base for climatic maps. The earth is parceled out into blocks of different types of plants in a corresponding manner to the way different climatic conditions are distributed. The distinctive appearance of each formation results from the adaptations of local plants to common climatic and other environmental conditions. Thus, it is correct to say that vegetation-formation mapping does provide a basis for climatic mapping. If formations merge, climatic conditions should converge. Similarly, if formations diverge, climates should differ as well.

The distribution of animals is closely linked with the distribution of the vegetation in which they live and it is best to accept the formation boundaries and to include the animals in the formation descriptions. Biologists find it convenient to recognize a limited number of major formations called biomes. Thus the grassland of the western United States is one biome, and the nearby desert is a second biome. A grassland and a desert also exist in South America, but the species that comprise these biomes are almost entirely different from their counterparts in the United States. We speak of these biomes that resemble one another in physical appearance but differ in species composition as comprising a worldwide biome-type.

Describe the various land biomes that are usually encountered by a traveler going from the equator to the arctic polar ice cap.

Solution: In the equatorial regions of several continents (Africa, Asia, and South America) are tropical rain forests with enormous trees (See figure, problem 28-15). The interlocking canopy of leaves blocks out the sun and allows only dim light to penetrate. As a result, the ground is sparsely covered with small plants. The canopy also interrupts the direct fall of the plentiful rain, but water drips from it to the forest floor much of the time, even when it is not raining. It also shields the lower levels from wind and hence greatly reduces the rate of evaporation. The lower levels of the forst are consequently very humid. Temperatures near the forest floor are nearly constant.

The pronounced differences in conditions at different levels within such a forest result in a striking degree of vertical stratification. Many species of animals and epiphytic plants (plants growing on large trees) occur only in the canopy, while others occur only in the middle strata and still others occur only on the forest floor. Decomposition of fallen leaves and dead wood is so rapid that humus is even missing in spots on the floor. The diversity of life here is the greatest found anywhere on earth. In a single square mile of the richest tropical rain forests, hundreds of species of trees, hundreds more of birds, butterflies, reptiles and amphibians, and dozens of species of mammals make their habitat.

Huge areas in both the temperate and tropical regions of the world are covered by grassland biomes. These are typically areas with relatively low total annual rainfall or uneven seasonal occurrence of rainfall. This type of climate is unfavorable for forests but suitable for growth of grasses. Temperate and tropical grasslands are remarkably similar in appearance, although the particular species they contain may be very different. In both cases, there are usually vast numbers of large herbivores, which often include the ungulates (hoofed animals). Burrowing rodents or rodentlike animals are also often common.

North of the tropics, in the temperate regions of Europe and eastern North America, are deciduous forests. Small creepers and epiphytes such as lichens and mosses may be found, but they are not so conspicuous an in the rain forests. There is richer vegetation on the ground below, which is often covered by a carpet of herbs. Such a forest may be called temperate deciduous forest. In those parts of the temperate zone where rainfall is abundant and the summers are relatively long and warm, the climax communities are frequently dominated by broad-leaved trees, whose leaves change color in autumn then fall off in winter and grow back in the spring.

North of the deciduous forest of temperate North America and Eurasia is a wide zone dominated by coniferous (cone-bearing) forests sometimes referred to as the boreal forests. This is the taiga biome. Instead of being bushy-topped, the trees are mostly of the triangular Christmas-tree shape. The trees of the boreal forest are evergreens, cone-bearing and needle-leaved consisting mostly of spruce, fir, and tamarack. They branch over most of their height and are relatively close-packed. The land is dotted by lakes, ponds and bogs. The winters of the taiga are very cold and during the warm summers, the subsoil thaws and vegetation flourishes. Moose, black bears, wolves, lynx, wolverines, martens, squirrels, and many smaller rodents are important in the taiga communities. Birds are abundant in summer.

The territory of the boreal forest is hundreds of miles wide, but eventually it meets the completely treeless vegetation of the tundra. The tundra is the most continuous of the earth's biomes, forming a circular band around the North pole, interrupted only narrowly by the North Atlantic and the Bearing Sea. The vegetation of the tundra resembles grassland but it is actually made up of a mixture of lichens, mosses, grasses, sedges and low growing willows and other shrubs. There are numerous perennial herbs, which are able to withstand frequent freezing and which grow rapidly during the brief cool summers, often carpeting the tundra with brightly colored flowers. A permanent layer of frozen soil lies a few inches to a few feet beneath the surface. It prevents the roots of trees and other deep growing plants from becoming established, and it slows the drainage of surface water. As a result, the flat portions of the tundra are dotted with shallow lakes and bogs, and the soil between them is exceptionally wet. Reindeer, lemmings, caribou, arctic wolves, foxes and hares are among the principal mammals, while polar bears are common on parts of the tundra near the coast. Vast numbers of birds, particulary shorebirds and waterfowl (ducks, geese, etc.), nest on the tundra in summer, but they are not permanent inhabitants and migrate south for the winter. Insects are incredibly abundant. In short, the tundra is far from being a barren lifeless land; but the number of different species is quite limited.

The succession of biomes between the equator and the Arctic is roughly mirrored in the Southern Hemisphere, although the absence of large landmasses makes the pattern less complete. There are also some other biomes that are more irregularly scattered about the world, but which have a distinctive form. These are the biomes of the desert regions and the sclerophylous bushlands.

The bushlands, common in the chaparral of California, maquis of the Mediterranean and the mallee heath of Australia, are made up of different species of plants, but have much in common. They have gnarled and twisted, rough and thorny plants. The leaves tend to be dark and are hairy, leathery, or thickly cutinized. All grow in places with hot dry summers and relatively cool moist winters. A climatic pattern such as this is rare and is always accompanied with vegetation of this form. In places where rainfall is very low grasses cannot survive as the dominant vegetation and desert biomes occur. Deserts are subject to the most extreme temperature fluctuations of any biome. During the day they are exposed to intense sunlight, and the temperature of both air and soil may rise very high. But in the absence of the moderating influence of abundant vegetation, heat is rapidly lost at night, and a short while after sunset, searing heat has usually given way to bitter cold. Some deserts, such as parts of the Sahara, are nearly barren of vegetation, but more commonly there are scattered drought-resistant shrubs and succulent plants that can store much water in their tissues. Most desert animals are active primarily at night or during the brief periods in early morning and late afternoon when the heat is not so intense. Most of them show numerous remarkable physiological and behavioral adaptations for life in their hostile environment.

We have thus seen that one moving North or South on the earth's surface may pass through a series of different biomes. The major biome types have been mentioned with the exception of the comparatively unproductive biomes such as the polar ice caps. Ecologists often differ in their classifications of the major patches of the biosphere because the biomes are strictly defined by the elements put into them. The borders of the biomes shift as we add or substract species. The point to remember is that the biomes seldom exist as sharply defined patches. They have broad borders, and the species that comprise them have weakly correlated geographical ranges.

• PROBLEM 30-26

Why isn't a biome map of the earth a true representation of the vegetation formations found on land?

Solution: The naturalist travelers of the eighteenth and nineteenth centuries saw strikingly different kinds of vegetations parceled out into formations. They drew maps showing the extent of each formation. However, drawing a map involves drawing boundaries. Where there are real, distinct boundaries on the ground this is an easy task. Such boundaries include those between sea and land, along deserts or beside mountain ranges. But often there are no real boundaries on the ground. In fact, most of the formation boundaries of the earth are not distinct. Real vegetation types usually grade one into another so that it is impossible to tell where one formation ends and another begins.

The apparent distinctness of the vegetation formations, when viewed from afar, is largely an optical illusion as the eye picks out bands where individual species are concentrated. A vegetation map-maker could plot the position of a tree line well enough, if he had a few reports from places with distinct treelines which he could plot and link up. However, the gradual transitions, such as between the temperate woods and tropical rain forests would be much more difficult to plot. So map-makers draw boundaries using what seems to be the middle of the transitions. When the map is finished, the area is parceled out by blocks of formations of plants. A map like this shows the vegetation of the earth to be more neatly set into compartments than it really is.

• PROBLEM 30-27

Why do ecologists not attempt to divide the ocean into biome types like those on land? How do they instead distinguish the different parts of the ocean?

Solution: Because of the vastness and relative uniformity of the ocean, and the comparatively rapid mixing of the organisms within it, the various regions of the ocean cannot be divided into distinct biome types as are terrestrial regions. Even though the ocean is characterized by many different types of habitats, these habitats are all interconnected, and some organisms may move freely from one to the other. It seems more appropriate, therefore, to describe an ocean by zonation rather than by biome types.

The oceans and seas cover 71 per cent of the earth's surface, but, except for the shallow water of the margins, they are far less complex in structure and productivity than the land. The continental margin gradually slopes seaward to depths of 150 to 200 meters and then slopes steeply to depths of 3000 meters or more. This slope is known as the continental shelf. The continental shelf then gives way to the abyssal ocean floor.

The accompanying figure depicts the zonation in an ocean. Waters over the continental shelf comprise the neritic zone, and these beyond the edge of the shelf comprise the main part of the ocean basin, called the oceanic zone. The floor of the ocean basin, called the abyssal plain, ranges from 3000 to 5000 meters in depth and may be marked by such features as sea mounts, ridges and trenches. The oceanic zone is further distinguishable into four zones. The upper part of the open ocean into which enough light can penetrate is called the euphotic zone. Below this zone is the bathyal zone, which ranges to a depth of perhaps 2,000 meters. Below the bathyal zone, the waters over the abyssal plain (ocean floor) form the abyssal zone. Trenches and valleys of the ocean floor form the tidal zone, which is generally below 6,000 meters.

The euphotic zone, usually about 100 meters in thickness, has enough penetrated sunlight effective for photosynthesis. Below this is a transition zone where some light penetrates. From the transition zone down to the ocean floor, total darkness prevail and this region constitutes the aphotic zone. The exact boundary between the lighted zone and the deep lightless zone at any given locality depends upon the intensity of the sunlight, which in turn depends upon the latitude, and upon the turbidity of the water. Usually no light penetrates below 600 meters.

The physical environment of the ocean is most diverse in the shallow water along its margin. Here in the neritic (coastal) zone exists the greatest variation in temperature, salinity, light intesity and turbulence. Also, here, sunlight can penetrate to the bottom of the water body. For this reason, here occurs the most luxuriant forms of both plant and animal life. The edge of the ocean, which rises and falls with the tide, is the intertidal zone or littoral zone. It extends out into the water to the depths at which the water is no longer stirred by tides and waves. The region after the intertidal zone is called the subtidal zone and farther on is the lower neritic region. In general, the density of living organisms in the open ocean is much less than that found in the neritic region, but because of its large area and volume, the total biomass in the open ocean is greater.

• PROBLEM 30-28

Ecologists refer to the ocean as the marine ecosystem. How are the major marine communities classified and where are the regions of their habitation?

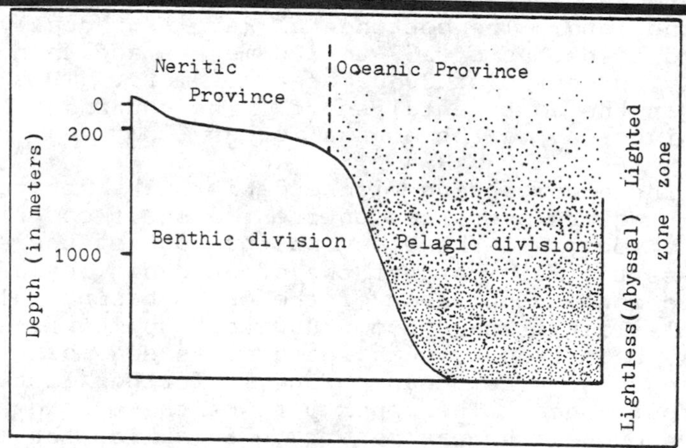

OCEANIC ECOSYSTEM (Special local communities such as coral reefs and intertidal zones not included in diagram)

Solution: The ocean is a very complex ecosystem with many different kinds of communities. Temperature, salinity, availability of light, and depth all affect the nature of these communities and present major barriers to the free movement of organisms that have specific adaptability to certain environmental conditions. The major currents, which keep the waters of the ocean in continous circulation, also affect the nature of marine communities, as do waves and tides. Another important phenomenon, called upwelling, occurs where the winds move surface waters away from the steep coastal slopes, bringing cold water to the surface from the depths which is rich in such nutrients as phosphates and nitrates; the regions of such upswellings are often among the most productive regions of the ocean.

The major marine habitats are shown in the accompanying diagram. Marine organisms in general are divided into two groups, depending upon whether they live in the open water or upon or beneath the bottom. Those organisms that live suspended or swimming in the open waters are said to be pelagic. Bottom dwellers, called benthic organisms, may live on the surface or beneath the surface of the ocean floor. The benthic organisms include both attached organisms and those organisms that move in or on the substratum. Bottom dwellers are generally distinct for each fo the neritic, or coastal, regions, and the type found in each region is largely dependent on the type of ocean floor, whether it is sand, rock, or mud. Many animals are adapted for living in the sand and this group includes representatives of virtually every phylum of animals.

Pelagic organisms are of two types. The microscopic plants and animals that are unable to move against the current are collectively called the plankton. The larger, free swimming animals that are able to move at will are called nekton. The planktonic plants constitute the phytoplankton. The phytoplanktonic algae are the most important producers throughout the ocean. The phytoplankton are eaten by a great array of invertebrate animals, ranging in size from protozoans to large medusae. These latter are the primary consumers. These, in turn, are consumed by other, larger invertebrates and vertebrates.

As would be expected, plankton attains its greatest density in the upper lighted zone where there is sufficient light for photosynthesis to occur. In productive waters, planktonic organisms may occur in such enormous numbers that the water appears turbid. The animals that are permanent inhabitants of the lower dark zones are carnivorous, suspension, or detritus (dead organic matter) feeders and depend ultimately on the photosynthetic activity of the microscopic algae in the upper, lighted regions. Therefore, vertical distribution of marine organisms is largely controlled by the depths of light penetration. Light sufficient for photosynthesis to exceed respiration penetrates only a short distance below the surface, with this distance being dependent upon the turbidity of the water. Below this is a region where some photosynthesis can occur, but here the production of organic matter is less than the loss through respiration.

• PROBLEM 30-29

What kind of marine life is found in the different regions of the ocean and what are these organisms' adaptations to their environment?

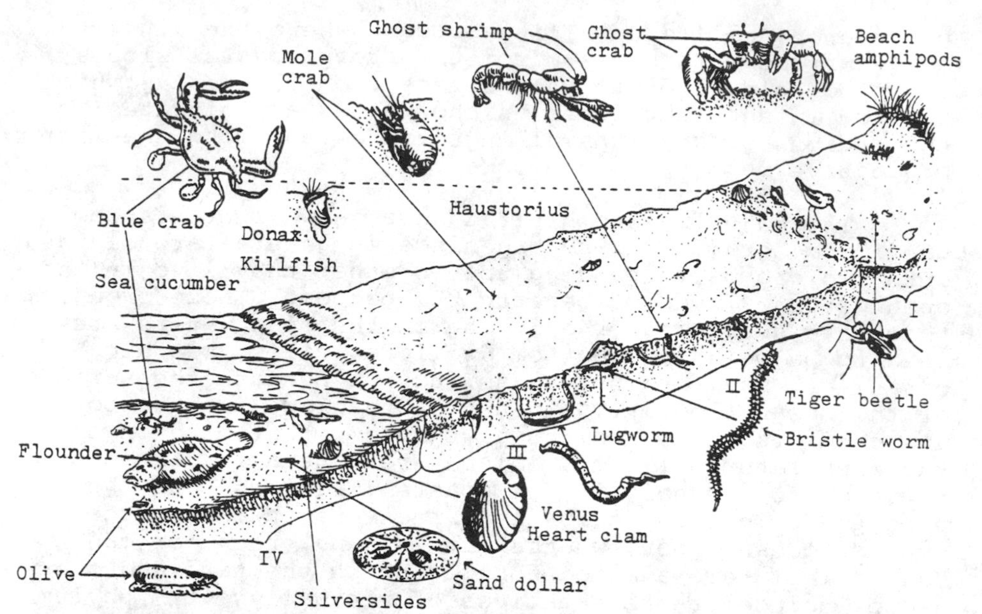

Figure 1. Life on a sandy ocean beach along the Atlantic coast. Although strong zonation is absent, organisms still change on a gradient from land to sea. (I) Supratidal zone: ghost crabs and sand fleas; (II) Flat beach zone: ghost shrimp, bristle worms, clams; (III) Intratidal zone: clams, lugworms, mole crabs; (IV) Subtidal zone: the dashed line indicates high tide.

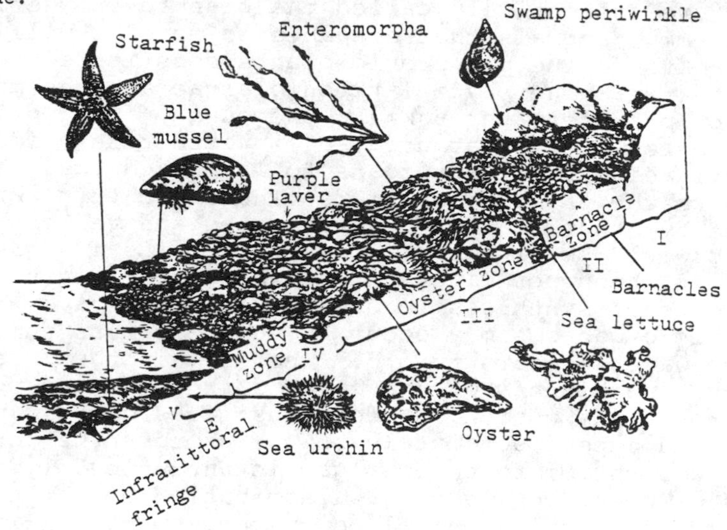

Figure 2. Life on a rocky shore, mid-Atlantic coastline. (I) Bare rock with some black algae and swamp periwinkle; (II) Barnacle zone: (III) oyster zone, oysters, Enteromorpha, sea lettuce, and purple laver; (IV) muddy zone: mussel beds; (V) infralittoral fringe: starfish and so on. Note absence of kelps.

Solution: The neritic, or coastal, waters support a greater population of marine life forms than do those of the open ocean. This abundance results from the rich supply of nitrates, phosphates, and other nutrients dumped into coastal waters by rivers and streams or brought to the surface by upwellings and turbulence. These substances are required by the producers who are the photosynthetic organisms that form the base of the food chain for animal life. Also, the ocean floor receives much more energy from sunlight here and as a result there is a great luxury of both plants and animals in the neritic waters.

More is known about the communities of the intertidal and subtidal regions than about those of the lower neritic region. On a typical sandy shore (Fig. 1) the dominant animals in the intertidal region might include ghost shrimps, burrowing isopods, burrowing amphipods, mole crabs, polychaete worms, and beach clams, and the dominant forms in the subtidal region might include hermit crabs, sand dollars, borrowing shrimp, ascidians, and copepods.

In contrast, on a rocky shore (Fig. 2) the dominant animal forms in the intertidal region might include barnacles, oysters, and mussels, with such forms as sea anemones, sea urchins, and corals in the subtidal region. On mud bottoms, various kinds of burrowing clams and snails are found. Many of the bottom dwellers are adapted to feed on plankton and dead organism.

The organisms of the intertidal region are subjected to alternating periods of exposure to the air and to the strong action of waves. They are adapted to these environmental factors in a number of ways. For example, the seaweeds have tough outer coverings and pliable bodies. Such animals as crabs, barnacles, mollusks and starfish have hard calcerous shells, whereas others like the sea anemones have a leathery outer covering. Also, many animals are able to burrow in the sand to escape exposure to air.

Neritic pelagic communities include a great variety of fish species, larger crustaceans, turtles, seals, whales, and others. Most of the fish are plankton feeders - the adult fish feed mostly on zooplankton and are thus secondary consumers. Some fish, such as sharks, are dangerous predators.

The euphotic zone of the oceanic region is inhabited by planktonic organisms and a variety of nektonic forms, including many kinds of fish. Mackeral and herring are among the commercial fish found in this oceanic region. Whales may also be present, as well as many kinds of birds. The bathyal and abyssal zones are incomplete ecosystems since there can be no producers where light cannot reach. The food of the inhabitants consists partly of the dead bodies of plants and animals that rain down from above, and partly of each other. A variety of species of fish, crustaceans, echinoderms, worms and mollusks are found in these zones and on the ocean floor. Many of the inhabitants posses light-producing structures which are significant adaptation in this darkened habitat.

The deep sea forms are adapted to withstand great pressure by keeping the pressure uniform inside their bodies and on the outside. In the hadal zone at the great depths of the Pacific, organisms are subjected to pressures greater than 1000 atmospheres. At a depth of 10,500 meters, sea cucumbers, sea anemones, bivalve mollusks and crustaceans have been found. This is close to the deepest part of the ocean yet discovered, which is 10,860 meters.

SHORT ANSWER QUESTIONS FOR REVIEW

Answer

To be covered when testing yourself

Choose the correct answer.

1. The ecological unit composed of organisms and their physical environment is the (a) niche, (b) population, (c) ecosystem, (d) community. — c

2. The orderly change from one ecological community to another in an area is called (a) convergence, (b) climax, (c) dispersal, (d) succession. — d

3. An edaphic climax (a) is when the succession of communities is stopped short of the expected climax community, (b) is when the climax community changes into a previous type of community, (c) is when the climax community is destroyed and replaced by a totally new type of community, (d) none of the above. — a

4. Organisms which break down the compounds of dead organisms are called (a) phagotrophs, (b) parasites, (c) saprotrophs, (d) producers. — c

5. Which must be present in an ecosystem if the ecosystem is to be maintained? (a) producers and carnivores, (b) producers and decomposers, (c) carnivores and decomposers (d) herbivores and carnivores. — b

6. The relationship between fungi and the algae in lichens is known as (a) mutualism, (b) parasitism (c) commensalism, (d) saprophytism. — a

7. All ecosystems have three basic living components. Which one of the following is not necessarily found in all ecosystems? (a) "producer" plants (b) animal "consumers" (c) decomposers (d) parasites and commensalists. — d

8. A sequence of species related to one another as predator and prey is a(n) (a) trophic level, (b) ecosystem, (c) food chain, (d) climax. — c

9. Which of the following pairs represents ecological equivalents? (a) Squirrel and rattlesnake (b) house cat and lion (c) seagull and codfish (d) wild horse and zebra. — d

10. Choose the statement that best describes the climax stage of an ecological succession. (a) It is usually populated only by plants, (b) it remains until there are severe changes in the environment, (c) it represents the initial phases of evolution, (d) it changes rapidly from season to season. — b

Answer

To be covered when testing yourself

11. The climax organism growing above the tree line on a mountain would be the same as the climax organism found in the (a) taiga, (b) tundra, (c) tropical forest, (d) desert.

b

12. The character of an ecosystem is determined by the environmental factor that is in shortest supply. This is (a) the law of minimum, (b) Borty's law, (c) the law of diminishing returns, (d) Dollo's law.

a

13. Which of the following is a non-renewable natural resource? (a) Forests (b) Field crops (c) Iron (d) Animal furs.

c

14. The "10 percent rule" (a) refers to the percentage of similar species that can coexist in one ecosystem, (b) refers to the average death total of all mammals before maturity, (c) is the percent of animals not affected by DDT, (d) refers to the level of energy production present in a given trophic level and used for production by the next higher level.

d

15. Choose the correct statement: (a) The earth is a closed system for materials, (b) in an ecosystem energy is usually not in balance, (c) all energy in an ecosystem is not ultimately lost as heat, (d) food relations are in chains, not in webs.

a

Fill in the blanks.

16. The steady replacement of one species by another as the environment changes due to the actions of organisms is known as _____.

ecological succession

17. The phenomenal population growth has been a result of the striking drop in the early ______.

death rate

18. Six important terrestrial biomes are recognized. They are: _____, _____, _____, _____, _____ and _____.

desert, grassland, equatorial forests, deciduous forests, coniferous forests and tundra

19. If a host and symbiont can exist outside of the symbiotic relationship, the association is _____.

facultative

Answer

To be covered when testing yourself

20. The three physical subdivisions of the biosphere are the _____, _____ and _____.

    hydrosphere, lithosphere, atmosphere

21. D.D.T. is a pollutant which is slowly _____ while metals such as mercury are _____ pollutants.

    degradable, non degradable

22. The law of minimum indicates that a _____ environmental factor can determine the character of a biome.

    single

23. _____ flows into and out of the world ecosystem at a fast rate, while _____ are a fixed component.

    energy, physical materials

24. _____ is the misplacement of resources by the alteration of biogeochemical cycles.

    pollution

25. _____ refers to studies of individual organisms, or populations of single species, and their interactions with the environment.

    autecology

26. _____ refers to studies of various groups of organisms which are associated to form a functional unit of the environment.

    Synecology

27. Within a given area, a group of organisms belonging to the same species is called a _____.

    population

28. A group of populations living in a given area is considered a _____.

    community

29. The community and the physical environment considered together is a(an) _____.

    ecosystem

30. Two species of organisms that occupy the same or similar ecological niches in different geographical locations are termed _____.

    ecological equivalents

Determine whether the following statements are true or false.

31. Biomes that resemble one another in physical appearance but differ in species composition belong to the same biome type.

    True

32. The tropical rain forest exhibits a great diversity of niches because it is a very young biome type.

    False

	Answer To be covered when testing yourself
33. The increase in the world population today is partially due to an increased birth rate.	False
34. Insects and related arthropods are man's chief competitors for the world's food sypply.	True
35. Each developmental stage in an ecological succession which results in a climax community is known as a sere.	True
36. Food chains seldom have more than five levels because the top carnivores are too sparse and contain too few calories to make predation worthwhile.	True
37. Both mutualism and commensalism may lead to parasitism.	True
38. A land biome is identified primarily by its dominant animals.	False
39. An ecosystem is self-sustaining if equal numbers of plants and animals are present.	False
40. Whether a land ecosystem supports a climax community of a deciduous forest or a grassy prairie depends chiefly upon longitude but not latitude.	False
41. The technique of pest control by male sterilization depends upon the fact that the females of susceptible insects mate only once.	True
42. When mercury or DDT are introduced into a food pyramid the level in which the greatest concentration is found is the producers.	False
43. A population of frogs protected from all predation would increase indefinately.	False
44. Large mammals are not common in the desert.	True
45. A major ecological grouping of different plant and animal species is a population.	False
46. The gastrointestinal tract can be considered an ecosystem.	True

CHAPTER 31

ANIMAL BEHAVIOR

TYPES OF BEHAVIORAL PATTERNS

• **PROBLEM** 31-1

What is behavior? Are behavior patterns solely inherited or can they be modified?

Solution: The term behavior refers to the patterns by which organisms survive and reproduce. The more an organism must actively search the environment to maintain life, the more advanced are its behavior patterns. Thus a deer shows more complex behavior patterns than does a planaria. However, even sedentary organisms such as sea coral display forms of behavior necessary for survival. Behavior is not limited to the animal kingdom; plants display simple behavior, such as the capture of prey by the venus flytrap.

Behavior has a genetic or hereditary basis that is controlled by an organism's DNA. The anatomical aspects of both the nervous and endocrine systems are chiefly determined by the genetic composition. These two systems are responsible for most behavioral pheneomena. In general, an organisms's behavior is principally an expression of the capabilities of its nervous system, modified to various extents by its endocrine system.

Behavioral development is often influenced by experience. This modification of behavioral patterns through an organism's particular experiences is called learning. For example, some birds transmit their species-specific song to their offspring solely by heredity, while others, such as the chaffinch, must experience the singing of the song by other chaffinches in order for them to acquire the song.

Both inheritence and learning are therefore fundamental in the determination of behavioral patterns. Inheritance determines the limits within which a pattern can be modified. The nervous system of a primitive animal can limit the modification of existing behavior patterns, allowing only simple rigid behavioral responses. In highly developed animals such as man, there are fewer purely inherited limits, permitting learning to play a significant role in determining behavior.

• **PROBLEM 31-2**

What is instinct?

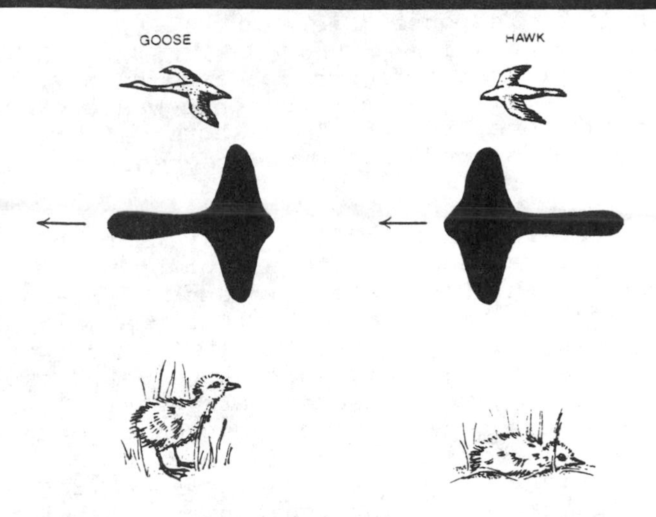

Ducklings show fear and crouch in grass at the sight of a hawk flying overhead, but not at the sight of a flying goose. Same responses are evoked by the sight of a cardboard silhouette passing overhead. When silhouette is moved as shown at left, ducklings respond as if it were a goose. When silhouette is reversed and moved as shown at right, ducklings respond as if it were a hawk.

Solution: The term "instinct" is difficult to define since its meaning has been so broadly applied. Instinct has been used to designate behavior that is stereotyped; that is, behavior which is rigidly executed and therefore highly predictable when the appropriate stimuli are presented. For example, when the eggs of herring gulls and other ground-nesting birds accidentally fall out of the nest area, the parent gulls use their bills to roll them back into the nest. This is a stereotyped response and can be repeated by purposely displacing the eggs. The eggs serve as releasing stimuli to elicit the egg-rolling behavior. When investigators placed abnormally large wooden models of eggs adjacent to the gulls' own displaced eggs, the gulls preferred to roll the larger models back to the nest. Any such stimuli having a quality that is preferred over that of the natural stimulus encountered in the life of an animal - in this case abnormally large size - is referred to as a supernormal stimulus.

Instinct is often believed to consist solely of innate responses which are determined by set neural pathways of the brain and not altered by experience. However, recent study has shown that learning may play a role in stereotyped instinctive behavior. For example, ducks and other ground-nesting birds either cry out alarm calls or crouch in fear when a bird of prey flies overhead (see Figure). It was discovered that the basic shape of the flying bird is the releasing stimulus for fear behavior; particular anatomical features are not crucial. Thus, when an investigator passed different shapes of cardboard silhouettes over the heads of ducklings, only the bird shape having a head shorter than its tail evoked the fear response (See Figure). When this same silhouette was passed over the ducklings in the reverse direction, so that the figure appeared to have a long head and short tail, no fear response was evoked. In the first instance, the bird's silhouette represented a hawk, while the second silhouette represented another duck or goose.

This fear response was initially thought to be a strictly inherited response to a sign stimulus. Further study, however, has indicated elements of learning in this behavior. Ducklings are initially afraid of all overhead flying objects. Eventually they stop reacting to ones they see repeatedly, such as other ducks and falling leaves. Hawks, however, are generally scarce, and the ducklings never become accustomed to their peculiar shape. Learning therefore plays a role in what seems to be a totally innate stereotyped response. Hence, it is difficult to make a distinction between instinct and other kinds of behavior, since some instincts can be shown to contain learned elements.

• PROBLEM 31-3

Discuss the role of a stimulus that elicits a behavior pattern.

Solution: Innate behavior patterns occur in response to certain stimuli detected by an animal's sensory receptors. Of the myriad of stimuli encountered in any situation, an animal responds only to a limited number of stimuli. These stimuli, which elicit specific responses from an animal, are referred to as sign stimuli. The sign stimuli involved in intra-specific communication are called releasers. The intensity of the stimuli necessary to evoke a behavior pattern is inversely proportional to the animal's motivation to perform that behavior. Motivation is the internal state of an animal which is the immediate cause of its behavior. The motivational aspects of behavior can be characterized by terms such as "drives", "goal-oriented behavior", and "satiation". Motivation is partly determined by hormones, influences from the brain, and previous experiences. The animal must possess neural mechanisms selectively sensitive to particular releasing stimuli (or sign stimuli). When the releasing or sign stimuli are appropriate to stimulate an animal and the animal is sufficiently motivated, a behavior pattern is initiated. The intensity of the stimuli necessary to evoke a behavior pattern is inversely proportional to the animal's motivation to perform that behavior. Thus, an animal highly motivated to perform a given behavior will require a less intense stimuli to invoke that behavioral response than an animal which is not as highly motivated.

• **PROBLEM 31-4**

What are the typical periods in the unfolding of a behavioral act?

Solution: Most behavior is goal-oriented; that is, the final behavioral act is directed at some specific object or state. There are usually three phases to a behavioral act. The first phase is called appetitive behavior. The animal orients its behavior toward some specific goal which is determined by its physiological needs. A hungry animal searches for food; a sexually mature animal searches for a mate. It is frequently difficult to identify the nature of the appetitive behavior until the goal is observed. The exploratory behavior of a wolf looking for a mate is very similar to its behavior when looking for prey.

Once the appetitive behavior has enabled the animal to reach its goal, another type of behavioral act -the second phase- ensues. The sign stimuli, which in the above examples could be the sight of food or the presence of a mate, evoke the appropriate response called the consummatory act. Eating is the consummatory act of feeding behavior; copulation, the consummatory act of sexual behavior.

The last phase is called quiescence. After the consummatory act is performed, an animal in this phase is not likely to be responsive to the same sign stimuli a second time. It usually slows down or stops its appetitive behavior; thus, a well-fed animal no longer searches for prey but may pursue some other goal. An animal in the quiescent phase also will not perform additional consummatory acts when presented with the same goal. If, for example, a recently-fed animal is presented with food it usually will not eat. Eventually, the quiescence phase is terminated and appetitive behavior gradually resumes.

ORIENTATION

• **PROBLEM 31-5**

What is the difference between a kinesis and a taxis?

Solution: Both kineses and taxes are types of animal orientation. Orientation is the process by which the animal directs the consummatory act at an appropriate object (the goal). It is involved in those forms of appetitive behavior in which the animal moves in a constant direction in search of a goal. Highly developed forms of orientation are involved in the migration and homing behavior of birds. Kineses and taxes are two simpler forms of orientation.

A kinesis is a form of orientation in which the animal does not necessarily direct its body toward the stimulus. The stimulus merely causes the animal to speed or slow down its movements, resulting in the eventual displacement of the animal toward or away from the stimulus. It is an undirected type of orientation. The movement of the woodlouse is a good example of a kinesis. A woodlouse is a crustacean that lives in damp environments beneath stones or pieces of wood. When the woodlouse finds itself dehydrating, it begins to move around. This increase in locomotor activity may cause the woodlouse to chance upon a moist spot. Here, the dampness causes the woodlouse to slow down and it now establishes a new home. The woodlouse did not direct itself toward the moist spot, but simply encountered it by random movement.

In contrast, a taxis is a type of orientation in which the animal directs its body toward or away from the stimulus. The type of stimulus is usually denoted by a prefix to the word taxis: a geotaxis is movement regulated by gravity; a phototaxis is movement guided by light; a chemotaxis is movement guided by chemical substances. Movement toward the stimulus is a positive taxis whereas movement away from the stimulus is a negative taxis. For example, the directed movement of cockroaches away from light is termed negative phototaxis. Another common example of a taxis is the directed movement of the bee flying to its hive or source of pollen. The bee maintains a straight course by moving at a constant angle to the sun, therefore exhibiting a phototaxis.

The term taxis should be distinguished from tropism: Tropism indicates growth of an organism in a particular direction. The phototropism of plants is a typical tropism.

• **PROBLEM 31-6**

Describe two ways in which taxes can be guided.

Solution: A taxis is a directed type of orientation in which an animal moves toward or away from a stimulus. But exactly how does the animal detect which direction to move in relation to the stimulus? One simple way is by testing the intensity of the stimulus at different short intervals of time: The animal simply takes momentary readings of the stimulus and directs itself toward the strongest reading. If it accidentally moves away from the source, it changes direction until a higher intensity of the stimulus is sensed. For example, dogs locate the source of an odor by repeatedly sniffing as they move. They periodically change their direction so as to continue smelling a high intensity of the odor.

A second, more complex taxis guidance system is the instantaneous reading of the stimulus using two receptor organs. Many animals move in response to a stimulus by positioning themselves so that equal intensities of the stimulus are sensed by the two organs. They are then

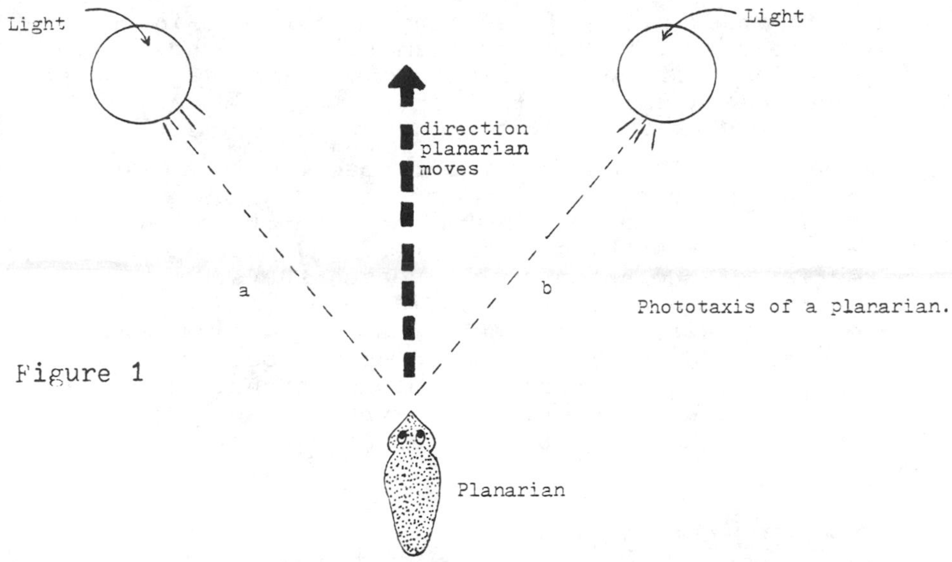

Figure 1

Phototaxis of a planarian.

precisely aligned to move either toward or away from the stimulus, depending on whether they are attracted or repelled by it. For example, an animal can orient toward a light source by moving its body, or simply its head, until an equal intensity of illumination is sensed in both eyes. This quidance system can be demonstrated by covering one eye, for instance, the right eye, of a positively phototactic animal. Since the animal senses all the light in its left eye, it moves toward the left. However, its next instantaneous reading directs the animal to the left again, since a balance of illumination was made impossible. This causes the animal to turn to

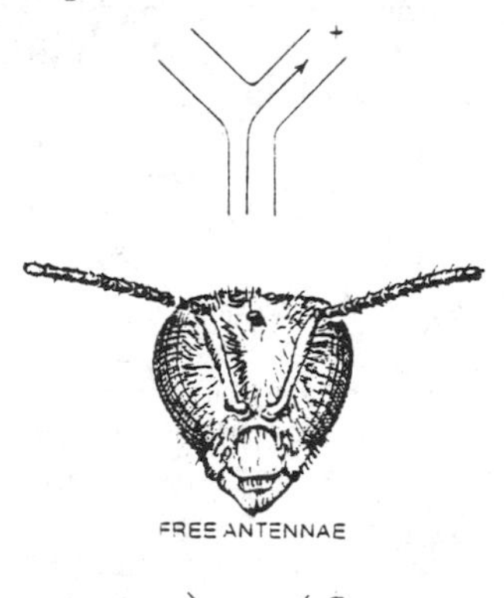

Chemotaxis in the honeybee can be demonstrated by crossing its antennae and gluing them in place. When put in a Y-maze, bees whose antennae have been left in their natural position (top) readily find their way to an odor source (+). Bees with crossed antennae enter wrong channel (bottom).

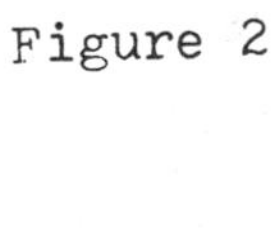

Figure 2

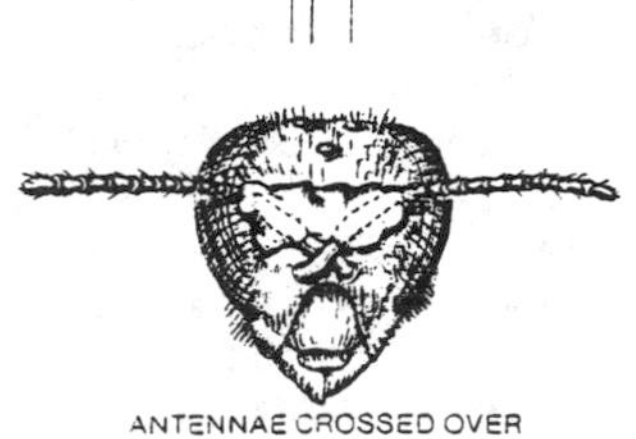

the left indefinitely, resulting in movement in conter-clockwise circles! Another example is the phototaxis of a planarian. If two equally bright lights are located at equal distances from the worm, it will move toward a pt. midway between the two lights (see Figure 1)

The orientation of the honeybee toward an odor is also one involuing two simultaneous readings of a stimuli. The honeybee tests the strength of an odor simultaneously with both antennae, and then turns in the direction of the antenna that sensed the stronger odor. If an investigator crosses the antennae and glues them into this position, the bee reads the odor as coming from the opposite direction. Should such a bee be placed in a Y-maze with an odor source in one channel, the bee will detect the source as originating from the other channel and move in that direction. (See Figure 2)

COMMUNICATION

• PROBLEM 31-7

Discuss the role of pheromones in chemical communication.

Solution: In general, communication between organisms can be defined as action on the part of one individual that alters the behavior of another individual. Communication need not necessarily be visual or auditory, nor must it always be confined to the present. If it is chemical, it obviously has the ability to span time. Specific chemicals involved in social communication are called pheromones, by analogy with hormones. Pheromones are small quantities of specific chemical substances released by an organism which act upon another organism at a distance from their point of release and in very specific ways.

Pheromones can be classified in two groups: those that possess releaser effects, which elicit immediate behavioral responses, and those that have primer effects, which work by altering the physiology and subsequent behavior of the recipient.

Releaser pheromones are used by a variety of animals for different purposes, such as the attraction of mates, individual recognition, or trail or territorial marking. The females of many moth species secrete pheromones to attract males over large areas. The area within which the substance is dense enough to be detected by a male moth is called the active space. When a sexually active male flies into the active space, it flies upwind until it reaches the pheromone-emitting female. This particular pheromone is very potent since only one molecule of it is needed to stimulate a receptor on the male's antennae. The sugar glider,a squirrel-like marsupial, uses its own glandular secretions to mark its territorial boundaries. Male sugar gliders also mark their mates with pheromone, for recognition and in order to warn other males that their mate is claimed. Ants use pheromones to lay down a trail from a recent food source to their nest. The pheromone, which is laid from the tip of their stinger, evaporates to form an active space. This enables other ants to rapidly find the food source.

Primer pheromones change the physiology of the recipient by activating certain endocrine glands to secrete hormones. These hormones subsequently cause changes in the behavioral patterns of the recipient. For example, social insects such as ants, bees and termites use primer pheromones in caste determination. In termites the king and queen secrete primer pheromones which are ingested by workers, preventing them from developing reproductive capabilities. In mammals. the estrous cycle of female mice can be initiated by the odor of a male mouse. In addition, the odor of an unfamiliar male mouse can block the pregnancy of a female that has been newly impregnated by another male. This enables the strange mouse to mate with the female. However, if the olfactory bulbs of the female are removed, the pregnancy block will not occur. This signifies that pheromones, unlike hormones, have an external rather than internal source.

• PROBLEM 31-8

Explain how visual displays are important in communicating an animal's "mood."

Solution: For animals having a well-developed visual sense, visual displays serve as signals to convey their potential behavior. Such displays can convey an animal's readiness to mate, to attack, or to retreat. During territorial contests and sexual encounters, cichlids, which are African fresh water fish, transmit their moods by displaying distinctive coloration or patterns. (See Figure 1) The particular coloration suggests the behavior that is likely to ensue.

Another example of a visual display is the courtship behavior of ducks. The odd movements of male ducks in spring signal to the female that they are eager to mate. These visual displays also synchronize the sexual physiology of both male and female and make the female more receptive to the male's advances. Each species has its own display pattern, ensuring that a female will only mate with a male of its own species.

Visual displays occur in other than reproductive behavior. They can be seen in the everyday behavior of cats and dogs. The greeting display, which involves the erection of the tail in cats and the wagging of the tail in dogs, conveys the animal's receptive mood. When dogs and cats are displaying antagonistic behavior, the teeth are bared, the hairs on the neck and back are raised, the ears are laid back, and the body raised. This conveys

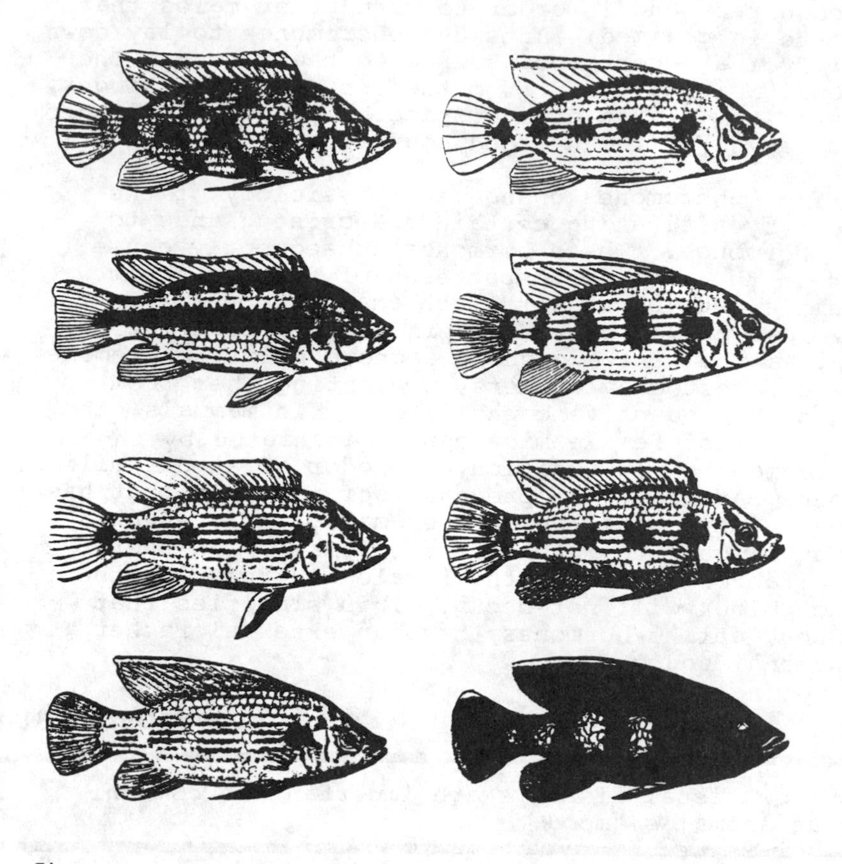

Figure 1.

African fresh-water fish (Hemichromis fasciatus) can change its body coloring rapidly. The color changes are displays that express eight different moods.

Figure 2. Agonistic displays of Black-headed Gull. (A) The first response given by a male on his territory when another male approaches is the "Long Call," in which the body is tilted downward, the wing butts are lifted, the head is thrust foward, and a characteristic call is given. (B) If the other male continues to approach, the defending bird may move to meet the intruder at the boundary of his territory, where he gives the "Upright" display, lifting his still-folded wings, stretching his neck upward, and pointing his bill downward. (C) If the intruder performs counterthreat displays, the defender may then adopt the "Choking" posture, tilting his body head down in an almost vertical position and moving his head in a series of quick up-and-down jerks. (D) "Facing Away" is an appeasement display in which a gull turns his head so that the other bird cannot see the beak and eyes or the black facial mask.

the animal's readiness to attack. A defeated or submissive animal displays appeasement behavior: the fur is sleeked back, the body lowered with legs bent, the tail is down or tucked between the legs, and the head turned away from the victor. Such a display shows the animal's willingness to retreat. In general, attack behavior usually involves directing the face toward the antagonist and making the body appear as large as possible through hair-raising and posturing. The antagonistic displays of a black-head gull are shown in figure 2. Retreat or appeasement behavior involves turning the face away and making the body appear small and vulnerable. Such behavior sometimes serves to prevent actual fighting between animals, as one of the contestants usually backs down. For instance, a defeated rat will roll over on his back, which usually causes the aggressor to stop his attack.

• PROBLEM 31-9

How is a human's perception of a flower different from that of a bee?

Solution: It is known that pigeons can sense the earth's magnetic field and bats can detect ultrasonic signals, both of which humans cannot detect. It is therefore obvious that humans do not have all the sensory skill possessed by other animals. Another fascinating example of animal sensory perception, is the detection of ultraviolet light by the honeybee. Certain flowers which appear to us to be plainly colored, actually contain highly contrasting patterns in the ultraviolet range. If one should look at these flowers through a transformer that shifts the ultraviolet energies to the visible spectrum, one could see how a bee perceives these flowers (See photograph). The patterns, completely invisible to the human eye, serve as guides to the bee in reaching the location of nectar and pollen.

The two photographs show the different views of the same flower as seen by a human eye (A) and and a bee (B). The bee is sensitive to ultraviolet light which is invisible to human beings.

• PROBLEM 31-10

How does a worker bee communicate to other bees the distance and direction of food from the hive?

Solution: After one bee has discovered a food source, many bees soon appear. The first bee must somehow communicate to the others the location of the new food source. It has been determined that when a worker bee discovers food, it returns to the hive, feeds several other bees, and performs a dance on the surface of the honeycomb. If the food source is near the hive, the bee performs a "round" dance. This consists of a repeating pattern of circling, first to the right and then to the left:

Figure 1.
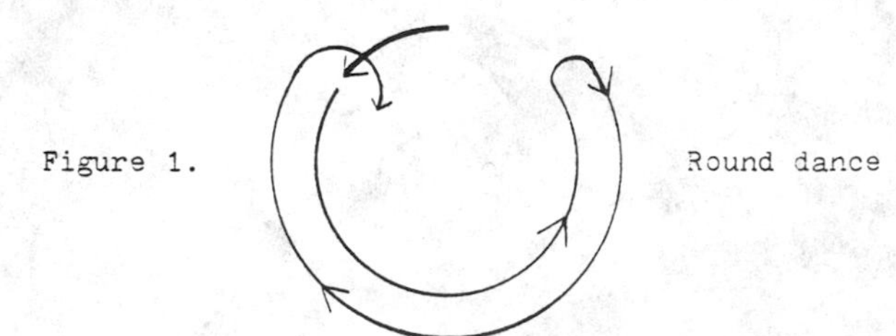
Round dance

The round dance excites the hive mates in the dancer's vicinity and they soon leave the hive in search of the food. This dance does not indicate direction or exact distance. It simply indicates that the food source is a short distance away.

With increasing distance of the source from the hive, the round dance is gradually transformed into a "waggle" dance, which gives information about distance and direction. In the waggle dance, the bee repeatedly runs in a figure-eight pattern, wagging its abdomen from side to side as it traverses the "waist" of the figure eight:

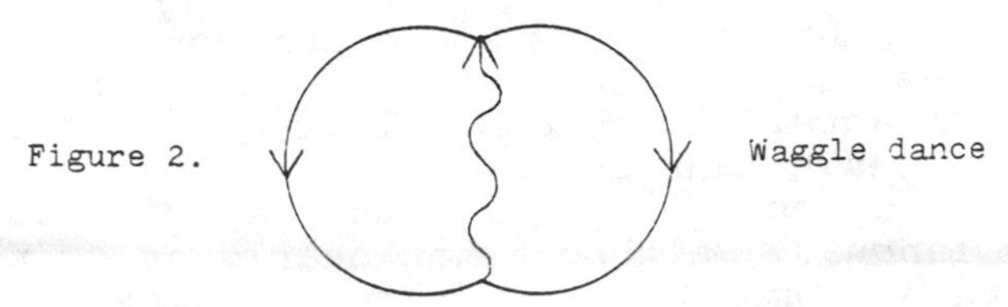
Figure 2. Waggle dance

The number of turns per unit time is inversely proportional to the distance of the food source. The faster the bee runs (or turns), the closer the food source. For honey-bees, an estimate of distance is based on the amount of energy expended traveling the round trip. When conditions exist such as headwinds, uphill flights, clipped wings, or added weight to the bees, excessive distances will be reported.

The waggle dance also indicates direction. The location of the food relative to the sun is indicated by the direction of the straight portion of the waggle dance. For instance, if the bee runs straight up the vertical comb, the food lies in the direction of the sun; if it runs down, the food is found in the opposite direction from the sun. A straight run at an angle indicates that the food source is to be found at that angle to the sun. Therefore a run 30° to the left of vertical indicates a food source 30° to the left of the sun. (See figure 3) If the bee should perform the waggle dance on a horizontal surface outside the hive, it simply dances in the direction of the food source relative to the sun.

Figure 3.

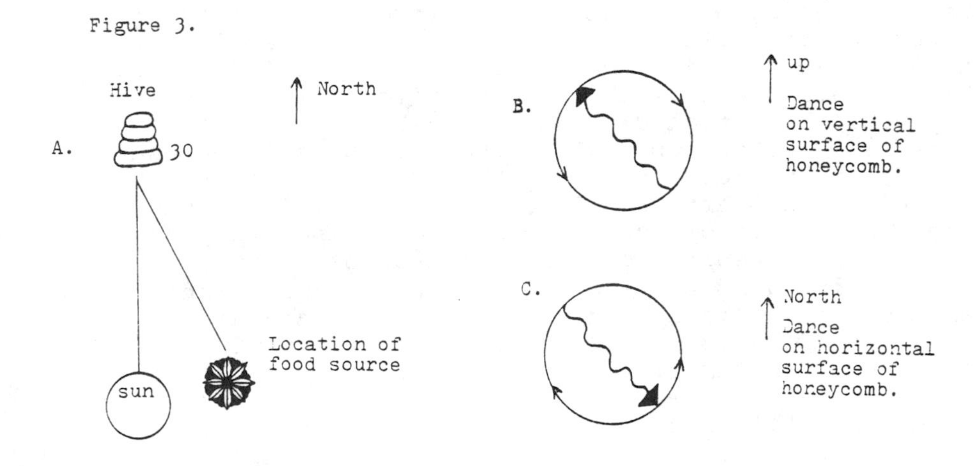

Waggle dance of bees:
B and C show the same dance, performed inside and outside the hive respectively, for locating the same food source in A.

• PROBLEM 31-11

What functions are served by sound communication in animals?

Solution: Communication by sound can convey a variety of messages necessary to the survival and reproduction of the participants. Sound communication is of special interest to us as humans because it serves as the fundamental basis of our language. Other species use forms of auditory language, although not nearly as complex or sophisticated as human speech. Mate attraction, species recognition, defense, and territorial claim are examples of functions of sound communication in animals.

The female Aedes mosquito creates a buzzing sound with her wings, which serves to attract the male mosquito. The sound waves cause the male's antennae to vibrate, stimulating sensory hairs. This allows the male to locate the female and copulate with her. An interesting adaptation for this communication is that in the sexually immature male mosquito, the antennae hairs lie close to the antenna shaft, causing near deafness. This prevents sexually immature males from responding to females. When the male mosquito reaches maturity, the antennae hairs become erect, allowing him to be receptive to the female's characteristic mating sound. The frequency of the buzzing sound is specific for each species of mosquito; thus, sexually mature males will respond only to females of their own species. The mating calls of frogs are also species-specific. In contrast to the Aedes mosquito, however, it is the male frog that attracts the female by calling.

Bird songs are a good example of sound communication being used for territorial defense and species recognition. A territory may be defined as the area defended by an animal. This area centers around the animal's breeding ground, nest site, and sources of food or other needs. The use of sound to defend a territory is exemplified in the following experiment. If silent models of wood thrushes are set up within the territory of a given thrush, they would be attacked by that thrush. If, however, a loudspeaker is set up next to the models, and the characteristic species song of another species played, the thrush will not attack. This is because males attack only other males of their own species, since they are the most threatening to their territories.

Singing also plays a role in the establishment of territories. A male bird chooses an area and sings loudly in order to warn away other males. During the spring, when boundaries are being established, the male thrushes that sing most loudly and forcefully are those who successfully acquire the largest territories.

• PROBLEM 31-12

What are some of the cues used by animals in migration?

Solution: Some animals have the extraordinary ability

to find their way home after making long-distance journeys. They must somehow "remember" the way home. Several cues which such animals use in their migrations have been established.

One cue used by many animals is scent. Salmon are known to be born in fresh-water streams and shortly thereafter journey to the sea. After they mature, they swim back upstream to spawn. But they almost invariably swim back up the same stream in which they were born. It has been determined that salmon have an acute sense of smell, and can remember the odor due to chemical composition of the stream in which they were hatched. Salmon whose olfactory tissues have been incapacitated cannot locate their native stream.

Birds are known to undertake long-distance migratory journeys. Each spring many birds fly North to obtain food and to rear their young. As winter approaches, food becomes scarce and they fly South again. One species, the golden plover, migrates from northern Canada to southern South America. Birds make these long journeys by relying on celestial clues. At migration time, if a caged bird is allowed to see the sun, it will attempt to fly in the direction of its migratory route. If it is overcast, the movements are undirected. Some birds fly at night and use the relative positions of the stars. The physiological state of such birds can be altered by exposing them to different artificial day lengths. Thus some birds will enter the spring migratory condition and others the fall migratory condition. When exposed to the artificial night sky of a planetarium, the "spring" birds try to fly North while the "fall" birds try to fly South.

In studying the homing behavior of pigeons and other birds, it was discovered that birds are able to sense the earth's magnetic field. When tiny magnets were attached to homing pigeons in order to cancel the effects of the earth's magnetic field, the pigeons became disoriented and could not make their way home. Experiments are being done to see if other cues are also used, such as barometric pressure and infra-sound (low-frequency sound waves).

• PROBLEM 31-13

What is the adaptive importance of display in aggressive behavior?

Solution: It is necessary to distinguish aggressive behavior between members of the same species from violent predatory behavioral patterns, which are usually directed at members of different species. Aggressive behavior is usually exhibited between same-species males in defending their territory or establishing their status in a social order. The critical factor in aggressive behavior is that most of the fighting consists of display.

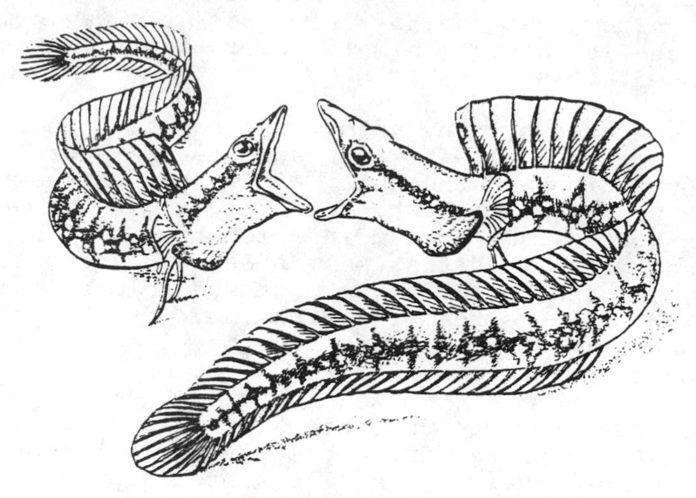

Aggresive display between fishes. Male pike blennies (Chaenopsis ocellata) defend territories on the sea floor with highly ritualized exchanges that include lowering of the gill case, raising the fins, gaping of the jaws, and changes of color. Winner is determined by size and aggressiveness.

Display consists of ritualized, highly-exaggerated movements or sounds which convey the attack motivation of the contestants. In attempting to appear as formidable as possible, the animal often changes the shape of certain body parts in an attempt to make itself appear as large as possible.

The adaptive significance of display is that the two contestants are rarely seriously hurt. Aggression is a type of bluff. The contestant's goal is to repel his opponent to a point outside of his territorial boundary or to convince his rival that a certain female is "taken." Aggressive behavior is therefore highly ritualized, with each movement conveying a certain message. It seldom results in bodily harm, even if physical contact is made. The animals can tell without serious fighting which one would be the probable winner if they were to actually fight. It would be biologically deleterious if during every aggressive contest between rivals, the loser were to be killed. Display behavior allows the "would-be loser" to escape and establish a territory or acquire a mate elsewhere. Survival and reproduction are therefore preserved.

• PROBLEM 31-14

How does communication serve in predator avoidance?

Solution: Both recognition and avoidance of predators can be considered forms of communication between species. One method by which animals avoid detection is called cryptic appearance. In this method, the animal uses its color or shape to resemble objects in the environment in

An example of Batesian mimicry. Top: The Monarch butterfly, a distasteful species. Bottom: The Viceroy, a species that mimics the Monarch. Species in the group to which the Viceroy belongs ordinarily have a quite different appearance.

which it lives. Many insects are shaped to resemble inedible objects, such as twigs. Some animals can even change their colors to resemble the surface on which they are temporarily resting.

Warning appearance is another form of inter-species communication. It is used by some animals to warn potential predators that they are dangerous or poisonous, thus serving to prevent attack. The bold red-and-black bands of the venomous coral snake and the black-and-yellow stripes of the hornet are familiar examples.

Mimicry is another method used to avoid attack by predators. It is used by animals that are not naturally protected; that is, they do not actually possess any characteristics harmful to a predator. The mimic, however, resembles some other species that does have dangerous characteristics and warning signals. In the form of mimicry called Batesian mimicry, the predator is unable to differentiate between the dangerous model and the harmless mimic. It thus leaves both alone. For example, the Monarch butterfly is distasteful to birds, who therefore avoid it. It serves as the model for the Viceroy butterfly, the edible mimic who closely resembles the Monarch in color, shape and behavior. Having the same warning appearance as the Monarch the Viceroy butterfly is protected. (See Figure)

A second form of mimicry, called Mullerian mimicry, involves the evolution of two or more inedible or unpleasant species to resemble each other. Since each species serves as both model and mimic, greater protection is afforded to them because their repellent qualities are more frequently advertised.

• PROBLEM 31-15

Compare the navigational system of dolphins and bats with that of the fish Gymnarchus (electric eel)

Solution: Most navigational systems in organisms consist of detecting stimuli originating from another object. However some animals bounce their own signals off the object and use the echoes of these signals to navigate.

Although dolphins use a variety of sounds to communicate many kinds of information, they also use sound to navigate and locate food. They possess a type of sonar system in which high-frequency sounds are emitted. The dolphins can then detect the echoes after the sound waves strike objects. The dolphin uses these echoes to determine the object's location by noting the direction of the echoes and how long they take to return. This type of navigation, called echolocation, is also used by bats. Bats are able to fly rapidly through deep caves in near total darkness without bumping into objects. The bat does this by emitting very rapid clicking sounds and detecting the echoes from obstacles. This is very similar to echolocation in dolphins except that the bat's sounds are at a higher frequency. Most bat frequencies are 50 to 100 kilohertz per second; these ultrasounds are out of human range (up to 20 kilohertz per second). Bats, like dolphins, use echolocation to locate food, mainly insects. However, some moths have evolved ears capable of detecting the clicks of bats and can thus dodge the bat's detection system.

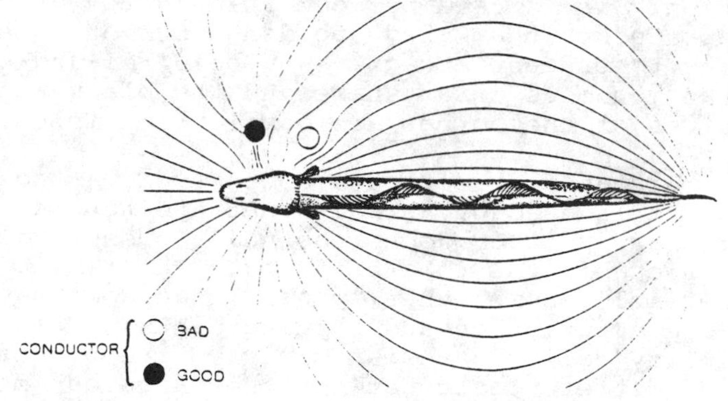

Electric eel (Gymnarchus) generates an electric field around its body. Foreign objects, whether good or bad conductors, distort field. Eel detects changes with receptors in its head.

The electric eel Gymnarchus also has a sophisticated form of energy-emitting navigation. However, the signals used are electrical, not auditory (See Figure). The eels generate these electrical impulses from special organs in the tail. A dipole field is thus formed between the head (which is positive) and the tail (which is negative). Objects which enter the electric field surrounding the body, distort the field. The eel detects these changes using special sensory organs located in the head. By using this form of sensory perception the eel can navigate in murky waters.

HORMONAL REGULATION OF BEHAVIOR

• PROBLEM 31-16

How has the role of internal body secretions in behavior been demonstrated in silk worms?

Solution: Although much of the stimuli which trigger specific behavioral responses have an external origin, many responses occur as a result of internal stimulation. Much of this internal stimulation is initiated by hormones.

The silkworm larva must spin its cocoon at a certain time in life, just following its larval stage. The stimulus for the spinning behavior comes from the disappearance of a specific substance, juvenile hormone. This is secreted by the corpora allata, a pair of glands located behind the brain of the silkworm. The secretion of juvenile hormone prevents the spinning of the cocoon. As the larva grows, the level of juvenile hormone drops, allowing for the spinning reaction to occur. No external sign stimuli is necessary, only internal hormonal levels. The cocoon-spinning behavior of the larvae of silk worms is an example of a behavior under the control of hormones.

ADAPTIVE BEHAVIOR

• PROBLEM 31-17

Black-headed gulls remove broken eggshells from their nests immediately after the young have hatched. Explain the biological significance of this behavior.

Solution: Behavior that contributes to the survival and reproduction of the animal is adaptive. Since those animals which demonstrate adaptive behavior survive longer and successfully raise more offspring, it is more likely that the behavior patterns of these animals will be continued throughout further generations.

The eggshell - removing habit of Black-headed gulls is such adaptive behavior. The gulls do not remove only broken shells, but also any conspicuous object placed in the nest during the breeding season. The more conspicuous the object, the more likely it is to be removed. The gulls seem to be discarding conspicuous objects as a defense mechanism against visual predators. When investigators placed conspicuous objects along with eggs in nests, these nests were robbed of eggs (by other gulls) more often than nests having only eggs. Thus eggshell-removing behavior is significant in that it reduces the chances of a nest being robbed, thus enhancing the survival of offspring. The gulls that evoke this behavioral pattern will successfully raise more offspring than those who do not, and thus this behavioral pattern will tend to be passed on.

COURTSHIP

• PROBLEM 31-18

During the courtship of the common tern, the male presents a fish to another tern. Account for this behavior and discuss why courtship in general may be necesary for reproduction.

Solution: Courtship is an important means of communication between some animals. Courtship involves a multitude of precopulatory behavior patterns which serve a specific purpose for the animal that performs them. The courtship of the common tern, a type of gull, exemplifies the purposes of courtship behavior.

During the mating season of the common tern, the males will catch a fish and present it to another tern. If the recipient tern is a sexually receptive female, she will probably accept it. The male tern then bows to her and prepares to make a nest in the sand where they will mate. If the male offers the fish to another male, the recipient will reject it with agressive behavior. Why would the male offer the fish to another male? In the common tern, both sexes look so similar that this courtship behavior is the only way of differentiating males from females. Thus, one function of courtship behavior is sexual identification.

Courtship is necessary for reproduction in some cockroaches and other selected insects. It serves to decrease aggressive tendencies long enough for the two animals to mate. In these insects, the male secretes a substance on its back which the female feeds on during copulation. This diverts her attention so mating may occur.

Courtship is also necessary for some animals to produce eggs. A female lyrebird will only lay eggs if she can see a male displaying his feathers and singing. However, eggs can also be laid if the neck of the female is tickled in a way to simulate the tickling done by the male during courtship, even if no male is around.

Finally, courtship is necessary to establish species identification. Many birds have species-specific songs and courtship displays which only have significance for others of the same species. Such behavior also serves to ward away other males of the same species from the displaying male's territory.

If courtship behavior did not exist, much energy would be wasted by animals trying to mate with others of the same sex or of different species.

• PROBLEM 31-19

In a certain species of flies, the male presents the female with a silken balloon before mating. Explain the evolutionary derivation of this behavior.

Solution: Animals with beneficial adaptive behavior patterns usually survive longer and successfully raise more offspring than those without. These behavior patterns will often be continued through successive generations. As they are passed on, they evolve. Thus, new behavior patterns can be seen to arise from ancestral behavior patterns. The presentation of a silken balloon to the female fly (family Empididae) before mating is the result of the evolution of a behavior pattern.

Different species of the fly family, Empididae, exhibit slightly different courtship behaviors. These different behaviors may represent the different stages of the same behavior as it evolved from the ancestral behavior pattern. The male of the most primitive Empididae species mates with the female without giving her anything. This often results in the female capturing and eating the male.

In more advanced species, the male will first capture an insect and present it to the female. While the female is occupied with eating the insect, the male mates with her. In even further advanced species, the male captures an insect, wraps it in silk and presents this to the female. The female attempts to open the silken wrapping while the male mates with her. In the most advanced species of Empididae the male presents an empty balloon of silk to the female. The female proceeds to unravel the silken balloon while the male mates with her. At this stage in the evolution of this behavior, the silken balloon (not the insect) is the important element.

By studying different species, a clue to the derivation of behavior patterns can be obtained. It is important to realize that many seemingly odd behaviors performed by animals have an evolutionary origin and have helped that animal adapt to its environment.

• PROBLEM 31-20

When an adult mouse or bird is castrated, its sexual behavior virtually disappears. Compare the effects of castration of highly developed primates with the less developed animals.

Solution: In simple behavior, animals respond to a particular stimulus in the environment with an appropriate response. However, few animals respond so automatically; stereotyped behavior can be modified. Sometimes the animal does not respond; sometimes it responds differently than expected. Its actions often depend on other controlling factors, such as hormone levels and previous experience. This question involves the use of these two factors in modifying sexual behavior.

Hormones modify behavior by affecting the level of an animal's motivational state. They alter the animal's physiology, which can then alter behavioral patterns. For example, the breeding term in birds is initiated by physiological changes in the reproductive organs which produce the sex hormones. These sex hormones cause the development of special breeding plumage which subsequently affects reproductive behavior. The male hormones also induce courtship displays like bowing and cooing upon sight of the female. The female responds to these courting displays by the release of its own reproductive hormones which affect egg production. Eventually nest building occurs and copulation results. In each step of this reproductive behavior, hormones lead to behavior acts which cause further release of other hormones directing further behavior.

However, as animals have evolved, there is less hormone-directed behavior, and more learned responses. Learning is the change in behavior as a result of experience. Only higher animals with larger and more complex brains demonstrate the process of learning. Learning affects man's behavior, very little of which is stereotyped. In man, hormones play a minor role in modifying behavior; learning plays a major role.

When a mouse is castrated (testes removed), it can no longer produce any male reproductive hormones. As in the bird, these hormones are necessary for the reproductive behavior of the mouse. Removal of the hormones thus causes the sexual behavior of the mouse to decline and eventually disappear. The mouse apparently cannot learn its sexual behavior. In higher primates such as man, castration does not affect the ability to perform the sexual act, provided the male is sexually mature and experienced. This is because the behavior has been learned through experience.

LEARNING AND CONDITIONING

• PROBLEM 31-21

A women feeds her new cat daily with canned cat food. As she opens the cans with a noisy electric can opener, she calls the cat. The cat comes in response, to its name, and the woman feeds it. Why does the cat eventually come running when it hears the electric can opener?

Solution: Many animals can learn to associate two or more stimuli with the same reward or punishment. This behavior, called associative learning, was first demonstrated by Pavlov, a Russian psychologist. When Pavlov placed meat powder in a dog's mouth, the dog salivated. Pavlov then added a new stimulus, the ticking of a metronome, at the same time the meat was given. After a number of such pairings, the dog was only permitted to hear the metronome, without any presentation of meat. It still salivated. The dog associated the ticking of the metronome with the presentation of food, and hence salivated even when no food was given. Associative learning is called classic reflex conditioning, when the response is a reflex, such as salivation.

The cat and can opener example is a variation of Pavlov's experiment. When the woman calls the cat by its name, it responds by running to her. The cat is rewarded or positively reinforced with food. A short time before the cat is fed, it hears the sound of the can opener. After a while the woman stops calling the cat. However, it still comes running when it hears the sound of the can opener. The cat has associated the sound of the can opener with the presentation of food, and it responds to that sound by running to the woman even when she is silent. One would expect this associative learning to continue as long as the cat is still given food (reward). If the cat is no longer fed after responding to the sound of the can opener, it will eventually stop responding to the can opener. The lack of reward will extinguish the behavior.

• PROBLEM 31-22

A young scientist is watching a group of new-born ducklings. On the first day, he observes them to crouch in fear at the sight of falling leaves. He also sees them "practice" flying. Over the next few days he observes that they do not react to the falling leaves and can now fly expertly. He concludes that the ducklings have learned to ignore falling leaves and have also learned to fly. Why is he partly incorrect?

Solution: The scientist is incorrect because he uses the term "learning" to describe both processes. Learning is the adaptive change of behavior as a result of experience The animal modifies its behavior to respond to stimuli in ways that promote its survival and reproduction; that is, in ways that adapt it to its environment.

One of the simplest forms of learning is called habituation. Habituation is the gradual decline in response to repeatedly presented stimuli not associated with reward or punishment. Young ducklings initially show fear and crouch when presented with any overhead flying objects, such as falling leaves. But the ducklings cease to respond to the leaves since experience has shown them that they are insignificant. (no associated reward or punishment)

The duckling, however, is not "learning" to fly. One of the factors that complicates the study of learning is maturation. Maturation is the automatic disclosing of a behavioral process due to physilogical development. The flight of birds is a prime example of maturation. As the scientist observed the flying attempts of the ducklings, he termed it "learning", thinking that the ducks flew better (change in behavior) because of this apparent "practice" (experience). But flying can be achieved solely by maturation, with no practice involved. This has been shown by experiments with pigeons. Pigeons were raised in narrow tubes that prevented them from moving their wings and practicing flying. However, when the pigeons were released at an age when they would have already been flying, they were able to fly as well as control pigeons raised under normal conditions. It is not practice that enables the flight of a young bird to improve; it is progressive maturation. The development of the bird's nervous and muscular systems enables it to improve its flying. The bird has not "learned" to fly since experience is unnecessary. Hence the young scientist is incorrect in this respect.

• PROBLEM 31-23

Explain how a "Skinner box" can be used in operant conditioning.

Solution: In operant conditioning, a particular behavior is rewarded or punished when it occurs, thereby increasing or decreasing the probability that it will occur again. The Skinner box (named after B. F. Skinner, a behavioral psychologist) is a principal means of studying this type of learning. The typical Skinner box is a small cage with a bar and food dish.

When a rat or other small animal is placed in the box, its exploratory behavior might cause it to press the bar accidentally. If the rat is rewarded by being given a food pellet, it is more probable that it will press the bar again. This is called positive reinforcement. It is critical that the reward be given within a second or so after the response in order for conditioning to be effective. The reward is given whenever the rat presses the bar. The rat eventually learns to associate bar-pressing with food. Repetition produces a degree of learning beyond which there is no improvement.

One can eliminate the bar pressing behavior by eliminating the reinforcement. This is called extinction, which is the decay of the original learned behavior in the absence of the reward. The rate of extinction depends on the extent and schedule of positive reinforcement. Increased repetition makes a response more resistant to extinction. Also, reinforcement at temporal intervals (giving a reward for every fourth bar press) is more difficult to extinguish than continuous reinforcement (reward for every bar press). It is also possible to condition the rat not to press the bar through the use of punishment. By electrically shocking the rat concurrently with bar pressing behavior, the rat associates the two and learns not to press the bar.

Operant conditioning is sometimes called trial-and-error learning when it occurs in nature. Young chicks will initially peck at any small object; eventually they peck only at edible objects. When the chick pecks at inedible objects, it gets no reward, so this behavior is extinguished. It learns to peck only at edible objects since they act as a positive reinforcement.

• PROBLEM 31-24

Using conditioning, how can one determine the sensitivity of a dog's hearing to sounds of different frequencies?

Solution: Dogs can usually discriminate between sounds of different frequencies or pitches (measured in cycles per second, cps). To determine how well they discriminate, one can use operant conditioning. By rewarding or punishing a particular act when it occurs, the probability that it will occur again is either increased or decreased. For example, one can expose the dog to a sound frequency of 1000 cps. The dog is then rewarded, whenever it responds to a sound of that frequency by barking. Soon the dog barks whenever it hears a sound of 1000 cps. In the same manner, the dog can be conditioned to raise its paw whenever it hears sounds of another frequency. By making this frequency closer and closer to 1000 cps, and observing the dog's response, either a bark or a paw lift, one can determine the smallest differences in frequency between which the dog can discriminate. A point is eventually reached beyond which the dog is unable to discriminate between a "bark" pitch and a "lift-paw" pitch. Dogs have been found to be capable of discriminating between two sounds that differ by only two cycles per second.

Generalization is the response that is contradictory to discrimination. This occurs when an animal conditioned to one stimulus responds in the same manner to a similar, though different stimulus. For example, a bird conditioned to peck at a dark blue spot may peck at a violet spot when the blue one is unavailable.

• PROBLEM 31-25

In the popular nursery rhyme "Mary Had a Little Lamb", her lamb would follow Mary wherever she went. Discuss a possible biological explanation.

Solution: The bond established between Mary and her lamb might have resulted from a form of learning called imprinting. The concept of imprinting was first formulated by Konrad Lorenz, an Austrian ethologist (ethology is the study of behavior). Lorenz discovered that new-born geese will follow the first moving object they see, especially if it makes noise. They form a strong and lasting bond to it, adopting the object as their parent. Usually, the first moving object the bird sees is its mother. Thus imprinting serves to establish a bond between the young and their mother in nature. This type of response has important survival values, such as keeping the young near their parent and establishing species recognition and interaction. Imprinting is characterized by having a short time period during which it can occur. This critical period lasts only about 32 hours in ducklings.

Lorenz demonstrated imprinting by allowing goslings (young geese) to see him as the first moving creature. They then followed Lorenz as if he were their parent.

Young birds of most species become independent when they attain sexual maturity. If birds are reared by a mother of a different species, they attempt to mate with birds of the foster mother's species when sexually mature.

Although imprinting was first demonstrated in birds, it is known to occur in sheep, goats, deer and other mammals. Orphan lambs reared by humans follow them and show little interest in other lambs. That is why "everywhere that Mary went, her lamb was sure to go."

• PROBLEM 31-26

A hungry chimpanzee is able to solve the problem of getting a banana that is out of reach by stacking boxes on top of one another. Explain.

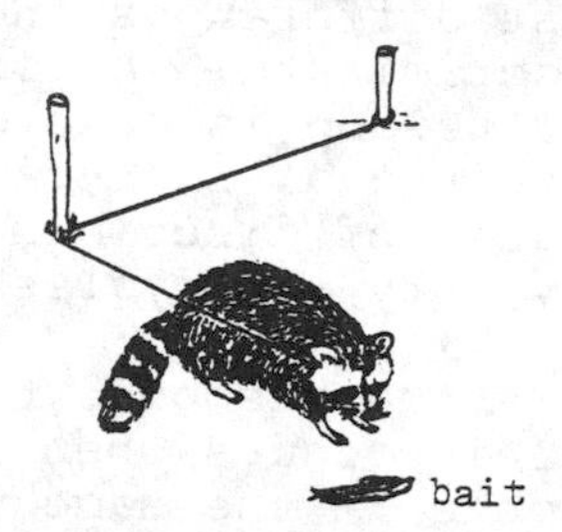

A detour problem for the raccoon in which the animal must first go away from the bait in order to reach it.

Solution: This type of behavior is called insight learning. Considered to be the highest form of learning, it is most prevalent in primates and man. Insight is the ability to respond correctly on the first and subsequent trials to problems not encountered before. The animal is able to apply its prior learning experience in other situations to a new problem and solve the problem without having to use a trial-and-error process.

In the situation given, the chimpanzee used insight to solve a problem it had never encountered before. Insight is the ability to incorporate concepts and principles with past experience in order to solve problems. During his survey of the problem, the chimp might mentally or visually internalize a trial-and-error process. By mentally integrating several elements originally learned separately, the chimpanzee can figure out the solution without doing the actual trial-and-error process and learning by mistake. In humans, insight learning is often referred to as reasoning. Dogs and cats demonstrate lack of insight or reasoning by being unable to perform well on detour tests. Detour tests are the simplest techniques used to investigate problem-solution capacities in animals. In these tests, an animal is blocked from a direct approach to food which he can see and smell by a barrier. In order to get to the food he must first walk away from the food and detour around the barrier (Fig. 1). Monkeys and chimpanzees are usually successful initially without having to rely on a trial and error process.

CIRCADIAN RHYTHMS

• PROBLEM 31-27

A biology student from Nebraska, vacationing at a Virginia beach, observes that the level of activity of a fiddler crab changes with the cycle of the tides. When he brings some crabs back with him to Nebraska, he notices that a similar activity cycle is shown, yet there are no ocean beaches for miles. Explain.

Solution: The fiddler crab lives on the beaches of the eastern coast of the United States. These crabs are very active at low tide. At high tide, they are inactive and usually burrow in the sand so they will not be swept away to sea. Since there are approximately two low tides and two high tides every 24 hours, the crabs undergo a cycle of activity. The biology student might expect this cycle of activity to stop when the crabs are in Nebraska. The fact that it does not suggests that the actual presence of the tides is inconsequential to the crab's level of activity. That is, the sensing of the water is not critical.

The crucial determinant of the level of activity of the crab is not the presence of the tides, but the gravitational pull of the moon which brings about the tides. The fiddler crab has a sensory mechanism which can detect the moon's gravitational pull, and its activity cycle varies according to this pull. However, the fiddler crab in Nebraska does not have the identical cycles of activity that it showed in Virginia. Since the time of the "rising" of the moon is slightly different in Nebraska, the crab will gradually reset its activity cycle. This cycle would correspond to the times of the tides were Nebraska located on an ocean.

• PROBLEM 31-28

Fiddler crabs are a dark brown color during the day and change to a pale brown color at night. Is it possible to reverse this pattern and make the crabs turn dark at night and light during the day? How?

Solution: Fiddler crabs undergo a daily cycle of color change. However, this is not just a simple adjustment to conditions of light and dark. When the crabs are brought into a room, which is kept dark continuously, they still turn dark brown at dawn and light brown at sunset. These crabs have an internal sense of time, called a "biological clock."

It is possible to reset this biological clock by artificially producing periods of light and darkness. For example, we can keep the crabs in a room in which a bright light is turned on at sunset and shut off at sunrise. After several days of this treatment, the crabs begin to turn dark brown at sunset and pale brown at sunrise. Even when the crabs are exposed to constant light or constant darkness, the 24 hour rhythm continues. But this cycle is 12 hours out of phase with the true day, showing that the crabs' internal clock has been reset. Rhythms of this sort, with approximately 24-hour periods, are called circadian rhythms.

The biological clock of the fiddler crab can also be reset by exposing it to ice water. This type of experimental manipulation seems to stop the clock. When it is taken out of the ice water, the clock restarts, but it has lost time. This manipulation seems to support the belief that biological clocks are controlled by metabolic rhythms, since metabolism slows down under very cold conditions.

SOCIETAL BEHAVIOR

• PROBLEM 31-29

Why would one call a flock of birds a "society"?

Solution: A society is a group of animals that belongs to the same species and is organized in a cooperative manner. The animals are usually bound together by reciprocal communication, which leads to the cooperative behavior.

A flock of birds can be considered an elementary form of society called a motion group. Motion groups include schools of fish and herds of mammals. Birds within a flock communicate with one another in order to stay together as they move from place to place. The cooperative behavior displayed by a flock of birds enables them to more efficiently detect and avoid predators. It is more likely that a flock of birds will detect a predator than a solitary bird. Many birds have special signals, such as alarm cries and wing movements, that seem to alert the entire group to the presence of a predator.

Membership in a group also affords protection once the predator attacks the group. For example, starlings usually fly in a wide, scattered pattern, but assemble quickly in tight formation when attacked by a falcon, their natural predator. The falcon usually attacks by swooping at high speed into the group. By assembling into a tight group, the starlings make it difficult for the falcon to hit one bird without colliding into others and possibly injuring itself.

Other cooperative behavior patterns which may be performed by a flock of birds are the discovery of food and the coordination of reproductive activities.

• PROBLEM 31-30

When a group of chickens from different flocks are placed together there is much more fighting than is observed in flocks which have been together for some time. Explain.

Solution: The group of chickens fight when first placed together in order to establish dominance-subordination relationships within the new flock. The relationships are formed through a series of aggressive encounters between each pair of chickens. Chicken A may be larger, stronger, or just more temperamental than all the chickens it encounters and thus may win all its fights. It becomes the dominant bird, while all the others become subordinate to it. Chicken B may be dominant over all the birds except A. For example, A may peck at B without B pecking back. Yet B may peck at C, D, E, etc. This social hierarchy is sometimes called a pecking order. Pecking orders can get complex. For example, chicken Z may be particularly strong in its encounter with X, yet it may have lost its fight with Y, who is subordinate to X. This pecking order can be diagrammed as follows:

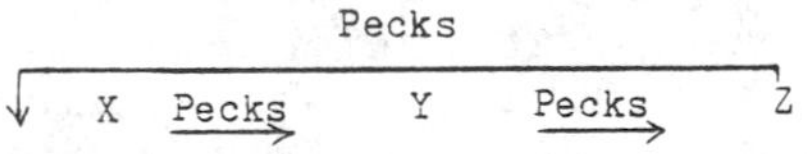

Pecks (Z → X)

X —Pecks→ Y —Pecks→ Z

Among the various factors that determine an animal's position in a hierarchy are: age, sex, size, strength, temperament, physiological condition (hormones), and seniority in the group. Once established, a "peck-right system," common in chickens, becomes rather fixed and rigid, with very few fighting encounters. Thus, an old, weak chicken may still be dominant because the subordinate chickens do not retaliate when pecked. However, in a "peck-dominance system," common in pigeons, subordinate birds often retaliate and thus are more apt to discover any weakness of the dominant pigeons. This system thus promotes more social mobility than the "peck-right system."

The adaptive value of the establishment of social hierarchies is the promotion of order and stability to the relationships in the group. The less fighting that occurs, the more time can be allocated to finding food, avoiding predators, and mating.

• PROBLEM 31-31

Discuss some of the elements found in advanced animal societies.

Solution: A society is a group composed of animals of the same species and organized in a cooperative manner. The simplest animal society could therefore be combinations of parents and their offspring. The communication between adults and young serves to hold this elementary society together. Another form of a simple society is a flock of birds, which is an example of a motion group. The members communicate to keep the group together as it travels in search of food. They cooperate in the mutual protection of members against predators.

An element of a more advanced society is leadership. A leader usually regulates the activities of the other members and directs the group as it moves. Wolves, elephants, and primates usually have a dominant individual who watches over the group's activities and leads them from place to place.

Another important element of advanced societies is the territory, an area occupied by the group and defended by aggression or warning messages. However, most animals move in a larger area in search of food. This area, called the home range, is often shared with the group's neighbors. Territorial behavior involves either the aggressive exclusion of outsiders or the continual vocal signalling that warns outsiders away. The songs of birds, crickets, and frogs serve as such warning messages. Territory may also be advertised by chemical signals. Rabbits are known to place characteristically scented fecal pellets at their territorial boundaries. The distinctive odor warns intruders that the territory is occupied. The attempts of domesticated dogs to urinate on various land marks and smell the urine and feces of other dogs are just manifestations of this territorial behavior.

A third characteristic of many advanced societies is a dominance hierarchy. The peck order, which is a set of dominance-subordination relationships, is an example of this. Dominance hierarchies exist in societies of wasps, bumblebees, hermit crabs, birds and mammals. The concept of a dominance hierarchy is different from that of leadership. In a dominance hierarchy, a subordinate animal may be dominant to other members.

In a society characterized by leadership, the leader is the only member who exhibits dominance. All these elements of advanced societies provide better organization and control for the group, allowing it to exist in its environment more successfully.

• PROBLEM 31-32

Explain the main differences between insect societies and primate societies.

Solution: The major characteristic differentiating primate societies from insect colonies is personal recognition of its members. Within primate societies and most complex vertebrate societies, every member knows every other member as an individual. Often each member bears some particular relationship to every other member. Parent-offspring relationships are tightly binding and persist a long time. The young may also be partly cared for by non-parental members who act as "aunts" and "uncles." Eventually the young primate leaves these personal relationships and establishes new bonds with its peers.

A weak division of labor exists in primate societies. Some members specialize in defense, others in foraging. However, each member is physiologically similar to every other member of the same sex. Therefore, this specialization of duties is not based upon physiologically different castes.

Insect societies, such as those of the ants, wasps, and bees, exist without personal relationships. No individual bonds are established during the short lifespan of the members. The colonies themselves are very large. Only very limited dominance hierarchies exist, whereas in primates, status is extremely important. Insects show little or no socialization and do not play. Young primates undergo long periods of socialization and often play. The queen of the insect colony is not recognized personally by the other members, but by means of pheromones. If one were to place these pheromones on a small piece of wood, the other members would react to it as if it were the queen. In contrast to primate societies, strong divisions of labor exist within insect societies. These divisions are based upon physiologically different castes. Some members are specialized to perform in reproduction (the queens), in labor (the workers), or in colony defense (the soldiers). This highly efficient means of organization is sacrificed in primate societies for the sake of individual freedom and reproduction.

• PROBLEM 31-33

It is easy to understand how selfish behavior can be explained by natural selection. But how can one account for altruism?

Solution: Selfish behavior is exemplified in the formation of dominance hierarchies. The high-ranking animals eat before the other animals when food is scarce and take preference in the selection of mates and resting places. Consequently, these individuals are generally healthier, live longer and reproduce more frequently. Natural selection favors these strong animals, in that they produce more offspring having their superior traits. Selfish behavior is expected to be transferred to future generations since it leads to the more successful survival of the animal.

Altruism is self-sacrificial behavior that benefits other members in a group. An altruistic animal reduces its own fitness, and therefore its chances of survival and reproduction, in order to increase the success of another animal or of the group as a whole. For example, a male baboon remains to fight a leopard while the troop escapes. Natural selection would not favor the continuation of this costly behavior since those who do show it are less likely to survive and reproduce.

One would therefore expect the genes for altruism to be lost, but they are not. The reason for this lies in the fact that the male baboon - only risks its life for baboons in his own troop, not for baboons in other troops. His behavior therefore benefits only those baboons related to itself; these baboons probably share common genes, among which are those for altruistic behavior. The altruistic behavior thus preserves the genes in the other members that are shared with the altruistic member by common descent. For the altruistic genes to be favored in natural selection, the benefit that the related members receive must outweigh the loss of the altruistic animal. Although the baboon might die and lose its ability to pass on its genes, the baboon's brother can compensate for the loss. To do so, it must have more than twice as many offspring as a result of its brother's altruistic behavior. (Since about ½ of the genes of a given animal are shared by its brother or sister.) Similarly, since first cousins share ¼ of their genes, on the average, compensation would involve quadrupling the survivor's offspring.

Altruism directed at a variety of relatives, from offspring to cousins, results in a summation of the benefits so that the altruistic gene(s) may be preserved and the related behavior may evolve. Thus, the existence of altruism can be accounted for by natural selection.

To demonstrate the compensation by a relative for the loss of an altruistic relative, we will assume that altruism is determined by one allele, A.

Frequency of A

P_1	1 A		
F_1	1/2 A	1/2 A	brother - brother
	↓	↓	
F_2	1/4 A	1/4 A	cousin - cousin
	↓	↓	
F_3	1/8 A	1/8 A	

If one offspring of F_1 generation dies, its brother can compensate by doubling its offspring, (F_2). This will double the frequency of A in F_2 ($2 \times 1/4$) to give 1/2, the frequency that A would have had in F_1 if the brother had not died. Similarly, if one altruistic offspring in the F_2 generation dies, its cousin (also in F_2) can compensate by quadrupling its offspring, F_3. Four times the frequency of A in F_3 ($4 \times 1/8$) gives 1/2, which would be the frequency of A in F_1 if the cousin had not died. It is important to note that compensation must always be referred back to the F_1 generation, since that is the only way to prevent a loss of the altruistic gene.

SHORT ANSWER QUESTIONS FOR REVIEW

Answer

To be covered when testing yourself

Choose the correct answer.

1. The chemical substances used for communication between individuals of the same species are called (a) hormones, (b) androgens, (c) inducer substances, (d) pheromones. — d

2. Reactive behavior is (a) automatic, (b) stereotyped, (c) endogenously controlled, (d) all of the above. — d

3. The phase of motivated behavior that will not be influenced by further stimuli is the (a) appetitive phase, (b) satiation phase, (c) consummatory phase, (d) elicitation phase. — b

4. Most arthropods and vertebrates have an individual distance which is (a) the minimum distance between two animals which does not elicit aggressive behavior, (b) the home range or area of daily travel, (c) the territory or area an animal defends, (d) the animal's niche. — a

5. Courtship patterns (a) stimulate the female to lay eggs, (b) help the male to recognize a receptive female, (c) help animals of a particular species to recognize each other, (d) all of the above are true, in some species at least. — d

6. A harmless animal that resembles some other species that is dangerous to a predator is an example of (a) Batesian Mimicry, (b) Mullerian Mimicry, (c) cryptic appearance, (d) aggressive behavior. — a

7. A particular act is rewarded (or punished) when it occurs, therefore increasing (or decreasing) the probability of the act being repeated. This is an example of (a) habituation, (b) operant conditioning, (c) a conditioned reflex, (d) classic reflex conditioning. — b

8. A wood louse becomes restless after it becomes dry. In response, it moves around randomly until it finds a new, moist home. This is an example of (a) Taxis, (b) Kinesis, (c) biological communication, (d) all of the above. — b

9. Aggressive behavior in animals (a) consists mostly of encounters between members of different species (excluding predation), (b) occurs most frequently in contests over food, (c) usually consists of nonviolent displays which avoid serious injury, (d) occur equally in both sexes. — c

Answer

To be covered when testing yourself

10. Choose the correct statement. (a) Young turkeys respond instinctively to the shape of predators. (b) Young turkeys initially fear all flying objects and eventually learn which ones are harmless. (c) All instincts are genetically innate and unalterable. (d) Experience has no effect on developing behavior of lower animals. — b

11. A releaser produces (a) aggressive behavior, (b) a complex series of of reactions not always directly related to the stimulus, (c) an abnormal behavior, (d) a group of reactions related to a learned experience. — b

12. Migration in birds is quided in part by (a) celestial navigation, (b) temperature changes, (c) olfactory cues, (d) day-length changes caused by latitude. — a

13. Which of the following can be considered as a society? (a) A swarm of flies attracted to a rotting fruit, (b) a collection of various insects placed in a jar, (c) a pack of dogs chasing their prey in relays, (d) a group of male crickets attracted to the same female. — c

14. Adaptive significance for "territoriality" most likely includes all of the following except (a) population control, (b) establishment of mating pairs, (c) efficient use and spacing of environmental resources, (d) familiarity with the area. — d

15. Choose the correct statement. (a) Learned patterns are retained by the anterior half of a worm that has been bisected. (b) Some research indicates that untrained planaria fed trained planaria acquire some of the trained behavior. (c) DNA is believed to be the part of the trained worm carrying information when fed to the untrained worm. (d) All of the above are correct. — b

Fill in the blanks.

16. A taxis is an automatic, usually orientational response which involves the _____ and _____ systems (in animals that possess them). — nervous, muscular

17. A persistant change in behavior resulting from experience is a definition of _____. — learning

18. The first phase of motivated activity is known as the phase of _____. — elicitation

	Answer To be covered when testing yourself
19. The searching phase of motivational activity is called _____ behavior.	appetitive behavior
20. The control center for motivated behavior is generally considered to be the _____.	hypothal-amus
21. An important characteristic of intraspecific fighting is that it _____ (often/rarely) results in physical damage.	rarely
22. Through evolutionary selection, many acts of aggression and courtship have become _____.	ritua-lized
23. A normal activity performed out of context is called a _____ activity.	displace-ment
24. The phenomenon whereby a young animal becomes "attached" to the first moving object it sees and reacts to it as it would towards its mother is called _____.	imprint-ing
25. The chemical signals used for intraspecific communication are called _____.	pheromones
26. The ability of an organism to orient itself with the sun's changing position is known as _____.	sun compass orienta-tion
27. In Pavlov's experiments with dogs, the sound of the bell may be classified as a _____.	conditioned stimulus
28. As _____ increases in a population, the frequency of fighting increases.	popu-lation density
29. The study of _____ is the study of animal behavior under conditions that are as nearly natural as possible.	ethology
30. Injecting _____ into sexually immature animals will induce sexual behavior prematurely.	sex hormones
Determine whether the following statements are true or false.	
31. Exaggerated stimuli are not preferred by organisms.	False
32. When organisms are shifted from a rhythmically varying environment to a constant environment, their circadian rhythms persist.	True
33. In organisms transported to different geographical locations, the biological clocks eventually reset.	True

Answer

To be covered when testing yourself

34. Bees are guided to pollen and nectar by the same flower patterns and colors visible to man. — False

35. Bird songs are transmitted from generation to generation entirely by heredity. — False

36. Distance and direction of a food source are communicated by the duration and angle from the vertical of the straight run in the bee's waggle dance. — True

37. An organism must be in the appropriate physiological state for a sign stimulus to evoke the proper consummatory act. — True

38. In general, insect societies are less complex and rigid than those of vertebrates. — False

39. The male sex hormone is the hormone that promotes aggressiveness in either sex. — True

40. Cognitive behavior does not depend to any great extent on operant conditioning. — False

41. The hypothalamus does not influence feeding and drinking behavior, sexual behavior, sleep, maternal behavior or behavioral responses to temperature change. — False

42. In order for insight learning to occur, an animal must remember previous experiences involving various stimuli and adapt the memory to solve new problems. — True

43. Hormonal control of the levels of motivational states is precise and quickly adjusted. — False

44. Practice flights of young pigeons are not required for them to fly properly. — True

45. Imprinting can occur at any age in an animal. — False

INDEX

Numbers on this page refer to PROBLEM NUMBERS, not page numbers

Numbers on this page refer to PROBLEM NUMBERS, not page numbers

Numbers on this page refer to PROBLEM NUMBERS, not page numbers

Numbers on this page refer to PROBLEM NUMBERS, not page numbers

Numbers on this page refer to PROBLEM NUMBERS, not page numbers

Numbers on this page refer to PROBLEM NUMBERS, not page numbers

Numbers on this page refer to **PROBLEM NUMBERS**, not page numbers

Numbers on this page refer to PROBLEM NUMBERS, not page numbers

Numbers on this page refer to PROBLEM NUMBERS, not page numbers

Numbers on this page refer to PROBLEM NUMBERS, not page numbers

Numbers on this page refer to PROBLEM NUMBERS, not page numbers

Numbers on this page refer to PROBLEM NUMBERS, not page numbers

Numbers on this page refer to PROBLEM NUMBERS, not page numbers

Numbers on this page refer to PROBLEM NUMBERS, not page numbers

Numbers on this page refer to PROBLEM NUMBERS, not page numbers

Numbers on this page refer to PROBLEM NUMBERS, not page numbers

Numbers on this page refer to PROBLEM NUMBERS, not page numbers

Numbers on this page refer to PROBLEM NUMBERS, not page numbers

Numbers on this page refer to PROBLEM NUMBERS, not page numbers

Numbers on this page refer to PROBLEM NUMBERS, not page numbers

Numbers on this page refer to PROBLEM NUMBERS, not page numbers

APPENDIX

CLASSIFICATION OF LIVING THINGS

KINGDOM MONERA

DIVISION SCHIZOMYCETES. Bacteria*

CLASS MYXOBACTERIA. *Myxococcus, Chondromyces*

CLASS SPIROCHETES. *Leptospira, Cristispira, Spirocheta, Treponema*

CLASS EUBACTERIA. *Staphylococcus, Escherichia, Salmonella, Pasteurella, Streptococcus, Bacillus, Spirillum, Caryophanon*

CLASS RICKETTSIAE. *Rickettsia, Coxiella*

CLASS BEDSONIA. *Chlamydia*

DIVISION CYANOPHYTA. Blue-green algae. *Gloeocapsa, Microcystis, Oscillatoria, Nostoc, Scytonema*

KINGDOM PLANTAE

DIVISION EUGLENOPHYTA. Euglenoids. *Euglena, Eutreptia, Phacus, Colacium*

DIVISION CHLOROPHYTA. Green algae

CLASS CHLOROPHYCEAE. True green algae. *Chlamydomonas, Volvox, Ulothrix, Spirogyra, Oedogonium, Ulva*

CLASS CHAROPHYCEAE. Stoneworts. *Chara, Nitella, Tolypella*

DIVISION CHRYSOPHYTA

CLASS XANTHOPHYCEAE. Yellow-green algae. *Botrydiopsis, Halosphaera, Tribonema, Botrydium*

CLASS CHRYSOPHYCEAE. Golden-brown algae. *Chrysamoeba, Chromulina, Synura, Mallomonas*

CLASS BACILLARIOPHYCEAE. Diatoms. *Pinnularia, Arachnoidiscus, Triceratium, Pleurosigma*

DIVISION PYRROPHYTA. Dinoflagellates. *Gonyaulax, Gymnodinium, Ceratium, Gloeodinium*

DIVISION PHAEOPHYTA. Brown algae. *Sargassum, Ectocarpus, Fucus, Laminaria*

DIVISION RHODOPHYTA. Red algae. *Nemalion, Polysiphonia, Dasya, Chondrus, Batrachospermum*

DIVISION MYXOMYCOPHYTA. Slime molds

CLASS MYXOMYCETES. True slime molds. *Physarum, Hemitrichia, Stemonitis*

CLASS ACRASIAE. Cellular slime molds. *Dictyostelium*

CLASS PLASMODIOPHOREAE. Endoparasitic slime molds. *Plasmodiophora*

CLASS LABYRINTHULEAE. Net slime molds. *Labyrinthula*

DIVISION EUMYCOPHYTA. True fungi

CLASS PHYCOMYCETES. Algal fungi. *Rhizopus, Mucor, Phycomyces*

CLASS OÖMYCETES. Water molds, white rusts, downy mildews. *Saprolegnia, Phytophthora, Albugo*

CLASS ASCOMYCETES. Sac fungi. *Neurospora, Aspergillus, Penicillium, Saccharomyces, Morchella, Ceratostomella*

CLASS BASIDIOMYCETES. Club fungi. *Ustilago, Puccinia, Coprinus, Lycoperdon, Psalliota, Amanita*

DIVISION BRYOPHYTA

CLASS HEPATICAE. Liverworts. *Marchantia, Conocephalum, Riccia, Porella*

CLASS ANTHOCEROTAE. Hornworts. *Anthoceros*

CLASS MUSCI. Mosses. *Polytrichum, Sphagnum, Mnium*

DIVISION TRACHEOPHYTA. Vascular plants

Subdivision Psilopsida. *Psilotum, Tmesipteris*

Subdivision Lycopsida. Club mosses. *Lycopodium, Phylloglossum, Selaginella, Isoetes, Stylites*

Subdivision Sphenopsida. Horsetails. *Equisetum*

Subdivision Pteropsida. Ferns. *Polypodium, Osmunda, Dryopteris, Botrychium, Pteridium*

Subdivision Spermopsida. Seed plants

CLASS PTERIDOSPERMAE. Seed ferns. No living representatives

CLASS CYCADAE. Cycads. *Zamia*

CLASS GINKGOAE. *Ginkgo*

CLASS CONIFERAE. Conifers. *Pinus, Tsuga, Taxus, Sequoia*

CLASS GNETEAE. *Gnetum, Ephedra, Welwitschia*

CLASS ANGIOSPERMAE. Flowering plants

SUBCLASS DICOTYLEDONEAE. Dicots. *Magnolia, Quercus, Acer, Pisum, Taraxacum, Rosa, Chrysanthemum, Aster, Primula, Ligustrum, Ranunculus*

SUBCLASS MONOCOTYLEDONEAE. Monocots. *Lilium, Tulipa, Poa, Elymus, Triticum, Zea, Ophyrys, Yucca, Sabal*

* There is no generally accepted classification for bacteria at the class level. The system used here is based on R. Y. Stanier, M. Doudoroff, and E. A. Adelberg, *The Microbial World* (3rd ed.; Prentice-Hall, 1970), though these authors do not formally designate their categories as classes. Other systems usually recognize more classes, some of them very small.

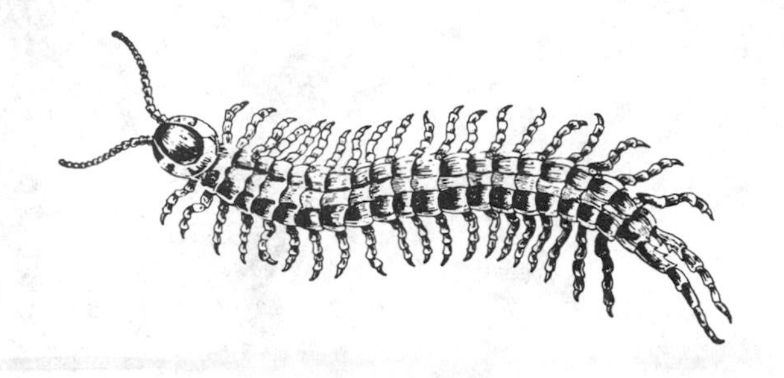

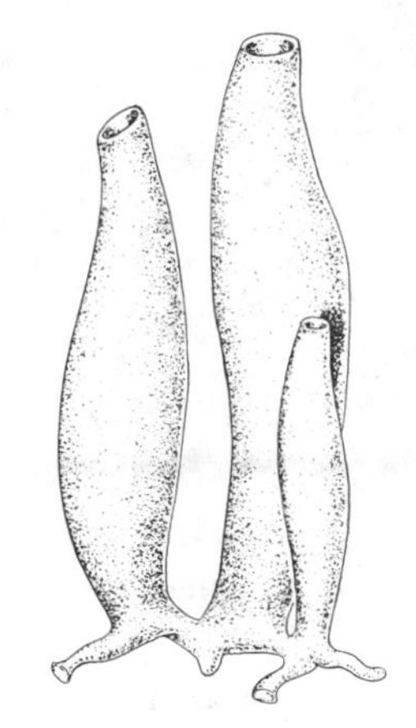

KINGDOM ANIMALIA

SUBKINGDOM PROTOZOA

PHYLUM PROTOZOA. Acellular animals

Subphylum Plasmodroma

CLASS FLAGELLATA (or Mastigophora). Flagellates. *Trypanosoma, Calonympha, Chilomonas* (also *Euglena, Chlamydomonas,* and other green flagellates included in Plantae as well)

CLASS SARCODINA. Protozoans with pseudopods. *Amoeba, Globigerina, Textularia, Acanthometra*

CLASS SPOROZOA. *Plasmodium, Monocystis*

Subphylum Ciliophora

CLASS CILIATA. Ciliates. *Paramecium, Opalina, Stentor, Vorticella, Spirostomum*

SUBKINGDOM PARAZOA

PHYLUM PORIFERA. Sponges

CLASS CALCAREA. Calcareous (chalky) sponges. *Scypha, Leucosolenia, Sycon, Grantia*

CLASS HEXACTINELLIDA. Glass sponges. *Euplectella, Hyalonema, Monoraphis*

CLASS DEMOSPONGIAE. *Spongilla, Euspongia, Axinella*

SUBKINGDOM MESOZOA

PHYLUM MESOZOA. *Dicyema, Pseudicyema, Rhopalura*

SUBKINGDOM METAZOA

SECTION RADIATA

PHYLUM COELENTERATA (or Cnidaria)

CLASS HYDROZOA. Hydrozoans. *Hydra, Obelia, Gonionemus, Physalia*

CLASS SCYPHOZOA. Jellyfishes. *Aurelia, Pelagia, Cyanea*

CLASS ANTHOZOA. Sea anemones and corals. *Metridium, Pennatula, Gorgonia, Astrangia*

PHYLUM CTENOPHORA. Comb jellies

CLASS TENTACULATA. *Pleurobrachia, Mnemiopsis, Cestum, Velamen*

CLASS NUDA. *Beroe*

SECTION PROTOSTOMIA

PHYLUM PLATYHELMINTHES. Flatworms

CLASS TURBELLARIA. Free-living flatworms. *Planaria, Dugesia, Leptoplana*

CLASS TREMATODA. Flukes. *Fasciola, Schistosoma, Prosthogonimus*

CLASS CESTODA. Tapeworms. *Taenia, Dipylidium, Mesocestoides*

PHYLUM NEMERTINA (or Rhynchocoela). Proboscis worms. *Cerebratulus, Lineus, Malacobdella*

PHYLUM ACANTHOCEPHALA. Spiny-headed worms. *Echinorhynchus, Gigantorhynchus*

PHYLUM ASCHELMINTHES

CLASS ROTIFERA. Rotifers. *Asplanchna, Hydatina, Rotaria*

CLASS GASTROTRICHA. *Chaetonotus, Macrodasys*

CLASS KINORHYNCHA (or Echinodera). *Echinoderes, Semnoderes*

CLASS NEMATODA. Round worms. *Ascaris, Trichinella, Necator, Enterobius, Ancylostoma, Heterodera*

CLASS NEMATOMORPHA. Horsehair worms. *Gordius, Paragordius, Nectonema*

PHYLUM ENTOPROCTA. *Urnatella, Loxosoma, Pedicellina*

PHYLUM PRIAPULIDA. *Priapulus, Halicryptus*

PHYLUM ECTOPROCTA (or Bryozoa). Bryozoans, moss animals

CLASS GYMNOLAEMATA. *Paludicella, Bugula*

CLASS PHYLACTOLAEMATA. *Plumatella, Pectinatella*

PHYLUM PHORONIDA. *Phoronis, Phoronopsis*

PHYLUM BRACHIOPODA. Lamp shells

CLASS INARTICULATA. *Lingula, Glottidia, Discina*

CLASS ARTICULATA. *Magellania, Neothyris, Terebratula*

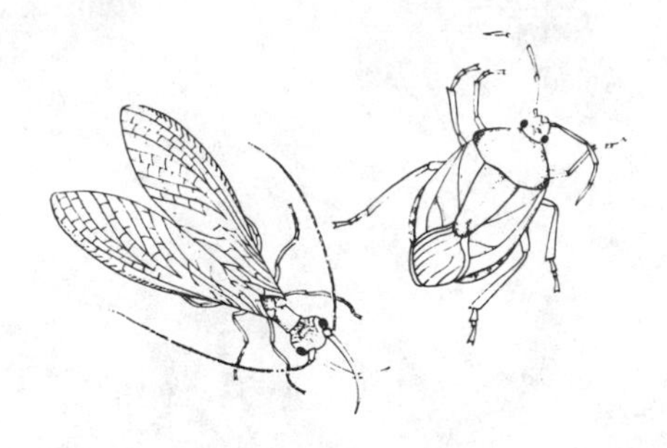

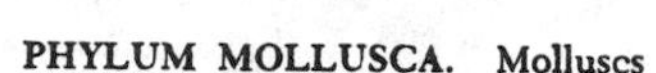

PHYLUM MOLLUSCA. Molluscs

CLASS AMPHINEURA

SUBCLASS APLACOPHORA. Solenogasters. *Chaetoderma, Neomenia, Proneomenia*

SUBCLASS POLYPLACOPHORA. Chitons. *Chaetopleura, Ischnochiton, Lepidochiton, Amicula*

CLASS MONOPLACOPHORA. *Neopilina*

CLASS GASTROPODA. Snails and their allies (univalve molluscs). *Helix, Busycon, Crepidula, Haliotis, Littorina, Doris, Limax*

CLASS SCAPHOPODA. Tusk shells. *Dentalium, Cadulus*

CLASS PELECYPODA. Bivalve molluscs. *Mytilus, Ostrea, Pecten, Mercenaria, Teredo, Tagelus, Unio, Anodonta*

CLASS CEPHALOPODA. Squids, octopuses, etc. *Loligo, Octopus, Nautilus*

PHYLUM SIPUNCULIDA. *Sipunculus, Phascolosoma, Dendrostomum*

PHYLUM ECHIURIDA. *Echiurus, Urechis, Thalassema*

PHYLUM ANNELIDA. Segmented worms

CLASS POLYCHAETA (including Archiannelida). Sandworms, tubeworms, etc. *Nereis, Chaetopterus, Aphrodite, Diopatra, Arenicola, Hydroides, Sabella*

CLASS OLIGOCHAETA. Earthworms and many fresh-water annelids. *Tubifex, Enchytraeus, Lumbricus, Dendrobaena*

CLASS HIRUDINEA. Leeches. *Trachelobdella, Hirudo, Macrobdella, Haemadipsa*

PHYLUM ONYCHOPHORA. *Peripatus, Peripatopsis*

PHYLUM TARDIGRADA. Water bears. *Echiniscus, Macrobiotus*

PHYLUM PENTASTOMIDA. *Cephalobaena, Linguatula*

PHYLUM ARTHROPODA

Subphylum Trilobita. No living representatives

Subphylum Chelicerata

CLASS EURYPTERIDA. No living representatives

CLASS XIPHOSURA. Horseshoe crabs. *Limulus*

CLASS ARACHNIDA. Spiders, ticks, mites, scorpions, whipscorpions, daddy longlegs, etc. *Archaearanea, Latrodectus, Argiope, Centruroides, Chelifer, Mastigoproctus, Phalangium, Ixodes*

CLASS PYCNOGONIDA. Sea spiders. *Nymphon, Ascorhynchus*

Subphylum Mandibulata

CLASS CRUSTACEA. *Homarus, Cancer, Daphnia, Artemia, Cyclops, Balanus, Porcellio*

CLASS CHILOPODA. Centipeds. *Scolopendra, Lithobius, Scutigera*

CLASS DIPLOPODA. Millipeds. *Narceus, Apheloria, Polydesmus, Julus, Glomeris*

CLASS PAUROPODA. *Pauropus*

CLASS SYMPHYLA. *Scutigerella*

CLASS INSECTA. Insects

ORDER COLLEMBOLA. Springtails. *Isotoma, Achorutes, Neosminthurus, Sminthurus*

ORDER PROTURA. *Acerentulus, Eosentomon*

ORDER DIPLURA. *Campodea, Japyx*

ORDER THYSANURA. Bristletails, silverfish, firebrats. *Machilis, Lepisma, Thermobia*

ORDER EPHEMERIDA. Mayflies. *Hexagenia, Callibaetis, Ephemerella*

ORDER ODONATA. Dragonflies, damselflies. *Archilestes, Lestes, Aeshna, Gomphus*

ORDER ORTHOPTERA (including Isoptera). Grasshoppers, crickets, walking sticks, mantids, cockroaches, termites, etc. *Schistocerca, Romalea, Nemobius, Megaphasma, Mantis, Blatta, Periplaneta, Reticulitermes*

ORDER DERMAPTERA. Earwigs. *Labia, Forficula, Prolabia*

ORDER EMBIARIA (or Embiidina or Embioptera). *Oligotoma, Anisembia, Gynembia*

ORDER PLECOPTERA. Stoneflies. *Isoperla, Taeniopteryx, Capnia, Perla*

ORDER ZORAPTERA. *Zorotypus*

ORDER CORRODENTIA. Book lice. *Ectopsocus, Liposcelis, Trogium*

ORDER MALLOPHAGA. Chewing lice. *Cuclotogaster, Menacanthus, Menopon, Trichodectes*

ORDER ANOPLURA. Sucking lice. *Pediculus, Phthirius, Haematopinus*

ORDER THYSANOPTERA. Thrips. *Heliothrips, Frankliniella, Hercothrips*

ORDER HEMIPTERA (including Homoptera). Bugs, cicadas, aphids, leafhoppers, etc. *Belostoma, Lygaeus, Notonecta, Cimex, Lygus, Oncopeltus, Magicicada, Circulifer, Psylla, Aphis*

ORDER NEUROPTERA. Dobsonflies, alderflies, lacewings, mantispids, snakeflies, etc. *Corydalus, Hemerobius, Chrysopa, Mantispa, Agulla*

ORDER COLEOPTERA. Beetles, weevils. *Copris, Phyllophaga, Harpalus, Scolytus, Melanotus, Cicindela, Dermestes, Photinus, Coccinella, Tenebrio, Anthonomus, Conotrachelus*

ORDER HYMENOPTERA. Wasps, bees, ants, sawflies. *Cimbex, Vespa, Glypta, Scolia, Bembix, Formica, Bombus, Apis*

ORDER MECOPTERA. Scorpionflies. *Panorpa, Boreus, Bittacus*

ORDER SIPHONAPTERA. Fleas. *Pulex, Nosopsyllus, Xenopsylla, Ctenocephalides*

ORDER DIPTERA. True flies, mosquitoes. *Aedes, Asilus, Sarcophaga, Anthomyia, Musca, Chironomus, Tabanus, Tipula, Drosophila*

ORDER TRICHOPTERA. Caddisflies. *Limnephilus, Rhyacophila, Hydropsyche*

ORDER LEPIDOPTERA. Moths, butterflies. *Tinea, Pyrausta, Malacosoma, Sphinx, Samia, Bombyx, Heliothis, Papilio, Lycaena*

SECTION DEUTEROSTOMIA

PHYLUM CHAETOGNATHA. Arrow worms. *Sagitta, Spadella*

PHYLUM ECHINODERMATA

CLASS CRINOIDEA. Crinoids, sea lilies. *Antedon, Ptilocrinus, Comactinia*

CLASS ASTEROIDEA. Sea stars. *Asterias, Ctenodiscus, Luidia, Oreaster*

CLASS OPHIUROIDEA. Brittle stars, serpent stars, basket stars, etc. *Asteronyx, Amphioplus, Ophiothrix, Ophioderma, Ophiura*

CLASS ECHINOIDEA. Sea urchins, sand dollars, heart urchins. *Cidaris, Arbacia, Strongylocentrotus, Echinanthus, Echinarachnius, Moira*

CLASS HOLOTHUROIDEA. Sea cucumbers. *Cucumaria, Thyone, Caudina, Synapta*

PHYLUM POGONOPHORA. Beard worms. *Siboglinum, Lamellisabella, Oligobrachia, Polybrachia*

PHYLUM HEMICHORDATA

CLASS ENTEROPNEUSTA. Acorn worms. *Saccoglossus, Balanoglossus, Glossobalanus*

CLASS PTEROBRANCHIA. *Rhabdopleura, Cephalodiscus*

PHYLUM CHORDATA. Chordates

Subphylum Urochordata (or Tunicata). Tunicates

CLASS ASCIDIACEA. Ascidians or sea squirts. *Ciona, Clavelina, Molgula, Perophora*

CLASS THALIACEA. *Pyrosoma, Salpa, Doliolum*

CLASS LARVACEA. *Appendicularia, Oikopleura, Fritillaria*

Subphylum Cephalochordata. Lancelets, amphioxus. *Branchiostoma, Asymmetron*

Subphylum Vertebrata. Vertebrates

CLASS AGNATHA. Jawless fishes. *Cephalaspis,* Pteraspis,* Petromyzon, Entosphenus, Myxine, Eptatretus*

CLASS PLACODERMI. No living representatives

CLASS CHONDRICHTHYES. Cartilaginous fishes. *Squalus, Hyporion, Raja, Chimaera*

CLASS OSTEICHTHYES. Bony fishes

SUBCLASS SARCOPTERYGII

ORDER CROSSOPTERYGII (or Coelacanthiformes). Lobe-fins. *Latimeria*

ORDER DIPNOI (or Dipteriformes). Lungfishes. *Neoceratodus, Protopterus, Lepidosiren*

SUBCLASS BRACHIOPTERYGII. Bichirs. *Polypterus*

SUBCLASS ACTINOPTERYGII. Higher bony fishes. *Amia, Cyprinus, Gadus, Perca, Salmo*

CLASS AMPHIBIA

ORDER ANURA. Frogs and toads. *Rana, Hyla, Bufo*

ORDER URODELA. Salamanders. *Necturus, Triturus, Plethodon, Ambystoma*

ORDER APODA. *Ichthyophis, Typhlonectes*

CLASS REPTILIA

ORDER CHELONIA. Turtles. *Chelydra, Kinosternon, Clemmys, Terrapene*

ORDER RHYNCHOCEPHALIA. Tuatara. *Sphenodon*

ORDER CROCODYLIA. Crocodiles and alligators. *Crocodylus, Alligator*

ORDER SQUAMATA. Snakes and lizards. *Iguana, Anolis, Sceloporus, Phrynosoma, Natrix, Elaphe, Coluber, Thamnophis, Crotalus*

CLASS AVES. Birds. *Anas, Larus, Columba, Gallus, Turdus, Dendroica, Sturnus, Passer, Melospiza*

CLASS MAMMALIA. Mammals

SUBCLASS PROTOTHERIA

ORDER MONOTREMATA. Egg-laying mammals. *Ornithorhynchus, Tachyglossus*

SUBCLASS THERIA. Marsupial and placental mammals

ORDER MARSUPIALIA. Marsupials. *Didelphis, Sarcophilus, Notoryctes, Macropus*

ORDER INSECTIVORA. Insectivores (moles, shrews, etc.). *Scalopus, Sorex, Erinaceus*

ORDER DERMOPTERA. Flying lemurs. *Galeopithecus*

ORDER CHIROPTERA. Bats. *Myotis, Eptesicus, Desmodus*

ORDER PRIMATES. Lemurs, monkeys, apes, man. *Lemur, Tarsius, Cebus, Macacus, Cynocephalus, Pongo, Pan, Homo*

* Extinct.

ORDER EDENTATA. Sloths, anteaters, armadillos. *Bradypus, Myrmecophagus, Dasypus*

ORDER PHOLIDOTA. Pangolin. *Manis*

ORDER LAGOMORPHA. Rabbits, hares, pikas. *Ochotona, Lepus, Sylvilagus, Oryctolagus*

ORDER RODENTIA. Rodents. *Sciurus, Marmota, Dipodomys, Microtus, Peromyscus, Rattus, Mus, Erethizon, Castor*

ORDER CETACEA. Whales, dolphins, porpoises. *Delphinus, Phocaena, Monodon, Balaena*

ORDER CARNIVORA. Carnivores. *Canis, Procyon, Ursus, Mustela, Mephitis, Felis, Hyaena, Eumetopias*

ORDER TUBULIDENTATA. Aardvark. *Orycteropus*

ORDER PROBOSCIDEA. Elephants. *Elephas, Loxodonta*

ORDER HYRACOIDEA. Coneys. *Procavia*

ORDER SIRENIA. Manatees. *Trichechus, Halicore*

ORDER PERISSODACTYLA. Odd-toed ungulates. *Equus, Tapirella, Tapirus, Rhinoceros*

ORDER ARTIODACTYLA. Even-toed ungulates. *Pecari, Sus, Hippopotamus, Camelus, Cervus, Odocoileus, Giraffa, Bison, Ovis, Bos*

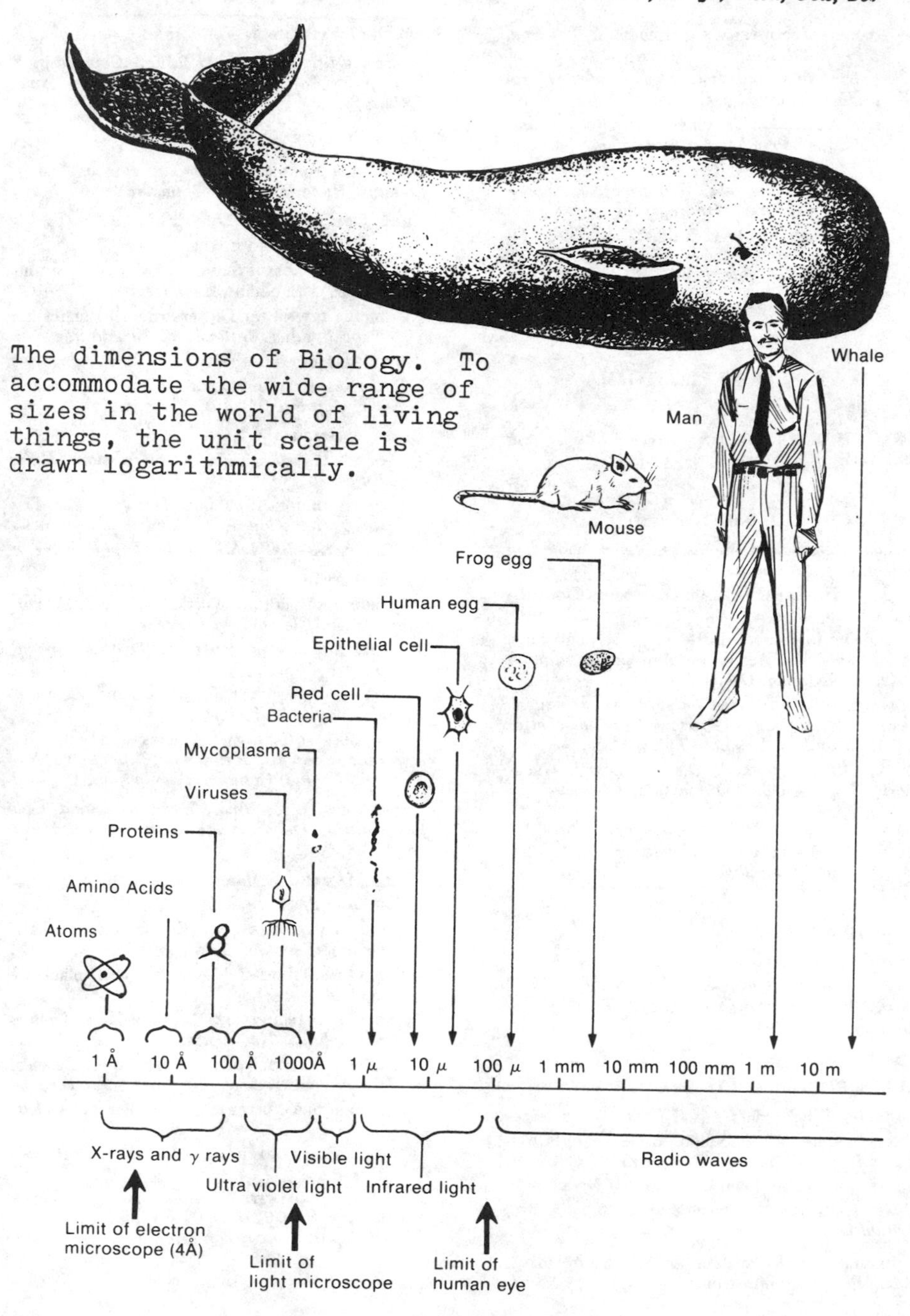

The dimensions of Biology. To accommodate the wide range of sizes in the world of living things, the unit scale is drawn logarithmically.

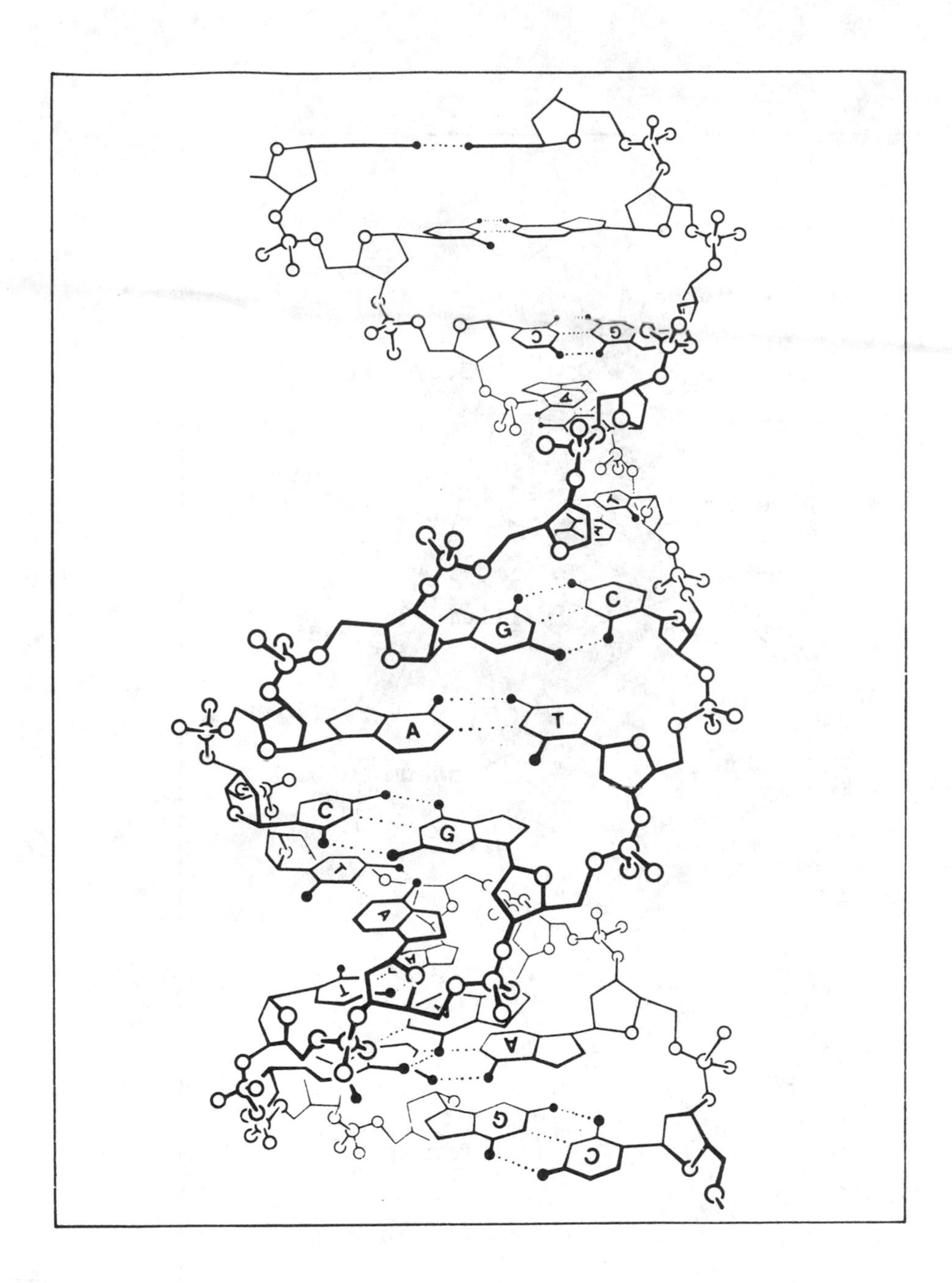

The DNA double helix. The purine (G= Guanine, A= Adenine) and pyrimidine (T= Thymine, C= Cytosine) base pairs are connected by hydrogen bonds. When these bonds are broken, the DNA can unwind and replicate (as in mitosis) or act as a template for mRNA synthesis. The ribose sugar and phosphate groups act as a backbone, helping to maintain the sequential order of the nitogenous bases. There are ten base pairs in one turn.

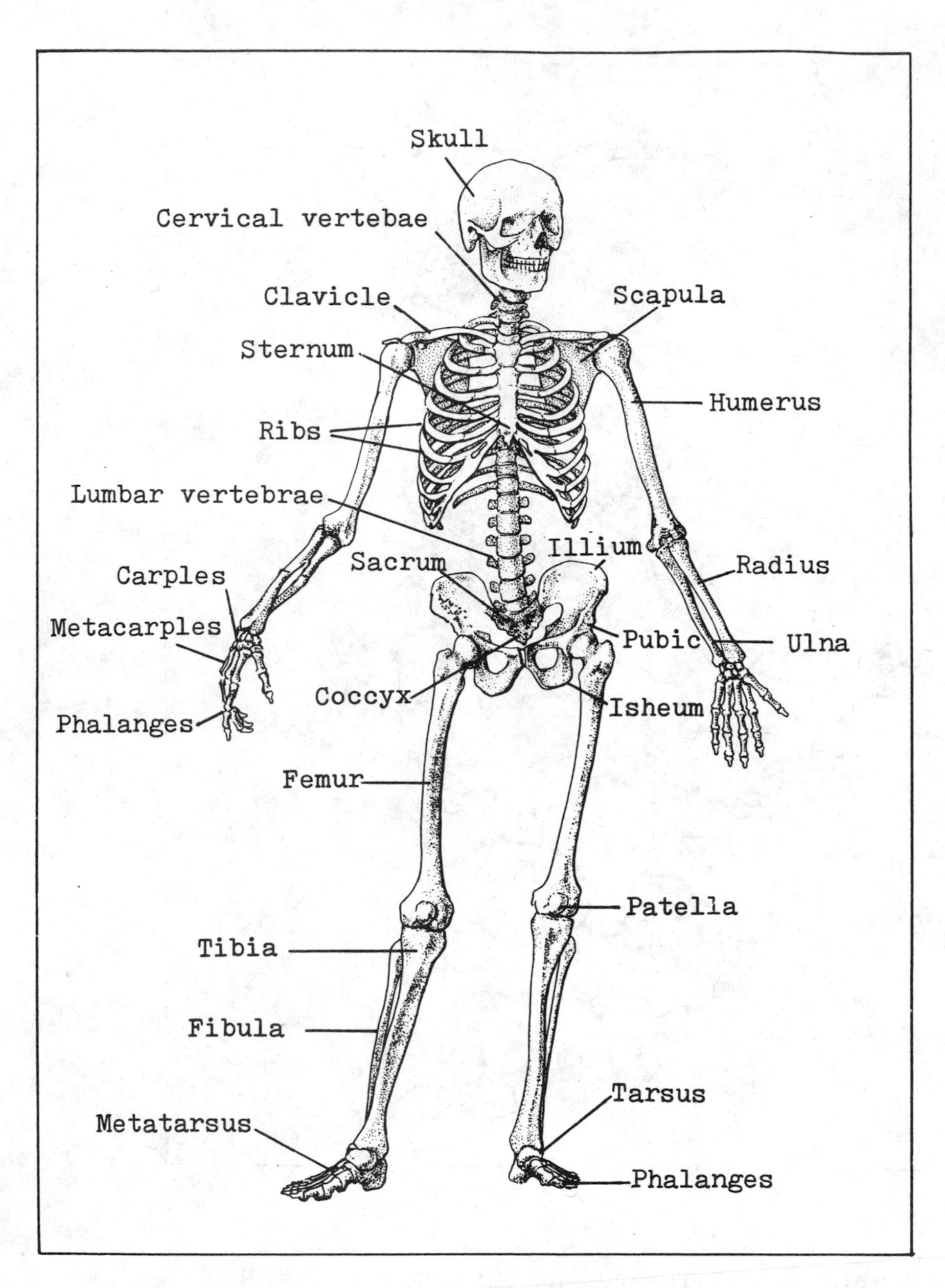

The human skeleton with major bones indicated.